Selected Papers on

Polarization

Books in SPIE's Milestone Series
Brian J. Thompson, *General Editor*

Selected Papers on:	*Editor*
CO_2 Lasers (1990)	James D. Evans
Polarization (1990)	Bruce H. Billings
Digital Image Processing (1990)	Mohan M. Trivedi
Holographic Particle Diagnostics (1990)	Chandra S. Vikram
Optical Shop Metrology (1990)	Daniel Malacara
Image Tubes (1990)	Illes P. Csorba
Coherence and Fluctuations of Light (1990)	Leonard Mandel, Emil Wolf
Acousto-Optics (1990)	Adrian Korpel
Natural Optical Activity (1990)	Akhlesh Lakhtakia
Radiometry (1990)	Irving J. Spiro
Visual Communication: Technology and Applications (1990)	T. Russell Hsing, Andrew G. Tescher
Quasielastic Light Scattering by Macromolecular, Supramolecular, and Fluid Systems (1990)	Benjamin Chu
Optical Fibers in Medicine (1990)	Abraham Katzir
Surface-Enhanced Raman Scattering (1990)	Milton Kerker
Infrared Fiber Optics (1990)	James A. Harrington
Fiber Optic Gyroscopes (1989)	Robert B. Smith
Deposition of Optical Coatings (1989)	Michael Ray Jacobson
Scattering in the Atmosphere (1989)	Craig F. Bohren
Optical Computing (1989)	H. John Caulfield, Gregory Gheen
Light Scattering (1988)	Milton Kerker
Optomechanical Design (1988)	Donald C. O'Shea
Infrared Design (1985)	R. Barry Johnson, William L. Wolfe
Laser Scanning and Recording (1985)	Leo Beiser

Forthcoming Milestones:

X-Ray Optics	Richard B. Hoover
Free-Space Laser Communications	David L. Begley
Interferometry	P. Hariharan, W. M. Steel
Thin Film Design	Michael Jacobson
Ellipsometry	Rasheed M. A. Azzam
Turbulence in a Refractive Medium	Ed Andreas
Laser Damage in Optical Materials	Roger M. Wood
Geometrical Aspects of Scattering	Philip L. Marston
Laser Design	Hugo Weichel
Optical Tolerancing	Gary Wiese

SPIE Milestone Series
Volume MS 23

Selected Papers on

Polarization

Bruce H. Billings, *Editor*
International Technology Associates, Inc.

Brian J. Thompson
General Editor, SPIE Milestone Series

SPIE OPTICAL ENGINEERING PRESS

A Publication of SPIE—The International Society for Optical Engineering
Bellingham, Washington USA

Library of Congress Cataloging-in-Publication Data

Selected papers on polarization / [compiled by] Bruce H. Billings.
p. cm. — (SPIE milestone series ; v. MS 23)
Includes index.
ISBN 0-8194-0494-2. — ISBN 0-8194-0495-0 (pbk.)
1. Polarization (Light) I. Billings, Bruce H., 1915–
II. Society of Photo-optical Instrumentation Engineers.
III. Series.
QC442.S44 1990
535.5′2—dc20 90-45397
CIP

ISBN 0-8194-0494-2 (hardbound)
ISBN 0-8194-0495-0 (softbound)

Published by
SPIE—The International Society for Optical Engineering
P.O. Box 10, Bellingham, Washington 98227-0010 USA
Telephone 206/676-3290 (Pacific Time) • Telex 46-7053

This book is a compilation of outstanding papers selected from the world literature on optical and optoelectronic science, engineering, and technology. SPIE Optical Engineering Press acknowledges with appreciation the authors and publishers who have given their permission to reprint material included in this volume. An earnest effort has been made to contact the copyright holders of all papers reproduced herein.

Printed in the United States of America

Introduction to the Series

There is no substitute for reading the original papers on any subject even if that subject is mature enough to be critically written up in a textbook or a monograph. Reading a well-written book only serves as a further stimulus to drive the reader to seek the original publications. The problems are, which papers and in what order?

As a serious student of a field, do you really have to search through all the material for yourself, and read the good with the not-so-good, the important with the not-so-important, and the milestone papers with the merely pedestrian offerings? The answer to all these questions is usually yes, unless the authors of the textbooks or monographs that you study have been very selective in their choices of references and bibliographic listings. Even in that all-too-rare circumstance, the reader is then faced with finding the original publications, many of which may be in obscure or not widely held journals.

From time to time and in many disparate fields, volumes appear that are collections of reprints that represent the milestone papers in the particular field. Some of these volumes have been produced for specific topics in optical science, but as yet no systematic set of volumes has been produced that covers connected areas of optical science and engineering.

The editors of each individual volume in this series have been chosen for their deep knowledge of the world literature in their fields; hence, the selection of reprints chosen for each book has been made with authority and care.

On behalf of SPIE, I thank the individual editors for their diligence, and we all hope that you, the reader, will find these volumes invaluable additions to your own working library.

Brian J. Thompson
University of Rochester

November 1990

This book is dedicated to
Dr. Edwin Land,
the best teacher I have ever had.

—Bruce H. Billings

Selected Papers on

Polarization

Contents

Section Three

Mathematical Analysis and Measurements of Polarization States

Section Four

Depolarizers

Section Five

Polarization Filters

Section Six

Polarization and Nature

Preface

I had two reasons for accepting the invitation to prepare this milestone volume on polarization.

First was nostalgia. Before World War II, I was Employee No. 80 at Polaroid Corporation, where my interest in polarized light began. It has continued ever since. Many of the early papers reproduced here were sent to me by my friends. In going through my own archives, it is interesting to find that some became Nobel laureates.

My second reason was that the project would offer me the opportunity to learn about the discoveries made in the last few decades. Many of these I had not taken time to try to comprehend. Although I don't want to lecture on degenerate four-wave mixing, at least I think I now have some understanding of it.

Unlike degenerate four-wave mixing, knowledge and application of polarization as such goes back several hundred years. The authors by and large were not Americans. It is interesting to remember how few people specialized in science before World War II. My undergraduate class in college had 1100 students and only 25 of us majored in physics. After a few years, only two of us were working in the field.

Many of our professors had received their degrees in Europe. In those days, we were required to know French and German with a high degree of fluency. I am ashamed to say that, although I had lectured in French and German, I flunked the German exam for my Ph.D. studies and had to take it a second time. Because so many of the major discoveries in polarization were made in Europe, a large number of papers are in French or German. How can I leave out the Malus paper on polarization of light on reflection?

Polarized light is with us constantly and its applications cover every field of science. We wear polarizing glasses to eliminate reflected glare from roads or water. Polarized light from the sky has been used for polar navigation. Determination of stress and mechanical systems is determined with plastic models and polarized light.

The purity of various organic chemicals such as sugar is measured with polarized light in polarimeters. Very high speed light modulators operate with polarized light using the Kerr and Pockel's effects.

Liquid crystals with polarizers are used in a large number of wristwatches in which the time is presented directly in numbers. Fiber optic systems are now being made in so-called "polarization-maintaining fibers" and fibers have been developed in which the birefringence can be altered electrically.

Almost each of these polarization areas could merit a Milestone reprint volume of its own and, in fact, a reprint volume is being put together on polarization in chemistry and another one exists on fibers and of course on optical fibers and their polarization characteristics.

To compress this area into one volume has been a fascinating, yet difficult, task. I have not really memorized the book. Accordingly, I expect to use it myself.

Thousands of papers exist on polarization. An enormous number could really be called milestone or breakthrough papers in the field. Accordingly, I am sure that many have been left out. I am also sure that some reader will be amused that I had the vanity to insert a few of my own papers. However, in glancing at other Milestone volumes, I found I was not alone.

I owe special thanks to many people who helped with the preparation of this volume. Most important is my wife, Fannie, without whose patience and many hours of effort it would have been impossible to finish this volume. I also owe thanks to Sam Tennant, president of The Aerospace Corporation, who gave me permission to call on the corporate library, and to Shirley McElroy of the library for her help in getting reprints of many of the articles in the volume.

Bruce H. Billings
International Technology Associates, Inc.

November 1990

Section One
Polarizers

The first polarizer used as such was a calcite crystal. This was described in Latin by Bartholinus in 1670 and was further investigated by Christaan Huygens in 1678.

Both of these articles are already reproduced in a book by Swindell.[1] As an optical physicist, one of my most exciting moments was to pick up a perfect calcite rhomb on a trail leading up to a calcite mine in California. This rhomb splits a beam of natural light into two separate beams that are oppositely polarized. Accordingly, the crystal can be used as a polarizer or as an analyzer for polarized light. Until 1828, this was the method used. Brewster, Malus, and other area pioneers in polarization used the simple crystal of calcite.

In William Nicol's 1828 paper, he invented the prism made of calcite that separates the two beams. We have included his paper in this section. His incredible invention is still used today. His device was so perfect that more than 100 years elapsed before the next new breakthrough in polarizers was made by Dr. Edwin H. Land. Land's invention had an enormous impact on the field of polarization. It made possible a tremendous number of advances in technology. The pocket television receiver and the lap-top computer are just a few of the many examples.

---

[1] William Swindell, *Polarized Light—Benchmark Papers in Optics,* Volume 1, Dowden, Hutchinson & Ross, Inc., Pennsylvania, 1975.

Reprinted from *Edinburgh Journal of Philsophy*, Vol. 6, 83–84 (1828).

On a Method of so far increasing the Divergency of the two Rays in Calcareous-spar, that only one Image may be seen at a time. By WILLIAM NICOL, Esq. Lecturer on Natural Philosophy. Communicated by the Author.

THE following simple method of constructing a prism of calcareous-spar, so that only one image may be seen at a time, will, perhaps, prove interesting to those who are in the habit of examining the optical properties of crystallised bodies by polarised light.

Let a rhomboid of calcareous-spar one inch long be reduced in breadth and thickness to three-tenths of an inch; let the obliquity of its terminal planes be increased about three degrees; or, in other words, let the angles formed by the terminal planes, and the adjoining obtuse lateral edges, be made equal to 68°, by operating on the terminal planes: these planes may now be polished. The rhomboid is then to be divided into two equal portions, by a plane passing through the acute lateral edges, and nearly touching the two obtuse solid angles. The sectional plane of each of the two halves must now be made to form exactly an angle of 90° with the terminal plane, and then carefully polished. The two portions are now to be firmly cemented together by means of Canada balsam, so as to form a rhomboid similar to what it was before its division.

If a ray of common light fall on the end of such a rhomboid in a direction parallel to the lateral edges, the two rays into which it is divided, in passing through the spar, will deviate so far from each other, that only the ordinary image will be seen. That image, too, will appear exactly in its true position, and free from colour. The range of the ordinary ray will be found considerably greater than the whole field of vision, as may easily be seen by making the rhomboid revolve on an axis parallel to the longer diagonal of the terminal planes. There is a tinge of blue where the ordinary ray vanishes on one side, and a tinge of orange, accompanied by a number of extremely minute obscurely coloured fringes, where it terminates on the other. If the rhomboid revolve beyond the fringes, the ordinary image will disappear, and the extraordinary image come into view. The latter, however, from the great obliquity of the incident light, occupies a smaller range, and is less distinct than the other. The ordinary ray passing out of the rhomboid in a direction parallel to its lateral edges, is therefore the best adapted for analytical purposes; and as calcareous-spar, when pure, and free from flaws, is not only transparent, but perfectly colourless, a rhomboid of that substance, of the above construction, developes the coloured rings of crystallized bodies with a degree of brilliancy not to be equalled by a plate of tourmaline, or perhaps by any other substance.

With the view of rendering the structure of the analyzing rhomboid more easily understood, I have supposed a piece of the spar to be divided into two equal portions. Such a division, however, would be a difficult task; but if two similar pieces of spar be taken, it will be found a very easy matter to remove one-half of each of them, either by grinding, or by the action of a file. The pieces of spar should not be much less than an inch long, and they need not be longer than 1.4 inch. If the latter dimension be adopted, the breadth and thickness will require to be about .48 of an inch.

In cementing the two pieces together, it will be proper to let the pointed end of the one project a little over the terminal plane of the other. By so doing, a more firm contact is obtained at the edges; and when the cement is sufficiently indurated, the whole of the projecting parts may easily be removed, according to their cleavages. The lateral planes should be left quite rough, to prevent the reflection of extraneous light.

UNITED STATES PATENT OFFICE

2,041,138

PROCESS OF FORMING IMPROVED LIGHT POLARIZING BODIES

Edwin H. Land, Wellesley Farms, Mass., assignor to Sheet Polarizer Company, Inc., Union City, N. J., a corporation of New Jersey

Original application March 10, 1930, Serial No. 434,833. Divided and this application December 24, 1935, Serial No. 55,941

11 Claims. (Cl. 18—57)

This invention relates to a process of forming improved light polarizing bodies.

The invention has for its object generally an improved process for forming improved polarizing bodies of the character indicated which accomplish plane polarization of transmitted light with substantially a minimum of scattering.

More especially the object of the invention is to provide a process for manufacturing a light polarizing body made from a transparent suspending medium with a mass of polarizing particles dispersed therein and oriented and retained therein with their polarizing axes in substantial parallelism.

A further object is to provide such an improved process for forming polarizing bodies with viscous material containing minute polarizing particles by orienting the particles within the viscous material through the utilization as the orienting agent of the differential action between cohesion and adhesion when pouring or flowing the particle-containing, viscous material.

A still further object of the invention is to provide a process for orienting asymmetric particles suspended in a suitable medium by subjecting the medium and the suspended particles to a pouring action.

Other objects of the invention will in part be obvious and will in part appear hereafter.

The invention accordingly comprises the several steps and the relations of one or more of these steps with respect to each of the others which are exemplified in the following detailed disclosure and the scope of the application of which will be indicated in the claims.

This application is a division of my copending application Serial No. 434,833, filed March 10, 1930, for Refracting polarizing body and process of making the same.

For a fuller understanding of the nature and objects of the invention reference should be had to the following detailed description taken in connection with the accompanying drawing in which the figure is a fragmentary sectional view showing a section of a polarizing body which has been flowed or otherwise deposited on a suitable support, all on a greatly enlarged scale, and constructed in accordance with the present invention.

In the preferred embodiment of the present invention a polarizing body is preferably made in the form of a film or sheet by pouring or flowing past an edge a relatively viscous medium which contains a large mass of very small light polarizing bodies, such, for example, as dichroic transparent crystals. By pouring or flowing a viscous medium of this character it is seen that the crystals dispersed therein are subjected to a mechanical field of force which orients them. This mechanical field is that resulting from the differential action of the forces of cohesion and adhesion which act upon the crystals so that they turn until their long axes are substantially parallel to the direction of flow. The crystals employed should preferably be very small and asymmetric in shape. There is hereinafter set forth a method of forming such crystals of herapathite (sulphate of iodo-quinine) which have been found suitable for use in the process of this invention in the formation of light polarizing sheets.

It is intended that after the viscous medium has been poured or flowed and the crystals suspended therein oriented, the mass should ordinarily be hardened or set so as to retain the crystals in oriented position.

It is to be understood that the crystals employed should preferably be of very small size, for example, they should have preferably one dimension shorter than the wave lengths of light to be transmitted by the polarizing sheet. Such crystals are ordinarily invisible under the microscope. They may be termed colloidal or microscopic. Where such crystals are employed the scattering of light transmitted by the finished polarizing sheets is materially reduced. A microscopic examination of such a set viscous medium containing such minute particles shows it to be uniform and clear and substantially no scattering of transmitted light is detected.

In order to obtain films or sheets in ribbon form containing such crystals set in their oriented position from a pouring or flowing operation, the suspending medium employed should have substantially the maximum viscosity compatible with flow. Under such circumstances and where the medium carrying the suspended particles is poured or flowed as, for example, past an edge, smooth ribbons or sheets are obtained free from air-bubbles and blemishes, the pouring operation being adapted to supply ribbon-like polarizing bodies rapidly and in commercial quantities.

A viscous suspension adapted for pouring or flowing in accordance with the present invention is readily prepared by adding to a suitable supporting medium such, for example, as a nitro-cellulose solution or a cellulose acetate solution, a jelly-like mass containing minute asymmetric particles, such as colloidal herapathite crystalline material.

Such a jell may be obtained in the following manner: Take 1.5 grams of quinine bisulphate and dissolve the same in 50 cc. of methyl alcohol which may then be brought to a boil and stirred, preferably with an iron stirring rod. While being stirred, 0.525 gram of iodine as a 20% solution in alcohol are added. By forming the solution of herapathite in this manner the saturation is increased from 50% to 75% above normal. The stirring may be continued while a jell forms and until the mass has cooled. The herapathite is rapidly precipitated out of the solution as a jelly of interwoven sub-microscopic needles. Since this jelly is incorporated in a viscous suspending medium, such for example as a cellulosic medium, the needles are readily distributed therethrough. The jelly clots may be gradually pulled apart and the suspension finally appears perfectly clear. The process outlined above for the formation of the jelly gives needle-like particles of herapathite having their optical axes corresponding with their physical axes so that when the particles are aligned physically their optical or polarizing axes are also oriented.

Where a nitro-cellulose solution is employed as the suspending medium for the cool mass of colloidal herapathite it should preferably be one having a relatively high ratio of nitro-cellulose to solvent. Such a solvent may be made, for example, by dissolving a quantity of the finest grade of highly nitrated cotton in as little solvent as will dissolve it. A suitable solvent may be amyl- or butyl-acetate. The solution of nitro-cellulose is highly adhesive and when the mass of herapathite is added, it disrupts the plates into which the needle-like crystals of herapathite are tending to gather. As a consequence there can thus be produced a relatively highly dispersed mass of minute polarizing particles in a viscous menstruum which, when poured from a bowl or plate or otherwise subjected to flow, as for example, to flow past an edge, forms a ribbon-like sheet containing crystals which are oriented so as to have their polarizing axes in substantial parallelism and which when set has the polarizing properties desired.

Commercial polarizing bodies may preferably be made in the form of a relatively thin sheet or film comprising the suspending medium and the minute particles dispersed therein. If desired, the polarizing body may itself be permanently or detachably fixed to a suitable support, preferably transparent, as for example, to a plate of glass or to a sheet of celluloid. Such a support may be desirable with conditions where it is found that the polarizing body itself may require some form of protection.

In the drawing such a construction is shown. 10 denotes the polarizing body which is represented as a film of a transparent medium having properly oriented therein minute particles of the desired character. As shown, this film is supported or superimposed upon any ordinary transparent body 11 which may be glass.

Where the polarizing body is poured or flowed on to a support in a still somewhat fluid condition it may be desirable to interpose between the polarizing body and the support a layer of a preferably transparent body 12 which may have preferably substantially the same viscosity as that of the polarizing body.

If the material 12 is kept in a somewhat fluid condition until the polarizing body has been distributed thereon, the entrapment of bubbles beneath the layer of the polarizing body may be avoided. This coating 12 may be of any suitable material, as for example a thin layer of fluid celluloid or dibutyl phthalate. It may, if desired, be adhesive, serving to unite the polarizing body with the transparent support. This latter quality is not essential and the polarizing body may be permitted to set or harden while in contact with the coating 12 and later be removed from contact therewith.

In the specification and claims reference to a set suspending medium or to a hardened suspending medium will be understood as implying merely such a change in the characteristics of the medium as will prevent movement of the particles therein from their oriented positions. It will be further understood that while the suspending medium has been described as a cellulose in a solvent, this description has been intended to cover the dispersion of the cellulosic material in the so-called solvent.

It is to be understood also that any suitable dispersing material may be employed as the dispersing medium for the polarizing particles and other solvents may be employed for said dispersion mediums provided, however, that the solvent should be a non-solvent for the polarizing particles carried in the suepension and the solvent and the suspending medium should be light transmitting.

It is also to be understood that while herapathite is mentioned as the preferred material from which the polarizing particles are obtained, any other suitable material may be employed. The periodides of the sulphates of cinchonine, cinchonidine or quinidine are known to be of the same class as herapathite.

Since certain changes in carrying out the above process may be made without departing from its scope, it is intended that the matter contained in the above disclosure shall be broadly interpreted and not in a limiting sense.

It is also to be understood that the following claims are intended to cover all of the generic and specific features of the invention herein described, and all statements of the scope of the invention which, as a matter of language, might be said to fall therebetween.

Having described my invention, what I claim as new and desire to secure by Letters Patent is:

1. The process of producing polarizing bodies which comprises preparing a suspending medium, introducing thereinto a mass of polarizing particles which are asymmetric in shape, dispersing said particles throughout said medium, orienting the polarizing axes of said particles by flowing said medium past an edge, and retaining said particles in oriented position by setting said medium.

2. The process of producing polarizing bodies which comprises preparing a suspending medium comprising a cellulosic material and a non-solvent of herapathite, preparing a suspension of herapathite, introducing said suspension into said medium, dispersing said herapathite throughout said medium, orienting the polarizing axes of said herapathite particles to substantial parallelism, and setting said medium to retain said particles in oriented position.

3. The process of producing polarizing bodies which comprises providing a fluid suspending medium, introducing into said medium a mass of asymmetric polarizing particles, orienting the polarizing axes of said particles by flowing said medium, depositing said medium upon a fluid support, and setting or hardening said medium while in contact with said support to retain the polariz-

ing axes of said particles in substantial parallelism.

4. The process of producing polarizing bodies, which comprises providing a fluid suspending medium, introducing into said medium a mass of asymmetric particles of heraphathite, orienting the polarizing axes of said particles to substantial parallelism, by flowing said medium past an edge, depositing said medium upon a fluid support, and setting or hardening said medium while in contact with said support to retain the polarizing axes of said particles in substantial parallelism.

5. The process of producing polarizing bodies, which comprises providing a fluid suspending medium of a cellulosic material and a suitable solvent, introducing a multiplicity of minute polarizing particles into said medium, orienting the polarizing axes of said particles by flowing said medium, depositing said medium upon a fluid support, and setting or hardening said medium while in contact with said support to retain the polarizing axes of said particles in substantial parallelism.

6. The process of producing polarizing bodies, which comprises providing a fluid suspending medium of nitro-cellulose and ethyl acetate, introducing and dispersing in said medium a mass of asymmetric herapathite particles, orienting the polarizing axes of said particles by flowing said medium, depositing said medium upon a fluid support, and setting or hardening said medium while in contact with said support to retain the polarizing axes of said particles in substantial parallelism.

7. The process of producing polarizing bodies, which comprises forming a suspension of minute particles of herapathite, preparing a suspending medium comprising nitro-cellulose and an acetate non-solvent of herapathite, distributing the herapathite suspension throughout the suspending medium, flowing the suspending medium to orient the polarizing axes of the particles therein, and setting the suspending medium to retain said particles in oriented position.

8. The process of preparing polarizing bodies, which comprises preparing a suspension of minute particles of herapathite in an alcohol, preparing a suspending medium comprising nitro-cellulose, incorporating and dispersing the herapathite suspension in the suspending medium, flowing the suspending medium past an edge to orient the particles, and setting the suspending medium to retain the particles in oriented position.

9. The process of producing polarizing bodies which comprises providing a fluid suspending medium, introducing into said medium a mass of needle-like polarizing particles, orienting the polarizing axes of said particles by flowing said medium, depositing said medium upon a fluid support, and setting said medium while in contact with said support to retain the polarizing axes of said particles in substantial parallelism.

10. The process of producing polarizing bodies which comprises providing a fluid suspending medium, introducing into said medium a mass of needle-like polarizing particles, orienting the polarizing axes of said particles to substantial parallelism by flowing said medium past an edge, and setting said medium immovably to retain said polarizing particles in substantial parallelism.

11. The process of producing polarizing bodies which comprises providing a suspending medium, introducing into said medium a mass of needle-like polarizing particles, orienting the polarizing axes of said particles to substantial parallelism by flowing said medium, and setting said medium to retain said particles in oriented position.

EDWIN H. LAND.

May 19, 1936. E. H. LAND 2,041,138

PROCESS OF FORMING IMPROVED LIGHT POLARIZING BODIES

Original Filed March 10, 1930

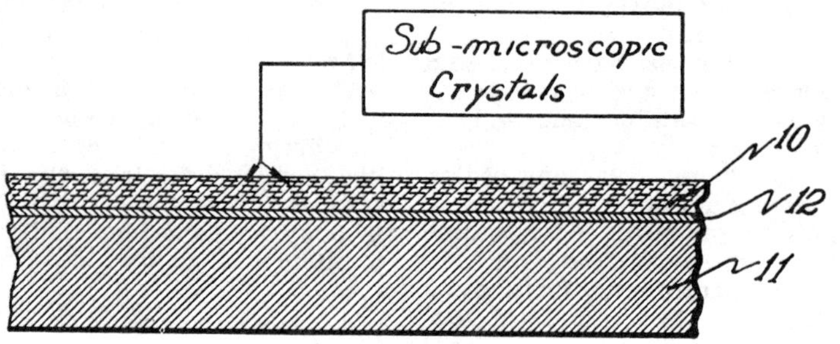

Edwin H. Land
INVENTOR

BY Brown & Jones
ATTORNEYS

UNITED STATES PATENT OFFICE

EDWIN H. LAND, OF NORWICH, CONNECTICUT, AND JOSEPH S. FRIEDMAN, OF BROOKLINE, MASSACHUSETTS, ASSIGNORS TO NORWICH RESEARCH, INC., OF NEW YORK, N. Y., A CORPORATION OF NEW YORK

POLARIZING REFRACTING BODIES

Application filed April 26, 1929. Serial No. 358,288.

This invention relates to polarizing refracting bodies, and particularly to polarizing bodies made from suspensions containing polarizing crystals.

The invention has for its object generally an improved polarizing body which may be produced in a convenient and efficient manner, and which is readily adapted to manufacturing procedure.

More specificially the invention has for its object the provision of an improved refracting body of solid material containing properly orientated polarizing bodies such as crystals which are dichroic and substantially transparent.

Other objects of the invention will in part be obvious and will in part appear hereinafter.

The invention accordingly comprises the article possessing the features, properties and the relation of elements, which are exemplified in the following detailed disclosure, and the scope of the application of which will be indicated in the claims.

For a fuller understanding of the nature and objects of the invention, reference should be had to the following detailed description taken in connection with the accompanying drawing, in which:

Fig. 1 is a view mainly diagrammatic, showing elementary apparatus adapted for carrying out the process of the present invention;

Fig. 2 is a fragmentary view of the medium showing its condition before treatment in accordance with the present invention;

Fig. 3 is a similar fragmentary view showing the condition of the medium after treatment;

Fig. 4 is a schematic view showing means employing a magnetic field of force for carrying out the process of the invention; and

Fig. 5 is a similar view showing means employing an electric field of force for carrying out the process of the invention.

In accordance with the present invention, a refracting polarizing body is produced by forming a colloidal suspension containing a mass of relatively small polarizing bodies which are oriented in a field of force so that the polarizing axes of the polarizing bodies are in alignment, or substantially parallel, whereby a body having permanent refracting and polarizing properties is obtained when the colloidal mass is set or hardened.

The polarizing bodies here preferably employed are relatively small crystals which are dichroic and transparent, and can be produced commercially in large quantities with substantial uniformity of size and other properties. It is generally desirable to employ colorless crystals, for example, small crystals of herapathite (sulphate of iodoquinine).

Herapathite crystals commonly occur in two modifications, generally designated alpha and beta varieties. Care should be taken to use only one variety in the suspension when prepared for mechanical treatment, as the geometric disposition of the polarizing axes in the two varieties is different. Crystals of herapathite of but one variety, however, result from any standard method of preparation.

The crystals which are preferred, when the orientation is to be accomplished by mechanical means as hereinafter set forth, are relatively small and needle-shaped, having one dimension longer than the others, for example, crystals with a width of 600 $\mu\mu$ (micromicrons) and a longitudinal dimension of 1 to 2 μ (microns).

A crystal-containing suspension is made by providing a suspending medium of proper viscosity, in which the crystals are introduced, the medium serving as a protecting agent adapted to prevent any tendency of the suspended crystals to flocculate. Suitable colloids for this purpose are the cellulose esters, particularly cellulose nitrate. The crystal-forming reaction may, with advantage, be made to take place in the colloidal mass itself. This is particularly true where it is desired to form needle-like crystals.

The following is an example of the manner in which a crystal-containing suspension of a character adapted to produce long crystals is prepared:

To 5 grams of nitrocellulose of low nitration dissolved in enough butyl acetate to make it plastic is added 1 gram of bisulphate of quinine, which has been dissolved in 2 cc. of methyl alcohol. This is worked to insure substantially uniform distribution of the bisulphate of quinine throughout the mass and is brought down substantially to its original plasticity. To this plastic material is then added 2.5 cc. of a 20% iodine solution dissolved in methyl alcohol. When this has been properly worked so as to bring about the crystal-forming reaction, it will be found to have substantially uniformly dispersed throughout the mass relatively fine needle-like crystals of herapathite.

The crystals thus dispersed throughout the colloidal mass have initially substantially all orientations of their polarizing axes, and form, in the first instance, only an opaque mass. In order to convert this mass into a transparent polarizing body, a quantity of the crystal-containing suspension supported in any suitable manner, and subjected to a field of force to which it is susceptible or to a system of constraints whereby differential velocities are produced throughout the mass by friction-applying means, as shown in Fig. 1.

Referring specifically to Fig. 1, 10 denotes a colloidal mass containing polarizing crystals having their polarizing axes disposed at random. This mass is shown as extruded between die-members 11 and 12, which apply frictional force to the sides of the extruded mass and thereby cause, by the stroking action, an orientation of the long crystals so that their geometrical and polarizing axes all point substantially in the same direction.

The extruded colloidal mass containing the oriented polarizing bodies is then allowed to become set by drying so that the resultant body is relatively hard and permanently has the property of polarizing light. When set, the polarizing particles are immovably embedded in and protected by the dried suspending medium in which they are positioned.

The die-members and the mass 10 in Fig. 1 may be deemed to show the parts greatly magnified as the die-members in practice are spaced apart a distance not much greater than the least dimension of the polarizing bodies (plus any support for the colloid that may be passed through the die-members). The polarizing crystals which have one dimension materially greater than the others, are thus susceptible to a mechanical application of force when stroked, which effects their desired orientation.

In Fig. 2 is shown schematically a strip of the colloidal suspension containing the polarizing bodies before being subject to a field of force, while Fig. 3 similarly shows the condition after such subjection, for example, after being stroked or extruded; the stippling or broken lines in Fig. 2 being employed to indicate that polarizing axes of the polarizing bodies have all orientations. In Fig. 3, the broken lines indicate that these axes have been brought into the substantial alignment or parallelism which imparts polarizing properties to the body. The condition of the medium shown in Figs. 2 and 3 may be regarded as enlarged views, showing rectangular sections taken in the plane of the paper from Fig. 1; Fig. 2 being taken from the portion of the medium shown at 10, while that in Fig. 3 lies in the plane of the paper and is taken from the portion 13.

Fields of force may be employed to accomplish the crystal orientation desired equally well, for example, magnetic and electric, since crystals in general have properties which make them susceptible to such fields of force. The unequal permeabilities of crystals respectively along the major crystal axes, make them susceptible to orientation by magnetic forces, while the similar phenomenon with respect to electric permeabilities along the crystal axes make them susceptible to electric fields. Small crystals of herapathite which are relatively geometrically symmetric, are readily oriented by suitable magnetic and electric fields.

Colloidal suspensions containing relatively small geometrically symmetric crystals which are particularly susceptible to magnetic and electric fields, are produced by grinding the crystals which have been previously obtained from solutions by supersaturation in any convenient manner, for example, by the Jörgensen method as described in "Journal für Praktische Chemie", Series 2, vol. 14, p. 230. This grinding appears to secure substantially uniform small size and distribution of the crystals in the colloidal mass. The following is an example of the manner in which such a crystal-containing suspension is made:

Two stock solutions are first prepared, as follows:

Solution No. 1, comprising ethyl acetate 80% and butyl acetate 20%; and

Solution No. 2, comprising 420 cubic centimeters of solution No. 1, to which is added 100 grams of one-half-second cellulose nitrate.

To make a convenient amount of the desired suspension, one takes 800 cubic centimeters of solution No. 1 and adds thereto 300 cubic centimeters of solution No. 2, and then introduces 2 grams of herapathite crystals. When made, this mass is ground by suitable means, for example, by a pebble-grinding machine, for a period of from 84 to 96 hours. At this juncture, it is desirable to introduce about 100 cubic centimeters of solution No. 2 and to continue the grinding for another period, for example, five hours or more. When the grinding is finished, the crystal-containing suspension may be further conditioned by adding a so-called "plasticizing agent", for example, di-butyl phthalate. The quantity of this latter agent that is added to a mass of the proportions given above is taken as substantially 30% by weight of the cellulose-nitrate introduced.

A suitable arrangement for applying a magnetic field of force to such a crystal-containing suspension, is shown in Fig. 4, where 20 and 21 denote windings or coils of electromagnets, which have opposed core members 22 and 23 of magnetic materials, which preferably have relatively high permeabilities; the electromagnets shown being of the open-magnetic circuit type. The windings of electromagnets are energized by an electric circuit which is here symbolically depicted as having a source of E. M. F. at 24, and conductors 25, 26 and 27, connected in series and controlled by a switch 28.

The electromagnets here employed are of a character adapted to produce a relatively intense magnetic field between the core members 22 and 23. A suitable field for this purpose has a flux density of at least 10,000 gauss. A frame 30 is shown between the opposed core members which is adapted to support a crystal-containing suspension prepared in accordance with the present invention for treatment in the magnetic field. After the insertion of the support 30, the field is energized by closing the switch 28, this field being applied until the colloid has dried or set. If desired, the colloid and its support may be moved for progressive treatment through the field.

In Fig. 5 there is shown schematically an arrangement of means for accomplishing the desired orientation of the polarizing crystals by an electric field of force. Here 31 and 32 denote respectively the plates of air-condenser between which an intense electrostatic field is arranged to be produced and employed as an orienting agent in accordance with the present invention. This condenser is energized from any suitable source of E. M. F., preferably a high potential alternating current source, since a direct-current source would tend to cause migration of the polarizing crystals from one portion of the suspension to another, which is not desired, this being due, apparently, to the fact that the crystals carry small charges so that they act like slow-moving ions. A source of alternating current is indicated at 33 and has electrical conductors 34 and 35 connecting the same to the plates 31 and 32. An impedance for regulating the intensity of the impressed E. M. F. is shown at 36.

A circuit-controlling means is also shown at 37. The support in this case which is adapted to contain the colloidal material for treatment, is shown at 40, the colloid containing the polarizing crystals in suspension. Here the orientation is accomplished by means of the change in the relatively different electrical permeativities of the polarizing crystals along their respective crystal axes. An alternating current of too low a frequency should be avoided.

In all the examples above described as methods of applying the field of force to orient the polarizing crystals in the colloidal suspension, the crystals are seen to have a physical property which is susceptible to the field of force to be applied. In the case of mechanical constraints the crystals have a geometrical dimension which is acted upon by the force applied frictionally to produce the desired orientation. When the magnetic field is employed, the crystals appear to move in the suspension so that their axis of maximum permeability lies along the lines of magnetic flux. In like manner, in the electric field, the crystals move so that their polarizing axes are substantially aligned.

Since certain changes in carrying out the above process, and certain modifications in the articles which embody the invention may be made without departing from its scope, it is intended that all matter contained in the above description or shown in the accompanying drawing shall be interpreted as illustrative and not in a limiting sense.

Having described our invention, what we claim as new and desire to secure by Letters Patent, is:

1. A polarizing refracting body, comprising a set transparent suspending medium of colloidal origin, and a plurality of granular crystalline bodies having their polarizing axes oriented to be in substantial parallelism and dispersed substantially uniformly through said medium and immovably embedded therein, the set suspension being adapted to retain its polarizing properties independent of external support.

2. A polarizing refracting body, comprising a set transparent suspending medium of colloidal origin, and a plurality of relatively small transparent dichroic crystals having their polarizing axes oriented into substantial parallelism and dispersed substantially uniformly through said medium and immovably embedded therein, the set suspension being adapted to retain its polarizing properties independent of external support.

3. A polarizing refracting body, comprising a set transparent suspending medium of colloidal origin, and a plurality of relatively fine crystals of herapathite having their polarizing axes oriented into substantial parallelism and dispersed substantially uniformly throughout the said medium and immovably embedded therein, the set suspension being adapted to retain its polarizing properties independent of external support.

4. A polarizing refracting body, comprising a set transparent suspending medium containing a cellulose ester, and a plurality of transparent dichroic crystals having their

polarizing axes oriented into substantial parallelism and dispersed substantially uniformly throughout said ester and immovably embedded therein, the set suspension being adapted to retain its polarizing properties independent of external support.

5. A polarizing refracting body, comprising a set transparent suspending medium containing cellulose nitrate, and a plurality of crystals of herapathite having a dimension in the neighborhood of 600 $\mu\mu$ dispersed substantially uniformly throughout said medium and immovable embedded therein and having their polarizing axes oriented into substantial parallelism, the set suspension being adapted to retain its polarizing properties independent of external support.

6. A polarizing refracting body, comprising a set transparent suspending medium containing cellulose nitrate, and a plurality of relatively fine needle-like crystals of herapathite having their polarizing axes oriented into substantial parallelism and dispersed substantially uniformly throughout said medium and immovably embedded therein, the set suspension being adapted to retain its polarizing properties independent of external support.

7. A polarizing body, comprising a set suspending medium and a plurality of bodies immovably embedded therein having their polarizing axes oriented to be in substantial parallelism, the set suspension being adapted to retain its polarizing properties independent of external support.

8. A polarizing body comprising a set suspending medium having relatively small transparent polarizing bodies dispersed and immovably embedded therein with their polarizing axes oriented to be in substantial parallelism, the set suspension being adapted to retain its polarizing properties independent of external support.

9. A polarizing body comprising a set suspending medium having herapathite particles dispersed and immovably embedded therein with their polarizing axes oriented to be in substantial parallelism, the set suspension being adapted to retain its polarizing properties independent of external support.

EDWIN H. LAND.
JOSEPH S. FRIEDMAN.

July 18, 1933. E. H. LAND ET AL 1,918,848

POLARIZING REFRACTING BODIES

Filed April 26, 1929

Fig. 4.

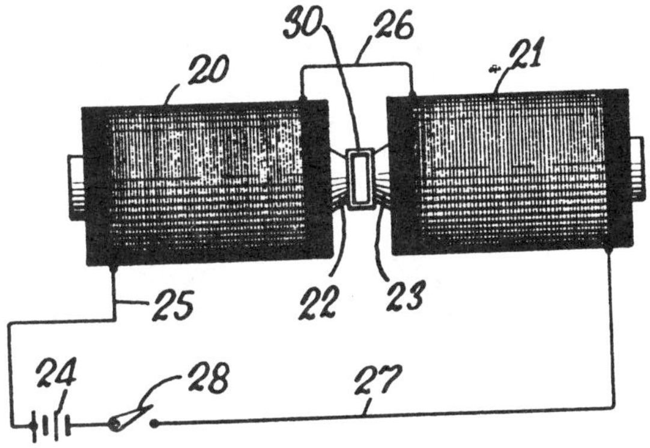

Fig. 2.

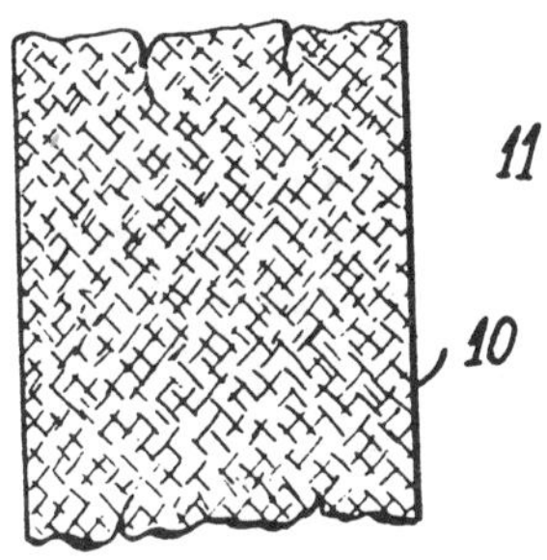

Fig. 1.

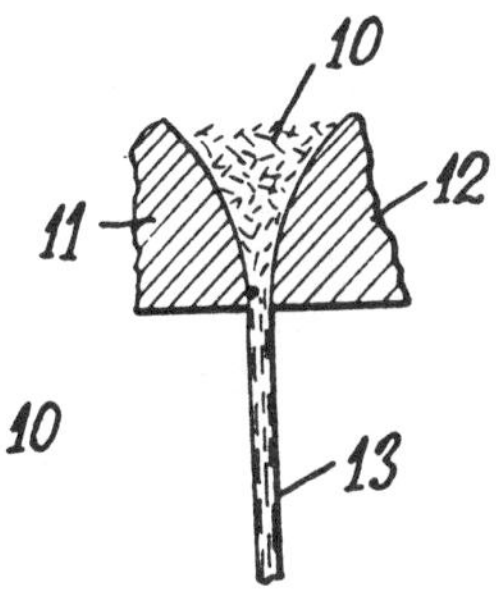

Fig. 3.

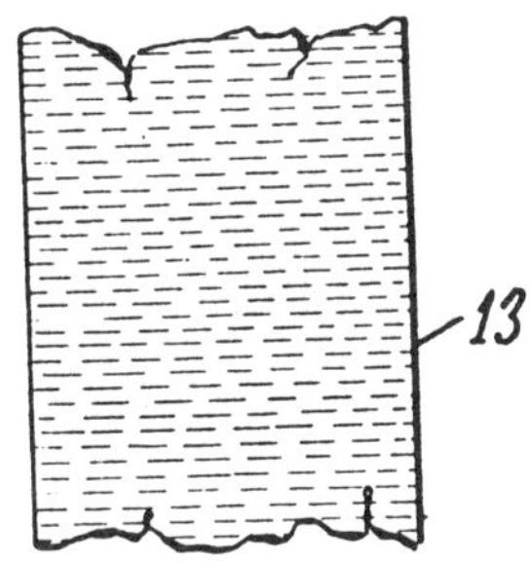

Fig. 5.

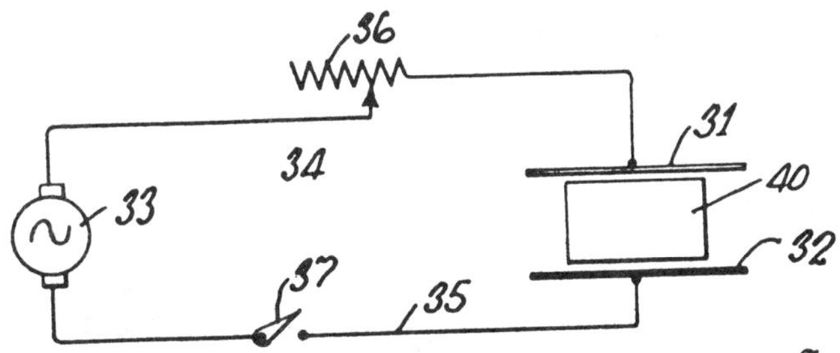

Inventor

Edwin H. Land and Joseph S. Friedman

By their Attorneys

Warfield & Watson

Dichroism and Dichroic Polarizers

EDWIN H. LAND AND C. D. WEST
Polaroid Corporation, Cambridge, Mass.

Historical Introduction

The phenomenon of linear dichroism was first observed in the mineral tourmaline by Biot in 1815. Since that time numerous observations, but relatively few quantitative measurements, of linear dichroism have been made on the widest variety of materials; these include, besides natural and artificial crystals, dichroic products obtained by the staining, dyeing or coloring of crystals, fibers, and other biological tissues, films, glass surfaces, and the like. This subject, although faithfully cultivated by a few devotees, never aroused a wide interest in the literature of science and hardly passed beyond a preliminary descriptive stage of development. The data of dichroism remained diffusely scattered through the journals of physics, chemistry, crystallography and mineralogy, biology, and dye and textile technology. This may explain why examples of later rediscoveries of the results of much of the earlier work are fairly numerous in the literature of this subject.

The phenomenon of circular dichroism was first reliably observed in alkaline chromium tartrate solution by another Frenchman, Cotton, in 1895. This property has been carefully measured for a wide variety of optically active molecules in the liquid and vapor states and in solutions; and these measurements have contributed substantially to the development of the theory of optical activity in asymmetric molecules, as related in several books on this subject.

The first satisfactory light polarizer was found in 1828 by W. Nicol, who devised the type of calcite prism which bears his name. The superiority of this type of polarizer needs no other testimony than its leadership of the field for over one hundred years; and it is still the preferred polarizer, when suitably constructed, for giving a completely polarized beam over the spectral range for which calcite transmits the extraordinary ray without absorption, namely about 0.23 — 3.0μ. The limitations of this class of polarizer are equally well recognized and have effectively prevented their application outside the laboratory. These are an angular aperture limited by the optical constants of calcite, and a linear aperture limited by the scarcity of large pieces of calcite of optical quality; great thickness; and occasionally the astigmatism of images transmitted through such prisms.

It was probably recognized quite early that physically the most satisfactory substitute for a calcite polarizer would be some kind of dichroic polarizer. To be specific, tourmaline sections were employed in the simple form of polariscope known as the tourmaline tongs, and also as polarizers for microscopes. The Englishman Herapath in 1852 discovered the intense dichroism of crystals of a sulfate periodide of the cinchona alkaloid quinine. He described in detail his attempts to grow and mount these crystals, which were named herapathite after him, for use as polarizers; and later worked with a similar crystal from cinchonidine alkaloid, a relative of quinine. Later attempts to realize Herapath's conception were described by Zimmern and Coutin in France (1926); [1] and about 1936 A. M. Marks in the United States began commercial production of macrocrystalline cinchonidine herapathite polarizers, while the Carl Zeiss firm in Germany was doing the same with quinine herapathite polarizers developed by F. Bernauer. In 1935 he [2] had previously outlined general methods for making large area polarizers, but none of these apparently was ever described in any detail. E. H. Land had already been occupied with the same problem for several years, and in February 1932 he had described and demonstrated before the Physical Colloquium of Harvard University a form of dichroic polarizer that was entirely new in that its structure was microcrystalline rather than macrocrystalline. In 1934 Land was in commercial production of this type of polarizer, using quinine herapathite as the crystal component.

The last six years have witnessed a rapid application of dichroic polarizers in a diversity of popular and scientific uses, and an equally rapid development of new types of dichroic polarizers. The most notable of these are not crystalline in the ordinary sense of this term, but would better be described as molecular in structure.

In this brief account we shall treat the general phenomenon of linear dichroism and its measurement, describe the structures of several kinds of useful dichroic polarizers, and point out some of their practical applications.

Dichroism: Definition and Measurement

Absorption. For ordinary homogeneous absorbing media, the absorption of light is conveniently expressed by the optical density of each wave length: $\frac{J}{J_0} = k = 10^{-d}$, or $\log \frac{J_0}{J} = \log \frac{1}{k} = d$. When different thicknesses of the same medium are compared, the law becomes $\log \frac{J_0}{J} = d = Kt$. When the volume concentration of the absorbing agent may be varied as well as the thickness, then in the ideal case the law becomes $\log \frac{J_0}{J} = d = Kt = ect$. It is sometimes more convenient to determine the area concentration c' than the volume concentration c, and in this case by a suitable choice of units $ect = ec'$. The proportionality constants K and e are constants of the absorbing agent, and like d they vary with wave length; when they are employed, the units for c and t must always be stated. We shall take t in cm, c in g per liter, and c' in g per 1000 cm^2; thus our e is a specific extinction coefficient. For ordinary media it is indifferent whether natural, linear, or circular polarized light is used to measure the absorption constants.

Dichroism. It is sometimes overlooked that there exist homogeneous media whose absorption of natural light does not follow the simple thickness law. If we let natural light of intensity J_0 fall on such a medium, and measure successively the emergent intensities J_1 and J_2 after traversing thicknesses t_1 and $t_2 = 2t_1$, we find after making any necessary corrections that $\frac{J_2}{J_0}$ is not equal to $\left[\frac{J_1}{J_0}\right]^2$ as required by $\frac{J}{J_0} = 10^{-Kt}$, but is somewhat greater than $\left[\frac{J_1}{J_0}\right]^2$. If now we analyze the emerging light we find that it has experienced a change in its vibration state; instead of being natural light, it is partially polarized, either circularly or linearly as the case may be. Accordingly, caution must always be exercised in calculating the results of a transmission measurement with natural light to an absorption coefficient if the specimen has this property to an appreciable degree.

Such media are called dichroic, and are characterized by unequal absorptions for two oppositely polarized beams of incident light; in one case the two beams of light are linear-polarized at right angles to each other; in the other case the two beams are circular-polarized in opposite senses. For such media *two* absorption constants are required for a complete description; for example, for linear dichroism $d_{\parallel}$ and $d_{\perp}$, or $K_{\parallel}$ and $K_{\perp}$ or $e_{\parallel}$ and $e_{\perp}$. Then the conventional mathematical expression for dichroism is $\Delta d = d_{\parallel} - d_{\perp}$, or ΔK, or Δe.

The constant $d_{\parallel} = \log \frac{1}{k_{\parallel}} = K_{\parallel} t$ of a dichroic specimen denotes the optical density of the specimen for linear-polarized light vibrating parallel to one of the two vibration directions of the specimen, and similarly for $d_{\perp}$. For a linear dichroic specimen $d = \log \frac{1}{k} = Kt$ may be defined and measured in two ways; it is either the optical density for natural or circular-polarized light, or it is the optical density for linear-polarized light vibrating at 45° to the vibration directions of the specimen.

The optical constants, for natural light, of a combination of ideal polarizer plus a dichroic specimen characterized by the constants $d_{\parallel}$, d, and $d_{\perp}$, are found as follows: since the d for an ideal polarizer is $\log 2 = 0.301$, the constants $d_{\parallel}$, d, $d_{\perp}$ are each to be increased by 0.301 to give the constants of the combination; and similarly the k constants are to be multiplied by 0.5 to give the constants of the combination.

Linear dichroism is in fact more complicated than this, for in the most general linear dichroic crystal the absorption constants $K_{\parallel}$ and $K_{\perp}$ vary with the orientation of the crystal section being measured, with respect to the orthogonal principal axes X, Y, Z, of the triaxial absorption ellipsoid. In the dichroic materials treated here, we shall avoid ambiguity by considering only what amounts to crystal sections of fixed orientation, and only that light which traverses such sections at normal incidence.

Detection of Dichroism. While a suitably mounted calcite cleavage rhomb was formerly employed as a dichroscope for detecting linear dichroism, it is now easier to construct an equivalent device by juxtaposing two linear film polarizers (plate P in Fig. 1). If this divided field plate be superposed with a quarter-wave sheet (plate

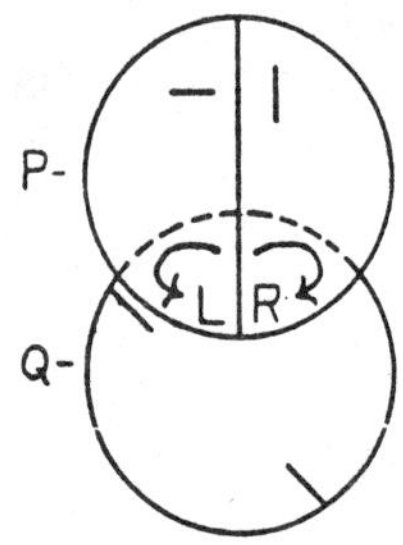

FIGURE 1.

Q), the combination serves as a circular dichroscope when the polarizer layer is held next to the eye, and as a linear dichroscope when it is turned over and the quarter-wave layer is next to the eye of the observer. With dichroscopes of this kind, in fact in the general use of dichroic polarizers, it is a great convenience to have the vibration direction or sense of the transmitted light plainly indicated by a reference mark.

For detecting linear dichroism in extra-visual portions of the spectrum, C. D. West has made use of the interference bands (Müller bands) given by a thin crystal section such as quartz cut parallel to the optic axis. This section is placed in the diagonal position between a known polarizer, such as a Nicol prism, and the unknown specimen, in front of the slit of a prism or grating spectrograph; and varying exposures are made with a light source giving a continuous spectrum. Then the interference bands, which occur at approximately equal intervals on a frequency scale, are registered only in those portions of the spectrum for which the unknown specimen is dichroic.

The Savart plate in conjunction with a Nicol prism is the familiar device for detecting traces of linear polarization in a light source. A convenient and easily made equivalent device results, when a film polarizer whose vibration direction is marked is cemented to a basal section of a strongly birefringent crystal such as calcite. The appearance of isochromatic circles will reveal the presence of polarization, while the azimuth at which the black isogyres join in the center will indicate the vibration direction of the polarized component.

Measurement of Absorption Constants of Linear Dichroic Media. The absorption constants of a dichroic medium can be measured with any kind of a photometer; but in practice the spectral variation of these constants, as measured with a spectrophotometer, is of primary interest.

It is to be recognized that the measurement of the absorption constants of strongly dichroic samples such as the present polarizing layers is equivalent to the measurement of relatively high and relatively low optical densities; if done with accuracy, these are difficult tasks by any spectrophotometric methods. In contrast to the more usual practice with isotropic liquid solutions, the operator here is generally not free to choose for measurement density ranges over which a given instrument yields results of maximum accuracy.

With a non-polarizing spectrophotometer one measures k, the transmission of a single dichroic layer; k_2, the transmission of two superposed identical layers in the parallel position; and k_x, the transmission of the same two layers crossed. In the measurement of k and k_2, polarization effects in the instrument which would affect the results must be guarded against. The three transmissions should satisfy the relations

$$k_2 > k_x,$$
$$2k^2 = k_2 + k_x.$$

The constants k_2 and k_x are related to the transmissions of a single layer for each of two oppositely polarized rays, $k_{\|}$ and $k_{\perp}$, by the equations (when $k_x = 0$, $k_{\perp} = 2k = \sqrt{2k_2}$)

$$k_2 = \tfrac{1}{2}(k_{\|}^2 + k_{\perp}^2), \qquad k_{\perp} = \frac{\sqrt{2}}{2}(\sqrt{k_2 + k_x} + \sqrt{k_2 - k_x}),$$

$$k_x = k_{\|} \cdot k_{\perp}; \qquad k_{\|} = \frac{\sqrt{2}}{2}(\sqrt{k_2 + k_x} - \sqrt{k_2 - k_x}).$$

Any non-polarizing spectrophotometer may be adapted to measure directly the constants of dichroic layers by inserting a fixed polarizer, preferably a calcite polarizer, in the optical system. By such means one can determine $k_{\|}$ and $k_{\perp}$ with only a single specimen, which obviously must be rotated through a right angle between the two measurements; k also may be measured in the diagonal position. This method is suggested for the photoelectric spectrophotometers which have in recent years become readily available, and which cover portions of the ultraviolet and infrared spectra as well as the visible.

Polarization spectrophotometers are also available for measurements over the visible spectrum. Such subjective instruments can be employed to measure independently four different constants of a dichroic specimen, namely k, $k_{\|}$, $k_{\perp}$, and the ratio of the last two or $R = \frac{k_{\|}}{k_{\perp}}$. For the measurement of the first three of these, the dichroic specimen covers only one window of the photometer, whereas in measuring R the specimen must be large enough to cover both windows. In terms of density the corresponding constants are d, $d_{\|}$, $d_{\perp}$, and $\Delta d = d_{\|} - d_{\perp}$. With such an instrument the logical practice is to measure whichever pair of these constants can be obtained with the greater accuracy, for example, Δd and $d_{\|}$, and calculate the other two constants therefrom. Incidentally, these four constants may be readily measured to a fair approximation for a not too strongly absorbing dichroic specimen (*e.g.*, a piece of commercial dyed cellophane) by means of an improvised photometer consisting of the dichroscope plate described above, in combination with an analyser and a divided circle (such as the stage of a polarizing microscope) for reading off the angular settings.

From a measured R the degree of polarization may be calculated as $V = \frac{1-R}{1+R}$, where V is the per cent of linear polarized light, $1 - V$ the per cent of unpolarized light in the total light transmitted.

The k, V and R constants are related to $k_{\parallel}$ and $k_{\perp}$ by the equations:

$$k_{\parallel} = k(1 - V) = \frac{2kR}{1+R}, \qquad k = \tfrac{1}{2}(k_{\parallel} + k_{\perp}),$$

$$k_{\perp} = k(1 + V) = \frac{2k}{1+R}; \qquad V = \frac{k_{\perp} - k_{\parallel}}{k_{\perp} + k_{\parallel}}.$$

All calculations are simplified by freeing the measured densities ($d_{\parallel}$, $d_{\perp}$, d) or transmissions from reflection losses. All the measurements in this paper have been corrected in this manner.

The effect of adding an isotropic absorber to a dichroic absorber is to decrease both the transmission k and the dichroic ratio $d_{\parallel}/d_{\perp} > 1$, while the dichroism Δd, the ratio R and the degree of polarization V characteristic of the dichroic absorber are left unchanged.

The average density $\bar{d}$ of a dichroic medium will enter into some of our calculations. It is defined as

$$\begin{aligned} \bar{d} &= \tfrac{1}{3}(d_x + d_y + d_z) \text{ for biaxial media;} \\ &= \tfrac{1}{3}(2d_{\perp} + d_{\parallel}) \quad \text{for uniaxial media;} \\ &\sim \tfrac{1}{3}d_{\parallel} \text{ when } d_{\perp} \sim 0. \end{aligned}$$

This is distinctly different from the density for natural light of a section of a dichroic medium

$$\begin{aligned} d = -\log k &= -\log \tfrac{1}{2}(k_{\parallel} + k_{\perp}) \\ &\sim \log 2 + d_{\perp} \text{ when } k_{\parallel} \sim 0. \end{aligned}$$

The spectral curves of $d_{\parallel}$, $d_{\perp}$, $\bar{d}$ and d for a strongly dichroic medium are illustrated in Fig. 7; it is seen here how the pronounced absorption band shown by the $d_{\parallel}$ and $\bar{d}$ curves vanishes completely when the data are calculated to the d curve for this medium. As has already been stated, the constants $d_{\parallel}$, d, and $d_{\perp}$ are measures of the intensity of the emergent light when a linear dichroic specimen is traversed by a beam of linear-polarized light vibrating at the respective azimuths 0, 45° and 90° to one of the vibration directions of the specimen. For a better idea of the relations between these three constants the reader is referred to Fig. 12.

Presentation of Results. In comparing a wide variety of dichroic media it is desirable to reduce measured quantities to a uniform system of constants. In the past the practice has been almost unanimous to report values of $d_{\parallel}$ and $d_{\perp}$, or absorption constants differing from these by a simple factor. But in describing the current dichroic polarizers one notes a tendency to give other constants, for example k_2 and k_s (Strong, 1936)[3] or k and V (Grabau, 1937).[4] This last system is doubtless quite satisfactory from the point of view of the person who expects to use a dichroic preparation as a polarizer; but for the purposes of the present treatment we prefer to retain the older $d_{\parallel}$, $d_{\perp}$ system. One advantage here is that not only the dichroism, Δd, but also the dichroic ratio $\frac{d_{\parallel}}{d_{\perp}}$ is easily obtained in this system; and this ratio is seen to be, for a given absorbing agent, a constant invariant with respect to changes in concentration and thickness, and thus it serves as a convenient figure of merit, other things being equal. This ratio was early employed in reporting quantitative measurements of dichroism by Pulfrich (1882),[5] and it was more recently resurrected by Preston (1931).[6]

Example of Linear Dichroism. Flow Dichroism of a Hydrosol. Previous measurements of flow dichroism and flow birefringence of colloidal solutions of dyes and of vanadium pentoxide in water were given in a valuable paper on linear dichroism of Zocher and Jacoby (1927).[7] These measurements give only the difference of the absorption, Δd, but not the value of either $d_{\parallel}$ or $d_{\perp}$ for the systems studied.

West has found that the strong blue color resulting from mixing extremely dilute solutions of α-naphthoflavone and iodine, first observed by Barger and Starling (1915)[8] and lately employed as an iodometric indicator, is characterized by a remarkable flow dichroism and flow birefringence not previously reported. This flow dichroism is positive, which means that the component of the incident light vibrating parallel to the streamlines is the more strongly absorbed; but the flow birefringence is negative (the slow ray is transverse to the streamlines), as would be expected if it were governed by the dichroic absorption band in the far red.

A glass cell was constructed in which the solution could be made to flow with a slow nearly uniform motion about a closed path between stationary walls. Measurements were made of $d_{\parallel}$ and $d_{\perp}$ for the liquid in motion, and of $\bar{d}$ for the stationary system, by means of a Gaertner polarization spectrophotometer for the visual spectrum.

The accompanying curves (Fig. 2) show the course of the measured $\bar{d}$, $d_{\parallel}$, and $d_{\perp}$ curves; and the $\Delta d = d_{\parallel} - d_{\perp}$ and $\bar{d}$ calc. $= \frac{1}{3}(d_{\parallel} + 2d_{\perp})$ curves calculated therefrom. It is seen that the calculated $\bar{d}$ points fall on the measured $\bar{d}$ curve to a good approximation; they would be expected to coincide exactly for a system in uniform uniaxial flow, a condition somewhat difficult to realize experimentally.

Dichroic Polarizers: Structure and Properties

The following list is representative of recent patents disclosing dichroic polarizers or processes of preparing them.*

Linear dichroism is a general property of absorbing anisotropic matter. It is not difficult to state the characteristics of dichroic materials potentially suitable for preparation of dichroic polarizers. Such materials should have:

(1) An absorption band covering the portion of the spectrum it is desired to polarize, ordinarily the visible spectrum 400 – 700 mμ.

(2) An adequately high dichroism throughout this band.

(3) Stability and permanence under the conditions to which it will be exposed in use—or generally, exposure to heat, to visible and ultraviolet radiation, to humidity, and to mechanical stresses.

(4) Freedom from scattering or diffusion of the transmitted component.

Under (2), we might arbitrarily take a dichroic ratio of about 7 as the threshold

* While this paper is intended to represent the state of the art as of January, 1943, this list includes patents up to June, 1944.

E. D. Bailey and M. M. Brubaker. U. S. 2,246,087 (1941).
R. P. Blake. U. S. 2,256,108 (1941).
C. H. Brown. U. S. 2,224,214 (1940); 2,287,598 (1942).
E. Käsemann. U. S. 2, 236,972 (1941).
L. A. Keim. U. S. 2,274,706 (1942); 2,340,476 (1944).
E. H. Land. U. S. 1,918,848 (1933) (Land and Friedman); 1,951,664 (1934); 1,956,867 (1934); 1,989,371 (1935); 2,011,553 (1935); 2,041,138 (1936); 2,078,254 (1937); 2,165,973 (1939); 2,173,304 (1939) (Land and Rogers); 2,178,996 (1939); 2,237,567 (1941); 2,281,100 (1942); 2,289,712 (1942) (Land and West); 2,289,713 (1942); 2,289,-714 (1942); 2,298,058 (1942); 2,306,108 (1942) (Land and Rogers); 2,319,816 (1943); 2,328,219 (1943); 2,343,775 (1944); 2,346,766 (1944).
H. Lapp. German 674,840 (1939).
A. M. Marks. U. S. 2,104,949 (1938); 2,167,899 (1939); 2,199,227 (1940); 2,344,514 (1944).
K. Meyer and others. German 675,217 (1939); 681,237 (1939); 681,347 (1940).
L. Pollack. U. S. 2,286,569 (1942); 2,346,784 (1944).
H. G. Rogers. U. S. 2,255,940 (1941); 2,284,590 (1942).
O. Vierling and P. G. Gänswein. U. S. 2,344,117 (1944).
G. Wilmanns and W. Schneider. U. S. 2,176,516 (1939).
H. Zocher. U. S. 1,873,951 (1932).

of usefulness. A polarizer with $k_\parallel = 0.01$, $d_\parallel = 2.0$; $k_\perp = 0.50$, $d_\perp = 0.30$; and $k = 0.25$, $d = 0.61$, $V = 96.1$ per cent, would approximate this figure. It is of interest that colored tourmalines so far reported have a dichroic ratio of this order of magnitude, irrespective of their color. An ideal polarizer would have, for comparison, $k_\parallel = 0$, $d_\parallel = \infty$; $k_\perp = 1.00$, $d_\perp = 0$; $k = 0.50$, $d = 0.30$, $V = 100$ per cent, giving an infinite dichroic ratio.

Linear dichroic polarizers are conveniently treated in three classes according to structure: macrocrystalline, microcrystalline, and those associated with oriented linear high polymers. In the following, dichroic specimens of these classes will be illustrated by $d_\parallel$, $d_\perp$ curves, all taken in these laboratories by means of the Gaertner spectrophotometer (Martens type) previously mentioned, and all corrected for reflection losses (Figs. 2-18); these are chosen to show the nature of the absorption bands

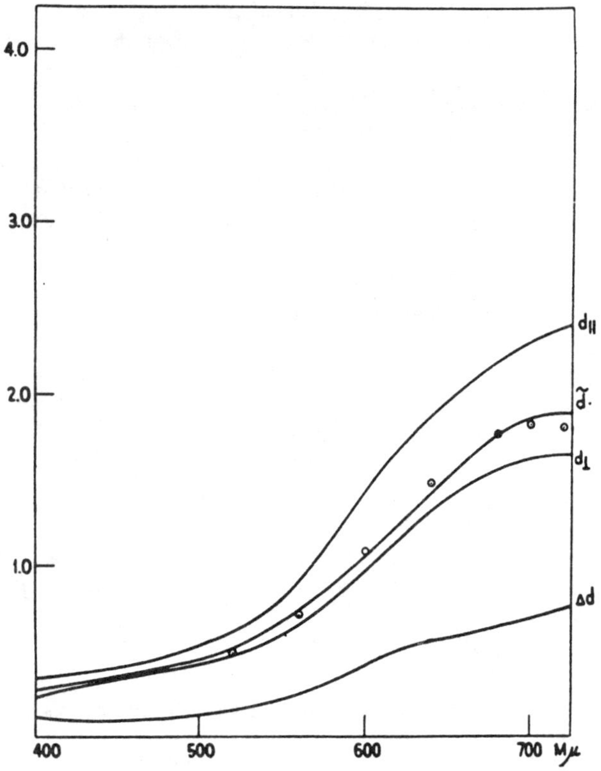

FIGURE 2. Flow dichroism of α-naphthoflavone product with iodine. The curve of average density $\bar{d}$ is for the fluid at rest, the circles are values of $\bar{d}$ calculated from $d_\parallel$ and $d_\perp$ for the fluid in motion. Cell thickness 2 cm.

in question rather than to be representative of the commercial polarizers which are or may be prepared from these materials.

In considering these curves it is to be kept in mind that optical densities greater than 3 are in general subject to increasingly large errors of measurement as the density increases above 3.

We wish to mention, among other members of the research laboratory of Polaroid Corporation, especially W. F. Amon, Jr., who prepared many photometric samples, and A. S. Makas, who took many spectrophotometric curves, which were the necessary preliminary to the discussion of the following section.

Macrocrystalline Polarizers. These are prepared by cutting, or by growing, sections of dichroic crystals of suitable orientation and thickness to give useful polarizers. When the section is thin or mechanically weak, it is cemented between transparent protecting covers such as glass plates.

Tourmaline Polarizers. This relatively abundant silicate mineral crystallizes in the trigonal system, and when colored always shows stronger absorption of the ordinary ray (negative dichroism). As with all uniaxial crystals, the maximum dichroism is exhibited by sections parallel to the optic axis, whereas sections perpendicular to the axis are isotropic. In the past the absorption constants $K_{\parallel}$ and $K_{\perp}$ have been repeatedly measured through the visible spectrum for specimens of various colors, and while they are found to vary quite widely, the dichroic ratio is substantially constant. This suggests that the absorbing agent is of the same nature in all tourmalines, but is present in varying concentrations. No explanation of the dichroism has yet been undertaken; it is recognized only that colored tourmalines generally contain cations of the transition metals, for example manganese in rose tourmaline and iron in the brown or black varieties. The dichroism of a black tourmaline is shown in the accompanying density curves (Fig. 3).

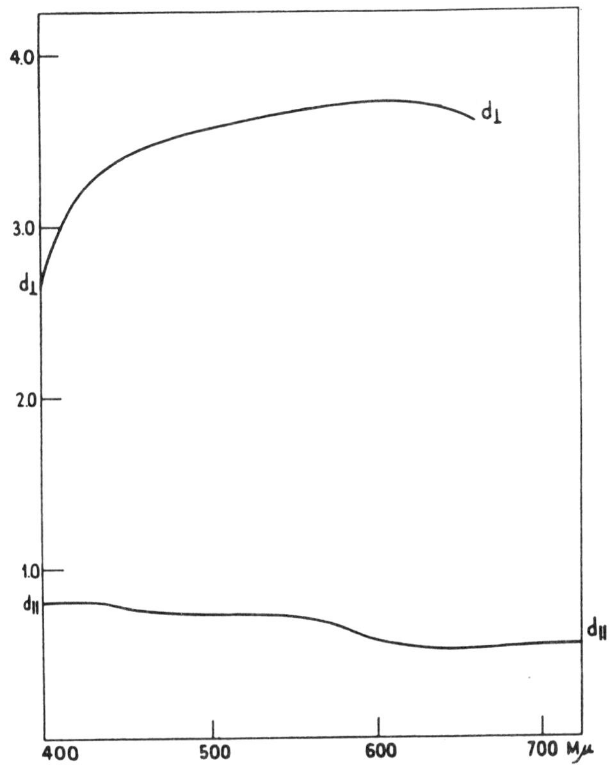

FIGURE 3. Dichroism of black tourmaline, section about 0.2 mm thick, cut 24° from parallelism with the optic axis, for light traveling perpendicular to the optic axis.

A mineral which, when colored, has a very strong uniaxial positive dichroism colorless-deep blue, but whose absorption constants have not been measured, is the rare silicate, benitoite, $BaTiSi_3O_9$ or $BaO{\cdot}TiO_2{\cdot}3\ SiO_2$. It is interesting to speculate whether the absorption here might be due to the presence of another transition metal, trivalent titanium, which is known in water solutions to have a narrow absorption band with a peak at 744 mμ.

Quinine Herapathite Polarizers. This quinine sulfate tri-iodide crystallizes in thin pinacoidal plates belonging to the orthorhombic system, which absorb one vibration in their plane and transmit the other two vibrations (positive dichroism). Its formula, according to Jörgensen (1876),[9] is $4C_{20}H_{24}N_2O_2 \cdot 3H_2SO_4 \cdot 2HI \cdot 2I_2 \cdot xH_2O$, and it has been observed in these laboratories to begin to melt at about 165°. The crystallography of this material has been described by West (1937).[10] The absorption constants $K_\parallel$ and $K_\perp$ of this crystal are not yet accurately known, since $d_\parallel$ and $d_\perp$ have not been measured for specimens of known thickness; however, a thickness of about 0.0002 inch (5 microns) is probably enough to give a good polarizer. Plates that are too thin have red extinction, while those that are too thick have red-brown transmission when viewed against a white light source. The absorption curves for this crystal are illustrated by the data for a Zeiss Herotar filter (Fig. 4).

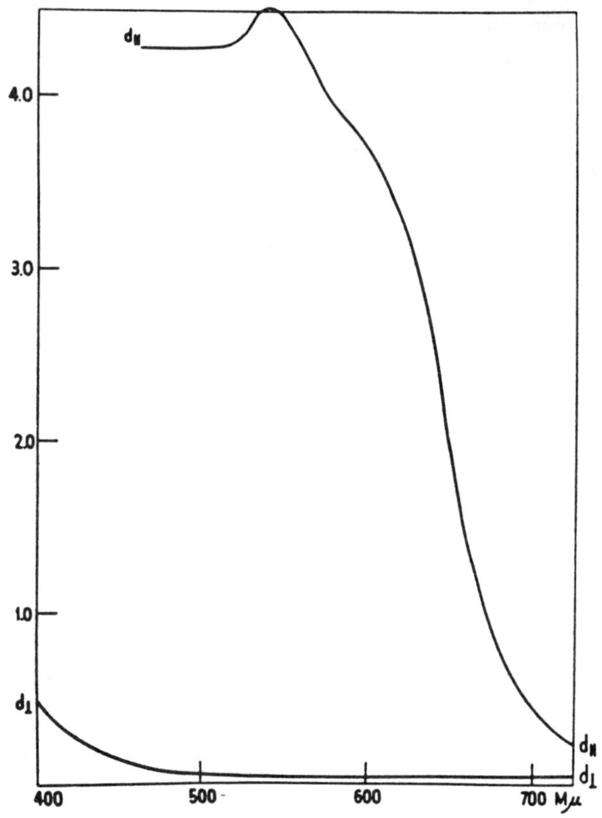

FIGURE 4. Dichroism of a Zeiss Herotar filter (quinine herapathite).

Cinchonidine Herapathite Polarizers. The crystal habit and optical properties of cinchonidine herapathite are similar to but not identical with those of quinine herapathite; it grows in orthorhombic pinacoidal plates, which begin to melt at about 154°. The formula given by Jorgensen (1876),[11] $12C_{20}H_{24}N_2O \cdot 9H_2SO_4 \cdot 8HI \cdot 12I_2 \cdot 8H_2O$, requires revision, since the formula now accepted for cinchonidine is $C_{19}H_{22}N_2O$. This crystal is superior to herapathite in that its positive dichroic absorption band persists into the near infrared, while that of herapathite cuts off in the visible far red; but as in the preceding example the absorption constants have not yet been published for this crystal. Presumably they will be of the same order as the constants of herapathite. Measurements on a Marks polarizing plate may be taken as representative of the dichroism of this crystal (Fig. 5).

Microcrystalline Polarizers. Polarizers of this type, invented by Land, are made by first preparing a suspension of dichroic crystal needles in a viscous medium, and then subjecting the suspension to a uniform flow process which will yield a sheet containing the crystal needles all oriented parallel to a common direction, namely the direction of the streamlines in the flow process. After the flow process the sheet is hardened, as by the escape of the volatile solvent used in preparing the viscous medium. Practically the process differs from the example of flow dichroism described in a previous section in respect to the higher concentrations, viscosities, and velocity gradients employed. Suitable crystal needles may have one of two main kinds of dichroism: they may absorb light vibrating parallel to the needle length, and transmit the two light vibrations perpendicular to this direction (positive dichroism), or they may transmit the parallel vibrations and absorb the two perpendicular vibrations (negative dichroism); a third possibility would be offered by a crystal needle which transmitted the parallel vibration and absorbed only one of the two perpendicular vibrations. In practice, only crystal needles of the first class, those characterized by positive dichroism, have so far been employed. If one starts with a strongly dichroic crystal species, the quality of the resulting polarizer depends chiefly on the perfection of the orientation attained, and this is conditioned by two factors: first, all the crystals in the suspension should have the correct shape and should fall in a certain size range; second, the flow conditions should be correctly chosen to set them all in parallelism. There are further obvious precautions to be observed in making these polarizers: the viscous medium should not have an appreciable solvent power for the crystal; temperatures must be avoided which melt, decompose or dissolve the crystals; and mechanical stresses must be avoided which break the crystals.

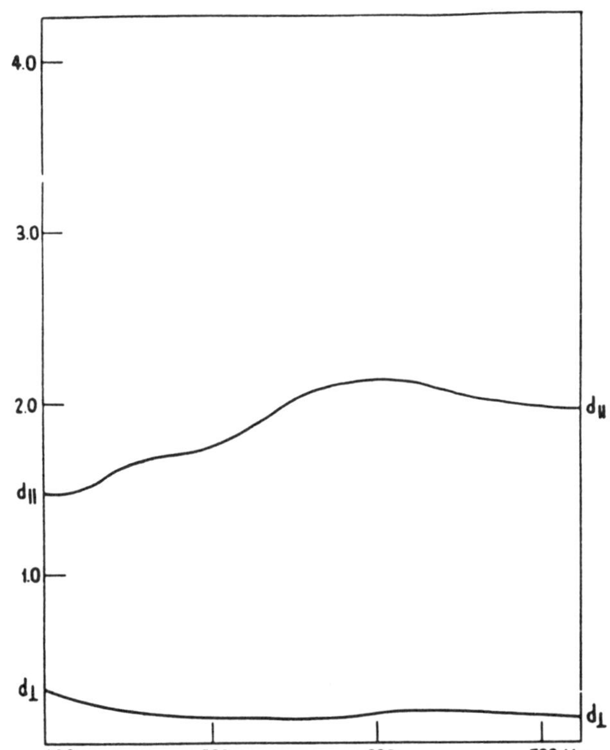

FIGURE 5. Dichroism of a Marks Polarizing Plate (cinchonidine herapathite).

West has observed that a section of Polaroid J-sheet cut across the needle axes is found to be not isotropic but weakly dichroic, which indicates a departure of its optical properties from strict uniaxiality. If we denote the film normal by X, the direction of the needle axes in the plane of the film by Z, and the transverse direction in the plane of the film by Y, then the absorptions are in the order $Z \gg Y > X$. Consistent with this is the observation that when a film is immersed in mineral oil and tilted about Y, two positions (corresponding to the optic axes of a biaxial dichroic crystal) are found where the dichroism vanishes. The angle between these two positions measured in the XZ plane over X might be termed the "monochroic angle." The simplest interpretation is that the long axes of the needles have a spatial distribution such that they are more nearly parallel to the plane of the film (YZ plane) than they are to the XZ plane.

Dichroic polarizers of this class are often characterized by a slight scattering of the transmitted light; this scattered light is far from having a uniform distribution, rather it reaches a maximum in what we have just called the XY plane. This subject was studied by Farwell (1938).[12] While this sometimes undesirable property would be minimized by having the refractive index of the medium as close as possible to the refractive index of the crystals, practically it is found easier to control it by keeping the largest crystals below a limiting size. It would be even more effectively eliminated, as Land has suggested, by using negative dichroic crystals if such were available.

The conclusion from experience so far is that microcrystalline polarizers are technically easier to produce than macrocrystalline polarizers, over any useful range of areas.

Useful dichroic polarizers have been made by this process from both quinine and cinchonidine herapathites. The preparation of quinine herapathite suspensions for this purpose was described in the Land patents and in a recent paper by Godina and Faerman (1941).[13] The commercial polarizer known as Polaroid J-sheet is of this type and consists of oriented quinine herapathite needles in a cellulose acetate film. It is of interest that overheating under suitable conditions will shift the dichroic absorption band of this film towards the blue. The effect of this shift is to change the extinction color of crossed films from purple to red, when viewed against a white light source. Absorption curves for a dichroic film made by the Polaroid J-sheet process before and after such heating are shown in the accompanying figure (Fig. 6).

Data are also given for a dichroic film prepared from microcrystalline cinchonidine herapathite (Fig. 7). A general similarity of Figs. 7 and 5 and likewise of Figs. 6 and 4, is to be observed.

A third microcrystalline dichroic polarizer was developed by Polaroid Corporation in 1942 from an acid periodide of 3, 4, 5, 6 dibenzacridin. The somewhat lower dichroism of this polarizer (see Fig. 8) as compared to the two preceding ones, is to some extent offset by the higher melting point of this crystal, which is in the neighborhood of 200°. The colored sorption compound of dibenzacridin with iodine was described in a paper by Kermack, Slater and Spragg (1930).[14]

Dichroic Polarizers Associated with Oriented Linear High Polymers. Linear high polymeric materials initially in an isotropic state are quite generally capable of being put into a uniformly oriented state by subjection to a relatively large uniform tensile strain or deformation under suitable conditions. The deformation is generally at least partly recoverable, likewise under suitable conditions. The process of orienting a linear high polymer may be formulated as a rather complex case of combined elastic and viscous deformation, the resistance to which is quite large; much larger, apparently than the viscous resistance encountered in the manufacture of the microcrystalline polarizing films treated in the preceding section.

Evidences of the oriented state of a linear high polymer are the anisotropy of mechanical properties (increase of modulus and tensile strength in the direction of the original tension); the anisotropy of optical properties (birefringence which cannot be explained in terms of internal stresses, uniaxial about the direction of the original tension); and frequently but not always the property of giving an x-ray fiber diffraction pattern, when the primary beam travels perpendicular to the direction of the original tension. Some natural fibers such as ramie (cellulose) and silk

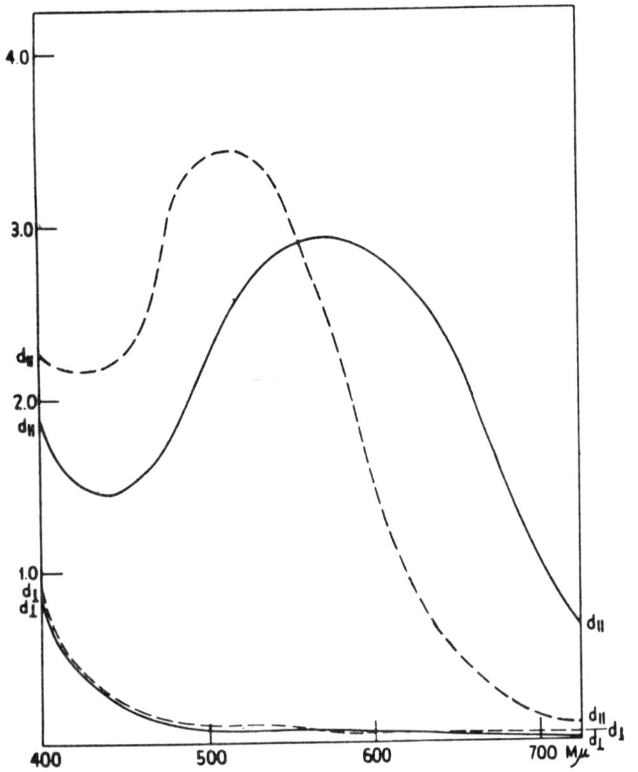

Figure 6. Dichroism of a quinine herapathite microcrystalline polarizer before (full lines) and after (broken lines) heat treatment.

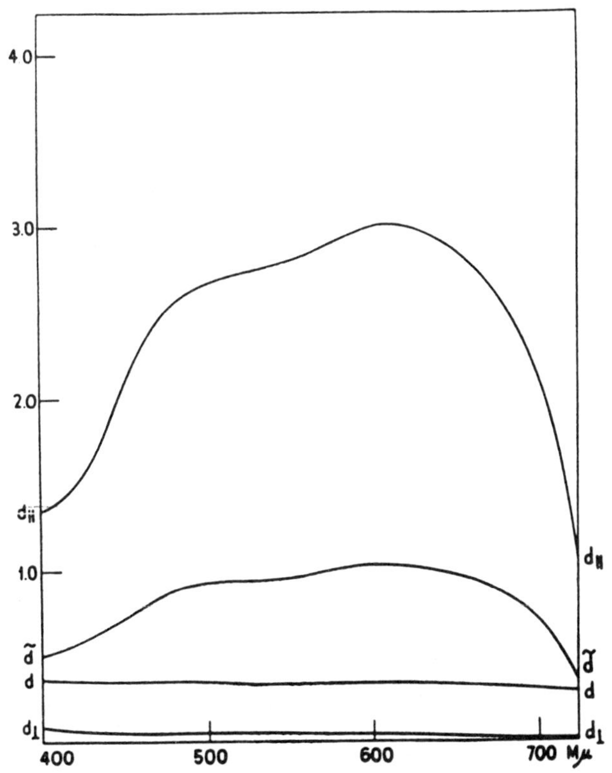

Figure 7. Dichroism of a cinchonidine herapathite microcrystalline polarizer. Curves of average density $\tilde{d}$ and density for natural light d are also shown.

represent quite perfectly oriented states of linear high polymers, but these are not very suitable for preparing dichroic polarizers.

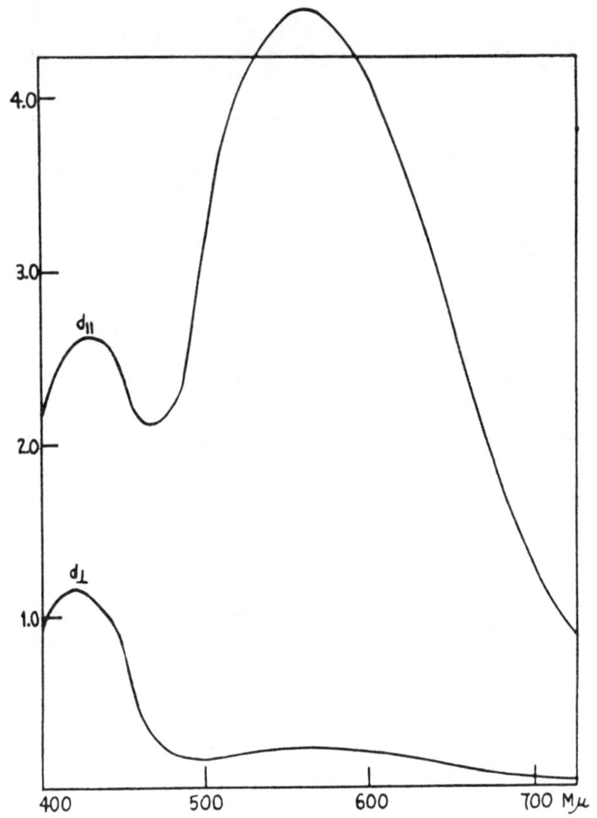

FIGURE 8. Dichroism of a dibenzacridine periodide polarizer.

A polymer which Land has found to be well suited for preparing the present class of dichroic polarizers is the synthetic product polyvinyl alcohol $(-CH_2-CHOH-)_x$, manufactured commercially by hydrolysis of polyvinyl acetate $(-CH_2-CH(OOCCH_3)-)_x$. This polymer, which has an extensibility of the same order as that of rubber, is formed into sheets which are oriented by uniform tensile deformations. The fully oriented sheet has a definitely fibrous structure and is characterized by high tensile strength in the fiber direction, of the order of 10^5 psi; by uniaxial positive birefringence ($\varepsilon = 1.560$, $\omega = 1.526$, $\varepsilon - \omega = 0.034$ are representative saturation values); and by a sharp x-ray fiber diffraction pattern. The atomic structure of the oriented polyvinyl alcohol fiber as worked out by x-ray methods by Mooney (1941)[15] has a plane zigzag of carbon atoms (*trans, trans,* . . .) with hydroxyls attached to alternate carbons (1, 3 diol groups) and all lying on the same side of the plane of the carbon chain. As is true of cellulose films, the transparency of polyvinyl alcohol films extends over considerable ranges into the ultraviolet and infrared. The weak ultraviolet absorption band reported for polyvinyl alcohol at 280 mμ by Marvel and Denoon (1938)[16] does not exist, according to our measurements.

As with the microcrystalline polarizers, all the useful dichroic polarizers so far made from oriented high polymers are characterized by positive dichroism.

Dye Polarizers. Preston (1933)[17] found that dichroic products resulted from the application of direct dyes to films or fibers of oriented cellulose, and published quantitative data on the dichroism; he also referred to earlier observations of this effect.

Other dichroic polarizers can also be made from oriented polyvinyl alcohol films, and these are found to have useful dichroic ratios.

An outstanding characteristic of such preparations is that the light absorption by each dye molecule (more strictly, ion) does not change very much, whether it is measured in solution, in the unoriented film, or in the oriented film. This observation suggests that the dye goes onto the oriented film in single molecules, and that the dichroism of the film generally may be taken as a direct measure of the dichroism of the individual dye molecules. In making this comparison, one measures the e of the solution in the customary manner for comparison with the average extinction coefficient $\tilde{e}$ of the dichroic film, as we have already done in the example of flow dichroism. This line of investigation is a relatively easy one, first because it is possible to cast the dye directly into the polyvinyl alcohol film by the aid of mutual solvents; secondly, because it is possible to calculate the e's of the dye in the film from the measured d's through the area concentration of the dye, taken as the product of the weight per cent of the dye in the film times the area density of the film: $c' =$ (g dye per g of dyed film) $\times$ (g dyed film per 1000 cm^2); then $\tilde{e} = d/c' = \frac{1}{3c'}(2d_{\perp} + d_{\parallel})$. In this way the labor of a chemical analysis becomes unnecessary.

Another indication that dichroic direct dyes on oriented high polymers are present as single ions, and not as polymeric forms, is afforded by the observation that x-ray diffraction photographs of such preparations show only amorphous scattering even when the dye is present in high concentrations. Our unpublished data confirm Valko (1941)[18] on this point.

Figs. 9 and 10 illustrate the constancy of the absorption by the Congo Red anion; curves are also given for dichroic preparations of both the red and the blue (acid) forms of this dye.

The curves for Congo Red are representative of the generality of direct dyes: the absorption bands of single dyes are too narrow to cover the entire visible spectrum satisfactorily at once. For this purpose one may combine two or more dyes in separate parallel films, or in the same oriented film (Fig. 11). Also it is seen that

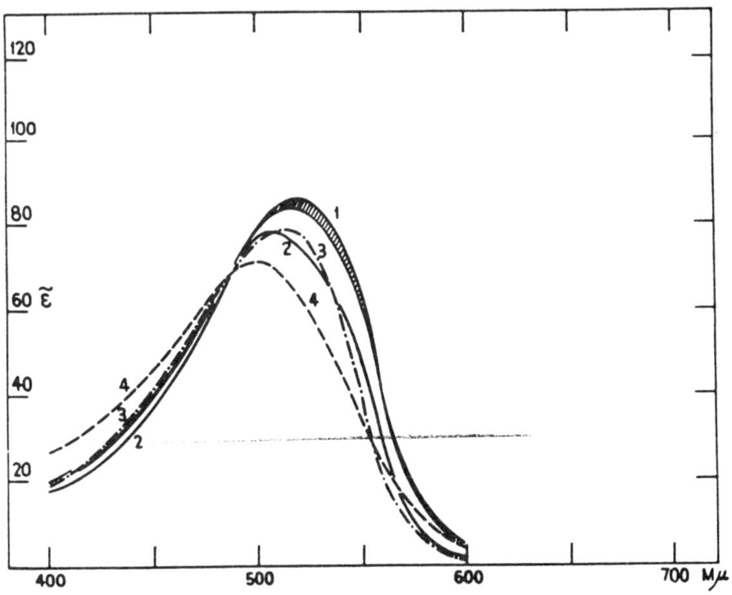

Figure 9. Absorption curves of Congo Red in varying environments. (1) Two dichroic preparations on polyvinyl alcohol, $c' = 0.00566$ and 0.01187 respectively. (2) Isotropic preparation on unoriented polyvinyl alcohol, $c' = 0.01155$. (3) Methyl "Cellosolve" solution. (4) Water solution. For (3) and (4), $c = 0.0174$, thickness = 1 cm.

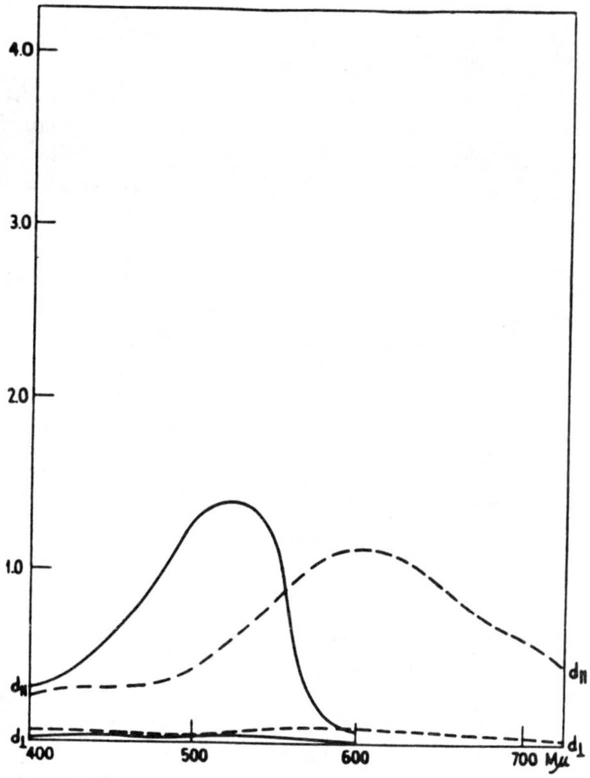

Figure 10. Dichroism of Congo Red on oriented polyvinyl alcohol, red (full lines) and blue (dotted lines) forms.

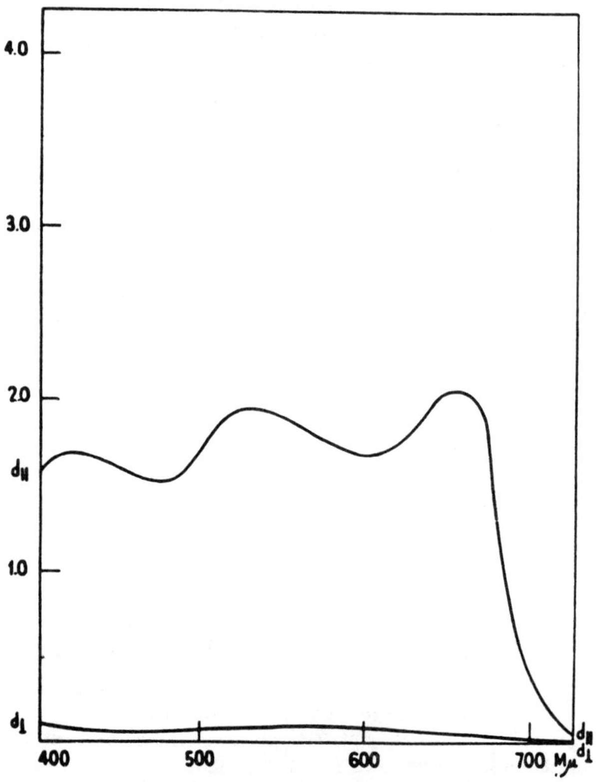

Figure 11. Dichroism of a mixture of dyes on oriented polyvinyl alcohol: yellow (Color Index No. 622), red (C. I. 278), and blue (C. I. 518).

quite striking effects are observed when two dichroic films, made with dyes whose absorption bands do not overlap, are crossed and examined with a dichroscope. Absorption curves of such a combination are shown in Fig. 12. When the two dyes in

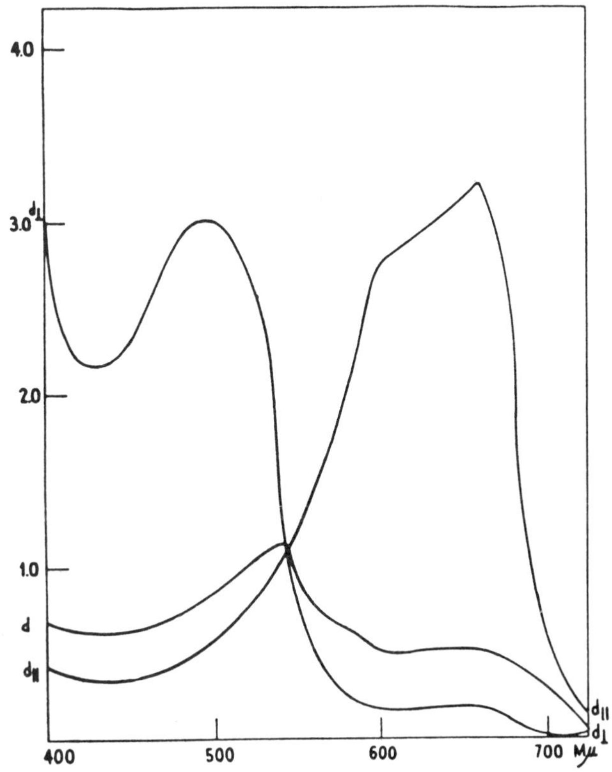

FIGURE 12. Dichroism of crossed dyed films of polyvinyl alcohol, orange (C. I. 374) and blue (C. I. 518).

a crossed combination of this kind are complementary, such combinations when taken in pairs have the interesting property of extinguishing white light when rotated to the crossed position.

It should not be inferred from the foregoing that only direct dyes give useful dichroism on oriented polyvinyl alcohol. The basic dye methylene blue, in which the dye ion is a cation, is an example of a dye with a dichroic ratio of about ten on the oriented polymer.

For dyes having disc-shaped molecules one would expect to find a maximum dichroic ratio not greater than 2, namely $\frac{d_\parallel}{d_\perp} = \frac{e_x}{\frac{1}{2}(e_x + e_z)}$, which equals 2 when $e_z \sim 0$ (z denotes the normal to the molecule plane, x a direction in the molecule plane). This is confirmed by observations of such dye cations and anions as crystal violet, aurin, and picric acid on oriented polyvinyl alcohol.

From the complete absence of dichroism of other dyes on oriented polyvinyl alcohol, one concludes that their structures are substantially spherically symmetrical. An example is the coördination compound of trivalent iron, Naphthol Green B.

From evidence of this sort one is justified in drawing conclusions concerning proposed structural formulas for dye molecules. The stilbene dye Color Index 620 has a dichroic ratio of about 15; therefore, the ring structure assigned to it seems highly improbable. Similarly, the observation that aniline black applied to oriented polyvinyl alcohol in a variety of ways always exhibits negligible dichroism apparently precludes the beautifully linear structures ascribed to this molecule in its various stages of oxidation.

The present method also has some value in helping to identify the several absorption maxima of a dye molecule with geometrical directions or axes in the molecule. Our experience has been that when a given molecule has positive dichroism for its strongest absorption maximum in the visible, then a second maximum will generally exhibit dichroism of the same sign. For example, when polyvinyl alcohol sheet containing ¼ per cent by weight of the much studied basic dye 1, 1′-diethyl-2, 2′-cyanine chloride is stretched and its dichroism is measured, twin maxima having the same dichroism (dichroic ratio about 5) and the same sign are found at 527 and 495 mμ; from which we conclude that both these maxima are identified with electronic transitions along the long direction of the planar molecule. Such a conclusion is at variance with the views of Scheibe, and of Lewis and Calvin, who ascribe the 495 maximum to an electronic transition transverse to the long direction of the molecule and in its plane.

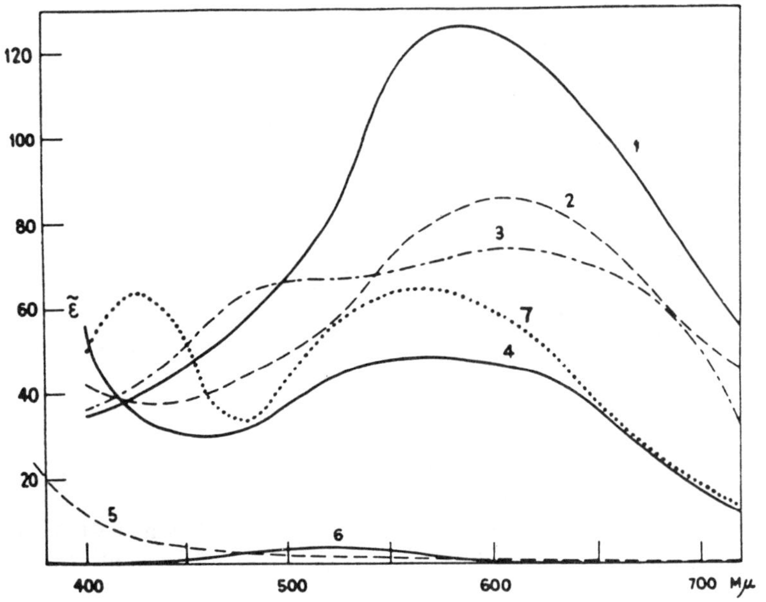

FIGURE 13. Absorption curves of iodine in varying environments. (1) Starch-iodine (2.54 mg I, 50 mg starch per 100 cc water; $c' = 0.0254$). (2) Iodine in preparation of Fig. 14, $c' = 0.012$ by titration. (3) Iodine in preparation of Fig. 7, $c' = 0.014$ calc. (4) Iodine in preparation of Fig. 6, $c' = 0.02115$ calc. (5) Iodine in aqueous KI, after Brode, *J. Am. Chem. Soc.*, **48**, 1877 (1926). (6) Iodine in hexane, after Groh, *Z. physik. Chem.*, **149**, 153 (1930). (7) Iodine in preparation of Fig. 8, $c' = 0.02625$ calc.

Iodine Polarizers. The ability of iodine to give dichroic dark colors on oriented cellulose has been known for some time. That a water solution of polyvinyl alcohol gives a blue coloration with iodine, when the concentrations are relatively high as compared for example with the starch-iodine system, was shown by Staudinger (1927)[19] and by Gallay (1936).[20] Land has prepared films of extraordinarily high dichroism from iodine and the oriented polymers of vinyl alcohol and vinyl butyral; and this reaction seems to be a fairly general property of oriented polyvinyl oxy-compounds, represented by the formula $(-CH_2-CHOR-)_x$, where OR is taken to indicate hydroxyl, acetal, ketal, ether or ester groups, etc., or mixed polymers thereof.

These colored products differ distinctly from the foregoing products made with dye molecules. For all simple iodine solutions, whether brown or violet, are characterized by relatively weak absorption in the visible spectrum, and especially in the red, whereas iodine in association with water solutions of suitable polymeric materials gives very intense absorption bands in the visible, as for example the well-known blue color with starch solution or the red color with dextrine solution. The state of the iodine in these water solutions is evidently similar to the state of the iodine in the dichroic preparations from oriented films of cellulose or polyvinyl alcohol, but must be radically different from the state of the iodine in the brown or violet solutions in simple solvents, as shown by an examination of the quantitative absorption curves (Fig. 13).

A vast amount of work has been done and published in the past with the object of elucidating the structure of this strongly absorbing state of iodine, which is called forth by the presence of suitable high polymers and water; but even the recent activity in this field has not given the final word.

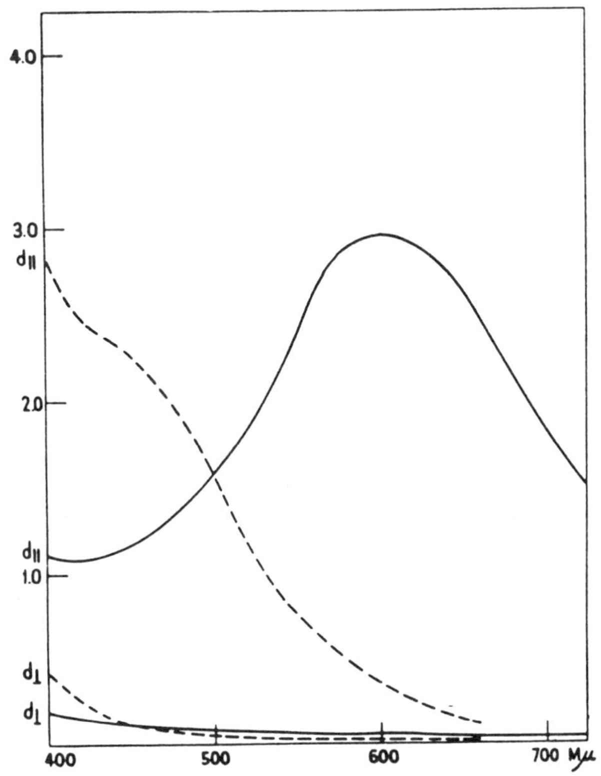

FIGURE 14. Dichroism of iodine on oriented polyvinyl alcohol, blue (full lines) and brown (broken lines) forms.

West has obtained definite evidence, not yet published, from the x-ray diffraction study of dichroic films of iodine on oriented polyvinyl alcohol that the iodine is present in a *polymeric* form, namely as independent long strings of iodine atoms all lying parallel to the fiber axis, with a periodicity in this direction of about 3.10Å. A slightly smaller interatomic distance is known for the tri-iodide ion in crystals.

The commercial polarizer known as Polaroid H-sheet is of this type.

When oriented polyvinyl alcohol film is stained with iodine in the absence of water, a dichroic brown product results. An x-ray photograph of one such preparation indicates the presence of polymeric iodine here also.

Absorption curves of brown and blue dichroic preparations of iodine on oriented polyvinyl alcohol are shown in Fig. 14.

Polyvinylene Polarizers. Land and Rogers found that when an oriented transparent film of polyvinyl alcohol is heated in the presence of an active dehydration catalyst such as HCl, the film darkens slightly and acquires a strong positive dirchroism. It is also observed that the darkened film, which has undergone only a slight decrease in weight of the order of 1 per cent in the process, can still be strongly swelled but no longer dissolved in hot water.

The chemistry of this change in light absorption obviously involves a partial dehydration of the transparent polyol $(-CH_2-CHOH-)_x$ to form an unsaturated compound or polyene $(-CH=CH-)_x$. The polyene must be pictured as a conjugated polyene, with the conjugation extending through long chains, in order to provide an acceptable explanation for the absorption of visible light. To account for the insolubility of the product, it seems quite probable that some dehydration takes place between chains; this would have the effect both of interrupting the conjugation and of forming a three-dimensional structure from the original linear polymer.

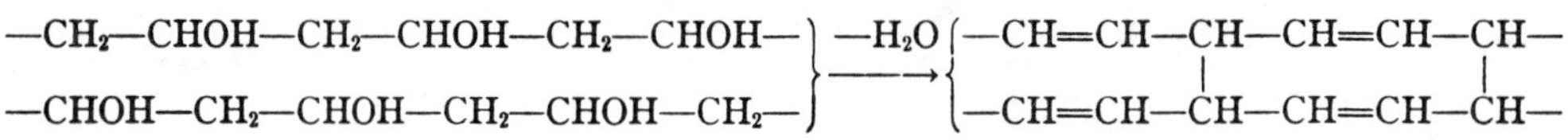

There might also be some oxygen bridge formation by elimination of water from hydroxyl pairs in adjacent chains.

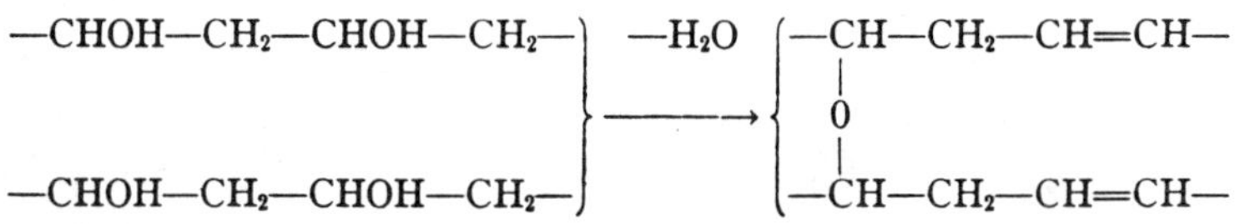

As required by this conjugated polyene structure, which was first proposed by West, the product is found to be bleached by common oxidizing agents with relative ease, but it is very stable indeed when oxidizing agents are excluded.

A quantitative determination of the number of double bonds in a photometered specimen of this polarizer would allow a comparison with the known absorption of light by the low molecular polyenes, a subject extensively studied by Kuhn and his associates (1935,[21] 1938[22]). Qualitatively the foregoing picture is consistent with both Kuhn's experimental work and Mulliken's theoretical treatment of this subject. We quote from a summary by the latter (1939):[23] "The longest wave length electronic transition of appreciable intensity in absorption (in a conjugated polyene $CH_2(=CH-CH=)_nCH_2$) is then $N \rightarrow V_1$. This shifts to longer wave lengths with increasing n. The calculations showed that if the polyene chain has the most elongated possible form (*trans, trans, trans, . . .*), the $N \rightarrow V_1$ transition should be far stronger than all the other $N \rightarrow V$ transitions together, and its strength should increase with the length of the chain. This calculated result is paralleled by experimental data on the polyenes, especially the carotinoids. The calculations showed that for less elongated (more *cis*-like) molecules, the intensity of $N \rightarrow V_1$ should be reduced in favor of $N \rightarrow V_2$, $N \rightarrow V_3$, and so on. A further point of interest was that the $N \rightarrow V_1$ transition in molecules of the most elongated form should be polarized approximately along the long axis of the molecules."

When oriented polyvinyl alcohol films are heated in the absence of active dehydration catalysts, dichroic yellow, orange or brown products result; the structure of these products has not been more closely investigated. The absorption curves for such a preparation are also shown in Fig. 15.

To give a better picture of the spatial relationships involved in the foregoing linear polymer polarizers, the approximate atomic dimensions of polyvinyl alcohol, of the Congo Red molecule and the iodine and polyvinylene polymers, are shown in the accompanying figures (Fig. 16).

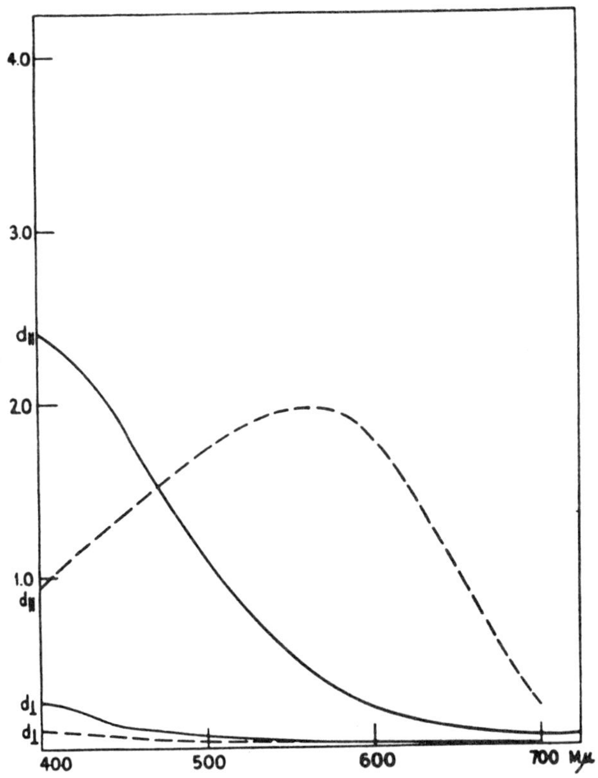

FIGURE 15. Dichroism of oriented polyvinylene (dotted lines) and of yellow product from polyvinyl alcohol (full lines).

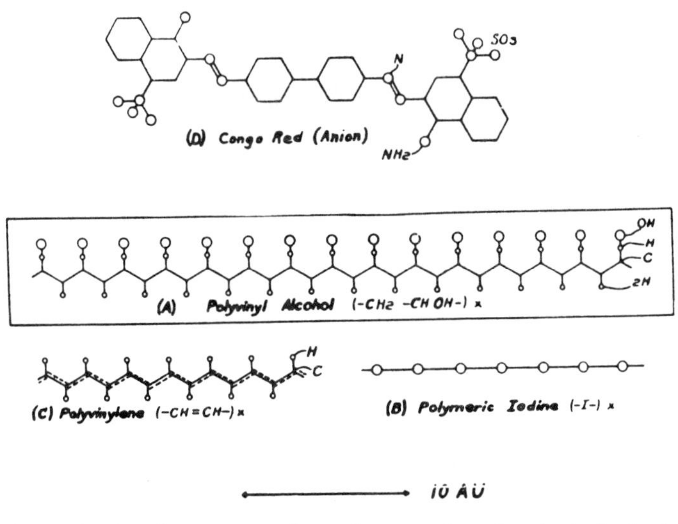

FIGURE 16. Schematic drawings, approximately to scale, of polyvinyl alcohol, Congo Red, polyiodine and polyvinylene molecules.

Metallic Polarizers. When metallic salts are incorporated in oriented linear high polymers and are subsequently reduced to the metals, the product is often characterized by a strong dichroism. The preparation and the structure investigation of such products has been described at some length by Frey-Wyssling (1937),[24] although the writers are not aware that quantitative measurements of dichroism have yet been published. For example, Frey-Wyssling finds for dichroic preparations made by depositing gold or mercury on oriented cellulose, that the gold is microcrystalline with its well known cubic structure and the mercury is liquid; the metallic particles are extremely small, widely separated rods oriented parallel to the cellulose fibers. Particle dimensions of the order of 10mμ diameter, 100mμ in length, are mentioned by Frey-Wyssling. Thus these particles are plainly much larger than the dichroic molecules treated earlier in this section. Furthermore, in contrast to all the polarizers so far treated, the oriented absorbing particles here are not of themselves dichroic, since every cubic crystal or liquid particle is of necessity an optically isotropic system.

An earlier review of this subject was given by Berkman, Boehm and Zocher (1926) [25]; these writers show how some of the optical properties of dichroic metal preparations may be predicted from O. Wiener's theory, but point out the failure, important from practical considerations, of this theory to relate the optical properties of a given preparation to the dimensions of the metallic particles. Figs. 17, 18, and 19

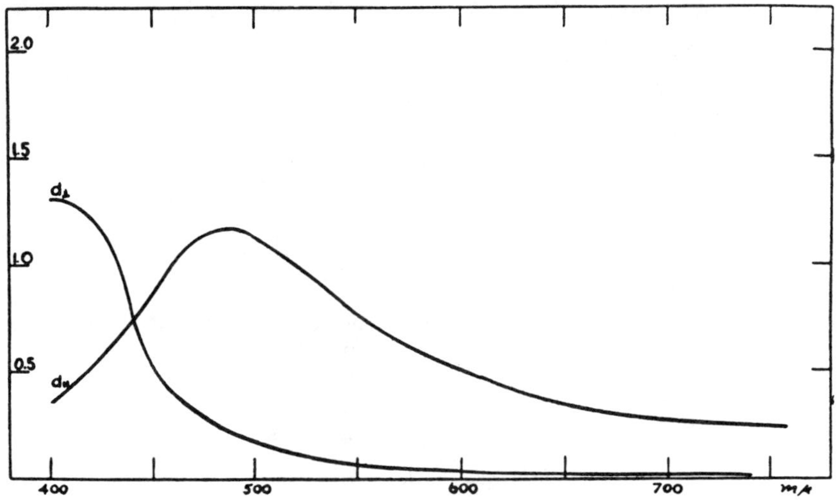

FIGURE 17. Dichroism of silver on oriented polyvinyl alcohol.

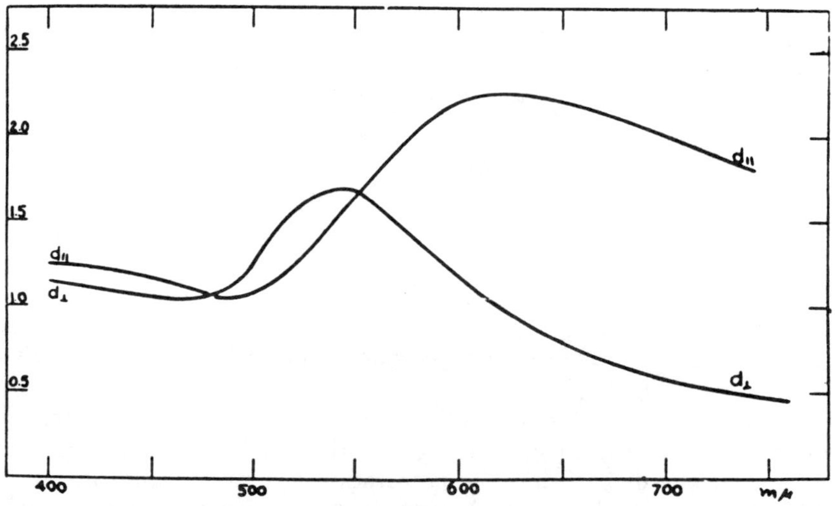

FIGURE 18. Dichroism of gold on oriented polyvinyl alcohol.

give data for dichroic preparations made from silver, mercury and gold on oriented polyvinyl alcohol. It is of interest that for the gold preparation, which has a red-green dichroism, the curves show a suggestive resemblance to some of Lange's (1928) [26] absorption curves for gold hydrosols of measured particle diameters; and namely the $d_{\parallel}$ curve resembles in shape Lange's curve for the 56mμ hydrosol, while the $d_{\perp}$ curve is close to that of the 7mμ hydrosol, even if their relative heights are out of order. It is seen that these dimensions are of the same order as those obtained by Frey-Wyssling by entirely different methods.

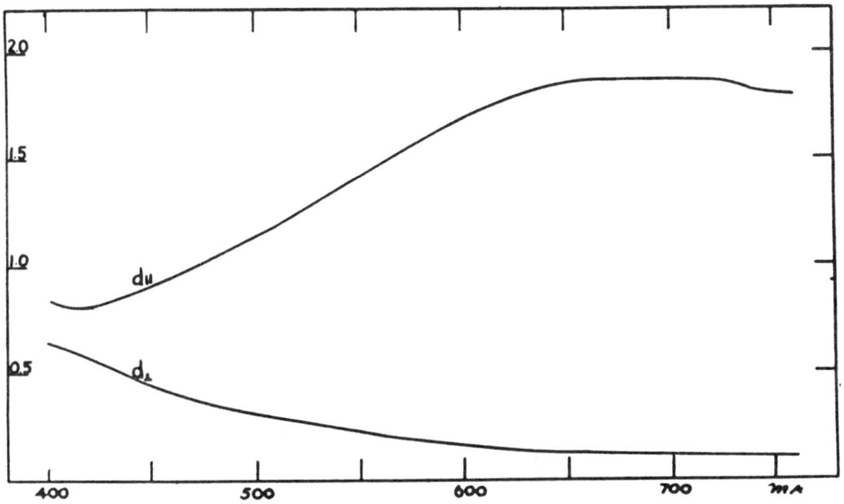

FIGURE 19. Dichroism of mercury on oriented polyvinyl alcohol.

Circular Polarizers. It appears doubtful that useful polarizers will ever be made from homogeneous circular dichroic materials, because the dichroism here is so far of such a low order of magnitude. The circular polarizers widely used in polariscopes designed for photoelastic investigations are customarily made by combining linear dichroic polarizers with quarter-wave birefringent films. However, if anyone ever discovers a useful circular dichroic polarizing material, this could obviously be combined with a quarter-wave film to make a useful linear polarizer also.

Summary. The useful artificial dichroic polarizers treated here, setting an arbitrary threshold of usefulness at a dichroic ratio of about 7:1, all have linear positive dichroism, and are made either by growing dichroic macrocrystals (herapathite), or by orienting dichroic microparticles. These particles may be dichroic microcrystalline needles (herapathite), dichroic simple molecules (dyes), or dichroic polymeric molecules (iodine, polyvinylene). The preferred methods of orienting the absorbing microparticles all involve some kind of a flow process in a viscous fluid medium.

Some Practical Applications of Dichroic Polarizers

Introduction

Since the applications of dichroic polarizers have in recent years been repeatedly written up in readily available sources, as shown by the following list of general papers, a brief recapitulation will suffice here.

W. E. Forsythe and E. Q. Adams, "Polarized Light." *Gen. Elec. Rev.*, **42**, 346-52 (1939).
H. Freundlich, "Polaroid Films." *Chemistry & Industry*, **56**, 698-9 (1937).
M. Grabau, "Polarized Light Enters the World of Everyday Life." *J. Applied Phys.*, **9**, 215-25 (1938).
—, "Introduction to Polarized Light and Its Applications." 46 p., Cambridge, Mass., Polaroid Corporation, 1940.

M. Haase, "New Polarization Filters Using Dichroic Crystals." *Z. tech. Physik,* **18,** 69-72 (1937).
H. A. Levey, "New Polarizing Media and Their Applications." *Ind. Eng. Chem., News Ed.,* **17,** 527-8 (1939).
E. Nähring, "Polarization Filters," *Phot. Ind.,* **38,** 599-601; 629-30 (1940).
A. Pollard, "Polarization of Light and Some Technical Applications." *Nature,* **138,** 311-14 (1936).
S. Rösch, "Some Properties and Applications of Dichroic Polarizers." *Z. Instrumentenk.,* **58,** 181-92 (1938).
L. W. Taylor, "Physics, the Pioneer Science." Chapter 38, pp. 549-75, Polarized Light. Boston, Houghton Mifflin Co., 1941.

For our purposes we may consider applications of dichroic polarizers for the man in the street or in the home, and applications for the man in the laboratory or in the lecture hall. In the former class, the numerous uses of dichroic polarizers are characterized throughout by novelty, even though some of the underlying ideas were proposed before the advent of practical dichroic polarizers, and by a practically exclusive concern with visual processes, either directly or at second hand through photography. In the latter class of applications, dichroic polarizers have widely replaced calcite polarizers for many laboratory uses. In some of the more expensive precision polarizing instruments, the instrument maker has been understandably slow in making the change. The period of active exploration in both classes of applications is by no means closed. Military applications, a third class which has now outstripped the other two, are not treated here.

Dichroic Polarizers in Applied Science

As Aids in the Visual Process. The widest application of dichroic polarizers so far has been in aiding the visual process. This end is achieved by cutting out, by means of one or more polarizers between the eye and the light source, some of the visual radiation, so that one may see more clearly or more comfortably with what radiation is left. This field of applied physical optics is primarily subject to physiological, and to some extent to psychological considerations.

There are three general ways of using polarizers to aid vision:

(1) A single polarizer is used over the eye, to cut out undesired light partially polarized by natural reflection and scattering.

(2) A single polarizer is used over the source for the same purpose as (1).

(3) The polarizers are used in pairs to cut out undesired light, and either superposed in space, or else separated in space, so that one of the pair goes with the eye, the other with the source.

Sun Glasses. By observation as with a dichroscope it is quickly found, first, that when sunlight is reflected or scattered from flat surfaces, it thereby generally becomes partially polarized with the vibrations normal to the plane of incidence predominating; secondly, that this component of the light is a hindrance rather than an aid in perceiving detail in or possibly under the flat surface. The degree of polarization varies widely with the angle of incidence, reaching a maximum for angles intermediate between normal and grazing incidence, and with the nature of the flat surface.

The function of the sun glasses is to absorb the component of the light vibrating normal to the plane of incidence, and to transmit only the component vibrating in the plane of incidence. Since flat surfaces are more often horizontal, for example the surface of a body of water, the pavement of a road, etc., polarizing sun glasses commonly have a fixed orientation so as to transmit only vertical vibrations.

It is of interest that when the sun is in the zenith on a clear day the skylight just above the horizon is partially polarized by scattering with the horizontal vibrations predominating, so that polarizing sun glasses "selectively absorb," so to speak, this horizon light as well as the ground light.

Reading Lamps. The light scattered or reflected from printed pages, by which we read, is also found to be partially polarized, and here too perception of detail under the optimum conditions is greatly aided when only vibrations in the plane of incidence are admitted to the eye. As would be expected, some improvement may be noted when one reads with sun glasses under high levels of illumination. It is generally more efficient and more convenient to employ a polarized source of light, that is, a lamp provided with a polarizer and designed so that its radiation is incident on the printed page at an angle to give a maximum effect. The improvement is especially noticeable with material printed on glossy paper.

Variable Density Filters. Under some conditions it is desired to attain a wider range of illumination control than is possible by absorbing a single component of the light. There is then the possibility of using dichroic polarizers in pairs, and either superposed, or separated in space.

Large polarizers have been used superposed in pairs to provide variable daylight illumination through windows, smaller polarizers similarly in variable-density sun glasses. The illumination in both cases is varied by rotating one polarizer of a pair with respect to the other.

When the problem arises of completely eliminating the mirror reflection from any flat surface, regardless of the angle of incidence of the light and the nature of the mirror, the solution is found in polarizing the light source in or normal to the plane of incidence, and in viewing the reflected ray with an analyser crossed with the polarizer. Then the flat surface can be seen only by that light, if any, which is depolarized as by scattering or by defects in the mirror. By suitable rotations of one polarizer with respect to the other more or less of the mirror component of the reflected light may be transmitted to the eye.

Automobile Headlights. Here the polarizers are to be used in crossed pairs to eliminate dazzle, with the components of the crossed pairs necessarily separated in space. The polarizers have a fixed orientation so that the analyzer in front of a driver's eyes is always automatically parallel to the polarizers over the headlights on his car and on other cars driving in his direction, while it is always crossed with the polarizers over the headlights of all approaching cars. This is effected by orienting the vibration directions of all polarizers at a uniform angle of 45° from the horizontal, always measured in the same sense.

Of all the applications of dichroic polarizers this probably has been the one most thoroughly discussed in the patent and technical literature.[27-34]

From the manufacturing point of view, the polarized headlight that fills the rather exacting requirements of this application is best made by starting with a headlamp of the sealed beam type and applying to the outer surface of its lens a thin layer containing dichroic polyvinylene.

Photographic Images. Many of the principles which apply in improving visual images through the use of polarizing filters can be utilized equally well in photography. Here, however, the situation is somewhat simpler in that ordinary cameras have only one objective, and filter holders which can be set at any desired azimuth are readily available for most cameras. In using a polarizing filter one must take into consideration the curve of its transmission for natural light as a function of wave length, since the sensitivity curve of the photographic layer may be quite different from the sensitivity curve of the eye.

There is already a considerable literature on the photographic uses of polarizing filters. Filters over the objective are employed to cut out vibrations perpendicular to the plane of incidence when there is undesired illumination from reflecting surfaces,—horizontal mirrors such as bodies of water, vertical mirrors such as window panes, etc. Such filters too are advantageously employed to darken a background that is illuminated by light partially linear-polarized by scattering, such as the skylight 90° from the sun on a clear day; their use in underwater photography has been proposed on the same principle.

In photography too there is the already mentioned possibility of using polarizers in pairs—together as with both components over the objective or over the source, or more commonly separated as with one polarizer over the source and one over the objective, to vary the illumination over wider ranges than is possible with a single rotatable polarizer. Polarizers have been used in pairs over the objectives of motion picture cameras for taking fades and lap dissolves.

As Aids in Viewing Stereoscopic Pairs of Images. Polarizing devices have already been widely used for viewing stereoscopic pairs of images, either directly or after projection. Three different systems may be cited here to illustrate some of the possibilities.

In all these systems, the observer employs polarizing spectacles to admit the one image of the stereo pair to the eye to which it belongs.

Projection of juxtaposed stereo images through twin projectors.[35, 36] This system is the exact analog of the anaglyph system, except that polarizing filters are employed in place of complementary (*e.g.*, red and green) color filters.

Projection or viewing of vectograph stereo images.[37, 38] In this strikingly novel application, due to Land and Mahler, each image of the stereo pair is formed directly as a positive image in a polarizing layer; for the high lights, $d_{\parallel} = d_{\perp} = 0$, while for the shadows $d_{\perp} = 0$, $d_{\parallel} > 0$. For viewing or projection, the two images are superposed with their axes crossed; thus only a single ordinary projector is required in contrast to current practice for the preceding case. The vectograph may also be viewed directly, and either as a diapositive, or by backing with a suitable non-depolarizing diffusing layer. An aluminum lacquer has been used for this purpose as well as for preparing projection screens for reflecting polarized images without depolarization.

In addition to their uses in stereoscopic viewing or projection, vectographic images have other novel applications.

A third system,[39] also due to Land, has been used for stereoscopic viewing of the large diapositives encountered in x-ray radiographic work.

Display Devices. The full range of interference colors obtainable with doubly refracting materials of the normal dispersion of birefringence, when viewed between polarizers against a standard white light source, has been analytically determined, tabulated and plotted in a valuable paper by Buchwald (1940).[40] This writer starts with the I.C.I. (International Commission on Illumination) data, using the illuminant B and gypsum as bases for calculation, and ends up with the first five orders of interference colors (gypsum thicknesses 0-0.30 mm) tabulated according to the color triangle coördinates x, y, and the dominant wave length λ and saturation σ derivable therefrom. The color curves for the first three orders of such a gypsum wedge between crossed and parallel polarizers respectively are shown as spirals around the white point in Figs. 20 and 21; the colors of the fourth and fifth orders between crossed polarizers are shown in Fig. 22; and the brightness, Hx, for the first five orders between crossed polarizers is shown in Fig. 23 (the brightness for parallel polarizers is $H\| = 100 - Hx$). These data for $(x, y, H)_x$ and $(x, y, H)_{\|}$ could advantageously be presented in three-dimensional figures. Fig. 24 shows the transition of the color in a 0.10-mm section of gypsum to its complementary as the analyzer is rotated from the crossed position ($\beta = 90°$) to the parallel position ($\beta = 0°$).

Buchwald's results would be of scientific interest only, were it not for the fact that some common transparent sheet materials such as cellophane are characterized by birefringence and normal dispersion thereof. This means that any point on Buchwald's color curves can be reproduced by selecting a suitable piece of birefringent film and viewing it between dichroic polarizers. Still other colors are obtainable by preparing transparent films having abnormal dispersion of birefringence.

An application of such polarization colors, either alone or in combination with ordinary isotropic colors or the dichroic colors described earlier, has been made in various display devices. Common to many of these is the play of colors resulting from continuous rotation of one or more members of the optical system, ordinarily one of the polarizers.

Dichroic Polarizers in Scientific Research

The papers of the following two lists have in common that they all include some mention of dichroic polarizers. Somewhat in the manner of *Science Abstracts,* the

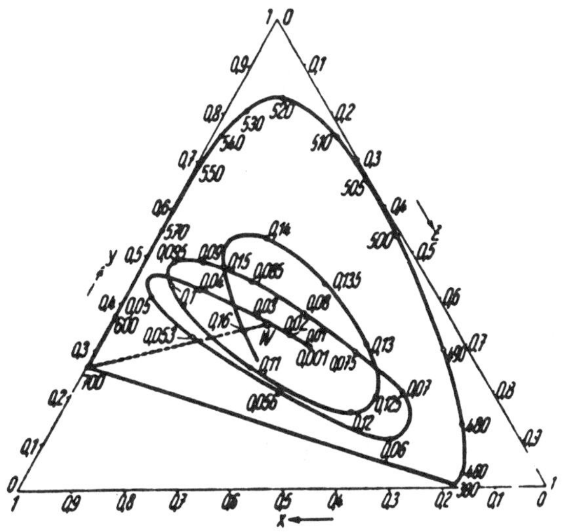

FIGURE 20. Metric of normal linear-birefringent interference colors, reproduced from E. Buchwald. Gypsum sections 0 to 0.30 mm thick, I.C.I. illuminant B. Crossed polarizers, orders I-III.

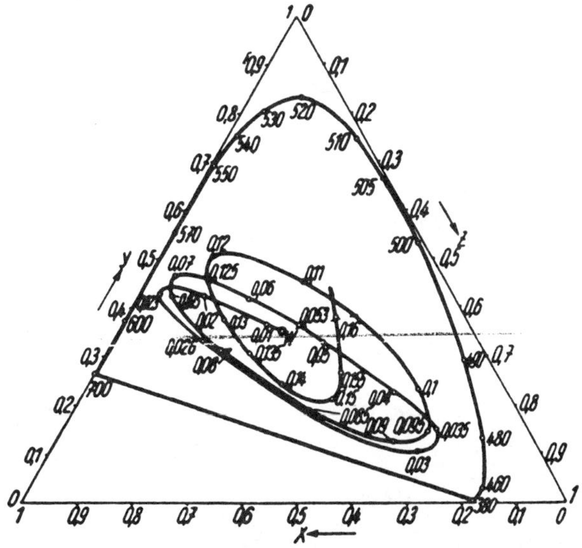

FIGURE 21. Same as Figure 20, with parallel polarizers, orders I-III.

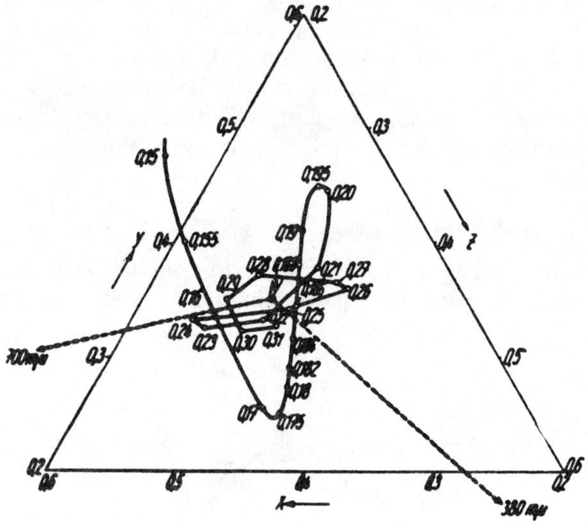

FIGURE 22. Same as Figure 20, with crossed polarizers, orders IV and V.

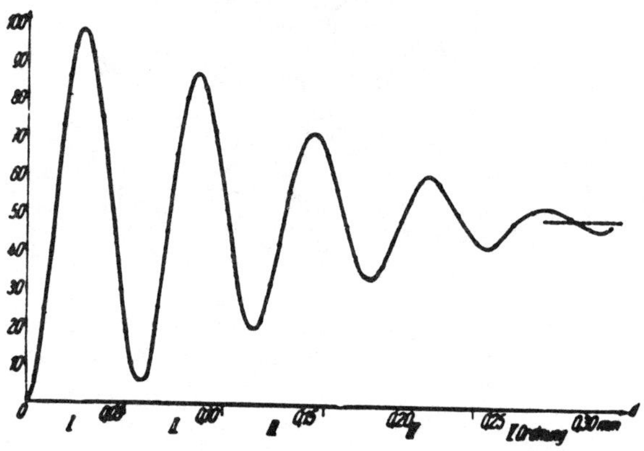

FIGURE 23. Same as Figure 20, with crossed polarizers, brightness for orders I-V.

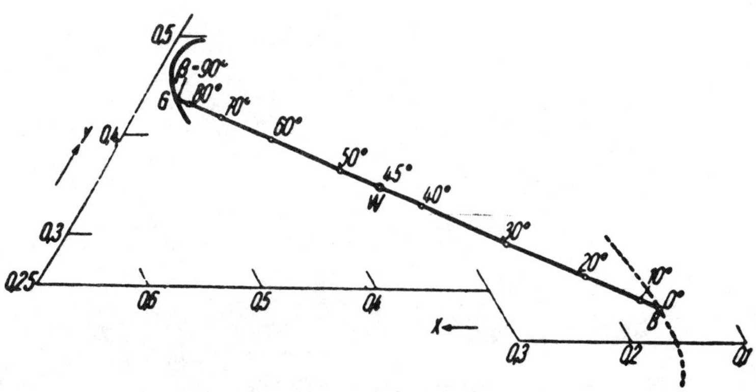

FIGURE 24. Same as Figure 20. Transition of interference color of 0.10 mm section to its complementary as analyzer is rotated from crossed ($\beta = 90$) to parallel position ($\beta = 0$).

first is by subjects and the second deals with apparatus and instruments. While these lists are not offered as exhaustive, it is hoped that they will be fairly representative of the uses to which dichroic polarizers have been put in the laboratory up to January 1943.

For colloid chemists, there are several fields of direct interest in which polarization effects are observed and measured:

(1) The study of the size and shape of colloid particles from the state of polarization of the light scattered thereby. S. Bhagavantam (1940) has treated this subject in his book "Scattering of Light and the Raman Effect."

(2) The study of certain fluid systems by the optical anisotropy of flow. This term is used broadly to include both birefringence and dichroism, and to include matter in states of both actual flow and "frozen" flow. A valuable review of streaming birefringence has been written by J. T. Edsall (1942), as cited below.

(3) The study of very thin films by methods involving the reflection of polarized light.

List of Subjects

Mechanics:

Photoelasticity of solids under steady stress

Frocht, M. M., "Photoelasticity," Vol. I. New York, Wiley, 1941.

Weller, R., "Three-Dimensional Photoelasticity Using Scattered Light," *J. Applied Phys.*, **12**, 610 (1941).

Photoelasticity of vibrating systems

Ferry, J. D., "Photoelastic Method for Transverse Vibrations in Gels," *Rev. Sci. Instruments*, **12**, 79 (1941).

—, "Rigidities of System Polystyrene-Xylene and Their Dependence on Temperature and Frequency," *J. Am. Chem. Soc.*, **64**, 1323 (1942).

Photoviscosity of fluids in steady flow

Astronomical optics:

Allen, C. W., "Polarization of Solar Corona," *Monthly Notices Roy. Astron. Soc.*, **101**, 281 (1941).

Meteorological optics:

Byram, G. M., "Polarization of Atmospheric Haze," *Science*, **94**, 192 (1941).

Physiological optics:

Bannon, R. E., "Technique of Measuring and Correcting Aniseikonia," *Am. J. Optometry*, **18**, 145 (1941).

Gehrcke, E., "New Phenomena in Physiological Optics," *Physik. Z.*, **41**, 540 (1940).

Land, E. H., and Hunt, W. A., "Use of Polarized Light in Simultaneous Comparison of Retinally and Cortically Fused Colors," *Science*, **83**, 309 (1936).

Physical optics:

Optical anisotropy in various systems

Biological materials

Tartar, V., and Richards, O. W., "Use of Polarized Light in Biological Science," *Collecting Net*, **12**, 4 p. (1937)

Crystals, etc.

Burbage, E. J., and Anderson, B. W., "Analysis of Movements of Shadow-Edges on the Refractometer in the Case of Biaxial Gemstones," *Mineralog. Mag.*, **26**, 246 (1942).

West, C. D., "Optical Properties and Polymorphism of Paraffins," *J. Am. Chem. Soc.*, **59**, 742 (1937).

Fluids in flow

Edsall, J. T., and Mehl, J. W., "Effect of Denaturing Agents on Myosin. II. Viscosity and Double Refraction of Flow," *J. Biol. Chem.*, **133**, 409 (1940).

Farwell, H. W., "Double Refraction in Colloidal Systems," *J. Applied Phys.*, **8**, 416 (1937).

Hauser, E. R., and Dewey, D. R., "Visual Studies in Flow Patterns," *J. Phys. Chem.*, **46**, 212 (1942).

Kraemer, E. O., ed., "Advances in Colloid Science." Vol. 1, p. 269. Article by Edsall, J. T.: "Streaming Birefringence and Its Relation to Particle Size and Shape," New York: Interscience Publishers, Inc., 1942.

Langmuir, I., "Role of Attractive and Repulsive Forces in Formation of Tactoids, Thixotropic Gels, Protein Crystals and Coacervates," *J. Chem. Phys.*, **6**, 873 (1938).

Lauffer, M. A., "Optical Properties of Solutions of Tobacco Mosaic Virus," *J. Phys. Chem.*, **42**, 935 (1938).

Lauffer, M. A., and Stanley, W. M., "Stream Double Refraction of Virus Proteins," *J. Biol. Chem.*, **123**, 507 (1938).

Stanley, W. M., "Biophysics and Biochemistry of Viruses," *J. Applied Phys.*, **9**, 148 (1938).

Glass

Edgerton, H. E., and Barstow, "Further Studies of Glass Fracture with High Speed Photography," *J. Am. Ceram. Soc.*, **24**, 131 (1941).

Schardin, H., Elle, D., and Stroth, W., "Velocity of the Fracture Process in Glass and Artificial Resins," *Z. tech. Physik*, **21**, 393 (1940).

Plastics

Farwell, H. W., "Double Refraction and Change of Length of Certain Plastics," *J. Applied Phys.*, **10**, 109 (1939).

—, "Tension and Birefringence in a Vinylite Plastic," *J. Applied Phys.*, **11**, 271 (1940).

Middlehurst, D., and Weller, R., "Polarizing Properties of Cellophane," *Rev. Sci. Instruments*, **11**, 108 (1940).

Spence, J., "Optical Anisotropy and the Structure of Cellulosic Sheet Materials," *J. Phys. Chem.*, **43**, 865 (1939).

Electro-optics

Mueller, H., "Electro-optical Field Mapping," *J. Optical Soc. Am.*, **31**, 286 (1941).

Mueller, H., and Sakmann, B. W., "Electro-optical Properties of Colloids," *J. Optical Soc. Am.*, **32**, 309 (1942).

Reflection

Laurence, J., "Reflection Characteristics with Polarized Light," *J. Optical Soc. Am.*, **31**, 9 (1941).

Scattering, Raman

Bernstein, H. J., and Martin, W. H., "Depolarization of Raman Lines and Structure of Complex Ions," *Trans. Roy. Soc. Can., Sect. III*, **32**, 43 (1938).

Nedungadi, T. M. K., "Effect of Crystal Orientation on Raman Spectrum of Sodium Nitrate," *Proc. Indian Acad. Sci.*, **10**, 197 (1939).

Scattering, Rayleigh or Tyndall

Gehman, S. D., and Field, J. E., "Colloidal Structure of Rubber in Solution," *Ind. Eng. Chem.*, **29**, 793 (1937).

Hoover, C. R., Putnam, F. W., and Wittenberg, E. G., "Depolarization of Tyndall Light of Bentonite and Iron Oxide Sols," *J. Phys. Chem.*, **46**, 81 (1942).

Thin Films

Bishop, F. L., Jr., "Surface Layers of Sheet Glass," *Phys. Rev.*, **62**, 295 (Abstract) (1942).

Blodgett, K. B., and Langmuir, I., "Built-up Films of Barium Stearate and Their Optical Properties," *Phys. Rev.*, **51**, 964 (1937).

Pfund, A. H., "Haidinger Interferences in Silvered Mica," *J. Optical Soc. Am.*, **32**, 383 (1942).

Photochemistry:

Rama Char, T. L., "Photochemical Stability of Mixtures of Vanadic and Tartaric Acids. II . . . Induced Optical Activity by Circularly Polarized Light," *J. Indian Chem. Soc.*, **18**, 563 (1941).

List of Apparatus and Instruments

Color filters

Lyot, B., "A Monochromatic Filter Especially Practical for Investigations of the Sun," *Compt. rend.*, **212**, 1013-17, (1941).

Öhman, Y., "A New Monochromator," *Nature*, **141**, 157 (1938).

Pettit, E., "The Interference Polarizing Monochromator." Pub. of the Astronomical Soc. of the Pacific, **53**, 313 17-27 (1941).

Compensators and retardation plates

Piecewicz, C. T., "Plastic Wedge Compensator," *Rev. Sci. Instruments,* **13,** 296 (1942).

Tuzi, Z. and Oosima, H., "Artificial Quarter Wave Plate and Its Theory," *Sci. Papers Inst. Phys. Chem. Research (Tokyo),* **905,** 72 (1939).

West, C. D., "Substitute for the Quartz Wedge Used with the Polarizing Microscope," *Am. Mineral.,* **23,** 531 (1938).

Dichroscopes, divided field devices

Pfund, A. H., "Polaroid Half Shade Analyzer," *J. Optical Soc. Am.,* **26,** 453 (1936).

Thibault, N. W., "A Simple Dichroscope," *Am. Mineral.,* **25,** 88 (1940).

Ellipticity and retardation, measurement of

Ingersoll, L. R., "Improved Spectroradiometric Method and Apparatus for Measuring Small Polarization Ellipticities," *J. Optical Soc. Am.,* **27,** 411 (1937).

Kent, C. V., and Lawson, J., "Photoelectric Method for the Determination of Elliptically Polarized Light," *J. Optical Soc. Am.,* **27,** 116 (1937).

v. Look, Th., "New Method for Measuring Retardations; the Temperature Dependence of Birefringence of Quartz, Beryl and Adular," *Z. angew. Mineral.,* **1,** 229 (1938).

Microscopes

Plato, W., "Determination of the Optical Crystallographic Properties of Polished Crystalline Stones," *Z. wiss. Mikroskop.,* **57,** 137 (1940).

West, C. D., "Polarizing Accessories for Microscopes," *J. Chem. Education,* **19,** 66 (1942).

Optometry and ophthalmology, instruments for

American Optical Company, "Eikonometer," *Rev. Sci. Instruments,* **12,** 282 (1941).

Bausch and Lomb Company, "Ortho-Fusor," *Am. J. Optometry,* **19,** 39 (1942).

Photometers, densitometers, spectrophotometers, etc.

Adam Hilger, Ltd., "Photoelectric Color Comparator (for Reflected Light)," *Rev. Sci. Instruments,* **12,** 337 (1941).

Moon, P., and Laurence, J., "Construction and Test of a Goniophotometer," *J. Optical Soc. Am.,* **31,** 130 (1941).

Öhman, Y., "New Method of Standardization in (Photographic) Spectrophotometry," *Stockholms Obs. Ann.,* **13,** 6 16 p. (1940).

Sweet, M. H., "Physical Density Comparator," *J. Optical Soc. Am.,* **28,** 349 (1938).

Polarimeters and Saccharimeters

Estey, R. S., "The Use of the Spencer Polarimeter," 32 p. Buffalo, Spencer Lens Company, 1941.

Karush, F., "Photoelectric Polarimeter," 16 p. Thesis, Univ. of Chicago, 1941.

Marion, A., "Adapting Polarizing Microscopes for Use as Polarimeters," *Ind. Eng. Chem., Anal. Ed.,* **12,** 777 (1940).

Spengler, O., and Hirschmüller, H., "New Photoelectric Saccharimeter," *Z. Wirtschaftsgruppe Zuckerind.,* **90,** 26 (1940).

Polariscopes, for crystals, etc.

Arshinov, V. V., "Pocket Mineralogical or Polarizing Magnifier," *Am. Mineral.,* **23,** 287 (1938).

Buckley, H. E., "Demonstration of Optical Interference Figures," *Nature,* **143,** 801 (1939).

Quirke, T. T., "Direct Projection of Optical Figures," *Am. Mineral.,* **23,** 595 (1938).

Rösch, S., "Polariscope for Physical Chemistry, Mineralogy and Examination of Gems," *Fortschr. Mineral. Krist. Petrog.,* **22,** li-lii (1937).

Schaub, B. M., "On the Use of Polaroid for Photographing Large Thin Sections in Crossed Polarized Light," *Am. Mineral.,* **21,** 384 (1936).

Zuloaga, G., "Photomicrography with an Enlarger," *Phototechnique,* p. 53 (Nov. 1941).

Polariscopes, for glass products

Douglas, R. W., and Breadner, B. L., "Strain Viewer for Glass Articles," *J. Sci. Instruments,* **17,** 187 (1940).

Hull, A. W., and Burger, E. E., "Simple Strain Tester for Glass Seals," *Rev. Sci. Instruments,* **7,** 98 (1936).

Swicker, V. C., "The Polariscope as a Glass Factory Instrument," *Glass Ind.,* **23,** 13 (1942).

Polariscopes, photoelastic

Föppl, L., and Müller-Lufft, E., "Stress-Optical Apparatus with Polarization Filters," *Arch. tech. Messen,* **100,** T125-6 (1939).

Frocht, M. M., "Photoelasticity," Vol. I, p. 382-9. New York, Wiley, 1941.

Tuzi, Z., and Nisida, M., "New Apparatus for Photoelastic Stress Type. Quarter Wave Plate of Large Field," *Sci. Papers Inst. Phys. Chem. Research (Tokyo)*, **31,** 676, 99 (1937).

Refractometers

West, C. D., "Measurement of Refractive Indices of Resins and Plastics," *Ind. Eng. Chem., Anal. Ed.,* **10,** 627 (1938).

Sextants, telescopes, etc.

Hardy, A. C., "Automatic Telescope Control," *J. Optical Soc. Am.,* **30,** 654 (Abstract) (1940).

Hulburt, E. O., "Sextant with Improved Filters," *J. Optical Soc. Am.,* **26,** 216 (1936). **26,** 216 (1936).

Spectrographs for detecting polarization of spectral lines

Cleveland, F., and Murray, M. J., "Depolarization Measurements on Raman Lines by an Easy Accurate Method," *J. Chem. Phys.,* **7,** 396 (1939).

Edsall, J. T., and Wilson, E. B., Jr., "Simple Method of Determining Polarization of Raman Lines," *J. Chem. Phys.,* **6,** 124 (1938).

Glockler, G., and Baker, H. T., "Polarization Measurements of Raman Lines," *J. Chem. Phys.,* **10,** 404 (1942).

Spectrometers, goniometers, etc.

Anderson, B. W., "Refractive Index Measurement by Brewster's Angle," *Gemnologist (London)*, **10,** 61 (1941).

Pfund, A. H., "Brewsterian Angle and Refractive Indices," *J. Optical Soc. Am.,* **31,** 679 (1941).

References

1. Zimmern, A., and Coutin, M., *Compt. rend.,* **182,** 1214 (1926).
2. Bernauer, F., *Fortschr. Mineral. Krist. Petrog.,* **19,** 22 (1935).
3. Strong, J., *J. Optical Soc. Am.,* **26,** 256 (1936).
4. Grabau, M., *J. Optical Soc. Am.,* **27,** 420 (1937).
5. Pulfrich, C., *Z. Krist. Mineral. Petrog.,* **6,** 142 (1882).
6. Preston, J., *J. Soc. Dyers Colourists,* **47,** 312 (1931).
7. Zocher, H., and Jacoby, F., *Kolloidchem. Beihefts,* **24,** 365 (1927).
8. Barger, G., and Starling, *J. Chem. Soc.,* **107,** 411 (1915).
9. Jörgensen, S., *J. prakt. Chem.,* **14,** 213 (1876).
10. West, C., *Am. Mineral.* **22,** 731 (1937).
11. Jörgensen, S., *J. prakt. Chem.,* **14,** 356 (1876).
12. Farwell, H., *J. Optical Soc. Am.,* **28,** 460 (1938).
13. Godina, D., and Faerman, G., *J. Applied Chem. (U.S.S.R.)*, **14,** 362 (1941).
14. Kermack, Slater and Spragg, *Proc. Roy. Soc. Edinburgh,* **50,** 243-61 (1930).
15. Mooney, R., *J. Am. Chem. Soc.,* **63,** 2828 (1941).
16. Marvel, C., and Denoon, C., *J. Am. Chem. Soc.,* **60,** 1045 (1938).
17. Preston, J., see reference 6. See also Silverman, S., *Rev. Sci. Instruments,* **12,** 212 (1941).
18. Valkò, E., *J. Am. Chem. Soc.,* **63,** 1433 (1941).
19. Staudinger, H., Frey, K., and Starck, W., *Ber.,* **60,** 1782 (1927).
20. Gallay, W., *Can. J. Research,* **14B,** 105 (1936).
21. Hausser, K., Kuhn, R., and others, *Z. physik. Chem.,* **29,** 363 (1935).
22. Kuhn, R., *J. Chem. Soc.,* p. 605 (1938).
23. Mulliken, R., *J. Chem. Phys.,* **7,** 570 (1939).
24. Frey-Wyssling, A., *Protoplasma,* **27,** 372-411; 563-571 (1937).
25. Berkman, S., Boehm, J., Zocher, H., *Z. physik. Chem.,* **124,** 83 (1926).
26. Lange, B., *Z. physik. Chem.,* **132,** 27-46 (1928).
27. Christoph, W., and Neugebauer, H., *Z. Tech. Physik.,* **20,** 257 (1939).
28. Chubb, L., *Trans. Illum. Eng. Soc. (N.Y.)*, **32,** 505 (1937).
29. Gamble, A., *Bull. soc. franc. élec.,* **7,** 511 (1937).
30. Gibbs, H., *J. Franklin Inst.,* **228,** 719 (1939).
31. Land E., *J. Franklin Inst.,* **224,** 269 (1937).
32. Land, E., *Univ. Mich. Official Pub.,* **42,** No. 42 (1940): "Michigan-Life Conference on New Technologies in Transportation," pp. 151-68.
33. Roper, V., and Scott, K., *Illum. Eng. Soc. (N.Y.)*, **36,** 1205 (1941).
34. Sauer, H., *Z. Ver. deut. Ing.,* **82,** 201 (1938)

35, 36. Norling, J., *J. Soc. Motion Picture Eng.,* **33,** 612 (1939); **37,** 516 (1941).

37. Land, E., *J. Optical Soc. Am.,* **30,** 230 (1940).
38. Dudley, B., *Phototechnique,* **3,** 30 (May, 1941).
39. Stamm, R., *Am. J. Roentgenol. and Radium Therapy,* **45,** 744 (1941).
40. Buchwald, E., *Ann. Physik,* **38,** 325 (1940).

Some Aspects of the Development of Sheet Polarizers*

EDWIN H. LAND
Polaroid Corporation, Cambridge, Massachusetts
(Received October 10, 1951)

This paper reviews the 25 years of activity, by the author and his co-workers, in the development of synthetic sheet polarizers. The early work during the nineteenth century is described briefly, and then the various stages of the modern development in the author's laboratory are chronicled. A description is given of the nature and the optical properties of the currently-available sheet polarizers of the Polaroid *J*, *H*, *K*, and *L* types, and of the quantitative methods used in characterizing them. The reasons are given for developing special polarizers for each of a variety of applications such as optical instruments, vectographs, and headlights.

INTRODUCTION

I SHOULD like to begin by describing some work that I did about twenty-five years ago and which should have been reported at that time. But the days slipped by and the months slipped by and now twenty-five years have slipped by, and, although my co-workers and I have reported from time to time on some of the applications of that work, I was surprised to find that the work itself has not yet been reported to the Optical Society of America.† I want to take this occasion to review with you the field to which a number of us have devoted a large part of our professional careers, in order to see what our early purposes were, and to what extent we have fulfilled them.

In about 1926 I became interested in the question of whether or not one could make a material in the form of a synthetic sheet which could polarize light in the same way that tourmaline does, only more efficiently—a material that might have the desirable properties of a Nicol prism and yet be free from some of its undesirable properties such as small area and limitation in angular aperture.

HERAPATH'S WORK

In the literature there are a few pertinent high spots in the development of polarizers, particularly the work of William Bird Herapath,[1] a physician in Bristol, England, whose pupil, a Mr. Phelps, had found that when he dropped iodine into the urine of a dog that had been fed quinine, little scintillating green crystals formed in the reaction liquid. Phelps went to his teacher, and Herapath then did something which I think was curious under the circumstances; he looked at the crystals under a microscope and noticed that in some places they were light where they overlapped and in some places they were dark. He was shrewd enough to recognize that here was a remarkable phenomenon, a new polarizing material.

Doctor Herapath spent about ten years trying to grow these green crystals large enough to be useful in covering the eyepiece of a microscope. He did get a few, but they were extremely thin and fragile—for it is very hard to grow them. We find that they were not successfully grown as large single crystals until at the Zeiss Works, Bernauer[2] succeeded about ten or twelve years ago in growing some that were several inches in diameter.

Herapath's work caught the attention of Sir David Brewster, who was working in those happy days on the kaleidoscope. Brewster thought that it would be more interesting to have interference colors in his kaleidoscope than it would to have just different-colored pieces of glass. The kaleidoscope was the television of the 1850's and no respectable home would be without a kaleidoscope in the middle of the library. Brewster, who invented the kaleidoscope, wrote a book[3] about it, and in that book he mentioned that he would like to use the herapathite crystals for the eyepiece. When I was reading this book, back in 1926 and 1927, I came across his reference to these remarkable crystals, and that started my interest in herapathite.

AMBRONN'S WORK

In the work of Herapath, iodine was seen to be definitely associated with polarizers. Iodine appears in another place in the early literature, in the work of Ambronn,[4] who noticed that when he stained membranes of cells, living cells of animal tissue, he obtained an elevation of the absorption coefficient of the iodine. As he treated certain natural tissues, they became darker, and he observed that these darkened tissues were dichroic. In a sense he "missed the boat" on making a polarizer. In fact he said flatly that he saw no prospect of making a polarizer by using this phenomenon, a statement which I find very hard to understand. Ambronn was a competent worker, but he was

* Presented as an invited paper in the Symposium on Polarization at the March, 1951, meeting of the Optical Society of America, Washington, D. C.

† The development of the *J* polarizer was reported to the Harvard Physics Colloquium on February 8, 1932, under the title "A new polarizer for light, in the form of an extensive synthetic sheet." See reference 6 for many references to polarizers and their applications.

[1] W. B. Herapath, Phil. Mag. (4th ser.) **3**, 161 (1852).

[2] F. Bernauer, Fortschr. d. Mineralogie **19**, 22 (1935).

[3] D. Brewster, *The Kaleidoscope, Its History, Theory, and Construction* (John Murray, London, 1858), second edition.

[4] H. Ambronn, Ann. Physik u. Chem. **34**, 340 (1888).

one of those persons who are more interested in conducting scientific experiments than in the utilization of the results obtained. If Ambronn had been working in America, the outcome might have been different!

One of the concepts that has been emphasized thus far is the important place of iodine in the history of the development of synthetic polarizers. Herapathite is a crystalline form of quinine sulfate per-iodide.

USE OF MICROSCOPIC CRYSTALS

My immediate inspiration was the work of Herapath. It was apparent that, since his ten years of effort had not sufficed to make a large polarizing crystal, this was probably a hard way to make a polarizer. I wondered if, instead of trying to grow one large crystal, one might not use a multiplicity of small crystals, all similarly oriented. However, there are several difficulties: (1) A number of little crystals will scatter light if they are not carefully prepared; (2) There is a problem of orientation; (3) How does one prepare very small crystals; and (4) In what form does one handle them? In visualizing this problem—how would one go about making a synthetic sheet containing oriented herapathite crystals—the first elementary concept is not to make big crystals, but to make microscopic ones. If one is then able to turn all of the microscopic crystals the same way, the remaining difficulty is that it is not sufficient to have them microscopic, but because they will scatter light, they must be submicroscopic.

THE FIRST SYNTHETIC POLARIZER

In my then youthful innocence it seemed to me that these problems could be solved in a rather short time, perhaps in a few months. I have preserved as an exhibition piece the first synthetic polarizer which I made. This first polarizer was made by grinding herapathite crystals in a ball mill for a month; the mill contained a solution of nitrocellulose lacquer.

Then came the most exciting single event in my life. The suspension of herapathite crystals was placed in a small cell—a cylinder of glass about a half-inch in diameter and a quarter-inch in length. The cell was placed in the gap of a magnet which could produce about 10,000 gauss. Before the magnet was turned on, the Brownian motion caused the particles to be oriented randomly so that the liquid was opaque and reddish black in color. When the field was turned on—and this was the big moment—slowly and somewhat sluggishly the cell became lighter and quite transparent; when we examined the transmitted light with a Nicol prism, it went from white to black as the prism was turned. This first polarizer experiment was a success, but in the twenty-five years between then and now, there have been many technical details which required solution.

In making a solid polarizer from a liquid polarizer, the following steps were taken: The same suspension was placed in a test tube and a plastic sheet was dipped into it; the test tube as a whole was placed in the magnetic field and then the tube was withdrawn so as to leave a coating of the dispersion of polarizing crystals on the sheet; the sheet was then allowed to dry in the magnetic field, and a few moments later we had a polarizer.

OTHER METHODS OF ORIENTATION

One can also orient the particles with an electrostatic rather than a magnetic field; however, I would like to mention a quite different orientation method. A special sheet of rubber was procured, a sheet about eight inches square, and this sheet was stretched in one direction to a length of several feet. A very viscous suspension of crystals was coated on the sheet in the stretched condition. The sheet was then permitted to contract, and was then stretched in the perpendicular direction. The stretch ratio obtained in this way is the square of the stretch ratio that can be obtained by a simple stretching operation.

All of this, while exciting at the time and indicative of the potential of this approach, did not provide a good way to make polarizers in large quantities. We abandoned use of electrostatic and magnetic fields and put about a year's work into making a colloidal dispersion of submicroscopic needles of herapathite, this dispersion being formed into a viscous mass which was then extruded between long narrow slits. As the sheet was extruded, the needles were oriented parallel to one another. This is the method employed in making the *J* polarizer.

SIZE OF THE CRYSTALS

These needles were thought to be about one micron long and a small fraction of a micron in diameter. It was quite important to have the diameter that small, because this minimized the light scattered by this two-phase system. In the recent decade it became feasible to study these needles with the electron microscope, and thus a good estimate could be made of their actual size. The technique was this: The polarizing sheet made with the small crystals of quinine sulfate per-iodide was treated with silver nitrate to form silver iodide in the place where the herapathite crystals had been. This was done in the solid state, the liquid being introduced without dissolving the sheet. Then the plastic sheet itself was dissolved and the crystals of silver iodide were examined with the electron microscope. Figure 1 is one of the pictures obtained in this manner. In this figure the needles are indeed of about the size originally suspected; these needles were prepared from the Polaroid *J* polarizer. An interesting fact which will be discussed in another part of this presentation is that the sheet is still a polarizer when one replaces the herapathite with metallic silver.

THE PRINCIPAL ABSORPTION COEFFICIENTS

Dichroic polarizers have two different principal absorption coefficients which are effective for normal incidence, so that the absorption of normally incident linearly polarized light is markedly different for vibration directions parallel to the two principal axes of the polarizer. In a good dichroic polarizer the two absorption coefficients must be very different; in an ideal polarizer one linear component is transmitted without absorption and the other is absorbed completely. Thus an ideal polarizer involves an infinite ratio for the two absorption coefficients. As will appear later, we are theoretically unable to obtain an infinite ratio, but in good modern polarizers the ratio can run as high as 100 or more. It is not necessary for the ratio of absorption coefficients to be constant throughout the visible spectrum; indeed, there are some applications where one does not want the ratio to be constant. It is desirable, however, in an ordinary polarizer that the one absorption coefficient be so high throughout the visible spectrum that substantially none of the light of that orientation passes through the sheet, and that the other absorption coefficient be so low that substantially all of the light goes through. Such a polarizing material is remarkable in that its transmittance for unpolarized light is quite constant throughout the spectrum, with the result that a good visual polarizer is sometimes a better neutral-density filter than filters which are made with nonpolarizing absorbing materials.

QUANTITATIVE DESCRIPTION OF POLARIZERS

It is useful to be able to describe a polarizer in several different ways. The first way is in terms of the two principal transmittances. In Fig. 2, for one density-type of *H* polarizer, k_z is the transmittance for linearly-polarized incident light whose electric vector is parallel to the "machine direction" z, and similarly k_y is the transmittance when the electric vector of the incident light is perpendicular to the machine direction. This method of describing a polarizer permits one to show in a graph the value of k_y with fair precision, but the actual value of k_z is not indicated well. Figure 3 shows the same information described now in terms of the principal optical densities. In Fig. 3 the transmitted component is at the bottom and the nearly extinguished component appears as the upper curve. In the density representation the behavior of the z component is shown well, but the transmittance for the y component is not shown very accurately.

THE STRETCH RATIO

It is of interest to inquire as to what actually determines the density ratio of the polarizing sheet—i.e., the ratio of the principal optical densities. There are two parameters; obviously one is the efficiency of the crystals (the intrinsic density ratio of the dichroic absorbing material), and the other is the degree of orientation of the crystals.

The degree of orientation is measured in terms of the stretch ratio; the method of determining the stretch ratio in the laboratory is very simple. A circle is drawn on the sheet; then after stretching, one measures the axial ratio of the ellipse into which the circle has been deformed. Figure 4 shows the relation between the density ratio of the final polarizer and the stretch ratio, on the basis of the assumption that the dichroic crystals are needle-like and that the absorption axis is parallel with the needle axis.[5] In Fig. 4 it is apparent that even if the density ratio of the crystals is infinite, that is, even if the crystal itself is a perfect polarizer, the sheet cannot have a density ratio greater than about 1.3 times the stretch ratio. We have found, however, that this limitation on the density ratio which holds for the microcrystalline *J* sheet does not apply to the more modern *H* and *K* polarizers.

THE SCATTERING POLARIZER

We have developed an entirely different kind of polarizer, which very few people indeed have ever

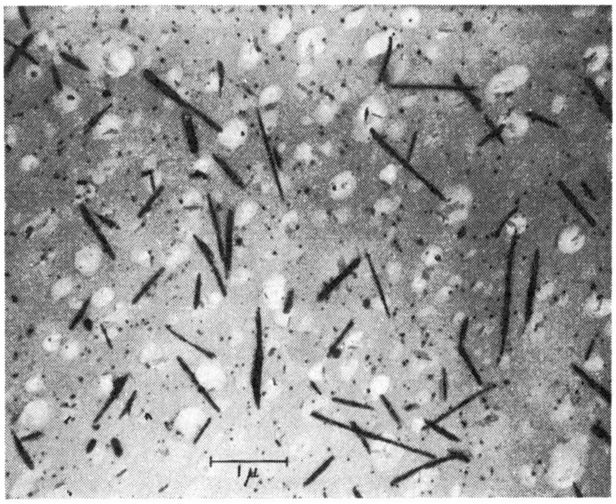

FIG. 1. Silver iodide replicas of (disoriented) herapathite needles in the *J* polarizer. (Electron micrography by C. E. Hall, Massachusetts Institute of Technology, 1949.)

seen, and for which there seems to be no obvious use. In this polarizer we use crystals of a doubly-refracting material instead of dichroic crystals, and we deliberately make the crystals with a size several times the wavelength of light. Furthermore, the refractive index of the plastic matrix is made equal to one of the two principal indexes of the birefringent crystals. What is then obtained is quite a beautiful product: when the sheet is examined with an analyzer, it is as transparent as glass for one position of the analyzer, but when the analyzer is turned through 90°, the sheet is turbid and highly diffusing. It is a variable diffuser, and one would think that there would be a use in optical instrumentation for such a component.

[5] R. Clark Jones, J. Opt. Soc. Am. **35**, 803(A) (1945).

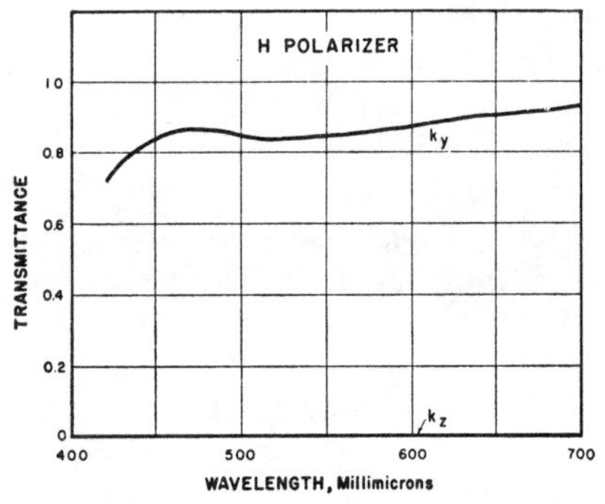

FIG. 2. Principal transmittances *versus* wavelength, for a typical *H* polarizer. (Transmittance values have been corrected for reflection at the surfaces.)

THE METAL POLARIZERS

Still different are the metal polarizers, like the one whose properties are illustrated in Fig. 5 which utilizes very small needles of tellurium, all oriented parallel with one another. This polarizer works, probably not so much because of the dichroism of the individual needles as because the absorption of light by the lattice or grating of needles depends on the direction of the electric vector with respect to the needle axis. The tellurium crystals were prepared and oriented in a way quite different from that used in making the *J* polarizer. Instead of being dispersed mechanically in a plastic matrix before extrusion, the reagents were introduced into a different plastic, namely polyvinyl alcohol. Either before or after the sheet is stretched, the tellurium needles are precipitated chemically. Figure 6 is an electron micrograph which indicates the actual size of these needles. Because of the method of preparation of the pellicle, the needles are disoriented and are much fewer in number per unit area than in the original polarizer. The preceding sentence applies also to Fig. 1.

We have also prepared metal polarizers of silver, gold, and mercury; the optical density curves of these have been published.[6] The particles of all three of these metals are optically isotropic.

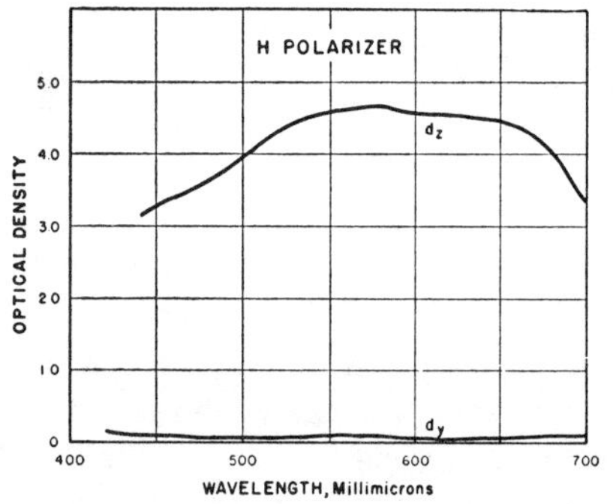

FIG. 3. Principal optical densities *versus* wavelength, for the typical *H* polarizer of Fig. 2. (The data have been corrected for surface reflection.)

[6] E. H. Land and C. D. West, "Dichroism and Dichroic Polarizers." *Colloid Chemistry*, J. Alexander, Ed. (Reinhold Publishing

POLYVINYL ALCOHOL

A new material of considerable importance in the field of polarizers is polyvinyl alcohol, which can be fabricated into a beautiful, water-clear sheet which lends itself to stretching when heated. In the stretched state it is the matrix for a series of important and useful polarizers.

The *H* polarizer is made by absorbing iodine in a stretched sheet of polyvinyl alcohol. The various dye polarizers are made by absorbing dyes in stretched polyvinyl alcohol. These dye polarizers, which we call *L* polarizers, have the property of polarizing only portions of the spectrum and of transmitting both components in the remainder of the spectrum. A very important new polarizer is the *K* polarizer, whose dichromophore is polyvinylene. The *K* polarizer is made by dehydrating stretched polyvinyl alcohol.

Figure 7 indicates the approximate relative shape and size of the molecular structures of polyvinyl alcohol,

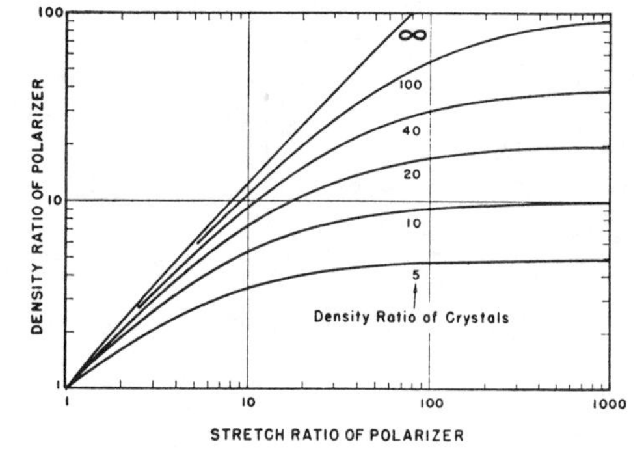

FIG. 4. Optical density ratio of a microcrystalline polarizer (calculated) *versus* the stretch ratio of the polarizer sheet, for needle-shaped crystals of different intrinsic density ratios. The crystals are assumed to be uniaxial in optical properties; the unique axis is the axis with the greater absorption coefficient, and is parallel with the geometrical axis of the needle.

polyvinylene (*K* polarizer), and the polymeric iodine of the *H* polarizer.

Blake, Makas, and West showed that when dichromophores of the *H* and *K* types are combined in a single layer of polyvinyl alcohol, contrary to expectation a new dichromophore termed *HR* is formed which has a definite infrared absorption maximum at about 1.5μ. This dichromophore makes possible the preparation of efficient dichroic polarizers for the broad spectral range 0.75 to 2.8μ. (See U. S. Patent 2494686, 1950.)

THE VECTOGRAPH

We have found one rather interesting application of the iodine-polyvinyl alcohol polarizer in the vectograph, and this is one of the applications which has

Corporation, New York, 1946), Vol. 6, pp. 160–190. This paper contains much original data, and lists many references to articles and patents concerning polarized light, including those of Käsemann and Marks. Dreyer's work is first described in U. S. Patent 2400877, 1946.

been reported previously to the Optical Society.[7] Prior to this it was not realized that three-dimensional pictures could be obtained if the two images of the stereoscopic pair were made in terms of percent polarization. We then undertook to develop methods for obtaining good "polarization anaglyphs." If two vectographs are superposed with their stretch directions at right angles, one will see one image and then the other, if one views the pair with an analyzer which is rotated slowly. In order to prepare such images, the same procedure is used as that employed in making color separations by the dye transfer process, except that iodine is used instead of a dye. The procedure involves making what is called a wash-off relief, in which the thickness of the gelatin is proportional to the image density that is desired. The wash-off relief is then dipped into an iodine bath and is then placed in contact with a sheet of stretched polyvinyl alcohol. Images formed in this way with

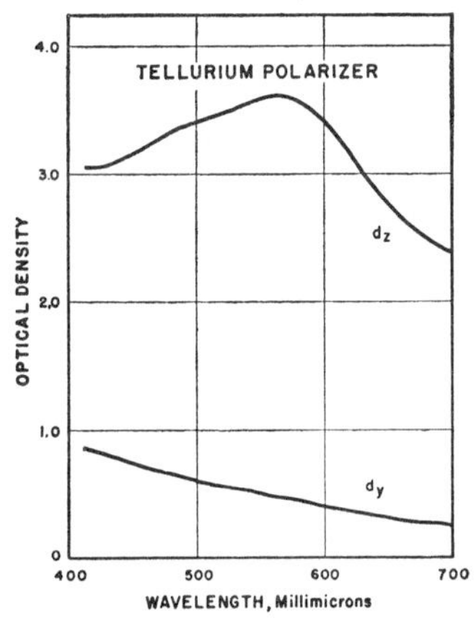

FIG. 5. Principal optical densities *versus* wavelength, for a tellurium polarizer. (The data have been corrected for surface reflection.)

iodine in polyvinyl alcohol are really images formed in terms of degree of polarization.

In particular, the two images one observes can be members of a stereoscopic pair, so that one can use an ordinary lantern-slide projector or an ordinary movie projector with a nondepolarizing screen, to obtain pictures which are threedimensional when viewed with polarizing spectacles. A method has also been worked out by which one can make vectograph prints of ordinary size that can be held in the hand.

By using the dye polarizers mentioned above, instead of the iodine polarizers, vectographs can be prepared in color. This provides the only method of projecting full-color stereoscopic pictures in standard motion-picture and lantern-slide projectors without any alteration of the instrument. No polarizers are used on the projector, so that all the light from the source is utilized. This application illustrates a requirement for

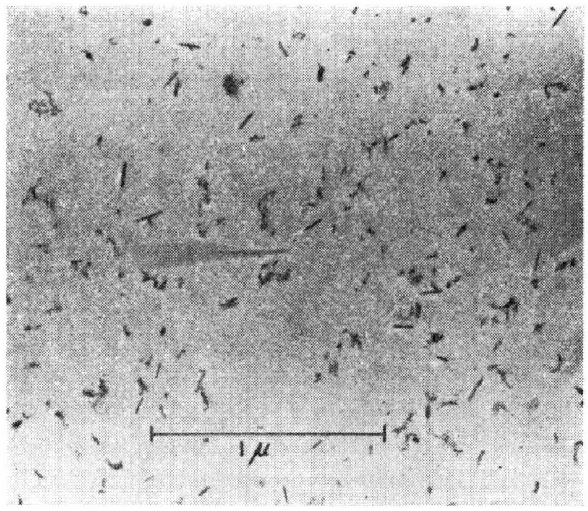

FIG. 6. Electron micrograph of tellurium needles in polyvinyl alcohol, precipitated *in situ* in the oriented state. (The matrix was re-dissolved in making the pellicle, which fact accounts for the disorientation.) It is interesting to speculate that needles like these may possibly account for the polarizing action of interstellar space.

polarizers which have high absorption for one component over only a portion of the visible spectrum.

THE *H* POLARIZER

The *H* polarizer is the one that is now in most common use. Its principal transmittances and principal optical densities were shown in Figs. 2 and 3 respectively, and are shown also in Fig. 8, which is a combination graph designed to exhibit both k_y and k_z effectively. The optical density d_z is high in the middle of the spectrum where the eye is most sensitive; even at 680 mμ the optical density is greater than three. Also, the transmittance k_y is quite high and stays high throughout the visible spectrum. This polarizer looks neutral in color, when viewed in unpolarized light, and it is remarkably free of light scattering because the dichromophore is of molecular dimensions. The extinction color is ordinarily a deep blue, but a very bright light is required to see the extinction color in the denser varieties of the *H* polarizer.

THE *K* POLARIZER

Figure 8 also shows the optical behavior of a typical *K* polarizer. The density d_z of this polarizer is greater than that of the typical *H* polarizer just discussed for

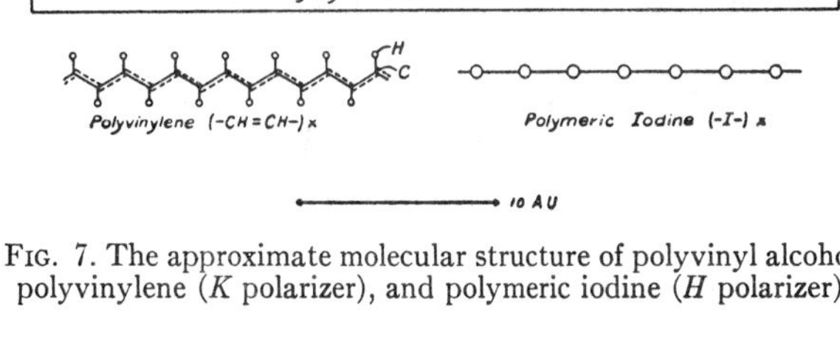

FIG. 7. The approximate molecular structure of polyvinyl alcohol, polyvinylene (*K* polarizer), and polymeric iodine (*H* polarizer).

[7] E. H. Land, J. Opt. Soc. Am. **30**, 230 (1940); see also p. 184 of reference 6 regarding the work of Land and Mahler.

all wavelengths shorter than about 650 mμ, but the density d_z is less than that of the *H* polarizer for longer wavelengths. As a result, the *K* polarizer has a red extinction color instead of the blue extinction of the *H* polarizer.

The *K* polarizer is particularly interesting for two reasons: first, the chemical simplicity of the simple straight-chain hydrocarbon, polyvinylene, employed as the dichromophore; second, its extraordinary stability with respect to heat and radiation.

The *K* polarizer was developed especially for use on automobile headlamps in connection with our system for eliminating the glare of opposing headlamps in night driving while still providing drivers with excellent

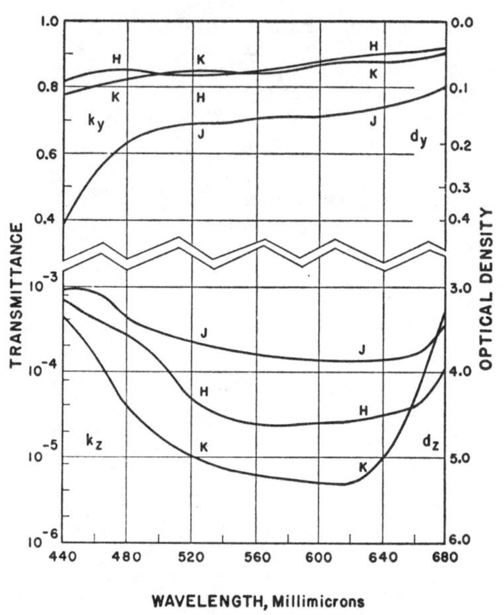

FIG. 8. The optical behavior of typical *J*, *H*, and *K* polarizers *versus* wavelength. The curves represent transmittances if referred to the ordinate scale on the left, and represent optical densities if referred to the scale on the right. The principal transmittance k_y is plotted on a linear ordinate scale in the upper part of the diagram, and k_z, the transmittance of the highly absorbed component, is plotted on a logarithmic scale in the lower part. (The data have been corrected for surface reflection.)

visibility of the road ahead. In this system a polarizer is laminated to the front surface of the headlamp; the requirements placed upon this polarizer are severe: it must stand the intense heat of the radiation during the periods when the car is stationary with the lights turned on, and it should last for the life of the car, even though its owner subjects it to continuous weathering in the tropics.

The motivation back in the early days when I was making the polarizers in the magnetic field was threefold: First, there was the tremendous excitement of trying to make a polarizer the size of a large piece of window glass such as you see here on the stage; second, there was a good deal of scientific interest in such things as Kerr cells, the Faraday effect, and polarized-light microscopy, and all of these interests required polarizers with an optical performance that could be met at that time only by the Nicol prism, which of course had a severely limited aperture. Incidentally, Kerr cells were a lively item in the early television research; Alexanderson[8] was using them in his researches, and it looked as though Kerr cells might play an important part in television. But the third and most important motivation was, and still is, the field of polarizing headlights.

HEADLIGHTS

The glareless headlight system involves placing on the headlights polarizers with their axes tilted in the vertical plane at about 45° from the horizontal. A similar polarizer is incorporated in the windshield or otherwise placed before the driver, and its axis is substantially parallel with the polarizers on the headlights. If you now consider two such cars, and if you imagine that one is turned around so that it faces the other, you see that the headlight polarizers of the approaching car are "crossed" with the one before the driver's eyes. The performance of this arrangement is such that the driver's viewer selects the opposing headlamps from among all other objects in the field of view and extinguishes them. Meanwhile, the same viewer causes little reduction in the driver's visibility of his own roadway because of the parallel relationship between the viewer and his own headlight polarizers. With this system in universal use, every driver will be able to drive at night with light at least comparable to the *upper* beam of your present headlamps, and approaching cars will seem to have headlights whose brightness is equal to only a fraction of that of the present *lower* beams.

At the present time the status of our glareless headlight system might be summed up as follows: We have available for the system polarizers of such stability and high efficiency that little would be gained by making them better; and 125-watt headlamps have been developed by General Electric for meeting the requirements of the system. These headlamps, three times as powerful as present day headlamps, are, when in use in the polarizing system, about 50 percent brighter than today's. With this matched system, including the polarizing viewer, large-scale quantitative tests involving careful measurements[9] have been conducted. These tests show that through the viewer one has sufficient light for at least as good visibility as one has today with the high beam for open road driving, and

[8] E. F. W. Alexanderson and R. D. Kell, U. S. Patent 1783031, 1930.

[9] V. J. Roper, Traffic Engineering **19**, 151 (1949), and E. H. Land and L. W. Chubb, Jr., Traffic Engineering **20**, 265 and 384, 399 (1950).

that this good visibility is maintained while a similarly-equipped car is being met and passed.

CONCLUSION

In closing I would like to acknowledge the assistance received from the many colleagues, co-workers, and assistants who have been associated in this program. Many of them are known from their own publications, and others are named in the references that have been cited. One of the great pleasures in reviewing this field has been the recalling to mind of these many happy relationships.

Practical Methods of Making and Using Multilayer Filters [1]

MARY BANNING
Institute of Optics, University of Rochester, Rochester, New York
(Received February 3, 1947)

The general method of making an evaporated multiple-layer interference filter from two dielectrics is described for any desired wave-length range. The proper thickness of each layer to give maximum reflection in the desired region and at the desired angle of incidence is first computed, and next the visible reflectivity of such a film at normal incidence. This visible reflectivity is used as the gauge to determine the proper thicknesses of the films and the point at which the evaporation of each layer should be stopped. Practical difficulties encountered in making these filters are also discussed with various ways of surmounting them.

Several examples of special filters for particular purposes are reviewed in detail, including infra-red and polarizing filters.

A GREAT deal of material has been published during the last year on methods of computing the spectral reflection and transmission of multilayer films.[2–5] However, little has yet been said about the making of such filters and their practical applications. The present article describes methods of making multilayer filters and easy "rules of thumb" by which the proper film thicknesses may be computed. For lack of sufficient personal experience, no attempt is made to describe metallic-dielectric films such as are used to increase metallic reflection,[6] or to form selective transmission filters,[7] although the same problems occur in these cases. The exact theoretical treatment of both metallic and dielectric multilayer films is given in a forthcoming article by H. D. Polster, whose work coincided with a large part of that described here.

The first highly reflecting films of zinc sulfide were made by Pfund in 1933,[8] and the first "nonreflecting" films of low index fluorides evaporated in 1935 by Strong;[9] in 1939, Cartwright and Turner tried a combination of low and high index films and produced selective reflection.[10] Since that time many laboratories have been working on this problem, but because of the war little has been published except for a short account from the RCA Laboratories.[11]

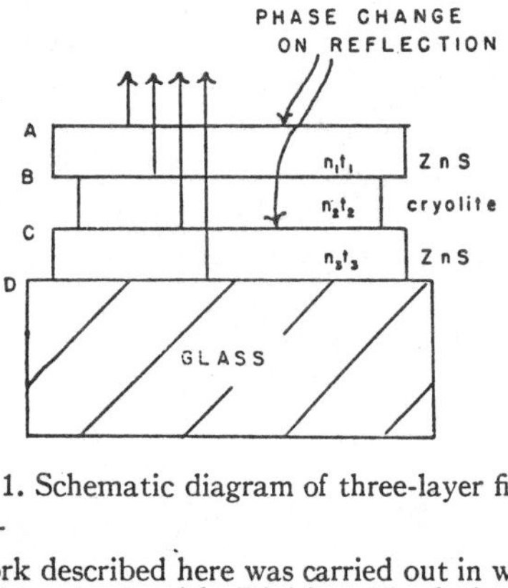

FIG. 1. Schematic diagram of three-layer filter.

SIMPLE FILTERS

Development of multilayer filters at the University of Rochester arose accidentally, in September, 1942, from the need of producing a reflex plate that would reflect approximately 50 percent of the incident light and transmit the rest. A single coating of zinc sulfide, evaporated to an optical thickness of a quarter wave-length for green light, did not give sufficient reflection; if a plate of glass was coated on both sides, the reflection was nearly satisfactory, but this necessitated exposing one of the soft coatings to rough handling and atmospheric conditions. Two plates coated on one side and mounted with their coated faces together were difficult to align per-

[1] The work described here was carried out in whole or in part under contract with Division 16 of the Office of Scientific Research and Development, at the University of Rochester.
[2] P. King and L. B. Lockhart, J. Opt. Soc. Am. 36, 513 (1946).
[3] D. L. Caballero, J. Opt. Soc. Am. **36**, 710A (1946).
[4] R. L. Mooney, J. Opt. Soc. Am. **36**, 256 (1946).
[5] H. D. Polster, J. Opt. Soc. Am. 350A (1946).
[6] A. F. Turner, J. Opt. Soc. Am. **36**, 711A (1946).
[7] Such filters are now being commercially produced by the Farrand Optical Company, Inc., New York City.
[8] A. H. Pfund, J. Opt. Soc. Am. **24**, 99 (1934).
[9] J. Strong, J. Opt. Soc. Am. **26**, 73 (1936).
[10] C. H. Cartwright and A. F. Turner, Phys. Rev. **55**, 1128A (1939).
[11] G. L. Dimmick, J. Soc. Mot. Pic. Eng. **38**, 36 (1942).

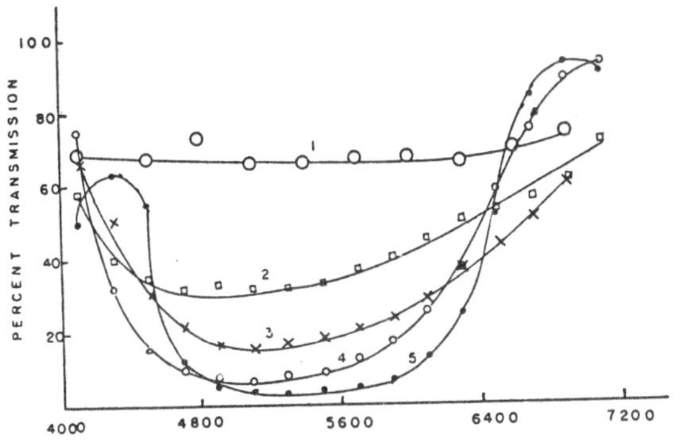

FIG. 2. Effect of addition of layers, first-order interference. (1) 1 ZnS layer; (2) 2 ZnS and 1 cryolite; (3) 3 ZnS and 2 cryolite; (4) 4 ZnS and 3 cryolite; (5) 5 ZnS and 4 cryolite.

fectly parallel and double images resulted, while cementing the two together decreased the reflection too much.

At the suggestion of J. W. Evans, a low index dielectric layer of cryolite was evaporated on top of the high index layer and followed by a second high index film, thus "substituting" for the intermediate air layer. It was soon found that the thickness of the low index layer was as critical as that of the high index layers and that best results were obtained when all three had an optical thickness of a quarter wave-length. Such a film showed 60 percent reflection for the middle portion of the visible spectra.

The explanation of this is very simple. Figure 1 shows a schematic three-layer filter. According to a convention adopted here, high index films are shown with greater length than low index films. Light reflected from all four interfaces, A, B, C, and D, must be in phase for maximum reflection. A phase change of $\lambda/2$ will take place whenever light is reflected from a high index medium into a lower index medium, or at surfaces A and C in Fig. 1. The ray reflected at A suffers a phase change of $\lambda/2$. The ray from B goes $2n_1t_1$ farther than that from A and thus differs in phase by $2n_1t_1$ plus $\lambda/2$; now if n_1t_1 equals $\lambda/4$, the two rays differ by λ and are in phase. The ray from C changes phase by $2n_1t_1$ plus $2n_2t_2$ plus $\lambda/2$ at reflection and is, therefore, $2n_2t_2$ plus $\lambda/2$ out of phase with that from A; to be in phase n_2t_2 must be $\lambda/4$. Similarly n_3t_3 must be $\lambda/4$.

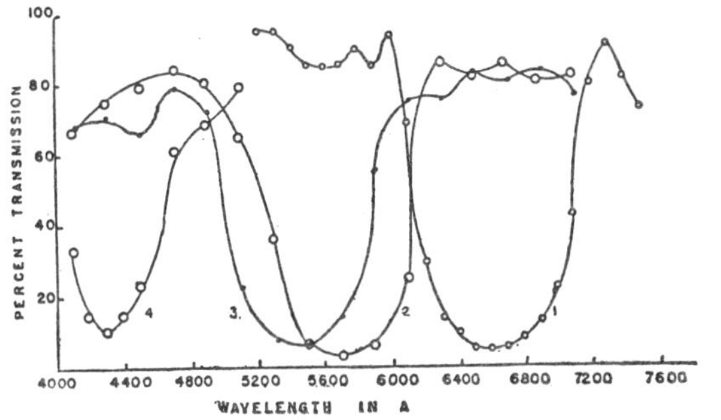

FIG. 3. Second-order filters of seven layers. (1) ZnS to second magenta; (2) ZnS to second yellow; (3) ZnS to second white; (4) ZnS to second blue, first bright blue.

In the case mentioned above, that of a reflex plate using visible light, the wave-length for which this must hold is approximately 5400A. The effect of adding more and more layers is shown in Fig. 2, with the maximum reflection becoming greater and the width of the reflection band becoming narrower. The graphs are shown as transmission values. The reflection is obtained by subtracting these values from 100 percent, as the absorption loss is negligible. This filter is the simplest to make, as the color changes are quite definite and easy to see. It is the fundamental *first-order* filter, with the optical thickness of all layers a quarter wave-length for green light. It is not advisable to put down more than four sulfide layers (with three of cryolite), as the total thickness becomes so great that the film tends to crackle off. This is even more pronounced with layers of greater thickness, as described below.

Reference to Fig. 1 shows that if all the films are made twice as thick ($nt=\lambda/2$), the ray from

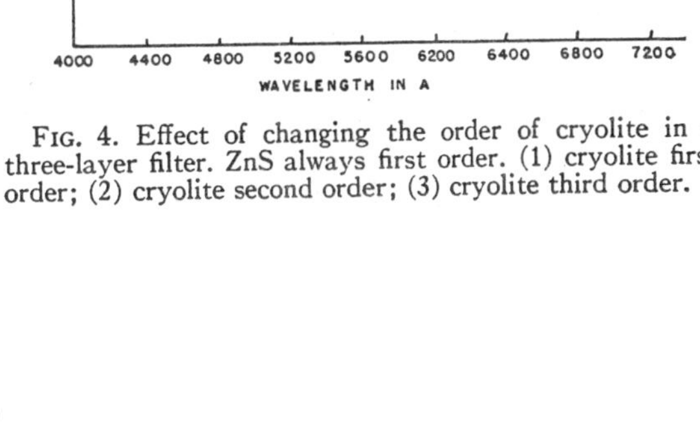

FIG. 4. Effect of changing the order of cryolite in a three-layer filter. ZnS always first order. (1) cryolite first order; (2) cryolite second order; (3) cryolite third order.

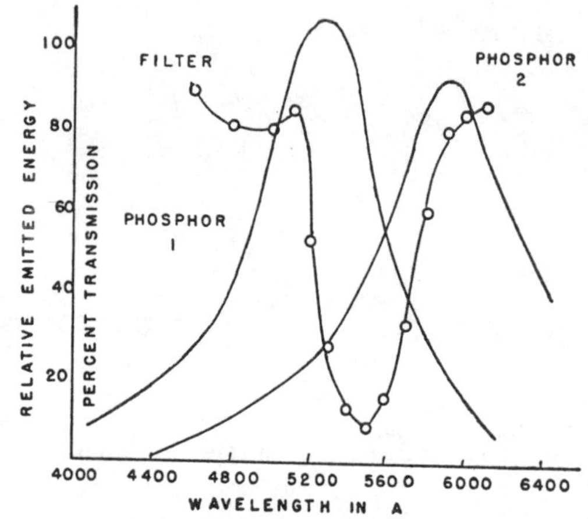

FIG. 5. Multilayer filter designed to distinguish between the emission of two phosphors.

A will tend to cancel that from *B*, while that from *C* will cancel the one from *D*. Thus there will be very little reflection of green light. However, light of twice this wave-length, about 1.1 microns, will find the optical thickness equal to $\frac{1}{4}$ of its wave-length, and will be strongly reflected. This filter has been found useful as a reflex plate for stimulating and observing infrared phosphors which have a maximum sensitivity to radiation of 1.1 microns, and emit in the green. Now if the films are made even thicker so that $nt=\frac{3}{4}\lambda$, the four rays will again be in phase and again green light will be strongly reflected. This is called a *second-order* filter, several of which (of 7 layers) are shown in Fig. 3, for light of various wave-lengths. It can be seen that the reflection band is much narrower than is the one of a first-order filter. Going to the third order, or $\frac{5}{4}\lambda$-thickness, gives an even sharper band, useful when only a short spectral range is to be considered. In some cases it is desirable to combine cryolite layers of first order with second-order zinc sulfide layers, or *vice versa*. The effect of changing the cryolite order while holding that of the sulfide constant is shown for a three-layer filter in Fig. 4.

DEPOSITION OF LAYERS

Before taking up more complicated filters, the method of deposition should be mentioned. It is not at all necessary to break the vacuum in an evaporator during the making of a filter; although this does no harm to the filter, the total time consumed is much greater than if all layers are deposited in the same vacuum. A small test piece, shielded from evaporating materials except for a small area, can be placed beside the actual sample and rotated slightly after each layer to present a clean surface for observation. The cryolite and zinc sulfide are placed in crucibles side by side on the bottom of the evaporating chamber and the sample centered above them.

If the sample is large, it can be rotated during evaporation about an axis perpendicular to its face, and the crucibles offset from the center to provide an even coating. The proper crucible

TABLE I. Colors of different thickness of zinc sulfide and cryolite films as seen by reflection at normal incidence.

Color: ZnS	Color: Cryolite	Optical Thickness for Green Light
bluish white	yellow	
white	magenta	$\lambda/4$, first order maximum
yellow	blue	
magenta	white	$\lambda/2$, first order minimum
blue	yellow	
greenish white	magenta	$3\lambda/4$, second order maximum
yellow	blue	
magenta	greenish white	λ, second order minimum
blue	yellow	
green	magenta	$5\lambda/4$, third order maximum

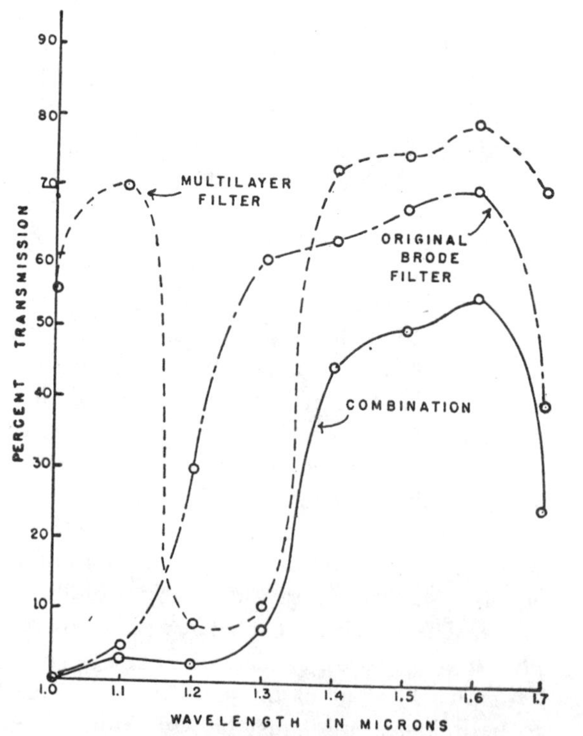

FIG. 6. Multilayer filter designed to sharpen the cut-off of another infra-red filter.

positions are found by successive approximations. Circular filters of 7-inch diameter have been made in this manner, showing only 1 percent variation in reflectivity over their surfaces. This can be done with crucibles only nine inches below the level of the plate, in a 12-inch diameter bell jar. A small hole drilled in the center of the disks allows access to a test piece supported just above the large sample.

The evaporators used here have large flat glass tops for viewing purposes.[12] Thicknesses of the layers are judged by watching, at as near to normal incidence as possible, the changing colors reflected from the sample or test piece. A fluorescent lamp hung over the chamber provided a good white light source. Both the back of the sample and the inside of the glass top are coated with mineral oil to prevent deposition on these surfaces in the form of a reflecting film and consequent confusion in viewing the reflected light.

As the evaporation of zinc sulfide proceeds, it first shows a bluish white reflection, then as it gets thicker it reflects a brilliant white (optical thickness of $\lambda/4$ for green light), yellow, magenta, blue, greenish white, etc. The third "white" is a definite green, as even a single layer shows a much narrower reflection band in the third order. Finally the color changes alternately from pink to green as the orders progress. Cryolite starts with a faint yellow, hard for a novice to see, then turns magenta ($\lambda/4$ for green light), blue, white, deep yellow, deep magenta, blue, green, etc. Table I shows the colors seen at various thicknesses for both the sulfide and fluoride. If a plate is to be used at an angle of incidence other than normal, the colors should be viewed *at that angle*, since at high angles they will differ considerably from those shown at normal incidence.

MORE COMPLICATED FILTERS

If a wide band filter reflecting yellow light is desired, zinc sulfide is deposited to the first yellow reflection, and cryolite to the first blue. A narrower band as shown in Fig. 3 can be made with the second yellow and the second blue. Other visible light filters can be made by following the same law; zinc sulfide to the color

[12] M. Banning and F. W. Paul, Rev. Sci. Inst. **15**, 152 (1944).

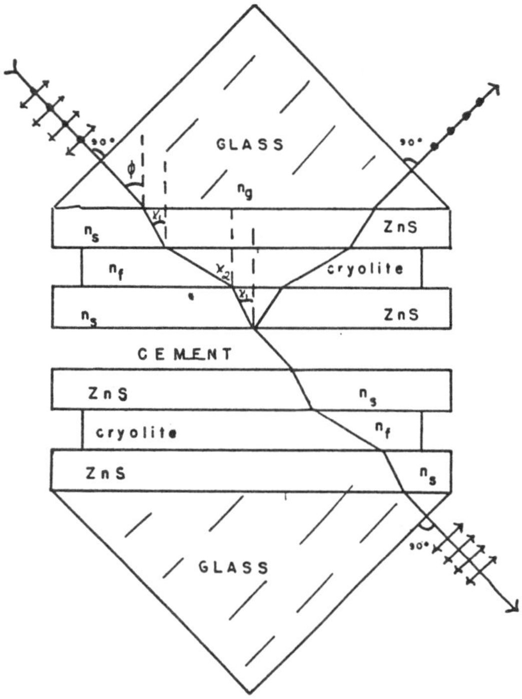

FIG. 7. Schematic diagram of a polarizing beam splitter.

desired to be reflected and cryolite to the complementary color.

For example, a problem arose involving two different phosphors with separate but close emission peaks. The complete emission bands overlapped to such an extent that distinction between the two was difficult. A filter was desired that would cut out the common region between the two bands and at the same time preserve as much as possible of the rest of the emission. Figure 5 shows the two emission bands and the curve of the filter designed to distinguish them. A green reflection was desired, so white sulfide and magenta cryolite were indicated. Since a very sharp cut-off was needed, it was decided to use third-order cryolite and first-order sulfide. The filter curve shown is the result of the first attempt.

Filters such as these cannot be made for the ultraviolet except for the region very close to the visible spectrum, as zinc sulfide begins to have an appreciable absorption around 3650A. However, the infra-red region is still open to exploration, the only limit imposed is that of the necessary thickness of the layers. Since "colors" in the infra-red cannot be detected easily during evaporation it is necessary to calculate what color of visible light will be reflected at the proper thickness for the infra-red, and watch for this. Although the resulting filter may not be perfect the first time, it is usually close enough to predict what minor changes must be made. A specific example of this occurred when it became desirable to sharpen up the cut-off of another kind of filter, a gel filter made at Ohio State University by Dr. Wallace Brode. As shown in Fig. 6, this was opaque from 0.6 to 1.0 micron and then slowly rose in transmission to 60 percent from 1.3 to 1.6. However, it was to be used for a purpose in which radiation of 1.3 microns was not desired but that of 1.4 microns was necessary.

This called for a second-order filter with maximum reflection at 1.3μ. Since second-order films at 1.3μ are very thick, it was decided to compromise on first-order sulfide and second-order cryolite. To find the wave-length reflected by the sulfide in the visible, the following formula is used:

$$nt = \lambda/4 = 3x/4,$$

where λ is the wave-length in the infra-red and x is the wave-length in the visible. In this case the region around 4300A, or second-order blue, is reflected. Thus the sulfide must be evaporated until it reflects second-order blue.

The cryolite is to be second order in the infra-red; to find what color it will reflect in the visible, the appearance of the sulfide, if evaporated to second order at 1.3 microns, is first calculated. In this case the formula is

$$nt = 3\lambda/4 = 5x/4 = 7y/4 = 9z/4.$$

Supplementary maxima are seen to be at 7800, 5570, and 4300A of third, fourth, and fifth orders, respectively. This would, with sulfide, probably appear to be the fourth blue just before the fourth green. The *complementary* cryolite color is the fourth yellow. The curve obtained with second blue sulfide and fourth yellow cryolite is shown in Fig. 6 along with the resultant curve of the two filters.

Another method for making these high order filters is to compute the color of the sulfide, and then find the cryolite color by trial and error; a good guess is usually approximately right. Once the proper colors are found, they are not at all hard to duplicate. A second person, once shown the colors, can reproduce a filter so that the curves made by two different operators are nearly indistinguishable.

If the filter is to be used at an angle of incidence other than normal, and if for some reason it cannot be watched at this angle during evaporation, a simple calculation gives the color it will reflect at normal incidence. In this case the optical thickness is no longer nt but $nt\cos\theta$, where θ is the angle of refraction in the film. θ_1, the angle of refraction in the sulfide, and θ_2, the angle in the fluoride, are computed from the known angle of incidence and refractive indices by Snell's law. Thus

$$nt = \frac{\lambda}{4\cos\theta} = \frac{\lambda'}{4},$$

where λ is the wave-length reflected at the angle of incidence and λ' is the wave-length reflected normally ($\lambda' > \lambda$).

POLARIZING BEAM SPLITTERS

In January, 1943 we observed marked polarization by reflection from multilayer films on glass plates. In March, 1943 a request was made by Dr. S. M. MacNeille, of Eastman Kodak Company, for the development of polarizing beam splitters made by the deposition of multilayer filters on glass prisms which, as he pointed out, would permit light to enter the film at the polarizing angle. These have been made and found to work quite satisfactorily. Films are deposited on the hypotenuse faces of two 45–90 prisms which are then cemented to form a cube, as shown in Fig. 7. Half of the incident light is reflected and half transmitted, each part being plane polarized to more than 98 percent, over a small angle of incidence.

In order for this to be true, the incident light must strike each interface of the multiple film at the polarizing angle, and the total reflectivity of the film must be such that all of the light polarized perpendicular to the plane of incidence is reflected. It is also desirous that the light path be normal to the entrance and exit faces of the cube.

The index of refraction of the glass, as well as those of the two evaporated substances, is very important in assuring that the reflection takes place at the critical angle. As shown in Fig. 7, n_g, n_s, and n_f are the indices of refraction of the glass, sulfide, and fluoride, respectively; while φ, x_1 and x_2 are the angles of incidence at the glass-sulfide, the sulfide-fluoride and the fluoride-sulfide interfaces. From Brewster's and Snell's law the following equation must hold for the polarizing angle.

$$n_g^2 = 2n_f^2 n_s^2/(n_f^2 + n_s^2).$$

Assuming that zinc sulfide has an index of 2.3 and cryolite of 1.25 (evaporated close to 10^{-3} mm Hg), the index of the glass prisms as computed from this formula must be 1.55. Another factor to be taken into account is the V-number of the glass. Zinc sulfide has a high dispersion with a V-number of 17. Thus in order to make certain that all wave-lengths reach the sulfide-fluoride interfaces at approximately the polarizing angle, the dispersion of the glass must be carefully chosen. It can easily be proved that for the ideal case, assuming the dispersion of cryolite is negligible:

$$V_g = n_g(n_f^2 + n_s^2)V_s(n_g - 1)/n_s(2n_f^2 - n_g^2)(n_s - 1).$$

Substituting in the equation gives $V_g = 48.5$. Fortunately, either extra light flint or light barium crown will do for the purpose, with the first alternative being by far the best.

Once the best indices of materials and glass (the same process can be repeated for different evaporated materials) are chosen, the proper thickness of the two films remains to be found. A broad reflection maximum, or first-order green, is desired, to cover as much of the visible spectrum as possible. Because total reflection will take place at the hypotenuse faces before the prisms are cemented together, it is impossible to view the prisms at the polarizing angle while evaporating. Therefore, it is desirable to know what color will be reflected at approximately normal incidence if a film of proper thickness is deposited. This is done by computing the angles x_1 and x_2 and using the cosine formula given above. Roughly, the high index film will appear a faint first yellow and the low index film the first greenish white. With extra light flint, zinc sulfide, and cryolite, the ratios of the undesired to desired polarized components are shown in Table II, for both beams and three different colors, taken normal to the cube faces. This beam splitter was made with three zinc sulfide layers on each prism. The polarization for white light remains greater than 98 percent over 5 degrees on each side of the normal entrance.

TABLE II. Ratio of undesired to desired components of sample polarizing prism.

	Red	Green	Blue
Reflected beam	.0035	.0065	.0103
Transmitted beam	.0138	.0031	.0545

There is no reason why equally good polarizers cannot be made for the infra-red.

Although only zinc sulfide and cryolite are mentioned here, it is solely for the purpose of simplicity. Any high or low index materials will, in general, act the same way. The advantage of sulfide-cryolite films is that they are easy to evaporate and give a large index difference, insuring a large reflection at each interface. However, they are soft and sensitive to atmospheric changes.

The author wishes to express her special gratitude to Dr. J. W. Evans for his constant encouragement and advice on all this work, and to Mr. Robert Hills, Jr., for his assistance during most of the development.

©1989 Optical Society of America. Reprinted from *Applied Optics*, Vol. 28(14), 2847–2853 (July 1989).

Modified MacNeille cube polarizer for a wide angular field

Jacques Mouchart, Jacqueline Begel and Eugène Duda

Well known relationships exist between thicknesses and indices of a stack of optical layers which permit the definition of very good polarizers of light. One of the most famous types of polarizer is the MacNeille cube. In this case the hypotenuse face of an isosceles right angled prism is coated with a polarizing deposit and then cemented to the hypotenuse face of an identical prism. This polarizer can be used over a large spectral range located in the *s* vibration stop band. Unfortunately, the insertion losses for the *p* vibration are large when the angular field is larger than ±2°. Using very simple optical considerations, we design suitable coating structures that allow the use of cube polarizers over a wide angular range, typically ±10° in air, when $T_p > 0.97$ and $T_s \leq 10^{-3}$. However, the spectral range is reduced. Diagrams are given to calculate the thickness of the layers according to the substrate and the indices of the evaporated materials. The prism angle is also determined to have a symmetric angular field in air. Such polarizers are suitable for semiconductor lasers because they can be used without a collimating lens in spite of their large divergence. Good optical characteristics up to $T_p \approx 0.95$ and $T_s \approx 10^{-4}$ over the range of ±5° have been measured for these polarizers manufactured in the Laboratoires de Marcoussis.

I. Introduction

MacNeille cube polarizers are well known in optics.[1,2] Their function is to split the *s* and *p* vibrations of an incident beam in two perpendicular directions, perpendicular and parallel to the plane of incidence. Very good polarizers are obtained when the Banning condition[2] exists between the indices of the substrate and the materials used for the evaporation of the thin films. The polarizing coating consists of a stack of $\lambda_0/4$ alternate high and low index layers. It can be used over a large spectral range which depends on the chosen materials. Usually the range is between $0.8\lambda_0$ and $1.3\lambda_0$. However, the beam must be collimated, the angle of incidence at the entrance face must be normal to within ±2°. When the incidence increases, the insertion losses are higher.

Used for years, they are easy to manufacture. In special applications associated with monochromatic laser sources, it is interesting to use a cube polarizer, in spite of the fact that it does not match well with its surroundings. Its large spectral range is not used, while its small angular range is penalizing. We researched the conditions necessary to spread out the angular field while keeping good optical characteristics. With this result, the spectral range is reduced.

After a short summary concerning the properties of the MacNeille polarizer, we give the reason it is unsuitable for large incidences. Using very simple optical considerations we design coating structures able to be used over a large range of incidence. Some results are given, measured on polarizers manufactured in our laboratory.

II. MacNeille Polarizer

A. Principle

The principle of a polarizer with multidielectric layers is as follows. The reflectance of a beam on a dielectric coating at a skew angle of incidence changes with the *s* or *p* vibration. One can find geometrical configurations for which $R_s \approx 1$ and $T_p \approx 1$, R_s being the reflectance for the *s* vibration and T_p the transmittance for the *p* vibration.

The MacNeille polarizer is a glass cube cut along the diagonal at 45°. The two diagonal faces are cemented allowing the beam to pass, one of them coated with a polarizing deposit[3,4] (see Fig. 1). We shall suppose afterward that the indices of the cement and substrate are the same.

The formula of the optical coating $(\mathrm{LH})^N$ or $(\mathrm{LH})^N\mathrm{L}$, $(\mathrm{HL})^N$, $(\mathrm{HL})^N H$ is composed with a stack of two dielectric layers L and H, called the base period, of low and

The authors are with Centre de Recherches de la Compagnie Générale d'Electricité, Laboratoires de Marcoussis, Route de Nozay, 91460 Marcoussis, France.

Received 2 May 1988.
0003-6935/89/142847-07$02.00/0.

high indices n_L and n_H repeated N times. The optical thicknesses $(nt)_L$ and $(nt)_H$ are in $\lambda/4$ configuration for an incidence θ_{sub} of 45° in the glass (corresponding to $\theta_0 = 0$ at the entrance face in air).

The angles θ_0 and θ_D (at the diagonal face) are interlinked by the relation

$$\theta_D = \arcsin\left(\frac{\sin\theta_0}{n_{sub}}\right) + \theta_{sub}.$$

The necessary conditions to obtain a polarizing deposit are that they be located in the stop band of the base period for the s vibration, have a high reflectance R_s (this determines the number N of the base periods), be located in the passband of the base period for the p vibration, and have a reflectance R_p as low as possible.

1. p Vibration

To obtain a MacNeille polarizer, the choice of materials is such that for $\theta_{sub} = 45°$, the following relation must be achieved between the admittances y_p for the p vibration:

$$\left(\frac{n}{\cos\theta}\right)_L = \left(\frac{n}{\cos\theta}\right)_H, \tag{1}$$

where θ is the angle of the beam in the layer.

For $(LH)^N$, $T_p = 1$ when the admittances $(y_p)_L$ and $(y_p)_H$ are equal. If we suppose that the indices of the materials are independent of the wavelength (which is often true in the IR) we have $T_p \# 1$ for this incidence θ_{sub}, whatever the wavelength, when the p admittance of the substrate $(y_p)_{sub}$ is such that $(y_p)_{sub} \# (y_p)_L = (y_p)_H$.

According to Snell's law,

$$(n \sin\theta)_L = (n \sin\theta)_H = (n \sin\theta)_{sub}, \tag{2}$$

and $\theta_{sub} = 45°$, we obtain the Banning relation between indices:

$$n_{sub}^2 = \frac{2n_L^2 n_H^2}{n_L^2 + n_H^2}. \tag{3}$$

2. s Vibration

For the s vibration, the admittances for the layers, $(n \cos\theta)_L$ and $(n \cos\theta)_H$, are different. There is an index jump at each interface between two layers, and then the reflectance R_s exists. Because the layers are in a $\lambda_0/4$ configuration, the wavelength λ_0 is at the center of the stop band. The coating is a polarizing one for each wavelength located in the stop band. The reflectance R_s increases with the number N of the base periods.

Figure 2 shows the theoretical transmittances T_p and T_s vs the wavelength λ when $\theta_0 = 0$. The design of the coating (coating 1) is $(LH)^{11}$ with $n_{sub} = 1.70$ $n_L = 1.45$ and $n_H = 2.20$ and the optical thicknesses $(nt)_L = 0.447$ μm and $(nt)_H = 0.298$ μm. The coating is centered at $\lambda_0 = 1$ μm.

B. Influence of the Incident Angle

The consequence of a change in the angle of incidence θ_D is a mismatch for the p admittances of the layers, Eq. (1) being no longer true. A p stop band

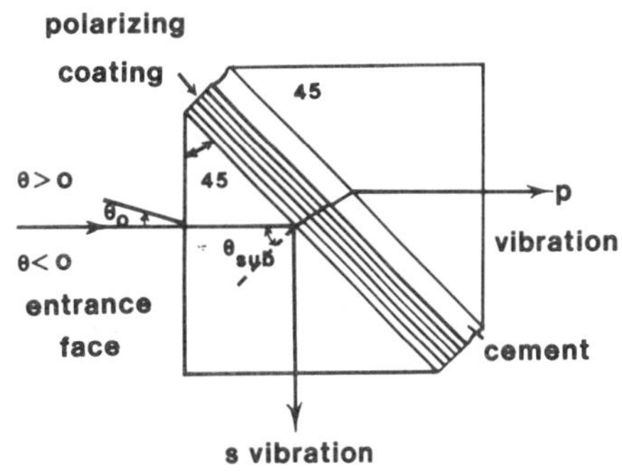

Fig. 1. MacNeille cube polarizer.

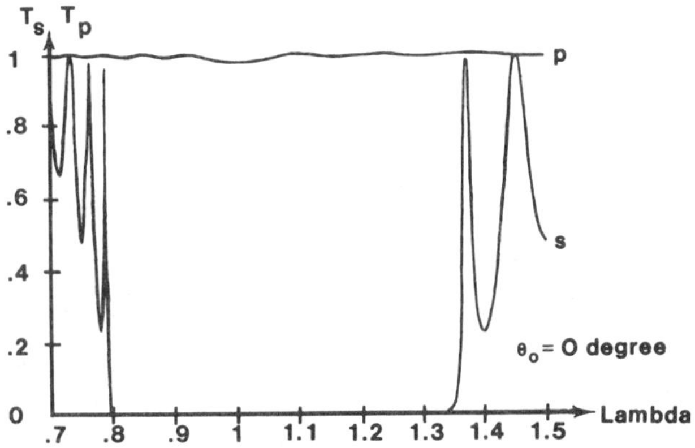

Fig. 2. MacNeille cube polarizer centered at $\lambda = 1$. Transmittances T_p and T_s vs wavelength for the incidence $\theta_0 = 0$ in air ($\theta_{sub} = 45°$ in the substrate). The design of the coating is $(LH)^{11}$ with $n_{sub} = 1.70$, $n_L = 1.45$, $(nt)_L = 0.447$ μm, $n_H = 2.20$, $(nt)_H = 0.298$ μm.

centered near λ_0 appears. Rigorously, the layers are no more in the $\lambda_0/4$ configuration. However, they are in about the same configuration when the variation of the incidence is small. There is a residual reflectance R_p in the middle of the stop band whose value is in a first approximation for a coating $(LH)^N$:

$$R_p = \left(\frac{1 - z^{2N}}{1 + z^{2N}}\right)^2 \tag{4}$$

with $z = (y_p)_L/(y_p)_H$ and the entrance and exit media identical. A variation in the angle of incidence involves a fall in the p transmittance T_p. The s admittances are also changed. The reflectance R_s is modified as well as the range of the stop band.

The cube is still a polarizer when used in transmission, but it is not a polarizer when used in reflection, because a mixing appears between the s and p vibrations. Figure 3 shows the p admittances of the L and H layers as well as the substrate as a function of the angle incidence θ_0 on the entrance face. Curves are for coating 1. Figure 4 is similar to Fig. 2: The transmittances T_p and T_s are given as a function of the wavelength for $\theta_0 = -2°$. Thus the MacNeille cube polarizer in which all the optical thicknesses are in the $\lambda/4$ configuration is not suitable for large angular field use.

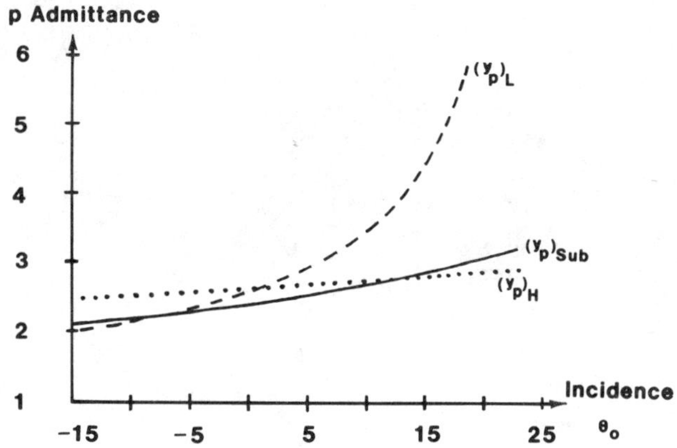

Fig. 3. MacNeille cube polarizer. Values of the p admittances y_p for the substrate and the L and H layers vs incidence θ_0 in air (in degrees).

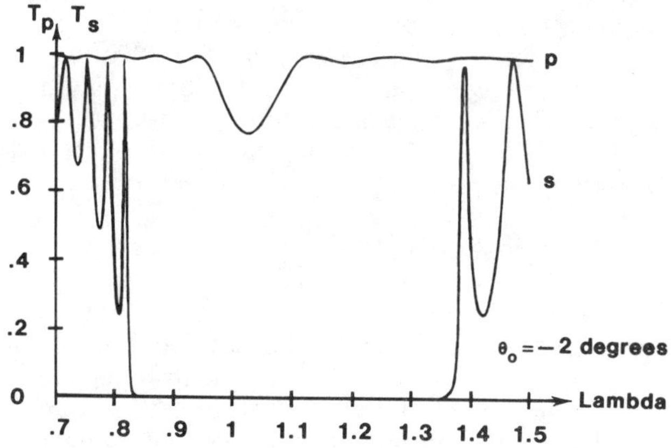

Fig. 4. MacNeille cube polarizer. Same as for Fig. 2 but with incidence $\theta_0 = -2°$ in air.

III. Large Field Polarizer

A. Principle

To understand the optical functioning of a polarizing coating, the best way is to consider a stack of symmetrical bases instead of bases with two different layers as before. We consider a symmetrical base composed of three layers, with optical thicknesses not necessarily $\lambda/4$. The formula of the coating is then $(L/2\ H\ L/2)^N$.

We shall introduce two parameters $\sum$ and P related to the optical thicknesses $(nt)_{L/2}$ and $(nt)_H$ by the following relations:

$$\sum = 2(nt)_{L/2} + (nt)_H.$$

$\sum$ is the sum of the optical thicknesses in the base, and

$$P = \frac{2(nt)_{L/2}}{2(nt)_{L/2} + (nt)_H}.$$

P is the ratio between the low index optical thickness and the sum of the optical thicknesses in the base.

Knowledge of $\sum$ and P is sufficient to calculate the elements of the base $(nt)_{L/2}$ and $(nt)_H$:

$$(nt)_{L/2} = \sum P/2 \qquad (nt)_H = \sum(1 - P).$$

In a first approximation, small changes in thicknesses of the layers with $\sum$ constant affect the spectral centering of the base period only slightly.

1. *p Vibration*

The aim of this study is to improve the transmittance T_p when the incidence increases. For a position located in the passband of the symmetrical base, the base is equivalent to a single layer whose oblique admittance E, the Herpin admittance, is varying with wavelength and incidence.[4] For the p vibration and a three-layer symmetrical base, the admittance is given by

$$E_p = (y_p)_L \left(\frac{u - v}{u + v}\right)^{1/2} \tag{5}$$

with $u = \sin 2\phi_{L/2} \cos\phi_H + 1/2(z + 1/z) \times \cos 2\phi_{L/2} \sin\phi_H$,

$v = 1/2\,(z - 1/z) \sin\phi_H$,

$\phi_{L/2} = 2\pi/\lambda\ (nt \cos\theta)_{L/2}$, and

$\phi_H = 2\pi/\lambda\ (nt \cos\theta)_H$.

First, we shall assume that Eqs. (1) and (3) are satisfied for $\theta_D = 45°$. For this incidence $z = 1$: the Herpin admittance E_p is equal to both $(y_p)_L$ and $(y_p)_H$ and is very near the admittance of the substrate $(y_p)_{sub}$. Notice that here the value of E_p is independent of the thicknesses of the layers.

Every shift in the angle of incidence changes the values of $(y_p)_{sub}$, $(y_p)_L$, $(y_p)_H$ and consequently the values of E_p, a mismatch between admittances appears. The inequality $(y_p)_L \neq (y_p)_H$ involves $z \neq 1$, and thus E_p varies with a change in the thickness of the base.

The difference between $(y_p)_{sub}$ and E_p creates a residual reflectance which depends on the number N of the bases. The maximum reflectance is given by the envelope reflectance defined in the passband of the base.[5] Because the entrance and exit media are the same, the maximum is given by

$$R_p = \left(\frac{1 - q^2}{1 + q^2}\right)^2 \qquad 1/q = (y_p)_{sub}/E_p.$$

The existence of R_p is one reason which limits the angular field of the polarizer.

The problem now is to vary the thicknesses to minimize the mismatch between $(y_p)_{sub}$ and E_p over the whole angular field. Thus owing to the symmetric base design one can modify the admittances to spread out the field of use. For example, we indicate in Fig. 5 the variations of the admittances $(y_p)_{sub}$ and E_p vs incidence. The indices of the materials are:

$$n_{sub} = 1.70 \qquad n_L = 1.45 \qquad n_H = 2.20$$

and the following thicknesses:

Coating 2: $(nt)_{L/2} = 0.2235\ \mu m$, $(nt)_H = 0.298\ \mu m$, with $\sum = 0.745\ \mu m$ and $P = 0.60$, corresponding to the design of a $(\lambda_0/8, \lambda_0/4, \lambda_0/8)$ base with $\lambda_0 = 1$. This coating has a $\lambda_0/4$ configuration except for the first and last layers, which are of $\lambda_0/8$ thickness.

Coating 3: $(nt)_{L/2} = 0.25225\ \mu m$, $(nt)_H = 0.2405\ \mu m$, with $\sum = 0.745\ \mu m$ and $P = 0.677$, for which the Herpin admittance is matched more nearly.

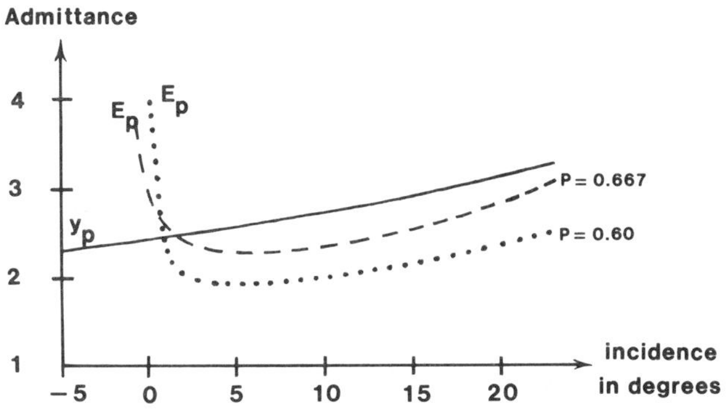

Fig. 5. Use of a symmetrical base. Values of the p admittance of the substrate $(y_p)_{\text{sub}}$ and of the p Herpin equivalent admittance E_p as a function of incidence θ_0 for two different values of the parameter P.

This reasoning is not changed when Eqs. (1) and (3) are not satisfied. The intersection of the representative curves of $(y_p)_{\text{sub}}$ and E_p (Fig. 5) is obtained now for incidences other than $\theta_0 = 0$. A change in the angle θ_{sub} between the entrance face and diagonal of the cube allows centering the angular field.

2. s Vibration

A modification of the thickness $(nt)_{L/2}$ and $(nt)_H$ involves also a change in the reflectance R_s as well as in the angular field of the s stop band. The reflectance R_s and the width of the stop band are maximal when the thickness are in the $\lambda_0/4$ configuration, that is,

$$(nt\cos\theta)_{L/2} = \lambda_0/8 \qquad (nt\cos\theta)_H = \lambda_0/4.$$

Thus an arrangement must be found between the transmittance T_p and the reflectance R_s.

B. Choice of the Thicknesses

At first sight, it is not easy to find the optimal thickness of the base, because the optimization must be made on both parameters $\sum$ and P. In practice, we proceed as follows: $\sum$ is always kept constant, that is, the spectral centering of the coating is not changed significantly. At first the optimization is made for P and then for the wavelength λ. When P is taken as a parameter, solutions are found by calculation of the diagram of angular solutions for given indices of the layers and substrate. Four different limiting conditions are shown in this diagram:

(1) the s stop band of the base;

(2) the p passband of the base;

(3) the insertion losses R_p;

(4) the angle of incidence θ_0 in air which corresponds to an angle of 90° in the lower index (layers or substrate).

Figure 6 shows the diagram calculated for $\lambda_0 = 1\ \mu$m and valid when $n_{\text{sub}} = 1.70$, $n_L = 1.45$, and $n_H = 2.2$, the insertion losses R_p are imposed to be <0.03. The solution which has a larger angular field ($0 \leq \theta_0 \leq 23°$) corresponds to $P = 0.667$.

Figures 7 and 8 show diagrams valid for wavelengths of 1.05 and 1.1 μm. In a first approximation, one can see a translation of the diagrams followed by a slight

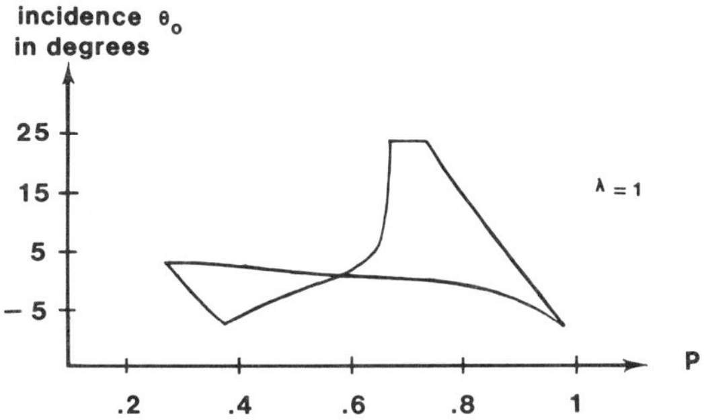

Fig. 6. Research for large field polarizer. Diagram of the angular existence of solutions vs the P parameter for $\lambda = 1$. The indices of materials are $n_{\text{sub}} = 1.70$, $n_L = 1.45$, and $n_H = 2.20$.

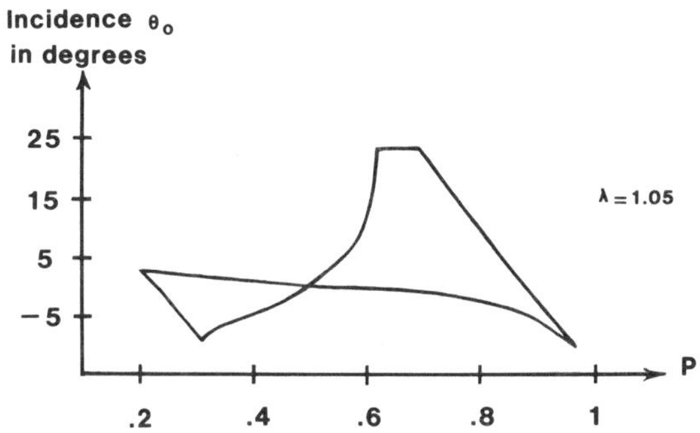

Fig. 7. Same as for Fig. 6 but for $\lambda = 1.05\ \mu$m.

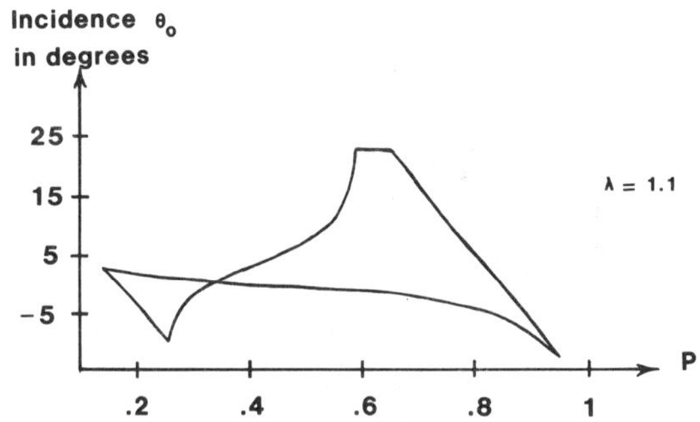

Fig. 8. Same as for Fig. 6 but for $\lambda = 1.10\ \mu$m.

deformation. With the above example optimization is obtained when $P \approx 0.667\ \mu$m and $\lambda = 1.05\ \mu$m. Centered at $\lambda_0 = 1$, this solution is

$$(nt)_{L/2} = 0.2483\ \mu m \qquad (nt)_H = 0.2483\ \mu m.$$

C. Easy Determination of Solutions

As the $\sum$ and P parameters can vary greatly with the chosen materials, we have introduced the variables $\sum_0$ and P_0 equal to $\sum$ and P when the layers are in a $\lambda_0/4$ configuration at the oblique incidence (45° in the substrate). These variables $\sum_0$ and P_0 are taken as reference values, so

$$\sum = \sigma \sum_0 \qquad P = \rho P_0,$$

σ and ρ being the relative characteristics of the base with respect to $\lambda_0/4$ layers. For $\lambda_0 = 1\ \mu$m, the optical thicknesses $(nt)^*_{L/2}$ and $(nt)^*_H$ of the $\lambda_0/4$ layers are

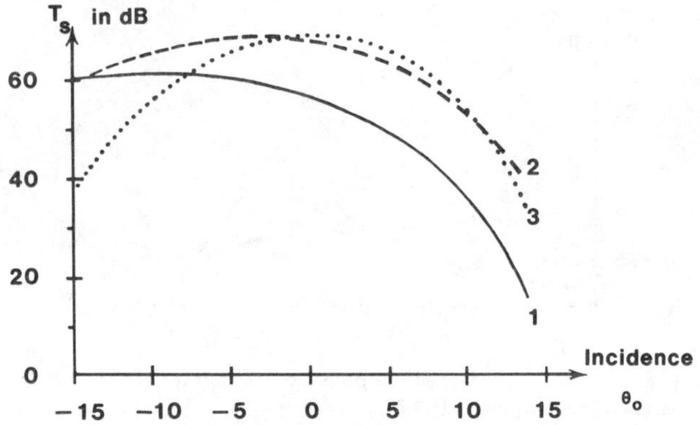

Fig. 9. Large field polarizer. Transmittance T_s in decibels vs incidence θ_0 for three different values of the parameter P. The design of the coating is $(L/2\ H\ L/2)^{11}$ with $P = 1$ (curve 1), $P = 1.15$ (curve 2), $P = 1.30$ (curve 3), and $\lambda = 1$.

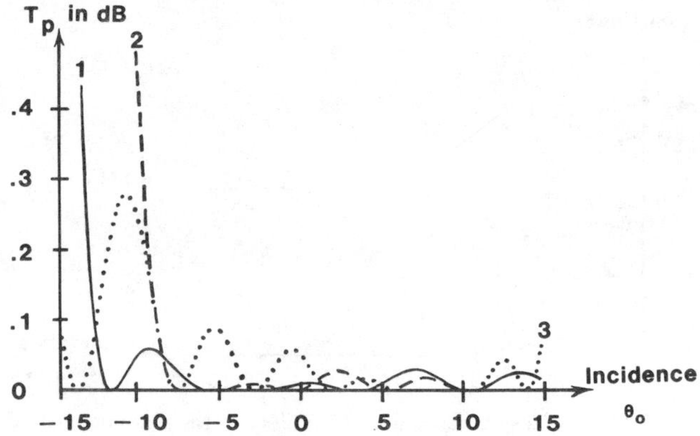

Fig. 10. Large field polarizer. Transmittance T_p in decibels vs incidence θ_0. Same coatings as for Fig. 9.

$$(nt)^*_{L/2} = \frac{0.125\ \mu}{\left[1 - \frac{(n \sin\theta)^2_{sub}}{n_L^2}\right]^{1/2}}, \qquad (nt)^*_H = \frac{0.25\ \mu}{\left[1 - \frac{(n \sin\theta)^2_{sub}}{n_H^2}\right]^{1/2}}.$$

Then $\sum_0 = 2\,(nt)^*_{L/2} + (nt)^*_H$ and $P_0 = 2\,(nt)^*_{L/2}/\sum_0$.

Consequently, a base defined by $\sigma = 1$ and $\rho = 1$ corresponds to a coating, the layers of which are in a $\lambda_0/4$ configuration, except for the first and last layers of thickness $\lambda_0/8$.

Practically, for the majority of materials used, the optimal centering of a $\lambda_0/4$ deposit (with, however, the first and last layers of $\lambda_0/8$) is located near $\lambda_0 = 1.15\ \mu$m and not at $\lambda_0 = 1\ \mu$m. To take account of this spectral shift, we found that the σ and ρ parameters are linked by the empirical relation:

$$1/\sigma \approx 1.92 - 0.77\rho.$$

In that case, the base is entirely characterized by the single parameter ρ. Thus the value of ρ can be adjusted to find the largest angular field for the use of the polarizer, which depends on the minimum values imposed for the transmittance T_ρ and the reflectance R_s.

For example, Figs. 9 and 10 show the transmittances T_p and T_s (expressed in decibels) vs the incidence θ_0 ($\lambda_0 = 1\ \mu$m for three different ρ values. The design of the coating is $(L/2\ H\ L/2)^{11}$ with $n_{sub} = 1.70$, $n_L = 1.45$, $n_H = 2.20$, and

ρ		$(nt)_{L/2}$	$(nt)_H$
$\rho = 1$	which involves	$(nt)_{L/2} = 0.1944\ \mu$m	$(nt)_H = 0.2596\ \mu$m
1.15		$0.2485\ \mu$m	$0.2237\ \mu$m
1.30		$0.3162\ \mu$m	$0.1789\ \mu$m.

The angle between the entrance face and diagonal is 50°, so that the angular field is approximately symmetrical.

The blocking rate for the s vibration (60 dB) is lower in practice, because the scattering in the layers as well as birefringence of the substrate must be taken into account. The insertion losses for the p vibration (<0.05 dB) are generally higher and are produced by a small lack of efficiency in the antireflective coating on the entrance and exit faces of the cube and by a small mismatch between the indices of the substrate and cement.

D. Fabrication and Measurement

We have manufactured small (2.5 mm) and large (20 mm) cube polarizers. The small polarizers are used with a semiconductor laser emitting at $\lambda = 1.55\ \mu$m. The optical configuration needs a 5° angular field polarizer.

With the materials used,

substrate:	$n_{sub} = 1.70$	(D2050 Sovirel glass),
layer:	low index $n_L = 1.45$	(SiO_2),
	high index $n_H = 1.95$	(substance 1 Merck),

the angular field is symmetrical when the diagonal face is at 45°. The design of the polarizing coating is $(L/2\ H\ L/2)^{11}$ with

$$(nt)_{L/2} = 0.303\ \mu\text{m} \qquad (nt)_H = 0.43\ \mu\text{m}.$$

The cube polarizers are cemented with an optical cement, the index of which is 1.57. The entrance and exit faces are coated with an antireflection coating centered at $\lambda = 1.55\ \mu$m; the lack of efficiency is estimated to be 0.5%/face.

Figures 11 and 12 show the theoretical transmittances T_p and T_s (in decibels) as a function of incidence θ_0 for the 1.50-, 1.55-, and 1.60-μm wavelengths.

The measurement is made using an automatic testing system shown in Fig. 13. The light source S is a BH diode laser emitting at $\lambda = 1.531\ \mu$m a linearly polarized beam. This beam is collimated by a Selfoc lens L. The linear vibration is changed to a circular vibration by a Babinet B compensator used in quarterwave. A

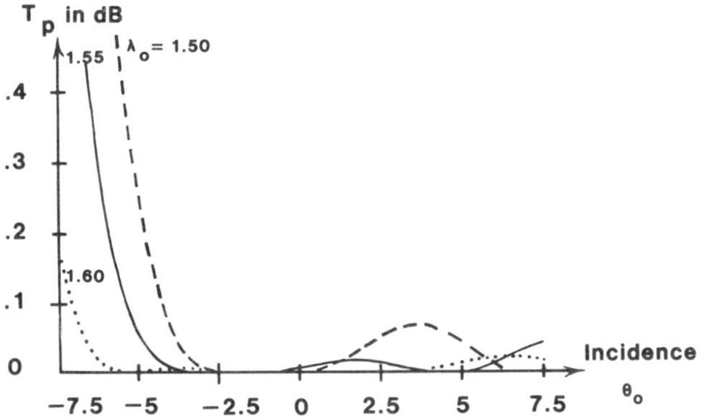

Fig. 11. Large field polarizer. Theoretical transmittance T_p as a function of incidence for three wavelengths: 1.50–1.55 and 1.60 μm. The design is $(L/2\,H\,L/2)^{11}$ with $n_{sub} = 1.70$, $n_L = 1.45$, and $n_H = 1.95$. Optical thicknesses: $(nt)_{L/2} = 0.303\ \mu m$, $(nt)_H = 0.43\ \mu m$.

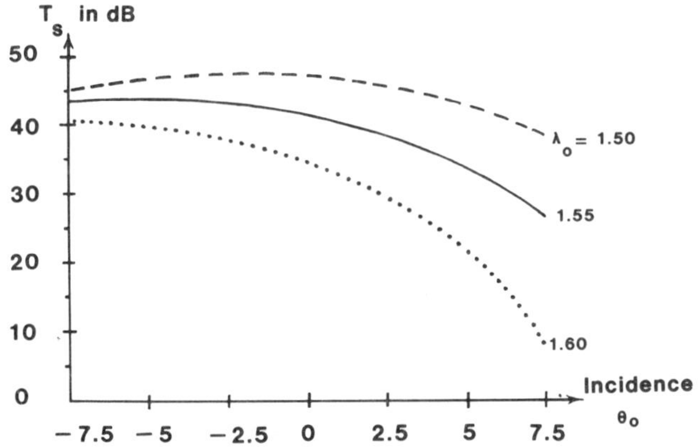

Fig. 12. Large field polarizer. Theoretical transmittance T_s in decibels as a function of incidence θ_0. Same coating as for Fig. 11.

Glazebrook polarizer G adjustable by rotation (360° range) perpendicular to the beam gives a linear vibration. The polarizer to test P is placed behind the Glazebrook polarizer. The beam is then focused on a InGa/As detector D. The measurement is made in synchronous detection; the frequency of modulation of the beam (1 kHz) is obtained by a chopper C.

Figures 14 and 15 show the transmittances T_p and T_s in decibels measured on a large polarizer in the meridian plane (+) and in a plane perpendicular (×) for incidences of −6 and +6°. The blocking rate for the s vibration is ~40 dB in accordance with theory. (For example, for two crossed Glazebrook polarizers the measurement is ~60 dB.) The insertion losses for the p vibration are between 0.2 and 0.3 dB.

The optical characteristics for small polarizers are lower. We measure for incidences varying between ±5°

insertion losses for the p vibration: between 0.4 and 0.8 dB,

blocking rate for the s vibration: between 30 and 37 dB.

Measurement made on a cube similar to a polarizer but without a polarizing coating gives the following results:

insertion losses: 0.3 dB;

birefringence, measured between two crossed Glazebrook polarizers at 54 dB.

The appearance of a natural oxidation on the glass and difficulty in cleaning the surfaces of the small polarizers before coating are probably the reasons for the decrease in the performance of these small polarizers.

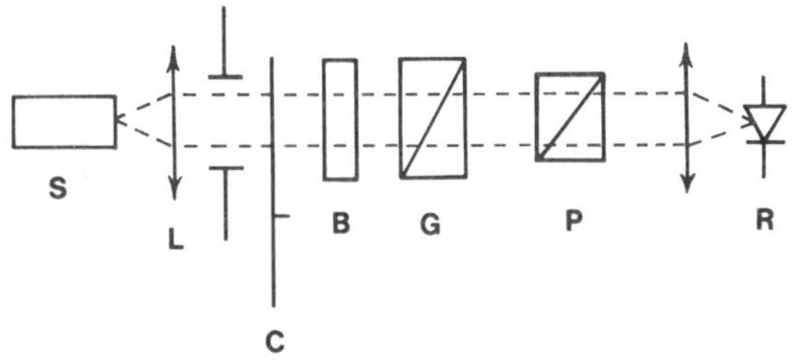

Fig. 13. Automatic test system: S, light source emitting at λ = 1.531 μm; L, Selfoc; C, chopper; B, Babinet compensator; G, Glazebrook polarizer; P, polarizer to test; R, receptor.

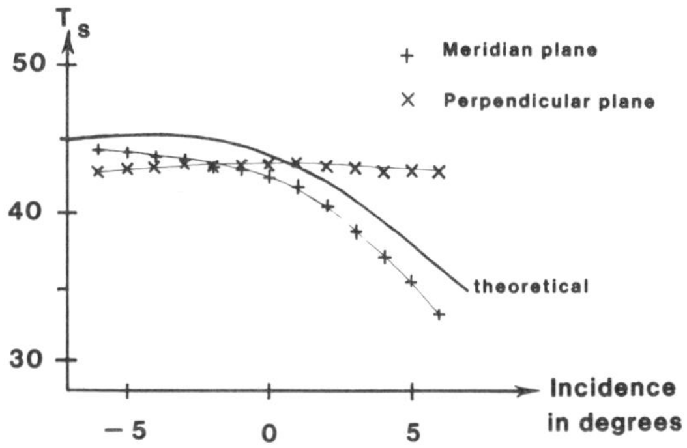

Fig. 14. Measured values of transmittance T_p in decibels (see Fig. 11) in the meridian plane (+) and perpendicular (×). These values are compared with the theory for the meridian plane (full line curve).

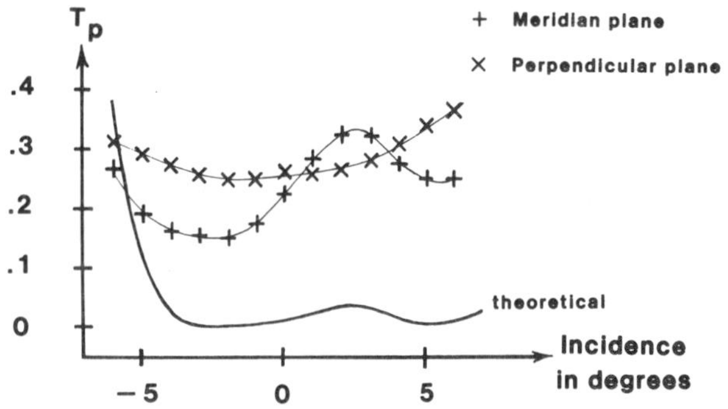

Fig. 15. Measured values of transmittance T_s in decibels. Same coating as for Fig. 14.

IV. Conclusion

MacNeille cube polarizers are efficient for small angular fields. With the introduction of symmetric base periods, the angular field can be increased to about ±10° with classical materials when they are used with monochromatic light sources such as lasers.

A method is presented for determination of the thicknesses of the base where the optimization is made using a single parameter. Experimental results obtained on polarizers manufactured at the Laboratoires de Marcoussis are in good agreement with theoretical values.

The authors take the opportunity to thank Jacques Augé and Jean-Jacques Guérin who took the measurements and Julian Cave for help during the translation of this paper.

References

1. S. M. MacNeille, "Beam Splitter," U.S. Patent 2,403,731 (9 July 1946).
2. M. Banning, "Practical Methods of Making and Using Multilayer Filters," J. Opt. Soc. Am. **37**, 792–797 (1947).
3. H. A. MacLeod, *Thin Film Optical Filters* (Macmillan, New York, 1986).
4. R. P. Netterfield, "Practical Thin Film Polarizing Beam Splitters," Opt. Acta **24**, 69–79 (1977).
5. J. Mouchart, "Revêtements optiques à structure périodique. I: Fonctions caractéristiques," Appl. Opt. **20**, 4201–4206 (1981).

The Wire Grid as a Near-Infrared Polarizer

GEORGE R. BIRD AND MAXFIELD PARRISH, JR.
Research Division, Polaroid Corporation, Cambridge 39, Massachusetts
(Received January 22, 1960)

Wire grids have been fabricated with microscopic elements and periodicity. These wire grids behave as linear polarizers over large portions of the infrared spectrum, confirming predictions from electromagnetic theory. The method of fabrication of grids with periods as small as 463 mμ is described. Limits on the performance of a wire-grid polarizer are set by fineness of period, optical constants of the metal, spectral transmission of the supporting substrate, and by skin-depth effects. Within these limits, excellent polarizer performance may be obtained over a very wide band of wavelengths, and these new wire-grid polarizers do, in fact, cover the range 2–15 μ and beyond. Appreciable polarization is observed at near-infrared and visible wavelengths. Wire grids may be used in convergent beams of radiation or in restricted geometrical arrangements, unlike many pile-of-plate or prism polarizers. These wire-grid polarizers have proved useful in measuring the imperfections of other types of infrared polarizers.

WIRE-GRID POLARIZERS

A. History

THE wire grid is actually an antique optical element. David Rittenhouse made and demonstrated the first known diffraction grating in 1786 in Philadelphia.[1] His transmission grating was a grid of hairs (which might be considered as the finest wires then available) positioned by the threads of two screws. Fraunhofer began his classic work in spectroscopy in 1810 with a wire-grid diffraction grating. Heinrich Hertz first used a wire grid as a polarizer in 1888 to test the properties of the newly discovered radio waves.[2] He found that a wire grid with 1-mm copper wires spaced 3 cm apart reflected or transmitted 66-cm radio waves as the wires were turned parallel or perpendicular to the electric vector of the radiation. In other words, the wire grid behaved as a dielectric sheet except when turned so that the radiation field could induce current to flow along the wires.

Hertz's results with radio waves were extended to the far infrared by duBois and Rubens[3] in 1911. Noble-metal wires of 25-μ diam and larger were used, and the experimenters used broad wavelength bands centered about 24, 52, 100, 108, and 314 μ. Some of the details of the behavior of wire grids were obscured by the broad bands of wavelengths used, but the implication was clear that finer wire grids would polarize at still shorter wavelengths.

The most recent and definitive investigation of grids of macroscopic wires is a study by Peters and Pursley[4] in which both 3-cm microwaves and 80- to 500-μ infrared radiation were used. The finest wire used was 50-μ diam, and this was mounted with 100 μ and larger periods. A modern grating spectrograph was used to provide far-infrared radiation of high spectral purity. Peters and Pursley demonstrated that the optical properties of a system were preserved as wavelength, period, and wire diameter were identically scaled up or down in size. Correlations were obtained with the theoretical treatment by Ignatowsky[5] for ordinary wires and by

[1] (a) D. Rittenhouse, Trans. Amer. Phil. Soc. **2**, 201 (1786); (b) For a more readily available account of 1(a), see T. D. Cope, J. Franklin Inst. **214**, 99 (1932); (c) The writers are indebted to Professor Brooke Hindle of New York University for detailed information on Rittenhouse's work.

[2] H. Hertz, *Electric Waves* (Macmillan and Company, Ltd., London, 1893), p. 177.

[3] H. duBois and H. Rubens, Ann. Physik **35**, 243 (1911).

[4] C. W. Peters and W. K. Pursley, J. Opt. Soc. Am. **42**, 877(A) (1954). (a) W. K. Pursley, Doctoral thesis, University of Michigan, 1956.

[5] W. v. Ignatowsky, Ann. Physik, **18**, 495 (1905); **23**, 875 (1907); **25**, 116 (1908); **26**, 1032 (1908); **44**, 369 (1914).

Baldwin and Heins[6] for strips. Peters and Pursley varied the angle of incidence of radiation on the grating, and showed that the transmittance was insensitive to changes of angle up to 30° from the normal, unless the ratio of wavelength to period was near unity. Since only good conductors were used with far-infrared or microwave radiation, it was not necessary to consider resistive loss in their work. If poor conductors are used, or near-infrared and visible radiation is involved, losses become important. These effects have been treated theoretically by Lewis and Casey.[7]

B. Grids of Microscopic Wires

A single mechanical problem has prevented the obvious extension of wire-grid techniques to the visible and near-infrared spectrum. Wires less than 0.001 in. or 25-μ diam are drawn only with the greatest difficulty, and are generally so fragile that forming a multiple array is a formidable task. Techniques do exist for producing individual wires as small as 1-μ diam (the Wollaston process for bolometer wire, for example[8]), but these are clearly inadequate for preparation of a grid of reasonable area. What has been lacking is a process for preparing a complete grid of microscopic wires on a supporting substrate in a single operation. We have solved this problem, and are thus able to examine the behavior of wire grids in the near-infrared and visible regions.

The solution to this problem is to use a transmission grating as a form, and to deposit metal preferentially on some portions of the individual grooves. The best and most easily controlled method for selective deposition is a vacuum evaporation at nearly grazing incidence to the grating surface.[9] This forms an array of metallic elements on part of one face of each groove. The individual elements run the full length of the grating (neglecting occasional defects) and are considerably smaller in width than the period of the grating. Thus, a wire grid of reasonable area, having microscopic elements and period, is formed in a single operation. Provided that certain qualifications are met by the metal and the material of the transmission grating, this assembly should behave exactly like the coarser wire-grid polarizers constructed by earlier workers. Since the metallic elements have been scaled down in size, the polarizing properties of the new grids occur in the infrared and visible spectrum.

It is fortunate that diffraction grating technology has progressed to a high level in recent years, for the original grating used to form transparent replicas has a profound effect on the finished polarizers made from the replicas.

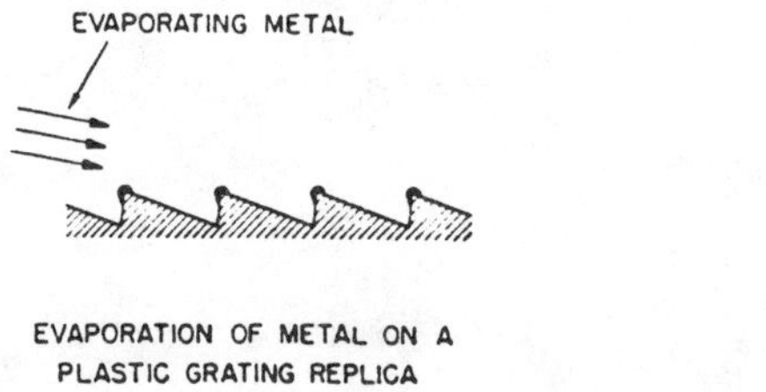

FIG. 1. Evaporation of metal on a plastic grating replica to form a wire grid. The source is a fine tungsten coil at some distance from the replica. The effective size of the source is somewhat larger than the coil because of residual gas pressure and collisions between metal atoms during evaporation. To limit deposition to the steep faces of the grooves, the source is positioned well below the plane defined by the broad faces of the grooves.

The period of the finished wire grid is set by the period of the original grating. Equally important, the groove shape of the original grating is controlled by setting the ruling diamond, and this shape is copied or blazed in the transparent replica. The detailed form of the finished metal deposit is in part determined by the shape of the grooves. We have found the best groove shape to be a blaze near 20°, i.e., one face inclined about 20° from the face of the grating, and the opposite face at about the complement of this angle.

If two gratings with reasonably good groove shape are available, one naturally chooses the grating with the smaller period, since polarizers made from this grating will have a shorter wavelength limit to their satisfactory polarization properties. The most satisfactory results to date have been obtained with the Bausch and Lomb 2160 lines/mm rulings, with blaze at 18° 22′. In addition to possessing a good groove shape, these gratings have the smallest period ever ruled with uniformity over a large area. The Bausch and Lomb ruling group has produced 10 800 lines/mm rulings, but only in very small areas. These were intended only for x-ray spectroscopy, and have a shallow unblazed groove poorly suited for controlled evaporation of metallic elements. It is intriguing to consider what might be done with a grating of this period with a favorable groove shape, but the production of such an object is a formidable task.

The metal to be deposited on the grating is chosen for two properties. It must be highly conductive for the wavelengths involved, which is equivalent to a requirement of high reflectivity for the bulk metal. Second, the metal must be resistant to corrosion, either through lack of reactivity or through formation of a protective oxide skin. Gold and aluminum have been used with satisfactory results.

The deposition of metal is carried out in a conventional bell jar apparatus. The configuration is as shown in Fig. 1, with the evaporating filament positioned well below the plane defined by the broad faces of the replica. Large-area polarizers are prepared by defining a deposition area with baffles and passing the replica across this area repeatedly. Thus, all portions of the replica

[6] G. L. Baldwin and A. E. Heins, Math. Scand. 2, 103 (1954).
[7] E. A. Lewis and J. P. Casey, J. Appl. Phys. 23, 605 (1952).
[8] C. G. Montgomery, *Technique of Microwave Measurements* (McGraw-Hill Book Company, Inc., New York, 1947), pp. 157–160.
[9] Patent application has been filed on this method of preparing grids of microscopic wires, and on other, less direct methods.

FIG. 2. Electronmicrograph of the wire elements from a gold grid. The metallic elements shown here are from a section of the gold grid described in the text and in Fig. 4. The gold strips were lifted from the Kel-F replica with a layer of Scotch Tape cement, the cement and strips placed on a conventional electron microscope sample grid, and the cement leached out with organic solvent. The resulting sample has strips less uniform than the original grid and set at a different angle, but the photograph will illustrate the continuity and general form of the microscopic wire grid. Total magnification is 11,000×. The black areas correspond to the metal elements of the grid. The electron microscope replica was prepared by E. S. Emerson. He was also able to make visual observation of the original grid on $(CF_2CFCl)_n$, and found that the metal elements were much more uniform than those appearing in the figure. The plastic was so rapidly degraded by the very high beam current required that no photographs could be obtained.

receive the same amount of metal from the same average angle. The amount of metal deposited is monitored by observing light transmission through a flat piece of glass just beside the replica. The actual shape of the deposit is unknown, but may be rather like a comma in cross section, as shown in Fig. 1. An electronmicrograph of the elements from a 2160 lines/mm gold grid at 11 000× magnification, taken by E. S. Emerson, is shown in Fig. 2.

The plastic substrate does not participate directly in the optical behavior of the grid, but there are certain stringent requirements to be met. First of all, the plastic must be suitable for solvent casting or for some other method giving detailed replication of the master grating surface. Second, the plastic must be transparent to the radiation to be polarized—in this case the visible and infrared spectrum. There is no single plastic which is transparent throughout the infrared, but two polymers, polyethylene and poly-chlorotrifluorethylene, do have broad transmission bands. Excellent replicas are obtained with $(CF_2CFCl)_n$, and this plastic is highly transparent out to 7 μ. Polyethylene is generally superior for broader coverage of the infrared spectrum, having only three narrow absorption bands at 3.5, 6.8, and 13.8 μ.

The formation of the plastic replica is actually the most difficult step in fabrication of a wire-grid polarizer, since the master grating should be recovered intact for further replications. Excellent replicas of $(CF_2CFCl)_n$ have been prepared by casting low-molecular-weight polymer from ketone solvents, but the casting master is attacked by the solvent after several casts. Polyethylene is not readily solvent-cast, so we have used hot pressing techniques for this plastic. The master gratings available to us have a plastic base, and are not suitable for hot pressing, so a metallic pressing master was prepared by vacuum evaporation and electroplating on a $(CF_2CFCl)_n$ cast replica, and this pressing master was used to emboss 3-mil polyethylene sheet. The polyethylene embossed replicas are thus third-generation replicas from the master (which is in turn a manufacturer's replica of a ruled original). The embossed polyethylene replicas made to data have been markedly inferior in quality of replication to the $(CF_2CFCl)_n$ cast replicas, but still good enough to demonstrate the ultimate superiority of polyethylene for broad-wavelength coverage. The search for better methods of polyethylene replication is being continued.

PERFORMANCE OF WIRE GRID POLARIZERS

The performance of two wire grid polarizers is shown in Figs. 3 and 4, and transmission data at selected wavelengths are given in Table I. The first of these polarizers has a relatively thick deposit of aluminum on a $(CF_2CFCl)_n$ replica, and the second has a thin deposit of gold on the same substrate. Both cast replicas are derived from a Bausch and Lomb replica grating, and have 52×52 mm ruled area and one metallic element per 463 mμ.

The polarizing properties of these grids cover such a broad band of wavelengths that three spectrometers and three reference polarizers were used to measure their optical properties. These were a Cary model 11M double-beam spectrometer and Glan-Foucault polarizing biprism for visible measurements, a Perkin-Elmer model 112 single-beam spectrometer with double-passed LiF prism and PbS detector combined with a dense Polaroid HR near-infrared sheet polarizer for the range 0.8–2.7 μ, and a Perkin-Elmer model 21 double-beam spectrometer with NaCl prism for the range 2.5–15 μ. Beyond 2.5 μ the wire grids were superior to any other infrared polarizer available to the authors, so the thick aluminum grid was used to test the thin gold grid. This is possible because the density difference $(\log k_1 - \log k_2)$ of the aluminum grid is much greater than that of the gold grid. In all cases, the wire grid was mounted perpendicular to the direction of the incident beam, and only slightly convergent beams ($N.A. \leq 0.10$) were used.

The values shown on the graphs, k_1 and k_2, are the transmittances for plane-polarized light with electric vector perpendicular and parallel, respectively, to the metallic elements. No correction for reflection has been made, but the reflection losses corresponding to a single

TABLE I. Principal transmittance values of the wire-grid polarizers as shown in Figs. 3 and 4 at selected wavelengths.

λ (microns)		0.50	0.70	1.0	1.5	2.0	3.0	4.0	5.0	6.0
Thick aluminum	k_1	5.8%	23.8%	27.6%	31.6%	39.6%	51.8%	60.1%	68.6%	68.6%
wire grid	k_2	3.8%	2.2%	1.1%	0.5%	0.3%	...	<0.2%	...	<0.2%
Thin gold	k_1	40.7%	44.7%	52.0%	72.4%	80.2%	86.1%	86.6%	88.8%	86.4%
wire grid	k_2	29.2%	28.8%	13.7%	6.4%	3.5%	2.0%	0.9_4%	0.70%	0.5_6%

film of the plastic are shown by a dashed line. Note that the transmittances are shown on a log scale to give a clear indication of k_2 values less than 1%. If the ratio of densities of a wire grid is desired at any particular wavelength, it may be obtained by taking the ratio of linear distances from the dashed line to the k_2 and k_1 lines at that wavelength.

The general features appearing on these plots are as follows:

(1) At 400–500 mμ a crossover appears in the k_1 and k_2 values for the aluminum polarizer, and a series of irregularities appears for the gold grid. This is the region

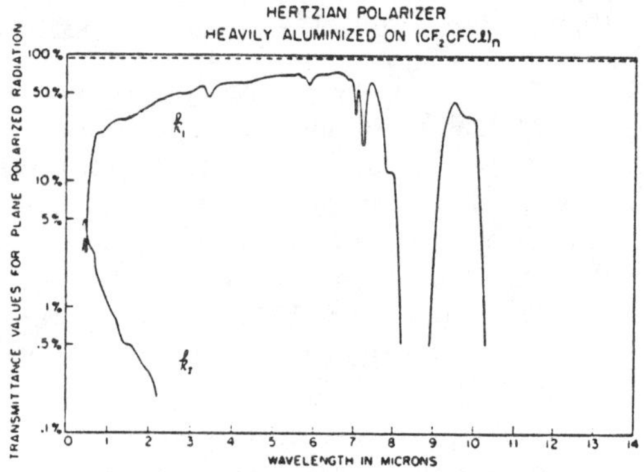

FIG. 3. Principal transmittance values of a wire-grid polarizer thickly aluminized. k_1 = maximum transmittance for plane-polarized radiation (no reflection correction); k_2 = minimum transmittance for plane-polarized radiation (polarizer rotated 90° from setting for k_1). Dashed line = reflection correction for $(CF_2CFCl)_n$. This polarizer is a 5.2×5.2 cm sq with one thick aluminum element per 463 mμ.

in which the grid begins to function as a diffraction grating, with the first order being propagated along the face of the grid at ~463 mμ. Both theoretical calculations and the observations of Peters and Pursley on macroscopic grids confirm this crossover of k_1 and k_2 for a conductive grid in the region where $\lambda = d$ (wavelength = period).

(2) At 650 mμ a maximum appears in the k_1 and k_2 values for the gold grid, but is absent for the aluminum grid. This is associated with the well-known transmission of visible light by thin gold films, and represents a departure from metallic conduction.

(3) In the region of 1–2 μ the performance of both polarizers improves monotonically with increasing wavelength. This, too, is in accord with theoretical predictions and observations on grids of macroscopic wires. The small sawtooth pattern in this region is associated with thin film interference effects in the plastic cast.

(4) In the neighborhood of 2 μ both of these grids become very useful polarizers having high and increasing density ratios. The gold grid is the better polarizer for most infrared applications where light intensity is at a premium. The aluminum grid has a much greater density difference ($\log k_1 - \log k_2$) and is thus useful for applications where leakage of the undesired component of polarization is to be minimized. This low leakage is obtained at the cost of decreased transmittance of the desired component. Beyond 2.2 μ the aluminum grid has a greater density difference than any polarizer available to the authors, and k_2 values could not be measured. There are good theoretical grounds for assuming that the monotonic decrease of k_2 continues beyond 15 μ and probably to 100 μ. This is shown in part by the monotonic decrease of k_2 for the gold grid in the region of 2–7 μ, as determined by testing the gold grid against the aluminum grid. Beyond 7 μ, k_2 values for the gold grid are comparable to instrumental noise, and so are not shown.

(5) At 7.0 μ the first of a series of absorption bands of $(CF_2CFCl)_n$ appears. Both grids are essentially useless in the ranges 7.7–9.2 and 10.0–11.0 μ. Beyond 11.0 μ the plastic is partially transparent, and useful polarization properties are again obtained, as shown on Fig. 4. It is the existence of these absorption bands in $(CF_2CFCl)_n$ that makes replication in polyethylene desirable.

No tests of the wire grids have been made beyond 15-μ wavelength, and only the roughest estimates can be

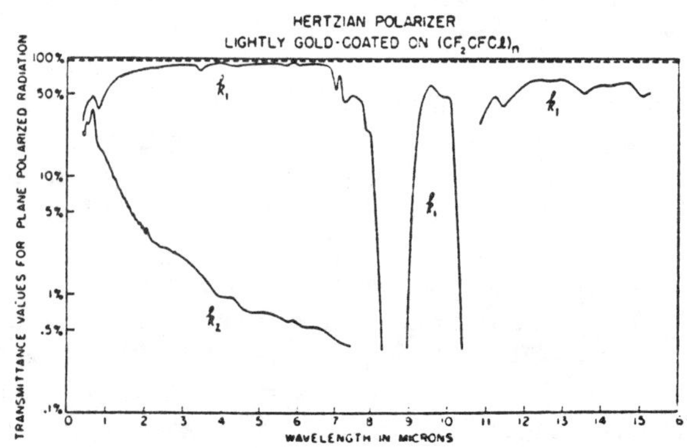

FIG. 4. Principal transmittance values of a wire-grid polarizer thinly gold coated. This polarizer is identical with the aluminum grid shown in Fig. 3 except that the grid has one thin gold element per 463 mμ.

made of the long-wavelength limit of high polarization for these grids. If the grid is considered as a continuous metal sheet with respect to long-wavelength k_2 transmission, skin-depth effects will become important at some very long wavelength, and k_2 (the leakage transmission) will then begin to increase gradually with increasing wavelength. The onset of skin conduction may become significant in the range from 100-μ to 1-cm wavelength.[10] If polarization at these longer wavelengths is desired, a grid of macroscopic wires may be mounted on a frame and used as a microwave polarizer.

No reflection measurements have been made on the existing wire grids, but it is assumed that at $\lambda > d$ the untransmitted component of polarization is largely reflected. This opens the possibility of using wire grids as polarizing beam splitters.

Only qualitative tests of the effect of oblique incidence on the polarizers have been made. These agreed with the careful study of oblique incidence by Pursley,[4a] who found that the transmittances of the grid are nearly independent of oblique incidence up to 30° off-normal. This near-independence holds except when, with $\lambda < 2d$, rotation of the grating causes the first diffracted order to appear. The wire grid is certainly suitable for use in converging optics of $N.A. = 0.5$ or higher.

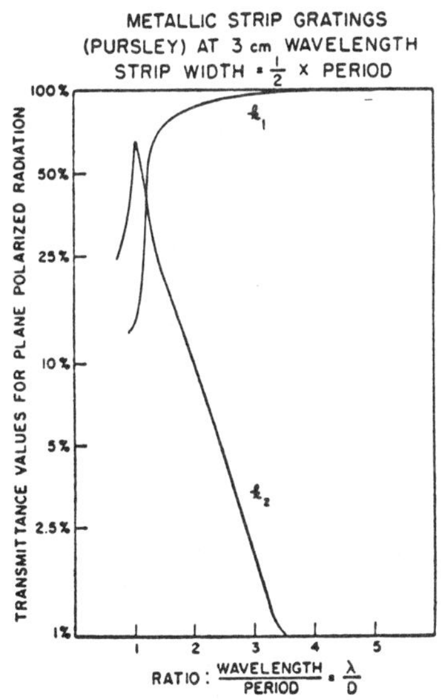

FIG. 5. Principal transmittance values of grids of macroscopic metal strips for 3-cm microwaves. This figure is comparable to Figs. 3 and 4, but wavelength is fixed at 3.2 cm and k_1, k_2 values are obtained from a series of gratings of varying period made from strips of aluminum foil. This figure is a reproduction of Fig. 20 of footnote reference 4(a) shifted from linear to semilog scale. We are indebted to Professor C. W. Peters for permission to reproduce these data. To compare these microwave results with Fig. 3, consider 463 mμ as $\lambda/D = 1$ in Fig. 3. Most of the quantitative difference may be attributed to resistive losses in aluminum in the visible and near infrared.

[10] *Note added in proof*: Measurements by F. W. Behnke of Perkin-Elmer Corporation show no detectable leakage transmission in the 15 to 25-μ region.

CORRELATION WITH THEORY AND WITH GRIDS OF MACROSCOPIC WIRES

A direct theoretical treatment of these grids of microscopic wires seems out of the question at the moment for two reasons. The actual cross section of the deposited metal elements is unknown, and is probably not a shape amenable to reasonable mathematical treatment. It certainly is neither a round wire nor a flat strip. Also, the metallic elements are continuous, but by no means uniform, as shown in the electromicrograph. Even so, the qualitative correlation with Pursley's results on strip gratings is striking. His results for the special case of perfectly conducting thin strips obscuring half the area of the polarizer are shown in Fig. 5, adjusted to a log scale similar to the transmittance data on the grids of microscopic wires. Pursley's results are in agreement with the theory of Baldwin and Heins and of Lamb.[11]

COMPARISONS WITH OTHER INFRARED POLARIZERS

All broad-band infrared polarizers, other than the wire grid, work on the principle of selective reflection of multiple plates. AgCl, Se, and Ge have been used in transmission, and germanium plates[12] have been used in reflection. A polarizer employing a rhomb of germanium with two faces in contact with mercury[13] has recently been discussed. All of these share a number of objections: they are large and awkward, difficult to rotate, and cannot be used in sharply converging beams of radiation. The silver chloride pile-of-plates polarizer is the most commonly used, and for special purposes it may be designed to fit into a small space and have a high degree of polarization.[14] However, an analysis of the use of this type of polarizer with spectrometers has demonstrated an unfortunate limitation.[15] A pile-of-plates polarizer designed for high degree of polarization cannot be rotated in relation to the spectrometer slit without drastic decrease in polarization. None of the foregoing limitations apply to the wire-grid polarizer. It is a planar structure, it is easily rotated, and it can accept sharply converging beams. The principal limitations on it are set by the absorption bands of the substrate and the fineness of the ruling period.

There is at least one sheet polarizer available for the near-infrared region and this is the Polaroid HR near-infrared polarizer.[16] This is a polyvinyl alcohol-polyvinylene-iodine complex with a strong absorption centered about 1.5 μ. This dichromophoric system has an absorption band broad enough to give satisfactory polarizing properties from 0.7 to 2.8 μ with varying thicknesses. The "tail" of this band extends out to 4 μ

[11] H. Lamb, Proc. London Math. Soc. 29, 523 (1898).

[12] D. F. Edwards and M. J. Bruemmer, J. Opt. Soc. Am. 49, 860 (1959).

[13] N. J. Harrick, J. Opt. Soc. Am. 49, 376, 379 (1959).

[14] A. S. Makas and W. A. Shurcliff, J. Opt. Soc. Am. 45, 998 (1955).

[15] G. R. Bird and W. A. Shurcliff, J. Opt. Soc. Am. 49, 235 (1959).

[16] R. P. Blake, A. S. Makas, and C. D. West, J. Opt. Soc. Am. 39, 1054 (1949).

and beyond, but is obscured by the OH and CH stretching bands of PVA from 2.8–3.5 μ. A composite graph of k_1, k_2 for a thin HR polarizer has been compiled by A. S. Makas from data on several unlaminated HR polarizers of various densities. This is shown in Fig. 6. The HR polarizer used to test the wire grids in the range 0.8–2.5 μ was a much denser unit than the "composite" shown here, having very similar dichroic ratios but greater densities and density differences. With the combination of a grid of macroscopic wires, a 463-mμ period wire grid, dense and thin HR polarizers, and dense Polaroid H sheet, complete spectral coverage with sheet polarizers would be available from infinite wavelength to 350 mμ.

CONCLUSION

Wire grids with microscopic elements are shown to be very useful infrared polarizers. They are compact, easily rotatable, and may be used with sharply converging beams. The most remarkable feature of the wire grid is the extreme breadth of the wavelength band of good polarizing properties. Existing wire grids made from the best available diffraction gratings cover the range from 2–15 μ and probably on to 100 μ with excellent properties. Polarization decreases as wavelength decreases, so that only limited use is likely with wavelength less than four times the period of the grid: a wavelength of about 2 μ for present grids. The other principal limitation on existing wire grids is the absorption of the plastic substrate.

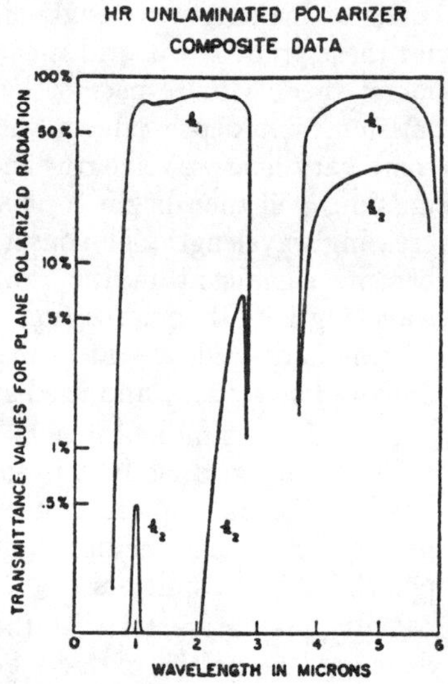

FIG. 6. Principal transmittance of polaroid HR near-infrared polarizer. k_1 and k_2 are as in Figs. 3–5, and represent values compiled by A. S. Makas from a number of HR samples having a range of dichromophore concentrations.

ACKNOWLEDGMENTS

Replication techniques for $(CF_2CFCl)_n$ plastic were developed by James Bailey, and all replicas described in this paper were made by him. The low-angle shadowing operations were conducted by Carl Basso. Electron and optical microscope studies on the replicas and grids were made by Eugene S. Emerson, and many helpful discussions were held with him. David Richardson of the Bausch and Lomb Company provided helpful advice and information on replicas.

A Survey of the Theory of Wire Grids*

TOVE LARSEN†

Summary—**The paper gives a survey of the literature concerning the electromagnetic properties of wire grids. As an introduction to the literature survey, a short description of the properties and applications of wire grids is given. Finally some particular grid configurations are mentioned.**

I. Introduction

THIS SURVEY is written from an electromagnetic point of view and all but a few papers concerning optical grids have been omitted. The survey does not claim to be complete. Papers, the contents of which are known by the author only from summaries in the *Electrical Engineering Abstract* or other abstracts are marked with an asterisk.

A description of the properties and applications of grids is given as an introduction to the literature survey. This is followed by the survey, which is ordered chronologically and divided into two parts: early papers, and currently used papers from a later period. Finally is given a description of some particular grid configurations.

Most attention is paid to the theoretical papers dealing with the simple grid with wires of circular cross section. The formulas found by the various authors for the reflection coefficient of such a gird are given and compared. For the rest of the papers describing experimental investigations and examinations of grids with noncircular wires and special grid configurations only brief descriptions are given, as the results in most of these cases cannot be stated briefly. The notation used in the formulas is shown in Fig. 1. The time factor is $e^{-i\omega t}$ and the Giorgi unit system is used.

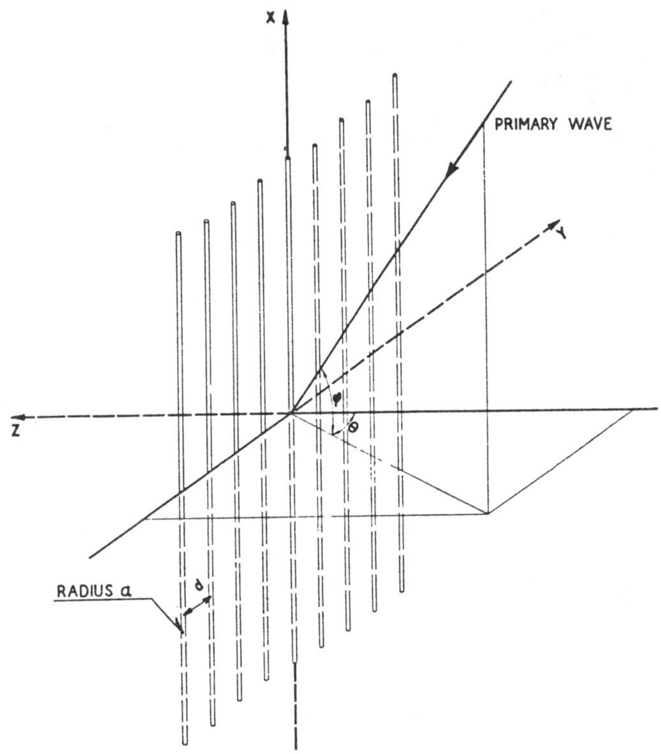

	AIR	WIRES
DIELECTRIC CONSTANT	ε_0	$\varepsilon_1 = \varepsilon_r \varepsilon_0$
PERMEABILITY	μ_0	$\mu_1 = \mu_r \mu_0$
CONDUCTIVITY	$\sigma_0 = 0$	σ_1
PROPAGATION CONSTANT	$k_0 = \omega\sqrt{\varepsilon_0 \mu_0}$	$k = \omega\sqrt{\mu_1 \varepsilon_1 (1 + i\frac{\sigma_1}{\omega \varepsilon_1})}$
INTRINSIC IMPEDANCE	$\zeta_0 = \sqrt{\frac{\mu_0}{\varepsilon_0}}$	$\zeta_1 = \sqrt{\frac{\mu_1}{\varepsilon_1 (1 + i\frac{\sigma_1}{\omega \varepsilon_1})}}$

Fig. 1—Plane grating with a plane wave at arbitrary incidence.

II. Properties and Applications of Grids

The grids which will be discussed in this paper may be defined as plane metallic systems which have in one direction a periodic structure with a period called the grid constant, which may be of the same order of magnitude as the wavelength of the incident field. The material properties of the grid are constant in the direction of the wires. The influence of such a grid depends primarily on the polarization of the incident wave and of the ratio between the grid constant and the wavelength. This dependence will be mentioned briefly in what follows together with the definition of the quantities commonly used to describe the properties of grids. Some applications of grids with a few references to papers describing these applications are also briefly mentioned.

A: Polarization

The grid usually has the greatest influence when the electric vector of the incident wave is polarized in the direction of the wires. This is called the Hertz effect; in certain cases, however, the grid may effect a wave polarized perpendicular to the wires more severely than a wave polarized parallel to the wires; this is called the Dubois effect. The type of transmission that takes place for a given grid depends upon the ratio between the wire diameter and the wavelength and upon the material of which the grid is made.

In radio engineering the Hertz effect is of most in-

* Received by the PGMTT, November 9, 1960; revised manuscript received, February 2, 1962.
† Laboratory of Electromagnetic Theory, Technical University, Copenhagen, Denmark.

terest. It was exactly this effect which Hertz [1] utilized in 1889 in his experiments. Around the end of the century there were several theoretical investigations dealing with the electromagnetic fields near grids (see Section III-A). Formulas for the reflection coefficient of a grid were derived for both the case where the polarization of the incident wave is parallel and where the polarization is transverse to the wires. It turned out that usually the reflection coefficient was vanishingly small when the polarization of the incident wave was perpendicular to the wires. In fact, in most of the recent work on the subject (see Section III-B) it is assumed that a wave with a polarization perpendicular to the wires of the grid passes the grid uninfluenced. In computing the field around a grid one often expresses the total field as the incident field as it would be if the grid were not present plus a scattered field radiated from the grid in both directions by a current in the wires of the grid induced by the component of the field parallel to the wires. The reflected wave in front of the grid is identical with the scattered field, whereas the transmitted wave behind the grid is equal to the sum of the scattered field and the incident wave (assuming losses in the grid are negligible). Since the scattered field can have a polarization different from the polarization of the primary field, it is possible to produce an elliptically polarized field behind the grid. This fact can be exploited in the construction of antenna systems for circularly or elliptically polarized fields; such antennas have been examined for example by Andreasen [2], who computed the field reflected from two parallel grids being so arranged that the wires of the first grid formed an angle with the wires of the second grid, and by Aagesen [3], who has investigated the field from a reflector consisting of a grid placed parallel to a metal sheet.

B. Diffraction Pattern

A qualitative description of the diffraction pattern around a grid may easily be given for an electromagnetic as well as for an optical transmission grating, the simple mechanism being that the periodic elements of the grid (current-carrying wires or slits with light) radiate elementary cylindrical waves in all directions. The phase of these radiated secondary waves will increase from one wire or slit to the next with the amount $k_0 d \sin\theta$, where $k_0 = 2\pi/\lambda$ is the propagation constant of the incident wave, d the grid constant and θ the angle of incidence of the primary field (this field is assumed not to change in the direction along the wires or slits, $\phi = 0°$). The secondary waves will arrive with the same phase to planes, the normals of which form the angle ψ_n with the positive normal to the grid, where ψ_n is determined from (see Fig. 2)

$$d \sin\theta - d \sin\psi_n = n\lambda, \qquad n = 0, \pm 1, \pm 2 \cdots. \tag{1}$$

By the positive normal to the grid we understand the normal pointing from the front to the back side of the grid (seen from the generator). The angle ψ_n should be counted negative when the corresponding direction is to the same side of the normal to the grid as the direction of propagation of the incident wave, positive when it is to the opposite side.

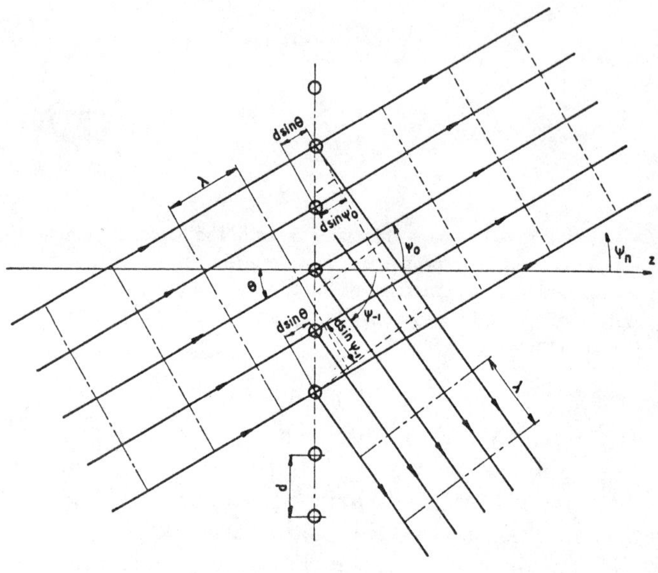

Fig. 2—Diffraction pattern around a grid, side waves are shown for $z>0$ only. The symmetrical waves are present for $z<0$. ($\lambda/d = 1.3$).

The secondary waves propagating from the grid and outwards will form a sequence of plane waves propagating in directions determined by the angles ψ_n.

From (1) we find

$$\sin\psi_n = \sin\theta + n\frac{\lambda}{d}, \qquad n = 0, \pm 1, \pm 2, \cdots. \tag{2}$$

In order to obtain a real solution for ψ_n to this equation we must have

$$\left|\sin\theta + n\frac{\lambda}{d}\right| \le 1. \tag{3}$$

The number of real solutions ψ_n is therefore finite. For those values of n for which the above inequality is not satisfied we find infinitely many solutions to (2). These solutions, however, are all complex. It may be demonstrated that these complex values of ψ_n correspond to waves propagating along the grid surface, but being attenuated exponentially in the direction perpendicular to the gird. When the wavelength relative to the grid constant increases, more and more of the real side waves will change to attenuated waves. A very illustrative description of the side waves is given by MacFarlane [4].

There will always be at least two real solutions to (2), namely for $n = 0$ and $\psi_0 = \theta$ or $\psi_0 = \pi - \theta$, *i.e.*, there will always be a transmitted wave which continues in the direction of the incident wave and a reflected wave with a reflection angle equal to the angle of incidence. If it is desired that only these two zero-order waves shall be present, we must have

$$\frac{\lambda}{d} > 1 - \sin\theta. \tag{4}$$

In optics it is exactly the side waves that are of interest since by using a suitably designed grid we may split a light-wave in waves corresponding to the various colors, as the angle ψ_n becomes different for different wavelengths. This was the purpose of the first practical grids (or diffraction gratings), which were constructed by Fraunhofer in 1823 in connection with his investigations in the field of spectral analysis.

In radio technology it is mostly the directly reflected wave and not the side waves that is of interest, since the grids are often used as a substitute for solid metal plates for reflecting purposes; for example, grid constructions have been used as reflector surfaces in antenna systems and in ground wire systems for antennas. An experimental study of grid reflectors for antennas has been made for example by Gresky [5] and ground wire systems for antennas have been treated for example by Abbott [6], Monteath [7] and by Knudsen and Larsen [8].

However as the lower limit of the wavelengths used in radio technology becomes shorter and approaches the infrared, the diffraction pattern of grids becomes of interest. A spectrometer for millimeter waves has for example been described by Coates [*9], and Klein, *et al.* [*10] describes how it is possible to generate millimeter waves by letting a magnetron oscillate with higher harmonics and then filtering the waves with desired wavelength by using a diffraction grid.

C. *Grid Impedance, Transmission and Reflection Coefficients*

The purpose of most of the electromagnetic investigations of grids has been to find the complete field distribution around the grid in the case of a plane incident wave. However, some authors have found it practical to characterize the influence of the grid upon the field exclusively by a single quantity, the grid impedance, or by the reflection and the transmission coefficients. These quantities are of most value for the directly transmitted and reflected wave, but they have been defined also in the case where side waves occur.

The grid impedance has been defined in two different ways. Wessel [11] and Hornejäger [12] (see Section III-B) defined their grid impedance Z_w per unit length of wire as the ratio between the electric field strength parallel to the wires and the current of a single wire, and used this definition also when side waves occur. MacFarlane [4] and Wait [13] (see Section III-B) calculated the reflection coefficient of the grid and found that it had the same form as the reflection coefficient for a transmission line shunted with an impedance. Accordingly, they defined their grid impedance Z_g as the shunt impedance of the equivalent transmission line.

The connection between the two impedances is given by

$$Z_w d = \frac{Z_0}{2} + Z_g, \tag{5}$$

where Z_0 is the characteristic impedance of the equivalent transmission line, which in the general case has the form

$$Z_0 = \zeta_0 \frac{\cos\phi}{\cos\theta}, \tag{6}$$

where ζ_0 is the characteristic impedance of free space.

The voltage reflection and transmission coefficients r_v and t_v are defined as the ratio between the field strength of the reflected and the transmitted wave to the field strength of the incoming wave (same polarization).

The power reflection and transmission coefficients r_p and t_p are defined from these quantities by:

$$r_p = |r_v|^2, \tag{7}$$

$$t_p = |t_v|^2. \tag{8}$$

For a lossless grid with no side waves we have

$$r_p = 1 - t_p. \tag{9}$$

For some applications it is sufficient to attribute to the grid surface an infinitely large conductivity in the direction of the wires, and the conductivity zero in the transverse direction. For example, this approximate description has been used in the investigations by Andreasen [2] and by Aagesen [3] referred to above, and it is also used to describe a helix in a traveling-wave tube, the helix being approximated by a tube with an infinitely thin wall having an infinitely large conductivity in the direction of the wires, and the conductivity zero in the transverse direction (see for example Chu and Jackson [14]).

However, in computations regarding ground wire systems the simple description of the impedance is insufficient. In such cases the equivalent shunt impedance Z_g has been used.

The impedance properties of the grid for electromagnetic waves may be utilized technically for example for matching purposes (described for example by Jones and Cohn [15] for a dielectric lense), for filters (for example described by Lewis and Casey [*16]) and for artificial dielectrics (described for example by Kaprielian [17]).

III. The Development of the Theory of Grids

A. *Investigations from an Early Period*

As was mentioned in Section II-B Fraunhofer was the first physicist to make use of grids. This was done in his experiments in spectral analysis around 1823. Since the purpose of this paper is a description of investigations of the electromagnetic properties of grids, and since the literature on grids published from the time of Fraunhofer up to the end of the century is difficult to trace, in what follows only references will be made to papers published in 1889 or later, 1889 being the year when Hertz [1] showed that grids are of interest also for waves having a wavelength essentially larger than the wavelength of light.

The first attempt to a quantitative explanation of Hertz' experiments was made by J. J. Thomson [18] in 1893. He considered a plane grid of infinite extent having a grid constant smaller than the wavelength and with the polarization of the normally incident wave parallel to the wires of the grid. Thomson found that at a certain distance from the grid the intensity of the reflected wave is the same as in the case of reflection from a metallic surface, but that the phase differs with a quantity depending upon the diameter of the wires and the grid constant.

In 1898 Lamb [19] published a paper in which he computed the reflection and the transmission coefficient for the grid as well in the case when the polarization of the incident wave is parallel to the wires as in the case when the polarization is perpendicular to the wires (perpendicular incidence). These computations were based on the potential and the stream functions for the stationary field around the grid. Lamb investigated partly a grid consisting of strips, and partly a grid consisting of circular rods, the grid constant d being in both cases smaller than the wavelength λ. In contradiction to Thomson he found in accordance with later theories and measurements that the reflection coefficient depends upon the wire diameter as may be seen from the following expression found by Lamb for the power reflection coefficient r_p for a grid consisting of circular rods with radius a.

1) Electric vector parallel to the direction of the wires

$$r_p = \frac{1}{1 + \left(\frac{2d}{\lambda} \ln \frac{d}{2\pi a}\right)^2} . \tag{10}$$

2) Electric vector perpendicular to the direction of the wires

$$r_p = \frac{\left(\frac{2\pi^2 a^2}{\lambda d}\right)^2}{1 + \left(\frac{2\pi^2 a^2}{\lambda d}\right)^2} . \tag{11}$$

The last formula shows that when $a \ll d$, the reflection coefficient will be very small, *i.e.*, a wave polarized perpendicular to the direction of the wires passes the grid almost uninfluenced as shown experimentally by Hertz.

In 1906 Schaefer and Laugwitz [20] tried to verify J. J. Thomson's results experimentally (at this time they were not acquainted with Lamb's work). They measured and compared the phase difference occurrir in the reflection from a metal plate and in the reflectic from a grid, and they found that it did not agree at a with Thomson's results. Whereas Thomson had four a pronounced dependence of the grid constant, Schaef and Laugwitz's experiments showed that there w practically no dependence.

In 1907 G. H. Thomson [21] also carried out mea urements to check the theories of J. J. Thomson as we as Lamb's theories. Using a different method he o tained results which agreed well with Lamb's theo but not with the theory of J. J. Thomson. Throug G. H. Thomson's paper, Schaefer and Laugwitz becan aware of Lamb's investigation, and their paper fro 1906 was followed by a second paper [22] in which the showed that their measurements, too, were in agreeme with Lamb's theory.

The influence on the transmission through the grid b the material of the wires in the grid was investigate experimentally by Schaefer and Laugwitz [23] in 190 At this time there were no theoretical investigatio available with which the measurements could be con pared, since all theories developed up until that tim were based on the assumption that the wires were pe fectly conducting. Another assumption upon which a theories known to that time are based is that the radiu of the wires is much smaller than the grid constant an the wavelength. However, the smaller the radius is, th larger is the influence of the finite conductivity of th wires. For this reason Schaefer and Laugwitz mac measurements on grids with very thin wires of variou metals. They measured the power transmission co efficient for these grids and found a certain dependenc of the grid material.

At the same time as the above theories for explainin Hertz's grid experiments were developed and measur ments made, some papers on optical and acoustic grids were published. A few of these papers will b mentioned here.

In optics there was some interest in the influence o grids on the polarization of light. In fact, as early a 1861 Fizeau had made experiments with polarized ligh incident upon a screen with slits, and Hertz's exper ments with grids revived the interest in this field. I 1911 du Bois and Rubens [24] gave a review of all inve tigations made in this field up until this time, and ne investigations were made by themselves. An interest wa manifested in investigating whether the light had th same property as Hertz's radio waves, namely, that th effect of the grid is larger when the incident wave is po larized parallel to the wires than it is when the polariza tion is in the transverse direction. In 1910 Schaefer an Reiche [25] called the two effects the Hertz effect an the Dubois effect (with reference to a previous work o du Bois), respectively, and they investigated theore ically when these two effects will occur. In their compu tations they assumed that the wire diameter is small a compared to the wavelength, and that the grid constan is so large as compared to the wavelength that the computation of the field can be made on the basis of the computation of the field for a single cylinder. In a following paper from 1911 Schaefer and Reiche [26] elaborated on their theory, still assuming that the grid constant is much larger than the wavelength so that the mutual influence of the wires can be neglected in computing the field.

In acoustics Lord Rayleigh in particular worked on problems regarding the scattering of sound waves around various types of bodies including grids. His investigations form the basis for Lamb's theory of grids from 1898.

All the viewpoints mentioned so far were united in a very comprehensive theoretical investigation by Ignatowsky [27] from 1914. He was the first to consider an angle of incidence different from zero (the plane of incidence being perpendicular to the wires) and an arbitrary cross section as well as an arbitrary material for the wires. Later authors have usually found Ignatowsky's investigation too general and therefore too difficult to interpret clearly, for which reason his formulas have seldom been used. (However in a paper from 1958 Meecham, *et al.* [28] have made numerical computations based directly on Ignatowsky's formulas.)

In 1914 Arkadiew [29] made some experimental investigations of the reflection from grids consisting of only 4 wires. Both ferromagnetic and nonferromagnetic materials were used, but as was the case for Schaefer and Laugwitz in 1907 no theoretical investigations were available at that time with which to compare the experimental results.

This theory was developed in 1920 by Gans [30] and in an appendix [31] to the first paper in 1921. He considered grids with circular wires the radius of which were much smaller than the grid constant, and this again much smaller than the wavelength. In the first paper he assumed that $\mu_r=1$, but in the second paper he found that the formulas were valid also for $\mu_r\neq 1$. Only considering the case of perpendicular incidence he investigated partly the case where the polarization of the incident wave is transverse to the wires, and partly the case where it is parallel to the wires. In the last mentioned case he found the following expression for the voltage reflection coefficient:

$$r_v = \frac{-1}{1 - i\dfrac{2d}{\lambda}\left(\ln\dfrac{d}{2\pi a} + \tau_0\right)}, \tag{12}$$

where

$$\tau_0 = \frac{\mu_r}{k_1 a}\,\frac{J_0(k_1 a)}{J_0'(k_1 a)}\,. \tag{13}$$

Here $J_0(k_1a)$ and $J_0'(k_1a)$ are the zero order Bessel function and its first derivative of argument (k_1a). The formula (12) is seen to agree with (10) calculated by Lamb, when $\tau_0=0$, which will be the case for perfectly conducting wires.

Gans made numerical computations in the case of the electric vector being parallel to the wires using the same constants as were used by Schaefer and Laugwitz in their experimental work from 1907. However, it turned out that his results did not agree at all with the measurements. In 1924 Schaefer [32] found a very simple explanation of this fact, as he showed by repeating the experiments that the discrepancy was due to a printing error in the paper from 1907, where the figures for the reflection and the transmission coefficients were changed. After correction of this error the measurements agreed with the theory.

Arkadiew [33] in 1924, too, compared his experimental results for the nonferromagnetic wires with the theory of Gans and found a rather good agreement. In 1926 [34] he extended Gans's theory to include the case of ferromagnetic wires with a complex permeability and compared the results with his experiments from 1914. He found that the formulas found by Gans are valid also when the permeability of the wire material is complex.

B. Recent Investigations

Since the publications of the investigations mentioned in the last section, the last of these being from 1926, little work seems to have been done in the theory of grids for several years. In 1939 the problem was dealt with again, and since the end of the Second World War several contributions have been given to the theory of grids.

With reference to an experimental investigation by Esau, Ahrens and Kebbel [35], Wessel [11] made in 1939 a calculation of the transmission through a grid, and he investigated theoretically as well as numerically the equivalent grid impedance Z_w (for the definition of this impedance see Section II-C). Wessel based his investigation on the method mentioned in Section II-A using the current in the wires of the grid in his computations. He assumed the wires to have infinite conductivity and a circular cross section the radius of which is much smaller than the wavelength. The direction of propagation of the incident wave was assumed to be perpendicular to the plane of the grid. Wessel considered cases where the grid constant varies from being twice the wavelength to values much smaller than the wavelength. His numerical computations showed good agreement with the corresponding experimental investigations.

For the case $d<\lambda$ (no side waves) Wessel found the following expression for the power reflection coefficient (Durchlässigkeit D defined by Wessel equal to t_p):

$$r_p = \frac{1}{1 + \left(\dfrac{\omega L}{R_\infty}\right)^2}, \tag{14}$$

where

$$Z_w = R_\infty - i\omega L$$
$$= \frac{\zeta_0}{2d} - i\frac{\zeta_0}{\lambda} \cdot \left\{\ln\frac{d}{2\pi a} + \sum_{n=1}^{\infty}\left\{\frac{1}{\sqrt{n^2 - \left(\frac{d}{\lambda}\right)^2}} - \frac{1}{n}\right\}\right\}, \quad (15)$$

which for $d \ll \lambda$ (the sum negligible) agrees with the expression (10) given by Lamb.

For $d > \lambda > d/2$ Wessel found for the equivalent grid impedance

$$Z_w = R - i\omega L = \frac{\zeta_0}{2d}\left[1 + \frac{2}{\sqrt{1 - \left(\frac{\lambda}{d}\right)^2}}\right]$$
$$- i\frac{\zeta_0}{\lambda}\left\{\ln\frac{d}{2\pi ae} + \sum_{n=2}^{\infty}\left[\frac{1}{\sqrt{n^2 - \left(\frac{d}{\lambda}\right)^2}} - \frac{1}{n}\right]\right\}, \quad (16)$$

i.e., the first term in the sum has turned from an imaginary value to a real value corresponding to the first order side wave turning from an attenuated wave to a propagating wave.

In 1946 MacFarlane [4], like Lamb [19], used a quasi-stationary method of computation, the computation of the transmission coefficient and the impedances in the time dependent case being based on the stationary solution for the field around the grid. However, MacFarlane made an extension of Lamb's investigation in that he investigated the case where the primary wave, polarized parallel to the wires, has an arbitrary angle of incidence. MacFarlane found in this case the following expression for the voltage reflection coefficient under the assumption that $a \ll d$ and that the condition (4) (no side waves) is fulfilled

$$r_v = \frac{-1}{1 - i2\frac{Z_g}{Z_0}}, \quad (17)$$

where Z_g and Z_0 are the impedances defined in Section II-*C* given by

$$Z_0 = \frac{\zeta_0}{\cos\theta} \quad (18)$$

$$Z_g = \zeta_0\frac{d}{\lambda}\left[\ln\frac{d}{2\pi a} + F\left(\frac{d}{\lambda}, \theta\right)\right], \quad (19)$$

where F is given by

$$F = \frac{1}{2}\sum_{n=1}^{\infty}\left\{\left[\left(\frac{d}{\lambda}\sin\theta + n\right)^2 - \left(\frac{d}{\lambda}\right)^2\right]^{-1/2} + \left[\left(\frac{d}{\lambda}\sin\theta - n\right)^2 - \left(\frac{d}{\lambda}\right)^2\right]^{-1/2} - \frac{2}{n}\right\}. \quad (20)$$

This expression is seen to agree with the one found by Wessel, when $\theta = 0°$. F may be regarded as a correction term, which is negligible when $d \ll \lambda$. Curves of F as a function of θ with d/λ as a parameter are given by MacFarlane.

MacFarlane [36] utilized in 1946 like Booker [37] in 1947 Babinet's principle for further grid investigations. MacFarlane's investigations were related to diaphragms in transmission lines and waveguides, whereas Booker made a general investigation of the impedance concept.

In 1948 Hornejäger [12] extended Wessel's investigations to include an angle of incidence different from zero (the incident wave being polarized parallel to the wires), and he considered any value of the ratio between the grid constant and the wavelength.

He found for the power transmission quotient of the direct transmitted wave

$$t_p = \frac{\left(\frac{R}{R_\infty} - 1\right)^2 + \left(\frac{\omega L}{R_\infty}\right)^2}{\left(\frac{R}{R_\infty}\right)^2 + \left(\frac{\omega L}{R_\infty}\right)^2}, \quad (21)$$

where R_∞ is the real part of the grid impedance Z_w, when the inequality (4) is fulfilled (no side waves). R_∞ is given by

$$R_\infty = \frac{\zeta_0}{2d\cos\theta}. \quad (22)$$

The quantities R and ωL are real and imaginary parts of the grid impedance Z_w in the general case:

$$Z_w = R - i\omega L$$
$$= \frac{\zeta_0}{2d}\left[\frac{1}{\cos\theta} + G\right] - i\frac{\zeta_0}{\lambda}\left[\ln\frac{d}{2\pi a} + \frac{\lambda}{2d}M\right], \quad (23)$$

where G and M are expressions which take different forms according to the relative value of λ/d and θ (depending on the number of real side waves). In the case of no side waves [(4) fulfilled] G and M take the values

$$G = 0 \quad (24)$$

$$M = \sum_{n=1}^{\infty}\left\{\left[\left(n\frac{\lambda}{d} + \sin\theta\right)^2 - 1\right]^{-1/2} + \left[\left(n\frac{\lambda}{d} - \sin\theta\right)^2 - 1\right]^{-1/2} - \frac{2}{n}\frac{d}{\lambda}\right\}. \quad (25)$$

In this case the above expression for the transmission coefficient agrees with the reflection coefficient found by MacFarlane.

In 1949 Miles [38] investigated a grid of strips with an arbitrary value of d/λ, but only for perpendicular incidence. By setting up an integral equation and solving it by variational principles he investigated acoustic waves as well as electromagnetic waves polarized parallel to and perpendicularly to the wires. Originally only infinitely thin strips were considered, but later the investigations were extended to strips of finite thickness.

In 1951 Shmoys [39] formulated an integral equation for computing the scattering matrix for an infinitely large grid, consisting of wires having an infinite conductivity and an arbitrary cross section for the case where the plane of incidence for the incident wave is perpendicular to the wires, the direction of polarization of the incident wave being perpendicular to or parallel to the wires. In the last mentioned case he used a variational method and found an approximate expression for the scattering matrix for a grid, the wire radius and the grid constant of which are much smaller than the wavelength.

In the "Waveguide Handbook" from 1951 Marcuvitz [40] presents formulas and curve sheets for the equivalent, normalized parameters for various grid configurations all of which have been found by an integral equation method. In all the cases considered here the plane of incidence is perpendicular to the wires. The following cases have been treated: grids of infinitely thin strips with either the magnetic or the electric field strength parallel to the wires, grids with a circular or a rectangular cross section in the case where the magnetic field strength is parallel to the wires, and grids with elliptical, circular, and rectangular cross section in the case where the electric field strength is parallel to the wires.

In 1951 Grillini [*41] has published results of an experimental investigation of grids of strips.

In 1952 a new extension of Wessel's and Hornejäger's investigations was made by Lewis and Casey [42] who considered the case where the wires have a finite conductivity. They limited their investigation to the case $d<\lambda/2$, thereby excluding the possibility of side waves, and they gave numerical results for the reflection and the transmission coefficients for various values of the conductivity of the wires for the case where the electric field strength is parallel to the wires but where the direction of propagation of the incident wave is arbitrary. They found that the effect of losses in the wires is to increase the grid impedance Z_w as if the single wire internal impedance per unit length Z_i were added in series. Lewis and Casey used the value of the internal wire impedance given by Ramo and Whinnery [43]:

$$Z_i = -\frac{k_1 J_0(k_1 a)}{2\pi a \sigma_1 J_0'(k_1 a)}, \tag{26}$$

where $k_1 = \sqrt{i\omega\mu_1\sigma_1}$, *i.e.*, it is assumed that $\sigma_1 \gg \omega\epsilon_1$. For high frequencies and not too small wire radii this expression reduces to

$$Z_i \cong \sqrt{\frac{\omega\mu_1}{2\sigma_1}}\,\frac{1-i}{2\pi a}. \tag{27}$$

Curves of the real and imaginary part of Z_i have been calculated by Lewis and Casey.

They elaborated on the formula for Z_i and the formulas given by Hornejäger to find explicit expressions for the numerical value and phase angle of the reflection and transmission coefficient.

However, in order to compare the value of the reflection coefficient with the results found by other authors we write down the complex reflection coefficient found by Lewis and Casey [formula (15) in their paper]

$$r_v = \frac{-1}{1 + Z_i' - iX_p'}, \tag{28}$$

where Z_i' is the normalized internal wire impedance

$$Z_i' = \frac{d\cos\theta}{60\pi} Z_i = \frac{2d\cos\theta}{\zeta_0} Z_i, \tag{29}$$

and where

$$X_p' = \frac{2d\cos\theta}{\lambda}\left[\ln\frac{d}{2\pi a} + F\right], \tag{30}$$

F being the correction factor found by MacFarlane. This expression for the reflection coefficient is seen to agree with those found by MacFarlane and Hornejäger, when $Z_i = 0$. It is also seen to agree with the results found by Gans, when $\theta = 0$ and $\lambda \gg d (F=0)$ as we have

$$r_0 = i\frac{Z_i\lambda}{\zeta_0}. \tag{31}$$

Using an integral equation method Müller [44] in 1953 investigated the diffraction around a strip grid, and in a following paper [45] from the same year he demonstrated how his results could be used in investigating discontinuities in waveguides.

Extending Wessel's, Hornejäger's, and Lewis' and Casey's investigations to include a completely arbitrary direction of incidence and polarization and omitting any assumptions regarding the material of the grid wires Wait [13] in 1955 made a very general investigation of the diffraction through a grid with circular wires. Wait discussed the case where the grid-constant is much smaller than the wavelength so that no side waves occur and he derived an expression for the equivalent grid impedance defined on the basis of an equivalent transmission line. Wait found the following expression for the voltage reflection coefficient.

$$r_v = \frac{-1}{1 + 2\dfrac{Z_g}{Z_0}}, \tag{32}$$

where Z_g and Z_0 are the impedances defined in Section II-C.

$$Z_0 = \frac{\zeta_0 \cos\phi}{\cos\theta} \tag{33}$$

$$Z_g = -i\frac{d}{\lambda}\zeta_0 \cos^2\phi\left[\ln\frac{d}{2\pi a} + F\left(\frac{d\cos\phi}{\lambda}, \theta\right)\right] + dZ_i, \tag{34}$$

where

$$F\left(\frac{d\cos\phi}{\lambda}, \theta\right) = \frac{1}{2}\sum_{n=1}^{\infty}\left\{\left[\left(n+\frac{d}{\lambda}\cos\phi\sin\theta\right)^2 - \left(\frac{d}{\lambda}\cos\phi\right)^2\right]^{-1/2} + \left[\left(n-\frac{d}{\lambda}\cos\phi\sin\theta\right)^2 - \left(\frac{d}{\lambda}\cos\phi\right)^2\right]^{-1/2} - \frac{2}{n}\right\}. \tag{35}$$

Wait defines the internal wire impedance in accordance with Schelkunoff [46] as

$$Z_i = \frac{\zeta_1 I_0(-ik_1 a)}{2\pi a I_1(-ik_1 a)}, \tag{36}$$

where I_0 and I_1 are modified Bessel functions. This expression is equal to (26) used by Lewis and Casey, when $\sigma_1 \gg \omega\epsilon_1$ so that $k_1 = \sqrt{i\omega\mu_1\sigma_1}$.

Whereas Lewis and Casey found that the internal wire impedance Z_i should be added in series to the grid impedance Z_w, Wait found that dZ_i should be added in series to the equivalent shunt grid impedance Z_g. This is in accordance with the relation (5).

It is seen that the term F is the same as the F found by MacFarlane when $\cos\phi = 1$. In fact the curves of F as a function of θ with d/λ as a parameter calculated by MacFarlane may be used also to compute the value of F found by Wait, when the parameter is put equal to $d\cos\phi/\lambda$.

In 1955 and 1956 Yampolskii [*47], [48] published two papers on grid theory investigating the reflection coefficient from a wire grid when the electric vector is parallel to the wires and perpendicular to the wires, respectively. In the last mentioned case he found for the voltage reflection coefficient when $\lambda \gg d \gg a$

$$r_v = \frac{3\pi^2 a^2}{d\lambda}; \tag{37}$$

this expression is seen to agree with (11) found by Lamb except for the factor 3 where Lamb found a factor 2.

In an unpublished thesis Pursley [*49] in 1956 has described microwave and infrared light measurements on grids consisting of wires of aluminum and brass.

An experimental investigation of the transmission through inclined wire gratings including some simplified theoretical considerations was given by Snow [50] in 1956, the main emphasis, however, being on the side wave pattern.

By using a variational method Primich [51] in 1957 computed the field around a grid of strips with a perpendicularly incident primary wave and for arbitrary values of d/λ. He compared his results with corresponding measurements and found a reasonable agreement.

A semi-infinite grid has been investigated by Fel'd [*52] (1958), who found that the difference between the field in the case of the infinite grid and in the case of the semi-infinite grid is vanishing except around the first four conductors of the grid.

As was mentioned above on the basis of Ignatowsky's theory, Meesham, *et al.*, [28] in 1958 made numerical computations of the transmission through a grid with $d/\lambda > 1$ for a perpendicularly incident wave as well in the case where the polarization is parallel to as where it is perpendicular to the wires. The computations have been made on the basis of rather complicated series expressions; for this reason only a limited number of numerical results have been obtained. These results have been compared with Pursley's experimental results, and a reasonable agreement was observed.

In a thesis [53] from 1958 and later in a paper [54] Hansen has used an integral equation method for dealing with the problem of the diffraction of a plane, perpendicularly incident wave through a finite number of infinitely long slits in an infinitely large screen. He gives numerical results for the transmission coefficient in the case where a plane wave polarized in the direction perpendicular to the slits hits a screen with one, two, four or six slits, and he compares these results with the results for strip-grids with infinitely many slits or strips.

In 1959 Decker [55] published results of measurements of the transmission through a grid as a function of the angle of incidence and the frequency. The incident wave was polarized parallel to the wires. Decker compared his results with computations made on the basis of the formulas derived by Wait.

In several papers [56]–[58] (1959) and in a thesis [59] (1960) Særmark has treated the problem of diffraction around a number of strips having the same longitudinal direction but being otherwise arbitrarily oriented with respect to each other. In these investigations he uses a new extension of the addition theorem for Mathieu functions [60]. He also considers the case of several parallel slits in a plane screen. Numerical computations show that when the electric field strength is parallel to the slits, the transmission coefficient for four slits or more does not deviate essentially from the corresponding coefficient for a grid with infinitely many slits (except at resonance $\lambda = d$). Hansen found that this is not the case when the electric field strength is transverse to the slits.

In 1959 Skwirzynski and Thackray [61] have made

me calculations closely related to the work by natowsky. General but involved formulas are given r an arbitrary angle of incidence (incoming wave larized parallel to the wires) and approximate formus suitable for numerical computations are given in e case of normal incidence for a grid with circular ires. These formulas are valid for $d<\lambda/4$ but no rerictions are placed on d in the lower limit as is the se for all the expressions for reflection coefficients, etc., valuated by the authors mentioned in this section. umerical calculations of the transmission coefficient r various values of wire radius and distance between ires are given, the lower value of d being equal to vice the wire radius. These results were compared with xperimental results by Goodall and Jackson [62] 1959) and good agreement was obtained.

IV. Some Recent Investigations of Special Grids Configurations

. Grids with Wires of a Special Material

1) Ferromagnetic Grids: Kozinets [*63] has in 1946 nd Epelboim [*64] has in 1947 elaborated on Arkaiew's investigations of ferromagnetic grids.

2) Grid Wires of Two Different Materials: In 1956 aramonov [65] investigated both theoretically and xperimentally the transmission through a grid conisting of circular conducting wires uniformly coated vith a dielectric (ice); it turns out that the reflection ecreases considerably with increasing thickness of the ielectric coating. The computed curves are in good greement with the experimental data.

3) Grid Wires with Lumped Elements: A new type f grids was invented by Trentini during the Second World War (published 1953 [66]). He inserted usual etwork elements periodically along the grid wires vhereby he was able to obtain an arbitrary impedance or an incident electromagnetic wave. The first grid of his type was applied as an absorbing coating for metal valls in the meter wave range. Later such grids have ound application, for example: as matching elements, n filters, and in electromagnetic lenses.

In 1954 Franz [67] published a theoretical investigaion of an idealized model of Trentini's absorption grid vhere he considered an infinite grid consisting of very hin wires with a uniformly distributed complex impedance placed parallel to a plane sheet with infinite onductivity. His investigation also includes the cases vhere the conducting sheet is coated with an absorbing r dielectric material.

B. Two or More Parallel Grids

The experimental work by Esau, Ahrens, and Kebbel [35] which was carried out in connection with Wessel's theoretical investigations in 1939 also contained measurements on several parallel grids. Franz [68] first of all treated this case theoretically in 1949. Using a simplified form of Wessel's method of computation he investigated an arbitrary number of infinite grids with the same grid constant placed parallel to each other with parallel wires. He found the transmission coefficient for various systems of grids as a function of the angle of incidence, the distance between neighboring grids, and their position relative to each other. He found a fair agreement with Esau, Ahrens, and Kebbel's measurements.

Lewis and Casey [*16] described in 1951 how two parallel grids may be used as an interference filter for microwaves.

In 1952 Groves [69] investigated theoretically as well as experimentally the transmission through two parallel grids the wires of which form an arbitrary angle with each other. He found the power transmission coefficient as a function of the distance between the grids, the parameters of the grids, and the angle which the grid wires form with each other. Groves limited himself to investigating grids with a grid constant much smaller than the wavelength and to an arbitrarily polarized wave with the direction of propagation being perpendicular to the grids. He based his theoretical investigations on Wessel's work, and he found good agreement between measured and theoretical values.

The above mentioned investigation [66] from 1953 by Trentini also describes measurements on two and three parallel grids. In 1955 Trentini [*70] further investigated the transmission through two parallel grids theoretically as well as experimentally. He was particularly interested in investigating under what circumstances the power transmission was a maximum.

In 1956 Fel'd [*71] made a theoretical investigation of the reflection from and the transmission through two grids when the incident wave was polarized parallel to the wires, whereas the angle of incidence was arbitrary.

In connection with work on antenna systems for circularly or elliptically polarized waves Andreasen [2] in 1956 investigated two parallel wire grids, the wires of which form an aribitrary angle with respect to each other.

Several parallel grids or lattices of parallel, cylindrical, conducting rods are of interest for the construction of artificial dielectrics. Twersky [72]–[75] has in several papers from 1950–52 considered the problem of scattering of waves around an arbitrary number of parallel cylinders. He investigated the accuracy of Schaefer and Reiche's [26] method of computation from 1911 which can be used for optical grids where the distance between the single wires is so large as compared to the wavelength that the mutual influence can be neglected. Twersky first extended this method by taking into account the mutual influence between neighboring cylinders, and he later extended the investigation so that he found a completely general expression for the field around an arbitrary number of parallel cylinders of arbitrary cross section. Sakurai [*76] in 1950 and Kaprielian [17] in 1956 have investigated applications of such systems of cylinders for artificial dielectrics.

C. Grid Parallel to the Plane Interface Between Two Media

In 1954 Wait [77] investigated the reflection from a grid parallel to a conducting plane for an arbitrary angle of incidence for the primary field. In this connection he also considered the possibility of constructing an absorption grid by increasing the resistance of the wires, a problem which has been investigated by Franz [67] in 1954. As has been mentioned above also Aagesen [3] in 1957 treated the problem of a grid parallel to a conducting surface with a view to obtaining a polarization transforming reflector.

In 1957 Wait [78]–[80] treated in three papers the grid problem which is of greatest interest for the theory of ground wire systems, namely the computation of the field around a grid placed parallel to the interface between two dielectric media. Wait shows that under certain circumstances the two media and the grid may be considered equivalent to a composite transmission line being shunted with a certain impedance defined as the equivalent grid impedance. This description is valid for 1) oblique incidence, when the electric vector is always parallel to the wires, 2) normal incidence for any polarization, 3) perfectly reflecting interface for any angle of incidence and polarization, and 4) oblique incidence, when the magnetic vector is always perpendicular to the wires. Wait's impedance formulas for the grid have been derived for a grid placed in front of the interface between the two media, but through a simple interchange of symbols his formulas can also be used in the case most often met with in the investigation of ground wire systems, a grid buried in the ground and parallel to the surface of the ground. A numerical investigation of Wait's formulas has been made by Larsen [81] in 1960.

A special configuration of a grid placed parallel to the interface between two media is a grid situated inside a dielectric slab, which for the case of a normally incident plane wave was investigated by Yampol'skii [*82] in 1958.

Sivov [83] has in 1961 made an investigation of a grid with different dielectric materials on either side of the plane of the grid. Both circular and rectangular cross sections of the wires were treated, the magnetic field strength being parallel to the wires.

V. Conclusion

A review has been given of the literature pertaining to the electrical properties of wire grids. A great many papers have been written on this subject most of them investigating the case of a plane wave propagating towards a plane grid consisting of an infinite number of equal, equidistant and parallel wires. Most papers deal with grids with wires of circular cross section, but also a number of papers describing strip grids have appeared and a few concerning grids with wires of arbitrary cross section.

Various special grid configurations have been treated in the literature too, for example, ferromagnetic wir grids, dielectric coated wire grids, grids with lumped element wires, more parallel grids with parallel wire and with crossed wires, and grids parallel to conductin and dielectric interfaces.

Papers with the main emphasis on the scattered sid waves have not been included in this survey.

VI. Acknowledgment

This investigation was carried out at the Laborator of Electromagnetic Theory at the Technical Universit of Denmark by means of a grant from *Den polyteknisk Læreanstalts ingeniørvidenskabelige fond og G. A. Hage manns Mindefond* and by means of support from th Air Force Cambridge Research Center of the Air Re search and Development Command, United States Ai Force. The author wishes to express her thanks to Pro H. L. Knudsen for having suggested this investigatio and to Dr. J. R. Wait for critical comments.

Literature

[1] H. Hertz, "Über Strahlen elektrischer Kraft," *Ann. Physik*, vo 36, pp. 769–783; 1889.

[2] M. G. Andreasen, "Reflection from and transmission throug two parallel wire grids," *Acta Polytech. Scand.*, Elec. Engrg. Ser vol. 7, no. 2; 1956.

[3] J. Aagesen, "Polarization-transforming plane reflector f microwaves," *Acta Polytech. Scand.*, Elec. Eng. Ser., vol. pp. 1–27; 1957.

[4] G. G. MacFarlane, "Surface impedance of an infinite parall wire grid at oblique angles of incidence," *J. Inst. Elec. Engrs* pt. 3 A, vol. 93, pp. 1523–1527; 1946.

[5] G. Gresky, "Die Wirkungsweise von Reflektoren bei kurze elektrischen Wellen," *Jahrb. drahtl. Tel.*, vol. 32, pp. 149–16 November, 1928.

[6] F. R. Abbott, "Design of optimum buried-conductor RF groun system," Proc. IRE, vol. 40, pp. 846–852; July, 1952.

[7] G. D. Monteath, "The effect of the ground constants, and of a earth system, on the performance of a vertical medium wav aerial," *Proc. IEE*, pt. C, vol. 105, pp. 292–306; March, 195

[8] H. L. Knudsen and T. Larsen, "Wirtschaftlich optimale D mensionierung von Erddrahtsystemen für ausgedehnte Ante nensysteme," *Wiss. Z. Hochschule Elektrotechnik*, Ilmenau DD vol. 6, pp. 41–53; 1960.

[*9] R. J. Coates, "A grating spectrometer for millimeter waves *Rev. Sci. Instr.*, vol. 19, pp. 586–590; September, 1948.

[*10] J. A. Klein, J. H. N. Laubser, A. H. Nethercot and C. Townes, "Magnetron harmonics at millimeter wavelength," *R Sci. Instr.*, vol. 23, pp. 78–82; February, 1952.

[11] W. Wessel, "Über den Durchgang elektrischer Wellen dur Drahtgitter," *Hochfrequenztech. Elekt. Akust.*, vol. 54, pp. 62–7 August, 1939.

[12] R. Hornejäger, "Über die Beugung elektromagnetischer Well an einem Drahgitter," *Ann. Physik*, vol. 4, pp. 25–35; Januar 1948.

[13] J. R. Wait, "Reflection at arbitrary incidence from a parall wire grid," *Appl. Sci. Research*, vol. B4, pp. 393–400; 1955.

[14] L. I. Chu and J. D. Jackson, "Field theory of travelling-wa tubes," Proc. IRE, vol. 36, pp. 853–865; July, 1948.

[15] E. M. T. Jones and S. B. Cohn, "Surface matching of dielectr lenses," *J. Appl. Phys.*, vol. 26, pp. 452–457; April, 1955.

[*16] E. A. Lewis and J. P. Casey, "Metal grid interference filter *J. Opt. Soc. Am.*, vol. 41, p. 360; May, 1951.

[17] Z. A. Kaprielian, "Electromagnetic transmission characterist of a lattice of infinitely long cylinders," *J. Appl. Phys.*, vol, 2 pp. 1491–1502; December, 1956.

[18] J. J. Thomson, "Recent researches on electricity and magne ism," Oxford Univ. Press, England, pp. 425–428; 1893.

[19] H. Lamb, "On the diffraction in transmission of electric waves a metallic grating," *Proc. London Math. Soc.*, vol. 29, pp. 52 544; 1898.

[20] C. L. Schaefer and M. Laugwitz, "Über die bei Reflexion ele trischer Wellen an Hertzscher Gittern auftretende Phase verluste," *Ann. Physik*, vol. 21, pp. 587–594; 1906.

[21] G. H. Thomson, "Über den Durchgang Hertzscher Well durch Gitter," *Ann. Physik*, vol. 22, pp. 365–390; 1907.

[22] C. L. Schaefer and M. Laugwitz, "Zur Lambschen Gittertheorie," *Ann. Physik*, vol. 23, pp. 399–600; 1907.

[23] C. L. Schaefer and M. Laugwitz, "Abhängigkeit des Verhaltens Hertzscher Gitter vom Leitvermögen," *Ann. Physik*, vol. 23, pp. 951–956; 1907.

[24] H. du Bois and H. Rubens, "Polarization ungebeugter langwelliger Wärmestrahlen durch Drahtgitter," *Ann. Physik*, vol. 35, pp. 243–276; 1911.

[25] C. L. Schaefer and F. Reiche, "Ein Beitrag zur Theorie der Gitterinversion," *Ann. Physik*, vol. 32, pp. 577–588; 1910.

[26] ——, "Zur Theorie der Beugungsgitters," *Ann. Physik*, vol. 35, pp. 817–859; 1911.

[27] W. von Ignatowsky, "Zur Theorie der Gitter," *Ann. Physik*, vol. 44, pp. 369–436; 1914.

[28] W. C. Meecham, S. T. Choh and D. W. Mooman, "Calculation of the reflection of radiation from saw-tooth surfaces and the reflection and transmission of radiation by parallel cylinder systems," Final rept., Air Force Office of Scientific Research Contract No. AF 49(638)-26, University of Michigan, Ann Arbor; January, 1958.

[29] W. Arkadiew, "Über die Reflexion elektromagnetischer Wellen an Drähten," *Ann. Physik*, vol. 45, pp. 133–146; 1914.

[30] R. Gans, "Das Verhalten Hertzscher Gitter," *Ann. Physik*, vol. 61, pp. 447–454; 1920.

[31] ——, "Bemerkungen zu meiner Arbeit: Das Verhalten Hertzscher Gitter," *Ann. Physik*, vol. 66, pp. 427–428; February, 1922.

[32] C. L. Schaefer, "Das Verhalten Hertzscher Gitter (Bemerkungen zu einer Abhandlung des Herrn R. Ganz)," *Ann. Physik*, vol. 74, pp. 275–284; May, 1924.

[33] W. Arkadiew, "Reflexion elektromagnetischer Wellen an Hertzschen Gittern," *Ann. Physik*, vol. 75, pp. 426–434; 1924.

[34] W. Arkadiew, "Die Reflexion Hertzscher Wellen an ferromagnetischer Drahtgittern," *Ann. Physik*, vol. 81, pp. 649–665; November, 1926.

[35] A. Esau, E. Ahrens und W. Kebbel, "Über die Durchlässigkeit von Drahtgittern für elektrische Wellen," *Hochfrequenztech. Elekt. Akust*, vol. 53, pp. 113–115; April, 1939.

[36] G. G. MacFarlane, "Quasi-stationary field theory and its application to diaphragms and functions in transmission lines and wave guides," *Proc. IEE*, pt. 3A, vol. 93, pp. 703–719; 1946.

[37] H. G. Booker, "The elements of wave propagation using the impedance concept," *J. IEE*, pt. 3, vol. 94, pp. 171–202; May, 1947. See especially pp. 187–191.

[38] J. W. Miles, "The diffraction of a plane wave through a grating," *Quart. Appl. Math.*, vol. 7, pp. 45–64; May, 1949.

[39] J. Shmoys, "Diffraction of electromagnetic waves by a plane wire grating," *J. Opt. Soc. Am.*, vol. 41, pp. 324–328; May, 1951.

[40] N. Marcuvitz, "Waveguide Handbook," M.I.T. Rad. Lab. Ser., McGraw-Hill Book Co., New York, N. Y., no. 10, pp. 280–289; 1951.

[*41] L. Grillini, "An experiment using plane gratings with microwaves," *Nuovo Cimento*, vol. 8, pp. 952–959; December, 1951. (In Italian.)

[42] E. A. Lewis and J. Casey, "Electromagnetic reflection and transmission by gratings of resistive wires," *J. Appl. Phys.*, vol. 23, pp. 605–608; June, 1952.

[43] S. Ramo and J. R. Whinnery, "Fields and waves in modern radio," John Wiley and Sons, Inc., New York, N. Y., p. 212; 1947.

[44] R. Müller, "Eine strenge Behandlung der Beugung elektromagnetischer Wellen am Streifengitter," *Z. Naturforsch.*, vol. 8a, pp. 56–60; January, 1953.

[45] ——, "Reflexion elektromagnetischer Wellen an inhomogenen Grenzflächen mit einer zum E-Vektor senkrechten periodischen Struktur," *Arch. elekt. Übertragung*, vol. 7, pp. 492–500; October, 1953.

[46] S. A. Schelkunoff, "Elektromagnetic Waves," D. Van Nostrand Company, Inc., New York, N. Y., p. 264; 1944.

[*47] V. G. Yampol'skii, "Oblique incidence of a plane wave on a wire lattice," *Radiotekhnika*, vol. 10, no. 9, pp. 39–48; 1955. (In Russian.)

[48] ——, "Reflection of a plane wave from wire mesh at normal polarization," *Radiotekhnika*, vol. 11, no. 11, pp. 33–37; 1956. (In Russian.)

[*49] W. K. Pursley, Ph.D. thesis, The University of Michigan, Ann Arbor; 1956.

[50] O. J. Snow, "Transmission characteristics of inclined wire gratings," IRE Trans. on Antennas and Propagation, vol. AP-4, pp. 650–654; October, 1956.

[51] R. J. Primich, "Some electromagnetic transmission and reflection properties of a strip grating," IRE Trans. on Antennas and Propagation, vol. AP-5, pp. 176–182; April, 1957.

[*52] Y. N. Fel'd, "The diffraction of an electromagnetic wave by a semi-infinite mesh," *Radiotekh. Elektron.*, vol. 3, no. 7, pp. 882–889; 1958. (In Russian.)

[53] E. B. Hansen, "The Diffraction of a Plane Wave Through Two or More Slits in a Plane Screen," Licentiate degree, Laboratory of Electromagnetic Theory, Technical University, Denmark; 1958. (In Danish.)

[54] ——, "The diffraction of a plane wave through two or more slits in a plane screen," *Appl. Sci. Research*, vol. B8, pp. 73–83; 1959.

[55] M. T. Decker, "Transmission and reflection by a parallel wire grid," *J. Res. NBS*, vol. 63D, pp. 87–90; July-August, 1959.

[56] K. Særmark, "Scattering of a plane monochromatic wave by a system of strips." *Appl. Sci. Research*, vol. B7, pp. 417–440; 1959.

[57] ——, "Acoustic forces and torques on a system of strips," *Appl. Sci. Research*, vol. B8, pp. 13–28; 1959.

[58] ——, "Transmission coefficient for a system of parallel slits in a thin, plane screen," *Appl. Sci. Research*, vol. B8, pp. 29–34; 1959.

[59] ——, "On some two dimensional diffraction problems," Ph.D. dissertation, Dansk Videnskabs Forlag, Copenhagen, Denmark; 1960.

[60] ——, "A note on addition theorems for Mathieu functions, *Z. Math. Phys.*, vol. 10, pp. 426–428; 1959.

[61] J. K. Skwirzynski and J. C. Thackray, "Transmission of electromagnetic waves through wire gratings (theory)," *Marconi Rev.*, vol. 22, pp. 77–90; 1959.

[62] E. G. A. Goodall and J. A. C. Jackson, "Transmission of electromagnetic waves through wire gratings (experimental), *Marconi Rev.*, vol. 22, pp. 91–98; 1959.

[*63] O. J. Kozinets, "Reflection of electric waves on the wires of a Hertz screen parallel to the magnetic vector," in "Problems of Ferromagnetism and Magnetodynamics," Moscow, pp. 89–92; 1946. (In Russian.)

[*64] J. Epelboim, "On the behaviour of a ferromagnetic wire grating in the centimetre-wave band," *J. Phys. Radium*, vol. 8, pp. 251–256; August, 1947. (In French.)

[65] V. K. Paramonov, "The effect of deposits on the electrical properties of wire mesh surfaces," *Radiotekhnika*, vol. 11, no. 9, pp. 12–20; 1956. (In Russian.)

[66] G. von Trentini, "Gitter als Schaltelemente elektrischer Wellen im Raum," *Z. Angew. Phys.*, vol. 5, pp. 221–231; June, 1953.

[67] W. Franz, "Theorie des Trentinischen Absorptionsgitters," *Z. angew. Phys.*, vol. 6, pp. 449–456; 1954.

[68] ——, "The transmission of electric waves through wire grids," *Z. angew. Phys.*, vol. 1, pp. 416–423; June, 1949.

[69] W. E. Groves, "Transmission of electromagnetic waves through pairs of parallel wire grids," *J. Appl. Phys.*, vol. 24, pp. 845–854; July, 1953.

[*70] G. von Trentini, "Maximum transmission of electromagnetic waves by a pair of wire gratings, *J. Opt. Soc. Am.*, vol. 45, pp. 833–885; October, 1955.

[*71] Y. N. Fel'd, "The incidence of electromagnetic waves on a pair of infinite grids," *Dokl. Akad. Nauk. SSSR*, vol. 107, pp. 71–74; 1956. (In Russian.)

[72] V. Twersky, "On the non-specular reflection of plane waves of sound," *J. Acoust. Soc. Am.*, vol. 22, pp. 539–546; September, 1950.

[73] ——, "Multiple scattering of radiation by an arbitrary configuration of parallel cylinders," *J. Acoust. Soc. Am.*, vol. 24, pp. 42–46; January, 1952.

[74] ——, "Multiple scattering of radiation by an arbitrary planar configuration of parallel cylinders and by two parallel cylinders," *J. Appl. Phys.*, vol. 23, pp. 407–414; April, 1952.

[75] ——, "On a multiple scattering theory of the finite grating and the Wood anomalies," *J. Appl. Phys.*, vol. 23, pp. 1099–1118; October, 1952.

[*76] T. Sakurai, "Electromagnetic phenomena on ribbon elements gratings," *J. Phys. Soc. Japan*, vol. 5, pp. 203–207; July-August, 1950.

[77] J. R. Wait, "Reflection from a wire grid parallel to a conducting plane," *Canad. J. Phys.*, vol. 32, pp. 571–579; September, 1954.

[78] ——, "The impedance of a wire grid parallel to a di-electric interface," IRE Trans. on Microwave Theory and Techniques, vol. MTT-5, pp. 99–102; April, 1957.

[79] ——, "On the theory of reflection from a wire grid parallel to an interface between homogeneous media I," *Appl. Sci. Research*, vol. B6, pp. 259–275; 1956.

[80] ——, "On the theory of reflection from a wire grid parallel to an interface between homogeneous media II," *Appl. Sci. Research*, vol. B7, pp. 355–360; 1958.

[81] T. Larsen, "Numerical investigation of the equivalent impedance of a wire grid parallel to the interface between two media," *J. Res. NBS*, vol. 66D, pp. 7–14; January-February, 1962.

[*82] V. G. Yampol'skii, "The diffraction of plane waves by a wire grid situated inside a dielectric slab," *Radiotekh. Elektron.*, vol. 3, pp. 1516–1518; December, 1958. (In Russian.)

[83] A. N. Sivov, "Incidence of a plane electromagnetic wave on a flat grating (case of the H-vector parallel to the conductors)," *Radiotekh. Elektron.*, vol. 6, pp. 58–66; January, 1961. (In Russian.)

A NEW TIME PHENOMENON IN PHOTOGRAPHIC EMULSIONS.

By F. Weigert.

Received 11*th March*, 1938.

In the course of experiments on film-dichroism, the technique of which will be published separately, it has been observed that photo-dichroism can be produced by natural light as well as by the plane polarised light used in our previous researches. The use of natural light is an improvement, not only because it is more convenient, but also because the photo-dichroism it produces is more closely related to the exciting wave-length than is the case with plane polarised light.

The production of photo-dichroism by natural light [1] resembles in certain respects the Tyndall effect in which the scattered light is partially plane polarised laterally to the direction of a primary beam of natural light; and photo-dichroism can be detected in an exposed film if the light with which it is observed passes through it in a different direction to that of the exciting light. The effect is best obtained when the two beams are directed perpendicularly to each other.

Light sensitive films suitable for producing photodichroism are composed of minute dichroic particles with their axes oriented at random. In previous papers we have called these particles "light sensitive micelles." They are changed into "photomicelles" preferentially by light vibrating parallel to their axes.[2] A beam of natural light passing through the film produces therefore photomicelles whose axes are oriented perpendicularly to the beam. On observing the film in the direction of the beam no photo-dichroism is obtained because the azimuths of the axes have all possible values. But vertically to this direction the planes which contain the axes of the photomicelles appear on edge and photodichroism is revealed.

In our experiments we used a fine-grain silver chloride gelatin emulsion containing a slight excess of silver ion which was coated on glass plates. Light from two carbon arcs was focused (from opposite sides) on the same place of the film for studying the photo-dichroic effects of differently coloured radiations acting simultaneously or successively. One of these coloured lights must contain radiations of such short wave-lengths as to produce some free silver by photolysis of silver chloride crystals, since both silver halide and silver are indispensible for the production of photo-dichroism. The plates covered with the gelatin film were fixed (inclined at 45° to the vertical) on a small stand which could be placed in a trough with plane end plates filled with benzene, so that a horizontal beam would pass through the film at 45°. After exposure of the film the plate was turned through 180° in its own plane. This operation enabled us to observe on edge the planes which contain the axes of the photomicelles if any were produced by the exposure to natural light. The trough with the stand was then placed between the polariser and analyser of a polarimeter in order to determine the photodichroism and its distribution throughout the spectrum. The most striking results were obtained by the combined

[1] F. Weigert, *Ann. Physik* (4), 1920, **63**, 693; A. E. Cameron and A. M. Taylor, *J. Opt. Soc. Am.*, 1934, **24**, 327.

[2] F. Weigert and J. Matulis, *Koll. Beihefte*, 1933, **38**, 384; F. Weigert, *Trans. Faraday Soc.*, 1935, **31**, 359.

action of blue and red light. The " blue " light was produced by using a filter of 11 M. ammonia to which copper sulphate was added until no green or red light was transmitted. The resulting radiation consisted of blue, violet and near ultra-violet. The red filter was made from two dark cobalt glasses and a red filter. Only light of longer wave-length than 650 mμ was transmitted.

In experiments in the same series the total time of exposure to blue and red light was kept constant. In the typical example demonstrated by Fig. 1 this time was 1 minute for blue as well as for red.

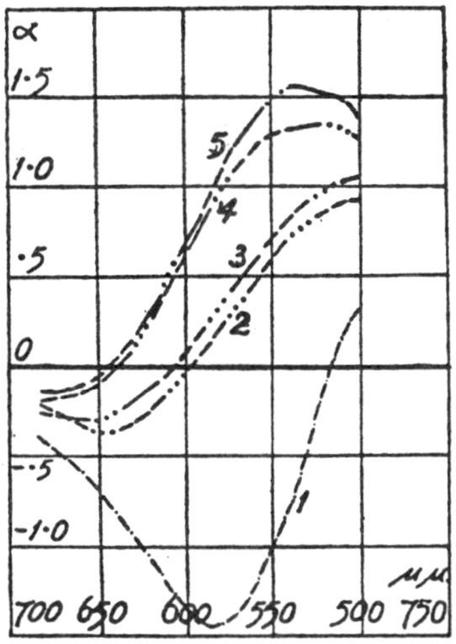

FIG. 1.—Photodichroism produced by the combined action of blue and red natural light on silver chloride gelatin.

(1) —.—.—.— blue and red acting simultaneously, Exp. time 1 min.
(2) ---...---... blue, red, blue, etc., Exp. time 2 mins.
(3) ---...---- blue, red, dark, blue, etc., Exp. time 6 mins.
(4) ...------... red, blue, dark, red, etc., Exp. time 6 mins.
(5) -----...— blue, dark, red, dark, blue, etc., Exp. time 12 mins.

Curve 1 shows the photodichroism after a simultaneous exposure of the plate to blue and red natural light for 60 secs. Curves 2, 3, 4 and 5 were obtained by a subdivision of the total time of exposure of 60 secs. into 12 periods of 5 secs. in the following order :

Exp. (2) 5 secs. blue, 5 secs. red, 5 secs. blue, etc. Total exposure time = 120 secs.

Exp. (3) 5 secs. blue, 5 secs. red, 20 secs. dark, 5 secs. blue, etc. Total exposure time 360 secs.

Exp. (4) 5 secs. blue, 20 secs. dark, 5 secs. red, 5 secs. blue, etc. Total exposure time 360 secs.

Exp. (5) 5 secs. blue, 25 secs. dark, 5 secs. red, 25 secs. dark, 5 secs. blue, etc. Total exposure time 720 secs.

The general feature of the graphs can be described as exhibiting negative photodichroism in the red part of the spectrum which becomes positive in the green part. This kind of distribution of photodichroism throughout the spectrum has already been discussed [3] when dealing with the action of plane polarised red light on photochlorides.

The photochloride used in the previous experiments was produced by a pre-exposure of silver chloride to blue light. In our new experiment (1) both coloured lights acted simultaneously and the shape of curve 1 is an example of strong " colour adaptation " to red light.

In comparison with curve 1 curves 2 to 5 show a pronounced decay of the negative values in red and an increase of the positive values in green after alternating exposures to blue and red light. Hence the whole curves shift towards the red end of the spectrum.

From (1), (2) and (5) it is evident that the time which has elapsed between a blue and red exposure is the reason for the shifting of the curves. The tendency to produce negative values of photo-dichroism diminishes with increasing time intervals between both exposures.

The most striking results were obtained from experiments 3 and 4, each of which lasted 6 minutes. They are distinguished from each other only by the order in which the two colours were arranged in each pair of exposures, separated from the next pair by a dark interval of 20 seconds.

[3] *Cf.* F. Weigert and E. Elvegård, *Z. physik. Chem. B.*, 1929, **4**, 239.

If a blue exposure immediately preceded the red light, curve 3 was nearly identical throughout with curve 2 which was produced within 2 minutes with no interval at all between blue and red. If red light immediately preceded the blue exposure the shape of curve 4 was about the same as that of curve 5 which lasted 12 minutes and in which each single colour was separated from the next by an interval of 25 seconds.

We see that when a dark interval succeeds an exposure to *blue* light the trend of the curves is changed, whereas a dark interval after a red exposure has no perceptible influence. Some matter must be introduced into the system by the action of blue light which is required for the production of negative values by red light and which loses its activity during the dark interval following the blue exposure. This matter is of course most efficient during simultaneous exposures in exp. 1.

This is most simply explained by the production of a sensitiser for red due to the action of the blue light; this can only be free silver and it is obvious that the diminution of its activity results from coagulation of the silver atoms produced by the photolysis. This conception accords well with familiar ideas on the effect of light in photographic emulsions, but it seems somewhat unexpected that this coagulation is a slow process which can be revealed by the simple methods used in these experiments.

Analogous experiments with orange and green light as the long wavelength component of the combined radiation gave less definite results than red light in combination with blue light. The graphs indicated no remarkable differences after a simultaneous exposure to both radiations and after an alternating exposure in periods of 5 seconds. The reason for this behaviour is the production of free silver even in pure green and orange light which does not occur in red light to a perceptible degree. In yellow light and in green light of 530 and 507 mμ which were purified by means of Christiansen filters, minima of negative photo-dichroism appeared at the correct places of the spectrum, which means clear colour adaptations. But also in pure blue light with its maximum of energy at about 450 mμ, in the violet with a maximum at 410 mμ, and even in the ultra-violet with a maximum at 390 mμ which were purified by colour filters the production of photo-dichroism was evident. In these cases no actual minima at the right parts of the spectrum could be measured with the instruments at our disposal, but the slope of the curves allowed us to extrapolate the expected shape. They will be reproduced in a separate paper.

The mechanism governing all these phenomena of sensitising and auto-sensitising actions (which are directly connected with the orientation of the light vibrations) I have already suggested,[2] and it is confirmed by the new time phenomenon. According to this conception a photomicelle is produced by the transfer of an electron from a silver atom to a silver ion which belongs to the silver halide, and which is properly oriented with respect to the light vibration. Red light requires silver atoms acting as electron donors which must be introduced by photolysis by an additional short wave-length light, as in our experiments. But the radiations of shorter wave-lengths than red produce their own donors either by photolysis of the silver halide or by means of a "silver body," according to Fajans.

In all cases the silver atoms effective in the production of photomicelles lose something of their activity as donors of electrons by coagulation to larger silver crystals and aggregates. This loss of activity can be revealed by the time phenomenon which was described above.

To correlate these experiments with the mechanism of the formation of the latent photographic image further experiments in natural light will be necessary. Some facts already known indicate that the relationships are very close.

In work on developed photo-dichroism on Ilford lantern plates produced by mixed radiations, Stiebel discovered very pronounced differences in the graphs when blue and red polarised light struck the film simultaneously or alternatingly.[4] Furthermore, Shidei found exceptionally strong deviations from the reciprocity law in his experiments on developed photo-dichroism.[5]

These failures of photographic emulsions to behave normally may be due to a slow decay in activity of an auto-sensitiser produced during the exposure. Thus the formation of the latent image by a given amount of light energy is the more inhibited the longer the time of exposure.

Summary.

The photodichroism produced on an oblique film of silver chloride gelatin by the combined action of beams of natural light of blue and red colour when plotted against the spectrum was the more positive as the length of the dark time interval elapsing between a blue and red exposure was increased. A dark interval following a red exposure had no perceptible influence. It is suggested that this is due to the production, by the action of the blue light, of a sensitiser for red which slowly loses its activity.

The author is greatly indebted to the " Society for the Protection of Science and Learning " for the grant of a Research-Fellowship and to the University of Glasgow for the generous hospitality given to him in the Chemistry Department. The determinations of photodichroism were carried out with an excellent polarimeter and monochromator. I am very grateful to the " Deutsche Forschungsgemeinschaft," Berlin, for having lent the instruments to me, and for the permit to use them in Glasgow.

[4] F. Weigert and F. Stiebel, *ibid.*, 1931, **13**, 289.
[5] F. Weigert and J. Shidei, *ibid.*, 1930, **9**, 344.

On the Use of a Pile-of-Plates Polarizer—The Transmitted Component

J. L. Weinberg

Formulas for the transmitted component of a pile-of-plates polarizer are derived in terms of the angle of incidence, the index of refraction, and the number of (transparent) plates for singly and multiply reflected rays. Numerical results are discussed for one, two, five, and ten plates and for indices of refraction of 1.5170, 1.5180, 1.5725, 1.6210, and 2.0060.

As part of a recent photoelectric study of the zodiacal light,[1] we have extended the calculation of the polarization produced by a pile of plane-parallel plates.

The reflected component has a higher degree of polarization, p, per surface than the transmitted component, but it has the disadvantage that it is not parallel to the original beam. Its use frequently requires the use of mirrors or other methods of changing the direction of the light path, and unwanted polarization is introduced. The transmitted component has only the slight disadvantage of a small degree of polarization per plate.

In view of the lack of a unified treatment of the polarization properties of a pile-of-plates at angles of incidence other than Brewster's angle, we will formulate here the general problem (for transparent plates) and discuss some of the numerical results.

The reflection coefficients of a dielectric surface were originally derived by Fresnel,[2] and they yield the orthogonal components of both the reflectivity, $\mathcal{R}$, and the transmissivity, $\mathfrak{J}$:

$$\mathcal{R}_\perp = \sin^2(\theta_i - \theta_t)/\sin^2(\theta_i + \theta_t), \tag{1a}$$

$$\mathcal{R}_\parallel = \tan^2(\theta_i - \theta_t)/\tan^2(\theta_i + \theta_t), \tag{1b}$$

$$\mathfrak{J}_\perp = \sin 2\theta_i \sin 2\theta_t/\sin^2(\theta_i + \theta_t), \tag{1c}$$

$$\text{and } \mathfrak{J}_\parallel = \sin 2\theta_i \sin 2\theta_t/[\sin^2(\theta_i + \theta_t)\cos^2(\theta_i - \theta_t)], \tag{1d}$$

where

θ_i = angle of incidence,

θ_t = angle of refraction, and

$\perp,\parallel$ = those components of intensity which are perpendicular and parallel, respectively, to the plane of incidence.*

These formulas will be used to obtain the degree of polarization of the transmitted component which is composed of the directly transmitted rays diluted by either singly or multiply reflected rays.

Transparent Plates and Singly Reflected Rays

It is well known that, after passing through one boundary of a plate, the ratio of the transmitted intensities of the two components is

$$\mathfrak{J}_\perp/\mathfrak{J}_\parallel = \cos^2(\theta_i - \theta_t). \tag{2}$$

The degree of polarization of the component transmitted through one boundary of a plate is defined as

$$p \equiv |(\mathfrak{J}_\parallel - \mathfrak{J}_\perp)/(\mathfrak{J}_\parallel + \mathfrak{J}_\perp)|. \tag{3}$$

After passing through both boundaries of a plate, the ratio of the transmitted intensities is

$$(\mathfrak{J}_\perp/\mathfrak{J}_\parallel)^2 = \cos^4(\theta_i - \theta_t) < 1. \tag{4}$$

Extension to m plates ($2m$ surfaces) is straightforward, and the degree of polarization becomes

$$p(m,n,\theta_i) \equiv \left|\frac{\mathfrak{J}_\parallel^{2m} - \mathfrak{J}_\perp^{2m}}{\mathfrak{J}_\parallel^{2m} + \mathfrak{J}_\perp^{2m}}\right| = \frac{1 - \cos^{4m}[\theta_i - \theta_t(n)]}{1 + \cos^{4m}[\theta_i - \theta_t(n)]}, \tag{5}$$

where n is the index of refraction. Equation (5) is applicable when only the directly transmitted component reaches the detector undiluted by multiply reflected radiation. If the majority of the multiply

The author was with the High Altitude Observatory, Boulder, Colorado. His present address is Hawaii Institute of Geophysics, University of Hawaii, Honolulu, Hawaii.

Received 19 December 1963.

* The notation follows that of Born and Wolf.[3]

Table I. The Degree of Polarization for Piles of One, Two, Five, and Ten Transparent Glass Plates (for Singly and Multiply Reflected Rays) for $n = 1.5170$ (e.g., Borosilicate Crown or BSC-2 or BK-7)

θ_i (deg)	θ_t (deg)	Singly reflecting 1	Singly reflecting 2	Singly reflecting 5	Singly reflecting 10	Multiply reflecting 1	Multiply reflecting 2	Multiply reflecting 5	Multiply reflecting 10
00	00	0	0	0	0	0	0	0	0
05	03.29	0.000887	0.00177	0.00443	0.00887	0.000851	0.00157	0.00321	0.00492
10	06.57	0.00358	0.00716	0.0179	0.0358	0.00343	0.00635	0.0130	0.0199
15	09.82	0.00818	0.0163	0.0408	0.0816	0.00784	0.0145	0.0296	0.0453
20	13.03	0.0148	0.0297	0.0740	0.147	0.0142	0.0263	0.0536	0.0821
25	16.18	0.0238	0.0476	0.118	0.234	0.0228	0.0422	0.0858	0.131
30	19.24	0.0354	0.0708	0.175	0.340	0.0340	0.0626	0.127	0.193
35	22.22	0.0502	0.100	0.246	0.464	0.0480	0.0882	0.178	0.268
40	25.07	0.0686	0.136	0.330	0.596	0.0654	0.120	0.238	0.355
45	27.78	0.0914	0.181	0.429	0.724	0.0859	0.157	0.307	0.450
50	30.33	0.120	0.236	0.538	0.834	0.113	0.202	0.382	0.544
55	32.68	0.154	0.302	0.652	0.915	0.144	0.252	0.457	0.626
56.61[a]	33.39	0.167	0.325	0.688	0.934	0.155	0.269	0.479	0.648
60	34.81	0.197	0.380	0.761	0.964	0.181	0.306	0.522	0.682
65	36.69	0.249	0.470	0.855	0.988	0.223	0.359	0.566	0.701
70	38.28	0.313	0.570	0.924	0.997	0.268	0.404	0.582	0.681
75	39.55	0.388	0.675	0.967	0.999	0.314	0.435	0.565	0.627
80	40.48	0.477	0.777	0.989	0.999+	0.355	0.444	0.522	0.554
85	41.05	0.576	0.865	0.997	0.999+	0.384	0.429	0.461	0.473
90	41.24	1.00	1.00	1.00	1.00	0.394	0.394	0.394	0.394

[a] Brewster's angle ($\tan^{-1} n$).

Table II. The Degree of Polarization for Piles of One, Two, Five, and Ten Transparent Glass Plates (for Singly and Multiply Reflected Rays) for $n = 1.5180$ (e.g., Picture Glass)

θ_i (deg)	θ_t (deg)	Singly reflecting 1	Singly reflecting 2	Singly reflecting 5	Singly reflecting 10	Multiply reflecting 1	Multiply reflecting 2	Multiply reflecting 5	Multiply reflecting 10
00	00	0	0	0	0	0	0	0	0
05	03.29	0.000889	0.00178	0.00444	0.00889	0.000853	0.00158	0.00322	0.00493
10	06.57	0.00359	0.00718	0.0179	0.0359	0.00344	0.00637	0.0130	0.0199
15	09.82	0.00819	0.0164	0.0409	0.0818	0.00786	0.0145	0.0296	0.0454
20	13.02	0.0149	0.0297	0.0742	0.148	0.0143	0.0264	0.0537	0.0822
25	16.16	0.0239	0.0477	0.119	0.234	0.0229	0.0423	0.0860	0.131
30	19.23	0.0355	0.0710	0.176	0.341	0.0340	0.0627	0.127	0.193
35	22.20	0.0503	0.100	0.246	0.465	0.0481	0.0884	0.178	0.268
40	25.05	0.0687	0.137	0.331	0.597	0.0656	0.120	0.238	0.355
45	27.76	0.0916	0.182	0.430	0.725	0.0871	0.158	0.308	0.450
50	30.31	0.120	0.236	0.539	0.835	0.113	0.202	0.382	0.545
55	32.66	0.155	0.302	0.653	0.916	0.144	0.252	0.457	0.627
56.62[a]	33.38	0.168	0.326	0.689	0.935	0.156	0.270	0.480	0.648
60	34.78	0.198	0.380	0.762	0.964	0.181	0.306	0.522	0.683
65	36.66	0.250	0.470	0.856	0.988	0.223	0.359	0.567	0.702
70	38.25	0.313	0.571	0.925	0.997	0.269	0.405	0.582	0.681
75	39.52	0.389	0.676	0.968	0.999	0.315	0.436	0.566	0.628
80	40.45	0.478	0.778	0.989	0.999+	0.356	0.444	0.522	0.554
85	41.02	0.577	0.866	0.997	0.999+	0.384	0.429	0.462	0.474
90	41.20	1.00	1.00	1.00	1.00	0.395	0.395	0.395	0.395

[a] Brewster's angle ($\tan^{-1} n$).

Table III. The Degree of Polarization for Piles of One, Two, Five, and Ten Transparent Glass Plates (for Singly and Multiply Reflected Rays) for $n = 1.5725$ (e.g., Light Barium Crown or Light Flint)

θ_i	θ_t	Singly reflecting				Multiply reflecting			
(deg)	(deg)	1	2	5	10	1	2	5	10
00	00	0	0	0	0	0	0	0	0
05	03.18	0.00101	0.00202	0.00506	0.0101	0.000965	0.00176	0.00350	0.00522
10	06.34	0.00408	0.00817	0.0204	0.0408	0.00389	0.00711	0.0141	0.0210
15	09.47	0.00932	0.0186	0.0466	0.0929	0.00888	0.0162	0.0322	0.0480
20	12.56	0.0169	0.0338	0.0843	0.167	0.0161	0.0294	0.0583	0.0868
25	15.59	0.0271	0.0541	0.135	0.264	0.0258	0.0471	0.0932	0.138
30	18.54	0.0402	0.0804	0.199	0.382	0.0383	0.0697	0.138	0.203
35	21.39	0.0569	0.113	0.277	0.515	0.0540	0.0981	0.192	0.282
40	24.13	0.0776	0.154	0.370	0.651	0.0735	0.133	0.257	0.373
45	26.72	0.103	0.204	0.476	0.776	0.0973	0.174	0.330	0.471
50	29.15	0.134	0.264	0.590	0.875	0.126	0.222	0.409	0.568
55	31.39	0.173	0.336	0.703	0.941	0.160	0.276	0.487	0.653
57.55[a]	32.46	0.196	0.377	0.758	0.963	0.180	0.305	0.523	0.687
60	33.42	0.220	0.419	0.807	0.977	0.200	0.333	0.554	0.711
65	35.19	0.276	0.514	0.889	0.993	0.245	0.389	0.600	0.732
70	36.70	0.344	0.615	0.946	0.998	0.294	0.436	0.615	0.713
75	37.90	0.424	0.719	0.978	0.999+	0.342	0.467	0.598	0.661
80	38.78	0.515	0.814	0.993	0.999+	0.384	0.475	0.554	0.587
85	39.31	0.615	0.893	0.998	0.999+	0.413	0.460	0.493	0.505
90	39.49	1.00	1.00	1.00	1.00	0.424	0.424	0.424	0.424

[a] Brewster's angle ($\tan^{-1} n$).

Table IV. The Degree of Polarization for Piles of One, Two, Five, and Ten Transparent Glass Plates (for Singly and Multiply Reflected Rays) for $n = 1.6210$ (e.g., Dense Flint or DF-3)

θ_i	θ_t	Singly reflecting				Multiply reflecting			
(deg)	(deg)	1	2	5	10	1	2	5	10
00	00	0	0	0	0	0	0	0	0
05	03.08	0.00112	0.00224	0.00560	0.0112	0.00106	0.00192	0.00372	0.00542
10	06.15	0.00452	0.00904	0.0226	0.0452	0.00428	0.00773	0.0150	0.0219
15	09.19	0.0103	0.0206	0.0515	0.103	0.00976	0.0176	0.0342	0.0498
20	12.18	0.0187	0.0373	0.0931	0.185	0.0177	0.0319	0.0619	0.0901
25	15.11	0.0299	0.0598	0.148	0.291	0.0283	0.0511	0.0988	0.144
30	17.97	0.0444	0.0866	0.219	0.417	0.0420	0.0756	0.146	0.211
35	20.72	0.0627	0.125	0.304	0.556	0.0592	0.106	0.203	0.292
40	23.36	0.0853	0.169	0.403	0.694	0.0803	0.143	0.271	0.385
45	25.86	0.113	0.224	0.514	0.813	0.106	0.188	0.347	0.486
50	28.20	0.147	0.288	0.630	0.902	0.137	0.238	0.429	0.585
55	30.35	0.189	0.364	0.742	0.957	0.174	0.295	0.510	0.672
58.33[a]	31.70	0.220	0.421	0.808	0.978	0.201	0.334	0.557	0.715
60	32.29	0.239	0.452	0.839	0.985	0.216	0.355	0.579	0.732
65	33.99	0.299	0.549	0.912	0.996	0.264	0.413	0.625	0.755
70	35.43	0.370	0.651	0.960	0.999	0.314	0.461	0.642	0.738
75	36.58	0.453	0.751	0.985	0.999+	0.364	0.493	0.625	0.687
80	37.41	0.546	0.841	0.996	0.999+	0.408	0.501	0.581	0.614
85	37.92	0.646	0.912	0.999	0.999+	0.438	0.485	0.519	0.531
90	38.09	1.00	1.00	1.00	1.00	0.449	0.449	0.449	0.449

[a] Brewster's angle ($\tan^{-1} n$).

Table V. The Degree of Polarization for Piles of One, Two, Five, and Ten Transparent Silver Chloride (AgCl) Plates (for Singly and Multiply Reflected Rays) for $n = 2.0060$

θ_i	θ_t	Singly reflecting				Multiply reflecting			
(deg)	(deg)	1	2	5	10	1	2	5	10
00	00	0	0	0	0	0	0	0	0
05	02.49	0.00192	0.00384	0.00960	0.0192	0.00173	0.00287	0.00478	0.00614
10	04.96	0.00774	0.0155	0.0386	0.0772	0.00696	0.0116	0.0192	0.0247
15	07.41	0.0176	0.0352	0.0877	0.174	0.0158	0.0263	0.0438	0.0562
20	09.82	0.0317	0.0634	0.157	0.307	0.0285	0.0475	0.0789	0.101
25	12.16	0.0506	0.101	0.248	0.467	0.0455	0.0755	0.125	0.160
30	14.43	0.0746	0.148	0.357	0.634	0.0670	0.111	0.183	0.234
35	16.61	0.104	0.206	0.481	0.781	0.0936	0.155	0.254	0.323
40	18.69	0.141	0.276	0.610	0.889	0.126	0.206	0.336	0.424
45	20.64	0.184	0.356	0.732	0.953	0.164	0.267	0.426	0.533
50	22.45	0.236	0.447	0.835	0.984	0.209	0.335	0.522	0.642
55	24.10	0.297	0.546	0.910	0.996	0.261	0.408	0.615	0.741
60	25.58	0.367	0.647	0.958	0.999	0.319	0.482	0.696	0.817
63.50[a]	26.50	0.422	0.716	0.978	0.999+	0.362	0.532	0.740	0.850
65	26.86	0.446	0.744	0.984	0.999+	0.381	0.552	0.754	0.859
70	27.93	0.534	0.831	0.995	0.999+	0.445	0.609	0.781	0.862
75	28.78	0.627	0.900	0.999	0.999+	0.505	0.645	0.774	0.829
80	29.40	0.721	0.949	0.999+	0.999+	0.556	0.656	0.736	0.767
85	29.78	0.808	0.978	0.999+	0.999+	0.590	0.641	0.676	0.688
90	29.90	1.00	1.00	1.00	1.00	0.602	0.602	0.602	0.602

[a] Brewster's angle ($\tan^{-1} n$).

reflected rays could be prevented from reaching the detector, the degree of polarization would correspond very closely to that predicted by Eq. (5). Several techniques for disposing of these rays are described by Bird and Shurcliff.[4]

Transparent Plates and Multiply Reflected Rays

The correct intensity formulas for m plates were derived by Provostaye and Desains,[5] Stokes,[6] and others:

$$r_{2m} = \frac{2m\mathcal{R}}{1 + (2m - 1)\mathcal{R}}, \tag{6}$$

and

$$t_{2m} = \frac{1 - \mathcal{R}}{1 + (2m - 1)\mathcal{R}}. \tag{7}$$

We are concerned here only with the degree of polarization of the transmitted component, and this can be obtained by considering the orthogonal components of Eq. (7). If we make use of Eqs. (1a) and (1b), these components become

$$t_{2m,\perp} = \frac{\tan^2(\theta_i + \theta_t) - \tan^2(\theta_i - \theta_t)}{\tan^2(\theta_i + \theta_t) + (2m - 1)\tan^2(\theta_i - \theta_t)}, \tag{7a}$$

and

$$t_{2m,\parallel} = \frac{\sin^2(\theta_i + \theta_t) - \sin^2(\theta_i - \theta_t)}{\sin^2(\theta_i + \theta_t) + (2m - 1)\sin^2(\theta_i - \theta_t)}. \tag{7b}$$

The degree of polarization in this case is defined as

$$p_{2m} \equiv \left|\frac{t_{2m,\parallel} - t_{2m,\perp}}{t_{2m,\parallel} + t_{2m,\perp}}\right|. \tag{8}$$

Equation (8) is applicable when no absorption occurs in the plates and *all* the transmitted rays (the directly transmitted component plus those rays which arise from multiple reflections) reach the detector. At Brewster's angle, Eq. (8) reduces to the familiar form

$$p_{2m} = \frac{m}{m + [2n/(n^2 - 1)]^2}. \tag{9}$$

The degree of polarization [according to Eq. (8)] is plotted against the angle of incidence in Fig. 1 for m equal to 1, 2, 5, and 10, and n equal to 1.518. It is interesting to note that, in the case of multiple reflections, the degree of polarization attains a maximum at an angle of incidence which approaches the polarizing angle as the number of plates is increased (see also, Stokes[6]). The depolarizing effects of multiple reflections, which take place between the surfaces of the plates, are clearly seen by comparing the results of Eq. (8) with those of Eq. (5) which are also shown in Fig. 1.

Conclusions

In Tables I, II, III, IV, and V we list the degree of polarization (to three significant figures) for the singly and multiply reflecting cases and indices of refraction of 1.5170, 1.5180, 1.5725, 1.6210, and 2.0060, respectively. The results are tabulated for the similar cases of 1.5170 and 1.5180 to illustrate the effect on p of a small change in the index of refraction. From a further comparison of Tables I to V we note that: (1) the depolarization associated with multiple reflections becomes increasingly more important as either

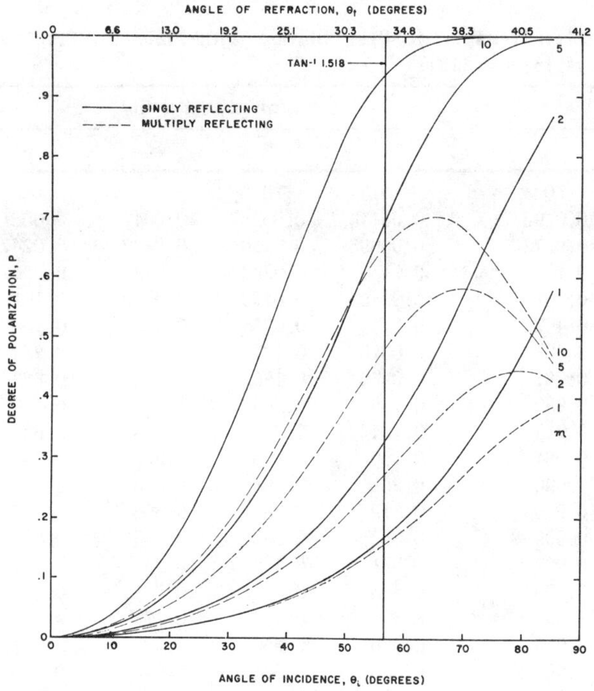

Fig. 1. The degree of polarization for piles of one, two, five, and ten transparent glass plates (singly and multiply reflected rays) for $n = 1.518$.

the number of plates or the angle of incidence (or both) is increased, (2) the degree of polarization increases with increasing index of refraction, and (3) for a given index of refraction the degree of polarization for any number of plates tends to a single value as grazing incidence is approached.

In practice, neither Eq. (5) nor Eq. (8) is strictly applicable even for transparent plates. Several investigators have measured (at Brewster's angle) a greater degree of polarization than can be accounted for by Eq. (9). Conn and Eaton[7] have pointed out that in some optical systems only a small number of the multiply reflected rays may reach the detector, and that, in such cases, the degree of polarization will be larger than that given by Eq. (9).

In a subsequent paper we will consider the degree of polarization of the reflected component and the effects of absorption.

This research was carried out as part of a polarimetric study of the zodiacal light which was supported by the National Aeronautics and Space Administration under contracts.

References

1. J. L. Weinberg, dissertation, "Photoelectric Polarimetry of the Zodiacal Light at λ5300," University of Colorado (1963).
2. A. J. Fresnel, Oevres Complètes (Paris) **10,** 640 (1866).
3. M. Born and E. Wolf, *Principles of Optics* (Pergamon, London, 1959), p. 42.
4. G. R. Bird and W. A. Shurcliff, J. Opt. Soc. Am. **49,** 235 (1959).
5. F. de la Provostaye and P. Desains, Ann. Chim. Phys. (3rd series) **30,** 159 (1850).
6. G. G. Stokes, Proc. Roy. Soc. (London) **11,** 545 (1862).
7. G. K. T. Conn and G. D. Eaton, J. Opt. Soc. Am. **44,** 553 (1954).

ERRATA

equation (7a): $t_{2m,\|\perp}$ should read $t_{2m,\perp}$

equation (7b): $t_{2m,\|\|}$ should read $t_{2m,\|}$

equation (9): should read

$$p_{2m} = \frac{m}{m + [2n/(n^2 - 1)]^2}$$

last line should read "...contracts NsG-15-59 and NsG-135-61."

Vol. 57(2), 202–205 (July 1986).

Polarization *Pendellösung* and the Generation of Circularly Polarized X Rays with a Quarter-Wave Plate

J. A. Golovchenko, B. M. Kincaid, R. A. Levesque,[a] A. E. Meixner,[b] and D. R. Kaplan
AT&T Bell Laboratories, Murray Hill, New Jersey 07974

A continuously tunable x-ray compensator that works in conjunction with synchrotron-radiation sources has been developed. The phenomena of polarization *Pendellösung* has been observed during its normal operation and the generation of circularly polarized x rays is verified through an observation of the magnetic Compton effect.

Several recent investigations have dealt with the generation of polarized x rays by use of dynamical x-ray diffraction techniques.[1–4] Besides the inherently interesting x-ray–crystal interactions demonstrated in these studies the prospect of extending the classical long-wavelength manifestations of crystal optics, (i.e., optical birefringence, rotary power, etc.) into the hard-x-ray region presents several new challenges and opportunities for experimental and theoretical physicists. An application of particular current interest lies in the ability to obtain spin-dependent momentum profiles of magnetic materials by examination of the Compton scattering of circularly polarized x rays.[5–7] Until now the only source of spin-polarized photons for such experiments has been radioactive decay and a new strong source, e.g., synchrotron radiation, would have a significant influence on such studies. However, present synchrotron-radiation sources provide mainly linearly polarized light. The possibility of providing circular polarization through the use of either helical wigglers[8] or tandem crossed wigglers[9] is currently under discussion for the next generation of storage rings. External optics which would serve the same purpose without perturbing electron orbits or the expense of large magnetic structures would clearly be desirable.

We describe here a recent series of experiments using a synchrotron-radiation source and x-ray compensator plate designed to isolate polarization *Pendellösung* in the plate and generate circularly polarized x rays at arbitrary energies. The *Pendellösung* or oscillation effect we observe is somewhat different in several respects from the one commonly known in diffraction physics[10] and it is probably best introduced here simply by a description of the experimental setup for observing it.

Figure 1 is a schematic illustration of the experiment. X rays from the Cornell high-energy synchrotron source operating at 5.2 GeV and 12 mA in the colliding-beam mode were collected by an asymmetrically cut silicon crystal oriented to diffract from the (220) planes in the Bragg geometry. A few square millimeters of the white beam were intercepted. The asymmetry was such as to increase the angular acceptance of the incident beam and reduce the angular divergence of the Bragg-diffracted emergent beam. Completely glacing incidence was obtained at ~18 keV, which determined the highest energy of our experiments. Renninger has described the collimating properties of such an element.[11] The collimated monochromatic beam thus obtained is then directed onto another silicon crystal also oriented to diffract from the (220) planes but now in the transmission Laue geometry, i.e., the diffracted beam propagates into the crystal. This crystal serves as the x ray com-

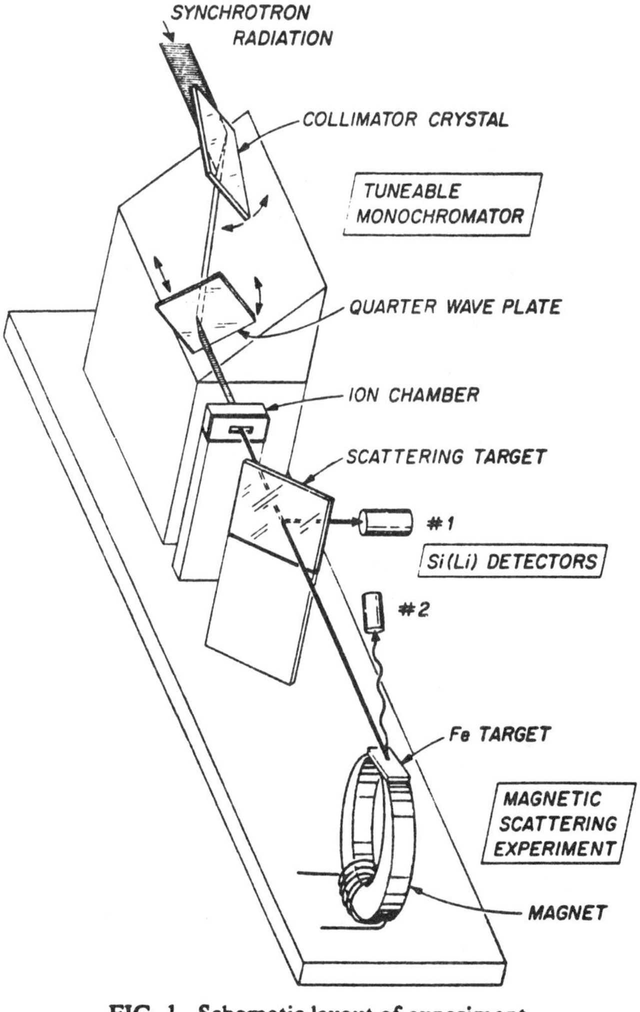

FIG. 1. Schematic layout of experiment.

pensator and quarter-wave plate. The (220) planes are normal to the x-ray entry surface and the thickness of the plate was chosen to be 1.39 mm. The two crystals were mounted in a piezoelectric, feedback-stabilized, mechanical-linkage system[12] capable of keeping the two sets of crystal planes parallel to <0.1 arcsec while the energy selected from the synchrotron is varied by changing the Bragg angles. Furthermore, the design yields an exit beam whose offset with respect to the incident beam is independent of the chosen beam energy. Finally, the entire assembly was rotated about the incident-synchrotron-beam direction such that the plane formed by incident and diffracted beams was near 45° degrees to the vertical.

The x-ray beam from the compensator plate was allowed to fall upon a 1-mm-thick piece of Plexiglas after passing through a slit and ion chamber used for normalization purposes. A lithium-drifted silicon Si(Li) detector observed those x rays Compton scattered 90° in the horizontal plane by the Plexiglas. The counting rate in the Si(Li) detector, normalized to the beam incident on the Plexiglas, as a function of beam energy chosen by the monochromator-polarizer system is plotted in Fig. 2. Strong oscillations, with increasing amplitude for increasing energy are noted. The horizontal scale in the plot is chosen in accord with Eq. (1) (see below) which results in the oscillatory behavior of the data being represented quite accurately by a damped sinusoidal wave. The reason for this will be described shortly.

The origin of the above oscillations, which we refer to as polarization *Pendellösung*, may be understood as follows. For the case of a single strong diffraction in

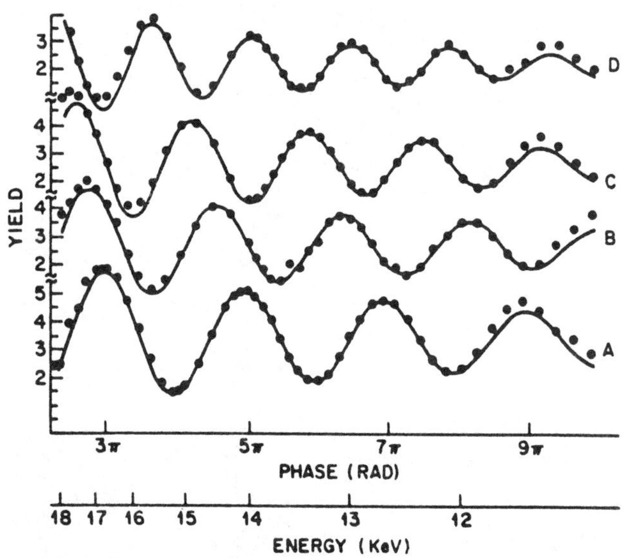

FIG. 2. Normalized yield in Si(Li) detector No. 1 vs energy of monochromator containing x-ray compensator. Curves *A–D* refer to the different compensator thicknesses. The phase assignment in the abscissa applies only to curve *A*.

the Laue geometry, the eigensolutions to the Maxwell equations inside the crystal can be classified into two possible states of linear polarization. One of these, the solutions of the so-called sigma type, consists of running Bloch waves each composed of coupled plane waves, whose electric field vector is perpendicular to the plane formed by the incident- and diffracted-beam directions. For this transverse electric case, the polarization direction is parallel to the diffracting plane. The other class of polarization solutions, called pi, is transverse magnetic and consists again of coupled plane waves in the incident and diffracted directions but now the electric vector of each of these lies in the plane formed by the two directions. Therefore, the polarization can be considered as perpendicular to the diffracting plane.

For each state of polarization at fixed x-ray energy there are two kinds of x-ray Bloch-wave states. One, the so-called alpha-type solution, has standing-wave character perpendicular to the diffracting planes such that the field intensity has the same periodicity and a phase which places its nodes directly on the planes. The other type solution, called beta, is identical except that its antinodes are on the diffracting planes.

We now imagine an external plane wave impinging on the crystal at the exact Bragg condition with a polarization vector at 45° to the diffracting planes. In order to match the internal- and external-wave solutions at the entry surface a linear combination of equal amounts of sigma and pi waves must be excited in the crystal, with each of these having equal amounts of alpha and beta character. The Poynting vector for each of these Bloch waves at the exact Bragg condition lies along the crystal planes so they all propagate in the same direction into the crystal.

We note that the alpha solutions have their nodal planes on the diffracting planes, which is where the majority of the strong x-ray absorption takes place and hence these waves should propagate significantly farther through the crystal than the beta solutions, which have their maximum intensity on the planes. Hence, following the suggestion of Hart,[2] our compensator plate has been cut thick enough so that only the alpha solutions survive to the back surface. Therefore, normal *Pendellösung*, which is basically a beating between alphalike and betalike solutions will have nothing to do with the oscillations with which this paper deals.

The effective index of refraction for the surviving two alpha solutions is slightly different (because the two solutions sample slightly different electron densities) leading to different phase velocities for the two waves. At the back surface of the plate, plane waves will be formed whose polarization state will depend upon the accumulated phase difference between the two internal eigenfunctions. Using the dynamical theory to account quantitatively for the effect one nds that the phase difference between the two soluons at the back of the plate is given by

$$\delta = 4\pi^3 \left(\frac{mc^2}{\alpha E} \right)^3 \frac{|F_G|}{\cos\theta_B} \frac{r_e^4}{d^2 v} D, \tag{1}$$

·here E is the x-ray energy, α the fine-structure conιant, r_e the classical electron radius, v the crystal nit-cell volume, d the Miller-plane spacing, D the ιrystal-plate thickness, F_G the crystal structure factor, ınd θ_B the Bragg angle. For $\delta = 2\pi n$ with n an integer follows that the final state of polarization is the same s the initial state, i.e., plane polarized in the horizonıl direction. For $\delta = 2\pi(n + \frac{1}{2})$ the final state of poırization should be plane polarized in the vertical irection. Finally for $\delta = 2\pi(n + \frac{1}{4})$ or $2\pi(n - \frac{1}{4})$ we xpect left- and right-hand circularly polarized x rays, espectively, and hence the compensator becomes a uarter-wave plate.

Figure 2, curve A, shows the data for the 90° Compэn yield in the horizontal plane plotted as a function f beam energy and the derived phase δ from Eq. (1). 'he fitted curve corresponds to an exponentially amped sinusoidal function of δ with period 2π. The ıinima occur at $\delta = 2\pi n$ as they should for horizontal-/ polarized x rays. The maxima at $\delta = 2\pi(n + \frac{1}{2})$ epresent vertical polarization. The states of circular olarization should occur halfway between the maxima nd minima; verification of this state of polarization /ill be demonstrated shortly through the observation f magnetic Compton scattering. First, however, we ote that it is possible to independently determine the umber of "quarter waves" in the plate by changing ıe plate thickness and recording the shifted polarizaon *Pendellösung* fringes. This was accomplished exerimentally by rotating the plate about an axis perendicular to the diffracting planes. This changes only ıe effective plate thickness seen by the beam. Thus 'ig. 2, curve B, obtained with a rotation of approxiıately 22°, Fig. 2, curve C, with 32°, and Fig. 2, curve), with 45°. Notice that at an energy of approximately 4 keV increasing the plate thickness by 41% results in ne additional polarization cycle. This agrees precisely /ith the phase assignment of $\delta = 5\pi$ for the peak in urve A at 14 keV and constitutes an independent ıeasurement of the oscillation length inside the plate.)ur final comments on the data in Fig. 2 deal with the egree of modulation in the oscillatory signals. Clearly ıe degree of polarization seems to be improving as ıe beam energy increases. The main reason for this eature lies in the fact that the asymmetric collimating rystal has its highest asymmetry, and hence its ighest degree of angular collimation, at the highest nergies in Fig. 2. Eigenfunctions in the quarter-wave late excited by incident rays slightly off the exact ıragg condition do not have perfect nodes on the planes for the sigma polarization state and the difference between the effective index of refraction for sigma and pi polarization states is reduced. This results in a modified *Pendellösung* length as a function of incidence angle to the plate and a smearing of results at low energies as a result of beam divergence. In addition, no special precautions were taken in these preliminary experiments to minimize air scattering near the Si(Li) detector No. 1 and multiple scattering in the Plexiglas target, both of which contribute to reduced modulation in Fig. 2.

A magnetic-Compton-scattering experiment was performed to demonstrate directly the presence of circularly polarized x rays. A beam energy of 17.5 keV was chosen (nearly twice the critical energy where the incident synchrotron radiation is essentially completely horizontally polarized). The plate was adjusted to a thickness of $\frac{11}{4}$ *Pendellösung* waves (see curve A in Fig. 2), and the transmitted x rays were scattered from a magnetized iron target as shown in Fig. 1. The flux onto the sample was $\sim 3 \times 10^9$ photons/sec, consistent with the expected 15% conversion efficiency from plane to circular polarization for the compensator plate. The Compton-scattering rate at a scattering angle of 145° was monitored with a solid-state detector with the target and its immediate vicinity bathed in a helium atmosphere to avoid air-scattering background. Figure 3 shows the spectrum observed in the Si(Li) detector after a total of 9 h of data accumulation with the sample magnetized alternatively along and against the spin direction of the incident x rays. The difference spectrum is also shown in the figure. The cross section is seen to differ by 0.5% for the two cases.[13]

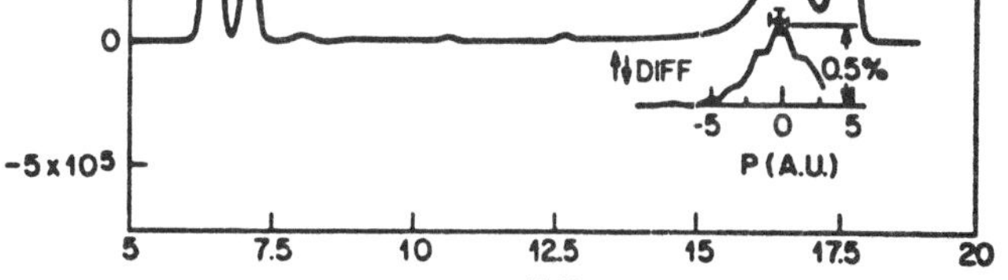

FIG. 3. Spectrum in Si(Li) detector No. 2 from magnetic-Compton-scattering experiment. Also shown is the difference spectrum described in the text. Vertical error beams are due to counting statistics and horizontal to detector resolution. The Fe fluorescence signals are spin independent and are used for accurate normalization in constructing the difference spectrum.

The theoretically expected value[14] is 0.75% for a 100% circularly polarized beam. The local abscissa for the difference spectrum in Fig. 3 has been converted to projected momentum as is customary in Compton-profile studies of electron-momentum distributions.[15] Unfortunately the poor relative resolution obtained by use of a Si(Li) detector at such low energies precludes our obtaining useful information on the contribution of conduction electrons to the magnetic moment of magnetized iron. This would require high resolution in the region below 1 a.u. which we have not obtained. It will therefore be necessary to perform similar experiments at higher energies (>80 keV) where momentum resolution is greater or to develop higher-resolution detectors for use at lower energies.

The difference between the magnetic-scattering theory and this experiment is due to incomplete polarization of the beam. A density-matrix analysis of the compensator-prepared x-ray beam using the above two scattering experiments indicates that at the highest energy studied here the beam was 83% circularly polarized. Several of the reasons for lack of perfect polarization have already been discussed. In addition, strains in the crystal elements and alignment errors can contribute to a deterioration of performance. We hope to substantially improve the situation in future experiments.

In conclusion, we have succeeded in isolating the phenomenon of polarization *Pendellösung* using an x-ray compensator plate in conjunction with continuous and linearly polarized radiation from a synchrotron source. The effect may be used to provide experimenters with a flexibility in choosing various states of photon polarization that is not currently available.

We gratefully acknowledge the contributions of th Cornell High-Energy Synchrotron Source staff to thi work.

(a)Present address: Lawrence Livermore National Labor tory, Livermore, CA 94550.

(b)Present address: AT&T Bell Laboratories, Whippan NJ 07981.

1P. Skalicky and C. Malgrange, Acta Crystallogr., Sect. **28**, 501 (1972).

2M. Hart, Philos. Mag. B **38**, 41 (1978).

3M. Hart and A. R. D. Rodriquez, Philos. Mag. B **40**, 1 (1979).

4O. Brummer, Ch. Eisenschmidt, and H. R. Hoche, Z. N turforsch. Teil A **37**, 524 (1982).

5P. M. Platzman and N. Tzoar, Phys. Rev. B **2**, 35 (1970).

6N. Sakai and K. Ono, Phys. Rev. Lett. **37**, 351 (1976).

7N. Sakai and K. Ono, J. Phys. Soc. Jpn. **42**, 770 (1977).

8B. M. Kincaid, J. Appl. Phys. **48**, 2684 (1977).

9K. J. Kim, Nucl. Instrum. Methods **219**, 425 (1984).

10L. Azaroff, R. Kaplow, N. Kato, R. Weiss, A. Wilso and R. Young, *X-Ray Diffraction* (McGraw-Hill, New Yor 1974). Note the *Pendellösung* effect shown in Fig. 2 of Re 4 is not of the kind we discuss here.

11M. Renninger, Z. Naturforsch. Teil A **16**, 1110 (1961).

12J. A. Golovchenko, R. A. Levesque, and P. L. Cowa Rev. Sci. Instrum. **52**, 509 (1981).

13Changing the sign of the circular polarization by adjus ing the compensator changed the sign of the effect as e pected.

14M. Bernardini, P. Brovetto, S. Debenedetti, and S. Fe roni, Nuovo Cimento **7**, 416 (1958).

15*Compton Scattering*, edited by B. Williams (McGraw-Hil New York, 1977).

Section Two

Polarization State Shifters

Polarized light can be shifted from one state to another in a tremendous number of ways. As a result, this section has more papers than other parts of this volume and also important papers may have been left out.

Polarization state shifters along with polarizers are used in all of the applications of polarized light. My favorite state shifter is the turkey wrapping found in any grocery store. To strip this plastic from the turkey body seems more difficult each year. The price is not just for the turkey. It includes this marvelous optical material, which can be a half-wave plate or a multi-wave plate. In the early days, the plastic around a cigarette package was almost a perfect half-wave plate. During my lectures on polarized lights, I would ask the audience if I could borrow a pack of cigarettes and then with a pair of crossed polarizers in front of the projector illustrate how the light could be turned on and off.

Kerr effect modulators and liquid crystals all have made an impact on the field of science.

Reprinted from *Philosophical Magazine*, Series XIX, 295–317 (November 1845).

XLIX. *Experimental Researches in Electricity.—Nineteenth Series.* By MICHAEL FARADAY, *Esq., D.C.L., F.R.S., Fullerian Prof. Chem. Royal Institution, Foreign Associate of the Acad. Sciences, Paris, Cor. Memb. Royal and Imp. Acadd. of Sciences, Petersburgh, Florence, Copenhagen, Berlin, Göttingen, Modena, Stockholm, &c. &c.**

§ 26. *On the magnetization of light and the illumination of magnetic lines of force*†.

¶ i. *Action of magnets on light.*

2146. I HAVE long held an opinion, almost amounting to conviction, in common I believe with many other lovers of natural knowledge, that the various forms under which the forces of matter are made manifest have one common origin; or, in other words, are so directly related and mutually dependent, that they are convertible, as it were, one into another, and possess equivalents of power in their ac-

* From the Philosophical Transactions for 1846, Part I., having been read November 20, 1845.

† The title of this paper has, I understand, led many to a misapprehension of its contents, and I therefore take the liberty of appending this explanatory note. Neither accepting nor rejecting the hypothesis of an æther, or the corpuscular, or any other view that may be entertained of the nature of light; and, as far as I can see, nothing being really known of a ray of light more than of a line of magnetic or electric force, or even of a line of gravitating force, except as it and they are manifest in and by substances; I believe that, in the experiments I describe in the paper, light has been magnetically affected, *i. e.* that that which is magnetic in the forces of matter has been affected, and in turn has affected that which is truly magnetic in the force of light: by the term magnetic I include here either of the peculiar exertions of the power of a magnet, whether it be that which is manifest in the magnetic or the diamagnetic class of bodies. The phrase "illumination of the lines of magnetic force" has been understood to imply that I had rendered them luminous. This was not within my thought. I intended to express that the line of magnetic force was illuminated as the earth is illuminated by the sun, or the spider's web illuminated by the astronomer's lamp. Employing a ray of light, we can tell, *by the eye*, the direction of the magnetic lines through a body; and by the alteration of the ray and its optical effect on the eye, can see the course of the lines just as we can see the course of a thread of glass, or any other transparent substance, rendered visible by the light: and this was what I meant by *illumination*, as the paper fully explains.—December 15, 1845. M. F.

tion*. In modern times the proofs of their convertibility have been accumulated to a very considerable extent, and a commencement made of the determination of their equivalent forces.

2147. This strong persuasion extended to the powers of light, and led, on a former occasion, to many exertions, having for their object the discovery of the direct relation of light and electricity, and their mutual action in bodies subject jointly to their power†; but the results were negative and were afterwards confirmed, in that respect, by Wartmann‡.

2148. These ineffectual exertions, and many others which were never published, could not remove my strong persuasion derived from philosophical considerations; and, therefore, I recently resumed the inquiry by experiment in a most strict and searching manner, and have at last succeeded in *magnetizing and electrifying a ray of light, and in illuminating a magnetic line of force.* These results, without entering into the detail of many unproductive experiments, I will describe as briefly and clearly as I can.

2149. But before I proceed to them, I will define the meaning I connect with certain terms which I shall have occasion to use:—thus, by *line of magnetic force*, or *magnetic line of force*, or *magnetic curve*, I mean that exercise of magnetic force which is exerted in the lines usually called magnetic curves, and which equally exist as passing from or to magnetic poles, or forming concentric circles round an electric current. By *line of electric force*, I mean the force exerted in the lines joining two bodies, acting on each other according to the principles of static electric induction (1161, &c.), which may also be either in curved or straight lines. By a *diamagnetic*, I mean a body through which lines of magnetic force are passing, and which does not by their action assume the usual magnetic state of iron or loadstone.

2150. A ray of light issuing from an Argand lamp, was polarized in a horizontal plane by reflexion from a surface of glass, and the polarized ray passed through a Nichol's eye-piece revolving on a horizontal axis, so as to be easily examined by the latter. Between the polarizing mirror and the eye-piece, two powerful electro-magnetic poles were arranged, being either the poles of a horse-shoe magnet, or the contrary poles of two cylinder magnets; they were separated from each other about two inches in the direction of the line of the ray, and so placed, that, if on the same side of the polarized ray, it might pass near them; or, if on contrary sides, it might go between them, its direction being always parallel, or nearly so, to the magnetic lines of force (2149.). After that, any transparent substance placed between the two poles, would have passing through it, both the polarized ray and the magnetic lines of force at the same time and in the same direction.

* Experimental Researches, 57, 366, 376, 877, 961, 2071.

† Philosophical Transactions, 1834. Experimental Researches, 951–955.

‡ *Archives de l'Electricité*, ii. pp. 596–600.

2151. Sixteen years ago I published certain experiments made upon optical glass*, and described the formation and general characters of one variety of heavy glass, which, from its materials, was called silicated borate of lead. It was this glass which first gave me the discovery of the relation between light and magnetism, and it has power to illustrate it in a degree beyond that of any other body; for the sake of perspicuity I will first describe the phænomena as presented by this substance.

2152. A piece of this glass, about two inches square and 0·5 of an inch thick, having flat and polished edges, was placed as a *diamagnetic* (2149.) between the poles (not as yet magnetized by the electric current), so that the polarized ray should pass through its length; the glass acted as air, water, or any other indifferent substance would do; and if the eye-piece were previously turned into such a position that the polarized ray was extinguished, or rather the image produced by it rendered invisible, then the introduction of this glass made no alteration in that respect. In this state of circumstances the force of the electro-magnet was developed, by sending an electric current through its coils, and immediately the image of the lamp-flame became visible, and continued so as long as the arrangement continued magnetic. On stopping the electric current, and so causing the magnetic force to cease, the light instantly disappeared; these phænomena could be renewed at pleasure, at any instant of time, and upon any occasion, showing a perfect dependence of cause and effect.

2153. The voltaic current which I used upon this occasion, was that of five pair of Grove's construction, and the electro-magnets were of such power that the poles would singly sustain a weight of from twenty-eight to fifty-six, or more, pounds. A person looking for the phænomenon for the first time would not be able to see it with a weak magnet.

2154. The character of the force thus impressed upon the diamagnetic is that of *rotation*; for when the image of the lamp-flame has thus been rendered visible, revolution of the eye-piece to the right or left, more or less, will cause its extinction; and the further motion of the eye-piece to the one side or other of this position will produce the reappearance of the light, and that with complementary tints, according as this further motion is to the right- or left-hand.

2155. When the pole nearest to the observer was a marked pole, *i. e.* the same as the north end of a magnetic needle, and the further pole was unmarked, the rotation of the ray was right-handed; for the eye-piece had to be turned to the right-hand, or clock fashion, to overtake the ray and restore

* Philosophical Transactions, 1830, p. 1. I cannot resist the occasion which is thus offered to me of mentioning the name of Mr. Anderson, who came to me as an assistant in the glass experiments, and has remained ever since in the Laboratory of the Royal Institution. He has assisted me in all the researches into which I have entered since that time, and to his care, steadiness, exactitude, and faithfulness in the performance of all that has been committed to his charge, I am much indebted.—M. F.

the image to its first condition. When the poles were reversed, which was instantly done by changing the direction of the electric current, the rotation was changed also and became left-handed, the alteration being to an equal degree in extent as before. The direction was always the same for the same *line of magnetic force* (2149.).

2156. When the diamagnetic was placed in the numerous other positions, which can easily be conceived, about the magnetic poles, results were obtained more or less marked in extent, and very definite in character, but of which the phænomena just described may be considered as the chief example: they will be referred to, as far as is necessary, hereafter.

2157. The same phænomena were produced in the silicated borate of lead (2151.) by the action of a good ordinary steel horse-shoe magnet, no electric current being now used. The results were feeble, but still sufficient to show the perfect identity of action between electro-magnets and common magnets in this their power over light.

2158. Two magnetic poles were employed end-ways, *i. e.* the cores of the electro-magnets were hollow iron cylinders, and the ray of polarized light passed along their axes and through the diamagnetic placed between them: the effect was the same.

2159. One magnetic pole only was used, that being one end of a powerful cylinder electro-magnet. When the heavy glass was beyond the magnet, being close to it but between the magnet and the polarizing reflector, the rotation was in one direction, dependent on the nature of the pole; when the diamagnetic was on the near side, being close to it but between it and the eye, the rotation for the same pole was in the contrary direction to what it was before; and when the magnetic pole was changed, both these directions were changed with it. When the heavy glass was placed in a corresponding position to the pole, but above or below it, so that the *magnetic curves* were no longer passing through the glass parallel to the ray of polarized light, but rather perpendicular to it, then no effect was produced. These particularities may be understood by reference to fig. 1, where *a* and *b* represent the first positions of the diamagnetic, and *c* and *d* the latter positions, the course of the ray being marked by the dotted line. If also the glass were placed directly at the end of the magnet, then no effect was produced on a ray passing in the direction here described, though it is evident, from what has been already said (2155.), that a ray passing *parallel* to the magnetic lines through the glass so placed, would have been affected by it.

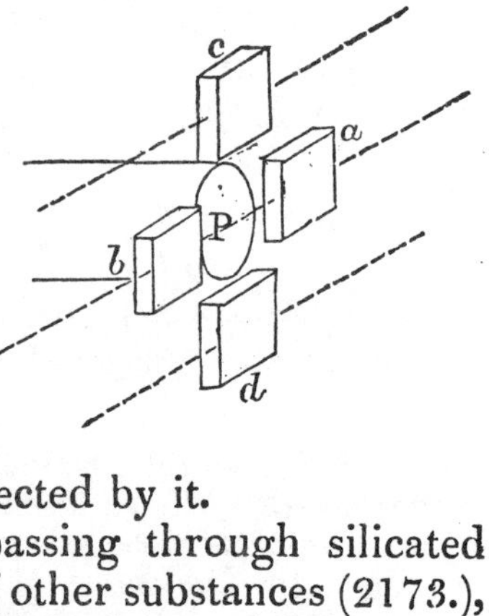

2160. Magnetic lines, then, in passing through silicated borate of lead, and a great number of other substances (2173.), cause these bodies to act upon a polarized ray of light when the lines are parallel to the ray, or in proportion as they are parallel to it: if they are perpendicular to the ray, they have no action upon it. They give the diamagnetic the power of rotating the ray; and the *law* of this action on light is, that if a magnetic line of force be *going from* a north pole, or *coming* from a south pole, along the path of a polarized ray coming to the observer, it will rotate that ray to the right-hand; or, that if such a line of force be coming from a north pole, or going from a south pole, it will rotate such a ray to the left-hand.

2161. If a cork or a cylinder of glass, representing the diamagnetic, be marked at its ends with the letters N and S, to represent the poles of a magnet, the line joining these letters may be considered as a magnetic line of force; and further, if a line be traced round the cylinder with arrow heads on it to represent direction, as in the figure, such a simple model, held up before the eye, will express the whole of the law, and give every position and consequence of direction resulting from it. If a watch be considered as the diamagnetic, the north pole of a magnet being imagined against the face, and a south pole against the back, then the motion of the hands will indicate the direction of rotation which a ray of light undergoes by magnetization.

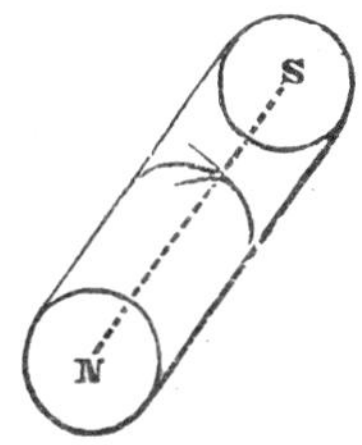

2162. I will now proceed to the different circumstances which affect, limit, and define the extent and nature of this new power of action on light.

2163. In the first place, the rotation appears to be in proportion to the extent of the diamagnetic through which the ray and the magnetic lines pass. I preserved the strength of the magnet and the interval between its poles constant, and then interposed different pieces of the same heavy glass (2151.) between the poles. The greater the extent of the diamagnetic in the line of the ray, whether in one, two, or three pieces, the greater was the rotation of the ray; and, as far as I could judge by these first experiments, the amount of rotation was exactly proportionate to the extent of diamagnetic through which the ray passed. No addition or diminution of the heavy glass on the *side* of the course of the ray made any difference in the effect of that part through which the ray passed.

2164. The power of rotating the ray of light *increased* with the intensity of the magnetic lines of force. This general effect is very easily ascertained by the use of electro-magnets; and within such range of power as I have employed, it appears to be directly proportionate to the intensity of the magnetic force.

2165. Other bodies, besides the heavy glass, possess the same power of becoming, under the influence of magnetic force, active on light (2173.). When these bodies possess a rotative power of their own, as is the case with oil of turpentine, sugar, tartaric acid, tartrates, &c., the effect of the magnetic force is to add to, or subtract from, their specific force, according as the natural rotation and that induced by the magnetism is right- or left-handed (2231.).

2166. I could not perceive that this power was affected by any degree of motion which I was able to communicate to the diamagnetic, whilst jointly subject to the action of the magnetism and the light.

2167. The interposition of copper, lead, tin, silver, and other ordinary non-magnetic bodies in the course of the magnetic curves, either between the pole and the diamagnetic, or in other positions, produced no effect either in kind or degree upon the phænomena.

2168. Iron frequently affected the results in a very considerable degree; but it always appeared to be, either by altering the direction of the magnetic lines, or disposing within itself of their force. Thus, when the two contrary poles were on one side of the polarized ray (2150.), and the heavy glass in its best position between them and in the ray (2152.), the bringing of a large piece of iron near to the glass on the other side of the ray, caused the power of the diamagnetic to fall. This was because certain lines of magnetic force, which at first passed through the glass parallel to the ray, now crossed the glass and the ray; the iron giving two contrary poles opposite the poles of the magnet, and thus determining a new course for a certain portion of the magnetic power, and that across the polarized ray.

2169. Or, if the iron, instead of being applied on the opposite side of the glass, were applied on the same side with the magnet, either near it or in contact with it, then, again, the power of the diamagnetic fell, simply because the power of the magnet was diverted from it into a new direction. These effects depend much of course on the intensity and power of the magnet, and on the size and softness of the iron.

2170. The electro-helices (2190.) without the iron cores were very feeble in power, and indeed hardly sensible in their effect. With the iron cores they were powerful, though no more electricity was then passing through the coils than before (1071.). This shows, in a very simple manner, that the phænomena exhibited by light under these circumstances, is directly connected with the magnetic form of force supplied by the arrangement. Another effect which occurred illustrated the same point. When the contact at the voltaic battery is made, and the current sent round the electro-magnet, the image produced by the rotation of the polarized ray does not rise up to its full lustre immediately, but increases for a couple of seconds, gradually acquiring its greatest intensity; on breaking the contact, it sinks instantly and disappears apparently at once. The gradual rise in brightness is due to the *time* which the iron core of the magnet requires to evolve all that magnetic power which the electric current can develope in it; and as the magnetism rises in intensity, so does its effect on the light increase in power; hence the progressive condition of the rotation.

2171. I cannot as yet find that the heavy glass (2151.), when in this state, *i. e.* with magnetic lines of force passing through it, exhibits any increased degree, or has any specific magneto-inductive action of the recognized kind. I have placed it in large quantities, and in different positions, between magnets and magnetic needles, having at the time very delicate means of appreciating any difference between it and air, but could find none.

2172. Using water, alcohol, mercury, and other fluids contained in very large delicate thermometer-shaped vessels, I could not discover that any difference in volume occurred when the magnetic curves passed through them.

2173. It is time that I should pass to a consideration of this power of magnetism over light as exercised, not only in the silicated borate of lead (2151.), but in many other substances; and here we perceive, in the first place, that if all transparent bodies possess the power of exhibiting the action, they have it in very different degrees, and that up to this time there are some that have not shown it at all.

2174. Next, we may observe, that bodies that are exceedingly different to each other in chemical, physical, and mechanical properties, develope this effect; for solids and liquids, acids, alkalies, oils, water, alcohol, æther, all possess the power.

2175. And lastly, we may observe, that in all of them, though the degree of action may differ, still it is always the same in kind, being a rotative power over the ray of light; and further, the direction of the rotation is, in every case, independent of the nature or state of the substance, and dependent upon the direction of the magnetic line of force, according to the law before laid down (2160.).

2176. Amongst the substances in which this power of action is found, I have already distinguished the *silico-borate of lead* (2151.) as eminently fitted for the purpose of exhibiting the phænomena. I regret that it should be the best, since it is not likely to be in the possession of many, and few will be induced to take the trouble of preparing it. If made, it should be well-annealed, for otherwise the pieces will have considerable power of depolarizing light, and then the particular phænomena under consideration are much less strikingly observed. The *borate of lead*, however, is a substance much more fusible, softening at the heat of boiling oil, and therefore far more easily prepared in the form of glass plates and annealed; and it possesses as much magneto-rotative power over light as the silico-borate itself. *Flint-glass* exhibits the property, but in a less degree than the substances above. Crown-glass shows it, but in a still smaller degree.

2177. Whilst employing crystalline bodies as diamagnetics, I generally gave them that position in which they did not affect the polarized ray, and then induced the magnetic curves through them. As a class, they seemed to resist the assumption of the rotating state. *Rock-salt* and *fluor-spar* gave evidence of the power in a slight degree; and I think that a crystal of alum did the same, but its ray length in the transparent part was so small that I could not ascertain the fact decisively. Two specimens of transparent fluor, lent me by Mr. Tennant, gave the effect.

2178. Rock-crystal, four inches across, gave no indications of action on the ray, neither did smaller crystals, nor cubes about three-fourths of an inch in the side, which were so cut as to have two of their faces perpendicular to the axis of the crystal (1692, 1693.), though they were examined in every direction.

2179. *Iceland spar* exhibited no signs of effect, either in the form of rhomboids, or of cubes like those just described (1695.).

2180. *Sulphate of baryta*, *sulphate of lime*, and *carbonate of soda*, were also without action on the light.

2181. A piece of fine clear *ice* gave me no effect. I cannot however say there is none, for the effect of water in the same mass would be very small, and the irregularity of the flattened surface from the fusion of the ice and flow of water, made the observation very difficult.

2182. With some degree of curiosity and hope, I put gold-leaf into the magnetic lines, but could perceive no effect. Considering the extremely small dimensions of the length of the path of the polarized ray in it, any positive result was hardly to be expected.

2183. In experiments with liquids, a very good method of observing the effect, is to inclose them in bottles from $1\frac{1}{2}$ to 3 or 4 inches in diameter, placing these in succession between the magnetic poles (2150.), and bringing the analysing eye-piece so near to the bottle, that, by adjustment of the latter, its cylindrical form may cause a diffuse but useful image of the lamp-flame to be seen through it: the light of this image is easily distinguished from that which passes by irregular refraction through the striæ and deformations of the glass, and the phænomena being looked for in this light are easily seen.

2184. Water, alcohol, and æther, all show the effect; water most, alcohol less, and æther the least. All the fixed oils which I have tried, including almond, castor, olive, poppy, linseed, sperm, elaine from hog's lard, and distilled resin oil, produce it. The essential oils of turpentine, bitter almonds, spike lavender, lavender, jessamine, cloves, and laurel, produce it. Also naphtha of various kinds, melted spermaceti, fused sulphur, chloride of sulphur, chloride of arsenic, and every other liquid substance which I had at hand and could submit in sufficient bulk to experiment.

2185. Of aqueous solutions I tried 150 or more, including the soluble acids, alkalies and salts, with sugar, gum, &c., the list of which would be too long to give here, since the great conclusion was, that the exceeding diversity of substance caused no exception to the general result, for all the bodies showed the property. It is indeed more than probable, that in all these cases the water and not the other substance present was the ruling matter. The same general result was obtained with alcoholic solutions.

2186. Proceeding from liquids to air and gaseous bodies, I have here to state that, as yet, I have not been able to detect the exercise of this power in any one of the substances in this class. I have tried the experiment with bottles 4 inches in diameter, and the following gases: oxygen, nitrogen, hydrogen, nitrous oxide, olefiant gas, sulphurous acid, muriatic acid, carbonic acid, carbonic oxide, ammonia, sulphuretted hydrogen, and bromine vapour, at ordinary temperatures; but they all gave negative results. With air, the trial has been carried, by another form of apparatus, to a much higher degree, but still ineffectually (2212.).

2187. Before dismissing the consideration of the substances which exhibited this power, and in reference to those in which it was superinduced upon bodies possessing, naturally, rotative force (2165. 2231.), I may record, that the following are the substances submitted to experiment: castor oil, resin oil, oil of spike lavender, of laurel, Canada balsam, alcoholic solution of camphor, alcoholic solution of camphor and corrosive sublimate, aqueous solutions of sugar, tartaric acid, tartrate of soda, tartrate of potassa and antimony, tartaric and boracic acid, and sulphate of nickel, which rotated to the right-hand; copaiba balsam, which rotated the ray to the left-hand; and two specimens of camphine or oil of turpentine, in one of which the rotation was to the right-hand, and in the other to the left. In all these cases, as already said (2165.), the superinduced magnetic rotation was according to the general law (2160.), and without reference to the previous power of the body.

2188. Camphor being melted in a tube about an inch in diameter, exhibited high natural rotative force, but I could not discover that the magnetic curves induced additional force in it. It may be, however, that the shortness of the ray length and the quantity of coloured light left, even when the eyepiece was adjusted to the most favourable position for darkening the image produced by the naturally rotated ray, rendered the small magneto-power of the camphor insensible.

¶ ii. *Action of electric currents on light.*

2189. From a consideration of the nature and position of the lines of magnetic and electric force, and the relation of a magnet to a current of electricity, it appeared almost certain that an electric current would give the same result of action on light as a magnet; and, in the helix, would supply a form of apparatus in which great lengths of diamagnetics, and especially of such bodies as appeared to be but little affected between the poles of the magnet, might be submitted to examination and their effect exalted: this expectation was, by experiment, realized.

2190. Helices of copper wire were employed, three of which I will refer to. The first, or *long helix*, was 0·4 of an inch internal diameter; the wire was 0·03 of an inch in diameter, and having gone round the axis from one end of the helix to the other, then returned in the same manner, forming a coil sixty-five inches long, double in its whole extent, and containing 1240 feet of wire.

2191. The second, or *medium helix*, is nineteen inches long, 1·87 inch internal diameter, and three inches external diameter. The wire is 0·2 of an inch in diameter, and eighty feet in length, being disposed in the coil as two concentric spirals.

The electric current, in passing through it, is not divided, but traverses the whole length of the wire.

2192. The third, or *Woolwich helix*, was made under my instruction for the use of Lieut.-Colonel Sabine's establishment at Woolwich. It is 26·5 inches long, 2·5 inches internal diameter, and 4·75 inches external diameter. The wire is 0·17 of an inch in diameter, and 501 feet in length. It is disposed in the coil in four concentric spirals connected end to end, so that the whole of the electric current employed passes through all the wire.

2193. The long helix (2190.) acted very feebly on a magnetic needle placed at a little distance from it; the medium helix (2191.) acted more powerfully, and the Woolwich helix (2192.) very strongly; the same battery of ten pairs of Grove's plate being employed in all cases.

2194. Solid bodies were easily subjected to the action of these electro-helices, being for that purpose merely cut into the form of bars or prisms with flat and polished ends, and then introduced as cores into the helices. For the purpose of submitting liquid bodies to the same action, tubes of glass were provided, furnished at the ends with caps; the cylindrical part of the cap was brass, and had a tubular aperture for the introduction of the liquids, but the end was a flat glass plate. When the tube was intended to contain aqueous fluids, the plates were attached to the caps, and the caps to the tube by Canada balsam; when the tube had to contain alcohol, æther or essential oils, a thick mixture of powdered gum with a little water was employed as the cement.

2195. The general effect produced by this form of apparatus may be stated as follows:—The tube within the long helix (2190.) was filled with distilled water and placed in the line of the polarized ray, so that by examination through the eye-piece (2150.), the image of the lamp-flame produced by the ray could be seen through it. Then the eye-piece was turned until the image of the flame disappeared, and, afterwards, the current of ten pairs of plates sent through the helix; instantly the image of the flame reappeared, and continued as long as the electric current was passing through the helix; on stopping the current the image disappeared. The light did not rise up gradually, as in the case of electro-magnets (2170.), but instantly. These results could be produced at pleasure. In this experiment we may, I think, justly say that a ray of light is electrified and the electric forces illuminated.

2196. The phænomena may be made more striking, by the adjustment of a lens of long focus between the tube and the polarizing mirror, or one of short focus between the tube and the eye; and where the helix, or the battery, or the substance experimented with, is feeble in power, such means offer assistance in working out the effects: but, after a little experience, they are easily dispensed with, and are only useful as accessories in doubtful cases.

2197. In cases where the effect is feeble, it is more easily perceived if the Nichol eye-piece be adjusted, not to the perfect extinction of the ray, but a little short of or beyond that position; so that the image of the flame may be but just visible. Then, on the exertion of the power of the electric current, the light is either increased in intensity, or else diminished, or extinguished, or even re-illuminated on the other side of the dark condition; and this change is more easily perceived than if the eye began to observe from a state of utter darkness. Such a mode of observing also assists in demonstrating the rotatory character of the action on light; for, if the light be made visible beforehand by the motion of the eye-piece in one direction, and the power of the current be to *increase* that light, an instant only suffices, after stopping the current, to move the eye-piece in the other direction until the light is apparent as at first, and then the power of the current will be to *diminish* it; the tints of the lights being affected also at the same time.

2198. When the current was sent round the helix in one direction, the rotation induced upon the ray of light was one way; and when the current was changed to the contrary direction, the rotation was the other way. In order to express the direction, I will assume, as is usually done, that the current passes from the zinc through the acid to the platinum in the same cell (663. 667. 1627.): if such a current pass under the ray towards the right, upwards on its right side, and over the ray towards the left, it will give left-handed rotation to it; or, if the current pass over the ray to the right, down on the right side, and under it towards the left, it will induce it to rotate to the right-hand.

2199. The LAW, therefore, by which an electric current acts on a ray of light is easily expressed. When an electric current passes round a ray of polarized light in a plane perpendicular to the ray, it causes the ray to revolve on its axis, as long as it is under the influence of the current, in the *same direction* as that in which the current is passing.

2200. The simplicity of this law, and its identity with that given before, as expressing the action of magnetism on light (2160.), is very beautiful. A model is not wanted to assist the memory; but if that already described (2161.) be looked at, the line round it will express at the same time the direction both of the current and the rotation. It will indeed do much more; for if the cylinder be considered as a piece of iron, and not a piece of glass or other diamagnetic, placed between the two poles N and S, then the line round it will represent the direction of the currents, which, according to Ampère's theory, are moving round its particles; or if it be considered as a core of iron (in place of a core of water), having an electric current running round it in the direction of the line, it will also represent such a magnet as would be formed if it were placed between the poles whose marks are affixed to its ends.

2201. I will now notice certain points respecting the degree of this action under different circumstances. By using a tube of water (2194.) as long as the helix, but placing it so that more or less of the tube projected at either end of the helix, I was able, in some degree, to ascertain the effect of length of the diamagnetic, the force of the helix and current remaining the same. The greater the column of water subjected to the action of the helix, the greater was the rotation of the polarized ray; and the amount of rotation seemed to be directly proportionate to the length of fluid round which the electric current passed.

2202. A short tube of water, or a piece of heavy glass, being placed in the axis of the Woolwich helix (2192.), seemed to produce equal effect on the ray of light, whether it were in the middle of the helix or at either end; provided it was always within the helix and in the line of the axis. From this it would appear that every part of the helix has the same effect; and, that by using long helices, substances may be submitted to this kind of examination which could not be placed in sufficient length between the poles of magnets (2150.).

2203. A tube of water as long as the Woolwich helix (2192.), but only 0·4 of an inch in diameter, was placed in the helix parallel to the axis, but sometimes in the axis and sometimes near the side. No apparent difference was produced in these different situations; and I am inclined to believe (without being quite sure) that the action on the ray is the same, wherever the tube is placed, within the helix, in relation to the axis. The same result was obtained when a larger tube of water was looked through, whether the ray passed through the axis of the helix and tube, or near the side.

2204. If bodies be introduced into the helix possessing, naturally, rotating force, then the rotating power given by the electric current is superinduced upon them, exactly as in the cases already described of magnetic action (2165. 2187.).

2205. A helix, twenty inches long and 0·3 of an inch in diameter, was made of uncovered copper wire, 0·05 of an inch in diameter, in close spirals. This was placed in a large tube of water, so that the fluid, both in the inside and at the outside of the helix, could be examined by the polarized ray. When the current was sent *through* the helix, the water within it received rotating power; but no trace of such an action on the light was seen on the outside of the helix, even in the line most close to the uncovered wire.

2206. The water was inclosed in brass and copper tubes, but this alteration caused not the slightest change in the effect.

2207. The water in the brass tube was put into an *iron* tube, much longer than either the Woolwich helix or the brass tube, and quite one-eighth of an inch thick in the side; yet when placed in the Woolwich helix (2192.), the water rotated the ray of light apparently as well as before.

2208. An iron bar, one inch square and longer than the helix, was put into the helix, and the small water-tube (2203.) upon it. The water exerted as much action on the light as before.

2209. Three iron tubes, each twenty-seven inches long and one-eighth of an inch in thickness in the side, were selected of such diameters as to pass easily one into the other, and the whole into the Woolwich helix (2192.). The smaller one was supplied with glass ends and filled with water; and being placed in the axis of the Woolwich helix, had a certain amount of rotating power over the polarized ray. The second tube was then placed over this, so that there was now a thickness of iron equal to two-eighths of an inch between the water and the helix; the water had *more* power of rotation than before. On placing the third tube of iron over the two former, the power of the water *fell*, but was still very considerable. These results are complicated, being dependent on the new condition which the character of iron gives to its action on the forces. Up to a certain amount, by increasing the development of magnetic forces, the helix and core, *as a whole*, produce increased action on the water; but on the addition of more iron and the disposal of the forces through it, their action is removed in part from the water and the rotation is lessened.

2210. Pieces of heavy glass (2151.), placed in iron tubes in the helices, produced similar effects.

2211. The bodies which were submitted to the action of an electric current in a helix, in the manner already described, were as follows:—Heavy glass (2151. 2176.), water, solution of sulphate of soda, solution of tartaric acid, alcohol, æther, and oil of turpentine; all of which were affected, and acted on light exactly in the manner described in relation to magnetic action (2173.).

2212. I submitted *air* to the influence of these helices carefully and anxiously, but could not discover any trace of action on the polarized ray of light. I put the long helix (2190.) into the other two (2191. 2192.), and combined them all into one consistent series, so as to accumulate power, but could not observe any effect of them on light passing through air.

2213. In the use of helices, it is necessary to be aware of one effect, which might otherwise cause confusion and trouble. At first, the wire of the long helix (2190.) was wound directly upon the thin glass tube which served to contain the fluid. When the electric current passed through the helix it raised the temperature of the metal, and that gradually raised the temperature of the glass and the film of water in contact with it, and so the cylinder of water, warmer at its surface than its axis, acted as a lens, gathering and sending rays of light to the eye, and continuing to act for a time after the current was stopped. By separating the tube of water from the helix, and by other precautions, this source of confusion is easily avoided.

2214. Another point of which the experimenter should be aware, is the difficulty, and almost impossibility, of obtaining a piece of glass which, especially after it is cut, does not depolarize light. When it does depolarize, difference of position makes an immense difference in the appearance. By always referring to the parts that do not depolarize, as the black cross, for instance, and by bringing the eye as near as may be to the glass, this difficulty is more or less overcome.

2215. For the sake of supplying a general indication of the amount of this induced rotating force in two or three bodies, and without any pretence of offering correct numbers, I will give, generally, the result of a few attempts to measure the force, and compare it with the natural power of a specimen of oil of turpentine. A very powerful electro-magnet was employed, with a *constant* distance between its poles of 2½ inches. In this space was placed different substances; the amount of rotation of the eye-piece observed several times and the average taken, as expressing the rotation for the ray length of substance used. But as the substances were of different dimensions, the ray lengths were, by calculation, corrected to one standard length, upon the assumption that the power was proportionate to this length (2163.). The oil of turpentine was of course observed in its natural state, *i. e.* without magnetic action. Making water 1, the numbers were as follows:—

Oil of turpentine . . .	11·8
Heavy glass (2151.) . .	6·0
Flint-glass	2·8
Rock-salt	2·2
Water	1·0
Alcohol	less than water.
Æther	less than alcohol.

2216. In relation to the action of magnetic and electric forces on light, I consider, that to know the conditions under which there is no apparent action, is to add to our knowledge of their mutual relations; and will, therefore, very briefly state how I have lately combined these forces, obtaining no apparent result (955.).

2217. Heavy glass, flint-glass, rock-crystal, Iceland spar, oil of turpentine, and air, had a polarized ray passed through them; and, at the same time, lines of electro-static tension (2149.) were, by means of coatings, the Leyden jar, and the electric machine, directed across the bodies, parallel to the polarized ray, and perpendicular to it, both in and across the plane of polarization; but without any visible effect. The tension of a rapidly recurring, induced secondary current, was also directed upon the same bodies and upon water (as an electrolyte), but with the same negative result.

2218. A polarized ray, powerful magnetic lines of force, and the electric lines of force (2149.) just described, were combined in various directions in their action on heavy glass (2151. 2176.), but with no other result than that due to the mutual action of the magnetic lines of light, already described in this paper.

2219. A polarized ray and electric currents were combined in every possible way in electrolytes (951–954). The substances used were distilled water, solution of sugar, dilute sulphuric acid, solution of sulphate of soda, using platinum electrodes; and solution of sulphate of copper, using copper electrodes; the current was sent along the ray, and perpendicular to it in two directions at right angles with each other; the ray was made to rotate, by altering the position of the polarizing mirror, that the plane of polarization might be varied; the current was used as a continuous current, as a rapidly intermitting current, and as a rapidly alternating double current of induction; but in no case was any trace of action perceived.

2220. Lastly, a ray of polarized light, electric currents, and magnetic lines of force, were directed in every possible way through dilute sulphuric acid and solution of sulphate of soda, but still with negative results, except in those positions where the phænomena already described were produced. In one arrangement, the current passed in the direction of radii from a central to a circumferential electrode, the contrary magnetic poles being placed above and below; and the arrangements were so good, that when the electric current was passing, the fluid rapidly rotated; but a polarized ray sent horizontally across this arrangement was not at all affected. Also, when the ray was sent vertically through it, and the eye-piece moved to correspond to the rotation impressed upon the ray in this position by the magnetic curves alone, the superinduction of the passage of the electric current made not the least difference in the effect upon the ray.

¶ iii. *General considerations.*

2221. Thus is established, I think for the first time *, a true, direct relation and dependence between light and the magnetic and electric forces; and thus a great addition made to the facts and considerations which tend to prove that all natural forces are tied together, and have one common origin (2146.). It is, no doubt, difficult in the present state of our knowledge to express our expectation in exact terms; and, though I have said that another of the powers of nature is, in these experiments, directly related to the rest, I ought, perhaps, rather to say that another form of the great power is distinctly and directly related to the other forms; or that the great power manifested by particular phænomena in particular forms, is here further identified and recognised, by the direct relation of its form of light to its forms of electricity and magnetism.

* I say, for the first time, because I do not think that the experiments of Morrichini on the production of magnetism by the rays at the violet end of the spectrum prove any such relation. When in Rome with Sir H. Davy in the month of May 1814, I spent several hours at the house of Morrichini, working with his apparatus and under his directions, but could not succeed in magnetising a needle. I have no confidence in the effect as a *direct* result of the action of the sun's rays; but think, that when it has occurred it has been secondary, incidental, and perhaps even accidental; a result that might well happen with a needle that was preserved during the whole experiment in a north and south position.

January 2, 1846.—I should not have written "for the first time" as above, if I had remembered Mr. Christie's experiments and papers on the Influence of the Solar Rays on Magnets, communicated in the Philosophical Transactions for 1826, p. 219, and 1828, p. 379.—M. F.

2222. The relation existing between *polarized* light and magnetism and electricity, is even more interesting than if it had been shown to exist with common light only. It cannot but extend to common light; and, as it belongs to light made, in a certain respect, more precise in its character and properties by polarization, it collates and connects it with these powers, in that duality of character which they possess, and yields an opening, which before was wanting to us, for the appliance of these powers to the investigation of the nature of this and other radiant agencies.

2223. Referring to the conventional distinction before made (2149.), it may be again stated, that it is the magnetic lines of force *only* which are effectual on the rays of light, and they *only* (in appearance) when parallel to the ray of light, or as they tend to parallelism with it. As, in reference to matter not magnetic after the manner of iron, the phænomena of electric induction and electrolysation show a vast superiority in the energy with which electric forces can act as compared to magnetic forces, so here, in another direction and in the peculiar and correspondent effects which belong to magnetic forces, they are shown, in turn, to possess great superiority, and to have their full equivalent of action on the same kind of matter.

2224. The magnetic forces do not act on the ray of light directly and without the intervention of matter, but through the mediation of the substance in which they and the ray have a simultaneous existence; the substances and the forces giving to and receiving from each other the power of acting on the light. This is shown by the non-action of a vacuum, of air or gases; and it is also further shown by the special degree in which different matters possess the property. That magnetic force acts upon the ray of light always with the same character of manner and in the same direction, independent of the different varieties of substance, or their states of solid or liquid, or their specific rotative force (2232.), shows that the magnetic force and the light have a direct relation: but that substances are necessary, and that these act in different degrees, shows that the magnetism and the light act on each other through the intervention of the matter.

2225. Recognizing or perceiving *matter* only by its powers, and knowing nothing of any imaginary nucleus, abstract from the idea of these powers, the phænomena described in this paper much strengthen my inclination to trust in the views I have on a former occasion advanced in reference to its nature*.

2226. It cannot be doubted that the magnetic forces act upon and affect the internal constitution of the diamagnetic, just as freely in the dark as when a ray of light is passing through it; though the phænomena produced by light seem, as yet, to present the only means of observing this constitution and the change. Further, any such change as this must

* A speculation, &c. Philosophical Magazine, 1844, vol. xxiv. p. 136.

belong to opake bodies, such as wood, stone, and metal; for as diamagnetics, there is no distinction between them and those which are transparent. The degree of transparency can at the utmost, in this respect, only make a distinction between the individuals of a class.

2227. If the magnetic forces had made these bodies magnets, we could, by light, have examined a transparent magnet; and that would have been a great help to our investigation of the forces of matter. But it does not make them magnets (2171.), and therefore the molecular condition of these bodies, when in the state described, must be specifically distinct from that of magnetized iron, or other such matter, and must be *a new magnetic condition*; and as the condition is a state of tension (manifested by its instantaneous return to the normal state when the magnetic induction is removed), so the *force* which the matter in this state possesses and its mode of action, must be to us a *new magnetic force* or *mode of action* of matter.

2228. For it is impossible, I think, to observe and see the action of magnetic forces, rising in intensity, upon a piece of heavy glass or a tube of water, without also perceiving that the latter acquire properties which are not only *new* to the substance, but are also in subjection to very definite and precise laws (2160. 2199.), and are equivalent in proportion to the magnetic forces producing them.

2229. Perhaps this state is a state of *electric tension tending to a current*; as in magnets, according to Ampère's theory, the state is a state of *current*. When a core of iron is put into a helix, every thing leads us to believe that currents of electricity are produced within it, which rotate or move in a plane perpendicular to the axis of the helix. If a diamagnetic be placed in the same position, it acquires power to make light rotate in the same plane. The state it has received is a state of tension, but it has not passed on into currents, though the acting force and every other circumstance and condition are the same as those which do produce currents in iron, nickel, cobalt, and such other matters as are fitted to receive them. Hence the idea that there exists in diamagnetics, under such circumstances, a tendency to currents, is consistent with all the phænomena as yet described, and is further strengthened by the fact, that, leaving the loadstone or the electric current, which by inductive action is rendering a piece of iron, nickel, or cobalt magnetic, perfectly unchanged, a mere change of temperature will take from these bodies their extra power, and make them pass into the common class of diamagnetics.

2230. The present is, I believe, the first time that the molecular condition of a body, required to produce the circular polarization of light, has been artificially given; and it is therefore very interesting to consider this known state and condition of the body, comparing it with the relatively unknown state of those which possess the power naturally: especially as some of the latter rotate to the right-hand and others to the left; and, as in the cases of quartz and oil of turpentine, the same body chemically speaking, being in the latter instance a liquid with particles free to move, presents different specimens, some rotating one way and some the other.

2231. At first one would be inclined to conclude that the natural state and the state conferred by magnetic and electric forces must be the same, since the effect is the same; but on further consideration it seems very difficult to come to such a conclusion. Oil of turpentine will rotate a ray of light, the power depending upon its particles and not upon the arrangement of the mass. Whichever way a ray of polarized light passes through this fluid, it is rotated in the same manner; and rays passing in every possible direction through it *simultaneously* are all rotated with equal force and according to one common law of direction; *i. e.* either all right-handed or else all to the left. Not so with the rotation superinduced on the *same* oil of turpentine by the magnetic or electric forces: it exists only in one direction, *i. e.* in a plane perpendicular to the magnetic line; and being limited to this plane, it can be changed in direction by a reversal of the direction of the inducing force. The direction of the rotation produced by the natural state is connected invariably with the direction of the ray of light; but the power to produce it appears to be possessed in every direction and at all times by the particles of the fluid: the direction of the rotation produced by the induced condition is connected invariably with the direction of the magnetic line or the electric current, and the condition is possessed by the particles of matter, but strictly limited by the line or the current, changing and disappearing with it.

2332. Let *m*, in fig. 3, represent a glass cell filled with oil of turpentine, possessing naturally the power of producing right-hand rotation, and *a b* a polarized ray of light. If the ray proceed from *a* to *b*, and the eye be placed at *b*, the rotation will be right-handed, or according to the direction expressed by the arrow-heads on the circle *c*; if the ray proceed from *b* to *a*, and the eye be placed at *a*, the rotation will still be right-handed *to the observer*, *i. e.* according to the direction indicated on the circle *d*. Let now an electric current pass round the oil of turpentine in the direction indicated on the circle *c*, or magnetic poles be placed so as to produce the same effect (2155); the particles will acquire a further rotative force (which no motion amongst themselves will disturb), and a ray coming from *a* to *b* will be seen by an eye placed at *b* to rotate to the right-hand more than before, or in the direction on the circle *c*; but pass a ray from *b* to *a*, and observe with the eye at *a*, and the phænomenon is no longer the same as before; for instead of the new rotation being according to

Fig. 3.

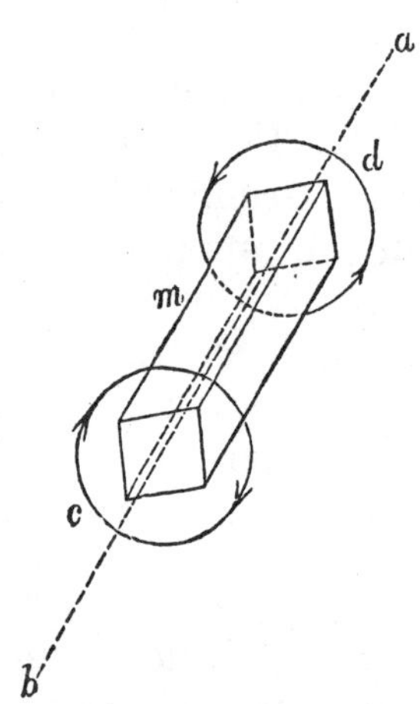

the direction indicated on the circle *d*, it will be in the contrary direction, or to the observer's left-hand (2199). In fact the induced rotation will be added to the natural rotation as respects a ray passing from *a* to *b*, but it will be subtracted from the natural rotation as regards the ray passing from *b* to *a*. Hence the particles of this fluid which rotate by virtue of their natural force, and those which rotate by virtue of the induced force, cannot be in the same condition.

2233. As respects the power of the oil of turpentine to rotate a ray in whatever direction it is passing through the liquid, it may well be, that though all the particles possess the power of rotating the light, only those whose planes of rotation are more or less perpendicular to the ray affect it; and that it is the resultant or sum of forces in any one direction which is active in producing rotation. But even then a striking difference remains, because the resultant in the same plane is not absolute in direction, but relative to the course of the ray, being in the one case as the circle *c*, and in the other as the circle *d*, fig. 3; whereas the resultant of the magnetic or electric induction is absolute, and not changing with the course of the ray, being always either as expressed by *c* or else as indicated by *d*.

2234. All these differences, however, will doubtless disappear or come into harmony as these investigations are extended; and their very existence opens so many paths, by which we may pursue our inquiries, more and more deeply, into the powers and constitution of matter.

2235. Bodies having rotating power of themselves, do not seem by that to have a greater or a less tendency to assume a further degree of the same force under the influence of magnetic or electric power.

2236. Were it not for these and other differences, we might see an analogy between these bodies, which possess at all times the rotating power, as a specimen of quartz which rotates only in one plane, and those to which the power is given by the induction of other forces, as a prism of heavy glass in a helix, on the one hand; and, on the other, a natural magnet and a helix through which the current is passing. The natural condition of the magnet and quartz, and the constrained condition of the helix and heavy glass, form the link of the analogy in one direction; whilst the supposition of currents existing in the magnet and helix, and only a tendency or tension to currents existing in the quartz and heavy glass, supplies the link in the transverse direction.

2237. As to those bodies which seem as yet to give no indication of the power over light, and therefore none of the assumption of the new magnetic conditions; these may be divided into two classes, the one including air, gases and vapours, and the other rock crystal, Iceland spar, and certain other crystalline bodies. As regards the latter class, I shall give, in the next series of these researches, proofs drawn from phænomena of an entirely different kind, that they do acquire the new magnetic condition; and these being so disposed of for the moment, I am inclined to believe that even air and gases have the power to assume the peculiar state, and even to affect light, but in a degree so small that as yet it has not been made sensible. Still the gaseous state is such a remarkable condition of matter, that we ought not too hastily to assume that the substances which, in the solid and liquid state, possess properties even general in character, always carry these into their gaseous condition.

2238. Rock-salt, fluor-spar, and, I think, alum, affect the ray of light; the other crystals experimented with did not; these are equiaxed and singly refracting, the others are unequiaxed and doubly refracting. Perhaps these instances, with that of the rotation of quartz, may even now indicate a relation between magnetism, electricity, and the crystallizing forces of matter.

2239. All bodies are affected by helices as by magnets, and according to laws which show that the causes of the action are identical as well as the effects. This result supplies another fine proof in favour of the identity of helices and magnets, according to the views of Ampère.

2240. The theory of static induction which I formerly ventured to set forth (1161, &c.), and which depends upon the action of the contiguous particles of the dielectric intervening between the inductric and the inducteous bodies, led me to expect that the same kind of dependence upon the intervening particles would be found to exist in magnetic action; and I published certain experiments and considerations on this point seven years ago (1709—1736). I could not then discover any peculiar condition of the intervening substance or diamagnetic; but now that I have been able to make out such a state, which is not only a state of tension (2227), but dependent entirely upon the magnetic lines which pass through the substance, I am more than ever encouraged to believe that the view then advanced is correct.

2241. Although the magnetic and electric forces *appear* to exert no power on the ordinary or on the depolarized ray of light, we can hardly doubt but that they have some special influence, which probably will soon be made apparent by experiment. Neither can it be supposed otherwise than that the same kind of action should take place on the other forms of radiant agents as heat and chemical force.

2242. This mode of magnetic and electric action, and the phænomena presented by it, will, I hope, greatly assist hereafter in the investigation of the nature of transparent bodies, of light, of magnets, and their action one on another, or on magnetic substances. I am at this time engaged in investigating the new magnetic condition, and shall shortly send a further account of it to the Royal Society. What the possible effect of the force may be in the earth as a whole, or in magnets, or in relation to the sun, and what may be the best means of causing light to evolve electricity and magnetism, are thoughts continually pressing upon the mind; but it will be better to occupy both time and thought, aided by experiment, in the investigation and development of real truth, than to use them in the invention of suppositions which may or may not be founded on, or consistent with fact.

Reprinted from *The London, Edinburgh, and Dublin Philosophical Magazine and Journal of Science,* Series IV, 337–348 (November 1875).

XL. *A new Relation between Electricity and Light: Dielectrified Media Birefringent.* By JOHN KERR, *LL.D., Mathematical Lecturer of the Free-Church Training College, Glasgow*.*

THE thought which led me to the following inquiry was briefly this:—that if a transparent and optically isotropic insulator were subjected properly to intense electrostatic force, it should act no longer as an isotropic body upon light sent through it. Faraday was often occupied with expectations of this kind; and he has mentioned in his memoir on the Magnetization of Light, and elsewhere in his 'Researches,' how he experimented in this very direction, upon electrolytes as well as dielectrics, at different times and in many ways, but always without success†. As far as I remember, I have not read or heard of an attempt in this field by any other naturalist. I proceed to offer a few notes of some recent experiments of my own. The investigation is not so complete as I should wish it to be; but it has been carried forward as far as my limited time and means would allow. At present I confine myself to solid dielectrics, reserving the case of liquids for a second paper. The principal results given in this first paper are stated apart, in articles 11, 17, 23.

1. *Dielectric of Plate Glass.*—A piece of good plate glass, $\frac{3}{4}$ inch thick, is formed roughly, before it leaves the shop, into a rectangular block 6 inches long and 2 wide. In this and subsequent operations, the original polish of the plate is carefully preserved. Two holes, about $\frac{1}{10}$ inch wide, are drilled into the block from its opposite ends; they lie exactly as in continuation of each other, in a line parallel to the longest edges of the block, and midway between opposite faces; and they terminate in well-rounded bottoms at the centre of the block, with a short extent (a quarter inch or less) of clear glass between them. Two fine pillars of glass rise from a stand on the table, distant an inch or

* Communicated by the Author.

† Faraday's 'Experimental Researches,' 2216, p. 951, or Maxwell's 'Treatise on Electricity,' vol. ii. p. 399.

two from each other. The block is placed across the pillars (at about a foot from the table), its plate-faces vertical, and the line of borings horizontal; and in this position it is tied securely to the pillars by coils of silk thread. Two thick wires of copper, sheathed in gutta percha, have their bared extremities inserted in the borings, down to the ends. As a provision against the strongest electric action applied in any case, these wires are coated very deeply with lac or sealing-wax at their junction with the glass, and an inch or more outwards. The whole surface of pillars and block is well varnished with lac—except a small space which is left clean upon each of the plate-faces, to allow distinct vision through the centre of the block.

When the dielectric has been thus prepared, its transparency is all that can be desired. Objects bright or faint are seen horizontally through the central parts of the plate (between and around the ends of the borings) quite as well as through a fine window.

2. The electricity is obtained from a Ruhmkorff's induction-apparatus, which gives a spark of 20 to 25 centimetres. The dielectric just described stands upon the table close to the inductorium. The outer ends of the wires from the dielectric are screwed into the knobs of the secondary coil. From the same knobs two wires are led to the other side of the instrument, and are connected with two insulated metallic balls, which act as spark-terminals. The distance of these balls, or the actual spark-length in air, is under the observer's control at every instant.

The ends of the secondary coil are separated thus at one place by so many inches of air, and at another place by a quarter inch of glass. When the primary circuit is closed in the usual way, through the oscillating rheotome of the instrument, the air between the spark-terminals is broken by a sensibly incessant discharge, while the glass between the induction-terminals is traversed by a strong electric force. By simple increase or diminution of the distance between the spark-terminals, the intensity of electric strain thus produced at the centre of the glass block may be raised or lowered at once, and in any degree, as the observer pleases. I may mention that on several occasions, when the instrument was working at full power and the spark-terminals were drawn 7½ or 8 inches apart, strong discharges burst across between the two induction-wires, at the rate of about one per second, without cessation of the principal discharge. The insulation was so good that the dielectric was not in any way damaged. The discharges took place through the air in dense white sparks, from end to end of the block, a distance of 7 or 8 inches. In these circumstances, the part of the glass block between the induction-terminals must have been subjected to a strain little short of the utmost it could bear. Electric forces of such intensity were hardly ever applied in the experiments, were not indeed required.

3. The polariscope consists of two Nicol's prisms. A flat paraffin flame, presented edgeways, is used as a source of light. Next to the lamp and close to it comes the first Nicol; then at a distance of 2 feet or more comes the dielectric of plate glass; then at a like distance comes the second Nicol. The pieces are

so arranged that the observer, looking horizontally through the polariscope, and keeping the first Nicol at the centre of the field of vision, sees the flame through the centre of the dielectric, midway between the induction-terminals. The light crosses the dielectric at right angles to the plate-faces, and therefore at right angles to the lines of force.

4. *Neutralizing Plate.*—Every very thick plate of glass which I have yet worked with, exerts at most of its points a slight depolarizing action upon transmitted light. When such a plate is inserted between the two Nicols without compensation, any thing like perfect extinction is generally unattainable, and the sensibility of the polariscope is lost.

The principal section of the first Nicol being in any desired position, and that of the second Nicol perpendicular to it, the dielectric (still unexcited though connected with the coil) is inserted properly between the Nicols, and the light reappears well in the polariscope.

The neutralizing plate, a piece of glass about 6 inches square, taken from the same original plate as the dielectric, is then placed upon a stand immediately in front of the second Nicol, and is moved by trial into such a position that the restored light is again extinguished perfectly by a very small rotation of the analyzer. The apparatus is now ready for work.

5. The highest powers applied in the experiments are sufficiently indicated thus:—battery in the primary circuit, a series of six Grove's or Bunsen's elements; corresponding spark-length 9 or 10 inches; actual spark-length, or thickness of air between the spark-terminals, 6 inches, rarely 7; thickness of glass between the induction-terminals $\frac{3}{16}$ inch. But powers a good deal lower give effects distinct enough.

6. *First experiment.*—The pieces are arranged in the order and manner just described (2, 3, 4). The spark-terminals are fixed at a distance of 5 or 6 inches; the polarizing Nicol is laid with its principal section at 45° to the horizon; and the analyzer is turned, with the help of the neutralizing plate, into the position of perfect extinction. No piece of the apparatus is now touched, except the commutator, till the end of the experiment.

Looking through the polariscope, the observer closes the primary circuit. In about 2 seconds the light begins to reappear through the dielectric, at its old place between the induction-terminals, very faintly at first; but it brightens continuously for 10 seconds, 20, even 30, till it is almost brilliant. When the primary circuit is now broken by the commutator, the light fades away continuously, at first rapidly, then more slowly, to perfect extinction. The time that elapses (in the latter part of the experiment) between the opening of the primary circuit and perfect extinction in the polariscope depends very noticeably upon the intensity and duration of the electric action, increasing as these increase.

7. The light thus restored by electric action cannot be extinguished again, at any stage of the experiment, by any rotation of the analyzer either way.

8. *Second experiment.*—The polarizing Nicol is laid with its principal section either horizontal or vertical; the analyzer is turned into the corresponding position of perfect extinction; all the other arrangements and the procedure are as in the first experiment. There is now no regular effect obtained in the polariscope. In many cases, indeed, even when the strongest electric action (5) has been kept up for 20 or 30 seconds, any recovery of the light is very doubtful, rather a guess than a perception.

Small effects do sometimes present themselves; but they are trifling and irregular in comparison with those obtained in the first experiment. They are probably due to known causes, such as imperfection of adjustments, irregularities of molecular structure in the dielectric, possibly also slight changes of temperature. If a small and irregular allowance be made for one or more such disturbing influences, there is now no effect in the polariscope.

9. *Third experiment.*—Distance of the spark-terminals small (say, 2 inches), the arrangements otherwise as in the first experiment. The electric action is kept up for a minute or more, till the intensity of the restored light is certainly constant. The spark-terminals are then separated all at once to a distance of 6 inches, and in a second or two there is an evident increase of effect in the polariscope. Simple arrangements will be described soon, which exhibit this increase of effect as a reappearance from extinction.

10. *Fourth experiment.*—The same as the first, except that the primary current, instead of being constantly in one direction, is regularly reversed by the commutator at successive equal intervals of time (say, every second). The optical effect is as good as in the first experiment, if not better.

11. *Summary.*—When plate glass is intensely dielectrified, and traversed by polarized light in a direction perpendicular to the lines of force, it exerts a partially depolarizing action upon the light, giving an effect which is much more than merely sensible in a common polariscope. There is a good regular effect when the plane of polarization is at 45° to the lines of force, no regular effect when the plane of polarization is parallel or perpendicular to the lines of force. Electric force and optical effect increase together. The optical effect of a constant electric action takes a certain time (apparently about 30 seconds in my observations) to reach its full intensity, which it does by continuous increase from zero; and it falls again slowly to zero after the electric force has vanished. There is as good an effect with a rapid succession of contrary (Ruhmkorffian) electrizations as with a continued (Ruhmkorffian) electrization in one direction.

12. *Optical Compensator.*—Not having a regular instrument of this kind, I supply its place by a simple slip of glass held in the hands and subjected to varying stress. The action of strained glass upon transmitted light has been exactly determined by experiment. Compressed glass acts as a negative uniaxal crystal with its axis parallel to the line of compression; stretched glass acts as a positive uniaxal with its axis along the line of tension.

Illustrative optical experiment.—All the pieces placed as in the first experiment (2, 3, 4), the plane of polarization at 45° to the horizon, the extinction in the polariscope perfect, and the dielectric always unexcited. Two additional pieces are introduced into the course of the beam—say, between the dielectric and the neutralizing plate. The first is a small square of thin plate glass, held edgeways in a vice with its surfaces perpendicular to the beam, and feebly compressed in the direction of its length, which is horizontal. When this piece is inserted, the light is well restored from extinction. The second piece is the compensator—a rectangular slip of plate plass, shaped like a common microscopic slide, but generally larger. It is held by the two hands in front of the neutralizing plate, with its surfaces perpendicular to the beam, and its long edges horizontal; it is gently bent by the hands, the axes of the couples applied being perpendicular to the plate-faces, so that (say) the upper parts of the slip are extended horizontally, and the lower parts compressed; and it is lowered or raised so that the light, after traversing the first piece, is transmitted to the analyzer through the upper or lower parts of the second piece. Through the extended parts the light is weakened, and, with a right degree of tension, extinguished perfectly, and with a greater tension restored again; through the compressed parts it is always strengthened.

Generally, when the directions of stress in the two pieces are parallel, compression and tension counteract each other; two compressions or two tensions reinforce each other. When the lines of stress are at right angles, two compressions or two tensions counteract each other; compression and tension strengthen each other.

The action of a strained piece in the polariscope is most distinct when the direction of the stress lies, as it does here, midway between the principal sections of the two crossed Nicols. There is no action when the direction of stress lies in either of these planes.

There are several variations of the illustrative experiment that ought to be noticed here for their bearing on what follows. The first piece may be stationed on the other side of the dielectric; it may even be removed altogether, and the compression or tension applied to the dielectric plate itself in a horizontal direction perpendicular to the beam; the results are then the same as formerly. It appears thus, without reference to theory, that horizontal compression of the first piece has always to be compensated in the same way (that is, by horizontal tension of the second piece), whether the first action is applied before the dielectric, or behind it, or in it.

The illustrative experiment takes its simplest form when dielectric and neutralizing plate are both removed. It appears thus that horizontal compression of a first piece has to be compensated in the same way (that is, by horizontal tension of a second piece), whether the other two mutually balanced pieces, the unexcited dielectric and the neutralizing plate, are present or absent. Accordingly, and without reference to theory, in proceeding to characterize the birefringent action of the excited dielectric by means of the compensating slip, I assume that the mutually balanced actions of unexcited dielectric and neutralizing plate are without effect, and therefore to be left out of account.

13. The compensating slips used in the following experiments were of different sizes. But there was one which I came at last to employ almost exclusively, as I found it well adapted to the whole range of effects examined. It was a rectangular piece of very good plate, $\frac{3}{16}$ inch thick, 2 inches wide, and 10 long. It had no sensible action in the polariscope while unstrained.

14. *Fifth experiment.*—A compensating slip is introduced between the dielectric and the analyzer; all the other arrangements and the procedure are as in the first experiment (6). The plane of polarization being at 45° to the horizon, and the initial extinction in the polariscope perfect, the electric force is applied, and the light is well restored. When the compensating slip is now raised into the course of the beam, and stretched horizontally, the light is weakened, and, with a proper degree of tension, extinguished perfectly. As the electric action is continued an increasing force has to be applied to the compensator to produce extinction; but the force is always of the same kind—tension exactly or nearly parallel to the lines of electric force, or compression in a direction exactly or nearly perpendicular.

After a few trials I got this phenomenon to present itself with perfect regularity. The compensator is chosen carefully (13), and inequalities of temperature avoided. A strong electric action is kept up for any convenient time, 5 or 6 seconds to 30 or 40, till the light is well restored. The slip is introduced as in the illustrative experiment (12), and carefully bent, so that the part of it crossed by the light is stretched horizontally by a continuously and slowly increasing force. In these circumstances, and constantly, the light fades away from a very considerable brightness to an extinction as pure as I have ever got in a polariscope. After this it is almost superfluous to say that, when the compensating slip is compressed in a direction parallel to the lines of force, the light restored by electric action is always distinctly strengthened.

15. *Sixth experiment.*—Distance of the spark-terminals small (say, 3 inches), the other arrangements as formerly. The electric action is kept up for a minute or more, till the light in the polariscope is certainly constant. The neutralizing plate is then moved about, and the analyzer turned (through a small angle in any case), till the light is again well extinguished. All the optical pieces being left untouched, the primary circuit is broken; the light reappears in a few seconds, and increases continuously up to a certain permanent intensity; but the compensator, to produce extinction at any stage of this increase, has to be compressed in a direction parallel to the lines of force. And this is evidently as it ought to be; for what the compensator has now to do is to reinforce the failing action of the dielectric, an action which has been proved equivalent optically to compression along the lines of force (14).

16. *Seventh experiment.*—Distance of the spark-terminals small, $1\frac{1}{2}$ or 2 inches. As in the preceding experiment (15), a constant effect of electric action is obtained in the polariscope, and the light again extinguished by the neutralizing plate, which is then left untouched, as well as the other optical pieces, till

the end of the experiment. The electric action being still kept up, the distance of the spark-terminals is suddenly increased up to 5 or 6 inches; the light soon reappears and is allowed time to come out distinctly. When the primary circuit is now broken, the light fades away to extinction, and afterwards reappears; but before and after this passing extinction the light has contrary characters, as tested by the compensator, being extinguished in the former case by tension parallel to the lines of force, and in the latter case by compression.

17. From the last three experiments, or simply from the first of them, interpreted by the illustrative optical experiment (12), we infer that the dielectrization of plate glass is equivalent optically to a compression of the glass along the lines of electric force. Dielectrified glass acts upon transmitted light as a negative uniaxal with its axis parallel to the lines of force.

18. *Eighth experiment.*—All the preliminary arrangements are as in the fifth experiment (14). A strong electric action (spark of 6 inches) is kept up without ceasing for 20 minutes, and all the optical pieces are left untouched for an hour, the induction-terminals being connected with each other through the secondary coil from first to last. One thing very noticeable under such conditions is, the length of time which the light takes to fade away to extinction. In the present experiment there is a distinct, though faint, effect in the polariscope, even as long as 30 or 40 minutes after the external electric action has ceased. The effect has a constant character from first to last: the light restored by electric action is always extinguished perfectly by a right degree of tension of the compensator in a direction parallel to the lines of force. With the standard compensator (13), there is not a very great effort required for extinction at any time during the whole hour of observation.

19. *Ninth experiment.*—The analyzer is mounted in such a way that it may be moved in different directions at right angles to the beam, without rotation; the other arrangements are as in the fifth experiment. When a good effect has been obtained through the centre of the electric field, the second Nicol is moved about so as to receive the light through different parts of the dielectric.

Keeping first to the perpendicular bisector of the line joining the terminals, the intensity of the optical effect diminishes as the distance from the centre of the field increases: and this is particularly noticeable at first, while the effect is rising; but the action has always the same character, is always neutralized by horizontal tension of the compensator. As far as my means would allow, I have assured myself of the fact that, as soon as there is a good virtual compression of the dielectric at the centre of the field, there is an effect of the same kind, a virtual compression in the same direction, beginning to manifest itself all along the equator, but more faintly as the distance from the centre of the field increases. Out of this equatorial line, and well back from the wires, the direction of apparent compression of the dielectric changes from point to point of the field, and is at some points vertical, perpendicular to the line joining the terminals. Only at points very close to the induction-wires does the compensator fail decisively to extinguish or greatly weaken the restored light.

20. In connexion with the preceding facts, I may mention that under certain conditions, in the fifth experiment, the compensator does not extinguish the restored light all at the same time, but produces a very dark broad band, which descends from the outer parts of the field towards the centre as the tension of the compensator increases. This happens, for instance, though not very regularly, when the electric force is very intense from the beginning and the optical effect is just beginning to show itself; it happens also when the lines of force are inclined to the horizon, so that the fine vertical band of flame seen through the centre of the dielectric crosses the equator of the field at about 45°. In these cases there appears to be an unusually rapid variation of birefringent action through the part examined of the electric field.

I have now done with the dielectric of plate glass. Half a dozen other solids were tried; and I will conclude with a short account of the only two of them which gave results worth mentioning. The great difficulty was to get a sufficiently strong superficial insulation, the masses being too small. In only one case, that of Iceland spar, was the dielectric perforated by discharge without giving a clear effect; but I think the crystal had received a predisposing flaw in the operation of boring.

21. *Dielectric of Resin.*—This is a piece very similar to the dielectric of plate glass in form, size, and adjuncts. A quantity of clear amber-resin is kept at a gentle fusion for some time till it is free from air-bubbles; it is then poured into a suitable mould and left to cool. Two thick stocking-wires of steel, previously fixed in the mould, along one line through its ends, remain imbedded as induction-terminals in the solid resin, leaving less than $\frac{1}{4}$ of an inch of clear dielectric between them at the centre of the block. The polarized light enters and leaves the central part of the block through small squares of thin plate glass, which are in optical contact with the resin and parallel to each other.

This dielectric, even the best specimen of it, is far inferior to the dielectric of plate glass. It gives evidence of permanent and irregular strain in the neighbourhood of the terminals; it exerts a pretty strong photogyric action in the polariscope, separating the blue and red by a small angle; it is also imperfectly transparent for very faint light; but its chief defect is, that it allows a spark-discharge over its surface, a length of 7 inches, before the distance of the spark-terminals has much exceeded $2\frac{1}{2}$ inches. With all these deficiencies, the dielectric of resin gives a definite and regular effect; and the action is contrary to that of glass.

All the arrangements are essentially as in the fifth experiment (14). The block is tied to two fine insulating pillars of glass; the induction-wires of the dielectric are connected with the knobs of the secondary coil; and the light is seen through the centre of the block, midway between the terminals. A pretty good initial extinction is obtained in the polariscope between the blue and red; the light is then well restored by electric action, and the compensator is introduced between the dielectric and the second Nicol. By tension of the compensator in a direction parallel to the lines of force, the restored light is constantly and distinctly strengthened; and by compression parallel to the lines of force, it is regularly and greatly weakened, though never quite extinguished.

To test the value of these indications, I repeated the illustrative optical experiment (12) with the block of resin put in place of the large block of plate glass; and the results were satisfactory. I varied the experiment also by manipulating the block of resin itself as a compensator against the small square of compressed glass, the resin being simply pulled or pushed gently at both ends; and the results were equally distinct. The lines of stress being parallel, compression of glass and compression of the resin always reinforced each other, compression of glass and extension of the resin always counteracted each other, but never down to perfect extinction. It is true that these last results might be due more or less to the thin plates of glass which limit the resin. From all the observations, I infer that dielectrization of resin is optically equivalent to tension of the resin along the lines of force.

22. *Dielectric of Quartz.*—This is a plate perpendicular to the axis, made as for ordinary experiments in the polariscope, thickness 3 millims., length 20. Two fine holes are drilled into the plate as in the dielectric of plate glass, their bottoms flat, and with $\frac{1}{16}$ of an inch of clear crystal between them. Wires of copper are inserted in the borings, and are fixed along with the crystal to simple bearings made of glass rods; and the whole piece is coated very deeply with fused lac, a narrow window being left at the centre of the plate. The action of the crystal in the polariscope is well neutralized by a contrary plate of equal thickness; the other arrangements are as in the fifth experiment. With my rough apparatus the adjustments are so troublesome that I have executed only one short series of observations, of which I append all the notes preserved.

Distance of spark-terminals $\frac{1}{2}$ inch: a faint but clear restoration of the light from almost perfect extinction; the compensator not working distinctly; the insulation perfect.

Distance 1 inch: the light clearly restored by electric action, and then well extinguished by tension of the compensator parallel to lines of force. The insulation is now failing, sparks passing occasionally over the surface of the lac near the crystal.

Distance $\frac{3}{4}$ inch: the insulation still defective; the light well restored by electric action and then well weakened by horizontal tension, and strengthened by compression. From what I have seen in some other experiments, I think that the effects in this case and the preceding may have been produced wholly or partially by slight changes of temperature, due to occasional spark-discharge over the surface of the crystal.

Distance again $\frac{1}{2}$ inch: the insulation again good. In this case it is noticed that, before the electric force is applied, while the body of the light is barely perceptible in the polariscope, there is a short length of it from the apex downwards (due exactly through the centre of the electric field) which is perfectly extinguished. By electric action this upper part of the light is restored very clearly, and is then as clearly extinguished by tension of the compensator parallel to the lines of force.

23. Upon the whole, though better experimental results are desirable, I consider it proved that dielectrified quartz (like glass) acts upon transmitted light as if compressed along the lines of force, while dielectrified resin (unlike glass) acts as if extended along the lines of force.

24. *Theory.*—Faraday's views as to the constitution and function of dielectrics apply here very aptly.

When the induction-terminals are charged, the particles of the dielectric throughout the field are electrically polarized, and tend accordingly to arrange themselves end to end, and to cohere in files along the lines of force, just as iron filings do in a magnetic field. As far as this tendency of the polarized particles towards a file arrangement along the lines of force takes effect, there is a new molecular structure induced in the dielectric.

If we neglect the influence of ordinary strains transmitted from point to point of the solid, and assume, as a good first approximation warranted by facts (19), that the change of molecular arrangement at each point is determined solely or principally by electric force at the point, we cannot easily suppose the new structure in a dielectric, originally isotropic, to be any thing else than uniaxal, symmetrical at each point with reference to the line of force through the point. And, even in the case of an æolotropic body, we may assume, as a simple and sure approximation to the truth, that the effect of electric force is to superinduce a uniaxal structure upon the primitive structure.

The uniaxal structure thus induced by dielectrization has been experimentally detected and characterized by birefringent action in three cases. As a matter of fact, it appears to be negative in glass and quartz, but positive in resin.

The electric force has probably a certain resistance to overcome, something analogous to coercive force in the case of magnetism. A sensible time is therefore required for the development of the uniaxal structure by electric action, and for its disappearance after the electric action has ceased. Under an intense and long-sustained electric force, the new structure of the dielectric may assume the character of a very stiff and perhaps permanent set, analogous to permanent or subpermanent magnetism (18). We shall see afterwards, as might indeed be expected, that there is nothing similar to this in the phenomena presented by dielectrified liquids.

Contrary electrizations rapidly succeeding one another exert contrary actions of electric polarization, but conspiring actions of molecular arrangement: they are therefore as effective as a continued electrization in one direction; and a Ruhmkorff's coil is as effective as an electrical machine of equal strength.

I have made some experiments, and have had a good many reflections, bearing on other explanations of the phenomena; and I think it not unlikely that strains due to the mutual actions of intensely charged shells of the dielectric, or strains due to changes of temperature, may have something to do with the facts. But in the meantime I offer the preceding remarks as a sketch of what appears to me to be the only probable theory.

Glasgow, September 22, 1875.

The Electro-Optic Effect in Uniaxial Crystals of the Type X H_2PO_4. I. Theoretical

BRUCE H. BILLINGS
Research Laboratory, Baird Associates, Inc., Cambridge 38, Massachusetts
(Received June 22, 1949)

When voltage is put across a crystal of the type X H_2PO_4, the index ellipsoid in the crystal is rotated and the length of the axes changes. Two electro-optic coefficients are necessary to describe the effect in the most general plate of the crystal. The necessity for two coefficients is shown from consideration of the crystal symmetry. The behavior of Z-cut and X-cut plates are treated in detail. For Z-cut plates with the voltage in the Z direction the crystal becomes biaxial with the plane of the optic axes at 45° to the X and Y crystallographic axis. The retardation along the normal to the plate is directly proportional to the voltage and independent of the thickness. The characteristics of the polarization interference pattern for excited X and Z plates are discussed.

INTRODUCTION

THERE are many places where an electric light valve or shutter has useful applications. One of the most obvious forms of light valve is the Kerr cell. This cell contains a liquid between a pair of parallel plate electrodes. When an electric field is applied, the liquid becomes birefringent. If the cell is placed between crossed polarizers, light will not pass through the combination until voltage is applied. There are many difficulties associated with the ordinary liquid Kerr cell. This liquid cell uses nitrobenzene which has a high Kerr constant only when it is extremely pure. The operation of the cell tends to decompose the material and the most successful cells have been those in which the nitrobenzene is continuously redistilled as the cell is used. In addition, the liquid has a disagreeable odor and is intensely poisonous.

A solid electro-optic shutter avoids many of these difficulties. As a result, such a device becomes a practical solution for many different problems and should be a useful scientific tool.

The Kerr effect in most solids is extremely small. In certain crystals, however, there is a large electro-optic effect which requires voltage no higher than those required in the liquid cell. This effect varies with first power of the electric field rather than with the square of the field as does the normal Kerr effect.

During the last few years several papers[1,2,3,4] have appeared discussing the electro-optic effect in tetragonal scalenohedral crystals of the type X H_2PO_4. Now that some of these crystals such as $NH_4H_2PO_4$ or ADP[5] are available as light valves, it seems appropriate to give a more complete treatment of the useful effect than is available in any of these papers.

THEORETICAL

The birefringence of a crystal can be described in terms of its index ellipsoid. The index ellipsoid has axes which are equal to the three crystal indices. If a plane is passed through the center of this ellipsoid at right angles to the direction in which a ray of light is going, it will cut the ellipsoid in an ellipse whose major and minor axes are the two indices associated with that particular direction in the crystal. In the case of a biaxial crystal the equation for this ellipsoid can be written

$$a^2x^2+b^2y^2+c^2z^2=1 \tag{1}$$

where a, b and c are the three reciprocal indices of

[1] G. D. Gottschall, J. Soc. Mot. Pict. Eng. **51**, 13 (1948).
[2] Hans Jaffe, Phys. Rev. **73**, 95 (1948).
[3] B. Zwicker and P. Scherrer, Helv. Phys. Acta **16**, 214 (1943).
[4] B. Zwicker and P. Scherrer, Helv. Phys. Acta **17**, 346 (1944).
[5] This crystal is grown under the trademark "PN" by the Brush Development Company, 3405 Perkins Avenue, Cleveland, Ohio.

refraction. In a uniaxial crystal two of the coefficients become equal and the equation reduces to

$$o^2(x^2+y^2)+e^2z^2=1 \tag{2}$$

where o is the ordinary reciprocal index and e the extraordinary reciprocal index. The electro-optic effect in crystals is a change in crystalline birefringence which occurs when a crystal is placed in an electric field. The first general treatment of the effect was given by Pockels.[6,7] He showed the number of constants to be expected in the different crystal classes and made experimental measurements in the case of quartz, $NaClO_3$ and rochelle salts. This change can be described in terms of a change of the orientation and the dimensions of the index ellipsoid. The numerical constants involved in this change can be considered as small corrections to be applied to the index ellipsoid constants. In order to write them it is necessary to put the equation of the index ellipsoid in its most general form. Crystallographers commonly refer the faces of a crystal to a cartesian coordinate system. The directions of these crystallographic axes are prescribed for each crystal class. Instead of referring the ellipsoid to its own principal axes it is referred instead to these crystallographic symmetry axes. The equation of the ellipsoid then becomes

$$1=a_{11}x^2+a_{22}y^2+a_{33}z^2+2a_{23}yz+2a_{31}zx+2a_{12}xy. \tag{3}$$

The new constants, a_{11}, a_{12}, etc., are called the polarization constants. If the a's are all known, it is possible

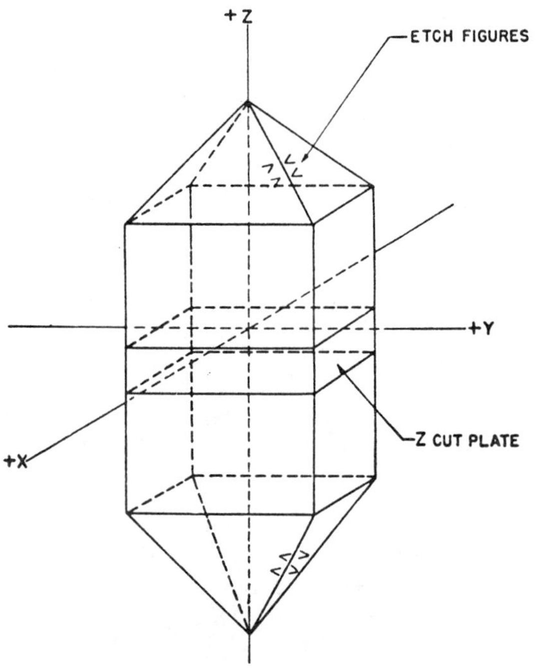

FIG. 1. Sketch of an $NH_4H_2PO_4$ crystal with its crystallographic axes labelled.

[6] F. Pockels, Abhandlungen der Gesellschaft der Wissenschaften zu Gottingen, 39, 1 (1893).

[7] F. Pockels, *Lehrbuch der Kristalloptik* (Leipzig, 1906).

to find the orientation of the axes of the ellipsoid with respect to the crystallographic axes.

The direction cosines connecting the two sets of axes may be described in the matrix

$$\begin{array}{c|ccc} & X' & Y' & Z' \\ \hline X & \alpha_1 & \alpha_2 & \alpha_3 \\ Y & \beta_1 & \beta_2 & \beta_3 \\ Z & \gamma_1 & \gamma_2 & \gamma_3 \end{array} \tag{4}$$

where the primed axes are the crystallographic axes, the unprimed axes are the axes of the index ellipsoid, and α_1 is the cosine of the angle between x and x', etc. When an electric field is applied, the polarization constants can be written

$$\begin{aligned} a_{11}-a_0^2&=r_{11}E_x+r_{12}E_y+r_{13}E_z \\ a_{22}-b_0^2&=r_{21}E_x+r_{22}E_y+r_{23}E_z \\ a_{33}-c_0^2&=r_{31}E_x+r_{32}E_y+r_{33}E_z \\ a_{23}&=r_{41}E_x+r_{42}E_y+r_{43}E_z \\ a_{31}&=r_{51}E_x+r_{52}E_y+r_{53}E_z \\ a_{12}&=r_{61}E_x+r_{62}E_y+r_{63}E_z \end{aligned} \tag{5}$$

where the r_{ij} are the electro-optic coefficients and E_x, E_y, E_z are the components of the field along the three crystallographic axes. These equations have been written for a crystal in which the crystallographic axes coincide with the polarization axes. They differ from the equations of Pockels by referring to the field instead of the electrical polarization. The constants r_{ij} are thus related to Pockels' constants e_{ij} by the equation

$$r_{ij}=\frac{K_j-1}{4\pi}e_{ij}$$

where K_j is the dielectric constant in the direction in which the field is applied. The symbol r_{ij} is used in accordance with Cady's[8] suggestion that the Mueller[9] convention be used. If no field were applied the constants would reduce to the three coefficients given in Eq. (1).

The most striking characteristic of a crystal is its symmetry. This symmetry is described by a set of operations sometimes called "covering operations" which can be performed on the crystal without altering its appearance or physical behavior. For example, if a cube is rotated 90° about an axis normal to any face the cube will not appear to have been moved. If the cube is isotropic with respect to all its other physical properties, the operations which leave its external appearance unchanged will be the characteristic symmetry operations. If the cube is made of wood which has a grain perpendicular to one face, the tensile strength would not be the same in the case of the 90° rotation about an axis parallel to this face. In general the physical characteristics of a crystal possess the same or higher symmetry as the symmetry of its form. This is,

[8] W. G. Cady, *Piezoelectricity* (McGraw-Hill Book Company, Inc., New York, 1946).

[9] H. Mueller, Phys. Rev. 58, 805–811 (1940).

of course, true only for an ideal crystal of any material.

Since many readers are perhaps rusty in their crystallography, it seems apropos to carry through the actual reduction of the constants in Eq. (4). In Fig. 1 is a sketch of an $NH_4H_2PO_4$ crystal with its crystallographic axes labeled. At first glance the crystal appears to possess a center of symmetry, i.e., it looks as though every point in the crystal could be reflected through the center to an equal distance on the other side and the resulting crystal would be identical. This turns out not to be the case. It is most simply demonstrated by etch figures. If one of the end faces is slightly moistened and dried, a series of little pits appear. These pits are triangular in shape and are oppositely oriented on opposite faces. There are two main symmetry operations which can be performed on the crystal. The first is a rotation through $\pi/2$ about the z axis followed by an inversion through the center. This, of course, does not mean that the crystal has a center of symmetry because the inversion must always be preceded by the $\pi/2$ rotation about z. The second operation is a rotation of π about either the x or y axis. Crystals whose characteristics are unchanged after these operations are said to belong in the class V_d or $42m$. If these operations are now performed on Eq. (5) the constants must remain unchanged. The operation of rotating through an angle π about the x axis leaves x unchanged and changes y to $-y$ and z to $-z$. The field E_x is unchanged whereas E_y becomes $-E_y$ and E_z becomes $-E_z$. When these substitutions are made in Eq. (5) and Eq. (3) the coefficients a_{11} through a_{12} must change sign in such a way that Eq. (3) is unchanged. The effect on the electro-optic coefficients may be simply shown by writing Eq. (5) in matrix form. For further aid the multiplier of the a_{ij}'s in Eq. (3) is placed in the a_{ij} vector.

$$\begin{vmatrix} a_{11}(x^2) \\ a_{22}(y^2) \\ a_{33}(z^2) \\ a_{23}(yz) \\ a_{31}(xz) \\ a_{12}(xy) \end{vmatrix} = \begin{vmatrix} r_{11} & r_{12} & r_{13} \\ r_{21} & r_{22} & r_{23} \\ r_{31} & r_{32} & r_{33} \\ r_{41} & r_{42} & r_{43} \\ r_{51} & r_{52} & r_{53} \\ r_{61} & r_{62} & r_{63} \end{vmatrix} \cdot \begin{vmatrix} E_x(x) \\ E_y(y) \\ E_z(z) \end{vmatrix}.$$

If the substitutions $x \to x$, $y \to -y$, $z \to -z$ are made in this matrix equation, it is apparent that a_{11}, a_{22} and a_{33} will not change signs. Since E_y and E_z do change sign, this means that all the r_{ij}'s in the first three rows are zero with the exception of r_{11}, r_{21}, and r_{31}. The operation of rotating the crystal through an angle π about the y axis changes the sign of x and z but leaves y unchanged. This operation makes these the remaining three coefficients in the first three rows vanish. The final array of elements in the matrix becomes

$$\begin{vmatrix} 0 & 0 & 0 \\ 0 & 0 & 0 \\ 0 & 0 & 0 \\ r_{41} & 0 & 0 \\ 0 & r_{41} & 0 \\ 0 & 0 & r_{63} \end{vmatrix}. \tag{7}$$

The constants a_{ij} thus reduce to

$$a_{11}=a_{22}=o^2 \quad a_{23}=r_{41}E_x$$
$$a_{33}=e^2 \quad a_{31}=r_{41}E_y \tag{8}$$
$$a_{12}=r_{63}E_z.$$

FIELD IN Z DIRECTION

The crystal section which is most useful as a shutter is the Z-cut or basal plate with field applied in the Z direction and the light traveling in the same direction. For light thus parallel to the optic axis the plate has no retardation and appears isotropic. The analysis below shows that in this direction the electrically induced retardation is also highest for a given voltage.

For this case the fields E_x and E_y become zero and the only non-zero coefficients are thus a_{11}, a_{22}, a_{33} and a_{12}.

These values can be inserted into Eq. (3) to give

$$e^2z^2+o^2(x^2+y^2)+2r_{63}E_zxy=1. \tag{9}$$

From this equation it is apparent that the Z and Z' axes coincide. This shows that $\gamma_1=\gamma_2=\alpha_3=\beta_3=0$ and $\gamma_3=1$.

To find the remaining cosines and hence the angle by which the ellipsoid is rotated, Eq. (9) is transformed to the coordinate system which has its axes along the ellipsoid axes. This equation connecting the two sets can be written down directly from the matrix (4) which represents the direction cosines between them.

$$x'=\alpha_1x+\beta_1y$$
$$y'=\alpha_2x+\beta_2y.$$

Since the rotation is in a plane

$$-\alpha_2=\beta_1; \quad \alpha_1=\beta_2.$$

These equations can now be put in Eq. (9) and the cross terms set equal to zero so that the axes will coincide with the ellipsoid axes.

$$2r_{63}C(\alpha_1^2-\alpha_2^2)xy=0. \tag{10}$$

The expression for the angle then becomes

$$\cos 2\alpha=0 \quad \alpha=\pm 45°. \tag{11}$$

Since this rotation is independent of the field it is possible at this stage of the analysis to say what will be the appearance between polarizers of a z-cut slab of PN when a field is applied in the z direction. The crystal becomes biaxial and the characteristic uniaxial fringe pattern of circles is replaced by a series of ovals. The long axis of the oval lies at 45° to the crystallographic axis. The length of the long axes of the ovals in the pattern will change as the field is changed. The direction, however, remains constant. To find how the axis length changes with the field it is necessary to compute the actual value of the crystal indices.

It is necessary now to find the new reciprocal indices, a, b and c. These are related to the polarization constants by formulas which can be written directly from Eq. (3) and the matrix (4)

$$\begin{aligned}a^2&=a_{11}\alpha_1^2+a_{22}\alpha_2^2+a_{33}\alpha_3^2+2a_{23}\alpha_2\alpha_3\\&\quad+2a_{31}\alpha_3\alpha_1+2a_{12}\alpha_1\alpha_2\\b^2&=a_{11}\beta_1^2+a_{22}\beta_2^2+a_{33}\beta_3^2+2a_{23}\beta_2\beta_3\\&\quad+2a_{31}\beta_3\beta_1+2a_{12}\beta_1\beta_2\\c^2&=a_{11}\gamma_1^2+a_{22}\gamma_2^2+a_{33}\gamma_3^2+2a_{23}\gamma_2\gamma_3\\&\quad+2a_{31}\gamma_3\gamma_1+2a_{12}\gamma_1\gamma_2.\end{aligned}\tag{12}$$

Since the only rotation is ϕ_z in the xy plane,

$$\begin{aligned}a^2&=a_{11}\cos^2\phi+a_{22}\sin^2\phi+2a_{12}\sin\phi\cos\phi\\&=o^2\cos^2\phi+o^2\sin^2\phi+2r_{63}E_z\sin\phi\cos\phi\\b^2&=a_{11}\sin^2\phi+a_{22}\cos^2\phi-2a_{12}\sin\phi\cos\phi\\&=o^2\sin^2\phi+o^2\cos^2\phi-2r_{63}E_z\sin\phi\cos\phi\\c^2&=a_{33}=e^2\end{aligned}\tag{13}$$

since $\phi_z=-45°$

$$\begin{aligned}a^2&=o^2-2r_{63}E_z\frac{1}{\sqrt{2}}\cdot\frac{1}{\sqrt{2}}=o^2-r_{63}E_z\\b^2&=o^2+2r_{63}E_z\frac{1}{\sqrt{2}}\cdot\frac{1}{\sqrt{2}}=o^2+r_{63}E_z.\end{aligned}\tag{14}$$

One can write

$$\begin{aligned}c^2&=e^2=\frac{1}{\epsilon^2}=\frac{1}{\alpha^2}\\a^2&=\frac{1}{\gamma^2}\quad b^2=\frac{1}{\beta^2}.\end{aligned}\tag{15}$$

Since the only field is in the Z direction, E_z can be replaced by V/d when V is the applied voltage and d is the plate thickness. The value of r_{63} then includes the dielectric constant.

The retardation which will appear along the normal to the basal section after the voltage is applied can be written:

$$\Gamma=\frac{d}{\lambda}\left(\frac{1}{a}-\frac{1}{b}\right).\tag{16}$$

This can be rewritten:

$$\begin{aligned}\Gamma&=\frac{d}{\lambda}\frac{a^2-b^2}{(a+b)ab}\\&=\frac{d}{2\lambda}(a^2-b^2)\omega^3.\end{aligned}\tag{17}$$

If Eq. (13) is inserted in this final equation, one gets the equation:

$$\Gamma=\frac{r_{63}V\omega^3}{\lambda}=pV.\tag{18}$$

This shows that in this particular cut of crystal the retardation along the normal is a function of the voltage rather than the field strength. In other words, the retardation for a given voltage is independent of the thickness of the plate.

The value for r_{63} becomes:

$$r_{63}=\lambda\Gamma/\omega^3V\tag{19}$$

The value of this constant has been determined for $NH_4H_2PO_4$ or "PN" under a variety of conditions by Dr. Hans Jaffe of The Brush Development Company.

The value of the new indices can now be calculated from Eq. (13):

$$\gamma=\frac{1}{a}=\frac{1}{\left(o^2-\frac{\lambda\Gamma}{\omega^3d}\right)^{\frac{1}{2}}}\tag{20}$$

or

$$\gamma=\omega+\frac{\lambda\Gamma}{2d}.\tag{21}$$

Similarly,

$$\beta=\omega-\frac{\lambda\Gamma}{2d}.\tag{22}$$

One of the striking things about these two index values is the small change which occurs with the application of voltage. A half-wave retardation requires nearly 10,000 volts for PN and this retardation is the result of index changes of only a few parts in a million for plate thicknesses of the order of a centimeter.

Although this value seems small, it is larger in ADP than in other crystals which have been previously studied. The coefficient r_{63} in ADP is 8.2×10^{-7} whereas the largest coefficients for quartz and rochelle salts are $r_{11(\text{quartz})}=4.67\times10^{-8}$ cm/kv and $r_{52(\text{Rochelle})}=-36.0\times10^{-8}$ cm/kv.

When voltage is applied to the crystal it becomes biaxial with an axial angle Ω. The retardation along the normal which was expressed in Eq. (14) can be written in terms of this axial angle. The expression for the retardation as a function of angle of incidence in a Z-cut slab of biaxial crystal can be written:

$$\frac{\Gamma}{kd}=\sin\theta_1\sin\theta_2/\cos r\tag{23}$$

where θ_1 and θ_2 are the angles made by the ray with the two axes; r is the angle of incidence of the ray; d is the thickness of the slab, k is a constant. When $r=0$, this equation becomes:

$$\Gamma=kd\sin^2\Omega\tag{24}$$

where Ω is the angle between the optic axes and the Z axis. In combination with Eq. (14) this becomes:

$$kd\sin^2\Omega=pV\tag{25}$$

where p is a constant and V is the voltage across the crystal plate.

It is apparent from these equations that the value of the axial angle is inversely proportional to the square root of the thickness and directly proportional to the square root of the voltage.

FIELD IN X OR Y DIRECTIONS

Since the equations are symmetrical, the effect produced by a field in the X direction is the same as a field in the Y direction. Accordingly, this section will be limited to a discussion of the effect of a field along X. In this case the only non-zero coefficients are a_{11}, a_{22}, a_{32} and a_{23}. These values can be inserted into Eq. (3) to give

$$o^2(x^2+y^2)+e^2z^2+2r_{41}E_xyz=1. \tag{26}$$

From this equation it is apparent that the X and X' axes coincide and the only rotation is in the yz plane. From an analysis similar to that used in the case of the basal section it can be shown that the axes are rotated through an angle

$$\tan 2\xi=\frac{2r_{41}E_x}{e^2-o^2}. \tag{27}$$

This differs from the previous case in that the angle is dependent on the field and that it is quite small. Its size can, of course, only be stated with a knowledge of r_{41}. This coefficient will be discussed in a later paper. For $NH_4H_2PO_4$ it is of such a size that the rotation is only a few minutes.

The reciprocal indices can be found from Eq. (11) which can be rewritten

$$\begin{aligned} b^2&=o^2\\ a^2&=o^2\cos^2\xi+e^2\sin^2\xi+2r_{41}E_x\cos\xi\sin\xi\\ c^2&=o^2\sin^2\xi+e^2\cos\xi-2r_{41}E_x\cos\xi\sin\xi. \end{aligned} \tag{28}$$

As in the first case this represents a biaxial crystal. To go from Eq. (12) to Eq. (28) it is necessary to interchange a and b. This is because b is by definition intermediate between a and c. If the field had been applied in the Y direction, no such interchange would have been necessary. Since the axial plane is always at right angles to b and since the direction of b is independent of the field, the plane of the axis is again independent of the field as in the Z case. By inspection of Eq. (26) it is apparent that the section of the ellipsoid perpendicular to Z is a circle, and the position of this circular section is independent of the field. This indicates that one of the optic axes of the crystal coincides with the original optic axis and is independent of the field. The other axis moves from this first axis as the field is applied. The axial angle which is the angle between the ellipsoid axis and the optic axis is exactly ξ. This can also be checked by using the equation

$$\sin^2\Omega=b^2-c^2/a^2-c^2. \tag{29}$$

Observation in the Z direction thus does not give a useful effect when field is applied in the X direction. Furthermore, the axial angle of the new biaxial crystal is only a few minutes as opposed to several degrees which is obtained when the field is applied in the Z direction.

In the Z-cut case the retardation was zero in the Z direction until the field was applied in this direction. In the X-cut case the retardation in the X direction is already large. For practical use the change in retardation for a reasonable voltage must be large. The retardation without field is:

$$\Gamma=(\epsilon-\omega)d/\lambda. \tag{30}$$

The change in retardation may be calculated to be

$$\Delta\Gamma=\tfrac{1}{2}r_{41}{}^2\frac{V^2}{d}\,\frac{\omega^2\epsilon^2(\omega^3+\epsilon^3)}{\omega^2-\epsilon^2}. \tag{31}$$

This turns out to be of the order of 10^{-3} for a PN crystal 1-cm thick with 25 kv applied in the x direction. Even in an isotropic medium such low induced birefringence would not be particularly useful.

CONCLUSIONS

The conclusions of these analyses can be summarized as follows:

For the field in the Z direction:

1. Uniaxial crystals of the form X H_2PO_4 become biaxial.
2. The plane of the optic axes is independent of the magnitude of the voltage and is inclined at 45° to the crystallographic axis.
3. For rays parallel to the field and at a given wavelength the retardation through the plate depends only on the voltage and is independent of the plate thickness.

For a field in the X direction:

4. The crystal again becomes biaxial.
5. The plane of the axes is independent of the magnitude of the voltage and is perpendicular to the X axis.
6. The retardation change through the plate for rays parallel to the field depends on the square of the applied voltage and the inverse first power of the plate thickness.
7. If the r_{41} and r_{63} coefficients are of the same order of magnitude a 1-cm plate of ADP gives a retardation change of the order of 10^{-3} for 25 kv applied voltage.

The problem of measuring the coefficients and of determining experimentally the properties of the specific crystal ADP will be discussed in the next paper.

In conclusion the author wishes to express appreciation to Miss Elizabeth Weichel of this laboratory for her help during the preparation of this paper.

The Electro-Optic Effect in Uniaxial Crystals of the Type XH_2PO_4. II. Experimental

BRUCE H. BILLINGS
Research Laboratory, Baird Associates, Inc., Cambridge 38, Massachusetts

(Received June 22, 1949)

In this paper it is shown experimentally that when a voltage is applied to the Z direction to a basal section of ammonium dihydrogen phosphate the crystal becomes biaxial. The normal to the plate is the bisector of the axes and the plane of the axes is inclined at 45° to the crystallographic axes. The location of the axial plane is independent of the magnitude of the voltage. The retardation along the normal is directly proportional to the voltage and is independent of the thickness of the basal section. Photographs are shown of the interference pattern obtained with polarized light when voltage is applied to the crystal.

The behavior of the crystal plate with an alternating voltage is also described. Between polarizers the crystal can operate as a light valve both for short pulses as in a camera shutter or for complex sound waves as in a sound-on-film device. It is shown that with circular polarized light 74 percent sinusoidal modulation introduces only 3 percent harmonic distortion.

THE previous paper[1] discussed some theoretical aspects of the electro-optic effect in tetragonal scalenohedral crystals of the type XH_2PO_4. It was shown that the only large effects take place in a basal section with the field in the Z direction. For the field in the Z direction and for those members of the family in which a measurable retardation could be produced, the following conclusions were drawn:

(1) Uniaxial crystals of the form XH_2PO_4 become biaxial.
(2) The plane of the axes is independent of the magnitude of the voltage and is inclined at 45° to the crystallographic axis.
(3) At a given wave-length the retardation along the normal through the plate depends only on the voltage and is independent of the plate thickness.

The experimental techniques used in checking these conclusions will be discussed and the agreement between theory and experiment demonstrated in the case of the particular crystal ammonium dihydrogen phosphate[2] or ADP. This paper will also mention some of the problems connected with the use of an ADP basal section as an electro-optic shutter.

Perhaps the most important problem in connection with the use and measurements of the ADP electro-optic effect is the application of electrodes. Since the electric field must be applied parallel to the direction in which the light travels, the electrodes must be transparent. The electrode problem is made even more difficult by the physical characteristics of the crystal. Not only is the material brittle but it is water soluble and slightly hygroscopic.

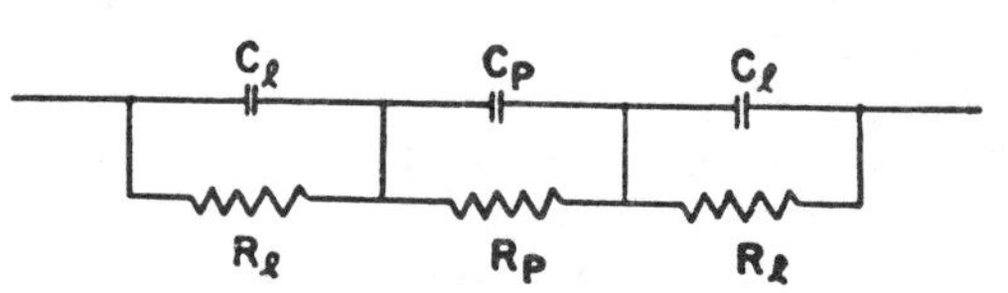

FIG. 1. Equivalent circuit of the laminated ADP crystal. C_l and C_p are the capacities of the laminating compound and the crystal. R_l and R_p are the resistance of the laminating compound and the crystal.

For the measurements which were made by Dr. Hans Jaffe beginning in 1942 at the Brush Development Company, glycerin was originally used as a conducting electrode. This had the disadvantages associated with the use of liquids and though the resulting electrodes were satisfactory electrically they were not by nature permanent.

At the Polaroid Corporation in 1943 evaporated gold was used as an electrode material. The gold layer which was used had a resistance of the order of 1000 ohms per square. Contact was made to this evaporated gold layer through an evaporated gold ring which was much heavier and had a resistance of only a few ohms per square. The disadvantage of this scheme is the loss of light at the gold surface. No matter how thin the evaporated layer, the reflection loss is still so great that more transparent electrodes would be desirable. An attempt to achieve greater transparency was made by Brian O'Brien, Jr. and Lt. Cmdr. G. A. Long at the University of Rochester in 1944. They used an evaporated grid of gold and relied on the surface conductivity of the PN to make the field uniform.

Most recent electrodes have been made using the Pittsburgh Plate Glass Nesa material. This material is a conducting layer of stannous oxide which is formed on glass. The material has no absorption in the visible part of the spectrum and for resistances of the order of 10,000 ohms per square the layers are so thin that the reflection losses are not much greater than the losses of the glass itself.

The optical quality of the Nesa is much better in thin coats than in very thick coats. One characteristic is haziness and non-uniformity that is apparent in thicknesses having resistances of a few 100 ohms/square. In the thicknesses giving 15,000 to 25,000 ohms/square this haziness disappears and the coat seems more uni-

[1] Bruce H. Billings, J. Opt. Soc. Am. 39, 797 (1949).
[2] Made by The Brush Development Company, Cleveland, Ohio, under the trademark "PN."

FIG. 2. ADP plate laminated between plates of Nesa-coated glass.

form. In both these and thick coats the glass is often severely strained. Selected pieces may be found which are strain-free. The surfaces are not flat and seem to depart from flatness by as many as 5–10 fringes in a 2-square.

The Nesa electrode introduces problems which are avoided with the use of glycerin or evaporated gold. In these other electrodes the crystal is in intimate contact with the conducting material. With the glass Nesa electrodes there is a thin film of non-conducting material between the crystal and the electrodes. If it

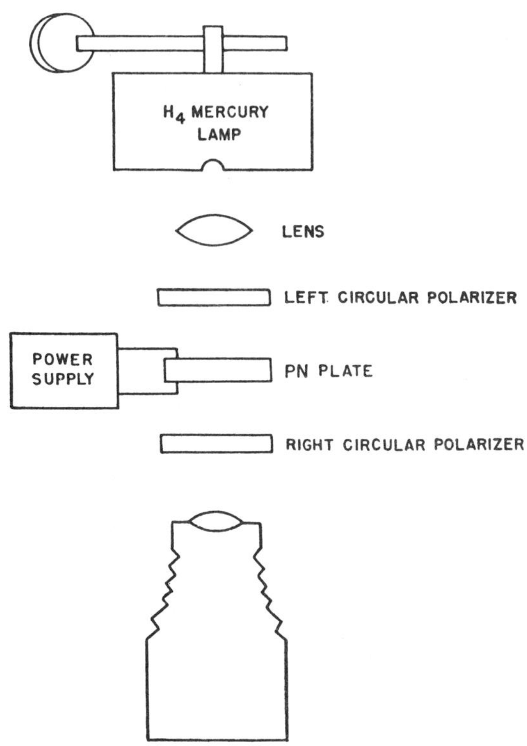

FIG. 3. Sketch of arrangement used to study the interference pattern of the ADP crystal with voltage applied.

were possible to get Nesa-coated plates which were optically flat, optical contact might be achieved between the Nesa and the crystal. Unfortunately, the method of preparing the conducting layer on the glass is of such a drastic nature that the glass is distorted. This makes optical contact out of the question.

Because of the extreme fragility of the PN crystal, an obvious technique is to fill the space between the crystal and the electrodes with a laminating compound such as Canada balsam or butyl-methacrylate. The glass electrode would then serve as a protective mount for the crystal. In this situation the voltage which actually appears across the crystal is a function of the

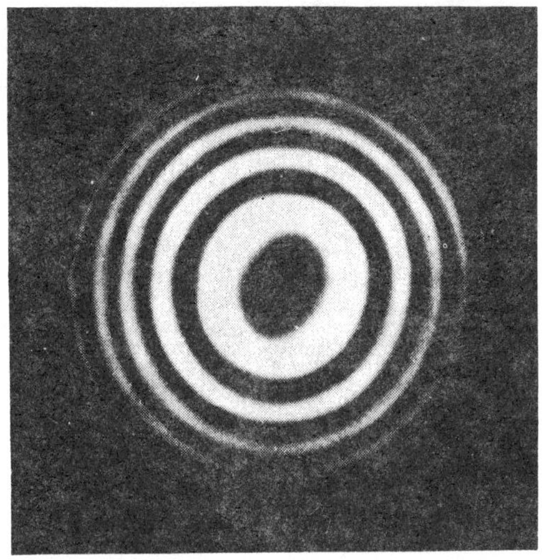

FIG. 4. Pattern observed at 546 mμ with basal section of ADP between crossed circular polarizers. Plate 0.31 cm thick.

characteristics of the laminating layer. The complete equivalent circuit of the ADP-Nesa glass sandwich is shown in Fig. 1. This circuit is applicable only at frequencies below about two megacycles. Below 10 kc the d.c. values of the constants can be used. Since in practice the Nesa plates have low resistance contacts along one edge only, the equivalent circuit at higher frequencies cannot be considered as made up of lumped elements. The new equivalent circuit will have the electrical properties of a transmission line. A detailed examination of the frequency response of the electro-optic effect in ADP and the significance of the electrodes at very high frequencies will be considered in a later paper.

At zero frequency the potential is divided according to the resistances of the ADP and laminating layers and is independent of the capacities. At a.c. frequencies the potential across the ADP can be written by inspection of the equivalent circuit and is:

$$V=\frac{V_0}{1+2(R_l/R_p)\left(\dfrac{1+j\omega R_pC_p}{1+j\omega R_lC_l}\right)}, \tag{1}$$

FIG. 5. Pattern observed at 546 mμ with basal section of ADP 0.31 cm thick between crossed circular polarizers. 9600 volts on plate.

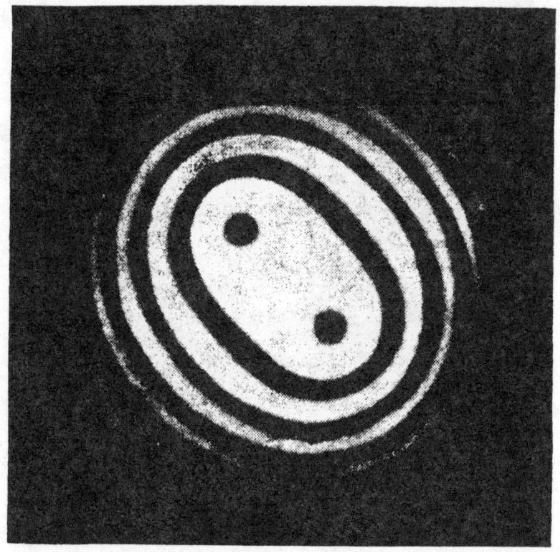

FIG. 6. Pattern observed at 546 mμ with basal section of ADP 0.31 cm thick between crossed circular polarizers. 4500 volts on plate.

for a condenser dielectric the quantity f_x, defined by

$$\frac{1}{2\pi R_x C_x}=\frac{1.80\times 10^{12}}{\rho_x K_x}=f_x, \tag{2}$$

where ρ and K are the volume resistivity (ohms-cm) and dielectric constant, respectively, is characteristic of the material and independent of the dimensions.

Equation (1) can be rewritten:

$$\left|\frac{V}{V_0}\right|^2=\frac{R_p^2(f_l^2+f^2)}{f_l^2(R_p+2R_l)^2+f^2\left(R_p+2R_l\frac{f_l}{f_p}\right)^2}. \tag{3}$$

According to this equation there are three different types of frequency response for the laminated ADP crystal.

Case I. $f_l=f_p$.

In this case the a.c. response is the same as the response to d.c. This case is unlikely to occur.

Case II. $f_l>f_p$.

In this case the response as a function of frequency drops for frequencies larger than:

$$f=f_l\frac{R_p+2R_l}{R_p+2R_l\frac{f_l}{f_p}}, \tag{4}$$

and then becomes constant for frequencies greater than $f=f_l$.

Case III. $f_l<f_p$.

This case is most likely to be realized in practice. Here the response climbs at $f>f_l$ and then becomes constant for frequencies greater than:

$$f=f_l\frac{R_p+2R_l}{R_p+2R_l\frac{f_l}{f_p}}. \tag{5}$$

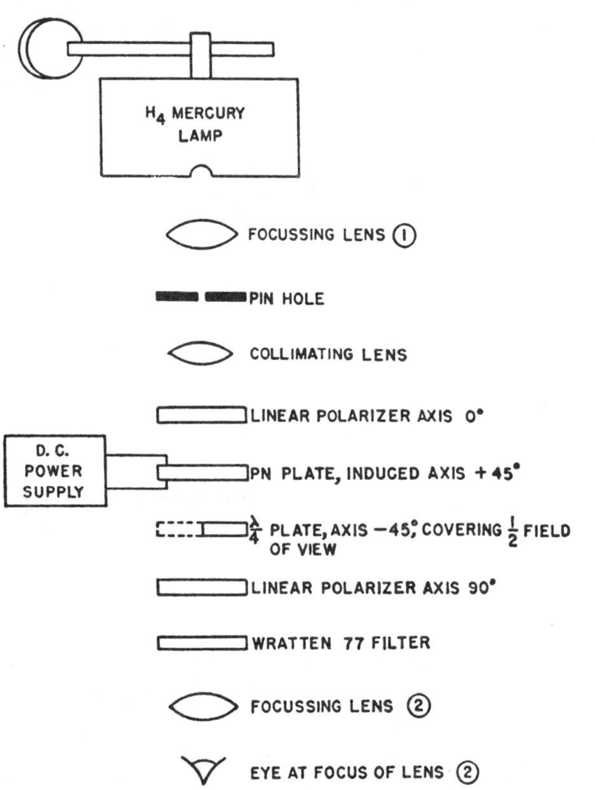

FIG. 7. Arrangement used for showing voltage required to produce the same normal retardation in basal sections of ADP of different thicknesses.

The values for ADP and for a typical laminating plastic such as plasticized polyvinyl butyral are:

$$\text{PN:} \quad f_p \sim 16 \quad \text{c.p.s.}$$
$$\text{Plastic:} \; f_l \sim \; 3.2 \; \text{c.p.s.}$$

For this particular laminating material the crystal will not respond adequately to frequencies below 3.2 cycles per second.

The d.c. response of such a lamination may be very poor. The crystal has a resistivity of 7800 megohm-cm and a typical plate will have a resistance of 300 megohms. A typical laminating layer will have a resistance of 500 megohms or higher. With such a combination only about a quarter of the applied voltage will appear across the crystal.

The improvement is so great for frequencies higher than the characteristic frequency of the lamination that for any a.c. work such a lamination is quite satisfactory.

One interesting consequence of the large crystal resistance is the slow rate of discharge if no low resistance shunt is present in the power supply. If d.c. is applied and the power supply disconnected from the crystal, the retardation remains unchanged. It is thus necessary in high speed pulse application to make provision for discharging the crystal at the end of a pulse as well as for charging it at the beginning.

Another set of problems arise because of the high voltage necessary to produce a half-wave retardation. For green light approximately 9500 volts is required to give one half-wave. The d.c. breakdown potential of a crystal 0.4 mm thick is above 25 kv/mm, but much lower values are found for a.c. potentials.[3]

For thicknesses of 0.080 in. which are used in the Baird Associates, Inc., cells, 9500 volts is well below the d.c. breakdown figure. It is necessary, however, to take special precautions to avoid breakdown around the

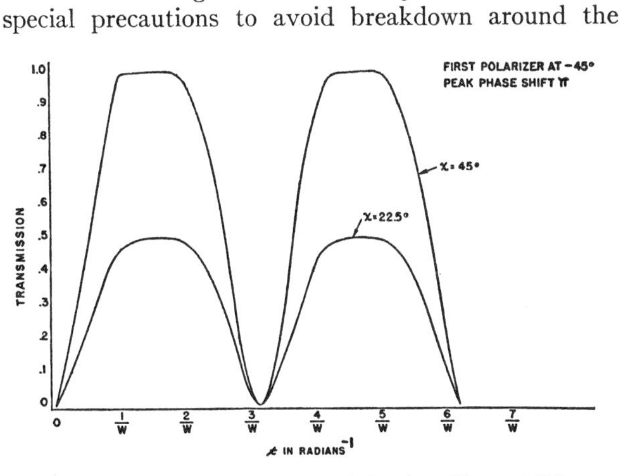

FIG. 8. ADP basal section transmission in collimated light as a function of time when a steady alternating current is applied. The crystal is between linear polarizers. One polarizer is at $-45°$ to the plane of the induced axis, the other at $+45°$ to the plane of the induced axis, the other at $+45°$ and $22\frac{1}{2}°$ to the plane of the induced axes. Voltage peak gives $\frac{3}{2}$-wave retardation.

[3] Personal communication from Dr. Hans Jaffe of The Brush Development Company, Cleveland, Ohio.

TABLE I. Voltage to produce retardation of approximately $\frac{1}{8}\lambda$ in crystals of different thicknesses.

Thickness	Voltage
0.08 in.	2000±200
0.25 in.	2000±200

TABLE II. Half-angular field to 1 percent and 0.1 percent transmission points for PN crystals of different thicknesses.

Thickness	Intensity	
	1.0%	0.1%
0.080 in.	2°20′	44′
0.250 in.	1°20′	26′

edge of the crystal. Accordingly, the laminations are made so that the crystal projects well beyond the electrodes.

To check the conclusions of the analysis of the previous paper it is necessary to observe the interference pattern when voltage is applied to the crystal. These observations can be made either visually or with photographs. In general they can be made only with d.c. on the crystal and, from the data given above, a plastic lamination has too high a resistivity. Accordingly, the low resistance laminations were made with oleic acid which was suggested by Dr. Elkan R. Blout of Polaroid Corporation.

A typical ADP plate laminated between Nesa plates is shown in Fig. 2. In Fig. 3 is a sketch of the arrangement used in studying the behavior of the crystals as voltage is applied. The retardation through the plate is described by the equation:

$$\Gamma = \frac{d}{\lambda} \frac{\omega^2 - \epsilon^2}{\epsilon^2 \omega} \sin^2 i, \tag{6}$$

where λ is the wave-length, d the crystal thickness, ω and ϵ the ordinary and extraordinary indices, i is the angle of incidence, and Γ the retardation. When the

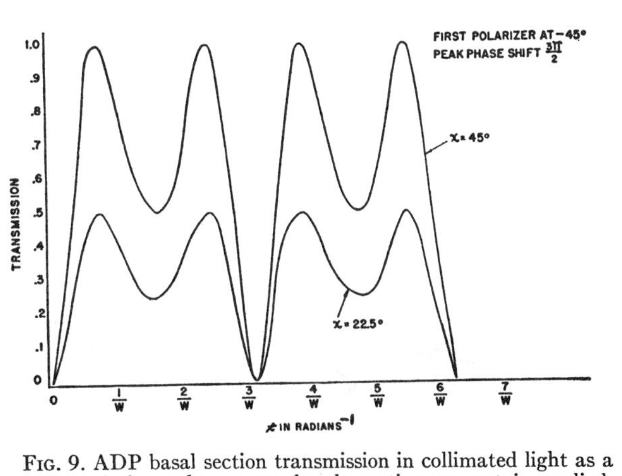

FIG. 9. ADP basal section transmission in collimated light as a function of time when a steady alternating current is applied. The crystal is between linear polarizers. One polarizer is at -45 to the plane of the induced axis, the other at $+45°$ and $22\frac{1}{2}°$ to the plane of the induced axes. Voltage peak gives $\frac{3}{2}$-waves retardation.

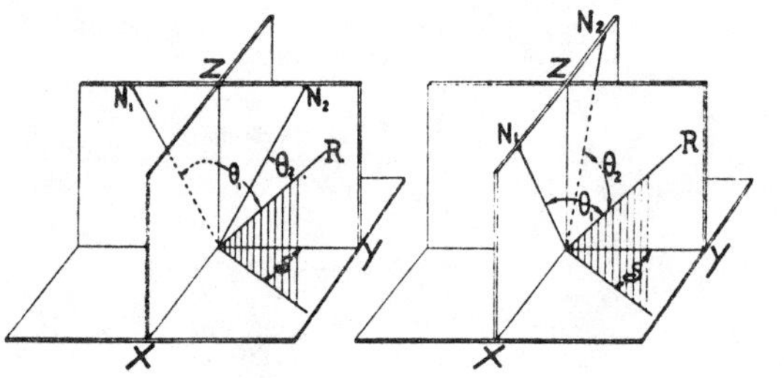

FIG. 10. Sketch showing that the angles between an oblique ray and optic axis are changed when the plane of the axis is rotated 90 degrees.

plate is placed between crossed circular polarizers it is opaque for integral values of Γ. The interference pattern thus consists of a series of rings and is shown in Fig. 4. It is typical of a basal section of any uniaxial crystal between crossed circular polarizers.

The first conclusion reached in the theoretical discussion in paper I was that the crystal would become biaxial with the axial plane at 45° to the crystallographic axis. The crystal section shown in Fig. 2 is approximately square and has edges which are parallel to the crystallographic axes. Accordingly, the biaxial interference pattern formed when voltage is applied should be at 45° to the edges of the crystal plate. This turns out to be the case. In Fig. 5 is shown the pattern obtained when 9600 volts is put across the crystal. In Fig. 6 is the pattern obtained with 4500 volts. The crystallographic axes were here kept parallel to the edge of the photograph. The optic axes remained in the same plane as the voltage was varied.

An experimental test was also made on the effect of crystal thickness on the voltage required for a given retardation.

The crystals were laminated with Nesa glass and oleic acid. The measurements were made with an arrangement devised by Mr. Sage of this laboratory. This is shown in Fig. 7. The crystal to be tested was oriented so that its induced axes would be at 45° to a linear polarizer. On the opposite side of the crystal was placed a strip of plastic of about $\frac{1}{4}$ λ-retardation at 546 mμ with its fast axes parallel to the slow axis of the crystal. Beyond the retardation sheet was placed a second polarizer at right angles to the first. Collimated light from an H4 mercury arc was passed through the system and examined visually through a lens. A Baird Associates, Inc., interference filter was placed in the system to restrict measurements to a wave-length of 546 mμ. The voltage on the crystal was then adjusted until the light passing through the retardation sheet and crystal was of the same intensity as the light which passed through the crystal alone. At this point

$$\Gamma_c = \frac{\Gamma_s}{2}, \tag{7}$$

when Γ_c is the retardation through the crystal and the retardation of the retardation sheet. With this arrangement it was not necessary to know the actual retardation of the crystal. Balance in the light beams is achieved at one value of the retardation and for this to be independent of thickness it must take place at the same voltage for crystals of different thickness. The results of the measurements are shown in Table I and it is apparent that the predicted independence of retardation and thickness is realized experimentally.

ANGULAR FIELD OF VIEW

When the crystal is used in an instrument as a light valve, there are several other characteristics that are involved in its behavior. One of the most important is the angular field of view. Since the crystal with zero applied voltage is uniaxial, the retardation is zero only for light parallel to the optic axis. For light inclined to the axis the retardation is finite and linearly polarized light is rendered sufficiently elliptical to penetrate the second polarizer. Superficially it might appear that the angular field of view under such conditions should be zero. Actually the angular field is finite as a practical limit can be set on the amount of light to be passed when the shutter is closed.

The equation which describes the light transmitted through a basal section of a uniaxial crystal between crossed circular polarizers can be written:

$$I = \frac{1}{2}\sin^2\left(\pi \frac{d}{\lambda} \frac{\omega^2 - \epsilon^2}{2\omega\epsilon^2} \sin^2 i\right). \tag{8}$$

If the light leak at the edge of the field is set at, for

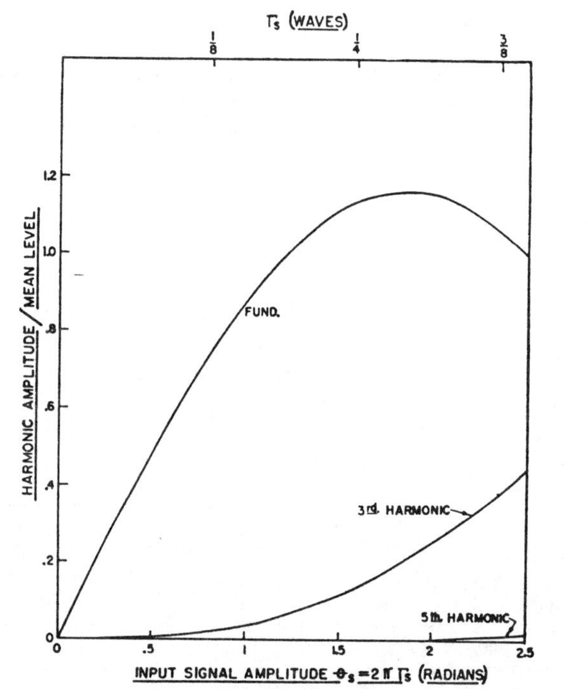

FIG. 11. Relative value of fundamental third harmonic and fifth harmonic in transmitted light when crystal is modulated at various amplitudes with a sine wave.

instance, 1 percent of the center transmission, this equation gives a value for the field of view as a function of the crystal thickness. It shows that the field of view is inversely proportional to the square root of the crystal thickness. In Table II is shown values of the field for different thickness systems with an allowance of 0.1 percent and of 1 percent light leak.

It might appear that any light leak would be intolerable especially in most photographic applications. This is not true if a simple mechanical shutter is put in series with the crystal shutter. The mechanical shutter is opened just as the crystal shutter is to be used. The tolerable light leak will then depend on the brightness of the scene and the ratio of the time of opening of the crystal shutter to the time of opening of the mechanical shutter.

The field of view of the open shutter is less critical than the field of the closed shutter. Since the whole area which is usefully closed becomes transparent when the crystal is open, no investigation needs to be made here of this case.

OPERATION WITH ALTERNATING CURRENT

We shall consider first the case in which the crystal is placed between crossed linear polarizers. The modulation voltage will be assumed to be sinusoidal. With appropriate frequency response curves it will then be possible to show the behavior of the crystal for any modulation function. The crystal can be considered as a simple retardation plate between crossed polarizers. The intensity of the light through the combination can be written:

$$I=\tfrac{1}{2}[1-\cos(\delta_0 \sin\omega t)]\sin^2(2\chi), \tag{9}$$

where χ is the angle between induced crystal axes and the first polarizer, ω is the angular frequency of the a.c. voltage, and δ_0 is the retardation of the crystal at peak voltage. A plot of this equation as a function of the time t is shown in Fig. 8. This is plotted with a value of for δ_0. If δ_0 is more than π, the peaks are split. A plot of this splitting is shown in Fig. 9. This figure is drawn for the case then the peak phase shift is $3\pi/2$.

If the light goes through the crystal at other than normal incidence, the pattern is more complex. When the field changes sign the optic axes open in a plane at 90° to the original direction. The retardation depends on the angles between the ray direction and the two optic axes. These angles will be different when the plane of the axes is shifted 90°. The relation between the ray and the axes is shown in Fig. 10. When the crystal is placed between crossed polarizers in an alternating field, the intensity will thus be different for the two parts of the cycle.

APPLICATIONS

From these various experiments both with direct current and with alternating current it is possible to see many of the possible applications of the electro-optic crystal. The crystals can be used as a light valve for putting sound on film. This has been discussed by Gotschall.[4]

The device has several advantages over the conventional sound on film light valve. It is easily adaptable to portable operations and is essentially non-microphonic.

One problem in connection with this application is the question of the linearity of the crystal. The change in intensity of the light which passes through the PN must follow exactly the change in voltage on the crystal. The analysis below was carried out by Mr. R. O'B. Carpenter of this laboratory. The intensity along the normal to the crystal between parallel polarizers can be written:

$$I/I_0=\tfrac{1}{2}\cos^2\pi\Gamma, \tag{10}$$

where Γ is the retardation of the crystal. This retardation is induced by the voltage. In addition to this electro-optic retardation, a retardation can be produced by putting in series with the crystal a fixed retardation sheet.

Consider that such a fixed plate of retardation Γ_0 is in series with the crystal and that a sinusoidal voltage is applied to the PN. In this case

$$I/I_0=\tfrac{1}{4}[1+\cos 2\pi(\Gamma_0+\Gamma_s \sin\omega t)], \tag{11}$$

where Γ_s is the peak retardation produced in the modulation elements by the applied voltage and ω is the angular frequency of this voltage. Equation (11) can be expanded to

$$I/I_0=\tfrac{1}{4}[1+\cos\theta_0 \cos(\theta_s \sin\omega t)-\sin\theta_0 \sin(\theta_s \sin\omega t)], \tag{12}$$

where $\theta=2\pi\Gamma$.

This can be rewritten in terms of Bessel functions as

$$\begin{aligned}I/I_0=\tfrac{1}{4}[1&+\cos\theta_0(J_0(\theta_s)+J_2(\theta_s)\cos 2\omega t\\&+J_4(\theta_s)\cos 4\omega t+\cdots)\\&-\sin\theta_0(J_1(\theta_s)\sin\omega t+J_3(\theta_s)\sin 3\omega t+\cdots].\end{aligned} \tag{13}$$

This gives the amplitudes of the various harmonics as a function of the operating point and signal amplitude.

At $\theta_0=\pi/2(\Gamma_0=\tfrac{1}{4}$ wave) all even harmonics disappear. At $\theta_0=0$ all odd harmonics including the fundamentals disappear. In Fig. 11 is plotted the amplitude of the various harmonics relative to maximum transmission for the proper operating point $\theta_0=\pi/2$. With a maximum signal of $\tfrac{1}{8}$ wave the percentage modulation is 74 percent and the third harmonic is less than 3 percent of the fundamental—all other harmonics being negligible.

The crystal also can be used as a high speed camera shutter.[5] This application is restricted by the narrow angular field of view. For very short exposure times film speeds are such that only the illumination of

[4] G. D. Gotschall, J. Soc. Mot. Pict. Eng. **51**, 13 (1948).
[5] Roy Gunter, J. Opt. Soc. Am. **38**, 419 (1948).

special flash discharge tubes is sufficiently intense. A high speed shutter might appear unnecessary. Professor Edgerton has pointed out, however, that the light pulse from these lamps has a tail which is troublesome for certain problems. A crystal shutter could effectively remove this tail.

Besides being used as a high speed shutter the crystal appears useful in the lower speed range because of the precision with which the exposure can be set.

In the tunable birefringent filter[6] this crystal is one of the most promising tuning elements and makes possible extremely rapid shifts of the filter passband.

In conclusion, the author wishes to express appreciation to Mr. Stanley J. Sage, Mr. Edgar E. Barr, and Mr. George Economou of this laboratory for their help during the preparation of this paper.

---

[6] B. H. Billings, J. Opt. Soc. Am. **37**, 738–746 (1947).

United States Patent Office

2,768,557
Patented Oct. 30, 1956

1

2,768,557

UNIAXIAL CRYSTAL ELECTRIC LIGHT VALVE COMPENSATED FOR DIVERGENT LIGHT

Walter L. Bond, New Providence, N. J., assignor to Bell Telephone Laboratories, Incorporated, New York, N. Y., a corporation of New York

Application October 6, 1952, Serial No. 313,336

2 Claims. (Cl. 88—61)

This invention relates in general to electro-optic apparatus and more particularly to light valves comprising birefringent electro-optically active crystalline materials.

A ray of light from a small dimensional source passing through a basal cut uniaxial birefringent crystal positioned between crossed polarizers, produces a concentric ring pattern on an interposed screen, such that the rings are crossed by a pair of dark radial bands. If the crystal is electro-optically active, as is for example ammonium dihydrogen phosphate, the application of an electric field along the optic axis, in the direction of crystal thickness, distorts the pattern, changing the circles into ellipses, and the dark cross into a pair of hyperbolae slightly separated at the bases. Thus, a small aperture in the center of the screen passes no light in the absence of the electric field, but does pass light when the field is applied displacing the dark cross. An electro-optical light valve operating on the principle described above is disclosed in W. P. Mason Patent 2,467,325, April 12, 1949.

The useful aperture size in such a light valve is limited by the size of the first ring of the pattern cast on the screen. Although this ring size can be increased by reducing the thickness of the crystal, such a procedure may be disadvantageous, since a certain minimum thickness (of the order of a quarter of an inch in amomnium dihydrogen phosphate) is required to sustain the applied voltage without fracture.

Accordingly, a general object of the present invention is to provide improved characteristics in a light valve of the type described. More particular objects of the invention are to provide in a light valve of the type described a significant increase in the allowable divergence of the incident polarized light beam and a decrease in the chromatic dispension thereof.

A ray of light passing through a basal cut uniaxial crystal breaks up into a radially vibrating component and a tangentially vibrating component which travel with different velocities. On emerging from the crystal these components recombine. Assuming that the crystal is positioned between a crossed polarizer and analyzer, if the vector sum of the component rays lies along the extinction direction of the analyzer, the intensity upon the screen is zero.

The tangentially vibrating component is known as the ordinary ray, designated ω, below its velocity of propagation is the same for all directions of propagation. The radially vibrating ray is known as the extraordinary ray, designated ϵ, because its velocity of propagation varies with the dierction of propagation, being equal to the ordinary velocity for rays traveling along the optic axes.

Unilateral birefringent crystals are divided into two classes, designated optically positive and optically negative. For the positive uniaxial crystals, the ordinary velocity is greater than the extraordinary velocity, whereas for negative uniaxial crystals, the reverse is true.

In accordance with the present invention the divergence and chromatic aberration in a light valve, of the general type described in United States Patent 2,467,325, supra, comprising an optically active birefringent Z-cut crystal interposed between crossed-polarizers with its major faces disposed transversely to the intercepted beam, is compensated by interposing in tandem with the initial crystal in the path of the transmitted beam, one or more additional crystals of opposite optical sign and proper thickness. By this device, the first ring falling on the screen can be made very large; and hence the allowable divergence of the incident polarized beam is markedly increased. If, for example, the first, electro-optic crystal is ammonium dihydrogen phosphate, which is an optically negative uniaxial crystal, a positive uniaxial crystal having a thickness computed in accordance with the techniques described hereinafter should be chosen to compensate it. Such a compensating crystal should be clear and colorless, and should not rotate the plane of polarization. Its refractive indices should not differ too greatly from those of ammonium dihydrogen phosphate.

For the purposes of the present exposition, zinc fluosilicate ($ZnSiF_6 \cdot 6H_2O$) has been chosen because, although its indices are not quite the same as those of ammonium dihydrogen phosphate, it is one of the few positive uniaxial crystals for which the literature gives refractive indices for more than one color of light.

Since the source is not monochromatic to compensate for chromatic aberration as will be described hereinafter, an additional crystalline component comprising potassium periodate (KIO_4) is added.

In one embodiment of the invention, rectangular Z-cut crystals of ammonium dihydrogen phosphate, potassium periodate, and zinc fluosilicate, each having its major surfaces perpendicular to its Z axis and having respective thicknesses in a Z direction of 3.5 millimeters, 1.947 and 7.6 millimeters, are placed together in tandem. A pair of electrode films, transparent to permit the passage of light, are evaporated onto opposing X—Y surfaces of the ammonium dihydrogen phosphate crystal and connected to an electrical source for applying a controlling field.

It will be apparent from the discussion hereinafter that a light valve comprising the composite crystal unit indicated has superior characteristics, both as to the allowable divergence of the incident polarized beam, and the decrease in chromatic dispersion. Such characteristics may make possible the application of the light valve of the present invention to various types of switching operations.

Other objects, advantages, and features of the present invention will be apparent from a study of the specification hereinafter and the attached drawing, in which

Fig. 1 shows an electro-optical light valve of the general type disclosed in W. P. Mason Patent 2,467,325;

Fig. 2 shows a variation in the pattern appearing on the screen included in the light valve of Fig. 1 of the drawings;

Fig. 3 is a cross-sectional diagram for definition of terms in the formulae set forth in describing the present invention;

Fig. 4 is a plot for sodium light of phase-divergence curves I, II, and III for compensators including respectively ammonium dihydrogen phosphate alone, zinc fluosilicate crystal alone, and a composite of the two using sodium light;

Fig. 5 shows a series of three curves C, D, and F for different spectral lines indicating the color response of a uniaxial crystal of ammonium dihydrogen phosphate compensated by a crystal of zinc fluosilicate;

Fig. 6 shows a series of three curves for the same spectral lines as Fig. 5, using a uniaxial crystal of ammonium dihydrogen phosphate compensated with a crystal of potassium periodate;

Fig. 7 shows a series of three curves using the same spectral lines as indicated in Figs. 5 and 6, using a triple-element crystal compensator including uniaxial crystals of ammonium dihydrogen phosphate, zinc fluosilicate, and potassium periodate;

Fig. 8 shows the proportions of the compensated crystal element including ammonium dihydrogen phospate, potassium periodate, and zinc fluosilicate, the thicknesses of which are computed in accordance with the formulae disclosed;

Fig. 9 shows a series of three curves for different spectral lines treated above showing the divergence and color response of a compensator in accordance with Fig. 8;

Figs. 10, 11, and 12 are showings of the crystalline habits respectively of ammoninum dihydrogen phosphate, zinc fluosilicate, and potassium periodate; and

Fig. 13 is a showing in cross section of a divergence and color compensated light valve assemblage in accordance with the present invention.

The electro-optical light valve of Fig. 1 is of the general type indicated in W. P. Mason Patent 2,467,325.

Referring in detail to Fig. 1, a beam of light is supplied from a source **1**. Aligned transversely in the path of the beam from the source **1** are a polarizer **2** and an analyzer **3**, which may be of any of the forms well known in the art, such as Nicol prisms, or plates of material known commercially as "Polaroid." The polarizer **2** and analyzer **3** are set for extinction of the beam passing through their thicknesses. A screen **4** is coaxially aligned with the plates **2** and **3**, in a parallel plane beyond the analyzer. Interposed betwen plates **2** and **3** is a uniaxial birefringent crystal such as, for example, ammonium dihydrogen phosphate, the major surfaces of which are substantially parallel to each other and normal to the Z axis.

Elecertodes **7** and **8**, which serve to apply an electrical stress from the alternating-current source **6**, are adhered to opposing major surfaces of the crystal **5**. These electrodes **7** and **8** are designed to permit the passage of light, and hence, may assume the form of transparent metallic films, e. g. gold 50 to 100 angstroms thick, evaporated on the crystal surfaces. The process for preparing such evaporated metallic electrodes is described in detail in Chapter IV of "Procedures in Experimental Physics" by John Strong, Prentice Hall, 1939. Alternatively the electrodes may assume the form shown in Fig. 5 of Mason Patent 2,467,325, supra, comprising a pair of grids of conductive material such as metallic mesh, or wires, or a metallic paint secured on the major faces of the crystal element **5**. Electrodes **7** and **8** are connected by lead wires to the electrical control source **6**.

In the absence of an applied electrical field, a beam of light from the source **1** passing through the crossed polarizing plates **2** and **3** along the optic axis of the ammonium dihydrogen phosphate crystal **5** produces on the screen **4** a pattern of concentric interference rings **4***a* of the type well known in the art, superposed on which is an additional pattern in the nature of a dark cross **4***b* concentric with the pattern of rings. When a field is applied from the electrical source **6** along the optic axis of the crystal **5**, the ring pattern **4***a* is changed in the manner shown in Fig. 2 of the drawings, becoming elliptical, and the arms of the dark cross **4***b* are separated forming a pair of hyperbolae spaced apart by a bright spot.

As pointed out in the introductory paragraphs of the specification, a light valve providing a greater angle of divergence for the incident beam would be possible if the size of the rings were increased. In accordance with the present invention, this is accomplished by placing in tandem with the uniaxial birefringent crystal **5**, which in the present illustrative example is ammonium dihydrogen phosphate, an electro-optically negative crystal, and one or more compensating electro-optically positive crystals. Since zinc fluosilicate ($ZnSiF_6 \cdot 6H_2O$) is one of the few positive uniaxial crystals for which the literature gives values of the refractive indices for more than one color of light, it has been found to be a useful material for the purposes of the present invention.

For ammonium dihydrogen phosphate the International Critical Tables, volume VII, page 19 gives the following values:

λ	ω	ϵ
.486 F	1.5314	1.4847
.589 D	1.5246	1.4792
.656 C	1.5212	1.4768

where λ is the wave length in microns, ω is the index of refraction for the ordinary ray, ϵ is the index of refraction for the extraordinary ray and the letters F, D, and C are well-known Fraunhofer designations. On the other hand, for zinc fluosilicate, P. Groth's Chemische Kristallographie, Engelmann, 1906, volume I, page 599 gives the following values:

λ	ω	ϵ
.486 F	1.3860	1.3992
.589 D	1.3824	1.3956
.656 C	1.3808	1.3938

An approximate theory can be built around M. Bertin's formula (Preston, Theory of Light, 4th edition, The MacMillan Co., 1912, pages 408 and 410). Consider a uniaxial crystal illustrated diagrammatically in enlarged cross secction in Fig. 3 of the drawing. If a ray with an angle of incidence i is refracted so that it passes at angle θ to the optic axis, which is perpendicular to the major faces of the crystal, the phase difference of the two components upon emerging is:

$$\varphi=\frac{Bl}{\lambda} \tag{1}$$

where B is the difference of refractive indices for this direction of propagation (i. e. the birefringence), l is the path length and λ is the wave length in air, φ being the phase difference in waves. Obviously

$$l=\frac{t}{\cos\theta}$$

Bertin's formula states that

$$B\doteq B_0 \sin^2\theta \tag{2}$$

where

$B_0=\epsilon-\omega$ (for a beam directed along the optic axis)

whence

$$\varphi\doteq\frac{B_0 t \sin^2\theta}{\lambda\cos\theta}$$

But

$$\omega\sin\theta=\sin i \tag{3}$$

whence

$$\varphi\doteq\frac{B_0 t}{\lambda\omega}\frac{\sin i}{\sqrt{\omega^2-\sin^2 i}} \tag{4}$$

For ammonium dihydrogen phosphate in sodium light (the Fraunhofr D line), $B_0=1.4792-1.5246=-.0454$, $\omega=1.5246$, $\lambda=0.5893$ microns. Hence, if the thickness is 3530 microns (3.53 millimeters) the phase difference becomes

$$\varphi=\frac{-178.3\sin^2 i}{\sqrt{2.324-\sin^2 i}}$$

Evaluating this for angles of incidence $i=2, 4, 6 \ldots 16$ degrees, curve I of Fig. 4 is obtained.

Let us now consider an ammonium dihydrogen phosphate crystal or thickness t_a, ordinary ray refractive index

ω_a, birefringence $\epsilon_a-\omega_a=B^a{}_0$; and also a zinc fluosilicate crystal of thickness t_z, ordinary index ω_z birefringence $\epsilon_z-\omega_z=B_0{}^z$. The resultant phase difference of the combination is the sum of the phase differences of the components, that is:

$$\Phi=\varphi_a+\varphi_z \qquad (5)$$

or

$$\Phi \doteq \frac{B_0{}^a t_a \sin^2 i}{\lambda\omega_a\sqrt{\omega_a{}^2-\sin^2 i}}+\frac{B_0{}^z t_z \sin^2 i}{\lambda\omega_z\sqrt{\omega_z{}^2-\sin^2 i}} \qquad (6)$$

which is zero if

$$\frac{B_0{}^a t_a}{\omega_a\sqrt{\omega_a{}^2-\sin^2 i}}+\frac{B_0{}^z t_z}{\omega_z\sqrt{\omega_z{}^2-\sin^2 i}}=0 \qquad (7)$$

For small angles of incidence i, Equation 7 is a good approximation if the thickness ratio is

$$\frac{t_a}{t_z}=-\frac{B_0{}^z\omega_a{}^2}{B_0{}^a\omega_z{}^2} \qquad (8)$$

For ammonium dihydrogen phosphate and zinc fluosilicate in sodium light, Equation 8 gives

$$\frac{t_a}{t_z}=\frac{-.0132\times 1.5246^2}{-.0454\times 1.3824^2}=0.353$$

Whence if $t_z=10$ millimeters, then t_a should be 3.53 millimeters, the value chosen for curve I, Fig. 4. By using the approximate formula, Equation 6, the approximate relationship between φ and i can now be computed. Substituting these values in Equation 6:

$$\Phi=\left(\frac{-.0454\times 3530}{1.525\sqrt{2.323-\sin^2 i}}+\frac{.0132\times 10000}{1.382\sqrt{1.911-\sin^2 i}}\right)\frac{\sin^2 i}{0.5893}$$

$$\Phi=\left(\frac{-178.3}{\sqrt{2.323-\sin^2 i}}+\frac{162.0}{\sqrt{1.911-\sin^2 i}}\right)\sin^2 i$$

This equation is represented by curve II, Fig. 4. It is apparent that while the first ring of the uncompensated crystal comes at about 5.2 degrees from the axis, the compensated one comes at about 36 degrees. The phases are reversed in the two curves; that is, in the uncompensated curve there is one more ordinary wave length in the crystal than extraordinary wave length at $i=5.2$ degrees; while at $i=36$ degrees in the compensated crystal, there is one more extraordinary wave length than ordinary wave length.

Equation 6 can be solved for $\Phi=0$ at some arbitrary value of i, to make Φ smaller over a selected range. For example, if $\Phi=0$ at $i=20$ degrees, the curve will fall very slightly below zero at $i=10$ degrees, rise to zero at $i=20$ degrees, then rise more steeply than curve II. This curve has a very small Φ values out to $i=30$ degrees.

Inasmuch as curve II is only valid for a single frequency, the next step is to investigate the changes in compensation due to a change of wave length of the light. Using the crystal data for the Fraunhofer F line, $\lambda=0.486\mu$, Φ values are computed as before, to obtain curve III, Fig. 4. Hence, it is apparent that the divergence compensation does depend on color. To achromatize the divergence compensator, the Bertin's approximate equation is dropped, and a more accurate equation, which is an approximation superior to Bertin's equation is adopted. This follows:

$$\Phi=\frac{\omega t}{\lambda}\left(\sqrt{1-\frac{\sin^2 i}{\epsilon^2}}-\sqrt{1-\frac{\sin^2 i}{\omega^2}}\right)=\frac{(\epsilon-\omega)(\epsilon+\omega)t\sin^2 i}{\lambda\left(\sqrt{1-\frac{\sin^2 i}{\omega^2}}+\sqrt{1-\frac{\sin^2 i}{\epsilon^2}}\right)\epsilon^2\omega} \qquad (9)$$

Let

$$f=\frac{(\epsilon-\omega)(\epsilon+\omega)}{2\lambda\omega\epsilon^2} \text{ and } \delta=\tfrac{1}{4}\left(\frac{1}{\omega^2}+\frac{1}{\epsilon^2}\right) \qquad (10)$$

then

$$\varphi=ft\sin^2 i+ft\delta\sin^4 i+\ldots \qquad (11)$$

where f and δ, for any given material, depends on ω, ϵ, and λ. These constants are computed for three crystals below.

Crystal		λ (microns)	ω	ε	f	δ
$NH_4H_2PO_4$						
Ammonium dihydrogen phosphate (tetragonal, colorless indices 1.525 and 1.479).	F	.4861	1.5314	1.4847	−.04293	.220
	D	.5893	1.5246	1.4792	−.03469	.222
	C	.6563	1.5212	1.4768	−.03058	.223
$ZnSiF_6\cdot 6H_2O$						
Zinc fluosilicate (Hexagonal prisms, colorless).	F	.4861	1.3860	1.3992	.013951	.258
	D	.5893	1.3824	1.3956	.011556	.259
	C	.6563	1.3808	1.3938	.010251	.260
KIO_4						
Potassium periodate (tetragonal, colorless).	F	.4861	1.6346	1.6651	.022850	.184
	D	.5893	1.6205	1.6479	.017272	.188
	C	.6563	1.6151	1.6416	.015114	.189

(The KIO_4 data is from the International Critical Tables, volume VII, page 27.)

In the foregoing table, λ represents wavelength in microns, ω and ϵ, the indices of refraction for the ordinary and extraordinary rays respectively, and f and δ are defined in Equations 10.

Whence, to compensate ammonium dihydrogen phosphate by means of $ZnSiF_6\cdot 6H_2O$, for small angles of incidence i, for light corresponding to the sodium D line, one may use the equation:

$$f_a t_a+f_z t_z=0$$

or

$$-.03469t_a+.011556t_z=0$$

so that

$$t_z=\frac{.03469}{.011556}t_a$$

or

$$t_z=3.0t_a$$

This value differs somewhat from that derived from Bertin's formula, being more accurate than the old value. Assume that $t_a=3500$ microns, $t_z=10500$ microns.

Whence one has now:

$$\varphi_D=(-.03469\times 3500\times .011556\times 10500)\sin^2 i+(-.03469\times 3500\times .222+.011556\times 10500\times .259)\sin^4 i$$

Whence

$$\varphi_D=.08\sin^2 i+4.37\sin^4 i$$

which is plotted as curve D, Fig. 6.

For the F line, one has

$$\varphi_F=(-.04293\times 3500+.013951\times 10500)\sin^2 i+(-.04293\times 3500\times .220+.013951\times 10500\times .358)\sin^4 i$$

or

$$\varphi_F=-3.77\sin^2 i+19.38\sin^4 i$$

This is curve F, Fig. 5 of the drawing. Curve C is drawn similarly for the C line.

If the indices of the compensating crystal had been higher than those of ammonium dihydrogen phosphate instead of lower, the F curve might have been above the C curve instead of below. For example, for KIO_4, for the D line to be compensated for small angles of incidence, one finds:

$$-.03469t_a+.017272t_k=0$$

or

$$t_k=\frac{.03469}{.017272}t_a=2.01t_a$$

Let $t_a=3500$ microns, then $t_k=7030$ microns.

Whence

$$D\varphi=(-.03469\times3500\times.222+.017272\times7030\times.188Q\sin^4 i$$

$$\varphi_D=-4.1\sin^4 i$$

which shows the reversal, as φ_D is negative, whereas φ_D was positive for the previous case. For the F line one has:

$$\varphi_F=(-.04293\times3500+.022850\times7030)\sin^2 i+(-.04293\times3500\times.220+.022850\times7030\times.184)\sin^4 i$$

$$\varphi_F=10.7\sin^2 i-3.5\sin^4 i$$

which is plotted as curve F, Fig. 6.

At angle of incidence, $i=20°$, the F value for this last pair is 3.1 times as large as for the first combination, but opposite in sign. Hence, if the two thicknesses are divided by 3.1 and added to the first pair, the composite gives good compensation at $i=20°$. This gives

$$\left(3500+\frac{3500}{3.1}\right)=$$

4630 microns of ammonium dihydrogen phosphate, 10500 microns of $ZnSiF_6{\cdot}6H_2O$

and

$$\frac{7030}{3.1}=2270 \text{ microns of } KIO_4$$

To keep the electric breakdown conditions comparable with previous conditions, each thickness should be multiplied by

$$\frac{3500}{4630}=0.755$$

to give 3500 microns of ammonium dihydrogen phosphate, 7930 microns of $ZnSiF_6.6H_2O$, and 1715 microns of KIO_4. For this triple system, the phase relations are:

$$\Phi_D=(-.03469\times3500+.011556\times7930+.017272\times1715)\sin^2 i+(-.03469\times3500\times.222+.011556\times7930\times.259+.017272\times1715\times.188)\sin^4 i$$

or

$$D\varphi=-.155\sin^2 i+2.350\sin^4 i$$

$$\varphi_F=(-.04293\times3500+.013951\times7930+.02285\times1715)\sin^2 i+(-.04293\times3500\times.220+.013951\times7930\times.258+.02285\times1715\times.184)\sin^4 i$$

$$\varphi_F=-.446\sin^2 i+2.685\sin^4 i$$

$$\varphi_C=(-.03058\times3500+.010251\times7930\times.015114\times1715)\sin^2 i+(-.03058\times3500\times.223+.010251\times7930\times.260+.015114\times1715\times.189)\sin^4 i$$

$$\varphi_C=-0.0\sin^2 i+2.1\sin^4 i$$

These equations are respectively plotted for progressively increasing angles of incidence as curves D, F, and S in Fig. 7. It is seen that the compensation is so good in the major part of the visible spectrum, that the errors do not exceed a quarter wave unless the angle of incidence i is greater than 35 degrees.

Another more accurate method for computing the relative thicknesses of the crystal plates to be added in tandem as follows: If plates of ammonium dihydrogen phosphate, zinc fluosilicate, and potassium periodate of thicknesses t_a, t_z, and t_k respectively are placed in tandem, the phase difference between emergent ordinary and extraordinary rays for the C, D, and F Fraunhofer lines of light is represented by substitutions in Equation 11 in the manner carried out in previous computations:

$$\begin{aligned}\Phi_C=&-.03058t_a\sin^2 i-.03058\times.223t_a\sin^4 i\\&+.010251t_z\sin^2 i+.010251\times.260t_z\sin^4 i\\&+.015114t_k\sin^2 i+.015114\times.189t_k\sin^4 i\\\Phi_D=&-.03469t_a\sin^2 i-.03469\times.222t_a\sin^4 i\\&+.011556t_z\sin^2 i+.011556\times.259t_z\sin^4 i\\&+.017272t_k\sin^2 i+.017272\times.188t_k\sin^4 i\\\Phi_F=&-.04293t_a\sin^2 i-.04293\times.220t_a\sin^4 i\\&+.013951t_z\sin^2 i+.013951\times.258t_z\sin^4 i\\&+.02285t_k\sin^2 i+.02285\times.184t_k\sin^4 i\end{aligned}$$

If the D line is compensated for small values of i, $\sin^4 i$ in the Φ_D equation may be neglected, and the remaining term set equal to zero.

$$\Phi_D=(-.03469t_a+.01156t_z+.017272t_k\sin^2 i=0$$

whence

$$-.03469t_a+.011556t_z+.017272t_k=0$$

Now, assume $\Phi_F=\Phi_C$, for $i=20°$. This gives the equation:

$$-.001481t_a+.000445t_z-.0009234t_k=0 \qquad (12)$$

Solving Equations 2 and 3 for t_z and t_k in terms of t_a gives

$$t_z=2.171t_a, \quad t_k=0.556t_a$$

whence if $t_a=3500$ microns, then $t_z=7600$ microns, and $t_k=1947$ microns. These values, placed in the equations for Φ_C, Φ_D and Φ_F, show the degree of compensation of Φ for both in angle and color:

$$\Phi_C=0.30\sin^2 i+1.95\sin^4 i$$
$$\Phi_D=2.11\sin^4 i$$
$$\Phi_F=0.24\sin^2 i+2.48\sin^4 i$$

These equations are respectively plotted for progressively increasing angles of incidence as curves C, D and F in Fig. 9 of the drawings. It is seen that the phase is wrong by a quarter wave, only if i exceeds 30 degrees, and that this holds for all colors from blue ($\lambda=.4861$ microns) to red ($\lambda=.6563$ microns).

The compensation for dispersion and chromatic aberration achieved in the manner described hereinbefore could probably be achieved with a single compensator, instead of two, were there available in the art a crystal having the same δ value as ammonium dihydrogen phosphate, uniaxial, colorless, and without rotation. However, since such a crystal is not at present available, a composite compensator must be employed for providing the wide-angle, color-corrected light valve described.

For preparing such a compensator, crystal plates *5a*, *5c*, and *5b* as indicated in Fig. 8, of the requisite thicknesses, are cut from mother crystals of ammonium dihydrogen phosphate, zinc fluosilicate and potassium periodate, of the forms despectively indicated in Figs. 10, 11, and 12 of the drawings. The ammonium dihydrogen phosphate and potassium periodate crystallize in tetragonal form, whereas the sodium fluosilicate crystallizes in hexagonal form, each with the Z axis running the length of the crystal. Each of the plates, *5a*, *5b*, and *5c* is a so-called Z-cut crystal, the major faces being normal to the Z (optic) axis of the crystal. The two major surfaces of each of the crystals are made parallel and are polished in a manner well known in the art.

For example, if the dimensions are chosen in accordance with the method of computations last described, Z-cut crystal elements are cut with the following thickness dimensions:

5a—ammonium dihydrogen phosphate—3.5 millimeters.
5b—potassium periodate—1.947 millimeters.
5c—zine fluosilicate—7.6 millimeters.

These three elements **5***a*, **5***b*, and **5***c*, are assembled with the elements previously described with reference to Fig. 1 of the drawings to form a divergence-color compensated light valve, which is indicated in cross section in Fig. 13 of the drawings. This comprises the source **1'** which is focused centrally on an opening in a baffle **15**. A lens **16** centrally located on the other side of the baffle **15**, receives beam passing through the opening and serves to focus the beam on the optical assembly including the crystal plates **5***a*, **5***b*, and **5***c*. These need not be contiguous, but may be spaced at random distances between the polarizing plates **2'** and **3'**, the ammonium dihydrogen phosphate crystal **5***a* being located closest to polarizing plate **2'**. The surfaces of the ammonium dihydrogen phosphate crystal are coated with transparent films **7'** and **8'** of gold, similar to the electrode films **7** and **8** described with reference to Fig. 1 of the drawings. As in Fig. 1, the electrodes **7'** and **8'** are connected to the electrical control source **6'**. The beam emerging from the analyzer **3'** is collimated by means of the lens **17** which focuses it on a utility surface **4'** which may, for example, be part of a photoelectric circuit.

It is to be understood that it is the relative, and not the absolute thicknesses of plates **5***a*, **5***b*, and **5***c* which is of critical interest in the embodiments of the present invention.

What is claimed is:

1. A divergence and color compensated light valve comprising a plurality of crystalline elements respectively including ammonium dihydrogen phosphate, zinc fluosilicate, and potassium periodate positioned in tandem, each of said crystalline elements having two substantially plane faces substantially perpendicular to a common optic axis, wherein the ratio between the respective thicknesses of the elements of zinc fluosilicate and ammonium dihydrogen phosphate approximates 2.17, and the ratio between the respective thicknesses of the elements of potassium periodate and ammonium dihydrogen phosphate approximates 0.56.

2. A divergence and color compensated light valve comprising in combination a plurality of crystalline elements including ammonium dihydrogen phosphate, zinc fluosilicate, and potassium periodate arranged in a tandem group, each of said elements cut with its major dimensions substantially perpendicular to a common optic axis, wherein the ratio between the thicknesses of zinc fluosilicate and ammonium dihydrogen phosphate approximates 2.17 and the ratio between the thicknesses of potassium periodate and ammonium dihydrogen phosphate approximates 0.55, translucent conducting electrodes attached to opposing surfaces of said ammonium dihydrogen phosphate element for impressing a controlling electrical field thereacross, polarizing elements arranged at opposite ends of said group, a light source, a collimating lens directing light from said source through a first one of said polarizing elements in the direction of the optic axis common to said crystalline elements, and means for utilizing light emerging from said crystal elements through the second one of said polarizing elements.

References Cited in the file of this patent

UNITED STATES PATENTS

2,591,701	Jaffe	Apr. 8, 1952
2,607,272	Bond	Aug. 19, 1952
2,705,903	Marshall	Apr. 12, 1955

Oct. 30, 1956 W. L. BOND 2,768,557

UNIAXIAL CRYSTAL ELECTRIC LIGHT VALVE COMPENSATED FOR DIVERGENT LIGHT

Filed Oct. 6, 1952 3 Sheets-Sheet 1

FIG. 1

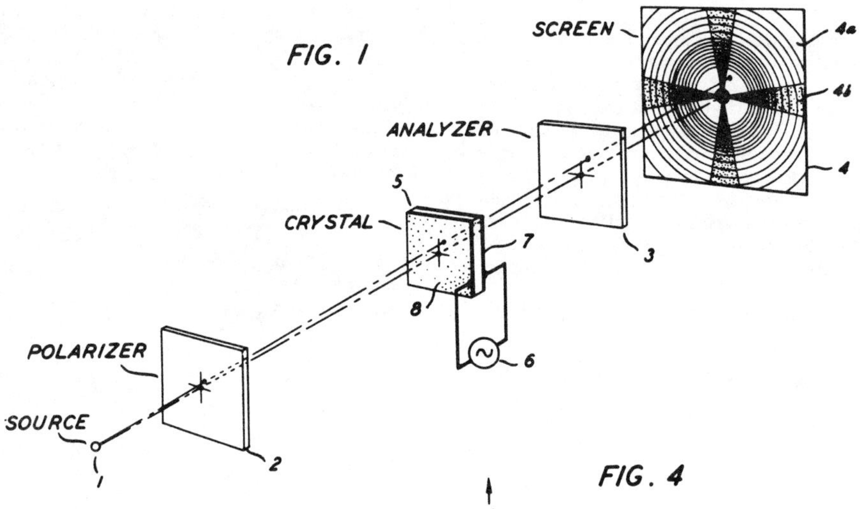

FIG. 2

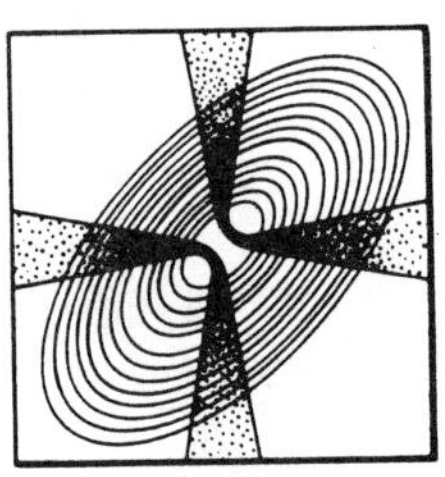

FIG. 3

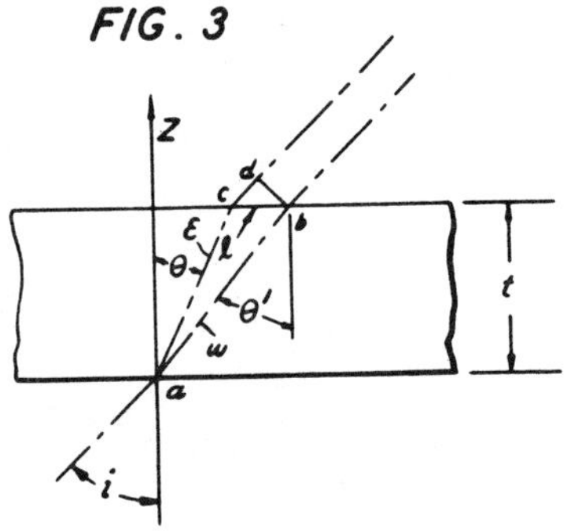

FIG. 4

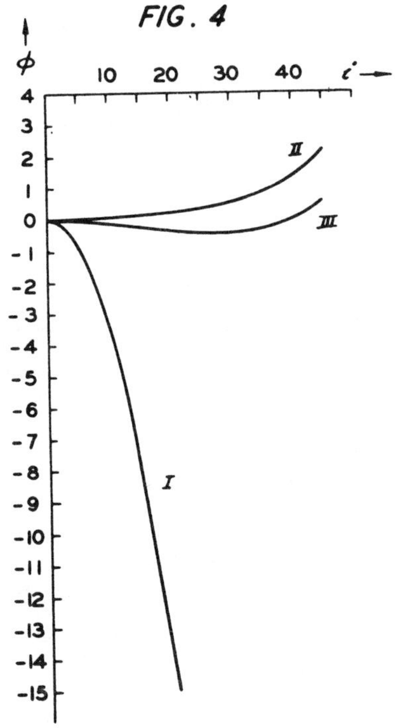

INVENTOR
W. L. BOND
BY N. S. Ewing
ATTORNEY

Oct. 30, 1956 W. L. BOND 2,768,557

UNIAXIAL CRYSTAL ELECTRIC LIGHT VALVE COMPENSATED FOR DIVERGENT LIGHT

Filed Oct. 6, 1952 3 Sheets-Sheet 2

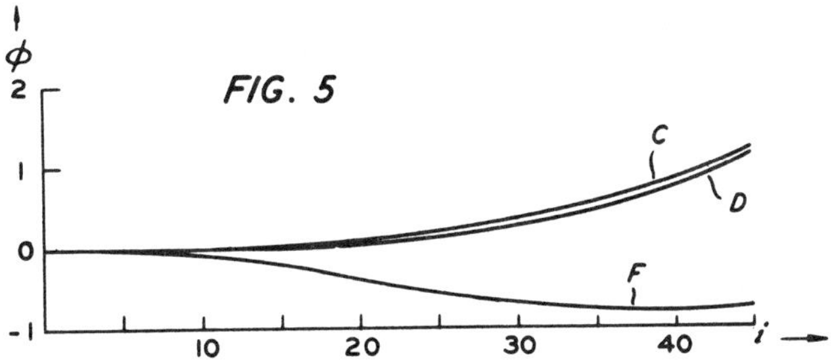

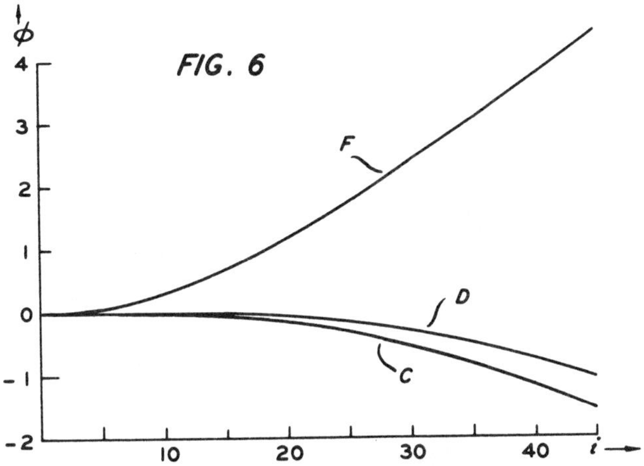

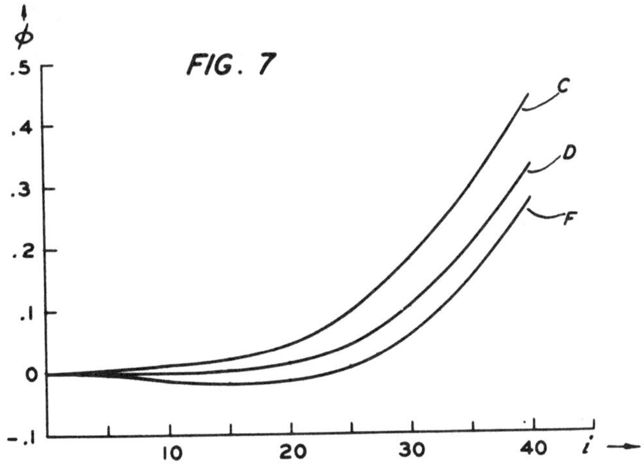

INVENTOR
W. L. BOND
BY N. S. Ewing
ATTORNEY

Oct. 30, 1956 — W. L. BOND — 2,768,557

UNIAXIAL CRYSTAL ELECTRIC LIGHT VALVE COMPENSATED FOR DIVERGENT LIGHT

Filed Oct. 6, 1952 — 3 Sheets-Sheet 3

FIG. 8

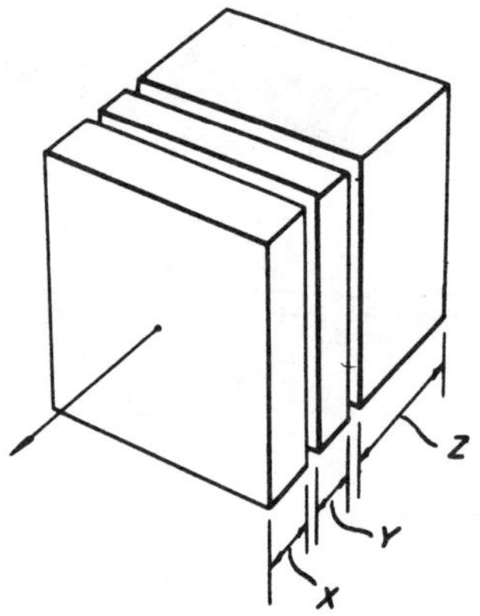

FIG. 9

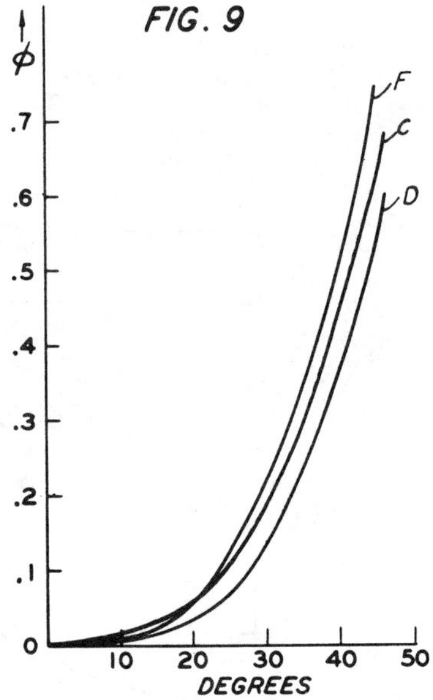

FIG. 10

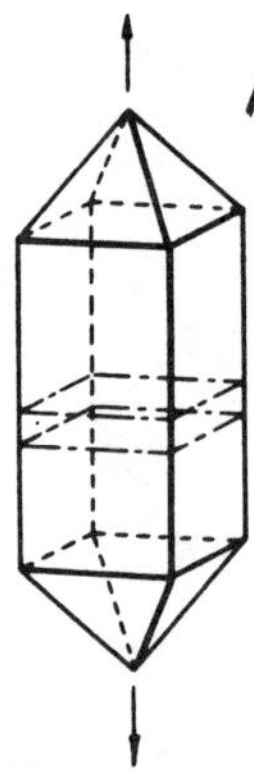

FIG. 11

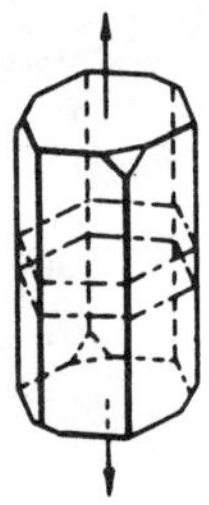

FIG. 12

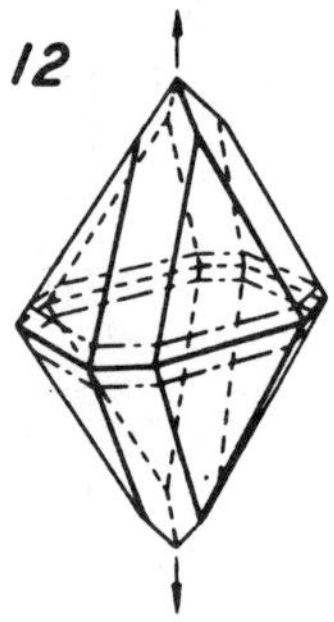

FIG. 13

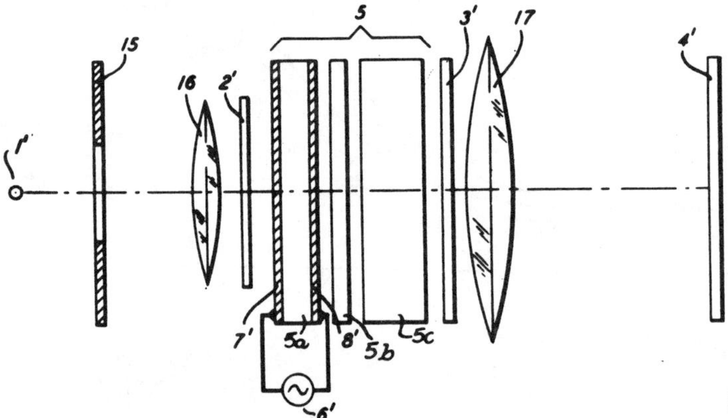

INVENTOR
W. L. BOND
BY
N. S. Ewing
ATTORNEY

Microwave Modulation of Light in Paramagnetic Crystals*

N. Bloembergen, P. S. Pershan, and L. R. Wilcox
Gordon McKay Laboratory, Harvard University, Cambridge, Massachusetts
(Received August 3, 1960)

The considerations of Dehmelt and several other workers about the modulation of light by radio-frequency signals in atomic vapors are extended to paramagnetic solids. It is shown that these materials, driven near a microwave resonance at low temperatures may be used both to create and to detect modulation of light at microwave frequencies. Experimental design criteria are discussed at the hand of two numerical examples, modulated circular dichroism in ruby and modulated Faraday rotation in a broad class of ionic rare earth compounds. Some possible applications of microwave modulated light are reviewed.

I. INTRODUCTION

THE modulation of light has a long history. As early as 1881 Righi[1] obtained fluctuations in an interference pattern by a periodic mechanical rotation of one in a pair of crossed Nicols. Rupp[2] produced unresolved side bands on light by operating a Kerr cell near 10^8 cps and detected their passage through a vapor cell absorbing only the central component. More recently Dehmelt[3] proposed that a precessing magnetization of atoms in alkali vapors should give a modulation of the absorption coefficient for circular polarized light propagating in the plane of the precessing component. Such a time dependent magnetization may be created by polarization of the alkali atoms (e.g., by optical pumping) and driving the resonance between two Zeeman levels with a radio-frequency magnetic field.

The successful outcome of such an experiment at 550 kc/sec on Na vapor has been reported by Bell and Bloom.[4] Bloom has announced a similar experiment on a higher frequency transition in K vapor. Series[5] has reported related experiments on modulation in Hg vapor where the precessing magnetization is induced in the excited state.

It is well known that large precessing components of magnetization can be produced by exciting the spin resonance in paramagnetic crystals at low temperatures. It is the purpose of this paper to investigate what happens to a circularly polarized light beam which passes through a crystal in which a precessing magnetization is established by an exciting microwave field.

Several optical experiments and their interplay with the spin resonances in ruby have recently been reported.[6–9] The analogy with corresponding experiments in alkali vapors has been pointed out. In paramagnetic crystals the polarization is established by thermal relaxation processes and need not be established by optical pumping. The characteristics of an experiment in ruby analogous to the Bell and Bloom experiment in Na vapor will be analyzed in Secs. II and III.

Kastler[10] has pointed out that the Faraday rotation in paramagnetic crystals could be decreased by saturation of the microwave resonance(s) in the ground-state multiplet. Opechowski[11] has given a detailed theory of the effect, which can be described simply by saying that microwave saturation changes the effective magnetic temperature. Daniels and Wesemeyer[12] have demonstrated the disappearance of optical Faraday rotation in

* The research reported in this paper was made possible through support extended Cruft Laboratory, Harvard University, jointly by the Navy Department (Office of Naval Research), the Signal Corps of the U. S. Army, and the U. S. Air Force.

[1] A. Righi, J. phys. **2**, 437 (1883).
[2] E. Rupp, Z. Physik **47**, 72 (1927).
[3] H. G. Dehmelt, Phys. Rev. **105**, 1924 (1957).
[4] W. E. Bell and A. L. Bloom, Phys. Rev. **107**, 1559 (1957).
[5] G. W. Series, Proceeding of the Rochester Conference on Coherence, June, 1960 (unpublished).
[6] I. Wieder, Phys. Rev. Letters **3**, 468 (1959).
[7] S. Geschwind, R. J. Collins, and A. L. Schawlow, Phys. Rev. Letters **3**, 545 (1959).
[8] J. Brossel, S. Geschwind, and A. L. Schawlow, Phys. Rev. Letters **3**, 548 (1959).
[9] T. H. Maiman, Phys. Rev. Letters **4**, 564 (1960).
[10] A. Kastler, Compt. rend. **232**, 953 (1951).
[11] W. Opechowski, Revs. Modern Phys. **25**, 264 (1953).
[12] J. M. Daniels and H. Wesemeyer Can. J. Phys. **36**, 405 (1958).

neodymium ethyl sulphate when the transition between the two Zeeman levels in the ground-state Kramers doublet of Nd^{+++} is saturated.

A variation of this experiment, in which the direction of the incident light is at right angles to the dc magnetic field, but parallel to the microwave magnetic field is discussed in Sec. IV. An observable microwave modulation of the plane of polarization would result.

II. THEORY OF MODULATED CIRCULAR DICHROISM IN RUBY

Consider the geometry of Fig. 1(a) in which circular polarized light is incident along the z axis, parallel to the c axis of the crystal. A homogeneous magnetic field is applied at right angles, parallel to the x axis. The crystal is placed in a microwave cavity with a strong magnetic radio-frequency component in the z direction. The incident light is assumed to contain black-body radiation in a frequency interval, which covers completely the structure of the R_1 line alone, or the structure of the R_2 line alone. The incident light will be treated classically and will be considered as a small perturbation. The Hamiltonian of the problem is

$$\mathcal{H}=\mathcal{H}_{cr}+\mathcal{H}_0+\mathcal{H}_{rf}+V_L,$$

where $\mathcal{H}_{cr}$ is the crystalline field potential including the spin orbit coupling. It determines the splittings between the quartet ground state ${}^4A_2$ and the doublet states $\bar{E}({}^2E)$ and $2\bar{A}({}^2E)$. The notation of Sugano and

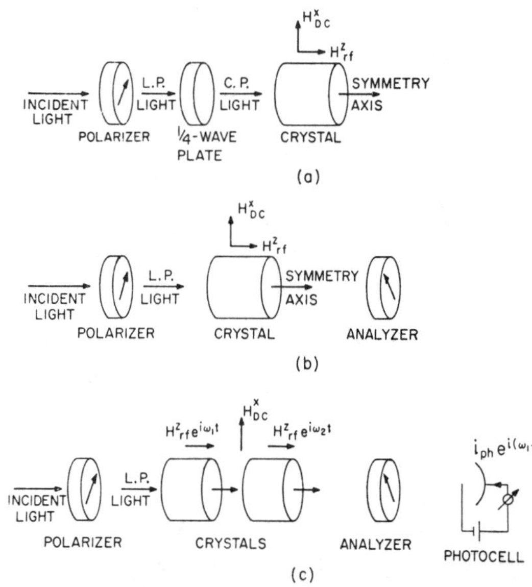

FIG. 1(a). Experimental geometry for microwave modulation of magnetic circular dichroism of light. (b). Microwave modulation of light by optical Faraday rotation of the plane of polarization or by circular dichroism. The light passes through crossed polarizer and analyzer, between which a crystal with a radiofrequency component of magnetization is inserted. (c). Heterodyne detection of microwave modulation of light. The photocurrent is modulated at an intermediate beat frequency.

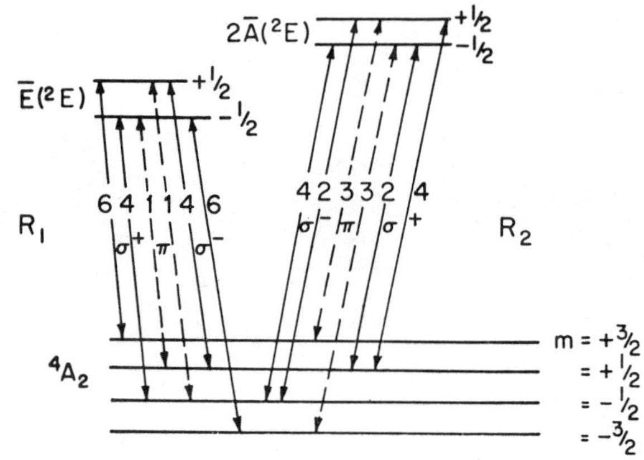

FIG. 2. Relative transition probabilities and polarizations of the R lines in ruby according to Sugano and Tanabe.[13] The optic axis is chosen as the axis of quantization.

Tanabe[13–15] has been adopted. In this representation the crystal field is diagonal. The spin quartet will be labeled by the magnetic quantum number along the z-axis $m=+\frac{3}{2}, +\frac{1}{2}, -\frac{1}{2}$, and $-\frac{3}{2}$. In the presence of a strong magnetic field in the x direction these states are of course not proper states of the system. The Zeeman energy $\mathcal{H}_0$ is not diagonal in this representation.

The set of states in which $\mathcal{H}_{cr}+\mathcal{H}_0$ is diagonal will be denoted by μ, and can be obtained from the m states by a unitary transformation U with matrix elements defined by

$$\psi_\mu=\sum_m U_{m\mu}\psi_m. \tag{1}$$

The excited states of two Kramers doublets are denoted by $2\bar{A}^{\pm}$ and $\bar{E}^{\pm}$, where $+$ and $-$ indicate the quantization with respect to the z axis. Again $\mathcal{H}_0$ is not diagonal in this representation. The Zeeman splitting of the excited doublets will in general be different from the ground-state splittings. The perpendicular spectroscopic splitting factor of the excited states is small. It will henceforth be assumed that no two splittings are equal. A microwave field H_{rf} in the z direction will therefore be near resonance for only one transition. Its frequency will be chosen close to a resonant frequency $\omega_{\mu\mu'}$ in the ground-state quartet.

Sugano and Tanabe[13] have calculated the matrix elements of the light operator in the R_1 and R_2 doublet for three modes of polarization; σ^+ is right circular, σ^- is left circular and π is polarization parallel to the z axis. The matrix elements are simplest in the m representation with quantization along the z axis. The squares of the nonvanishing σ components are reproduced in Fig. 2.

Although all matrix elements of $\mathcal{H}$ in the 8×8 representation can be written down explicitly, a com-

[13] S. Sugano and Y. Tanabe, J. Phys. Soc. (Japan) **13**, 880 (1958).

[14] S. Sugano and Y. Tsujikawa, J. Phys. Soc. (Japan) **13**, 899 (1958).

[15] Y. Tanabe and H. Kamimura, J. Phys. Soc. (Japan) **13**, 394 (1958).

plete solution of the equation of motion for the density matrix,[16]

$$i\hbar\dot{\rho}=(\mathcal{H}\rho-\rho\mathcal{H}), \tag{2}$$

is not feasible in the present case. Complete solutions for the case of a free atom under the simultaneous influence of a strong radio-frequency field and optical resonance radiation have been described by Cagnac, Barrat[17] and others.[18] The presence of the crystalline field and spin-orbit coupling, inhomogeneous broadening and relaxation mechanisms dictate the use of a perturbation procedure.

Consider the 4×4 submatrix $\rho_{\mu\mu'}(\mu,\mu'=1,2,3,4)$ of the ground state quartet in the μ representation in which $\mathcal{H}_{cr}+\mathcal{H}_0$ is diagonal. A microwave field of saturating strength, but smaller than the line width, is applied near the resonant frequency of the transition $\mu=1\rightarrow 2$.

The diagonal elements of the density matrix can be obtained from the usual relaxation rate equations for the populations with $\rho_{11}=\rho_{22}$. The problem is then reduced to a two-level spin resonance with total population $\rho_{11}+\rho_{22}$. The solution for the two by two density matrix is equivalent to the classical solution of Bloch for a precessing magnetization.[16] The off-diagonal element ρ_{12} precessing in phase with the driving microwave field can therefore be determined from the theory of Redfield.[19] The dispersive component of the magnetization does not saturate, if H_{rf} is smaller than the linewidth. The value of the real part of the susceptibility is essentially the same as the unsaturated value. Redfield gives more details (compare Fig. 6 of his paper), but for our purposes the above assumption is adequate. It is particularly good, if most of the magnetic broadening is due to unlike spins (Al^{27} nuclei). For inhomogeneous broadening, such as may be caused by a distribution of crystalline field parameters, Portis[20] arrived earlier at the same conclusions for the dispersive part of the magnetization.

Redfield's assumption about the relaxation in the precessing field and return to thermal equilibrium will be taken over, even though the presence of other levels complicates the relaxation mechanism. This is permissible because his results for the dispersive component do not depend critically on details of spin-lattice relaxation.

With microwave field $H_{rf}\cos\omega t$ in the z direction, the expectation value of the z component of the precessing magnetization is

$$\langle M_z\rangle=|M_{12}{}^z|^2\hbar^{-1}N(\rho_{11}{}^0-\rho_{22}{}^0)\times(\omega-\omega_{12})T_2{}^2\{1+(\omega-\omega_{12})^2T_2{}^2\}^{-1}H_{rf}\cos\omega t, \tag{3}$$

for

$$\gamma^{-1}(T_1T_2)^{-\frac{1}{2}}\ll H_{rf}(\gamma T_2)^{-1}. \tag{4}$$

The difference in populations per unit volume between levels 1 and 2 in thermal equilibrium is given by $N(\rho_{11}{}^0-\rho_{22}{}^0)$, where N is the total number of ions per unit volume. The coefficient of $H_{rf}\cos\omega t$ is equal to the Bloch expression for the unsaturated real part of the susceptibility in a linearly polarized field.[21] There is no out-of-phase component in Eq. (3), since the absorptive component is completely saturated in the regime of Eq. (4). The values of the off-diagonal matrix elements of the density matrix and magnetic moment operator are closely related. From the expression

$$\langle M_z\rangle=(\rho_{12}M_{21}{}^z+\rho_{21}M_{12}{}^z)N,$$

and Eq. (3) it can be inferred that

$$\rho_{12}=\tfrac{1}{2}M_{12}{}^z\hbar^{-1}(\omega-\omega_{12})T_2{}^2\{1+(\omega-\omega_{12})^2T_2{}^2\}^{-1}\times(\rho_{11}{}^0-\rho_{22}{}^0)H_{rf}\exp(i\omega t). \tag{5}$$

In the Heisenberg representation $M_{12}{}^z$ and ρ_{12} have to be multiplied by the time factor $\exp(-\omega_{12}t)$.

The number of light quanta absorbed is equal to the transitions out of the ground state quartet caused by the light perturbation. Note that all terms in the Hamiltonian, except the incident light field, leave the partial trace over the ground state quartet invariant. Consider a time interval $t-t_0\ll\omega^{-1}$. In this interval the variation of the density matrix due to the presence of a microwave driving field at the frequency ω can be ignored. In such a short time interval the perturbation by light can be treated independently from the perturbation by the microwave field. The solution of the equation of motion,

$$i\hbar(\partial/\partial t)_{\text{light}}\rho=V\rho-\rho V,$$

is obtained by standard perturbation theory,

$$\begin{aligned}\rho_{\mu\mu}(t)&=\left\{\exp\left[-i\hbar^{-1}\int_{t_0}^{t}V(t')dt'\right]\times\rho(t_0)\exp\left[+i\hbar^{-1}\int_{t_0}^{t}V(t'')dt''\right]\right\}_{\mu\mu}\\&=\rho(t_0)-i\hbar^{-1}\int_{t_0}^{t}dt'[V_{\mu\alpha}(t')\rho_{\alpha\mu}(t_0)-\rho_{\mu\alpha}(t_0)V_{\alpha\mu}(t')]\\&\quad+\hbar^{-2}\int_{t_0}^{t}\int_{t_0}^{t}dt'dt''\{V_{\mu\alpha}(t')\rho_{\alpha\beta}(t_0)V_{\beta\mu}(t'')\}\\&\quad-\tfrac{1}{2}\hbar^{-2}\int_{t_0}^{t}\int_{t_0}^{t}dt'dt''\{V_{\mu\alpha}(t')V_{\alpha\mu'}(t'')\rho_{\mu'\mu}(t_0)\\&\quad+\rho_{\mu\mu'}(t_0)V_{\mu'\alpha}(t')V_{\alpha\mu}(t'')\}.\end{aligned} \tag{6}$$

It is understood that summations over all excited states α and β, and over μ' are carried out. The initial time t_0 should be considered as a variable parameter. Since at t_0 all elements of the density matrix connecting with an excited level vanish, only the contribution of the last term will have to be considered. This statement is

[16] U. Fano, Revs. Modern Phys. **29**, 74 (1957).
[17] B. Cagnac and J. P. Barrat, Compt. rend. **249**, 534 (1959).
[18] J. P. Barrat, J. phys. radium **20**, 541 and 633 (1959).
[19] A. G. Redfield, Phys. Rev. **98**, 1787 (1955).
[20] A. M. Portis, Phys. Rev. **91**, 1071 (1953).
[21] F. Bloch, Phys. Rev. **70**, 460 (1946).

equivalent to saying that the intensity of the incident light is not capable of producing a noticeable pumping effect. The light absorption is proportional to

$$\sum_{\mu}\{\rho_{\mu\mu}(t)-\rho_{\mu\mu}(t_0)\}$$

$$=-\hbar^{-2}\sum_{\mu,\mu',\alpha}\int_{t_0}^{t}\int_{t_0}^{t}dt'dt''V_{\mu\alpha}(t')V_{\alpha\mu'}(t'')\rho_{\mu'\mu}(t_0). \quad (7)$$

If the usual assumption is made that the vector potential is constant over the volume of one ion, the matrix elements of the light perturbation can be written explicitly in the Heisenberg representation of $\mathcal{H}_0+\mathcal{H}_{cr}$,

$$V_{\mu\alpha}(t)=(e/mc)\mathbf{p}_{m\alpha}\cdot[\mathbf{A}(t)\exp(i\langle\omega_R\rangle t)+\mathbf{A}^*(t)\exp(-i\langle\omega_R\rangle t)]\exp(-i\omega_{\mu\alpha}t). \quad (8)$$

Here $\mathbf{p}$ is the momentum operator in the Schrödinger representation. The light wave is represented by a narrow band Gaussian random process, centered at an average absorption frequency $\langle\omega_R\rangle$ in the R_1 or R_2 line structure. The complex amplitude of the vector potential is a random function of time. Its character is described by a set of correlation functions $\langle A_i^*(t')A_j(t'')\rangle=\langle A_i^*(0)A_j(\tau)\rangle$, where i and j denote components of polarization. The correlation function becomes negligibly small for $\tau=t''-t'>\tau_c=\Delta\omega^{-1}$, where τ_c is the correlation time and $\Delta\omega$, the bandwidth of the light. The interval of integration is chosen so that $\tau_c<t-t_0<\omega^{-1}$. This is possible if the bandwidth of the light is large compared to the microwave frequency, i.e., if the light covers the entire structure of the absorption line.

When Eq. (8) is substituted into Eq. (7), one should also add the explicit Heisenberg time factor to $\rho_{\mu\mu'}(t_0)$. Introduction of new variables of integration $T=t'+t''$ and $\tau=t'-t''$ leads to a light absorption proportional to the time,

$$\sum_{\mu}[\rho_{\mu\mu}(t)-\rho_{\mu\mu}(t_0)]$$

$$=-(e^2/\hbar^2m^2c^2)\sum_{\mu,\mu',\alpha}\sum_{i,j^*}p_i(\mu\alpha)p_j(\alpha\mu')\rho_{\mu\mu'}(t_0)$$

$$\times\exp(-i\omega_{\mu\mu'}t)\frac{1}{2}\int_{2t_0}^{2t}\exp\{i(\omega_{\mu\alpha}+\omega_{\alpha\mu'})T\}dT$$

$$\times\int_{-\infty}^{\infty}\langle A_i^*(0)A_j(\tau)\rangle\exp\{i(\langle\omega_R\rangle-\tfrac{1}{2}\omega_{\mu\alpha}-\tfrac{1}{2}\omega_{\mu'\alpha})\tau\}d\tau.$$

The exponential function in the last integral can be put equal to unity, because the correlation function approaches zero before the exponent becomes appreciable, $\{\langle\omega_R\rangle-\frac{1}{2}(\omega_{\mu'\alpha}+\omega_{\mu'\alpha})\}\tau_c\ll1$.

The next to last integral becomes simply,

$$\frac{1}{2}\int_{2t_0}^{2t}\exp\{i(\omega_{\mu\alpha}+\omega_{\alpha\mu'})T\}dT=(t-t_0)\exp(i\omega_{\mu\mu'}t_0),$$

since the time interval $t-t_0$ can be chosen small compared to any microwave period, $(t-t_0)\omega_{\mu\mu'}\ll1$. In this manner one obtains a rate of light absorption proportional to

$$\sum_{\mu}\dot{\rho}_{\mu\mu}(t_0)=-(e^2/\hbar^2m^2c^2)\sum_{\mu,\mu',\alpha}\sum_{i,j}p_i(\mu\alpha)p_j(\alpha\mu')\rho_{\mu\mu'}(t_0)$$

$$\times\int_{-\infty}^{+\infty}\langle A_i^*(0)A_j(\tau)\rangle d\tau. \quad (9)$$

The last integral is proportional to the intensity of the incident light $\mathcal{I}(\langle\omega_R\rangle)$, which has been assumed constant over the absorption band. The absorption rate derives an explicit time dependence from $\rho_{\mu\mu'}(t_0)$ which is given by Eq. (5) when $\mu=1$, $\mu'=2$. The light absorption rate is modulated at the microwave driving frequency ω. All Heisenberg time factors have canceled each other.

The expression (9) is invariant under a change in representation. Since the matrix elements for the momentum, or electric dipole moments, are given by Sugano and Tanabe[13] for a representation quantized along the c axis, it is convenient to make the transformation from the ground-state designation of $|\mu\rangle$ and the excited state, of $|\alpha\rangle$ to the new representation where the ground state is $|m\rangle$ and the excited state is $|a\rangle$. Equation (9) is to be rewritten as

$$\sum_{\mu}\rho_{\mu\mu}(t_0)=\sum_{m}\dot{\rho}_{mm}(t_0)$$

$$=-(e^2/\hbar^2m^2c^2)\sum_{m,m',a}\sum_{i,j}p_i(ma)p_j(am')\rho_{mm'}(t_0)$$

$$\times\int_{-\infty}^{\infty}\langle A_i^*(0)A_j(\tau)\rangle d\tau. \quad (10)$$

The elements

$$\rho_{mm'}=\sum_{\mu,\mu'}\rho_{\mu\mu'}U_{m\mu}^*U_{m'\mu'} \quad (11)$$

derive an explicit time dependence $\exp(i\omega t)$ from $\rho_{\mu\mu'}(t_0)$, if the microwave driving field is applied at a frequency ω near $\omega_{\mu\mu'}$. The transformation matrix U has been tabulated[22] for various magnitudes and orientations of the dc magnetic field. The interest here is primarily in the case that the applied field is at right angles to the c axis and that the Zeeman splitting is large compared to the zero-field splitting. In this case the transformation matrix corresponds to the four-dimensional $(S=\frac{3}{2})$ representation of a rotation by 90° around the y axis,

$$U_{m\mu}=D_{m\mu}^{3/2}(0,\pi/2,0). \quad (12)$$

Rose[23] has given a clear account of the Wigner coefficients and his notation is followed here.

A similar argument can still be given if the incident light were strictly monochromatic, but the excited final states α of the ion were diffuse. In this case all transitions

[22] See, e.g., J. Weber, Revs. Modern Phys. **31**, 681 (1959), Appendix.

[23] M. E. Rose, *Elementary Theory of Angular Momentum* (John Wiley & Sons, Inc., New York, 1959), Chap. 4.

$\mu \rightarrow \alpha$ would still be excited. In ruby, however, not only the ground levels m of ^4A, but also the excited doublets (^2E) are extremely sharp. If the incident light could only excite the transition from one state $\mu \rightarrow \alpha$, then the light absorption would be proportional to $\rho_{\mu\mu}$, which is independent of time. For such narrow line excitation the preceding argument breaks down, because in a suitable time interval of integration $t-t_0 > \tau_c > \omega^{-1}$ the variation of matrix elements $\rho_{mm'}$ with the microwave precession would not be negligible.

III. EXPERIMENTAL DESIGN CONSIDERATIONS AND NUMERICAL EXAMPLE

Consider the specific case with left circular polarized light incident along the c axis, covering all transitions $\mu \rightarrow \alpha$ in the R_2 structure. The light Hamiltonian for this polarization is given by Eq. (8) with $\mathbf{A}=\mathbf{A}_0(\hat{x}+i\hat{y})$ where $\hat{x}$ and $\hat{y}$ are unit vectors in the x and y axis, respectively. The only nonvanishing matrix elements according to Fig. 2 occur for $m=m'=-\frac{1}{2}$, $(a=2A^+|p_+|m=-\frac{1}{2}) = (m=-\frac{1}{2}|p_-|a=2A^+) = \sqrt{2}(a=2A^-|p_+|m=-\frac{1}{2}) = \sqrt{2}(m=-\frac{1}{2}|p_-|a=2A^-)$, where the notation $p_+=p_x+ip_y$ and $p_-=p_x-ip_y$ has been introduced. The absorption coefficient is proportional to $\rho_{-\frac{1}{2},-\frac{1}{2}}$, which is modulated at the microwave frequency ω according to Eqs. (5), (10), and (11). Left circular polarized light in the R_1 lines can make transitions from the $m=-\frac{3}{2}$ and $m=+\frac{1}{2}$ level with relative strengths 3:2. The absorption coefficient is proportional to $3\rho_{-\frac{3}{2},-\frac{3}{2}}+2\rho_{\frac{1}{2},\frac{1}{2}}$. In this case the relative phase of the time variation in these elements is important. Similar consideration can be given for right circular polarized light.

Consider a specific numerical example. A pink ruby crystal at 1.6°K is placed in a dc magnetic field of about 10 500 oersteds perpendicular to the c axis in the x direction, so that the microwave resonance frequency between the two lowest levels $\mu=-\frac{3}{2} \rightarrow -\frac{1}{2}$ occurs at 24 kMc/sec. Since the Zeeman energy is large compared to the zero field splitting, the magnetic quantum number μ along the x axis is nearly perfect and the transformation to the m representation is very nearly that given by Eq. (12). The ruby is placed with the c axis parallel to the axis in a right-cylindrical cavity, driven in the TE_{011} (wave meter) mode. The microwave magnetic field is predominantly along the c axis and is assumed to have an amplitude $H_{rf}=1$ oersted. With a cavity volume of about 1 cc and an unloaded $Q_0=30\,000$, this corresponds to a power dissipation in the walls of 10 milliwatts. This power consumption can easily be supplied by standard klystroms and can be conveniently dissipated by liquid helium at 1.6°K. It corresponds to a pump speed of 0.35 l/sec at the cryostat. The estimate for Q_0 is considered conservative for a pure copper cavity at low temperature, where the resistivity is mainly determined by the magnetoresistance and the anomalous skin effect.

The power dissipation by the magnetic ions can always be made negligible compared to the wall losses, if a sufficiently high degree of saturation is reached. The maximum power transferred by the spins is determined by their spin-lattice relaxation time and is roughly $(\Delta N_0)T_1^{-1}h\nu$. With $T_1=10^{-2}$ sec and $\Delta N_0=10^{18}$ (see below) in a dilute ruby sample (0.02% Cr) of 0.2 cc volume one obtains 0.7 milliwatts. It is clear from these considerations that a long spin-lattice relaxation time and a very high quality factor are favorable in an experiment to obtain the largest possible precessing component of magnetization for a given power consumption. A dilute pink specimen is chosen because it has the narrowest microwave line width of about $T_2^{-1}=1.8\times10^8$ sec^{-1} according to Strandberg ($\Delta\nu=60$ Mc/sec between points of half maximum absorption). The optical density (opacity) at the R lines is still sufficiently large for such a specimen. The optical absorption coefficient[24] is about 5 cm^{-1} for the population of the levels occurring in our example.

Since the condition (4) is satisfied by our choice of parameters, Eq. (5) may be applied. Since $\exp(-h\nu/kT)=0.5$ with our choice of temperature and frequency, the equilibrium populations are $\rho_{11}{}^0=0.54$, $\rho_{22}{}^0=0.27$, $\rho_{33}{}^0=0.13$, $\rho_{44}{}^0=0.06$. Advantage of the Boltzmann factor is taken if the resonance between the two lowest levels is saturated, $\rho_{11}{}^0-\rho_{22}{}^0=0.27$. The matrix element $(1|M_z|2)$ between the $\mu=-\frac{3}{2}$ and $\mu'=-\frac{1}{2}$ state is $g_{||}\beta\frac{1}{2}\sqrt{3}$. Note that $g\beta h^{-1}H_{rf}T_2$ is just the ratio between the half linewidth at half maximum and the microwave amplitude, both measured in gauss. This ratio is 0.1 in this numerical example. The field or frequency is, of course, adjusted off resonance to maximize the dispersive component; $(\omega-\omega_{12})T_2=1$. One finds $\rho_{21}=\rho_{12}=\frac{1}{8}\sqrt{3}\,0.1\times0.26\exp(i\omega t)$. The diagonal elements depend somewhat on the details of the relaxation mechanism but a reasonable choice for the saturated state is $\rho_{11}=\rho_{22}=0.4$, $\rho_{33}=0.14$, $\rho_{44}=0.06$. Equations (11) and (12) then yield

$$\rho_{m,m=-\frac{1}{2},-\frac{1}{2}}(t_0)=\tfrac{1}{8}\sqrt{3}\{\rho_{12}(t_0)+\rho_{21}(t_0)\} +\tfrac{3}{8}\rho_{11}+\tfrac{1}{8}\rho_{22}+\tfrac{1}{8}\rho_{33}+\tfrac{3}{8}\rho_{44}, \tag{13}$$

$$\rho_{-\frac{1}{2},-\frac{1}{2}}(t)=0.24(1+0.01\cos\omega t).$$

The modulation depth is one percent. In the same manner one finds that the absorption coefficient for the right circular polarized R_2 line is proportional to

$$\rho_{+\frac{1}{2},\frac{1}{2}}(t)=0.24(1-0.01\cos\omega t).$$

It turns out that the case of circularly polarized R_1 light is less favorable, because the modulation of the $m=-\frac{3}{2}$ and $m=+\frac{1}{2}$ levels have opposite phase. The modulation depth under the same conditions is only 0.2% in this case. Similarly, driving of the $\mu=-\frac{1}{2} \rightarrow +\frac{1}{2}$ transition leads to smaller modulation depths of 0.33% for the R_2 lines; but to a larger depth of 1.33% for the R_1 line. This larger value is obtained in spite of the fact that the population $\rho_{33}{}^0-\rho_{22}{}^0$ is only half of $\rho_{22}{}^0-\rho_{11}{}^0$. It

[24] F. Varsanyi, D. L. Wood and A. L. Schawlow, Phys. Rev. Letters 3, 544 (1959).

has its origin in the constructive interference of the modulation of $\rho_{-\frac{1}{2},-\frac{1}{2}}$ and $\rho_{+\frac{1}{2},+\frac{1}{2}}$ which are now in phase.

The intensity in a circularly polarized light beam is related to the amplitude of the vector potential A_0 by

$$(\mathcal{I}_R)=(2\pi\nu_L^2/c)\int_{-\infty}^{\infty}\langle A_0(0)A_0(\tau)\rangle d\tau. \quad (14)$$

The absorbed light energy/cm² sec from a beam of left circular polarized R_2 light in a segment dz is equal to the decrease in beam intensity,

$$+d\mathcal{I}=Nh\nu_L\rho_{-\frac{1}{2},-\frac{1}{2}}{}^0dz. \quad (15)$$

With the use of Eqs. (10) and (13) this becomes

$$d\mathcal{I}=-\frac{|p_+|^2e^2}{\nu_L\hbar cm^2}N\mathcal{I}\langle\nu_R\rangle 0.24(1+0.01\cos\omega t)dz. \quad (16)$$

Here $e^2|p_+|^2/4\pi^2\nu_L{}^2m^2=e^2|x+iy|^2$ is the square of the electric dipole moment operator connecting the $m=-\frac{1}{2}$ ground level with the $2A^+$ level.

It is not immediately possible to integrate Eq. (16) for large thickness z, because the spectral quality of the incident white light will change after traversing a finite thickness. In a dilute ruby with well-resolved Zeeman components a rather complex situation will result. In a darker ruby, with sufficient overlap of components the expression $N\mathcal{I}(\nu)$ may be replaced by $N\mathcal{I}g(\nu)$, where $\mathcal{I}$ is now the total light intensity and $g(\nu)$ is a normalized shape function, describing the combined spectral response of all ions. This distribution does not change with the distance z. Cross relaxation processes in the spin system will keep it constant. An absorption coefficient β for incident monochromatic light at frequency ν can then be defined,

$$\beta=-(\partial\mathcal{I}/\partial z)/\mathcal{I}$$
$$=4\pi^2\nu_L\hbar^{-1}c^{-1}e^2|x+iy|^2g(\nu)N\rho_{-\frac{1}{2}-\frac{1}{2}}(t). \quad (17)$$

The modulation depth s is introduced by

$$\rho_{-\frac{1}{2},-\frac{1}{2}}=C(1+s\cos\omega t),\quad \beta=\beta_0(1+s\cos\omega t). \quad (18)$$

The transmitted intensity after a thickness z is modulated according to

$$\mathcal{I}=\mathcal{I}_0\exp(-\beta_0z)(1-s\beta_0z\cos\omega t), \quad (19)$$

valid for $s\beta_0z\ll 1$. The oscillating part is maximum for $\beta_0z=1$ and then the modulation depth of the intensity is also s, valid for $s\ll 1$. In practice the modulation depth will be considerably smaller, because the incident light will contain wavelengths for which $\beta_0z\ll 1$.

The detection of light modulation has thus far been based on the photoelectric effect which is proportional to the light intensity. For frequencies higher than 10^8 cps the photocathode has to become an integral part of the microwave circuitry. Forrester[25] and co-workers have established that the variation in the emitted photocurrent can follow light fluctuations at 10 kMc/sec. It is probable that the photocurrent from a cathode forming the wall of a suitably constructed K-band cavity could re-excite microwave power. It would, however, require considerable development and it would involve the simultaneous testing of two effects, microwave modulation of magnetic circular dichroism and of the photoelectric emission at K-band frequencies.

It is therefore proposed here to utilize the same effect for generation and detection of light modulation. Since the modulation depth is so small, it is also advantageous to use a light bridge scheme to avoid the shot effect of the unmodulated photon current. Consider therefore the case of R_2 light, linearly polarized perpendicular to the c axis as shown in Fig. 1(b). Decompose into a right and left circular wave. These waves will propagate according to $\exp(2\pi in_Lz/\lambda)$ and $\exp(2\pi in_Rz/\lambda)$, where λ is the vacuum wavelength and the complex index of refraction is

$$n_L=n_0+i(\lambda/2\pi)(\beta_0/2)(1-s\cos\omega t),$$
$$n_R=n_0+i(\lambda/2\pi)(\beta_0/2)(1+s\cos\omega t). \quad (20)$$

After having traversed a distance z, the light will be elliptically polarized and the linear component at right angles to the original incident polarization will have an amplitude

$$|A_L|-|A_R|=2A_0\sinh(\tfrac{1}{2}\beta_0zs\cos\omega t)\exp(-\tfrac{1}{2}\beta_0z).$$

The intensity observed through a crossed polarizer and analyzer, with the dichroic ruby in between, will be, for $\beta_0zs\ll 1$,

$$\mathcal{I}_\perp=\mathcal{I}_0\exp(-\beta_0z)(\beta_0zs)^2\tfrac{1}{2}(1+\cos 2\omega t). \quad (21)$$

The signal is a maximum for $\beta_0z=2$, $s\ll 1$. It can be considered as the output of an autodyne or square law rectifier. This becomes evident if one considers two ruby crystals in the same orientation, but driven with different phases, as shown in Fig. 1(c). The factor $\frac{1}{2}(1+\cos 2\omega t)$ in Eq. (21) is then replaced by $\cos\phi_1\cos\phi_2$.

The difference in phase may be created by passing the light beam twice through the same crystal with the aid of mirrors. In this way the time interval between the two light passages is measured, which is a basic observation in the determination of the velocity of light.

If the phases are different because two different microwave driving frequencies are used, $\phi_1=\omega_1t$ and $\phi_2=\omega_2t$, an intermediate frequency beat at $\omega_2-\omega_1$ in the photoelectric current will result, which is detectable with a conventional photomultiplier. This signal would be a more specific indication of the contemplated effect from an experimental point of view than the dc signal given by Eq. (21). In either case further differentiation against spurious effects can be obtained by audiofrequency modulation of the microwave power, frequency, or the dc magnetic field H_0. The signal is admittedly weak, since Eq. (21) for $\beta_0z=2$ gives

$$\mathcal{I}_\perp=0.27\mathcal{I}_0s^2,$$

[25] A. T. Forrester, R. A. Gudmundsen, and P. O. Johnson, Phys. Rev. **99**, 1691 (1955).

where $\mathcal{I}_0$ is the intensity incident on the crystal in the frequency interval of the absorption line and the modulation depth $s\sim 10^{-2}$. Nevertheless the back-ground light in the crossed analyzer can be kept with some care at the same low level, and the modulation depth is of the order of unity. It then becomes a matter of detecting a beam of intensity which is a few parts in 10^5 of the incident beam. This is perfectly feasible. If necessary, the dark current in the photocell can be reduced by cooling to eliminate its associated shot noise.[26] The remaining fluctuations are then essentially those inherent in the statistics of counting individual photons.

If intermittent operation is used by applying, e.g., a 90° pulse to the microwave spin resonance, much larger values of s can be obtained. The equations given in this section are of course equally valid for this case. During the pulse and in the free precession immediately following the microwave signal, the light signal will be more than two orders of magnitude larger than assumed above.

A variation of these experiments is obtained when the light source itself is modulated. This can be done in ruby by magnetic resonance in the excited state. The fluorescent ruby light source should have a higher Cr concentration so that the emitted light spectrum is broadened sufficiently to cover the entire absorption spectrum of a second ruby crystal.

The magnetic resonance condition in these paramagnetic crystals at very low temperature should be corrected for shape demagnetization effects in the same manner as is customary for ferromagnetic materials.

IV. MODULATION OF FARADAY ROTATION IN CRYSTALS WITH RARE EARTH IONS

The dispersive companion effect to circular dichroism is Faraday rotation of the plane of polarization of linear polarized light. Condon[27] has reviewed this relationship for optically active molecules, but it holds equally well for paramagnetic rotation. $\boldsymbol{\phi}$ and $\boldsymbol{\psi}$ denote two vectors in the direction of the incident light. The magnitude of $\boldsymbol{\phi}$ corresponds to the angle of rotation per unit length and the magnitude of $\boldsymbol{\psi}$ corresponds to the ellipticity per unit length, acquired by incident linear polarized light,

$$\boldsymbol{\phi}+i\boldsymbol{\psi}=8\pi^2\nu_L{}^2Nh^{-1}e^{-1}\epsilon^{-\frac{1}{2}} \times\sum_{fg}\frac{i\mathbf{p}_{fg}\times\mathbf{p}_{gf}}{\nu_L{}^2-\nu_{gf}{}^2+2\pi i\nu_L\Gamma_{gf}}\rho_{ff}. \quad (22)$$

Here $\mathbf{p}_{gf}$ is the matrix element of the electric dipole moment operator between a state f, occupied with a probability ρ_{ff} and an optically excited state g, separated by an energy interval $h\nu_{gf}$. The formula is valid for substances which are optically isotropic with dielectric constant ϵ in the absence of a magnetic field, if the rotation per wavelength and the ellipticity are small, $|\phi|\lambda\ll 1$ and $|\psi|\ll 1$. It contains the example of the preceding section as a special case. Take $\nu=\nu_{fg}=\nu_{ma}$ and consider only the z component of one term. Note the relationship,

$$\begin{aligned}2i\{(p_x)_{ma}(p_y)_{am}-(p_y)_{ma}(p_x)_{am}\}\\ =(p_x-ip_y)_{ma}(p_x+ip_y)_{am}\\ -(p_x+ip_y)_{ma}(p_x-ip_y)_{am}\\ =e^2\{|(x-iy)_{ma}|^2-|(x+iy)_{ma}|^2\}.\end{aligned} \quad (23)$$

The inverse linewidth Γ^{-1} is equal to $\pi^2 g_{\max}(\nu)$ and the relationship $|\psi|=\frac{1}{4}(\beta_L-\beta_R)$ is readily established by comparison of Eqs. (17), (22), and (23).

The interest is now centered on transparent regions of the crystal, where the term $2\pi i\nu_L\Gamma_{fg}$ in the denominator is negligible compared to $\nu_L{}^2-\nu_{fg}{}^2$. Equation (22) then reduces to a well-known expression first given by Kramers.[28] One might think that the effects off-resonance should be even smaller than the result of the preceding section. In that case, however, the optical matrix element is very small and is due to higher order admixtures of the spin-orbit coupling. In that way a dependence of the spin orientation in the ground state (spin-only) quartet could be obtained.

If the orbital angular momentum is not quenched, much larger effects which are still appreciable in the transparent regions of the crystal result. For the present purpose the interest will therefore be centered on crystals containing ions of the rare earth group in "non-S" states, in which microwave magnetic resonances are observable.

Consider first the case of free ions. The occupied states f are characterized by principal quantum number n, total angular momentum j. The summation over f reduces to a summation over the magnetic quantum number m. The excited states g are characterized by $n'j'm'$. The frequency ν_{fg} can be assumed to be independent of m and m', $\nu_{gf}=\nu_{n'j'}-\nu_{nj}$, since the Zeeman splittings are small compared to the optical frequencies. The dependence of the numerator on m and m' is, according to the Wigner-Eckart theorem and Eq. (23), contained in a term with the Clebsch-Gordan coefficients[29] $\{C^2(j1j';m,m'=m+1)-C^2(j1j';m,m'=m-1)\}\propto m$.

The paramagnetic Faraday rotation around the z axis in a $2j+1$-manifold of given total angular momentum is therefore proportional to the z-component of magnetization, and can be expressed as a partial trace,

$$\phi_z\propto\sum_m m\rho_{mm}=\mathrm{Tr}_m\{M_z\rho\}=\langle M_z\rangle. \quad (24)$$

Consider next the influence of a crystalline field. Following the argument of Van Vleck,[30] one assumes that no other j' states are admixed to the ground-state multiplet. The crystal field splitting is assumed to be

[26] J. M. W. Milatz and N. Bloembergen, Physica **9**, 449 (1946).

[27] E. U. Condon, Revs. Modern Phys. **9**, 432 (1937).

[28] H. A. Kramers, Proc. Acad. Sci. (Amsterdam) **33**, 959 (1930); also *Collected Papers*, p. 522. A factor 4π appears to be missing from his Eq. (34).

[29] Compare reference 23, p. 46 and pp. 85 ff.

[30] J. H. Van Vleck and M. H. Hebb, Phys. Rev. **46**, 17 (1934).

small compared to the separation between the ground-state multiplet and excited states. The new ground-state multiplet can be denoted by $nj\mu$, the new excited states by $n'\zeta'\mu$. The new ground-state multiplet is connected to the old one by the unitary transformation

$$\psi_\mu = \sum_m \delta_{nj:n'j'} U_{\mu m} \psi_m.$$

Furthermore, it is assumed that the variation in the frequency denominators due to the crystalline field is negligible,

$$\nu_L^2 - \nu^2(n'\zeta'\mu'; nj\mu) = \nu_L^2 - \nu^2(n'j', nj).$$

With these assumptions the trace relationship (24) remains valid on the application of a crystal field.

This argument of Van Vleck remains *a fortiori* valid if a microwave resonance field is applied near the frequency $\nu_{\mu\mu'}$, causing transitions between the states μ and μ' in the ground state multiplet. The unitary transformation associated with the microwave resonance is described by a two-by-two matrix, $U_{\mu\mu',\mu''\mu'''}$ and is similar to the transformation of a spin $S=\frac{1}{2}$ quantized along H_0, to a rotating coordinate system and subsequent quantization along the direction of the effective field.

The trace relationship (24) still holds. The modulation of the plane of polarization is proportional to the modulation of magnetization along the direction of the incident light. This latter quantity is, however, given by Eq. (3) with $\mu=1, \mu'=2$. If the dc paramagnetic Faraday rotation is known, its modulation can be calculated,

$$\phi_z = (\langle M_z \rangle / M_s) \phi_{\text{sat}}. \tag{25}$$

Here M_s is the saturation magnetization and ϕ_{sat} is the saturation value of the rotatory power.

Consider for example a crystal of neodymium ethylsulphate with linearly polarized light incident along the axis of symmetry, which is taken in the z direction. A dc magnetic field is applied along the x axis of about 5500 gauss which splits the ground state Kramers doublet by 0.5 cm^{-1}, corresponding to a resonant frequency $\nu_0=15$ kMc/sec.

The crystal is placed along the axis of a cylindrical cavity driven in the TE_{011} mode. The microwave field is along the z axis with an amplitude of 1 gauss. The geometry is therefore quite similar to that used in the preceding example, Fig. 1(b).

The saturation value of the optical paramagnetic Faraday rotation has been determined by Becquerel, de Haas, and van den Handel[31] as 114°/mm, for a saturation magnetization $\frac{1}{2}g_{\parallel}\beta N$. If the crystal is kept at 1.6°K, $\rho_{11}^0 - \rho_{22}^0 = \tanh(h\nu_0/2kT) = 0.22$.

In the undiluted salt there is no resolved structure for H_0 perpendicular to the axis of symmetry. The half line width between half maximum power points ΔH is about 150 oersteds.[32] This width, which may be equated roughly to $(g_\perp \beta h^{-1} T_2)^{-1}$, is due to the magnetic interactions between Nd^{+++} ions.

The anisotropic g factor has also been measured by Bleaney,[33] $g_{\parallel}=3.53$ and $g_\perp=2.07$. Note the advantage of the large $g_{\parallel}$, when one substitutes for M_{12}^z in Eq. (3). One finds for $(\omega-\omega_{12})T_2=1$,

$$\langle M_z \rangle = -\frac{1}{4}\frac{g_{\parallel}}{g_\perp}\frac{H_{\text{rf}}}{\Delta H}(\rho_{11}^0 - \rho_{22}^0) M_s \cos\omega t. \tag{26}$$

With Eq. (25) one finds for the modulation amplitude ϕ_M of the plane of polarization, after passage through a 5-mm thick crystal, about 20 minutes of arc. If a magnetically dilute ethylsulphate crystal is taken (3% Nd, 97% La), the line width ΔH is about 7.5 oersted and caused mostly by the local field from the protons. One gains a factor 20 in inverse width. At the same time one loses a factor 30 in saturation magnetization. One has to multiply also by a factor 0.8, representing the abundance of the even isotopes. The odd isotopes have a resolved hfs and do not contribute to the precessing magnetization. The rotation amplitude would be 12′ of arc in 5 mm thickness. Nevertheless, a dilute specimen is preferable because the microwave resonance is more clearly defined, the power dissipation per unit volume is much reduced and the ratio of modulated to dc rotary power is increased.

In dilute (3% Ce, 97% La) ethylsulphate with $g_{\parallel}=0.95$, $g_\perp=2.18$ and a saturation rotation 430°/mm, Eqs. (25) and (26) yield about the same value for ϕ_M under the same conditions.[34,35] In either case a microwave field of one oersted fulfills the condition (4). The spin-lattice relaxation time in the Nd salt has recently been measured by observation of the recovery of dc Faraday rotation after saturation. Daniels[36] finds $T_1 \sim 10^{-2}$ sec. The relaxation time in Ce salts drops exponentially at low temperature. The process takes place via excited doublets.[37] Although the relaxation time in the ethylsulphate is not known accurately, it is expected to become longer than a millisecond below 2°K. The power absorbed by the spins from the microwave field is of the order of a milliwatt/cc, and there is no absorption of light in this case.

The expected rotation of about a quarter of a degree is just near the limit of resolution of a pair of crossed Nicol prisms. Let $\langle\phi_0\rangle$ and $\langle\phi_0^2\rangle$ denote the mean and mean square deviation from perfectly crossed polarizers.

[31] J. Becquerel, W. J. deHaas, and J. van den Handel, Physica **5**, 753 (1938).

[32] B. Bleaney, R. J. Elliott, and H. E. D. Scovil, Proc. Phys. Soc. (London) **A64**, 933 (1951).

[33] B. Bleaney, H. E. D. Scovil, and R. S. Trenam, Proc. Roy. Soc. (London) **A223**, 15 (1954).

[34] K. D. Bowers and J. Owen, *Reports on Progress in Physics* (The Physical Society, London, 1955), Vol. 18, p. 364.

[35] J. Becquerel, W. J. deHaas, and J. van den Handel, Physica **5**, 857 (1938).

[36] J. M. Daniels and K. E. Riethoff, Can. J. Phys. **38**, 604 (1960).

[37] C. B. P. Finn, R. Orbach and W. P. Wolf, Proceedings of the Seventh International Conference on Low-Temperature Physics, Toronto, 1960 (unpublished).

The light field after passage through polarizer, modulating crystal and analyzer is represented by

$$A_0 \cos\omega_L t \sin[\phi_M \cos\omega t + \langle\phi_0\rangle] \approx A_0 \cos\omega_L t(\phi_M \cos\omega t + \langle\phi_0\rangle), \quad (27)$$

since $\phi_M \approx 0.5\times10^{-2} \ll 1$ and $\langle\phi_0\rangle \ll 1$.

The intensity of the signal in terms of the incident intensity $\mathcal{I}_0$ is

$$\mathcal{I} = \mathcal{I}_0\phi_M^2 \times \tfrac{1}{2}(1+\cos 2\omega t) + 2\mathcal{I}_0\phi_M\langle\phi_0\rangle \cos\omega t + \mathcal{I}_0\langle\phi_0^2\rangle. \quad (28)$$

The modulation can be detected in the same manner as discussed in the preceding section. It can also be detected by spectral analysis. If the incident light consists of a sharp spectral line with width $\Delta\omega$ smaller than the microwave frequency ω, side bands at $\omega_L \pm \omega$ can be resolved. The intensity in each side band is $\frac{1}{4}\phi_M^2 \mathcal{I}_0 \sim 3\times10^{-6}\mathcal{I}_0$. This should be compared with the intensity of the central component $\mathcal{I}_0\langle\phi_0^2\rangle$ which is of the order of $10^{-5}\mathcal{I}_0$ for crossed Nicols. A spectrometer of moderate resolution (0.1 cm^{-1}) should suffice. A resonant vapor absorption cell would also be suitable to eliminate the central carrier. The resonance *D* light from an alkali vapor discharge may, e.g., be passed through an absorption cell containing the same alkali vapor. The first term on the right-hand side in the expression (28) for the intensity arises from beats of the side bands with themselves and each other. The second term represents beats of the side bands with the central carrier. It would be detectable as modulation of the photoelectric effect. This effect assumes a maximum value for $\langle\phi_0\rangle = 45°$, as can be verified immediately from the left-hand side of Eq. (27) when the approximation $\langle\phi_0\rangle \ll 1$ is not made.

The analogy of the side bands discussed here with Raman lines is only superficial. There is no definite phase relationship between the Raman lines and the incident light, as the incident and scattered light and the molecular vibrations have random phases. In our case the coherent microwave precession of the spins produces a well defined phase relationship between each light component in the central line and the side bands.

An important advantage of the Faraday rotation is the absence of light absorption, in contrast to the absorptive nature of circular dichroism. The incident light may cover a broad frequency interval. Considerably larger total light intensities can be handled. The thickness of the active crystal is not limited by absorption.

In noncubic crystals the aperture is severely limited, because of rotation of the plane of polarization by the anisotropic dielectric tensor. The interference with the Faraday rotation has been described by Kramers[28] and Becquerel.[38] For this reason the incident light has been taken parallel to the optic axis in our examples. For a light ray making a small angle θ with the optical symmetry axis, the angle ϕ of the plane of polarization will be changed to $-\phi$ by birefringence in a distance $d = \lambda\epsilon_\perp/\theta^2(\epsilon_\parallel - \epsilon_\perp)$ where λ is the wavelength in the crystal for light moving parallel to the axis. The aperture of the light beam and the thickness of the crystal are therefore limited by the relation

$$d\theta^2 \ll \lambda\epsilon_\perp/(\epsilon_\parallel - \epsilon_\perp).$$

Since the relative anisotropy of the refraction index in the ethylsulphates is about 0.5% or $(\epsilon_\parallel - \epsilon_\perp)/\epsilon_\perp = 10^{-2}$, the apex angle θ is limited to about 3° for a 5-mm thick crystal. The thickness of the crystal is of course also limited by the phase of the microwave field. It should not be more than half a guide wavelength thick.

A crystal with cubic symmetry and an isotropic index of refraction would offer an important advantage. The alignment of the crystal with respect to light beam and magnetic field would be much less critical and larger apertures could be used. A CaF_2 crystal with rare earth ions as impurities is an example of such a material. It is not necessary that the individual ions are at sites of cubic symmetry. Microwave resonances from cubic and noncubic sites have been reported.[39–41]

V. OTHER METHODS AND APPLICATIONS

Although the calculated effects are small, they should be detectable in the laboratory and could be used in a determination of the velocity of light.[42] The light beam would be reflected from the mirror and pass the same paramagnetic crystal a second time. As the distance between mirror and crystal is varied, the modulation depth of the light, for, e.g., intensity in the side bands, would pass alternatively through maxima and zero. The crystal would act like a very fast shutter, about two orders of magnitude faster than the fastest Kerr cell, for a small fraction of the light. This small fraction can readily be identified by the modulation and crossed polarizer techniques described in this paper.

The magnitude of the effect can be enhanced considerably in several ways over the numerical results obtained in the preceding section. The magnitude of the precessing magnetization, and therefore ϕ_M, can be enhanced by an order of magnitude in microwave pulses. A light source could be pulsed at the same time, so that the product $\mathcal{I}_0\phi_M$ or $\mathcal{I}_0\phi_M^2$ in Eq. (28) is increased by factor 10^3 or more. The rotation per unit length also increases generally toward the shorter optical wavelengths. The efficiency of the photoelectric detector increases at the same time. Incident light from the near infrared to the near ultraviolet can readily be used in these experiments. Significant increases in the modulation depth can be obtained according to Eq. (22) in a wavelength interval on either side of narrow optical

[38] J. Becquerel, J. phys. radium **9**, 337 (1928).

[39] J. M. Baker, B. Bleaney, and W. Hayes, Proc. Roy. Soc. (London) **B247**, 141 (1958).

[40] J. M. Baker, W. Hayes, and D. A. Jones, Proc. Phys. Soc. (London) **73**, 943 (1959).

[41] M. Dvir and W. Low, Proc. Phys. Soc. (London) **75**, 136 (1960).

[42] E. Bergstrand, *Encyclopedia of Physics* (Springer-Verlag, Berlin, Germany, 1956), Vol. 24, p. 1.

absorption lines in the spectrum of the rare earth ion. No Faraday rotation experiments appear to have been made in the vicinity of such absorption lines.

In the absence of significant light absorption the thickness of the crystal may be increased. As in the case with other magnetic resonance applications, extension to higher frequencies into the millimeter range is straightforward if sufficient microwave driving power is available. The increased value of $h\nu/kT$ will also enhance the modulation angle ϕ_M. If the dimension of the crystal becomes comparable to or larger than the microwave wavelength, a traveling wave geometry is indicated in which the light wave and microwave travel with the same phase velocity. It is at this time a matter of speculation whether the modulated photoemission current and frequency multiplication effects inherent in Eqs. (25)–(28) can be used to advantage at these high frequencies. Phase control of microwave signals at different points in space by means of a modulated light beam is another possibility.

The corresponding effect in ferromagnetic materials would offer the advantage of a large magnetization at a more convenient operating temperature. In particular the exchange narrowed resonance in yttrium iron garnet (YIG) appears attractive. Dillon[43] has briefly mentioned the possibility of a combination of microwave and optical phenomena in garnets. Nevertheless, the maximum obtainable effect in YIG is rather smaller than in the paramagnetic crystals cited above. The transverse rotating component of magnetization assumes a maximum value $(\Delta H_k M_s/4\pi)^{\frac{1}{2}}$ for a critical microwave field strength $H_{\mathrm{crit,rf}}=\Delta H(\Delta H_k/4\pi M_s)^{\frac{1}{2}}$, at which other spin waves with finite wavenumber become excited through nonlinear coupling effects.[44] In these expressions M_s is the saturation magnetization, ΔH_k the true line width due to finite lifetime, ΔH is observed width, which includes inhomogeneity broadening. The maximum ratio of transverse to saturation magnetization, $\langle M_z\rangle/M_s$ in Eq. (25), is about 0.05 at a microwave field strength of a few tenths of an oersted in carefully polished garnet. Although larger ratios exist in principle in ferromagnetic materials with larger damping ΔH_k, the microwave power required for maximum precession angle is prohibitive in such cases.

The dc Faraday rotation in YIG is 200°/mm at 7000 A. Since the absorption coefficient is 1300 cm^{-1} at this wavelength, the maximum useable thickness is only 0.02 mm. Nothing would be gained by going to other wavelengths, since the rotary power and absorption coefficient are roughly proportional. It follows from Clogston's theory[45] of ferromagnetic rotary power and the pseudo-vector transformation property of magnetic rotary power[28] that Eq. (25) remains valid in this ferromagnetic case. The plane of polarization would be modulated by $0.02\times200\times0.05=12$ minutes of arc. Although this effect is of the same order of magnitude as for the ethylsulphates, it would be much harder to detect in the presence of magnetostriction causing optical birefringence and of circular dichroism. The possibilities for enhancement discussed above for paramagnetic crystals do not exist in this case.

Ultrasonic vibrations with concomitant variations in the optical index of refraction by acoustical strain have been used extensively in quartz at lower frequencies.[42] Extension of ultrasonic techniques to the microwave frequency range has recently been achieved. It has not yet been possible[46] to set up a microwave sound field in a well-defined mode over a rather large volume containing many acoustical wavelengths. This would be a prerequisite for optical modulation by this technique, which also requires very low temperatures because of ultrasonic attenuation.

It may therefore be concluded that an experimental investigation of microwave modulation of light in paramagnetic crystals is well warranted, especially in cubic crystals containing rare earth ions. Such a program is currently under way in our laboratory.

Note added in proof. The solid state Kerr effect[47] which occurs in dihydrogenphosphates may be modulated at microwave frequencies. The displacements of the hydrogen atoms in this crystal can follow rapid variations in an electric field. It turns out accidentally that a given energy density of electromagnetic microwave radiation produces about the same rotation per unit length of the plane of polarization in dihydrogenphosphate at room temperature as in the paramagnetic crystals at liquid helium temperature.

[43] J. F. Dillon, J. Appl. Phys. **29**, 539 (1958). See footnote on p. 541.

[44] See, e.g., H. Suhl, J. Phys. Chem. Solids **1**, 207 (1957); J. J. Green, thesis, Harvard University, 1959 (unpublished).

[45] A. M. Clogston, J. phys. radium **20**, 151 (1959).

[46] E. H. Jacobsen, *Proceeding of Conference on Quantum Electronics* (Columbia University Press, New York, 1960), p. 468.

[47] R. O. Carpenter, J. Opt. Soc. Am. **40**, 225 (1950).

Vol. 58(4), 563–566 (April 1987).

Efficient electro-optic modulator for optical pumping of Na beams

J. F. Kelly and A. Gallagher[a)]

Joint Institute for Laboratory Astrophysics, University of Colorado and National Bureau of Standards, Boulder, Colorado 80309-0440

(Received 20 November 1986; accepted for publication 5 January 1987)

An electro-optic modulator using $LiTaO_3$ is described which yields 34% of the carrier intensity in each of the first-order sidebands with rf phase modulation frequencies $f_m \sim 1.0$ GHz and $\lesssim 1.0$ W input power. The modulator makes use of a lumped resonator with $Q \sim 200$ to obtain efficient production of the sidebands. It is shown that the device can be scaled for operation at ~ 2 GHz. Applications of this modulator include optical pumping of the lighter alkali atoms, *FM* sideband spectroscopy, and laser phase/frequency stabilization using rf modulation techniques.

INTRODUCTION

There are a number of applications, notably laser cooling techniques, where it is useful to prepare a beam of spin-polarized alkali atoms in the highest $F,|M|$ state of the ground manifold of hyperfine states.[1,2] Preparation of a beam of such atoms, to produce a "two-level" system, typically requires two laser frequencies to empty out the lower F,M states, which are initially populated and further repopulated by leakage due to imperfect circular polarization and other effects.[2,3] Past techniques employed either two separate cw lasers[4] or one standing wave laser cavity operated simultaneously at two mode frequencies to pump both sets of F levels. The same goal has been achieved by rf phase modulating an electro-optic (EO) crystal to obtain first-order sidebands separated by the appropriate hyperfine splitting.[3]

We report here an analogous technique for optically pumping a beam of Na into the $F,|M| = 2,2$ levels. In this technique we modulate at *half* the hfs frequency and use the first-order sidebands only. [By modulating an EO crystal at $\omega_m = \text{hfs}/2$ with sufficient rf power, one effectively obtains the two necessary cw frequencies with power $I_{\pm}^{\max}(\omega_0 \pm \omega_m) \approx 0.34\, I_0$ using one actively stabilized laser with power I_0.] The construction of a lumped EO resonator of moderate $Q(\sim 200)$, which can produce the maximum first-order sidebands with modest rf powers ($\lesssim 1$ W) is described. For our applications a modulation frequency of ~ 886 MHz is required, but we show that the device could be scaled to $f_m \sim 2$ GHz.

The principles of lumped and traveling wave EO modulators are well known[5] so we shall give only a first-order discussion of the principles in Sec. I. No attempt is made to analyze the EM fields of our device since experience has shown that the resonant frequency is sensitive to the geometry of rf coupling and parasitic reactances of the mounting/shielding box and feed-in cable. A more complete analysis of a similar resonant cavity modulator is given in Ref. 6. Since it is possible to tune the cavity by minor geometry adjustments, exacting construction and design principles are not needed. We believe this design has some practical advantages over previously reported designs because it is easy to construct and operate and less expensive in terms of the rf technology. The cavity is based on a split-ring resonator design used successfully in magnetic resonance work.[7] The layout of the resonator and inductive coupling used to excite the cavity are shown in Fig. 1.

The device we describe uses $LiTaO_3$ as the modulating medium. A prototype was also successfully built with $LiNbO_3$; however, it suffered photorefractive damage with light levels $\gtrsim 20$ mW/mm^2. The damage was reversible by heating the crystal above $\sim 170\,^\circ$C. The $LiTaO_3$ crystal shows no evident optical damage with extended use at light levels $\gtrsim 150$ mW/mm^2.

I. PRINCIPLES OF APPLICATION

The technique of generating rf sidebands on a laser field is based on the equivalence of phase modulation and frequency modulation given by[8]

$$E(t) = E_0 \sin(\omega_0 t + \eta \sin \omega_m t)$$

$$= E_0 \sum_{n=-\infty}^{\infty} J_n(\eta) \sin(\omega_0 + n\omega_m)t\,. \qquad (1)$$

Applying a sinusoidal change of phase $\eta \sin \omega_m t$ at modulation frequency ω_m on a pure sinusoidal input field yields an output field consisting of sidebands with amplitudes equal to the nth-order Bessel functions $J_n(\eta)$. For the case of light linearly polarized along the ith direction of an EO crystal, the modulation index η_{ij} is calculated from the linear electro-optical coefficients r_{ij} by the relation[8]

$$\eta_{ij} = \pi L n_i^3 r_{ij} E_j / \lambda_0\,. \qquad (2)$$

Here L is the length of the crystal, n_i is the index of refraction for the ith direction of light polarization, E_j is the applied electric field of modulation along the jth axis of the crystal, and λ_0 is the vacuum wavelength of the light field. Maximum first-order sidebands occur when $\eta \approx 1.84$ which yield $J_1^2(1.84) \approx 0.34$, $J_0^2(1.84) \approx 0.100$, and $J_2^2(1.84) \approx 0.100$. It is also possible to generate a sizable second-order sideband, which may be advantageous for some applications. One could obtain two fairly strong laser sources separated by several multiples (3–4) of the modulation frequency.

An estimate of the rf power needed to sustain this modulation in $LiTaO_3$ and $LiNbO_3$ can be derived from the data given in Table I.[10] For comparison, the maximum first-order modulation η_{33} in $LiTaO_3$ is produced by applying a 43-V peak rf field across a 1-mm-sq bar with 25 mm length. A traveling wave resonator made from this crystal (with $\sim$50-Ω impedance) must transmit about 16 W of rf power to obtain this modulation index. A lumped resonator with $Q \approx 100$ can sustain the same modulation index using <1 W of rf to compensate for dielectric and skin losses in the cavity. For a practical comparison, our 3 (x axis)$\times$3 (z axis)$\times$25 (y axis) mm^3 $LiTaO_3$ crystal used $\sim$1.5-W rf power while the traveling wave resonator of Ref. 3, with dimensions $0.6\times0.5\times25$ mm^3, used $\sim$4–6 W of rf power to achieve the same index of modulation. This smaller crystal should yield nearly the same f_0 in our modulator and require $\ll 1$ W, but breakage is a serious concern and it was not tried. Longer crystals will use less rf power, but will have reduced bandwidth due to phase-velocity mismatch.[6]

A first-order analysis of the cavity shown in Fig. 1 assumes the resonator can be treated as a lumped $\mathscr{L}_{RC}$ tank circuit. This assumption seems reasonable since the high dc dielectric constant of $LiTaO_3$ ($\epsilon/\epsilon_0 \approx 43$) should concentrate the electric field lines within the gap filled by the EO crystal. If this is true then the gap can be treated as an ideal capacitor with lumped capacitance $\epsilon wL/d$. Here ϵ is the dielectric constant of the crystal at the modulation frequency ω_m. It follows, also, that the circulating current distribution along the inside cylinder wall is uniform, so the lumped inductance of the loop (with $L>r$) is nearly that given by an ideal cylindrical current sheet $\mu_0\pi r^2/L$. This approximation is expected to hold when the gap is small compared to the loop circumference, i.e., $d \ll 2\pi r$. The resonant frequency of this series $\mathscr{L}C$ circuit is thus given as

$$f_0 = \frac{1}{2\pi}(\mathscr{L}C)^{-1/2} = \frac{1}{2\pi}\left(\frac{c}{r}\right)\left(\frac{d}{\pi w K}\right)^{1/2}, \tag{3}$$

where $K = \epsilon(\omega_m)/\epsilon_0$ and c is the speed of light. It is noted that the operating frequency is independent of crystal length L. For the case of a $LiTaO_3$ crystal with $W/d = 1$ and modulated along the z axis, a resonant frequency of 860 MHz, appropriate for Na optical pumping, is obtained with $r \sim 0.49$ cm. The dc value of K is assumed. (Its value at 1 GHz is unknown and may be a substantial cause of error in calculating f_0.)

Power is coupled into the resonator by mutual inductance with a single-loop drive coil (Fig. 1). The effect this induction coil has on the resonant frequency of the modulator cavity has been ignored in Eq. (3) since the circuit is operated in the limit of low coupling coefficient κ between

TABLE I. Physical properties of $LiTaO_3$ and $LiNbO_3$ taken from Ref. 9.

	ϵ_3(dc)[a]	n_3[b]	r_{33}[c]	$\tan\delta_l$[d]
$LiTaO_3$	0.38	2.188	30.3	5×10^{-4}
$LiNbO_3$	0.25	2.217	30.8	N.A.

[a] in 10^{-9} f/m.
[b] at 0.60 μm.
[c] in 10^{-12} m/V at 0.633 μm and $f_m \sim 75$ MHz.
[d] for z axis at 1 GHz and 25 °C.

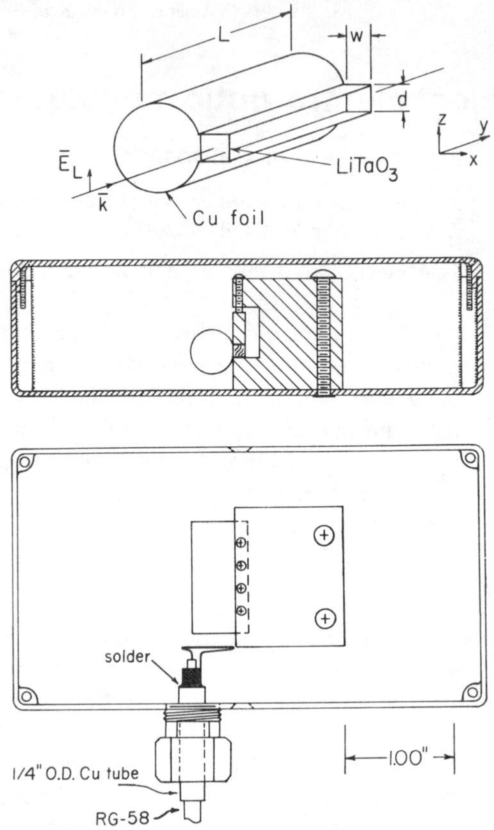

FIG. 1. Schematic of Cu foil resonator and scale drawing of operational modulator for 860 MHz. See text for details.

drive coil and resonator. Maximum rf power transfer occurs when $\kappa Q \sim 1$ so the limit of weak coupling can be achieved if the Q of the cavity is large. We have observed that the transfer of power for the circuit shown in Fig. 1 is large enough to store the necessary rf energy without much effort given to matching the circuit to the rf source impedance. One could use swept frequency reflection measurements to match the circuit and source for more critical applications.

An estimate of the Q value of this circuit is difficult to make because it depends on several fabrication variables. The three notable loss mechanisms are (a) skin resistance which depends on the surface preparation of the metal loop; (b) dielectric loss in the EO crystal; and (c) rf losses to the surrounding structure which supports the cavity. We do not attempt to specify the last effect except to note that we were careful to construct the supporting structure from low dielectric loss plastics. The dielectric loss of the crystal is calculated given the loss tangent $\tan\delta_l$. The total power $P^{(b)}$ dissipated in the crystal is[11]

$$P^{(b)} = (V^2/2)G .$$

Here G is the effective parallel conductance of the crystal given by

$$G = \omega_m C \tan\delta_l$$

The $Q^{(b)}$ of the circuit with only this loss is

$$Q^{(b)} \equiv 2\pi\frac{\text{Energy stored per cycle}}{\text{Energy lost per cycle}}$$

$$= \frac{\pi\omega_m C V^2}{P^{(b)}} = 2\pi/\tan\delta_l . \tag{4}$$

Assuming $\tan \delta_l \approx 5\times10^{-4}$ at 1 GHz for $LiTaO_3$ yields $Q > 10^4$. The Q value which would result from the finite conductance σ of the metal foil is calculated similarly and gives[7]

$$Q = r/\delta ,$$

where δ is the skin depth and r the radius of the coil. The former quantity is given by

$$\delta = \left(\frac{2}{\mu_0\omega_m\sigma}\right)^{1/2} .$$

Quoting Ref. 7, at 1 GHz the conductivity of Cu, 0.5×10^6 $(\Omega\ \text{cm})^{-1}$, results in a skin depth δ of 2.4 μm. Thus, a highly polished Cu loop (scratch depth $< 1\ \mu$m) should result in $Q^{(a)} \gtrsim 10^3$. Generally, our resonators had an order of magnitude smaller $Q(\sim 200)$ as measured by

$$Q = \omega_0/\Delta\omega(\text{FWHM}) .$$

II. CONSTRUCTION AND OPERATION OF MODULATOR

The $LiTaO_3$ crystal used in our work is a $3\times3\times25\ \text{mm}^3$ bar with the x and z axes along the shorter dimensions. The y faces are broadband AR coated with a multilayer coating. A more cost-effective AR coating might be a single $\lambda/4$ layer of MgF_2 since $n(MgF_2) \approx [n(LiTaO_3)]^{1/2}$.[12] A final caveat concerning AR coatings: A coating vendor should be made aware that $LiTaO_3$ has a very large pyroelectric coefficient; large temperature changes in time will produce large electric fields along the crystal which may result in discharging.[12]

The resonator shown in Fig. 1 was formed from 0.005-in. Cu foil. Two tabs ($3\times25\ \text{mm}^2$) were initially bent at right angles to form a shallow C channel. A layer of indium was soldered to each of the tab faces to improve electrical contact of the foil and crystal. The cylinder of the resonator was then formed around a $\frac{13}{32}$-in. hardened and polished steel pin gauge taking care not to scratch the Cu surface unduly.

The surfaces of the indium coatings were then "dressed up" as follows: A hard, polished spacer (a sapphire window works very well) of the requisite thickness was inserted between the tabs (the indium is facing inside the gap) and this assembly was squeezed in a machinists' vise with well-polished jaws. The soft indium flowed to form a mirror finish on each tab; the excess flow over the edges was trimmed off. Several interations of this process can be performed to obtain a very thin layer of indium ($\lesssim 0.0005$ in.).

The assembly of crystal and foil resonator was clamped together with a low dielectric loss plastic vise as shown in Fig. 1, and mounted in an Al shield box with $2\frac{1}{4}$-in.-W$\times 4\frac{1}{4}$-in.-L$\times 1\frac{1}{8}$-in.-H. The primary induction coil consisted of a simple loop termination of the inner wire of RG-58 coax cable to its sheath; the diameter of the loop was nearly the same as the resonator coil. Before adding this loop, a $\frac{1}{4}$-in.-o.d. Cu tube sleeve was fitted over the outer plastic sheath of the RG-58 cable and this was passed through a compression fitting mounted to the Al box. The feed-in position can be fixed securely by finger-tightening this fitting. In this way, one can smoothly adjust the mutual inductance between the drive and resonator loops to obtain the best coupling.

The resonant frequency f_0 of the EO cavity and drive coil can be measured with swept frequency reflection measurements, however, we opted to measure the modulator f_0 by placing a second loop antenna near the cavity to pick up the resonance condition. The weak coupling limit will apply to the pick-up coil also, so the measured f_0 is only slightly different than the actual frequency of operation. [Figure 2 shows the additional port on the power splitter for measuring reflected power; this is a suggested improvement (Ref. 12) for ease of implementation.] With the dimensions noted, we measured the initial resonant frequency to be near 860 MHz. This compares well with a calculated $f_0 \sim 806$ MHz using Eq. (2) with $K = K_{dc} \approx 43$. However, the match is probably coincidental, as we show later.

Figure 2 shows the operational setup of our device. The sideband intensity is measured directly with a 2-GHz optical spectrum analyzer. The voltage controlled oscillator frequency is monitored with a 1-GHz counter while tuning for maximum optical pumping for Na (≈ 886 MHz). The first-

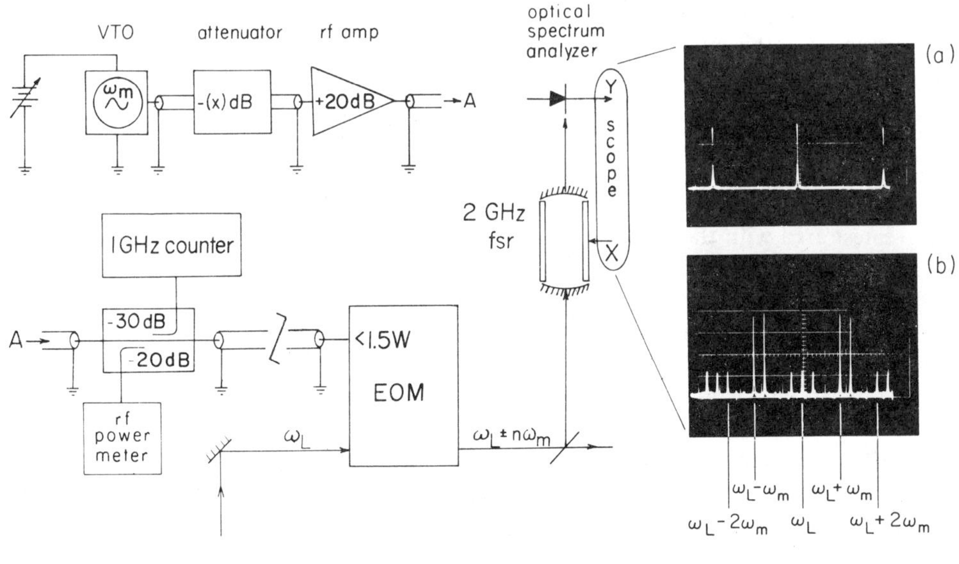

FIG. 2. Schematic of modulator technology. VTO = varactor tuned oscillator. The oscillograph trace shown in (a) is for the case of no modulation. The spectrum orders are separated by 2 GHz. Trace (b) shows the case with modulation index $\eta \sim 1.8$. The following components, which provide a low-cost solution for the rf technology, were used: VTO (0.6–1.0 GHz with 10-mW output power) Avantek VTO-8060, rf amplifier (10–10 000 MHz, $\gtrsim 1.0$ W output) Mini-Circuits ZHL-2-8, frequency counter (50 Hz–1.0 GHz) Digimax Instruments Corp. (The names of the various instruments have been specified for technical completeness and should not be considered an endorsement of particular products.)

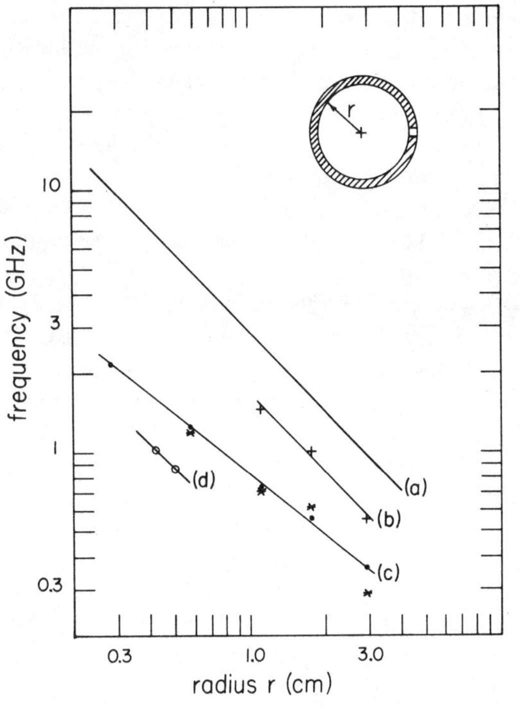

FIG. 3. Frequency response of split ring resonator: (a) response based on Eq. (3) with $w = 0.112 \times d = 0.127$ cm^2 and airspaced ($K = 1$); (b) measured response of same with 25-mm L Cu ring and airspace; (c) measured response of same with $LiTaO_3$ crystal added; (d) measured response of operational foil resonator based on Fig. 1 with $r = 0.516$ cm.

order sideband amplitudes are then maximized by adjusting the attenuation between the oscillator and amplifier stages.

One drawback of this resonator design is its temperature-dependent resonant frequency. The resonator takes an hour to equilibrate to a higher operating temperature (~ 50 °C) as it dissipates energy. This heating alters the performance of the device by shifting the resonant frequency downward. The Q of the cavity is reduced also, since tan δ_l increases with temperature.[10] (It also probably alters η because ϵ and r_{ij} are temperature-dependent parameters, but data are not available concerning their temperature dependence.)

This is not a critical problem in our experiment since we allow the modulator to reach equilibrium; any small changes after this are not important since the intensity of the sidebands is sufficient to achieve optical saturation. For this purpose the modulator's initial resonant frequency is readjusted to a higher f_0 which then settles to the desired operating frequency. f_0 can also be adjusted by deforming the cylindrical coil of the resonator to an elliptical cross section. The largest inductance, hence lowest f_0, results when the coil is circular since this form encloses the greatest area with the shortest boundary. (The ratio of magnetic flux to current density is maximal.) A consequence of this procedure is the reduction of Q, but this is tolerable in our applications since $Q \sim 200$ at 860 MHz was reduced to $Q \sim 156$ at 890 MHz. External cooling/temperature stabilization should eliminate this problem for more critical applications.

It should be practical to scale this lumped EO modulator to higher frequencies, subject to phase-matching considerations as the crystal length becomes a significant fraction of the rf wavelength.[6] In order to test the scaling properties of this device, a number of split-ring resonators were fabricated from OFHC copper as shown in the inset of Fig. 3. A crystal with dimensions 1.1 (x axis) $\times$ 1.3 (z axis) $\times$ 25 (y axis) mm^3 was used in this resonator design. The resonant frequencies of these cavities, shown in Fig. 3, were obtained using the second technique of frequency measurement discussed above. The Q's of these cavities, measured by $\omega_0/\Delta\omega$(FWHM), were ~ 100. These frequency measurements were performed with the cavities sitting ~ 7.5 cm above a metal plane on a dielectric block, but otherwise unshielded.

The difference between these results and the operational device are due to many factors such as the linkage geometry of the coils and external reactances. Placement of external metal objects within 0.5 m of an unshielded cavity shifted f_0 down ~ 3–5 MHz, which implies an rf shielding box will pose sizable perturbations. In lieu of this, Eq. (3) should only be used as a crude reference. The results of Fig. 3 also imply K at 1 GHz is significantly smaller than quoted values of $K_{dc} \approx 43$. A more precise evaluation of K and r_{33} (and their temperature dependence) should be performed for frequencies $\gtrsim 1$ GHz.

Nevertheless, the results of Fig. 3 imply that these devices can produce efficient rf frequency modulation in the range of 0.3–2.2 GHz. The efficiency of these devices makes them useful for applications requiring a large modulation index η at a fixed frequency. Two such applications are laser phase and frequency stabilization techniques requiring rf phase modulation,[13] and frequency-modulation spectroscopy.[14]

This work has been supported in part by the Fundamental Interactions Branch of the Department of Energy. The authors thank J. Hall for many fruitful suggestions and equipment loans in the course of developing this modulator, and his valuable criticisms of this manuscript. We also thank D. Z. Anderson for loaning us the $LiTaO_3$ crystal used in this work.

[a] Staff member, Quantum Physics Division, National Bureau of Standards.

[1] D. E. Pritchard, E. L. Raab, V. Bagnato, C. E. Wieman, and R. N. Watts, Phys. Rev. Lett. **57**, 310 (1986).
[2] R. Blatt, W. Ertmer, and J. L. Hall, Prog. Quantum Electron. **8**, 237 (1984).
[3] W. Ertmer, R. Blatt, J. L. Hall, and M. Zhu, Phys. Rev. Lett. **54**, 996 (1985).
[4] S. Chu, L. Hollberg, J. E. Bjorkholm, A. Cable, and A. Ashkin, Phys. Rev. Lett. **55**, 48 (1985), and references therein.
[5] See, e.g., I. P. Kaminow, *An Introduction to Electrooptic Devices* (Academic, New York, 1974), and references therein.
[6] T. F. Gallagher, N. H. Tran, and J. P. Watjen, Appl. Opt. **25**, 510 (1986).
[7] W. N. Hardy and L. A. Whitehead, Rev. Sci. Instrum. **52**, 213 (1981).
[8] See, e.g., I. P. Kaminow, *An Introduction to Electrooptic Devices* (Academic, New York, 1974), p. 213.
[9] See, e.g., I. P. Kaminow, *An Introduction to Electrooptic Devices* (Academic, New York, 1974), p. 167.
[10] Data supplied by Crystal Technology, 1035 E. Meadow Circle, Palo Alto, CA 94303.
[11] See, e.g., I. P. Kaminow, *An Introduction to Electrooptic Devices* (Academic, New York, 1974), p. 84.
[12] J. L. Hall, Joint Institute for Laboratory Astrophysics, Boulder, CO 80309-0440 (private communication).
[13] R. W. P. Drever, J. L. Hall, F. V. Kowalski, J. Hough, G. M. Ford, A. J. Munley, and H. Ward, Appl. Phys. B **31**, 97 (1983).
[14] J. L. Hall, L. Hollberg, T. Baer, and H. G. Robinson, Appl. Phys. Lett. **39**, 680 (1981), and references therein.

TRANSPARENT ELECTROOPTIC PLZT CERAMICS-A PROCESS REVIEW

Gene Haertling
Motorola Inc.
Albuquerque, NM

Abstract

In the last 16 years since the first transparent ferroelectric electro-optic materials were developed in the PLZT (lead lanthanum zirconate titanate) compositional system, considerable progress has been made toward understanding the basic phenomena associated with these materials and developing various processing methods which are optimum for good chemical uniformity and high optical transparency. This review describes the various processes which have been developed for powder preparation of the PLZT's (mixed oxides, chemical co-precipitation), their fabrication (hot pressing, sintering) and the factors controlling optical transparency and performance (composition, microstructure, internal stress). Overall results point out that the majority of control over the final properties of the ceramic is critically dependent on the combined effects of powder processing and final densification.

1. INTRODUCTION

Ceramics have long been known for their desirable structural (high modulus), electrical (dielectric and electromechanical (piezoelectric) properties; however, it has only been within the last 16 years that ceramics have been shown to also possess quite useful and, indeed, optimum optical and electro-optic properties.

The event which was most instrumental in bringing this about was the successful development of the first truly transparent PLZT (lead lanthanum zirconate titanate) ferroelectric materials in the fall of 1969 and reported in 1971.[1] At that time, research work was primarily devoted to defining the phase diagram of the PLZT system and investigating processing methods for achieving higher transparency and better chemical homogeneity. Subsequently, more of the total effort has gone into (1) understanding the many interactive phenomena (photoferroelectric, ferroelastic, electrically induced phases) which are operative in these complex materials and either limit or enhance their use and (2) identifying markets, designing prototypes and developing devices for various electrooptic applications such as shutters, modulators, displays and image storage devices.

Since a general discussion of the total status of the PLZT materials would be beyond the scope of this paper, it was decided to confine this review to the portion of the PLZT technology pertaining to processing; i.e., powder preparation, fabrication and characterization. A brief mention of applications will also be included.

2. MATERIALS

The PLZT's were first produced in transparent form as a result of efforts to increase the optical translucency of the conventional PZT (lead zirconate titanate) ceramics through the addition of chemical modifiers such as Ba, Sr, Sn, Bi and La. Of those tested, La was found to be unique and significantly better than all of the others in reducing light scattering, thus promoting transparency. It is now known that lanthanum oxide is effective largely because of (1) its ability to reduce the distortion (anisotropy) of the oxygen octahedral (ABO_3) unit cell thereby approximating a more scatter-free cubic structure, (2) its high solubility in the PLZT perovskite structure yielding an extensive series of homogeneous, single phase compositions and (3) its ability to produce a significant number of lattice vacancies on the A and B sites via charge compensation, resulting in the enhancement of the densification process, and the promotion of a controlled and highly uniform microstructure.

2.1 Phase Diagram

The PLZT ceramics do not, in fact, consist of a single composition; but rather, constitute a series of homogeneous solid solutions with varying characteristics which can be tailor-made to suit specific applications. A general formula describing all of the compositions in the PLZT system is given by:

$$Pb_{1-x}La_x(Zr_yTi_{1-y})_{1-x/4}O_3 \qquad (1)$$

where La^{+3} ions replace Pb^{+2} ions (ions of similar size) on the A sites of the octahedral structure and charge balance is maintained by the creation of lattice site vacancies on the B sites. In reality, both A and B sites possess vacancies when stoichiometric balance is achieved, usually resulting in the expulsion of some free PbO at elevated temperatures. This free PbO has been credited with playing a key role in total densification via liquid phase sintering.

The room temperature phase diagram of the PLZT system[2] is given in Figure 1. Several features of the

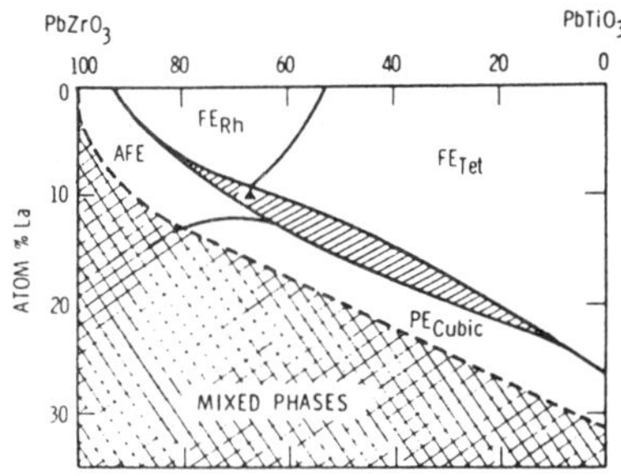

Fig. 1 Room Temperature PLZT Phase Diagram

phase diagram are worth noting; i.e., (1) complete solid solution exists between the two end members and La in small-to-moderate concentrations (<35 atom %), (2) increasing La decreases the stability of the ferroelectric phases (rhombohedral and tetragonal) in favor of the non-ferroelectric cubic and antiferroelectric phases, (3) La additions reduce the Curie point in a linear manner to below room temperature at the rate of approximately 37°C per atom % La, and (4) the crosshatched area existing along the border of the ferroelectric phases denotes a region of diffuse, metastable ferroelectric phases that can be electrically induced with an electric field. Compositions within the crosshatched region are of special interest because they possess unusually high quadratic electrooptic (Kerr) properties. A preferred composition in the PLZT system; i.e., 9/65/35 (La/Zr/Ti), is indicated by the triangular marker in Figure 1.

3. POWDER PROCESSING

The PLZT materials have been successfully prepared in transparent form by either of two general methods; i.e., (1) the mixed oxide (MO) process or (2) a chemically prepared (CP) co-precipitation technique of which there are many varieties. Table 1 summarizes the

TABLE 1

POWDER PREPARATION PROCESSES FOR PLZT CERAMICS

TYPE	DESCRIPTION	YEAR
MIXED OXIDES	WET BLENDING OF OXIDES	1971
CHEM. PREP.	CO-PRECIPITATION OF TBZ AND TBT WITH LaAc & PbO	1972
CHEM. PREP.	CO-PRECIPITATION OF ALL ALKOXIDES	1972
CHEM. PREP.	CO-PRECIPITATION OF TBT AND NITRATES W/AMMONIA	1974
CHEM. PREP.	CO-PRECIPITATION OF NITRATES AND CHLORIDES W/AMMONIA	1976
CHEM. PREP.	ALCOHOL DEHYDRATION OF OF CITRATES	1981
HYDRO-THERMAL	Ti and Zr GELS REACTED WITH La & Pb at 300°C	1984
CHEM. PREP.	CO-PRECIPITATION OF NITRATES AS OXALATES	1985

various methods of preparing PLZT powders that have been reported over the years.[3-9] As can be seen, the CP processes dominate and consist of either organic or inorganic precursors. The inorganic method is potentially more economical, however, the organometallic alkoxide process is the only one, thus far, that has been successfully used in production for a number of years.

Historically, the PLZT's were first prepared from the mixed oxides, but this soon gave way to the CP approach as it became evident that greater levels of chemical uniformity and optical transparency were required than could be achieved (at that time) with the mixed oxides. As the quantities of PLZT material became larger and production costs more critical, there has been a recent trend back to the use of the MO process. In all liklihood, both processes will continue to be used; that is, MO for the less demanding and more cost-conscious applications and CP for the more stringent applications involving the military and the like.

A flow sheet describing the essential steps for both the MO and CP processes is given in Figure 2. It is quite evident that there are many steps which are common to both methods. The essential difference between the two occurs in the powder forming stage, early in the process. For the MO technique, this very simply consists of wet milling (slurry form) the individual oxides or other compounds such as the carbonates, acetates or nitrates which decompose into the oxides during calcining (a high temperature solid state chemical reaction) at 800-900°C. In the CP process, the starting materials are usually solutions which are mutually soluble in each other, thus producing an atomically homo-

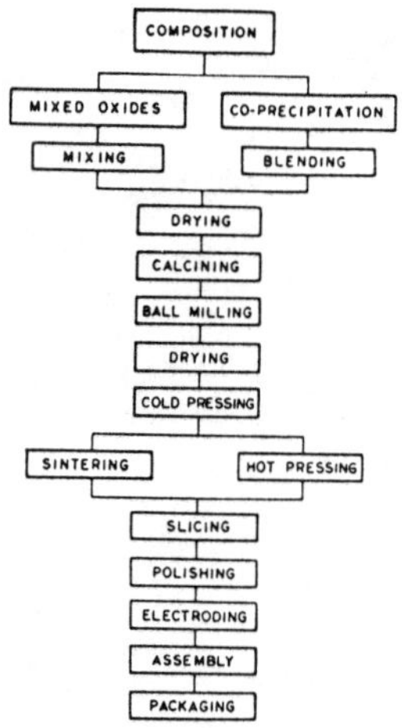

Fig. 2 Flow Sheet for MO and CP Processes

geneous precursor solution which is precipitated (pH change) while blending with another solution such as ammonia gas or ammonium hydroxide to produce a very fine particle size material that is subsequently separated (filtered) from the liquid and dried. Because the particle size of the CP powders are usually much finer than the MO's (0.03-0.1 um vs. 1 um), the CP powders are more reactive and are calcined at a lower temperature of approximately 500°C for 1 hour. Ball milling of the calcined material is necessary for both types of powders in order to produce the required chemical and optical homogeneity. This is a very critical step in the process since too little milling will not produce the necessary homogeneity and over milling will increase the liklihood of contamination, leading to optical haziness and eventual opacity. A common practice is to use a rubber or plastic lined mill with high density media and a non-polar, non-flammable milling liquid such as Freon TF. Depending on the particular power characteristics, milling times may vary from 8-16 hours.

The milled powders are dried thoroughly, mechanically broken up and homogeneized in a V-blender, and then stored in clean, air-tight containers for further processing.

4. FABRICATION

In addition to composition and powder preparation, densification of the powder into a fully dense slug is the third area of processing which is extremely critical to achieving optical transparency. Referring again to the flowsheet in Figure 2, the PLZT powder is first consolidated into a given shape by cold pressing (without binders) either uniaxially or isostatically. This pre-formed slug is further consolidated and densified into its final form using a sintering technique, a hot pressing method or combinations thereof. Although both basic methods will produce transparent material, the process which has been found to be the most reliable in volume production is that of hot pressing; i.e., the technique whereby external pressure is applied to the material during sintering at an elevated temperature. However, this area, like that of powder preparation, is seeing a change to less costly methods of production, and there is now a trend toward conventional sintering of the PLZT's.

A typical hot pressing setup is shown in Figure 3. Experience has shown that a simple, uniaxial,

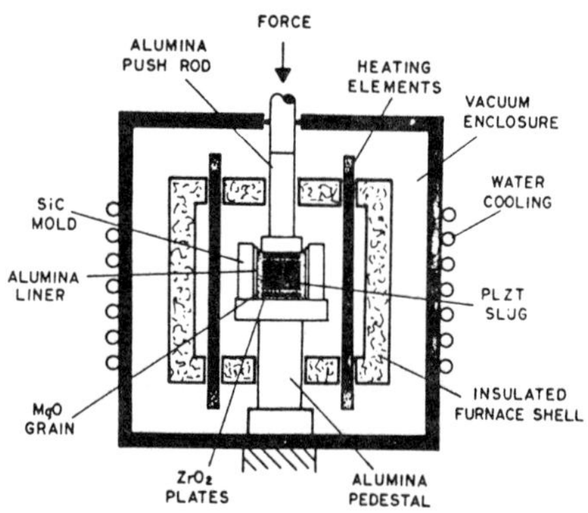

Fig. 3 A Typical Hot Pressing Setup

single-ended hot pressing arrangement is both reliable and economical in producing consistently high optical quality material. The pre-pressed slug is placed into an alumina-lined, silicon carbide mold and completely surrounded with a thin layer of refractory grain such as magnesia or zirconia in order to transmit the pressure more evenly and to facilitate the removal of the slug after hot pressing. A push rod is placed on top of the slug and a modest amount of pressure is applied for alignment purposes. During heat up of the furnace, a vacuum is drawn on the slug by means of a water cooled chamber surrounding the furnace. Oxygen is backfilled into the chamber at 700°C, full pressure is applied, and the furnace temperature is raised to its final value. Typical hot pressing conditions are 1250°C for 18 hours at 2000 psi. After hot pressing, the slug is extracted from the mold, cleaned of refractory grain, annealed and polished for optical evaluation. Examples of PLZT quality, sizes and shapes are given in Figure 4.

Figure 5 shows a slug and push rod assembly which was kept intact after extraction from the mold and then polished. The right hand portion of the figure shows a cross section of the assembly with the mold liner still intact and the slug (as hot pressed) still in place. The remaining half of the hot pressed slug was polished and displayed below the assembly.

The conventional setup for atmosphere sintering the PLZT's is shown in Figure 6. This technique is an adaption of the process normally used in the industry to sintering PZT transducer materials.[10] Flowing oxygen is used to provide the proper oxidizing atmosphere along with the barrier atmosphere powder to prevent voltilization of PbO form the PLZT.

Fig. 4 Examples of PLZT Quality, Sizes and Shapes

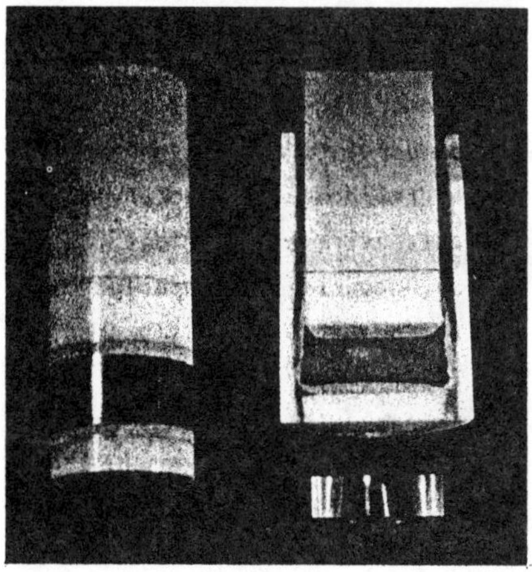

Fig. 5 Hot Pressed PLZT Slug and Pushrod Assembly

Typical conditions for sintering are 1250°C for 50 hours. A two-step sintering cycle is often used in order to obtain higher transparency, although recently reported results do not find this necessary.[11] As previously mentioned, the sintering method can generally produce good material but reproducibility and reliability are still significant problems. Examples of PLZT, first sintered by Snow,[12] are shown in Figure 7.

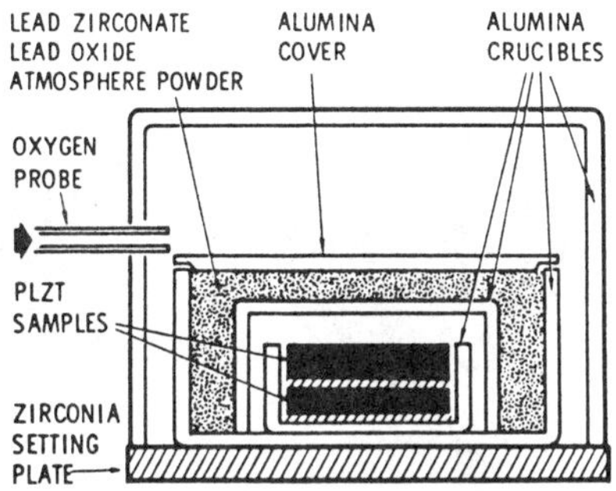

Fig. 6 Setup for Atmosphere Sintering PLZT

Fig. 7 Atmosphere Sintered PLZT Material

The fabrication of the PLZT materials has also undergone a continual evolution from one technique to another.[13-19] Table 2 traces this history of fabrication methods for the PLZT materials from the earliest hot pressing techniques to the more recent atmosphere sintering methods. As one can see, several single and two-stage processes have been utilized in an effort to facilitate fabrication, reduce costs and maximize material quality. Just which one of these processes will eventually dominate in high volume production is yet to be determined.

TABLE 2

FABRICATION TECHNIQUES FOR PLZT CERAMICS

TYPE	DESCRIPTION	YEAR
HOT PRESSING	UNIAXIAL HP OF OXIDES IN AIR	1971
OXYGEN HP	UNIAXIAL HP OF CP MAT'L IN O_2	1972
ATM. SINTERING	SINTERING OF CP MAT'L IN O_2	1972
ATM. SINTERING	ATM. SINTERING CP MAT'L IN O_2, HEAT TREATMENT IN AIR	1973
VACUUM HP	UNIAXIAL HP OF OXIDES IN VAC.	1974
ISOSTATIC HP	SINTERING OF OXIDES IN O_2, ISOSTATIC HP IN N_2	1975
VAC. SINTERING	SINTERING CP MAT'LS IN VAC.	1976
VACUUM HP/ ATM. SINTERING	UNIAXIAL HP OXIDES IN VAC. ATM. SINTERING	1976
VAC. SINTERING/ ATM. SINTERING	SINTERING OXIDES IN VAC. ATM. SINTERING	1976
VAC.-OXYGEN HOT PRESSING	PREHEAT OXIDES IN VAC. UNIAXIAL HP OF CP MAT'LS IN O_2	1977
ATM. SINTERING (LIQUID PHASE)	ATM. SINTERING OF OXIDES IN O_2, LARGE EXCESS OF PbO	1978
ATM. SINTERING	ATM. SINTERING OF CP MAT'L IN O_2	1985

5. PROPERTIES

All of the materials which are to be used for electrooptic devices must undergo careful inspection and evaluation in order to determine their properties in conformity to the specific applications. Table 3 summarizes all of these properties normally evaluated. The majority of these pertain to the physical condition of the material as it relates to homogeneity and transparency. In addition, several of these properties are evaluated under polarized light since defects are most noticeable under crossed polarizers, and this is the configuration most often used in devices.

TABLE 3

PROPERTIES NORMALLY EVALUATED IN PLZT CERAMICS

TYPE		DESCRIPTION
PHYSICAL		POROSITY GRAIN SIZE MICROSTRUCTURE
	EVALUATED UNDER POLARIZED LIGHT	MOTTLING CROSSHATCH STARS STRESS
ELECTRICAL		DIELECTRIC CONSTANT REMANENT POLARIZATION INDUCED POLARIZATION
OPTICAL		TRANSMITTANCE HAZE (SURFACE)
ELECTROOPTIC		HALF-WAVE VOLTAGE REMANENT SPACE CHARGE

For inspection purposes, it is common practice to also insert a quarter-wave plate in between the crossed polarizers, thus producing circularly polarized light which is very sensitive to subtle inhomogeneities and stress imperfections in the PLZT material. Figure 8 illustrates such a setup for viewing three slugs with different types of defects and one slug which meets all specifications. Of these defects, mottling (gross chemical inhomogeneity) and random stars (localized stress centers) are the most prevalent and troublesome. Both are usually caused by inadequate powder preparation, however, improper hot pressing can also be a contributing factor. Overall stress in the slug is a defect

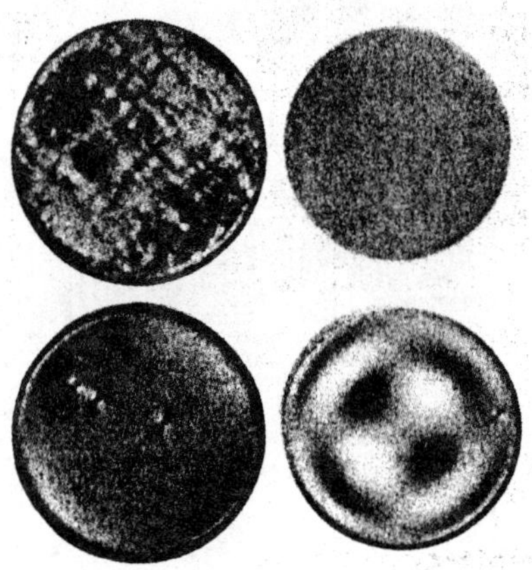

Fig. 8 Typical Defects Which Occur in PLZT Slugs as Viewed Under Circularly Polarized Light. Counterclockwise (Starting in Upper Left), these are (1) Mottling, (2) Stars, (3) Radial Stress and (4) A Good, Well Annealed Slug

which occurs in all hot pressed slugs to some degree, but fortunately, it can be and is routinely eliminated by a post-anneal process.

5.1 Microstructure

The PLZT materials are characterized by a very uniform microstructure of controlled grain size ranging from about 2 um to 15 um depending on the conditions of fabrication. A typical grain size (average grain diameter) is 8 um, as illustrated in the photomicrographs of Figure 9. A distinctive feature of the PLZT's as seen in the figure, is their fully dense, pore-free structure which possesses

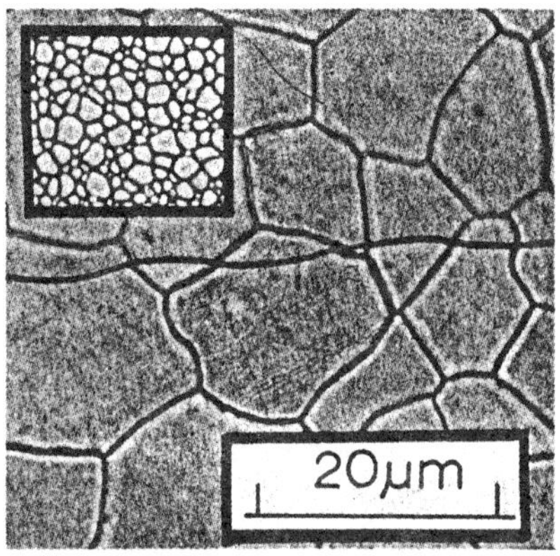

Fig. 9 A Thermally Etched PLZT Microstructure Illustrating Uniform Grain Size, Clean Grain Boundaries and Crack Behavior. The Inset Shows the Microstructure at a Lower Magnification

only one homogeneous crystalline phase and very clean, narrow grain boundaries. All of these attributes are necessary in order to maximize transparency and minimize light scattering from pores, grain boundaries and second phases. The inset view of Figure 9 shows the microstructure on a larger scale where more grains are in the field of view. Also shown in Figure 9 is a crack which has propagated through the part, illustrating the cohesiveness of the grains. It may be noted that the crack is not inclined to follow the grain boundaries - a condition very different from most polycrystalline ceramics and more characteristic of glasses.

5.2 Optical Properties

Lanthanum oxide, as a chemical modifier in the PLZT system, is very unique in its ability to produce high transparency at modest hot pressing or sintering conditions. Thin (0.375 mm) polished plates characteristically transmit between 67 to 69% of the incident light. This increases to about 98% when broadband antireflection coatings are applied to the major surfaces of the PLZT. Surface reflection losses are approximately 18% per surface (n = 2.5) at 550 nm wavelength (green) but will vary as a function of wavelength.

A characteristic transmission curve for a 9/65/35 PLZT wafer throughout the optical and near infrared is given in Figure 10.

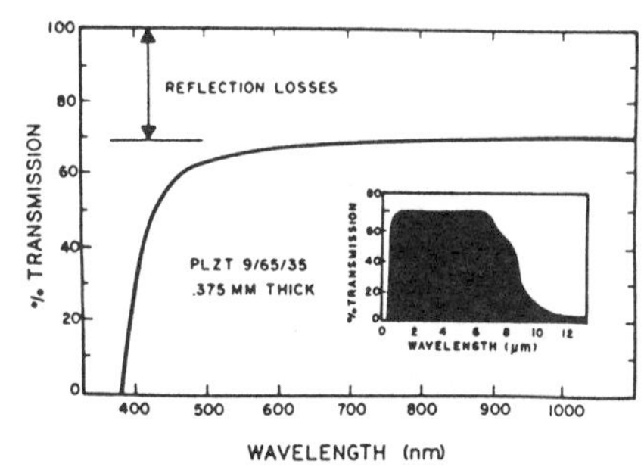

Fig. 10 Optical Transmission Characteristics of Hot Pressed PLZT 9/65/35

As noted, the material is transparent over most of the region. Absorption begins to occur in the IR at about 6.5 um, but it does not become appreciable until 10 um. Beyond 12 um, and out to at least 40 um, the material is completely absorbing. On the short wavelength end of the spectrum, the transmission curve will typically begin to "roll off" at about 450 nm (0.45 um) and will be totally absorbing at 370 nm, the value which is usually quoted as the absorption edge of the PLZT materials.

5.3 Electrooptic Properties

A quantitative and convenient method of judging the magnitude of the electrooptic effect in the PLZT materials may be made by determining the half-wave voltage (voltage required to produce 90° rotation of the linearly polarized light) of the material, and from this, calculating a quadratic R coefficient according to the relationship:

$$R = -\frac{1}{2} \frac{\Gamma}{n^3 t E^2} \qquad (2)$$

where Γ = total optical retardation in nanometers, n = index of refraction, t = activated thickness of the plate and E = applied voltage/electrode gap. For the PLZT ceramics, the R values are exceptionally high and exceed those of most other crystalline materials.[1] Indeed, it is primarily these high electrooptic coefficients which make the PLZT materials especially attractive in electrooptic applications.

The half-wave voltage measurement is usually performed in a shutter configuration as shown in Figure 11. The polished and electroded PLZT plate of known thickness and

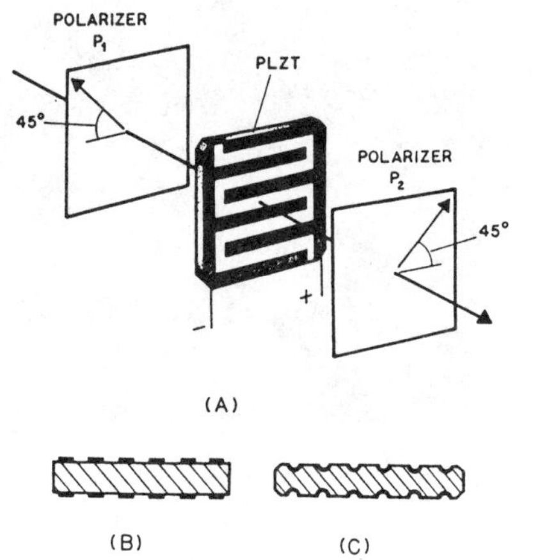

Fig. 11 A Typical Shutter Configuration (A) Showing Cross Section Views of Surface (B) and Grooved (C) Electrodes

electrode gap is sandwiched between two crossed polarizers and this assembly is situated between a light source and a detector. As voltage is applied, an electrooptic birefringence (retardation) is created in the PLZT which is quadratically related to the applied voltage and is manifested in an opening up of the shutter, thus allowing light to pass as long as the voltage is applied. Removing the voltage automatically closes the shutter. The voltage corresponding to the first maximum in light transmission is known as the half-wave voltage. This half-wave voltage (and thus, R coefficient) is a fundamental characteristic of the material, itself; however, since it is usually determined with either surface or grooved electrodes which do not provide a true, transverse electric field, the values determined are specific to the particular type of device measured.

Most of the devices presently utilizing the PLZT materials in electrooptic applications are constructed as shown in Figure 11. Since the electrodes are so situated as to block part of the light passing through the device, it is important to minimize the electrode width and maximize the gap. Of course, making the gap larger also increases the voltage of operation, so there is a trade-off to be made in the design of the particular device. Grooved electrodes are often used because they lower the voltage (for a given gap width) by providing a better electric field penetration into the PLZT. Operating voltages (half-wave) may range from about 650V for a 1 mm (40 mil) gap device to 325V for a 0.5 mm (20 mil) device and 200V for a 0.05 mm (2 mil) device. Figure 12 illustrates the behavior of a typical PLZT shutter when operated to half-wave voltage and beyond. Applying voltage higher than half-wave results in the generation of colors, thus making it a voltage variable color filter.

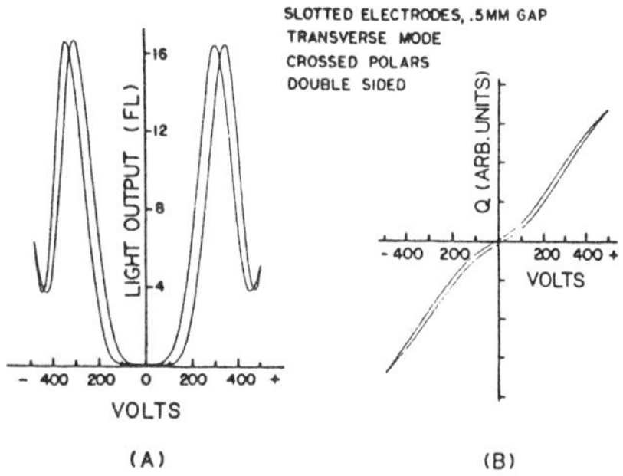

Fig. 12 Light Output Characteristics as a Function of Voltage (A) for a 9/65/35 (9065) PLZT shutter. The Hysteresis Loop of This Device is Shown in (B)

6. APPLICATIONS

The transparent PLZT ceramic materials continue to display high potential for many electrooptic applications. They do, indeed, possess desirable characteristics such as high transparency, fast response (1 usec), wide viewing angle and multi-color capability; however, this must be weighed against the disadvantages of high operating voltage (incompatible with IC's) and the lack of multiplexing capability. It must also be recognized that the generally limited availability of low cost PLZT in larger sizes has hampered the design and use of this material in some electrooptic devices. This situation should improve in the coming years as several new manufacturers make it available in the marketplace. At present, most of the material is utilized in military applications; however, recent new applications in displays and spatial light modulators (linear gate arrays) promise to significantly change this situation.

7. REFERENCES

1. Haertling, G.H. and Land, C.E., J. Am. Ceram. Soc., 54, 1 (1971)
2. Land, C.E., Thacher, P.D. and Haertling, G.H., Applied Solid State Science, (R. Wolfe, ed.), Vol. 4, Academic Press, 1974, p. 180.
3. Haertling, G.H. and Land, C.E., Ferroelectrics, 3, 269 (1972)
4. Brown, L.M. and Mazdiyasni, K.S., J. Am. Ceram. Soc., 55, 541 (1972)
5. Thomson, J., Bull. Am. Ceram. Soc., 53, 421 (1974)
6. Murata, M., Wakino, K., Tanaka, K. and Hamakawa, Y., Mat. Res. Bull., 11, 323 (1976)
7. Yin, Z.W., Ferroelectrics, 35, 161 (1981)
8. Kutty, T.R. and Balachandran, R., Mat. Res. Bull, 19, 1479 (1984)
9. Song, B.M., Kim, D.Y., Yamamura, H., Shirasaki, S.I. and Tanada, M., Japan. J. Appl. Phys., 24, Suppl. 24-2, 439 (1985)
10. Murray, T.F. and Dungan, R.H., Ceramic Industry, 82, 74 (1964)
11. Tanaka, K., Okabe, S., Murata, M. and Wakino, K., Japan. J. Appl. Phys., 24, Suppl. 24-3, 107 (1985)
12. Snow, G.S., J. Am. Ceram. Soc., 56, 479 (1973)
13. Okazaki, K., Igarashi, H., Nagata, K. and Hasegawa, A., Ferroelectrics, 7, 153 (1974)
14. Hardtl, K., Bull. Am. Ceram. Soc., 54, 201 (1975)
15. Pottharst, J. and Schichl, H., Ferroelectrics, 10, 191 (1976)
16. Okazaki, K., Ohtsubo, I. and Toda, K., Ferroelectrics, 10, 195 (1976)
17. Nagata, K., Schmitt, H., Stathakis, K. and Muser, H.E., Proc. 3rd. Int'l. Meeting on Modern Ceramics, Rimini, Italy, 233 (1976)
18. Dungan, R.H. and Snow, G.S., Bull. Am. Ceram. Soc., 56, 781 (1977)
19. James, A.D. and Messer, P.F., J. Brit. Ceram. Soc., 77, 152 (1978)

8. BIOGRAPHY

Gene Haertling is Vice President of the Technical Staff and Manager of the Ceramic R & D department in the Communications Group of Motorola. Until 1973, he was employed by Sandia National Laboratories, engaged in research and development of ferroelectric ceramic materials for electronic and electrooptic applications. He left Sandia in 1973 to form a new company, Optoceram, Inc., where he served as President. Optoceram was acquired by Motorola in 1974. Gene received his B.S. in ceramic engineering from the University of Missouri at Rolla in 1954 and his Ph.D. in ceramic engineering from the University of Illinois in 1961. He is a Fellow of both the American Ceramic Society and IEEE, and is a member of the National Institute of Ceramic Engineers, Sigma Xi, the Society for Information Display and the Materials Research Society.

Hot-Pressed (Pb, La)(Zr,Ti)O_3 Ferroelectric Ceramics for Electrooptic Applications

G. H. HAERTLING and C. E. LAND

Sandia Laboratories, Albuquerque, New Mexico 87115

Transparent ferroelectric ceramic materials suitable for a variety of electrooptic applications were found in the quaternary (Pb,La)(Zr,Ti)O_3 system. These PLZT materials are prepared from mixed oxides and hot-pressed typically at 1100°C for 16 h at 2000 psi. Modifying the lead zirconate-titanate system with lanthana linearly reduces the Curie point with increasing lanthana. Transmission measurements in the visible and infrared show that these materials exhibit a nearly constant response from the absorption edge of 0.37 μm to ≈6 μm. The highest transmission values, essentially 100% (neglecting reflection losses of ≈18%) for thin polished plates, were noted for compositions containing 8 at.% La or more. Specific compositions within the system display electrooptic memory or conventional linear or quadratic electrooptic effects; on the basis of the magnitude of the electrooptic effects, they compare quite favorably with single crystals.

I. Introduction

SINGLE-CRYSTAL ferroelectric materials have been studied intensively for applications in electrooptic devices. The optical transparency of single crystals makes them particularly suitable for such applications; however, other factors, including limited physical size, variable optical homogeneity, and occasionally unpredictable chemical compositional control, detract from their desirability. Ferroelectric ceramics are easier to fabricate in a variety of sizes and shapes with good compositional control, but their optical transparency is commonly so low as to make them unusable in optical applications. Improved fabrication methods such as hot-pressing have enhanced the optical transparency to a degree such that electrooptic devices using ceramic materials now appear feasible.

Studies by Land,[1,2] Land and Thacher,[3,4] Land and Haertling,[5] Maldonado and Meitzler,[6] and Haertling[7] are the first reported on ferroelectric ceramic materials for electrooptic applications. Land[1] and Land and Thacher[3,4] described the general phenomena of light scattering and optical retardation in ferroelectric lead zirconate-titanate ceramics and suggested methods for applying the effects. Land[2] and Land and Haertling[5] described the effects of Ba and Sn additives in the lead zirconate-titanate system. Maldonado and Meitzler[6] used one of the switching schemes (180° switching in the plane of the plate), in conjunction with a new voltage-controlled switching mode, to study the behavior of ferroelectric ceramics as light gates. Haertling[7] reported on the fabrication and properties of a variety of Bi-doped lead zirconate-titanate compositions which were evaluated in the studies of optical scattering and retardation just mentioned.

The present paper describes physical, electrical, and electrooptic properties of a series of ferroelectric compositions in the (Pb,La)(Zr,Ti)O_3 solid-solution (PLZT) system which are significantly more transparent in ceramic form than are other known ferroelectrics. The characteristics of this system include (1) high transparency essentially independent of grain size, (2) ease of fabrication, and (3) electrically variable optical retardation, including electrooptic memory and linear or quadratic modulation capabilities.

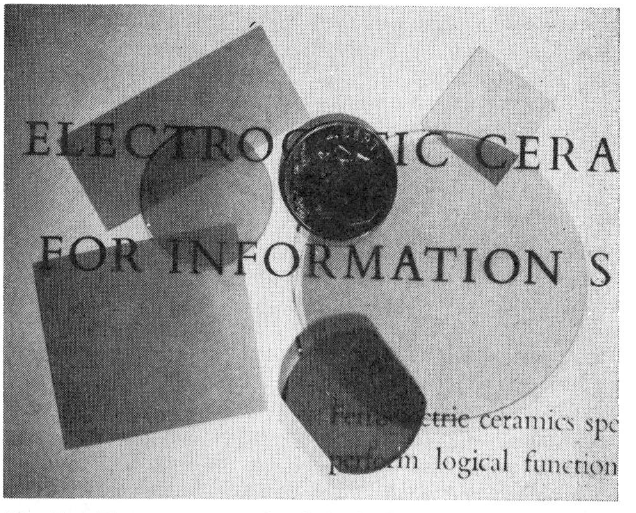

Fig. 1. Transparency of polished plates of PLZT. Plates clockwise from inclined plate 1¾ in. in diameter are 0.3, 2.5, 0.5, 0.5, 0.35, and 0.75 mm thick.

Although there is no information in the literature on the PLZT system as such, a somewhat similar system, $PbZrO_3$-$PbTiO_3$-$LaFeO_3$, was investigated by Ikeda and Okano,[8,9] who found that (1) at least 20 mol% $LaFeO_3$ enters into solid solution with lead zirconate-titanate, (2) the addition of $LaFeO_3$ enhances the region of stability of the antiferroelectric phase, and (3) increasing $LaFeO_3$ favors the tetragonal ferroelectric phase over the rhombohedral ferroelectric phase. A similar study of the same system by Berlincourt[10] essentially substantiated the findings of Ikeda and Okano. In both of these investigations, the materials, as fabricated, were optically opaque as a result of the use of iron oxide as a primary ingredient. The results of these investigations are similar to those obtained by introducing La into solid solution with lead zirconate-titanate except that modification by La alone produces optical transparency. Examples of the transparency of PLZT compositions are shown in Fig. 1.

II. Experimental Procedure

(1) Formulation

All compositions in the PLZT system were prepared according to the general formula:

Presented at the 72nd Annual Meeting, The American Ceramic Society, Philadelphia, Pa., May 4, 1970 (Electronics Division, Nos. 2–E–70 and 3–E–70). Received August 1, 1970; revised copy received October 3, 1970.

Supported by the United States Atomic Energy Commission.

$$Pb_{1-x}La_x(Zr_yTi_z)_{1-x/4}O_3 \qquad (1)$$

where the concentration of La (x) was varied from 2 to 30 at.%, and the ratio of Zr to Ti (y/z) ranged from 100/0 to 0/100, i.e. $PbZrO_3$ and $PbTiO_3$ were end compositions. This formula assumes that all the A (Pb,La) sites of the ABO_3 perovskite structure are filled and that charge compensation is brought about by vacancies in the B (Zr,Ti) sites; one B site vacancy is created for every 4 at.% La.

An alternate formula,

$$Pb_{1-3x/2}La_x(Zr_yTi_z)O_3 \qquad (2)$$

is commonly used in compositions in which trivalent dopants such as La are used in Pb site substitutions. It assumes that electrical neutrality is maintained by the creation of Pb vacancies; one Pb vacancy is created for every 2 at.% La present.

In comparing the above formulas, it would at first appear that the creation of (Zr,Ti) vacancies as proposed in formula (1) is not as probable as the creation of Pb vacancies postulated in formula (2), especially considering the high volatility of Pb ions above 1000°C. On the other hand, the creation of A or B site vacancies does not require complete volatilization of the excess ions because they may exist interstitially as a separate undetectable entity. Eror and Smyth,[11] in their work on oxygen stoichiometry of La-doped $BaTiO_3$, proposed that oxidized A site vacancies (BaO) do not necessarily represent even a separate entity but rather an ordered structural accommodation; these vacancies quite probably are undetectable by X-ray analysis.

The present work was not aimed at establishing the validity of one formula or mechanism over the other. Nevertheless, formula (1) is favored over formula (2) because (1) consistently yielded transparent material, whereas (2) consistently did not; conventional atmosphere firing experiments revealed that formula (1) material lost weight (PbO) in firing, whereas formula (2) material consistently gained weight from the PbO-enriched atmosphere, indicating that formula (2) is lead-deficient; formula (1) is probably lead-rich, but this condition is conducive to higher densification rates in hot-pressing, as noted previously[12]; deliberate addition of excess PbO to formula (2) material resulted in higher transparency.

(2) Material Preparation

Specimens were prepared from high-purity (>99.5%) small-particle-size (<2 μm) PbO,* La_2O_3,† ZrO_2,‡ and TiO_2.§ A 500-g batch of each composition selected was formulated and processed by techniques developed in a previous investigation of hot-pressed Bi-doped lead zirconate-titanate materials for electrooptic applications.[7] Cold-pressed slugs $^1/_4$ in. in diameter by $^3/_4$ in. high were hot-pressed at 1000° to 1300°C for 1 h to 10 days at 300 to 10,000 psi. Typical hot-pressing conditions were 1100°C for 16 h at 2000 psi.

(3) Physical and Electrical Measurements

The physical and electrical measurement techniques were similar to those described in Ref. 7. Bulk density was measured by the liquid (trichlorotrifluoroethane) displacement technique. Grain sizes were determined from polished and etched specimens; in some cases, considerable difficulty was experienced with the common HF/HCl etch because the etched domain patterns interfered with the delineation of the grain boundaries. In such cases, polished plates were thermally etched at 1300°C for $\approx$1 h in a PbO-rich atmosphere; this technique revealed grain boundaries but not domain patterns. Since 1300°C is higher than most of the hot-pressing temperatures, the possibility of grain growth during thermal etching was investigated; none was detected.

X-ray diffraction patterns were determined with a spectrometer using $CuK\alpha$ radiation through an Ni filter. The specimens were hot-pressed disks which were lapped flat on a 320-grit wheel. Unit cell measurements were determined from

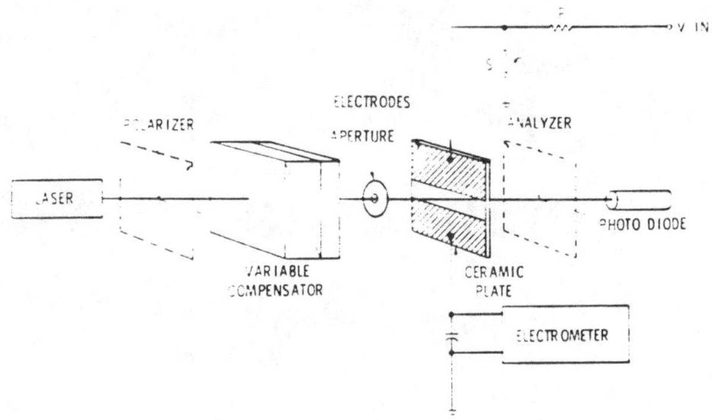

Fig. 2. Experimental setup for Γ measurements.

several front- and back-reflection lines; the accuracy of these measurements is estimated to be ±0.005 Å.

(4) Optical and Electrooptic Measurements

(A) Transmittance vs Wavelength: The transmittance of unpoled polished plates was measured as a function of wavelength from $\lambda=0.2$ to 2.6 μm using two spectrophotometers‖ and from $\lambda=2.5$ to 15 μm using an infrared spectrophotometer.¶

(B) Effective Birefringence: The effective birefringence,[3] $\overline{\Delta n}=\overline{n_e-n_o}$, was determined directly from measurements of the phase retardation, Γ, of the components of a linearly polarized monochromatic light wave incident on the ceramic plate[5,6]; n_e and n_o refer to the extraordinary and ordinary refractive indices, respectively. When Γ and the light path length, l (in the plate), are expressed in the same units of distance, $\overline{\Delta n}=\Gamma/l$.

The arrangement used to obtain data for Γ as a function of remanent polarization, P_r, or applied electric field, E, is shown in Fig. 2. The light source is a 6328 Å He-Ne laser; the beam is initially polarized so that the E field vector of the light is oriented at 45° with respect to the direction of the applied electric field in the ceramic plate. The polarized beam, after passing through the variable compensator,** is reduced in size by the aperture; it then passes through the gap between the electrodes on the surface of the ceramic plate. Light emerging from the plate is processed by the analyzer, crossed with respect to the polarizer, and detected by a photodiode.††

A voltage was applied to the electrodes to produce an electric field in the gap equal to about $3E_c$, where E_c is the coercive field of the material. When the field was removed by closing switch S, the ceramic remained poled at saturation remanence, P_R. The polarity of the voltage was reversed, switch S was opened momentarily, and the gap was switched to some new state of polarization, P_r. For each remanent polarization state achieved by momentarily opening switch S, the retardation was measured by adjusting the variable compensator to obtain zero net retardation for the ceramic plate-compensator combination. The remanent polarization of the

---

*Type 100Y, Hammond Lead Products, Inc., Hammond, Ind.

†Code 528, American Potash & Chemical Corp., New York, N. Y.

‡Cp grade, Tizon Chemical Corp., Flemington, N. J.

§Type 2650, Whittaker, Clark & Daniels, Inc., New York, N. Y.

‖Model 14RI, Cary Instruments, Monrovia, Calif., and Model DK-2, Beckman Instruments, Inc., Fullerton, Calif.

¶Model 21, Perkin-Elmer Corp., Norwalk, Conn.

**Model L-135 Soliel Babinet, Gaertner Scientific Corp., Chicago, Ill.

††4207, Hewlett-Packard Co., Palo Alto, Calif.

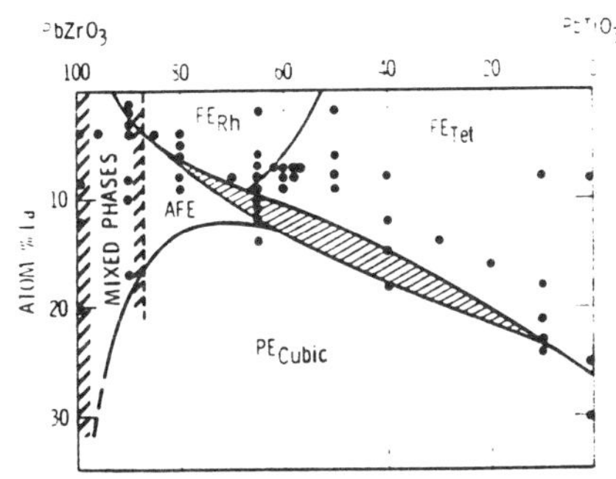

Fig. 3. Room-temperature phase diagram of PLZT system.

material in the gap was proportional to the electrometer reading.

The arrangement shown in Fig. 2 was also used to obtain data for Γ as a function of electric field; Γ was measured with the electric field applied (*S* open).

III. Results and Discussion

(1) Phase Diagram

The room-temperature phase diagram of the PLZT system is given in Fig. 3. Small-signal dielectric data, X-ray analysis, and large-signal hysteresis loop characteristics were combined to establish phase boundaries; the points on the diagram represent actual compositions prepared. The phase diagram is shown only up to 30 at.% La since higher La contents produce a cubic nonferroelectric phase at room temperature for all ratios of $PbZrO_3$ to $PbTiO_3$. It should be noted that the ordinate is expanded twofold in relation to the $PbZrO_3/PbTiO_3$ axis; this was done so that phase boundaries could be more easily identified and located.

Several features of the phase diagram are quite noticeable: (1) small variations in La content produce rather large changes in areas of phase stability, (2) increasing La content extends the antiferroelectric phase (AFE) over the ferroelectric (FE) phase, and (3) increasing La content favors the tetragonal ferroelectic phase over the rhombohedral. These results are similar to those reported by Ikeda and Okano[8] and by Ikeda[9] for the ternary system $PbZrO_3$-$PbTiO_3$-$LaFeO_3$. The principal difference between this system and the present quaternary system is the opposite curvature of the phase boundary between the AFE and FE phases; in the ternary system, the boundary is convex toward the FE phase, whereas in the present system, it is concave.

Modifying pure $PbTiO_3$ with La systematically reduced the Curie point by solid-solution formation; ≈26 at.% La was required to reduce the Curie point to room temperature. When pure $PbZrO_3$ was modified with La, increasing the La concentration did not noticeably alter the Curie point as indicated by small-signal dielectric measurements. X-ray analysis revealed that mixing of phases rather than solid solution occurred at the high $PbZrO_3$ end; the phases were identified as $PbZrO_3$ and $La_2Zr_2O_7$. This mixed-phase region extended from the $PbZrO_3$ end-member to a Zr/Ti ratio of 90/10.

A moderately broad transition region (cross-hatched area) exists between the FE-AFE and FE-cubic phases. The exact nature of this region has not been well defined; it may consist of mixed phases which result, in part, from local variations in chemical composition on a microscopic scale. Compositions within this region possess diffuse or smeared-out phase transitions much like those reported by Goodman[12] for barium zirconium metaniobate, by Bokov and Myl'nikova[14] for lead magnesium niobate, and by Isupov[15] for the general phenomena of broadened ferroelectric transitions. Goodman considered such diffuse transitions as higher than second order, whereas Bokov and Myl'nikova and Isupov suggested that they resulted from compositional microheterogeneities brought about by random composition fluctuations in the ionic lattice. Berlincourt,[16] on the other hand, reported the presence of mixed phases, extensive enough to be detected by X-ray analysis, in the boundary regions of the $PbZrO_3$-$PbTiO_3$-$LaFeO_3$ system.

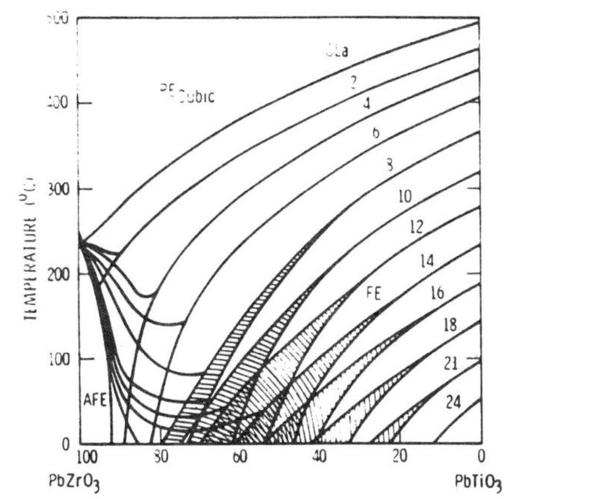

Fig. 4. Phase diagram of PLZT system showing constant La concentration levels.

The location of the phase boundary between the FE (rhombohedral) and FE (tetragonal) phases was established from X-ray, electrical, and optical results. This boundary also is indistinct; it has a finite width of 3 to 4% change in Zr/Ti ratio.

The phase diagram as a function of temperature for various levels of La (Fig. 4) illustrates the systematic change in Curie points for the complete system as the La content is increased. The extension of the AFE phase at the expense of the FE phase is evident, as well as the extension of the transition region to higher temperatures for varying Zr/Ti ratios. There is also a reversal of the slope of the phase boundary between the AFE and FE phases as the La concentration is increased from 2 to 4 at.%

Figure 5 illustrates the Curie point lowering for increasing La concentration in two series of compositions; the effect is essentially linear. Deviations from linearity were noted for those compositions which had Curie points near or below room temperature.

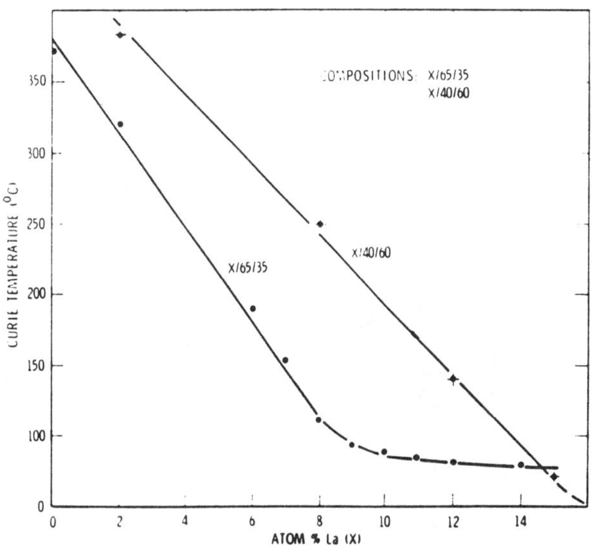

Fig. 5. Curie point lowering with increasing La content.

Table I. X-Ray Data for PLZT Compositions

Composition (La/Zr/Ti)	Phase	a_0 (Å)	c_0 (Å)	$\sqrt[3]{vol}$	Unit cell vol (Å^3)	Theor. density (g/cm^3)
2/65/35	Rh			4.093	68.57	7.976
6/65/35	Rh			4.092	68.52	7.900
8/65/35	Rh			4.091	68.47	7.863
9/65/35				4.091	68.47	7.845
10/65/35				4.091	68.47	7.824
12/65/35				4.090	68.42	7.784
8/80/20	Tet/Cubic			4.116	69.73	7.872
8/70/30	Rh			4.116	68.32	7.932
8/65/35	Rh			4.091	68.47	7.863
8/60/40	Tet/Rh	4.078	4.086		67.95	7.871
8/58/42	Tet	4.076	4.088		67.92	7.854
8/53/47	Tet	4.051	4.081		66.97	7.912
8/50/50	Tet	4.059	4.086		67.32	7.892
8/40/60	Tet	4.029	4.074		66.13	7.874
8/10/90	Tet	3.971	4.034		63.61	7.854
7/65/35	Rh			4.096	68.72	7.856
7/62/38	Rh			4.086	68.22	7.882
7/58/42	Tet	4.066	4.086		67.55	7.897
4/100/0	Mixed	(4.16)	(4.12)		(71.30)	
20/100/0	Mixed	(4.16)	(4.14)		(71.65)	
17/90/10	Mixed			(4.129)	(70.39)	
18/10/90	Tet	3.958	3.985		62.43	7.787
30/0/100	Cubic			3.944	61.35	7.551

NOTE: Values in parentheses are for major $PbZrO_3$ phase; Rh = rhombohedral, Tet = tetragonal.

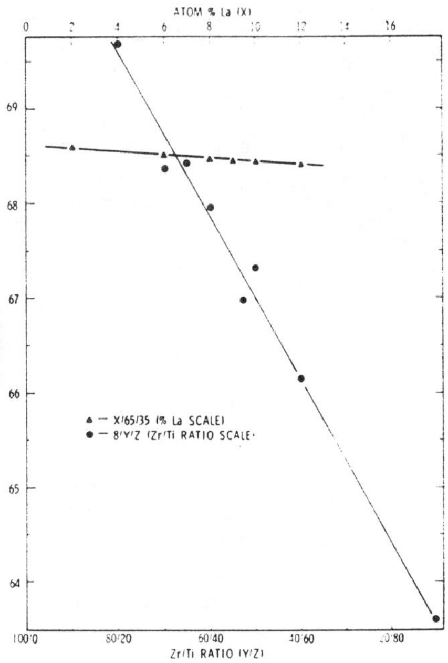

Fig. 6. Variation of PLZT unit cell volume with Zr/Ti ratio and La concentration.

X-ray results for selected series of compositions in the system are summarized in Table I. Selected unit cell volume data from this table are plotted in Fig. 6 as a function of La content and of Zr/Ti ratio; unit cell volume decreases linearly as a function of decreasing Zr/Ti ratio and increasing La content. Both effects support the view that extensive solid solution exists in this system.

(2) Physical Properties

The microstructure of a typical PLZT composition (Fig. 7) is fully dense and highly uniform. Most of the specimens had a small amount (<0.1%) of residual porosity which led to some light scattering within the material: this effect was most noticeable in plates 0.5 mm or more thick. The pores were ≈0.5 to 5 μm in diameter. Polished plates 0.25 mm or less thick appear transparent to the naked eye despite the residual porosity.

Bulk density decreases linearly with increasing La content (Fig. 8) for all Zr/Ti ratios investigated; however, the incremental decrease in density/at.% La was less for the lower Zr/Ti ratios (nearer $PbTiO_3$). Comparison of these density curves with the theoretical densities given in Table I shows that the measured densities are lower than those anticipated from the optical quality and residual porosity of the material. The difference is most noticeable at the larger concentrations of La, e.g. PLZT 12/65/35 had a measured density of 7.730 g/cm^3 and a theoretical density of 7.784 g/cm^3, yielding a relative density of only 99.3%. This value, compared with an anticipated relative density >99.8%, suggests that the formula weight of the composition, as calculated from the formula, is too high and that some material is lost by volatilization during hot-pressing. Quantitative wet chemical analysis should confirm the possibility of volatilization.

(3) Electrical Properties

Relative dielectric constant as a function of temperature for compositions with 65/35 and 40/60 Zr/Ti ratios is shown in Fig. 9. The peaks in the dielectric constant are reduced in height and in temperature as the La content is increased. For the series with 65/35 Zr/Ti ratios, a fairly consistent and

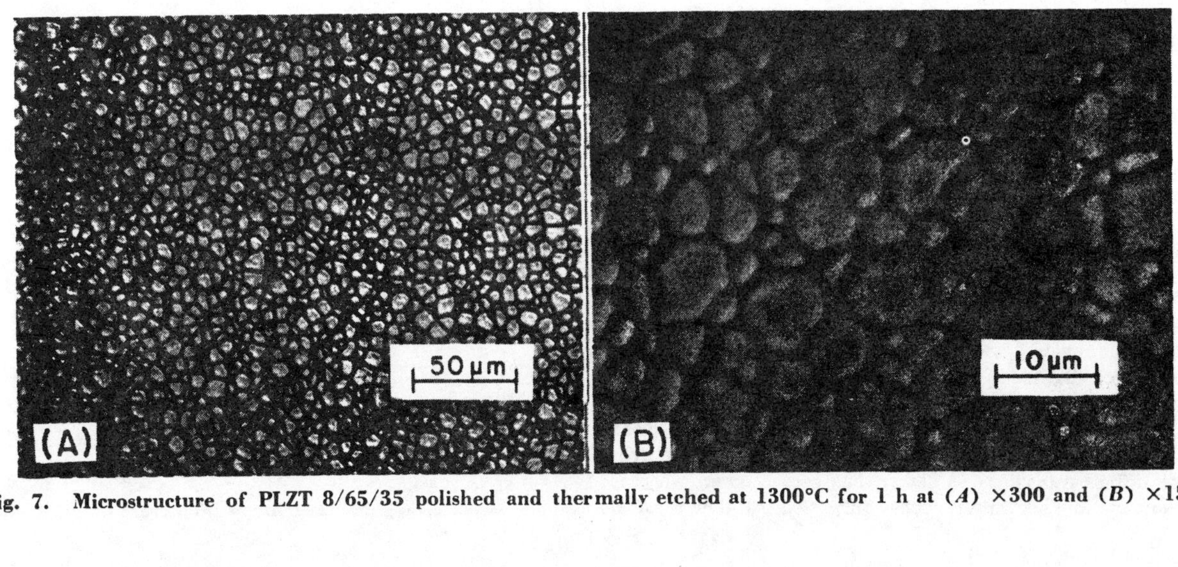

Fig. 7. Microstructure of PLZT 8/65/35 polished and thermally etched at 1300°C for 1 h at (*A*) ×300 and (*B*) ×1500

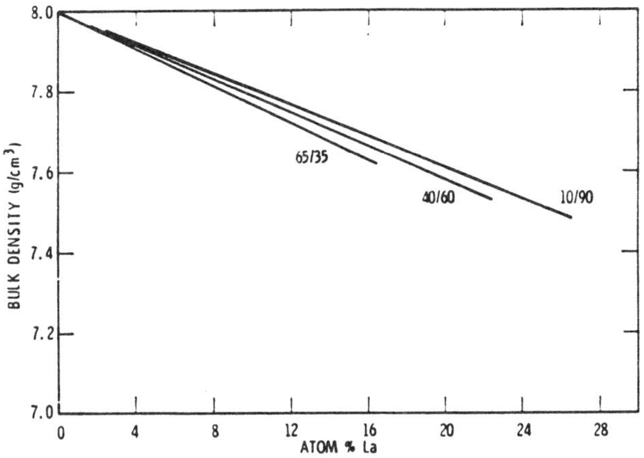

Fig. 8. Bulk density as a function of La content.

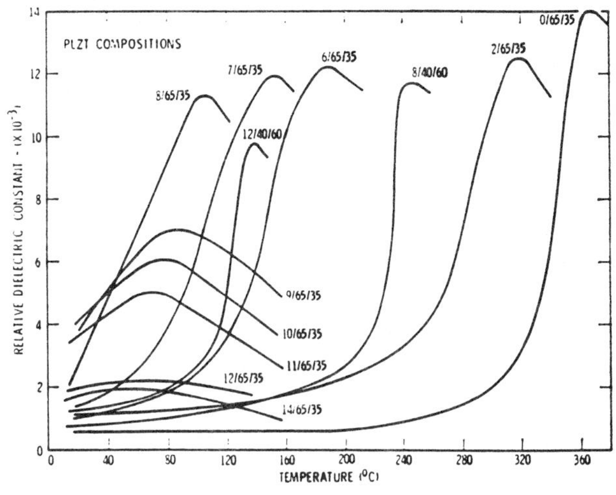

Fig. 9. Dielectric constant as a function of temperature.

gradual reduction in the Curie point occurs up to La concentrations of 8 at.% or less. Higher concentrations of La reduce the height of the dielectric constant peak, making it more diffuse, without significantly changing the temperature of the peak. This change in behavior between 8 and 9 at.% La may indicate that mixed FE/cubic phases are present on a microscopic or macroscopic scale.

Loss tangent was measured as a function of temperature on the compositions of Fig. 9. In general, losses for all materials were 2.5% or lower at room temperature; this value increased to 7.0% at 80°C for the 8/65/35 composition and 5.0% at 40°C for the 9/65/35 composition, the highest values recorded in the entire system. At 100°C and above, the losses decreased to ≈1.0% and remained constant until dc conductivity became predominant, usually at ≈200°C.

Room-temperature dielectric constants for several compositions are plotted directly on the phase diagram in Fig. 10(*A*). As a rule, $PbZrO_3$-rich and $PbTiO_3$-rich compositions are char-

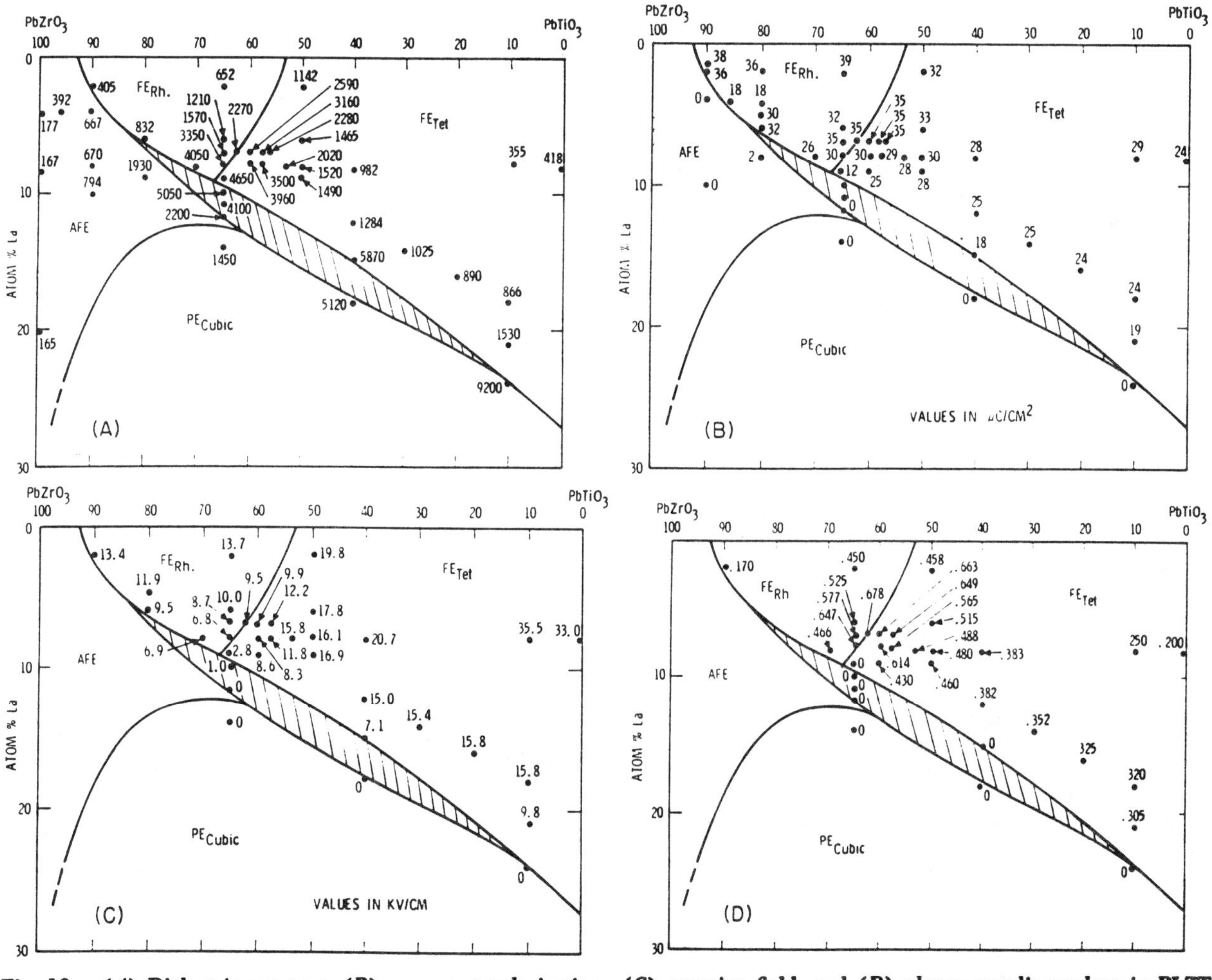

Fig. 10. **(*A*) Dielectric constant, (*B*) remanent polarization, (*C*) coercive field, and (*D*) planar coupling values in PLZT system.**

acterized by low dielectric constants and loss tangents of ≈1% or less. Maxima in dielectric constants occur along the FE-cubic and the FE (tetragonal)-FE (rhombohedral) boundaries.

Figure 10(*B*) illustrates the variation in saturation remanent polarization (P_R) throughout the phase diagram. Values are fairly constant in the system, even to Zr/Ti ratios as low as 10/90. Significant reductions, of course, occur at the FE-AFE and FE-cubic phase boundaries.

Electric switching field values (E_c) for the system (Fig. 10(*C*)), unlike remanent polarization, change more gradually and over a wider range in a given area of phase stability. For example, E_c for compositions with a constant Zr/Ti ratio of 65/35 varies regularly from 13.7 to 6.8 kV/cm within the FE (rhombohedral) phase region; E_c values range from 2.8 to 35.5 kV/cm in the FE (tetragonal) region of stability.

Values of electromechanical planar coupling (k_p) as a function of the phase system are given in Fig. 10(*D*). Compositions of constant 65/35 Zr/Ti ratio exhibit an increasing coupling factor to a peak value of 0.647 for the 8/65/35 composition; an abrupt decrease occurs at 9/65/35. Altering the Zr/Ti ratio from 65/35 to 10/90 at a constant 7 at.% La also first increases the coupling and then decreases it. The maximum planar coupling factor found in the PLZT system is 0.678 for the 7/62/38 composition, which is located on the phase boundary between the FE (rhombohedral) and FE (tetragonal) phases.

The manner in which planar coupling and dielectric constant vary as a function of La content for a constant Zr/Ti ratio of 65/35 is shown in Fig. 11(*A*). Coupling decreases to zero quite abruptly at about 8 to 9 at.% La; dielectric constant peaks in the region of transition between the FE and cubic phases. The effective optical birefringence, $\overline{\Delta n}$, for this compositional series, also shown, like coupling, decreases essentially to zero at 9 at.% La.

Figure 11(*B*) illustrates the variation of the same properties as a function of the Zr/Ti ratio at a constant 7 at.% La. The dielectric constant peaked at a Zr/Ti ratio of 58/42, whereas planar coupling and birefringence were maximum at 62/38. This result suggested a possibility of mixed phases along the FE (rhombohedral)/FE (tetragonal) phase boundary. A finite boundary width of ≈4 at.% change in Zr or Ti is not unreasonable, especially in a highly modified system such as the present one. X-ray analysis confirmed the existence of mixed phases. Similar results for niobia-doped $PbZrO_3$-$PbTiO_3$ materials were reported by Berlincourt.[16]

(4) *Optical and Electrooptic Properties*

(A) General: By far the most significant property of the PLZT solid-solution system is the transparency of compositions containing 8 at.% or more La. Polished plates of these materials 0.25 mm or less thick exhibit essentially 100% transmittance, neglecting reflection losses, in the visible and near infrared (Fig. 12). The curve on the upper scale of Fig. 12 shows that the fundamental absorption edge is near λ=0.37 μm for the material containing 8 at.% La, compared to the previously reported value of λ=0.34 μm for a similar material containing 2 at.% La.[4] The refractive index at 6328 Å is about 2.5, accounting for the reflection losses shown in Fig. 12.[4]

In addition to the significantly increased transparency of the PLZT ceramics, these materials are extremely versatile with respect to device properties, as illustrated by the areas enclosed by dashed lines on the room-temperature phase diagram (Fig. 13). Area ABC includes compositions which have good electrooptic memory characteristics[3,4]; area BDEC includes compositions with high coercivity which have linear electrooptic characteristics at saturation remanence; area EFGH includes compositions which have slim-loop[17] quadratic electrooptic characteristics. The points on the phase diagram represent compositions which were used to obtain data on electrooptic properties.

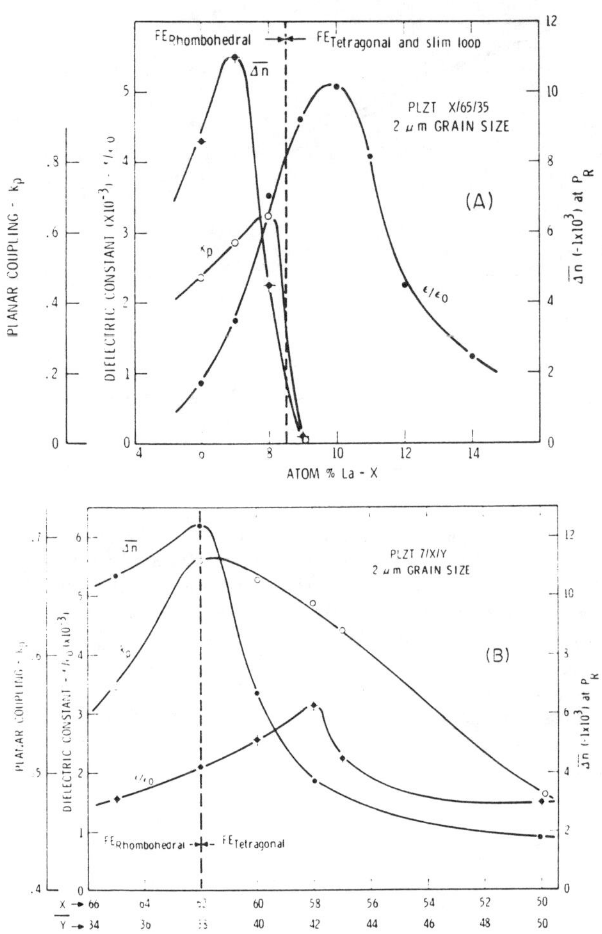

Fig. 11. Selected electrical and optical properties as functions of (*A*) La content for 65/35 Zr/Ti ratio and (*B*) Zr/Ti ratio for La content of 7 at.%.

(B) Memory Materials: Phase retardation, Γ, was measured as a function of remanent polarization for the compositions in area ABC of Fig. 13. From these data, effective birefringence, $\overline{\Delta n}$, vs normalized remanent polarization, $P = P_r/P_R$, was calculated. Figure 14 shows results for PLZT 8/65/35 with 2 μm grain size. The gap was first poled to

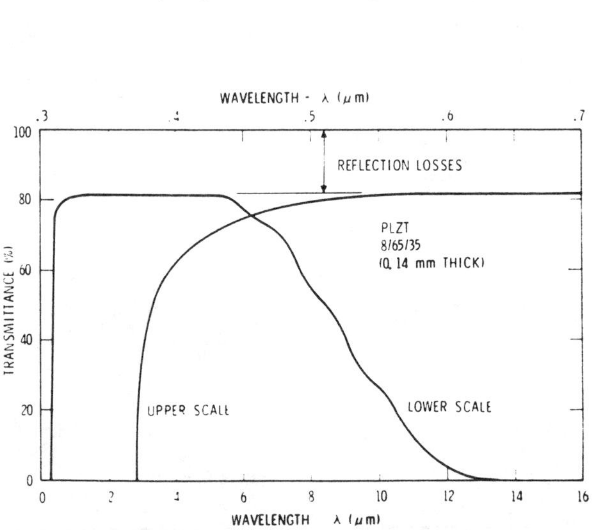

Fig. 12. Transmittance as a function of wavelength for PLZT 8/65/35. Reflection losses ≈18%.

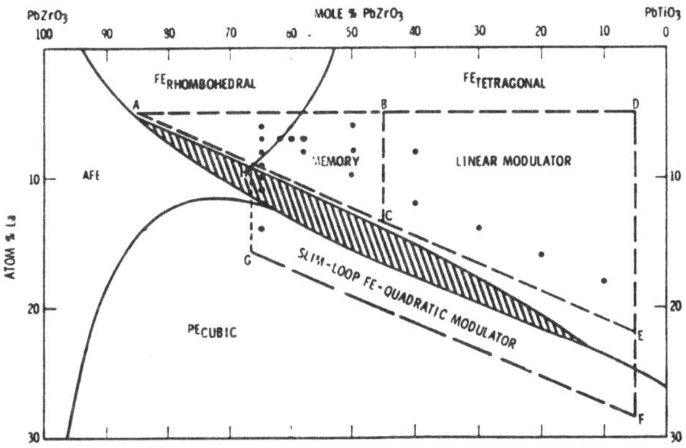

Fig. 13. Phase diagram of PLZT system showing compositional areas particularly applicable to memory, linear modulator, and quadratic slim-loop modulator devices.

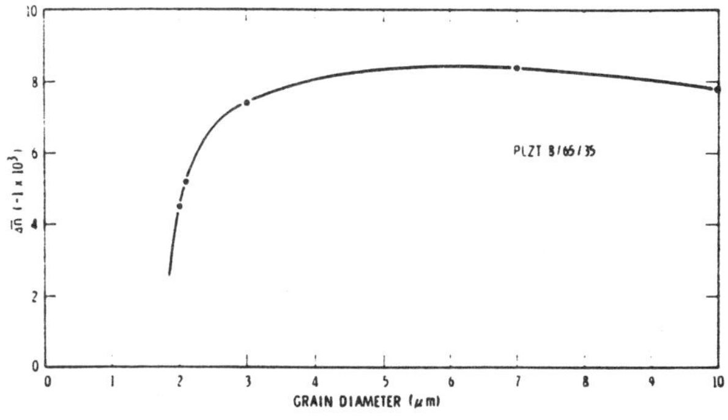

Fig. 15. Effective birefringence $\overline{\Delta n}$ (P_R, 0) measured at saturation remanence P_R vs grain size for PLZT 8/65/35.

positive saturation remanence ($P=1.0$) by applying an electric field $E=3E_c$; it was then switched incrementally from $P=1.0$ to negative saturation remanence, $P=-1.0$. The polarity of the switching voltage was reversed, and the gap was switched incrementally from $P=-1.0$ to $P=1.0$. The hysteresis is considerably larger and more asymmetrical than that observed for PLZT 2/65/35.[4] The asymmetry of the $\overline{\Delta n}$ vs P characteristic always depends on the polarity of the original poling field; the characteristic was repeatable within the accuracy of the measurement over several switching cycles. It is significant that the range of $\overline{\Delta n}$ values for 2-μm PLZT 8/65/35 is 100%, whereas for PLZT 2/65/35[4] it is only about 45% of its maximum value at $P=1.0$. PLZT 8/65/35 is the only electrooptic ceramic so far observed to have $\overline{\Delta n}=0$ for a state of remanent polarization; this result indicates that the material is optically isotropic on a macroscopic scale at a P near zero. The value of $\overline{\Delta n}$ at $P=1.0$ and $\lambda=6328$ Å is -0.0044 for PLZT 8/65/35, compared to -0.012 for PLZT 2/65/35.[4]

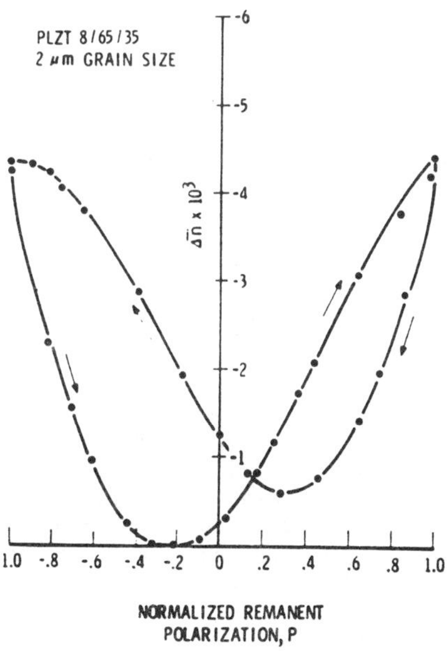

Fig. 14. Effective birefringence $\overline{\Delta n}$ (P_r, 0), measured at P_r with $E=0$, as a function of normalized remanent polarization P for PLZT 8/65/35. Arrows indicate switching direction

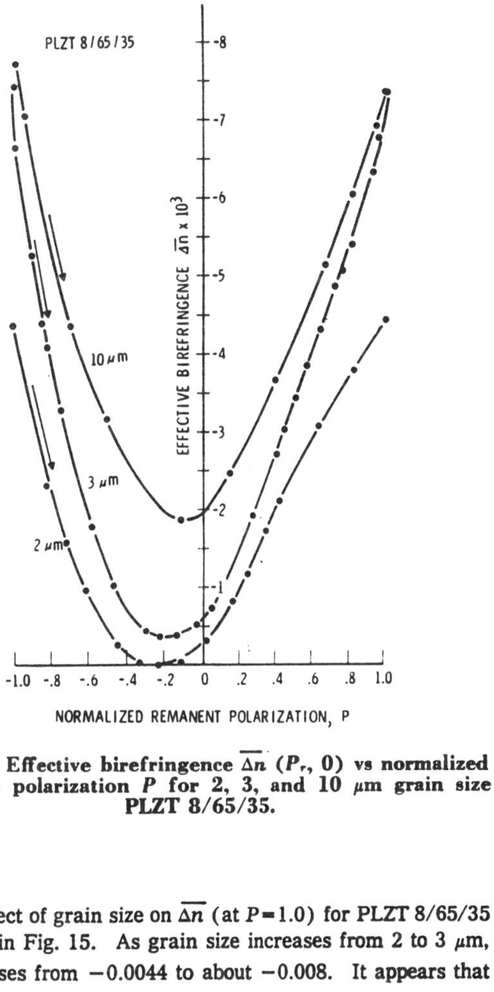

Fig. 16. Effective birefringence $\overline{\Delta n}$ (P_r, 0) vs normalized remanent polarization P for 2, 3, and 10 μm grain size PLZT 8/65/35.

The effect of grain size on $\overline{\Delta n}$ (at $P=1.0$) for PLZT 8/65/35 is shown in Fig. 15. As grain size increases from 2 to 3 μm, $\overline{\Delta n}$ increases from -0.0044 to about -0.008. It appears that $\overline{\Delta n}$ ($P=1.0$) remains nearly constant as grain size is further increased to 10 μm. The smaller value of $\overline{\Delta n}$ ($P=1.0$) of PLZT 8/65/35 compared to that of PLZT 2/65/35 may be explained, in part, by the decreasing Curie points of the materials with increasing La content.

Varying grain size also affects the range of $\overline{\Delta n}$ vs P (Fig. 16). The range of $\overline{\Delta n}$ for 2-μm grains is ≈0.004; for 3-μm grains, it is ≈0.007.

(C) Linear Electrooptic Materials: For many electrooptic devices, it is desirable that $\overline{\Delta n}$ be a linear (odd) function of E near $E=0$. For example, light modulators which must operate without a biasing field require such a relation between $\overline{\Delta n}$ and E. "Hard" (i.e. high-coercivity) tetragonal PLZT composi-

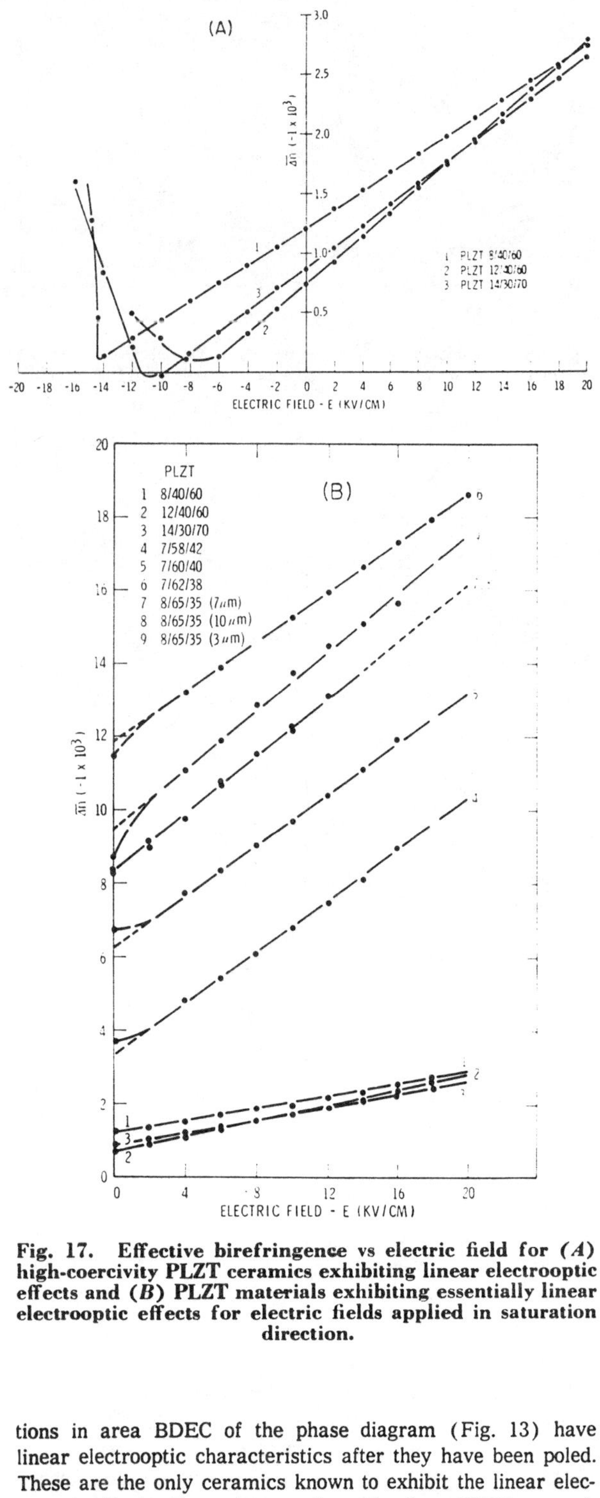

Fig. 17. Effective birefringence vs electric field for (*A*) high-coercivity PLZT ceramics exhibiting linear electrooptic effects and (*B*) PLZT materials exhibiting essentially linear electrooptic effects for electric fields applied in saturation direction.

tions in area BDEC of the phase diagram (Fig. 13) have linear electrooptic characteristics after they have been poled. These are the only ceramics known to exhibit the linear electrooptic effect. Examples are shown for three PLZT compositions in Fig. 17(*A*). These materials were initially poled to positive saturation remanence, P_R, using a field $E=2E_c$ ($\approx$30 kV/cm). Effective birefringence was then measured at P_R with various values of increasing and decreasing positive and negative dc electric field applied. The $\overline{\Delta n}$ vs E curves remain linear for positive applied fields exceeding 20 kV/cm and for negative applied fields which do not exceed the domain switching threshold, which occurs from −6 to −14 kV/cm for the compositions of Fig. 17(*A*) and is indicated by the abrupt change of slope in the curves.

Some of the softer tetragonal materials, as well as the larger-grained rhombohedral materials, exhibit nearly linear

Table II. Electrooptic Coefficient, r_c

Material	$r_c \times 10^{10}$ (m/V)
$LiNbO_3$	0.17
$LiTaO_3$	0.22
$Ba_2NaNb_5O_{15}$	0.36
$Ba_2(K_{0.9}Na_{0.1})Nb_5O_{15}$	0.52
KD_2PO_4	0.52*
PLZT 8/40/60	1.02
PLZT 14/30/70	1.12
$(Sr_{0.5}Ba_{0.5})Nb_2O_6$	2.10
PLZT 7/62/38	4.43
PLZT 8/65/35 (10 μm)	5.23
PLZT 8/65/35 (3 μm)	6.12
$(Sr_{0.75}Ba_{0.25})Nb_2O_6$	14.00

*$r_c=2r_{63}$ measured at $\lambda=0.546$ μm.

electrooptic characteristics for electric fields applied in the saturation direction. Examples of these are shown in Fig. 17(*B*), in which the characteristics of the hard compositions of Fig. 17(*A*) are replotted for comparison. The $\overline{\Delta n}$ vs E characteristics of all the softer materials deviate from linearity near $E=0$, except for those of the 7- and 10-μm rhombohedral materials. This result may indicate that all the electrooptic characteristics become more linear for larger grain sizes. The maximum $\overline{\Delta n}$ at saturation remanence occurs for the composition PLZT 7/62/38, which is on the morphotropic rhombohedral-tetragonal phase boundary. Compositions on either side of the phase boundary have reduced $\overline{\Delta n}$; the reduction is more pronounced in the tetragonal phase. This reduction of $\overline{\Delta n}$ in the tetragonal phase tends to confirm the results obtained by Fushimi and Ikeda[18] on single-crystal lead zirconate-titanate compositions.

A convenient and physically significant comparison of the linear electrooptic effect in PLZT ceramics with those of several linear electrooptic crystals may be made by examining their respective transverse electrooptic coefficients, r_c.[19–20]

$$r_c=\left[\left(\frac{n_3^3}{n_1^3}\right)r_{33}{}^T-r_{13}{}^T\right]=-\frac{2\Gamma}{n_1^3 lE_3} \tag{3}$$

$$\Gamma=\overline{\Delta n}l \tag{4}$$

Therefore,

$$\overline{\Delta n}=-(1/2)n_1^3 r_c E_3 \tag{5}$$

In Eqs. (3) through (5), n_1 and n_3 are the refractive indices normal and parallel, respectively, to the applied field, E_3; Γ is the phase retardation; and l is the light path length through the electrooptic medium. The electrooptic coefficients $r_{33}{}^T$ and $r_{13}{}^T$ are defined in Refs. 19 and 20. Equation (5) shows that r_c is simply a measure of the slopes of the curves in Fig. 17, but it is, additionally, related to the index ellipsoid of the electrooptic material.[19] Values of r_c for several electrooptic crystals and some of the PLZT ceramics of Fig. 17(*B*) are listed in Table II. The values for the PLZT compositions are exceptionally high compared with those for all except the (Sr,Ba)Nb_2O_6 crystals.

(D) Slim-Loop Quadratic Electrooptic Materials: The compositions within area EFGH of the phase diagram (Fig. 13) are called slim-loop ferroelectric materials because of their low or zero remanence and their nearly anhysteretic polarization vs electric field characteristics.[13] These compositions are paraelectric at room temperature. When an electric field is applied, the ferroelectric (rhombohedral or tetragonal) phase is field-enforced, and ferroelectric anisotropy increases with increasing electric field. As a result of this unique behavior, the materials have a classical Kerr quadratic electrooptic effect. In the absence of an electric field, the materials are optically isotropic; with a field E applied, they become birefringent, and their birefringence increases as the product of a constant and

E^2. The $\overline{\Delta n}$ vs E characteristics for PLZT slim-loop compositions are shown in Fig. 18(A). The dramatic change in electrooptic sensitivity of these compositions as at.% La varies from 9 to 11 is illustrated in Fig. 18(B), where $\overline{\Delta n}$ is plotted as a function of E^2.

A convenient method of comparing the quadratic electrooptic effects in slim-loop PLZT materials with those of other quadratic electrooptic materials is to first calculate the transverse quadratic electrooptic coefficients, R and g[20]: R is a measure of the slope of $\overline{\Delta n}$ vs E^2 curve (Fig. 18(B)); g coefficients relate $\overline{\Delta n}$ and polarization P via the quadratic electrooptic effect. The transverse g coefficient defined in Eq. (12) should be essentially constant for all slim-loop PLZT and (Pb,Ba)(Zr,Ti)O_3 (PBZT) materials (see Ref. 21).

$$R=(R_{33}-R_{13})=-\frac{2\Gamma}{n_1^3 l E_3^2}\quad(9)$$

Combining Eqs. (4) and (6) gives

$$\overline{\Delta n}=-(1/2)n_1^3 R E_3^2 \quad (7)$$

Similarly,

$$g=(g_{33}-g_{13})=-\frac{2\Gamma}{n_1^3 l P_3^2} \quad (8)$$

$$\overline{\Delta n}=-(1/2)n_1^3 g P_3^2 \quad (9)$$

If the relation

$$P_3=\varepsilon_0\left(\varepsilon_{33}^T-1\right)E_3\approx\varepsilon_0\varepsilon_{33}^T E_3 \quad (10)$$

is used, then

$$\overline{\Delta n}\approx-(1/2)n_1^3 g(\varepsilon_0\varepsilon_{33}^T E_3)^2 \quad (11)$$

and

$$g\approx R/(\varepsilon_0\varepsilon_{33}^T)^2 \quad (12)$$

It is apparent that R may be calculated directly from Eq. (7) and the curves of Fig. 18(B); g can then be calculated using Eq. (12) if the free static dielectric constants $\varepsilon_0\varepsilon_{33}^T$ of the PLZT compositions are known. The results of such calculations for the materials of Fig. 18 are given in Table III. The values of g in Table III agree well with those of several PBZT compositions and that of a PLZT composition reported by Thacher.[21] The value of R for PLZT 9/65/35 is 9.1, compared to a maximum of 3.2 for a PBZT composition reported in Ref. 21. Electric fields required to produce equal retardations in similar plates of the two materials would have a ratio of $\approx 1/\sqrt{3}$, the larger field being required by the PBZT material.

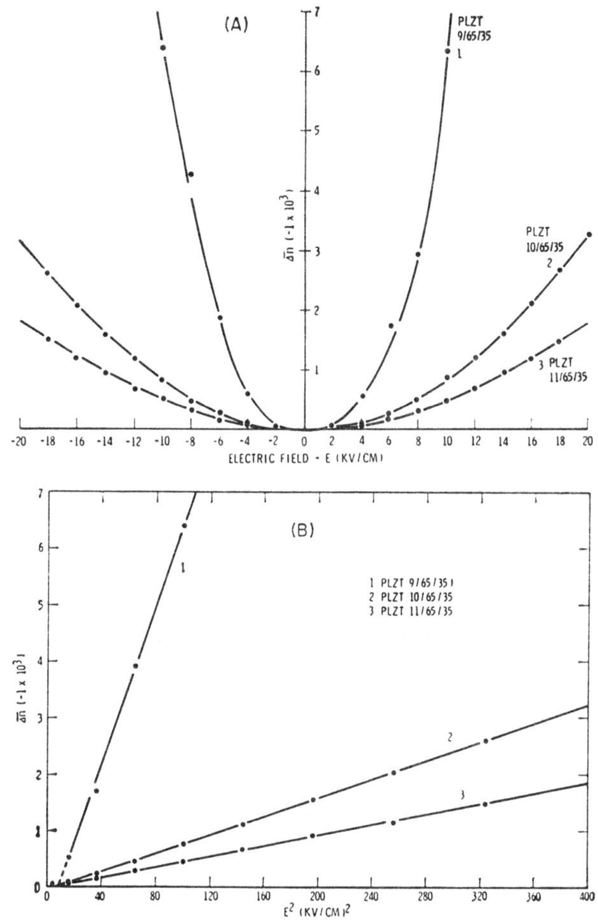

Fig. 18. Effective birefringence vs (A) E and (B) E^2 for PLZT slim-loop ceramics exhibiting quadratic electrooptic effects.

Table III. Electrooptic R and g Coefficients for PLZT Compositions

Composition (La/Zr/Ti)	$R\times10^{16}$ (m^2/V^2)	g (m^4/C^2)
9/65/35	9.116	0.018
10/65/35	1.073	.010
11/65/35	0.602	.014

(5) *Applications*

(A) *General:* Several types of devices using the electrically controlled birefringence of lead zirconate-titanate ceramics have been described.[1-6] These devices include light gates or shutters, light modulators, optical memories, displays, and spectral filters. In addition, Meitzler *et al.*[22,23] recently announced a new class of image storage and display devices.

Ceramic variable birefringence devices may be divided into categories according to the type of light source used:

(1) Devices which use a monochromatic light source vary the intensity of light transmitted by a polarizer-ceramic plate-analyzer network. These devices include light gates, modulators, optical memories, image stores, and displays.

(2) Devices which use white light vary the wavelength of light transmitted by a polarizer-ceramic plate-analyzer network. These devices are spectral filters which may be used as discrete single or multistage devices[4] or in linear or two-dimensional arrays.

Materials requirements for these device categories are discussed in the following sections.

(B) *Devices Using Monochromatic Light Source:* A range of control of phase retardation equal to ½ the light wavelength is desirable in nearly all devices in this category. The reason for this requirement is clear when the case of a light gate or shutter[6] is considered. If the light gate is set initially for extinction (OFF condition), the phase retardation, Γ, of the electrooptic material must be switched by $\lambda/2$ to obtain maximum light transmission through the analyzer (full ON condition). Since $\Gamma=\overline{\Delta n}l$, the change in retardation available using a given device depends on both the change in $\overline{\Delta n}$ and the light path length.

In devices requiring memory, the maximum available change in $\overline{\Delta n}$ is an intrinsic property of the material selected. In PLZT 2/65/35 material,[4] the maximum change in $\overline{\Delta n}$ is large (0.01), but light scattering is so pronounced that devices with light path lengths greater than 0.1 mm are not usually practical. Half-wavelength changes in phase retardation (where $\lambda/2$ does not exceed 1 μm) are, therefore, possible in single-stage devices, but two-stage networks are marginal, and multistage networks are not feasible.[4] On the other hand, the maximum available change in Δn for PLZT 8/65/35 is only ≈ 0.007, but light path lengths exceeding 0.5 mm are possible because of the greatly increased transparency of this material, especially at visible wavelengths. Thus, devices may be designed which produce half-wavelength changes in Γ with small incremental changes in P, and multistage networks are quite feasible. The substantially reduced light scattering in PLZT materials containing 6 at.% or more La provides improved extinctions for the OFF state of devices and, therefore, greatly improved ON to OFF ratios.

In devices using the conventional linear and quadratic electrooptic effects, the PLZT materials are clearly superior to previously reported electrooptic ceramics. The electrooptic coefficients, both linear and quadratic, of PLZT are the highest yet observed for ceramics and most single crystals, and their high transparency allows increased path lengths, leading to significant reductions in half-wavelength voltages and appreciably higher ON to OFF ratios.

(C) Devices Using White Light Source: Spectral filters for visible wavelengths require a minimum change in Γ of about 0.9 μm/filter stage. Also, these devices require reasonably uniform transmittance vs wavelength response from 0.4 to 0.7 μm. Previously reported electrooptic ceramics (e.g. PLZT 2/65/35) met the first requirement for a single-stage spectral filter, but scattering dispersion in the visible greatly attenuated the shorter wavelengths (0.4 to 0.5 μm).[4] The increased transparency and improved transmittance vs wavelength response in the visible of PLZT ceramics with 6 at.% or more La (Fig. 12) make multistage (narrow bandwidth) spectral filters quite feasible.

(D) Image Storage and Display Devices: The strain-biasing techniques of Meitzler *et al.*[22,23] make possible longitudinal electrooptic effect devices using ferroelectric ceramics. In these devices Δn is varied by electric fields parallel to the light propagation direction using transparent In_2O_3 electrodes. Thus, the size of electrooptically active areas of ceramic plates needs no longer be limited by plate thickness or high voltage requirements. This development greatly increases the scope of possible ceramic electrooptic devices.

IV. Conclusions

(1) The quaternary (Pb,La)(Zr,Ti)O_3 system (PLZT) represents a class of transparent ferroelectric materials readily adaptable to a variety of electrooptic applications.

(2) Hot-pressed ceramics of compositions in the PLZT system were prepared from mixed oxides and hot-pressed in alumina dies at typically 1100°C for 16 h at 2000 psi.

(3) Lanthana modification of the lead zirconate-titanate solid solutions (1) extends the AFE phase at the expense of the FE (rhombohedral) phase, (2) favors the FE (tetragonal) phase over the FE (rhombohedral) phase, (3) lowers the Curie point linearly, and (4) lowers the bulk density linearly.

(4) Room-temperature dielectric constants in the PLZT system were maximum along the phase boundary between the FE and cubic phases.

(5) Planar coupling values reached a maximum along the boundary between the FE (rhombohedral) and FE (tetragonal) phases. Coupling also increased with increasing La content until the Curie point was reduced to room temperature or below. A maximum value of 0.678 was measured for PLZT 7/62/38 (La/Zr/Ti).

(6) Remanent polarization was fairly constant throughout the PLZT system. Coercive field decreased with increasing Zr/Ti ratio and increasing La content.

(7) The transmittance vs wavelength response of PLZT ceramic plates is essentially flat from the absorption edge near 0.37 μm to ≈ 6 μm. The highest transmittance values (essentially 1.00, neglecting reflection losses of ≈ 0.18) were noted for plates up to 0.25 mm thick which contained 8 at.% La or more.

(8) The range of variation of birefringence with remanent polarization for PLZT 8/65/35 depended on grain size. The range for 2-μm grains was ≈ 0.004; for 3-μm grains, it was ≈ 0.007, the maximum range measured.

(9) Linear electrooptic effects were observed in high-coercive-field tetragonal compositions and in lower-coercive-field tetragonal and rhombohedral compositions having a grain size of 3 μm or larger. The linear transverse electrooptic coefficients, r_c, for the PLZT ceramics were larger than those reported for most single crystals, except for (Sr,Ba)Nb_2O_6 compositions.

(10) Quadratic electrooptic effects were observed in slim-loop PLZT compositions near the PE-FE (rhombohedral or tetragonal) phase boundary. These materials are optically isotropic with electric-field-enforced FE anisotropy. A wide range of quadratic electrooptic R coefficients is available with slight variations in La content. Electrooptic g coefficients are consistent with those previously measured for (Pb,Ba)(Zr,Ti)O_3 compositions.[21]

(11) Single-stage ceramic electrooptic devices with increased light path lengths (>0.5 mm) and multistage devices, such as narrow-band spectral filters, are possible because of the increased transparency of PLZT ceramics with 6 at.% or more La.

(12) Ceramics for memory and linear or quadratic modulator applications can be custom-designed because of the versatility of the PLZT system.

Acknowledgments

The writers express sincere appreciation to S. F. Duliere for the X-ray measurements, to I. D. McKinney for preparation of all samples for electrooptic measurements, and to P. D. Thacher and P. D. Wilcox for critical reading and discussion.

References

[1] C. E. Land, "Ferroelectric Ceramic Electrooptic Storage and Display Devices"; presented at the IEEE Electron Devices Meeting, Washington, D. C., Oct. 18–20, 1967; Sandia Laboratories Research Rept. SC–R–67–1219, Oct. 1967.

[2] C. E. Land, "Electrooptic Ceramic Storage and Display Devices," *Oyo Butsuri*, **39** [1] 18–29 (1970).

[3] C. E. Land and P. D. Thacher, "Ferroelectric Ceramic Electrooptic Materials and Devices," *Proc. IEEE*, **57** [5] 1969.

[4] P. D. Thacher and C. E. Land, "Ferroelectric Electrooptic Ceramics with Reduced Scattering," *IEEE Trans. Electron Devices*, **16** [6] 515–21 (1969).

[5] C. E. Land and G. H. Haertling, "Optical Properties of Ferroelectric Ceramics," *J. Phys. Soc. Jap., Suppl.*, **28**, 96–99 (1970).

[6] J. R. Maldonado and A. H. Meitzler, "Ferroelectric Ceramic Light Gates Operated in a Voltage-Controlled Mode," *IEEE Trans. Electron Devices*, **17** [2] 148–57 (1970).

[7] G. H. Haertling, "Hot-Pressed Ferroelectric Lead Zirconate Titanate Ceramics for Electro-Optical Applications," *Amer. Ceram. Soc. Bull.*, **49** [6] 564–67 (1970).

[8] Takuro Ikeda and Tatsuya Okano, "Piezoelectric Ceramics of Pb(Zr,Ti)O_3 Modified by $A^{1+}B^{5+}O_3$ or $A^{3+}B^{1+}O_3$," *Jap. J. Appl. Phys.*, **3** [2] 63–71 (1964).

[9] Takuro Ikeda, "Antiferroelectric Phase in the System $PbTiO_3$-$PbZrO_3$-$LaFeO_3$," *ibid.*, [8] 493–94.

[10] D. A. Berlincourt, "Research on Piezoelectric Materials and Phenomena," Sandia Laboratories Summary Rept. 323160, 1965.

[11] N. G. Eror and D. M. Smyth; pp. 62–73 in Chemistry of Extended Defects in Non-Metallic Solids. North-Holland Publishing Co., Amsterdam, 1970.

[12] G. H. Haertling, "Grain Growth and Densification of Hot-Pressed Lead Zirconate-Lead Titanate Ceramics Containing Bismuth," *J. Amer. Ceram. Soc.*, **49** [3] 113–18 (1966).

[13] G. Goodman, "Ferroelectric Behavior in Barium Zirconium Metaniobate," *ibid.*, **43** [2] 105–13 (1960).

[14] V. A. Bokov and I. E. Myl'nikova, "Electrical and Optical Properties of Single Crystals of Ferroelectrics with a Diffused Phase Transition," *Sov. Phys.-Solid State*, **3** [3] 613–23 (1961).

[15] V. A. Isupov, "Causes of Phase Transition Broadening and Nature of Dielectric Polarization Relaxation in Some Ferroelectrics," *ibid.*, **5** [1] 136–40 (1963).
[16] D. A. Berlincourt, "Properties of Lead Titanate Zirconate Ceramics," Tech. Rept. 3, Sandia Corp., Rept. 30135, 1957.
[17] P. D. Wilcox and C. A. Hall, "Dielectric Properties of Relaxator-Like Ceramics in the (Pb,Ba)(Zr,Ti)O_3 System"; presented at the Fall Meeting of the Basic Science and Electronics Divisions, The American Ceramic Society, St. Paul, Minn., Sept. 18, 1968 (Paper No. 21–BE–68F); for abstract see *Amer. Ceram. Soc. Bull.*, **47** [8] 754 (1968).
[18] Shoichi Fushimi and Takuro Ikeda, "Optical Study of Lead Zirconate-Titanate," *J. Phys. Soc. Jap.*, **20** [11] 2007–12 (1965).
[19] I. P. Kaminow and E. H. Turner, "Electrooptic Light Modulators," *Proc. IEEE*, **54** [10] 1374–90 (1966).
[20] I. P. Kaminow; pp. 183–96 in Ferroelectricity. Edited by Edward Weller. American Elsevier Publishing Co., New York, 1967.
[21] P. D. Thacher, "Electrooptic *g* Coefficients of Pb-Containing Oxygen Octahedra Ferroelectrics: Ceramic (Pb,Ba)(Zr,Ti)O_3," *J. Appl. Phys.*, **11** [12] 4790–97 (1970).
[22] A. H. Meitzler, J. R. Maldonado, and D. B. Fraser, "Fine-Grain, Ferroelectric Ceramics in Image Storage and Display Devices"; presented at the 72nd Annual Meeting of The American Ceramic Society, Philadelphia, Pa., May 4, 1970 (Electronics Division, No. 4–E–70); for abstract see *Amer. Ceram. Soc. Bull.*, **49** [4] 412 (1970).
[23] J. R. Maldonado and A. H. Meitzler, "Strain-Biased Ferroelectric Picture Device"; presented at the IEEE Electron Devices Research Conference, Seattle, Wash., June 1970; submitted to *Proceedings of Institute of Electrical and Electronics Engineers.*

Helical Kerr Cell

E. F. DAWSON AND N. O. YOUNG
Ericsson Telephones Ltd., Instrument Division, New Basford, Nottingham, England
(Received March 27, 1959)

A cell having a pair of helical electrodes and containing material exhibiting the Kerr electrostatic effect is shown to give a pure optical rotation proportional to the fourth power of the electrostatic field. Experiments in support of the theory are described.

IT has long been known[1–3] that a skewed arrangement of retardation plates distributed through 180° will give an approximately pure optical rotation. The Pockels calculation shows that rotation $\bar{W}$ is related to the angular retardation Δ_s per sheet and the angle ω between the optic axes of successive birefringent sheets.

For the case where $\omega=\pi/m$, m being the number of sheets, Pockels gives the relation

$$\bar{W}=(m/8)\Delta_s^2\cot(\pi/m). \quad (1)$$

In general, the light emerging from such systems is elliptically polarized. Jones[4] shows how the amplitudes of the electric vectors in such systems may be calculated using matrix algebra and verifies that for low ellipticity the plates should be skewed through an integral number of semicircles.

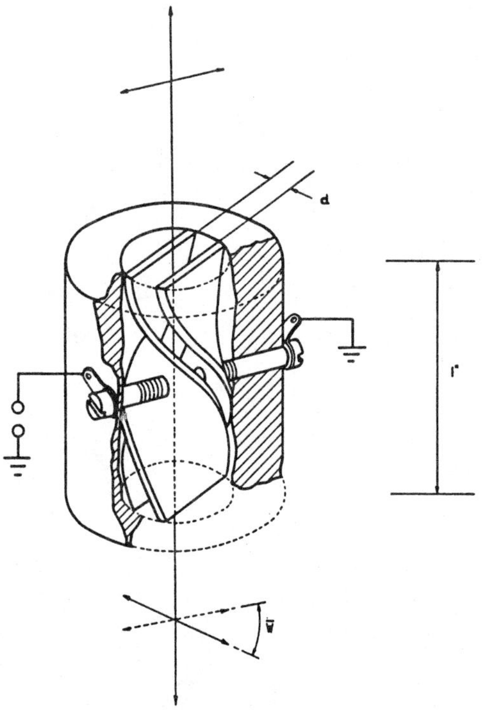

FIG. 1. Helical Kerr cell.

[1] E. Reusch, Pogg. Ann. **138**, 628 (1869).
[2] L. Sohncke, Math. Ann. **9**, 505 (1876).
[3] F. Pockels, *Lehrbuch der Kristall Optic* (Teubner, Leipzig, 1906), pp. 289–290.
[4] R. C. Jones, J. Opt. Soc. Am. **34**, 488–503 (1941).

Evans[5] investigated the Solc color filter, which also is an example of the use of skewed retardation plates, although the angle of distribution in this case is not 180°. In one form of Solc filter, Evans employs the methods of Jones to verify that the optimum angle of distribution is 90° and to compare Solc with Lyot filters.

The desirability of simple methods of standardizing sensitive polarimeters such as the E.T.L.–N.P.L. Automatic Polarimeter* led Young[6] to the investigation of helically symmetrical films of birefringent material. Using a special "off normal" evaporating technique with rotating specimen holders, it was demonstrated that significant optical rotation for this purpose could be achieved. Ellipticity is a matter of some difficulty, in general, and from experiment the best distribution angle appeared to be a multiple of 360°, although it is probably more correct to assume that an exact multiple of 180° would be sufficient.

The resemblance of the helically deposited films to Reusch rotators appears to be analogous to the similarity between "distributed" and "lumped" transmission systems in communications. Evidently such systems may be described mathematically in terms of differential equations[7] or of matrix algebra[8] by using Sylvester's theorem.[9] Distributed parameters may be introduced by allowing the number of elements to approach infinity in the latter method.

Of the various devices showing birefringence and retardation the Kerr cell is of special interest since it has an electrically controllable retardation.

The development of the foregoing ideas has shown that a 180° skewed set of Kerr cells is analogous to the Reusch rotator while a single Kerr cell having a helical electrode geometry, Fig. 1, is analogous to the Young helical film rotator.

A helical Kerr cell having birefringence distributed continuously in azimuth appeared to be the more interesting case and such a cell was built. Experiments

[5] J. W. Evans, J. Opt. Soc. Am. **48**, 142–145 (1958).
* Developed at Ericsson Telephones Ltd., from a prototype made at the National Physical Laboratories, Teddington, Middlesex.
[6] N. Young and J. Kowal, Nature **183**, 104–105 (1959).
[7] C. W. Oseen, Arkiv. Math. Astron. Fys. **21A**, pt. 11, No. 25 (1929).
[8] R. C. Jones, J. Opt. Soc. Am. **38**, 671–684 (1948), **46**, 126–131 (1956).
[9] Frazer, Duncan, and Collar, *Elementary Matrices* (Cambridge University Press, London, 1938), p. 78.

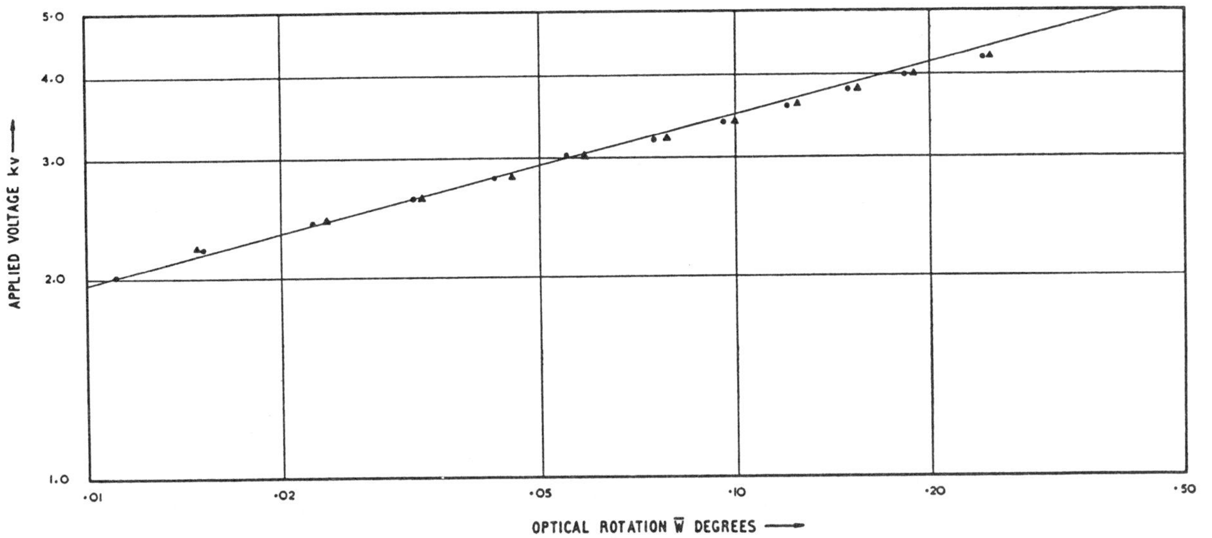

FIG. 2. Optical rotation *vs* applied voltage showing experimental points and a line drawn according to the theory.

showed that its activity $\bar{W}$, was proportional to the fourth power of the applied voltage,

$$\bar{W} \propto (V/d)^4, \tag{2}$$

where d is the electrode separation. This formula is obtained by applying the usual Kerr-effect retardation formula,

$$\Delta_s = 2\pi\beta\delta(V/d)^2 \times 1.12 \times 10^{-5} \text{ rad}, \tag{3}$$

where β is the Kerr constant, δ is the length of an elementary electrode, and the factor $1.12\times10^{-5} = (1/299.7)^2$ is used to convert from esu to practical units.

Introducing the over-all electrode length $Z = m\delta$ and inserting Δ_s in (1), the optical rotation of m skewed Kerr cells becomes,

$$\bar{W}_m = (m/8)[2\pi\beta(V/d)^2(Z/m)\times1.12\times10^{-5}]^2 \times\cot(\pi/m). \tag{4}$$

By allowing $m\rightarrow\infty$, the distributed helix gives a rotation

$$\bar{W} = 6.27\times10^{-11}\pi\beta^2Z^2(V/d)^4 \text{ rad}. \tag{5}$$

Figure 2 shows the rotation measured by an E.T.L.-N.P.L. Automatic Polarimeter Type 143A in which the specimen was a helical-electrode Kerr cell.

The over-all length of the electrodes was 2.53 cm, and they were skewed through 180°. The mean electrode spacing was 0.48 cm, and the cell was filled with nitrobenzene of Kerr constant $\beta = 220\times10^{-7}$. The calculated rotation is 0.053° for 3 kv compared with the experimental value of 0.056° for this voltage. The discrepancy is small considering the rapid variation of rotation with electric field. At the lower applied voltages it was found that small bubbles tend to congregate in the optical path and it is believed that this accounts for some of the discrepancies since the accuracy of measurement with the automatic polarimeter at 0.01° is high. With larger voltages on the Kerr cell the bubbles are swept clear and the optical path is less obstructed.

It is felt that improvement in the present cell design and the use of high purity nitrobenzene would eliminate this difficulty and also allow the cell to be operated up to, say, 8 kv where a rotation of 3° would be observed.

Mechanical rotation of the cell while maintaining its polarizing voltage constant showed that the helical Kerr cell was very nearly a pure rotator. The peak-to-peak variation of optical rotation with azimuth is a measure of residual birefringence. At a polarizing voltage of 4 kv the cell showed a peak-to-peak variation about 3% of the mean optical rotation. We ascribe this mainly to imperfections in electrode geometry and electrostatic field fringe effects.

ACKNOWLEDGMENTS

The writers are grateful to Mr. J. Kowal for carrying out many of the measurements and to Dr. W. T. Welford of the Imperial College for discussions on the use of matrices in optics problems. We also thank Dr. J. H. Mitchell, Director of Research, Ericsson Telephones Limited, for permission to publish this paper.

Piezo-Optical Birefringence Modulators: New Use for a Long-Known Effect

James C. Kemp
Department of Physics, University of Oregon, Eugene, Oregon 97403
(Received 6 March 1969)

A general description is given of a long-neglected use for the piezo-optic or photoelastic effect (stress-induced birefringence). An acoustic vibration, such as a fundamental extensional mode in a bar or long thin plate, is set up in a block of isotropic transparent material, such as glass or fused silica; the vibration is sustained by a transducer. The resulting modulated birefringence can be used in a variety of ways, notably to produce a beam of alternately left- and right-circularly polarized light for circular-dichroism measurements. Strains of the order 10^{-5} are required, considerably below the breakage point for most materials. Because advantage is taken of the high Q of the vibrational modes, typically 10^3 to 10^4, very small transducer power is needed, usually less than 1 W. The literally enormous useful angular aperture, of the order 50° total cone angle, makes the device far superior to Pockels or Kerr cells for many applications. Reference is made to current practical realizations and to present and future uses of the device.

Index Headings: Birefringence; Photoelasticity; Modulation of beams; Polarization; Glass; Quartz; Ellipsometry; Magneto-optics; Optical systems.

If a light source is viewed through a pair of polarizers adjusted for extinction, i.e., with their passing axes crossed, and if a piece of glass or other normally isotropic transparent material is inserted between the polarizers, it is well known that strain in the material will destroy the extinction. A pattern of light and dark appears, revealing the strain contours. This is the piezo-optical or photoelastic effect. In a material that is otherwise isotropic and uniform, uniaxial strain (i.e., compression or distension) along one direction, say the x direction, makes the refractive index n_x differ from the index n_y, where y is perpendicular to the strain axis. This birefringence results, as in a wave plate (e.g., $\lambda/4$ plate, etc.), in a path difference or retardation $\phi=2\pi t(n_x-n_y)/\lambda$ between the x and y polarization components of a light beam of wavelength λ which has traversed a path length $\Delta z=t$ through the material. Any ray traversing a region with such a strain and a strain direction such that $\phi=(N+\frac{1}{2})\pi$ ($\lambda/2$-plate action) produces a spot of full brightness in the pattern, i.e., same brightness as would appear with the polarizers set parallel; while the pattern is completely dark where $\phi=N\pi$. These extremes strictly apply only when monochromatic light is used, although with white light the zero-strain contours remain completely dark.[1]

The long-standing use of this effect is in detecting mechanical strain in transparent bodies. Equally attractive should be the converse application: the production of optical retardation, i.e., wave-plate action, by applied stress. It is rather incredible that such a use has been almost totally neglected—or at the very least scarcely exploited—until just recently. The simplest such application is to make a wave plate, by statically stressing a block of glass, fused silica, or similar substance. Very modest force, easily attained with a small **C**-clamp or even finger pressure, is enough to produce $\lambda/4$ retardation at visible wavelengths, in a matchbox-size block of almost any of the common optical materials. The writer (and no doubt others) has considered using **C**-clamped blocks of CaF_2 for uv-passing $\lambda/4$ plates; a vernier dial on the pressure screw would give a convenient retardation adjustment.

A further step beyond the use of static stress is to make use of mechanical vibration, and more specifically of self-vibrations, i.e., of standing sound waves set up in the block of material. We consider this for two reasons: (1) In many applications there is a natural advantage in working with a modulated rather than with a constant birefringence; and more critically, (2) we benefit from the mechanical Q of the material. The applied forces needed to sustain a given amplitude of strain vibration, thus a given amplitude of modulated birefringence, are much less, by the factor Q, than those needed for an equivalent static strain. This is because the internal dynamical inertia of the material comes into play. If there were no acoustic losses (infinite Q), an acoustic oscillation would persist indefinitely, and a freely suspended block would generate modulated birefringence by itself, of more or less arbitrary amplitude. Practical optical materials have Q in the range 10^2 to 10^4, for vibrations in the audio to 100-kHz range.

Schematically, the piezo-optic modulator may appear as in Fig. 1. A simple extensional (longitudinal) acoustic standing wave is set up in a bar of transparent material. The fundamental mode, $l=\lambda_s/2$, where λ_s is the sound wavelength, is assumed to be used. We sketch the bar as unconstrained, i.e., freely suspended; the coupled transducer needed to sustain the oscillation is not shown. The extensional displacements along the bar are given by $\delta=\delta_0\sin(2\pi x/\lambda_s)\sin(\omega t)$; the corresponding strain is proportional to $\partial\delta/\partial x$, i.e., to $\cos(2\pi x/\lambda_s)$, and is maximum at the bar's center. The birefringence (n_x-n_y) is proportional to the strain and varies in time as $\sin(\omega t)$.

Two or more general sorts of applications are suggested in Fig. 1, involving the addition of polarizers. If light polarized at 45° to the bar axes, with flux I_0,

[1] H. J. Jessop, in *Encyclopedia of Physics, Vol. VI*, S. Flügge, Ed. (Springer-Verlag, Berlin, 1958). This is a review and includes references to books on mechanical engineering applications.

passes through the modulator, it takes on a time-varying, elliptical polarization. If the peak retardations are $\pm\lambda/4$, the polarization of the transmitted light varies between left- and right-circular polarizations: This modulated circular polarization can be used very effectively in measuring circular dichroism,[2] i.e., the differential absorption of left- vs right-circularly polarized light, an intrinsic property of some substances and a general property of all media when a magnetic field is present (magnetic circular dichroism).

If a second polarizer, P' in Fig. 1, is placed after the modulator bar, with axis inclined at $-45°$ to the bar axis and 90° relative to the input-polarizer (P) axis, an optical switch or irradiance modulator results. In the arrangement shown, for monochromatic light the time-varying irradiance of the exit wave can be shown to be

$$I_2(t)=I_0[\sin(\phi_0 \sin\omega t)]^2=I_0\{\tfrac{1}{2}[1-J_0(\phi_0)] + J_2(\phi_0)\cos(2\omega t)+J_4(\phi_0)\cos(4\omega t)+\cdots\},$$

where ϕ_0 is the peak retardation in radians (magnitude at maximum stress). The analysis is identical to that applying to Pockels-cell modulators. The time-averaged transmittance $\bar{I}_2/I_0$, specifically the term $\tfrac{1}{2}[1-J_0(\phi_0)]$ above, shows characteristic Bessel-function fluctuations as a function of ϕ_0—note Fig. 2(b). A maximum mean transmittance of about 70% is obtained with $\phi_0 \simeq 3.8 \sim \pi$, that is, when the retardation amplitude is slightly greater than half-wave. When broad-band (e.g., white) light is used, the Bessel-function fluctuations of $\bar{I}_2/I_0$ are more or less smoothed out, and the effective time-averaged transmittance increases roughly monotonically, reaching slightly more than 0.5 when ϕ_0 for the mean wavelength is about π and then levelling off to 0.5.

Two modes of use for this flux modulation come to mind:

(a) If the rapid modulation at the sound-vibration frequency and its harmonics is ignored, we may use the device as a simple switch: we merely turn the bar excitation on or off. With the bar quiescent, $\phi_0=0$, and $I_2=0$ since the polarizers are crossed. Excitation of the bar destroys the extinction, on the average. Apart from simple on–off shuttering, the power fed to the transducer that drives the bar can be envelope-modulated, producing an arbitrary modulation of the light flux. But in such a mode of use, the birefringence modulator is not a particularly fast device, at least in the form discussed here. The response time is the ringing time of the mechanical mode of the bar, $\tau=2Q/\omega=Q/(\pi f)$. For practical designs (see below) τ falls in the range 10^{-2} to 10^{-1} sec.

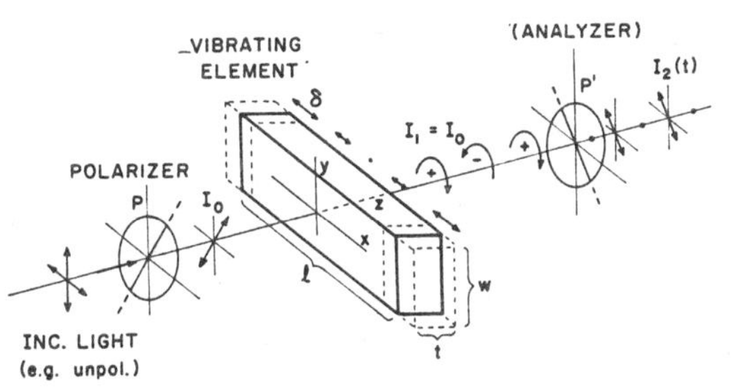

FIG. 1. The piezo-optical birefringence modulator in rudimentary form. A simple extensional vibration is set up in a transparent bar, sustained by an acoustic transducer (not shown). Incident light I_0 linearly polarized at 45° to the bar axis is rendered alternately right- and left-circularly polarized by the alternating birefringence, if the peak birefringence corresponds to $\lambda/4$ retardation; this output (I_1) can be used in circular-dichroism measurements. Addition of P', or of a stationary $\lambda/4$ plate followed by P', permits flux modulation at frequencies 2ω or ω, respectively.

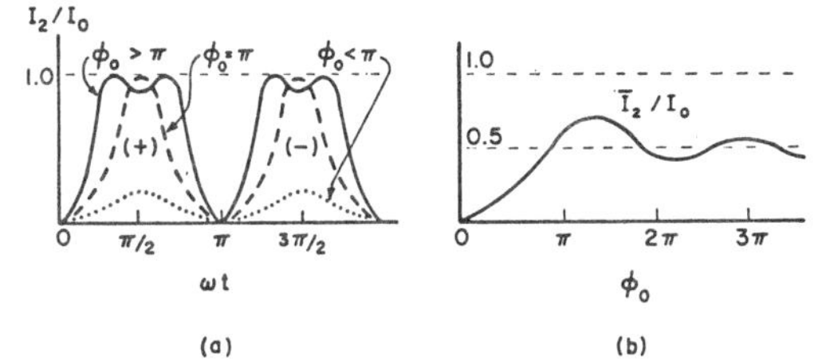

FIG. 2. The output flux I_2 of Fig. 1, as a function of time (a); the time-averaged mean flux as a function of the retardation amplitude ϕ_0 is shown in (b). The (+) and (−) peaks in (a) correspond to the senses of elliptic or circular polarizations in the beam incident on P' in Fig. 1, for the case $0<\phi_0<\pi$.

(b) As a strictly time-periodic modulator, the device produces an almost pure sinusoidal waveform, provided that the retardation amplitude is less than π—note Fig. 2(a). For $\phi_0>\pi$, there is increasing harmonic content. Actually, the crossed-polarizer setup in Fig. 1 yields only double-frequency (2ω) modulation, plus higher even harmonics. Fundamental-frequency (ω) modulation is easily achieved by inserting on ordinary $\lambda/4$ plate between the modulator and P', and near-perfect $\sin(\omega t)$ output is then obtained when $\phi_0 \simeq \pi/2$[(2)].

COMPARISON WITH OTHER DEVICES

Other devices with applications in common with those of the piezo-optic modulator are the Kerr cell and the Pockels cell; and, for certain purposes, the ordinary static wave plate.[3] The most important advantage of the piezo-optic device is the relatively enormous angular aperture.

A. Angular-Aperture Considerations

Exact treatment of the refraction through an anisotropic but homogeneous medium is the central topic of crystal optics.[4] In the birefringence modulator, we deal with an almost isotropic medium, having a small uniaxial birefringence: $n_x \neq n_y = n_z$, where x is the stress

[2] M. Billardon and J. Badoz, Compt. Rend. **262**, 1672 (1966); **263**, 139 (1966).

[3] A review of modern polarization devices is given by W. A. Shurcliff, *Polarized Light* (Harvard Univ. Press, Cambridge, 1962). For the state of the art in electro-optic devices see R. S. Ploss, Opt. Spectry. **3**, (1)63 (1969).

[4] A. Sommerfeld, *Optics* (Academic Press Inc., New York, 1964); Ch. IV.

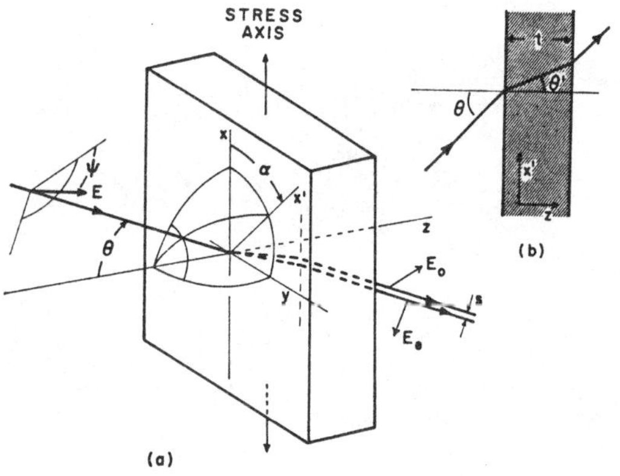

FIG. 3. Geometry for angular-aperture considerations. In the stressed plate of isotropic material, with peak stress-induced retardation of the order one wave, the double-refraction separation s between extraordinary and ordinary exit waves is of the order one wavelength, and is therefore greatly exaggerated in the figure. An upper limit for the retardation increase with θ is given by the simple path increase, $\sec\theta$ or $\sec\theta'$ in (b).

axis and z is the normal to the stressed plate face—see Fig. 3. This fact is the reason for the large usable angular aperture.

Consider a general off-axis ray with an arbitrary linear polarization (**E**-vector in sketch) incident on the plate face. It is well known that such a ray is doubly refracted into two linearly polarized component rays within the medium; the rays emerge parallel, but displaced and with a phase difference ϕ (retardation). The latter is a function of both of the incidence angles, i.e., of α (azimuth) and θ (polar angle).

Now in our case we can show that double refraction is negligible and not important. The retardation is of order $(\Delta n/n)t/\lambda$. This is to be of order unity (e.g., π, for half-wave retardation). But the double-refraction separation s between the emergent rays is of the order $(\Delta n/n)t$. Thus s is of the order of one optical wavelength; on a macroscopic scale, the ordinary and extraordinary rays are coincident.

The variation of ϕ with the incidence angles is due to two effects: one, the simple $\sec\theta$ dependence of the geometrical path length on θ; and two, the variation of the effective birefringence seen by different off-axis rays. If the birefringence is small, i.e., if $|\Delta n|/n \ll 1$, the second effect has the form

$$\Delta n(\theta,\alpha)/n \simeq [\Delta n(0,0)/n](1-\sin^2\theta \cos^2\alpha).$$

Thus both effects are of the same order. The net retardation is given by $\phi(\theta,\alpha) \cong \phi(0,0) \sec\theta(1-\sin^2\theta \cos^2\alpha)$. (Here we have not included simple Snell's law refraction. If this is done, θ is replaced by θ', where $\sin\theta = n \sin\theta'$). A general estimate of the useful angular aperture is therefore given by $\Delta\phi/\phi \simeq (\sec\theta - 1) \simeq \frac{1}{2}\theta^2$. If a 10% retardation spread is tolerated, we obtain $\theta_{max} = 25°$, corresponding to a full cone angle of 50°. This is a truly generous specification!

In contrast, the much smaller angular aperture (1 or 2 deg) of the standard Pockels cell is due to the fact that the latter is made of an intrinsically birefringent crystal, e.g., KDP. The electric field-induced birefringence, analogous to that induced by stress in the piezo-optic device, is again $n_x - n_y$ and is of the same order of magnitude as in the above discussion. But the intrinsic birefringence $n_z - n_x$, affecting all but precisely axial rays, is so large that for θ as large as a few degrees, retardation due to this effect becomes overwhelming and the interference pattern varies in a rapidly accelerating fashion with θ. In principle this could be made small by using very thin plates (small t), but this is practically impossible, if only because an enormous driving voltage would be needed. (The Pockels effect also exists in a certain class of cubic crystals, which are not intrinsically birefringent. But apparently no crystal of this type, with uv transparency and other favorable properties, is yet at hand.)

An analogous comparison can be made with ordinary (stationary) wave plates, which are normally made of birefringent crystals, e.g., quartz or mica, with intrinsic birefringence $\Delta n = n_z - n_{x,y}$. Such plates are sometimes made not of the primary thickness, i.e., with retardation $\lambda/4$ or $\lambda/2$, etc., but many times thicker, i.e., with retardation $(N+\frac{1}{4})\lambda$, $(N+\frac{1}{2})\lambda$, etc., where N is a large integer. This is a practical necessity when plates with $N=0$ would be too thin to make. All plates for the uv are made of such thick elements (mica is limited to $\lambda \gtrsim 3600$ Å). They may take the form of pseudo single-wave plates, a sandwich of two plates each of thickness $\sim N\lambda$ but with their optic axes crossed, made so that retardation for axial rays is cancelled to within $\lambda/4$ or $\lambda/2$, etc.; but far off-axis rays are again subject to retardation path differences of order $\Delta nN\lambda$, of the same order as through a single thick plate. These all, therefore, have very limited angular aperture, usually less than 10°, for the same basic reason as does the KDP Pockels cell. Our piezo-optic device is thus attractive for general wave-plate use, even apart from the high-frequency-modulation feature. As already mentioned, the device can take the trivial form of a statically stressed plate, if desired.

Aperture limitations of the Kerr cell are less fundamental in nature. But to avoid excessive drive voltages, we understand that the constriction through the cell electrodes restricts the angular aperture to some 10°. The Kerr cell has other well-known disadvantages such as limited wavelength range.

B. Spectral Range

The optical element in the piezo-optical modulator can be made of any transparent material having reasonable mechanical properties. Fused silica has a very high Q (about 10^4), and is adequate through the middle uv to about 1800 Å. CaF_2 and LiF would no doubt work very well, extending the range toward 1200 Å. The excitation requirement (strain amplitude) decreases with λ, so that lower Q's and/or smaller piezo-optic constants could be tolerated in the uv. In the infrared, we can consider using NaCl elements.

C. Limitations

The single disadvantage of the piezo-optic modulator is that it is really a time-periodic, single-frequency device, assuming it takes the form of a high-Q mechanical resonator. It is thus not appropriate for fast-transient or single-pulse modulation, outside of the bandwidth limitation $\Delta f = f/Q$. On the other hand, if nonresonant operation is considered, the response time is ultimately limited to the propagation time of sound waves across the required spatial aperture of width d, i.e., to a time d/v where v is the sound velocity. Such an ultimate limitation was pointed out by R. W. Dixon, in connection with some proposed related devices based on the acousto-optic interaction.[5]

POWER AND DRIVE CONSIDERATIONS

For the basic freely suspended $\lambda_s/2$ vibrating bar (Fig. 1), we can estimate the transducer drive power needed. We assume the simple longitudinal displacement $\delta(x,t) = \delta_0 \sin(2\pi x/\lambda_s) \sin(\omega t)$. The stored energy W is the total kinetic energy at the velocity peak,

$$W = \frac{1}{2}\int\int\int \rho u_{\max}{}^2 dx dy dz = \tfrac{1}{4}\rho\omega^2\delta_0{}^2 wtl$$

where $u = \partial\delta/\partial t$ is the velocity of particles within the bar, of magnitude $\omega\delta$, and $u_{\max}{}^2 = [\omega\delta_0 \sin(2\pi x/\lambda_s)]^2$; ρ is the density and w, t, and l are the dimensions. Since $l = \lambda_s/2$ we will write $\omega = 2\pi v_s/\lambda_s$ where v_s is the sound velocity.

The retardation is $\phi = 2\pi t\Delta n/\lambda_0$, where $\Delta n = n_x - n_y$ and λ_0 is the free-space optical wavelength. The fractional birefringence $\Delta n/n$ is proportional to the strain $\sigma = \partial\delta/\partial x = \pi\delta/l$; we write $\Delta n/n = r\sigma$, where r is a strain-optical constant. This is related in our context to conventional strain-optical constants[6] by $r = n(p_{11} - p_{12})/2$. For common substances these constants are of the order of a fraction of unity.

The required drive power is $P = \omega W/Q$. If we now specify a peak retardation $\phi_0 = m\pi$ (i.e., m half-waves), requiring a strain amplitude $\sigma_0 = m\lambda_0/(2rtn)$, we obtain

$$P = \frac{\pi}{16}\,\frac{w\rho v_s{}^3\lambda_0{}^2 m^2}{tn^2r^2Q}.$$

Oddly, the bar length l, and thus the vibration frequency ω, do not appear in the power formula, except insofar as Q depends on ω. This is because the specified retardation calls for a certain stress amplitude. The stored energy can also be written as the square of this stress times the bar volume. Doubling the bar length doubles the the energy; but this halves the frequency, and because $P = \omega W/Q$, P remains the same. For many materials including fused quartz, Q is constant over the kHz and MHz range.[7] Thus the frequency choice is governed by other factors, such as the desired useful area of the plate faces.

As an example, consider fused-silica or glass units, which the writer has found convenient. These were 2.5- to 10-cm long, resonating at f from about 20 to 80 kHz. Appropriate values of the strain-optical constant r, from data in Ref. 6, are 0.10 for typical glasses and 0.15 for fused silica. Consider the data:

$$\left.\begin{array}{l} w = 12\text{ mm},\ t = 6\text{ mm}\ (w/t = 2) \\ \rho = 2.3\text{ gm/cm}^3;\ v_s = 5\times10^5\text{ cm/sec} \\ n = 1.5;\ \lambda_0 = 5000\text{ Å};\ r \cong 0.1 \end{array}\right\} P \cong 1000\text{m}^2/Q(\text{W})$$

For half-wave ($m = 1$) peak retardation, if a glass element is used, for which $Q \sim 10^3$, we need $P \sim 1.0$ W; while with fused silica, for which $Q \sim 10^4$, we need only 0.1 W. Very modest requirements! Experimental designs (see the next section) have borne out the fair accuracy of such estimates.

With regard to spatial aperture, recall that the retardation falls off as $\cos(\pi x/l)$ from the central displacement node of the bar. In the above sample design, with $l = 8$ cm($f \simeq 30$ kHz), a spatial aperture of approximate dimensions 12×12 mm is obtained, over-which retardation is uniform to within about 10%.

PRACTICAL REALIZATION

The most obvious way to drive the optical plate is with a piezoelectric transducer crystal.[7] The driving-power estimate of the preceding section of course applies only to a simple $\lambda_s/2$ bar, and does not include losses or acoustic-impedance mismatch associated with the coupling to the transducer; with good coupling the total power input to the transducer will be of the order of twice the power P above. The transducer can be attached in any of several ways, on the ends or along the edges of the bar or block.

At the present writing, two papers on piezo-optic birefringence modulators have appeared, the first by Billardon and Badoz,[2] who must be recognized as the inventors, and the second by Mollenauer *et al.*[8] Several developments are in progress, or remain unreported. The devices described in Refs. 2 and 8 take the form of composite resonators, in which only a fraction of the

[5] R. W. Dixon, IEEE J. Quant. Electr., **QE-3**, 85 (1967).

[6] K. Vedam and S. Ramaseshan, in *Progress in Crystal Physics* R. S. Krishnan, Ed. (S. Viswanathan, Madras, India, 1958). Some photoelastic data are given in terms of the stress-optical constants q_{ij}, essentially the p_{ij} times elastic constants. Data for glasses and fused quartz are compiled in *Landolt-Bornstein Tables*, (Springer-Verlag, Berlin, 1962), Vol. II, Part 8, pp. 3–542; these are generally in terms of a stress-optical constant C_0, related for practical purposes to constants we use by $\Delta n/n = r\sigma = (C_0 Y/n)\sigma = (n/2) \times (p_{11} - p_{12})\sigma$, where Y is Young's modulus and σ is a uniaxial strain.

[7] W. P. Mason, *Piezoelectric Crystals and Their Application to Ultrasonics* (D. Van Nostrand Co., Inc., New York, 1950); Ch. XV.

[8] L. F. Mollenauer, D. Downie, H. Engstrom, and W. B. Grant, Appl. Opt. **8**, 661 (1969).

total mass of vibrating material is in the optical element; the other elements are steel stub-bars and the transducers. The latter are of the ceramic type (PZT, lead zirconium titanate). In such composite resonators, the effective Q and the frequency are more or less determined by the parts other than the optical element.

A natural and efficient driving arrangement is to use the well-known sympathetic-resonance principle. The transducer crystal, most logically consisting of single-crystal quartz, since this matches the acoustic properties of fused silica, is cut to have the same fundamental-mode frequency as the optical element. The bond between the two parts can then be quite weak, because it supports no reactive stresses. The writer has found that an X-cut crystal-quartz bar, with comparable dimensions to that of a fused silica optical element and matched precisely to the latter in frequency, can in fact merely be laid loosely on top of the optical element, this being supported from below at the node, and sufficient coupling can be thus obtained. Or a thin bond of relatively soft cement, such as silicone rubber, can be used, joining the two bars side by side (along one long edge of each part) or end to end. Using such a weak bond avoids zero-excitation stress in the optical element. In composite resonators in which sympathetic resonance is not used,[2,8] the cement must be epoxy or a similar type as strong as the optical-element material itself, since it supports first-order reactive stresses. Such cement inevitably introduces some degree of residual stress and therefore undesirable residual birefringence in the optical element.

We have also experimented with an electromagnetic drive, basically the mechanism of the dynamic loudspeaker. Very light single-turn coils made of aluminum foil, carrying ac at the bar frequency, are cemented to the ends of a $\lambda_s/2$ bar (Fig. 1); the ends protrude into the fields of permanent magnets. Owing to a basic electromechanical mismatch, this system is very inefficient. But $\lambda/4$ retardation of visible light was achieved in a fused silica bar, dimensions $10\times6\times12$ mm, with about 5 W of drive power. The arrangement is useful for experimental studies, because the transducer bond is trivial (the foil loops can be put on with Duco cement) and because it is a virtually unperturbed, single-element resonator. We have used it to measure the frequencies of bars of silica, glass, and other materials to be used in modulators of the sympathetic resonance type discussed above. (The frequencies of such bars are given only approximately by $v_s/2l$, being somewhat dependent on the width and thickness.)

With high Q, it is advantageous to operate the birefringence modulator as an oscillator, rather than to drive it from an independent source. This is easily done by making the unit function as the frequency-determining element in a positive feedback loop.[2] The writer has built a portable oscillating modulator, giving $\lambda/2$ modulation in the visible, using a single battery-driven transistor.

APPLICATIONS

A few areas are suggested as follows.

(1) Circular dichroism spectroscopy, including magnetic circular dichroism. This application was the original incentive for developing this device.[2] The invention permits an improvement of the sensitivity of such measurements by over two orders of magnitude, when advantage can be taken of the large angular aperture. Related spectroscopic measurements possible with the apparatus are linear dichroism and rotary dispersion.

(2) Light modulation and switching. This is subject to limitations pointed out above. An electrical shutter might be desired having modest response time, say 0.01 sec, without moving parts—or at least none that the eye can see. (We have in mind the simple on-off switching mode discussed earlier.) As a purely time-periodic modulator, the device can modulate an arbitrarily intense wide-aperture source, up to 100%, at practical rates up to 1 MHz. Thus, photographic applications are possible, in situations where stroboscopic sources are inadequate.

(3) Astronomy. We do not know to what extent circular-polarization measurements have been exploited in astronomy. The piezo-optic modulator could detect minute degrees of circular polarization, probably as small as 10^{-6} under favorable conditions, revealing the existence of weak stellar or interstellar magnetic fields. The instrumental set-up for such measurements could be quite simple: A telescope, a modulator unit, one polarizer, a photomultiplier, and a lock-in amplifier–detector.

(4) Laser developments. It appears that the device cannot readily be made to work as a Q-switch in lasers, since the requirement in that case is modulation with a very small duty cycle, i.e., short on-pulse relative to the interval. One method of producing an effectively short duty cycle would be to arrange that laser threshold occur very near the top of the transmission waveform of the modulator [note Fig. 2(a)].

Vol. 13(9), 701–703 (September 1988).

Binary polarization modulator: a simple device for switching light polarization between orthogonal states

R. M. A. Azzam

Department of Electrical Engineering, University of New Orleans, Lakefront, New Orleans, Louisiana 70148

Received May 9, 1988; accepted June 8, 1988

A simple binary polarization modulator is described that periodically switches the polarization of a monochromatic light beam between orthogonal linear, circular, or elliptical states. It consists of a synchronously rotating disk whose planar surface is divided into annuli. Each annulus is segmented by radial lines into sectors, and each sector is coated by a thin film that reflects light of only one of two orthogonal states. Multiple annuli may be used for multiple-wavelength operation. Although a single layer is adequate to achieve the desired effect, multilayers would be required for higher performance standards.

The state of polarization of a monochromatic light beam can be modulated using a variety of electro-optic, magneto-optic, or photoelastic devices. These modulators are typically driven by an analog (e.g., sinusoidal) input, and the output polarization varies through a continuum of states (as represented, e.g., by all points on a closed contour on the Poincaré sphere). This is certainly the case of the widely used photoelastic modulator, in which a resonant standing stress wave in a transparent solid gives rise to a sinusoidal time-varying birefringence and retardation.[1,2] Photoelastic modulators are now incorporated into instruments that measure natural or induced linear or circular birefringence and dichroism[3,4] in light-scattering nephelometers[5,6] and in ellipsometers.[7,8]

This Letter describes a simple binary polarization modulator (BPM) that periodically switches the polarization of a light beam between orthogonal linear, circular, or elliptical states. The BPM is ideal for measuring linear and circular dichroism,[9] circular intensity differential scattering,[10] and related effects, because it considerably simplifies the output signal of the photodetector. It also has important possible applications in lidar and laser target designation and reconnaissance.[11]

The BPM, which may also be called a polarization flipper, is shown in Fig. 1. It consists of a circular disk that rotates at a uniform angular speed ω. The disk has an annulus that is segmented by radial lines into N sectors of the same size. The disk (or annulus) consists of a flat and smooth dielectric, semiconductor, or metallic substrate, which is coated by a transparent thin film of certain refractive index N_1 and thickness that alternates between specific values d_A and d_B over adjacent sectors A and B of the substrate. The light beam, whose polarization is to be switched between orthogonal states, is incident at a specific angle ϕ on the segmented annulus and is linearly polarized at an azimuth α from the plane of incidence.

For simplicity, we consider here the case of a dielectric substrate of high refractive index N_2. To switch the reflected polarization between the p and s linear states (parallel and perpendicular to the plane of incidence, respectively), we should have

$$\phi = \phi_B = \tan^{-1} N_2, \tag{1}$$

$$N_1 = \sqrt{2} \sin \phi_B = \sqrt{2}N_2/(N_2^2 + 1)^{1/2}, \tag{2}$$

$$d_A = 0.3536(\lambda/N_1), \tag{3}$$

$$d_B = 0. \tag{4}$$

ϕ_B is the substrate Brewster angle, Eq. (2) indicates wave refraction in the film at 45°, and d_A is the quarter-wave thickness at ϕ_B. These relations were derived previously in connection with the division-of-wave-front beam splitter[12] and now assume added importance in the context of the BPM. To Eqs. (1)–(4) we add

$$R_s = [(N_2^2 - 1)/(N_2^2 + 1)]^2, \tag{5}$$

$$R_p = [R_s/(2 - R_s)]^2. \tag{6}$$

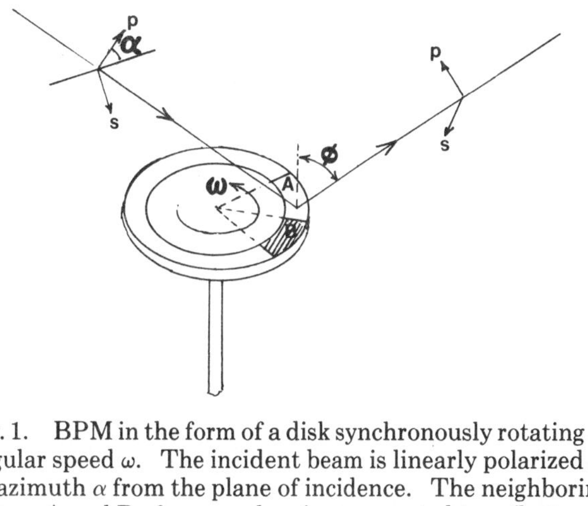

Fig. 1. BPM in the form of a disk synchronously rotating at angular speed ω. The incident beam is linearly polarized at an azimuth α from the plane of incidence. The neighboring sectors A and B of an annular ring are coated to reflect only one of two orthogonal polarizations (e.g., p and s).

To equalize the reflectances of the coated and uncoated sectors of the dielectric substrate, the azimuth α of incident linearly polarized light is chosen such that

$$(\cos^2 \alpha)R_p = (\sin^2 \alpha)R_s$$

or

$$\alpha = \tan^{-1}(R_p/R_s)^{1/2}. \tag{7}$$

This choice of azimuth eliminates the intensity modulation of the reflected light, a condition that can be sensitively detected as a null of the ac output current of a polarization-insensitive photodetector that is placed to intercept the reflected beam. The reflected beam, therefore, is purely polarization modulated between the binary p and s states, as is schematically indicated in Fig. 2. In Fig. 2 the dashed vertical lines define brief transition intervals during which the light spot on the disk is split by the demarcation line between adjacent sectors. One might argue that the instant the light spot is split equally the reflected light is effectively unpolarized. With a small spot size, the transition interval can be made a negligible fraction of the dwell time on one sector. The modulation frequency of the BPM is given by

$$f_{\mathrm{BPM}} = (1/2)Nf_{\mathrm{mech}}, \tag{8}$$

where f_{mech} is the mechanical rotational frequency (rotations per second) of the disk and N is the number of sectors. Frequencies in the kilohertz range can be attained by increasing N.

As a specific example, consider a Ge substrate that is transparent over a wide IR spectral range (3–13 μm) with refractive index $N_2 \cong 4$. From Eqs. (1)–(7), we obtain $\phi = \phi_B = 75.96°$, $N_1 = 1.372$, which corresponds to a fluoride (e.g., ThF_4) thin-film coating material, $d_A = 0.2577\lambda$, $d_B = 0$, $R_p = 40.63\%$, $R_s = 77.85\%$, and $\alpha = 35.85°$.

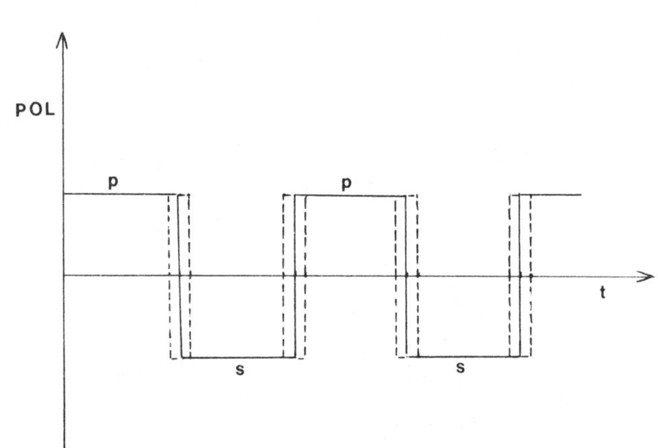

Fig. 2. Schematic representation of the periodic modulation of polarization of the light reflected by the BPM. p and s identify the linear states parallel and perpendicular to the plane of incidence, respectively. Switching between orthogonal circular or elliptical states is also possible.

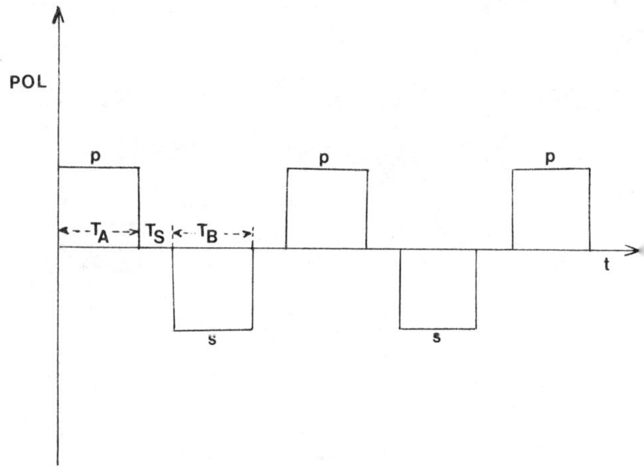

Fig. 3. Spaced orthogonally polarized successive light pulses in a periodic pulse train. Spacing is achieved simply by separating sectors A and B in Fig. 1 by a nonreflecting gap.

When the film and substrate exhibit weak dispersion (i.e., N_1 and N_2 are essentially constant independent of λ), which is possible in the IR, several modulators can be multiplexed (integrated) on the same substrate. This is done by creating a number of successive annuli on the disk surface, with the thickness of the film on the coated sectors of a given annulus determined by the corresponding wavelength through Eq. (3). In this case, the incident beam should be translated radially along the disk surface as the wavelength is changed; however, the angle of incidence remains the same.

To achieve high reflectance, or to operate in different spectral ranges, including the visible, an absorbing (e.g., metallic) substrate is required. The principle of the BPM remains as described above, except that both sectors are coated and Eqs. (1)–(6) are replaced by more complicated ones. This is discussed in Ref. 13.

The BPM can be modified to produce p and s orthogonally polarized light pulses that are separated by an interval T_s as shown in Fig. 3. This is done by separating sectors A and B in Fig. 1 by a (blackened rough) nonreflecting gap. The relative pulse width T_A, T_B, and T_s can be easily controlled to suit a particular application. It is worthwhile to compare the simplicity of this BPM with a scheme that uses Q switching in a system[11] developed for laser target designation and reconnaissance to generate a beam of light with spaced orthogonally polarized pulses.

The beam direction and axis can be maintained by two steering highly reflecting mirrors placed after the BPM such that the plane of incidence is one and the same for all three reflections. The incident linear polarization azimuth α is determined by Eq. (7), except that R_p and R_s are now the net, three-bounce reflectances for the p and s linear polarizations.

The BPM is also versatile in that it can be designed to switch the polarization of a light beam between the right- and left-circular states. This is important in circular birefringence or optical rotation, circular dichroism, and circular intensity differential scattering, as stated before. With a dielectric substrate of high refractive index, this effect can be accomplished at a suitably high angle of incidence by the correct choice of film index N_1, coating thicknesses d_A and d_B, and azimuth α of incident linearly polarized light. For example, for a Ge substrate of refractive index 4 (the same example taken for the p–s switching), we may take $\phi = 70°$. From results already published elsewhere,[14] we find that $N_1 = 1.3831$ (a fluoride thin-film coating material). The associated thicknesses required to reflect the right- and left-circular polarization states, when the incident light is linearly polarized at a 45° azimuth from the plane of incidence, are $d_A = 1.9559$ μm and $d_B = 3.2667$ μm, respectively, assuming an IR (CO_2 laser) beam of wavelength $\lambda = 10.6$ μm. The uniform reflectance of the coated Ge disk is 26.62%.

Further generalization to produce alternating light pulses of orthogonal elliptical states is also possible in view of more-recent work.[15]

It should be evident that single-layer coatings can be replaced with multilayers to enhance various aspects of device performance, such as angular insensitivity, high reflectance, and a wider spectral range.

In summary, we have described a chopperlike[16] device that operates as a polarization flipper to switch a monochromatic light beam between orthogonal polarization states. This simple and versatile binary polarization modulator promises to be useful in a wide range of applications, especially in analytical instruments that employ the interaction of polarized light with matter.

References

1. M. Billardon and J. Badoz, C. R. Acad. Sci. **262,** 1672 (1966).
2. J. C. Kemp, J. Opt. Soc. Am. **59,** 950 (1969).
3. See, e.g., L. A. Nafie, T. A. Keiderling, and P. J. Stephens, J. Am. Chem. Soc. **98,** 2715 (1976).
4. See, e.g., F. A. Modine and R. W. Major, Appl. Opt. **14,** 761 (1975).
5. A. J. Hunt and D. R. Huffman, Rev. Sci. Instrum. **44,** 1753 (1973).
6. R. J. Perry, A. J. Hunt, and D. R. Huffman, Appl. Opt. **17,** 2700 (1978).
7. S. N. Jasperson and S. E. Schnatterly, Rev. Sci. Instrum. **40,** 761 (1969).
8. B. Drevillon, J. Perrin, R. Marbot, A. Violet, and J. L. Dalby, Rev. Sci. Instrum. **53,** 969 (1982).
9. P. J. Stephens, Proc. Soc. Photo-Opt. Instrum. Eng. **88,** 75 (1976).
10. M. F. Maestre, C. Bustamante, J. Subirana, T. Hays, and I. Tinoco, Jr., Nature **298,** 773 (1982).
11. V. A. Misek, U.S. Patent 4,333,008 (June 1, 1982).
12. R. M. A. Azzam, Appl. Opt. **23,** 1296 (1984).
13. R. M. A. Azzam, J. Opt. Soc. Am. A **2,** 189 (1985).
14. R. M. A. Azzam, Opt. Commun. **55,** 297 (1985).
15. R. M. A. Azzam, J. Opt. Soc. Am. A **5,** 1576 (1988).
16. A conventional mechanical chopper may be transformed into a polarization modulator by fitting the apertures of its wheel with polarizing optical elements. This brute-force approach suffers from a serious limitation. The principal axes of a polarizing element (e.g., a sheet polarizer or wave retarder) rotate during the gating time of a light pulse, making the polarization continuously variable instead of being strictly constant. Therefore, *binary* polarization modulation is not strictly produced. This effect is absent in the reflective thin-film BPM of this Letter.

Reprinted from *Acta Physicochimica U.R.S.S.*, Vol. IV(1), 91–98 (1936).

Die Einwirkung des elektrischen Feldes auf die smektische Meso-Phase

Von *V. Freedericksz* und *A. Repiewa*

Bekanntlich besteht die charakteristische Struktur der smektischen Meso-Phase aus einzelnen Gruppen, die aus Molekülen derartig gebildet sind, dass die Molekülachsen einer solchen Gruppe sich längs den Erzeugenden einer Serie von Kegeln lagern, deren Spitzen die

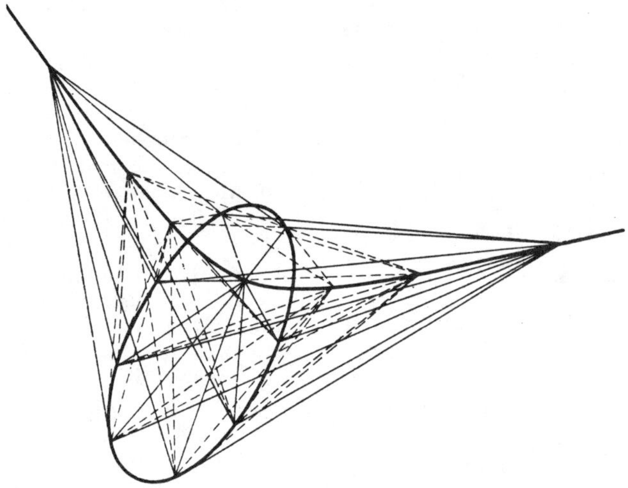

Fig. 1. Modell der fokal-konischen Struktur.

aufeinanderfolgenden Punkte an einem der Zweige einer Hyperbel einnehmen, die Grundflächen aber auf der Ellipse ruhen, und umgekehrt, die Spitzen an der Ellipse liegen und die Grundflächen sich auf die Hyperbel stützen (Fig. 1). Ellipse und Hyperbel sind miteinander durch folgende Beziehungen verknüpft:

1. Die Ebenen der Ellipse und der Hyperbel sind senkrecht zueinander.

2. Der Scheitel der Ellipse liegt im Fokus der Hyperbel und umgekehrt. Was die Dimensionen der Gruppen und die Exzentrizität der Kegelschnitte betrifft, so können sie verschieden sein. Zu bemerken ist, dass der Grenzfall $e_e = 0$, wenn die Ellipsenachsen $a = b = r = 0$ ist, eine homogene parallele Orientierung der optischen Achsen ergibt. Der zweite Grenzfall $e_e = 1$ gibt gleichfalls eine parallele Orientierung der Achsen, die im Vergleich zum ersten Fall eine Wendung von 90° aufweist.

Wird die smektische Phase an einem gewöhnlichen mikroskopischen Präparat zwischen gekreuzten Nikol-Prismen beobachtet, so erscheinen die einzelnen fokal-konischen Gruppen scharf ausgeprägt und lassen deutlich singuläre Linien hervortreten, wo die Moleküle ihre Achsenrichtung scharf ändern, d. h. Ellipse und Hyperbel (Fig. 1 u. 7).

Bei natürlichem Licht bleiben aber Ellipse und Hyperbel allein sichtbar (Fig. 6). Im gewöhnlichen, genügend dicken Präparat können die konischen Gruppen verschiedene Lagen einnehmen. Fig. 1 zeigt geschlossene Ellipsen mit den den Ellipsen entsprechenden Teilen der Hyperbeln, die in der nahe dem Präparat befindlichen konjugierten

Fokalebene des Mikroskops liegen, und langgestreckte Hyperbeläste mit den ihr entsprechenden Ellipsenteilen, die wiederum in der Nähe der genannten Fokalebene des Mikroskops liegen. Durch eine Veränderung der Fokussierung am Mikroskop lässt sich die Form einer ganzen fokalkonischen Gruppe verfolgen. Es sei bemerkt, dass die Grenzformen $e = 0$ und $e = 1$ in dem betrachteten Falle gewöhnlich nicht vorkommen.

Wir untersuchten, wie das elektrische Feld auf Lage und Struktur der fokal-konischen Gruppen einwirkt.

Zu diesem Zweck werden zwischen den beiden Gläsern des Präparats zwei Platin-Elektroden von 0,1 bis 1 mm Dicke eingeführt. Der Abstand zwischen ihnen betrug $^1/_2$ — 1 mm. Die Spannung wurde ihnen von einem Transformator zugeführt. Die Beobachtungen wurden an solchen Substanzen vorgenommen, die eine typisch fokale konische Struktur besitzen: *p*-Azoxybenzoesäureäthylester und *p*-Azoxyzimtsäureäthylester.

Beide Präparate verhalten sich im elektrischen Felde analog. Das zweite erfordert zwar eine geringere Feldspannung, dafür aber eine höhere Temperatur für das Zustandekommen der smektischen Phase. Daher war es günstiger, das erste Präparat zu beobachten und zu photographieren.

Wir beginnen unsere Untersuchungen mit hohen Temperaturen, wo die Zähigkeit der Substanz der Einwirkung des elektrischen Feldes weniger Hindernisse entgegensetzt.

1. Der Übergang aus der flüssigen amorphen Phase in die smektische ohne elektrisches Feld ergibt eine ungeordnete Verteilung der fokal-konischen Gruppen (Figg. 2 und 7, Plate 3).

Im relativ schwachen elektrischen Felde (ungefähr bis 100 V pro mm) lagern sich die beim Abkühlen der amorphen Flüssigkeit entstehenden stäbchenförmigen Tropfen der „Smektik" ihrer Länge nach senkrecht zu den Kraftlinien (Figg. 3, u. 4, Plate 3). Die diesen Stäbchen entwachsenden konischen Gruppen orientieren sich in der Weise, dass die Ellipsen sich längs den Kraftlinien, die Hyperbeln sich senkrecht zu denselben richten (Figg. 5 und 6, Pate 3. Die Gruppen haben eine sehr gestreckte Form, d. h. ihre Exzentrizität ist nahezu oder gleich Null, und ihre Dimensionen sind bedeutend grösser als ohne Feld. Es muss bemerkt werden, dass ein Präparat, welches der Einwirkung des elektrischen Feldes ausgesetzt war, auch nachher, ohne Feld im allgemeinen sehr grosse Gruppen aufweist (Fig. 7, Plate 3).

Die Schlussfolgerung auf die geringe Exzentrizität bei den Fokalgruppen, die sich im elektrischen Felde bilden, lässt sich daraus ziehen, dass die Projektionen ihrer Ellipsen durch das Kreuzen ihrer Hyperbeln, wie auf den Photo-Aufnahmen ersichtlich, in gleiche Teile geteilt werden; also haben sich die Ellipsen in Kreise verwandelt. Ausserdem sind die Hyperbeln als ziemlich lange gerade Linien wahrzunehmen, besitzen also folglich eine geringe Krümmung. Lassen wir die vom Felde orientierten stäbchenförmigen Tropfen sehr langsam abkühlen, so können wir sehen, wie die aus ihnen entstandenen fokalen Gruppen ihre Ellipsen zusammenziehen, letztere schliesslich ganz verschwinden und durchweg homogene orientierte Gebiete ergeben, in denen die optischen Achsen der Moleküle senkrecht zu den Kraftlinien gerichtet sind, d. h. der Fall $e = 0$ $a = b = r$ artet in $r = 0$ aus (Fig. 5).

2. Beim Abkühlen der amorphen Flüssigkeiten im elektrischen Felde hoher Spannung (Tausende V pro mm) lagern sich die entstehenden stäbchenförmigen Tropfen der „Smektik" ihrer Länge nach deutlich parallel zu den Kraftlinien (Fig. 8, Plate 4) und verbleiben einige Zeit regungslos. Darauf, sowie mit Sinken der Temperatur und im Beginn des Entstehens der konischen Fokalgruppen tritt augenblicklich eine überaus heftige Bewegung ein, bei der es nicht möglich ist, irgend etwas zu unterscheiden.

Schaltet man die Spannung in dem Augenblick aus, wo das Präparat genügend mit Stäbchen angefüllt, die turbulente Bewegung aber noch nicht eingetreten ist, und lässt man durch Sinken der Temperatur konische Gruppen entstehen, so lagern sich jetzt ihre Ellipsen den Kraftlinien parallel, die Hyperbeln dagegen — senkrecht. Die dabei entstehenden Gruppen sind auch sehr gross und haben, ebenso wie im vorhergehenden Fall, die Neigung, sich in homogen orientierte Gebiete umzuwandeln. Ihre Exzentrizität ist gleichfalls gering (Fig. 9, Plate 4).

Unterhalten wir bei hoher Spannung eine Temperatur, die sich dem Übergangspunkt der „Smektik" zur amorphen Phase nähert, so können wir sehr effektvolle Explosionen der grösseren Stäbchen beobachten, die momentan in kleinere zerfallen, wobei diese letzteren ihre dem Felde parallele Orientierung beibehalten.

3. Wenn wir die dem stäbchenförmigen Zustand der „Smektik" entsprechende Temperatur genau einhalten und ziemlich rasch die Feldspannung von hoch zu niedrig oder umgekehrt wechseln, so können wir sehen, wie die Stäbchen sich um 90° wenden.

4. Die smektische Phase, die aus der amorphen bei mittlerer Feldspannung (ca. 500 V pro mm) entsteht, ergibt keine verlängerten Stäbchen. Ihre Tropfen haben hier eine mehr abgerundete Form und bleiben nicht regungslos, sondern sie drehen sich rasch und unausgesetzt um sich selbst.

5. Wir wollen nun die Veränderungen, die im Felde mit hoher Spannung bei fortgesetztem Steigen der Temperatur des Präparates stattfinden, näher verfolgen.

Beginnen wir mit dem festen Zustand.

Das schmelzende Kristallpräparat bewahrt anfangs im smektischen Zustand die Spuren derjenigen Orientierung, die es im festen Zustande besass (Fig. 10 und 11, Plate 4). Dank der Zähigkeit des Präparats hat das elektrische Feld keinen Einfluss auf dasselbe. Beim weiteren Erwärmen beginnt eine schwache Bewegung, die darin besteht, dass unter den ungeordnet gelagerten konischen Gruppen neue entstehen, die die früheren verdrängen. Diese neuen Gruppen bilden wohlgeordnete Reihen, die aus Gruppen von gleichen Dimensionen bestehen (Fig. 12, Plate 4, und 13, s. S. 95). Die fokalen Gruppen sind überall von den Ellipsen längs den Kraftlinien orientiert. Sämtliche Gruppen bewegen sich unausgesetzt, indem sie sich längs ihrer Reihe drehen. Mit zunehmender Temperatur wächst die Geschwindigkeit der Bewegung an, indem sie die Richtung längs dem Felde beibehält. Die Grenzen zwischen den Reihen werden sehr scharf und geradlinig, darauf wird die Bewegung turbulent. In der Nähe der Übergangstemperatur zum amorphen Zustand hört die heftige Bewegung augenblicklich auf. Die Überreste der smektischen Flüssigkeit, umringt von der amorphen, nehmen die Form von langen Stäbchen an, die sich längs den Kraftlinien des Feldes erstrecken und fast regungslos verbleiben, während sie allmäh-

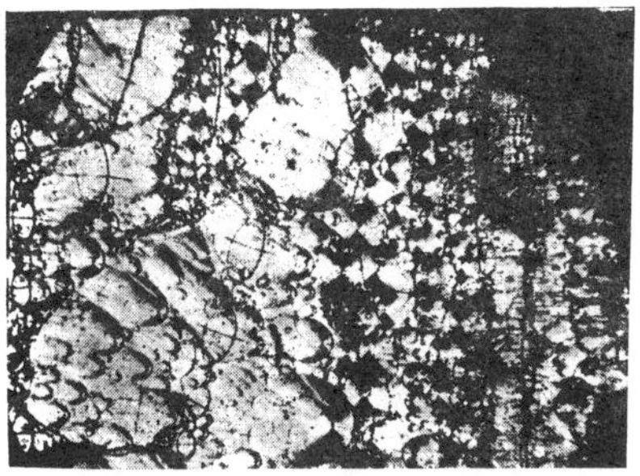

Fig. 2. Verschiedene Lagen der fokal-konischen Gruppen (ohne Analysator, mit einem Polarisator).

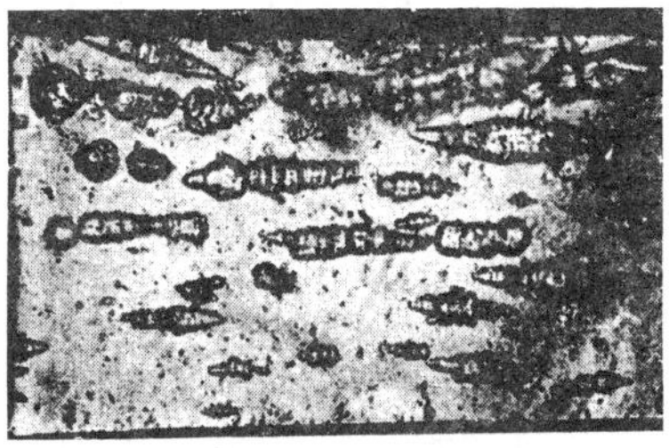

Fig. 3. Stäbchenförmige Tropfen der smektischen Phase im elektrischen Felde schwacher Spannung.

Fig. 4. Grosse Stäbchen, aus der Verschmelzung der kleinen bei schwachem Felde entstanden.

Fig. 5. Fokal-konische Gruppen, die aus den durch schwaches Feld orientierten stäbchenförmigen Tropfen hervorgehen.

Fig. 6. Fokal-konische Gruppen im schwachen elektrischen Felde orientiert (ohne Nikols).

Fig. 7. Grosse konische Gruppen im Präparat nach vorausgehender Einwirkung des Feldes (zwischen gekreuzten Nikols).

Fig. 8. Stäbchenförmige Tropfen der „Smektik" im starken Felde orientiert.

Fig. 9. Fokal-konische Gruppen, die sich aus den durch ein starkes elektrisches Feld orientierten Stäbchen entwickeln. Links ist das homogen orientierte Gebiet zu sehen.

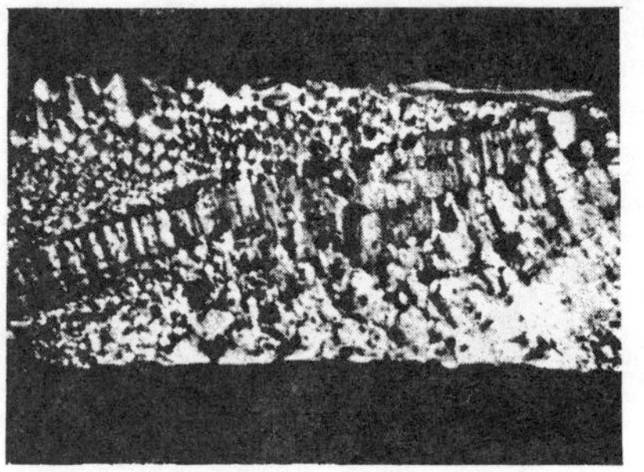

Fig. 10. Erste Phase.

Fig. 11. Smektische Phase bei ihrem Entstehen aus der festen.

Fig. 12. Orientierte Reihen der konischen Gruppen, die durch ein starkes elektrisches Feld in Bewegung gesetzt waren, durch Ausschalten der Spannung zum Stehen gebracht.

Fig. 13. Grenzen zwischen den Reihen konischer Gruppen - schwarze Linien durch Pfeile bezeichnet.

lich zu amorpher Flüssigkeit schmelzen, mitunter explodieren. Beim Abkühlen der amorphen Phase wiederholt sich der Prozess in umgekehrter Reihenfolge.

Wie aus den angeführten Beobachtungen ersichtlich, lässt sich die Einwirkung des elektrischen Feldes auf die smektische Phase nicht auf eine einfache Orientierung der bipolaren oder der dielektrisch-anisotropen Moleküle oder deren Aggregate zurückführen, wie dies in Bezug auf das Magnetfeld wohl möglich war[1]. Hier wird die Sache durch induzierte Polarisation, durch die im elektrischen Felde eintretende mechanische Bewegung der smektischen Flüssigkeit, durch Anisotropie der Zähigkeit und der Oberflächenspannung und durch andere derartige Ursachen erschwert.

Die bei der Bewegung der smektischen Phase im elektrischen Felde beobachteten Erscheinungen lassen die Vermutung aufkommen, dass die Zähigkeit in den Richtungen, die den Ebenen der Ellipsen parallel sind, geringer ist als längs den Hyperbeln. Oder vielmehr, dass es den konischen Gruppen leichter ist, sich in der den Ellipsen parallelen Richtung fortzubewegen (zu rollen), als in beliebiger anderer Richtung. Es ist möglich, dass eben diese Ursache die fokal-konischen Gruppen bei Bewegung mit mässiger Geschwindigkeit veranlasst, sich reihenweise aufzustellen und auf diese Weise die Bewegung der „Smektik" dem Felde entlang zu erleichtern, wodurch die von Friedel[2] nachgewiesenen Regeln von der Angrenzung der Gruppen befolgt werden.

Die Grenzen zwischen den Reihen der fokalen Gruppen, die in Fig. 12 und 13 als schwarze Linien zu sehen und den Kraftlinien parallel gerichtet sind, haben viel Ähnlichkeit mit denen von Naggiar[3] untersuchten Linien, die bei der turbulenten Bewegung der nematischen Phase beobachtet wurden. Vermutlich sind sie gleichen Ursprungs.

Besonderes Interesse erregt die verschiedene Orientierung der stäbchenförmigen Tropfen bei schwachem und bei starkem Felde. Möglicherweise findet im schwachen Feld eine Orientierung der Dipole statt, bei hoher Spannung dagegen wird der Effekt der induzierten Polarisation superponiert.

Dass die Molekülgruppen der „Smektik" bei hohen Spannungen stark polarisiert sind, beweisen die Explosionen der stäbchenförmigen Tropfen in der Nähe der Übergangstemperatur zur isotropen Phase, wenn der gegenseitige Zusammenhang der Moleküle erschlafft.

Es muss hervorgehoben werden, dass die smektische Phase im elektrischen Feld einen überaus reichhaltigen Stoff zur Erforschung ihres Wesens liefert, und schwerlich darf so ohne weiteres nur auf Grund von Beobachtungen bei niedriger Temperatur, wo die Zähigkeit noch gross ist, der Schluss gezogen werden, die dielektrische Anisotropie der „Smektik" sei negativ, wie dies Zocher und Birstein[4] tun. Sie haben allerdings die verschiedene Orientierung der Stäbchen in der „Smektik" bei schwachem und bei starkem Feld bloss bei Äthoxybenzalaminozimtsäureäthylester, das zwei Meso-Phasen — eine nematische und eine smektische — besitzt, wahrgenommen.

Zusammenfassung

Untersucht werden *p*-Azoxybenzoesäureäthylester und *p*-Azoxyzimtsäureäthylester.

In der Nähe der Übergangstemperatur zum amorphen Zustand ist eine verschiedene Orientierung der stäbchenförmigen Tropfen der smektischen Phase im schwachen (bis 100 V pro mm) und im star-

[1] V. Freedericksz u. A. Repiewa, Z. Physik. **42**, 532 (1927).
[2] G. Friedel, Ann. d. Physik., XVIII, 311 (1922).
[3] V. Naggiar, C. R. **200**, 903 (1935), Mars.
[4] H. Zocher u. V. Birstein, Z. physik. Chem., **142**, 186 (1929).

ken (über 1000 V pro mm) elektrischen Feld wahrgenommen worden. In Feldern von mittlerer Spannung besitzen die Tropfen der „Smektik“ nicht Stäbchenform, sondern eine mehr abgerundete Form und sind unausgesetzt in Drehbewegung begriffen. In Feldern von hoher Spannung sind öfters Explosionen der stäbchenförmigen Tropfen beobachtet worden. Aus den vom Felde orientierten Stäbchen haben sich beim Ausschalten des Feldes und beim Sinken der Temperatur grosse spindelförmige fokal-konische Gruppen entwickelt, die je nach schwacher oder starker Spannung eine entsprechend um 90° gedrehte Orientierung besassen. Bei stetiger Temperaturänderung von der festen bis zur amorphen Phase wurde im Felde hoher Spannung eine Aufstellung der konischen Gruppen in regelmässige, orientierte Reihen (die Ellipsen parallel den Kraftlinien, die Hyperbeln senkrecht zu ihnen), und eine Bewegung derselben, die erst langsam in der Richtung der Reihen, nachher stürmisch und turbulent verlief, wahrgenommen. In der Nähe der Übergangstemperatur zur amorphen Phase hörte die Bewegung auf, die Tropfen der „Smektik“ dehnten sich zu langen Stäbchen und lagerten sich ihrer Länge nach parallel dem Felde.

Photo-Aufnahmen der beobachteten Erscheinungen sind beigegeben.

Die Einwirkung des elektrischen Feldes auf die smektische Meso-Phase darf nicht auf eine blosse Orientierung der dielektrisch anisotropen Moleküle zurückgeführt werden, wie es für das Magnetfeld möglich war. Hier wirkten gleichzeitig die Orientierung der Dipole, die induzierte Polarisation, die durch das elektrische Feld hervorgerufene, mechanische Bewegung der Flüssigkeit, die Anisotropie der Zähigkeit und der Oberflächenspannung und andere Ursachen mit.

Industrielles Institut,
Leningrad.

Eingegangen am
20. Oktober 1935.

Reprinted from *Zeitschrift fuer Naturforschung,* Vol. 20, 849–851 (1965).

Electric Field Effects in Cholesteric Liquid Crystals

JOHN H. MULLER

U.S. Army Engineer Research and Development Laboratories
Fort Belvoir, Virginia (USA)

(Z. Naturforschg. **20 a**, 849—851 [1965]; received 14 May 1965)

Several physical effects of strong electric fields upon a *cholesteric* liquid crystal have been found. While the alignment of nematic liquid crystals by electric and magnetic fields has been known for some time [1]; and various authors, particularly FRIEDEL (ref. [1], p. 344), have classified the cholesteric phase as an alternate form to the nematic phase; we are not aware of any electric field effects upon cholesteric liquid crystals reported in the literature.

It might seem natural to examine optical effects in electric fields transverse to the viewing direction, in analogy with other electro-optic effects. However, one of our group [2] reasoning from the following considerations reached a different conclusion.

a) The intensity of the selectively reflected light in the plane cholesteric texture depends strongly upon the substrate material. E. g. Mylar plastic oriented by rolling and known for its strongly polar surface induces more reflection than glass, a comparatively amorphous material.

b) The model of DE VRIES [3] explains the strong reflection band and the high optical rotatory power of opposite sign on either side of that band by a screw structure. Even though only normal incidence and reflection is considered by this model, it suggests that there exists a preferred axis parallel to, and a preferred plane normal to the screw axis.

c) Hence the most sensitive effect might be observed by a field directed along such a screw axis and parallel to the viewing direction, for only in this direction would the field effect upon the individual molecular dipole axes be cumulative rather than periodic along the screw structure. We have therefore examined several cholesteric compounds in polarized light in a field parallel to the viewing axis. In this letter we wish to report on effects observed in cholesterol nonanoate, hereafter designated "CN".

Observations were made with a polarizing microscope using incident illumination, a transparent upper electrode of NESA glass, and an opaque, blackened, lower electrode of anodized copper. The microscope analyzer was crossed with the polarization of specularly reflected light, and magnifications of 30 to 250 were used.

Both instantaneous and time average light intensity were monitored by a photomultiplier on a third tube of the trinocular polarizing microscope. Readout was by means of a meter or oscilloscope.

The thickness of the sample was controlled with the aid of a mylar spacing sheet usually 15 microns thick between the electrodes. A circular hole of 3 mm dia. was cut in the mylar to accomodate the sample.

Fields up to 2.4×10^7 V./M. were monitored directly by a voltmeter connected to the electrodes. Near dielectric breakdown, a current of 3 microamperes was observed, but almost all measurements were taken under smaller fields at which the current leakage was less than 10^{-7} amperes.

Impurity effects can be ruled out by the following: The cholesterol nonanoate, (CN) used was chromatographically purified and checked. The behaviour of the sample did not degrade or alter with time or continued exposure to high electric fields. The zero field behaviour of the samples throughout was the same as that of samples exposed only to glass. Additionally, CN is known to be chemically stable at the temperatures used. All observations except as noted were made on the selectively scattered light from the cholesteric mesophase.

The twelve following main features have been observed in CN. The first five of these have mainly to do with D. C. fields, the last seven with A. C. fields.

[1] M. G. FRIEDEL, Ann. Physique 18, 378 [1922].
[2] R. ENNULAT, private communication.
[3] H. L. DE VRIES, Acta Cryst. 4, 219 [1951].

1. Application of a strong D. C. field parallel to the viewing direction yields up to a five-fold increase in the intensity of reflected light.

2. When the liquid crystal is cooled from the clearing point with the field continuously applied, the intensity of reflected light is up to seven times greater than when the field is turned on only within the color temperature range.

3. The intensity of reflected light is an approximately linear function of applied voltage up to electrical breakdown of the material, providing, that the field is applied above the clearing point, and maintained while the material is cooled into the colored cholesteric range. Some slight saturation of the effect with field may take place, but is within experimental error.

4. A marked hysteresis is shown. The intensity does not return to its original value when the field is switched off, but remains nearly two-thirds of full intensity. Attempts to recover the original no-field intensity by use of A. C. fields are only partly effective. See below.

5. At fields near but below breakdown strength, approximately 2.4×10^7 V./M., perhaps ten to twenty small, dark points per square millimeter appear suddenly and just as rapidly disappear. Except for the color, the appearance under the microscope could be likened very much to a mud volcano, with bubbles bursting irregularly and repeatedly at the surface. Because we will deal with two other kinds of spots we will label these "type A".

Application of an A. C. field to CN gives a complicated frequency-dependant behaviour as follows: In general, the same type of behaviour occurs at higher frequencies for higher temperatures (blue); and at lower frequencies for lower temperatures (red). For a given behaviour, the A. C. field frequency is *more nearly* inversely proportional to visible color wavelength, rather than proportional to temperature.

6. Below 1 to 5 cycles per second, one sees primarily the D.C. behaviour with the added feature of switching transients. The transient consists of various dark spots of the order of 20 – 100 microns dia. suddenly appearing in the field of view and then fading out. Two sets of spots appear alternately, one for each sign of polarity change. Such spots move about to some extent, and are well-defined at much lower fields than those which appear near breakdown. Additionally, these spots appear synchronously with polarity reversal so that we will call them "type B".

7. Between 1 and 35 cps, depending upon temperature, there is a very broad minimum in time average intensity of reflected light as a function of frequency. At this minimum, about 25 percent of the scattered light intensity of the D. C. case remains. The broad minimum is also found both for 30 micron and for 8 micron samples; but the intensity decrement is smaller for the very thin case. As freqency rises in the 1 to 35 cycle range, the spots grow in size and number until the field of view is mainly occupied by spots of the order of 100 – 200 microns dia. nearly equal in size, and closely packed in an almost-hexagonal, irregular pattern. The edges of the dark spots have a definite color, characteristic of a lower temperature than the rest of the medium. These spots are in continual agitated movement. Since these spots have transformed continuously from the low-frequency regime, they are still "type B".

Also, for some samples there is a sharp temperature-independant minimum in the average reflected light intensity to as little as 13 percent of the D. C. case. This sharp minimum, however, seems to be dependant upon the gross dimensions of the samples, and furthermore, requires extended times of the order of ten minutes for the light intensity to fall after suitable temperature and frequency conditions have been initiated. No special visible structure seems to be associated with this minimum except for the already occurring spots.

8. As the frequency rises further in the 1 to 35 cycle range the spots move into irregular rows, and growing in size, amalgamate into heavy dark lines of the order of 20 to 100 microns wide and 100 to 500 microns long depending on temperature. It must be remarked that both these lines and the spots preceeding them are remarkably uniform in size in an isothermal field of view, varying at most 10 percent in thickness. Surprisingly, the average measured light intensity is starting to rise slowly again with the formation of these dark rows.

9. As the frequency increases further, now between 18 and 65 cycles, again depending upon temperature, the heavy bars begin to fade and grow thinner. The average light intensity rises until it returns to that of the D. C. case. The bars disappear completely, and other very thin lines appear. These give a threaded appearance reminiscent of a nematic liquid crystal.

10. In this regime, there can also be seen a few dark dots of about 10 microns apparent size. These are in continual irregular translation with velocities of the order of .5 mm/sec. A very high proportion of the motion is circular or spiral; often with several tens of consecutive orbits observed. There is no visual evidence that these spots are partaking of a gross vortex motion of the fluid, but it seems that these spots are following some sort of regular discontinuity within the liquid crystal. These spots are followed by a trail which fades out some 150 microns or so behind them. Such spots are sometimes "captured" at some fixed, visible, local discontinuity where they merge into the local detail; but they have not been seen to stop in the absence of a visible discontinuity. These dots are designated "type C".

11. The application of a strong A. C. field of 50 cps to a sample of CN which had been very rapidly cooled from above the clearing point, (especially if the "homeotropic"[4] texture remained into the usual color range) tended to generate large stable spiral structures within a minute. More than twenty complete turns could be counted in some instances; the nearly perfect spirals being up to 500 microns diameter and quite uniformly colored.

[4] G. W. Gray, Molecular Structure and the Properties of Liquid Crystals, Academic Press, New York 1962, p. 41.

Such spirals nucleated at the center of and were concentric with, maltese crosses of "spherulitic" structures observed growing in the homeotropic texture. Smaller spirals are observed otherwise in this and other materials, but we have never observed such extensive, rapid and repeatable development of these spirals as with high frequency A. C. electric fields.

12. Observation of the instantaneous intensity of reflected light indicated sharp spiked decreases in light intensity at twice the applied A. C. field frequency. These dips were approximately proportional to the average intensity and could be observed by the 1 P 21 photocell up to the clearing point, considerably warmer than the color temperature. The dips disappeared above 1000 cps applied A. C. frequency, but an A. C. component to the light reflected could be restored by applying a D. C. bias field equal to the peak A. C. field; the maximum instantaneous field remaining below breakdown. As one might suppose, this A. C. component was at the fundamental frequency, and could be observed up to 3000 cps.

Some regularities in the evidence suggest themselves.

1. The field seems to align the somewhat disoriented putative mesophase screw axes of DE VRIES' Theory[3] normal to the electrodes. The temperature dependance of reflection wavelength is not shifted, implying no change in the screw pitch.

2. The electric field effect depends, as in the nematic case, not only upon the individual molecular dipole moment, but also on the cooperative structure of the mesophase. This is shown by the relatively weak probable dipole moment of CN as compared to the supposedly strong dipole moments of certain other cholesteric compounds showing no detectable effect. Work in progress is expected to elucidate this matter.

3. Very crude calculations based on the observed time constants and estimated viscosity and density yield an electric dipole moment for CN of the order of .01 to 1 Debye molecule. The large uncertainties make agreement with a crude estimate based on molecular structure fortuitous.

4. In the colored cholesteric regime, the large spirals centered upon the "spherulitic" nuclei (ref. [1], pp. 364 to 373 and 386) lend more credence to a structure which has both an internal screw axis and a preferred plane of orientation parallel to the sample boundary, and which is consistent with the general picture presented, for instance by GRAY (ref. [4], pp. 47 – 54).

5. No magnetic effects have been observed, and due to the smallness of the electric currents passing into the electrodes, none are expected here. Work is presently in progress to determine whether strong magnetic fields can affect cholesteric liquid crystals.

6. Since the liquid crystal was separated from the massive copper stage only by a very thin oxide layer, the liquid crystal could not have been at a very different temperature from the stage. The small electrical power, usually less than fifty microwatts, should not affect the very well heat-sinked microscope stage. Furthermore, since a careful study showed no displacement of the intensity as a function of temperature, temperature effects due to local heating are extremely unlikely.

7. Rectification due to the copper oxide blackening on the stage was suspected, but was found to be much less important as measured by the D. C. component of the A. C. current than the D. C. unbalance in amplifier output. Nevertheless even this unbalance could be kept under one percent.

It is a pleasure to thank Dr. R. ENNULAT for suggesting this work and for many stimulating discussions and encouragement. Thanks are also due to Dr. W. ELSER and Dr. J. POHLMANN for much chemistry and information; and Mr. R. SHERMAN for many measurements.

Liquid-crystal laser optics: design, fabrication, and performance

S. D. Jacobs, K. A. Cerqua, K. L. Marshall, A. Schmid, M. J. Guardalben, and K. J. Skerrett

Laboratory for Laser Energetics, University of Rochester, 250 East River Road, Rochester, New York 14623-1299

Received January 29, 1988; accepted April 26, 1988

We describe the development of laser optics utilizing liquid crystals. Devices discussed constitute passive optical elements for both low-power and high-power laser systems, operating in either the pulsed or cw mode. Designs and fabrication methods are given in detail for wave plates, circular polarizers, optical isolators, laser-blocking notch filters, and soft apertures. Performance data in the visible to near infrared show these devices to be useful alternatives to other technologies based on conventional glasses, crystals, or thin films. The issue of laser damage is examined on the basis of off-line threshold testing and daily use in OMEGA, the 24-beam Nd:glass laser system at the Laboratory for Laser Energetics. Results demonstrate that long-term survivability has been achieved.

1. INTRODUCTION

The development of liquid-crystal optics began in 1979, when it became apparent that some optical-component requirements of large-aperture solid-state lasers could not easily be met by solid-crystalline materials. Since that time we have found that the flexibility offered by these anisotropic fluids permits them to be used as replacements for conventional laser-system hardware, as novel solutions to existing device problems, and as entirely new optical elements. In this paper we briefly describe the properties of nematic and chiral nematic (cholesteric) liquid crystals, reviewing the important link between chemical structure and optical properties. We give the procedures required for fluid processing, substrate preparation, and optical-element construction. Several device concepts are then described in a manner that includes design, fabrication, and performance data. Laser-damage resistance is treated as a separate issue, and we discuss results from both off-line threshold testing and daily use in OMEGA, the 24-beam Nd:glass laser system at the Laboratory for Laser Energetics (LLE).

2. LIQUID-CRYSTAL MATERIALS

Low-molecular-weight, monomeric liquid-crystal compounds have been employed for the results reported here. Bulk viscosities of the order of 25–60 centistokes (cS) at room temperature permit these fluids to be filtered, blended, and processed without difficulty. Liquid-crystal mesophases can occur at temperatures below −20°C (melting point or glass transition) and greater than 100°C (clearing point), permitting the construction of devices that operate over wide temperature ranges without the need for precise temperature control. Environmentally stable compounds are commercially available from several sources[1] in quantities of 1–100 g at a cost of dollars to tens of dollars per gram. An extensive tabulation of compounds and properties relevant to laser applications has recently been published.[2]

Nematic materials constitute the basic building blocks of our studies. They can be generically represented by the following structure:

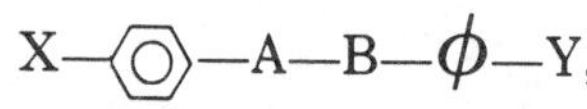

where A and B are bridging groups that bind two rings together (which may be benzene or cyclohexyl) and X and Y are terminal groups along the molecular long axis. The A–B linkage groups may be of the types –CH=N– (Schiff bases), –N=N(O)– (azoxy compounds), –N=N– (azo compounds), –COO– (esters), and biphenyls (direct ring-to-ring linkage). The group represented by ϕ in the above structure can consist of phenyl, cyclohexyl, biphenyl, or pyrimidine ring structures. The elongated nematic molecule is optically anisotropic and may be characterized by a uniaxial optical indicatrix. The birefringence of the molecule is positive, and its optical axis (extraordinary refractive index) is coincident with the molecular long axis.

Other liquid-crystal compounds that are important are referred to as chiral nematics. By adding percentage amounts of chiral twisting agents to the nematics described above, we can create optically active structures that exhibit the property of selective reflection.[3,4] Because many chiral twisting agents are composed of cholesteryl derivatives, the resulting liquid crystals are often referred to as cholesterics. A left-handed (LH) or a right-handed (RH) twist can be induced in nematics by choosing a suitable additive.

The major microscopic structural differences between nematics and cholesterics are shown in Fig. 1. In Fig. 1(a) the spontaneous parallel alignment of nematic molecules is depicted over a volume of roughly 40 nm^3. This order carries over into the cholesteric structure [Fig. 1(b)], but it is confined to a single molecular layer. A structure results that is characterized by the distance required for the molecular long axes of the nematic molecules in each layer to rotate by 360°. This distance is defined as pitch, P, of the cholesteric structure.

The optical axes for both liquid-crystal phases are along the vertical direction in the figure. For the case of nematics, this corresponds to the long direction of the molecule; for cholesterics, the optical axis is coincident with the helical axis of the structure.

Selective reflection is observed when optical radiation

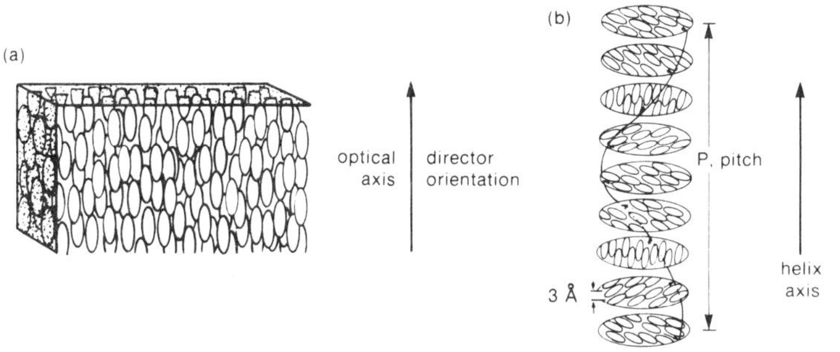

Fig. 1. Spontaneous organization of liquid-crystalline structures: (a) nematic phase, (b) cholesteric phase.

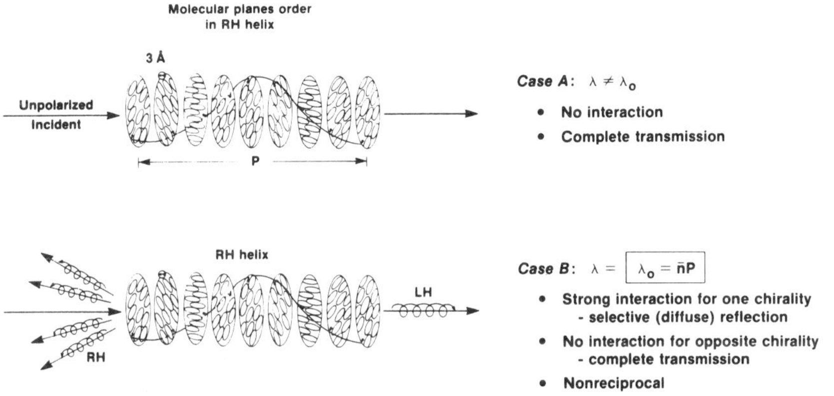

Fig. 2. Selective reflection effect in the cholesteric structure.

propagating along the cholesteric helix axis satisfies the relationship

$$\lambda = \lambda_0 = \bar{n}P, \qquad (1)$$

where $\bar{n}$ is the average refractive index of the compounds and P is the pitch. That wavelength, λ_0, that exactly satisfies Eq. (1) will interact with the cholesteric structure depending on its polarization state, whereas all other wavelengths and polarizations will pass through the helical structure without interacting. Figure 2 illustrates the interaction for unpolarized incident monochromatic radiation for case A ($\lambda \neq \lambda_0$) and case B ($\lambda = \lambda_0$). If, as in the figure, the cholesteric structure is assumed to be RH, then case B illustrates the following: the structure breaks incident optical radiation into LH and RH components, passing the portion whose helicity is opposite to that of the structure, and reflecting the same-handedness portion. There is no 180° phase shift when the optical radiation is reflected from the structure, and the effect is *not* absorptive.

3. LIQUID-CRYSTAL CELLS

Because of their low viscosities, liquid crystals must be contained in a sandwich configuration between a pair of circular glass substrates. High-optical-quality and survivability constraints are placed on liquid-crystal cells intended for insertion into laser systems.

A. Cell Fabrication

The internal refractive-index homogeneity and finished surface quality for a pair of glass substrates are chosen to be commensurate with laser-system requirements. Typical tolerances for transmitted wave-front quality, surface flatness, scratch dig, and wedge are described Subsections 4.A.1 and 5.B of this paper.

Spacers are utilized to define the liquid-crystal fluid thickness and to minimize wedge in assembled cells. Mylar foil tabs or thin-film islands deposited with vacuum coating techniques set the gap at between a few micrometers and tens of micrometers. Cells are filled by employing capillary action at atmospheric pressure.

Sealing is performed by forming a continuous epoxy bead in the bevel valley of the substrate–substrate interface. The type of epoxy,[5] rate of cure, size of bead, and its elasticity are determined in part by the cell diameter, the substrate composition, and its aspect ratio (typically 10:1). The sealing cycle must allow for in-process manipulation before final cure in order to permit wave-front and wedge requirements to be achieved. The long-term integrity of the epoxy seal against popping from thermal-expansion mismatch must be balanced against the desire to disassemble damaged or defective cells for refilling and reuse.

B. Macroscopic Alignment and Surface Anchoring

The most important aspects of device fabrication are cleanliness and substrate inner-surface treatment. The presence of particulate contamination can impede uniform filling of cells, alter the desired fluid-layer thickness, and/or act as a nucleating site for laser-induced damage. This latter issue is discussed in more detail in Section 5. The methods used for substrate inner-surface preparation define the type of liquid-crystal device, its optical quality, and its level of performance.

The molecular order depicted in Fig. 1 does not extend

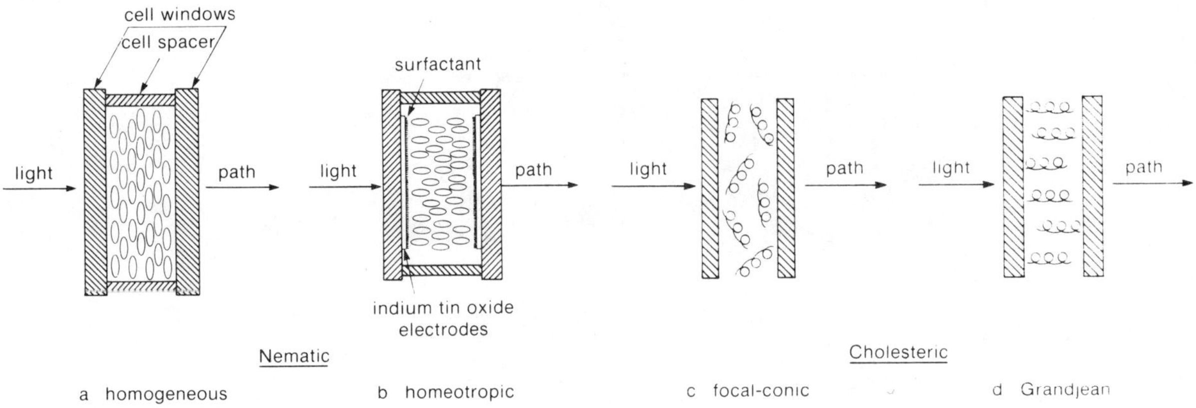

Fig. 3. Externally imposed alignment configurations for nematic (a and b) and cholesteric (c and d) compounds.

over macroscopic dimensions unless some external input is provided to create a preferred direction for the liquid crystal at the fluid–substrate interface. This imposed order extends from the interface into the bulk of the fluid for distances of tens of micrometers. A considerable amount of data has been published that addresses this alignment issue.[6] The various configurations are illustrated in Fig. 3. Each configuration is suited to a particular device application. For example, the homogeneous texture in Fig. 3a is the basis for optical retardation plates, whereas the Grandjean texture in Fig. 3d is employed for passive devices such as circular polarizers, isolators, and laser-beam apodizers. The orientation in Fig. 3b is useful in electro-optic and opto-optic applications and will not be discussed here.

Anchoring of liquid-crystal molecules in the homogeneous texture (Fig. 3a), both in the bulk of the fluid and over millimeters to centimeters of clear aperture, is achieved by roughening the inner cell surface. For bare glass this is commonly done by unidirectionally rubbing with a fine polishing compound (0.05-μm alumina or 0.1-μm diamond). One theory[6] holds that peaks and valleys are thereby created in the surface. A minimum free-energy condition is achieved for molecules that align parallel to each other in head-to-tail fashion along the linear grooves. It has recently been suggested,[7] however, that rubbing promotes homogeneous alignment by the breaking of sigma bonds in the SiO_2 glass former. The liquid-crystal molecules align parallel to these disrupted bonds. Homogeneous alignment may also be achieved with unidirectional buffing of a thin polymer layer previously deposited upon the inner substrate surface by spin coating. The higher degree of alignment achieved with unidirectional buffing is essential for the fabrication of liquid-crystal wave plates.

A large scattering effect is observed in cholesterics for optical radiation propagating in the focal–conic (Fig. 3c) configuration. Mechanical shearing of glass substrates during cell assembly ensures that the light propagates along the helical axis of the cholesteric texture in the Grandjean (Fig. 3d) configuration, reducing scatter losses to less than 2%. The shearing of bare-glass substrates without any prior rubbing is adequate for the construction of liquid-crystal polarizers/isolators.

4. PASSIVE-OPTICAL ELEMENTS

In this section we describe the principles of operation for several kinds of optical elements constructed from liquid-crystal fluids. Special fabrication issues not previously described are included where appropriate, and performance data are given.

A. Liquid-Crystal Wave Plate

Laser applications that require the manipulation of the state of polarization of optical radiation rely almost exclusively on quarter-wave and half-wave optical retarders. Materials most frequently considered for the fabrication of optical wave plates include mica, crystal quartz, synthetic sapphire, Polaroid poly(vinyl alcohol) (PVA), and DuPont poly(ethylene terephthalate) (PET) (or Mylar). Table 1 compares the various properties of these materials with those of nematic liquid crystals.

The major limitations to the use of natural or synthetic

Table 1. Properties of Various Wave-Plate Materials (λ = 1053 nm)

Material	Birefringence Δn	Quarter-Wave Thickness (μm)	Fabrication Technology	Aperture Limit (mm)	Comments
Natural mica	0.0039	67	Cleave zero order	150	Absorption, size
Natural quartz	0.0087	30	Polish multiple order	220	Availability, cost
Synthetic quartz	0.0087	30	Polish multiple order	150	Laser damage, size
Synthetic sapphire	0.008	33	Polish multiple order	100	Quality, cost
Polymer PET	0.10–0.23	1–3	Extrude zero order	None	Quality, availability
Polymer PVA	0.0076	37	Extrude zero order	None	Quality, availability
Liquid crystals	0.05–0.30	0.5–5.0	Pour zero order	>200	Flexibility

Table 2. Low-Birefringence Nematics for Wave-Plate Fabrication (λ = 1053 nm, T = 22 ± 0.1°C; Values Are ±0.00005)

Material[a]	n_0	n_e	$\bar{n}$	Δn
ZLI-3125	1.45602	1.50608	1.48105	0.05005
ZLI-2359[b]	1.45651	1.50807	1.48105	0.05156
ZLI-1800	1.46912	1.54057	1.50485	0.07146
ZLI-2244-100[b]	1.46521	1.53679	1.50100	0.07158
ZLI-1646	1.47008	1.54641	1.50824	0.07633

[a] See Table 7 for a description of these compositions.
[b] Measured at λ = 1.047 μm, 21.9°C.

crystals is the expense associated with obtaining these materials with good optical quality in circumstances when sizes greater than 100 mm are required. Mylar or Polaroid PVA are not adequate because of poor optical quality or the lack of available sheets with appropriate retardance.

The attractiveness of liquid crystals as wave plates is based on the availability of nematic compounds whose birefringence and spectral transparency can be tailored to any optical application. As indicated in Table 1, birefringence can be adjusted by compound blending from 0.05 to 0.30. The fabrication of zero-order wave plates with decreased angular sensitivity is therefore possible. Optical transparency spans the UV–IR range. Other than single-crystal CdS or tellurium, which are not available in sizes greater than 25 mm, few transmissive, birefringent materials exist for retarder applications in the IR.

1. *Device Fabrication*

Key to achieving superior performance and quality of liquid-crystal wave plates is strict control over material selection and assembly procedures. Early research at the LLE[8] to develop wave plates utilizing liquid crystals was plagued by regions of nonuniform retardation across the device aperture. This nonuniformity was due to inadequate substrate surface flatness as well as to regions of varying alignment quality. As will be discussed, uniform alignment resulting from proper substrate surface preparation, rigid gap-thickness control, and proper fluid selection and filling techniques are essential for producing satisfactory devices.

Liquid-crystal wave plates are assembled using a lengthy fabrication process. Before assembly, the fluid of choice is determined on the basis of its birefringence at the wavelength of use. All the devices fabricated at the LLE have been for use at 1053 nm. Table 2 shows potential nematic materials and their respective values of birefringence at λ = 1053 nm, ranked in order of increasing birefringence. Listed in the table are several of the lowest-birefringence nematics currently available from commercial sources. Low Δn materials are required for wave-plate applications, as zero-order cells of manageable thickness (3 μm or greater) for quarter-wave applications in the near IR cannot be made with a fluid whose Δn is greater than 0.1.

Refractive indices of liquid-crystalline materials are measured utilizing a temperature-controlled (±0.1°C) Abbé refractometer.[9] A Nd:YAG laser[10] modified with an YLF rod serves as the 1053-nm source for the experiment. The refractometer prism is cleaned with methanol and lens tissue using a unidirectional wipe, and liquid-crystal material is then applied dropwise to the prism surface. Several strokes with lens tissue are sufficient to provide adequate homogeneous alignment of the material, permitting observation of the fluid's two distinct lines (n_0, n_e) with an IR viewer through the refractometer eyepiece. Index values are repeatable to ±0.00005.

Selection of appropriate spacer thickness (t) is determined on the basis of fluid birefringence (Δn) and desired device phase shift (δ) from the relationship

$$t = \frac{\delta\lambda}{2\pi\Delta n}. \tag{2}$$

Because slight variations in gap thickness severely affect retardance uniformity and ultimate device quality, accurate knowledge of spacer thickness and substrate flatness are required. For our experiments, Mylar spacers of thickness 3.5–5 μm were used. Test cells are assembled utilizing 100-mm-diameter borosilicate glass substrates [12.5 mm thick, flat to $\leq\lambda/10$ peak to valley (P-V) at 632 nm across 90% of the hard aperture, and antireflection coated for 1053 nm].

Thorough cleaning of spacers and substrates is performed before assembly. All cell materials undergo a 15-min ultrasonic cleaning followed by thorough rinsing with ASTM Type I water. Once clean, all materials are blown dry with filtered nitrogen, treated with an antistat gun to remove static charge, and stored in a clean, dry container within a class 100 laminar-flow hood. All subsequent cell preparation and assembly take place within the class 100 environment.

Wave-plate inner-substrate surfaces are spin coated with an alignment layer of PVA. The 2% PVA aqueous solution is applied dropwise to the spinning substrate (250 rpm) and is permitted to disperse uniformly. Substrates are then cured at 120°C for 30 min on a programmable hot plate, allowing for adequate heating/cooling cycles to prevent thermal shock to the substrates. Once cooled, the coated substrates are rubbed unidirectionally with a felt-roller-type buffing machine, developed in house, to impart small microgrooves to the inner-substrate surface. These microgrooves aid in orienting the liquid-crystal molecules.

Following an antistat treatment, substrates are blown off and inspected to ensure complete removal of residual lint, which could lead to device contamination and eventual damage if not removed. Spacers are then placed on one clean, coated inner-substrate surface at three points, 120° apart, around the substrate's edge. A second substrate is contacted to the first, with the rub direction of its PVA coating antiparallel to its counterpart, maintaining proper position and flatness (no creases or crinkles) of the sandwiched spacers. The completed cell is then loaded into a variable load press and held under a 50–100-kg load for 15 min to ensure uniform gap thickness. The cell is tacked under pressure with 5-min epoxy at the spacers and is then permitted to cure under pressure for 30 min. Because the thickness uniformity of the fluid gap is perhaps the most important aspect of the device, this step was found to be essential in our achieving reproducible, uniform air gaps.

Accurate measurements of the air gap are made utilizing an interferometric technique,[11] which resolves t to within ~1%. Once an accurate mean gap thickness is known for a cell, some blending of the liquid-crystal fluids possessing

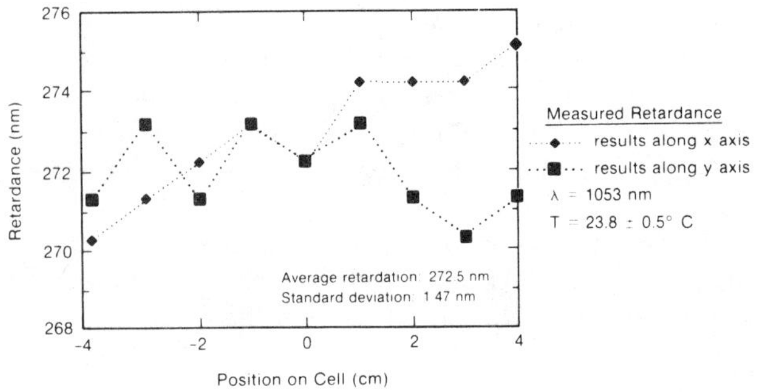

Fig. 4. Measured retardance across the x and y axes of 100-mm-diameter liquid-crystal quarter-wave plate shows good uniformity across the device aperture.

Table 3. Measured Gap Thickness and (Calculated Retardance) at Six Sites Across Retarder QWP-2-860611

Direction	Position on Cell (cm) −4	0	4
x	3.86 ± 0.10 μm (276 ± 7 nm)	3.93 ± 0.12 μm (281 ± 9 nm)	4.06 ± 0.22 μm (290 ± 16 nm)
y	3.91 ± 0.11 μm (280 ± 8 nm)	3.93 ± 0.12 μm (281 ± 9 nm)	3.85 ± 0.09 μm (276 ± 6 nm)

different birefringence values may be necessary to generate the target retardance (Γ) equal to the $t\Delta n$ product.

Cell filling is performed at room temperature by injecting degassed liquid-crystal material into the cell gap through a filtered syringe. (Degassing of fluids in their isotropic phase under vacuum for several hours before filling removes trapped gas in the material, which could lead to bubbles in the completed cell.) Slow incorporation of the fluid into the cell is found to minimize trapped air bubbles. Once sealed with degassed epoxy and cured, each device undergoes characterization for its retardance uniformity, contrast, and transmission.

2. Performance

Measured retardance data for one such device are shown in Fig. 4. Device QWP-2-860611 was designed to be a quarter-wave retarder at 1053 nm (Γ_{target} = 263 nm). Air-gap measurement data made before filling are shown in Table 3. From these data, mean gap values were used to select an appropriate fluid. Fluid ZL1-2244-100 was chosen for use without additional blending.

Several observations can be made from these results. Figure 4 shows that, although the uniformity of retardance across the device is excellent (±1.47 nm), the target retardance was exceeded by approximately 9 nm. Minor compound blending to fine tune the birefringence of the mixture would have improved the performance of the wave plate. There is also a discrepancy of several nanometers between calculated retardances based on actual gap thickness values (Table 3, in parentheses), and the measured values in Fig. 4. Recent work by Wu and Efron[12] suggests that this discrepancy occurs in thin (<3-μm) cells, and it is due to disordered alignment near the inner cell surfaces, which reduces effective birefringence. The measurements of Fig. 4 and the calculations of Table 3 do agree within the precision of the gap-thickness determination.

Interferometric measurements performed on assembled wave plates at λ = 632.8 nm have shown excellent transmitted wave-front uniformity. Device QWP-2-860611 exhibited 0.075 wave (P-V) of deformation with little wedge. This, we believe, is due to the high degree of flatness uniformity imparted to the cell during the pressing of the spacer gap before filling. Additionally, the average refractive index of the fluid is also a reasonable match to that of the substrates.

This technology allows for custom tuning of birefringence to bring retardance to within 1% of a target value. It is more flexible than other technologies, which require crystal cleaving or grinding and polishing. Because they are true zero-order elements with a minimum thickness of birefringent material, liquid-crystal waves plates are considerably less angularly sensitive than thicker, multiple-order solid-crystalline counterparts.

B. Liquid-Crystal Polarizer/Isolator

The selective reflection effect in a cholesteric forms the basis for its use as a polarizer/isolator.[2,13,14] A single, passive liquid-crystal element can perform *both* functions, provided that the laser system of interest is configured to propagate circularly polarized optical radiation. Figure 5 illustrates the most basic use of a liquid-crystal polarizer/isolator.

As a polarizer, the liquid-crystal cell exhibits a high level of transmittance for incident laser radiation whose chiral sense is *opposite* that of the element. Any backreflections of the passed radiation from misaligned optical surfaces or targets are blocked by the element because of the 180° phase shift in the handedness of the laser beam on reflection. This isolation feature is defeated, however, if passed radiation becomes depolarized on reflection or if passed radiation is intentionally converted to linear polarization before reflection from downstream components. To compare this technology with more conventional ones, a number of optical properties must be discussed.

1. Handedness and Tuning

Chiral twisting agents may be chosen from primarily commercial sources with LH or RH structures. They may be doped into nematics in amounts varying from a few percent to 20% to create LH and RH circular polarization elements. Figure 6 gives a few of the compounds that we have found useful in our work. Some of these materials, although they are polycrystalline powders or waxes at room temperature,

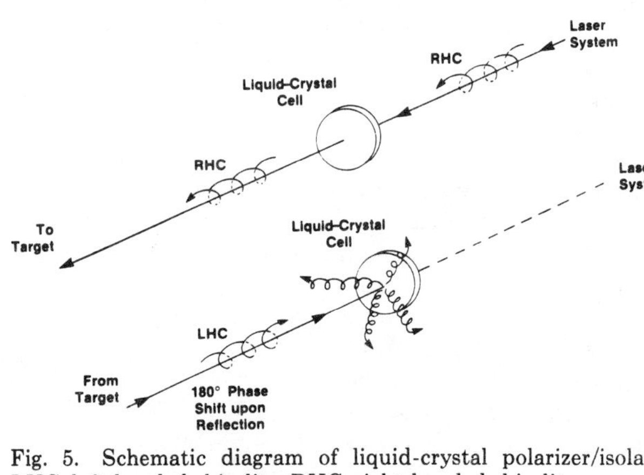

Fig. 5. Schematic diagram of liquid-crystal polarizer/isolator: LHC, left-handed chirality; RHC, right-handed chirality.

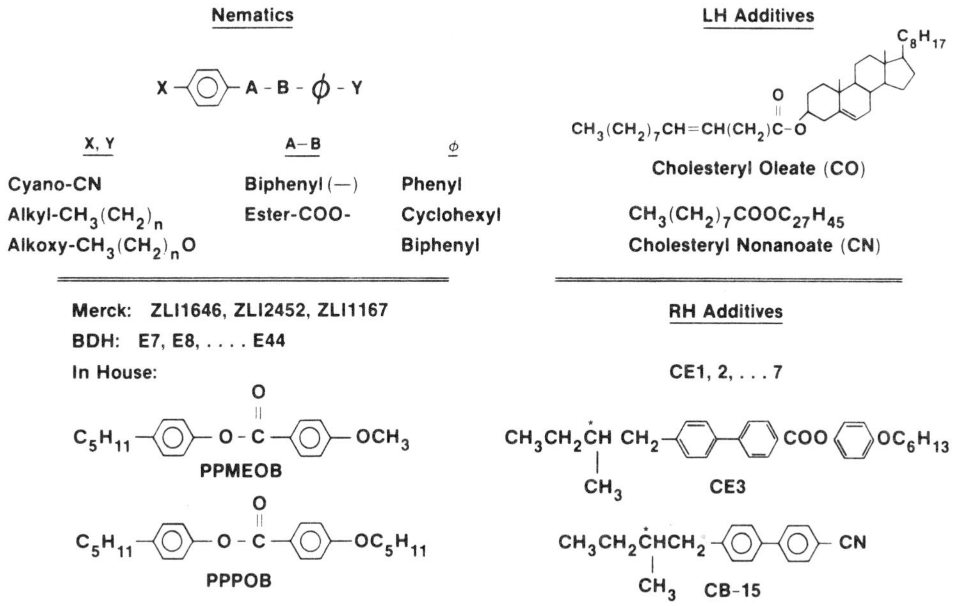

Fig. 6. Base nematics and chiral additives.

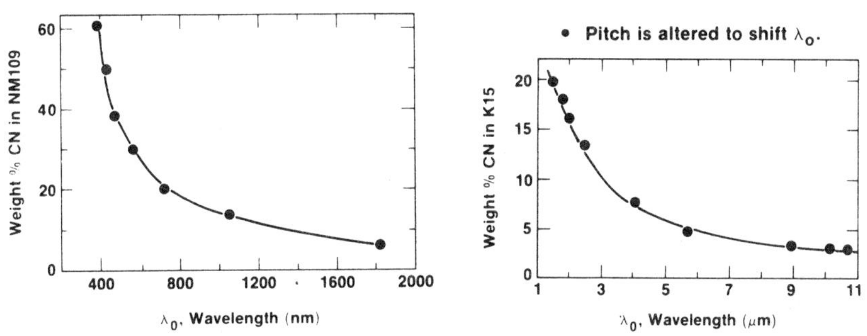

NM109: Mixture of cholesteryl oleate in nematic eutectic PPMEOB-PPPOB (ALX Corp).

CN: Cholesteryl Nonanoate (Kodak)
K15: Cyanoalkylbiphenyl Nematic (EM Laboratories)

Fig. 7. Wavelength tuning with composition.

form moderate-viscosity fluids after blending at elevated temperatures.

Liquid-crystal polarizers are wavelength selective, in accordance with Eq. (1). Because average refractive indices for most blends fall between 1.5 and 1.7, wavelength tuning is most readily accomplished by pitch contraction. Chiral twisting agents are added to shift λ_0 to the desired wavelength, as shown in Fig. 7.

Agents with tight intrinsic pitch are preferred because their incorporation in small amounts provides a large degree of pitch contraction without causing major changes to the thermo-optic properties of the stable nematics. For laser applications it is important to choose compound blends whose refractive indices and pitch are not strongly temperature dependent. Extensive tabulations for twisting agents, which include handedness (LH or RH), selective reflection peaks for blends, and temperature sensitivities, are given in Ref. 2.

2. Spectral Transmittance and Bandwidth

Molecular vibrations prevent complete transparency in the IR,[15] although liquid-crystal device concepts have been successfully pursued in transmission windows near $\lambda = 10.6$ μm.[16] The spectral transmission for liquid crystals in the near IR and visible is essentially free from absorption. Figure 8 shows two scans taken separately in a spectrophotometer for two polarizer elements. Apart from the 50% attenuation loss exhibited by each element near the reflection peak

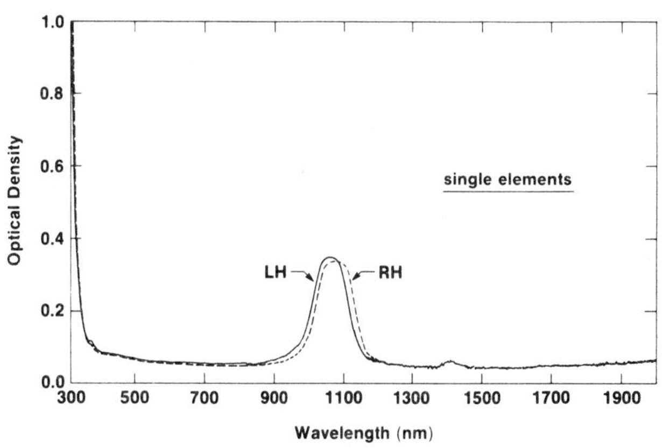

Fig. 8. Spectral transmission through liquid-crystal polarizers (LCP's): fluid path length, 11 μm.

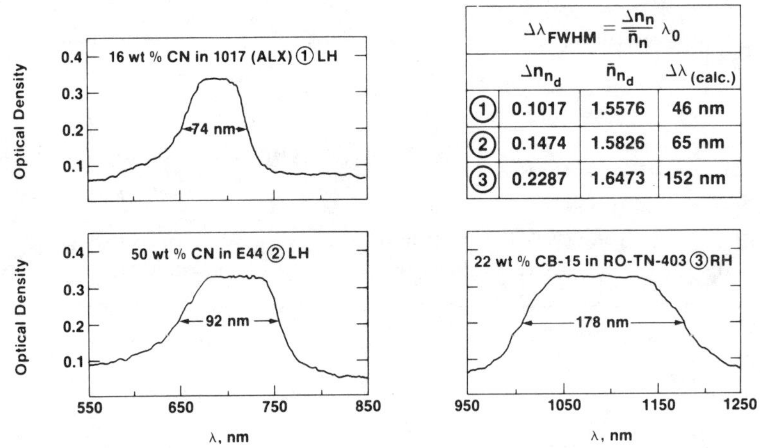

	$\Delta\lambda_{FWHM} = \frac{\Delta n_n}{\bar{n}_n}\lambda_0$		
	Δn_{n_d}	$\bar{n}_{n_d}$	$\Delta\lambda$ (calc.)
①	0.1017	1.5576	46 nm
②	0.1474	1.5826	65 nm
③	0.2287	1.6473	152 nm

Fig. 9. Bandwidth of selective reflection for three blends at two wavelengths.

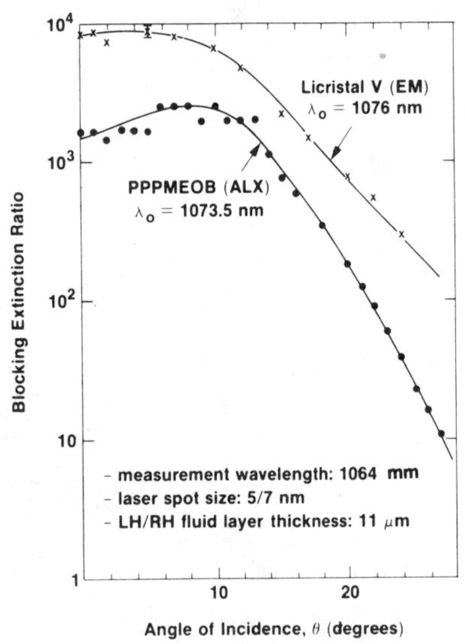

Fig. 10. Blocking extinction and angular acceptance for LCP's. Base nematic compound is identified in the figure.

of λ_0 = 1064 nm, there are no other attenuation features down to the onset of UV absorption in the borosilicate glass substrate.

The bandwidth for selective reflection (FWHM) is given approximately by[17]

$$\Delta\lambda_{FWHM} = \frac{\Delta n}{\bar{n}}\lambda_0. \qquad (3)$$

It is possible to broaden or to narrow the bandwidth at a given wavelength by choosing a compound with an appropriate birefringence. A natural broadening occurs as the selective reflection wavelength is shifted to the IR. Both the broadening and the narrowing of the bandwidth are illustrated in Fig. 9, where calculated bandwidth compares moderately well with measurements for λ_0 = 694 nm and λ_0 = 1064 nm.

3. *Contrast and Angular Acceptance*

Theory[18] and experiment[19] suggest that an optical interaction path length of at least ten pitch lengths is required to maximize selective reflection and the degree of polarization provided by the cholesteric liquid-crystal structure. We have confirmed, by experiment, that a fluid path length of 10–12 μm is optimum for constructing liquid-crystal elements for the near IR. Figure 10 shows that if the fluid alignment is good, LH and RH polarizer element pairs can be constructed with an extinction ratio approaching 10^4:1. Transmission losses from scattering are less than 2% per fluid layer. Increasing fluid-layer thickness results in no further improvement in blocking contrast and causes an increase in scatter loss.

The acceptance angle for liquid-crystal polarizers depends on the placement of the selective reflection peak with respect to the wavelength of interest and its bandwidth. As the angle of incidence increases from θ = 0 for a liquid-crystal element, the effective λ_0 shifts to *shorter* wavelengths according to the equation[17]

$$\lambda = \lambda_0 \left[\cos\left\{\sin^{-1}\left(\frac{1}{\bar{n}}\sin\theta\right)\right\}\right] \qquad (4)$$

This equation is similar to the one that describes the shift in transmission peak for a thin-film interference filter. Figure 10 shows how, by biasing λ_0 to 1076 nm, performance at λ = 1064 nm may be extended to ±20° from normal.

4. *Environmental Stability and Technology Comparison*

Sealed liquid-crystal cells are stable against humidity and moderate exposure to sunlight. Temperature-induced effects are mitigated to some degree by the thermal mass of the cell substrates, but shifts in peak wavelength do occur. Changes in selective reflection-peak wavelength for two

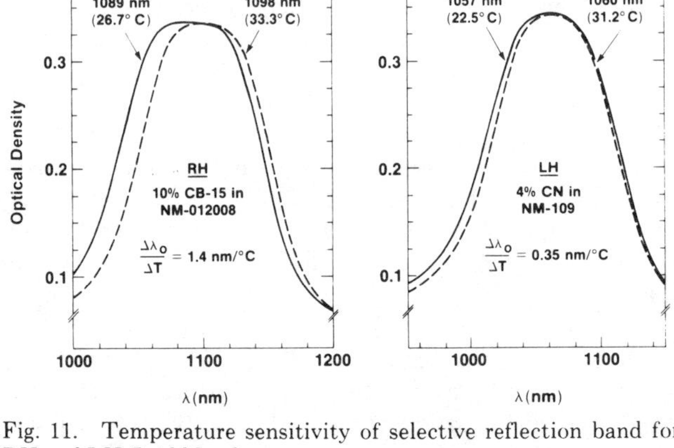

Fig. 11. Temperature sensitivity of selective reflection band for RH and LH fluid blends.

Table 4. Comparison of Polarizer Technologies

Attribute	Crystal Prism	Brewster Thin Film	Dielectric Cube	Dyed Plastic	Liquid Crystal
Optical quality at large apertures	No	Yes	No	No	Yes
High contrast with angular insensitivity	No	No	No	Yes	Yes
High transmission for passed polarization	Yes	?	Yes	No	Yes
Environmental stability	Yes	No	Yes	Yes	Yes
Laser-damage resistance	Some	Yes	Some	No	Yes
Backreflection protection	No	No	No	No	Yes

blends are shown in Fig. 11. Shifts for these and other blends are positive and are of the order of 0.3–1.5 nm/°C. We have not separated out the individual contributions of dn/dt and dP/dt to $d\lambda_0/dt$.

Liquid-crystal polarizer/isolators provide a useful alternative to conventional technologies. A brief comparison is provided in Table 4. Liquid crystals are unique for their ability to provide both polarization and isolation in a single, passive element.

C. Liquid-Crystal, Laser-Blocking Notch Filter

In various laser applications situations often arise that require selective optical filtration. Optical elements that exhibit extremely low transmittance at the specific laser wavelength λ_0 and high transmittance at other wavelengths are required to filter optical radiation incident, for example, upon UV diagnostics or IR thermal-imaging systems. Laser goggles for eye protection of personnel constitute another area where laser-blocking notch filters are useful.

Some of the design and performance goals for a laser-blocking notch filter operating at a wavelength of λ_0 are as follows: extinction at λ_0 greater than 10^3:1; transmittance away from λ_0 of greater than 85%; ease of alignment, or broad angular acceptance; a device that is compact, lightweight, and inexpensive; and adequate laser-damage resistance.

High in-band attenuation and high out-of-band transmittance are required to provide optical sensors with protection against lasers without sacrificing the ability to detect background radiation. Angular insensitivity to laser radiation and light weight are important for personnel protection devices such as laser goggles.

Absorbing glasses or dye/plastic filters, thin-film interference filters,[20] and holographic filters[21] are examples of materials technologies that have been used to construct notch filters. Liquid crystals offer an alternative method for constructing laser-blocking notch filters; this alternative method has been explored in our laboratory.

Notch filters may be constructed from cholesteric liquid crystals by stacking two liquid-crystal cells filled with fluids possessing opposite chiralities, which have been tuned to the same wavelength, λ_0.[22] The concept is shown in Fig. 12 (L is LH, R is RH). At the wavelength for selective reflection, the device performs as a nearly perfect pair of crossed circular polarizers, blocking incident laser radiation. Optical radiation at wavelengths away from λ_0 is transmitted through the filter with negligible attenuation. The spectral scan superimposed upon the device concept in Fig. 12 illustrates the results achieved for a ruby blocking filter. Out-of-band

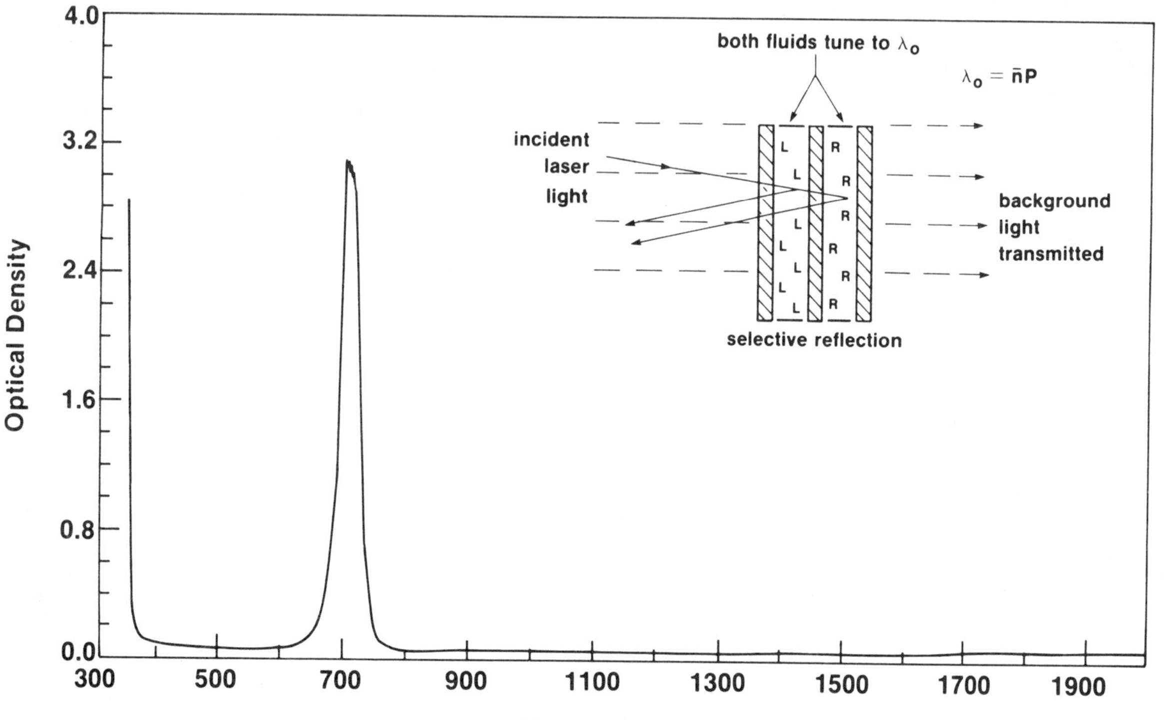

Fig. 12. Concept and proof of concept for liquid-crystal notch filter at λ = 694 nm.

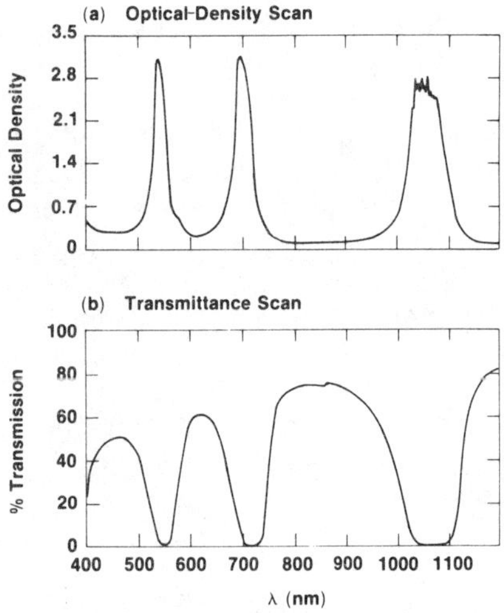

Fig. 13. Spectral performance for a triple-wavelength notch filter composed of six fluid layers.

transmission is uncluttered by the secondary attenuation features characteristic of other technologies.

Issues of transmittance, extinction, angular acceptance, and thermal sensitivity are similar for liquid-crystal notch filters and liquid-crystal polarizers. One important distinction concerns the quality of the substrate. The residual stress birefringence of the central spacer element in the notch filter, as depicted in Fig. 12, determines in part the ultimate blocking extinction of the composite filter. This element must be constructed of finely annealed glass (2 nm/cm), whereas the outer elements may be composed of polycarbonate if ballistics protection is required. Uniformity of extinction is a function of fluid alignment. We have measured ±0.25 optical-density units across a 38-mm-diameter notch filter tuned to λ_0 = 1064 nm. Multiple-wavelength blocking filters may be designed and constructed. Performance of a triple-wavelength notch filter is demonstrated in Fig. 13. The temperature sensitivity of the device (~2 nm/°C) limits its usefulness to temperature-controlled environments.

D. Liquid-Crystal Soft Aperture—Apodizer

High-quality, laser-beam apodization has been the goal of solid-state laser device research programs since the early 1970's. The ability to shape the spatial intensity profile of a low-to-moderate-power laser beam, before propagation through several stages of amplification, determines to a large extent the ultimate performance of high-peak-power and high-average-power systems.

The effect of a perfect apodizer device is illustrated schematically in Fig. 14. With this element we can create a quasi-flat-topped intensity profile from a Gaussian input. The soft edges of the apodizer minimize the occurrence of Fresnel diffraction rings in the output beam, permitting its propagation and amplification with a large fill factor and minimal small-scale self-focusing.

A perfect apodizer preserves the temporal shape of the input pulse and causes little distortion to the phase across the transmitted beam profile. Transmission in the device's clear aperture is close to 1.0, whereas transmission at the edge falls to less than 10^{-3}. Good circular symmetry of the apodizer is a requirement in most applications, as is an edge profile that approximates the transmittance function

$$I(r) = I_0 \exp[-(r/r_0)^N], \qquad N = 5\text{–}10, \tag{5}$$

where the radial clear-aperture parameter r_0 is selected on the condition that $I(r_1) = 10^{-3} I_0$, where $2r_1$ is the entry aperture of the amplifying device that follows the apodizer.[23] Finally, the device must exhibit a high degree of laser-damage resistance for high-peak-power applications and low absorptive losses for high-average-power applications.

The fact that no perfect apodizer has been created is evident from the literature. The following passive-device technologies have been pursued: shaped absorbing liquid cells,[24] saturable absorbers,[25] shaped solid absorbers,[24,26–28] graded photographic emulsions,[24,29,30] graded metallic films,[24,31] single and multilayer dielectric films,[24] total internal reflection devices,[23,24,32,33] radially birefringent elements,[8,34–36] and annular lenses.[37]

Electro-optic and magneto-optic concepts have also been proposed,[24] but none of these passive or active approaches has found broad acceptance. In fact, most high-peak-power solid-state laser systems, such as LLE's OMEGA, utilize a hard aperture and an image-relaying scheme invented by Hunt *et al.*[38] to maximize fill factor and to minimize diffraction effects.

We have developed a new apodizer device based on liquid-crystal technology[39] that has demonstrated properties resembling those of a perfect apodizer. Originating from our previous work with liquid-crystal polarizers and laser-blocking notch filters, our liquid-crystal soft aperture possesses the following characteristics[40]: passive and nonabsorptive with no aperture limitations, wavelength selectivity from UV to IR, no input polarization requirements, ease of fabrication in circular or other cross sections, high laser-damage resistance, and low wave-front distortion in the soft-edge region.

The concept of a liquid-crystal soft aperture follows directly from that of a laser-blocking notch filter. The selective reflection effect can be made radially varying if curved surfaces are employed on the central spacer glass, as shown in Fig. 15. The fluid path is therefore made negligible at the center of the filter and increases to tens of micrometers at the filter edge. Blocking extinction increases radially with fluid-layer thickness.

The cell geometry that is most useful for apodizer devices

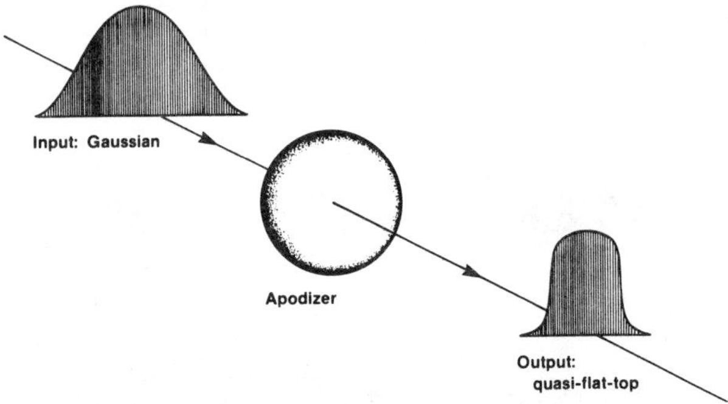

Fig. 14. Laser-beam apodizer. A perfect apodizer truncates the wings of a Gaussian input without introducing diffraction effects. Good optical phase quality and intensity in the center of the beam are preserved.

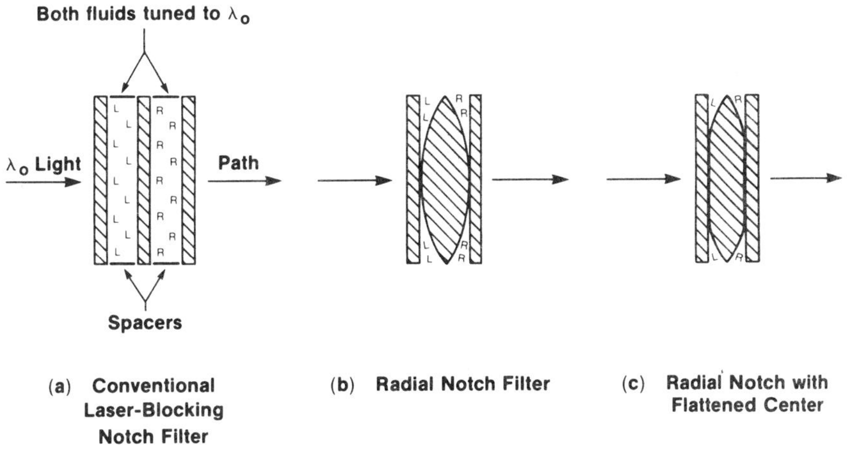

Fig. 15. Evolution of the concept for liquid-crystal soft aperture. By changing the shape of the central spacer element in a conventional laser-blocking notch filter, we can synthesize filters with radially varying transmission profiles. Blocking extinction, which originates from the selective reflection effect, varies from zero at element center to greater than 10^3 at the edge. Refractive-index matching between the fluids and the glass minimizes phase distortion.

is illustrated in Fig. 15(c). Here a biconvex central element has had optical flats polished onto the center of each curved surface. The flats determine the clear aperture of the apodizer, and the soft edge is created by the increase in LH and RH fluid-layer thickness along the device radius. Fundamental to the success of the configuration shown in Fig. 15(c) is the ability to adjust the average refractive index of each liquid-crystal layer to match that of the confining substrates to 1 part in 10^3 to 1 part in 10^4. This minimizes optical wave-front distortion across the device's clear aperture and well into the soft edge. The reflective nature of the effect promotes high laser-damage resistance and minimal thermally induced distortions at high average powers.

We have designed liquid-crystal soft apertures for use at λ = 1064, 800, 532 nm. The transmission profile for this type of apodizer at the soft edge is modeled by an equation of the form

$$I(r) = I_0 e^{-Qt(r)}, \tag{6}$$

where Q is a factor that describes the degree of extinction for circular polarization based on the quality of liquid-crystal alignment (in reciprocal micrometers) and $t(r)$ is the LH or RH fluid-gap thickness in micrometers. For flattened, spherical central spacer elements, the functional form for t is given as

$$t(\text{micrometers}) = [(r^2 - r_0^2)/4f(n-1)] \times 10^3 \qquad (r \geq r_0), \tag{7}$$

where r_0 is the radius of flattened clear aperture [as defined in Eq. (5)] in millimeters, r is the radial distance from the center of the device in millimeters, f is the focal length of the biconvex lens in millimeters, and n is the refractive index of lens material at λ_0.

The Q is determined empirically by constructing liquid-crystal notch filters in a wedged fluid-layer configuration. Transmission as a function of layer thickness is measured. Results for three separate wedges, tuned to λ = 1064, 800, 532 nm, are given in Fig. 16. Nominal Q-factor fits to the data are shown. The quality of the fit depends strongly on the quality of alignment imposed by inner-cell-surface preparation.

A computer program combines clear-aperture and edge-transmission-function specifications with Q factors to design apodizers. Table 5 shows the calculated results for soft-aperture elements at two wavelengths; the soft-aperture elements are designed to have an edge transmittance profile [refer to Eq. (5)] where N = 9.

Figure 17 gives measured and target-edge profiles for two soft apertures fabricated according to the design specifications in Table 5. Measurement conditions for these data are summarized in Table 6.

The dashed curves in Fig. 17 correspond to the calculated transmission profiles for each device. For cell SA-3, there is good agreement between the two curves; the optical density at the device's edge approaches 2.0. The good extinction at the edge of the apodizer is the result of good alignment in this region. Cell SA-4 also agrees quite well with its predicted profile, although it does not exhibit optimum extinction at its outer edge. The central-element curvature of this apodizer is steeper than that of the green cell, and it is

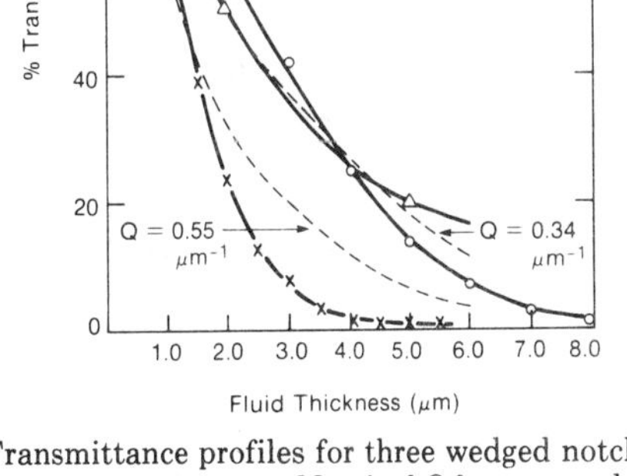

Fig. 16. Transmittance profiles for three wedged notch filter elements at λ = 1064, 800, 532 nm. Nominal Q factors are derived from the data for subsequent design calculations.

Table 5. Design Parameters for the Construction of Soft Apertures

(Element Identification) Design Wavelength (λ, nm)	Q Factor $[(\mu m)^{-1}]$	Edge Function (N)	Required Clear Aperture (mm)	Calculated Focal Length of Biconvex Element (mm)
(SA-3) 532	0.55	9	4	600
(SA-4) 800	0.34	9	4	400

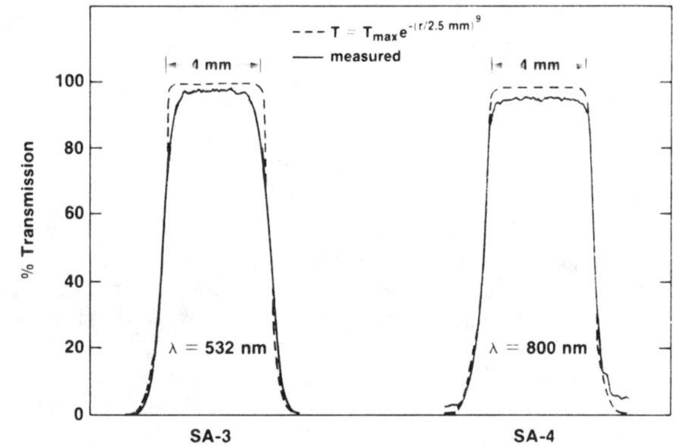

Fig. 17. Design and measured performance for two liquid-crystal soft apertures.

Table 6. Soft-Aperture Transmission-Profile Measurement Conditions

Feature	SA-3	SA-4
Laser	Doubled Nd:YAG, λ = 532 nm	Diode laser, λ = 795 nm
Polarization	Linear	Unpolarized or linear
Spot size	128 ± 40 μm (10–90% points)	190 ± 35 μm (10–90% points)
Measured center transmittance	97.5 ± 0.5%	96 ± 0.5%

believed that the anchoring of the liquid-crystal molecules to the glass surface was not sufficient to give maximum blocking in the area 2.5 mm from the center of the device. Both elements possessed AR-coated outside surfaces appropriate to their respective design wavelengths. Central transmittance losses of 1–2% per fluid layer are evident from the data in Fig. 17 and Table 6. These are scatter-related losses.

Extensive work is required to develop LH and RH blends whose selective reflection peaks coincide with the design wavelength, while simultaneously providing a refractive-index match to the borosilicate glass substrates. Satisfying this latter requirement helps to minimize wave-front distortion in the soft-edge region. Interferometry performed on devices SA-3 and SA-4 shows good wave-front quality. Figures 18(a) and 18(b) illustrate the measured wave-front quality of each as a function of radial distance from aperture center. By overlaying the measured transmission profiles (shown as dashed lines) for each cell onto their respective wave-front distortion plots, the resulting phase distortion at the $1/e^2$ (~13.5% transmittance) point may be evaluated. Cell SA-3 shows approximately λ/4 of distortion, whereas SA-4 is slightly better at λ/5. These values, however, are not indicative of the true quality of the devices. The data for both cells show a quadratic phase factor (defocus) that is due primarily to the fluids not being index matched at the measurement wavelength of 632.8 nm. Using index data and calculating expected defocus at this measurement wavelength, we conclude that the devices, if measured at their respective design wavelengths, would exhibit even better wave-front quality than that measured at λ = 632.8 nm. Figure 18 demonstrates that, with index matching to 1 part in 10^3 and temperature control to ±2°C (the temperature coefficient of the refractive index for our fluids is approximately $-3 \times 10^4/°C$), we can keep wave-front distortion below 0.25 wave at a point on the soft edge where transmittance drops to 20%. For comparison, glass apodizers exhibit

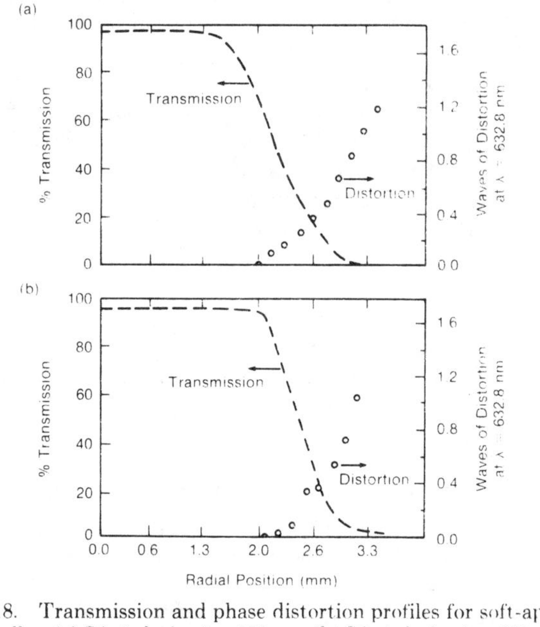

Fig. 18. Transmission and phase distortion profiles for soft-aperture cells: (a) SA-3, design λ = 532 nm; (b) SA-4, design λ = 800 nm.

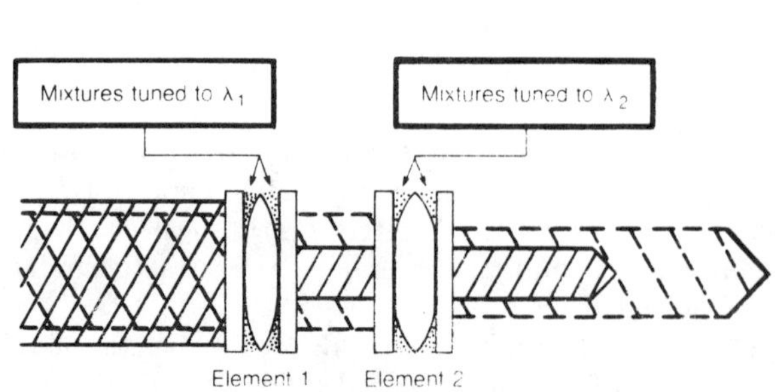

Fig. 19. Liquid-crystal soft-aperture flexibility. Because the selective reflection effect in liquid crystals is wavelength specific, liquid-crystal soft apertures may be stacked in tandem to operate on collinearly propagating laser beams whose wavelengths differ. Light at a given laser wavelength may be apodized with a unique clear-aperture size and soft-edge profile without affecting light at other wavelengths.

a wave-front distortion in excess of 1.2 waves at their 20% transmittance point.

The high laser-damage resistance and low absorption of the compounds used to construct liquid-crystal soft apertures should permit their use in high-peak- and high-average-power systems operating in the UV to near IR. The wavelength selectivity of the selective reflection effect permits apodization of one wavelength only in situations (see Fig. 19) when two or more wavelengths propagate collinearly. Device performance neither depends on nor does it alter the state of polarization of the incident laser beam.

5. LASER-INDUCED DAMAGE IN LIQUID CRYSTALS

We have studied the laser-damage issue for liquid-crystal elements in two ways. Small-scale testing with a well-characterized Nd:glass zigzag slab-laser system has been employed to construct laser-damage-resistance data for a number of nematic and chiral nematic compounds. N-on-one and one-on-one results at 1053 nm and 1 nsec have been obtained. In-system performance for liquid-crystal polarizers at an aperture of 100 mm has been evaluated for hundreds of elements in the OMEGA system during the past 18 months. Whole-beam data illustrate different features than those from off-line damage-testing experiments because they are more sensitive to isolated extrinsic impurities.

A. Laser-Induced Damage Threshold Experiments

Laser damage in optical solids is characterized on a microscopic scale by the formation of irreversible lattice modifications. Fluid-medialike liquid crystals offer the possibility of exhibiting *transient damage*, which self-heals between shots in pulsed applications. Gas bubble formation and dissolution have been observed in our damage-testing facility. Bubbles that do not dissolve have been seen to drift out of the cell's clear aperture if viscosity and fluid-layer thickness permit fluid motion that is due to gravity. Irradiation of liquid crystals at high fluences has led to the observation of pyrolysis and the formation of carbonaceous residues. The time duration of the transient damage events can be of the order of seconds, and this has forced us to devote considerable care to the methods used for establishing damage-fluence thresholds for liquid crystals.

We conducted laser–materials interaction experiments at λ = 1053 nm with a flash-lamp-pumped zigzag slab-phosphate glass-laser system. The 1-nsec output of this system is weakly focused with a 2.75-m lens to a spot size of 3-mm FWHM at the sample location. Transverse beam-intensity profiles are recorded on a shot-by-shot basis in the sample-equivalent plane by a charge-injection device camera with 244×248 pixel resolution. Total pulse energy is recorded by a calibrated pyroelectric detector. Using this pulse-energy information, we can calculate the maximum fluence (in joules per square centimeter) with a microcomputer for each shot, and the position of maximum fluence within the irradiated area of the sample is identified.

Two types of testing are conducted. One-on-one testing is performed, wherein photographs of the interaction site are taken at 100× magnification under dark-field illumination before and after irradiation by a single pulse. Damage is defined to exist if bubbles appear in the photographed region, regardless of whether they subsequently reabsorb into the fluid. Detection is limited to bubbles 5 μm in diameter or greater. Damage is defined not to exist if no bubbles are recorded, regardless of whether any bubbles were generated and reabsorbed before postirradiation photography.

N-on-one testing is performed by operating the zigzag slab laser at 1/5 Hz. Pulse-to-pulse energy fluctuations under N-on-one conditions are 8%. This low pulse-repetition rate gives us time to examine the interaction region for bubbles with a long-working-distance microscope after each shot. Unlike other materials for which subthreshold conditioning has been seen to raise N-on-one thresholds,[41] we see no such behavior in liquid-crystal fluids.

Considerable care must be devoted to test cell design and fluid filling if intrinsic rather than extrinsic damage thresholds are to be measured. Two borosilicate glass substrates separated by 100-μm spacers make up the test cell. The entrance substrate is a plane-parallel plate, whereas the exit substrate is in the form of a 30–60–90-deg prism. This configuration eliminates reflection of the laser pulse back into the fluid. After testing, a cell is disassembled, final-cleaned in ASTM Type I water, and dried in a class 100 laminar-flow hood before reassembly. A liquid-crystal fluid is commonly final filtered to 0.3 μm (0.2 μm in some cases) and injected onto the substrate–substrate interface through a glass syringe. Capillary action causes filling of the 100-μm gap. The accumulation of dust on the exposed fluid bead is minimized by performing the filling in the clean hood. No form of molecular alignment has been imposed on the liquid crystals for the damage-testing results reported here.

The selection of test fluids is driven by the application. Properties such as clearing-point temperature and miscibility with chiral additives are considered. Except for filtration, compounds are tested as received from commercial sources. Table 7 gives vendor-supplied physical properties data and our laser-damage test results for several nematic liquid crystals. [Testing has been performed on one chiral nematic blend used exclusively in fabricating liquid-crystal polarizers for OMEGA. The results for CB-15 in E7 were 6.0 ± 0.2 J/cm^2 (one-on-one) and 5.9 ± 0.4 J/cm^2 (N-on-one). In-system performance for this blend is discussed in Subsection 5.B.]

The data in Table 7 have been arranged in order of decreasing extraordinary refractive index. We have found no correlation between damage-threshold values and any of the listed physical properties of the evaluated compounds. The damage thresholds of these organics compare well with results (1053 nm, 1 nsec) reported by others for Teflon (one-on-one: 12.0 ± 1.3 J/cm^2) (Ref. 42) and chlorofluorocarbon oils (N-on-one: 5 J/cm^2).[43] The decrease in threshold when N-on-one testing is carried out for some of the compounds is a well-known phenomenon. Manenkov *et al.*[44] have studied this effect in a variety of organic compounds and attribute it to intensity-dependent photochemical reactions and thermomechanical effects in the vicinity of absorbing defects. Our own attempts to identify molecular fragments, resulting from laser-induced dissociation, have involved the use of Fourier-transform IR spectroscopy in specially prepared cells. We have not yet generated persuasive results.

In the absence of clear evidence for impurity- or photodissociation-driven damage in nematic liquid crystals, other mechanisms must be considered. The most apparent

Table 7. **Properties and Values of Laser-Damage Resistance for Commercial Liquid Crystals**

Commercial Liquid-Crystal Material[a]	Constituents[a,b]	Clearing Point (°C)	Specific Heat at 45°C (Cal/g deg)	Viscosity at 20°C (cS)	Optical Properties[a] at λ = 589.6 nm: Refractive Index (n_e)	Birefringence (Δn)	Damage Threshold at 1053 nm, 1 nsec: One on One (J/cm^2)	N on One (J/cm^2)
E7	Mixture of K, M, and T	60.5	0.465	40	1.746	0.22	15.1 ± 1.7	6.8 ± 0.7
ZLI-2140-000	Mixture of K, PCH, ME, D, and HD	63	0.446	35	1.660	0.15	6.8 ± 1.2	6.7 ± 1.0
ZLI-2140-000	Mixture of K, PCH, D, and HD	75	0.442	29	1.631	0.13	8.1 ± 0.3	7.7 ± 0.6
ZLI-1646	Mixture of PCH, HD, CCH, PCH-O, and CH	60	0.452	26	1.565	0.08	8.8 ± 0.8	8.6 ± 0.6
ZLI-2244-100	Mixture of CCH, CH, HD, OS, CBC, and D	92	0.457	31	1.555	0.07	10.5 ± 3.7	8.8 ± 0.8
ZLI-1800-000	Mixture of PCH, CCH, HD, CH, and D	60	0.478	29	1.555	0.08	10.5 ± 0.3	9.7 ± 0.1
K-15	K	35.3	0.446	29	1.535	0.18	9.6 ± 2.0	5.4 ± 1.3
ZLI-2359	Mixture of CCH, OS, and CH	68	0.464	28	1.519	0.05	8.9 ± 1.5	7.0 ± 1.2

[a] Information obtained from EM Chemicals, Hawthorne, New York.
[b] Identification of constituents:
K, 4-Cyano-4′-n-alkylbiphenyl
M, 4-Cyano-4′-n-alkoxybiphenyl
T, 4-Cyano-4″-n-alkylterphenyl
P, 4-Cyano-4′-n-alkylphenylcyclohexane
ME, 4-Cyanophenyl-4′-n-alkylbenzoate
D, 4-Cyanophenyl-4′-n-alkylcyclohexane carboxylate
HD, 4″-n-Alkylcyclohexylphenyl-4′-n-alkylcyclohexanecarboxylate
PCH-O, 4-n-Alkyl-4′-n-alkoxyphenylcyclohexane
CH, 4-n-Alkylcyclohexyl-4′-n-alkylbicyclohexylcarboxylate
CBC, 4′-(4-n-Alkylcyclohexyl)-4″-n-alkylbiphenylcyclohexane
OS, 4-n-Alkylcyclohexyl-4′-n-alkylcyclohexanecarboxylate
CCH, 4-Cyano-4′-(4″-n-alkylcyclohexyl)cyclohexane

among these mechanisms is nonlinear refractive-index-induced self-focusing. In nematic liquid crystals such an optical nonlinearity can arise from two physical causes: either from orientational effects based on the materials birefringence or from the polarizability of the molecule's electron distribution. The major difference between the two is their response times. Orientational nonlinearities have been observed for laser pulses as short as 5–20 nsec.[45] However, at the 800-psec-to-1-nsec pulse lengths of interest here, the orientational effects are most likely too slow to be effective. The electronic polarizability depends on the degree of conjugation in the molecules. The groups of molecules listed in Table 7 are systems with conjugated cores in which some of the electrons are completely delocalized. These highly delocalized electrons dominate the nonlinear susceptibility[46] at short time scales and thereby influence self-focusing and other nonlinear-optical effects.

The electronic nonlinear susceptibilities for nematic materials in their liquid-crystalline phases have not generally been measured. $\chi^{(3)}$ values for some biphenyls, measured in their isotropic phases, were reported by Madden *et al.*[47] The nonlinear refractive indices of other nematic systems, unrelated to the ones discussed here, were measured in their isotropic phases by Soileau *et al.*[48] However, data taken from materials in their isotropic phases cannot simply be extrapolated to their mesogenic phases. In the isotropic phase, where all molecular orientations are equally probable, the macroscopic nonlinear susceptibility $\chi^{(3)}$ isotropically averages over the microscopic, molecular susceptibilities. In the nematic phase, and for linearly polarized incident light, an anisotropic relation exists between the macroscopic $\chi^{(3)}$ and the molecular susceptibilities as established by Wong and Garito.[49] There is a significant difference in order between a cell in which the nematic alignment is enhanced by static, external fields or by organic surface layers and the order in an unaligned, nematic cell. The nonlinear-optical effects are expected to arise with different magnitudes among varying alignment conditions.

The damage-threshold data reported in Table 7 must therefore be interpreted with caution. In the unaligned cells used here, local variations in the magnitude of the macroscopic $\chi^{(3)}$ may potentially enhance self-focusing and, as a result, may lower the damage threshold. To supplement these data, experiments using homeotropically aligned, nematic cells are now under way.

B. Liquid-Crystal Polarizer Performance in OMEGA

Brewster-angle linear polarizers composed of alternating high- and low-index, thin-film layers, which have been deposited onto glass substrates, have been utilized in the 24-beam OMEGA laser system at the LLE since the laser was first constructed. Produced by standard electron-gun deposition technology in apertures ranging from 2 to 20 cm, these optical elements routinely achieve polarization levels of 500:1 at 1053 nm with moderate (3–5 J/cm^2) laser-damage resistance. They are essential to the operation of electro-optic switching devices such as Pockels cells. Three disadvantages of Brewster-angle polarizers include environmental instability, tendency to displace the laser beam, and physical space requirements.

In 1986 we replaced Brewster-angle linear polarizers in OMEGA with more than 60 LCP's varying in clear aperture from 38 to 100 mm. LCP's have none of the disadvantages associated with Brewster-angle polarizers and offer additional features not available with any other polarizer technology. This versatility has permitted us to incorporate the following features in the OMEGA system: propagation of circularly polarized laser radiation through the entire front end, minimizing stress birefringence induced by the laser rods and reducing nonlinear effects; reduced concern with backreflected ghosts, owing to the isolation properties of the LCP's; polarization cleanup before and after most amplification stages; and ease of installation and alignment.

Each LCP is assembled from a pair of borosilicate glass substrates, 8.5 mm thick by 100 mm in diameter. Optical specifications for the substrates, which have proved to be critical to device performance, are surface quality (better than 20/10 scratch-dig per MIL-0-13830A rev. L), surface figure (flat to λ/16 P-V at 1053 nm), wedge (less than 10 arcsec), and transmitted wave-front quality (better than λ/13 P-V at 1054 nm). All 100-mm LCP elements fabricated for use in OMEGA contain a blend of a nematic E7 doped with the RH chiral twisting agent CB-15. Wavelength tuning of the blend is performed in a spectrophotometer[50] equipped with a thermal control cell of in-house design to regulate temperature. The blend's peak position is adjusted so that each device may be installed in OMEGA and used at λ = 1053 nm at a temperature of 22 ± 3°C and at an angle of incidence of 8° from normal. Light rejected from a LCP is reflected within a 1–2° cone angle, as shown in the semilogarithmic plot of Fig. 20. It is important that this near-specular reflection be directed away from the laser-system optic axis.

1. Early In-System Performance

The first sets of LCP's installed in OMEGA in late 1986 and early 1987 were assembled using a diamond rub for molecular alignment. The inside surfaces of the substrates were rubbed for 1–3 min with 0.1-μm diamond paste under a load that varied from 5 to 25 kg. The substrates were cleaned by hand scrubbing with 0.05-μm alumina polishing compound, hand scrubbing with Micro detergent, rinsing with ASTM Type I water, and blowing dry with 0.2-μm filtered nitrogen. Assembly of each LCP with Mylar spacers was conducted in

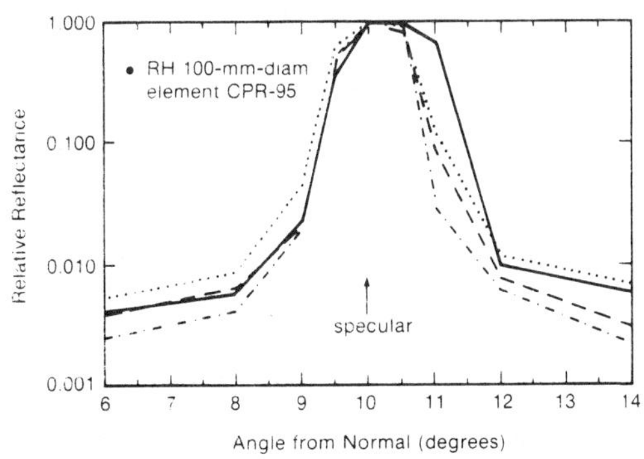

Fig. 20. Reflectance near specular for a LCP. The probe beam was 1 mm in diameter and RH polarized at a wavelength of 1052 nm, and the angle of incidence was 10° from normal. An integrating sphere detector fitted with a 1-mm aperture measured reflectivity as a function of angle around an arc centered at 130 mm from the polarizer. Scans taken at four different sites on a 100-mm-diameter element are given.

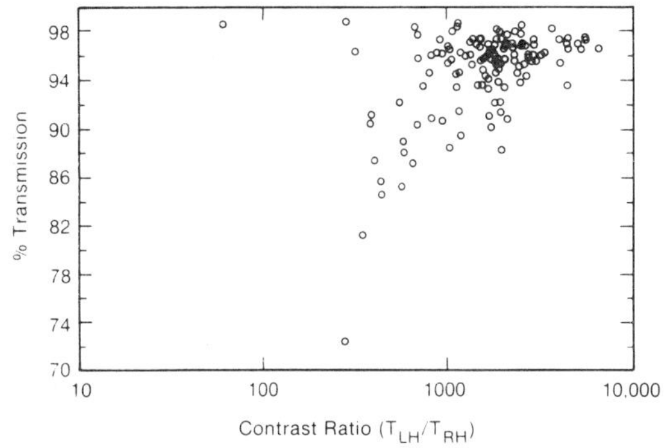

Fig. 21. Polarization characteristics of 133 diamond-rubbed, 100-mm-diameter, RH LCP's at 1052 nm. Transmission is for LH circularly polarized light. The contrast ratio is for LH/RH circularly polarized light.

Table 8. Incident Energy at LCP Positions

Position	J/cm² (1053 nm, 1 nsec)
Before first 64-mm amplifier in a beam line	0.3–0.5
Before second 64-mm amplifier in a beam line	0.6–1.2
Before the 90-mm amplifier in a beam line	2.0–3.8

a horizontal laminar-flow hood on a hot plate, where the cell was filled with the liquid-crystal fluid at a temperature above the fluid's isotropic point and was permitted to cool. Each cell was then given a mechanical shear of approximately 0.5 mm and was sealed with epoxy. The contrast and transmission of these LCP's were generally higher than those of the Brewster-angle dielectric thin-film polarizers previously used to control polarization in OMEGA. Test results showing contrast and transmission for these rubbed LCP's are given in Fig. 21.

Diamond-rubbed LCP's have performed well in the driver line and at the input 64-mm amplifiers in the beam lines. Most of these polarizers have sustained minimal or no damage since their installation in September 1985 for the driver line and in September 1986 for the beam lines. Most of the damage that has occurred at these positions has resulted from aberrant energy fluxes well above the normal maximum energy ranges listed in Table 8.

Damage was a problem at the position preceding the 90-mm amplifier in a beam line and was generally of two types shown in Fig. 22. Damage appeared to occur randomly within the clear aperture of the LCP and was highly variable as to the time interval between installation of the LCP in the system and the onset of damage. Thorough characterization of the beam lines was undertaken to examine the possibility that hot spots or ghosts were responsible for the observed damage, but no correlation was found. In an attempt to reduce damage susceptibility a number of modifications to the assembly process were made in early 1987.

2. Processing for Improved Performance

Current assembly protocol no longer includes the diamond rubbing step. This removes a potential source of contamination and glass damage. Substrates, if disassembled from a previous LCP, are reground and polished to remove any residual glass damage. Substrate cleaning takes place in a clean room, which maintains the air at fewer than 1000 0.5-

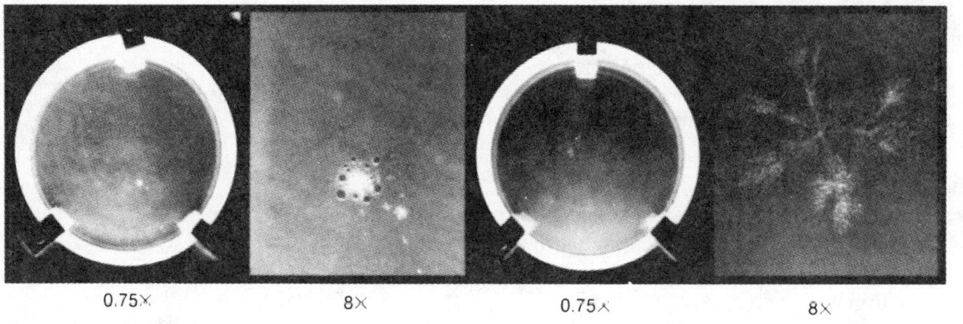

Fig. 22. LCP damage from in-system use: (a) a burst that shows a high amount of scatter and often involves bubbles in the liquid crystal and chips in the glass surface at the center of the damage site; (b) a web that has lines of scatter radiating from a center; bubbles, if present, are usually at the ends of the arms.

Table 9. Improved Type-I Water Specifications for LCP Cleaning

Specification	Value
Resistivity	>18 MΩ
Point of use filtration	0.1 μm
Silica	<0.02 mg/L
Bacteria	<1 colony forming unit/mL

μm particles per cubic foot (~28,000 cm^3). High-purity water is used, which is monitored and maintained to the specifications in Table 9. The cleaning process involves the following steps: 1-min ultrasonic cleaning in detergent, 5-min spray rinse, 1-min ultrasonic cleaning in wetting agent, 6-min spray rinse, IR drying.

The substrates are carried through the cleaning process in Teflon holders, and handling is minimized. Rinsing, drying, and assembly are done in a horizontal laminar-flow hood in the clean room, which maintains the air at less than 10 0.5-μm particles per cubic foot. Operator procedures for maintaining contamination-free conditions are critical at these steps to minimize potential damage sites in the resulting LCP's. An inspection step has been introduced to look for any deviations from surface-quality specifications and any substrate damage, visible cleaning residue, or particulate that remains after the cleaning process. These undesirable features are mapped, and if they are severe the substrate is rejected.

After the substrates and Mylar spacers are assembled into a cell, the cell is removed from the hood and placed on a hot plate in the clean room. This minimizes contamination of the assembly area by the hot plate and filling operations. The cell is heated to a temperature that is greater than the isotropic-point temperature and is then filled with the liquid-crystal fluid, using 0.5-μm point-of-use filtration. Care is taken to avoid dragging in any contaminants on the bevel. Cooling proceeds slowly to a point below the isotropic/liquid-crystal transition, at which point the LCP is transferred from the hot plate to the benchtop. A mechanical shear of approximately 1 mm is performed by hand, and the LCP is permitted to equilibrate to room temperature. Repetitive shears are then done in the direction of fill and crosswise to it, while the decrease in scatter in the LCP is visualized. The LCP is again permitted to equilibrate and can then be sealed.

The LCP's assembled by our clean-room assembly process show higher and more consistent contrast and transmission than those made with previous diamond-rub procedures, as shown in Fig. 23. Because the technique for assembly has improved, we have observed an increase in device damage resistance. System shots are now routinely taken at the energy levels of 2–3.8 J/cm^2, as shown in Table 8. When damage occurs, growth of the sites and rapid deterioration owing to generation of bubbles now proceed at a significantly slower rate. LCP's are generally left in OMEGA until bub-

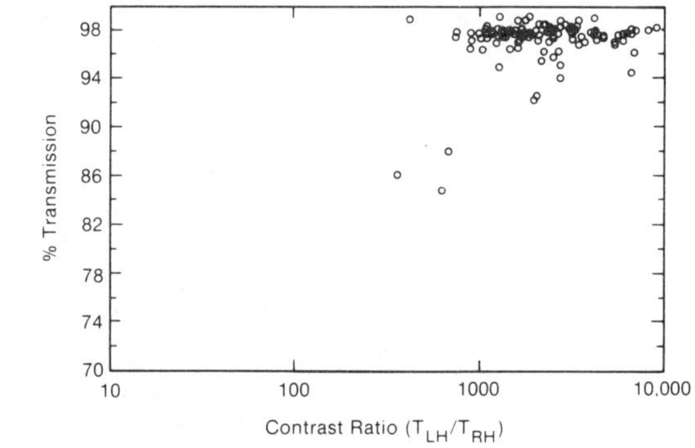

Fig. 23. Polarization characteristics of 120 100-mm RH LCP's assembled in 1987 by using the improved process. Transmission is for LH circularly polarized light. The contrast is for LH/RH circularly polarized light at 1053 nm.

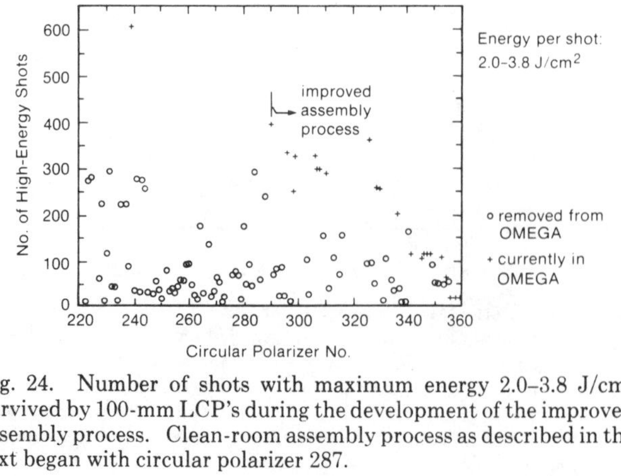

Fig. 24. Number of shots with maximum energy 2.0–3.8 J/cm^2 survived by 100-mm LCP's during the development of the improved assembly process. Clean-room assembly process as described in the text began with circular polarizer 287.

bles appear that are large enough or numerous enough to disrupt the beam. LCP lifetimes in OMEGA, according to these criteria, are shown in Fig. 24.

There has been little or no correlation between sites of damage that occur in OMEGA and sites identified and mapped during the assembly procedure. One exception, however, was noted: If a substrate suffered glass damage at a LCP damage site and the damage was not fully removed by grinding and repolishing, the new LCP was seen to become damaged at the previous damage site, usually rapidly. Recently a correlation has been noted between LCP damage at the 90-mm laser-rod position and the size of targets being irradiated in laser-fusion experiments. It is suspected that unconverted 1053-nm laser radiation, backreflected from large targets and amplified in the 90-mm rods, is contributing to damage in the LCP's. Methods to eliminate this additional source of laser-induced damage in LCP's are currently being studied.

6. SUMMARY

The goal of this paper has been to describe several new device applications for liquid crystals in laser systems. Extensive design, fabrication, and performance data have been provided to assist optical engineers in understanding the properties of these materials and their interactions with laser radiation. Liquid crystals offer unique opportunities for device development in laser systems in which aperture, function, and/or performance limit conventional materials.

ACKNOWLEDGMENTS

Much of the work reported here was carried out with the assistance of undergraduate students at the University of Rochester. The authors would like to acknowledge the contributions of following: R. Gingold and K. Thibodeau (liquid-crystal wave plates), T. Gunderman and B. Wagner (LCP's), M. Topp and P. Laverty (liquid-crystal apodizers), and A. Bevin (laser-damage testing). We would also like to acknowledge the technical assistance of R. Bahr, B. Castle, R. Hutchison, J. Kelly, T. Kessler, A. Lindquist, S. Morse, M. Shoup III, and D. J. Smith. The manuscript was skillfully prepared for us by L. Cardone and J. Steve.

This research was supported by the U.S. Department of Energy, Office of Inertial Fusion, under agreement DEFC08-85DP40200, and by the Laser Fusion Feasibility Project at the LLE, which has the following sponsors: Empire State Electric Energy Research Corporation, New York State Energy Research and Development Authority, Ontario Hydro, and the University of Rochester. Additional support was provided by the New York State Center for Advanced Optical Technology of the Institute of Optics, McDonald Douglas Astronautics Company, General Electric Laser Group in Binghamton, New York, and the U.S. Army Night Vision and Electro-Optics Laboratory.

REFERENCES AND NOTES

1. EM Chemicals, Inc., Hawthorne, New York; Hoffman-La-Roche, Nutley, New Jersey; Eastman Kodak Company, Rochester, New York; Aldrich Chemical Company, Milwaukee, Wisconsin.
2. S. D. Jacobs, "Liquid crystals for laser applications," in *Optical Materials Properties, CRC Handbook of Laser Science and Technology*, M. J. Weber, ed. (CRC Press, Cleveland, Ohio, 1986), Vol. IV, pp. 409–465.
3. H. L. deVries, "Rotatory power and other optical properties of certain liquid crystals," Acta Crystallogr. **4,** 219 (1951).
4. J. L. Fergason, "Liquid crystals," Sci. Am. **211,** 76 (1964).
5. One useful epoxy is 3M Scotch-Weld 2216 epoxy.
6. See, for example, J. Cognard, "Alignment of nematic liquid crystals and their mixtures," Mol. Cryst. Liq. Cryst. Suppl. **1,** 1 (1982).
7. J. L. Fergason, Optical Shields, Ltd., 1390 Willow Road, Menlo Park, California 94025 (personal communication, September 1987).
8. S. D. Jacobs, "Liquid crystals as large aperture wave plates and circular polarizers," in *Polarizers and Applications*, G. B. Trapani, ed., Proc. Soc. Photo-Opt. Instrum. Eng. **307,** 98 (1981).
9. Bellingham and Stanley, Kent, England, Model 60/HR Abbé refractmeter.
10. CVI Laser Corporation, Albuquerque, New Mexico, Model C-95 CVI YAG-MAX laser.
11. D. Kinzer, "An optical technique for measuring liquid crystal cell thickness," Mol. Cryst. Liq. Cryst. Lett. **5,** 147 (1985).
12. S. T. Wu and U. Efron, "Optical properties of thin nematic liquid crystal cells," Appl. Phys. Lett. **48,** 624 (1986).
13. S. D. Jacobs, J. A. Abate, K. A. Bauer, R. P. Bossert, and J. M. Rinefierd, "Liquid crystal isolator for fusion lasers," in *Digest of Conference on Lasers and Electro-Optics* (Optical Society of America, Washington, D.C., 1980), pp. 128–129.
14. S. D. Jacobs, K. A. Cerqua, T. J. Kessler, W. Seka, and R. Bahr, "Retrofit of a high-power Nd:glass laser system with liquid crystal polarizers," Natl. Bur. Stand. (U.S.) Spec. Pub. **727,** 15–22 (1986).
15. S. T. Wu, U. Efron, and L. D. Hess, "Infrared birefringence of liquid crystals," Appl. Phys. Lett. **44,** 1033 (1984).
16. J. G. Pasko, J. Tracy, and W. Elser, "Liquid crystal infrared modulator," in *Active Optical Devices*, J. Tracy, ed. Proc. Soc. Photo-Opt. Instrum. Eng. **202,** 82 (1979).
17. J. L. Fergason, "Cholesteric structure I. Optical properties," in *Liquid Crystals*, G. H. Brown, G. J. Dienes, and M. M. Labes, eds. (Gordon & Breach, New York, 1967), p. 89.
18. P. G. deGennes, *The Physics of Liquid Crystals* (Clarendon, Oxford, 1974).
19. T. J. Scheffer, "Twisted nematic display with cholesteric reflector," J. Phys. D **8,** 1441 (1975).
20. A. Thelen, "Design of optical minus filters," J. Opt. Soc. Am. **61,** 365 (1971).
21. J. R. Magarinos and D. J. Coleman, "Holographic mirrors," Opt. Eng. **24,** 769 (1985).
22. J. Adams, W. Haas, and J. Dailey, "Cholesteric films as optical filters," J. Appl. Phys. **42,** 4096 (1971).
23. I. K. Krasyuk, S. G. Lukishova, P. P. Pashinin, A. M. Prokhorov, and A. V. Shirkov, "Formation of the radial distribution of intensity in a laser beam by 'soft' apertures," Sov. J. Quantum Electron. **6,** 725 (1976).
24. V. R. Costich and B. C. Johnson, "Apertures to shape high power beams," Laser Focus **10**(9), 43 (1974).
25. A. Penzkofer and W. Frohlich, "Apodizing of intense laser beams with saturable dyes," Opt. Commun. **28,** 197 (1979).
26. Y. Asahara and T. Izumitani, "Process of producing soft aperture filter," U.S. Patent 4,108,621 (August 22, 1978).
27. G. Dubé, "Progress in glass components for neodymium lasers," in *Advanced Laser Technology and Applications*, L. Esterowitz, ed., Proc. Soc. Photo-Opt. Instrum. Eng. **335,** 10 (1982).
28. V. I. Kryzhanovskii, B. M. Sedov, V. A. Serebryakov, A. D. Tsvetkov, and V. E. Yashin, "Formation of the spatial structure of radiation in solid-state laser systems by apodizing and hard apertures," Sov. J. Quantum Electron. **13,** 194 (1983).
29. A. J. Campillo, B. Carpenter, B. E. Newnam, and S. L. Shapiro, "Soft apertures for reducing damage in high-power laser-amplifier systems," Opt. Commun. **10,** 313 (1974).
30. E. W. S. Hee, "Fabrication of apodized apertures for laser beam attenuation," Opt. Laser Technol. **7,** 75 (1975).
31. S. B. Arifzhanov, R. A. Ganeev, A. A. Gulamov, V. I. Redkorechev, and T. Usmanov, "Formation of a beam of high optical

quality in a multistage neodymium laser," Sov. J. Quantum Electron. **11,** 745 (1981).
32. G. Dubé, "Total internal reflection apodizers," Opt. Commun. **12,** 344 (1974).
33. J.-C. Diels, "Apodized aperture using frustrated total reflection," Appl. Opt. **14,** 2810 (1975).
34. S. B. Papernyi, V. A. Serebryakov, and V. E. Yashin, "Formation of a smooth transverse distribution of intensity in a light beam by a phase-rotating plate," Sov. J. Quantum Electron. **8,** 1165 (1978).
35. G. Giuliani, Y. K. Park, and R. L. Byer, "Radial birefringent element and its application to laser resonator design," Opt. Lett. **5,** 491 (1980).
36. J. M. Eggleston, G. Giuliani, and R. L. Byer, "Radial intensity filters using radial birefringent elements," J. Opt. Soc. Am. **71,** 1264 (1981).
37. B. J. Feldman and S. J. Gitomer, "Annular lens soft aperture for high power laser systems," Appl. Opt. **16,** 1484 (1977).
38. J. T. Hunt, J. A. Glaze, W. W. Simmons, and P. A. Renard, "Suppression of self-focusing through low-pass spatial filtering and relay imaging," Appl. Opt. **17,** 2053 (1978).
39. S. D. Jacobs and K. A. Cerqua, "Optical apparatus using liquid crystals for shaping the spatial intensity of optical beams having designated wavelengths," U.S. Patent 4,679,911 (July 14, 1987).
40. S. D. Jacobs, K. A. Cerqua, K. L. Marshall, T. J. Kessler, R. J. Gingold, P. J. Laverty, and M. Topp, "High power laser beam apodization using a liquid crystal soft aperture," in *Digest of Conference on Lasers and Electro-Optics* (Optical Society of America, Washington, D.C., 1986), paper THG2, p. 258.
41. J. Swain, S. Stokowski, D. Milam, and F. Rainer, "Improving the bulk laser damage resistance of KDP by pulsed laser irradiation," Appl. Phys. Lett. **40,** 350 (1982).
42. M. C. Staggs and F. Rainer, "Damage thresholds of fused silica, plastics and KDP measured with 0.6 μs, 355-nm pulses," Natl. Bur. Stand. (U.S.) Spec. Pub. **688,** 84 (1985).
43. *Lawrence Livermore National Laboratories Annual Report* (Lawrence Livermore Laboratory, Livermore, Calif., 1983), pp. 6–60.
44. A. A. Manenkov, G. A. Matyushin, V. S. Nechitailo, A. M. Prokhorov, and A. S. Tsaprilov, "Nature of the cumulative effect in laser damage to optical materials," Sov. J. Quantum Electron. **13,** 1580 (1983).
45. I. C. Khoo, R. R. Michael, and P. Y. Yan, "Optically-induced molecular reorientation in nematic liquid-crystals and nonlinear optical processes in the nanoseconds regime," IEEE J. Quantum Electron. **QE-23,** 267 (1987).
46. S. Risser, S. Klemm, D. W. Allender, and M. A. Lee, "Huckel model—calculations of polarizability and hyperpolarizability for conjugated molecules," Mol. Cryst. Liq. Cryst. **150B,** 631 (1987).
47. P. A. Madden, F. C. Saunders, and A. M. Scott, "Measurement of the nonlinear susceptibility of liquid crystal materials in the isotropic phase," Opt. Acta **33,** 405 (1986).
48. M. J. Soileau, E. W. van Stryland, and S. Guha, "Nonlinear optical properties of liquid crystals in the isotropic phase," Mol. Cryst. Liq. Cryst. **143,** 139 (1987).
49. K. Y. Wong and A. F. Garito, "Third-harmonic-generation study of orientational order in nematic liquid-crystals," Phys. Rev. **34,** 5051 (1986).
50. A Lambda 9 spectrophotometer from Perkin-Elmer, Inc., Norwalk, Connecticut, was used.

©1971 Masson S.A. Reprinted from *Nouvelle Revue d'Optique Appliquée*,
Vol. 2(4), 221–228 (1971).

CONVERTISSEUR D'IMAGES A CRISTAL LIQUIDE ET PHOTOCONDUCTEUR

par

M. FRAPPIER, G. ASSOULINE, M. HARENG et E. LEIBA

MOTS CLÉS :
Cristal liquide
Convertisseur

KEY WORDS :
Liquid crystal
Converter

RÉSUMÉ

Un milieu nématique soumis à un champ électrique et traversé par un courant électrique devient fortement diffusant. La juxtaposition d'une couche photoconductive au nématique permet de moduler la diffusion du cristal liquide par des variations d'éclairement. On réalise ainsi un « relais de lumière » qui, par sa simplicité, remplace avantageusement les autres procédés de projection d'informations sur grand écran utilisé actuellement (écrans à forte luminance ; systèmes à couche d'huile, etc...). Les caractéristiques du relais varient avec le choix du photoconducteur et du nématique : il présente l'intérêt de pouvoir être utilisé dans de nombreuses applications :

- *reprographie, hard copy,*
- *convertisseurs d'images infrarouge ou ultraviolette,*
- *projection d'informations sur grand écran.*

La sensibilité et la rapidité des photoconducteurs utilisés permet d'envisager la projection sur grand écran d'une image de télévision.

INTRODUCTION

Les cristaux liquides de structure nématique sont des substances qui permettent une modulation en transmission de la lumière. Ceci est dû au phénomène dit de « diffusion dynamique » [1] [2] sur lequel nous reviendrons plus en détail. Disons simplement qu'une couche nématique transmet normalement un faisceau de lumière visible. Si on la soumet à une tension électrique appliquée par l'intermédiaire de deux électrodes transparentes, comme l'indique la figure 1, le faisceau est fortement atténué.

Nous décrivons ici une structure constituée de la couche de nématique juxtaposée à une couche photoconductrice, l'ensemble étant placé entre deux électrodes transparentes (fig. 2).

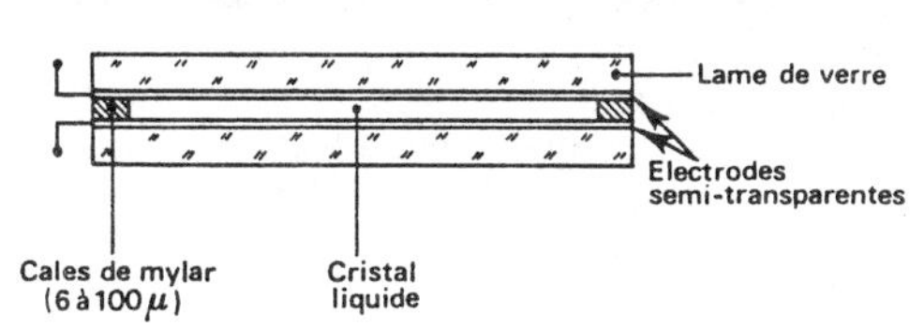

FIG. 1. — *Cellule à cristal liquide.*

FIG. 1. — *Liquid crystal cell.*

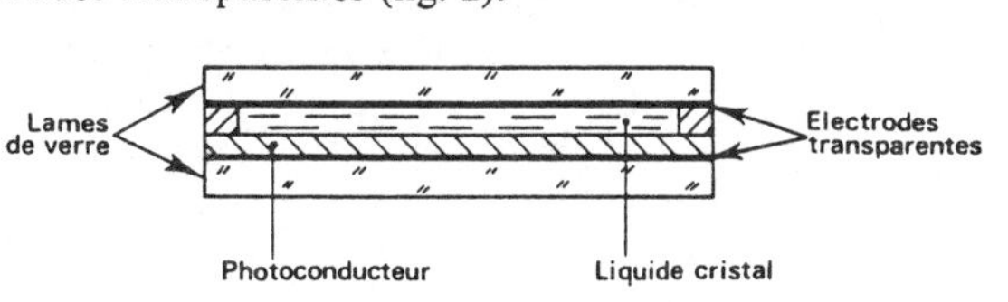

FIG. 2. — *Cellule à cristal liquide et photoconducteur.*

FIG. 2. — *Liquid crystal and photoconductor cell.*

Une tension continue V est appliquée aux électrodes. Dans l'obscurité la résistance du photoconducteur est très supérieure à celle du nématique. De ce fait la quasi-totalité de la tension *V* est appliquée au photoconducteur et le nématique, n'étant soumis qu'à une tension très faible, reste transparent. Si maintenant le photoconducteur est éclairé, sa résistance diminue au prorata de la lumière reçue et une partie plus ou moins grande de la tension se reporte sur le nématique modulant ainsi sa transparence. Par suite une image formée sur le photoconducteur s'inscrit en diapositive sur le nématique et peut être projetée.

La figure 3 représente un exemple d'application de ces cellules. L'image formée sur le photoconducteur provient d'un tube à rayons cathodiques. Elle s'inscrit en diapositive sur le sandwich et est simultanément projetée sur un grand écran. C'est une solution possible à la projection sur grand écran actuellement étudiée dans nos laboratoires.

Dans ce qui suit nous parlons brièvement des cristaux nématiques et du phénomène de « diffusion dynamique ». Puis nous donnons quelques notions sur les photoconducteurs et leur préparation en couches minces. Le troisième chapitre est consacré au sandwich complet et le quatrième à ses applications.

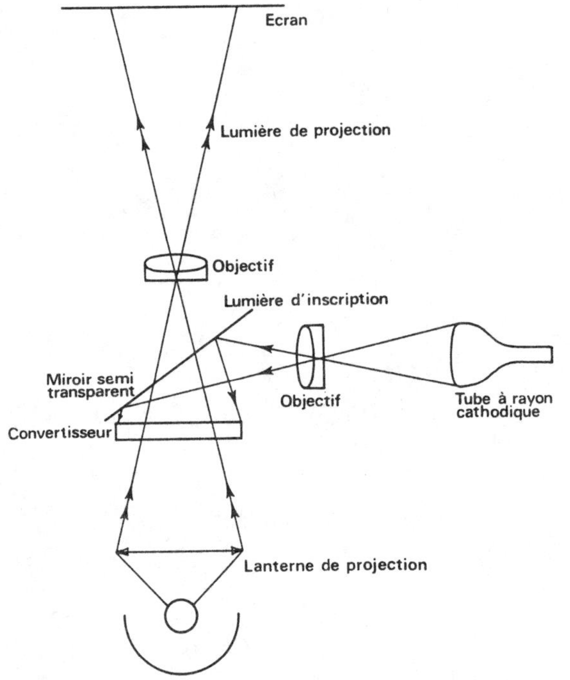

FIG. 3. — *Système de projection de télévision sur grand écran utilisant le convertisseur à cristal liquide et photoconducteur.*

FIG. 3. — *Apparatus for the large screen television projection utilizing the liquid crystal and photoconductor cell.*

I. — CRISTAUX LIQUIDES NÉMATIQUES ET PHÉNOMÈNE DE DIFFUSION « DYNAMIQUE »

Les cristaux liquides

Les cristaux liquides [3] sont des corps organiques constitués de molécules allongées, dont le comportement mécanique est celui des liquides mais dont le comportement optique s'apparente à celui des corps cristallins. Cette anisotropie est due à un arrangement des molécules constituant le corps.

On a classé les cristaux liquides en trois catégories se distinguant par l'ordre moléculaire. Ce sont la phase smectique, la phase cholestérique et la phase nématique. La figure 4 montre les trois structures. Nous ne nous intéressons ici qu'à la phase nématique caractérisée par une orientation commune des molécules. Elle peut être parallèle ou perpendiculaire au support : si on y applique un champ électrique ou magnétique l'orientation commune est parallèle à ce champ.

La phase « cristal liquide » n'existe qu'entre deux températures correspondant au passage à l'état solide en refroidissant et à l'état liquide isotrope en chauffant.

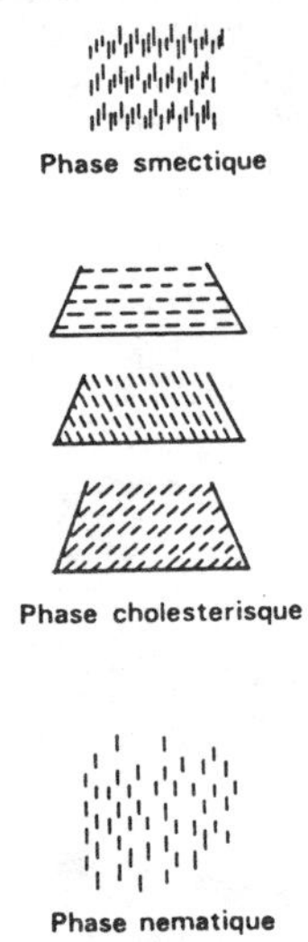

FIG. 4. — *Les trois structures des cristaux liquides.*

FIG. 4. — *The three liquid crystal structures.*

La diffusion dynamique

On appelle ainsi un phénomène découvert par un chercheur américain dénommé Heilmeier [1]. On place entre deux électrodes transparentes constituant les armatures d'un condensateur plan une couche nématique à la place du diélectrique (fig. 1). L'ensemble est transparent. Si on applique une tension aux électrodes le nématique devient fortement diffusant. L'origine physique de ce phénomène s'explique de la manière suivante : en absence de champ le milieu est ordonné ; en présence du champ, les ions sont mis en mouvement et cisaillent le liquide en entraînant ainsi de la matière. Lorsque le nombre d'ions et leur vitesse sont suffisants, l'écoulement n'est plus laminaire et il apparaît des tourbillons créant ainsi le désordre du milieu et par suite la diffusion de lumière. L'effet de diffusion est donc dû aux ions [4].

Courbes de diffusion en fonction de la tension. — Les mesures de la diffusion optique du nématique ont été réalisées à l'aide du dispositif décrit par la figure 5. Une lentille est placée entre la couche et un photomultiplicateur afin de recueillir le rayonnement diffusé dans la direction perpendiculaire à la couche. La diffusion se produisant surtout vers l'avant, l'électrode inférieure est parfaitement réfléchissante.

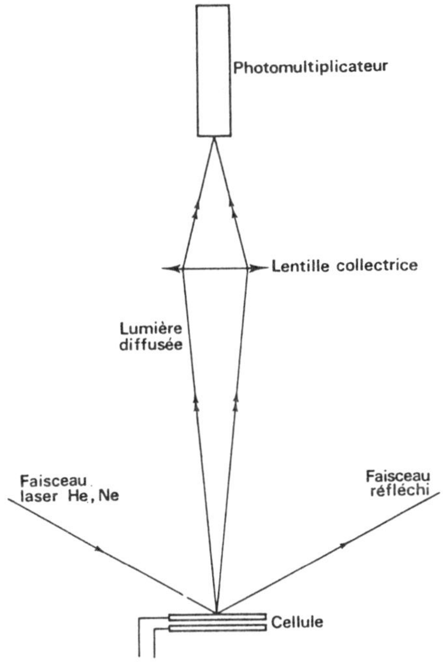

FIG. 5. — *Dispositif de mesure de la diffusion par réflexion.*

FIG. 5. — *Apparatus for the scattering measurements.*

En l'absence de champ, le courant de photomultiplicateur, correspond à la faible diffusion existante, est pris égal à l'unité.

La figure 6 représente la courbe de contraste relevée dans ces conditions. On appelle contraste le rapport des intensités lumineuses diffusées en présence et en absence de champ électrique données par le photomultiplicateur.

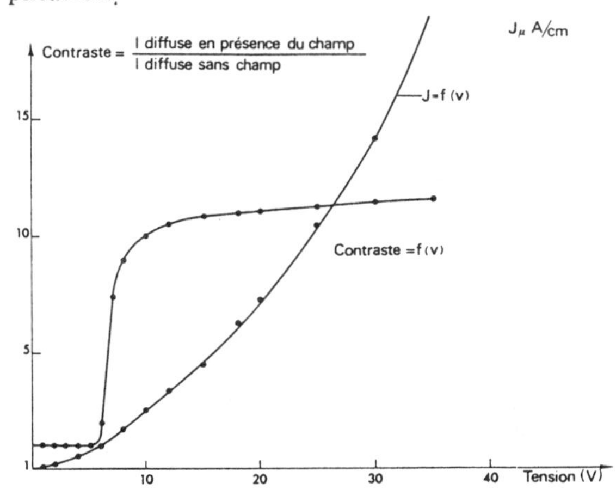

FIG. 6. — *Diffusion et courant en fonction de la tension. Epaisseur* 8 μ, $T = 62°$.

FIG. 6. — *Scattering and current vs the voltage.*

On constate que la diffusion induite par le champ électrique débute assez brusquement à une tension de 5,5 V. Cette diffusion débute toujours à la même tension quelle que soit l'épaisseur de la couche de nématique, il s'agit donc effectivement d'un seuil en tension.

Lorsqu'on augmente la tension le contraste augmente très rapidement et linéairement, puis au-delà d'une certaine valeur variable avec l'épaisseur de la couche, l'effet présente une saturation plus ou moins marquée. Le contraste maximum est indépendant de l'épaisseur de la cellule mais fonction du produit et des conditions de mesure, il peut être très élevé. La figure 6 donne également la variation du courant en fonction de la tension. On constate que le courant existe avant que ne se manifeste l'effet de diffusion et qu'il ne présente pas de saturation.

Transmission en fonction de la tension. — Ces mesures ont été faites à l'aide d'un dispositif semblable au montage de la figure 5, seulement ici le photomultiplicateur reçoit la lumière transmise par la cellule en absence et en présence de tension. La figure 7 représente la courbe d'atténuation pour différents produits. L'atténuation étant le rapport entre l'intensité I_0 reçue par le PM en absence de tension et l'intensité I avec une tension appliquée.

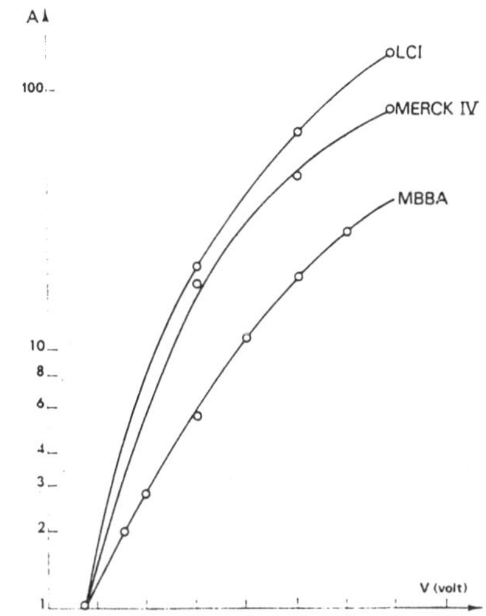

FIG. 7. — *Atténuation comparée des différents produits nématiques commerciaux : cellule* 20 μ, *excitation :* 50 Hz.

FIG. 7. — *Compared attenuation with various commercial nematic products — Cell* 20 μ — *Voltage* 50 Hz.

On constate que l'atténuation débute à une tension de l'ordre de 7 à 8 volts (tension de seuil) et augmente ensuite avec la tension. On constate également que cette atténuation dépend surtout du cristal liquide nématique, elle varie peu avec l'épaisseur de la cellule.

Temps de réponse et de relaxation de la cellule. — Le temps de montée d'une cellule est de l'ordre de la milliseconde. Ce temps est suffisant pour l'affichage. Pour la formation d'une image du type télévision il est nécessaire de disposer d'un élément supplémentaire capable de garder l'information pendant un temps nécessaire à la réponse du cristal liquide.

Le temps de relaxation varie avec le produit, sa résistivité et l'épaisseur de la cellule, il peut varier entre 50 et 500 ms.

Sur la courbe 8 sont représentés les temps de descente de trois produits différents.

Mélanges nématiques cholestériques. — Ces mélanges possèdent les propriétés de diffusion des nématiques et présentent l'avantage d'être mis en mémoire après avoir été excités par une tension continue.

Le cristal liquide à mémoire est obtenu en dopant du produit nématique par 5 à 15 % de produit cholestérique. La diffusion induite dans la cellule se maintient après suppression du champ électrique et ceci pendant un temps plus ou moins long suivant la composition du mélange. On peut obtenir des temps de mémorisation allant de quelques minutes à quelques semaines.

Le temps de réponse d'une cellule à mémoire est de 10 millisecondes.

L'effacement est obtenu par l'application d'une tension alternative, son temps est lui aussi fonction de la composition du mélange, il est d'autant plus long que la durée de mémorisation est longue. Par exemple une cellule qui tient la mémoire pendant 1 h s'efface en 1 seconde par l'application d'une tension efficace de 90 volts à une fréquence de 50 Hz.

Applications. — La phase nématique présente des possibilités d'applications étendues grâce au phénomène de diffusion et de variation de biréfringence avec le champ électrique.

L'application la plus proche est l'affichage numérique et alpha-numérique par segments et par points. Ces affichages offrent par rapport à leurs concurrents (nixies, diodes électroluminescentes) l'avantage de nécessiter des basses tensions compatibles avec les

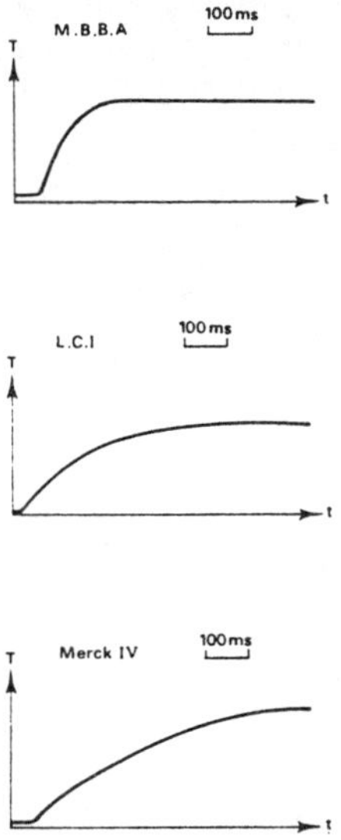

FIG. 8. — *Variation du temps de descente en fonction de la nature du produit (Courbe de transmission T en fonction du temps t).*

FIG. 8. — *Decay time for various products.*

circuits intégrés et d'être faiblement consommateurs (2 $\mu A/cm^2$ sous 20 V).

La deuxième application immédiate est l'atténuateur de lumière, utile en particulier pour la protection des caméras de télévision de nuit.

La variation de la biréfringence sous l'effet du champ permet de moduler la couleur transmise à partir d'un faisceau de lumière blanche.

II. — NOTIONS SUR LES PHOTOCONDUCTEURS

Photoconducteurs

La couche photoconductrice permet de moduler optiquement la tension appliquée au cristal liquide. On lui demande les caractéristiques suivantes :

— Se déposer en couches minces sous vide pour avoir une bonne qualité de surface et une homogénéité d'épaisseur ;

— Une grande résistivité : l'épaisseur de la couche photoconductrice est du même ordre de grandeur que celle du cristal liquide. La résistivité doit donc être plus élevée d'environ deux ordres de grandeur. Typiquement les cristaux liquides font 10^9 Ω.cm et le photoconducteur 10^{11} Ω.cm dans l'obscurité.

Il est en outre souhaitable d'avoir :

— Une bonne sensibilité ;

— Des constantes de temps à la montée et à la descente aussi courtes que possible ;

— Une transparence dans une partie importante du spectre visible.

Spectre de sensibilité en liaison avec les courbes de transmission

Un photoconducteur n'est évidemment sensible qu'aux longueurs d'ondes qu'il absorbe ; les rayonnements transmis ne l'impressionnent pas. Nous avons reporté sur la figure 9 les courbes de transmission de différents photoconducteurs en couches minces et les courbes de photoréponse correspondantes. Ainsi un photoconducteur tel que l'oxyde de zinc transmet dans le spectre visible et est sensible dans le proche ultra-violet. C'est le photoconducteur idéal pour notre application. Malheureusement sa sensibilité et ses constantes de temps sont médiocres. Au contraire, le sélénium qui est très sensible et a de courtes constantes de temps est tout à fait opaque et ne conviendrait que dans un montage en réflexion. Enfin, le sulfure de cadmium représente un bon compromis à tous points de vue et c'est celui que nous utilisons le plus actuellement.

Résistivité et sensibilité

Forte résistivité et sensibilité sont deux caractéristiques généralement antagonistes dans les photoconducteurs. Pour conserver une forte résistivité nous sacrifions donc un peu de la sensibilité, ce qui a au moins l'avantage de préserver en même temps la transparence et la parfaite clarté des couches.

Exemples de photoconducteurs et préparation

Nos études sont portées sur un grand nombre de matériaux dont : le sulfure de zinc, l'oxyde de zinc, le sulfure de cadmium, le sélénium et le sélénium-tellure. Pour l'instant nous ne parlerons que des trois derniers qui ont des performances satisfaisantes.

Les couches de sulfure de cadmium s'obtiennent par évaporation d'une poudre dopée sous une pression de 10^{-6} mm de Hg et sur support à 100 °C.

Le sélénium doit être très pur au départ et est déposé dans un très bon vide sur un support à 55 °C. Les couches de sélénium sont amorphes et on doit éviter de les laisser cristalliser ce qui est provoqué par une élévation de température supérieure à 60 °C.

Le sélénium-tellure est un composé obtenu en ajoutant 5 à 10 % de tellure au sélénium et en réalisant l'alliage avant évaporation. Les couches obtenues sont plus sensibles et moins fragiles : la température de cristallisation est supérieure à 100 °C.

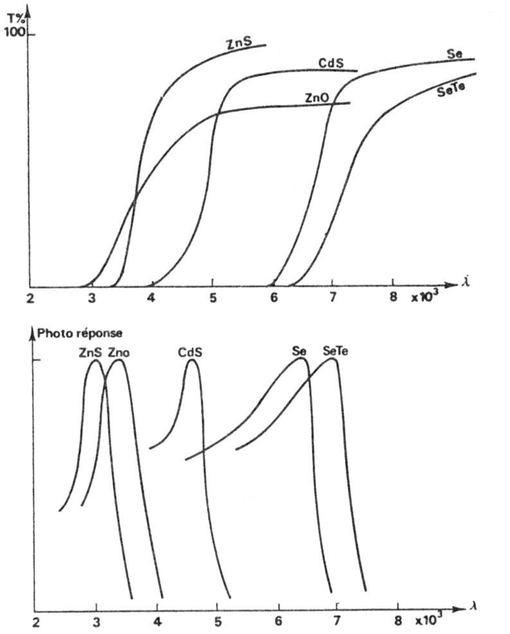

FIG. 9. — *Courbes de transmission et de sensibilité relative de quelques photoconducteurs en fonction de la longueur d'onde.*

FIG. 9. — *Curves of transmission and relative sensitivity of various photoconductors vs wavelengths.*

III. — CARACTÉRISTIQUES DU CONVERTISSEUR

Cellules au sélénium et sélénium-tellure

Ces couches sont obtenues par évaporation sous vide à une température voisine de 60° ; de couleur rouge, elles sont peu transparentes, leur épaisseur est de l'ordre de 4 µ.

Jusqu'à présent ce sont les couches de sélénium qui nous ont données de meilleurs résultats ; ce photoconducteur est sensible dans la moyenne partie du spectre visible.

En observant les courbes de transmission faites sur des échantillons nous remarquons que le sélénium absorbe la lumière jusqu'à une longueur d'onde de 6 200 Å, puis devient au-delà de cette valeur peu à peu transparent. Le sélénium est donc sensible dans la majeure partie du spectre visible, on ne peut donc pratiquement pas l'utiliser dans notre montage actuel. En effet nous ne pouvons pas utiliser la lumière blanche comme lumière de projection du fait que le photoconducteur est sensible à cette lumière. Si nous filtrons la région du spectre sensible qui excite le photoconducteur pour ne laisser passer que la lumière rouge, la projection en rouge n'est pas suffisamment lumineuse. Nous utiliserons donc de préférence ce photoconducteur en réflexion comme il sera décrit plus loin.

La figure 10 donne les courbes de transmission d'une cellule au sélénium, le photoconducteur étant dans l'obscurité puis sous un éclairement de 1 000 lux. La transmission est mesurée à l'aide d'un faisceau d'analyse pour lequel le sélénium est insensible. On constate que dans l'obscurité la transmission de la cellule reste supérieure à 80 % jusqu'à une tension de 300 volts. Par contre en présence de l'éclairement la cellule devient tout à fait opaque à partir de 40 volts.

Sur le même graphique on a porté les courbes de courant d'obscurité et de photocourant du sandwich.

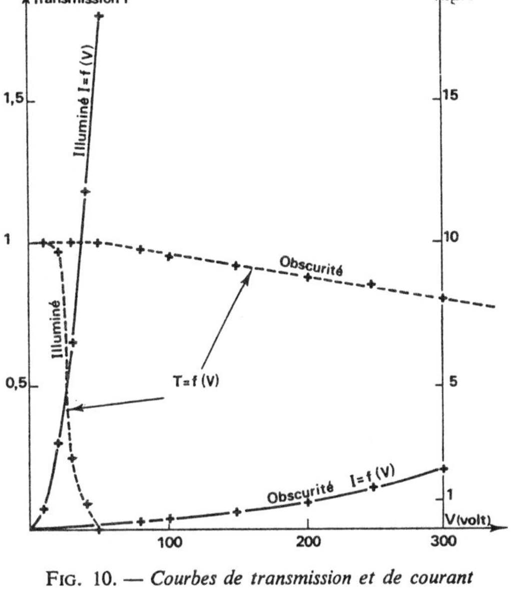

FIG. 10. — *Courbes de transmission et de courant d'une cellule au Sélénium.*

FIG. 10. — *Currvs of transmission and current of a Selenium cell.*

Ce courant dans l'obscurité reste inférieur à 2 µA jusqu'à 300 volts, par contre il croît très rapidement en présence de l'éclairement et atteint 20 µA à 50 volts.

La sensibilité des cellules au sélénium est améliorée en dopant ce matériau de 7 % de tellure.

Les temps de réponse de la cellule ont été faits à l'aide du montage suivant :

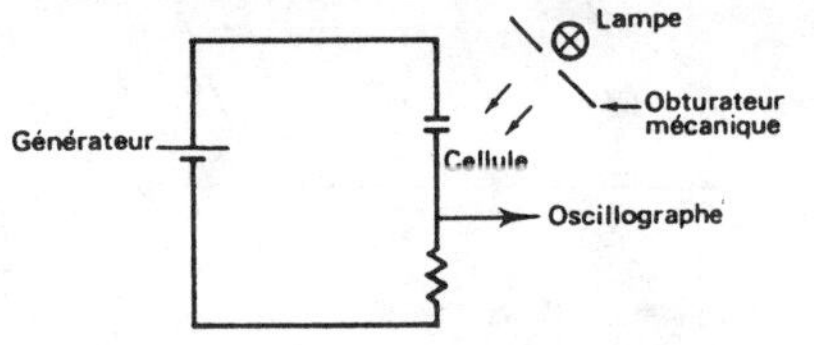

FIG. 11. — *Dispositif de mesure des temps de réponse des cellules.*

FIG. 11. — *Apparatus for the measuring of the response time of the cells.*

Les deux courbes figurant sur la figure 12 représentent les temps de montée et de descente d'une cellule au sélénium : les valeurs sont directement issues des courbes inscrites sur l'oscillographe.

En tenant compte du temps d'ouverture de l'obturateur (sensiblement 1 ms), nous pouvons conclure que le temps de montée est très court, bien inférieur à 1 ms (environ 40 µs) il est de 20 ms pour la descente.

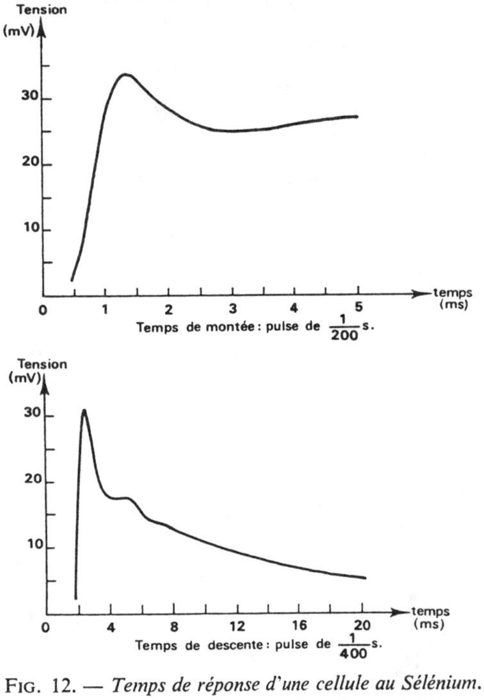

FIG. 12. — *Temps de réponse d'une cellule au Sélénium.*

FIG. 12. — *Response time of a Selenium Cell.*

Cellule au sulfure de cadmium

Les couches obtenues par évaporation sous vide à une température voisine de 100° sont claires et homogènes ; leur épaisseur est de l'ordre de 1 µ.

Le Cds absorbe toute la lumière jusqu'à une longueur d'onde voisine de 4 500 Å, au-delà de cette valeur il devient peu à peu transparent et est ainsi insensible à la lumière rouge.

Il présente donc l'avantage d'être sensible à la lumière blanche tout en laissant une marge suffisante du spectre visible qui nous permet de le projeter sur l'écran en lumière blanche filtrée jusqu'à 5 000 Å. La lumière de projection est alors suffisante pour nous permettre d'observer convenablement les détails de l'image de la cellule.

La résistivité chute sensiblement de 2 ordres de grandeur lorsque l'on passe de l'obscurité à un éclairement de 1 000 lux sous une tension de 30 volts :

$$(R_{\text{nématique}} + R_{pc})_{\text{obscurité}} = \frac{30}{0{,}025} 10^6 = 1\,200 \text{ M}\Omega .$$

$$(R_{\text{nématique}} + R_{pc})_{\text{illuminé}} = \frac{30}{1{,}6} 10^6 = 18{,}7 \text{ M}\Omega .$$

Le temps de montée du photoconducteur est de 10 ms, son temps de descente de 75 ms.

Les caractéristiques de fonctionnement d'une cellule au Cds sont représentées sur la figure 13 : la variation de transmission en fonction de la tension appliquée,

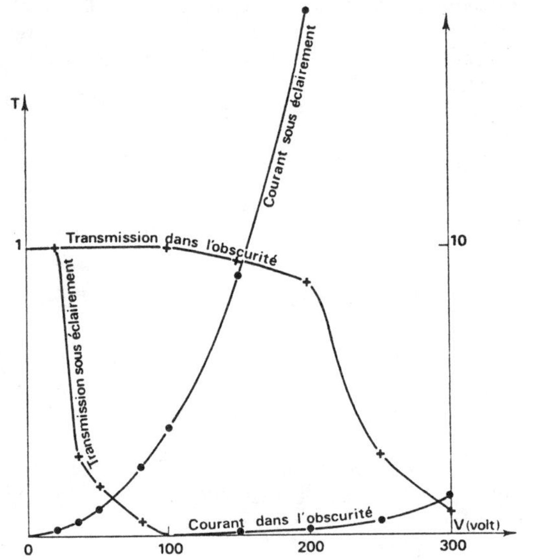

FIG. 13. — *Courbes de transmission et de courant d'une cellule au sulfure de Cadmium.*

FIG. 13. — *Curves of transmission and current of a Cadmium Sulfide Cell.*

la cellule étant dans l'obscurité puis en présence d'un éclairement de 1 000 lux. On constate que si dans l'obscurité la transmission de la cellule reste supérieure à 80 % jusqu'à une tension de 200 volts, elle devient tout à fait opaque à partir de 40 volts en présence de l'éclairement. Sur le même graphique sont portées les courbes de courant d'obscurité et de photocourant dans la cellule. Ce courant reste très faible même pour des tensions élevées ; il est inférieur à 1 µA à 300 volts dans l'obscurité ; par contre, il monte très rapidement à partir de 30 volts en présence de l'éclairement.

Objectifs

Cellules en réflexion. — Notre montage actuel utilise le sandwich photoconducteur nématique en transmission, cela nous satisfait en particulier avec le sulfure de cadmium qui est relativement transparent. Effectivement, la lumière de projection étant filtrée, ce photoconducteur nous permet malgré tout d'obtenir une projection sur l'écran suffisamment lumineuse pour distinguer les détails et avoir un contraste relativement bon.

Le sélénium est un photoconducteur dont la sensibilité est supérieure à celle du sulfure de cadmium. Cependant, ce matériau présente l'inconvénient de filtrer trop la lumière de projection, par suite la projection de la cellule sur l'écran est insuffisamment lumineuse. Notre objectif est d'utiliser ces cellules au

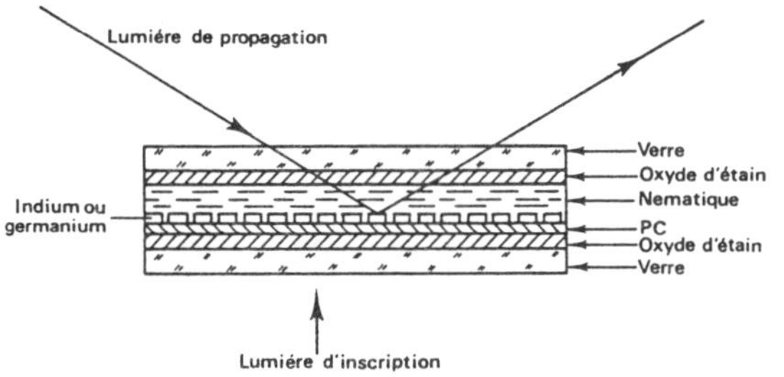

FIG. 14. — *Cellule en réflexion.*

FIG. 14. — *Reflection Cell.*

sélénium en réflexion. Dans ce but, nous étudions une structure semblable à celle de la figure 14.

Entre les deux couches photoconductrices et nématiques on a interposé une couche opaque et réfléchissante. La lumière d'inscription arrive alors sur le photoconducteur directement. La lumière de projection traverse le film de cristal liquide et est réfléchie sur la couche intermédiaire sans agir sur le photoconducteur.

Cellules à l'oxyde de zinc. — L'oxyde de zinc est un photoconducteur qui est insensible à la lumière blanche à partir d'une longueur d'onde de 4 000 Å il devient transparent. Ce photoconducteur qui est sensible dans l'ultra-violet va donc nous permettre d'utiliser comme lumière de projection, la lumière blanche.

Nous obtenons des couches très transparentes en laissant recuire les couches pendant quelques heures.

Les mesures effectuées sur quelques échantillons nous laissent espérer que l'oxyde de zinc serait le photoconducteur qui nous permettrait de réaliser un convertisseur dans l'ultraviolet.

IV. — APPLICATIONS

Description du montage actuel

La figure 3 représente le schéma général de montage d'utilisation des cellules.

L'image donnée par le tube à rayons cathodiques est réformée sur le photoconducteur par un système optique conventionnel. Nous appliquons la tension continue au sandwich ; l'image du tube à rayons cathodiques qui excite le photoconducteur s'inscrit sur le cristal liquide. Pour nous permettre de projeter cette image nous devons filtrer la lumière de projection afin que son rayonnement n'impressionne pas le photoconducteur. Si le nématique est à mémoire il n'est pas nécessaire que le photoconducteur soit insensible au faisceau de projection, on peut séparer la phase enregistrement de la phase projection. L'effacement de la cellule est obtenu par l'application d'une tension alternative pour les sandwichs à mémoire ; pour les cellules à nématique simple la suppression de la tension entraîne la disparition de la diffusion avec une rémanence de l'ordre de 50 ms.

Caractéristiques actuelles du relais de lumière. — La photographie représentée sur la figure 15 a été faite

FIG. 15. — *Mire enregistrée sur une cellule au* Cds.

FIG. 15. — *Test plate taken with one cell* Cds.

à l'aide d'une cellule au sulfure de cadmium et du montage général d'utilisation.

La photographie a été prise directement sur l'écran où est projetée l'image de la cellule enregistrée. Les dimensions de la cellule sont 10 × 10 cm et l'image occupe une surface de 4 cm de diamètre.

La résolution est de 5 paires de lignes au millimètre ; elle est limitée par l'épaisseur totale de la cellule et la diffusion dans le photoconducteur et le cristal liquide.

Nous pensons atteindre une résolution de 10 à 20 paires de lignes au millimètre.

Le seuil de sensibilité de la cellule est de 2.10^{-2} mj/cm^2 de lumière.

Reprographie Hard Copy

La brillance du tube à rayons cathodiques sur lequel est affiché des résultats provenant d'un calculateur par exemple est insuffisante pour réaliser la production directe pour les procédés classiques d'électrophotographie.

En inscrivant l'image sur le relais photoconducteur nématique on obtient ainsi un éclairement suffisant.

Pour la reprographie il est nécessaire que la projection se fasse en bleu ou en violet, les papiers existants étant généralement sensibles dans cette région du spectre. Le relais de lumière doit donc être adapté et par conséquent utiliser un photoconducteur du type oxyde de zinc ou sulfure de zinc.

Convertisseurs d'images infra-rouges et ultra-violettes

Le choix du photoconducteur détermine la région spectrale de fonctionnement de la cellule. On peut ainsi envisager de réaliser un convertisseur dans l'ultraviolet en utilisant l'oxyde de zinc ou le sulfure de zinc, ou un convertisseur dans le proche infra-rouge à l'aide de l'arséniure de gallium et de l'antimoniure d'indium.

Projection de la télévision sur grand écran

Les caractéristiques obtenues actuellement avec nos cellules laissent espérer que nous sommes près d'appliquer le relais photoconducteur-nématique à la projection sur grand écran de la télévision « grand public » ou « semi-professionnelle ».

Nous obtenons effectivement des temps de montée et de descente suffisants pour cette application et nous observons des demi-teintes de bonne qualité.

(Manuscrit reçu le 26 avril 1971).

BIBLIOGRAPHIE

[1] Heilmeier (G. H.) et al. — *P. I. E. E. E.*, juillet 1968.

[2] Assouline (G.) et Leiba (E.). — *J. Physique Supp.*, nov.-déc. 1969, p. 109.

[3] Gray (G. W.). — Molecular Structure and the Properties of Liquid Crystals. *Academic Press.*

[4] Assouline (G.), Hareng (M.), Leiba (E.). — « Electrical measurements in nematic liquid crystals ». A paraître dans *I. E. E. E.*

Electrochiral effects in nematic liquid crystal-chiral salt mixtures

Zhang Fuliang (*) and G. Durand

Laboratoire de Physique des Solides, Bât. 510, Université de Paris-Sud, 91405 Orsay, France

(Reçu le 25 mai 1988, accepté sous forme définitive le 11 janvier 1989)

Résumé. — Une cellule hybride d'un cristal liquide nématique est un instrument commode de mesure de torsion induite, puisque l'ancrage homéotrope peut se tordre sans contrainte. Nous observons la rotation de polarisation de la lumière qui se propage à travers une telle cellule hybride, dopée par des ions chiraux. Un champ électrique continu entraîne les ions dans, ou hors de, la zone où la torsion induite se voit optiquement. Il en résulte un effect électrochiral linéaire. Aux grands champs, l'effet quadratique de déformation de texture et les effets électrochimiques sont clairement séparés. Cette observation permet aussi des mesures de coefficient de diffusion, d'énergie d'ancrage, et d'effet flexoélectrique.

Abstract. — A hybrid nematic liquid crystal cell is a sensitive twisting power measuring device, since the homeotropic anchoring can be twisted without any torque. We observe the rotation of polarization for a light wave which propagates across such a hybrid nematic cell, doped with chiral ions. A DC electric field drags the ions inside or outside the optical twisting region of the hybrid cell. This results in a linear electrochiral effect. At large field, the quadratic effect from texture change and electrochemical effects are clearly separated. The diffusion constant, surface anchoring energy and flexoelectric constant are also measured.

1. Introduction.

Nematic liquid crystals are usually made of rod-like non-chiral molecules [1]. In absence of external field, they align spontaneously in a uniform texture without any twist. It is well known [2] that the dissolution of chiral molecules in a nematic transforms it into a cholesteric material, which aligns with a helical texture indicative of a spontaneous twist. Much work has been done on these mixed systems, to relate for instance the helical twisting power to the concentration of chiral molecules [1, 3, 4]. Up to now, it seems that almost no investigation has been made to study the helicity induced in nematics by chiral ions, instead of neutral chiral molecules. The reason is probably related to the few number of real salts which can be dissolved in usual non-polar nematic materials. In this work, we have studied the twisting power of chiral ions dissolved in a nematic liquid crystal, in presence of DC or AC electric field. We have used partly dissociated salts giving chiral ions in absence of field, and also

(*) On leave from Physics Department, Wuhan University, China.

undissociated chiral molecules, which can be ionized by electrolysis under the action of the DC field. To measure the twisting power, we have observed the twist of light polarization, propagating across an hybrid cell containing the chiralized nematic. Because we have worked on low concentration systems, we have been able to make the now classical measurement [5] of the mass diffraction coefficient of the chiral compounds dissolved in our cells. Finally, working with large AC field, we can estimate the anchoring energy of the nematic on the planar oriented plate of the hybrid cell.

2. A simple model for a twisted hybrid cell.

The expected texture of a hybrid cell containing a weakly twisted cholesteric material is shown in figure 1. The lower plate is treated to give a planar orientation along **x**. The upper plate, at a distance d, is treated to give an homeotropic alignement along **z**. At any point M, the molecular line is twisted by an angle $\varphi(z)$ in the (**x**, **y**) plane, and $\theta(z)$ compared to **z**. The problem is to calculate $\theta(z)$ and $\varphi(z)$ and to further estimate the optical twist of polarized light propagating along **z** through the cell. For simplicity we make the one constant curvature (K) approximation. A two constant model for a uniform pitch cholesteric has already been made [6]. We call **n** the director of the system. The cholesteric curvature free energy density writes as :

$$f = \frac{K}{2}\left[(\text{div}\,\mathbf{n})^2 + (\mathbf{n} \times \mathbf{curl}\,\mathbf{n})^2 + (\mathbf{n} \cdot \mathbf{curl}\,\mathbf{n} + q)^2\right] \quad (1)$$

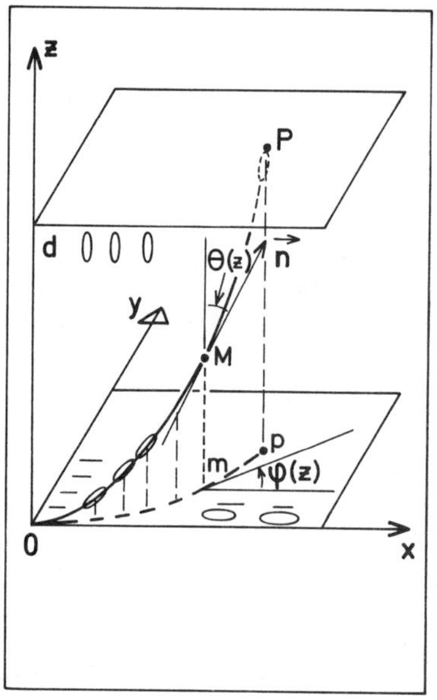

Fig. 1.

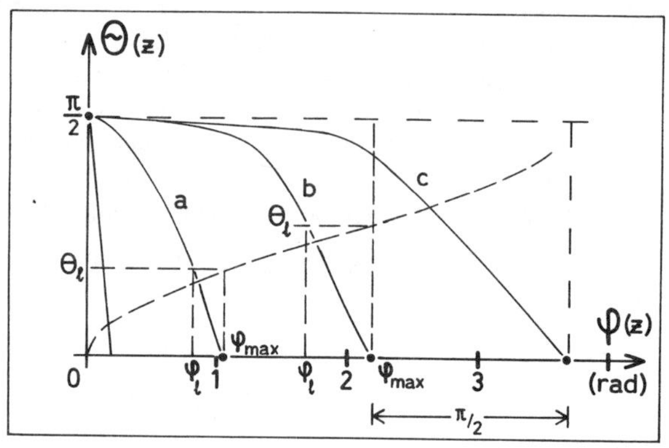

Fig. 2.

Fig. 1. — Geometry of molecular alignment in the hybrid cell. The molecular line starts from 0 at the planar lower plate and stops at P on the homeotropic upper plate. **n** is the director.

Fig. 2. — $\theta(z)$ *versus* $\varphi(z)$ in the cell, for small (a), medium (b) and large (c) total twist $\varphi_{\max}$. The dotted curve determines the limit of the Mauguin regime i.e. the effective optical twisting thickness of the cell between $\theta = \pi/2$ down to θ_ℓ. φ_ℓ is the measured optical twist.

where q is the spontaneous wave vector of the cholesteric helix ($q = 2\pi/P$, P = pitch). With our notations,

$$f = \frac{K}{2}\left[\left(\frac{d\theta}{dz}\right)^2 + \sin^2\theta\left(\frac{d\varphi}{dz}\right)^2 - 2q\sin^2\theta\frac{d\varphi}{dz}\right]. \tag{2}$$

Minimizing (2) compared to θ and φ gives :

$$\sin 2\theta\left[\left(\frac{d\varphi}{dz}\right)^2 - 2q\frac{d\varphi}{dz}\right] = 2\frac{d^2\theta}{dz^2} \tag{3}$$

$$\frac{d}{dz}\left[\sin^2\theta\left(\frac{d\varphi}{dz} - q\right)\right] = 0 \tag{4}$$

where q is now a function of z. Equation 4 means that the hybrid cholesteric texture transmits everywhere the same twist torque around z, from the upper to the lower plate. With our hypothesis of homeotropic anchoring, this torque must be zero, since molecules cannot transmit torque along **n** on the homeotropic plate (the nematic is assumed to be uniaxial). Equation (4) is then integrated as :

$$\varphi = \int_0^z q(z')\,dz'. \tag{5}$$

The total twist $\varphi_{max} = \varphi(d)$ of the texture from the lower planar plate ($z = 0$) to the upper homeotropic plate ($z = d$) is :

$$\varphi_{max} = \int_0^d q(z')\,dz' = \bar{q}d \tag{6}$$

φ_{max} does not depend on the $\theta(z)$ profile. It just measures the thickness averaged twisting power $\bar{q}$.

The integration of the θ equation (3) is straigthforward when q is uniform, i.e. $q = \bar{q} = q_0 > 0$ independent of z. Integrating (3), we can write :

$$\frac{d\theta}{dz} = -q_0(C - \sin^2\theta)^{1/2} \tag{7}$$

since $\theta(z)$ decreases from the planar to the homeotropic plate. Finally, θ is given by :

$$q_0 z = +\int_\theta^{\pi/2} \frac{du}{(C - \sin^2 u)^{1/2}}. \tag{8}$$

The integration constant $C > 1$ is fixed by the boundary condition :

$$q_0 d = \int_0^{\pi/2} \frac{du}{(C - \sin^2 u)^{1/2}} \tag{9}$$

$q_0 d$ is the total twist of the cell across d. We have plotted in figure 2 the behavior of $\theta(z)$ *versus* $\varphi(z)$ in the cell. For thick enough cells or small pitch systems ($q_0 d > 1$), $\theta(z)$ remains close to $\pi/2$, and drops only close to the homeotropic plate on a thickness $\sim P/4$ over which φ rotates of $\pi/2$. This corresponds to a situation where cholesteric layers pile up parallel to the planar plate ($\theta = \pi/2$). Only the last 1/4 layer is destroyed by the imposed θ bend.

For thin cells or large pitch systems ($q_0 d < 1$), θ decreases linearly to zero as : $\theta = \frac{\pi}{2}\left(\frac{d-z}{d}\right)$. Figure 2 is useful to understand the optical properties of the cell. Normally, for low concentration of chiral molecules, the pitch P is much larger than the wavelength λ of the light used to measure the twist. One is then in the Mauguin limit [2], where the light polarization follows adiabatically the twist, provided that :

$$p\,\Delta n(\theta) > \lambda \tag{10}$$

$\Delta n(\theta)$ is the molecular birefringence. The point is that, close to the homeotropic region, θ tends to zero and $\Delta n(\theta)$ accordingly. In this region, the bended nematic remains mechanically twisted, but is no more capable to twist the light polarization. To estimate the effective thickness of the hybrid cell which can induce an optical twist, we can use the approximate relationship,

$$\Delta n(\theta) \sim (n_e - n_o)\sin^2\theta \tag{11}$$

where n_e and n_o are the extraordinary and ordinary indices of the nematic material. Condition (10) gives then :

$$\sin^2\theta > \lambda/(P\,.\,(n_e - n_o))\,. \tag{12}$$

Calling θ_ℓ the limit value for θ_ℓ which can twist the light polarization, we can write :

$$\sin\theta_\ell \sim \left(\frac{\lambda\,\varphi_{\max}}{d(n_e - n_o)}\right)^{1/2} \tag{13}$$

where $\varphi_{\max} = q_0 d$. We have plotted $\theta_\ell(\varphi_{\max})$ in figure 2, for typical values : $d = 10\ \mu\text{m}$, $n_e - n_o = 0.2$, $\lambda = 0.546\ \mu\text{m}$ corresponding to our experiment. Only the *planar* region of the cell $\frac{\pi}{2} > \theta > \theta_\ell$ is in the Maugin limit, i.e. is capable of inducing an optical twist. The *homeotropic* region $\theta_\ell > \theta > 0$ will behave as a birefringent plate, the eigen axis of which are oriented along $\phi_{\max}$ and $\phi_{\max} + \pi/2$. In general, a linearly polarized light wave propagating along **z** and entering the cell at the planar plate with its electric field vibrating along the director **n**, should be elliptically polarized at the homeotropic output plate. The axis of the ellips should rotate by the angle $\varphi_\ell = \varphi(\theta_\ell)$ shown in figure 2. In this first order analysis, we are not interested in the polarization ellipticity, but simply in the polarization twist. As $\varphi_{\max}$ is linear in d, one can directly visualize the effective fraction d_ℓ/d of the cell which controls the light polarization twist. The observed twist is not $\varphi_{\max}$, but $\varphi_\ell = \varphi(d_\ell)$ shown in figure 2. For low $\varphi_{\max}$, i.e. for small q_0, φ_ℓ is close to $\varphi_{\max}$. The non-twisting part of the cell has then a relative thickness :

$$\frac{d - d_\ell}{d} = \frac{\varphi_{\max} - \varphi_\ell}{\varphi_{\max}} = \frac{2}{\pi}\left(\frac{\lambda\,\varphi_{\max}}{d(n_e - n_o)}\right)^{1/2} \tag{14}$$

which goes to zero as $\varphi_{\max}$. For large $\varphi_{\max} > \pi$, the twist induced by the cell is the one of the cholesteric layers close to the planar plate, and φ_ℓ tends toward $\varphi_{\max} - \pi/2$. We shall not discuss anymore this domain, since we shall restrict, in what follows, to the low $\varphi_{\max}$ regime.

We discuss now a situation where the total number of chiral molecules is fixed in the cell, with an inhomogeneous distribution along z, so that q depends now on z, with a fixed average $\bar{q}$. As previously discussed, $\varphi_{\max} = \bar{q}d$ will not depend on the $q(z)$ distribution. A measurement of φ_{max} would not give any information on $q(z)$. On the other hand, φ_ℓ depends on the $q(z)$ distribution. The Mauguin condition for the optical wave guide regime (Eq. (13)) remains valid locally, as :

$$\sin \theta_\ell \sim \left(\frac{\lambda q(z)}{n_e - n_o} \right)^{1/2} \tag{15}$$

provided we use the right tilt $\theta(q(z))$. In fact, in the situation we consider, the relative variation in space of the twist is small $\left(\frac{dq}{dz} < \frac{q}{d} \right)$, so that we can estimate $\theta_\ell(z)$ by using its calculated value for $q_0 = \bar{q}$, and also replace q by $\bar{q}$ in equation (15). The hybrid cell can then be visualized as constituted of two regions : the optical twisting quasi planar region $(\theta > \theta_\ell)$ close to the planar plate $(z < d_\ell)$ and the optical non twisting quasi homeotropic region $(\theta < \theta_\ell)$ close to the homeotropic plate $(d_\ell < z < d)$. An optical measurement of φ_ℓ gives an information on the total number of chiral molecules in the optical twisting region.

One now understands the principle of the linear electrochiral effect described in this experiment : assume that part of the twisting power of chiral molecules comes from chiral ions. Assume that on the upper and lower plates of the hybrid cell are placed transparent electrodes, across which one can apply an electric field **E** (Fig. 3) $[|\mathbf{E}| = V/d]$. We illuminate the sample by a light wave linearly polarized along **x**, propagating along **z**. At the output of the cell, the light polarization has rotated over an angle φ_ℓ defined by the available twisting power of the lower part of the cell $(z < d_\ell)$ in the wave guide regime. In absence of a field, the ion concentration is uniform along z. One can measure a mean twist φ_ℓ. In presence of field, one will drag the chiral ions, according to their sign, inside or outside of the twisting region. φ_ℓ should then depend linearly on E.

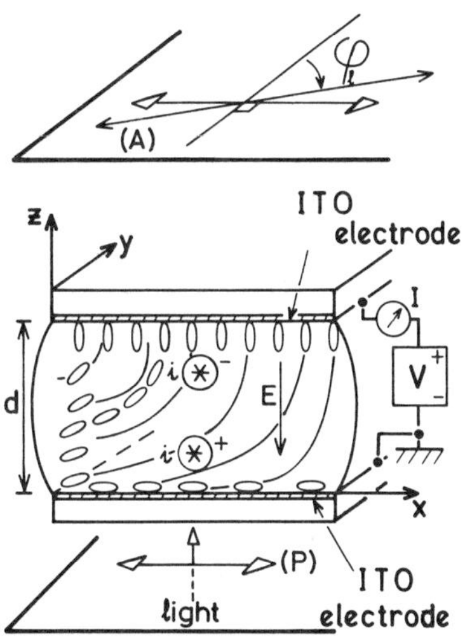

Fig. 3. — Experimental set up. P : polarizer, A : analyzer, V : electric generator. I ammeter. The field **E** drags the ions i in the twisting (lower) region $(i > 0)$ or in the non twisting (upper) region $(i < 0)$, resulting in a linear change in the optical rotation φ_ℓ across the cell.

In practice, most liquid crystals possess a dielectric anisotropy ε_a. In presence of a field, this causes a texture change quadratic in E. This quadratic effect will be superimposed on the previously predicted linear electrochiral effect. We can simply describe this quadratic dielectric effect. Assume $\varepsilon_a > 0$, i.e. the nematic director **n** will align along **E**, inducing an homeotropic texture for large E. The twisting power of the cell must decrease, because of the general decrease of θ. The new texture $\theta(z)$ in presence of E is obtained by adding in f the dielectric term $\frac{1}{2}\frac{\varepsilon_a}{4\pi}E^2 \sin^2\theta$. The φ equation (4) does not change since, because of its cylindrical symmetry, the coupling of E with the dielectric anisotropy does not depend on φ. Equation (3) becomes now :

$$\frac{2\,\mathrm{d}^2\theta}{\mathrm{d}z^2} = \sin 2\theta\left[\left(\frac{\mathrm{d}\varphi}{\mathrm{d}z}\right)^2 + \frac{1}{\xi^2} - 2\,q\,\frac{\mathrm{d}\varphi}{\mathrm{d}z}\right] \tag{16}$$

where the usual [1] coherence length ξ is given by :

$$\frac{K}{\xi^2} = \frac{\varepsilon_a E^2}{4\pi} \tag{17}$$

θ remains given by (8), where $q_0 = \bar{q}$ must be replaced by $(\xi^{-2} - q_0^2)^{1/2}$. The field reduces the φ, θ coupling, i.e. it reduces the tendency of θ to remain around $\pi/2$ close to the planar plate, since E unwinds the cholesteric layers. For $q_0\,\xi = 1$, $\theta = \frac{\pi}{2}\frac{d-z}{d}$ is the same as the one of a non-chiral cell. For larger field ($q_0\,\xi < 1$), the curvature of the $\theta(z)$ dependence changes its sign and θ goes more rapidly to zero (Fig. 4). The twisting region of the cell becomes thinner when E increases, and accordingly $\varphi_\ell(E) \sim \frac{\xi}{d}\,\varphi_\ell(0)$ goes also to zero. A more detailed description would necessitate a numerical computation, non essential for our purpose.

Another difficulty, in large electric fields, may be related to the finite anchoring strength of the plates. We have not to consider the azimuthal anchoring strength of the planar plate, since

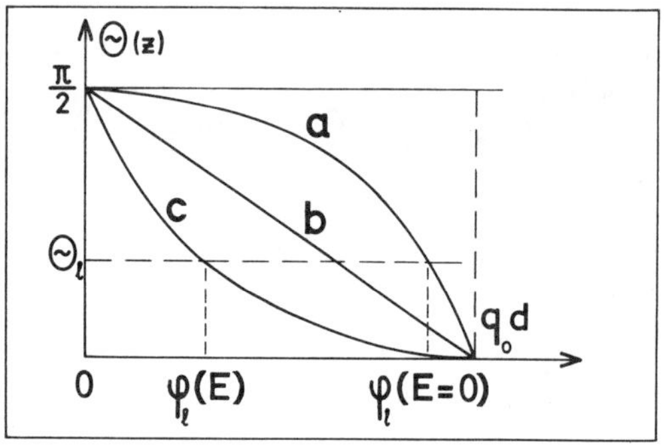

Fig. 4.

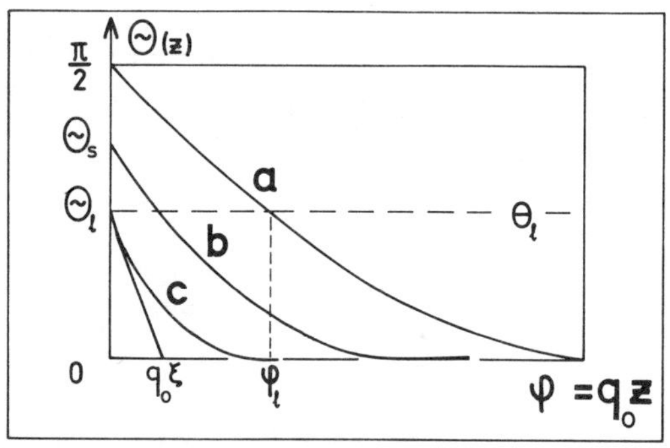

Fig. 5.

Fig. 4. — Variation of $\theta(z)$, with E^2, for a positive ($\varepsilon_a > 0$) dielectric anisotropy material. (a) $E = 0$. (b) medium field ($q_0\,\xi = 1$). (c) large field ($q_0\,\xi = < 1$). θ_ℓ, independent of E, is obtained from φ_{max} (see Fig. 2). The measured optical twist $\varphi_\ell(E)$ decreases continuously down to zero, for $E \to \infty$.

Fig. 5. — Variation of $\theta(\varphi = q_0\,z, E^2)$. The surface value of $\theta(0, E^2) = \theta_S$ is now allowed to be smaller than $\pi/2$, because of weak anchoring. The z dependence is restricted to a thickness ξ: (a) $\theta(E = 0)$. (b) intermediate situation $\theta(E^2)$. (c) zero optical twist situation $\theta_S = \theta_\ell$.

even in presence of field, there is no azimuthal torque around **z** in the free twisted texture. We have just to consider the zenithal anchoring strength of the planar plate. In large field, the $\pi/2$ variation of $\theta(z)$ is confined to a distance ξ (Fig. 5) resulting in a large surface torque $K\frac{\pi}{2\xi}$. This torque can change the $\theta = \theta_S$ orientation on the planar plate. Let us characterize as usual [1] the planar plate anchoring strength with a surface energy W_S:

$$W_S = \frac{1}{2}\frac{K}{L}\cos^2\theta_S \tag{18}$$

where L is the usual extrapolation length [1] ($L \sim 0.1$-$1\ \mu$m). We must write the balance of curvature and surface torques at the planar plate. From equation (16), we obtain a curvature surface torque $K\sin\theta_S/\xi$, neglecting the effect of twist, since $q_0\,\xi < 1$. This results in the torque equation:

$$\frac{K}{L}\sin\theta_S\cos\theta_S = \frac{K}{\xi}\sin\theta_S\,. \tag{19}$$

If $\xi < L$, the solution is $\theta_S = 0$, all of the cell is homeotropically aligned. If $\xi > L$, the solution of equation (19) is:

$$\cos\theta_S = L/\xi \tag{20}$$

or

$$\frac{\theta_S^2}{2} \sim \frac{\xi - L}{\xi} \sim L\left(\frac{\varepsilon_a}{4\pi K}\right)^{1/2}(E_S - E) \tag{21}$$

where E_S is the field for which $\xi = L$. We are not interested in the $\theta_S \to 0$ limit, but in the zero optical twist regime. Using equations (12) and (10), we obtain:

$$\sin^2\theta_S = 1 - L^2/\xi^2 = \frac{\lambda q_0}{n_e - n_o} \tag{22}$$

i.e. a field E_0 for zero twist, given by:

$$E_0 = \frac{1}{L}\left(\frac{4\pi K}{\varepsilon_a}\right)^{1/2}\left(1 - \frac{\lambda q_0}{n_e - n_o}\right)^{1/2}. \tag{23}$$

For typical values ($\lambda = 0.5\ \mu$m, $q_0 = 2\pi/60\ \mu\text{m}^{-1}$, $n_e - n_o = 0.2$) the last term is of order 1. This remains valid for small λq_0. Using $\varepsilon_a \sim 4\pi$ and $K \sim 10^{-6}$ cgs, one expects E_0 values in the range of $E_0 \sim V_0/d \sim 0.3\,L$ V/cm independent of q_0. In principle, the observation of E_0 allows for calculation of L. The $\theta(z)$ variation for $\theta_S \neq \pi/2$ is shown in figure 5. Apart for a thickness ξ close to the planar plate, the sample is expected to be homeotropic. Above E_0 the twisting power is expected to increase linearly in field.

Finally, we could also in principle suspect another linear electric field effect to occur in our system. The hybrid cell presents a flexoelectric polarization $\mathbf{P}_f$ [1], associated with the splay-bend texture. A DC electric field will couple linearly with $\mathbf{P}_f$, through the coupling energy density $-\mathbf{E}\cdot\mathbf{P}_f$. This situation has already been studied [7]. Following the work of the Bordeaux group [8], we can assume that $\mathbf{P}_f$ has mostly a quadrupolar origin, i.e. $\mathbf{P}_f = \frac{e}{2}\nabla\cdot(\bar{n}\bar{n} - \bar{\bar{I}}/3)$ where e is the sum $e_1 + e_3$ of the usual [1] flexoelectric coefficients. Assume now **E** uniform. The volume energy density integrates out as a surface term on the planar plate $+\frac{e}{2}E\cos^2\theta_S$. Because there is no flexoelectric polarization associated with a twist distorsion, this surface term, independent of φ, changes only the zenithal anchoring energy. We expect now a more complicated behavior. Equation (19) becomes :

$$\left(\frac{K}{L}+eE\right)\sin\theta_S\cos\theta_S=\frac{K}{\xi}\sin\theta_S=\left(\frac{K\varepsilon_a}{4\pi}\right)^{1/2}|E|\,. \tag{19'}$$

For small flexoelectric coefficient $e<\left(\frac{K\xi_a}{4\pi}\right)^{1/2}$, one expects now two differents critical fields to produce the complete homeotropic alignment ($\theta_S=0$). These fields are solution of :

$$\frac{\xi(E)}{L}=1-\text{sign}\,(E)\,.\,e\left(\frac{4\pi}{\varepsilon_a K}\right)^{1/2}.$$

For most liquid crystals, e is positive. A positive **E** (aligned along **z** from the planar plate to the homeotropic one) will then decrease ξ, i.e. increase the critical field. A negative E should decrease the critical field.

For large flexoelectric coupling $\left(e>\left(\frac{K\varepsilon_a}{4\pi}\right)^{1/2}\right)$, there is just one critical field, of same polarity as e which aligns homeotropically the cell. Of course, the flexoelectric coupling should become zero when using AC fields, of frequency larger than the relaxation frequency of the splay bend texture.

3. The experimental set up.

We have prepared a hybrid cell filled with room temperature 5CB (pentyl cyanobiphenyl) which undergoes a nematic → isotropic phase transition at $T_c=34.5$ °C. Our experiments are made at $T=19.5\pm1$ °C, i.e. 15° below T_c. For this material at $T=19.5$ °C, $n_e=1.739$ and $n_o=1.536$ for $\lambda=5\,461$ Å, the green mercury wavelength [9]. The elastic curvature constants are [10] K_1 (splay) $=6.2\times10^{-7}$ cgs, K_2 (twist) $=3.1\times10^{-7}$ cgs and K_3 (bend) $=8.8\times10^{-7}$ cgs. The dielectric anisotropy is [11] $\varepsilon_a=12$. The plates are ITO (indium tin oxyde) transparent electrodes. The upper one is coated with DMOAP silane [12], to induce homeotropic orientation. The lower one is obliquely evaporated with SiO, at an angle of 66° compared to the normal, which results in a planar orientation [13]. A linearly polarized light beam (Fig. 3) is send across the cell, with the optical **E** vector along **P**//**x** along the planar plate. The sample is observed through a polarizing microscope. An analyzer inside the microscope measures the rotation φ_ℓ of the output polarization, from the rotation of the light extinction angle. Accuracy and reproductibility are in the one degree range. Note that a positive algebraic φ_ℓ in the (**x**, **y**, **z**) frame corresponds to a negative φ_ℓ for the opticians. In what follows, we have kept the optical convention for the sign of φ_ℓ. For comparison, we give in table I the sign of the rotatory power in solution of the chiral compounds used to dope the hybrid cell.

To observe a possible electrochiral effect, we can apply an electric field $|\mathbf{E}|=V/d$ between the two electrodes. The electric generator can be an AC or DC source with adjustable polarity, in the range ± 30 V. The sample thickness d is 10 μm, obtained using silica sphere spacers. We can also measure the DC current I through the cell using a sensitive ammeter. The measured voltages are always referenced compared to the grounded lower planar plate. This means that in the **x**, **y**, **z** frame a positive V induces a negative **E**. For convenience, when reporting our observations, we have chosen for positive the sense of **E** along – **z**, so that **E** and V have the same sign.

To calibrate our set-up, we have made various samples with different concentrations of a chiral dopant, the cholesteryl oleyl carbonate (COC). We have first measured the pitch of a

Table I. — *Sign of optical rotatory power, in solution, of the various chiral compounds used in this work. (Optical convention).*

Compound	Solvent	Sign
Phenyllactic acid	water	−
Alanine ethyl ester chlorhydrate	water	+
Phenylalanine ethyl ester chlorhydrate	ethanol	+
Alanine chlorhydrate	water	+
N-benzyl quininium chloride and bromide	water	−
K alaninate	water	+
Krypted K phenyllactate	water	−
Cholesteryl octanoate	chloroform	−
2-Octanol	pure	−
Cholesteryl oleyl carbonate	chloroform	−

cholesteric mixture 5CB + COC for various mole concentration c of COC, using the Cano wedge [14] method. The two plates forming the wedge are SiO coated to give parallel planar orientation. We observe a periodic set of disclination lines parallel to the wedge corner. The

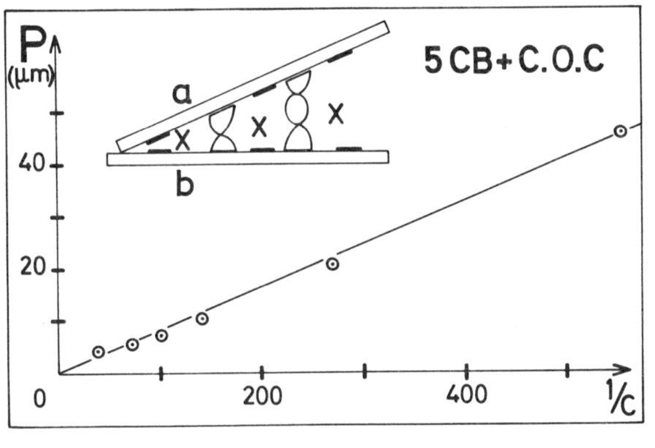

Fig. 6.

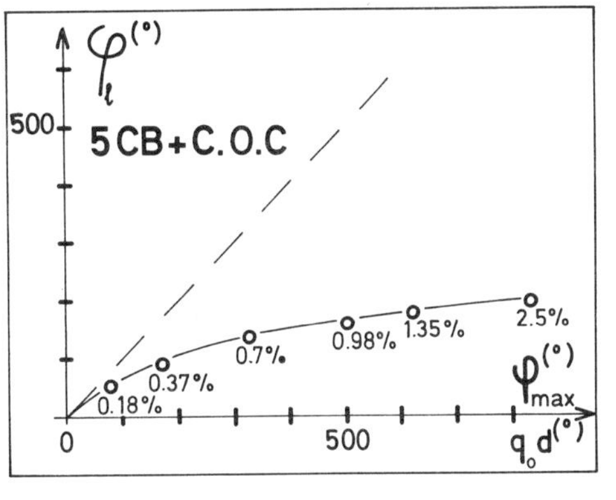

Fig. 7.

Fig. 6. — Pitch measurement with a Cano wedge. c mole concentration of the solution, P pitch (μm). a and b planar aligned plates. X disclination lines.

Fig. 7. — Optical twist φ_ℓ calibration *versus* the total twist $\varphi_{\max}$. The figures are the mole concentration of COC in 5CB.

change of wedge thickness between two adjacent lines is taken as $\frac{P}{2}$ (half the pitch). The result is shown in figure 6, which shows the classical c^{-1} dependence of P [1]. We have now measured the rotation φ_ℓ *versus* c, in our hybrid cell, for the 5CB-COC mixtures. We have plotted φ_ℓ *versus* $\varphi_{max} = q_0 d$, calculated from the measured pitch P, in figure 7. The data show clearly that φ_ℓ tends toward φ_{max} only for small concentrations typically in the range $\varphi_{max} < 90°$, as expected. In the following experiments we have always worked in this small rotation regime where the Mauguin criterium works and for which there is no 180° error possible for φ_ℓ.

4. The electrochiral effects.

4.1 POSITIVE CHIRAL IONS. — We have first tried to observe the electrochiral effect on mixtures of 5CB doped with a chiral salt. Obviously, the dissociation of a « salt » is not expected to be very large in a weakly polar material like 5CB. We have first used benzyl quininium bromide (BQB) at $C = 0.5\ \%$, which is expected to dissociate to give a *positive* chiral ion BQ^+ and a non-chiral negative bromide ion. We have measured the optical rotation φ_ℓ *versus* E, in the range $\pm$ 6 V/10 μm. We do observe (fig. 8) around $E = 0$ a linear variation of φ_ℓ, followed by sharp variations at $\pm V_1/d$ and finally a decrease at large E. The large $|E|$ decrease is easy to understand. On the same figure 8, we have shown the observed φ_ℓ variation *versus* the rms value of a high frequency ($f = 3$ kHz) E field, which results only into a dielectric quadratic effect. This quadratic effect is the expected E^2 decrease of the cell

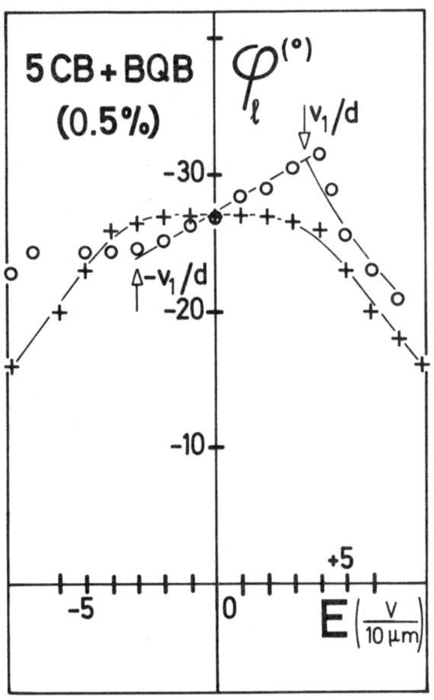

Fig. 8.

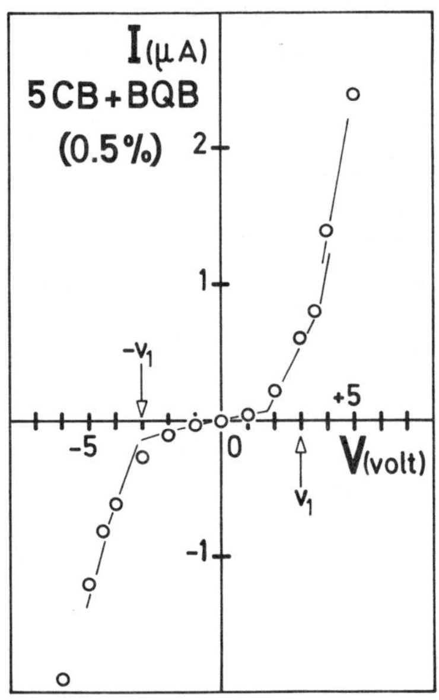

Fig. 9.

Fig. 8. — Electrochiral effect for a 5CB + BQB mixture. (O) : $\varphi_\ell(E)$ (DC field) ; (+) : $\varphi_\ell(E)$ (AC field, $f = 3$ kHz). $V_1/d \sim 3$ V/10 μm corresponds to V_1 in figure 9, and is the threshold for electrochemical reaction.

Fig. 9. — Current I *versus* DC voltage V across the 5CB + BQB cell of figure 8. V_1, corresponding to a large increase of current, is an electrochemical threshold.

twisting power, due to the induced molecular tilt. For large values of $|E|$, the DC field results also in the same quadratic decrease of φ_ℓ, when molecules align along **E**. The low field observed variation of $|\varphi_\ell|$ corresponds to a larger (smaller) twist for positive (negative) voltage on the upper plate. If positive chiral ions are present in the cell, they must concentrate close to the negative planar electrode, in the active region for twist of the hybrid cell, increasing the rotation $|\varphi_\ell|$ as observed. We then interpretate this linear variation as the expected electrochiral effect from positive BQ^+ ions. Note that φ_ℓ is negative for this mixture. The singularities observed at $\pm V_1/d$ with $V_1 \sim 3$ V can be understood if one looks at the measured electric current I through the cell (Fig. 9). I is linear (ohmic like) for low voltage, due to the presence of ions. Above $\pm V_1$, we observe a sharp increase of the current. V_1 corresponds probably to the onset of an electrochemical reaction which results in a large increase of the total ions density. The $\pm$ signs indicate that this reaction can happen on both electrodes. In figure 8, we do see that, below $-V_1/d$, the rotation increases and that above V_1/d, the rotation decreases sharply. This means probably that the electrochemical reaction decreases the density of chiral ions BQ^+ or create negative chiral ions. We cannot be more specific at this stage.

We have reproduced the experiment with a mixture of 5CB plus benzyl quininium chloride (BQC), which is also expected to give positive BQ^+ ions. We do observe the same linear effect for the low field variation of $\varphi_\ell(E)$ (Fig. 10), with the same sign for the slope, as expected. The electric current in the cell shows now sharp variations for $V = \pm V_1 \sim 1$ V (Fig. 11), for which the linear effect on φ_ℓ seems to increase slightly. It may be that at this voltage the density of BQ^+ chiral ions increases.

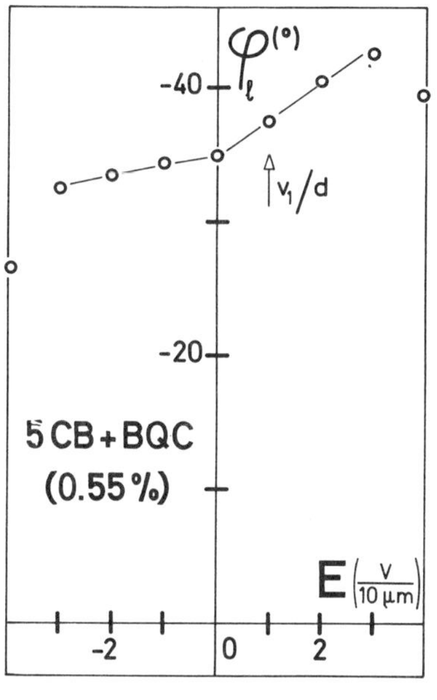

Fig. 10.

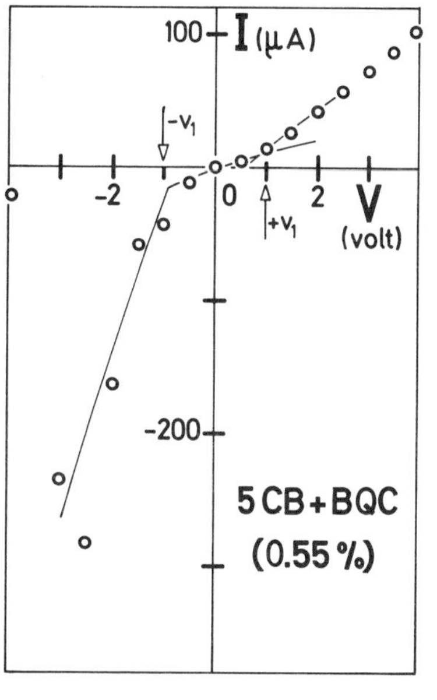

Fig. 11.

Fig. 10. — Linear electrochiral $\varphi_\ell(E)$ (DC) for a 5CB + BQC mixture. V_1/d corresponds to $V_1 \sim 1$ V/10 μm in figure 11.

Fig. 11. — $I(V)$ relationship for the 5CB + BQC, mixture of figure 10. $V_1 \sim 1$ V is an electrochemical threshold.

Many other compunds which could give positive chiral ions have been found poorly soluble in 5CB and give a weak effect. This is the case for instance of Alanine ethyl ester hydrochloride (AEEH). A saturated (unknown) concentration of this compound gives a zero field rotation $\varphi_e \sim -1°$. A positive E increases $|\varphi_e|$ as expected, but the effect is too weak to be quantitatively analyzed.

4.2 NEGATIVE CHIRAL IONS. — We have now made the same experiments on salts which can dissociate to give *negative* chiral ions. We have prepared a $c = 4.95\ \%$ solution of phenyllactic acid (PhLA) in 5CB. Our results are plotted in figure 12. For this compound, the twist is now optically positive. For low field, we do observe a linear variation of φ_ℓ, with a negative slope, as expected from negative chiral ions $(PhLA)^-$ which for positive voltage on the homeotropic upper plate, should get out from the efficient twisting region of the hybrid cell. We also find at $E = \frac{V_1}{d} = +2$ V/10 μm a sharp decrease of φ_ℓ. This is correlated with the current voltage curve of the cell (Fig. 13). $I(V)$ shows an ohmic region between -2 V and $+2$ V, and a large increase beyond these values. As φ_ℓ decreases more sharply above V_1/d, we can infer that electrolysis creates probably above V_1 a larger concentration of the negative chiral ions responsible for the twist.

We have reproduced the experiment with another related compound which is also expected to give negative chiral ions, the potassium phenyl lactate (KPhL). We have prepared a $c = 0.43\ \%$ solution of KPhL in 5CB. In absence of field, we measure $\varphi_\ell = +57°$. A positive field of 2 V/10 μ decrease φ_ℓ down to 55°, as expected for chiral PhL^- ions.

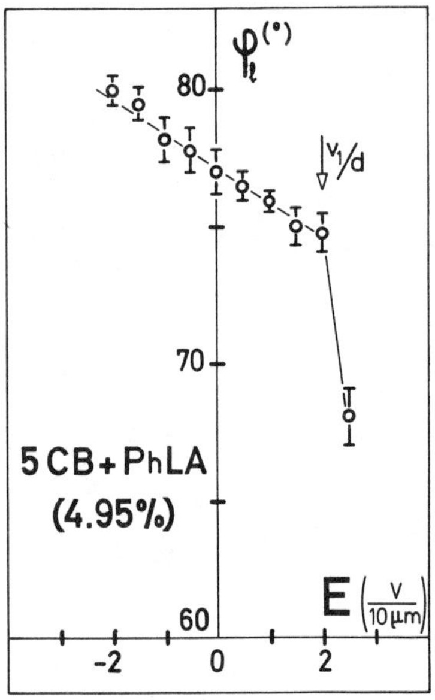

Fig. 12.

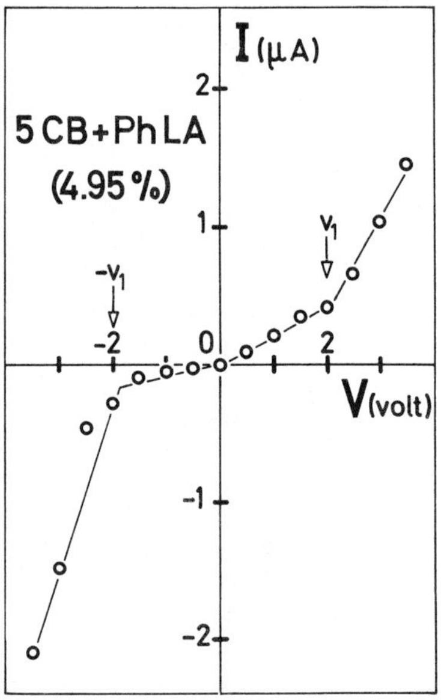

Fig. 13.

Fig. 12. — Linear electrochiral effect for a 5CB + PhLA mixture. $V_1/d \sim 2$ V/10 μm corresponds to V_1 in figure 14.

Fig. 13. — $I(V)$ curve for the 5CB + PhLA, mixture of figure 1. $V_1 = 2$ V is an electrochemical threshold.

4.3 UNDISSOCIATED CHIRAL DOPANTS. — For comparison, we have studied chiral compounds which are not expected to spontaneously dissociate, like chiral alcohols and esters. We have checked octanol-2 (OC-2), cholesterol (CH) and its esters. The behavior of a $c = 0.64\,\%$ mixture of CH in 5CB is shown in figure 14. For low field ($|E| < 2$ V/10 μm) φ_ℓ does not depend on E. Increasing the voltage drop across the cell above 2 V, one observes a sharp decrease of φ_ℓ for $E > 0$ and an increase for $E < 0$, which indicates the appearance of negative chiral ions. These variations are deduced in comparison with the purely dielectric quadratic effect demonstrated on the same figure, measured with a 3 kHz AC field. The observed φ_ℓ variation must be correlated with the $I(V)$ curve of this cell (Fig. 15). We do find an ohmic region between $\pm V_1 = \pm 2$ V. As there is no change in φ_ℓ in this region, we must accept the idea that the ions responsible for this conduction are non-chirals. Beyond ± 2 V, we observe a large increase of I, and a further one above $V_2 = 6$ V. We can say that V_1 and V_2 are the thresholds for electrolysis to produce chiral negative ions (CH^- ?).

Cholesteryl octanoate (CO) ($c = 0.4\,\%$) shows an equivalent behaviour (Fig. 16) : no variation of φ_ℓ for low field, an abrupt decrease of φ_ℓ above $E = 1$ V/10 μm and an increase of φ_ℓ (compared to the quadratic AC curve) for $E < -1$ V/10 μm. This also can be correlated with the non-linear behavior of the $I(V)$ curve (Fig. 17). The same behavior has been found for cholesteryl nonanoate (CN), cholesteryl decanoate (CD), cholesteryl propronate CP, cholesteryl oleylcarbonate (COC) and cholesteryl chloride (CC). Note that all these cholesterol compounds result in negative φ_ℓ for low concentration, as already reported [4].

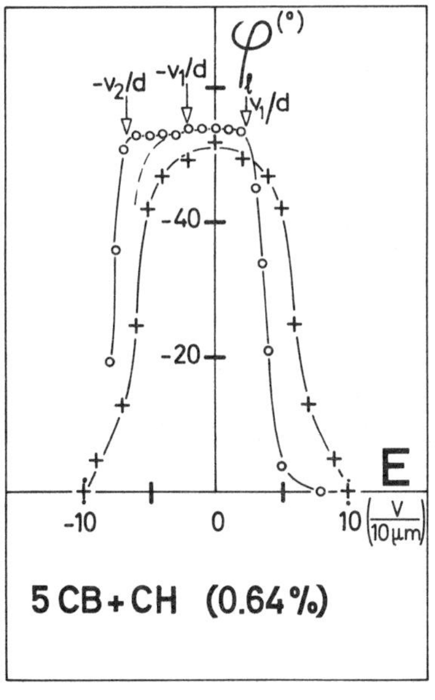

Fig. 14.

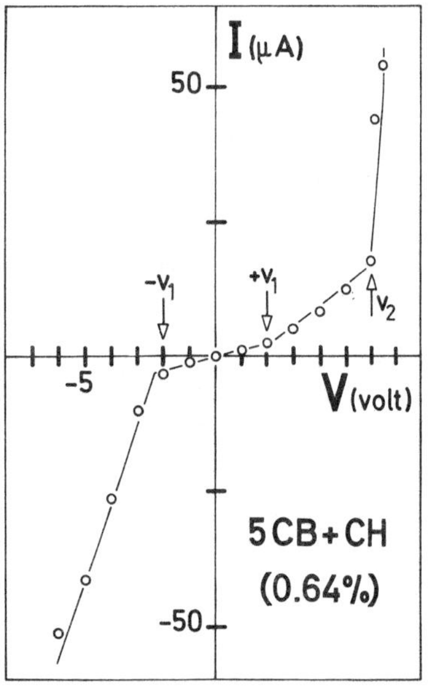

Fig. 15.

Fig. 14. — Electrochiral effect, for a 5CB + CH mixture. (O) : DC field ; (+) : AC field (3 kHz). No linear effect is observed at low field. $V_1/d \sim 2$ V/10 μm corresponds to V_1 in figure 15. $V_2 \sim$ (6-8) V/10 μm corresponds to V_2 in figure 15.

Fig. 15. — $I(V)$ curve for the 5CB + CH mixture of figure 14. $V_1 \sim 2$ V and $V_2 \sim 6$ V are electrochemical thresholds.

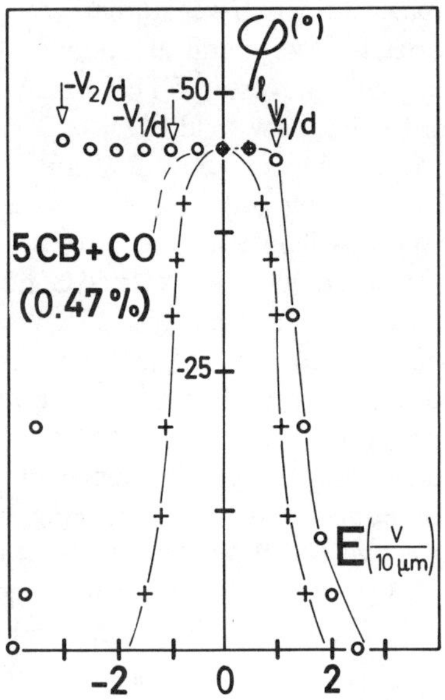

Fig. 16.

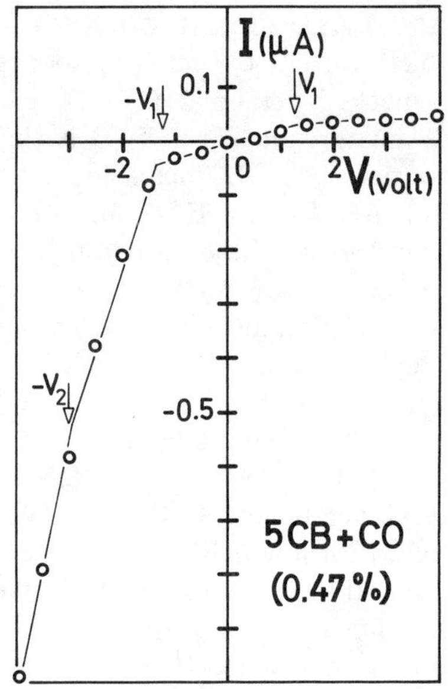

Fig. 17.

Fig. 16. — Electrochiral effect for a 5CB + CO, mixture. (○) : DC measurement ; (+) : AC measurement ($f = 3$ kHz). No linear electrochiral effect is observed. $V_1/d \sim 1$ V/10 μm and $V_2/d \sim 3$ V/10 μm correspond to V_1 and V_2 in figure 17.

Fig. 17. — $I(V)$ curve for the 5CB + CO mixture of figure 16. V_1, V_2 electrochemical thresholds.

All these compounds show the appearance of chiral negative ions from electrolysis, above a voltage threshold $V_1 \sim$ 1-2 V. Finally, the chiral octanol-2 (OC-2) at $c = 4.3$ % shows also a similar behavior but the weak initial rotation ($\varphi_\ell = -3°$) does not allow for a good enough accuracy.

4.4 Discussion. — We have not made any detailed calculation to modelize the expected change in ions concentration across the cell, *versus* the applied voltage V. Nevertheless, we can consider two limiting cases : assume first that the used compound is a perfect salt, i.e. its ionisation energy qV_1 in the nematic is lower, or comparable, to $k_B T$ (q ion charge). The compound should be completely dissociated at room temperature. The linear electrochiral effect should saturate for $V > k_B T/q \sim 25$ mV, since the chiral ions will segregate on one side of the cell. In case, now, where qV_0 is much larger than $k_B T$, no effect should be observed until $V \sim V_1$, for which an electrochemical reaction can be triggered. This last situation corresponds to the case of the cholesteryl esters described in part c), where we observe a simultaneous increase of the current and the onset of a large electrochiral effect for $V_1 \sim 2$ V. We have not observed the first situation. It is likely that the compounds we have used are not fully dissociated. We can just estimate that the maximum relative change of twist must compare with the relative density of ions, i.e. to the dissociation factor. From our data, we can estimate this factor at about 5/75 ~ 8 % for PhLA, 3/60 ~ 5 % for KPhL, 8/40 ~ 20 % for BQC and 5/30 ~ 16 % for BQB. The benzyl quininium derivative are the one which behave closer to real salts. In fact, we have observed that the instantaneous current through the cell is much larger than the measured stationnary current. This means that the used non-blocking electrodes are polarized. The potential drop across the cell will be localized closed to the electrodes, which explains also the observed absence of saturation for the electrochiral effect, since the effective voltage inside the cell is much smaller than the applied voltage V. A quantitative model of the charges and potential distribution across the cell cannot be made at the present stage, with just the integrated data of the electrochiral effect and the electric current.

5. Diffusion and anchoring energy measurements.

5.1 DIFFUSION. — When trying to make solutions of chiral salts with 5CB, we had some difficulties to dissolve the chiral molecules. One method was to prepare an hybrid cell of pure 5CB, and to place on the edge of the cell small crystallites of the compound to dissolve. The chiral molecules, at the constant saturation concentration on the edge of the cell, diffuse into the nematic cell along the **x** direction. The optical measurement of the twist allows for a measurement of the local concentration $c(x, t)$ at time t. We have then a method to measure the mean diffusion coefficient D of a chiral compound in the cell. This method is akin to a previously demonstrated experiment [5], with the advantage of a better spatial accuracy, and the drawback of the averaging process due to the splay bend of the hybrid texture. We demonstrate this method with a solution of PhLA in 5CB. Figure 18 shows microphotographic pictures of the cell, placed between the polarizer P (Fig. 3) and the analyzer. A turned from the extinction direction by 10°, 20°, ... up to 90°, 24 h after placing the crystallites on the left of the pictures. The horizontal scale represents 4 mm. From these pictures, we can draw the $\varphi_\ell(x, t_0)$ at $t_0 = 24$ h (Fig. 19). Taking the same data at times $t = 2$ h, 4 h, 8 h, we obtain in figure 20 an ensemble of curves $\varphi_\ell(x, t)$. φ_ℓ is the same at the edge of the sample, because it corresponds to the saturated concentration. We could analyse exactly the analytical profiles of these curves, which represent the classical smoothing of a constant concentration step function, by diffusion. In fact, the accuracy of the observation is not terribly good. It is simpler to derive D from the definition :

$$D = \frac{\partial c}{\partial t} \Big/ \frac{\partial^2 c}{\partial x^2} = \frac{\partial \varphi_\ell}{\partial t} \Big/ \frac{\partial^2 \varphi_\ell}{\partial x^2}$$

$\frac{\partial \varphi}{\partial t}$ and $\frac{\partial^2 \varphi}{\partial x^2}$ are measured in figure 20. This results in $D = 1.7 \times 10^{-7}$ cm²/s. This compares well with the data from reference [5] for cholesteryl esters in MBBA. We do not know of any measurements of cholesteryl esters diffusion constant in 5CB.

5.2 ANCHORING ENERGY. — We can now interpretate the large field behavior of the quadratic electrochiral effect as shown in figure 14 for instance, for the 5CB + CH mixture. Let us observe first the saturation effect of the AC field for which we do not expect any flexoelectric effect. The optical twist φ_ℓ goes linearly to zero for a finite $E_0 = 10$ V/10 μm. Using formula (23), we can estimate $L = 0.3$ μm. This value is a bit larger than the already measured $L = 0.1$ μm for strong planar anchoring [15]. It may be that chiral molecules, or ions, trapped at the interface, decrease the anchoring energy, or even change the orientation of minimum energy. This could explain the very large observed E_0 dispersion (2-20 V/10 μm) between cholesteryl esters and BQB respectively. One possible mechanism would be the following : surface ions, trapped on the electrode, create a surface field E_S normal to the electrode, over a thickness equal to the Debye screening length. The coupling of E_S with the positive anisotropy of 5CB would decrease the planar anchoring energy, since it orients the nematic director normal to the electrode. This mechanism has already been

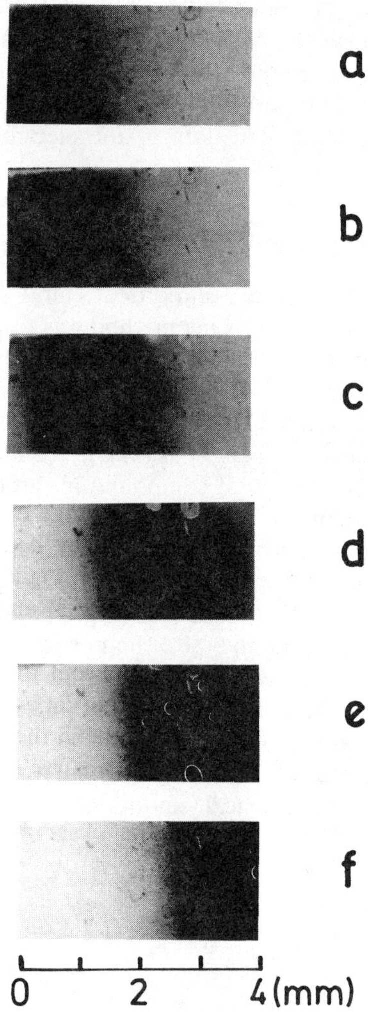

Fig. 18. — Optical appearance of the hybrid cell under microscope 24 h after placing crystallites of PhLA on the left edge. The x distance to the edge is indicated in mm. The black line corresponds to the region where the optical rotation is φ_ℓ : a) $\varphi_\ell = 90^\circ$; b) $\varphi_\ell = 80^\circ$; c) $\varphi_\ell = 60^\circ$; d) $\varphi_\ell = 30^\circ$; e) $\varphi_\ell = 20^\circ$; f) $\varphi_\ell = 10^\circ$. The x displacement of the dark line indicates a concentration gradient of chiral PhLA.

proposed [16] to explain the so-called « spontaneous Freedericksz ransitions ». More carefull observations would be necessary to clarify this point.

We now observe the large DC field saturation effect of φ_ℓ, also visible for instance in figure 14. Compared to the AC behavior, we do observe a decrease of $|E_0|$ for positive V, and an increase of E_0 for negative V, as expected for a flexoelectric contribution to the surface energy, with $e > 0$. The relative variation of E_0 is of the order of 30 %. This is reasonable with our model which predicts a relative change $\sim e\left(\dfrac{4\,\pi}{\varepsilon_a K}\right)^{1/2}$. With $\varepsilon_a \sim 12$ and $K \sim 0.7 \times 10^{-6}$ cgs, we find $e \sim 2.5 \times 10^{-4}$ cgs. This value is too low by a factor 2 or 3, from the one measured [3] in AC by the Bordeaux Group. This effect is probably due to the partial

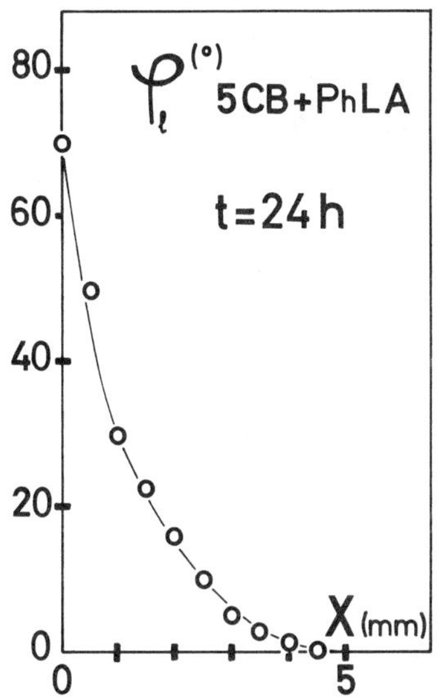

Fig. 19.

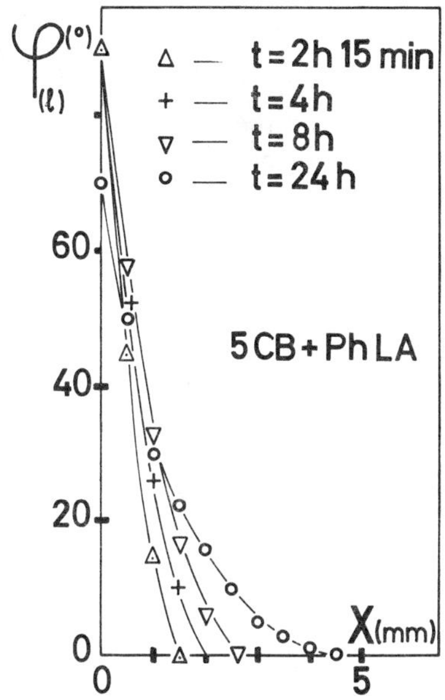

Fig. 20.

Fig. 19. — $\varphi_\ell(x, t = 24\ \text{h})$ for the 5CB + PhLA inhomogeneous sample of figure 18.

Fig. 20. — Same as figure 19, for various t, from $t = 2$ h to 24 h.

screening of the field by the ions close to the non-ohmic electrodes. It would also be important to know the associated non uniformity of E which would give a direct coupling with the electric quadrupole density of the nematic material. As always, the DC behavior of nematic cells remains essentially unexplained.

Conclusion.

A nematic hybrid cell is the simplest liquid crystal texture which allows to measure the twisting power of chiral compounds, since by definition the homeotropic plate cannot transmit twist torque, and then will not disturb a weak induced helicity. We have used this property to measure the twisting power of chiral molecules and salts in the nematic 5CB, in presence of DC or AC electric field. A small DC field drags the chiral ions in the region of large visible induced helicity (the planar side) or in the region where twist is not visible, because of molecular tilt (the homeotropic region). Although the total twist of the texture remains the same, because the optical wave guide condition of Mauguin is non-linear, this chiral ion drag results in a linear electrochiral effect. For large field, the quadratic effect of the dielectric anisotropy of the nematic changes the hybrid texture to an homeotropic texture and suppresses the linear effect. At large AC field, the twist vanishes for a finite field value, which gives a new method to measure surface zenithal anchoring energies. For DC field, the same observation gives a measurement of the flexoelectric constant of the nematic material. The twist in the hybrid cell can also be used to probe concentration changes in space and time, through diffusion process for instance. Our results on 5CB room temperature nematic liquid crystal demonstrate the existence of the polar electrochiral effect, for both positive and negative chiral ions. Benzyl quininium derivative seems to be 10 to 20 % dissociated : phenyllactic derivative can give 5-8 % ions. In its present form the effect is too weak to lead to practical applications. For this purpose one should use more dissociated chiral salts and zero dielectric constant anisotropy mixtures. Other geometries can also be considered, where the electric field is parallel to the plates for instance. This allows for a nice 2 dimensional mapping of the chiral agent concentration, but leads, in presence of chiral ions, to complicated flows and could be the subject of further work. Finally, we have estimated the mean diffusion constant of Phenyllactic acid in the cell [$D = 1.7 \times 10^{-7}\ \mathrm{cm^2/s}$ at $T = 20\ ^\circ\mathrm{C}$] and the anchoring energy on the planar plate (extrapolation length in the 0.1 μm to 1 μm range). The observed dispersion of this anchoring energy is not really understood.

Acknowledgments.

We thank Pr. H. Kagan for the supply of the benzyl quininium derivatives and a critical reading of the manuscript. We thank Dr. C. Germain for mixtures preparations. This work was supported by a CNRS grant to one of us (Z.F.L.).

References

[1] For general references, see for instance, DE GENNES P. G., The Physics of Liquid Crystals (Oxford University Press, Clarendon) 1974.
[2] MAUGUIN C., *Bull. Soc. Fr. Minér. Cristallogr.* **34** (1911) 3.
[3] BAESSLER H. and LABES M. M., *J. Chem. Phys.* **52** (1970) 631.
[4] KOZAWAGUCHI H. and WADA M., *Jap. J. Appl. Phys.* **14** (1975) 651-660.
[5] HAKEMI H. and LABES M. M., *J. Chem. Phys.* **61** (1974) 10, 4020.
[6] DOZOV I., PENCHEV I., *J. Phys. France* **47** (1986) 373.
[7] MADHUSUDANA N. V., DURAND G., *J. Phys. Lett. France* **46** (1985) L-195.
[8] PROST J., MARCEROU J. P., *J. Phys. France* **38** (1977) 315 ;
MARCEROU J. P., PROST J., *Mol. Cryst. Liq. Cryst.* **58** (1980) 259.
[9] KARAT P. P. and MADHUSUDANA N. V., *Mol. Cryst. Liq. Cryst.* **36** (1976) 54.
[10] MADHUSUDANA N. V. and PRATIBHA R., *Mol. Cryst. Liq. Cryst.* **89** (1982) 255.
[11] PATNA B. R. and SHASHIDAR R., *Mol. Cryst. Liq. Cryst.* **42** (1977) 116.
[12] KAHN F. J., *Appl. Phys. Lett.* **22** (1973) 386.
[13] URBACH W., BOIX M. and GUYON E., *Appl. Phys. Lett.* **25** (1974) 479.
GUYON E., PIÉRANSKI P. and BOIX M., *Appl. Phys. Lett.* **25** (1974) 479.
[14] CANO R. and CHATELAIN P., *C.R. Acad. Sci. Paris* **B 259** (1964) 252 ;
CANO R., *Bull. Soc. Miner. Cryst.* **91** (1968) 20.
[15] RIVIÈRE D., LEVY Y. and GUYON E., *J. Phys. Lett. France* **40** (1979) 215.
[16] BARBERO G. and DURAND G., *Liq. Cryst.* **2** (1987) 401.

©1988 SPIE. Reprinted from *Optical Computing '88,* Proc. Vol. 963, 165–172 (1988).

ELECTRO-OPTIC POLYMER WAVEGUIDE DEVICES

A.J. Ticknor, G.F. Lipscomb, R. Lytel, M.A. Stiller, and J.I. Thackara
Research and Development Division, LOCKHEED MISSILES AND SPACE COMPANY, INC.
D-9720, B-202, 3251 Hanover St., Palo Alto, CA 94304

1. INTRODUCTION

The promise of organic and polymeric materials for electro-optic and nonlinear optic applications has been advanced a great deal in recent years[1-4]. This interest arises from the fact that successful molecular design[5-7] has produced materials with useful nonlinear susceptibilities while maintaining compliant structural, mechanical, and processing properties[8-15]. This paper will discuss the application of such materials in the construction of electro-optic channel-waveguide devices. In section 2 we will discuss a technique for fabricating channel-waveguide devices in polymer materials. Section 3 will briefly cover the interest in high-speed electro-optic devices in these materials, and in section 4 we will describe a device architectures that can take advantage of the processing properties used to make polymer waveguides.

2. FABRICATING WAVEGUIDE DEVICES

Integrated-optic waveguide systems are invariably limited to a planar environment. In the predominant technology, titanium-indiffused lithium niobate, this planar environment is a thin surface region of the substrate. The qualities of the device region are inherently dependent upon the qualities of the substrate. In the polymeric materials the device region is a thin film, or as more appropriately realized, an autonomous film stack consisting of an inactive cladding layer on each side of the active film layer. The only function of the substrate is to mechanically locate and support the film stack, which is generally a few microns thick. Such a film stack could cover the surface of a relatively large substrate to emulate existing integrated-optic device structures, or it could be so thin and long as to not need a substrate and would be considered a fiber. In fact, polymeric waveguide technology can cover a domain continuously intermediate to planar waveguides and optical fibers.

There are several possible methods for constructing these film layers. The properties of the materials to be used determine the appropriate methods. The processes and characteristics we will describe are appropriate for the class of materials we are using, specifically guest-host and side-chain polymer sytems incorporating nonlinear moieties. The approach we typically use is as follows: spin-coat a buffer layer, such as UV-curing epoxy, onto a substrate with patterned electrodes; spin-coat the active-layer material onto the buffer layer; process the active layer to define the desired waveguide devices; and, if necessary, add an upper buffer layer and upper electrodes to finish the structure.

The spun-on polymer film is essentially amorphous and has no significant electro-optic response. The material must be processed to yield an electro-optic susceptibility as well as to laterally define the waveguide devices. Both of these goals can be accomplished through electric-field poling of the material[16-19]. If the material is heated above its glass transition temperature (50-150°C in materials we have used) and an electric field on the order of 100V/μ is applied, the dipoles of the nonlinear moiety will tend to align with the field, inducing a macroscopic second-order susceptibility. If the material is then cooled to room temperature with the poling field applied, the orientation freezes in place. The field can be removed and the susceptibility remains[7,20]. The degree of alignment achieved with this technique is around 10%, yielding electro-optic responses comparable with lithium niobate (14-53 pm/V as compared to 32 pm/V) in the more-active polymers. After poling, the film is an electro-optically active slab waveguide and can be used for slab waveguide devices like the phase modulator depicted in figure 1.

A summary of the properties of a simple, single-channel phase modulator is shown in table 1. The electro-optic material was an experimental nitro-aromatic side-chain polymer system, identified as HCC-1238 (formerly refered to as C-22), synthesized by the Hœchst-Celanese Research Division .

In addition to producing the electro-optic susceptibility, the poling process induces a birefringence in the material. Light polarized with its E-field vector parallel to the original poling field experiences a higher index of refraction than in the unpoled film. This is also the polarization for maximal electro-optic response in the film. This property is the basis of a channel-waveguide formation technique we refer to as the selective poling procedure (SPP). By using patterned electrodes to apply the poling field only where the channels are desired, electro-optically active channel waveguides are formed for the appropriate polarization. Figure 2 depicts a cross-sectional view of the forming of two channel waveguides for TM polarization by SPP. In part (a) of this figure, the channels are poled between the patterned electrodes on the substrate and an area electrode on the polymer. After poling, the area electrode is removed and a buffer layer and switching electrode are added to form the device depicted in part (b) of figure 2.

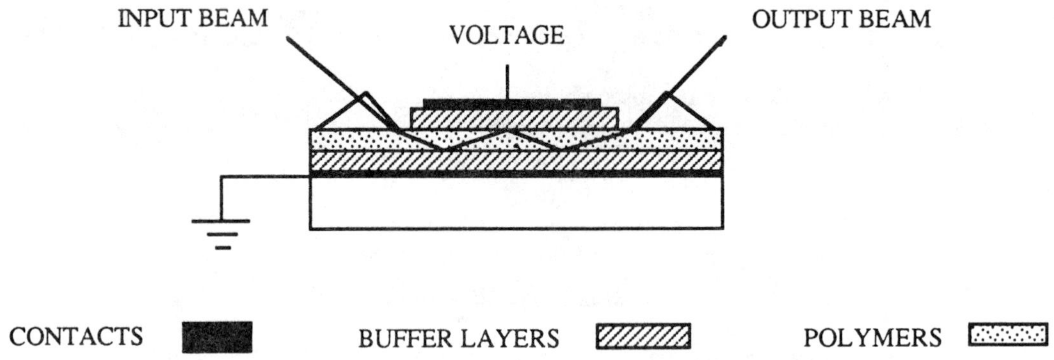

Figure 1. Typical Slab Waveguide Modulator and Components

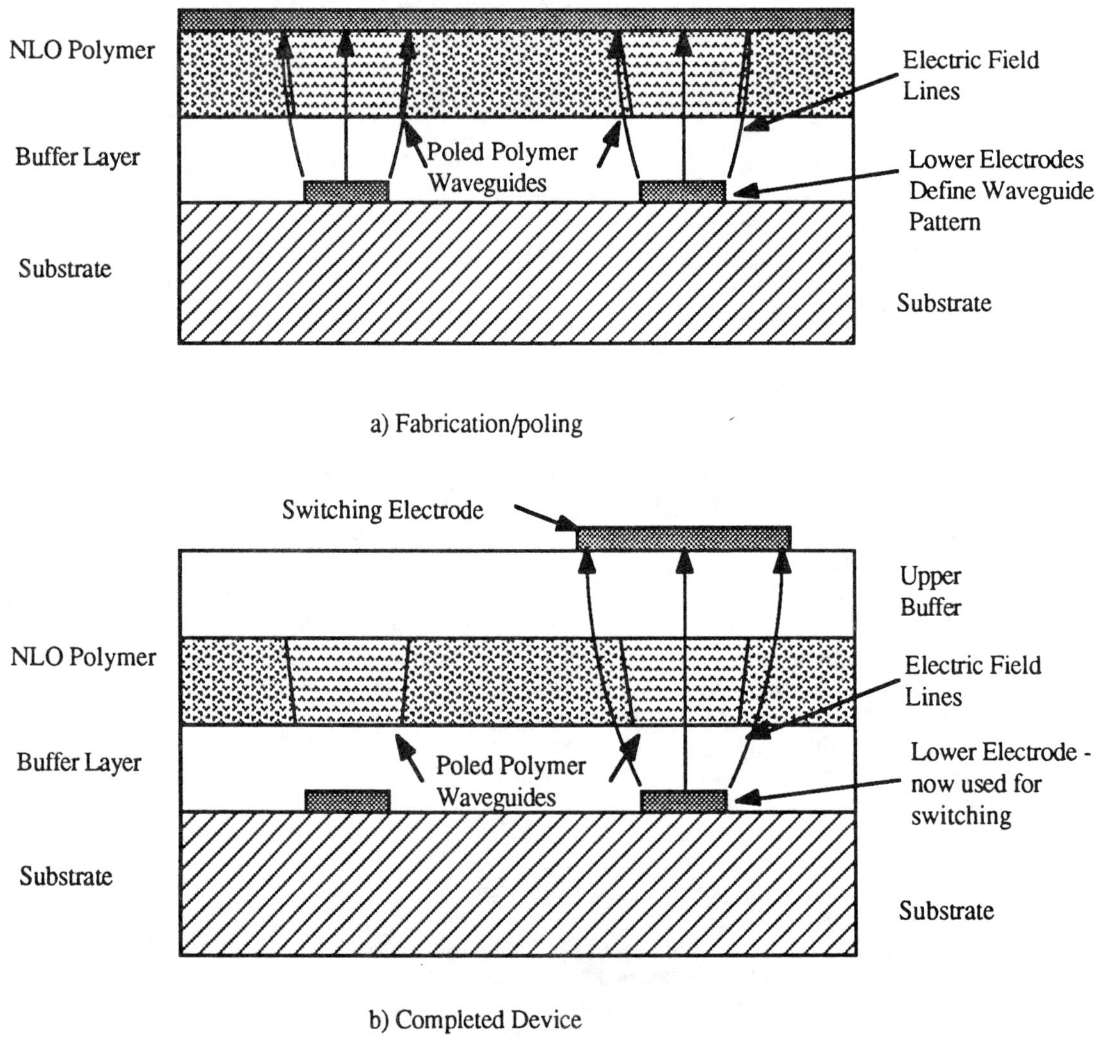

Figure 2. Selective Poling Procedure for Active Channel Waveguides

(wavelength = 0.83 μm)	HCC-1238
E-O COEFFICIENT (pm/V)	14.0
TM REFRACTIVE INDEX (POLED)	1.58
TM INDEX DIFFERENCE (POLED-UNPOLED)	0.005
WG LENGTH (cm)	2.5
MEASURED LOSS (dB/cm)	0.8
HALF-WAVE VOLTAGE (volts)	7

Table 1. Physical Parameters for a C-22 Slab Waveguides

The DC electric field between and around the poling electrodes is smoothed by fringing effects and does not mimic the fine, discontinuous features of the electrodes. Since the field serves to create the channel, the lateral boundaries of the waveguide are quite free of the roughness that causes scattering in guides produced by other methods, such as etched ribs or channel filling. This conclusion is supported by the observation that the propagation loss in a straight, poled channel was nearly the same as the propagation loss in the slab[18].

The maximum index rise for poling C-22 was observed to be about 0.005. The precise index profile for narrow poled guides has not yet been determined, but with the properties observed, it is apparent that 2-3 micron widths will produce single-mode guides with relatively tight confinement.

We have used SPP to construct several waveguide devices, such as phase modulators, Mach-Zehnder intensity modulators, and multi-mode directional couplers. The phenomenological performance of these devices revealed the differences in sensitivity to polarization and in the magnitude of the electro-optic response as compared to similar devices of more conventional construction, but otherwise they performed the same as would their counterparts made in $LiNbO_3$.

3. HIGH-SPEED DEVICES

An important area of integrated optics where organic materials promise to shine brightly is in high-speed traveling-wave devices. If an electro-optic material has substantial dispersion in the dielectric constant between optical and microwave frequencies, the light signal outruns the modulating signal fairly rapidly. This phenomenon results in a fundamental length-bandwidth product for a material. If the material experiences little loss of the modulating signal over the practical length of the device, then the interaction length is roughly proportional to the sensitivity of the device (ratio of modulation depth to signal level). A device with finite sensitivity must have a limited bandwidth to remain within the length-bandwidth product of the material. Because the polymers have only slight dispersion in the dielectric constant between the microwave and optical frequencies, they have length-bandwidth products of around 120 cm-GHz, compared to about 10 cm-GHz for lithium niobate. Since the electro-optic coefficients are similar, the polymer devices should be able to have around an order-of-magnitude greater bandwidth than otherwise equivalent devices. The actual microwave losses of the materials we have been using have not yet been determined. The microwave loss in other polymeric waveguide materials, such as poly(methyl methacrylate), is known to be small, and hence it is anticipated that the loss can be small in the more active polymers as well.

4. NOVEL DEVICES

The production of polymer waveguide devices offers mechanical flexibilities that lead to possible device architectures not practically realizable in conventional integrated-optic technology. An obvious difference is that with film stacks several device layers may be stacked onto a single substrate. Even if these layers are independent this would provide a much greater physical density for a collection of devices than could be achieved with a separate substrate for each layer, as is done with surface technology. If interactions between these layers are allowed and controlled, it is possible to have devices with more sophisticated operations than can be achieved in the planar environment of one layer per substrate.

Another novel property of poled-polymer waveguides is that the waveguide material is chemically the same as its lateral cladding. Lateral confinement is achieved by virtue of molecular alignment. If a section of a waveguide can be locally heated, such as by a short-wavelength-laser spot, the molecules can be disoriented and that section of the waveguide will be effectively erased. This may allow the "trimming" of individual devices to optimize their performance. It might also be possible to create new sections of waveguide by local heating with a field applied, but this would require a finer process and the maintainance of large temperature gradients for up to several minutes. Processes to accomplish this must be further developed and demonstrated before the "writing" of narrow waveguides by selectively heating the polymer can be deemed practical.

Most common integrated-optic devices are not apt to take best advantage of such properties. Optical signals are generally directed by bending waveguides off in new directions. This is not an appropriate method for directing signals between the device layers because of the complexity in establishing waveguides skew to the nominal orientation of the device planes. The common waveguide devices are also not generally susceptible to having their performance tuned by longitudinal trimming of the channels. Changing the length of the interaction areas would require reforming the input and/or output channels of the device. Lateral trimming of the channels would require the kind of conditions necessary to locally write a waveguide, and it has already been argued that this is probably less practical.

We have, however, designed some basic switching devices that are better suited to take advantage of the poled-polymer waveguide properties. These devices are based on the resonant coupling between parallel waveguides. Signals can be switched from the inputs among the various outputs by electro-optically perturbing the propagation constant (β) of the appropriate individual guides. This alters the eigenfunctions (amplitude distributions) and eigenvalues (phase speeds) of the guided eigenmodes of the parallel waveguide structure. The switching of the optical signals can be described as the controlling of the interference between these eigenmodes. Effective switching of these devices requires interfering eigenmodes with the appropriate amplitudes and phases. Changing β of the individual guides alters the eigenfunctions of the structure, redistributing the energy in each of the modes at the output. Unfortunately, making this perturbation within practical limits over the entire interaction region often provides unsatisfactory control to align the magnitudes and phases of the eigenmodes appropriately for efficient switching. To get more efficient switching we designed devices with both active and passive regions along the interaction length of the device. In the active region, β is perturbed to produce the appropriate distribution in the magnitudes of the eigenfunctions. In the passive regions, β is unaltered or varied very slightly to make small changes in the phase speeds without significantly altering the magnitudes in the eigenfunctions. These passive regions are used to appropriately align the phases of the eigenmodes to allow efficient operation of the device.

THE PSI-SWITCH

The first device of this type we will describe is a one-to-two multiplexer/demultiplexer we refer to as a psi-switch because the layout of the channels resembles the Greek capital letter Ψ. A schematic of the device is shown in figure 3. An optical signal is efficiently coupled between the single guide at one end and the desired one of the two guides at the other end. Switching is accomplished by addressing the electrode over the chosen guide with a voltage appropriate to induce an index change of approximately π/kl under the electrode, where k is the free-space propagation constant of the optical signal and l is the length of the electrode. If we define the resonance length of the three-guide structure as the distance required for a signal in the center guide to maximally couple into the two outer guides, then the interaction length of the psi-switch is approximately 2.7 resonance lengths and the active region (the electrodes) is one resonance length. Figure 4 shows beam-propagation-method simulations of the operation of a psi-switch as well as a cascade of psi-switches to produce a higher-order multiplexer/demultiplexer.

THE XI-SWITCH

The next resonantly-coupled device to be described is a two-by-two crossbar we call a xi-switch because of its resemblance to the Greek letter Ξ. It is depicted in figure 5. In this device, a central waveguide is placed between two extremely-weakly coupled waveguides. The central guide is nearly resonant with the outer guides and it acts as a catalyst for the coupling between the outer guides. The active region of the device consists of an electrode over half the central guide. Depending on the voltage applied to this electrode, the coupling can be enhanced, to produce the cross state, or it can be reversed, to produce the bar state. It often helps to also make very small perturbations in the index of the passive region of the central guide to "tweak" the phase distributions. Figure 6 shows a typical simulaton for this device as well as a cascade for a higher-order crossbar.

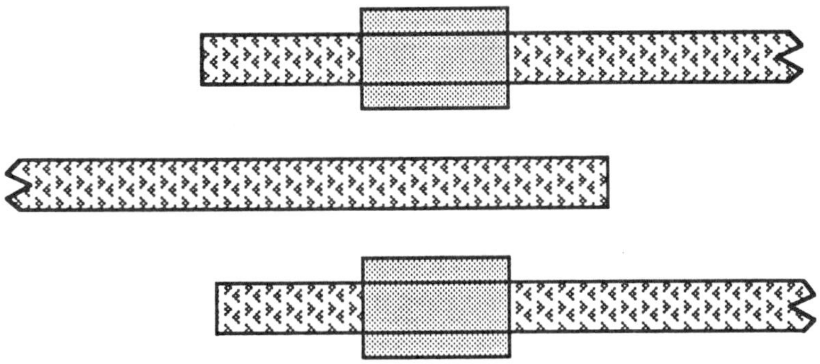

Schematic of a psi-switch: an optical signal is efficiently coupled between the center guide and one of the outer guides by applying an appropriate voltage to the electrode over the desired guide and not driving the other electrode.

Figure 3
Psi-Switch

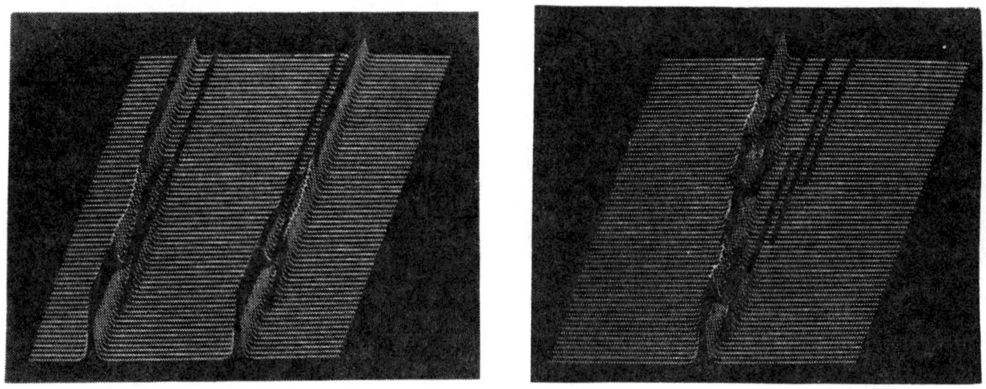

a b c

Demultiplexer simulations: single-stage psi-switch coupled to the left (a) and right (b) output channels and two-stage device coupled to the leftmost of four output channels (c). The dark stripes represent areas that have been poled and that are not electrically addressed for the simulation.

Figure 4
Demultiplexer Simulations

The central guide in the xi-switch carries no signal into nor out of the device. It does not interface with the guides of any other devices and needs not be of the same width as the outer guides. In fact, we often design this device with a fatter central guide of slightly lower index (to match the ß's) so we can further separate the outer guides and minimize the interaction in the absence of the catalyst. Minimizing this interaction is crucial to minimizing crosstalk in long cascades for high-order crossbars and the wider channel separation is better suited for coupling fibers to the inputs and outputs of a device.

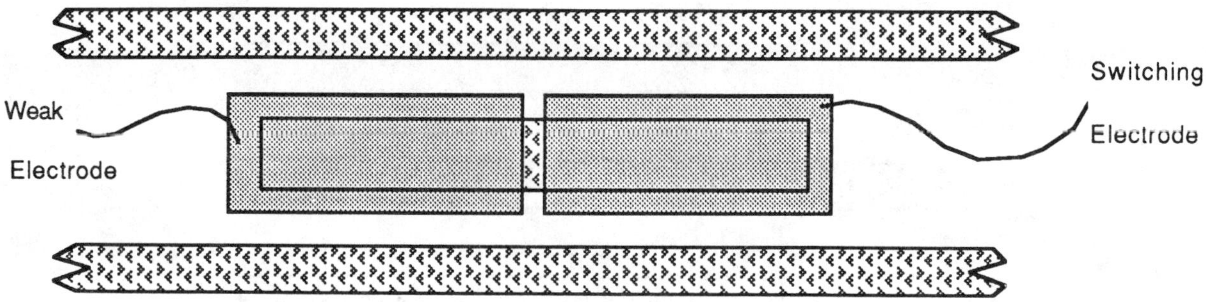

Schematic of a xi-switch: if no voltage, or only very small 'tuning' voltages are applied to the two electrodes, the center guide will 'catalyze' the coupling between the outer guides and produce the 'cross' state; if the switching electrode is driven to give an index difference of about π/kl, the interaction will be reversed and the 'bar' state will result.

Figure 5
XI-SWITCH

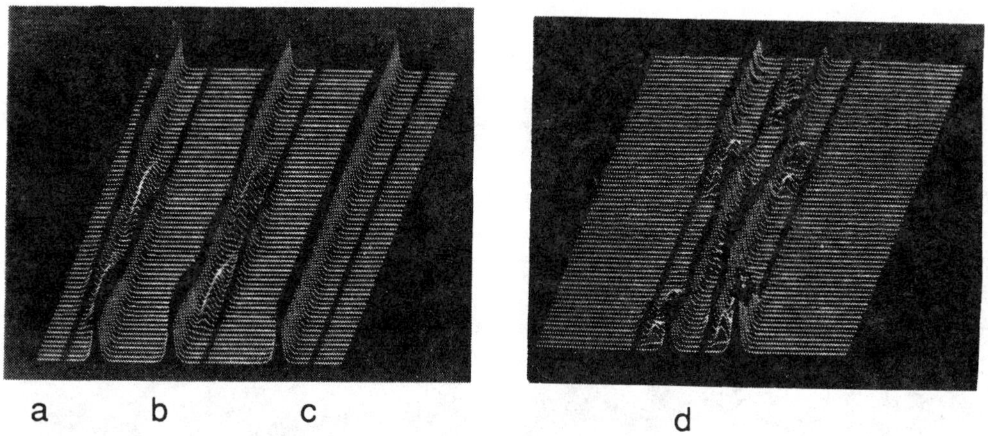

Xi-switch simulations showing the cross state (a), the bar state (b), and insignificant coupling over the same interaction length with the catalyst removed (c). Simulation d is of a four-level, six-catalyst cascade of xi-switches to form a fully-reconfigurable, four-by-four crossbar. In this simulation a large signal is coupled from the right-most channel to the left-most channel while a smaller signal is coupled from the inner left channel to the inner right channel.

Figure 6
Simulated Operations of a Xi-Switch

These parallel-waveguide devices are well suited to take advantage of the processing techniques developed for polymer waveguides. They can be used to switch signals between different layers of a stacked device as well as to direct them laterally in the same device layer. Changing the length of the interaction region to trim the device can be accomplished by "erasing" the ends of the guides at either end of the interaction region. This erasing would not require extreme precision in establishing steep temperature gradients because heat can diffuse into the buffer regions between the guides and not alter the guides not being trimmed. A serendipitous quality of these devices is that they also seem to have good crosstalk-rejection properties. The crosstalk must arise from coupling between guides that are well separated. If the device suffers thermal expansion or other slight variations from the optimum operating conditions, errant signal energy is coupled into abruptly ending waveguides and is radiated into (laterally) unguided energy. The first order defect of an imperfect device often arises as loss rather than crosstalk. This can be an advantage in many systems, such as several-channel multiplexers and crossbars or bidirectional devices, where crosstalk can result in unintelligible or unreliable signals.

5. SUMMARY

Many desirable optical properties are available in polymeric materials and the application of these materials to practical devices is advancing the potential of this field. We have demonstrated some new processing techniques applicable to these materials and used these techniques to produce functioning modulators and couplers. We have also described properties of these materials that lead to processing flexibilities not available in more conventional materials, and we have described two switching devices that can exploit these properties for improved performance.

6. ACKNOWLEDGEMENTS

The specific material discussed in this paper was supplied to us by Hoechst Celanese Research Division[5]. We wish to acknowledge Drs. J. Stamatoff, R. DeMartino, and H. Yoon for useful discussions.

7. REFERENCES

1. R. Lytel, G.F. Lipscomb, and J.I. Thackara, "Recent Developments in Organic Electro-optic Devices", in Nonlinear Optical Properties of Polymers, A.J. Heeger, J. Orenstein, and D.R. Ulrich, ed., Proc. Materials Research Society Vol. 109, 19 (1988).
2. R. Lytel and G.F. Lipscomb, "Nonlinear and Electro-optic Organic Devices", in Nonlinear Optical and Electro-active Polymers, P.N. Prasad and D.R. Ulrich, ed., Plenum Press, New York (1988), p. 415.
3. R. Lytel, G.F. Lipscomb, and J.I. Thackara, "Advances in Organic Electro-optic Devices", Proc. SPIE Vol. 824, pp.152-161 (1987).
4. R. Lytel, G.F. Lipscomb, P. Elizondo, B. Sullivan, and J. Thackara, "Optical Nonlinearities in Organic Materials: Fundamentals and Device Applications", Proc. SPIE 682, 125 (1986).
5 S.J. Lalama and A.F. Garito, "Origin of the Nonlinear Second-order Optical Susceptibilities of Organic Systems", Phys. Rev. A 20, 1179 (1979)
6. C.C. Teng and A.F. Garito, "Dispersion of the Nonlinear Second-order Optical Susceptibility of an Organic System: p-nitroaniline", Phys. Rev. Lett. 50, 350 (1983).
7. K.D. Singer, M.G. Kuzyk, and J.E. Sohn, "Second-order Nonlinear Optical Processes in Orientationally Ordered Materials: Relationships Between Molecular and Macroscopic Properties", J. Opt. Soc. Am. B4, 968 (1987).
8. R.N. Demartino, E.W. Choe, G. Khanarian, D. Haas, T. Leslie, G. Nelson, J. Stamatoff, D. Stuetz, C.C. Teng, and H. Yoon, "Development of Polymeric Nonlinear Optical Materials", in Nonlinear Optical and Electro-active Polymers, P.N. Prasad and D.R. Ulrich, ed., Plenum Press, New York (1988), p. 169.
9. D.J. Williams, "Nonlinear Optical Properties of Guest-Host Polymer Structures", in Nonlinear Optical Properties of Organic Molecules and Crystals, Vol. 1, D. Chemla and J. Zyss, ed., Academic Press, FLA (1986), p. 405.
10. G.F. Lipscomb, A.F. Garito, and R.S. Narang, "An Exceptionally Large Linear Electro-optic Effect in the Organic Solid MNA", J. Chem. Phys. 75, 1509 (1981).
11. D. Pugh and J.O. Morley, "Molecular Hyperpolarizabilities of Organic Materials", in Nonlinear Optical Properties of Organic Molecules and Crystals, Vol. 1, D. Chemla and J. Zyss, ed., Academic Press, FLA (1986), p. 193.
12. Nonlinear Optical Properties of Organic Molecules and Crystals, Vol. 1 and 2, D. Chemla and J. Zyss, ed. (Academic Press, FLA) 1986.
13. G.J. Bjorklund et. al, "Organic and Polymeric Materials", in Research on Nonlinear Optical Materials: An Assessment, Appl. Opt. 26, 227 (1987).
14. Nonlinear Optical Properties of Organic and Polymeric Materials , D.J. Williams ed., ACS Symposium Series 233 (American Chemical Society), 1983.
15. K.D. Singer and A.F. Garito, "Measurements of Molecular Second-order Optical Susceptibilities Using DC Induced Second Harmonic Generation", J. Chem. Phys. 75, 3572 (1981).

16. J. Thackara, M. Stiller, E. Okazaki, G.F. Lipscomb, and R. Lytel, "Optoelectronic Waveguide Devices in Thin-Film Organic Media", Conference on Lasers and Electro-optics, Baltimore, MD (1987), paper ThK29.

17. J. Thackara, M. Stiller, G.F. Lipscomb, A.J. Ticknor, and R. Lytel, "Poled Electro-optic Waveguide Formation in Thin-Film Organic Media", Appl. Phys. Lett. 52, 1031 (1988).

18. J. Thackara, M. Stiller, A.J. Ticknor, G.F. Lipscomb, and R. Lytel, "Poled Electro-optic Waveguide Devices in Thin-Film Organic Media", Conference on Lasers and Electro-optics, Anaheim, CA (1988), paper TuK4.

19. R. Lytel, G.F. Lipscomb, M. Stiller, J.I. Thackara, and A.J. Ticknor, "Organic Electro-optic Waveguide Modulators and Switches", Proc. SPIE 971 (1988, to be published).

20. C.S. Willand, S.E. Feth, M. Scozzafava, D.J. Williams, G.D. Green, J.I. Weinshenk, H.K. Hall, and J.E. Mulvaney, "Electric-Field Poling of Nonlinear Optical Polymers", p. 107; and K.D. Singer, J.E. Sohn, and M.G. Kuzyk, "Orientationally Ordered Electro-optic Materials", in Nonlinear Optical and Electro-active Polymers, P.N. Prasad and D.R. Ulrich, ed., Plenum Press, New York (1988), p. 189.

©1988 American Institute of Physics. Reprinted from *Applied Physics Letters*, Vol. 52(13), 1031–1033 (March 1988).

Poled electro-optic waveguide formation in thin-film organic media

J. I. Thackara, G. F. Lipscomb, M. A. Stiller, A. J. Ticknor, and R. Lytel
Lockheed Research and Development Division, Lockheed Missiles and Space Company, Inc., 3251 Hanover Street, O/97-20, B/202, Palo Alto, California 94304

(Received 30 November 1987; accepted for publication 26 January 1988)

We describe a novel technique for the fabrication of electro-optic (EO) waveguides in integrated optic device structures employing organic EO materials. The technique combines the poling and waveguide formation steps by utilizing patterned poling electrodes and the induced birefringence associated with the poling process. Several prototype waveguide devices fabricated using this procedure are reported.

The electro-optic (EO) waveguide is an essential structure in integrated optic devices. Devices fabricated from such waveguides include mode converters,[1] modulators,[2] Y-branch interferometers,[3,4] and directional couplers.[5] Ti in-diffusion in $LiNbO_3$ is now a mature technology for fabricating integrated optic devices.[6] Waveguides have also been fabricated in semiconductor materials such as GaAs[7] and InGaAs/InP multiple quantum well structures.[8] In recent years, organic and polymeric materials have emerged as a promising class of EO and nonlinear optical (NLO) materials.[9] These organic materials can be formed into thin films, making them very useful in the integrated optics field.

The processes used to form thin films from organic EO materials, such as spin or dip coating techniques, do not usually produce EO films since the bulk structure is still centrosymmetric. A molecular alignment process, such as electrically induced poling,[10] is needed to establish a noncentrosymmetric structure. To form three-dimensional waveguides in EO films, plasma etching[11] and channel filling[12] techniques have been utilized. In this letter, we describe a novel EO waveguide fabrication technique which combines the poling and waveguide formation steps by utilizing the induced birefringence associated with the poling process.

Two generic waveguide device structures are shown in cross section in Fig. 1. The first has electrodes above and below the film, while the second employs coplanar electrodes. In both cases, the EO layer is optically isolated from the electrodes by buffer or cladding layers. As shown in Fig. 1(a), before the poling step, the active molecules in what will be the EO layer are randomly oriented. After poling, the active molecules are partially ordered in areas defined by the electrodes, as shown in Fig. 1(b).

The partial ordering of the active molecules allows components of the second-order molecular nonlinearities to add, creating a region which exhibits the linear EO effect.[10] Since most organic nonlinear molecules possess an anisotropic microscopic linear polarizability, the poled region also becomes birefringent.[13] The poled regions are uniaxial, with n_e oriented along the direction of the local poling field. For materials in which poling induces positive birefringence, TM and TE waves propagating in vertical and transverse device structures, respectively, will experience a greater refractive index in the poled regions than in the surrounding unpoled regions, and so can be confined in the lateral dimension. Thus, by applying the poling fields using electrodes patterned to define the waveguide network, including both active and passive sections, no further patterning of the organic layer is needed to form the three-dimensional EO waveguide structures.

The devices can be fabricated in a number of ways depending on the type of EO material used and the final device structure required. For devices employing the vertical geometry, the upper electrode can be selectively removed over sections of the waveguides that must be inactive, or the entire upper electrode can be removed and a new electrode pattern deposited to address only the sections which must be active. For some EO layers, the upper electrode used during the poling/waveguide formation step can be deposited directly on the EO layer and then removed after the poling is accomplished. The devices can then be completed by depositing the upper buffer layer and a new upper electrode. It should also be possible in some cases to use electrodes with narrow (relative to the buffer layer thickness) gaps in them to pole the EO waveguides, and then use only certain sections to address the active portions of the device. In this later fabrication procedure, the poling/waveguide formation step would complete the device.

This technique of forming the EO waveguides in a single step has several advantages over other proposed fabrication procedures for organic waveguide devices. An obvious advantage is the simplification of the overall device fabrication

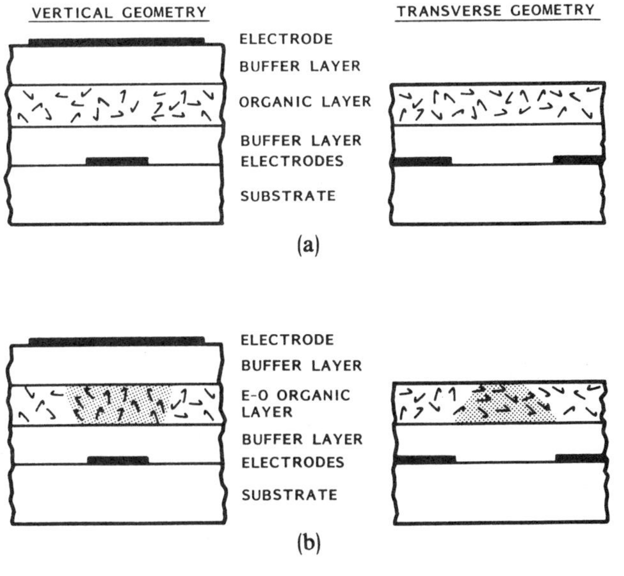

FIG. 1. Waveguide device structures (a) before and (b) after the poling/waveguide formation step.

process. A more important advantage is that the fringing of the poling fields in the buffer and EO layers tends to reduce the effects of electrode edge roughness on the roughness of the waveguide side boundaries. This is not the case when either fabricating channels in the substrate or etching the EO layer itself to form the waveguides. In fact, if the scale of the electrode edge roughness is small compared to the buffer layer thickness, the roughness of the side boundaries should be independent of the limitations of the photolithographic processes used to make the electrodes. This should result in low scattering losses from the waveguide side boundaries. Another advantage of this technique is that the waveguides that are formed guide only one polarization. One polarization therefore, can be selected from the incoming optical signal, facilitating the operation of subsequent devices, such as directional couplers or integrated interferometers, that operate most efficiently on light of a single polarization.[14]

Using this procedure, we have fabricated several prototype integrated optic devices in two different organic EO materials. The materials were PC6S and Celanese 22 (C22), both developed by the Hoechst–Celanese Research Company (HCRC). PC6S is a pendant side chain polymer that we have previously used as the EO layer in slab waveguide modulators,[15] and C22 is a new HCRC proprietary material.

The induced extraordinary index change in PC6S due to poling is large: from 1.64 to 1.70 for unpoled to poled regions at 830 nm. These indices were calculated from coupling angles into slab guides of PC6S from LaSF9 prisms ($n = 1.83$). The index change for C22 was calculated from measured phase changes on double pass reflections through poled and unpoled regions of a C22 film. The induced extraordinary index increase in C22 is approximately 0.005 (1.575–1.58) from unpoled to poled regions.

Figure 2 shows the results of a computation using a beam propagation method[16] for a normalized guide width of $5\lambda/n_{\text{eff}}$, where λ is the free-space wavelength and n_{eff} is the effective refractive index for light guided in the unpoled C22 layer. In Fig. 2(a) the indices in the poled and unpoled regions are the measured values for C22, while in 2(b) the

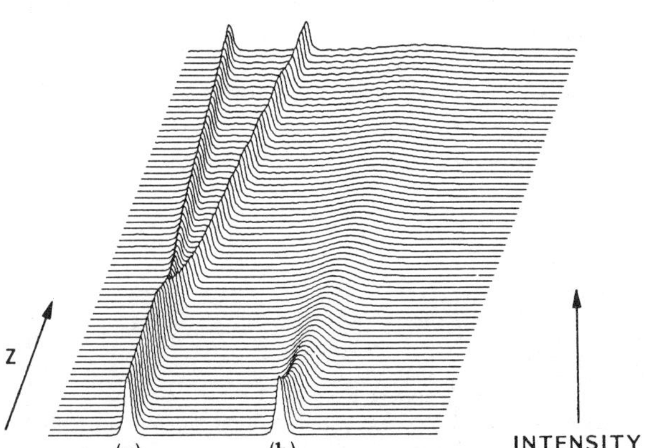

FIG. 2. Theoretical demonstration of beam confinement and splitting in a C22 film. (a) Optical signal launched in the z direction into a region selectively poled to define a Y-branch having a crossing angle of approximately 2.4° and widths of straight sections of $5\lambda/n_{\text{eff}}$. (b) Optical signal launched into an unpoled region demonstrating laterally unconfined propagation.

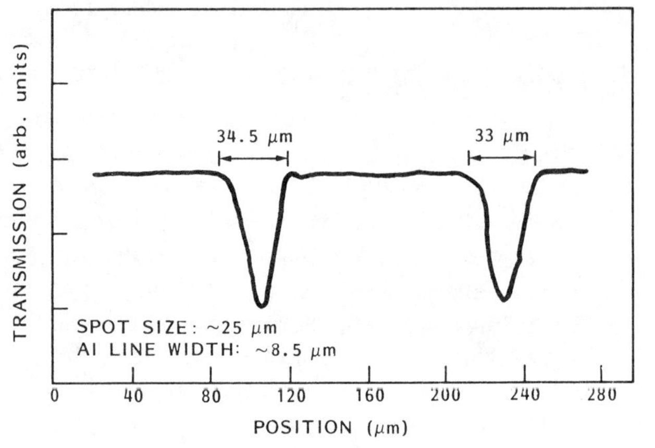

FIG. 3. Measured transmission ($\lambda = 632.8$ nm) through a PC6S structure containing two adjacent poled waveguides (each of vertical geometry) as a function of position. The beamwidth was approximately 25 μm. No birefringence is observed beyond the 8.5-μm-wide aluminum poling electrodes.

index is 1.575 uniformly. As expected, strong confinement is predicted.

A potential problem with this technique may be loss of resolution due to lateral charge spreading at the layer interfaces. This has not been a problem, however, in the structures we have fabricated. Figure 3 shows the results of an experiment in which a 25-μm-diam beam ($\lambda = 632.8$ nm) was scanned across two adjacent 8.5-μm-wide aluminum electrodes above which poled waveguide structures with PC6S as the EO layer had been fabricated. The aluminum electrodes partially blocked the beam as it passed over them; but no birefringence, as can easily be seen when scanning across poled slab waveguides formed on substrates coated with transparent conductors, was detected beyond the electrodes.

To fabricate the prototype devices, aluminum lower electrodes defining Y-branch interferometers and directional couplers were formed on glass substrates using standard deposition and photolithographic techniques. The dimensions of these electrodes were chosen to facilitate the evaluation of the poling/waveguide formation process and so were not scaled to optimize the performance of the completed devices made from the different materials. A 2–3-μm-thick UV-curable coating was deposited on each substrate to form the lower buffer layer. A 2-μm-thick layer of EO material was then deposited by spin coating.

Initial waveguide fabrication work was done with PC6S. A Y-branch interferometer was built by evaporating a gold

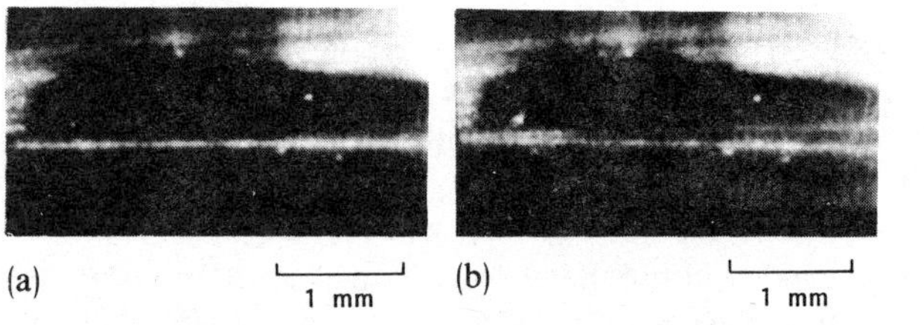

FIG. 4. Experimental verification of beam confinement and partial switching in a poled Y-branch interferometer in C22: (a) 120 V applied, and (b) 65 V applied. The remainder of the gold upper electrode can be seen above the waveguides.

electrode directly onto the PC6S, poling the waveguide pattern at 90 °C for 5 min with an electric field of 100 V/μm in the EO layer, etching off the electrode with dilute Aqua Regia, and finally bonding a piece of indium-tin-oxide-coated glass over one arm of the interferometer with a 3-μm-thick layer of epoxy. The epoxy also served as the upper buffer layer. Prism coupling was used to inject 830 nm light into the device. Guiding was observed along each arm, and modulation of the output beam was detected.

In addition to a Y-branch structure of the same basic design as that built using PC6S, a prototype directional coupler was fabricated using C22. The electrode that defined the final waveguide pattern consisted of two 7-μm-wide sections brought to a 13-μm-wide, 1-cm-long common section at a crossing angle of 3°. The common section was followed by a symmetric arrangement of two more 7-μm-wide sections. The overall length of the electrode was 3.8 cm. To build this device, a 3-μm-thick polysiloxane upper buffer layer was deposited onto the C22 layer by dip coating. A gold upper electrode was then deposited and the waveguide pattern was poled at 105 °C for 30 min with an electric field of 90 V/μm in the C22 layer. The device was completed by removing the gold electrode over the incoming and outgoing waveguides.

Partial switching of 830 nm light from one output arm to the other is shown in Fig. 4. Prism coupling was used to inject light into the lower input arm (not shown). An applied voltage of 120 V produced the bar state shown in Fig. 4(a), and an applied voltage of 65 V produced the partial cross state shown in Fig. 4(b). Complete switching was not observed, but this is not surprising due to the large crossing angle and the nonoptimum waveguide dimensions.

In conclusion, we have developed a novel technique for fabricating EO integrated optic devices in thin-film organic media. The technique should be applicable to many EO and NLO organic systems, produce waveguide structures with very smooth sidewalls, and simplify the overall device fabrication procedure.

The authors gratefully acknowledge the many helpful discussions with W. D. Eades and D. Armitage of the Lockheed Palo Alto Research Laboratories and the contributions of H. N. Yoon and R. N. DeMartino of the Hoechst–Celanese Research Company.

[1]R. C. Alferness and L. L. Buhl, Opt. Lett. **5**, 473 (1980).
[2]A. Neyer and W. Sohler, Appl. Phys. Lett. **35**, 256 (1979).
[3]T. R. Ranganath and S. Wang, IEEE J. Quantum Electron. **QE-13**, 290 (1977).
[4]F. J. Leonberger, Opt. Lett. **5**, 312 (1980).
[5]R. V. Schmidt and R. C. Alferness, IEEE Trans. Circuits Syst. **CAS-26**, 1099 (1979).
[6]R. C. Alferness and J. N. Walpole, IEEE J. Quantum Electron. **QE-22**, 803 (1986).
[7]K. Hiruma, H. Inoue, K. Ishida, and H. Matsumura, Appl. Phys. Lett. **47**, 186 (1985).
[8]U. Koren, B. I. Miller, T. L. Koch, G. D. Boyd, R. J. Capik, and C. E. Soccolich, Appl. Phys. Lett. **49**, 1602 (1986).
[9]A. F. Garito, K. D. Singer, and C. C. Teng, in *Nonlinear Optical Properties of Organic and Polymeric Materials*, edited by D. J. Williams (American Chemical Society, Washington, DC, 1983), p. 1.
[10]K. D. Singer, M. G. Kuzyk, and J. E. Sohn, J. Opt. Soc. Am. B **4**, 968 (1987).
[11]See, for example, D. L. Lee, *Electromagnetic Principles of Integrated Optics* (Wiley, New York, 1986), Chap. 7.
[12]S. Tomaru, M. Kawachi, and M. Kobayashi, Opt. Commun. **50**, 154 (1984).
[13]J. L. Oudar and J. Zyss, Phys. Rev. A **26**, 2016 (1982).
[14]H. S. Hinton, IEEE Commun. Mag. **25**, 16 (1987).
[15]J. I. Thackara, M. A. Stiller, G. F. Lipscomb, R. Lytel, and E. Okazaki, in CLEO Tech. Dig. Ser. 14 (Optical Society of America, Washington, DC, 1987), p. 260.
[16]See, for example, J. Van Roey, J. van der Donk, and P. E. Lagasse, J. Opt. Soc. Am. **71**, 803 (1981).

©1978 Optical Society of America. Reprinted from *Applied Optics*, Vol. 17(1), 52–56 (January 1978).

Polarization and magnetooptic properties of single-mode optical fiber

A. M. Smith

The polarization and magnetooptic properties of a sample of single-mode optical fiber have been investigated. The fiber acts as a linear retarder, and the degree of retardation is dependent on the external pressure applied to the fiber. The stress optic coefficient is found to be 8.72×10^{-10} N^{-1} m^2. The direction of linear polarization is rotated when a longitudinal magnetic field is applied to the fiber (Faraday effect). The Verdet constant is 1.56×10^{-2} min A^{-1}. The intrinsic specific linear retardation of the fiber is found to be less than 0.44 rad m^{-1} from the magnetooptic measurements. The fiber has been used in an experimental current measurement device.

I. Introduction

If polarized light is launched into a length of optical fiber, in general, the polarization parameters of the light emerging from the fiber are different from the original parameters. Some investigations of the polarization properties of multimode fibers,[1] which have a depolarizing effect, and of single-mode fibers,[2] have been reported. In addition papers have appeared on the polarization[3] and magnetooptic[4] properties of index-gradient fibers.

The present work describes the polarization and magnetooptic properties of high quality phosphosilicate single-mode optical fiber produced by the chemical vapor deposition technique.[5] The fiber has a step-index profile with a core of 7-μm diam and a cladding of 110-μm diam. The refractive index of the cladding is 1.458, the normalized frequency $V = 2$, and the numerical aperture is 0.048.

II. Experimental Apparatus

The experimental apparatus used for the polarization measurements is shown in Fig. 1. Linearly polarized light from an He–Ne laser is passed through a $\lambda/4$ plate and then a rotatable polarizing element. The orientation of the $\lambda/4$ plate is adjusted to give circularly polarized light so that the intensity of the linearly polarized light after the polaroid is independent of the polarization angle. The resulting beam is focused onto the entrance endface of a sample of single-mode optical fiber using a ×10 microscope objective. The small diameter of the core means that, inevitably, some light is launched into the cladding. This light is removed by passing the fiber through a liquid with refractive index greater than that of the fiber cladding. At the exit endface a second ×10 microscope objective is used to focus the emerging light. The focused beam is split into two angularly displaced beams with orthogonal linear polarizations by a Wollaston prism, and the intensities of the two beams I_1, I_2 are monitored by two silicon P-I-N photodiodes. These detectors are coupled to analog electronics which compute the function $\Gamma = (I_1 - I_2)/(I_1 + I_2)$. The Wollaston prism and detectors may be rotated to any angle relative to a known arbitrary direction.

For the magnetooptic measurements the fiber is passed through a solenoid which produces a known longitudinal magnetic field H over a length of fiber l_1. The region of the fiber where the magnetic field is applied is a distance l_0 from the entrance endface and a distance l_2 from the exit endface. The total length of the fiber is $l = l_0 + l_1 + l_2$.

III. Polarization Properties

A. Experimental Measurements

The polarizing element is set so that the output linear polarization is at an angle θ_0 to an arbitrary known direction, and the detector (consisting of the Wollaston prism and two photodiodes) is rotated so that the output of the analog electronics reaches a maximum value $P = (I_{max} - I_{min})/(I_{max} + I_{min})$ at a detector angle θ_1. Both θ_0 and θ_1 are measured in the plane of the fiber cross section (x,y plane) with respect to arbitrary directions. Figures 2 and 3 show the experimental data obtained on the variation of P and θ_1 with θ_0 for several different lengths of fiber l. The reference directions from which θ_0 and θ_1 are measured have been chosen so that the minima in Fig. 2 occur at $\theta_0 = 45°$ and $\theta_1 = 45°$ in each case. It was found that the reference directions were parallel.

The author is with Central Electricity Research Laboratories, Leatherhead, Surrey KT22 7SE, England.

Received 23 March 1977.
0003-6935/78/0101-0052$0.50/0.

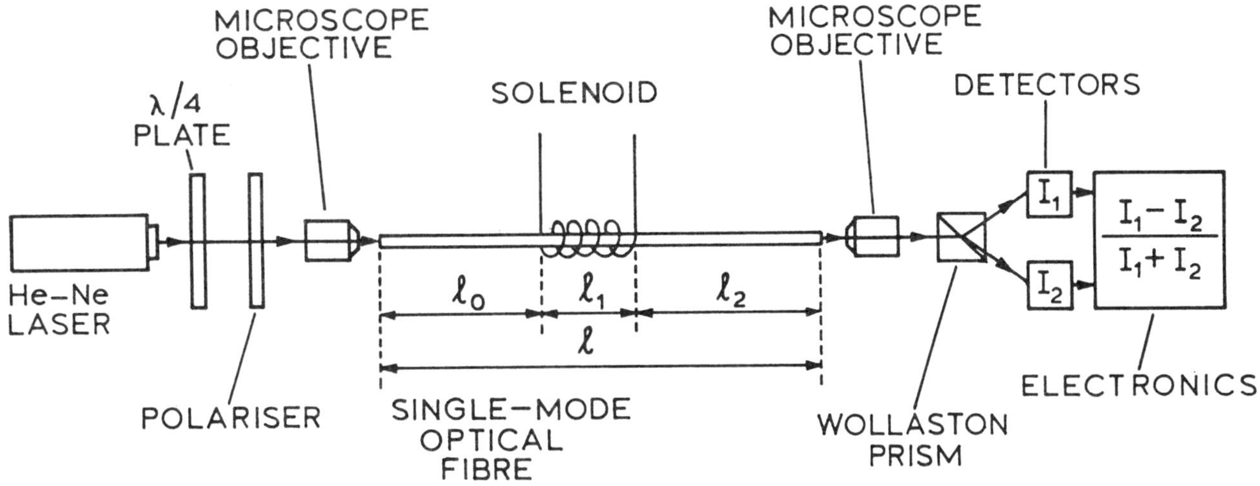

Fig. 1. Experimental apparatus.

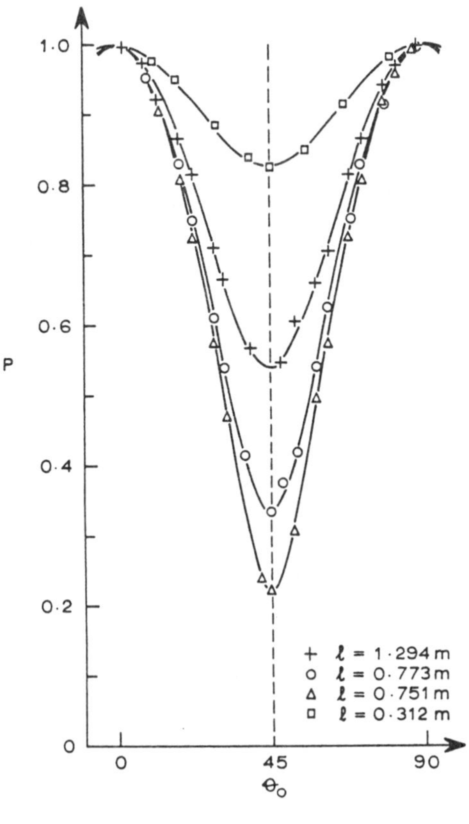

Fig. 2. Variation of intensity ratio P with input polarization angle θ_0.

B. Linear Retarder Model

The form of the experimental data indicates that the polarization properties of the single-mode optical fiber are similar to those of a linear retarder.

If x and y are the privileged directions of a linear retarder of retardation δ rad, the incident electric field components (E_xE_y) are modified to $(E_x'E_y')$ by the retarder, where, using Jones calculus,[6]

$$\begin{pmatrix} E_x' \\ E_y' \end{pmatrix} = \begin{bmatrix} \exp(j\delta/2) & 0 \\ 0 & \exp(-j\delta/2) \end{bmatrix} \begin{pmatrix} E_x \\ E_y \end{pmatrix}. \tag{1}$$

If the retarder is illuminated with linearly polarized light with amplitude $E_0 \exp(j\omega t)$ and polarization direction at an angle θ_0 to the x axis, and the light emerging from the retarder is monitored in two orthogonal directions X,Y (in the x,y plane) where X makes an angle θ_1 with x,

$$\begin{pmatrix} E_X \\ E_Y \end{pmatrix} = \begin{pmatrix} \cos\theta_1 & \sin\theta_1 \\ -\sin\theta_1 & \cos\theta_1 \end{pmatrix} \times \begin{bmatrix} \exp(j\theta/2) & 0 \\ 0 & \exp(-j\delta/2) \end{bmatrix} \begin{pmatrix} \cos\theta_0 \\ \sin\theta_0 \end{pmatrix} E_0 \exp(j\omega t). \tag{2}$$

In practice it is not the electric field components but the intensity which is monitored. This is proportional to the square modulus of the electric field components. Multiplying Eq. (2) by its complex conjugate yields

$$\begin{pmatrix} |E_X|^2 \\ |E_Y|^2 \end{pmatrix} = \begin{pmatrix} \cos^2\theta_1 \cos^2\theta_0 + \sin^2\theta_1 \sin^2\theta_0 + \tfrac{1}{2}\sin 2\theta_1 \sin 2\theta_0 \cos\delta \\ \sin^2\theta_1 \cos^2\theta_0 + \cos^2\theta_1 \sin^2\theta_0 - \tfrac{1}{2}\sin 2\theta_1 \sin 2\theta_0 \cos\delta \end{pmatrix} E_0^2. \tag{3}$$

The electronics used in this experiment effectively compute the function

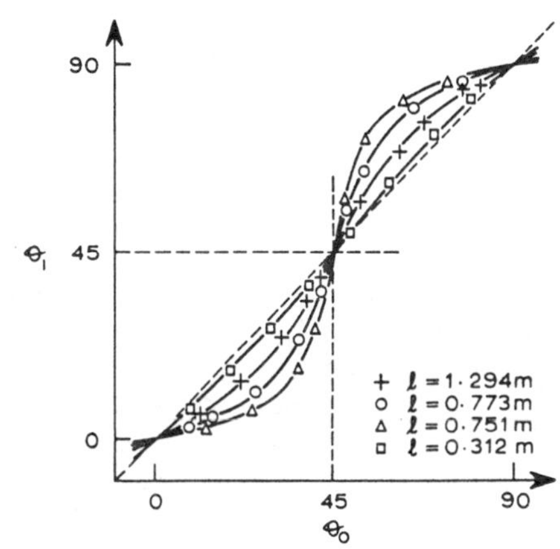

Fig. 3. Variation of output polarization angle θ_1 with input polarization angle θ_0.

$$\Gamma = (|E_X|^2 - |E_Y|^2)/(|E_X|^2 + |E_Y|^2). \tag{4}$$

Substituting using Eq. (3) yields

$$\Gamma = \cos 2\theta_1 \cos 2\theta_0 + \sin 2\theta_1 \sin 2\theta_0 \cos\delta. \tag{5}$$

For a given value of θ_0, Γ will be a maximum when $(\partial\Gamma/\partial\theta_1) = 0$. This leads to the condition

$$\tan 2\theta_1 = \tan 2\theta_0 \cos\delta. \tag{6}$$

The maximum value of Γ is denoted by P, where using Eqs. (5) and (6)

$$P = (\cos^2 2\theta_0 + \sin^2 2\theta_0 \cos^2\delta)^{1/2}. \tag{7}$$

Equations (6) and (7) relate P, θ_0, and θ_1 for the linear retarder model. The maximum and minimum of P are given by the conditions

$$\text{Maxima:}\quad P = 1 \text{ for } \theta_0 = m(\pi/2) \qquad m = \text{integer}$$
$$\text{Minima:}\quad P = |\cos\delta| \text{ for } \theta_0 = (2m+1)(\pi/4).$$

C. Comparison with Experimental Data

The definitions of θ_0, θ_1, and P used in Secs. III.A and III.B are equivalent, and hence the linear retarder model may be compared directly with the experimental data. For each length of fiber careful measurements were made of the minimum value of P (Fig. 2), and this value was taken as $|\cos\delta|$. Using Eqs. (6) and (7) theoretical curves of the variation of P and θ_1 with θ_2 were plotted for each value of $|\cos\delta|$, and these are shown in Figs. 2 and 3. It can be seen that the theoretical curves are an excellent fit to the experimental data which confirms the validity of the linear retarder model.

A single measurement of $\cos\delta$ cannot determine δ unambiguously because of the multivalued nature of the cosine function. If the phase retardation is a linear function of the fiber length,

$$\delta = \delta_s l, \tag{8}$$

where δ_s is the specific phase retardation in units of rad/m. Hence, by comparing the possible values of δ for several lengths of fiber it should be possible to find a value of δ_s. In practice it was not possible to find a unique value of δ_s. This discovery leads to one of two conclusions:

(1) The assumption that the specific phase retardation is a constant at all points along the fiber is incorrect.

(2) The intrinsic polarization properties are being swamped by induced polarization properties caused by pressure on the fiber, stress, or bending.

The first conclusion was supported by the work of Kapron and his fellow workers[2] who found that the specific phase retardation did vary substantially over short lengths of fiber (~2 m). However, their work was done on comparatively crude fiber, and the C.V.D. method of single-mode fiber preparation is expected to produce fibers with properties varying only over tens or hundreds of meters.

It was therefore concluded that at least part of the observed linear birefringence resulted from the effect of external stress of one type or another. In particular, it was found that the pressure caused by the clamps in the positioning devices at the ends of the fiber induced substantial birefringence.

In order to estimate this pressure induced linear birefringence, known pressures were applied to the fiber, and $\cos\delta$ was measured for each pressure. The variation of the specific linear retardation δ_s with pressure p was found to be linear for the moderate range of pressures applied, and $d\delta_s/dp = 4.07 \times 10^{-3}$ rad N^{-1} m.

This corresponds to a stress-optic coefficient[2] of 8.72×10^{-10} m^2 N^{-1}. (The stress-optic coefficient is defined as the rate at which the difference in the refractive indices of the retarder in its privileged directions changes with applied pressure.)

Even after the pressure induced by the fixing clamps on the fiber had been substantially reduced, it still appeared that the remaining birefringence was due at least in part to external effects acting on the fiber, and no accurate estimate of the specific retardation coefficient could be obtained directly.

IV Magnetooptic Properties

A. Superposition of the Faraday Effect and Linear Birefringence

The Faraday effect is a well known optical phenomenon. If an external magnetic field **H** is applied to a suitable medium such that there is a field component parallel to the direction of linearly polarized light in the medium, it is found that the direction of polarization of the emergent light has been rotated through an angle F, such that

$$F = V' \int_L \mathbf{H} \cdot d\mathbf{l}, \tag{9}$$

where L is the path traversed by the light in the medium and V' is a constant, which is a property of the medium, known as the Verdet constant. The Faraday effect is essentially a diamagnetic phenomenon, and consequently all dielectric materials exhibit this induced circular birefringence to some extent. In particular, it is found that glasses and silica, the materials from which optical fibers are constructed, have readily measurable Verdet constants, and so it is to be expected that fibers will exhibit the Faraday effect.

Hence if a longitudinal magnetic field is applied to a typical fiber with intrinsic linear birefringence there will be a superposition of the two birefringence effects, and, using Jones calculus[6] and the results of Ref. 7, the resultant polarization modification can be written in the form

$$\begin{pmatrix} E_x \\ E_y \end{pmatrix}_z = \begin{pmatrix} A & -B \\ B & A^* \end{pmatrix} \begin{pmatrix} E_x \\ E_y \end{pmatrix}_0, \tag{10}$$

where

$$A = \cos(\phi/2) + j\cos\chi \sin(\phi/2), \tag{11}$$

$$B = \sin\chi \sin(\phi/2), \tag{12}$$

$$(\phi/2)^2 = (\delta/2)^2 + F^2, \tag{13}$$

$$\tan\chi = F/(\delta/2). \tag{14}$$

F and δ are the Faraday rotation and phase retardation, respectively, of a fiber of length z. The x and y axes are chosen parallel to the privileged directions of the linear retarder as before.

If a length l_1 of fiber in the center of a fiber of length l is subjected to a longitudinal magnetic field, as shown in Fig. 1, the change in the polarization of the light may be written as

$$\begin{pmatrix} E_x \\ E_y \end{pmatrix}_l = \begin{bmatrix} \exp(j\delta_2/2) & 0 \\ 0 & \exp(-j\delta_2/2) \end{bmatrix} \begin{pmatrix} A & -B \\ B & A^* \end{pmatrix} \times \begin{bmatrix} \exp(j\delta_0/2) & 0 \\ 0 & \exp(-j\delta_0/2) \end{bmatrix} \begin{pmatrix} E_x \\ E_y \end{pmatrix}_0, \quad (15)$$

where δ_i is the phase retardation of a length l_i of fiber and A and B apply to a length l_1 of fiber.

In order to simplify the analysis of the output polarization state it is convenient to choose the input polarization such that

$$\begin{pmatrix} E_x \\ E_y \end{pmatrix}_0 = \begin{bmatrix} 0 \\ E_0 \exp(j\omega t) \end{bmatrix}. \quad (16)$$

This means that if no magnetic field is applied, the light emerging from the fiber is still completely linearly polarized in the same direction. Furthermore, it means that the final 2×2 matrix in Eq. (15) does not produce any change in the relative phase of the two field components and may be omitted.

Expanding Eq. (15) using Eq. (16) yields

$$\begin{pmatrix} E_x \\ E_y \end{pmatrix}_l = \begin{Bmatrix} -BE_0 \exp[j(\omega t + \delta_2/2)] \\ |A| E_0 \exp[j(\omega t + \delta_2/2 - \eta)] \end{Bmatrix},$$

where $\eta = \tan^{-1}[\cos\chi \tan(\phi/2)]$.

The detector is now rotated so that it detects the two polarization directions corresponding to $E_1 = (E_x + E_y)/\sqrt{2}$; $E_2 = (E_x - E_y)/\sqrt{2}$ and electronically produces an output proportional to $T = (|E_1|^2 - |E_2|^2)/(|E_1|^2 + |E_2|^2)$. Hence, after substitution

$$T = -2|A|B \cos(\delta_2 + \eta). \quad (17)$$

It can be seen that $|A|$, B, and η are all functions of the length of fiber in the magnetic field. Hence, if the magnetic field parameters (coil current, field length, and permeability) remain constant and $l_2 = (\delta_2/\delta_s)$ is varied, there will be a definite sinusoidal variation in the output T with period δ_s. It should be further noted that any induced birefringence caused by a pressure effect at the fiber's exit end-face will not change the character of this variation but may displace it a little since δ_2 will be made up of the intrinsic birefringence $\delta_s l_2$ and a constant pressure induced birefringence δ_p.

B. Experimental Measurements

A single-mode fiber was set up in the configuration described above, and the value of T was recorded for l_2 in the 2–45-cm range. (These figures represented the lengths that could be handled practically in the laboratory.) The length of fiber within the magnetic field l_1 was 0.295 m, and the field was 4.173×10^3 A m^{-1} in every case. It was found that over this length the output T remained a constant to within the experimental accuracy, the value being $T = (1.12 \pm 0.01) \times 10^{-2}$.

Neglecting any possible contribution from pressure induced birefringence, this puts an upper limit on the value of the specific birefringence coefficient: $\delta_s < 0.44$ rad m^{-1}; that is, a phase delay of 2π rad in >14.28 m.

In the approximation $F \gg (\delta_1/2)$, which is valid for the field strengths and lengths of fiber used, Eq. (17) reduces to

$$T = \sin 2F, \quad (18)$$

where F is the Faraday rotation angle defined in Eq. (9). Using this equation, the Verdet constant of the fiber material is found to be $V' = 1.56 \times 10^{-2}$ min A^{-1}. This compares well with the Verdet constant of fused silica which is 1.61×10^{-2} min A^{-1} at a wavelength of 632.8 nm.

The fact that the detector output is of the form of Eq. (18) indicates that the fiber could be used quite easily for magnetic field measurement. This application if discussed in Sec. VI.

V. Electrooptic Properties

Various materials exhibit polarization modification effects in the presence of transverse and longitudinal electric fields. However, these effects are normally associated with definite crystalline structure rather than with diamagnetism (cf. Faraday effect), and as the likelihood of any such structure in a glass or fused material is remote, it is probably that the fiber will not exhibit any detectable electrooptic effect.

The fiber was subjected to transverse electric fields of up to 2.5×10^4 V m^{-1} over ~15-cm length, and no polarization modification was detected. Also, longitudinal electric fields of up to 2.5×10^4 V m^{-1} over ~1-cm lengths of fiber were produced. These also failed to produce any measurable polarization modification.

VI. Experimental Current Measurement Device

Various devices have been described in the literature that make use of the Faraday effect in glassy materials to measure the magnetic field close to a conductor carrying a high current.[8,9] The current may be deduced from the Faraday rotation angle and the geometry of the device. These devices have the disadvantage that the free path propagating light beam used to interrogate the remote glass transducer can be attenuated or angularly shifted by adverse atmospheric conditions.[10] A current measurement device using single mode optical fiber both as a light guiding and transducing medium would not suffer from these difficulties. Such a device was set up in the laboratory, and a schematic drawing of it is shown in Fig. 4. Four turns of single-mode fiber were wound around a 12-cm radius former mounted on a standard 14-cm diam (400-kV) busbar. Currents in the 0–1000-A range rms at 50 Hz could be passed through the busbar. A cylindrical busbar produces a circumferential magnetic field when a current is passed through it so the fiber experiences a longitudinal magnetic field. The reason for the use of the former, rather than winding the fiber directly onto the busbar, is that the critical bend radius below which bending losses become severe is $\simeq 10$ cm for this fiber. The total length of fiber used was about 6 m.

The incident direction of linear polarization of the light entering the fiber was adjusted so that the output

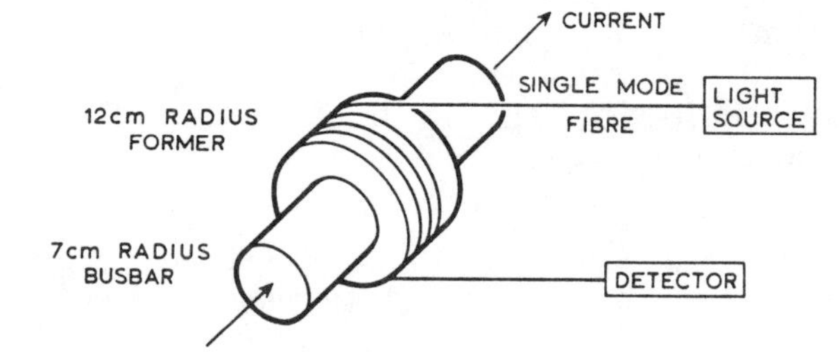

Fig. 4. Schematic diagram of a current measurement device using single-mode optical fiber.

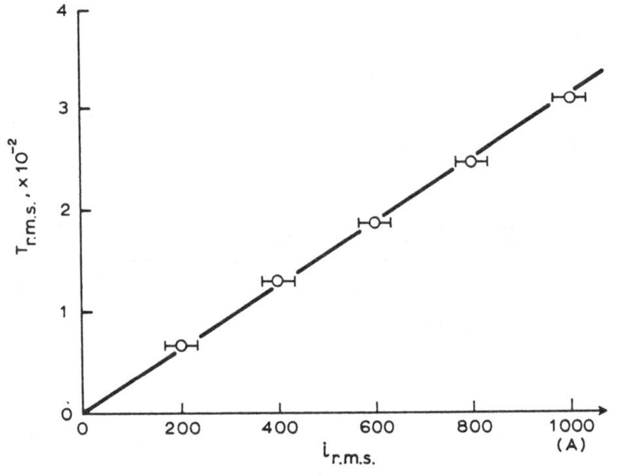

Fig. 5. Graph of detector output against busbar current.

polarization was linear when no current was flowing in the busbar. Then, the detector unit was rotated so that its output T (Sect. IV) was zero. The rms output was then measured for a variety of busbar currents, and the results obtained are shown in Fig. 5. The results indicated a linear variation of detector output T, with rms busbar current i, such that $T(\text{rms}) = (3.13 \pm 0.05) \times 10^{-5}\, i(\text{rms})$.

Using Eq. (18), the approximation for the output when no linear birefringence is present, a value of the Verdet constant of the material is found: $V' = (1.31 \pm 0.05) \times 10^{-3}$ min A^{-1}. This is lower than the previously obtained value, and this fact indicates that there is some linear birefringence. However, it is probable that part of this is induced by bending. The polarization rotation sensitivity is $(F/i) = 1.57 \times 10^{-5}$ rad A^{-1}.

Since $\sin x = x$ to within 1% for $x < 14°$ and to within 5% for $x < 31°$, it follows that currents up to 15 kA may be measured linearly to within 1% and up to 34 kA to within 5%.

The noise on the detector output due to the electronics was equivalent to a current of 9 A. Vibration, transmitted by the mechanical support given to the fiber, raised this figure to ~25 A. However, neither figure is believed to be representative of what would be possible in an operational system.

There will be an optimum number of turns of fiber for best sensitivity within minimum noise. However, at present no firm estimates can be made of how many turns this will be, but it will be large (>10).

VII. Conclusions

Experiments have been carried out to determine the polarization properties of a sample of single-mode optical fiber. It has been found that the specific linear birefringence of the fiber δ_s is less than 0.44 rad m^{-1} and that the main source of the observed linear birefringence is external pressure on the fiber. The stress optic coefficient was estimated to be 8.72×10^{-10} N^{-1} m^2.

When a longitudinal magnetic field was applied to the fiber a magnetooptic (Faraday) rotation of the direction of polarization was found to occur, and the Verdet constant was found to be 1.56×10^{-2} min A^{-1}, which is close to the accepted value for fused silica at the wavelength of the light used (632.8 nm). No field induced polarization effects were observed when transverse and longitudinal electric fields were applied to the fiber.

A simple device for the measurement of current was set up using four turns of fiber wound around a current carrying busbar. The output sensitivity of this device was 1.57×10^{-5} rad A^{-1}. Noise in the electronics limited the minimum measurable current to 9 A, and vibrational effects, which could occur under certain circumstance, introduced variations in the output equivalent to ~25 A. None of these figures is fundamental and may be altered by changing the length of fiber around the busbar and adjusting the electronics and fiber supporting system.

This work was carried out at the Central Electricity Research Laboratories, and the paper is published by permission of the Central Electricity Generating Board.

References

1. L. G. Cohen, Bell Syst. Tech. J. **50**, 23 (1971).
2. F. P. Kapron, N. F. Borelli, and D. B. Keck, IEEE J. Quantum Electron. **QE-8**, 222 (1972).
3. A. Papp and H. Harms, Appl. Opt. **14**, 2406 (1975).
4. H. Harms, A. Papp, and K. Kempter, Appl. Opt. **15**, 799 (1976).
5. W. A. Gambling, D. N. Payne, D. N. Hammond, and S. R. Norman, Proc. IEE **123**, 570 (1976).
6. R. C. Jones, J. Opt. Soc. Am. **31**, 488 (1941).
7. W. J. Tabor and F. S. Chen, J. Appl. Phys. **40**, 2760 (1969).
8. A. J. Rogers, Proc. IEE **120**, 261 (1973).
9. S. Saito, J. Hamosaki, Y. Fujü, K. Yokoyama, and Y. Ohno, IEEE J. Quantum Electron. **QE-3**, 589 (1967).
10. H. Hodara, Proc. Inst. Electr. Electron. Eng. **54**, 368 (1966).

An Improved Method for High Reflectivity Ellipsometry Based on a New Polarization Modulation Technique*

S. N. Jasperson† and S. E. Schnatterly
Palmer Physical Laboratory, Princeton University, Princeton, New Jersey 08540

(Received 10 October 1968; and in final form, 23 January 1969)

We report the development of a new ellipsometry technique which is especially well suited to the measurement of the dielectric constants of high reflectivity metals. Modulation methods are used to permit the measurement of a differential reflectance parameter which does not decrease in sensitivity in the high reflectivity limit as do the usual reflectance parameters measured with standard techniques. In addition the ellipsometer is faster than previous methods, being capable of recording the relevant data in a few minutes as the wavelength is swept continuously over a range of several thousand angstroms. The precision of the dielectric constants so measured is 1% in most cases. The polarization modulation technique, upon which the ellipsometer is based, is presented in detail, with emphasis on its wider applicability to any problem involving the study of dichroic phenomena.

INTRODUCTION

In recent years, optical measurements have contributed much information concerning the electronic structure of solids. Standard techniques when applied to the study of metals in the high reflectivity (Drude) region, however, have in general been of limited success in providing detailed information of the electronic behavior in terms of experimentally determined dielectric constants of these materials. The fundamental reason is the following: All previous techniques involve at least one measurement of the reflectance (R) or the ratio (ρ^2) of reflectances for light polarized parallel and perpendicular to the plane of incidence, and in the Drude region where the reflectance at all angles of incidence is nearly equal to one, R and ρ^2 become relatively insensitive measures of the dielectric constants.

Several modern techniques have been developed in an effort to at least minimize this difficulty. In 1955, Beattie and Conn[1] demonstrated that, in the high reflectivity limit, measurements of the ellipsometric parameters ρ and Δ (where Δ is the relative phase angle and ρ is defined above) were superior to two or more measurements of the reflectances R_s, R_p, or ρ^2 in uniquely determining the dielectric constants. Subsequently, they devised a new technique for extracting these parameters in terms of certain predesignated intensity measurements.[2] More recently, Archer[3] and Smith and Stromberg[4] have used a modern version of the original Drude null intensity method for determining these same parameters with a great deal of precision.

One disadvantage of these and nearly all existing techniques is their slowness. An interval of several minutes is required at every datum point, during which time several manual adjustments are performed to extract the desired information. This is undesirable since surface contamination can occur rapidly, even in a good vacuum, and one would like to be able to make measurements on atomically clean surfaces. An improvement in this respect has been the application of modulation methods to the development of a number of modified ellipsometric techniques. Although several simply use the modulation to facilitate manual adjustments,[5–7] the systems developed by Takasaki[8,9] and one described briefly by Winterbottom[10] have the advantage of being completely automated, with the result that measurements can be made electronically in a matter of seconds, rather than minutes. Presumably this would make possible the recording of data continuous in the wavelength, an impossibility with all other methods.

In the present paper, we report an approach to the high reflectivity ellipsometry problem which avoids the traditional difficulties through the use of a new method of polarization modulation. In brief, the polarization of a beam of light is continuously and periodically varied in such a way that information concerning the properties of a reflecting surface may be obtained by synchronously detecting and recording the magnitudes of the subsequent intensity modulations of the reflected and suitably analyzed beam. The system is relatively fast and perfectly suited to obtaining data continuous in the wavelength. To illustrate, this system has been used to record a complete set of data of 10 Å resolution over the visible spectrum in 2 h (approximately equal to the time required for one monolayer of impurities with sticking probability of unity to collect on the sample in a vacuum of 10^{-10} Torr)

* This work was supported in part by grants from the National Science Foundation and the Research Corporation.

† Present address: Department of Physics, University of Illinois, Urbana, Illinois.

[1] J. R. Beattie and G. K. T. Conn, Phil. Mag. **46**, 222 (1955).
[2] J. R. Beattie, Phil. Mag. **46**, 235 (1955).
[3] R. J. Archer, J. Opt. Soc. Amer. **52**, 970 (1962).
[4] L. E. Smith and R. R. Stromberg, J. Opt. Soc. Amer. **56**, 1539 (1966).
[5] J. F. Archard, P. L. Clegg, and A. M. Taylor, Proc. Phys. Soc. (London) **B65**, 758 (1952).
[6] G. K. T. Conn and G. K. Eaton, J. Opt. Soc. Amer. **44**, 546 (1954).
[7] T. E. Faber and N. V. Smith, J. Opt. Soc. Amer. **58**, 102 (1968).
[8] H. Takasaki, J. Opt. Soc. Amer. **51**, 463 (1961).
[9] H. Takasaki, J. Opt. Soc. Amer. **56**, 557 (1966); Appl. Opt. **5**, 759 (1966).
[10] A. B. Winterbottom, *Ellipsometry in the Measurement of Surfaces and Thin Films* E. Passaglia, R. R. Stromberg, and J. Kruger, Eds. (National Bureau of Standards, Washington, D. C., 1963).

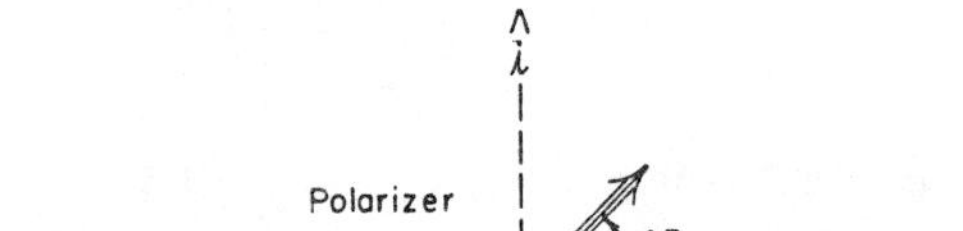

FIG. 1. Polarization modulator.

The primary advantage of this technique with respect to precision is that it enables the direct measurement of the quantity $(1-\rho^2)/(1+\rho^2)$ rather than ρ or ρ^2, in addition to a measurement of the phase angle Δ. This is of significance because the differential quantity does not become a less sensitive measure of the dielectric constants in the high reflectivity limit, as does ρ itself. To be specific, in those situations where $\rho \geq 0.9$, an uncertainty of 0.3% in $(1-\rho^2)/(1+\rho^2)$ results in less uncertainty in the dielectric constants than an uncertainty of 0.04% in ρ. For $\rho < 0.9$, the relative sensitivity of the new technique falls below that of the better static procedures, but its speed and scanning ability still make it useful for quickly locating interesting structure within a broad wavelength range.

The details of the experimental apparatus follow. In order to emphasize the fact that the technique of polarization modulation is useful in many experiments other than ellipsometry, that part of the system is discussed separately in the next section. A description of the complete ellipsometry system, including an excerpt from experimental data, is presented in Sec. II.

I. POLARIZATION MODULATION

The modulation of the polarization of a beam of light is accomplished by the high frequency (50 kHz) modulation of the optical properties of a stationary, normally isotropic block of fused quartz through which the light passes. Specifically, a sinusoidal, uniaxial strain induced along the $\hat{j}$ axis of a fused quartz block (Fig. 1) causes the phase velocity of light polarized along the $\hat{i}$ and $\hat{j}$ axes to differ in a periodic manner. If the incident light is linearly polarized so as to have nonzero components along both axes, the transmitted beam will consequently have the state of its polarization modulated.

The strain is induced by a piezoelectric transducer cemented to one end of the fused quartz block. The transducer is a $10\times20\times50$ mm block of crystal quartz, $-18°$ X cut, which resonates in a uniaxial mode directed along the $\hat{j}$ axis when a 50 kHz electric field is applied normally to the $\hat{i}$–$\hat{j}$ plane via conducting electrodes deposited on opposite faces of the crystal quartz. The fused quartz block has the same end dimensions, and its length is such that it, too, has a 50 kHz fundamental longitudinal resonance. The composite system consisting of the two blocks glued end-to-end rests on two knife edges milled into a Lucite support, one located under the center of each block. Since this constrains both centers to remain stationary, the natural resonant frequency of the unit is also 50 kHz. Moreover, the cement holding the blocks together reduces the Q of the resonant system little from that of the individual components because the boundary between the blocks is at a node in the standing strain wave.

During the initial experimentation, the modulator was always operated with its $\hat{j}$ axis horizontal, so that it could simply rest freely on the two knife edges, minimizing the power loss to the supports. Later, however, it was necessary to fasten the unit to the knife edges with epoxy to permit nonhorizontal orientations about the light beam. The presence of the epoxy decreased the resonant Q to half its original value. This decrease, however, was offset by the discovery that acoustic power, radiating directionally along the $\hat{j}$ axis from the ends of the modulator, could be reflected back into the unit, more than compensating for the power loss to the epoxy. Subsequently, metal plates, positioned so as to cause constructive interference, were mounted near both ends of the modulator.

The change in the index of refraction is a linear function of the strain,[11] which, in turn, is a sinusoidal function of the time. The relative phase difference between $\hat{i}$ and $\hat{j}$ components of the incident plane polarized wave is therefore

$$\delta = (2\pi d/\lambda)P \sin(\omega t), \qquad (1)$$

where d is the thickness of the block, λ is the wavelength of the light, and P is a number proportional to the peak strain amplitude. (In this work, the wavelength dependence of P was found negligible.) The strain amplitude is itself proportional to the product of the Q of the resonant system and the voltage amplitude of the applied ac signal. Because of the high Q (experimentally measured to be 20 000), only relatively small voltages are required to produce significant phase shifts. For example, a 20 V peak-to-peak signal produces a maximum relative phase shift of π for 4000 Å light.

For simplicity of notation, Eq. (1) is rewritten as

$$\delta = A \sin(\omega t), \qquad (2)$$

where A, the relative phase amplitude, is assumed to be effectively a function only of the ratio (V/λ), V being the peak-to-peak voltage applied to the transducer.

[11] M. Born and E. Wolf, *Principles of Optics* (The MacMillan Co. New York, 1964), p. 705.

For light initially polarized at 45° to the $\hat{\imath}$ and $\hat{\jmath}$ axes (Fig. 1), the amplitude of the incident wave may be written as

$$\mathbf{E}=E_0(\hat{\imath}+\hat{\jmath}), \tag{3}$$

which after passing through the modulator becomes

$$\mathbf{E}=E_0(\hat{\imath}+e^{i\delta}\hat{\jmath}), \tag{4}$$

where only the relative phase between components has been retained.

The effect of the modulation becomes clear when the transmitted beam is suitably analyzed and detected. If a linear analyzer of polarizing strength α (where α is defined as the difference between transmittances parallel and perpendicular to the transmission axis divided by their sum) is placed in the light beam following the modulator and oriented so that its transmission axis is parallel to that of the initial polarizer, the transmitted intensity is

$$I_L=I_0[1+\alpha\cdot\cos(\delta)], \tag{5}$$

If, on the other hand, a circular analyzer follows the modulator, the transmitted intensity becomes

$$I_C=I_0[1+\alpha\cdot\sin(\delta)], \tag{6}$$

where α is now defined in terms of transmittances for left and right circularly polarized light. Here, and throughout this paper, I_0 is defined to be an arbitrary constant of proportionality which contains all the time independent parameters of the optical and electronic detection system.

Since δ is itself a sinusoid, both $\sin(\delta)$ and $\cos(\delta)$ are seen to possess a complicated harmonic content which depends on the relative phase amplitude A. The strength of each harmonic is conveniently expressed in terms of Bessel functions of argument A according to the following expansions[12]:

$$\begin{aligned}\sin[A\,\sin(\omega t)]&=2J_1(A)\sin(\omega t)+2J_3(A)\sin(3\omega t)+\cdots\\ \cos[A\cdot\sin(\omega t)]&=J_0(A)+2J_2(A)\cos(2\omega t)+2J_4(A)\cos(4\omega t)+\cdots.\end{aligned} \tag{7}$$

Only terms of frequency 2ω and lower are of further interest. The properties of these expansions of relevance here are the following: (1) $J_1(A)$ assumes its maximum value for $A=105°$ (approximately $\frac{1}{4}\lambda$ retardation), whereas $J_2(A)$ is maximum for $A=175°$ (approximately $\frac{1}{2}\lambda$ retardation); and (2) the time average of $\sin(\delta)$ is strictly zero, whereas that of $\cos(\delta)$ is equal to $J_0(A)$, which is in general not zero.

By combining Eqs. (5) and (6) with Eq. (7), we may rewrite the expressions for the intensities as

$$\begin{aligned}I_L&=I_0\{1+\alpha[J_0(A)+2J_2(A)\cos(2\omega t)]\},\\ I_C&=I_0\{1+\alpha[2J_1(A)\sin(\omega t)]\}.\end{aligned} \tag{8}$$

[12] Andrew Gray and G. B. Billings, *A Treatise on Bessel Functions* (MacMillan and Co., New York, 1895).

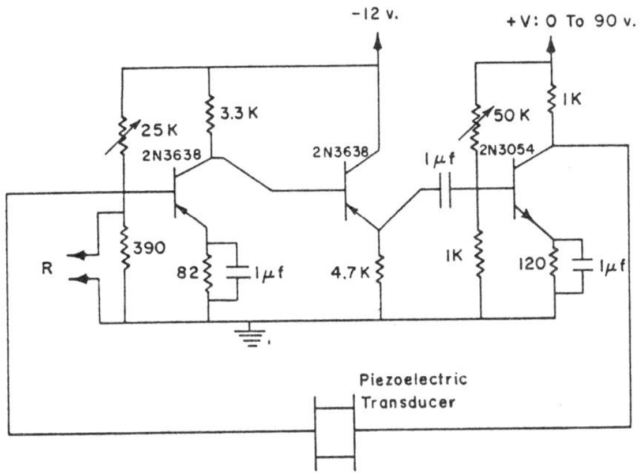

FIG. 2. Circuit diagram of oscillator used to drive the polarization modulator. The lock-in amplifier reference signal is taken from point R.

On the basis of these results, the way in which this modulation technique is applicable to the general study of dichroism can now be made explicit. If a beam of light with a suitable degree of modulation is transmitted through or reflected from a properly oriented linearly dichroic material, the resulting intensity will possess a 100 kHz frequency component, the strength of which is proportional to the amount of linear dichroism. If the light beam interacts similarly with a circularly dichroic material, the intensity will possess a 50 kHz component proportional in strength to the degree of circular dichroism.

Extending the analysis, one can verify that placement of a linear analyzer in the transmitted or reflected beam enables the extraction of information concerning the relative phase shift between components of orthogonal polarization. Thus, it is also possible to measure a second, independent parameter at each wavelength, as is required to specify directly the interaction between the light beam and the dichroic material.

Because polarization modulation is an ac technique which can subsequently exploit the advantages of phase sensitive detection, it is particularly well suited to the accurate measurement of small effects. To enumerate some of the possible applications of this technique in areas other than ellipsometry, it has been used in this laboratory to monitor the optical pumping of F centers in alkali halides (wherein transmission circular dichroism is measured) and in connection with magnetoreflectance experiments on metals (wherein the normal incidence paramagnetic Kerr effect is measured).

To return to the details of instrumentation, the polarization modulation unit is driven by a crystal controlled oscillator in which the piezoelectric transducer is itself the crystal. The transducer is a high impedance electrical element, except at its resonant frequencies, where it presents a purely resistive load of relatively small magnitude. When placed in the feedback loop of a simple ampli-

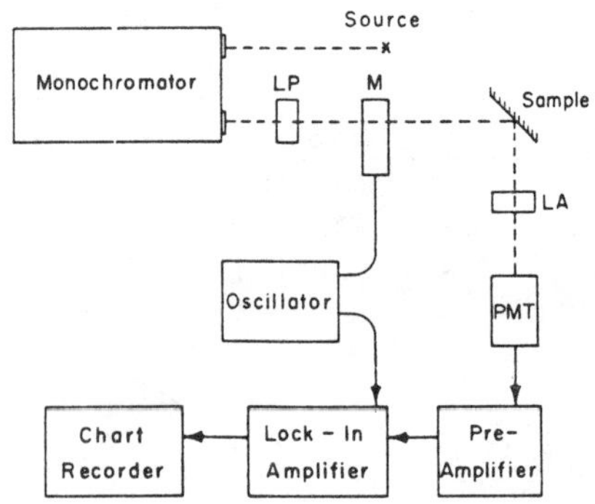

FIG. 3. Schematic diagram of the ellipsometry system. LP and LA are linear polarizers, M is the polarization modulator, and PMT is the photomultiplier tube.

fier, as shown in Fig. 2, it acts to phase shift and block all frequency components not on resonance. Because the modulator is such a stable high Q device, it is insensitive to the properties of the driving circuit, and the only requirements concerning the amplifier design are that its gain at 50 kHz be sufficient for oscillation and that the phase shift of its output be a multiple of 2π relative to the input.

The two variable resistors in the base circuits of the first and third stages allow the waveform at the output to be shaped to approximate a square wave. It does not matter to the optical detection system that higher harmonics are present in the strain; only 50 kHz strains contribute to intensity modulations detected at 50 and 100 kHz. The reference signal for the lock-in detector (see Fig. 3) may be taken from point R. Sufficient amounts of both 50 and 100 kHz components are present across this resistor to drive the reference channel without need for further amplification.

The square wave which drives the piezoelectric transducer has a magnitude V essentially equal to the voltage operating the last stage of the oscillator. Hence, by changing the voltage supply output, one may conveniently adjust the level of polarization modulation, since changes in V at fixed wavelength result in changes in $A=A(V/\lambda)$.

For certain applications it is necessary to be able to vary V as a continuous function of the wavelength so that V/λ, thus A, can be maintained at a constant value independent of the wavelength. Were this not done, the changing level of modulation would result in an additional wavelength dependent contribution to the signal which would make accurate quantitative work most difficult. In the present work, a voltage directly proportional to the wavelength is generated by passing a fixed current through a potentiometer which is geared to the wavelength drive of a grating monochromator. This voltage is then used to program the voltage supply which drives the output stage of the oscillator. Consequently, the voltage driving the modulator may be made to change with any desired rate vs wavelength by suitably adjusting the current through the potentiometer. The correct initial voltage is obtained by adjusting the resistance of the potentiometer once the correct current has been determined.

II. ELLIPSOMETRY

We now consider the specific application of polarizatio modulation to ellipsometry. With other methods, it i customary to measure the ellipsometric parameters (ρ,Δ) defined to be

$$\rho=r_p/r_s,\qquad \Delta=\delta_s-\delta_p, \tag{9}$$

where $r_s(r_p)$ is the reflectance amplitude and $\delta_s(\delta_p)$ th reflectance phase shift for light polarized perpendicula (parallel) to the plane of incidence. As defined above, ρ i a number between 0 and 1, and Δ is a positive angl lying in the first or second quadrant and equal to 180° a normal incidence. If it is assumed that the response of th reflecting medium can be specified by a local, positio independent dielectric function $\tilde{\epsilon}=\epsilon_1+i\epsilon_2$, it is a straight forward matter to express the theoretically interestin real and imaginary parts of $\tilde{\epsilon}$ in terms of the experimenta parameters and the angle of incidence, θ.[13]

In this work, however, it is not (ρ,Δ) which are measured but either of the pair of related quantities (N,S) or (N,C) where

$$N=(r_s^2-r_p^2)/(r_s^2+r_p^2)=(1-\rho^2)/(1+\rho^2),\qquad S=\sin(\Delta),\qquad C=\cos(\Delta). \tag{10}$$

The usual expressions for ϵ_1 and ϵ_2, now modified so as t be functions of these new parameters, are

$$\epsilon_1=\sin^2(\theta)+\sin^2(\theta)\tan^2(\theta)\frac{N^2-(1-N^2)S}{[1\mp(1-N^2)^{\frac{1}{2}}(1-S^2)^{\frac{1}{2}}]^2},\qquad \epsilon_2=2\sin^2(\theta)\tan^2(\theta)\frac{NS(1-N^2)^{\frac{1}{2}}}{[1\mp(1-N^2)^{\frac{1}{2}}(1-S^2)^{\frac{1}{2}}]^2} \tag{11}$$

$$\epsilon_1=\sin^2(\theta)+\sin^2\theta\tan^2(\theta)\frac{N-(1-N^2)(1-C^2)}{[1+C(1-N^2)^{\frac{1}{2}}]^2},\qquad \epsilon_2=2\sin^2(\theta)\tan^2(\theta)\frac{N(1-N^2)^{\frac{1}{2}}(1-C^2)^{\frac{1}{2}}}{[1+C(1-N^2)^{\frac{1}{2}}]^2}, \tag{12}$$

where the upper (lower) sign in the first pair of equation applies for Δ in the second (first) quadrant.

The optical and electronic systems used to measure th quantities N, S, and C are shown in block form in Fig. 3 Light from a high pressure xenon arc lamp is made mono chromatic (10 Å half-width band pass) by a grating mono chromator; linearly polarized (LP); polarization modu lated (M); obliquely reflected by the sample; linearl analyzed (LA) or not, depending upon the paramete being measured; and detected by a photomultiplier tub (PMT). The subsequent ac component of the PMT outpu current is amplified and fed into a lock-in amplifier whic

[13] For example, see Ref. 11, p. 620.

demodulates the signal and records it on a 25 cm potentiometric chart recorder.

The specific information which can be extracted from the detected light intensity depends upon the relative orientations of the elements of the optical system. We consider only the two most useful cases:

Case a. With the modulator oriented such that its $\hat{j}$ axis is at an angle of 45° with respect to the plane of incidence of the sample, the 100 kHz component of the reflected intensity is proportional to N.

Case b. With the modulator oriented such that its $\hat{j}$ axis lies in the plane of incidence and the linear analyzer (LA) oriented such that its transmission axis is at 45° to the plane of incidence, the 50 kHz component of the detected intensity is proportional to S, and the 100 kHz component is proportional to C. In both cases, it is assumed that the transmission axis of polarizer LP is aligned at 45° to the $\hat{i}$ and $\hat{j}$ axes of the modulator.

Since, as the wavelength is scanned, the magnitudes of the ac signal components would ordinarily be proportional to the lamp's spectral output, the transmittance of the optical elements of the system and the sensitivity of the PMT, as well as to the properties of the reflecting surface, it is necessary to provide some means of eliminating the spurious contributions from the recorded signal. In this work, correct normalization is accomplished by a modified PMT voltage supply which maintains the PMT average output current at a constant value, independent of the light intensity reaching the PMT, provided that the level of operation is within the capabilities of the PMT and its voltage supply. The ac components of the PMT output current are therefore proportional to the corresponding ac components of the light intensity divided by the dc average intensity, thus effectively canceling out all wavelength dependent contributions other than that of the reflecting surface.

A schematic diagram of the PMT voltage supply circuit is shown in Fig. 4. In normal operation, an internal control circuit in the high voltage power supply functions so as to establish a null voltage across the input terminals of an operational amplifier.[14] If a null voltage does not exist, the operational amplifier turns the high voltage pass element either on or off in an attempt to reestablish point K as a virtual ground.

To adapt the supply to the new mode of operation, the internal reference resistors in one arm of the regulation bridge are replaced by external potentiometers R_1 and R_2, through which the reference current of 1 mA then must flow. The voltage generated across the anode resistor by the PMT output current is applied, via a voltage follower for impedance matching, to point L between R_1 and R_2.

[14] Power Supply Handbook, Kepco, Inc., Flushing, N. Y. 11352, 1965.

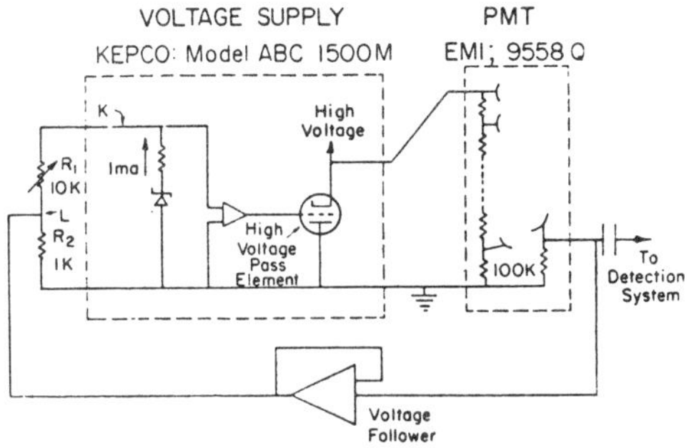

FIG. 4. Schematic diagram of the photomultiplier tube (PMT) regulation circuit which maintains the average PMT output current at a constant value, independent of the incident light level.

Point K can be a virtual ground only if the voltage applied to L is equal and opposite to the voltage drop across R_1 due to the reference current. The polarity of the feedback is such that the voltage supply will adjust the gain of the PMT until the PMT output current results in the correct voltage at L.

The speed of this system was experimentally determined by stepping the incident light intensity and observing the decay of the PMT current back to its preset value. The time constant of this decay was measured to be 0.02 sec, a response time fast enough to remove frequency components of less than 8 Hz, but slow enough to be completely insensitive to the 50 and 100 kHz signal components, which are capacitively coupled to the remainder of the detection system. Moreover, at a fixed wavelength the output signals remained constant to within $\frac{1}{2}\%$ as the voltage was driven through its effective range by a two orders of magnitude change in light level, indicating that changes in the ratio of ac to dc gain as a function of PMT voltage are not large enough to be of importance.

We are now in a position to express the magnitude of the detected signal in terms of the ellipsometric parameters of the reflecting surface. Starting with Eq. (4) as the representation for the light beam after having passed through the modulator and considering the effects of the remaining optical elements, it is a straightforward matter to obtain the following expressions for the light intensity reaching the PMT in each of the two cases mentioned above:

$$\begin{aligned}&\text{(case a)}, \; I_a = I_0\{1 + [(1-\rho^2)/(1+\rho^2)]\cos(\delta)\},\\ &\text{(case b)}, \; I_b = I_0\{1 + [2\rho/(1+\rho^2)]\cos(\delta-\Delta)\}.\end{aligned} \tag{13}$$

The detected and recorded signals are designated by the quantities n, s, and c, where each is essentially a measure of the corresponding upper case letters. Since the lock-in amplifier output is proportional to the ratio of the ac to dc components of the intensity, the definitions of these lower

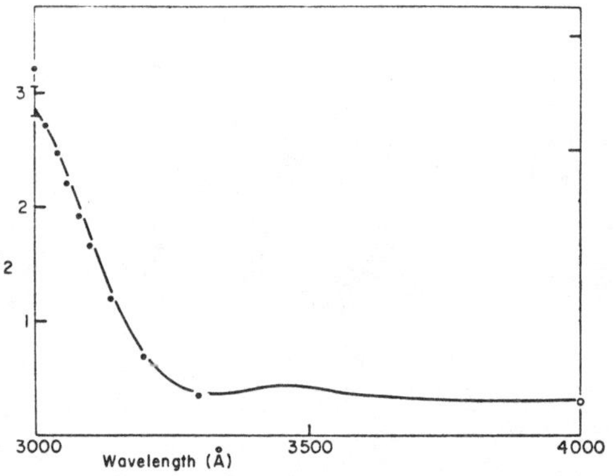

FIG. 5. ϵ_2 for silver in the range 3000 to 4000 Å calculated from data obtained with the new ellipsometer. Data obtained by Huebner (●) and Schulz (○) are shown for comparison.

case parameters are

$$n=I_0\frac{N\cdot J_2(A)}{1+N\cdot J_0(A)},$$
$$s=I_0\frac{(1-N^2)^{\frac{1}{2}}S\cdot J_1(A)}{1+(1-N^2)^{\frac{1}{2}}C\cdot J_0(A)}, \tag{14}$$
$$c=I_0\frac{(1-N^2)^{\frac{1}{2}}C\cdot J_2(A)}{1+(1-N^2)^{\frac{1}{2}}C\cdot J_0(A)}.$$

It is evident that these equations would simplify considerably if $J_0(A)$ were made to vanish. A fortuitous property of the Bessel functions is that at $A=138°$, where $J_0(A)=0$, both $J_1(A)$ and $J_2(A)$ are within 12% of their respective maximum values. Consequently, the same degree of modulation can be used in measuring both 50 and 100 kHz signals; the slight loss of signal level is more than compensated for by the comparative ease of data analysis and the fact that all measurements can be made with a single adjustment of the modulation level control circuit.

Calibration at any given wavelength is accomplished by comparing n, s, and c with their corresponding 100% values. To illustrate, the n measurement is calibrated by inserting a linear analyzer in the beam immediately preceding the sample with its transmission axis aligned perpendicularly to the plane of incidence. The sample then has no effect other than to deflect the light toward the PMT. The resulting 100 kHz measurement is

$$100\%\ n=I_0\cdot J_2(A) \tag{15}$$

[assuming that $J_0(A)=0$] and the ratio of these two n measurements is the ellipsometric parameter N.

Both phase measurements are calibrated by adding an additional phase to Δ, so that the angle argument becomes 90° in the case of an S measurement, and 0 or 180° in the case of C. This extra phase shift is introduced by a strained fused quartz block placed in the light beam between the modulator and the sample. The block is held in the jaws of a vise which is oriented so that the strain axis is either parallel or perpendicular to the plane of incidence. By carefully adjusting the pressure exerted by the vise, a maximum in the signal level can be found, which corresponds to the 100% value,

$$100\%\ s=I_0(1-N^2)^{\frac{1}{2}}J_1(A),$$
$$100\%\ c=I_0(1-N^2)^{\frac{1}{2}}J_2(A). \tag{16}$$

As in the case of N, the parameters S and C are simply the ratios of s and c to their respective 100% values.

The complete ellipsometric procedure is summarized as follows. In a preliminary set of measurements, the modulator level control circuit is adjusted so that A will be maintained at 138° throughout the wavelength range of interest. Then two successive scans over the same range are made in order to measure either the pair of quantities (n,s) or (n,c). The choice is dictated by whichever of the quantities S and $|C|$ is smaller, hence the more precise measure of Δ. After each scan the raw data are calibrated by measuring the appropriate 100% signal at one or more wavelengths, at which point the parameters (N,S) or (N,C) can be evaluated throughout the range. Finally ϵ_1 and ϵ_2 can be calculated by making the proper substitutions in Eqs. (11) or (12).

We have found the systematic errors in measurements of N, S, and C to be less than 1%, except in extreme situations where N falls below 0.01, and an approximate standard deviation for repeating values of N, S, and C to be 0.3%. Sources of error which were studied in the course of this work are summarized below:

(a) The fused quartz modulator block was found to possess a small static strain which introduced a spurious contribution to the phase measurement of approximately 2°. In a subsidiary measurement of S with no sample present, this spurious phase was determined to within a few minutes of arc at all wavelengths of interest, so that later data could be corrected accordingly. No other significant sources of spurious phase shifts were encountered.

(b) Corrections to Eqs. (13) and (14) must be made if the polarizing strength of the polarizers is different from one (i.e., $\alpha\leq0.996$, when 1% accuracy is desired). This is particularly important if α varies with wavelength, since the effect of such changes is to cause spurious structure in the detected signal of percentage magnitude equal to twice the percentage change in α. The condition $\alpha\neq1$ also makes necessary an additional correction term if the monochromator partially polarizes the initial beam.

(c) In preliminary measurements, it was found that the voltage required to maintain A equal to 138° was sufficiently linear in the wavelength between 3000 and 11 000 Å that perfect adjustment and stability of the modulator level control circuit would result in $J_0(A)\leq0.01$ over the entire range. This is a small enough departure from zero hat 1% measurements of the ellipsometric parameters can e obtained simply by ignoring terms in J_0 in Eqs. (14). n practice, however, the magnitude of J_0 was found to lrift an average of 0.02 in the course of 2000 Å scans. This drift was routinely compensated by taking duplicate neasurements at several discrete calibration points (where 1 was always readjusted so as to make $J_0(A)<0.01$) and uniformly shifting the continuous data on the chart reordings so as to coincide with these more accurate data t the corresponding wavelengths.

(d) The angular orientations of the optical elements vere uncertain by less than $\frac{1}{2}$°, hence causing no significant errors. The detection system was measured to be linear to %, and relative gain changes were calibrated to $\frac{1}{2}$%.

(e) Spurious contributions of phase coherent signals of undetermined origin were of magnitude 10^{-4}, much too mall to affect the phase measurements, but large enough o increase the systematic uncertainty in N for values less han 0.01. In the few situations where this might be a problem, multiple reflection from two or more similar urfaces could be used to increase the effective N to a nagnitude greater than 0.01.

(f) Since the f number of the optical system is finite, each experimentally measured parameter is in fact the veighted average of a spread of values corresponding to he spread of angles of incidence. Each system and each pectral range must be considered separately; however, luring the present work with an $f/8$ system, the percentage lifference between each parameter at the central angle of ncidence and the respective experimental parameter was calculated to be less than 1% in all situations encountered.

In the case of the relative phase measurement, an ndependent check of the accuracy exists, since both S and C can be measured equally well, and at any given waveength they should satisfy the identity $S^2+C^2\equiv 1$. Several neasurements of both S and C were made in instances where the magnitudes of both were approximately the same, and in every case, $|S^2+C^2-1|\leq 0.008$, indicating that the sum of both systematic and random uncertainties at fixed wavelengths can easily be made less than $\frac{1}{4}$%. A limit of 1% is quoted for scanning situations on the basis of the wavelength dependent systematic errors considered above.

These 1% uncertainties in (N,S) or (N,C) correspond to systematic uncertainties of approximately 3% in ϵ_1 and 1 to 5% in ϵ_2 in all regions where N is less than the smaller of S and $|C|$. Any uncertainty in the angle of incidence

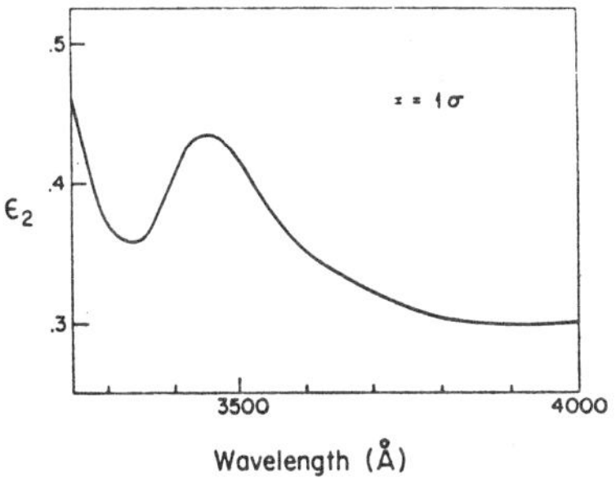

FIG. 6. An expanded view of the low absorption part of Fig. 5.

contributes to a rather large systematic uncertainty of at least 10% per degree in both ϵ_1 and ϵ_2, but since it is wavelength independent, it only decreases certainty in the absolute magnitude of the dielectric constants, while doing nothing to obscure the presence of structure in the data. The standard deviation of the scatter of ϵ_1 and ϵ_2 values about a smooth curve which best fits the data is 1% of the absolute magnitude, in agreement with the 0.3% random scatter in values of (N,S) or (N,C).

As an example of the use of this technique we present in Fig. 5 a graph of ϵ_2 for silver in the wavelength range 3000 to 4000 Å. Included for comparison are the results of Huebner *et al.*[15] and Schulz.[16] Figure 6 shows an expanded view of ϵ_2 in the wavelength range 3250 to 4000 Å. The error bar designates one standard deviation for the scatter of experimental ϵ_2 values about the smooth curve. The vertical position of the graph is uncertain by 10%, since the angle of incidence ($\theta=45°$) was measured only to the nearest degree.

The resonant absorption centered at 3450 Å was previously identified and reported by the authors,[17] and the results of further work pertaining to it are to be published elsewhere. The significance of these data for present purposes is that this absorption occurs in a wavelength range where the reflectance is rather large (0.9) and changing abruptly, thus tending to obscure the presence of structure in any discrete set of ρ measurements. On the other hand, the absorption was immediately obvious in the continuous recording of N, in spite of the changing background, and the resulting structure in ϵ_2 is approximately 25 times larger in magnitude than the noise level.

---

[15] R. H. Huebner, E. T. Arakawa, R. A. MacRae, and R. N. Hamm, J. Opt. Soc. Amer. 54, 1434 (1964).

[16] L. G. Schulz, J. Opt. Soc. Amer. 44, 357 (1954); L. G. Schulz and F. R. Tangherlini, J. Opt. Soc. Amer. 44, 362 (1954).

[17] S. N. Jasperson and S. E. Schnatterly, Bull. Amer. Phys. Soc. 12, 399 (1967).

Optically induced birefringence and dichroism in rigidly held dye molecules

M. Montecchi, M. Settembre, and M. Romagnoli

Fondazione Ugo Bordoni, Viale Europa 190, 00144 Rome, Italy

Received November 20, 1987; accepted June 29, 1988

Nonlinear-optical properties of Fluorescein-doped boric acid glass are investigated. By considering a linearly polarized pumping field we were able to obtain a complete expression of the dielectric susceptibility tensor. Utilizing a heterodyne detection polarimeter, we characterized the material at λ = 457.9, 476.5, and 632.8 nm. Large values of the nonlinear susceptibility are observed as well as considerable light-induced changes of the absolute values of the dielectric response. As an example, at $\lambda = 457.9$ nm and for a sample thickness $d = 100\ \mu$m, a concentration $C = 1.5 \times 10^{-3}$ M, and an incident intensity $I = 100$ mW/cm^2, the induced dichroism is $\Delta\alpha/\alpha = 30\%$, and the birefringence is $\Delta n/n = 3.2 \times 10^{-4}$.

INTRODUCTION

Currently there is great interest in the development of materials for applications in nonlinear optics. Of the possible nonlinearities, resonant media are of particular interest because of the reduced value of the intensity required to achieve a consistent change in the dielectric properties. The advantage provided by resonant, with respect to transparent, materials arises from their capacity for holding the memory of external perturbations for a longer time. However, a fundamental limit to the maximum amount of change is dictated by the finite number of centers available for the transitions, and, in fact, this kind of nonlinearity shows the phenomenon of saturation. The research in this field still requires the investigation of a large variety of materials in order to optimize the speed of response and the nonlinearity.

Many authors[1–3] suggest the utilization of dye molecules rigidly held in glassy compounds in order to improve the mechanism of preservation of the memory impressed by external perturbations. To our knowledge, Fluorescein-doped boric acid glass shows the lowest saturation intensity together with a high value of nonlinear susceptibility. Fluorescein is an organic molecule with a highly anisotropic shape; randomly dispersed in an isotropic host, it is particularly suitable for experiments and applications based on polarization effects.

In the sections that follow, the interaction of polarized light with Fluorescein-doped boric acid glass is theoretically treated. The complete expression of the susceptibility generated by a linearly polarized pump beam is obtained. An experimental apparatus for the direct measurement of the real and imaginary parts of the dielectric susceptibility is described, and a set of measurements for three different wavelengths is presented. The results discussed in the last part of the paper are encouraging: large values of nonlinear susceptibility are observed as well as considerable light-induced changes of the absolute values of the dielectric response. The experimental results indicate good agreement with the theoretical model, although, for intensities many times above the saturation intensity, thermal phenomena, which affect the normal decay routes, become relevant. As an example, at $\lambda = 457.9$ nm, for a sample thickness of $d = 100\ \mu$m, a concentration $C = 1.5 \times 10^{-3}$ M, and an incident intensity $I = 100$ mW/cm^2, the induced dichroism is $\Delta\alpha/\alpha = 30\%$, and the birefringence is $\Delta n/n = 3.2 \times 10^{-4}$.

THEORY

A simplified diagram of the energy levels for the Fluorescein molecule is given in Fig. 1. Actually, more excited states exist, but this is not relevant for the discussion that follows. Selection rules forbid radiative transitions between singlet and triplet states. The molecule can make a transition from the fundamental state of singlet S_0 to the first excited state S_1 by absorbing a photon of energy $h\omega_S$. The lifetime of S_1 is of the order of nanoseconds; from this state the molecule can return to S_0 or decay to T_0 by means of an energy-transfer process with a quantum yield Q.[4] Furthermore, the molecule can make a transition from T_0 to T_1, capturing a photon $h\omega_T$; otherwise, it can return back to S_0, either through S_1 by thermal activation or directly, giving rise to phosphorescence. The total probability of both transitions determines the average lifetime (τ) of the metastable state T_0. The value of τ is of the order of 1 sec and depends sensitively on the temperature.[4] This explains the extremely low value of the saturation intensity $I_S(\omega) = h\omega/[Q\sigma_S(\omega)\tau] \sim 2.5$ mW/cm^2 at λ = 457. 9 nm [where $\sigma_S(\omega)$ is the cross section of the transition $S_0 \rightarrow S_1$] and the phosphorescent behavior.

Our aim is to deduce the dielectric susceptibility of the Fluorescein in boric acid glass. As previously observed,[5] Fluorescein is a highly anisotropically shaped molecule, and therefore it is reasonable to assume that the electric dipole transition moments $\mu_S = \langle S_1|\mu|S_0\rangle$ and $\mu_T = \langle T_1|\mu|T_0\rangle$ are aligned in the direction of the molecular axis. Macroscopically the sample appears to be optically isotropic because of the random orientation of the dipoles in the fused boric acid. Initially the molecules having the axis direction inside an infinitesimal solid angle $d\Omega$ are considered; their number is proportional to $d\Omega/4\pi$. Extending the procedure adopted in Ref. 6 to a three-level system, we can obtain in a straightforward manner the dielectric response relative only to the transition $S_0 \rightarrow S_1$:

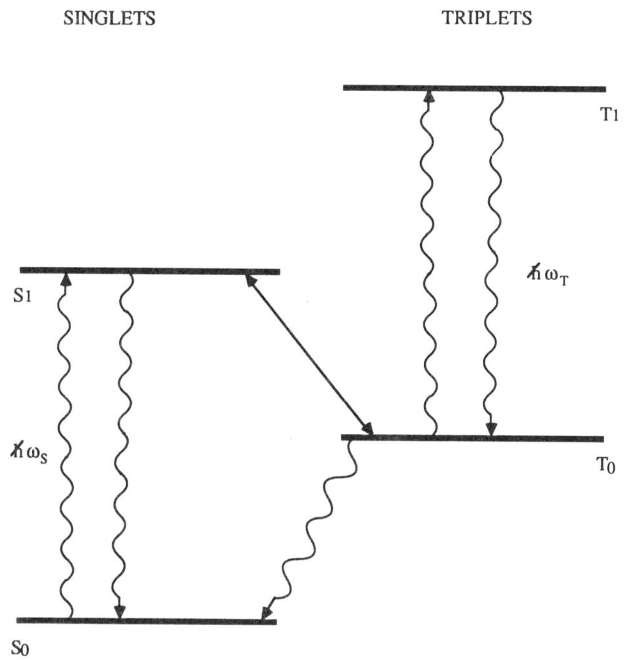

Fig. 1. Simplified diagram of Fluorescein molecule energy level and its main transitions. Selection rules forbid radiative transitions between singlet and triplet states. Because of its long lifetime, T_0 acts as a trap state.

$$d\chi_{\mu\mu}^{\rm Sing}(\omega) = d\Omega/4\pi[1 + I/I_S(\omega)\cos^2\alpha]^{-1}\chi_{\rm Sing}^{(1)}(\omega), \quad (1)$$

where I is the intensity of linearly polarized light at frequency ω, α is the angle between the electric dipole transition moment μ_S and the polarization vector,

$$I_S(\omega) = 2c\epsilon_0 h^2[1 + (\omega - \omega_S)^2 T_2{}^2]/(Q\tau T_2\mu_S{}^2) = h\omega/[Q\sigma_S(\omega)\tau] \quad (2)$$

(with T_2 being the transverse relaxation time) is the saturation intensity, and

$$\chi_{\rm Sing}^{(1)}(\omega) = \mu_S{}^2 T_2 N_0/(\epsilon_0 h)\{(\omega_S - \omega)T_2/[1 + (\omega - \omega_S)^2 T_2{}^2] - i/(1 + (\omega - \omega_S)^2 T_2{}^2)\} \quad (3)$$

is the linear susceptibility by the singlet transitions. In Eq. (1) the factor $[(1 + I/I_S(\omega)\cos^2\alpha]^{-1}$ is the portion of the total population density N_0 in S_0. For $I = 0$, all the molecules occupy the ground state; for $I \gg I_S$, S_0 is bleached, and the population completely fills the long-lived state T_0. From the lowest triplet state the molecules can still be excited to T_1, which is characterized by a lifetime of a few nanoseconds. That implies a saturation intensity I_T much higher than I_S. So, for $I \ll I_T$, the triplet contribution to the susceptibility is

$$d\chi_{\mu\mu}^{\rm Sing}(\omega) = (d\Omega/4\pi)[I/I_S(\omega)]\cos^2\alpha/[1 + I/I_S(\omega)\cos^2\alpha]\chi_{\rm Trip}^{(1)}(\omega), \quad (4)$$

where $\chi_{\rm Trip}^{(1)}(\omega)$ can be obtained from Eq. (3) by substituting the singlet parameters with those of the triplet.

In fact, when our model is followed, the only states significantly populated are S_0, with a population proportional to $1/(1 + I/I_S\cos^2\alpha)$, and T_0, with the remaining portion proportional to $I/I_S\cos^2\alpha/(1 + I/I_S\cos^2\alpha)$. The total susceptibility of the group of molecules oriented in the solid angle $d\Omega$ is

$$\begin{aligned} d\chi_{\mu\mu}(\omega) &= d\chi_{\mu\mu}^{\rm Sing}(\omega) + d\chi_{\mu\mu}^{\rm Trip}(\omega) \\ &= d\Omega/4\pi(\{1 + [I/I_S(\omega)]\cos^2\alpha\}^{-1}\chi_{\rm Sing}^{(1)}(\omega) \\ &\quad + [I/I_S(\omega)]\cos^2\alpha/\{1 + [I/I_S(\omega)]\cos^2\alpha\}\chi_{\rm Trip}^{(1)}(\omega)). \end{aligned} \quad (5)$$

Note that Eqs. (2) and (3) could lack validity when the band shapes are included; however, Eqs. (1), (4), and (5) are substantially correct under only the hypothesis of homogeneously broadened absorptions and codirectionality of μ_S and μ_T. In fact, although the glassy nature of this compound should provide an inhomogeneous broadening, the large size of the molecule has to be taken into account to justify the homogeneity of the interaction with the host.

To obtain the overall susceptibility, we must refer $\chi_{\mu\mu}$ to the laboratory frame and then average it on the solid angle. Because of these considerations,

$$\chi_{ij} = 1/(4\pi)\int_{4\pi} d\chi_{\mu\mu}\cos\alpha_{\mu j}\cos\alpha_{\mu i}, \quad (6)$$

where $\cos\alpha_{\mu j}$ and $\cos\alpha_{\mu i}$ are the director cosines of the μ axis relative to the i, j axes of the laboratory frame.

In what follows, a linearly polarized pump beam along the direction **z** is assumed. If we define

$$a = I/I_S(\omega), \qquad \chi_{\rm Sing}^R = {\rm Re}\{\chi_{\rm Sing}^{(1)}\},$$
$$\chi_{\rm Sing}^I = {\rm Im}[\chi_{\rm Sing}^{(1)}], \qquad \chi_{\rm Trip}^R = {\rm Re}[\chi_{\rm Trip}^{(1)}],$$
$$\chi_{\rm Trip}^I = {\rm Im}[\chi_{(1)}^{\rm Trip}], \qquad p = \chi_{\rm Trip}^R/\chi_{\rm Sing}^R,$$
$$q = \chi_{\rm Trip}^I/\chi_{\rm Sing}^I,$$

the dielectric response becomes

$$\begin{aligned} \chi_{zz} &= [\chi_{\rm Sing}^{(1)}/3]f(a) + [\chi_{\rm Trip}^{(1)}/3][1 - f(a)] \\ &= [\chi_{\rm Sing}^R/3][f(a)(1 - p) + p] + i[\chi_{\rm Sing}^I/3][f(a)(1 - q) + q], \end{aligned} \quad (7a)$$

$$\begin{aligned} \chi_{xx} = \chi_{yy} &= [\chi_{\rm Sing}^{(1)}/3]g(a) + [\chi_{\rm Sing}^{(1)}/3][1 - g(a)] \\ &= [\chi_{\rm Sing}^R/]3[g(a)(1 - p) + p] \\ &\quad + i[\chi_{\rm Sing}^I/3][g(a)(1 - q) + q], \end{aligned} \quad (7b)$$

where

$$f(a) = (3/a)[1 - \arctan(a)^{1/2}/(a)^{1/2} - (a)^{1/2}], \quad (7c)$$

$$g(a) = (3/2)[(1 + 1/a)\arctan(a)^{1/2}/(a)^{1/2} - 1/a] \quad (7d)$$

are the saturation functions.

When $a \ll 1$, $f(a) \sim 1 - (3/5)a$, $g(a) \sim 1 - a/5$, so

$$\begin{aligned} \chi_{zz} &= [\chi_{\rm Sing}^R/3][1 - (3/5)a(1 - p)] \\ &\quad + i[\chi_{\rm Sing}^I/3][1 - (3/5)a(1 - q)] \end{aligned} \quad (8a)$$

and

$$\begin{aligned} \chi_{xx} = \chi_{yy} &= [\chi_{\rm Sing}^R/3][1 - (a/5)(1 - p)] \\ &\quad + i[\chi_{\rm Sing}^I/3][1 - (a/5)(1 - q)]. \end{aligned} \quad (8b)$$

In the limit $a \to 0$, $\chi_{zz} = \chi_{xx} = \chi_{yy} = \chi_{\rm Sing}^{(1)}/3$ and the medium is isotropic, with χ determined exclusively by the singlet.

For $a \sim 1$, Eqs. (7) show that the material is anisotropic and that its anisotropy depends on a. For $a \gg 1$, $f(a) \to 0$ and $g(a) \to 0$, i.e., the singlet states are saturated and $\chi_{zz} =$

$\chi_{xx} = \chi_{yy} = \chi^{(1)}_{\mathrm{Trip}}/3$. In this situation the medium is again isotropic but with different susceptibility. The lifetime of the triplet excited states is of the order of a few nanoseconds, which requires an intensity $I \sim 10^9$ W/cm^2 to saturate even this transition; for this reason the triplet-state saturation is not considered.

PUMP-BEAM PROPAGATION

Seeking a solution for the pumping electromagnetic field (linearly polarized along **z** and propagating along **y**) in the form $\mathbf{E}_P = \mathbf{z}E_P(y)\exp[i(\omega t - ky)]$, with $E_P(y)$ slowly varying with respect to $\lambda = 2\pi/k$, we have

$$dE_P(y)/dy = -i\omega/(2nc)\chi_{zz}(|E_P|^2)E_P(y). \tag{9}$$

Multiplying Eq. (5) by $\epsilon_0 cE_P{}^*d/I_S$, where d is the sample thickness, and adding its complex conjugate, we obtain

$$da/d\xi = -a[f(a)(1-q)+q]\alpha_0, \tag{10}$$

where, as above, $a = I/I_S$, $\xi = y/d$, $\alpha_0 = -\omega\chi_{\mathrm{Sing}}{}^I d/(3nc)$. When we numerically solve the differential Eq. (10) for the initial value of the normalized incident intensity $a(\xi = 0) = a_0$, the evolution of $a(\xi)$ is known, and consequently so is the susceptibility.

PROBE-BEAM PROPAGATION

The induced change of susceptibility is probed by a secondary beam much weaker than the pump. This allows us to assume that the susceptibility is not affected by the probe field. Again, by using the slowly varying amplitude approximation, seeking solutions as $E_i(y) = E_{i0}\exp[-\alpha_i(y) + i\beta_i(y)]$ with $i = z, x$, separating the real from the imaginary part, and introducing the adimensional variable ξ, we have a set of four equations for the amplitudes and phases:

$$d\alpha_z/d\xi = \alpha_0/2[f(a)(1-q)+q], \tag{11a}$$

$$d\beta_z/d\xi = -\beta_0[f(a)(1-p)+p], \tag{11b}$$

$$d\alpha_x/d\xi = \alpha_0/2[g(a)(1-q)+q], \tag{12a}$$

$$d\beta_x/d\xi = -\beta_0[g(a)(1-p)+p], \tag{12b}$$

with $\beta_0 = \omega\chi_{\mathrm{Sing}}{}^R d/6nc$.

Once $a(\xi)$ has been derived from Eq. (10), Eqs. (11) and (12) are solvable by numerical integration. The integration is computed from $\xi = 0$ to $\xi = 1$. In $\xi = 0$, $\alpha_z = \alpha_x = \beta_z = \beta_x = 0$. The phases β_z and β_x could be separately measured but, because of either the heating phenomena or the vibrations of the setup, it is difficult to isolate uniquely the effect of the nonlinearity induced by the light. Thus, in the experiment, the measurable quantity is the relative phase difference $\Delta\beta(a_0) = \beta_z(\xi = 1, a_0) - \beta_x(\xi = 1, a_0)$ ruled by

$$d\Delta\beta/d\xi = -\beta_0(1-p)[f(a) - g(a)]. \tag{13}$$

The other measurable quantities are the transmissions for both the parallel and the orthogonal polarization with respect to the pumping field, $T_z(a_0) = \exp[-\alpha_z(\xi = 1, a_0)]$ and $T_x(a_0) = \exp[-\alpha_x(\xi = 1, a_0)]$. The transmissions, as a function of incident intensity, parametrically depend on α_0, I_S, and q. The parameter α_0 is connected to the linear transmission; in this case ($a_0 = I/I_S \ll 1$), $T_z = T_x = \exp(-\alpha_0)$.

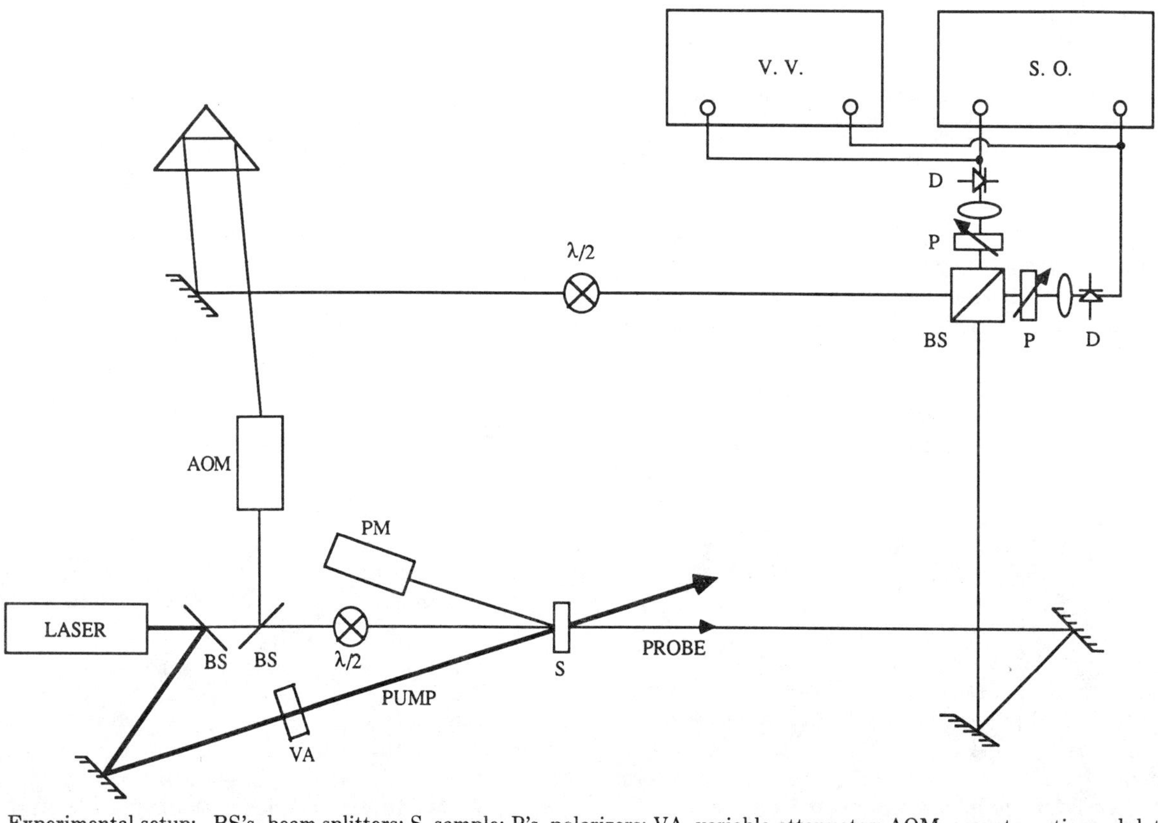

Fig. 2. Experimental setup: BS's, beam splitters; S, sample; P's, polarizers; VA, variable attenuator; AOM, acousto-optic modulator; D's, photodiodes; S.O., sampling oscilloscope; V.V., vector voltmeter; PM, power meter; λ/2's, half-wave plates.

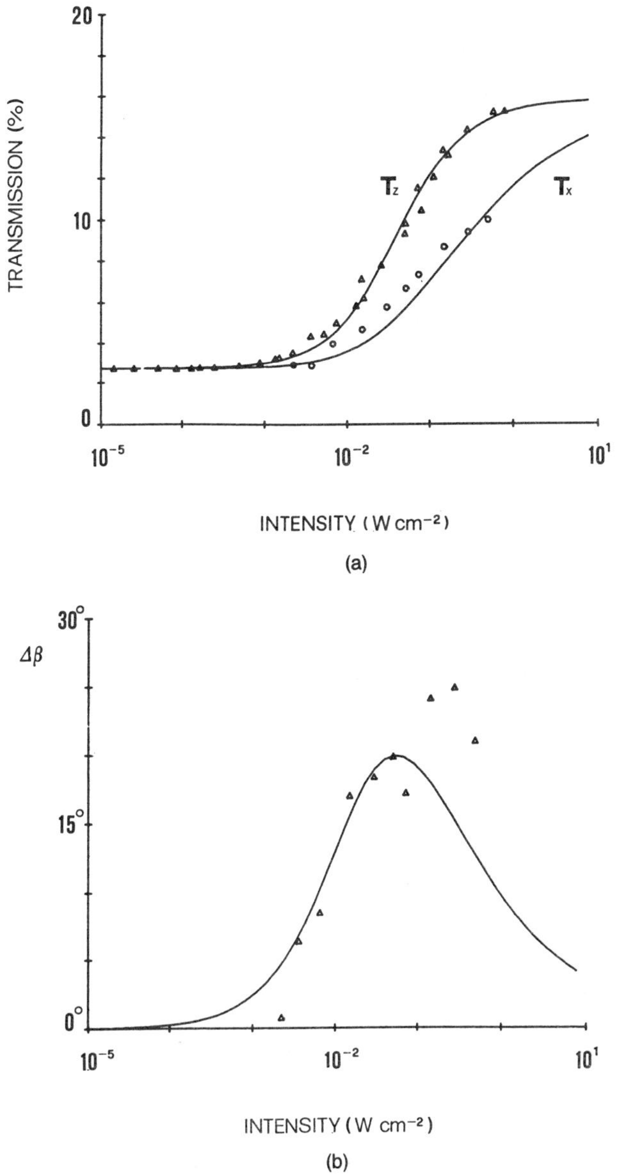

Fig. 3. (a) Transmission of a probe polarized along (T_z) and orthogonally (T_x) to the pump and (b) phase difference versus intensity I at $\lambda_{PUMP} = \lambda_{PROBE} = 457.9$ nm. The estimated parameters are summarized in Table 1 assuming that $n \sim 1.47$.

Conversely, at intensities such that the population of the singlet states ($a_0 = I/I_S \gg 1$) is bleached, $T_z = T_x = \exp(-\alpha_0 q)$. Once α_0, I_S, and q are determined, the remaining parameter $\beta_0(1-p)$ establishes the maximum amount of birefringence.

EXPERIMENT

The sample was prepared according to the technique described by Todorov *et al.*[7] The dye was Na Fluorescein ($C_{20}H_{10}Na_2O_5$) in boric acid; its absorption coefficient per unit length was $\alpha_0/d = 361$ cm^{-1} at $\lambda = 457.9$ nm for $C = 1.5 \times 10^{-3}$ M. The experiment was performed by measuring the change in the transmission and phase of a weak probe going through a region of the sample illuminated by the pump.

The experimental apparatus is shown in Fig. 2. The pumping beam was split from the output of a vertically polarized laser and crossed the sample at 13°. We checked that the spot size of the pump ($\sim 5.7 \times 10^{-2}$ cm^2) overlapped the spot size of the probe ($\sim 2 \times 10^{-2}$ cm^2). The pump intensity was adjustable over a wide range by means of calibrated neutral-density filters. The probe beam was polarized at 45° with respect to the pump beam; its vertical and horizontal components emerging from the sample were split and heterodyne detected by beating with the corresponding components of a 10-MHz frequency-shifted reference beam. The beat signals were sent to a sampling oscilloscope operating in the X–Y mode. This heterodyne technique permitted the visualization of the polarization ellipse. Simultaneously the signals were accurately measured with a vector voltmeter. However, for a better evaluation of α_0, low pump inten-

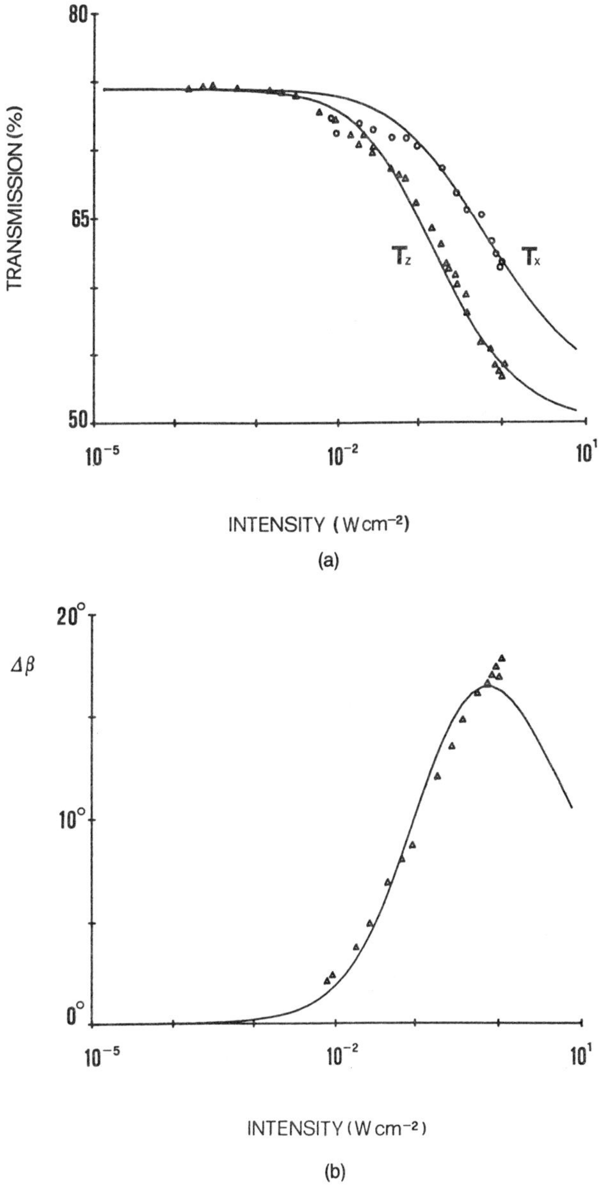

Fig. 4. Same as Fig. 3 but $\lambda_{PUMP} = \lambda_{PROBE} = 476.5$ nm.

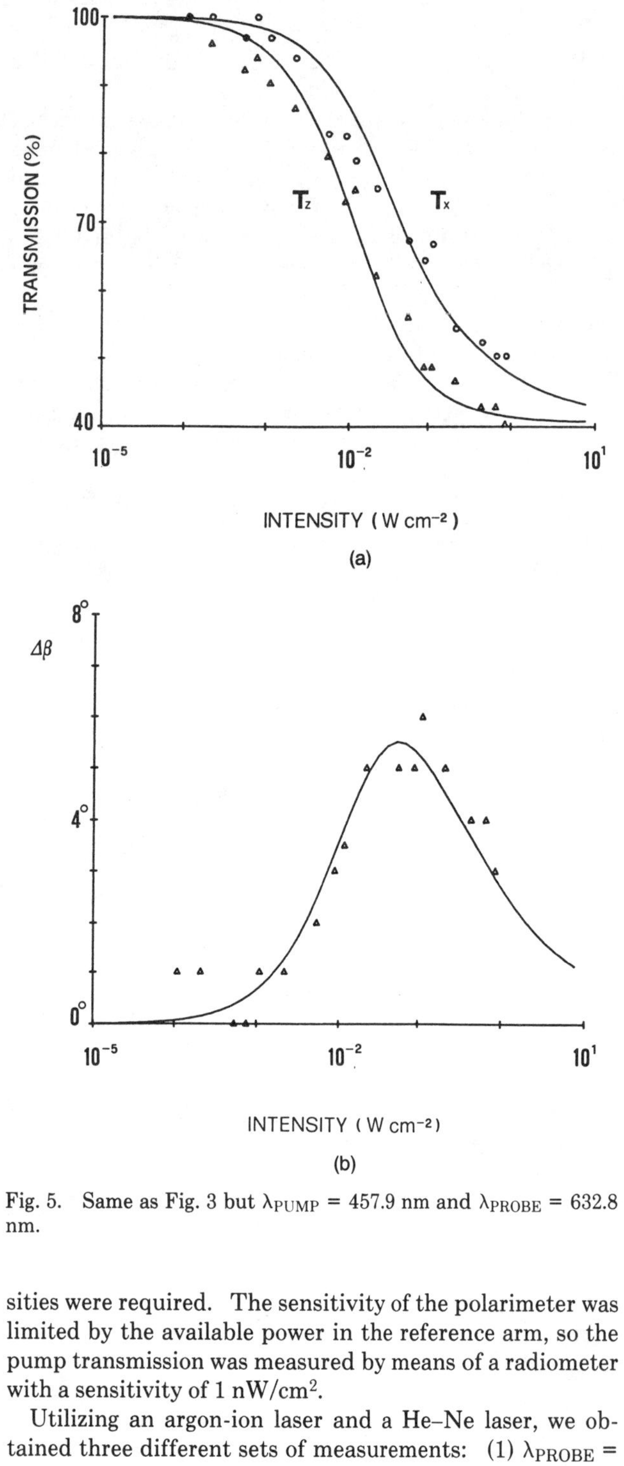

Fig. 5. Same as Fig. 3 but λ_{PUMP} = 457.9 nm and λ_{PROBE} = 632.8 nm.

sities were required. The sensitivity of the polarimeter was limited by the available power in the reference arm, so the pump transmission was measured by means of a radiometer with a sensitivity of 1 nW/cm^2.

Utilizing an argon-ion laser and a He–Ne laser, we obtained three different sets of measurements: (1) λ_{PROBE} = λ_{PUMP} = 457.9 nm, (2) λ_{PROBE} = λ_{PUMP} = 476.5 nm, (3) λ_{PROBE} = 632.8 nm, λ_{PUMP} = 457.9 nm. The fitting curves of the experimental points are shown in Figs. 3–5, and the estimated parameters are summarized in Table 1, for which we assume that n = 1.47. The first set is characterized by a decreasing absorption versus a growth of the medium saturation. This means that $|\mathrm{Im}[\chi^{(1)}_{\mathrm{Trip}}]|$ is less than $|\mathrm{Im}[\chi^{(1)}_{\mathrm{Sing}}]|$ ($q \sim 0.5$). On the contrary, the second set shows evidence of a gradual increase of absorption: $q \sim 2$. In both cases the singlet and triplet states actively contribute to the absorption. Conversely, for λ = 632.8 nm, only the triplet state affects the transmission (set 3). The first set shows the maximum birefringence ($\Delta\beta \sim 25°$).

Both of the values obtained for the saturation intensity are less than those reported by Kramer *et al.*[5] by approximately a factor of 4. We believe that such a difference is due to the close dependence of the lifetime τ on the amount of water lost by the boric acid during the preparation of the sample and on its cooling rate[4]: in fact, we have observed a longer lifetime. If the different absorption coefficients per unit length of the samples used by us and by Kramer *et al.* are taken into account, the order of magnitude of $\chi^{(3)}$ measured in both cases is comparable.

Despite the discrepancies appearing in the high-intensity range of Figs. 3(b) and 4(b), the theory satisfactorily overlaps the experimental points. Most likely, in this range the local rise in temperature (in particular, at $I \sim 50$ W/cm^2 the sample melts) reinforces the back intersystem crossing from T_0 to S_1. This phenomenon determines a conspicuous decrease of τ,[4] and consequently I_S grows. In such a way the saturation, and therefore the peak of $\Delta\beta$, shifts toward higher levels, and the observations do not follow the trend predicted by the model. To ensure the validity of the above hypothesis, we observed the time decay of the fluorescence peak (λ = 480.0 nm) after thermalizing the material by pumping it for a few minutes at λ = 457.9 nm ($I \sim 1$ W/cm^2). We show in Fig. 6 the logarithm of the normalized intensity of fluorescence. When the pump is turned off, the signal falls abruptly because of the fast depopulation of T_1. Afterward the signal decreases more slowly but not uniformly; in particular, its slope $(-1/\tau)$ increases. This indicates a progressive reduction of the contribution of the thermally activated transition $T_0 \rightarrow (S_1) \rightarrow S_0$ during the cooling down of the sample. Thus, initially, $\tau_{\mathrm{hot}} \sim 4.5 \times 10^{-1}$ sec, whereas at the end $\tau_{\mathrm{cold}} \sim 2$ sec. The value of τ_{hot} justifies the disagreement between the obtained values of I_S and those reported in Ref. (4); furthermore, the ratio $\tau_{\mathrm{cold}}/\tau_{\mathrm{hot}} = I_{S,\mathrm{hot}}/I_{S,\mathrm{cold}} \sim 4.5$ explains the observed deviations from the theory.

An accurate study of the dependence of the lifetime τ and the temperature on the pump intensity could be useful in

Table 1. Saturation Intensities and Dielectric Susceptibilities (in mks units) of Na Fluorescein-Doped Boric Acid Glass

Set	λ_{PUMP} (nm)	λ_{PROBE} (nm)	I_S (mW/cm^2)	χ^I_{Sing}	χ^I_{Trip}	q	$(\chi^R_{\mathrm{Sing}} - \chi^R_{\mathrm{Trip}})$
1	457.9	457.9	2.5	-1.2×10^{-2}	-6.1×10^{-3}	0.51	1.4×10^{-2}
2	476.5	476.5	85	-9.9×10^{-4}	-2.3×10^{-3}	2.35	1.1×10^{-2}
	457.9		2.5	-1.2×10^{-2}	-6.1×10^{-3}	0.51	
3							
		632.8		0	-4.0×10^{-3}	∞	2.7×10^{-3}

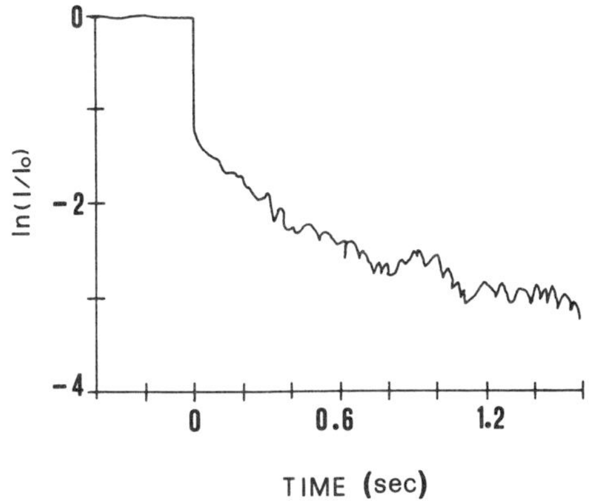

Fig. 6. Natural logarithm of the normalized intensity of fluorescence. The pump is turned off at $t = 0$ sec. The abrupt drop in fluorescence at $t = 0$ is due to the exhaustion of the fast transition $T_1 \rightarrow T_0$. Afterward, the signal decreases more slowly but not uniformly: $\tau_{hot} \sim 4.5 \times 10^{-1}$, whereas $\tau_{cold} \sim 2$ sec.

better understanding Fluorescein-doped boric acid glass, but it is not important for our purpose in this paper.

CONCLUSIONS

In conclusion, we have analyzed the propagation of a beam at three different wavelengths through a sample of Fluorescein-doped boric acid glass, optically pumped by a second beam. The measurements were taken with a polarimeter to permit simultaneous observation of the phase and absorption changes caused by the external pump. The material under investigation was highly nonlinear, although at relatively low intensities triplet absorption and eventually heating limit its nonlinearity. Nevertheless, the thermal effects may be overcome by pumping the medium with pulses shorter than the heating times. In this case the maximum amount of induced change is limited only by the triplet absorption, and the overall nonlinearity could be improved as estimated from the ratio $I_{S,hot}/I_{S,cold}$.

ACKNOWLEDGMENTS

This research was carried out at the Fondazione Ugo Bordoni under an agreement between the Fondazione Ugo Bordoni and the Istituto Superiore delle Poste e delle Telecomunicazioni. We wish to thank B. Daino for the useful discussions.

REFERENCES

1. T. A. Shankoff, "Recording holograms in luminescent materials," Appl. Opt. **8,** 2282–2284 (1969).
2. Y. Silberberg and I. Bar-Joseph, "Low power phase conjugation in thin films of saturable absorbers," Opt. Commun. **39,** 265–268 (1981).
3. H. Fujiwara and K. Nakagawa, "Phase conjugation in fluorescein film by degenerate four-wave mixing and holographic process," Opt. Commun. **55,** 386–390 (1985).
4. G. N. Lewis, D. Lipkin, and T. T. Magel, "Reversible photochemical processes in rigid media. A study of the phosphorescent state," J. Am. Chem. Soc. **63,** 3005–3018 (1941).
5. M. A. Kramer, W. R. Tompkin, and R. W. Boyd, "Nonlinear-optical interactions in Fluorescein-doped boric acid glass," Phys. Rev. A **34,** 2026–2031 (1986).
6. A. Yariv. ed., *Quantum Electronics*, 2nd ed. (Wiley, New York, 1975), pp. 149–155.
7. T. Todorov, L. Nikolova, N. Tomova, and V. Dragostinova, "Photochromism and dynamic holographic recording in a acid solution of Fluorescein," Opt. Quantum Electron. **13,** 209–214 (1981).

©1984 Optical Society of America. Reprinted from *Applied Optics*, Vol. 23(23), 4309–4312 (December 1984).

Polarization holography. 1: A new high-efficiency organic material with reversible photoinduced birefringence

T. Todorov, L. Nikolova, and N. Tomova

A new organic material for polarization holographic recording—azo-dye methyl-orange, introduced in polyvinyl alcohol (PVA)—is fabricated and investigated. It possesses all the good characteristics of the known polarization-sensitive materials—high sensitivity, reversibility, etc.—but excels them substantially in the magnitude of the photoinduced birefringence: $\delta n > 10^{-3}$. This makes possible the recording of reversible polarization gratings with diffraction efficiency over 30%. Depending on the conditions of production and subsequent mechanical and thermal treatments the layers of methyl orange/PVA may also have intrinsic birefringence. In this paper the results of experimental investigations into the properties of this new material are reported.

I. Introduction

Theoretical investigations in the last decade proved that the polarization of light can also be recorded and reconstructed by holographic means.[1] A new branch of holography emerged—polarization holography. The progress of polarization holography at the moment is hampered considerably by the performance of existing recording materials in which optical anisotropy is induced in accordance with the light polarization used. Polarization-sensitive media belong basically to one of the following groups: alkali halide crystals containing anisotropic color centers,[2,3] photochromic glasses or emulsions on the basis of silver halides (mainly AgCl) with a marked Weigert effect (photodichroism),[4,5] organic systems with metastable triplet levels,[6,7] and materials on the basis of dyes, undergoing photostructural changes, leading to a certain degree of stereoregularity in the new structure.[8,9] The main shortcoming of all these materials is the relatively weak photoinduced anisotropy, usually in the optical absorption (photodichroism) which is the reason for the low efficiency of the recorded polarization holograms (<1%). Most of the enumerated materials have low sensitivity and not all of them allow multiple use.

The authors are with Bulgarian Academy of Sciences, Central Laboratory of Optical Storage and Processing of Information, P.O. Box 95, Sofia 1113, Bulgaria.

Received 13 July 1984.
0003-6935/84/234309-04$02.00/0.

In this work results are presented from the investigations on a new polarization-sensitive material—a combination of an azo dye and polymer matrix, on which high-efficiency polarization holograms (>35%) can be recorded at relatively low recording light intensity, and which allows multiple erasure and repeated recording without apparent fatigue. Another feature of this new recording medium is that it can be prepared at will with intrinsic birefringence of the matrix (Δn) independent of the acting light. Producing samples with different Δn helped check experimentally the specific polarization properties of such holograms as predicted in Ref. 10.

II. Samples

The new anisotropic medium is in the form of layers of methyl orange introduced in a polymer matrix. Methyl orange is an azo dye in which, under the action of linearly polarized light, some degree of ordering of the molecules can be achieved leading to dichroism.[11] Our experiments showed that the molecules of methyl orange can be oriented reversibly and many times without apparent fatigue. The reverse process—destruction of the stereoregularity—depends on the matrix and in darkness it may last from a few seconds to several days, i.e., the material has different times of memory. This thermal process dictates the efficiency of the optically attained stereoregularity and the value of the induced anisotropy. Polyvinyl alcohol (PVA) turned out to be most suitable for our purposes. It is a well-known material[12] in which an optical axis along the direction of the force can be induced through mechanical deformation. Thus, the material becomes optically anisotropic—it acquires birefringence ($\Delta n = n_e - n_o$) over the whole visible region, the value of Δn depending on

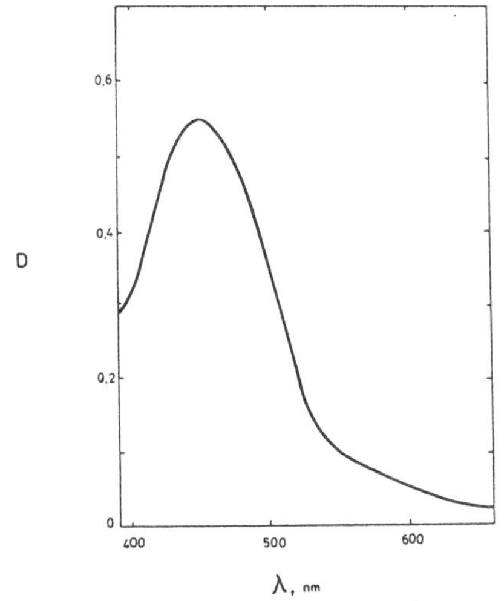

Fig. 1. Absorption spectrum of methyl orange/PVA layer at a dye concentration of 0.06 wt. %.

the regime of the mechanical and thermal treatment. We used samples of methyl orange/PVA in the form of ~100-μm thick films on a glass substrate. Depending on their preparation regime, they had different intrinsic birefringence and the phase difference between ordinary and extraordinary waves $\Delta\varphi = \varphi_o - \varphi_e$ after a 100-μm thick layer varied from 0 (no birefringence) to π ($\lambda/2$ plate).

III. Investigations into Photoinduced Anisotropy

Figure 1 shows the absorption spectrum of a 100-μm thick layer of PVA containing 0.06-wt. % methyl orange. Since PVA is transparent in the visible region, absorption is due entirely to the dye whose stable form is its transisomer.[11] In later experiments samples with greater dye concentrations were used and they ensured higher values of photoinduced anisotropy.

On absorbing the light of an Ar laser beam with a wavelength of 488 nm, a large part of the dye molecules undergoes spatial (trans-cis) isomerization. The cis-form of methyl orange is thermally unstable and a reverse thermal process of restoration of the transisomer begins. Also, if the light is linearly polarized the molecules tend to order themselves in such a way that the direction of their optical transition is perpendicular to the polarization direction of the light. Thus, the optical transmission for light, polarized along the polarization direction of the acting light, increases; for light polarized perpendicular to the polarization direction of the acting light, it decreases. Dichroism is induced. The kinetics of the anisotropic change in the optical transmission is shown in Fig. 2. Measurement was accomplished with a low-intensity light beam of the same wavelength. The intensity of the acting Ar laser light was 100 mW/cm^2. The moments of turning the Ar beam on and off are marked with arrows. After the light action is removed the sample begins to restore its original state and the ordering of the dye molecules is lost. The degree of restoration after the recording light is switched off and the rate of this process depend on the polymer matrix and in PVA it can be altered by thermal treatment. In

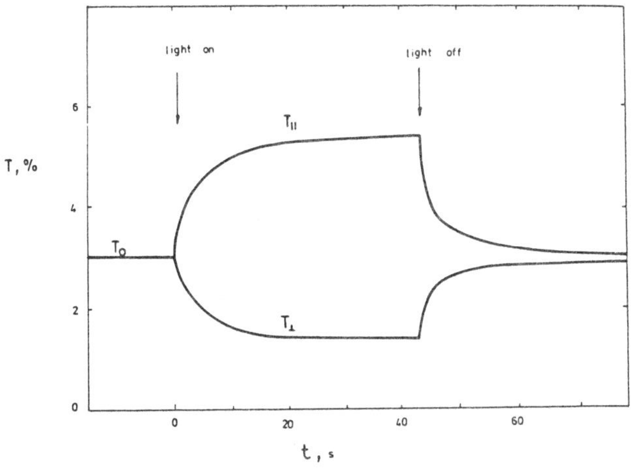

Fig. 2. Optical transmission change at λ = 488 nm under the action of an Ar^+ laser beam of the same wavelength and intensity of 100 mW/cm^2. $T_{\parallel}$ is the optical transmission for light, polarized parallel to the acting light polarization; $T_{\perp}$ is for perpendicularly polarized light. The moments of switching on and off of the acting light are marked with arrows.

the case corresponding to Fig. 2, such treatment was not carried out and the original optical transmission was restored almost completely in ~30 sec.

Along with dichroism, light birefringence is induced in the films under the action of Ar^+. Photoinduced birefringence, particularly for wavelengths outside the optical transmission, is of greater interest from a holographic point of view, since it leads to higher recording efficiency. We investigated birefringence at λ = 633 nm. For this purpose the samples were placed between two crossed polarizers P and A in the path of a He–Ne laser beam. They were irradiated with an Ar^+ laser beam, linearly polarized at ±45° with respect to the directions of transmission of P and A. The optical anisotropy, induced in the sample under the action of the blue light, causes a light signal to appear after the analyzer A. At λ = 633 nm this transmission is due solely to the photoinduced birefringence which is given by

$$I = I_0 \sin^2\left(\frac{\delta\varphi}{2}\right),$$

where I is the intensity of the transmitted light after A, I_0 is the intensity of the light passing through the pair of parallel polarizers,

$$\delta\varphi = \frac{2\pi}{\lambda}\,\delta n d,$$

d is the thickness of the sample (100 μm), and δn is the birefringence. The kinetic curve of such a process is plotted in Fig. 3. Here again switching the light on and off is marked with arrows.

The values of δn depend on the intensity of the Ar^+ beam, dye concentration, and preliminary treatment of the polymer. In optimal conditions $\delta\varphi$ of the samples exceeded $\pi/3$ ($\delta n > 10^{-3}$). Such large values for the birefringence make us think that the polymer matrix does not remain unchanged in the recording. The exact mechanism of this process is not clear but probably the

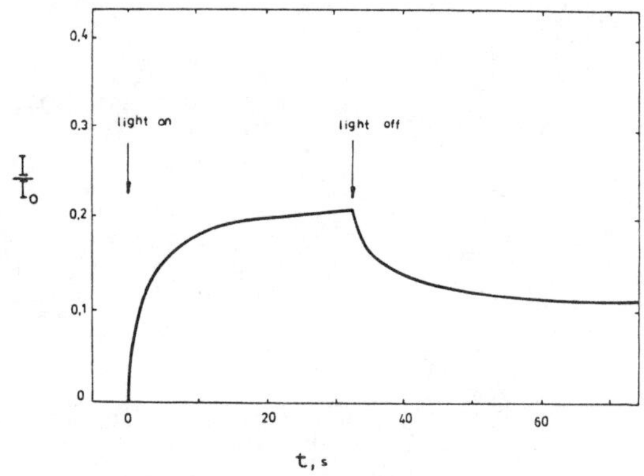

Fig. 3. Kinetics of the light signal change after passing through a pair of crossed polarizers, due to the photoinduced birefringence in a methyl orange/PVA layer placed between them.

PVA molecules are deformed in accordance with the photoinduced ordering of the dye molecules, and that is the reason for the large values of the induced birefringence.

The thermal erasure of the birefringence reached depends on the sample and may be slowed down substantially by preliminary thermal treatment. The sample, whose photoinduced change is shown in Fig. 3, was heated beforehand to 80°C for 30 min. After such treatment a large part (>0.5) of the photoinduced birefringence is retained for a long time after the acting light is removed. The ability of these layers to store information is several days.

Quite interesting is the induction of anisotropy in the samples with intrinsic birefringence achieved by mechanical tension of the polymer matrix. On studying the photoinduced birefringence these samples were positioned in such a way that their principal dielectric axes were at an angle of ±45° to the directions of transmission of P and A and the Ar^+ light was alternately polarized along one or the other axis. In this case the intensity of the light coming out of the analyzer was

$$I = I^0 \pm \delta I,$$

where I^0 is the intensity of the light transmitted due to the intrinsic birefringence of the layer, and δI is connected with the photoinduced birefringence. The sign depends on which axis is coincident with the Ar^+ beam polarization. The values of $\delta\varphi$ measured in this case are the same as those in the isotropic samples.

IV. Polarization Holographic Recording

We determined the basic holographic characteristics of the polymer layers described. The polarization holographic recording was accomplished by two plane waves with mutually orthogonal polarization at 488-nm wavelengths (Ar^+ laser). In this type of recording the resulting light field is not modulated by intensity but only by polarization. The induced optical anisotropy (dichroism or birefringence) is spatially modulated in accordance with the polarization modulation of the recording light field, i.e., a polarization holographic grating is recorded. As shown previously, simultaneous dichroism and birefringence is induced in the investigated layers. In spite of the relatively strong anisotropy in the optical transmission for $\lambda = 488$ nm (Fig. 2), the light with this wavelength diffracted from the recorded polarization grating is weak (diffraction efficiency $\eta <$ 0.1%), due to the large average optical density of the samples. Much larger is the diffraction efficiency for wavelengths outside the absorption region where strong birefringence is induced under the action of light, i.e., phase holographic recording is accomplished. Readout was done with a He–Ne laser beam at Bragg's angle and diffraction efficiency of the +1 order was measured.

The curve, illustrating the increase of η during the recording (Fig. 4, curve a) resembles the increase of δn, shown in Fig. 3. After ~20 sec η reaches saturation (in this case η_s = 20%) and remains constant as long as the recording continues.

When the recording light is turned off initially the diffraction efficiency falls to around one-third and remains practically unchanged for more than 24 h (curve

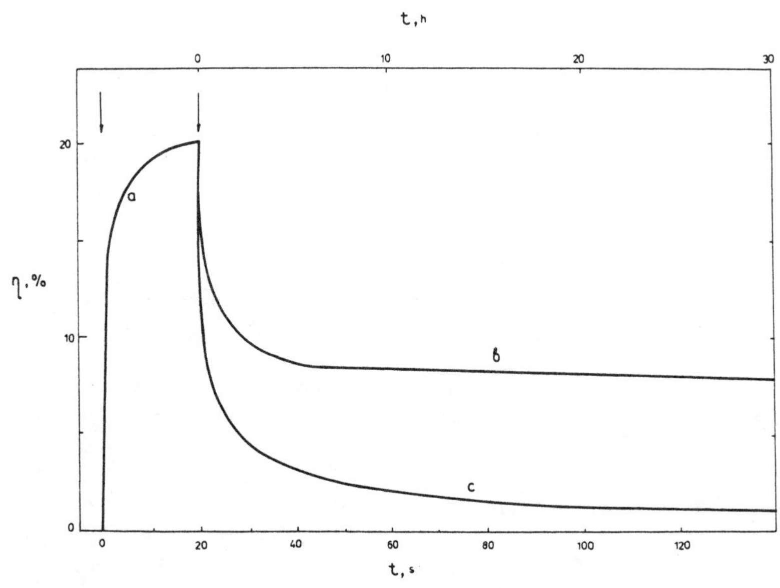

Fig. 4. Kinetics of the diffraction efficiency change in polarization holographic recording: curve a, during the recording; curve b, upper scale, erasure in darkness; curve c, lower scale, light erasure.

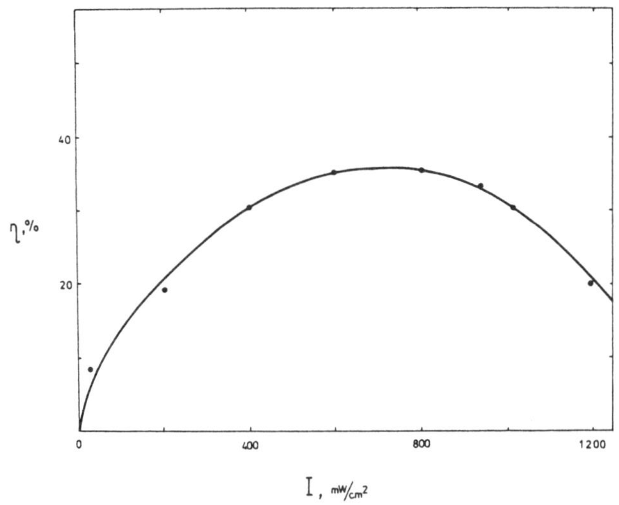

Fig. 5. Dependence of the photostationary value of the diffraction efficiency on the recording light intensity.

b, upper scale). As mentioned earlier, such memory effect is observed only in samples that have undergone preliminary thermal treatment; it is probably connected with residual deformation in the PVA matrix.

If, after reaching saturation of η during recording, only one of the recording beams is turned off, the decrease of η is much faster (curve *c*, lower scale). This is due to a new uniform re-ordering of the methyl orange molecules under the action of the light field, uniform in polarization. The optical erasure is still faster if it is accomplished with a light beam of greater intensity. (In the case corrresponding to Fig. 4, curve *c*, the erasing light intensity is half of the total recording light intensity.) A new holographic grating with the same diffraction efficiency can be recorded immediately on the same sample and this process of record–erase can be repeated many times with no fatigue.

The saturated value of the diffraction efficiency depends both on dye concentration and recording light intensity. This relationship is plotted in Fig. 5 for a sample with an optimal dye concentration of 0.6 wt. %, treated at 80°C beforehand. The maximum value of η ($\sim$35%) is obtained at a total intensity of the two recording waves I = 600–700 mW/cm^2.

V. Summary

The methyl orange/PVA layers are of high-efficiency material for polarization holographic recording, allowing multiple use and storage of information for several days. Also, this material makes it possible to obtain layers with intrinsic initial birefringence. Combined with the high diffraction efficiency, the material is suitable for investigating the properties of the various cases of recording polarization holographic gratings, studied theoretically in Ref. 10. The results of these investigations are reported in the second part of this work.[13]

References

1. Sh. D. Kakichashvili, "Method of Recording Phase Polarization Holograms," Kvantovaya Elektron. Moscow **1**, 1435 (1974).
2. H. Blume, L. Bader, and F. Lüty, "Bi-Directional Holographic Information Storage Based on the Optical Reorientation of F_A Centers in KCl:Na," Opt. Commun. **12**, 147 (1974).
3. L. Nikolova, T. Todorov, and P. Stefanova, "Polarization Sensibility of the Photodichroic Holographic Recording," Opt. Commun. **24**, 44 (1978).
4. N. F. Borelli, J. B. Chodak, and G. B. Hares, "Optically Induced Anisotropy in Photochromic Glasses," J. Appl. Phys. **50**, 5978 (1979).
5. J. M. C. Jonathan and M. May, "Anisotropy Induced in a Silver-Silver Chloride Emulsion by Two-Coherent and Perpendicular Light Vibrations," Opt. Commun. **29**, 7 (1979).
6. T. A. Shankoff, "Recording Holograms in Luminescent Materials," Appl. Opt. **8**, 2282 (1969).
7. T. Todorov, L. Nikolova, N. Tomova, and V. Dragostinova, "Photochromism and Dinamic Holographic Recording in a Rigid Solution of Fluorescein," Opt. Quantum Electron. **13**, 209 (1981).
8. T. Todorov, N. Tomova, and L. Nikolova, "High Sensitivity Material with Reversible Photoinduced Anisotropy," Opt. Commun. **47**, 123 (1983).
9. N. M. Burikin, N. N. Vsevolovod, T. V. Dyukova, E. Y. Korchemskaya, M. S. Soskin, and V. B. Taranenko, "Photoinduced Anisotropy of Bacteriorhodopsin," Ukr. Fiz. Zh. **28**, 1269 (1983).
10. L. Nikolova and T. Todorov, "Diffraction Efficiency and Selectivity of Polarization Holographic Recording," Opt. Acta **31**, 529 (1984).
11. A. M. Makushenko, B. S. Neporent and O. V. Stolbova, "Reversible Photodictioism and Photoisomerization of Complex Organic Molecules," Opt. Spektrosk. **31**, 741 (1971).
12. C. A. Finch, Ed., *Polyvinyl Alcohol, Properties and Applications* (Wiley, London, 1973).
13. T. Todorov, L. Nikolova, and N. Tomova, "Polarization Holography. 2: Polarization Holographic Gratings in Photoanisotropic Materials With and Without Birefringence," Appl. Opt. **24**, 0000 (1985).

©1984 Optical Society of America. Reprinted from *Applied Optics*, Vol. 23(24), 4588–4591 (December 1984).

Polarization holography. 2: Polarization holographic gratings in photoanisotropic materials with and without intrinsic birefringence

T. Todorov, L. Nikolova, and N. Tomova

Results are reported of experimental investigations on the diffraction efficiency and its polarization dependence of holographic phase polarization gratings in a photoanisotropic medium methyl orange/PVA with intrinsic birefringence. A comparative study of layers with different intrinsic birefringence was conducted. The conditions—initial birefringence and a type of polarization recording—were found to have high diffraction efficiency (>30%) and strong dependence on the direction of polarization of the linearly polarized readout beam. The results are in agreement with the theoretical considerations on such gratings.

I. Introduction

The feasibility of recording polarization holograms with an object and reference wave with mutually orthogonal polarizations in materials with photoinduced anisotropy was demonstrated for the first time by Kakichashvili.[1] In Ref. 2 we analyzed the various possibilities for recording polarization holographic gratings and showed that their diffraction efficiency η depends on the type of polarization interference pattern. In addition, at sufficiently large values of photoinduced anisotropy it is feasible to record polarization gratings with high efficiency—up to 25% for amplitude modulation and up to 100% for phase modulation. When the recording is accomplished with two orthogonal circularly polarized waves, η is strongly dependent on the ellipticity of the reconstructing wave; by varying it η can vary from 0 to η_{max}. We showed further that η may also depend on the polarization direction of the linearly polarized readout beam if the recording medium possesses intrinsic birefringence; in this case the diffraction efficiency is considerably higher than in recording with two parallel polarizations.[2] The intent of this work is to present experimental results verifying the described properties of polarization holographic gratings in photoanisotropic media.

The authors are with Bulgarian Academy of Sciences, Central Laboratory of Optical Storage and Processing of Information, P.O. Box 95, Sofia 1113, Bulgaria.

Received 16 July 1984.
0003-6935/84/244588-04$02.00/0.

II. Holographic Arrangement

In the experiments, layers of methyl orange introduced in a matrix of polyvinyl alcohol with a thickness of 100 μm were employed. Their properties were described in part 1 of this paper.[3] In some of the samples intrinsic birefringence was created by preliminary mechanical treatment.

The polarization holographic gratings were recorded on the setup in Fig. 1, using one of the lines ($\lambda = 488$ nm) of an Ar^+ laser (1). The setup enables the polarizations of the two recording waves R and S to be changed from linear in selected directions [using rotators (4) and (5) when the $\lambda/4$ plates (6) are removed: to circular (using the $\lambda/4$ plates).

III. Properties of Polarization Holographic Gratings in Samples without Intrinsic Birefringence

Since at 633 nm the samples have no optical absorption either before or after exposure, diffraction at this wavelength is entirely due to the phase modulation induced in the samples. This modulation is determined by the type of recording light field, i.e., the polarization of waves R and S.

If waves R and S have parallel polarizations an ordinary interference pattern results, i.e., the light field intensity is sinusoidally modulated. Because of the relatively strong absorption the modulation coefficient changes significantly with the depth of the recording medium; hence, the back surface of the medium is underexposed, the front surface is overexposed, and the diffraction efficiency is low.

When waves R and S have mutually orthogonal polarizations the intensity of the resultant light field is constant and only its polarization is periodically spatially modulated in accordance with the change of the

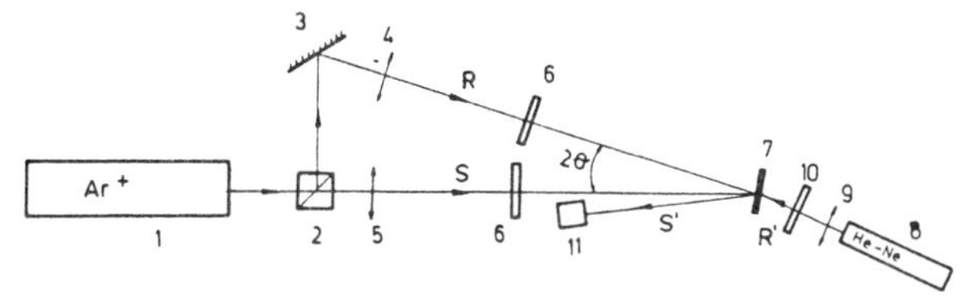

Fig. 1. Holographic setup: 1, Ar^+ laser; 2, beam splitter; 3, mirror; 4,5, polarization rotators; 6, λ/4 plates; 7, recording medium; 8, He–Ne laser; 9, polarization rotator; 10, λ/4 plate; 11, photosensor.

phase shift between waves R and S. The shape of this polarization modulation is shown in Fig. 2(a) when R and S have orthogonal linear polarizations, and in Fig. 2(b) when R and S have orthogonal circular polarizations. When the recording material does not possess intrinsic birefringence these polarization modulations are uniform in depth of sample and the light intensity diminishes slowly with the absorption. In holographic recording in a photoanisotropic material the polarization interference pattern is recorded as a spatially modulated optical anisotropy. Since the process involves reorientation of absorbing molecules,[3] the exposure curve reaches saturation. In such materials it is possible to record polarization holographic gratings with a uniform modulation index in the depth of the samples. In Fig. 2(b) this is obvious; in Fig. 2(a) the gratings are also practically uniform in depth for optical densities $D \leqslant 1$.[4]

As was shown in Refs. 2 and 5 for the case of recording with two orthogonal circularly polarized waves, the diffraction efficiency η is strongly dependent on the polarization of the reconstructing wave R', particularly on its ellipticity ε. Figure 3 illustrats the experimentally measured dependence η (ε). The readout wave ellipticity was controlled with a λ/4 plate. α denotes the angle between the fast axis of the λ/4 plate and the polarization direction of the incident laser beam on it. $\alpha = 45°$ corresponds to left-hand circular polarization of wave R' and $\alpha = -45°$ corresponds to right-hand circular polarization. As expected, the maximum value of η in the +1 order is reached when the polarization of the reconstructing wave coincides with that of the reference wave during the recording and falls to zero at orthogonal polarization of R' (curve a, Fig. 3). At the same time the change of η in the −1 order is described by a similar relation, shifted at 90°; when η_{+1} is at maximum, $\eta_{-1} = 0$ and vice versa. $\alpha = 0, 90°, 180°$ corresponds to linear polarization of the wave. At such polarization of R' the diffraction efficiency is $\eta = \frac{1}{2}\eta_{max}$ and is independent of the direction of linear polarization (curve b, Fig. 3; ψ is the angle between the polarization direction of R' and the vertical axis). The polarization of waves diffracted in the ±1 order are right-hand and left-hand, respectively, regardless of the polarization of R'.

In other equal recording conditions, if waves R and S have orthogonal linear polarizations η is substantially smaller (~11%). In this case the reconstructing wave polarization does not affect η but only the diffracted wave polarization.

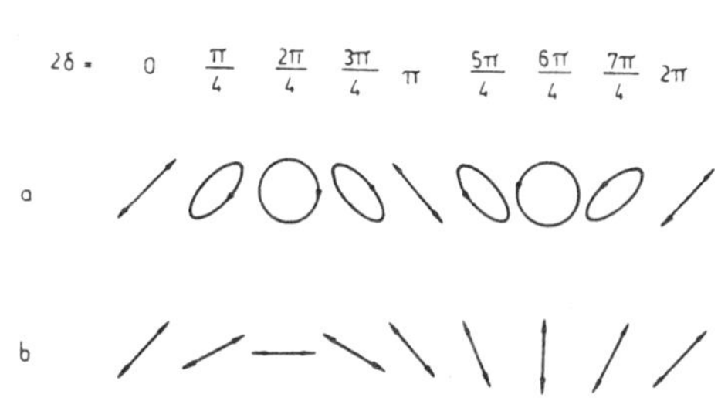

Fig. 2. Polarization interference pattern in recording when (a) the two recording waves have orthogonal linear polarizations; (b) the two recording waves have orthogonal circular polarizations. 2δ is the phase difference between the recording waves on the front surface of the sample.

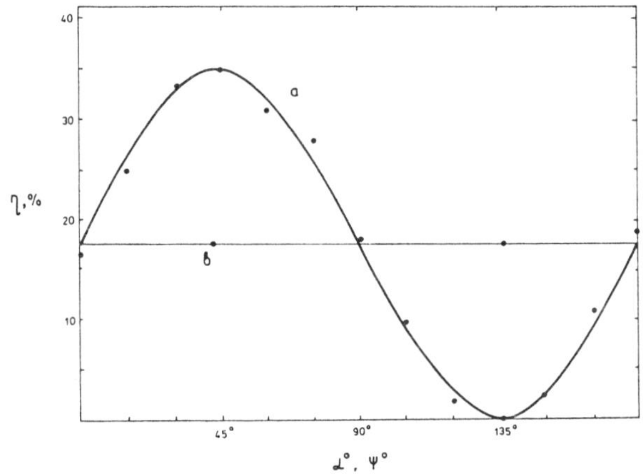

Fig. 3. Curve a, dependence of the diffraction efficiency on the angle α between the fast axis of the λ/4 plate (10 in Fig. 1) and the polarization direction of the readout beam after the polarization rotator (9 in Fig. 1); curve b, dependence of η on the polarization direction of the readout beam after the polarization rotator (9 in Fig. 1) at shifted λ/4 plate. ψ is the angle between the polarization direction of readout beam R' and the polarization direction of reference beam R in recording.

By way of comparison we should note that in holographic recording with parallel polarizations of R and S (same sample and intensity) the diffraction efficiency is ~1%.

IV. Polarization Holographic Gratings in Materials with Intrinsic Birefringence

If the recording medium has intrinsic birefringence before the recording (in our case when the polyvinyl alcohol matrix undergoes preliminary mechanical stretching), the phase difference between the two recording waves and their polarizations will vary with the depth of the sample; so will the polarization interference pattern. This change is shown in Fig. 4(a) when R and S are polarized in two mutually orthogonal directions at angles ±45° with the principal dielectric axes. As shown in Ref. 2 for this type of polarization recording, the diffraction efficiency η depends on the polarization direction of the linearly polarized reconstructing wave and reaches maximum when it coincides with that of the corresponding recording wave (here with the polarization of wave R). The ratio $m = \eta_{min}/\eta_{max}$ as well as the value η_{max} depends on the total phase difference between the two waves in the samples (ordinary and ex-

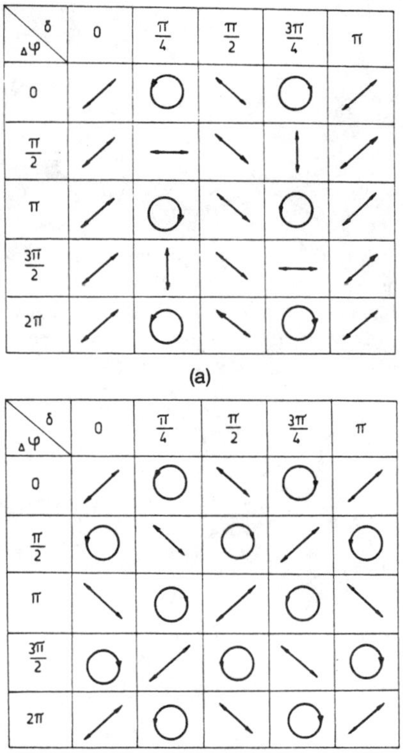

Fig. 4. Change of the polarization interference pattern in depth δ of the sample: (a) R and S are polarized at angles $\pm 45°$ to the principal dielectric axes; (b) R and S are polarized along the principal dielectric axes.

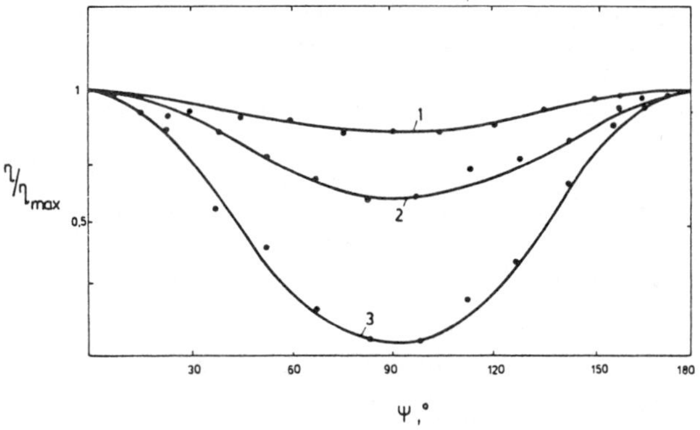

Fig. 5. Dependence of η on the polarization direction of the reconstructing wave in a sample with intrinsic birefringence (ψ is as in Fig. 3): curve 1, $\Delta\varphi = 30°$; curve 2, $\Delta\varphi = 60°$; curve 3, $\Delta\varphi = 140°$.

traordinary). Figure 5 shows the experimentally measured dependences of η, for the case corresponding to Fig. 4(a), on the polarization direction of the reconstructing wave R' at various values of the total phase difference $\Delta\varphi$. Here ψ denotes the angle between the polarization directions of the reconstructing wave R' and the corresponding recording wave R. The dependence of the ratio $m = \eta_{min}/\eta_{max}$ on the total phase difference $\Delta\varphi$ for curves such as those in Fig. 5 was calculated in Ref. 2. This relationship is plotted (solid lines) in Fig. 6 for the Fig. 4(a) polarization modulation. The same figure also illustrates the values of m, measured at different values of $\Delta\varphi$ in the range 0–π ($\lambda = 633$

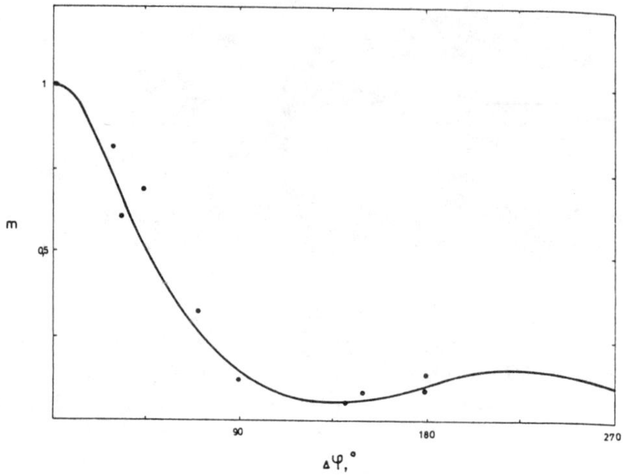

Fig. 6. Dependence of $m = \eta_{min}/\eta_{max}$ on the total phase difference $\Delta\varphi$ in recording with polarization modulation, as shown in Fig. 4(a).

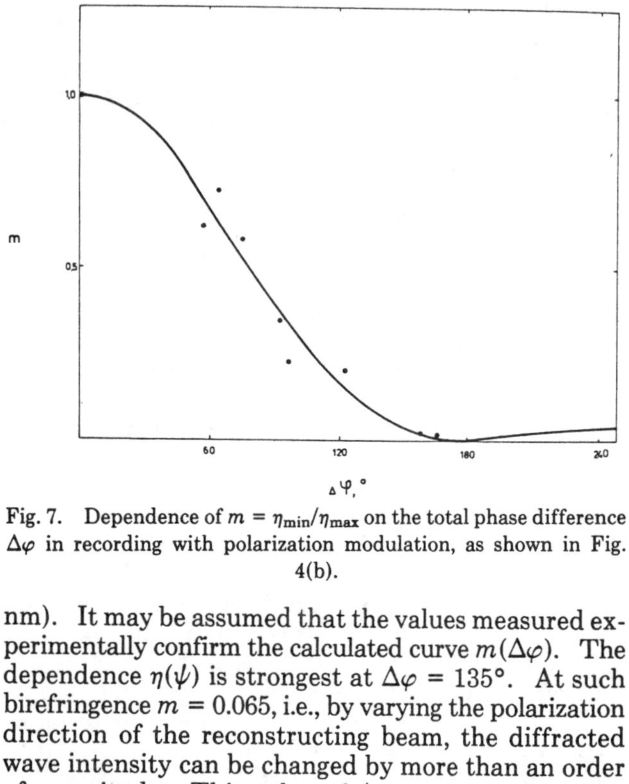

Fig. 7. Dependence of $m = \eta_{min}/\eta_{max}$ on the total phase difference $\Delta\varphi$ in recording with polarization modulation, as shown in Fig. 4(b).

nm). It may be assumed that the values measured experimentally confirm the calculated curve $m(\Delta\varphi)$. The dependence $\eta(\psi)$ is strongest at $\Delta\varphi = 135°$. At such birefringence $m = 0.065$, i.e., by varying the polarization direction of the reconstructing beam, the diffracted wave intensity can be changed by more than an order of magnitude. This value of $\Delta\varphi$ corresponds to the greatest absolute value of η_{max} obtained by us in this type of polarization recording with two orthogonal linear polarizations—$\eta = 28\%$. Figure 7 shows the analogous dependence for the case in Fig. 4(b) (the two recording waves are polarized along the principal dielectric axes). The relationship between the diffraction efficiency and polarization in this case is still stronger; it is described by the expression

$$\eta = \eta_0 \left[\cos^2\psi + \sin^2\psi \left(\frac{\sin\Delta\varphi}{\Delta\varphi}\right)^2\right], \quad (1)$$

where η_0 is the efficiency of a medium without intrinsic birefringence. As is evident from Eq. (1) and Fig. 7, m becomes 0 at $\Delta\varphi = 180°$ and is close to 0 for all values of $\Delta\varphi > 180°$. The presence of intrinsic birefringence in

Table I. Values of η_{max} and $m = \eta_{min}/\eta_{max}$ for Different Types of Recording

Type of recording (polarizations of R and S)	η_{max} (%)	m
(1) Parallel	1	1
(2) Orthogonal linear at 45° to the principal dielectric axes of the medium		
(a) $\Delta\varphi = 0°$	11	1
(b) $\Delta\varphi = 135°$	28	0.065
(3) Orthogonal linear, along the axes of the medium		
(a) $\Delta\varphi = 0°$	11	1
(b) $\Delta\varphi = 135°$	11	0
(4) Orthogonal circular, $\Delta\varphi = 0$	35	0

this case, however, does not lead to an increase in η_{max} and it remains the same as in the corresponding recording in a medium without birefringence (~11%).

The maximum values of η, obtained experimentally for the different cases of polarization recording in samples with equal thicknesses and optical densities are compared in Table I.

V. Summary

In conclusion we must point out that the new high-efficiency photoanisotropic medium we synthesized helped to obtain an experimental verification of the peculiar properties of polarization holographic gratings for the cases of recording media with and without intrinsic birefringence. The experiments revealed that the intensity of the +1 order diffracted wave is strongly dependent on the reconstructing beam polarization—on its ellipticity in the former case and on the polarization direction in the latter. In both cases the diffraction efficiency, corresponding to the optimal polarization of the reconstructing wave, substantially exceeds η in the recording with parallel polarizations of the two recording waves. The sensitivity of η to the polarization of the recording wave gives additional possibilities in using polarization holographic recording for raising the information capacity of the recording and solving problems in the field of optical image processing.

References

1. S. D. Kakichashvili, "Polarization Recording of Holograms," Opt. Spektrosk. **33**, 324 (1972); Opt. Spectrosc. **33**, 171 (1972).
2. L. Nikolova and T. Todorov, "Diffraction Efficiency and Selectivity of Polarization Holographic Recording," Opt. Acta **31**, 579 (1984).
3. T. Todorov, L. Nikolova, and N. Tomova, "Polarization Holography. 1: A New High-Efficiency Organic Material with Reversible Photoinduced Birefringence," Appl. Opt. **23**, 4309 (1984).
4. L. Nikolova and T. Todorov, "Volume Amplitude Holograms in Photodichroic Materials," Opt. Acta **24**, 1179 (1977).
5. M. Attia and J. M. C. Jonathan, "Anisotropic Gratings Recorded from Two Circularly Polarized Coherent Waves," Opt. Commun. **47**, 85 (1983).

SPATIAL FREQUENCY DEPENDENCE OF THE ENERGY TRANSFER IN TWO-WAVE MIXING EXPERIMENTS WITH BSO CRYSTALS

H. RAJBENBACH, J.P. HUIGNARD and B. LOISEAUX
Thomson-CSF, Laboratoire Central de Recherches, Domaine de Corbeville, B.P. no. 10, 91401 Orsay, France

Received 4 August 1983

Energy transfer between two beams interfering in an electrooptic $Bi_{12}SiO_{20}$ crystal is analyzed as a function of the spatial frequency of the recorded hologram. We report high coefficients in a wide range of low spatial frequencies ($\Lambda^{-1} \simeq 44$ mm^{-1}), when the drift length of the photocarriers becomes comparable to the fringe spacing. Exponential gain coefficients of $\Gamma = 7$ cm^{-1} have been reached for high intensity beams ratio ($\beta > 10^4$) and an optimized fringe displacement speed (moving grating recording). Application to coherent image amplification (×20) is shown.

1. Introduction

The recording of a phase volume hologram in photorefractive media leads to a stationary energy exchange between the two incident coherent light beams interfering in the crystal. This energy redistribution which has been observed in different electro-optical crystals [1–6] ($LiNbO_3$, $KNbO_3$, $BaTiO_3$, $Bi_{12}SiO_{20}$, ...) is due to a self-diffraction process of the reference beam in the dynamic phase grating photoinduced in the crystal volume. The beam coupling effect which is related to the phase shift ψ between the incident fringe pattern and the photoinduced index modulation brings up new applications such as coherent image processing and amplification, or self oscillations [7–10]. From diffraction considerations, it has been shown that the maximum of the energy transfer is reached for $|\psi|' = \pi/2$ [1,2]. In the photorefractive BSO (or BGO) crystals this condition is obtained by recording with an external applied electric field E_0 and moving the fringe pattern at a constant velocity V_0 across the crystal aperture [11] ("drift" recording mode with non local response). In a two wave mixing experiment such a displacement increases the amplitude of the $\pi/2$ phase shifted component of the index modulation and so induces the energy transfer from the reference beam to the low intensity probe beam. In this paper we report the dependence of the exponential gain coefficient Γ versus the spatial frequency of the written grating ($10 < \Lambda^{-1} < 400$ mm^{-1}), the incident beam intensity ratio, and the pump beam intensity. It is experimentally shown that for an optimized fringe displacement speed, very large values of the gain are obtainable, when the drift length L of the photocarriers becomes comparable to the fringe spacing Λ ($L = \mu\tau E_0$; μ and τ are respectively the mobility and life time of the photoexcited charges.)

In view of imaging applications, the photoinduced modulation versus the spatial frequency of the written grating in BSO was already considered in ref. [12] but the crystal response was first analyzed in terms of diffraction efficiency since all conditions for an efficient energy transfer where not satisfied.

2. Two wave mixing experiments at $\lambda = 568$ nm

The two beams (respectively reference and signal) interfere in the crystal so creating a phase volume grating (fig. 1). The incident beams polarizations are chosen parallel to the electro-optical axis X'. (Electric field applied in the 001 direction). For such conditions, incident and self-diffracted waves have the same state of polarization thus giving rise to a maximum of ener-

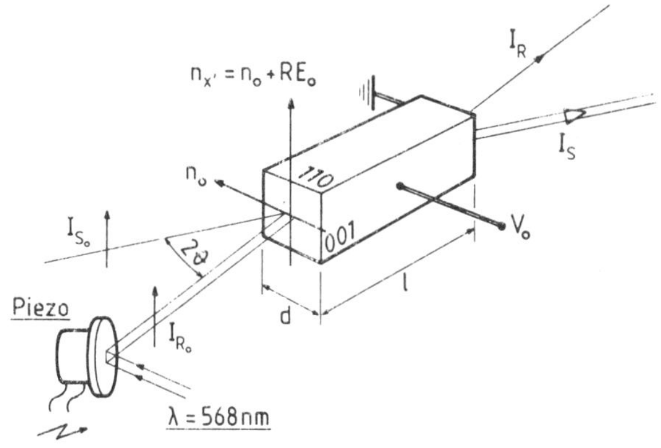

Fig. 1. Two wave mixing experiment in photorefractive BSO crystals. Electro-optic configuration for energy transfer. Inter-electrode distance d = 5 mm. Interaction length l = 10 mm.

gy transfer [4]. (Applying the voltage in the 110 direction induces a self-diffracted wave whose polarization is perpendicular to the one of the incident reference beam. Consequently self interference and soforth energy exchange cannot take place.) The non local response of the crystal is induced by moving the fringes at a constant speed v with a piezo mirror (fig. 1). This moving grating results from the coherent interference of the signal beam with the Doppler shifted reference beam due to the piezo mirror. The corresponding frequency translation is $\delta\omega = 4\pi v \sin\theta/\lambda$ [11–16]. The speed is adjusted such as at any time the index modulation is recorded with a spatial shift with respect to the incident moving fringe pattern. According to the formalism of the ref. [2], the intensity of the signal beam emerging from the crystal takes the form

$$I_S = I_{S_0} \exp(\Gamma - \alpha)\, l\,,$$

where I_{R0}, I_{S0} are the reference and signal beam intensities ($I_{S0} \ll I_{R0}$), α is the absorption coefficient at recording wavelength λ, $\beta = I_{R0}/I_{S0}$ is the incident beam intensity ratio, l is the crystal length, Γ is the exponential gain coefficient.

From the coupled waves equations applied to dynamic holographic medium, it results that the gain coefficient Γ in the steady state is given by the following relation [11]

$$\Gamma = \frac{4\pi\Delta n_s}{\lambda}\;\frac{Kv\tau}{1 + K^2v^2\tau^2}\,,$$

where $K = 2\pi/\Lambda$; Λ grating spacing. Δn_s is the maximum value of the index modulation which can be induced through the photorefractive effect; τ is the time constant to build-up the phase volume grating. The maximum of the energy transfer is so obtained for a fringe displacement speed $V_0 = \Lambda/2\pi\tau$ which corresponds to a frequency shift $\delta\omega_0 = 1/\tau$. According to the Kukhtarev model [2], Δn_s is a function of the applied electric field E_0, diffusion field E_D, fringe spacing Λ and trapping centers density. In order to induce appreciable values of the signal beam amplification, the experiments reported in this paper have been done with a crystal interaction length l = 10 mm and by choosing as a recording wavelength the yellow line of a single mode krypton laser (λ = 568 nm) which insures a low value of the absorption coefficient. ($\alpha \approx 0.6$ cm^{-1} including the reflection losses due to crystal interfaces, bulk absorption $\alpha_0 = 0.13$ cm^{-1}).

For such a wavelength, the typical recording-erasure cycle for an elementary grating ($\beta = 1$) is given in fig. 2. Typical writing-erasure energies S^{-1} are of the order of 100 mJ cm^{-2}, corresponding to a steady state diffraction efficiency η_s = 10% at a fringe spacing Λ = 6 μm. These values of energies are higher than those measured at the usual green line of the argon laser (S^{-1} = 500 J cm^{-2} at λ = 514 nm). This reduction of sensitivity seems principally due to a decrease of the quantum efficiency q of the photocarriers generation process at the considered wavelength. (q = 0.7 at λ = 514 nm; q = 0.1 at λ = 568 nm, from ref. [17]).

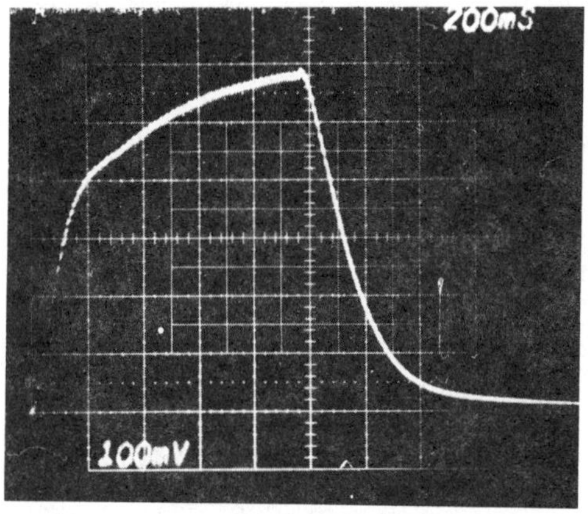

Fig. 2. Grating recording erasure cycle at λ = 568 nm. Incident beam intensities $I_{R0} = I_{S0}$ = 120 mW cm^{-2}. E_0 = 6 kV cm^{-1}. Fringe spacing Λ = 6 μm. Steady state diffraction efficiency η_s = 10% monitored with a low power He-Ne laser.

For the beam coupling experiments we chose a crystal orientation with respect to the reference beam and a direction of the fringes displacement such as the energy transfer is induced from the pump beam to the low intensity signal beam ($\beta = I_{R0}/I_{S0} \gg 1$).

The corresponding signal beam amplification will be experimentally characterized by the measurement of the parameter γ_0 defined as

$$\gamma_0 = \frac{I_S \text{ (with pump)}}{I_S \text{ (without pump)}} = \exp(\Gamma)\, l\,.$$

Figs. 3 show the dependence of γ_0 and Γ versus the spatial frequency of the written grating (3a) (explored spatial frequency range $10 < \Lambda^{-1} < 400$ mm^{-1}) and as a function of the applied electric field E_0 (3b). For each measurement, the fringes displacement speed is carefully adjusted such as the maximum of γ_0 is reached. The incident pump beam intensity is I_{R0} = 140 mW cm^{-2} and the intensity ratio of the two interfering waves is $\beta = 10^3$. These curves point out a strong increase of the energy transfer for $\Lambda^{-1} < 100$ mm^{-1} with a sharp maximum at $\Lambda_0^{-1} \approx 45$ mm^{-1}. The condition of signal beam amplification $\Gamma > \alpha$ is thus obtained for a wide range of the experimental conditions and for such beam ratio β a peak value of the exponential gain coefficient $\Gamma = 5$ cm^{-1} is reached for $E_0 = 10$ kV cm^{-1}. In the range of spatial frequencies considered on figs. 3 the diffusion field E_D is negligeable ($E_D \ll E_0$) and the condition $E_0 \ll E_q$ is also fulfilled, E_q = maximum steady state value of the photoinduced space charge field corresponding to a charge separation by a grating period [2], $E_q = (e/2\pi\epsilon_0\epsilon_r)N\Lambda$; E_q (V cm^{-1}) = $10^8\,\Lambda$ for a trapping density $N = 2 \times 10^{16}$ cm^{-3}. Therefore, the energy transfer is negligeable for a zero fringe displacement speed ($\Gamma \approx 0$ at $V = 0$). Fig. 4 represents the variations of Γ versus the incident beam ratio β corresponding to a spatial frequency of the grating near the optimum condition ($\Lambda_0^{-1} \approx 45$ mm^{-1}, $E_0 = 10$ kV cm^{-1}). This curve indicates an increase of the gain coefficient with the

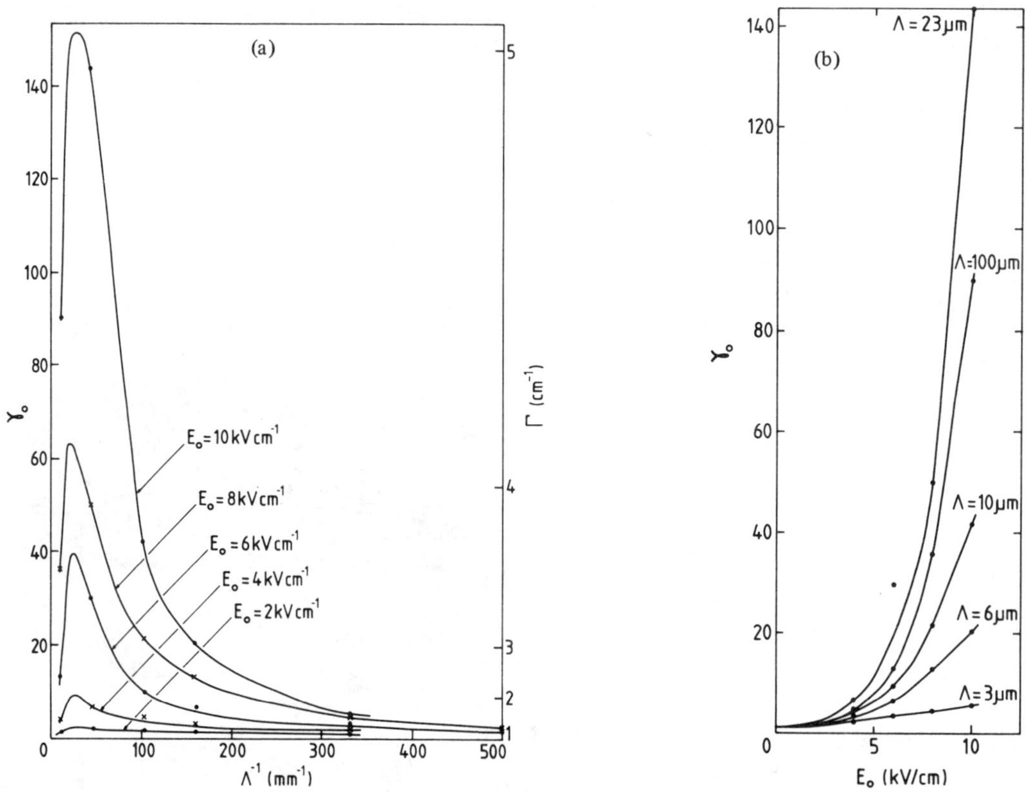

Fig. 3. Evolution of the gain coefficients γ_0, Γ versus the spatial frequency (a) and the applied electric field (b). Optimum fringe displacement speed $V_0 \approx \Lambda/2\pi\tau$. Incident beam intensities I_{R0} = 140 mW cm^{-2}; I_{S0} = 140 μW cm^{-2}. Beam ratio $\beta = 10^3$.

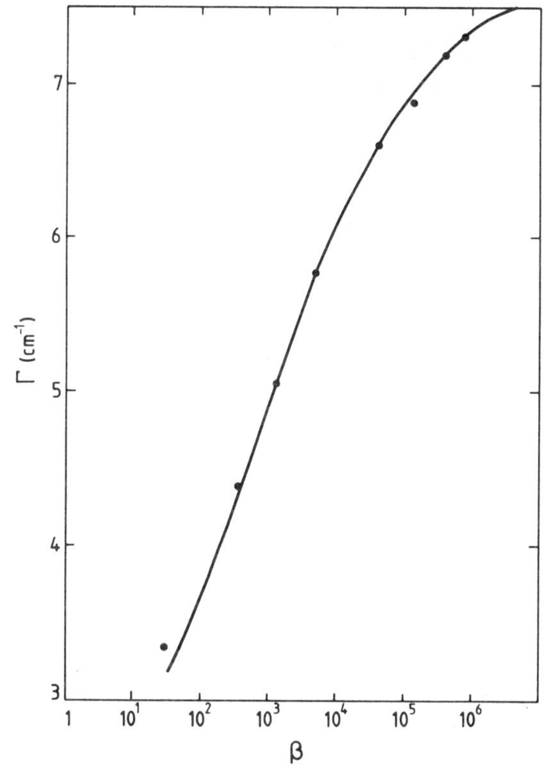

Fig. 4. Dependence of the exponential gain coefficient Γ on the incident beam ratio $\beta = I_{R0}/I_{S0}$. I_{R0} = 140 mW cm^{-2}, Λ = 22 μm, E_0 = 10 kV cm^{-1}.

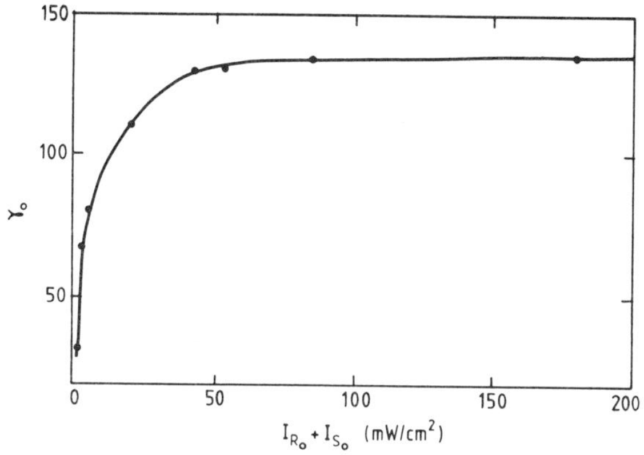

Fig. 5. Dependence of the gain coefficient γ_0 on the total incident beam intensity. $\beta = 2 \times 10^3$, Λ = 22 μm, E_0 = 10 kV cm^{-1}.

beam ratio β and a saturation value Γ = 7 cm^{-1} is reached for $\beta > 2 \times 10^6$. For a given beam ratio $\beta = 2 \times 10^3$, fig. 5 shows the evolution of the gain coefficient γ_0 versus the total incident intensity $I_{R0} + I_{S0}$. For the applied electric field E_0 = 10 kV cm^{-1}, this set of curves points out the experimental conditions ensuring a high signal beam amplification:

$$I_{R0} > 50 \text{ mW cm}^{-2}, \quad \Lambda_0^{-1} \sim 45 \text{ mm}^{-1},$$

$$\beta > 10^4 .$$

With such parameters, exponential gain coefficient values as high as Γ = 7 cm^{-1} are reached and taking account of the absorption and reflection losses ($\alpha \sim 0.6$ cm^{-1}) a signal beam amplification up to 6×10^2 was obtained. The time constant τ to buildup the energy exchanges is inversely proportional to the incident pump beam intensity. For I_{R0} = 120 mW cm^{-2}, τ is typically equal to 150 ms at λ = 568 nm and the corresponding fringe displacement speed is V_0 = 30 μm s^{-1} – related frequency shift due to the piezo mirror $\delta\omega/2\pi \sim 1$ Hz.

3. Discussions

A qualitative analysis of the dependence of the gain coefficient Γ versus the spatial frequency Λ^{-1} and incident beam ratio β points out the influence of a large transient energy transfer (TET) during the hologram formation in BSO crystals, when no fringe displacement exists ($V = 0$). Such a phenomena has been firstly reported and interpreted by Kukhtarev et al. in ferroelectric $LiNbO_3$ [18,19] and more recently by Günter in BGO crystals [20]. Description of the effect is the following: at the initial stage of the hologram recording the interference pattern and the photoinduced index modulation coincide and are normal to the crystal face. Then, self-diffraction of the beam tilts the illumination pattern and due to the crystal inertia for the index change it results a transient energy transfer and consequently a transient increase of the signal beam intensity.

Some specific characteristics of TET in BSO crystals are the following:

– TET takes place only for two beams having different intensities ($\beta \neq 1$) and is more pronounced as β increases.

– The peak value of the transmitted signal beam intensity increases as low spatial frequencies. For an applied electric field $E_0 = 10$ kV cm^{-1} and $\Lambda^{-1} < 200$ mm^{-1} the TET is becoming oscillatory due to transient spatial phase shift which may exceed $\pi/2$ in certain regions of the crystal. When no oscillation occurs, the time dependence of the transmitted signal beam intensity is given by the following relation [18–21]:

$$I_S = I_{S0} + (\tau l\Delta)^2 \times \frac{I_{S0} I_{R0}}{(I_{S0} + I_{R0})^2} (I_{R0} - I_{S0}) e^{-t/\tau} (1 - t/\tau - e^{-t/\tau}),$$

$$\gamma_0 \approx 1 + (\tau l\Delta)^2 \left(\frac{\beta - 1}{\beta + 1}\right) e^{-t/\tau} (1 - t/\tau - e^{-t/\tau}) .$$

For a given field E_0, Δ depends on the relative value of the photocarriers drift length L with respect to the fringe spacing Λ.

With the fringe pattern now moving at the resonance velocity V_0, we obtain a steady-state amplification of the signal beam intensity. For the large gain resulting from the moving grating (see fig. 6) we can formulate the following reasons: the moving grating gives rise to an additional phase shift due to the fringe displacement with respect to the time response of the crystal [11]. But it has also been shown by Stepanov et al. [22] that

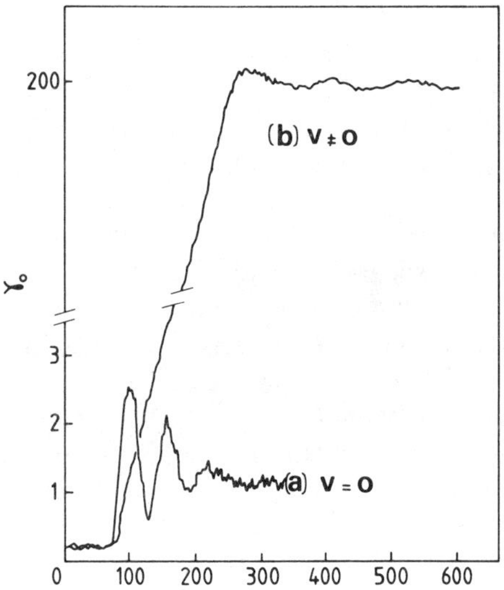

Fig. 6. (a). Transient oscillatory energy transfer on the signal beam for a fringe spacing $\Lambda = 35$ μm. No fringe displacement ($V = 0$); $I_{R0} = 120$ mW cm^{-2}, $\beta = 2 \times 10^3$, $E_0 = 10$ kV cm^{-1}. (b) Corresponding time evolution of the gain γ_0 due to the optimized fringe displacement speed.

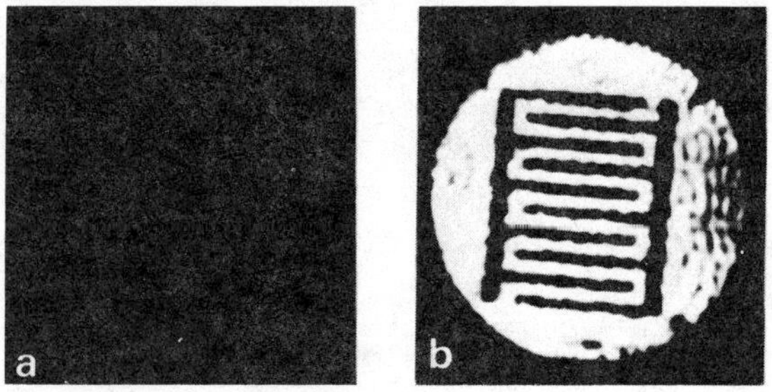

Fig. 7. Image amplification through energy transfer in two wave mixing experiments at $\lambda = 568$ nm. (a) Low intensity incident image. (b) ×20 amplified image ($I_S/I_{S0} = 20$), $E_0 = 6$ kV cm^{-1}.

illumination of a holographic grating in photorefractive crystals creates a charge pattern moving at velocity $V_r = L\tau^{-1}/(1 + K^2L^2)^{1/2}$. When now recording the hologram with a fringe pattern moving synchronously with the resonance velocity V_r, it results an increase of the steady state index modulation Δn_s [22] and consequently an increase of the gain coefficient (fig. 6). It thus appears that the theoretical description of the reported experiments should consider both the influences of the TET at $V = 0$ and the enhancement of the index modulation when recording with a moving fringe pattern.

4. Application to coherent image amplification

The large values of the gain permit the amplification of a low intensity complex signal wavefront. A binary photographic transparency is inserted on the signal beam path (fig. 1) and an amplified image due to energy transfer is received on a detector screen. Fig. 7 shows the experimental results, (a) is the low intensity signal beam, and (b) is the ×20 amplified image. ($I_S/I_{S0} = 20$, $E_0 = 6$ kV cm^{-1}.) The fringe displacement speed and reference-signal beams angles are adjusted such as the maximum gain corresponds to the hologram spatial carrier frequency $\Lambda_s^{-1} \sim 45$ mm^{-1}. Due to the bandpass characteristic of the amplifier (see fig. 3a) the difference in gain for the various spatial frequencies may be noticeable and limit the size of the image to be amplified and processed.

5. Conclusions

It has been experimentally shown that Γ is strongly dependent on the grating fringe spacing Λ applied electric field E_0 and intensity ratio β of the two interfering waves. Values of Γ as high as 7 cm^{-1} have been measured with optimized parameters at recording wavelength $\lambda = 568$ nm: $\Lambda_0 \sim 22\ \mu m$, $\beta = 10^6$, $E_0 = 10$ kV cm^{-1}, $I_{R0} > 50$ mW cm^{-2}. Beside this, a wide range of experimental conditions permits the gain to overpass the absorption and reflection losses. For the first time BSO gain coefficients approach the values already reported with ferroelectric crystals such as $LiNbO_3$, $BaTiO_3$ and SBN [6–8]. But contrary to $BaTiO_3$, these high values of Γ are now obtained in the low spatial frequency range ($\Lambda_0^{-1} < 100\ mm^{-1}$). These results in BSO can suggest new applications and improved performances of the crystal in the field of coherent image amplification and phase conjugation.

Acknowledgement

The authors are grateful to I. Ledoux, F. Gedin, L. Thépaut, D. Rak, G. Illiaquer and J.P. Herriau for very helpful contribution and discussion on this particular work. They also express their thanks to the D.R.E.T. for partial support of these studies.

References

[1] D.L. Staebler, in: Holographic recording materials, ed. H.M. Smith (Springer Verlag, 1977) p. 101.
[2] N.V. Kukhtarev, V.B. Markov, S.G. Odulov, M.S. Soskin and V.L. Vinetski, Ferroelectrics 22 (1979) 949.
[3] A. Krummins and P. Günter, Appl. Phys. 19 (1979) 153.
[4] A. Marrakchi, J.P. Huignard and P. Günter, Appl. Phys. 24 (1981) 131.
[5] J. Feinberg, D. Heinann, A.R. Tanguay and R.L. Hellwarth, J. Appl. Phys. 51 (1980) 1297.
[6] B. Fischer, M. Cronin-Golomb, J. White and A. Yariv, Optics Lett. 6 (1981) 519.
[7] J.P. Huignard and A. Marrakchi, Optics Lett. 6 (1981) 622.
[8] J. Feinberg and R.W. Hellwarth, Optics Lett. 5 (1980) 519.
[9] M. Cronin-Golomb, B. Fischer, J. Nilson, J. White and A. Yariv, Appl. Phys. Lett. 41 (1982) 219.
[10] R. Jain and G. Dunning, Optics Lett. 7 (1982) 420.
[11] J.P. Huignard and A. Marrakchi, Optics Comm. 38 (1981) 249.
[12] J.P. Huignard, J.P. Herrieu, G. Rivet and P. Günter, Optics Lett. 5 (1980) 102.
[13] M. Petrov, T. Pencheva and S. Stepanov, J. Optics 12 (1981) 287.
[14] J.P. Herriau, J.P. Huignard and P. Aubourg, Appl. Optics 17 (1978) 1851.
[15] A.R. Tanguay, J. Opt. Soc. Am. 12 (1981) 1832.
[16] J. Lam, Appl. Phys. Lett. 42 (1983) 155.
[17] R. Sprague, J. Appl. Phys. 46 (1975) 1673.
[18] N. Kukhtarev, V. Markov and S. Odulov, Optics Comm. 23 (1977) 338.
[19] N. Kukhtarev, V. Markov and S. Odulov, Sov. Phys. Tech. 25 (1980) 1109.
[20] P. Günter, Optics Comm. 41 (1982) 83.
[21] V.L. Vinetski, N. Kukhtarev, S. Odulov and M. Soskin, Sov. Phys. Usp. 22 (1979) 742.
[22] S.I. Stepanov, V. Kulikov and M. Petrov, Optics Comm. 44 (1982) 19.

Reprinted from *Comptes Rendus*, Vol. 7, 832–833 (1838).

MÉMOIRES LUS.

OPTIQUE. — *Sur l'absorption dans les milieux colorés biréfringents;* *par* M. BABINET.

(Commissaires, MM. Biot et Arago.)

(Extrait par l'Auteur.)

« L'absorption dans les milieux colorés biréfringents, diffère essentiellement du phénomène ordinaire auquel se rapportent les couleurs propres des corps. Là, comme dans beaucoup d'autres parties de l'optique moderne, les premières observations, celles qui constituent la découverte, sont dues à M. Arago. Il fit voir que dans les milieux colorés biréfringents, et notamment dans le sulfate jaunâtre de baryte, la lumière polarisée développait des teintes variables, suivant le sens où elle traversait le cristal. M. Biot et M. Brewster, mais surtout M. Biot, étudièrent ensuite le phénomène, mais sans en formuler la loi définitive. Dans une Note communiquée anciennement à l'Académie, j'avais annoncé que les cristaux colorés biréfringents, dits *négatifs* ou *répulsifs*, comme le spath d'Islande, la tourmaline, le saphir, absorbent en plus grande abondance le rayon ordinaire, tandis que les cristaux dits *positifs* ou *attractifs*, comme le quartz, le zircon, l'apophyllite ordinaire, éteignent en plus grande partie le rayon extraordinaire, et laissent par conséquent passer de la lumière partiellement polarisée, comme le rayon ordinaire. Il y a cependant quelques exceptions, comme pour le béryl, mais non pour l'émeraude. Enfin, dans le Mémoire sur les propriétés optiques des minéraux, le dichroïsme m'avait conduit à énoncer la loi du phénomène à la démonstration duquel se rapportent les expériences du présent Mémoire.

» Tout cristal biréfringent coloré, éclairé par de la lumière neutre, peut être considéré comme donnant passage à une certaine quantité de lumière colorée *non polarisée,* laquelle passe librement dans tous les sens, indépendamment de son axe ou de ses axes optiques (et par rapport à cette lumière, il est analogue à tout autre milieu coloré), et de plus, à une certaine quantité de lumière polarisée suivant la loi de M. Biot pour les cristaux bi-axes. Cette dernière quantité étant nécessairement variable quand on transmet de la lumière polarisée au travers du cristal, ou quand on analyse la lumière transmise avec un appareil biréfringent, il en résulte (quand cette lumière variable est elle-même colorée) une série nombreuse de nuances, depuis celle où la lumière non polarisée passe seule par l'extinction de la lumière polarisée, jusqu'à la nuance formée de l'ensemble de la lumière non polarisée et de la lumière polarisée tout entière. Ces deux cas extrêmes ont été seuls notés, sous le nom de *dichroïsme;* mais réellement les cristaux colorés biréfringents à un ou à deux axes, sont polychroïtes, et la règle suivante donnera la teinte d'un cristal quelconque :

» Prenez la teinte de la lumière non polarisée que transmet ce cristal en tous sens et la teinte polarisée variable suivant la direction de l'axe ou des axes qu'il transmet dans la direction donnée et pour l'azimuth donné, et la teinte résultante d'après les formules ordinaires de mélange des couleurs, sera la couleur actuelle du cristal. C'est ce qui résulte de quelques-unes des expériences exposées en détail dans le présent Mémoire.

» Ainsi, le phénomène de l'absorption chromatique des milieux polarisés, se trouve ramené à l'absorption ordinaire; mais celle-ci continue, avec la dispersion, le mélange des couleurs et la non-influence du mouvement de la Terre sur certains phénomènes d'optique, à former quatre grandes difficultés de la science, qui peut-être ne sont pas cependant insurmontables, puisque, de nos jours, avant Fresnel, il y avait encore une cinquième classe de phénomènes inexplicables, la diffraction. »

PHYSIQUE. — *Recherches expérimentales sur le passage de la chaleur d'un corps solide dans un autre corps solide; par* M. DESPRETZ.

(Commissaires, MM. Gay-Lussac, Pouillet.)

(Extrait par l'auteur.)

» Aucun physicien n'a publié de travail expérimental sur ce sujet. M. Poisson l'a considéré sous un point de vue théorique, dans un Mémoire *sur la Théorie mathématique de la Chaleur.*

» Le nouveau travail est propre à fournir des données que l'analyse mathématique réclamait, et à déterminer si la chaleur éprouve au contact d'un corps étranger un obstacle comme l'électricité en mouvement : on sait que ce dernier agent subit un changement de vitesse quand il change de conducteur.

» Pour résoudre la question, l'auteur a joint deux barres bout à bout par pression, au moyen d'une vis en bois : l'une de ces barres était en cuivre, l'autre en étain.

Reprinted from *Comptes Rendus*, Vol. 4, 758–766 (1837).

MÉMOIRES PRÉSENTÉS.

OPTIQUE MINÉRALOGIQUE. — *Précis d'un mémoire sur les caractères optiques des minéraux; par* M. BABINET.

(Commissaires, MM. Arago, Dulong.)

(Extrait par l'auteur.)

« Les caractères optiques des minéraux sont les propriétés diverses dont ils jouissent dans leur action sur la lumière, et qui peuvent servir à reconnaître ou à distinguer l'une de l'autre les nombreuses substances que nous offre la nature. Les cinq classes suivantes de phénomènes optiques servent principalement à caractériser les minéraux : 1° l'absorption sans polarisation et sans double réfraction; 2° l'absorption avec polarisation; 3° le dichroïsme ou polychroïsme; 4° les caractères analogues aux phénomènes de réseaux et de couronnes; 5° l'astérie et les phénomènes analogues; 6° la polarisation chromatique et ses applications.

I. *Absorption sans polarisation.*

» Toute substance non cristallisée, et, parmi les cristaux, tout cristal simplement réfringent ne transmet que de la lumière neutre et généralement, d'après M. Arago, toute extinction ou absorption de lumière ou de couleurs, non accompagnée de double réfraction, porte également sur les deux faisceaux polarisés en sens contraire, dont on peut supposer que toute lumière naturelle ou neutre est composée. Il ne s'agit donc ici que de l'absorption de telle ou telle couleur de préférence par une substance minérale colorée ou non. On peut citer pour type de cette absorption, celle qui a lieu dans les verres colorés. Pour l'observer, on emploiera les procédés suivants : 1° deux faces non parallèles du minéral forment un prisme au travers duquel on regarde la flamme d'une bougie. Par exemple, avec le rubis (corindon), l'extrémité violette du spectre sera supprimée, mais elle ne le sera pas avec le grenat; 2° illuminez le minéral avec la lumière monochromatique de l'alcool versé sur du sel ordinaire, le rubis sera presque noir et le grenat réfléchira encore une quantité notable de cette lumière jaune, qui correspond à la raie D de Fraunhofer, dans le spectre solaire; ou bien mettez l'échantillon sous la lumière d'un verre rouge, ou même, observez-le avec le verre rouge placé devant l'œil. Enfin, si vous avez à votre disposition un spectre solaire dans une chambre obscure, promenez le minéral dans les diverses nuances. Le saphir, la cordiérite (dichroïte), le disthène et la topaze bleue du Brésil ne se ressemblent plus alors, pour la teinte réfléchie ou transmise. Dans quelques cas, et notamment avec le rubis, on peut faire une expérience analogue à la suivante: regardez la flamme assez brillante de l'alcool salé avec une plaque liquide de sulfate de chrôme, comprise entre deux verres plans, et cette flamme sera complétement invisible.

II. *Absorption avec polarisation.*

» Tout cristal biréfringent coloré, absorbe inégalement les deux faisceaux polarisés de la lumière neutre. Certains cristaux de tourmaline, de cordiérite, d'épidote ne transmettent même que de la lumière polarisée. J'ai communiqué depuis plusieurs années à l'Académie, cette loi qui admet peu d'exceptions, savoir, que les cristaux biréfringents colorés, dits répulsifs ou négatifs (comme le spath coloré, la tourmaline, le corindon) laissent passer en plus grande abondance le faisceau extraordinaire, tandis que les cristaux, dits positifs ou attractifs (comme le quartz enfumé, le zircon, le gypse), transmettent en excès la lumière ordinaire. Ainsi, pour chaque espèce minérale, ou même pour chaque variété, la nature de la lumière transmise devrait être observée et notée ainsi que l'intensité de l'action polarisante. Par exemple, des amphiboles, des topazes bleues, parfaitement semblables pour la couleur, le poli, l'aspect, l'éclat, à des tourmalines polarisant complétement, ne donnent elles-mêmes qu'une très faible polarisation.

» Les quatre polariscopes dont j'ai fait usage sont dus à MM. Biot, Babinet, Arago et Savart. Le dernier surtout est éminemment applicable à l'optique minéralogique. On peut voir dans l'article que j'ai fourni à la traduction française de l'optique de M. Herschel, de quelle importance sont les appareils au moyen desquels on peut reconnaître les plus faibles traces de polarisation, depuis que M. Arago a ramené en grande partie toute la photométrie aux polariscopes.

III. *Dichroïsme.*

» Tout cristal qui polarise complétement la lumière, l'éteint aussi complétement suivant son axe ou suivant ses axes. Mais on remarque que pour une certaine couleur simple ou composée, plusieurs cristaux ne jouissent pas de la propriété polarisante et de la propriété d'absorption complète suivant l'axe. La tourmaline rouge, le rubis, laissent passer des rayons rouges non polarisés, et si l'on croise deux tourmalines rouges, cette même teinte se fait jour au travers du système, et les anneaux du spath interposé sont traversés par une croix rouge au lieu d'une croix noire qui aurait été produite avec deux tourmalines complétement polarisantes. Aussi, dans l'appareil de M. Herschell, les deux tourmalines que l'on veut croiser doivent être de teinte différente. Avec l'émeraude, les bandes noires du polariscope de M. Savart sont vertes; elles sont bleues avec l'aigue-marine et le saphir; elles sont rouges avec la sibérite et le rubis et manquent complétement quand on examine de la même manière le grenat et le spinelle. . . . .

. . . .

» On peut donc formuler comme il suit l'effet du dichroïsme : suivant l'axe ou les axes d'un cristal coloré biréfringent, on n'a que la couleur non soumise au *tamisage* polarisateur du cristal (si l'on peut se servir de cette expression) et *cette couleur n'est pas polarisée,* comme je l'ai très soigneusement observé, tandis que suivant toute autre direction, il se mêle à cette couleur non polarisée, qui passe dans tous les sens, la lumière ou teinte polarisée plus ou moins abondante que laisse passer l'action de l'axe ou des axes; en sorte que si par un procédé convenable de polarisation on supprime celle-ci, on retrouve la même teinte que suivant l'axe.

» On explique facilement par ce principe la coloration singulière observée par M. Brewster autour des axes de la cordiérite (autrement appelée dichroïte, iolite, saphir d'eau). J'ai fait tailler en sphère parfaite un très beau cristal de cette substance, qui dans un sens était du plus beau bleu de saphir, et dans l'autre d'un blanc jaunâtre mal défini. En observant cette sphère à la lumière ordinaire, on aperçoit deux taches bleues aux deux extrémités des axes, et suivant toute autre direction il venait se mêler à cette teinte bleue une proportion plus ou moins grande de la lumière polarisée suivant la loi de M. Biot pour les cristaux biaxes, que laissent passer les axes.

» On voit donc que ces mélanges en proportions variables doivent offrir des teintes multiples dont on n'a jusqu'ici noté que les extrêmes qui ont lieu suivant les axes et suivant une direction perpendiculaire au plan de ceux-ci; c'est un véritable polychroïsme et non un dichroïsme, comme on le dit ordinairement.

IV. — *Caractères analogues aux phénomènes de réseaux et de couronnes.*

» Tous les chatoiements superficiels, tels que ceux du corindon harmophane, de l'hyperstène, du labrador, de la nacre de perle, sont des phénomènes de réseaux dont j'ai donné la théorie dans les *Annales de Chimie et de Physique*. Les phénomènes de transmission de la nacre de perle, de l'agate irisée à couleurs récurrentes, sont aussi du même genre. Je ne parle pas ici des reflets intérieurs de la nacre, de l'opale, de la pierre de lune, de la cymophane, dont je n'ai pas l'explication. Quant aux chatoiements de surface et à ceux de transmission, les couleurs efficaces sont d'autant plus distantes du rayon réfléchi ou transmis régulièrement que les lames ou les stries de la substance sont plus fines. J'ai de beaux échantillons d'agate dont les couches sont espacées par soixantièmes de millimètre. Les couches de la nacre ne sont pas distantes de plus d'un centième de millimètre. Enfin, dans l'hyperstène et dans le labrador, la distance est encore beaucoup moindre, en sorte que les couleurs reflétées sont encore plus loin de la lumière directe et d'une teinte moins variable. C'est aussi le secret des singulières couleurs du colibri qui ne sont visibles que presque à l'opposite de la direction de la lumière incidente. On sait que dans ces cas, et sous l'incidence perpendiculaire, le produit de la distance des couches ou des stries par le sinus de l'angle de déviation de la couleur transmise est égal à l'intervalle fondamental des interférences. Soit a la distance de deux raies ou stries, α l'angle de déviation et λ l'intervalle fondamental des interférences, on a

$$a \sin \alpha = \lambda.$$

» Quant aux phénomènes analogues aux couronnes, on sait que si plusieurs fils égaux d'un diamètre d sont interposés entre l'œil et une bougie, on voit autour de celle-ci un ou plusieurs anneaux colorés tels que la distance angulaire α de chaque point de la circonférence du premier anneau ou couronne à la bougie est donnée par la relation suivante :

$$d \sin \alpha = \tfrac{1}{2} \lambda.$$

» Les autres anneaux ou couronnes ont des distances à la bougie qui sont successivement double, triple, quadruple de la distance de la première. Ceci peut servir à mesurer la grosseur des fibres ou cristaux élémentaires dans les cristaux fibreux, et indiquer en même temps si ces petits cristaux ou fibres sont de dimensions uniformes. Ainsi, avec le quartz œil-de-chat qui contient des filaments d'amianthe, avec le béril, avec l'asbeste fibreux compacte, on ne voit point de couronne, parce que les filaments n'ont point une dimension égale; mais avec le gypse fibreux, le quartz fibreux, le diopside, la witherite, la crokidolite, la couronne se voit et se mesure. Dans la première de ces espèces les fibres sont de un trentième à un quarantième de millimètre, et dans les échantillons les plus beaux, on pourrait exactement en fixer le diamètre en se servant de lumière homogène d'alcool salé pour laquelle je rappellerai que j'ai trouvé par les réseaux $\lambda = 0^{\text{milli.}},000\,588$. Ce qui donnerait d dans l'équation

$$d \sin \alpha = \tfrac{1}{2} \cdot 0^{\text{milli}},000\,588,$$

quand on aurait mesuré α. La mesure de cet angle s'obtient au reste sans difficulté au moyen de deux bougies dont on amène les deux couronnes à se toucher extérieurement en s'éloignant plus ou moins de ces bougies sur une ligne perpendiculaire au milieu de la distance qui les sépare.

V. — *Astérie et phénomènes analogues.*

» L'astérie à six branches du saphir et l'astérie du grenat, à quatre ou à six branches, suivant que les rayons parcourent l'axe d'un angle quadrièdre ou trièdre, est, dit M. Beudant, un fait dont nous n'avons point d'explication. Nous déduirons cette explication de l'existence de petites fibres ou solutions de continuité disposées dans le saphir parallèlement aux intersections des faces du prisme hexaèdre avec les bases de ce prisme qui, contrairement au prisme du quartz, est souvent terminé par des faces ou bases perpendiculaires aux arêtes du prisme, montrant ainsi une disposition moléculaire dont le résultat doit être une série de lignes ou de stries parallèles entre elles et aux arêtes de la base du prisme suivant trois directions diverses. Sur plusieurs saphirs on distingue au grand soleil deux ou même trois de ces systèmes de stries. La lumière qu'elles reflètent transversalement donne une étoile à six rayons qui doivent aboutir au milieu des faces du prisme, comme on l'observe en effet, et si l'on taille une lame de saphir et qu'on regarde une bougie au travers : on aperçoit (quelle que soit la place de l'œil; contrairement à l'explication d'Haüy) une étoile à six branches dont la lumière est le centre. Si cette théorie est vraie, en coupant le saphir astérie perpendiculairement à l'un ou à l'autre des systèmes de stries parallèles, on doit obtenir un cercle parhélique, c'est ce que j'ai vérifié en y sacrifiant de très belles astéries de saphir. On doit en dire autant de l'astérie du rubis qui n'est seulement qu'une autre couleur du corindon. Les astéries ne sont pas de la même teinte que le fond de la couleur de la pierre, comme cela doit être, puisqu'elles sont une réflexion spéculaire de la lumière transmise.

» L'œil-de-chat ne contient qu'une seule ligne de filaments qui sont de l'asbeste; aussi ne donne-t-il qu'une ligne lumineuse transversale aux filaments, et si on le taille perpendiculairement à ceux-ci, on a un cercle parhélique.

» Tous les cristaux fibreux, le gypse, la chaux carbonatée fibreuse, le zircon, le quartz, etc., donnent, en travers des filaments, une ligne astérique, et, dans le sens des filaments, un cercle parhélique.

» Le grenat, quand on le coupe perpendiculairement à l'axe d'un angle trièdre, donne une astérie à six branches dirigées aux angles (et non comme dans le saphir au milieu des côtés) de la lame hexaèdre qui résulte de cette section. Les filaments, ou fibres, ou solutions de continuité miroitantes sont à l'intersection de plans horizontaux (l'axe de l'angle trièdre étant vertical) avec les plans des faces de l'angle trièdre (ce qu'indiquent comme naturel les faces ou bases secondaires qui existent sur plusieurs échantillons), et les perpendiculaires à ces fibres vont droit aux angles de la lame hexaèdre (ce que l'inspection d'un cristal dodécaèdre montre de suite), mais de plus l'axe de cet angle trièdre n'est pas très éloigné de la direction des fibres qui appartiennent à un angle solide trièdre voisin, en sorte qu'on aperçoit, avec une lame hexaèdre, non-seulement l'étoile à six branches ou rayons, mais encore de plus un cercle parhélique qui passe par l'entrecroisement des branches de l'astérie où est aussi placée la bougie. J'avais pressenti par la théorie et avec quelque inquiétude l'existence de ce cercle parhélique dont j'ai vérifié avec grand plaisir l'existence sur tous les échantillons de grenat astérique qui ont été taillés convenablement. Les grenats astériques à quatre branches ne sont pas très rares. Sur 1000 à 1200 grenats pris au hasard dans le commerce on peut en trouver vingt ou trente qui jouissent de cette propriété; mais pour l'astérie à six rayons, je n'en ai trouvé qu'une sur au moins six mille grenats que j'ai essayés chez M. Mention, bijoutier. Si l'on considère l'angle solide quadriédre du grenat, on y reconnaît seulement deux systèmes de lignes ou de stries suivant l'intersection de plans parallèles aux faces de l'angle quadriédre (dont l'axe est ici censé vertical) avec des plans horizontaux dont l'existence est montrée par la troncature fréquente de cet angle quadriédre, et il en résulte deux systèmes de stries seulement, et, en travers, deux lignes lumineuses seulement allant aux angles de la lame carrée obtenue par la troncature artificielle de l'angle quadrièdre, comme le montre de suite l'aspect du dodécaèdre. Si, comme je l'ai fait, on détermine théoriquement un sens quelconque de fibres et que, perpendiculairement à cette direction, on fasse tailler deux faces parallèles, on a un cercle parhélique qui rend évidente l'existence du système de fibres indiqué par la théorie.

» Ce caractère minéralogique de fibres parallèles à des intersections de plans de structure ou de cristallisation, est par suite important en minéralogie, mais il est bien plus général qu'on ne peut l'imaginer au premier aperçu. J'oserais même dire qu'il est un des plus étendus et des plus commodes qu'on puisse consulter, et même, dans certaines circonstances, c'est un caractère unique. Les exemples suivants, que je pourrais multiplier à l'infini, prouveront suffisamment cette assertion.

» Je place une petite ouverture d'un millimètre ou deux devant une bougie, et, s'il est nécessaire, devant l'œil une carte percée d'un trou pour limiter l'ouverture de la pupille. La plupart du temps ces deux précautions accessoires sont inutiles. Cela posé :

» Un cristal hexaèdre de saphir blanc, de la plus belle eau, non soupçonnable de stries *à priori*, donne trois lignes blanches perpendiculaires aux faces du prisme hexaèdre, formant une véritable étoile à six rayons.

» Même observation pour un béryl et pour une émeraude, et pour une tourmaline, mais non pour un quartz hexaèdre.

» Pour un zircon, la croix sera à huit branches, formant des angles de 45° et non de 60°, et il en sera de même pour l'idocrase et pour tous les cristaux à base carrée.

» Si vous regardez par deux faces parallèles d'un prisme d'émeraude, et par des rayons perpendiculaires à l'axe, vous aurez une croix à quatre branches donnée par des filaments longitudinaux, résultant de l'intersection mutuelle des faces du prisme d'une part, et de l'autre, par les stries parallèles aux arêtes de la base. Le béryl ne donne point cette croix, par manque de ces derniers filaments; il ne donne qu'une ligne.

» Le disthène donne, suivant sa longueur, et en travers, comme l'émeraude, une croix rectangulaire.

» Le mica donne une grande variété d'apparences, dont je laisse aux minéralogistes l'étude compliquée. Le plus généralement, les lignes s'y coupent à 60°; mais il y en a qui donnent d'une manière très permanente des entrecroisements à 45° et à 90°.

» Les cristaux dont la forme primitive est plus compliquée donnent-ils des angles moins simples? C'est ce que je crois pouvoir affirmer d'après la baryte sulfatée, le gypse, le mica à deux axes, le feldspath. Mais je n'ai pas assez approfondi ce sujet pour énoncer ici autre chose que des présomptions. Nous avons projeté, M. Dufrénoy et moi, la revue optique de toutes les espèces minérales, en joignant aux notions que je viens de mentionner, toutes celles que la science doit aux physiciens qui ont contribué à enrichir l'optique minéralogique. Mais je pense que les principes que je viens de faire connaître peuvent, indépendamment d'applications plus complètes, servir de type à des travaux plus étendus.

VI. *Polarisation chromatique et ses applications dans la minéralogie.*

» La polarisation chromatique, que nous devons à M. Arago, est peut être le caractère le plus varié que fournisse l'optique à la minéralogie. Les anneaux colorés traversés d'une croix ou d'une ligne noire, les couleurs des lames suffisamment minces ou taillées dans le voisinage des axes, les hyperboles de compensation, l'aspect même des anneaux qui sont presque achromatiques dans l'apophyllite, et à couleurs inverses, dans l'idocrase (vésuvienne), ont été indiqués par les auteurs de minéralogie ou du moins dans celle de M. Beudant. Pour ne pas répéter ce qui est connu même moins généralement je mentionnerai les faits suivants : une lame d'émeraude donne des anneaux à croix noire, comme doit le faire un cristal uniaxe; une lame de béryl donne une ligne noire semblable au chiffre 8 qui défie jusqu'ici toute théorie. Ce cristal est autre chose encore qu'un cristal biaxe.

» Dans plusieurs cristaux, par exemple, le zircon, le béryl, le centre des branches n'est pas noir. Ce centre est analogue au centre des anneaux du quartz, et cependant il ne paraît pas y avoir rotation du plan de polarisation.

» Autre fait : la forme de plusieurs cristaux à base carrée ou hexaèdre semble indiquer un cristal à un axe et l'expérience donne deux hyperboles noires comme dans le nitre; exemples : le borate de magnésie électrique par la chaleur, l'arséniate de potasse de la chimie. Je pense que dans ce cas la substance a pu perdre de l'eau de cristallisation et passer à l'état de cristal à deux axes, comme je l'ai éprouvé sur le sulfate de quinine qui, frais préparé, donne une croix noire bien nette, et qui donne des hyperboles noires quand il est desséché.

» Je mentionnerai, en terminant, un fait observé par M. Norrenberg et par moi, mais publié d'après ses propres observations, par M. Müller. Si l'on fait passer au travers d'un cristal taillé parallèlement à l'axe un faisceau de lumière monochromatique polarisé et qu'on l'analyse par une tourmaline, à sa sortie, on obtient des hyperboles très faciles à mesurer et avec une épaisseur de cristal quelconque. Ces hyperboles dont les dimensions dépendent de la force biréfringente du cristal, sont encore un caractère facile à observer dans les cristaux et incomparablement plus approprié à la construction d'une table des forces biréfringentes que tout autre procédé fondé, soit sur l'écart des images doubles, soit sur l'emploi des compensations avec le quartz ou le gypse. L'axe de ces doubles hyperboles étant donné par l'angle qu'il soutend, on peut en conclure pour une lumière homogène, le rapport des deux constantes de la double réfraction ou celui des équivalents optiques maximum et minimum qui ont lieu perpendiculairement au plan des axes ou de l'axe du cristal.

» Si cependant ou peut casser en deux la lame biréfringente on obtient de suite, en croisant les deux fragments, les hyperboles de compensation et par la mesure de leurs axes le rapport des équivalents optiques des deux réfractions. Jusqu'ici l'on ne connaît ces éléments qui tiennent de si près à la constitution intime des corps, que pour une dixaine de cristaux, comme on peut le voir dans les traités d'optique de MM. Herschell et Brewster. »

GÉOLOGIE. — *Note sur l'existence, dans le département du Rhône, d'un relief orienté suivant le système de la chaîne du Pilat; par* M. FOURNET.

Cette note, qui fait suite à un mémoire du même auteur sur le sol des environs de Chessy et de l'Arbresle, est accompagnée d'une carte sur laquelle sont figurés les quatre principaux systèmes de reliefs du département du Rhône.

(Commission précédemment nommée.)

BOTANIQUE. — *Sur deux nouvelles espèces du genre* Oncidium, *de la famille des Orchidées, et sur quelques autres espèces peu connues du même genre; par* M. A. MUTEL.

Ce mémoire contient, avec les observations présentées précédemment par l'auteur, dans un mémoire portant le même titre que celui-ci, la description de l'*oncidium trifolium*, espèce que M. Mutel n'avait pas eu encore occasion de voir en fleurs à l'époque où il envoya son premier travail.

(Commission précédemment nommée.)

chromatic thin-film totally reflecting uarterwave retarders

Cojocaru, R. Dabu, V. Draganescu, T. Julea, and Nichitiu

National Center of Physics, P.O. Box MG-6, Bucharest, Romania.

Received 5 April 1988.
0003-6935/89/020211-02$02.00/0.

In many laser applications quarterwave plates are used to ange the polarization of light from linear to circular or om circular to linear. A quarterwave plate combined with linear polarizer is used in high-power laser systems to event reflected or scattered light from returning to the urce. Attempts have been made to replace the expensive ystalline quarterwave plates with reflective dielectric thin-yer films—quarterwave retarders—designed to produce a ° phase retardation between *s* and *p* polarizations. The nstraint of a high reflectivity approximately the same for th polarizations imposes a large number of layers which ve a low damage threshold at high fluences. Further fficulties in these designs are because the phase retarda-on varies rapidly with a change in wavelength or angle of cidence.

Spiller[1] reported designs of totally reflecting thin-film uarterwave retarders which are achieved with a prism as own in Fig. 1. The totally reflecting property of the prism self assures a reflectivity of 100% for both polarizations, d thus many fewer layers are required. As shown in Ref. 1 en a single-layer film of certain optical constants coated on e large surface of a right-angle prism can give a quarter-ave retarder having a reflectivity of 100% for both polariza-ons.

The purpose of this Letter is to analyze the design given by oiller. By using a numerical minimization method Spiller ried the refractive index and the number of layers until a ° phase retardation was obtained. It is well known that electrics used for coatings, for example, ZnS, MgF_2, SiO_2, O_2, have an almost constant refractive index over a large ectral range from visible to near infrared.[2] Then one can sign totally reflecting thin-film retarders that are achro-atic, as suggested in Ref. 1. We determined for several acks of layers of conveniently available optical constants ated on the large surface of a right-angle isosceles prism e fractional thickness of every layer to obtain a phase tardation of ~90° between the two polarizations over a range of wavelengths. Generally for a stack of k dielectric layers of real refractive indices $n_1, n_2, \ldots, n_k$ coated on a substrate of complex refractive index n_0 in an ambient medium of refractive index n_{k+1}, one uses at oblique incidence the recurrence formula for the reflected amplitude:[3]

$$r_{k+1,0} = \frac{r_{k+1,k} + r_{k,0}\exp(-i\Delta_k)}{1 + r_{k+1,k}r_{k,0}\exp(-i\Delta_k)}, \tag{1}$$

where $r_{k+1,k}$ are the Fresnel coefficients at the interface $k+1,k$ and Δ_k is the phase shift in the layer k.

The phase retardation in reflection is obtained from

$$\Delta\phi = \phi_p - \phi_s = \arg(r^p_{k+1,0}/r^s_{k+1,0}), \tag{2}$$

where $r^p_{k+1,0}$ and $r^s_{k+1,0}$ are the reflected amplitudes for the p and s polarizations, respectively.

We used a numerical minimization subroutine. The merit function for minimization is

$$Z(n,d) = \sum_{\lambda_i} (\Delta\phi - 90°)^2, \tag{3}$$

Table I. Designs of Totally Reflecting Achromatic Thin-Film Quarterwave Retarders Coated on the Large Surface of a Right-Angle Isosceles Prism[a]

Structure	Fractional thickness x ($x = d/\lambda_0$)	Wavelength range λ_0/λ with $\Delta\phi = (90 \pm 5)°$
Single-layer film		
ZrO_2	$x_1 = 0.168280$	0.95–1.06
TiO_2	$x_1 = 0.140540$	0.96–1.03
Double-layer film		
ZrO_2/SiO_2	$x_1 = 0.160720$	0.89–1.03
	$x_2 = 0.437820$	
Triple-layer film		
$SiO_2/ZrO_2/SiO_2$	$x_1 = 0.140200$	0.86–1.03
	$x_2 = 0.084213$	
	$x_3 = 0.446000$	
$SiO_2/TiO_2/SiO_2$	$x_1 = 0.156450$	0.85–1.00
	$x_2 = 0.048238$	
	$x_3 = 0.483960$	
Refractive indices:	$n_{SiO_2} = 1.46$; $n_{TiO_2} = 2.48$;	$n_{ZrO_2} = 2.1$; $n_{prism} = 1.509$

[a] Thin layers are counted from the air side toward the prism.

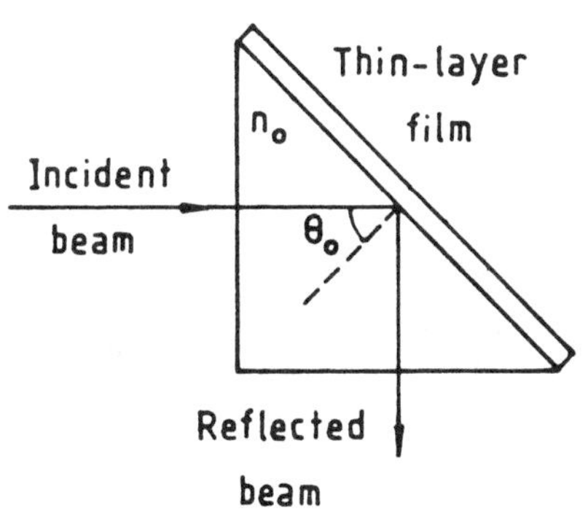

g. 1. Thin-film structure on the large surface of a right-angle sceles prism used to produce a totally reflecting quarterwave retarder.

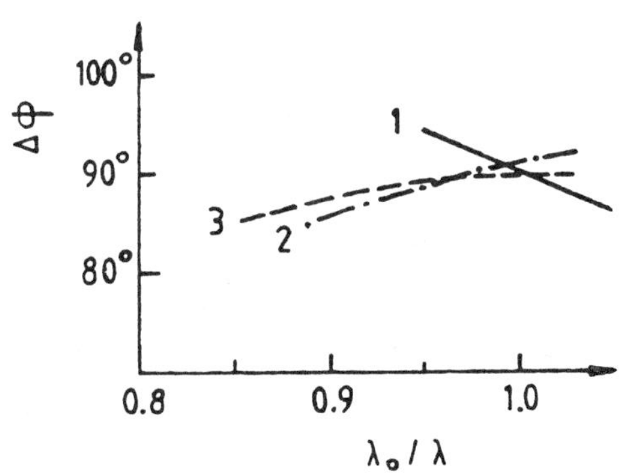

Fig. 2. Phase retardation $\Delta\phi$ in the $(90 \pm 5)\pi°$ range against λ_0/λ for three stacks of layers with parameters as given in Table I: 1, ZrO_2; 2, ZrO_2/SiO_2; 3, $SiO_2/ZrO_2/SiO_2$.

where d is the physical thickness of a layer and λ_i are the radiation wavelengths.

For several stacks of dielectric layers of constant refractive indices we looked for the fractional thickness $x = d/\lambda_0$ of every layer to obtain a phase retardation of the stack of $(90 \pm 0.001)T°$ at λ_0 and a continuous phase retardation $(90 \pm 5)T°$ over a wavelength range λ_0/λ. Results are presented in Table I for a right-angle isosceles glass prism of 1.509 refractive index. We especially considered hard coatings made of SiO_2 and ZrO_2 or TiO_2 which resist at high fluences.

Generally there are many solutions to the numerical minimization. We considered the lowest fractional thicknesses x corresponding to thinner layers which better satisfy the imposed conditions over a larger wavelength range λ_0/λ. In the case of a single-layer film the wavelength range is much shorter, as one can see in Fig. 2, the phase retardation having an almost linear dependence on λ_0/λ. The imposed conditions are better satisfied in the case of triple-layer films. These films with a SiO_2–air interface can be more resistant in ambient atmosphere and at high fluences, admittedly over a much larger wavelength range.

As a function of the desired wavelength range, the lifetime in an ambient atmosphere, and the level of fluence, one can choose either a single-, double-, or triple-layer structure.

References

1. E. Spiller, "Totally Reflecting Thin-Film Phase Retarders," Appl. Opt. **23**, 3544 (1984).
2. O. S. Heavens, *Thin Film Physics* (Methuen, New York, 1970).
3. M. Born and E. Wolf, *Principles of Optics* (Pergamon, New York

Achromatic Combinations of Half-Wave Plates

CHARLES J. KOESTER
Research Center, American Optical Company, Southbridge, Massachusetts
(Received August 7, 1958)

The plane of polarization of two or more wavelengths can be rotated by a predetermined amount through the use of two or more identical half-wave plates in series, the axes of which are oriented at predetermined angles with respect to the incident plane of polarization. For two wavelengths a rotation of 90° may be accomplished by the use of two plates with their slow axes at angles of $22.5°+\delta$ and $67.5°-\delta$, respectively. For three wavelengths a 90° rotation is obtained by using three plates at angles of $11.25°+\delta$, 45°, and $78.75°-\delta$ respectively. The quantity δ is a small angle usually less than 1° which determines the spectral range of achromatization. Identical half-wave plates are easily obtained by cutting a single splitting of mica or a plane parallel sheet of other birefringent material. The wavelength for which the plates have half-wave retardation is not critical. Rather than having a smaller angular aperture than the single half-wave plate, the three-element achromatic rotator has a larger angular aperture. The use of the stereographic projection of the Poincaré sphere for graphical solution of polarized light problems is discussed briefly.

INTRODUCTION

THE most frequent use for half-wave plates is to rotate the plane of polarization of incident light. In some instruments the angle of rotation is variable. The Laurent half-shade plate,[1] for example, is a half-wave plate placed partially over the polarizer, the transmission direction of which has an adjustable angle θ with respect to the axis of the wave plate. Thus the plane of polarization emerging from the half-wave plate has an angle $-\theta$ with respect to the half-wave plate axis. Another application in which the angle is variable is the polarization rectifier described by Inoué and Hyde.[2] In this case the light incident on the half-wave plate has a plane of polarization whose azimuth depends on the position of the ray within the aperture, and the half-wave plate gives each plane of polarization a rotation equal to twice the incident azimuth.

In many other applications, however, a half-wave plate is used to rotate plane polarized light of a fixed azimuth, ϕ, to another fixed azimuth, $\phi+\alpha$. To accomplish this the fast or slow axis of the wave plate is oriented with azimuth $\phi+\frac{1}{2}\alpha$. The condenser of AO Baker interference microscope[3] and the wide-angle compensator described by Francon and Sergent[4] are examples of this use, and in each the rotation angle, α, is 90°. In many cases where the required angular aperture is small and monochromatic light is used, the rotation could be accomplished by a circularly retarding plate, often called simply a rotatory plate, such as quartz cut perpendicular to the optic axis just as well as by the linearly retarding half-wave plate. Both the linearly retarding plates and the circularly retarding plates are wavelength sensitive.

Achromatic wave-plates using superimposed sheets with different dispersions of birefringence have been described by several authors.[5] Recently the internal reflection from a coated surface has been used by Lostis[6] to obtain a retardation of $\lambda/2\pm\lambda/50$ over the visible spectrum. Achromatic combinations of plates designed to produce or analyze circularly polarized light have been described by Destriau and Prouteau[7] and by Pancharatnam.[8] These employ two or three plates of the same birefringent material but differing thicknesses. Pancharatnam[9] has also described a combination of three plates of identical material which acts as a quarter-wave plate at two wavelengths.

This paper describes devices which accomplish the rotation of the plane of polarization through a fixed angle in an achromatic manner. The devices do not have all the other properties of half-wave plates. In particular they will not perform the task mentioned first, that of converting an incident plane polarized beam of arbitrary azimuth θ with respect to the wave-plate axis to an azimuth of $-\theta$. For this reason the name "achromatic rotators" is preferred for these devices rather than "achromatic half-wave plates."

GENERAL CONSTRUCTION

The achromatic rotators are made from two, three, or more half-wave plates which are identical in retardation and dispersion of birefringence. Thus they can be cut from a single splitting of mica, or a single sheet of other birefringent material. The wavelength, λ_0, for which they are half-wave plates should be near the center of the wavelength region of interest, but the exact value of λ_0 is not critical.

The orientations of the slow (or fast) axes of the plates and of the incident and final planes of polarization are shown in Fig. 1 for the case of 90° rotation.

---

[1] See, for example, A. C. Hardy and F. H. Perrin, *The Principles of Optics* (McGraw-Hill Book Company, Inc., New York, 1932), p. 610.
[2] S. Inoué and W. L. Hyde, J. Biophys. Biochem. Cytol. 3, 831 (1957).
[3] A. A. Lebedeff, Rev. opt. 9, 385 (1930), and F. H. Smith, Research (London) 8, 385 (1955).
[4] M. Francon and B. Sergent, Compt. rend. 241, 27 (1955); M. Francon, Optica Acta (Paris) 2, 182 (1955).
[5] C. D. West and A. S. Makas [J. Opt. Soc. Am. **39**, 791 (1949)], describe some achromatic plates and give references to earlier work.
[6] M. P. Lostis, J. phys. radium **18**, 51S (1957).
[7] M. G. Destriau and J. Prouteau, J. phys. radium **10**, 53 (1949).
[8] S. Pancharatnam, Proc. Indian Acad. Sci. **A41**, 130 (1955).
[9] S. Pancharatnam, Proc. Indian Acad. Sci. **A41**, 137 (1955).

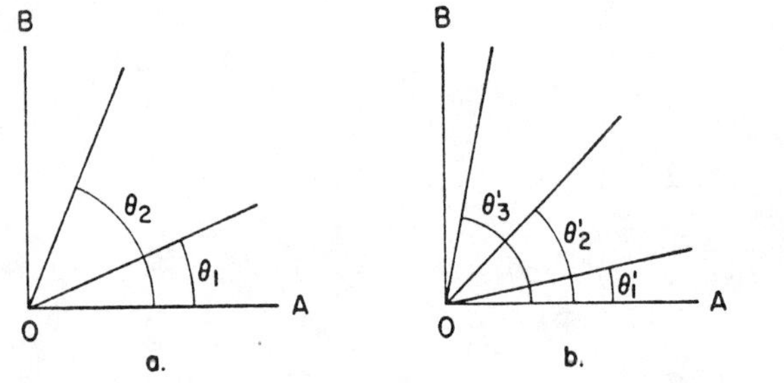

FIG. 1. (a) The two-element achromatic rotator. θ_1 and θ_2 are the azimuths of the slow axes of the two half-wave plates. (b) The three-element achromatic rotator. θ_1', θ_2', and θ_3' are defined as θ_1 and θ_2 above. OA is the azimuth of the plane of polarization of the incident light, OB is that of the emerging light.

In Fig. 1(a) it is easily seen that if $\theta_1=22.5$ and $\theta_2=67.5°$ then for the wavelength λ_0 the plane of polarization is rotated 45° by the first plate, another 45° by the second plate. This arrangement is not achromatic. If the angles are changed to

$$\theta_1=22.5°+\delta \tag{1a}$$

and

$$\theta_2=67.5°-\delta, \tag{1b}$$

where δ is a small angle, there will be two wavelengths, λ_1 and λ_2, having retardations Δ_1 and Δ_2, respectively, which will be rotated by 90°.

For $\delta=1°$ and 0.5°, and $\lambda_0=546$ mμ the values of Δ_1 and Δ_2 are given in Table I(a). From Δ_1 and Δ_2 the values of λ_1 and λ_2 are calculated for half-wave plates made of material with zero dispersion of birefringence.

If the half-wave plates have dispersion of birefringence, the exact wavelengths at which 90° rotation occurs will be different than the values given in Table I. But there will still be two wavelengths for which Δ_1 and Δ_2 are as given in the table, and at these wavelengths the rotation will be exactly 90°.

TABLE I. The achromatic 90° rotators.

a. Two-element

δ	Δ_1[a]	Δ_2[a]	λ_1[b]	λ_2[b]	$\beta(\Delta_0)$[c]
0.5°	158.8°	201.2°	619 mμ	489 mμ	$1.2\cdot10^{-3}$
1°	150.4°	209.6°	654	469	$4.9\cdot10^{-3}$

b. Three-element

δ	Δ_0[a]	Δ_1[a]	Δ_2[a]	λ_0[b]	λ_1[b]	λ_2[b]	$\beta(\frac{1}{2}\Delta_0+\frac{1}{2}\Delta_1)$[c]
0.25°	180°	147.2°	212.8°	546 mμ	668 mμ	462 mμ	$8.5\cdot10^{-5}$
0.75°	180°	138.4°	221.6°	546	710	444	$3.7\cdot10^{-4}$

c. Four-element

δ	Δ_1[a]	Δ_2[a]	Δ_3[a]	Δ_4[a]	λ_1[b]	λ_2[b]	λ_3[b]	λ_4[b]
0.5°	107°	154°	206°	253°	768 mμ	625 mμ	468 mμ	325 mμ

[a] For the two- and three-element rotators Δ_1 and Δ_2 were calculated by means of Eqs. (4) and (5), and were checked by constructions on the Poincaré sphere. The values agreed within ±0.5° for the two-element rotators and within ±3° for the three-element rotators. In the four-element case, Δ_1, Δ_2, Δ_3, and Δ_4 were obtained by means of the Poincaré sphere and are therefore only approximate.

[b] The wavelengths, λ_1, λ_2, ... were calculated assuming that the half-wave plate material had zero dispersion of birefringence and that $\lambda_0=546$ mμ.

[c] $\beta(\Delta)$ is defined as the ratio of light intensity transmitted when an analyser is placed after the rotators, first parallel to and then perpendicular to the polarizer, evaluated at the wavelength for which each component plate has a retardation of Δ. The quantity β can be calculated by matrix calculus[10] or from the Poincaré sphere.[11,12]

For the three-element rotator, the configuration of slow (or fast) axes is shown in Fig. 1(b). It can be seen that if $\theta_1'=11.25°$, $\theta_2'=45°$, and $\theta_3'=78.75°$, then for the wavelength λ_0 the plane of polarization is rotated 22.5° by the first plate, an additional 45° by the second plate, and 22.5° by the third plate, a total of 90°. If the angles are changed to

$$\theta_1'=11.25°+\delta \tag{2a}$$

$$\theta_2'=45° \tag{2b}$$

$$\theta_3'=78.75°-\delta \tag{2c}$$

the total rotation for wavelength λ_0 remains 90°. But in addition two other wavelengths, λ_1 and λ_2 are also rotated 90°. For $\delta=0.75°$ and 0.25° the values of Δ_1, Δ_2, λ_1, and λ_2 are given in Table I(b).

As with the two-element system, the smaller the value of δ the narrower the range of achromatization, but the greater the degree of achromatization within that range. In practice the value of δ is selected by turning the plates relative to each other until the best achromatization is obtained for the particular situation.

Achromatic combinations of four and even more plates are possible. The azimuth angles of the slow axes relative to the initial plane of polarization for the four-element rotator are

$$\theta_1''=7.5°+\delta, \tag{3a}$$

$$\theta_2''=30°, \tag{3b}$$

$$\theta_3''=60°, \tag{3c}$$

$$\theta_4''=82.5°-\delta. \tag{3d}$$

For $\delta=0.5°$ the values of Δ and of λ are given in Table I(c).

THEORY

The values Δ_1 and Δ_2 can be calculated by means of the matrix calculus described by Jones[10] The results for the two- and three-element rotators are

$$\text{2 element:}\quad \cos\Delta=1-[\sin^2 2\theta_1]^{-1}, \tag{4}$$

$$\text{3 element:}\quad \cos\Delta=1-[\sin 2\theta_1(1+\sin 2\theta_1)]^{-1}. \tag{5}$$

For understanding the achromatic combinations, the Poincaré sphere[11,12] is very useful. The principles are summarized here. Points on the sphere represent states of polarized light, the poles representing circular polarization and points on the equator representing linear polarization. A point of longitude 2ϕ and latitude γ represents elliptically polarized light with major axis azimuth of ϕ and a ratio of axes given by $b/a=\tan\gamma/2$. The northern hemisphere represents counter-clockwise rotation of the polarization vector. The action of a

[10] R. C. Jones, J. Opt. Soc. Am. **31**, 488 (1941).

[11] H. Poincaré, *Théorie mathématique de la lumière* II (Paris, 1892), Chap. 12.

[12] A good discussion of the Poincaré sphere is given by H. G. Jerrard, J. Opt. Soc. Am. **44**, 634 (1954).

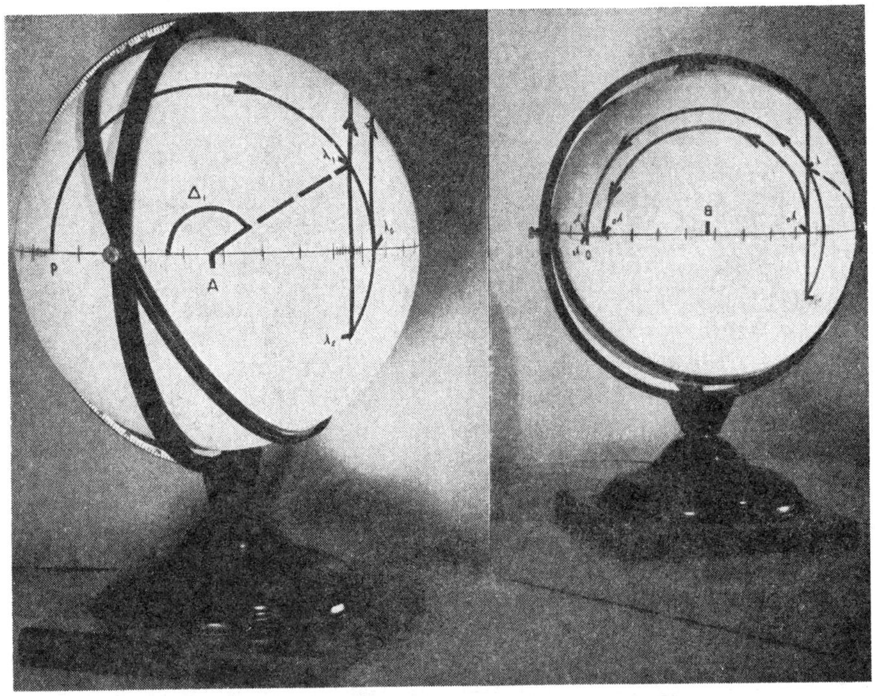

FIG. 2. A model of the Poincaré sphere. The necessary motions are obtained by rotation of the sphere about the vertical axis and the rotation of a perforated ring about a fixed horizontal axis. A pencil point inserted in a hole in the perforated ring will trace the action of a birefringent plate. In the right half of the picture a mirror is positioned to permit the side view of the sphere to be photographed simultaneously. Shown by heavy black lines are the actions of the two plates of an achromatic rotator with $\delta=1°$. The point P represents the plane of polarization of the incident light, A and B the slow axes of the two wave-plates, and Q the final plane of polarization of the wavelengths λ_1 and λ_2.

birefringent plate of retardation Δ and fast axis azimuth θ is represented on the sphere by a rotation of Δ about an axis passing through the center of the sphere and through the point on the equator at longitude 2θ. Thus the action of a half-wave plate upon linearly polarized light is represented on the sphere by rotation of 180° from the original point on the equator along the circumference of a small circle to the final point on the equator. Several such traces are shown in Fig. 2.

POINCARÉ SPHERE EXPLANATION OF ACHROMATIC ROTATORS

Figure 2 shows the construction for the two-element rotator, where $\delta=1°$. The action of the first plate is represented by rotation of the point P along the small circle about point A. For wavelength λ_1, the magnitude of the rotation is Δ_1, the end point being shown as λ_1. For λ_0 the magnitude of rotation is $\Delta_0=180°$, and for λ_2 it is Δ_2. The action of the second plate is represented by rotation about point B. By symmetry the points λ_1 and λ_2 are brought to the same point, Q. Thus the combination is achromatized for wavelengths λ_1 and λ_2.

The wavelength λ_0 also comes out linearly polarized, but with an azimuth of 86° instead of 90°. This gives an indication of the degree of achromatization.

The stereographic map of the Poincaré sphere is convenient for many polarized light problems. Ramachandran and Chandrasekharan[13] have described the method briefly. The mapping is accomplished by projecting a point from the sphere to the plane containing the equator, with the south pole as the center of projection. The equator, representing states of linearly polarized light, maps as a circle, called the equatorial or primitive circle and the southern hemisphere outside the primitive circle.

The stereographic map is particularly useful for polarized light problems because all circles on the sphere are mapped as circles or straight lines on the plane.[14] The constructions are considerably simplified by the use of stereographic coordinate paper, the Wulff net,[15] which contains the projections of a set of small circles with centers on the equator.

The stereographic projection is not limited to the particular configuration described in the foregoing. In general, any point on the sphere may be used as the center of projection and any plane perpendicualr to the diameter passing through the point can be the plane of the map. In his original treatment Poincaré[11] used a plane tangent to the sphere at the equator. However, there are several advantages in using the south pole and the equatorial plane. One is that all points representing linearly polarized light lie on the primitive circle. Another is that the actions of retardation plates are represented conveniently by the small circles on the Wulff net.

[13] G. N. Ramachandran and V. Chandrasekharan, Proc. Indian Acad. Sci. **A33**, 199 (1951). See also G. N. Ramachandran and S. Ramaseshan, J. Opt. Soc. Am. **42**, 49 (1952).

[14] More information on the stereographic projection is contained in the monograph: W. W. Flexner and G. L. Walker, *Military and Naval Maps and Grids* (Dryden Press, Inc., New York, 1942). A classical reference is S. L. Penfield, Am. J. Sci. **11**, 1, 115 (1901).

[15] The Wulff net is described in N. H. Hartshorne and A. Stuart, *Crystals and the Polarizing Microscope* (Edward Arnold and Company, London, 1952), second edition, p. 35. Wulff net coordinate paper is currently available from the University of Toronto Press, and transparent plastic protractors can be obtained from Ward's Natural Science Establishment, Inc., Rochester 9, New York, and from N. P. Nies, 969 Skyline Drive, Laguna Beach, California.

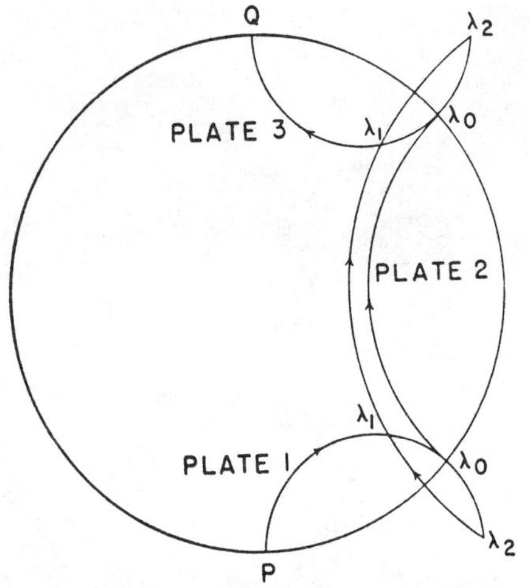

FIG. 3. The action of the three-element achromatic 90° rotator portrayed on a stereographic map of the Poincaré sphere. The complete circle is the equatorial circle. The essential features of stereographic projection are given in the text.

In Fig. 3 a stereographic map of the Poincaré sphere is used to illustrate the action of the three-element rotator. The value of δ is 0.75°, or $\theta_1 = 12.0°$. The action of the first plate on light of wavelength λ_1 is represented by a rotation of 138.4° along the small circle from P to λ_1. The action of the second plate is represented by a rotation of 138.4° along the small circle between the two points labeled λ_1. And the action of the third plate is symmetrical with that of the first, producing a rotation from λ_1 to Q. The construction for the other wavelengths is similar. Because the three plates are symmetrically oriented about the 45° position the azimuth of wavelength λ_0 is changed by 90° no matter what the value of δ.

PERFORMANCE

A single cleavage section of mica with no visible steps was cut into three pieces. The wavelength for which they were half-wave plates was $\lambda_0 = 545.9$ mμ. In the first test the pieces were mounted perpendicular to a collimated white light beam on divided circles over the table of a Gaertner polarizing spectrometer. The axis of each was aligned at the proper angle with respect to the polarizer, and the analyzer was turned parallel to the polarizer. The emerging light was passed through a monochromator and the resulting spectrum was viewed by eye. In the two-element case and especially in the three-element case the reduction in light intensity compared to that passed by a single half-wave plate rotator was marked, particularly outside the green spectral region.

For the second test the pieces of mica were mounted over holes in the centers of cardboard annuli each of which had 2° intervals marked on the periphery. The axes of the plates were aligned according to the requirements for the two- or three-element rotator. Then the assembly was placed between parallel Polaroid sheet polarizers and turned until the best extinction was obtained.

TABLE II. Orientations of plates for a desired rotation of α.[a]

	Single half-wave plate	Two-element rotator	Three-element rotator	Four-element rotator
Angles of optic axes of plates relative to initial plane of polarization	$\theta_1 = \frac{1}{2}\alpha$	$\theta_1 = \frac{1}{4}\alpha + \delta$ $\theta_2 = \frac{3}{4}\alpha - \delta$	$\theta_1 = \frac{1}{8}\alpha + \delta$ $\theta_2 = \frac{1}{2}\alpha$ $\theta_3 = \frac{7}{8}\alpha - \delta$	$\theta_1 = (1/12)\alpha + \delta$ $\theta_2 = \frac{1}{3}\alpha$ $\theta_3 = \frac{2}{3}\alpha$ $\theta_4 = (11/12)\alpha - \delta$
Number of wavelengths at which rotation is exactly α	1	2	3	4

[a] δ is a small angle, the magnitude of which determines the range and degree of achromatization. It can be adjusted by hand to give the best performance.

When this unit was placed immediately in front of a camera objective, a photograph of an illuminated white screen gave the pattern shown in Fig. 4. Each photograph is essentially a conoscopic view of one of the rotators viewed between parallel polarizers. Comparison of the photographs shows that for the tungsten source and panchromatic film used the two-element system is considerably more effective than the single half-wave plate on axis. The three-element rotator is more effective than either of these.

A somewhat unexpected result is that the three-element rotator has a greater angular aperture than either of the other systems. The photograph shows that the single plate system is effective to an average angle of about 10°, whereas the three-plate rotator is equally effective to at least 15°. Qualitatively the reason is that the axes of the waveplates are distributed over a range of azimuth angles. Therefore if an oblique ray receives more than the proper retardation in the first plate it will in general receive less than the proper retardation in the last. Because of the odd number of plates these effects tend to cancel. The five-element rotator probably has an even greater angular aperture. The use of a uniaxial crystal with the optic axis in the plane of the plate should also increase the angular aperture in each system.

The increased angular aperture would have a beneficial effect in devices such as the Francon and Sergent wide-angle compensator[4] and in the condenser of the AO Baker interference microscope.[3]

OTHER USES FOR ACHROMATIC ROTATORS

Rotations other than 90° can be accomplished achromatically by changing the angles between the several half-wave plates. Table II gives the values of θ_1, θ_2, etc., as functions of α, the desired rotation angle, and of δ, the small angle which determines the range of achromatization.

Using these equations a Laurent type half-shade plate could be constructed for use with white light for any desired half-shade angle, α.

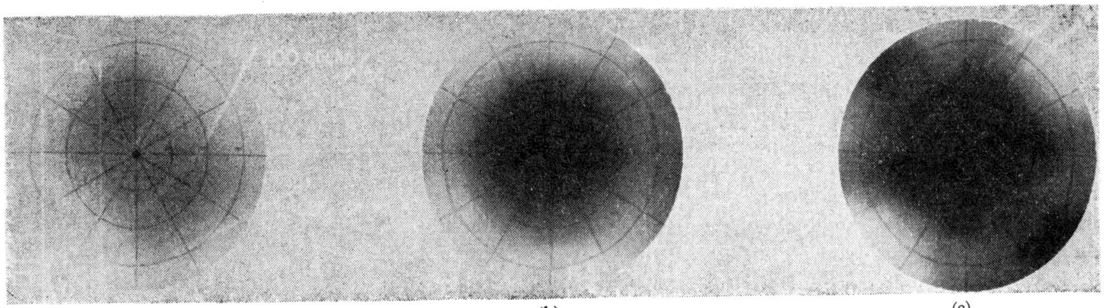

(a) (b) (c)

FIG. 4. Photographs taken through (a) a single wave plate, (b) a two-element achromatic 90° rotator, and (c) a three-element achromatic 90° rotator, each between parallel Polaroid sheet polarizers. The darker the area, the more effective is the rotator for the corresponding angular zone. The exposures were four seconds each on a single roll of Tri-X film which was developed for four minutes in D-19. The printing of the three negatives was done simultaneously on a paper of number 1 contrast. The circles were at 5°, 10°, and 15° to the axis of the system, measured from the camera objective. Half-wave plates of mica were used. In (a) the characteristic biaxial pattern can be seen. In (c) the pattern is complex, but the angular aperture is greater than in either (a) or (b).

If it is desired to rotate the plane of polarization of a certain wavelength λ_3 through an angle α, ordinarily a half-wave plate for λ_3 would be required. However, a combination of two identical half-wave plates for λ_0, where $\lambda_0 \neq \lambda_3$ can be constructed which will rotate the plane of polarization of λ_3 by α. This can be done experimentally, or if a rotation of $\alpha = 90°$ is desired, θ_1 can be found from Eq. (4). In the latter case the only restriction is that for the plates used $\Delta(\lambda_3)$ must be greater than or equal to 90°.

ACKNOWLEDGMENTS

The author is indebted to Dr. Gordon Walker for suggesting that constructions involving either great circles or small circles on a sphere can be carried out on a plane map by means of the stereographic projection. The encouragement given by Dr. Harold Osterberg is appreciated. Mr. Richard Weiss built the rings for the model of the Poincaré sphere and Mr. Herbert Willman performed some of the calculations for Table I.

ACHROMATIC COMBINATIONS OF BIREFRINGENT PLATES

Part II. An Achromatic Quarter-Wave Plate

BY S. PANCHARATNAM
(*Memoir No.* 71 *of the Raman Research Institute, Bangalore*)

Received March 5, 1955
(Communicated by Sir C. V. Raman)

1. INTRODUCTION

A COMBINATION of birefringent plates which can be used for producing circularly polarised light over a fairly large range of wave-lengths has been described in Part I. And it has there been pointed out that the arrangement cannot be properly described as an achromatic quarter-wave plate. A quarter-wave plate when rotated between crossed nicols will show two extinction positions; the achromatic combination described will show none. And to understand under what conditions a combination of plates will have the properties of a single birefringent plate—let alone a plate which is in addition achromatic—it is desirable to have a general method for combining the action of a pile of birefringent plates when their principal planes are not parallel. We shall follow a method differing slightly from that which may be found described by Pockels.[1]

2. GENERAL CONSIDERATIONS

The problem of designing an achromatic elliptic polariser is the problem of transforming a particular state of polarization represented by P_1 (on the Poincaré sphere) to another particular state P_1'. But, of an achromatic retardation plate we require that it transform *every* P to a corresponding P′ obtained by rotating the sphere by an angle δ about an equatorial diameter.

Passage of monochromatic polarised light through a succession of birefringent plates corresponds to a succession of rotations of the Poincaré sphere about a series of equatorial diameters. Any succession of rotations of the sphere can, however, be compounded into one single resultant rotation. Hence the action of a combination of birefringent plates corresponds to a rotation ϕ of the Poincaré sphere about some diameter EE′—which will in general be inclined to the equatorial plane; this means, of course, that the combination is equivalent to a single plate having elliptic birefringence—

E and E′ being the orthogonal elliptic vibrations propagated without change of form, and between which a phase retardation ϕ is incurred.

The condition that the combination should act as a purely birefringent plate (for a particular wave-length λ) is that the axis EE′ of the resultant rotation should be on the equator; so that, if 2ω denote the lattitude of E, the faster ellipse, $2\omega = 0$. And if the retardation of the plate is to have a required value δ, the magnitude of the resultant rotation ϕ should be equated to this value.

Both the magnitude ϕ as well as the axis EE′ of the resultant rotation will in general alter with the wave-length, since the magnitudes (though not the axes) of the component rotations are a function of wave-length. Hence, for the combination to be considered achromatic in the immediate neighbourhood of the wave-length λ, some more conditions have to be imposed; when the wave-length is increased by $\Delta\lambda$ the values of the composite phase retardation ϕ, as well as the latitude 2ω and longitude $2l$ of E, should remain unaltered at least to the first order of approximation.

Instead of following matrix methods, we shall use a well-known geometrical construction (illustrated in Fig. 1) for finding the resultant of any

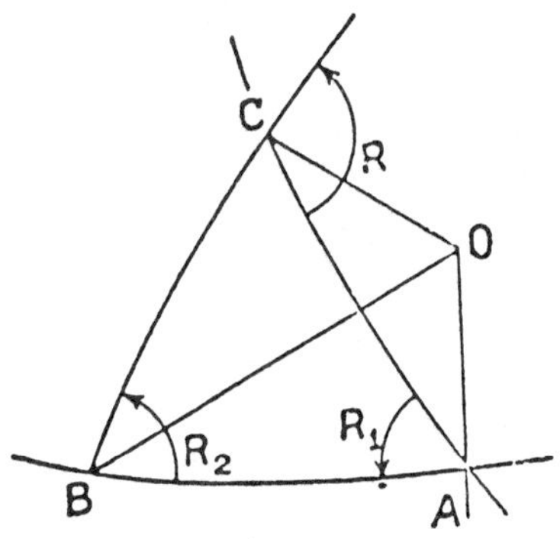

FIG. 1. Composition of Rotations: $2R_1 + 2R_2 = 2R$.

two rotations. If ABC be a spherical triangle described on a sphere whose centre is O then: *a rotation about AO through twice the internal angle at A, followed by a rotation about BO through twice the internal angle at B is equivalent to a rotation about CO through twice the external angle at C.*

3. A BIREFRINGENT COMBINATION WITH VARIABLE RETARDATION

We shall now show that a combination of three birefringent plates, the first and last of which have their corresponding principal planes parallel and their retardations identical, will behave as a single purely birefringent plate of retardation 2δ, whose fast vibration direction is inclined at an angle c_1 to that of the first plate, where 2δ and c_1 are given by:

$$\cos \delta = \cos 2\delta_1 \cos \delta_2 - \sin 2\delta_1 \sin \delta_2 \cos 2c \qquad (a)$$

and

$$\cot 2c_1 = \operatorname{cosec} 2c\,(\sin 2\delta_1 \cot \delta_2 + \cos 2\delta_1 \cos 2c) \qquad (b)$$

In the above relations $2\delta_1$ is the common retardation of the first and last plates, $2\delta_2$ that of the central plate, and c the angle between the fast vibration directions of the central plate and the other two.

Before proceeding to prove these relations, we may mention that by choosing 2 quarter-wave plates as the first and last plates, and a half-wave plate as the central plate, we can, by rotating the central plate, vary the retardation of the combination continuously from 0 to 2π; the principal planes of the combination will always be inclined at 45° to those of the first plate.

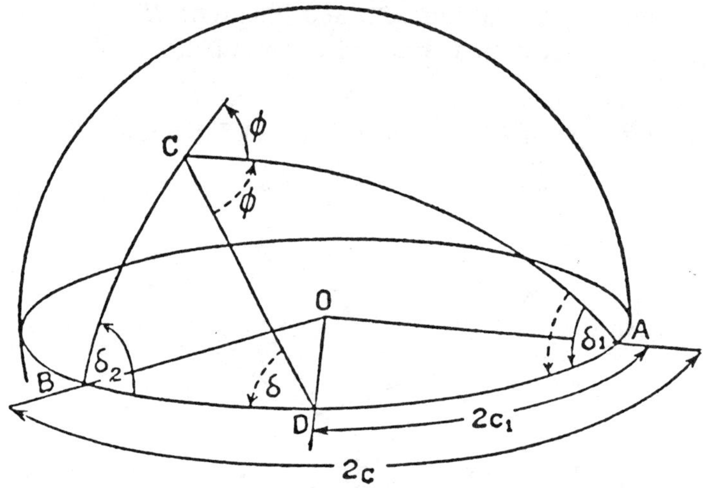

FIG 2.

Referring to Fig. 2, a rotation $2\delta_1$ about AO followed by a rotation $2\delta_2$ about BO, where A and B give the orientation of the fast axes of the first and middle plates respectively, is equivalent to a rotation 2ϕ about CO—where CAB $= \delta_1$, CBA $= \delta_2$ and $\pi -$ ACB $= \phi$. Let us draw an arc CD such that ACD $= \phi$, D being the point of intersection of this arc with the equator. Considering now the triangle ACD, the rotation 2ϕ about C (representing the combined action of the first 2 plates) followed by a rotation $2\delta_1$ about A (which now represents the action of the third plate) is equivalent to a rotation about D through twice the angle external to ADC. The combination is therefore equivalent to a birefringent plate of retardation 2δ whose fast axis is inclined at c_1 to that of the first plate—where $\delta = \pi - \text{ADC}$, and $2c_1 = \text{arc AD}$.

The expressions which we shall derive for δ and c_1 do not depend on assuming that D lies within AB.

From the triangle ADC, denoting arc CD by d

$$\cos \delta_1 = \cos \phi \cos \delta + \sin \phi \sin \delta \cos d \qquad (1)$$

From the triangle DBC

$$\cos \delta_2 = \cos 2\phi \cos \delta + \sin 2\phi \sin \delta \cos d \qquad (2)$$

Eliminating d by multiplying (1) by $2 \cos \phi$ and subtracting from (2), we obtain:

$$\cos \delta = 2 \cos \phi \cos \delta_1 - \cos \delta_2$$

The value of $\cos\phi$ to be substituted is obtained from the triangle ABC.

$$\cos \phi = \cos \delta_1 \cos \delta_2 - \sin \delta_1 \sin \delta_2 \cos 2c$$

Hence

$$\cos \delta = \cos 2\delta_1 \cos \delta_2 - \sin 2\delta_1 \sin \delta_2 \cos 2c$$

which is the required expression (a).

To get an expression for c_1, we use a well-known trigonometrical theorem, which gives from triangle ABC, on denoting arc AC by b

$$\cot 2c \sin b = \cot \text{ACB} \sin \delta_1 + \cos b \cos \delta_1$$

and from triangle ACD

$$\cot 2c_1 \sin b = \cot \text{ACD} \sin \delta_1 + \cos b \cos \delta_1$$

or

$$\sin b (\cot 2c_1 - \cot 2c) = 2 \sin \delta_1 \cot \phi$$

Substituting

$$\sin b = \sin \delta_2 \cdot \frac{\sin 2c}{\sin \phi}$$

we obtain,

$$\sin 2c (\cot 2c_1 - \cot 2c) = \frac{2 \sin \delta_1}{\sin \delta_2} \cos \phi$$

i.e.,

$$\sin 2c \cot 2c_1 - \cos 2c = \frac{2 \sin \delta_1}{\sin \delta_2} (\cos \delta_1 \cos \delta_2 - \sin \delta_1 \sin \delta_2 \cos 2c)$$

or

$$\sin 2c \cot 2c_1 = \sin 2\delta_1 \cot \delta_2 + \cos 2\delta_1 \cos 2c$$

which is the required expression (*b*).

4. Conditions for Achromatism

To make a combination of the type described above achromatic in the immediate neighbourhood of the wave-length λ we may differentiate the expressions on the right-hand side of the equations (*a*) and (*b*), and equate to zero. Since however our object is to attain achromatism over a finite range of wave-lengths we proceed in a different manner.

Let λ' and λ'' be two wave-lengths on either side of λ for which the retardations of each individual plate are multiplied by a factor $(1-f)$ and $(1+f)$ respectively. Values corresponding to these two wave-lengths will be denoted by corresponding single- and double-primed symbols.

We impose the conditions that the retardation $2\delta'$ and $2\delta''$ of the combination for these two wave-lengths be equal to one another and to 2Δ, the required retardation of the achromatic combination; also that the orientations of the fast axes of the combination for these two wave-lengths should be the same, *i.e.*, $c_1' = c_1''$. We then get three equations involving the three unknowns δ_1, δ_2 and c:

$$\cos \Delta = \cos 2\delta_1' \cos \delta_2' - \sin 2\delta'_1 \sin \delta_2' \cos 2c \qquad (1)$$

$$\cos \Delta = \cos 2\delta_1'' \cos \delta_2'' - \sin 2\delta_1'' \sin \delta_2'' \cos 2c \qquad (2)$$

$$\sin 2\delta_1' \cot \delta_2' + \cos 2\delta_1' \cos 2c = \sin 2\delta_1'' \cot \delta_2'' + \cos 2\delta_1'' \cos 2c \qquad (3)$$

From these three relations we may get two relations involving only the unknowns δ_1 and δ_2.

Thus eliminating $\cos 2c$ from (1) and (2)

$$\frac{\cos \delta_2' \cos 2\delta_1' - \cos \Delta}{\sin 2\delta_1' \sin \delta_2'} = \frac{\cos \delta_2'' \cos 2\delta_1'' - \cos \Delta}{\sin 2\delta_1'' \sin \delta_2''} \qquad (4)$$

For the second equation, we eliminate $\cos 2c$ as well as $\cos \Delta$ between the three equations. Thus equating the right-hand expression of (1) and (2), and then substituting the value of $\cos 2c$ from (3) we get:

$$\frac{\cos 2\delta_1' \cos \delta_2' - \cos 2\delta_1'' \cos \delta_2''}{\sin 2\delta_1' \sin \delta_2' - \sin 2\delta_1'' \sin \delta_2''} = \frac{\sin 2\delta_1' \cot \delta_2' - \sin 2\delta_1'' \cot \delta_2''}{\cos 2\delta_1'' - \cos 2\delta_1'}$$

which on simplification yields,

$$(1 - \cos 2\delta_1' \cos 2\delta_1'')\ (\cos \delta_2' + \cos \delta_2'')$$
$$= \sin 2\delta_1' \sin 2\delta_1'' \ \frac{\sin \delta_2' \cos \delta_2'' + \cos \delta_2' \sin \delta_2''}{\sin \delta_2' \sin \delta_2''}$$

Substituting

$$\delta_2' = (1 - f)\ \delta_2 \text{ and } \delta_2'' = (1 + f)\ \delta_2,$$

we get

$$(1 - \cos 2\delta_1' \cos 2\delta_1'')\ 2 \cos \delta_2 \cos f\, \delta_2 = \frac{\sin 2\delta_1' \sin \delta_1'' \cdot \sin 2\delta_2}{\frac{1}{2}(\cos f\, \delta_2 - \cos \delta_2)} \tag{5}$$

We have to find values of δ_1 and δ_2 which simultaneously satisfy (4) and (5). The task is considerably simplified when we note that equation (5) is identically satisfied for $\delta_2 = \pi/2$. The corresponding value of δ_1 may be found from (4), which now becomes:

$$\sin f \frac{\pi}{2} \cdot \sin 2\,(\delta_1' + \delta_1'') = \cos \Delta\,(\sin 2\delta_1'' - \sin 2\delta_1')$$

Substituting

$$\delta_1' = (1 - f)\ \delta_1 \text{ and } \delta_1'' = (1 + f)\ \delta_1,$$

this gives

$$\sin f\,.\,2\delta_1 = \frac{\sin f \cdot \frac{\pi}{2}}{\cos \Delta} \cdot \sin 2\delta_1 \tag{I}$$

An expression for $\cos 2c$ is got by eliminating Δ from (I) and (2).

$$\cos 2c = \frac{\sin f \frac{\pi}{2}}{\cos f \frac{\pi}{2}} \cdot \frac{\cos 2\delta_1' + \cos 2\delta_1''}{\sin 2\delta_1' - \sin 2\delta_1''}$$

or

$$\cos 2c = -\frac{\tan f \frac{\pi}{2}}{\tan f \cdot 2\delta_1} \tag{II}$$

For any required retardation 2Δ and for any choice of the arbitrary parameter f which determines the range of achromatism, the retardation $2\delta_1$ of the first and last plates of the achromatic combination may be found from I; since this is a transcendental equation it has to be solved by an iteration procedure. The orientation c of the central plate, which should be a half-wave plate, may then be obtained from II. The orientation c_1' of the principal planes of the combination may, if required, be calculated from

$$\sin 2c \cot 2c_1' = \sin (1 - f)\ 2\delta_1 \tan f \frac{\pi}{2} + \cos (1 - f)\ 2\delta_1 \cos 2c \tag{III}$$

The retardation and the orientation of the principal planes of the achromatic retardation plate, taken as a whole, will disperse slightly with wave-length. The combination will possess exactly the same retardation 2Δ and the same orientation c_1' of its fast vibration direction, for two wave-lengths; these wave-lengths, if we neglect the dispersion of birefringence, will be $(1+f)\lambda$ and $(1-f)\lambda$; and we may, somewhat arbitrarily, consider the range of achromatism as extending from $(1+\sqrt{2}f)\lambda$ to $(1-\sqrt{2}f)\lambda$.

We shall present the numerical solutions only for the case of an achromatic quarter-wave plate—for which we must substitute $2\Delta = 90°$. If we choose $f = \cdot 18$ the solution of I can be shown to be $2\delta_1 = 115° 42'$; so that from II, $\cos 2c = -\cdot 7639$ or $2c = 139° 48'$. The corresponding value of c_1' is got from III, which yields $\cot 2c_1' = \cdot 5489$ or $2c_1' = 61° 14'$.

On the other hand if $f = \cdot 1414$ we get

$$2\delta_1 = 115° 30', \; 2c = 140° 26', \; 2c_1' = 61° 30'$$

Thus an achromatic quarter-wave plate is obtained by superposing three plates of the same birefringent material (the dispersion of the birefringence of which need not be negligible). The central plate should be a half-wave plate for the wave-length λ in the middle of the spectral range to be covered. The first and last plates should have their principal planes in parallel orientation and their retardations of the same magnitude—equal to, say, $2\delta_1$ for the wave-length λ. With respect to the common fast vibration direction of the first and last plates, let the inclinations of the fast vibration direction of the central plate, and of the combination taken as a whole, be c and c_1' respectively: Then for covering the range of wave-lengths from about $1 \cdot 25\,\lambda$ to $\cdot 75\,\lambda$ the set of values, $2\delta_1 = 115° 42'$, $c = 69° 54'$ may be used—for which $c_1' = 30° 37'$. The achromatism is not high but can be increased by restricting the range to be covered. Thus for covering the region from $1 \cdot 2\,\lambda$ to $\cdot 8\,\lambda$ the values suitable are:

$$2\delta_1 = 115° 30', \; c = 70° 13',$$

for which

$$c_1' = 30° 45'$$

We may mention that the range over which these combinations may be considered achromatic would have to be altered if the dispersion of the birefringence is appreciable, but their essentially achromatic nature would not be changed. For retardation plates of mica, the dispersion of birefringence *is* negligible.[2]

Experimental Verification

Two plates of mica were cleaved, whose thicknesses were in the ratio of 180: 115·75, the former being nearly a half-wave plate for the D lines of sodium. The second plate was cut into two portions which were then superposed and cemented together at a corner—the principal planes of the two portions being made exactly coincident while the cement was still wet. After the cement was dry, the half-wave plate—with a drop of copal varnish on either side—was inserted *between* the other two plates at roughly the calculated angle. The combination was placed on a mirror and, with the aid of another mirror, white light from a point source was made to pass normally through a polaroid and the combination of plates and then back again—after which it reached the eye. The orientation of the central plate was then gradually altered while the varnish was still wet—and a position was found where the image of the point source could be almost completely extinguished by rotating the polaroid. It was also verified that the combination thus prepared showed two practically black extinction positions when rotated between crossed polaroids.

The author's thanks are due to Prof. Sir C. V. Raman for his kind interest in this work.

Summary

An achromatic quarter-wave plate is obtained by superposing three birefringent plates of the same material; the first and last should have the same retardation $2\delta_1$, their fast vibration directions being parallel to one another but inclined at a specific angle c to that of the central plate—of retardation π. The desired range of achromatism determines the optimum values of $2\delta_1$ and c (which, in turn, will determine the orientation of the effective principal planes of the combination). As an example, using mica retardation plates prepared for Hg 5461, the range from 4100 Å to 6800 Å is covered with $2\delta_1 = 115° 42'$ and $c = 69° 54'$.

Further, for a particular wave-length, a birefringent compensator of variable retardation (0 to 2π) is obtained by interposing a half-wave plate that can be rotated in its own plane, between two quarter-wave plates that have their fast vibration directions parallel.

The results follow from the Poincaré sphere by geometrically compounding successive rotations.

References

1. Pockels .. *Lehrbuch der Kristalloptik, Teubner,* 1906, 280.
2. Mathieu .. *Bull. Soc. Franc. Min.*, 1934, **57**, 233.

Vol. 39(9), 791–794 (September 1949).

The Spectral Dispersion of Birefringence, Especially of Birefringent Plastic Sheets

C. D. WEST AND A. S. MAKAS
Polaroid Corporation, Cambridge, Massachusetts

(Received June 6, 1949)

It is shown apparently for the first time how dispersion of birefringence (retardation, phase difference or ellipticity) can be measured through the visible spectrum by the use of a Martens type of polarization photometer and monochromator. Dispersion curves are given for several organic thermoplastic sheets including those with both normal and abnormal dispersion of birefringence, and possible uses for such sheets are suggested.

INTRODUCTION

PROBABLY the most complete account of spectral dispersion of birefringence of crystalline materials is found in a paper by Ehringhaus.[1] This writer found it useful to employ a function N, given by $N=(\Delta n(D)/(\Delta n(F)-\Delta n(C))$, to classify his material into two main groups: when $|N|$ is large, say greater than 30, the specimen is said to have normal dispersion, whereas when $|N|$ is small, say less than 30, the dispersion is abnormal. In this expression $\Delta n(D)$ denotes the birefringence n_1-n_2 of the crystal section for sodium light, etc. Otherwise stated, the birefringence of normal materials is essentially constant with wave-length between the C and F lines. A wedge cut from such a material and viewed in a polariscope exhibits a normal sequence of interference colors, whereas wedges cut from abnormal materials may exhibit very strange sequences of colors indeed.

A kind of abnormal dispersion of especial interest occurs when Δn is not constant, but varies so as to be directly proportional to wave-length, or $\Delta n=k\lambda$. In this case the *phase difference* δ introduced by passage of polarized light through the specimen is constant with the wave-length, and the material is suitable for the preparation of achromatic retardation plates. The N value necessary to give achromatic interference is thus

$$N=\frac{589}{486-656}=-3.46.$$

The relations employed in this work are

$$\Delta n=\frac{\Gamma}{t}=x\frac{\lambda}{t}=\frac{\delta}{2\pi}\frac{\lambda}{t},$$

where Γ is the path difference expressed in the same units as thickness t and wave-length λ; and x is the same thing expressed in fractions of a wave-length.

The Fresnel rhomb employing total internal reflections in an isotropic medium is well known to be an essentially achromatic device for introducing a 90° phase difference in a beam of polarized light; however its inconveniences have led to repeated attempts to find combinations of birefringent crystal plates having this desirable property, for example by Perucca,[2] Longchambon,[3] Gaudefroy,[4] and Ehringhaus.[5] Such combinations apparently have come into very little actual use.

It is known too that sheets and films of practically all rubber-elastic and thermoplastic organic high polymers are readily put into a birefringent state by stretching. While a number of studies have been published on the birefringence of thermoplastic materials, previous work on the spectral dispersion thereof is in effect limited to cellulose nitrate sheets.[6] In particular the measurements of Savur[6c] and of Seth and Harris[6e] show that their stretched Celluloid sheets (cellulose nitrate containing 24 percent camphor) had a birefringence that approached achromatism quite closely.

In the present brief survey we find that with suitable apparatus and samples it is quite as easy to measure dispersion of birefringence throughout the visible spectrum as it is to measure birefringence at a single wave-length.

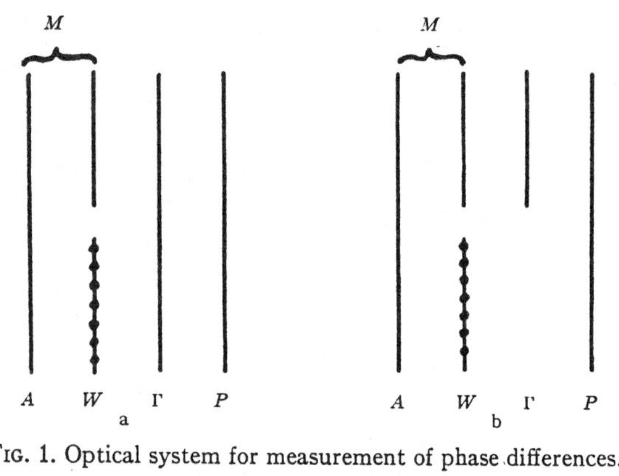

FIG. 1. Optical system for measurement of phase differences.

[1] A. Ehringhaus, Neues Jahrb. Mineral. Geol. Beilage B. 43, 557–618 (1920).

[2] E. Perucca, Atti Accad. Sci. Torino, **54**, 1013–34 (1918).

[3] L. Longchambon, Bull. Soc. Franç. Phys. No. **195**, 158 (1924).

[4] C. Gaudefroy, Comptes Rendus **189** 1289–91 (1929); Bull. Soc. Franç. Min. **52**, 147–60 (1929).

[5] A. Ehringhaus, U. S. Patent 1,926,716 (1933); Zeits. f. Krist. **75**, 179–95 (1930).

[6] (a) M. Wachtler, Kolloidchem. Beihefte **20**, 157 (1924); Zeits. f. tech. Physik **5**, 418 (1924). (b) Ramspeck, Ann. d. Physik **74**, 722 (1924). (c) Savur, Proc. Roy. Soc. (London) **A109**, 338 (1925). (d) Derksen, Katz, Hess, Trogus, Zeits. f. Physik. Chemie **149**, 371 (1930). (e) Harris and Seth, Proc. Phys. Soc. (London) **48**, 477 (1936).

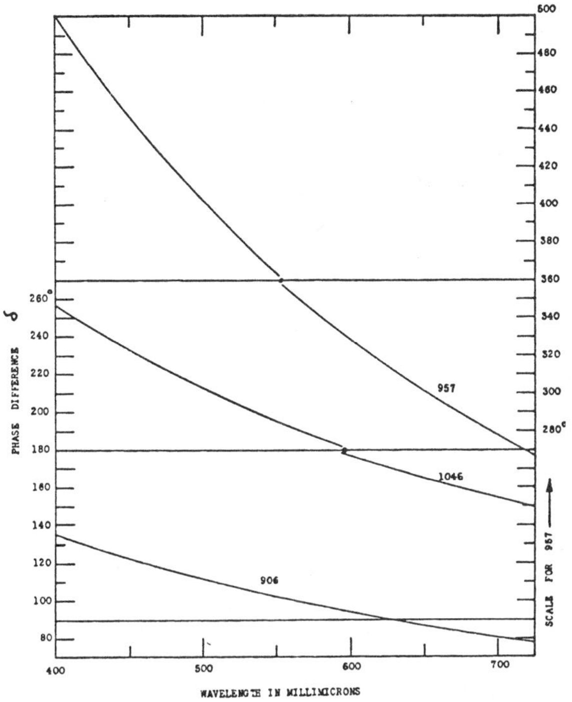

FIG. 2a. Curves of normal dispersion of phase difference.

MEASUREMENT OF PHASE DIFFERENCE

Since previous methods of measurement[1, 6e, 7] proved to be too laborious, after some consideration a simple optical system using a Gaertner polarizing spectrophotometer was found to be satisfactory. This system includes a Martens photometer, which is the preferred instrument for the measurement of dichroism as of sheet polarizers and the like.[8] The measurement of retardation Γ, phase difference δ, or ellipticity $\tan\vartheta = b/a$, where a high order of precision is not required, is illustrated in Fig. 1. In Fig. 1, W is the Wollaston prism with upper and lower windows W_u and W_e, A the analyzer of a Martens photometer. A is on a divided circle which reads zero when A is crossed with W_u, 90° when crossed with W_e, and 45° when the fields are balanced. P is a polarizer, for example a Polaroid polarizer, set exactly perpendicular to W_e, and therefore exactly parallel to W_u of the Wollaston. Γ is the retardation sheet covering W_u and W_e (Fig. 1a) and turned exactly 45° from its extinction position between P and W. These settings of P and Γ can be made very accurately. With the system thus adjusted, the light emerging from Γ is elliptic-polarized with the major axis of the ellipse parallel to the vibration direction of one of the Wollaston windows, for example W_u, and therefore perpendicular to W_e. If the ellipse axes are denoted by a and b, as is customary, then the intensities emerging from W_u and W_e are respectively a^2 and b^2. Now when the analyzer is turned to the azimuth α which balances the fields $J_u = J_e$, the intensity ratio is given by $\tan^2\alpha = J_e/J_u$, and in this way we arrive at the simple expressions $\tan^2\alpha = b^2/a^2$, $\tan\alpha = \tan\vartheta = b/a$, $2\vartheta = \delta$, etc. Or in words, the azimuth angle of the analyzer at balance is equal to the ellipticity angle of the vibration, or to one-half of the phase difference.

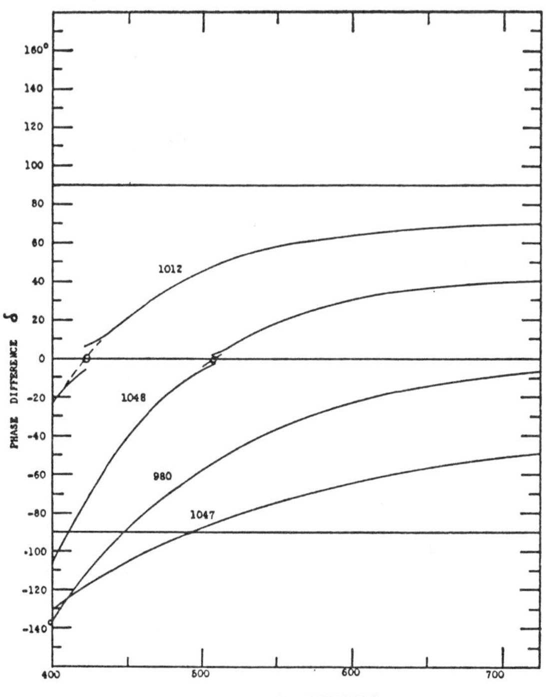

FIG. 2b. Abnormal dispersions.

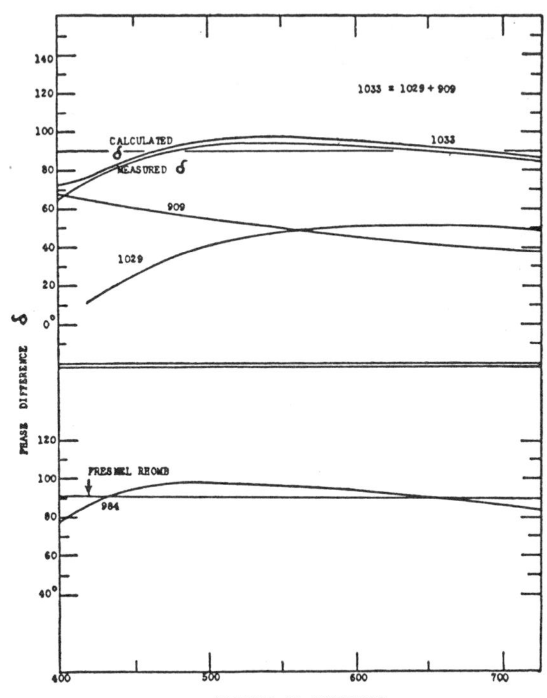

FIG. 2c. Achromatic dispersions.

[7] G. Szivessy, *Handbuch der Physik* (1928), Vol. 19, Chapter 28.

[8] M. Grabau, J. Opt. Soc. Am. **27**, 420–24 (1937).

This method does not distinguish between right and left ellipses.

By this method a good reproducibility of settings is found. For the purposes of this paper the results are plotted as dispersion curves of δ against λ from 400 to 760 mμ. The addition and subtraction of phase differences or retardations is readily demonstrated by this method (See Fig. 2). In Fig. 2 the sign of the ordinate is chosen to agree with the sign of birefringence, which in accordance with convention is taken as positive for $n_{\parallel}>n_{\perp}$, negative for $n_{\parallel}<n_{\perp}$, where $n_{\parallel}$ is the refractive index for light vibrating parallel to the direction of stretch.

In the regions where the dispersion curves cross the levels $\delta=0°$, 180°, 360°—there is an inaccuracy in the settings, as illustrated in Figs. 2a, b. This inaccuracy is due to the failure of the light intensity through W_u or E_e, as the case may be, to fall to zero, as it should under ideal conditions; rather the intensity passes through a minimum whose wave-length is readily found. This departure from ideal conditions is due in part to imperfect collimation, and in part to inter-reflexion; it prevents this method from being directly applicable to the accurate measurement of very small ellipticities, but does not materially affect the measurement of larger ellipticities.

It is seen that if the sample contains an isotropic light-absorbing material, the measurements by this method will not be affected by its presence. The advantage of the present system over previous ones for measuring ellipticities and phase differences is that it does not require a birefringent crystal plate whose constants have first to be determined. It would not be difficult to adapt this method for settings outside the visual spectrum by means of a suitable photoelectric cell circuit.

A less convenient and no more accurate system for making measurements of this kind is illustrated in Fig. 1b, where the retardation sheet covers only one window W_u, of the Wollaston, and is free of absorption. In this system the relation at balance is $\tan^2\alpha=b^2/(a^2+b^2)$, from which the corresponding value of $\tan\vartheta=b/a$ could be derived.

RESULTS

The results of this study are presented as dispersion curves (see Fig. 2, Tables I and II), for each curve the value of Ehringhaus' function N has been computed. In appraising these curves, it should be recalled that a small increase in thickness, or in stretch, of a given sheet has the effect of increasing δ by a small factor.

These results suggest that while many common films have normal dispersion of birefringence, others are abnormal and offer the possibility of preparing retardation plates of a wide range of N values, including those having a fair approximation to achromatism. An approach to achromatism may be found either in a single sheet of suitable composition, for example cellulose nitrate (984), or by combining two or more sheets with vibration directions parallel, such as cellulose acetate with cellulose nitrate or acetatebutyrate (909 parallel to 1029). Although it is easy to make substantial improvements in this way over normal materials such as mica, admittedly it is not easy to achieve the high degree of achromatism of the Fresnel rhomb.

In calculating achromatic combinations of such sheets, one could follow the analytical method of Gaudefroy,[2] but in practice it is simpler to employ the formula of Ehringhaus[3] for the N of a combination of two birefringent sections characterized by N values N_1 and N_2 respectively and by retardations for the D line of Γ_1 and Γ_2 respectively:

$$N=\frac{N_1N_2(\Gamma_1\pm\Gamma_2)}{N_2\Gamma_1\pm N_1\Gamma_2}=-3.46$$

for achromatism, where the + sign is used for the addition position, the − sign for subtraction. For example Gaudefroy calculated by his method that a 0.745-mm section of quartz crossed with a 0.038-mm section of calcite (both cut parallel to the axis, the thickness ratio being 1:0.0510) would give an achromatic phase difference of 180° through the visible spectrum. Employing Ehringhaus' N values of 33.67 and 23.68 respectively for quartz and calcite, and the respective birefringences for D light of 0.00911 and 0.17197, it is readily found by Ehringhaus' formula that

TABLE I. Description of Samples

Number	Sample	N	Γ_C	Γ_D	Γ_F
906	Stretched cellulose acetate	−45.0	158.5	157.0	155.0
1046	Mica	−42.6	299.0	298.0	292.0
957	Stretched polyvinyl alcohol	26.1	539.0	548.0	560.0
1048	Stretched cellulose nitrate (1) (36-mm film base)	−0.50	66.8	42.1	−17.6
1012	Stretched cellulose acetate butyrate	−1.48	122.5	102.4	53.4
980	Stretched cellulose nitrate (2) (Kodaloid)	0.67	−24.4	−42.6	−88.6
1047	Stretched cellulose triacetate	5.15	−103.7	−109.7	−125.0
909	Stretched cellulose acetate	−48.4	78.2	77.5	76.6
1029	Stretched cellulose nitrate (3) (unplasticized)	−1.96	92.3	81.4	50.8
1033	909 plus 1029	−3.73	164.6	152.4	123.7
984	Stretched cellulose nitrate (4)	−4.87	163.3	155.0	131.4
*	Fresnel rhomb (glass of $n_D=1.517$, $\nu=64.5$)	−3.50	164.0	147.2	122.0
*	Ideal	−3.46	164.0	147.2	121.5

* Calculated values.

TABLE II. Variation of N value with composition.

Number	Material	N values
—	Ramie*	26.0 at 20°. 24.4 at 50°.
978,976	Commercial Cellophane	122, 107
1053*a*, *b*	Cellulose acetate *E*-656 53.1% HOAc	−24, −41
1054*a*, *b*	Cellulose acetate *E*-309 56.3% HOAc	−8.1, −7.4
1047*a*, *b*	Cellulose triacetate 62.5% HOAc	5.2, 3.9

* Data of A. Frey-Wyssling, Helv. Chim. Acta 22, 981–8 (1939).

0.05097 mm of calcite is required to achromatize 1-mm of quartz. The equality of the thickness ratios shows the concordance of the two methods.

It may be pointed out that Ehringhaus, after setting up correct and useful equations, by a peculiar confusion of phase difference with path difference failed to appreciate that a N value of -3.46 is the criterion of achromatism (phase difference constant as wave-length varies); instead he only calculated several combinations of constant *path difference*, characterized by $N=\pm\infty$, which are of little practical or theoretical importance.

A comparison of curves 957 and 1048 shows that they will have similar interference colors between crossed polarizers in white light, and that accordingly, since 957 is sensitive to the addition or subtraction of a small retardation, 1048 will be similarly sensitive; it seems appropriate to call it a *zero-order sensitive tint.* Curve 1048 is in fact approximately the reflexion of curve 957 across the line $\delta=180°$. It is a matter of observation that a zero-order sensitive tint plate of type 1048 differs in two respects from a *normal first-order sensitive tint* of type 957: first, it is easier to prepare, once a suitable material has been found, in particular the interference color is relatively insensitive to thickness changes; and second, the interference color is very insensitive to the angle of incidence. Zero-order sensitive tints could also be calculated for a combination of two sheets having suitably different N values.

It will furthermore be apparent that the addition of a zero-order sensitive tint plate to a first-order one of which it is the reflexion, will give a combination with an achromatic retardation of one wave-length ($\delta=360°$) through the visible spectrum.

An abnormal first-order sensitive tint could be made from a sheet (or combination of sheets) whose $\delta(\lambda)$-curve is the reflexion of the normal first-order sensitive tint curve across the level $\delta=360°$. This can be called a *reversed first-order sensitive tint*, and would be exemplified by a suitable thickness of the material of curve 1029.

If we take as characteristic values for an ideal normal first-order sensitive tint $N=\pm\infty$, $\Gamma=\lambda^*=560$ mμ (a constant), then it is easy to show that the corresponding ideal zero-order sensitive tint has $\Gamma'=\lambda^*-\lambda$ (a variable with wave-length) and $N'=-0.17$; and similarly the corresponding ideal reversed first-order sensitive tint has $\Gamma''=2\lambda-\lambda^*$ (also variable with wave-length) and $N''=-1.82$.

The foregoing discussion could also be applied to the curves resulting when the ordinates are changed from phase difference δ to path difference Γ, and the abscissas remain wave-lengths. The level lines $\delta=0°, 180°, 360° \ldots$ parallel to the λ-axis become slanting lines $\Gamma=0, \lambda/2, \lambda, \ldots$ passing through the origin.

The purpose of Table II is to illustrate by measurements on duplicate samples how the N value varies as the composition changes from cellulose through acetone-soluble cellulose acetates to cellulose triacetate. These acetates are commercial preparations supplied by Eastman Kodak Company, Rochester, New York.

Note:—To the foregoing report, substantially as it was written in June 1944, there is little new to add. We have not worked out a good way to handle the false discontinuities on the $\delta(\lambda)$-curves where they cross the levels δ equal to an integral multiple of π. These probably result from various stray lights in the system plus depolarization effects arising in some samples.

Destriau and Prouteau,[9] have described an ingenious new way to realize a quasi-achromatic quarter-wave combination by superposing a quarter-wave with a half-wave layer of the same non-achromatic material, such as mica, at a suitable fixed oblique angle. A complete bibliography is included in this paper.

Some of the foregoing material is presented in the specification of the author's U. S. Patent 2,441,049 of May 1948, "Composite Achromatic Wave Plates."

[9] Destrian and Prouteau, J. de phys. et rad. 10, 53–55 (1949).

Section Three

Mathematical Analysis and Measurements of Polarization States

In this section, papers are included that describe devices that determine the degree of polarization of light and whether it is linear, circular, or elliptical.

Also in this section are some of the mathematical techniques that people have invented to allow the calculation of the state of polarization if the optical characteristics of the system elements are known. Jones calculus was one of the first to become widely used. Another was developed by Hans Mueller of MIT.

Mueller gave a course on polarized light and a group of us, including Jones, audited his class. He described the Stokes and Poincaré approach. When he got to the Jones calculus, he started the process of solving a fairly complex system. The board was covered with mathematics. Suddenly he realized he had made an error. He said, "Dr. Jones, what have I done wrong?" At that time, Jones looked about 17 years old. The class swiveled to see what surely, after Stokes and Poincaré, should be an elderly scholar. The shock was overwhelming. Jones quietly pointed out the error and the lecture continued.

A New Calculus for the Treatment of Optical Systems

I. Description and Discussion of the Calculus

R. CLARK JONES*

Research Laboratory, Polaroid Corporation, Cambridge, Massachusetts, and Research Laboratory of Physics, Harvard University, Cambridge, Massachusetts

(Received April 23, 1941)

The effect of a plate of anisotropic material, such as a crystal, on a collimated beam of polarized light may always be represented mathematically as a linear transformation of the components of the electric vector of the light. The effect of a retardation plate, of an anisotropic absorber (plate of tourmaline; Polaroid sheeting), or of a crystal or solution possessing optical activity, may therefore be represented as a matrix which operates on the electric vector of the incident light. Since a plane wave of light is characterized by the phases and amplitudes of the two transverse components of the electric vector, the matrices involved are two-by-two matrices, with matrix elements which are in general complex. A general theory of optical systems containing plates of the type mentioned is developed from this point of view.

INTRODUCTION

THE type of optical system with which this paper deals is not the more familiar type involving lenses, prisms, etc., but is rather the type which is composed of retardation plates, partial polarizers, and plates possessing the ability to rotate the plane of polarization. We shall therefore be concerned not with the directions of rays of light, but with the state of polarization and the intensity of the light as it passes through the optical system.

An example of a partial polarizer is a plate of tourmaline, or a sheet of light-polarizing film such as those sold under the trade-name "Polaroid." [1] For light incident perpendicularly on the plate, a partial polarizer may be characterized by two mutually perpendicular axes, with each of which is associated one of the two principal absorption coefficients. The retardation plate, or wave plate, is too well known to require description. In this paper we shall refer to a plate having the ability to rotate the plane of polarization as a *rotator*. A rotator is exemplified by a plate of quartz viewed along the optic axis, or by solutions of optically active molecules. We shall refer to the individual plates which make up the complete optical system as the elements of the system, or as the optical elements.

In the present paper it will be assumed that the light is always incident normally on the elements of the optical system.

The present paper is devoted to a description of the new calculus. When light passes through any of the three fundamental types of optical elements, the state of polarization and sometimes the intensity of the light will be changed. We shall find it possible to represent the effect of any optical element on the light as a linear operator acting upon the electric vector of the light wave. The operator is expressed in the convenient form of a two-by-two matrix, whose four matrix elements are in the general complex. From the associative property of matrices, it will then follow that the complete optical system may also be represented by a two-by-two complex matrix.

In Part II, the calculus is used to prove three theorems relating to optical systems of the type considered. The two theorems relating to systems containing partial polarizers are new, so far as the writer knows, and should be of some practical importance because of the recent development of a readily available and high quality polarizer in the form of Polaroid sheeting.

In Part III, the theory is applied to a rigorous treatment of the Sohncke theory of optical activity.

---

* National Scholar, Harvard University.

[1] Registered trade-mark of the Polaroid Corporation. For a description of Polaroid sheeting, see Martin Grabau, J. Opt. Soc. Am. **27**, 420 (1937).

Conventions and Notation

Consider a right-handed rectangular coordinate system, x, y and z. The elements of the optical system are considered to be arranged along the z axis, the z axis being perpendicular to the plane of the plates. The optical system will always be described as viewed from a point which is further out along the positive z axis than any of the elements; with this convention, the x and y axes have the relative position of the usual xy plane: the positive x axis may be superposed upon the positive y axis by rotating the x axis 90° counterclockwise.

Furthermore, let the two principal axes of the ith retardation plate or partial polarizer be indicated by x_i' and y_i'. We assume that the positive x_i' and y_i' axes have the same relative orientation as the positive x and y axes. We may now define the orientation of the ith element by stating the angle ω_i measured counterclockwise from the positive x axis to the positive x_i axis.

The light will be represented as plane waves progressing in either direction along the z axis. The state of polarization may be completely defined by stating the amplitudes and phases of the x and y components of the electric vector of the light wave. At any fixed point along the z axis, the components may be written in the usual complex form

$$\begin{aligned} E_x &= A_x \exp\left[i(\epsilon_x+2\pi\nu t)\right], \\ E_y &= A_y \exp\left[i(\epsilon_y+2\pi\nu t)\right], \end{aligned} \tag{1}$$

where A_x and A_y, ϵ_x and ϵ_y are real. If $\epsilon_x-\epsilon_y$ is an integral multiple of π, the light is plane polarized; otherwise, elliptically polarized.

Now let us consider the change in the character of a light wave as it passes through a retardation plate or a partial polarizer. As a matter of fact, in order to avoid giving a separate treatment of the two cases, we shall suppose that both the indices of refraction and the absorption coefficients are different along the two principal axes (x' and y') of the plate. If we know the x' and y' components ($E_{x'0}$ and $E_{y'0}$) of the electric vector as the light enters the plate, then the corresponding components of the light as it emerges from the other side of the plate are

$$\begin{aligned} E_{x'1} &= E_{x'0} \exp\left[-i(2\pi d/\lambda)(n_{x'}-ik_{x'})\right] \\ &= N_{x'}E_{x'0}, \\ E_{y'1} &= E_{y'0} \exp\left[-i(2\pi d/\lambda)(n_{y'}-ik_{y'})\right] \\ &= N_{y'}E_{y'0}. \end{aligned} \tag{2}$$

In these expressions, d is the thickness of the plate, λ is the wave-length of the light in vacuum, the n's are the principal indices of refraction, and the k's are the principal extinction coefficients. The extinction coefficient is a particular type of amplitude absorption coefficient. The N's are merely abbreviations for the exponential expressions. In the case of a retardation plate, the k's are the same and the n's different, whereas in the case of a partial polarizer, the n's are the same and the k's different. In practice, of course, it is difficult to secure a partial polarizer which is not also birefringent.

In general, however, we are interested in knowing the change in the x and y components of the light wave, rather than the change in the x' and y' components. If ω is the angle measured counterclockwise from the positive x axis to the positive x' axis, we have the relations

$$\begin{aligned} E_{x'} &= E_x \cos\omega + E_y \sin\omega, \\ E_{y'} &= -E_x \sin\omega + E_y \cos\omega. \end{aligned} \tag{3}$$

The elimination of the x' and y' components in (2) and (3) yields the result

$$\begin{aligned} E_{x1} &= (N_{x'}\cos^2\omega + N_{y'}\sin^2\omega)E_{x0} \\ &\quad +(N_{x'}-N_{y'})\sin\omega\cos\omega E_{y0}, \\ E_{y1} &= (N_{x'}-N_{y'})\sin\omega\cos\omega E_{x0} \\ &\quad +(N_{x'}\sin^2\omega + N_{y'}\cos^2\omega)E_{y0}. \end{aligned} \tag{4}$$

The equations (4) give the important relation between the x and y components of the entering and emergent light.

The Matrix Notation

The relations (4) may be put in a convenient and simple form by the use of a matrix notation.[2]

[2] In Parts I and II we shall need only the algebraic properties of matrices, as they are presented on pp. 348–352 of E. C. Kemble, *Fundamental Principles of Quantum Mechanics* (McGraw-Hill Book Company, Inc., New York, 1937). In Part III, we shall find it necessary to use also the transformation properties of matrices; see the reference just given, or V. Rojansky, *Introductory Quantum Mechanics* (Prentice-Hall, New York, 1938), pp. 285–340.

Let $\mathbf{M}$ be the two-by-two matrix*

$$\mathbf{M}\equiv\begin{pmatrix} m_1 & m_4 \\ m_3 & m_2 \end{pmatrix}, \tag{5}$$

where

$$\begin{aligned} m_1&=N_{x'}\cos^2\omega+N_{y'}\sin^2\omega,\\ m_2&=N_{x'}\sin^2\omega+N_{y'}\cos^2\omega,\\ m_3&=m_4=(N_{x'}-N_{y'})\sin\omega\cos\omega. \end{aligned} \tag{6}$$

Furthermore, let $\mathcal{E}_0$ and $\mathcal{E}_1$ be the one-column vectors

$$\mathcal{E}_0\equiv\begin{pmatrix} E_{x0} \\ E_{y0} \end{pmatrix};\quad \mathcal{E}_1\equiv\begin{pmatrix} E_{x1} \\ E_{y1} \end{pmatrix}. \tag{7}$$

The relations (4) are now equivalent to the vector equation

$$\mathcal{E}_1=\mathbf{M}\mathcal{E}_0. \tag{8}$$

The matrix $\mathbf{M}$ may itself be written in a simple form. Let $\mathbf{S}(\omega)$ be the rotation matrix

$$\mathbf{S}(\omega)\equiv\begin{pmatrix} \cos\omega & -\sin\omega \\ \sin\omega & \cos\omega \end{pmatrix} \tag{9}$$

and let

$$\mathbf{N}\equiv\begin{pmatrix} N_{x'} & 0 \\ 0 & N_{y'} \end{pmatrix}. \tag{10}$$

We then have

$$\mathbf{M}=\mathbf{S}(\omega)\mathbf{N}\mathbf{S}(-\omega) \tag{11}$$

or

$$\mathcal{E}_1=\mathbf{S}(\omega)\mathbf{N}\mathbf{S}(-\omega)\mathcal{E}_0. \tag{12}$$

We have thus been able to represent the effect of a retardation plate or a partial polarizer as a matrix operator which operates upon the vector describing the intensity and polarization of the entering light. The matrix operator has been factored into a product of two types of matrices, the first of which, $\mathbf{N}$, describes the optical element in a way independent of its orientation, and the second of which, $\mathbf{S}$, describes the orientation.

Multi-Element Systems

Suppose now that we have a series of n optical elements, represented by the n matrices $\mathbf{N}_1$, $\mathbf{N}_2$, $\cdots$, $\mathbf{N}_n$, with the respective orientations ω_1, ω_2, $\cdots$, ω_n, and suppose that the light passes through the elements in the order in which they are numbered. We then have

$$\begin{aligned} &\mathcal{E}_1=\mathbf{S}(\omega_1)\mathbf{N}_1\mathbf{S}(-\omega_1)\mathcal{E}_0=\mathbf{M}_1\mathcal{E}_0,\\ &\mathcal{E}_0=\mathbf{M}_2\mathcal{E}_1,\\ &\cdots\cdots\\ &\mathcal{E}_n=\mathbf{M}_n\mathcal{E}_{n-1}, \end{aligned} \tag{13}$$

where $\mathcal{E}_i$ is the light vector as it emerges from the ith element. By substituting each of the Eqs. (13) in the one following it, we have at once the relation between $\mathcal{E}_n$ and $\mathcal{E}_0$

$$\begin{aligned} \mathcal{E}_n&=\mathbf{M}_n\mathbf{M}_{n-1}\cdots\mathbf{M}_2\mathbf{M}_1\mathcal{E}_0\\ &=\mathbf{M}^{(n)}\mathcal{E}_0, \end{aligned} \tag{14}$$

where $\mathbf{M}^{(n)}$ is the two-by-two matrix

$$\mathbf{M}^{(n)}\equiv\mathbf{M}_n\mathbf{M}_{n-1}\cdots\mathbf{M}_2\mathbf{M}_1. \tag{15}$$

According to (6), each of the $\mathbf{M}_i$'s is a symmetrical matrix. In general, however, the product of two or more of the $\mathbf{M}$ matrices will be neither symmetrical nor anti-symmetrical. This criterion permits us to distinguish immediately between the matrices corresponding to a single element, and those corresponding to the superposition of two or more elements.

It is fruitful to examine the structure of $\mathbf{M}^{(n)}$ more closely. We have

$$\mathbf{M}^{(n)}=[\mathbf{S}(\omega_n)\mathbf{N}_n\mathbf{S}(-\omega_n)][\mathbf{S}(\omega_{n-1})\mathbf{N}_{n-1}\mathbf{S}(-\omega_{n-1})]\cdots[\mathbf{S}(\omega_2)\mathbf{N}_2\mathbf{S}(-\omega_2)][\mathbf{S}(\omega_1)\mathbf{N}_1\mathbf{S}(-\omega_1)]. \tag{16}$$

By the use of the easily proved relation:

$$\mathbf{S}(\omega_1+\omega_2)\equiv\mathbf{S}(\omega_1)\mathbf{S}(\omega_2), \tag{17}$$

the expression for $\mathbf{M}^{(n)}$ may be written in either of the following forms:

$$\mathbf{M}^{(n)}=\mathbf{S}(\omega_n)[\mathbf{N}_n\mathbf{S}(\omega_{n,\,n-1})\cdots\mathbf{N}_2\mathbf{S}(\omega_{2,\,1})\mathbf{N}_1\mathbf{S}(\omega_{1,\,n})]\mathbf{S}(-\omega_n), \tag{18}$$

$$=\mathbf{S}(\omega_1)[\mathbf{S}(\omega_{1,\,n})\mathbf{N}_n\mathbf{S}(\omega_{n,\,n-1})\cdots\mathbf{S}(\omega_{2,\,1})\mathbf{N}_1]\mathbf{S}(-\omega_1), \tag{19}$$

where

$$\omega_{i,\,j}\equiv\omega_j-\omega_i. \tag{20}$$

* We depart from the usual double-subscript notation for matrix elements solely for reasons of typographical convenience.

The matrices represented by the square brackets in Eqs. (18) and (19) depend only on the relative orientation of the elements of the optical system, and are independent of the orientation of the optical system as a whole. Thus, just as in the case of a single element, we have factored the operator representing the optical system into the product of two types of matrices, the first of which depends only on the nature of the optical elements and on their *relative* orientation, and the second of which describes only the orientation of the optical system as a whole with respect to the x, y axes.

So far we have excluded any consideration of rotators, which are plates possessing the ability to rotate the plane of polarization about the z axis. The formal change caused by the introduction of rotators into the system is very small, however. Suppose that a given rotator is able to rotate the plane of polarization through the angle $\bar{\omega}$ in the counterclockwise direction. The matrix operator corresponding to the rotator is then simply $\mathbf{S}(\bar{\omega})$. We shall not stop to prove this statement,* since its proof is elementary. If the rotator just considered is introduced between the ith and the $(i+1)$th element of the optical system considered in the last paragraph, the resulting optical system may still be represented in the form of Eqs. (18) and (19) if we redefine $\omega_{i+1,\,i}$ as

$$\omega_{i+1,\,i} \equiv \omega_i - \omega_{i+1} + \bar{\omega}. \tag{21}$$

The generalization to the case in which several rotators are placed in the optical system is obvious. If several rotators are placed in juxtaposition, they are equivalent to a single rotator, according to (17).

There is an important physical distinction between the behavior of retardation plates and partial polarizers on one hand, and the behavior of rotators on the other. With the coordinate system which we are using, it makes no difference in what direction the light passes through a retardation plate or partial polarizer. With the type of rotator represented by a crystal or solution possessing optical activity, however, the rotation changes sign if the light passes through the plate in the reverse direction. In determining the operator corresponding to a rotator, therefore, it is necessary to know in which direction the light passes through the plate.

* More precisely, the matrix of a rotator is $e^{i\phi}\mathbf{S}(\bar{\omega})$, where the phase factor takes into account the finite optical thickness of the plate. In this paper, however, rotators will be considered only in combination with retardation plates and partial polarizers, so that the phase factor may always be considered to be associated with the other elements of the optical system. For simplicity, we omit the phase factor.

Reversibility of the Optical System

It is, however, not necessary to make a recalculation if the light passes through the optical system in the opposite direction. Suppose that we have computed the matrix $\mathbf{M}^{(n)}$ for the case that the light passes through the system in the order in which the plates are numbered, and then suppose that the light is reversed so that it passes through the system in the opposite direction, passing first through the nth plate, and last through the first plate. In terms of the one-row vectors,

$$\bar{\varepsilon} \equiv (E_x\; E_y), \tag{22}$$

the relation between the vector $\bar{\varepsilon}_{\text{initial}}$ of the light entering the nth plate and the vector $\bar{\varepsilon}_{\text{final}}$ of the light emerging from the first plate is

$$\bar{\varepsilon}_{\text{final}} = \bar{\varepsilon}_{\text{initial}} \mathbf{M}^{(n)}, \tag{23}$$

where $\mathbf{M}^{(n)}$ is exactly the same matrix as that previously defined. This statement may be proved easily, first by showing that a relation of the form (23) holds for each element of the system, and then by eliminating the intermediate $\bar{\varepsilon}$ vectors.

The statement made in the last paragraph holds only if all of the rotators in the system are of the type represented by ordinary optical activity, and not of the type represented by the Faraday effect, in which case the direction of the rotation depends on the direction of the magnetic field, and not on the direction in which the light passes through the material. If any of the rotators are of the type represented by the Faraday effect, then the $\mathbf{M}^{(n)}$ which appears in (23) is not the same as the $\mathbf{M}^{(n)}$ which appears in (14), the difference being that the $\bar{\omega}$'s appearing in the $\mathbf{S}$ matrices which represent the Faraday rotators must have opposite signs in the two $\mathbf{M}^{(n)}$'s.

Discussion of the Matrices

When the optical system does not contain any partial polarizers but consists entirely of rotators and retardation plates, *all of the matrices are unitary.*‡ This fact may be proved by either of two methods.

First proof

The **N** matrices are all unitary, since their characteristic values are all of absolute value unity; the latter is a necessary and sufficient condition that a matrix be unitary. Furthermore, the **S** matrices are easily shown to be unitary, since the conjugate transpose of $\mathbf{S}(\omega)$ is $\mathbf{S}(-\omega)$, so that $\mathbf{S}^\dagger\mathbf{S}=1$. The general matrix $\mathbf{M}^{(n)}$ is the product of factors which are all unitary, and therefore $\mathbf{M}^{(n)}$ is unitary.

Second proof

This proof is not purely mathematical, but is based on the fact that an optical system composed only of rotators and retardation plates does not change the intensity of the light passing through it. The intensity of the light is proportional to the sum of the squares of the x and y components of the electric vector:

$$I \propto A_x{}^2+A_y{}^2=E^*{}_xE_x+E^*{}_yE_y=\bar{\varepsilon}^*\varepsilon=\bar{\varepsilon}\varepsilon^* \quad (24)$$

(cf. Eq. (1) for part of the notation). The intensity of the light is therefore proportional to the square of the length of the vector ε. Now a unitary matrix does not change the length of a vector, and conversely, if a matrix does not change the length of an arbitrary vector, then the matrix is unitary. Thus, since the optical system does not change the intensity of the light, the corresponding matrices must be unitary.

Conversely, if the first proof of the unitary character of the matrices is accepted, then the second may be regarded as a proof that an optical system composed only of retardation plates and rotators does not change the intensity of the light passing through it.

In nearly every case, we are not interested in the total phase change corresponding to the two axes of a retardation plate, but only in the phase difference of the two principal axes; we often refer to a retardation plate as a q wave plate, implying that the optical path difference for plane polarized light parallel to each of the two axes is q wave-lengths.

Since the two principal extinction coefficients of a retardation plate are both zero, we find by comparison of (2) with (10) that the matrix of a retardation plate whose axes are parallel to the x and y axes may always be written in the form $e^{i\phi}\mathbf{G}$, where

$$\mathbf{G}\equiv\begin{pmatrix} e^{i\gamma} & 0 \\ 0 & e^{-i\gamma} \end{pmatrix}. \quad (25)$$

If we are interested only in the phase difference, the phase factor $e^{i\phi}$ may be omitted, and we may consider **G** alone to be the diagonal representation of a wave plate.

If we choose to write the diagonal matrices of all the wave plates in the optical system in the form (25), then, since the determinant of an **S** matrix is unity, all of the matrices which occur as factors on the right-hand side of Eq. (16) will have determinants equal to unity, provided that the system contains only retardation plates and rotators. Then, since the determinant of a product of matrices is equal to the product of the determinants of the matrices, the determinant of $\mathbf{M}^{(n)}$ will have the value unity.

Our treatment of partial polarizers is similar. The diagonal form of the matrix corresponding to a partial polarizer may always be written in the form $e^{i\phi}\mathbf{P}$, where

$$\mathbf{P}\equiv\begin{pmatrix} p_1 & 0 \\ 0 & p_2 \end{pmatrix} \quad \begin{matrix} 0\leqslant p_1\leqslant 1 \\ 0\leqslant p_2\leqslant 1. \end{matrix} \quad (26)$$

Again, if we are not interested in the absolute phase of the two components of the light, but only in the relative phase, we may consider **P** alone to be the diagonal representation of a partial polarizer. If the plate is a perfect polarizer, that is to say, if the plate absorbs completely light parallel to one of its axes, then one of the two p's will be zero; in this case the determinant of **P** is zero. More generally, the determinant of **P** is real, with its value lying in the range zero to unity.

We thus find that if we use **G** to represent a retardation plate, and **P** to represent a partial polarizer, then the determinant of $\mathbf{M}^{(n)}$ for an optical system containing all three types of

‡ Let $\mathbf{M}^\dagger$ be the matrix obtained from **M** by first exchanging m_3 and m_4 and then replacing each of the matrix elements by its complex conjugate. The matrix **M** is unitary if $\mathbf{M}^\dagger\mathbf{M}=\mathbf{M}\mathbf{M}^\dagger=1$. The matrix $\mathbf{M}^\dagger$ is said to be the Hermitian adjoint, or the conjugate transpose of **M**.

elements will be real and non-negative, with its value in the range zero to unity.

It will be shown in Part II that the most general unitary matrix may be represented by an optical system consisting of only two elements, one a rotator and the other a retardation plate. Similarly, it will be shown that the general two-by-two complex matrix may be represented by an optical system containing not more than four elements, the only restriction being that the matrix may not be one which would increase the intensity of the light.

Elliptical Polarization

For convenience in reference, we state here the relations giving the shape and orientation of the ellipse representing the state of polarization of the light. Let the amplitude of the x and y components be A_x and A_y, and let $\delta=\epsilon_y-\epsilon_x$, as in Eq. (1). Further, let $\tan\theta$ equal the ratio of axes of the ellipse, and let ψ be the orientation of the major axis of the ellipse, measured counter-clockwise from the x axis. The following relations determine θ and ψ:[3]

$$\begin{aligned} \tan 2\psi &= \tan 2\alpha \cos \delta, \\ \cos 2\theta &= \sin 2\alpha |\sin \delta|, \end{aligned} \tag{27}$$

where

$$\tan \alpha = A_y/A_x, \quad 0 \leqslant \alpha \leqslant \tfrac{1}{2}\pi. \tag{28}$$

When $\sin\delta$ is positive, the ellipse is described in the clockwise direction, and *vice versa*.

[3] Max Born, *Optik* (Springer, Berlin, 1933), p. 23.

A New Calculus for the Treatment of Optical Systems

II. Proof of Three General Equivalence Theorems

HENRY HURWITZ, JR., *Research Laboratory of Physics, Harvard University, Cambridge, Massachusetts,*

AND

R. CLARK JONES,* *Research Laboratory, Polaroid Corporation, Cambridge, Massachusetts, and Research Laboratory of Physics, Harvard University, Cambridge, Massachusetts*

(Received April 23, 1941)

The general theory developed in Part I is used to prove three equivalence theorems about optical systems of the type under discussion. We prove that any optical system which contains only retardation plates and rotators is optically equivalent to a system containing only two plates—one a retardation plate, and the other a rotator. We then prove an exactly analogous theorem for systems containing only partial polarizers and rotators. Finally, it is proved that the most general optical system which contains any number of all three types of plates is optically equivalent to a system containing at most four plates—two retardation plates, one partial polarizer, and one rotator.

STATEMENT OF THE THEOREMS

THE power of the calculus described in Part I may be illustrated by using it to establish a few general theorems relating to optical systems of the type under consideration.

We shall prove the following three theorems:

I. For light of a given wave-length, an optical system containing any number of retardation plates and rotators is optically equivalent to a system containing only two elements—one a retardation plate, and the other a rotator.

II. For light of a given wave-length, an optical system containing any number of partial polarizers and rotators is optically equivalent to a system containing only two elements—one a partial polarizer and the other a rotator.

III. For light of a given wave-length, an optical system containing any number of re-

* National Scholar, Harvard University.

tardation plates, partial polarizers, and rotators is optically equivalent to a system containing four elements—two retardation plates, one partial polarizer, and one rotator. In a large and finite class of cases, the rotator is not necessary.

The method of proof of these theorems is essentially the same for each of them. We shall first determine the mathematical restrictions on the matrix representing the original optical system. Secondly, we shall write down the matrix corresponding to the substitute optical system. We shall then equate these two matrices. In each case, we shall find that the resulting equations are of such a form that they always admit a physically realizable solution for the parameters of the substitute optical system. The proof that the equations are always solvable will be obtained by actually solving the equations explicitly for the parameters of the substitute optical system. The proof will thus not only establish that the replacement is always possible, but it will also provide the explicit equations for computing the parameters of the substitute optical system corresponding to any given original optical system.

The first of the three theorems has been proved by H. Poincaré by an entirely different method (cf. below). So far as the writers have been able to learn, the remaining two theorems are new.

Proof of Theorem I

We saw in Part I that the matrix $\mathbf{M}^{(n)}$ representing any optical system composed of retardation plates and rotators was unitary. We shall demonstrate that any unitary matrix may be represented by a system consisting of one retardation plate and one rotator.

We shall write

$$\mathbf{M}^{(n)}=e^{i\alpha}\mathbf{U} \tag{1}$$

and we shall select the angle α so that the determinant of $\mathbf{U}$ is unity.

It is well known[1] that the general unitary matrix with two rows and columns with unit determinant may be written in the form:

$$\mathbf{U}\equiv\begin{pmatrix} e^{i\phi}\cos\theta & -e^{-i\psi}\sin\theta \\ e^{i\psi}\sin\theta & e^{-i\phi}\cos\theta \end{pmatrix}. \tag{2}$$

The real angles θ, ϕ and ψ may be determined from the matrix elements of $\mathbf{U}$ by the relations:

$$\begin{aligned} e^{2i\phi}&=u_1/u_2, & \sin\theta&=e^{-i\psi}u_3,\\ e^{2i\psi}&=u_3/u_4, & \cos\theta&=e^{-i\phi}u_4. \end{aligned} \tag{3}$$

For any given $\mathbf{M}^{(n)}$, it is thus possible to determine the values of α, θ, ϕ and ψ, although the determination is not unique; for example, if ϕ is a solution, then ϕ plus an integral multiple of π is also a solution.

The matrix of a retardation plate with the orientation ω is $\mathbf{S}(\omega)\mathbf{G}\mathbf{S}(-\omega)$, where $\mathbf{G}$ is defined by Eq. (25, I); the matrix of a rotator is $\mathbf{S}(\tilde{\omega})$. Whether the rotator is the first or second element of the system, the matrix of the combination will be of the form

$$\mathbf{V}\equiv\mathbf{S}(A)\mathbf{G}\mathbf{S}(B), \tag{4}$$

where the angles A and B are in general different.

If now for any values of θ, ϕ and ψ, we are always able to choose A, B and γ so that $\mathbf{V}=\mathbf{U}$, our theorem is proved. We multiply out explicitly the expression on the right of (4), so that the matrix elements of $\mathbf{V}$ appear as explicit functions of A, B and γ. We then set the four elements of $\mathbf{U}$ equal, respectively, to the four elements of $\mathbf{V}$, and after a little algebraic reduction, we find

$$\begin{aligned} \cos\phi\cos\theta&=\cos(A+B)\cos\gamma,\\ \sin\phi\cos\theta&=\cos(A-B)\sin\gamma,\\ \cos\psi\sin\theta&=\sin(A+B)\cos\gamma,\\ \sin\psi\sin\theta&=\sin(A-B)\sin\gamma \end{aligned} \tag{5}$$

as the conditions which must be satisfied in order that $\mathbf{V}=\mathbf{U}$.

From (5), we find as the explicit solution for A, B and γ:

$$\tan(A+B)=\frac{\cos\psi}{\cos\phi}\tan\theta=\frac{u_3-u_4}{u_1+u_2}, \tag{6}$$

[1] B. L. van der Waerden, *Die Gruppentheoretische Methode in der Quantenmechanik* (Springer, Berlin, 1932), p. 58.

$$\tan(A-B)=\frac{\sin\psi}{\sin\phi}\tan\theta=\frac{u_3+u_4}{u_1-u_2}, \quad (7)$$

$$\tan\gamma=\frac{\cos(A+B)}{\cos(A-B)}\tan\phi$$

$$=\frac{\sin(A+B)}{\sin(A-B)}\tan\psi. \quad (8)$$

It is evident that these equations always have a solution. The theorem is thus established. If we set

$$\mathbf{N}\equiv e^{i\alpha}\mathbf{G}, \quad (9)$$

then we have

$$\mathbf{M}^{(n)}=\mathbf{S}(A)\mathbf{N}\mathbf{S}(B). \quad (10)$$

We may at once show also that any matrix in the form **V** may be expressed in the form **U**. From (5) we have as the explicit solution for θ, ϕ and ψ in terms of A, B and γ:

$$\tan\phi=\frac{\cos(A-B)}{\cos(A+B)}\tan\gamma, \quad (11)$$

$$\tan\psi=\frac{\sin(A-B)}{\sin(A+B)}\tan\gamma, \quad (12)$$

$$\tan\theta=\frac{\cos\phi}{\cos\psi}\tan(A+B)=\frac{\sin\phi}{\sin\psi}\tan(A-B). \quad (13)$$

It is also evident that these equations always have a solution. We have thus shown that **U** and **V** are mathematically equivalent representations of the general unitary matrix with unit determinant. In the proof of Theorem III, we shall find it convenient to pass freely from one representation to the other.

Poincaré's Proof of Theorem I

Poincaré[2] has given a geometrical proof of Theorem I which appears at first sight to be quite different from ours. Actually, however, the two proofs are very closely related, as we shall now show.

Poincaré expresses the ratio of the two complex components of the light in the form

$$E_y/E_x=(A_y/A_x)\exp[i(\epsilon_y-\epsilon_x)]=\xi+i\eta. \quad (14)$$

In this expression, ξ and η together define completely the shape and orientation of the ellipse representing the state of polarization of the light. Thus each point on the ξ, η plane corresponds to a state of polarization, and conversely, each state of polarization corresponds to a point on the ξ, η plane.

Each point on the plane is considered to be projected stereographically upon a sphere of unit diameter to which the plane is tangent at the origin of coordinates. There is thus established a one-to-one correspondence between points on the sphere and points on the plane.

If now the sphere is subjected to a rotation about any diameter, and if we consider that the points on the sphere rotate rigidly with the sphere, then it is clear that all of the corresponding points on the plane (except two) will suffer a change of position. Thus a rotation of the sphere always may be said to correspond to a change in the state of polarization, and conversely, any change in the state of polarization may be related to some rotation of the sphere. Poincaré shows that the rotation corresponding to the action of a wave plate with retardation $\Delta=2\gamma$ is a rotation of angle Δ about the diameter which is perpendicular to the plane: he also shows that the rotation which corresponds to a rotator of angle $\bar{\omega}$ is a rotation of angle $2\bar{\omega}$ about the diameter which is parallel to the ξ axis.

Once the facts stated in the last paragraph have been established, the proof of Theorem I follows immediately, since it is physically obvious that any rotation of the sphere may always be compounded from two successive rotations, one about the diameter normal to the plane, and the other about the diameter parallel to the ξ axis.

Now we learn from the group theory of three-dimensional rotations[3] that any unitary two-by-two matrix with unit determinant may be associated uniquely with a rotation of a sphere. The elements of this matrix are referred to as the Cayley-Klein parameters of the rotation. If we express the matrix of the optical system of Theorem I in the form **U**, then the Eulerian angles specifying the rotation of Poincaré's sphere are linearly related to the

[2] H. Poincaré, *Théorie Mathématique de la Lumière* (Paris, 1892), Vol. II, Chapter XII.

[3] Carl Eckart, Rev. Mod. Phys. **2**, 305 (1930), pp. 341, 346.

angles ϕ, θ and ψ. This is the relation between Poincaré's geometrical proof and our proof by the use of matrices.

From a purely formal point of view, the relation between the proofs consists in the fact that the group of unitary two-by-two matrices with unit determinant is isomorphic to the group of three-dimensional rotations.

Proof of Theorem II

We saw in Part I that all of the matrices representing partial polarizers and rotators could always be so written that the matrix elements were real and the determinants real and non-negative, except for phase factors occurring as factors of the matrix as a whole. For an optical system composed only of partial polarizers and rotators, we may therefore set

$$\mathbf{M}^{(n)}=e^{i\alpha}\mathbf{R}, \tag{15}$$

where $\mathbf{R}$ is a matrix whose matrix elements are real and whose determinant lies in the range zero to unity.

The matrix of a partial polarizer with the orientation ω is $\mathbf{S}(\omega)\mathbf{P}\mathbf{S}(-\omega)$, where $\mathbf{P}$ is defined by Eq. (26, I). By the same argument that was used in the proof of Theorem I, the optical system composed of a rotator and a partial polarizer will always be represented by a matrix of the form

$$\mathbf{T}\equiv\mathbf{S}(A)\mathbf{P}\mathbf{S}(B), \tag{16}$$

where A and B are related in an obvious manner to ω and $\tilde{\omega}$.

In order to establish the theorem, we must show that any matrix $\mathbf{R}$ which satisfies the stated conditions may be written in the form $\mathbf{T}$. We multiply out explicitly the expression on the right of (16), so that the matrix elements of $\mathbf{T}$ appear as explicit functions of A, B, p_1, and p_2. We then set the four elements of $\mathbf{R}$ equal, respectively, to the four elements of $\mathbf{T}$, and after a little algebraic reduction, we find

$$\begin{aligned}(p_1+p_2)\sin(A+B)&=r_3-r_4,\\(p_1+p_2)\cos(A+B)&=r_1+r_2,\\(p_1-p_2)\sin(A-B)&=r_3+r_4,\\(p_1-p_2)\cos(A-B)&=r_1-r_2\end{aligned} \tag{17}$$

as the conditions which must be satisfied in order that $\mathbf{T}=\mathbf{R}$.

The quotient of the first two, and the quotient of the last two of the Eqs. (17) are

$$\tan(A+B)=(r_3-r_4)/(r_1+r_2), \tag{18a}$$

$$\tan(A-B)=(r_3+r_4)/(r_1=r_2). \tag{18b}$$

By the use of the trigonometric identity

$$\sin x\cos x\equiv\tan x/(1+\tan^2 x),$$

the product of the first two, and the product of the last two of the Eqs. (17) may be written

$$\begin{aligned}p_1p_2&=r_1r_2-r_3r_4\equiv\det\mathbf{R},\\p_1^2+p_2^2&=r_1^2+r_2^2+r_3^2+r_4^2.\end{aligned} \tag{19}$$

The Eqs. (18) always permit a solution for A and B, and, since the determinant of $\mathbf{R}$ is non-negative, the Eqs. (19) always permit a solution for p_1 and p_2 in which both p_1 and p_2 are positive. Although the Eqs. (19) always permit a solution for p_1 and p_2, they do not determine which is which; if $p_1=a$, $p_2=b$ is a solution of (19), then so is $p_1=b$, $p_2=a$. We must go back to the last two of the Eqs. (17) in order to determine which of the p's is the larger, and it is clear that the decision will depend on which of the two physically distinct solutions of (18b) has been chosen. Note also that only one of the two possible solutions of (18a) is admissible, since p_1+p_2 cannot be negative.

The theorem is thus established, provided only that we can show that neither of the p's is greater than unity. We proceed with this final part of the proof.

With every matrix $\mathbf{A}$ there may be associated a number $\Gamma(\mathbf{A})$ where Γ is the maximum value of the ratio $|\mathbf{A}\varepsilon|/|\varepsilon|$ with respect to any variation of the two components of ε. Such a maximum will always exist for a matrix $\mathbf{A}$ whose matrix elements are finite and do not depend on ε. For any two matrices $\mathbf{A}$ and $\mathbf{B}$, we obviously have the relation

$$\Gamma(\mathbf{AB})\leqslant\Gamma(\mathbf{A})\Gamma(\mathbf{B}). \tag{20}$$

Since a unitary matrix does not change the length of a vector, we have

$$\begin{aligned}\Gamma(\mathbf{U})&=1\\\Gamma(\mathbf{AU})&=\Gamma(\mathbf{UA})=\Gamma(\mathbf{A}),\end{aligned} \tag{21}$$

where $\mathbf{A}$ is any matrix and $\mathbf{U}$ is a unitary matrix.

The physical interpretation of Γ for a matrix which represents an element of an optical system is that Γ^2 is the maximum transmission factor of the given optical element. The value of Γ for such a matrix may therefore never be greater than unity. By repeated application of (20), we then find that the value of $\Gamma(\mathbf{M}^{(n)})$ may not be greater than unity. We clearly have $\Gamma(\mathbf{M}^{(n)})=\Gamma(\mathbf{R})$ and, since we have set $\mathbf{T}=\mathbf{R}$, we have also $\Gamma(\mathbf{M}^{(n)})=\Gamma(\mathbf{T})$. But by repeated application of (21), we have $\Gamma(\mathbf{T})=\Gamma(\mathbf{P})$, so that the value of Γ for $\mathbf{P}$ may not be greater than unity. The value of Γ for the matrix $\mathbf{P}$, however, is clearly the larger of the two p's, so that neither of the p's may be greater than unity. The proof of Theorem II is now complete.

Proof of Theorem III

In this section we shall show that any optical system composed of retardation plates, partial polarizers and rotators may always be replaced by an optical system composed of four elements; the four-element system consists of a partial polarizer placed between two retardation plates, with the addition of a rotator inserted at any of the four possible positions in the system.

We showed in Part I that, except for a phase factor, the matrix $\mathbf{M}^{(n)}$ representing the optical system could always be written so that the value of its determinant was real and in the range zero to unity. We therefore set

$$\mathbf{M}^{(n)}=e^{i\alpha}\mathbf{W}, \tag{22}$$

where $\mathbf{W}$ has a determinant which is real and which lies in the range zero to unity.

The matrix representing a partial polarizer placed between two retardation plates is

$$[\mathbf{S}(\omega_3)\mathbf{G}_1\mathbf{S}(-\omega_3)][\mathbf{S}(\omega_2)\mathbf{P}\mathbf{S}(-\omega_2)] \times[\mathbf{S}(\omega_1)\mathbf{G}_2\mathbf{S}(-\omega_1)]. \tag{23}$$

This matrix may be written in the form

$$\mathbf{\Phi}\equiv\mathbf{S}(A_1)\mathbf{G}_1\mathbf{S}(B_1)\mathbf{P}\mathbf{S}(A_2)\mathbf{G}_2\mathbf{S}(\mathbf{B}_2), \tag{24}$$

where the A's and B's are subject to the condition

$$A_1+B_1+A_2+B_2=0. \tag{25}$$

If now we add a rotator to the three-element system just described it is evident that it is always possible to choose the three ω's and the angle $\bar{\omega}$ of the rotator so that the A's and B's may all be chosen arbitrarily. That is to say, the addition of a rotator permits us to avoid the restriction (25).

We thus see that the four-element optical system just described may be represented by the matrix $\mathbf{\Phi}$, where the A's and B's are now all independent. The matrix $\mathbf{\Phi}$ is of the form $\mathbf{V}_1\mathbf{P}\mathbf{V}_2$, where $\mathbf{V}$ is the matrix defined by Eq. (4). We have shown that any matrix of the form $\mathbf{V}$ is mathematically equivalent to a matrix of the form $\mathbf{U}$, where $\mathbf{U}$ is defined by Eq. (2). Any matrix of the form $\mathbf{\Phi}$ may therefore be written in the form $\mathbf{U}_1\mathbf{P}\mathbf{U}_2$, *and conversely.*

In order to establish the theorem, we must now show that any matrix $\mathbf{W}$ can be written in the form

$$\mathbf{\Phi}\equiv\mathbf{U}_1\mathbf{P}\mathbf{U}_2. \tag{26}$$

This fact will be proved in the following paragraphs. In anticipation of this result, we should like to make an important observation: The matrices $\mathbf{W}$ and $\mathbf{\Phi}$ are not completely general, because they are restricted by the fact that the imaginary parts of their determinants are zero. They therefore contain eight minus one, or seven *independent* variables. The matrix $\mathbf{\Phi}$, however, contains eight variables (three in each of the $\mathbf{U}$'s and two in $\mathbf{P}$), so that these variables cannot be uniquely determined. We shall find that the extra degree of freedom afforded by the eight variables will often, but by no means always, permit us to dispense with the rotator; whenever the extra degree of freedom permits us to satisfy the relation (25), the rotator is unnecessary.

We now multiply out explicitly the product on the right in Eq. (26). We find

$$\mathbf{\Phi}=\begin{pmatrix} \exp[i(\phi_1+\phi_2)]p_1\cos\theta_1\cos\theta_2 & -\exp[i(\phi_1-\psi_2)]p_1\cos\theta_1\sin\theta_2 \\ -\exp[-i(\psi_1-\psi_2)]p_2\sin\theta_1\sin\theta_2 & -\exp[-i(\phi_2+\psi_1)]p_2\sin\theta_1\cos\theta_2 \\ \exp[i(\phi_2+\psi_1)]p_1\sin\theta_1\cos\theta_2 & -\exp[i(\psi_1-\psi_2)]p_1\sin\theta_1\sin\theta_2 \\ +\exp[-i(\phi_1-\psi_2)]p_2\cos\theta_1\sin\theta_2 & +\exp[-i(\phi_1+\phi_2)]p_2\cos\theta_1\sin\theta_2 \end{pmatrix}. \tag{27}$$

By examination of this matrix, we see that the ϕ's and ψ's occur only in four different combinations

$$\chi_1 \equiv \phi_1+\phi_2, \quad \chi_3 \equiv \phi_2+\psi_1,$$
$$\chi_2 \equiv \psi_1-\psi_2, \quad \chi_4 \equiv \phi_1-\psi_2. \tag{28}$$

These four relations are not linearly independent, in view of the relation

$$\chi_1+\chi_2=\chi_3+\chi_4.$$

We therefore choose to replace the ϕ's and ψ's by three new *independent* variables η, ξ_1, and ξ_2:

$$\phi_1+\phi_2=\eta+\xi_1, \quad \phi_2+\psi_1=\eta+\xi_2,$$
$$\psi_1-\psi_2=\eta-\xi_1, \quad \phi_1-\psi_2=\eta-\xi_2. \tag{29}$$

We have the following definitions of η, ξ_1, and ξ_2:

$$\begin{aligned} 2\xi_1 &\equiv \phi_1+\phi_2-\psi_1+\psi_2, \\ 2\xi_2 &\equiv -\phi_1+\phi_2+\psi_1+\psi_2, \\ 2\eta &\equiv \phi_1+\phi_2+\psi_1-\psi_2, \\ 2\sigma &\equiv \phi_1-\phi_2+\psi_1+\psi_2, \end{aligned} \tag{30}$$

where we have added a fourth variable σ in order that we may also solve the equations for the ϕ's and ψ's:

$$2\phi_1=\xi_1-\xi_2+\eta+\sigma, \quad 2\psi_1=-\xi_1+\xi_2+\eta+\sigma,$$
$$2\phi_2=\xi_1+\xi_2+\eta-\sigma, \quad 2\psi_2=\xi_1+\xi_2-\eta+\sigma. \tag{31}$$

We make the further changes of variable

$$q_1 \equiv e^{i\eta}p_1, \quad q_2 \equiv e^{-i\eta}p_2. \tag{32}$$

The q's are thus complex quantities.

With the changes of variable outlined in the last paragraph, the matrix $\mathbf{\Phi}$ may now be written in the simpler form

$$\mathbf{\Phi}=\begin{pmatrix} \exp(i\xi_1)(q_1\cos\theta_1\cos\theta_2-q_2\sin\theta_1\sin\theta_2) & -\exp(-i\xi_2)(q_1\cos\theta_1\sin\theta_2+q_2\sin\theta_1\cos\theta_2) \\ \exp(i\xi_2)(q_1\sin\theta_1\cos\theta_2+q_2\cos\theta_1\sin\theta_2) & -\exp(-i\xi_1)(q_1\sin\theta_1\sin\theta_2-q_2\cos\theta_1\cos\theta_2) \end{pmatrix}. \tag{33}$$

We now set the four matrix elements of $\mathbf{\Phi}$ equal to the four matrix elements of $\mathbf{W}$. The resulting four equations are equivalent to

$$\begin{aligned} (q_1+q_2)\sin(\theta_1+\theta_2) &= \exp(-i\xi_2)w_3-\exp(i\xi_2)w_4, \\ (q_1+q_2)\cos(\theta_1+\theta_2) &= \exp(-i\xi_1)w_1+\exp(i\xi_1)w_2, \\ (q_1-q_2)\sin(\theta_1-\theta_2) &= \exp(-i\xi_2)w_3+\exp(i\xi_2)w_4, \\ (q_1-q_2)\cos(\theta_1-\theta_2) &= \exp(-i\xi_1)w_1-\exp(i\xi_1)w_2, \end{aligned} \tag{34}$$

where the w's are the matrix elements of $\mathbf{W}$, and are in general complex.

We wish to show that these equations can always be solved by θ's and ξ's which are real, and by q's whose product is real and non-negative. The quotient of the first two, and the quotient of the last two of the above equations are

$$\begin{aligned} \tan(\theta_1+\theta_2) &= \frac{\exp(-i\xi_2)w_3-\exp(i\xi_2)w_4}{\exp(-i\xi_1)w_1+\exp(i\xi_1)w_2}, \\ \tan(\theta_1-\theta_2) &= \frac{\exp(-i\xi_2)w_3+\exp(i\xi_2)w_4}{\exp(-i\xi_1)w_1-\exp(i\xi_1)w_2}. \end{aligned} \tag{35}$$

In order that these equations be capable of solution by real values of the θ's, it is necessary that the imaginary parts of the expressions on the right side be zero. By rationalizing the denominators, we find that this condition is equivalent to

$$\text{I.P. }\{w_1{}^*w_3\exp[i(\xi_1-\xi_2)]-w_2{}^*w_4\exp[-i(\xi_1-\xi_2)]\}=0,$$
$$\text{I.P. }\{w_1{}^*w_4\exp[i(\xi_1+\xi_2)]-w_2{}^*w_3\exp[-i(\xi_1+\xi_2)]\}=0$$

or finally,

$$\xi_1-\xi_2=\arg\,(w_1w_3^*+w_2^*w_4),$$
$$\xi_1+\xi_2=\arg\,(w_1w_4^*+w_2^*w_3). \tag{36}$$

The last two equations serve to evaluate the ξ's, and with this choice of the ξ's, Eqs. (35) may always be solved for real values of the θ's.

We may now determine the q's from the Eqs. (34). In order, however, to show that the product q_1q_2 is real and non-negative, we form the product of the first two and the product of the last two of the Eqs. (34), and then subtract the two resulting equations. With the help of (35), we find the result reduces to

$$q_1q_2=w_1w_2-w_3w_4=\det \mathbf{W}. \tag{37}$$

The matrix $\mathbf{W}$ was obtained from $\mathbf{M}^{(n)}$ by requiring that the determinant of $\mathbf{W}$ be real and non-negative. After evaluating q_1 and q_2, we may determine the quantity η from the condition that p_1 and p_2 be individually real and non-negative. (Cf. Eq. (32).)

Up to this point we have shown that the equations obtained by setting $\mathbf{\Phi}$ equal to $\mathbf{W}$ can always be solved for the real quantities ξ_1, ξ_2, θ_1, θ_2, p_1, p_2 and η, in the order mentioned. If we assign an arbitrary real value to σ, the Eqs. (31) may always be solved for the ϕ's and ψ's. The proof that neither of the p's is greater than unity is exactly similar to the proof used in Theorem II. The present theorem is therefore established.

It remains to determine the condition that we be able to choose σ so that the equation

$$A_1+B_1+A_2+B_2=0 \tag{25}$$

is satisfied. If σ can be so chosen, the rotator is unnecessary. Now let us consider the equation

$$\tan\,(A_1+B_1)+\tan\,(A_2+B_2)=\text{constant}. \tag{38}$$

In order that we be able to choose the A's and B's so that they satisfy (25), it is necessary and sufficient that the constant in Eq. (38) be zero. The validity of the equation

$$\tan\,(A_1+B_1)+\tan\,(A_2+B_2)=0 \tag{39}$$

is therefore the necessary and sufficient condition that we be able to dispense with the rotator.

By the use of Eqs. (31) and Eq. (6), we find that (39) may be written:

$$\cos\psi_1\cos\phi_2\tan\theta_1+\cos\psi_2\cos\phi_1\tan\theta_2=0 \tag{40}$$

or

$$\cos\tfrac{1}{2}[-\xi_1+\xi_2+\eta+\sigma]\cos\tfrac{1}{2}[\xi_1+\xi_2+\eta-\sigma]$$
$$\times\tan\theta_1+\cos\tfrac{1}{2}[\xi_1+\xi_2-\eta+\sigma]\cos\tfrac{1}{2}[\xi_1-\xi_2+\eta+\sigma]\tan\theta_2=0. \tag{41}$$

By trigonometric reduction we find that the last equation is equivalent to

$$k_1\cos\sigma+k_2\sin\sigma+k_3=0, \tag{42}$$

where

$$k_1\equiv\cos\xi_1(\tan\theta_1+\tan\theta_2)$$
$$k_2\equiv\sin\xi_1(\tan\theta_1-\tan\theta_2) \tag{43}$$
$$k_3\equiv\cos\,(\eta+\xi_2)\tan\theta_1+\cos\,(\eta-\xi_2)\tan\theta_2.$$

The condition which must be satisfied in order that Eq. (42) be solvable for σ is

$$k_3^2\leqslant k_1^2+k_2^2. \tag{44}$$

This inequality is the condition which must be satisfied in order that it be possible to dispense with the rotator. As it is written, the condition is purely a mathematical one. Although the condition can be simplified slightly, the authors have not been able to attach any simple physical significance to the inequality expressed by Eq. (44).

A New Calculus for the Treatment of Optical Systems

III. The Sohncke Theory of Optical Activity

R. Clark Jones*

Research Laboratory, Polaroid Corporation, Cambridge, Massachusetts, and Research Laboratory of Physics, Harvard University, Cambridge, Massachusetts

(Received April 23, 1941)

Reusch and Sohncke have examined the properties of a system containing a large number n of identical retardation plates, each of which is rotated with respect to the one preceding it through the angle ω. The product of n and ω must be equal to $\mu\pi$, where μ is an integer. Under certain conditions this system is optically equivalent to a simple rotator. This system, which would be very difficult to examine by ordinary methods, is given a treatment which is completely rigorous and which is more general than any given heretofore.

Introduction

Reusch[1] has examined experimentally the optical properties of a system composed of n identical retardation plates, the corresponding axes of which are distributed uniformly about a whole number of semicircles. If the number of semicircles is denoted by μ, we may describe the system by stating that the axis of each plate is rotated through the angle $-\pi\mu/n$ with respect to the corresponding axis of the plate immediately preceding it. Reusch found experimentally that when the phase retardation of each of the plates was small, the complete optical system behaved approximately like a simple rotator. A few years after Reusch published his results, Sohncke[2] provided a rigorous mathematical treatment of the system composed of three retardation plates with their axes distributed through 180°—that is to say, with the relative angle 60°. In his book[3] on crystal optics, Pockels derives an approximate expression for the angle of the rotation produced by the system of n retardation plates distributed through 180°:

$$\bar{\omega}=\tfrac{1}{8}n\Delta^2 \cot(\pi/n). \tag{1}$$

In this expression, Δ is the phase retardation of each of the plates, and is assumed to be small. Just how small Δ must be, will be indicated in the treatment to be given.

We present here a rigorous treatment of the optical system described at the beginning of the previous paragraph. The results will be exact for arbitrary integral values of n and μ. By use of the transformation properties of matrices, we shall find an exact expression for the matrix $\mathbf{M}^{(n)}$ of the entire optical system. Since the optical system under consideration contains only retardation plates, it satisfies the conditions of Theorem I of Part II, and may therefore be replaced by a system composed only of one rotator and one retardation plate. We shall find exact and reasonably simple expressions (cf. Eqs. (18) and (20)) for the angle $\bar{\omega}$ of the equivalent rotator and the phase retardation $2\bar{\gamma}$ of the equivalent retardation plate, and we shall derive the condition (cf. Eq. (21)) which must be satisfied in order that the optical system behave approximately as a rotator. Our results will confirm Pockels' approximate result, Eq. (1).

Determination of the Matrix $\mathbf{M}^{(n)}$

With respect to the fixed axes x and y, let the orientation of the plate which the light enters first be zero. Then in general the orientation of the kth plate will be

$$\omega_k=-(k-1)\pi\mu/n. \tag{2}$$

Let us represent the retardation plate in the form $\mathbf{G}$ (cf. Eq. (25, I)). The phase retardation of the plate is then $\Delta=2\gamma$. According to Eqs. (19, I) and

* National Scholar, Harvard University.

[1] E. Reusch, Berliner Monatsber. (Gesammtsitzung, July 8, 1869) 530 (1869); Pogg. Ann. **138**, 628 (1869).

[2] L. Sohncke, Math. Ann. **9**, 504 (1876). See also Pogg. Ann. Ergänzungsband **8**, 16 (1878), and Zeits. f. Krist. **13**, 214 (1888).

[3] F. Pockels, *Lehrbuch der Kristalloptik* (Teubner, Leipzig, 1906), pp. 289–290.

(20, I), the matrix $\mathbf{M}^{(n)}$ of the complete optical system reduces to

$$\begin{aligned}\mathbf{M}^{(n)}&=\mathbf{S}(\pi\mu)(\mathbf{S}(\omega)\mathbf{G})^n\\&=(-1)^\mu(\mathbf{S}(\omega)\mathbf{G})^n,\end{aligned}\tag{3}$$

where ω is here minus the relative orientation of the plates

$$\omega\equiv\pi\mu/n.\tag{4}$$

The determination of $\mathbf{M}^{(n)}$ thus reduces to the problem of finding the nth power of a matrix which in this case is unitary.

Now there exists a simple method of finding the nth power of any matrix which can be diagonalized. Let $\mathbf{A}$ be any matrix which can be diagonalized, and let $\mathbf{T}$ be the matrix of the transformation which accomplishes the diagonalization:

$$\begin{aligned}\mathbf{D}&=\mathbf{TAT}^{-1},\\\mathbf{A}&=\mathbf{T}^{-1}\mathbf{DT},\end{aligned}\tag{5}$$

where $\mathbf{D}$ is the diagonal form of $\mathbf{A}$, and where $\mathbf{T}^{-1}$ is the reciprocal of $\mathbf{T}$:

$$\mathbf{T}^{-1}\mathbf{T}=\mathbf{TT}^{-1}=1.\tag{6}$$

The nth power of $\mathbf{A}$ is then

$$\begin{aligned}\mathbf{A}^n&=\mathbf{T}^{-1}\mathbf{DTT}^{-1}\mathbf{DT}\cdots\mathbf{T}^{-1}\mathbf{DT}\\&=\mathbf{T}^{-1}\mathbf{D}^n\mathbf{T}.\end{aligned}\tag{7}$$

Since the nth power of a diagonal matrix can be written down at once, the determination of $\mathbf{A}^n$ reduces to the problem of determining the transformation $\mathbf{T}$.

The writer has found by detailed calculation the following result. Let a matrix $\mathbf{A}$ be diagonalized by the transformation $\mathbf{T}$, and let d_1 and d_2 be the diagonal elements of the matrix $\mathbf{D}$ so obtained. Let now d_1 and d_2 be replaced by new diagonal elements d_1' and d_2', and let the resulting matrix be transformed back by the reciprocal of the transformation $\mathbf{T}$. We denote the matrix so obtained by $\mathbf{A}'$. The matrix elements of $\mathbf{A}'$ are then related to those of $\mathbf{A}$ by the relations

$$\begin{aligned}a_1'&=\frac{d_1'-d_2'}{d_1-d_2}a_1+\frac{d_2'd_1-d_1'd_2}{d_1-d_2},\\a_2'&=\frac{d_1'-d_2'}{d_1-d_2}a_2+\frac{d_2'd_1-d_1'd_2}{d_1-d_2},\\a_3'&=\frac{d_1'-d_2'}{d_1-d_2}a_3.\\a_4'&=\frac{d_1'-d_2'}{d_1-d_2}a_4.\end{aligned}\tag{8}$$

These relations involve only the old and new diagonal elements, and do not involve the transformation $\mathbf{T}$.

In application to Eq. (3), the matrix $\mathbf{A}$ is

$$\mathbf{A}=\mathbf{S}(\omega)\mathbf{G}=\begin{pmatrix}e^{i\gamma}\cos\omega & -e^{-i\gamma}\sin\omega\\ e^{i\gamma}\sin\omega & e^{-i\gamma}\cos\omega\end{pmatrix}.\tag{9}$$

The diagonal elements, or characteristic values, of this matrix are the roots of

$$(e^{i\gamma}\cos\omega-d)(e^{-i\gamma}\cos\omega-d)+\sin^2\omega=0\tag{10}$$

or

$$d_1=e^{i\chi},\quad d_2=e^{-i\chi},\tag{11}$$

where

$$\cos\chi=\cos\omega\cos\gamma.\tag{12}$$

For the sake of definiteness, we assume that χ is the positive solution of this equation. The matrix $\mathbf{D}^n$ has the diagonal elements $e^{in\chi}$ and $e^{-in\chi}$, so that if we set

$$d_1'=e^{in\chi},\quad d_2'=e^{-in\chi},\tag{13}$$

then the matrix $\mathbf{A}'$ is equal to $\mathbf{A}^n$.

By the indicated procedure, we find

$$\mathbf{M}^{(n)}=(-1)^\mu\mathbf{A}^n=(-1)^\mu\mathbf{A}',\tag{14}$$

where

$$\begin{aligned}a_1'&=\frac{\sin n\chi}{\sin\chi}\cos\omega e^{i\gamma}-\frac{\sin(n-1)\chi}{\sin\chi},\\a_2'&=\alpha_1'^*,\\a_3'&=\frac{\sin n\chi}{\sin\chi}\sin\omega e^{i\gamma},\\a_4'&=-a_3'^*.\end{aligned}\tag{15}$$

We find that these expressions for the matrix elements of $\mathbf{A}'$ yield the following expressions for the elements of $\mathbf{M}^{(n)}$, which we denote simply by m_1, m_2, m_3 and m_4:

$$\begin{aligned} m_1 &= \cos n(\chi-\omega)+i[\sin n(\chi-\omega)]\frac{\cos\omega\sin\gamma}{\sin\chi}, \\ m_2 &= m_1^*, \\ m_3 &= [\sin n(\chi-\omega)]\frac{\sin\omega}{\sin\chi}e^{i\gamma}, \\ m_4 &= -m_3^*. \end{aligned} \tag{16}$$

We should like to stress that these expressions are exact, no approximations having been used in their derivation.

Resolution of the Matrix $\mathbf{M}^{(n)}$

It is evident that the matrix $\mathbf{M}^{(n)}$ does not represent a pure rotation, since the matrix elements of a rotator are real. We wish to know, however, under just what conditions and in just what sense we may say that the matrix $\mathbf{M}^{(n)}$ is an approximation to a rotator. The most direct and elegant way to determine these conditions is to express the matrix $\mathbf{M}^{(n)}$ as the combination of a rotator and a retardation plate. We know by Theorem I of Part II that this may always be done.

We shall express the matrix $\mathbf{M}^{(n)}$ as the combination of a rotator $\mathbf{S}(\bar{\omega})$ and a retardation plate $\bar{\mathbf{G}}$ with the orientation $\bar{\omega}$:

$$\mathbf{M}^{(n)}=\mathbf{S}(\bar{\omega})[\mathbf{S}(\bar{\omega})\bar{\mathbf{G}}\mathbf{S}(-\bar{\omega})]. \tag{17}$$

From Eqs. (6–8, II), (12) and (16), we find

$$\begin{aligned} \tan\bar{\omega} &= \tan(A+B)=\frac{m_3-m_4}{m_1+m_2} \\ &= [\tan n(\chi-\omega)]\frac{\tan\omega}{\tan\chi}, \end{aligned} \tag{18}$$

$$\begin{aligned} \tan(\bar{\omega}+2\bar{\omega}) &= \tan(A-B)=\frac{m_3+m_4}{m_1-m_2} \\ &= \tan\omega, \end{aligned} \tag{19}$$

$$\begin{aligned} \tan\bar{\gamma} &= \frac{\sin(A+B)}{\sin(A-B)}\tan\gamma \\ &= \sin\bar{\omega}\frac{\tan\gamma}{\sin\omega}. \end{aligned} \tag{20}$$

We have found that the complete optical system is equivalent to the combination of a rotator with angle $\bar{\omega}$ and a retardation plate with the relative phase retardation $2\bar{\gamma}$. In order that the matrix $\mathbf{M}^{(n)}$ represent a rotator, it is therefore reasonable to require that $\bar{\gamma}$ must be small compared with $\bar{\omega}$, and in order to insure this inequality, it is by Eq. (20) necessary and sufficient to require that

$$\left|\frac{\tan\gamma}{\sin\omega}\right|\ll 1. \tag{21}$$

This condition may be more stringent than the condition that γ be small compared with unity, because Eq. (21) is equivalent to the condition

$$\gamma\ll\omega=\pi\mu/n. \tag{22}$$

The following two relations may be derived from the Eq. (12) defining χ:

$$\sin(\chi-\omega)=\sin\omega\cos\omega\cos\gamma \times\left[\left(1+\frac{\tan^2\gamma}{\sin^2\omega}\right)^{\frac{1}{2}}-1\right], \tag{23}$$

$$\frac{\tan^2\chi}{\tan^2\omega}=1+\frac{\tan^2\gamma}{\sin^2\omega}. \tag{24}$$

When the condition (21) is satisfied, by Eqs. (18), (23) and (24) we then have for the angle $\bar{\omega}$:

$$\begin{aligned} \bar{\omega} &\cong n(\chi-\omega)\cong n\sin(\chi-\omega) \\ &\cong \tfrac{1}{2}n\cot\omega\sin\gamma\tan\gamma\cong\tfrac{1}{2}n\gamma^2\cot\omega \end{aligned} \tag{25}$$

or

$$\bar{\omega}=O(n^2\gamma^2/2\pi\mu). \tag{26}$$

The approximately equal sign in these equations should be understood to mean that the ratio of the expressions on the two sides approaches unity as $\tan\gamma/\sin\omega$ approaches zero.

Comparing (22) with (26), we find that the rotation $\bar{\omega}$ is subject to the condition

$$\bar{\omega}\ll\tfrac{1}{2}\pi\mu. \tag{27}$$

We see from this relation that in order to obtain a rotation which is not infinitesimal, it is necessary that $\mu\gg 1$—that is to say, that the axes of the plates be distributed uniformly through a large number of semicircles.

It should be noted that whereas the orientations of the plates advance in the clockwise direction, the angle of the equivalent rotator is in the counter clockwise direction if $\cot \omega$ is positive.

Rotator Formed by Three Retardation Plates

The considerations presented so far in this part had at one time considerable theoretical interest in connection with the theory of crystal structure.[4] If one wishes to form a rotator from retardation plates alone, however, it may be done in a much simpler manner than might be inferred from the foregoing. In fact, it is possible to produce an optical system which behaves rigorously as a rotator with any given $\bar{\omega}$, by the use of only three retardation plates.

We may produce a rotator by inserting a retardation plate between two crossed quarter-wave plates in such a position that its axes are at 45° with respect to the axes of the quarter-wave plates. Let us write the matrix defined by Eq. (25, I) as $\mathbf{G}(\gamma)$. The statement we have just made then follows directly from the identity

$$\mathbf{S}(ab\gamma) \equiv \mathbf{G}(a\pi/4)[\mathbf{S}(b\pi/4)\mathbf{G}(\gamma) \times \mathbf{S}(-b\pi/4)]\mathbf{G}(-a\pi/4), \quad (28)$$

where

$$a = \pm 1, \quad b = \pm 1.$$

Since the relative phase retardation of the middle plate is 2γ, we see that the angle of the rotation is one-half of the phase retardation of the inserted plate.

By the use of two achromatic quarter-wave plates, it is possible in this manner to use a Soleil compensator as a variable rotator.[5]

There exists another theorem which may be considered as the converse of the one just proved. We may produce a retardation plate of any given retardation from two quarter-wave plates and a rotator. The rotator should be placed between the two crossed quarter-wave plates. The optical system then behaves as a retardation plate with its axes at 45° with those of the quarter-wave plates. This statement follows directly from the identity

$$\mathbf{G}(ab\bar{\omega}) \equiv \mathbf{S}(a\pi/4)[\mathbf{G}(b\pi/4)\mathbf{S}(\bar{\omega}) \times \mathbf{G}(-b\pi/4)]\mathbf{S}(-a\pi/4), \quad (29)$$

where

$$a = \pm 1, \quad b = \pm 1.$$

We see that the phase retardation of the optical system is twice the angle of the rotator.

With two achromatic quarter-wave plates, it is thus possible to use a variable rotator as a Soleil compensator.

According to Billet,[6] the physical statements corresponding to the identities (28) and (29) were known to Fresnel.

Conclusion

Part IV of this series, which is now in preparation, will contain, in addition to emendations to the general theory, derivations of the matrices which represent (1) plates or solutions exhibiting circular birefringence or circular dichroism, and (2) plates cut from crystals of such low symmetry (monoclinic and triclinic) that the principal axes of absorption are not parallel to the principal axes of refraction.

The writer is indebted to Mr. Edwin H. Land, President and Research Director of the Polaroid Corporation, for suggesting the problem of developing an operational calculus of the type presented here. He wishes to thank Dr. Cutler D. West, also of the Polaroid Corporation, for providing all of the references to the older optical literature, and he is indebted to his co-author of Part II, Dr. Henry Hurwitz, Jr., for helpful suggestions about Parts I and III. The author wishes to express his appreciation to Professor M. S. Vallarta, Department of Physics, Massachusetts Institute of Technology, for reading the first draft of the manuscripts and for making helpful suggestions.

[4] L. Sohncke, *Entwickelung einer Theorie der Krystallstructur* (Teubner, Leipzig, 1879). See particularly pp. 238–246.

[5] T. M. Lowry, *Optical Rotatory Power* (Longmans Green, New York, 1935), p. 194, Fig. 79(a).

[6] M. F. Billet, *Traité D'optique Physique* (Paris, 1859), Vol. 2, Chapter 20, Art. 1, pp. 328–351.

Magneto-Optic Rotation in Birefringent Media—Application of the Poincaré Sphere

G. N. Ramachandran and S. Ramaseshan
Department of Physics, Indian Institute of Science, Bangalore 3, India

(Received August 21, 1951)

It is pointed out that the concept of the Poincaré sphere appreciably simplifies the mathematical treatment of phenomena accompanying the passage of polarized light through a medium which exhibits birefringence, optical activity or both simultaneously. This is exemplified by using the Poincaré sphere to evolve techniques which could be used for determining the true Faraday rotation in the presence of birefringence. When birefringence is present, measurements made with the half-shade at the polarizer and analyzer ends are not equivalent. In either arrangement, the errors introduced as a result of birefringence are largely reduced by taking the mean of two measurements for opposite directions of the field. Formulae are also derived by which the magnitudes of the error can be calculated for the particular experimental set up, knowing the value of the birefringence. In certain cases, even this need not be known, and the true rotation can be determined purely from measurements of the apparent rotations for two different azimuths of the incident plane of polarization.

INTRODUCTION

THE Poincaré sphere[1] is a convenient geometric representation in which the state of polarization of a light beam can be denoted by a point on the sphere. Its utility arises from the fact that the phenomena accompanying the propagation of polarized light through a medium exhibiting birefringence, optical activity, or both can be determined by means of simple geometric constructions on the sphere. In spite of this, comparatively few applications appear to have been made of this concept in original investigations. In, two papers published in this journal,[2,3] the Poincaré sphere was used for obtaining elegant methods for the determination of the state of polarization of light propagated in a birefringent medium and in the design of an accurate apparatus for the analysis of elliptically polarized light. The Poincaré sphere has been applied to the case when both birefringence and optical rotation coexist by Becquerel[4] in his studies on the paramagnetic rotation in tysonit and by Bruhat and Grivet[5] for the determination of the rotatory power of quartz at right angles to the optic axis. In an investigation of the latter, made by Szivessy and Schweers,[6] the Poincaré sphere was not made use of, and one has only to compare the large number of complicated formulas in their paper with the simplicity of Bruhat and Grivet's treatment to appreciate the advantage of the concept of the Poincaré sphere.

As far as the authors are aware, no other investigations have been reported making use of this concept until quite recently. Ramaseshan and Chandrasekharan[7] made a brief study of the influence of birefringence on measurements of Faraday effect, and Ramachandran

[1] H. Poincaré, Traite de la lumiere, Paris 2, 165 (1892).
[2] F. A. Wright, J. Opt. Soc. Am. 20, 529 (1930).
[3] C. A. Skinner, J. Opt. Soc. Am. 10, 490 (1925).
[4] J. Becquerel, Communs. Phys., Lab. Univ. Leiden No. 91C (1928); 211a (1930).
[5] G. Bruhat and P. Grivet, J. phys. et radium 6, 12 (1935).
[6] G. Szivessy and C. Schweers, Ann. Physik 1, 891 (1929).
[7] S. Ramaseshan and V. Chandrasekharan, Current (India) Sci. 20, 150 (1951). S. Ramaseshan, Proc. Indian Acad. Sci. 34A, 32 (1951).

and Chandrasekharan[8] measured the photoelastic constants of sodium chlorate, a cubic crystal exhibiting optical activity. Both these investigations were conducted in this laboratory and they revealed to the authors the elegance and simplicity of the geometric methods using the Poincaré sphere compared to the algebraic manipulations that have to be made from the electromagnetic theory.

The present paper contains a very brief account of the Poincaré sphere proper, particularly in relation to the propagation of light in a medium exhibiting, simultaneously, both optical activity and birefringence. Some properties of the sphere, which do not appear to have been noted before, are also discussed. Since a medium exhibiting magneto-optic rotation behaves like one possessing optical activity (so long as the direction of a ray is not reversed in the medium), the Poincaré

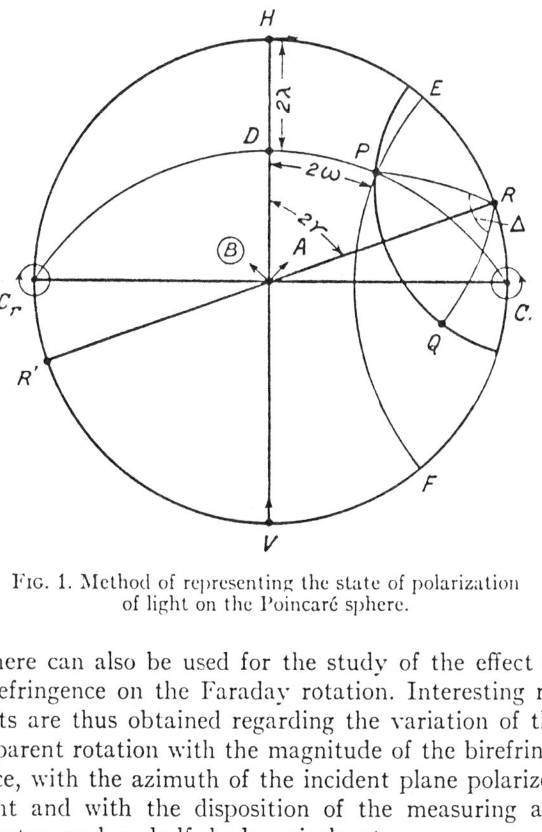

FIG. 1. Method of representing the state of polarization of light on the Poincaré sphere.

sphere can also be used for the study of the effect of birefringence on the Faraday rotation. Interesting results are thus obtained regarding the variation of the apparent rotation with the magnitude of the birefringence, with the azimuth of the incident plane polarized light and with the disposition of the measuring apparatus, such as half-shades, nicols, etc.

II. THE POINCARÉ SPHERE*

We shall always represent the sphere by its stereographic projection because suitable charts are available for making the calculations. The primitive circle is taken to be the "Greenwich meridian" or the great circle corresponding to longitudes 0° and 180°. All points on the upper hemisphere will be denoted by ordinary letters (e.g. A) while those in the lower hemisphere will be indicated by a circle around the letter.

A general point P (Fig. 1) represents a general elliptic vibration. The poles C_l and C_r represent left and right circularly polarized light, respectively. The equator $HAVB$ represents all plane vibrations, the point H of zero longitude standing for a horizontal vibration and a general point D, of longitude 2λ, representing a plane vibration whose direction is inclined at an angle λ to the horizontal. 2λ is taken to be positive for a counterclockwise rotation about C_l. Thus, the point A above has $\lambda=+45°$, while B, the correpsonding point below has $\lambda=-45°$. A general point P, with longitude 2λ and latitude 2ω, represents an elliptic vibration, whose major axis is at an angle λ to the horizontal and the ratio of whose axes b and a is given by

$$b/a=\tan\omega. \tag{1}$$

ω is taken to be positive measured from the equator towards C_l. Thus all left-rotating ellipses are on the hemisphere containing C_l and similarly all right-rotating ellipses are on the opposite hemisphere.

In a rotating birefringent medium, theory shows that two elliptic vibrations are propagated unchanged in type, along any direction of propagation. As in a non-optically active birefringent medium, these two vibrations are crossed with respect to each other. The two principal ellipses are propagated with different velocities so that plane polarized light incident on the medium emerges in general as elliptically polarized light. The state of polarization of light propagated in the medium can be obtained by the following simple construction. Let H and V be the two principal vibration directions of birefringence (i.e., in the absence of rotation) and let δ be the relative phase retardation introduced between the two per cm. Let ρ_0 be the true rotation† per cm (i.e., in the absence of birefringence). Draw the line RR' in the plane HC_lVC_r (Fig. 1) at an angle 2γ to HV where

$$\tan 2\gamma=2\rho_0/\delta_0. \tag{2}$$

Let

$$\Delta_0=(\delta_0^2+4\rho_0^2)^{\frac{1}{2}}. \tag{3}$$

Then, if P is the state of polarization of the incident light entering the medium of thickness t, then the state of polarization Q on emergence is obtained by rotating the sphere through an angle $\Delta=\Delta_0 t$ about the axis RR', which brings P to Q. The intermediate states of polarization lie on the small circular arc PQ.

The above construction follows from the electromagnetic theory of light propagation in an optically active birefringent medium.[9] Equation (3) is not strictly

[8] G. N. Ramachandran and V. Chandrasekharan, Proc. Indian Acad. Sci. **33A**, 199 (1951).

* For proofs of the statements made in this section, see reference 1.

† The rotation is taken to be positive when it is counter-clockwise. This is contrary to the convention used by physical chemists, but agrees with the mathematical definition of a counter-clockwise angle as being positive. See also Szivessy.[9]

[9] G. Szivessy, *Handbuch der Physik* (Julius Springer, Berlin, Germany), Vol. 20, Ch. 11.

valid, for according to theory $\Delta_0^2=\delta_0^2+4\rho_0^2+$ a small correction term, but the deviations from Eq. (3) are only of the order of 10^{-6} and may be neglected in any practical case.

In the special case of a purely birefringent crystal, $\rho_0=0$ and $2\gamma=0$ and the sphere is to be rotated about HV, the rotation being anticlockwise looking from H to V, if H is the faster ray, and conversely. The vibrations which are propagated unchanged are the horizontal and vertical plane vibrations. Similarly, for a purely rotating crystal, $2\gamma=\pi/2$ and the circular vibrations C_l and C_r are propagated unchanged. The rotation of the sphere is anticlockwise looking from C_l to C_r if the medium is left-rotating and conversely. Thus, if elliptically polarized light (say P of Fig. 1) is incident, only the major axis of the ellipse is rotated, but its eccentricity is unchanged. The states of polarization of the light lie on the small circle EPF. When both rotation and birefringence coexist, the ellipses that are propagated unchanged are those represented by R and R'. It is readily seen that the axes of both are parallel to H and V and that the ratio of their axes are, respectively, $\tan\gamma$ and $-\cot\gamma$ (since $\omega(R)=\gamma$ and $\omega(R)=\gamma-\pi/2$), so that their major and minor axes are interchanged and they have opposite sense of rotation.

The latitude 2ω and the longitude 2λ of a general elliptic vibration may be determined in various ways. One method is to use a quarter-wave plate for reducing the ellipse to a plane vibration and then measure the inclination of this to the axes of the quarter-wave plate. The azimuth of the axes gives λ, and the tangent of the angle between plane vibration and the major axis gives the ratio of the axes and hence ω. Obviously, because of the use of a quarter-wave plate, this method is applicable only to a particular wavelength. It can, however, be adapted such that any birefringent plate (not necessarily a quarter wave plate) is used as a compensator.[3] A simpler method would be the following.[8] One measures the azimuth of the major axis of the ellipse by using a biquartz half-shade and a linear analyzer.‡ A Babinet compensator is then introduced with its axes at 45° to the major axis and the shift s in the fringes is measured. If f is the fringe-width, then one has

$$2\omega=2\pi s/f. \tag{4}$$

III. DEFINITION OF TERMS USED LATER

It is necessary to evolve suitable terms for various quantities so as to simplify the discussion in the later sections. We shall use the term "principal directions" to denote the axes of the ellipses that are propagated unchanged in the medium. If the rotatory power of the medium becomes zero, these would correspond to the principal directions of the purely birefringent crystal thus produced. Similarly, "birefringence" means the difference in refractive indices of the two waves, when rotation is taken to be zero. The quantity Δ will be denoted by "composite phase difference," i.e., the phase difference between the two elliptic vibrations, while δ the corresponding phase difference caused by birefringence alone, $(=\delta_0 t)$ will be called "pure phase difference." When birefringence is present, the angle by which the major axis of the ellipse is rotated (ψ) would be different from the rotation (ρ) which it would have suffered if there were no birefringence. To distinguish between the two, ψ would be called the "apparent rotation" and ρ the "true rotation." The true rotation per unit distance traversed (ρ_0) is clearly the "rotatory power" of the medium.

Reverting to the Poincaré sphere, "light of polarization P" would mean that its polarization is represented by the point P. Similarly, a "polarizer P," is an apparatus which transmits light of polarization P when unpolarized light is incident on it, and an "analyzer P" is an apparatus which completely transmits light of polarization P, but only transmits a portion of light of any other state of polarization. For a general point P, these would be elliptic polarizers and elliptic analyzers.§ When P lies on the equator, or coincides with the poles of the Poincaré sphere, they become linear and circular polarizers and analyzers, respectively.

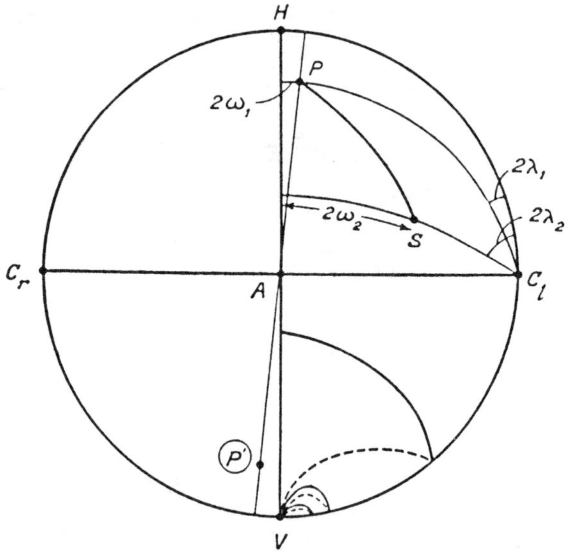

FIG. 2. Top half—Construction for determining the intensity of light transmitted by an elliptic analyzer. Bottom half—Locus of points representing the emergent light in Chauvin's experiment.

IV. A THEOREM ON THE POINCARÉ SPHERE

Suppose light of unit intensity having the polarization $S(\lambda_2, \omega_2)$ (Fig. 2) is incident on an elliptic analyzer $P(\lambda_1, \omega_1)$. It can be shown that the intensity trans-

‡ We shall use this term, instead of "nicol" or "polaroid" to denote an apparatus which transmits linearly polarized light. See next section on notation.

§ This is to be distinguished from the "elliptic analyzer" described by Ramachandran and Chandrasekharan.[8] Their set-up would be called crossed elliptic analyzers, according to our nomenclature.

mitted is $\cos^2(PS/2)$, where PS is the length of the great circular arc between P and S. As the proof of this is elementary, but tedious, only the main steps will be given here. The ratio of the axes of the ellipse S is $\tan\omega_2$ and the azimuth of the major axis is λ_2. The two axes thus have lengths $\cos\omega_2$ and $\sin\omega_2$, there being a phase difference of $\pi/2$ between the vibrations in the two directions. These can be resolved along the axes of the ellipse P. The effect of the analyzer would be to change the phase difference by $\pi/2$ and to transmit fractions $\cos\omega_1$ and $\sin\omega_1$ of the components along the major and minor axes, respectively. In this way, the intensity transmitted is found to be

$$\cos^2(\omega_1-\omega_2)\cos^2(\lambda_1-\lambda_2)+\sin^2(\omega_1+\omega_2)\sin^2(\lambda_1-\lambda_2). \quad (5)$$

This can be put in the form

$$\tfrac{1}{2}[1+\sin2\omega_1\sin2\omega_2+\cos2\omega_1\cos2\omega_2\cos2(\lambda_1-\lambda_2)] \quad (6)$$

$$=\tfrac{1}{2}(1+\cos PS)=\cos^2 PS/2, \quad (7)$$

from spherical trigonometry.

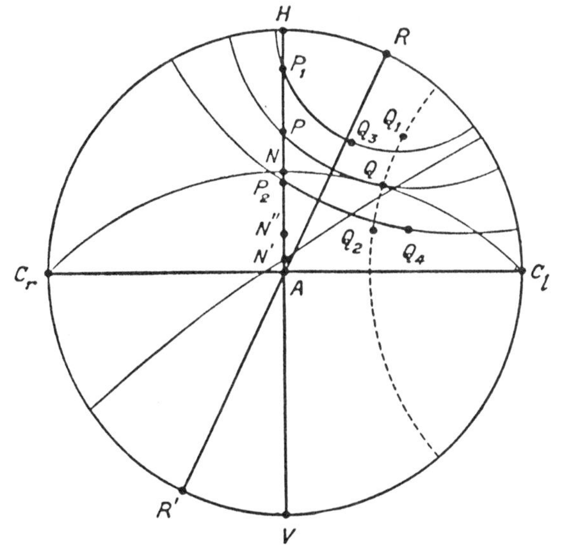

FIG. 3. The difference between putting the half-shade at the polarizer and the analyzer ends, illustrated on the Poincaré sphere.

An interesting consequence of this result is that the locus of points on the Poincaré sphere representing elliptic vibrations of which the elliptic analyzer P would transmit the same intensity, is a small circle with P as pole. In particular, it is seen that a linear analyzer of any azimuth would transmit half the intensity of circularly polarized light and a linear analyzer at 45° to H and V would transmit half the intensity of any elliptic vibration whose axes are vertical and horizontal. Further, an analyzer P would transmit zero intensity of light having a polarization represented by the antipodal point (P'). Thus, for every elliptic polarizer, there is a unique elliptic analyzer, which may be said to be "crossed" with respect to it, on the analogy of crossed nicols. Points R and R' in Fig. 1 thus represent ellipses which are "crossed" with respect to each other, as they should be because of their orthogonal property.

V. FARADAY EFFECT AND BIREFRINGENCE

A medium exhibiting magneto-optic rotation behaves similarly to one possessing natural optical activity. The only difference is that the sense of rotation is different for opposite directions of travel in the former case, while it is the same with optical activity. So long as one is interested in the phenomena accompanying light propagation only in a particular direction, the theory aforementioned can be applied *in toto* and used to evaluate the results when birefringence is also present. It has been known for a long time that the Faraday rotation of an isotropic solid, measured in the usual way with a half-shade at the analyzer end, appears to be less than its true value when the medium is birefringent.[10] Thus, Chauvin[11] measured the rotation in directions slightly inclined to the optic axis of calcite with the incident light polarized along a principal direction. With increasing magnitude of birefringence, he showed that the apparent rotation not only diminished in magnitude, but actually reversed in sign and also exhibited several reversals in sign. Chauvin explained these on the basis of the theory of light propagation in an optically active, birefringent crystal.

Obviously, therefore, measurements of the Faraday effect in a cubic crystal would be vitiated by the presence of accidental birefringence. Formerly, it has been the practice to choose for measurement as perfect a specimen as possible, judged by the least restoration of light under crossed polarizers, and to assume that the value obtained with this specimen is correct. As mentioned before, the errors introduced by birefringence in such measurements were considered by Ramaseshan and Chandrasekharan,[7] but not in great detail. The succeeding sections will deal with this problem, both in regard to the technique of measurement and the results obtained.

Before proceeding, we may indicate how the result of Chauvin can be simply explained from a geometric construction on the Poincaré sphere. Assuming the rotation ρ to be constant, and δ to increase continuously, the inclination 2γ of the axis of rotation continuously decreases while Δ increases, so that the state of polarization executes the spiral shown in the bottom half of Fig. 2. Obviously, the azimuth λ of the major axis first decreases from ρ, reverses sign, and then oscillates with successive reversal of sign, finally tending to zero, as it should for a purely birefringent crystal.

VI. EFFECT OF PUTTING THE HALF-SHADE AT THE POLARIZER AND ANALYZER ENDS

The apparatus used for the measurement of the Faraday effect is very similar to those used in polarimetry. In some polarimeters, the half-shade is put at

[10] H. Schutz, Handbuch Exp. Phys. 26, 48 (1936).
[11] M. Chauvin, Compt. rend. 102, 972 (1886).

the polarizer end and in others at the analyzer end, and different designs are used for the half-shade.‖ All the methods lead to identical results, so long as the medium is purely rotating, but when birefringence is present, they are different in effect, as will be shown as follows.

Three types of half-shades are in common use; (a) biquartz, (b) Lippich, and (c) Laurent. The last of these employs a half-wave plate and hence is useful only for one wavelength. In the Lippich type, the two halves are not of the same intensity, so that it is difficult to work out a general theory for it. In the special design of Jellet, this defect is remedied and effects obtained with it are identical with those with a biquartz half-shade. We shall therefore consider only the latter. If β is the half-shade angle its effects would be to rotate the azimuth of the incident light through angles of $+\beta/2$ and $-\beta/2$ in the two halves.

(i) Analyzer End

Let the principal directions be vertical and horizontal and let linearly polarized light (P, Fig. 3) be incident at an azimuth 2λ. Let the effect of the medium be to bring it to Q by rotation through an angle Δ about RR'. The effect of the biquartz would be to split Q into two ellipses Q_1 and Q_2 lying in the same latitude, but differing in longitude by 2β. From the construction of section 4, the locus of points representing analyzers for which the intensities of the two halves are equal is the great circle C_lNC_r bisecting the arc Q_1Q_2. Therefore, the setting of the plane analyzer when this happens is given by N, the intersection of this great circle with the equator. (Actually, the setting would not be N, but the antipodal point to N, i.e., 90° away, when the intensity is least, but no confusion would arise by calling it N. The corresponding setting for the incident light would be P itself according to this convention and not its antipodal point.) Clearly, N has the same longitude as Q, the midpoint of Q_1Q_2.

(ii) Polarizer End

In this case, the light incident on the medium from the two halves will have the states of polarizations P_1 and P_2 and the emergent light will have the polarizations Q_3 and Q_4 and by the same construction as before, the setting of the linear analyzer for equality in the two halves will be N', which is different from N. Indeed, it can be very far away from N, and the longitude of N' need not even lie in between those of Q_3 and Q_4.

This is illustrated in Figs. 4 and 5. We consider the case when $2\gamma=30°$, i.e., $2\rho/\delta=1/\sqrt{3}$ and the half-shade angle $\beta=20$. The composite phase retardation is 63°, and we assume that the mean azimuth of the half-shade is H. Fig. 4 gives the variation in intensity of the light transmitted by the linear analyzer at various azimuths.

‖ See Weissberger, *Physical Methods in Organic Chemistry* (Interscience Publishers, Inc., New York, 1945), Vol. 1, for details regarding various types of apparatus.

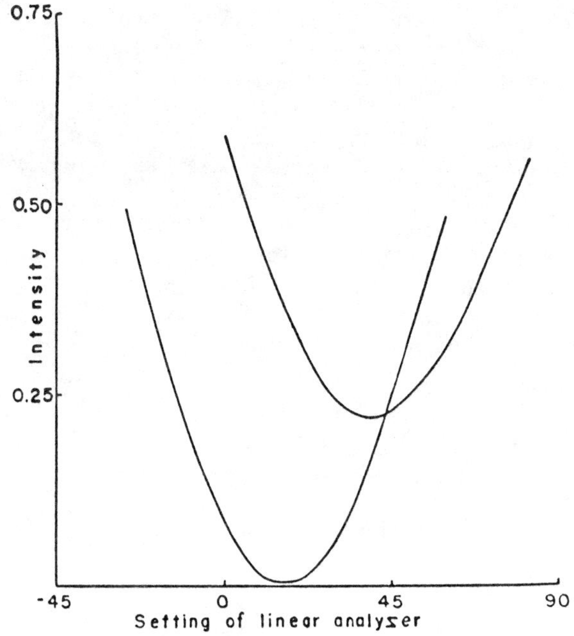

FIG. 4. Graph showing the variation of the intensity I_1 and I_2 of the two halves when the half-shade is put at the polarizer end and the linear analyzer is rotated.

It will be noticed that the minima for the two halves occur at about 20° and 40°, respectively, while the two are equal only at 44°. Figure 5 shows $\Delta I/I$ (i.e., $(I_1-I_2)/\frac{1}{2}(I_1+I_2)$) plotted against λ, and it will be seen that there is a sharp minimum at the azimuth of equality.

It is interesting to note that in both cases (i) and (ii) the setting N and N' of the analyzer is independent of the half-shade angle β. This follows from the construction that N (or N') should lie on the great circle bisecting Q_1Q_2 (or Q_3Q_4).

Considering now the effective rotation ψ in the two cases, the following formulas can readily be derived.

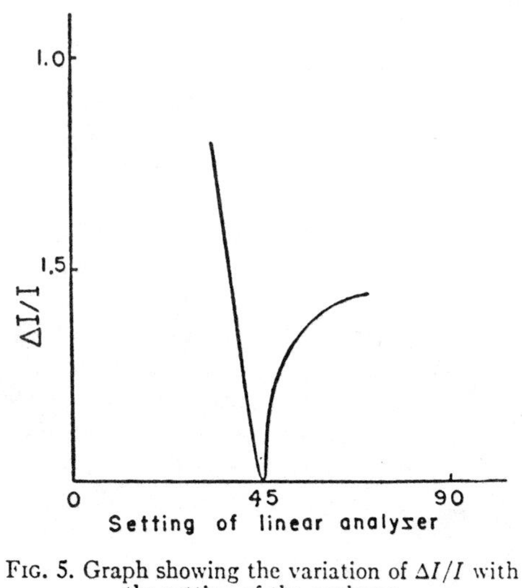

FIG. 5. Graph showing the variation of $\Delta I/I$ with the setting of the analyzer.

We shall not give the proof here. Let α be the azimuth of the polarizer in case (i) or the mean azimuth of the two beams in case (ii). Then we have

Case (i)

$$\alpha=0° \quad \tan 2\psi=\sin 2\gamma \sin\Delta/(\cos^2 2\gamma+\sin^2 2\gamma \cos\Delta), \quad (8)$$

$$\alpha=45° \quad \tan 2\psi=\sin 2\gamma \tan\Delta. \quad (9)$$

Case (ii)

$$\alpha=0° \quad \tan 2\psi=\sin 2\gamma \tan\Delta, \quad (10)$$

$$\alpha=45° \quad \tan 2\psi=\sin 2\gamma \sin\Delta/(\cos^2 2\gamma+\sin^2 2\gamma \cos\Delta.) \quad (11)$$

It will be noticed that the formulas for $\alpha=0$ and 45° have been interchanged in the two cases. We shall now prove that this is a particular case of the more general result, *viz.*, that the value of 2ψ in case (i) for an azimuth α of the incident light is the same as that in case (ii) with an azimuth $45+\alpha$ and vice versa.

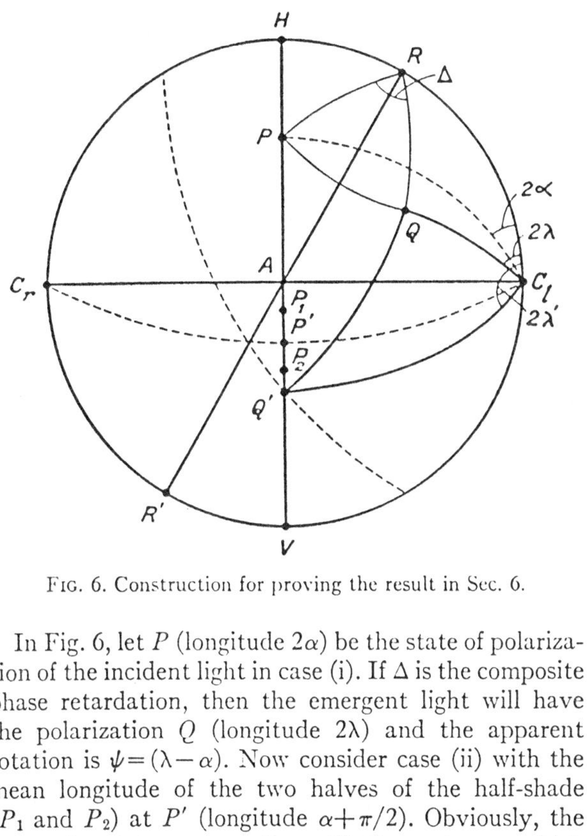

FIG. 6. Construction for proving the result in Sec. 6.

In Fig. 6, let P (longitude 2α) be the state of polarization of the incident light in case (i). If Δ is the composite phase retardation, then the emergent light will have the polarization Q (longitude 2λ) and the apparent rotation is $\psi=(\lambda-\alpha)$. Now consider case (ii) with the mean longitude of the two halves of the half-shade (P_1 and P_2) at P' (longitude $\alpha+\pi/2$). Obviously, the locus of points representing elliptic analyzers which would transmit equal intensities of the two halves is the great circle $C_lP'C_r$ *whose pole is* P. The effect of the passage of light through the medium is to rotate both P_1 and P_2 through an angle Δ about the axis RR'. The great circle $C_lP'C_r$, of equal intensity, would also therefore be rotated through the same angle Δ about the same axis. This great circle is best drawn by considering the fact that its pole (P) would also be rotated through the angle about RR' and would thus be brought to Q. The required great circle thus has for its pole the point Q, and its intersection with the equator gives Q', and the azimuth λ' of the plane analyzer for equality of intensity.

The following are clear from the diagram:

$$\text{arc}QQ'=\pi/2, \quad \text{arc}Q'C_l=\pi/2.$$

Hence

$$\angle Q'CQ=\pi/2 \quad \text{or} \quad 2\lambda'=2\lambda+\pi/2,$$

i.e.,

$$\lambda'=\lambda+\pi/4.$$

Since

$$\alpha'=\alpha+\pi/4,$$

we have

$$\psi'=\lambda'-\alpha'=\lambda-\alpha=\psi.$$

This interesting result, which was entirely unexpected, has in fact been verified by experiment. A rectangular block of optical glass (3×3×2 cm) was placed in a magnetic field and was subjected to different stresses. The light traveled along the thickness (2 cm) and the pressure was applied on the pair of faces 3×2 cm. In the absence of any strain, the glass exhibited a Faraday rotation of 17.40° for the field employed. The field was kept the same, and the glass was subjected to the successive loads shown in Table I. In each case, the birefringence introduced (δ) was directly measured with a Babinet compensator in the absence of the field (column 2). The succeeding columns contain the measured values of ψ, each of which is the average of two measurements made with the field reversed in between. This reduces the errors caused by nonuniformity of the strain (see Sec. 7). The results of the theory are verified to be true from the table. When ψ is of the order of 30° or more, the agreement is not satisfactory. This is a result of two causes: (a) the measurements themselves are not accurate because the ratio of the axes (b/a) of the ellipse approaches unity, and (b) a Lippich half-shade had to be used at the polarizer end owing to the nature of the experimental set up. In this, the two halves are not of equal intensity and the geometrical constructions outlined before are not exactly valid. It is, however, possible to work out the results from theory for this case. Thus, in Fig. 3, the setting with the Lippich would be N'', where $N''Q_3/N''Q_4$= the ratio of the intensities of the two halves. Obviously, no general result can be derived in this case as the results would depend on the value of the ratio, the half-shade angle, etc.

Table I also shows another result, mentioned earlier, that the apparent rotation *increases* with birefringence when $\alpha=45°$.

VII. EFFECT OF REVERSING THE FIELD ON THE MEASUREMENT OF APPARENT ROTATION

Equations (8) to (11) of the previous section show how the apparent rotation varies with birefringence when the incident light is polarized at 0° or 45° to the principal directions. In what follows, we shall consider only case (i) with the half-shade at analyzer end, since the formulas for case (ii) are readily deducible from these, utilizing the results of the previous section.

Suppose, linearly polarized light is incident on the medium at an arbitrary azimuth α, and let the major axis of the emergent ellipse have an azimuth λ_1. Then, it can be shown that

$$\cot 2\lambda_1 = \frac{\cos^2 2\gamma + \sin^2 2\gamma \cos\Delta - \sin 2\gamma \sin\Delta \tan 2\alpha}{\tan 2\alpha \cos\Delta + \sin 2\gamma \sin\Delta}. \quad (12)$$

ψ, the apparent rotation, is equal to $(\lambda_1 - \alpha)$ and can be calculated. Suppose now, the field is reversed, the birefringence remaining the same. Then the azimuth λ_2 of the emergent ellipse is given by

$$\cot 2\lambda_2 = \frac{\cos^2 2\gamma + \sin^2 2\gamma \cos\Delta + \sin 2\gamma \sin\Delta \tan 2\alpha}{\tan 2\alpha \cos\Delta - \sin 2\gamma \sin\Delta}. \quad (13)$$

It is common in Faraday effect measurements to determine the rotations for the two directions of the field and find the average which is $(\lambda_1 - \lambda_2)/2$. Now, from the aforenamed two equations,

$$\begin{aligned}\cot 2(\lambda_1 - \lambda_2) &= [(\cos^2 2\gamma + \sin^2 2\gamma \cos\Delta)^2 - \sin^2 2\gamma \sin^2\Delta \\ &\quad + \tan^2 2\alpha(\cos^2\Delta - \sin^2 2\gamma \sin^2\Delta)] \quad (14) \\ &\div 2 \sin 2\gamma \sin\Delta[\cos^2 2\gamma + \sin^2 2\gamma \cos\Delta \\ &\quad + \tan^2 2\alpha \cos\Delta].\end{aligned}$$

Although these equations are valid for any value of α, they find application particularly when α is near about 0° or 45°. This would cover the case when the polarizer is set parallel or at 45° to the principal directions, but when there is a slight mis-setting of either the polarizer or the specimen, or if the principal directions are not exactly the same at all points in the medium. To determine the order of magnitude of the errors introduced thereby in the apparent rotation, a few particular cases have been numerically evaluated and the results are shown in Tables II and III. The small deviation of α from 0° or 45° is made equal to 2°52′, so that $\tan 2\alpha = \frac{1}{10}$. ψ is the apparent rotation when α is exactly 0° or 45°; when it deviates from these, ψ_1 and ψ_2 are the apparent rotations for positive and negative fields, and $\bar{\psi}$ is the mean of the two ($\bar{\psi} = (\psi_1 - \psi_2)/2$).

Tables II and III show that the usual practice of taking the mean of the measurements for the two directions of the field largely eliminates the errors caused by slight mis-setting of the polarizer or specimen or as a result of variations in the strain axes in the specimen. The results have an important bearing on the method suggested by Ramaseshan[7] for determining photoelastic constants of isotropic solids by measuring the reduction in the apparent rotation due to strain. Since the strain directions may not be the same at all points in the medium, it would be expected that measurements made with one direction of the field alone would be highly erratic. In fact this was found to be so, but taking the mean, eliminated all the irregularities. The measurements of photoelastic constants are being reported elsewhere.

TABLE I. Effect of the position of the half-shade on measurements of apparent rotation.

Load in kg	δ in degrees	ψ in degrees: Analyzer end $\alpha=0°$	Analyzer end $\alpha=45°$	Polarizer end $\alpha=0°$	Polarizer end $\alpha=45°$
0	0.0	17.40	17.40	17.40	17.40
38.9	32.4	16.50	19.05	18.55	16.55
54.4	46.8	15.55	22.15	21.40	15.50
70.0	63.0	14.10	28.75	27.80	14.25
85.6	77.8	12.90	39.00	33.00	12.40
101.1	92.9	11.20	69.00	60.10	11.15

VIII. EVALUATION OF THE TRUE ROTATION IN THE PRESENCE OF BIREFRINGENCE

A general method of determining the true rotation (either natural or magneto-optic rotation) in the presence of birefringence is the following: Linearly polarized light is allowed to fall on the medium at an azimuth α to the principal directions and one measures the apparent rotation ψ by means of a half-shade at the analyzer end and the ratio of the axes ($\tan\omega$) of the emergent ellipse by a suitable method. Both δ and 2ρ can then be calculated. The formulas are:[8]

TABLE II. Effect of an error in setting on the apparent rotation, when the azimuth is near 0° ($\alpha = 2°52'$).

Δ	$2\gamma=30°$, $2\rho/\delta=1/\sqrt{3}$: ψ_1	ψ_2	$\bar{\psi}$	ψ	$2\gamma=60°$, $2\rho/\delta=\sqrt{3}$: ψ_1	ψ_2	$\bar{\psi}$	ψ
0	0°0′	0°0′	0°0′	0°0′	0°0′	0°0′	0°0′	0°0′
30	6°59′	−7°32′	7°15′	7°15′	12°16′	−12°57′	12°52′	12°51′
60	12°12′	−14°11′	13°12′	13°10′	24°53′	−25°22′	25°7′	25°6′
90	14°54′	−18°52′	16°53′	16°51′	36°48′	−37°14′	37°1′	36°57′
120	14°20′	−20°47′	17°33′	17°22′	50°7′	−49°37′	49°52′	49°43′
150	6°3′	−18°24′	12°13′	12°34′	68°7′	−65°4′	66°36′	66°21′
180	−8°30′	−8°30′	0°0′	0°0′	92°48′	−87°12′	90°0′	90°0′

$$\left.\begin{aligned}\tan 2\gamma &= [\cos 2\alpha - \cos 2\omega \cos 2(\alpha+\psi)]/\sin 2\omega \\ \cos\Delta &= 1-(1-\cos 2\omega \cos 2\psi)/(1-\cos^2 2\gamma \cos^2 2\alpha) \\ \delta &= \Delta \cos 2\gamma; \quad 2\rho = \Delta \sin 2\gamma.\end{aligned}\right\}. \quad (15)$$

This method may be used when neither δ nor 2ρ can be measured independently.

If, however, pure birefringence can be measured independently as in Faraday effect studies, then allowance may be made for it, and the true rotation can be deduced from a measurement of ψ alone.

If both 2ρ and δ are small, then it can be shown from Eqs. (8) and (9) that

TABLE III. Effect of an error in setting on the apparent rotation, when the azimuth is near 45° ($\alpha = 42°8'$).

Δ	$2\gamma=30°$, $2\rho/\delta=1/\sqrt{3}$: ψ_1	ψ_2	$\bar{\psi}$	ψ	$2\gamma=60°$, $2\rho/\delta=\sqrt{3}$: ψ_1	ψ_2	$\bar{\psi}$	ψ
0	0°0′	0°0′	0°0′	0°0′	0°0′	0°0′	0°0′	0°0′
30	7°44′	−8°20′	8°2′	8°3′	13°11′	−13°21′	13°16′	13°17′
60	19°4′	−21°31′	20°17′	20°27′	28°6′	−28°20′	28°14′	28°9′
90	45°29′	−44°38′	45°4′	45°0′	45°7′	−44°53′	45°0′	45°0′
120	73°20′	−65°57′	69°39′	69°33′	62°25′	−61°7′	61°46′	61°50′
150	86°14′	−77°46′	82°0′	81°57′	77°55′	−75°25′	76°40′	76°43′
180	94°17′	−85°42′	90°0′	90°0′	91°20′	−88°39′	90°0′	90°0′

(a) for the incident azimuth $\alpha=0°$ or $90°$

$$2\psi_0=2\rho(1-\delta^2/6)^7, \tag{16a}$$

and (b) for $\alpha=45°$,

$$2\psi_{45}=2\rho(1+\delta^2/3). \tag{16b}$$

The error in using these approximate formulas is less than 1 percent so long as δ and 2ρ do not exceed 30°.

It will be noticed from the forementioned formulas that the apparent rotation is *less* than the true rotation when $\alpha=0°$ or $90°$, but is *more* when $\alpha=45°$. It is interesting to see what would be the average value of 2ψ for all azimuths α of the incident light. Thus, from Eq. (12), $\tan 2\psi$ has the value given below for a particular value of 2α:

$$\tan 2\psi=\frac{\sin^2 2\gamma \sin\Delta-(1-\cos\Delta)\cos^2 2\gamma \tan 2\alpha+\sin 2\gamma \sin\Delta \tan^2 2\alpha}{\cos^2 2\gamma+\sin^2 2\gamma \cos\Delta+\cos\Delta \tan^2 2\alpha}.$$

The mean value of $\tan 2\psi$ over the range $-\pi/2$ to $+\pi/2$ of 2α can be shown to be the remarkably simple expression

$$(\tan 2\psi)_m=[\cot\Delta(\cot\Delta+\operatorname{cosec}\Delta \cot^2 2\gamma)]^{-\frac{1}{2}}. \tag{17}$$

With the same approximations as were used in deriving Eqs. (16a) and (16b), one obtains

$$2\psi_m=2\rho(1+\delta^2/12). \tag{18}$$

Although the deviation in the mean value of the apparent rotation is less than the deviation for either $\alpha=0$ or $\alpha=45°$, it does not vanish.

However, Eqs. (16a) and (16b) suggest a simple method of eliminating the effect of birefringence, *without measuring its value.* We have, from these,

$$(2\psi_0+\psi_{45})/3=\rho, \tag{19}$$

and this equation is correct to the third order in δ, the only terms that occur being δ^4 and higher powers.

The above formulas can be verified to be true from the data in Table I. The entries in the second row correspond to $\delta=32.4°$, which is near the limit of validity of the formulas. It is seen that the decrease in ψ for $\alpha=0$ is nearly half the increase for $\alpha=45°$ and their magnitudes are what are given by Eqs. (16a) and (16b). Further $(2\psi_0+\psi_{45})/3$ is 17.35°, which differs from the correct value 17.4° by less than 0.5 percent.

Another approximate formula, valid when 2ρ is small ($<30°$) and $\delta\gg 2\rho$ is

$$2\psi=2\rho \sin\delta/\delta, \tag{20}$$

if $\alpha=0$. For $\alpha=45°$, 2ψ increases indefinitely with increase of δ. The relation between the two is

$$\tan 2\psi=2\rho \tan\delta/\delta, \tag{21}$$

but this is not a convenient equation to use.

Equations (16) and (20) cover practically the whole range of values which occurs in Faraday effect studies, and can be used to deduce the true rotation from the measured apparent rotation.

A Comparative Survey of Some Possible Systems of Polarized Headlights

B. H. BILLINGS* AND E. H. LAND
Research Laboratory, Polaroid Corporation, Cambridge, Massachusetts

(Received May 5, 1948)

In a polarizing headlight system polarizers are placed on automobile headlights and in a viewer before the eyes of automobile drivers. These polarizers are so oriented that the viewer is crossed with the headlights of an approaching car. This blocks most of the light from approaching headlights. A polarizing system has other features which make it different from a non-polarizing system. The amount of the polarized component of the reflected light from an object illuminated by a car's headlights which is transmitted back through the viewer in the same car depends on the type and orientation of the polarizers. Thus objects have different brightness depending on the polarizing system being used. Similarly, the contrast between objects and their background will be different for different polarizing systems. In this paper some basic matrices of an optical calculus are derived. This calculus is then used to show the orientation of the headlights and viewer polarizers for all possible polarizing systems. For each of these systems the relative brightness of a specularly reflecting object is determined. Finally, equations are derived for the contrast of various types of objects with their background and a few special cases are calculated.

INTRODUCTION

THE use of polarizers for avoiding glare in automobile headlight systems has been known for some time. One of the first proposals was made by Short[1] in 1921. His scheme involved vertical and horizontal polarizers. With this system only the depolarized reflected component of the light from a driver's headlight is transmitted through his own analyzer. In the system proposed by Chubb[2] and by Land,[3] the windshield polarizer and headlight polarizers on a given car were parallel and at 45°. Both the depolarized and the polarized components were thus reflected back through the windshield.

This system was modified by Cooke[4] to include 1/4λ-plates in order to make circular polarizers. There is a brief discussion of elliptical polarizers for headlights in a paper by Christoph and Neugebauer[5] but this was in connection with the use of non-achromatic phase plates for making circular polarizers. Although there have been several papers[6] which have discussed polarization

* Now with Baird Associates, Inc., Cambridge, Massachusetts.

[1] F. Short, U. S. Patent 1,734,022, filed September 26, 1921.

[2] L. W. Chubb, U. S. Patent, 2,087,795, filed July 29, 1920.

[3] E. H. Land, U. S. Patent, 2,031,045, filed January 16, 1933.

[4] A. G. Cooke, British Patent 365,507, filed June 9, 1931.

[5] W. Christoph and H. E. J. Neugebauer, Zeits. f. Physik **20**, 257 (1939).

[6] L. W. Chubb, "Polarized Light for Motor Vehicle Lighting," I. E. S. Trans. **32**, 505 (1937); E. H. Land, "Polaroid and the Headlight Problem," J. Frank. Inst. **224**, 269 (1937); A. Gamble, "La lumière Polarisée son emploi dans les Projecteurs D'Automobiles," Bull. de la Société française des Électriciens **7**, 511 (1937); Henry T. Gibbs, "Polarized Light for Motor Vehicles," J. Frank. Inst. **228**, 719 (1939); Val Roper and K. D. Scott, "Seeing with Polarized Headlamps," I. E. S. Trans. **36**, 1205 (1941); E. H. Land, "Polarized Light in the Transportation Industries," The Michigan-Life Conference on New Technologies in Transportation (University of Michigan Official Publications) **42**, 151 (1940); H. Sauer, "Polarisiertes Licht

systems for night driving, they have invariably been concerned with one of the specific systems described above.

There are, however, an infinite number of polarized headlight systems all of which provide extinction of the approaching headlight. The headlight polarizer can be selected to produce any desired state of polarization. It can give linear polarized light inclined at any angle to the vertical. It can give elliptically polarized light with any desired value for the eccentricity of the ellipse and with the ellipse major axis inclined at any angle to the vertical. For each of these choices of headlight polarizers there is, of course, a viewer polarizer which will give extinction.

However, there are other important properties of a headlight polarizing system besides the property of extinction. The brightness of a given object and its contrast with its background may be different when viewed with different polarizing systems. The viewer may be able to eliminate a large fraction of horizontally polarized light which can be considered to be the major part of daytime reflected glare. The systems described above behave very differently with respect to these properties.

In order to determine this behavior it is necessary to describe the windshield polarizer orientation which corresponds to a given headlight polarizer orientation. This relationship may be visualized as follows: The axis of the windshield polarizer in an opposing car is at right angles to the headlight polarizer axis. In any given case the orientation of the windshield polarizer is then determined by a 180° rotation of this axis about the vertical. The windshield axis will then be in the same quadrant as the headlight axis and the angle between the windshield polarizer axis and the vertical will be the same as the angle between the headlight polarizer axis and the horizontal. This can be expressed mathematically as

$$\alpha+\beta=90°,$$

where α is the angle of the headlight polarizer from the vertical and β is the angle of the viewer polarizer both measured as seen from inside the car measured clockwise.

With knowledge of the angle of the headlight and viewer polarizer it is possible to calculate the apparent brightness of an object illuminated by the headlight. When the headlight and viewer polarizer are at 45° to the vertical the brightness is a maximum. At angles closer to the vertical or horizontal the brightness is less and is a minimum when the headlight polarizer is vertical or horizontal.

For any of these possible linear systems the transmission for horizontally polarized light can be calculated. The vertical viewer polarizer gives the best protection against daytime glare although it is poorest from the point of view of object brightness at night.

Although these various factors are not so spectacular as the reduction of the intensity of the opposing headlight by a factor of several thousand, they can nevertheless be important in particular cases. For example, some dark objects which have zero contrast with their background in the present non-polarizing system may have contrast as high as 50 percent in a system with linear polarizers at 45°. On the other hand some objects may have finite contrast in the non-polarizing system and have zero contrast in the 45° system. Also, the brightness of some objects may be zero in a 90° system whereas the same object would be quite bright in a 45° system.

In this paper the various possible polarized headlight systems will be discussed and their behavior as far as contrast, object brightness, and reduction of daytime glare are concerned.

The discussion will be theoretical in that it is based on perfect polarizers which are perfectly aligned. While the effects on the system of imperfect polarizers, misalignments, windshield tilt, etc., have also been carefully considered, we shall not attempt to treat such important practical details here.

THE MUELLER MATRICES

To discuss the various types of headlight systems, an optical calculus will be used which was developed by Perrin[7] and Soleillet[8] and put in matrix form by Mueller.[9] Its relationship to the optical calculus of R. Clark Jones[10] has been presented in a recent paper. This calculus has the

in der Kraftfahrzeug-Beleuchtung," Zeits. VDI **82**, 201 (1938).

[7] F. Perrin, J. Chem. Phys. **10**, 415 (1942).
[8] Paul Soleillet, Ann. de physique **12**, 23 (1929).
[9] H. Mueller, M.I.T. Course (8.26), Spring 1945.
[10] R. Clark Jones, J. Opt. Soc. Am. **37**, 107 (1947).

advantage that both polarizing and depolarizing systems can be represented. Totally polarized monochromatic light requires three parameters to represent its essential characteristics. Any three parameters could, of course, be used, but the most familiar ones are the amplitudes and phase difference of two linear components at right angles. The optical calculus uses in place of these three quantities the combination

$$\begin{aligned} M &= A_x^2 - A_y^2, \\ C &= 2A_xA_y\cos\phi, \\ S &= 2A_xA_y\sin\phi, \end{aligned} \qquad (1)$$

where A_x and A_y are the amplitudes of the components along the x and y axes and ϕ is the phase difference between the components.

It has been shown by Soleillet[8] that if a fourth variable I, the intensity of the light is added to the three variables above a linear relation can be written between the four components of the light going into an instrument and the four components of the light going out. These relationships can be written for depolarizing as well as polarizing systems. For totally polarized light in a non-depolarizing system only three of the variables are independent and we can write the equation

$$I^2 = M^2 + C^2 + S^2.$$

Since the equations connecting the incident and emergent light are all linear in I, M, C and S, the technique developed by Mueller can be used and the light can be represented as a vector in I, M, C, S space and the instrument as a matrix operating on the vector.

The Mueller matrices associated with an elementary instrument such as a wave plate at an arbitrary angle will be developed first. The product of the elementary matrices of a complex system will then give the matrix of the whole system.

The matrix of most of the elementary instruments can be written down from familiar relationships. For a partial polarizer whose axes are parallel to the x and y axes:

$$A_x' = P_xA_x \quad \text{and} \quad A_y' = P_yA_y \qquad (2)$$

where A_x and A_y are the x and y components of the light going into the polarizer, P_x and P_y are the transmission factors of the polarizers and A_x' and A_y' are the x and y components of light coming out of the polarizer.

In terms of the Perrin components, the x and y amplitudes can be written

$$A_x^2 = \tfrac{1}{2}(I+M), \quad A_y^2 = \tfrac{1}{2}(I-M). \qquad (3)$$

From this equation and Eq. (1)

$$M' = \tfrac{1}{2}P_x^2(I+M) - \tfrac{1}{2}P_y^2(I-M), \qquad (4)$$

where again primed components refer to light after it has gone through the polarizer. Similarly,

$$I' = \tfrac{1}{2}P_x^2(I+M) + \tfrac{1}{2}P_y^2(I-M). \qquad (5)$$

$$C' = P_xP_yC, \quad S' = P_xP_yS. \qquad (6)$$

The matrix for a partial polarizer with the axes in the x and y direction then becomes

$$\mathbf{P}_h = \tfrac{1}{2}\begin{vmatrix} P_x^2+P_y^2 & P_x^2-P_y^2 & 0 & 0 \\ P_x^2-P_y^2 & P_x^2+P_y^2 & 0 & 0 \\ 0 & 0 & 2P_xP_y & 0 \\ 0 & 0 & 0 & 2P_xP_y \end{vmatrix}. \qquad (7)$$

This discussion will be limited to perfect polarizers, i.e., $P_x = 1$ and $P_y = 0$.

The matrix now becomes:

$$\mathbf{P}_h = \tfrac{1}{2}\begin{vmatrix} 1 & 1 & 0 & 0 \\ 1 & 1 & 0 & 0 \\ 0 & 0 & 0 & 0 \\ 0 & 0 & 0 & 0 \end{vmatrix}. \qquad (8)$$

In order to describe the behavior of a polarizer at any angle α the matrix (8) must be transformed to arbitrary coordinate axes. This is done by the transformation equation.

$$\mathbf{P}(\alpha) = \mathbf{T}(-2\alpha)\mathbf{P}_h\mathbf{T}(2\alpha) \qquad (9)$$

where $\mathbf{T}(2\alpha)$ can be written

$$\mathbf{T}(2\alpha) = \begin{vmatrix} 1 & 0 & 0 & 0 \\ 0 & \cos 2\alpha & -\sin 2\alpha & 0 \\ 0 & \sin 2\alpha & \cos 2\alpha & 0 \\ 0 & 0 & 0 & 1 \end{vmatrix}. \qquad (10)$$

This is the matrix of a rotator, which is defined as an element which exhibits pure optical activity. α is defined as the angle which the polarizer transmission axis makes with the x axis. For the polarizing elements on an automobile the sign of α will be considered as positive if from the point of view of the car driver the angle is measured in a clockwise direction from the x axis.

Equation (9) can be expanded to give

$$\mathbf{P}(\alpha)=\frac{1}{2}\begin{vmatrix}1 & \cos 2\alpha & \sin 2\alpha & 0\\ \cos 2\alpha & \cos^2 2\alpha & \sin 2\alpha\cos 2\alpha & 0\\ \sin 2\alpha & \sin 2\alpha\cos 2\alpha & \sin^2 2\alpha & 0\\ 0 & 0 & 0 & 0\end{vmatrix}. \quad (11)$$

Similarly a matrix can be constructed for a wave plate of retardation δ. For a wave plate with axes parallel to the reference axes, this becomes:

$$\mathbf{R}(\delta)=\begin{vmatrix}1 & 0 & 0 & 0\\ 0 & 1 & 0 & 0\\ 0 & 0 & \cos\delta & -\sin\delta\\ 0 & 0 & \sin\delta & \cos\delta\end{vmatrix}. \quad (12)$$

With these matrices the various properties of any optical system and, in particular, the properties of a polarized headlight system can be calculated.

One of the first problems is the determination of the vector for completely unpolarized light. This can be accomplished with the phenomenological approach used by Perrin.[7] The action of a perfect polarizer on a general light vector (I, M, C, S) is represented by the equation

$$\mathbf{P}\begin{pmatrix}I\\ M\\ C\\ S\end{pmatrix}=\begin{pmatrix}I'\\ -\\ -\\ -\end{pmatrix} \quad (13)$$

which can be expanded to give

$$I'=\tfrac{1}{2}(I+M\cos 2\theta+C\sin 2\theta), \quad (14)$$

where θ is the angle of the polarizer to some predetermined axes.

In an actual measurement finite lengths of time are taken so that Eq. (14) should be rewritten

$$\bar{I}'=\tfrac{1}{2}[\bar{I}+\bar{M}\cos 2\theta+\bar{C}\sin 2\theta], \quad (15)$$

where

$$\bar{I}=\frac{1}{T}\int_0^T I\,dt \quad (16)$$

and T is the duration of the measurement. In order for I' to be independent of θ, as it is when the vector represents natural light, we must have $\bar{M}=\bar{C}=0$.

Similarly, the equation for light going through a quarter wave plate and a polarizer can be written

$$\bar{I}'=\tfrac{1}{2}[\bar{I}-\bar{M}\cos 2\theta-\bar{S}\sin 2\theta]. \quad (17)$$

Again I' is independent of θ and therefore $\bar{S}=0$. These same conclusions would have been reached by simply defining natural light as light which is unaffected by passing through any rotor or wave plate. In any case, the vector for natural light of unit intensity can be written

$$\begin{pmatrix}I\\ 0\\ 0\\ 0\end{pmatrix}. \quad (18)$$

LINEAR HEADLIGHT SYSTEM

In the general headlight system a series of linear polarizers and possibly retardation sheets are placed over all automobile headlights. The corresponding analyzer is placed before the automobile driver's eyes. We shall refer the axes of the various elements to the vertical. The elements on opposing cars will be reversed in order and rotated 180° about a vertical axis. If the axis of the viewer polarizer in a car is at +15°, the viewer polarizer in an opposing car must be at −15°.

The headlight polarizer will be represented by $\mathbf{P}_h(\epsilon)$ and the analyzer by $\mathbf{P}_w(\alpha)$ where ϵ and α are the angles of the axes with the vertical. The positive sign will be taken if the angles are measured in a clockwise direction from the vertical as viewed by a person inside the car. The corresponding retardation plates are $\mathbf{R}_h(\gamma,\delta)$ and $\mathbf{R}_w(\beta,\delta)$ where β and γ are the angles of the retardation axes with the vertical and δ is the retardation. Instead of beginning with the completely general case, we shall consider first the behavior of a system which employs only linear polarizers.

The positions of the polarizers on the headlight and windshield which will give extinction can be obtained from the equation

$$\mathbf{P}_w(\alpha)\mathbf{P}_h(-\epsilon)\begin{pmatrix}I\\ 0\\ 0\\ 0\end{pmatrix}=\begin{pmatrix}0\\ -\\ -\\ -\end{pmatrix}. \quad (19)$$

The expansion of these matrices gives

$$\tfrac{1}{4}[1+\cos 2(\alpha+\epsilon)]=0. \quad (20)$$

From this we can get the relationship

$$\alpha+\epsilon=\pm\frac{\pi}{2}. \tag{21}$$

This shows the possible arrangements of the polarizers in the headlight system. The equation says that on any car the angle of the headlight polarizer axis from the vertical is the same as the angle of the analyzer axis from the horizontal. This relationship was more simply deduced in the first part of this paper. For each of these possible arrangements, we can compute the transmission of the polarized or specular component of the light which is directly reflected back through the system. This can be done with the equation

$$I=\tfrac{1}{4}[1+\cos 2(\alpha-\epsilon)]. \tag{22}$$

If this result is combined with Eq. (20) a formula is obtained which holds only for those systems which give extinction. This can be written

$$I=\tfrac{1}{4}[1-\cos 4\epsilon]. \tag{23}$$

For the horizontal glare factor

$$\mathbf{P}_w(\alpha)\begin{Bmatrix}1\\-1\\0\\0\end{Bmatrix}=\begin{Bmatrix}I\\—\\—\\—\end{Bmatrix} \tag{24}$$

which can be expanded to give

$$I=\sin^2\alpha. \tag{25}$$

These curves are plotted in Fig. 1.

The contrast calculations are slightly more involved. The amount of light from the headlight reflected from an object back through the analyzer depends on the total reflectivity of the object and the amount by which the original percent polarization of the incident light is changed. An investigation of the polarized and unpolarized components reflected from various surfaces has been carried out by Lawrence.[11] If both of these quantities are known, a reflectivity matrix can be calculated which will show the behavior of the object in any headlight system. To simplify the calculation of this matrix, we will assume that the object is isotropic. This means that the change in percent polarization of light incident on this object is the same no matter what is the vibration direction of the light.

[11] Jacques Lawrence, J. Opt. Soc. Am. **31**, 9 (1941).

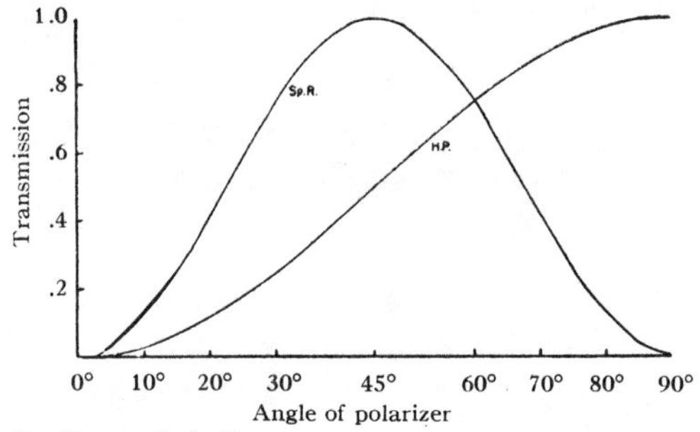

H.P. = Horz. polarization
SpR. = Specular reflection

Fig. 1. The relative amount of light from a polarizing headlight reflected specularly back through the automobile viewer and the amount of horizontally polarized light (daytime glare) transmitted by the automobile viewer as a function of the angle which the viewer axis is rotated from the vertical.

For most obstacles on the road this assumption is probably adequate. For certain road surfaces it may be dubious. This assumption of isotropy means that for all orientations the values of the elements in the reflection matrix are identical. This fact can be expressed in the equation

$$\mathbf{T}(2\theta)\boldsymbol{\varrho}\mathbf{T}(-2\theta)=\boldsymbol{\varrho} \tag{26}$$

where $\boldsymbol{\varrho}$ is the reflectivity matrix and $\mathbf{T}(2\theta)$ is the matrix of a rotator defined by Eq. (10). This equation is of exactly the same type as those used for finding the number of independent constants associated with a physical property of a crystal. In the case of a crystal, the angle θ depends on its symmetry. The solution of this equation can be written

$$\boldsymbol{\varrho}=\begin{vmatrix}\rho_{11} & 0 & 0 & \rho_{14}\\ 0 & \rho_{22} & \rho_{23} & 0\\ 0 & -\rho_{23} & \rho_{22} & 0\\ \rho_{14} & 0 & 0 & \rho_{44}\end{vmatrix}. \tag{27}$$

To find the significance of these remaining elements, we let the matrix, operate on various kinds of light. For example, in the case of unpolarized light

$$\begin{vmatrix}\rho_{11} & 0 & 0 & \rho_{14}\\ 0 & \rho_{22} & \rho_{23} & 0\\ 0 & -\rho_{23} & \rho_{22} & 0\\ \rho_{14} & 0 & 0 & \rho_{44}\end{vmatrix}\begin{vmatrix}1\\0\\0\\0\end{vmatrix}=\begin{vmatrix}\rho_{11}\\0\\0\\\rho_{14}\end{vmatrix}. \tag{28}$$

ρ_{11} thus turns out to be the natural reflectivity of the object and, also, since the object does not polarize $\rho_{14}=0$. With horizontal linearly polarized light

$$\begin{vmatrix} \rho_{11} & 0 & 0 & 0 \\ 0 & \rho_{22} & \rho_{23} & 0 \\ 0 & -\rho_{23} & \rho_{22} & 0 \\ 0 & 0 & 0 & \rho_{44} \end{vmatrix} \begin{vmatrix} 1 \\ 1 \\ 0 \\ 0 \end{vmatrix} = \begin{vmatrix} \rho_{11} \\ \rho_{22} \\ -\rho_{23} \\ 0 \end{vmatrix}. \tag{29}$$

Since the vibration direction of the polarized component of reflected light must be the same as the incident light

$$\rho_{23}=0. \tag{30}$$

Finally, with elliptically polarized light at 45°:

$$\begin{vmatrix} \rho_{11} & 0 & 0 & 0 \\ 0 & \rho_{22} & 0 & 0 \\ 0 & 0 & \rho_{22} & 0 \\ 0 & 0 & 0 & \rho_{44} \end{vmatrix} \begin{vmatrix} 1 \\ 0 \\ .7 \\ .7 \end{vmatrix} = \begin{vmatrix} \rho \\ 0 \\ .7\rho_{22} \\ .7\rho_{44} \end{vmatrix}. \tag{31}$$

Since again the polarized portion of the light must maintain its same vibration direction

$$\rho_{44}=\rho_{22}. \tag{32}$$

The elements ρ_{22} are the intensities of the polarized component. We will assume that the light from the automobile headlight is completely polarized. In this case

$$\rho_{22}=\rho_{11}P, \tag{33}$$

where P is the fractional polarization of the reflected light. ρ_{11} can now be replaced with the symbol ρ and the general matrix becomes

$$\varrho=\rho\begin{vmatrix} 1 & 0 & 0 & 0 \\ 0 & P & 0 & 0 \\ 0 & 0 & P & 0 \\ 0 & 0 & 0 & P \end{vmatrix}. \tag{34}$$

With this matrix it is now possible to calculate the contrasts between different objects in different headlight systems. The contrast will be defined as

$$\frac{dI}{I}=\frac{I\text{ object}-I\text{ background}}{I\text{ object}}. \tag{35}$$

When the background intensity is greater, it will be defined as

$$\frac{dI}{I}=\frac{I\text{ background}-I\text{ object}}{I\text{ background}}. \tag{36}$$

Consider first the contrast in a linearly polarized system. The headlight polarizer is at an angle ϵ and the windshield polarizer at an angle α. The intensity of the reflected light can be written

$$\mathbf{P}_w(\alpha)\varrho\mathbf{P}_h(\epsilon)\begin{Bmatrix} 1 \\ 0 \\ 0 \\ 0 \end{Bmatrix} = \begin{Bmatrix} I \\ - \\ - \\ - \end{Bmatrix}. \tag{37}$$

The contrast calculated from this equation is

$$\frac{dI}{I}=\frac{\rho_0[1+P_0\cos\,(\alpha-\epsilon)]-\rho_b[1+P_b\cos 2(\alpha-\epsilon)]}{\rho_0[1+P_0\cos 2(\alpha-\epsilon)]}, \tag{38}$$

where the subscript 0 refers to the object and the subscript b to the background. This contrast is to be compared with the contrast in an unpolarized system which can be written

$$\frac{dI}{I}=\frac{\rho_0-\rho_b}{\rho_0}. \tag{39}$$

For a headlight or viewer polarizer angle of $22\frac{1}{2}°$, Eq. (38) shows that the contrast is the same as the contrast in an unpolarized light system.

One of the most interesting things which is apparent from these last equations is the extreme change in contrast which is possible in going to a polarized light system.

Consider an object which reflects all of the light and is superimposed on a field which also reflects all the light. Suppose, furthermore, that half the light from the object remains polarized,

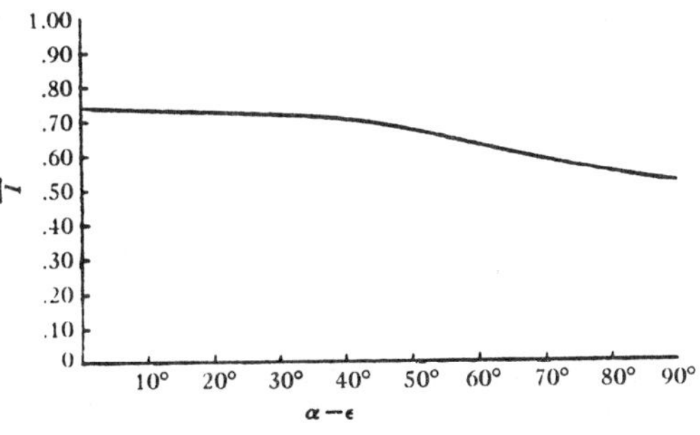

Object reflectivity = 2.6% % polarization = 58
Road reflectivity = 0.81% % polarization = 35

FIG. 2. Contrast of a coat of 2.6% reflectivity and 58% polarization against a macadam road of 0.81% reflectivity and 35% polarization plotted against the difference between the headlight polarizer angle and the viewer angle.

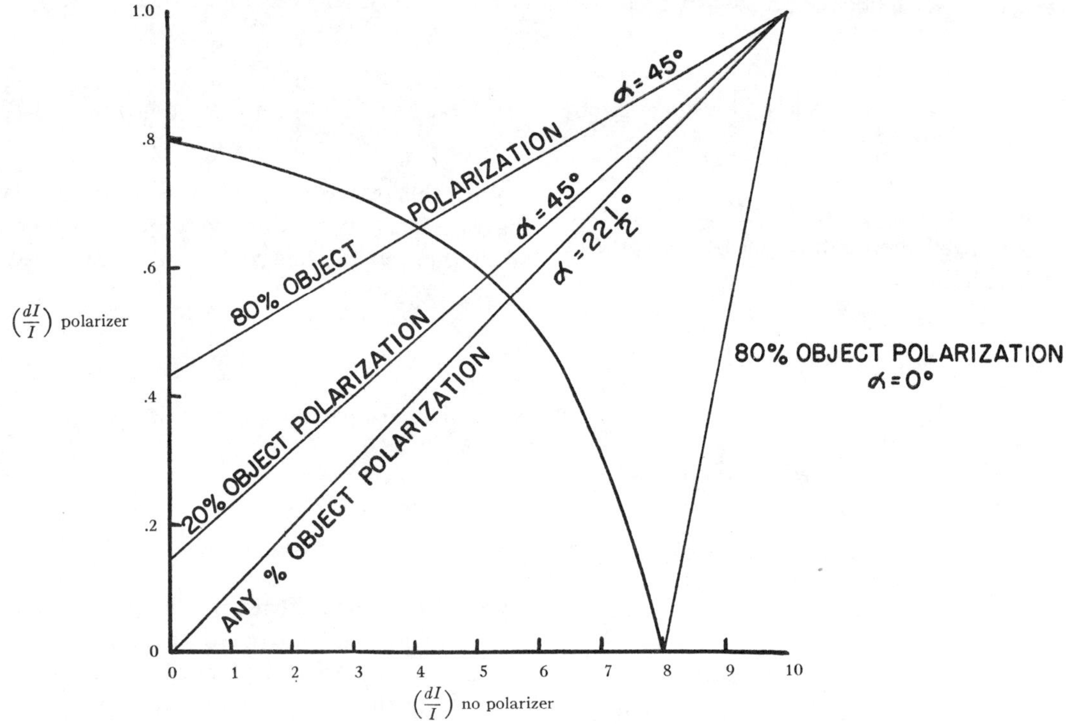

FIG. 3. The natural light contrast of an object of 5% reflectivity against various backgrounds plotted against the contrast in various linearly polarized headlight systems. [0% polarization of background; .05 = object reflectivity.]

whereas only a quarter of the field light is polarized. In this case the contrast is zero in unpolarized light. For a system with polarizers at 45° the contrast becomes

$$\left(\frac{dI}{I}\right)_{45}=\frac{1}{6}. \tag{40}$$

This contrast change is particularly spectacular when actually demonstrated in a driving test. Systems of this type were proposed for *IFF* during the war.

In order to see what contrast conditions are met in practice and thus to examine the differences in behavior of various headlight systems, it is necessary to reduce field brightness measurements to a form suitable for use in the equations.

Two measurements are necessary: (1) the brightness when illuminated with linearly polarized light from a headlight and viewed through a parallel polarizer; (2) the brightness when illuminated with linearly polarized light from a headlight and viewed with a crossed polarizer.

The equations connecting these two measurements with the polarization and reflectivity of the object can be written

$$\mathbf{P}_v\boldsymbol{\varrho}\mathbf{P}_v\begin{bmatrix}I_0\\0\\0\\0\end{bmatrix}=\begin{bmatrix}I_{vv}\\—\\—\\—\end{bmatrix},\quad \mathbf{P}_v\boldsymbol{\varrho}\mathbf{P}_h\begin{bmatrix}I_0\\0\\0\\0\end{bmatrix}=\begin{bmatrix}I_{vh}\\—\\—\\—\end{bmatrix}. \tag{41}$$

These can be condensed to

$$\frac{I_{vv}}{I_0}=\frac{\rho}{4}(1+P),\quad \frac{I_{vh}}{I_0}=\frac{\rho}{4}(1-P). \tag{42}$$

To illustrate the way the contrast varies in a typical situation a set of measurements were made by Mr. Dexter Cooper of these laboratories. The measurements were made with a polarized headlight system illuminating a man standing on a bituminous macadam road:

Measurement I.

Man's coat at 100 feet from headlights
$I_{vv}=0.30$ foot lambert
$I_{vh}=0.08$ foot lambert
Reflectance $=2.6\%=\rho$
Bituminous macadam road next to coat:
$I_{vv}=0.08$ foot lambert.

Measurement II.

Bituminous macadam road next to coat:
$I_{vv}=0.028$ foot lambert
$I_{vh}=0.0135$ foot lambert.

The first set of measurements was made with a Luckiesh brightness meter. The second set were made photographically. The photographic measurement is not so reliable as the meter measurement.

From the coat measurements and Eq. (42)

$$P_c = 0.58.$$

From the second road measurement:

$$P_r = 0.35.$$

From the first road measurement and the value for the coat reflectivity:

$$\rho_r = 0.81 \text{ percent.}$$

In Fig. 2 is shown a plot of the contrast of the coat against the road for a linear headlight system. In Fig. 3 is shown a plot of the contrast changes of a 5-percent reflectivity object against a road of varying reflectivities.

The data can be summarized as follows:

1. The largest part of the polarized reflected component reaches the driver's eyes in the 45° case. In other words, for the 45° case objects are brighter.
2. The most contrast is obtained for a non-depolarizing target with the 45° case.
3. For the $22\frac{1}{2}°$ case half the polarized reflected component reaches the driver's eyes.
4. The natural light contrast of objects is preserved in a $22\frac{1}{2}°$ system.

ELLIPTICAL SYSTEMS

The most general headlight system will consist of linear polarizers and retardation sheets so arranged as to give elliptically polarized light.

The Extinction Condition

In the general case exactly the same computations can be made as in the linear case. We are interested, first in the positions of the polarizers and retardation plates on our windshield and headlights which will give extinction. These can be obtained from the equation:

$$\mathbf{P}_w(\alpha)\mathbf{R}_w(\beta,\delta_1)\mathbf{R}_h(-\gamma,\delta_2)\mathbf{P}_h(-\epsilon)\begin{bmatrix}1\\0\\0\\0\end{bmatrix} = \begin{bmatrix}0\\-\\-\\-\end{bmatrix}. \quad (43)$$

The expansion of these matrices gives the equation:

$$\begin{aligned}1&+\sin 2(\beta+\alpha)\sin 2(\beta+\gamma)\\ &\times\cos 2(\gamma-\epsilon)\cos\delta_2(1-\cos\delta_1)\\ &+\cos 2(\beta+\alpha)\cos 2(\beta+\gamma)\\ &\times\cos 2(\gamma-\epsilon)(1-\cos\delta_1)\\ &+\sin 2(\beta+\alpha)\sin 2(\beta+\epsilon)\cos\delta_1\cos\delta_2\\ &+\sin 2(\beta+\alpha)\sin 2(\gamma-\epsilon)\sin\delta_1\sin\delta_2\\ &+\cos 2(\beta+\alpha)\cos 2(\beta+\epsilon)\cos\delta_1=0. \quad (44)\end{aligned}$$

This equation must be satisfied if the headlights are to be extinguished.

If the two wave plates are dissimilar and if their axes are not parallel to the polarizer axes, the polarizer axes' positions for extinction are not independent of δ_1 and δ_2 the value of the retardations. Since most wave plates are non-achromatic, the retardation δ is a function of frequency. This means that the value of the polarizer and retardation sheet extinction angles will be different for different frequencies. In other words all colors will not be extinguished simultaneously. If $\delta_1=\delta_2$, Eq. (44) can be rewritten

$$\begin{aligned}1&+\sin 2(\alpha-\beta)\sin 2(\gamma-\epsilon)[2\cos^2(\beta+\gamma)\cos^2\delta-1]\\ &+\cos 2(\alpha-\beta)\cos 2(\gamma-\epsilon)\cos 2(\beta+\gamma)\\ &-\sin 2(\beta+\gamma)\sin 2(\alpha-\beta-\gamma+\epsilon)\cos\delta=0. \quad (45)\end{aligned}$$

The real, non-trivial solution of this equation is given by the relationship

$$\begin{aligned}\alpha-\beta&=\gamma-\epsilon\\ \beta+\gamma&=\pm\frac{\pi}{2}\\ \alpha+\epsilon&=\pm\frac{\pi}{2}.\end{aligned} \quad (46)$$

This solution is particularly interesting as it does not involve δ. It shows that the system is "achromatized." In other words, if the light is extinguished for one color, it is extinguished for all colors.

In case the retardation δ is higher than one wave, we begin to get color in the specularly reflected component. In other words, the achromatization of the system does not extend beyond the headlight extinction. This color represents a loss of light. In the extreme case where δ is very large, the system will act for the polarized component of reflected light as though a 50 percent isotropic absorber were over both the headlight and the windshield. Because of this large light loss, we shall limit the discussion to low-order retardation plates.

Except for the extinction conditions a fixed value can be assigned to the retardation of a low-order plate and its variation with color ignored. By stipulating achromatism in the extinction Eq. (44) can be considerably simplified. A still further simplification can be made if we can show that all possible systems with different values for the retardation can be reproduced with a single value of δ.

Comparison between Elliptical Systems with Quarter-Wave Plates and Systems with Other Fractional Wave Plates

Any kind of elliptically polarized light can be produced with a quarter-wave plate and a linear polarizer. We need to show that the combination will behave the same for all the characteristics of an automobile system: day glare, polarized or specular reflection, etc., as the elliptical polarizer made of a retardation sheet other than a quarter wave.

In order to compare the $1/4\lambda$ system with the general system, one must know first the relationship between wave plate axes and polarizer axes in the two cases to produce elliptical light of identical shape and orientation. To do this we set up the relationship

$$\mathbf{Q}\left(\frac{\pi}{2}\right)\begin{bmatrix}1\\0\\0\\0\end{bmatrix} = \begin{bmatrix}I\\M\\C\\S\end{bmatrix} = \mathbf{Q}(\lambda)\begin{bmatrix}1\\0\\0\\0\end{bmatrix}, \tag{47}$$

where $\mathbf{Q}(\pi/2)$ is the matrix, $\mathbf{R}(\pi/2, \beta)\mathbf{P}(\alpha)$, of a polarizer followed by a quarter-wave plate, and $\mathbf{Q}(\lambda)$ is the matrix of a polarizer followed by a wave plate of any retardation.

From this matrix equation we get

$$\begin{vmatrix}q_{(\pi/2)11}\\q_{(\pi/2)21}\\q_{(\pi/2)31}\\q_{(\pi/2)41}\end{vmatrix} = \begin{vmatrix}q_{\lambda 11}\\q_{\lambda 21}\\q_{\lambda 31}\\q_{\lambda 41}\end{vmatrix} \tag{48}$$

which says simply that the elements in the first column in both matrices must be equal for them to have exactly the same effect on natural light.

It must now be demonstrated that with the two different elements arranged as analyzers in automobile windshields, the transmitted intensity is the same for any given kind of polarized light. In this case the retardation sheet comes first, and the matrix of the system is different. From the Rayleigh reciprocity law it can be shown that in this reverse instrument the elements of the first row have the same numerical value as the elements in the first column of the direct instrument. We can write

$$\mathbf{Q}^* = \begin{vmatrix}q_{11} & q_{21} & -q_{31} & q_{41}\\ — & — & — & —\\ — & — & — & —\\ — & — & — & —\end{vmatrix}, \tag{49}$$

where Q^* is the matrix of the retardation sheet and polarizer in reverse order. We want to show that

$$\mathbf{Q}^*\left(\frac{\pi}{2}\right)\begin{bmatrix}I\\M\\C\\S\end{bmatrix} = \mathbf{Q}^*(\lambda)\begin{bmatrix}I\\M\\C\\S\end{bmatrix} = \begin{bmatrix}I'\\—\\—\\—\end{bmatrix}. \tag{50}$$

Since the first column of the two direct matrices are the same, the first row of the two reverse matrices will also be the same, and the intensities of the light transmitted through the $\frac{1}{4}$-wave plate combination or the general retardation combination will be exactly the same.

It is thus apparent that a $\frac{1}{4}$-wave plate can be substituted for any other wave plate in the

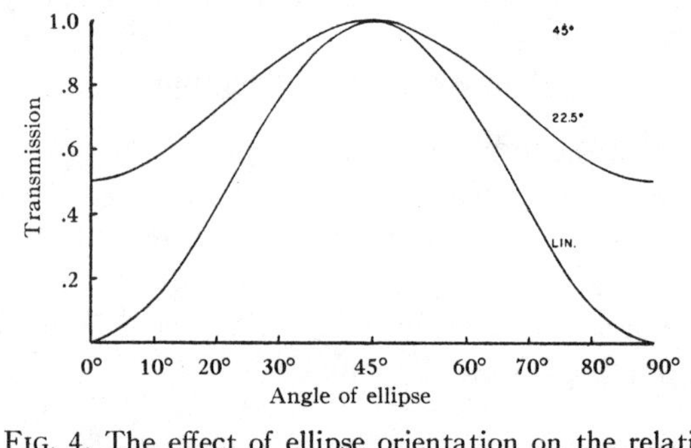

FIG. 4. The effect of ellipse orientation on the relative amount of light from headlights reflected specularly back through the viewer for elliptical systems in which the quarter-wave plate is at 45°, 22½° and 0° to the polarizer axis.

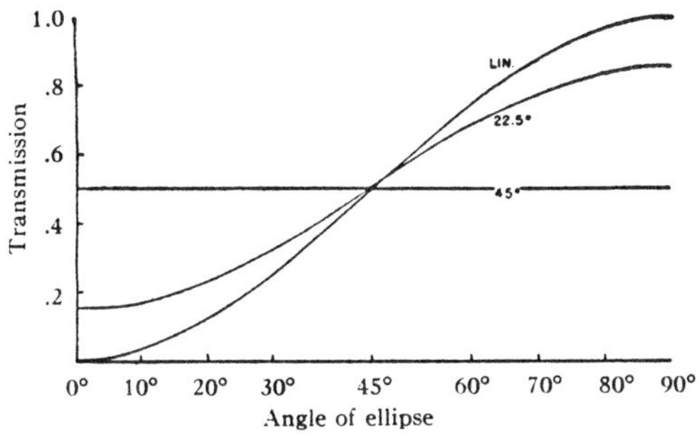

FIG. 5. The effect of ellipse orientation on the relative amount of horizontally polarized light transmitted by the viewer for elliptical systems in which the quarter-wave plate is at 45°, $22\frac{1}{2}$° and 0° to the polarizer axis.

mathematical analysis. If, however, a given kind of elliptical light should prove desirable, we will not be restricted to the use of a $\frac{1}{4}$-wave plate in order to obtain this elliptical light.

Since a quarter-wave sheet will do what any other retardation sheet will do, the main part of the following discussion will be limited to combinations of polarizers and quarter-wave plates.

A quarter-wave plate has the further advantage that when combined with a polarizer the plate axis coincides with the axis of the elliptical light which it produces.

Equation (46) shows the possible arrangements of the polarizers in the headlight system. For each of these possible arrangements we can compute the transmission of polarized reflected light from the driver's headlights back through his own windshield. This can be done with the matrix equation:

$$\mathbf{P}_w(\alpha)\mathbf{R}_w\left(\beta, \frac{\pi}{2}\right)\mathbf{R}_h\left(\gamma, \frac{\pi}{2}\right)\mathbf{P}_h(\epsilon)\begin{bmatrix}1\\0\\0\\0\end{bmatrix} = \begin{bmatrix}I\\—\\—\\—\end{bmatrix}. \quad (51)$$

Here we have let $\delta=\pi/2$ as the arguments above show this system is sufficiently general. From this we get the equation

$$I=\tfrac{1}{4}[1+\cos 2(\alpha-\beta)\cos 2(\gamma-\epsilon)\cos 2(\beta-\gamma) +\sin 2(\alpha-\beta)\sin 2(\gamma-\epsilon)]. \quad (52)$$

This result can be combined with Eq. (46) to give a formula which holds only for those systems which give extinction. This new equation can be written

$$I=\tfrac{1}{2}[1-\cos^2 2\gamma\cos^2 2(\gamma-\epsilon)]. \quad (53)$$

In Fig. 4 are shown curves of the intensity for various retardation plate angles and polarizer angles. From these curves it is apparent that various combinations differ widely with respect to their transmission of reflected light. The case $\gamma-\epsilon=45°$ represents circularly polarized light. Rotation of the headlight around an axis normal to the polarizer has no effect on the reflected light getting back through the windshield. When $\gamma-\epsilon=22\frac{1}{2}°$, the light is elliptically polarized and the transmission of the reflected light becomes a function of the position of the headlight polarized axes.

The amount of horizontally polarized light transmitted by the analyzer can be determined from the matrix equation

$$\mathbf{P}_\epsilon\mathbf{R}_\gamma\begin{bmatrix}1\\-1\\0\\0\end{bmatrix} = \begin{bmatrix}I\\—\\—\\—\end{bmatrix}. \quad (54)$$

This equation can be expanded to give

$$I=\tfrac{1}{2}[1-\cos 2\gamma\cos 2(\gamma-\epsilon)]. \quad (55)$$

A plot of this equation is shown in Fig. 5 for various orientations of the analyzer. Here again it is apparent that the various systems behave quite differently with respect to this factor. In a completely determined system we could consider

TABLE I. Characteristics of various headlight systems including system with 50 percent neutral filters over headlights and windshields.

Light produced by windshield polarizer	Vertical linear	45° linear	$22\frac{1}{2}$° linear	Circular	$22\frac{1}{2}$° ellipse at 45°	$22\frac{1}{2}$° ellipse at $22\frac{1}{2}$°	50% isotropic windshield
Transmission of horizontally polarized light	0	50	15	50	50	25	50
Transmission of specularly reflected light from headlights	0	100	50	100	100	75	50
Contrast of 7% reflectivity object against 3% reflectivity background. Object 25% depolarized. Background 50% depolarized	14	63	57	63	63	61	57
Contrast of 7% reflectivity object against 3% reflectivity background. Object 25% depolarized. Background 30% depolarized	49	58	57	58	58	58	57

Figs. 4 and 5 and find the "best" polarizing system for headlights.

This, of course, is calculated with the assumption of equal weights for the direct reflected factor and the horizontal glare factor which is something which can only be determined experimentally.

The calculation of contrast in elliptical systems is exactly the same as in linear systems. The reflected light is represented by the matrix equation

$$\mathbf{P}_w(\alpha)\mathbf{R}\left(\beta, \frac{\pi}{2}\right)\boldsymbol{\varrho}\mathbf{R}\left(\gamma, \frac{\pi}{2}\right)\mathbf{P}_h(\epsilon)\begin{bmatrix} I \\ 0 \\ 0 \\ 0 \end{bmatrix} = \begin{bmatrix} I \\ - \\ - \\ - \end{bmatrix}. \quad (56)$$

From this we get the contrast equation

$$\frac{dI}{I} = \frac{\rho_i - \rho_0 + (\rho_i P_i - \rho_0 P_0)[\cos^2 2(\alpha-\beta)\cos 2(\beta-\gamma) + \sin^2 2(\alpha-\beta)]}{\rho_i\{1 + P_i[\cos^2 2(\alpha-\beta)\cos 2(\beta-\gamma) + \sin^2 2(\alpha-\beta)]\}}. \quad (57)$$

The behavior of this equation is exactly the same as the equation for contrast in linear systems.

In order to show the difference between the various headlight systems, the material presented in the graphs has been summarized in Table I at the end of this section. One of the most interesting systems is that in which the viewer has a quarter-wave plate at $22\frac{1}{2}°$ to a polarizer at 0°. This gives elliptically polarized light with the ellipse axis at $22\frac{1}{2}°$ and with an ellipticity of $22\frac{1}{2}°$. The ellipticity is defined as $\tan^{-1}(b/a)$ where a and b are the major and minor axes of the ellipse. This system is also included in the table.

From the data in this paper it is apparent that ideally there are several equally useful polarizing headlight systems. For its current performance demonstrations, Polaroid Corporation is using with satisfactory results, linear polarizers at an inclination not shown in the table, namely, the viewer polarizer axis is inclined −35° from the vertical, this angle representing a counterclockwise rotation of his viewer as seen by the driver behind it.

In conclusion the authors wish to express appreciation to the members of the Polaroid Corporation for many helpful suggestions.

On a Calculus for Retardation Plates

J. G. COLLINS AND W. H. STEEL
C.S.I.R.O. Division of Physics, National Standards Laboratory, Sydney, Australia

(Received September 25, 1961)

THE choice of a suitable formalism within which to describe the effect of a plate of anisotropic material on a collimated beam of polarized light can greatly simplify investigations of multicomponent optical systems, such as birefringent filters. An example is the recent treatment by Evans[1] of the Šolc filter, using matrix calculus as introduced by Jones.[2]

Jones concentrated on a Cartesian representation of the state of polarization, but pointed out that some problems might be treated more simply in terms of other sets of orthogonal components. In particular, when the crystal plates are rotators, or exhibit circular dichroism, it is more convenient to express the light vector as the sum of two circularly polarized components.

During the design of a birefringent filter of the Lyot type,[3] it has been found that resolution into circularly polarized components also leads to a simpler treatment of retardation plates, even though here a description in terms of linear polarization seems more obvious at first sight. As any simplification is important when systems of many elements are investigated, we outline a treatment of retardation plates in terms of circular polarization.

If the electric vector $\mathbf{E}$ is specified by Cartesian components in a plane at right angles to the direction of propagation, the effect of a retardation plate, one of whose principal axes is at an angle ω to the x direction, has been given by Jones as

$$\begin{pmatrix} E_{x1} \\ E_{y1} \end{pmatrix} = \begin{pmatrix} \cos\omega & -\sin\omega \\ \sin\omega & \cos\omega \end{pmatrix} \begin{pmatrix} e^{i\gamma} & 0 \\ 0 & e^{-i\gamma} \end{pmatrix} \begin{pmatrix} \cos\omega & \sin\omega \\ -\sin\omega & \cos\omega \end{pmatrix} \begin{pmatrix} E_{x0} \\ E_{y0} \end{pmatrix}, \quad (1)$$

or $\mathbf{E}_1=\mathbf{S}(\omega)\mathbf{N}\mathbf{S}(-\omega)\mathbf{E}_0$. Here $\mathbf{E}_0$ and $\mathbf{E}_1$ are the incident and emergent light vectors, and 2γ is the phase difference for light linearly polarized along the two principal axes of the plate:

$$2\gamma=(2\pi/\lambda)d\delta n,$$

where d is the thickness of the plate and δn its birefringence, or difference of refractive index for these two directions of polarization.

This equation is made simpler for computation by the introduction of the transformation $\mathbf{T}$, which diagonalizes the rotation matrix $\mathbf{S}$ at the expense of producing a nondiagonal retardation matrix $\mathbf{N}$. We can write

$$\begin{aligned} \mathbf{TE}_1 &= \mathbf{TS}(\mathbf{T}^{-1}\mathbf{T})\mathbf{N}(\mathbf{T}^{-1}\mathbf{T})\mathbf{S}^{-1}(\mathbf{T}^{-1}\mathbf{T})\mathbf{E}_0 \\ &= (\mathbf{TST}^{-1})(\mathbf{TNT}^{-1})(\mathbf{TS}^{-1}\mathbf{T}^{-1})\mathbf{TE}_0. \end{aligned}$$

The required matrix is

$$\mathbf{T}=\frac{1}{\sqrt{2}}\begin{pmatrix} 1 & i \\ 1 & -i \end{pmatrix}, \quad \mathbf{T}^{-1}=\frac{1}{\sqrt{2}}\begin{pmatrix} 1 & 1 \\ -i & i \end{pmatrix},$$

and the transformed equation is

$$\begin{pmatrix} E_{x1}+iE_{y1} \\ E_{x1}-iE_{y1} \end{pmatrix} = \begin{pmatrix} e^{i\omega} & 0 \\ 0 & e^{-i\omega} \end{pmatrix} \begin{pmatrix} \cos\gamma & -i\sin\gamma \\ i\sin\gamma & \cos\gamma \end{pmatrix} \times \begin{pmatrix} e^{-i\omega} & 0 \\ 0 & e^{i\omega} \end{pmatrix} \begin{pmatrix} E_{x0}+iE_{y0} \\ E_{x0}-iE_{y0} \end{pmatrix},$$

or

$$\mathbf{E}_1'=\mathbf{S}'\mathbf{N}'(\mathbf{S}')^{-1}\mathbf{E}_0'. \quad (2)$$

When the matrix multiplications are explicitly carried out,

$$\mathbf{S}(\omega)\mathbf{N}\mathbf{S}(-\omega)=\begin{pmatrix} \cos\gamma+i\sin\gamma\cos2\omega & i\sin\gamma\sin2\omega \\ -i\sin\gamma\sin2\omega & \cos\gamma-i\sin\gamma\cos2\omega \end{pmatrix},$$

whereas

$$\mathbf{S}'(\omega)\mathbf{N}'\mathbf{S}'(-\omega)=\begin{pmatrix} \cos\gamma & -ie^{2i\omega}\sin\gamma \\ ie^{-2i\omega}\sin\gamma & \cos\gamma \end{pmatrix}.$$

In problems involving a series of dissimilar elements, the mathematical advantages of diagonalizing the rotation matrices are apparent, and the saving in computation increases with the number of elements.

The change in representation has a simple physical interpretation. In a Cartesian system, the light vector is resolved successively into components which are linearly polarized parallel to the principal axes of each element. Within the element the components are separately retarded, but not mixed. In the transformed system, the quantities $E_x\pm iE_y$ represent, respectively, right- and left-handed circularly polarized components, whose composition is varied within each element, but which merely undergo a relative phase shift when passing between elements whose principal axes have different orientations.

When the light is linearly polarized, the left- and right-handed circular components are represented by complex conjugates, and the matrix equations reduce to the calculation of a single component. This relation can be made to hold for all polarizations by artificially introducing a different imaginary j to represent the transformation to circularly polarized light (retaining i to represent phase) and taking the complex conjugate with respect to j only. Thus, if E_r and E_l are written for the right- and left-handed circular components $E_x\pm jE_y$, Eq. (2) can be written as

$$\begin{aligned} E_{r1}&=E_{r0}\cos\gamma-iE_{l0}e^{2j\omega}\sin\gamma, \\ E_{l1}&=E_{r1}^{*} \text{ (complex conjugate in } j). \end{aligned}$$

[1] J. W. Evans, J. Opt. Soc. Am. **48**, 142 (1958).
[2] R. C. Jones, J. Opt. Soc. Am. **31**, 488, 500 (1941); **32**, 486 (1942); H. Hurwitz and R. C. Jones, *ibid.* **31**, 495 (1941).
[3] W. H. Steel, R. N. Smartt, and R. G. Giovanelli, Australian J. Phys. **14**, 201 (1961).

On Calculating the Optical Constants from Reflection Coefficients

Richard Tousey
Tufts College, Medford, Massachusetts
(Received April 7, 1939)

Introduction

It is well known that the optical constants of a substance can be calculated from values of its specular reflection coefficient measured at several angles of incidence by means of formulae derived from the electromagnetic theory of light. This calculation is important since it offers, for the extreme ultraviolet region of the spectrum, one of the very few methods of obtaining the index of refraction, n, and the extinction coefficient, k. The other procedures which have been employed in this spectral region apply to very thin films only, since they require measurements of transmission. Such transmission methods, although very valuable because they permit the centers of absorption bands to be located easily, do not furnish an accurate determination of n and the absolute value of k since the thickness of the film, and also the correction for reflection from its two surfaces, are hard to measure with precision.

In order to obtain the optical constants from the reflecting power one employs Fresnel's reflection equations, generalized to an absorbing isotropic medium.[1] These are as follows:

$$R_s=\frac{a^2+b^2-2a\cos i+\cos^2 i}{a^2+b^2+2a\cos i+\cos^2 i},$$
$$R_p=R_s\cdot\frac{a^2+b^2-2a\sin i\cdot\tan i+\sin^2 i\cdot\tan^2 i}{a^2+b^2+2a\sin i\cdot\tan i+\sin^2 i\cdot\tan^2 i}, \tag{1}$$

where

$$2a^2=[(n^2-k^2-\sin^2 i)^2+4n^2k^2]^{\frac{1}{2}}+[(n^2-k^2)-\sin^2 i],$$
$$2b^2=[(n^2-k^2-\sin^2 i)^2+4n^2k^2]^{\frac{1}{2}}-[(n^2-k^2)-\sin^2 i].$$

R_s and R_p are the reflection coefficients for incident light vibrating in planes perpendicular and parallel, respectively, to the plane of incidence, i is the angle of incidence, n the index of refraction, and k the extinction coefficient. The complex index is $(n-ik)$. These equations can be written in the form

$$R=f(i, n, k, p),$$

where R is the reflection coefficient for light having a percentage polarization given by p. Suppose that a curve showing R as a function of i is obtained experimentally for a particular substance and wave-length. It would be possible, then, to determine the optical constants n and k of the material under investigation and also the percentage polarization of the incident light, p, by choosing values of n, k, and p such that the curve showing R as a function of i as computed from this formula would exactly fit the curve as determined from experiment.

The above procedure for obtaining the optical constants, but not the polarization, was suggested by Pfestorf.[2] Boeckner[3] used this procedure in the infra-red. But there the values of the optical constants were such that comparatively simple approximate formulae were avail-

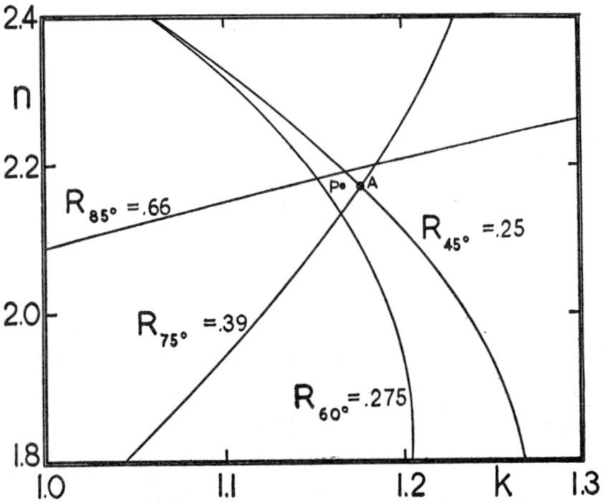

Fig. 1. Curves of constant reflection coefficient illustrating the application of the graphical method to a particular example.

[1] W. König, *Handbuch der Physik*, Vol. **20**, p. 242. n and k, as used in the present paper, are the same as König's n_0 and $(n_0\kappa_0)$.

[2] G. Pfestorf, Ann. d. Physik **81**, 906 (1926).

[3] C. Boeckner, J. Opt. Soc. Am. **19**, 7 (1929).

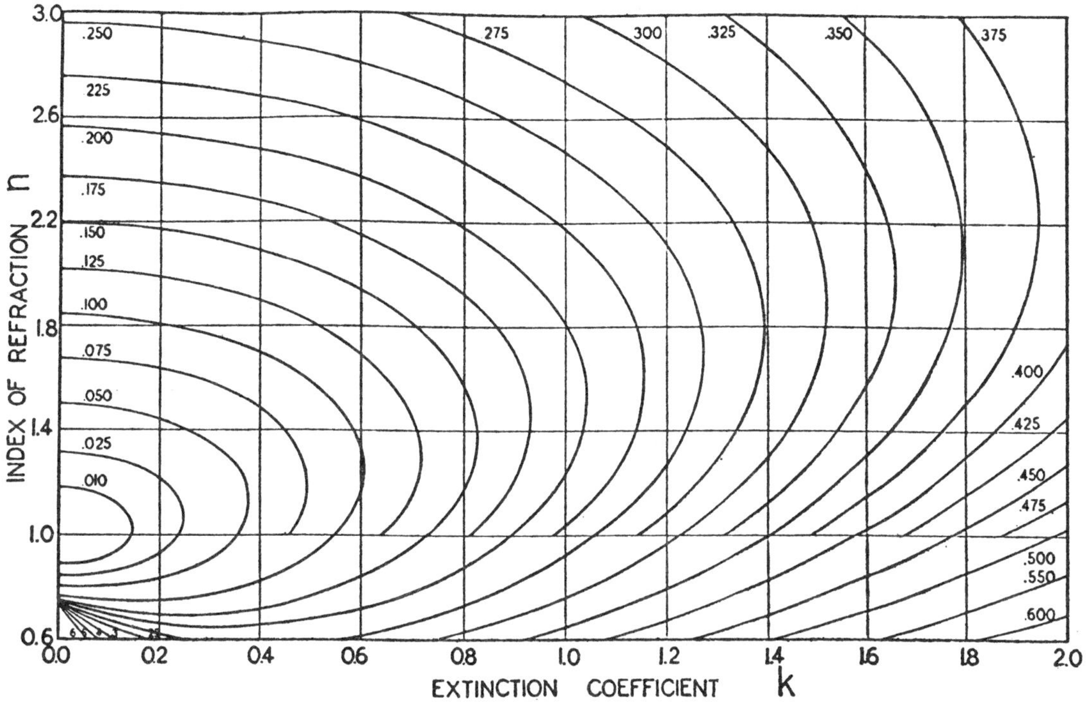

FIG. 2. Curves showing the relation between n and k for certain values of the reflection coefficient at 45°, for unpolarized incident light.

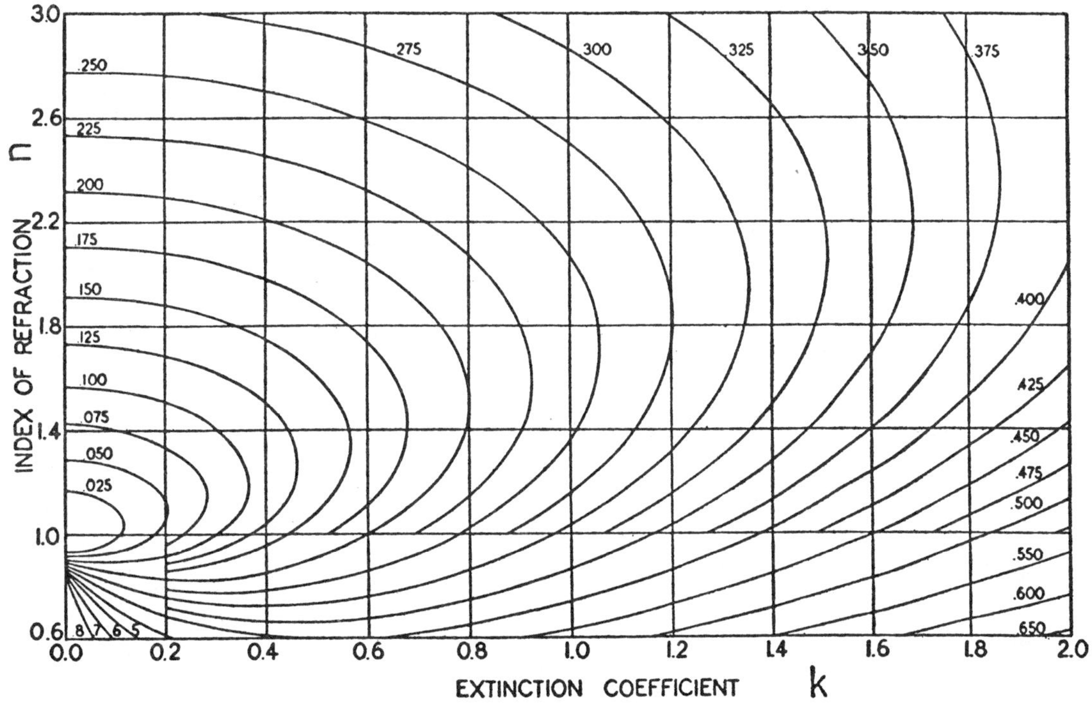

FIG. 3. Curves showing the relation between n and k for certain values of the reflection coefficient at 60°, for unpolarized incident light.

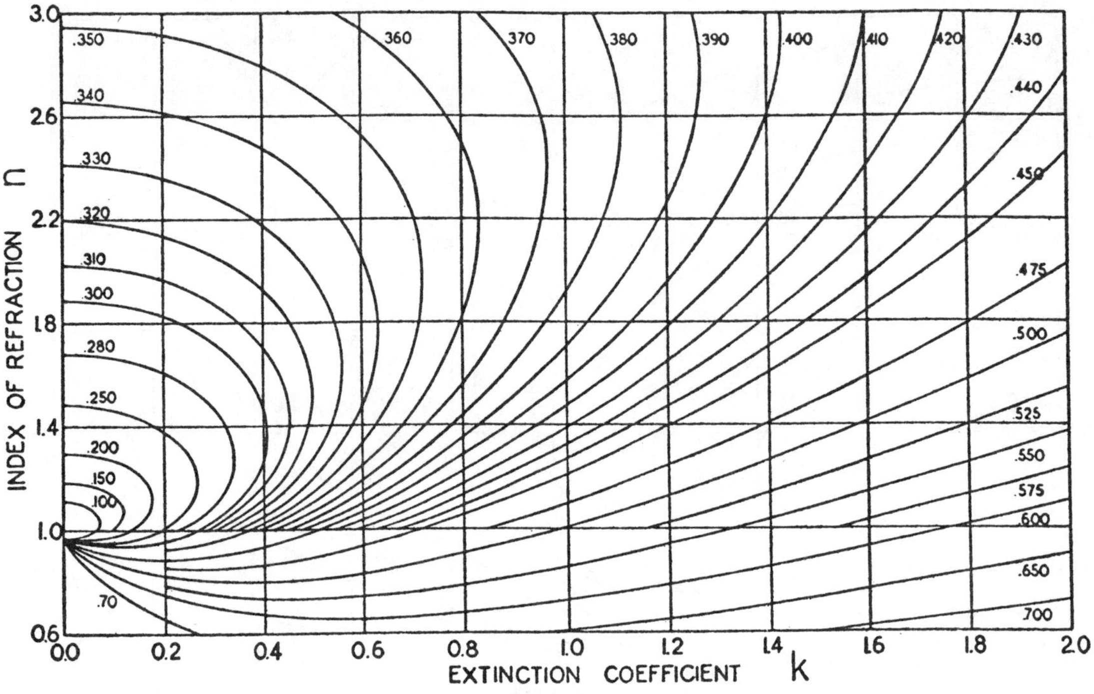

Fig. 4. Curves showing the relation between n and k for certain values of the reflection coefficient at 75°, for unpolarized incident light.

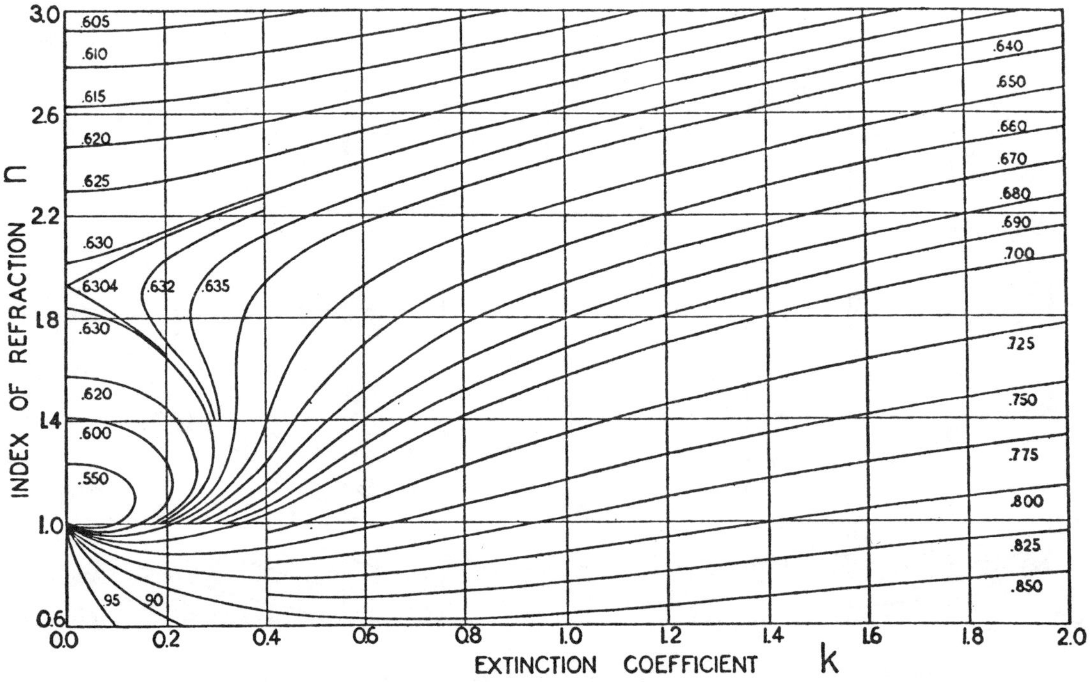

Fig. 5. Curves showing the relation between n and k for certain values of the reflection coefficient at 85°, for unpolarized incident light.

able. O'Bryan,[4] working in the extreme ultraviolet, applied the method to celluloid. In performing this calculation he was aided by direct measurements of the transmission through thin self-supporting films which, with estimates of the thickness of the films, gave approximate values of k with which to start. Further, he assumed the incident light to be unpolarized. In the case of the author's work on fluorite[5] in the extreme ultraviolet neither approximate formulae nor thin films were available and it was necessary to apply the general curve-fitting procedure. It is obvious from the complexity of the formulae that to carry out this process of curve-fitting simply by guessing and trying values of n and k for each particular case encountered would be very tedious indeed even though the problem be simplified by assuming that the state of polarization of the incident light is known. For this reason, a graphical method was developed which permits very rapid determination of the optical constants from the reflecting power data. It is the purpose of this paper to describe this graphical method and to present the curves necessary to handle the case where the incident light is unpolarized.

Graphical Method

Values of the reflection coefficient for two angles of incidence are sufficient to determine both n and k provided the state of polarization of the incident light is known. From Eqs. (1) the reflection coefficient R is then given in terms of the angle of incidence i and the optical constants of the material by an expression of the form $R=f(i, n, k)$. Let R_1 and R_2 be the reflection coefficients determined experimentally at the two selected angles i_1 and i_2. Two equations in the unknowns n and k can then be written:

$$R_1=f_1(n, k),$$
$$R_2=f_2(n, k).$$

The pair of values of n and k which can be obtained by solving these equations simultaneously, are the optical constants of the substance under investigation.

It is the solution of these equations which is performed graphically. Each equation defines a relation between n and k which can be plotted as a curve in the (n, k) plane. This is illustrated for a special case by any pair of the curves of Fig. 1. The point where these two curves intersect gives the values of n and k which satisfy the equations. It is a tedious matter, however, to plot these curves since the algebraic relation cannot be solved explicitly for either n or k. In order to avoid the necessity of calculating and plotting the curves every time new reflection coefficients are determined, families of such curves have been plotted to take care of the cases so far observed in the extreme ultraviolet. These families of "constant reflection coefficient curves" are reproduced in Figs. 2–5. They are calculated for the case of unpolarized incident light. The angles chosen at which to compute the reflection coefficients were 45°, 60°, 75°, and 85° incidence. Each family of curves refers to one of these angles and is plotted over a wide range of the parameter R. Thus each curve in a family gives those pairs of values of n and k which will yield the particular reflection coefficient belonging to this curve. A family for normal incidence is not included since its calculation is quite simple.

To show how these curves can be used let us take a particular case. Suppose reflection coefficients of 0.250 at 45° and 0.39 at 75° are determined experimentally for some substance at a particular wave-length. To find n and k for this case it is necessary only to select from the 45° and 75° families, interpolating if necessary, the two curves for $R_{45°}=0.250$ and $R_{75°}=0.39$ and find where they intersect when plotted together. This particular example is shown in Fig. 1, and we see that the point of intersection, A, gives $n=2.17$ and $k=1.175$. All this can be done very rapidly.

The accuracy of this graphical method of determining n and k depends mainly on the precision of the curves of Figs. 2–5. These curves have been carefully computed and plotted. It is believed that their locations as reproduced are nowhere in error by more than 0.3 mm, measured along the normal, in general however, by much less. This distance corresponds to about 0.005 in k or 0.01 in n. Throughout most of the range of n and k covered the experimental accuracy usually attainable in reflection coefficient measurements is not great enough to warrant larger and more accurate sets of curves. If more precise

[4] H. M. O'Bryan, J. Opt. Soc. Am. **22**, 739 (1932).
[5] R. Tousey, Phys. Rev. **50**, 1057 (1936).

values of n and k should be desired, the approximate values of n and k can easily be corrected, by computing the reflection coefficients from them by means of the general formula, and then by using the published curves as a basis for interpolation.

While reflection coefficients at only two angles are sufficient to determine n and k, and hence the whole curve of R against i, it is obvious that measurements of reflection coefficients at other angles should serve to increase the accuracy of this determination of n and k. As an illustration let us suppose that in the example previously given the following reflection coefficients are determined in addition to those for 45° and 75°; $R_{60°}=0.275$, $R_{85°}=0.66$. The curves for these values have been taken from the respective families and plotted on Fig. 1 along with those for $R_{45°}$ and $R_{75°}$. It is seen that they do not intersect in a single point as they should if no errors existed. But they do set limits to the likely values of n and k, and it is also possible to select "most probable" values of n and k such that the average difference, for all the angles used, between the computed reflection coefficients and the experimental ones is a minimum. In this case point P gives the "most probable" values of n and k; $n=2.17$, $k=1.165$.

Polarization

If the curves taken from the respective families do not intersect in a common point, and it is certain that this is not caused by either experimental error or a rough surface, then the degree of polarization must be different from that assumed in calculating the curves. One could then proceed by starting with other states of polarization and repeating the whole process until a common intersection is obtained. The reflection coefficient at normal incidence would be helpful in following this procedure since it is independent of the state of polarization.

The polarization of the incident light can be determined fairly easily, however, provided a substance which is transparent for the spectral range involved can be found. For this purpose reflection coefficient measurements must be made on this transparent substance for at least two angles of incidence. In this case $k=0$ and the general equations reduce to the form $R=f(i, n, p)$. From values of R for two angles, simultaneous equations are obtained just as before, but they are to be solved for n and p instead of n and k. The same graphical scheme for solution is very satisfactory for this purpose but since the curves are comparatively easy to compute, they are not included here. As before, measurements of R for additional angles serve to increase the accuracy of the result.

Thus from specular reflection measurements on any isotropic transparent substance the state of polarization of the incident light can be determined. Then curves such as those of Figs. 2–5 can be calculated for this amount of polarization and these should prove satisfactory for obtaining n and k for opaque materials provided that the same source of light is used. This procedure was employed by the author[5] in the extreme ultraviolet to check the fact that in his spectrograph the dispersed light was unpolarized.

The curves for unpolarized light as given in Figs. 2–5 can be used, however, regardless of the state of polarization of the incident light, provided it is possible to rotate the mirror surface under investigation about the line of the incident light as axis. If the reflection coefficients are determined at the same angle of incidence for two such positions of the mirror, 90° apart, the average value gives the reflection coefficient which would be obtained in case the incident light were unpolarized. The curves of Figs. 2–5 can then be used directly to obtain n and k from these average reflection coefficients.

If a common point of intersection still does not exist to within experimental error after the correct degree of polarization has been determined, or if the several curves for a transparent surface do not intersect in a common point, there must be present on the specimen some sort of surface layer, or its surface must not be perfectly smooth. An "ideal surface" was assumed in the derivation of the general reflection equations from the electromagnetic theory and if the test surface departs appreciably from this condition these equations cannot properly describe its behavior.

This work was begun while the author was at Harvard University. He is very grateful to Professor Theodore Lyman for encouragement in the preparation of this paper.

Intracavity polarimeter for measuring small optical anisotropies

Steven C. Read, Ming Lai, T. Cave, Stephen W. Morris, D. Shelton, A. Guest, and A. D. May

Department of Physics, University of Toronto, Toronto M5S 1A7, Canada

Received April 5, 1988; accepted May 26, 1988

An intracavity polarimeter for measuring small optical anisotropies is described. It is based on an optically pumped sodium dimer ring laser that supports two counterpropagating, orthogonally polarized modes. The optical anisotropy of an intracavity sample is detected by its effect on the beat frequency between the two modes. An analysis of the noise shows that a phase-retardation sensitivity of 6×10^{-8} rad can be achieved. A preliminary measurement of the Kerr effect in CO_2 is reported.

INTRODUCTION

A medium is said to be optically anisotropic when its properties depend on the polarization of the incident light. Optical anisotropies, such as birefringence, optical activity, and Faraday rotation, can occur naturally in a material or be induced by external forces or fields. These effects are conventionally measured by observing changes in the polarization of radiation that has traversed a sample.[1] In the case of field-induced effects, such measurements are often made relative to a standard such as nitrobenzene or carbon disulfide.[2] Typical reported[3–5] phase retardations for electric-field-induced birefringence (the Kerr effect) are of the order of 5×10^{-7} rad. It is of interest to improve this sensitivity to study small effects such as the optical activity associated with parity violation in atoms, χ^3 anisotropies, and optical anisotropies in thin films. It is also desirable to make absolute measurements of various anisotropies in systems simple enough to permit a quantitative comparison with theory. In this paper we describe a new polarimeter that is both absolute and sensitive. It is based on a dual-polarization laser with an intracavity cell.

The technique is simple in principle. The pioneering work of Doyle and White[6] showed that an anisotropic cavity can support two mutually orthogonal polarization modes. The difference in frequency between the two modes is a direct measure of the anisotropy of the cavity and can be determined by beating the two together with a square-law detector. If an anisotropic sample is introduced into the cavity or if an additional anisotropy is induced in a sample already present in the cavity, the *change* in the beat frequency is a measure of this extra anisotropy. This is the feature that we exploit in our polarimeter. In practice one must take into account effects of the gain medium, the residual anisotropy of the optical components, and whether the extra anisotropy alters the polarization modes of the original cavity. We have found that these complicating effects can be minimized by using an optically pumped sodium dimer ring laser[7–10] and have used such a laser to construct and test a polarimeter with a sensitivity of 6×10^{-8} rad. The potential exists to extend this sensitivity further by several orders of magnitude.

The paper begins with a discussion of design criteria and a description of the experimental apparatus. Next the sensitivity of the technique is estimated by analyzing fluctuations in the beat frequency that are due to noise. We conclude with a preliminary measurement of the dc Kerr effect in CO_2 and a summary.

APPARATUS

Figure 1 shows a schematic layout of the apparatus. Since the setup is rather complex, we preface the description of the apparatus with a discussion of the properties of an anisotropic cavity and optically pumped sodium that led us to the design shown in the figure.

Consider a sodium dimer ring laser that is longitudinally pumped by a single-frequency argon-ion laser. It is well known[11] that for many dimer lasers the gain results from a resonant Raman- or Λ-type transition, as shown in Fig. 2. In this figure levels 1 and 2 are occupied and unoccupied rovibrational levels of the ground electronic state, respectively. In our system, in which sodium is pumped by blue-green light, level 3 belongs to the excited *B* state. For such Raman processes, the amplification profile in the forward direction, i.e., in the direction of the pump beam, is sharply peaked, while that in the backward direction is broad and low.[11] Consequently, longitudinal optical pumping of a ring dimer laser can produce unidirectional laser operation in the same direction as the pump beam. For sodium, the effective Doppler width in the forward direction is comparable with the spectral free range of our 3.2-m ring cavity, and one can obtain a single longitudinal mode without additional frequency-selective intracavity elements. Furthermore, lasing can be restricted to the lowest-order transverse mode by properly matching the pump-beam profile to the TEM_{00} mode of the ring-laser cavity. In the following paragraph we examine the polarization of such a mode.

The residual optical anisotropy of the ring laser must be known if one is to predict the polarization modes of the bare laser, i.e., without the sample present. As is shown in Fig. 1, the windows of the sodium gain cell are perpendicular to the optic axis of the ring cavity, so that orthogonally polarized

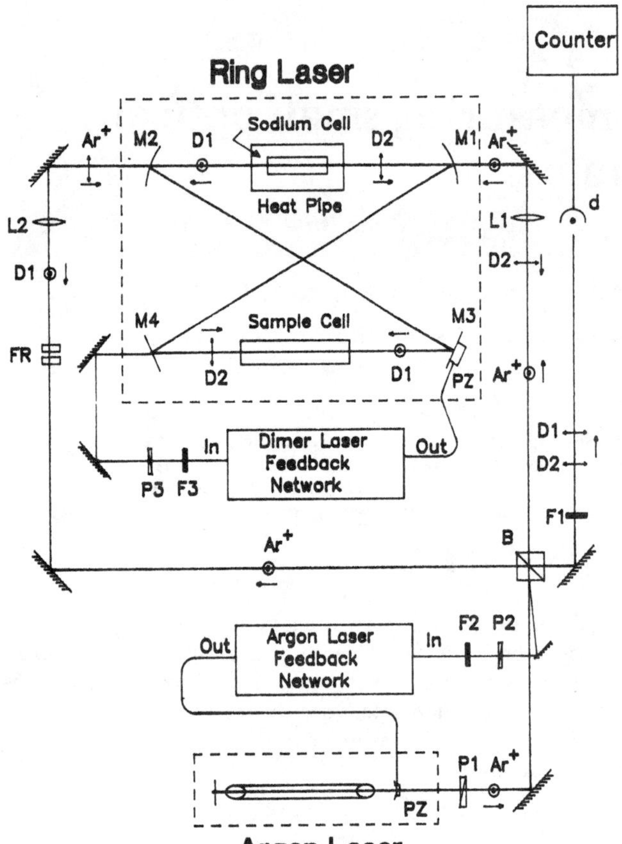

Fig. 1. Schematic diagram of the intracavity polarimeter. Abbreviations are defined in text.

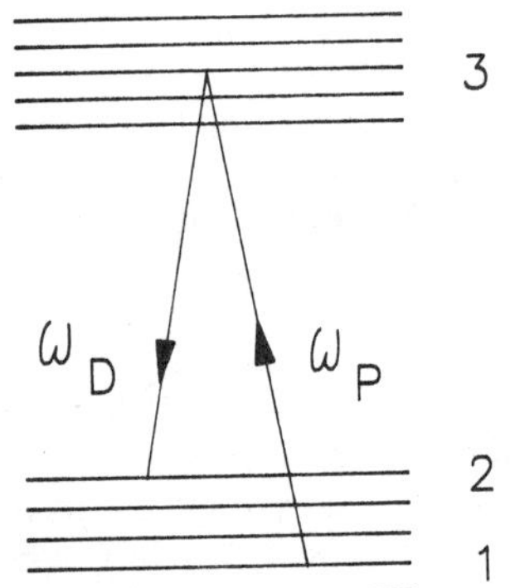

Fig. 2. Energy-level diagram of an optically pumped dimer laser. For our apparatus, ω_P is the frequency of the Ar^+ pump beam and ω_D is the frequency of the Na_2 laser.

modes will experience identical losses. The use of thin windows and the absence of étalons, optical diodes, etc. in the cavity ensures that residual strain birefringences play a negligible role in determining the polarization modes of the cavity. We find experimentally and have confirmed by calculations[7] that dominant source of anisotropy in the ring cavity is the phase shifts on reflection at the mirrors; *s*- and *p*-polarized beams experience different phase shifts for reflections at nonnormal incidence. Consequently, the modes of the bare laser are linearly polarized parallel and perpendicular to the plane of the ring. We have found that either of these polarization modes can be selectively excited by pumping with light of that polarization. This implies that the gain for light polarized parallel to the pump polarization is greater than for the orthogonal polarization. In summary, the pump beam can be adjusted such that a single-longitudinal, single-transverse, single-polarization mode with a single direction of circulation is excited in the ring laser. *We pump bidirectionally with orthogonally polarized beams whose polarizations are aligned with the modes of the ring cavity and thus obtain two dimer laser beams counterpropagating around the ring that themselves have orthogonal polarizations.*

Now consider what happens when a birefringent sample is inserted into the cavity or when a birefringence is induced in a sample by applying an external field. We shall describe a simple plane-wave theory for a ring laser with a cavity of length L and index n containing an intracavity sample cell of length l with principal indices n_1 and n_2. The cell is oriented in the cavity such that its principal axes are aligned with the polarization modes of the bare laser. This leaves the polarization modes the same as those of the bare laser. The cavity resonance condition for mode 1 is $2\pi m_1 = (2\pi\nu_1/c)[n(L - l) + n_1 l] + \phi_1$, where m_1 is the longitudinal mode number and ϕ_1 the total phase shift at the mirrors. Thus the frequency difference, or beat frequency, between the two polarization modes is given by

$$\nu_b = \nu_2 - \nu_1 = \frac{c(m_2 - \phi_2/2\pi)}{n(L - l) + n_2 l} - \frac{c(m_1 - \phi_1/2\pi)}{n(L - l) + n_1 l}. \qquad (1)$$

We are interested in cases in which the birefringence, $\Delta n = n_1 - n_2$, is so small that ν_b is much less than the spectral free range of the cavity. In that case the counterpropagating beams have the same longitudinal mode number, and Eq. (1) becomes

$$\nu_b - \nu_b^0 = f\nu\Delta n, \qquad (2)$$

where ν_b^0 is the bare-laser beat frequency, i.e., the beat frequency for $n_1 = n_2$. The filling factor, f, is l divided by the total optical length of the cavity. Equation (2) is the basic equation relating the measured change in the beat frequency to the optical anisotropy.

If dispersion of the gain medium can be neglected, fluctuations in the cavity length of the ring laser affect the frequency of both counterpropagating modes equally. This common mode rejection leads to a beat frequency that is insensitive to changes in the mean frequency of the laser arising from mechanical and thermal variations in the cavity length. Assuming a beat-frequency resolution of 1 Hz and a filling factor of the order of 1, it follows from Eq. (2) that the polarimeter is sensitive to birefringences of the order of 10^{-15}.

The above discussion was couched in terms of linearly polarized modes and birefringence. With the addition of quarter-wave plates to the cavity, one could obtain left-and or right-circularly polarized light in the region of the sample cell, and one could then observe either the Faraday effect or optical activity.

With this background, we turn now to consider the details shown in Fig. 1. The ring-laser cavity, formed by mirrors M1–M4 in a bow-tie configuration, contains both the sodium and the sample cell. Heating of the dimer cell was provided by a double heat pipe,[12] a design that yields a homogeneous temperature and density distribution of the sodium vapor. We find that optimum performance of the ring laser occurs with the argon buffer gas at a pressure of 3 to 5 Torr, which, for proper heat-pipe operation, requires a temperature in the range of 430 to 490°C. The bidirectional pumping was achieved by splitting the output beam, labeled Ar^+ in Fig. 1, of a single-frequency argon-ion laser with a 50/50 beam splitter, B. Lenses L1 and L2 were adjusted to match the Gaussian profiles of the pump beams to the profile of the TEM_{00} modes of the ring-laser cavity. The output polarization of the argon laser was originally perpendicular to the plane of the ring laser. The polarization of one pump beam was rotated by 90° by two Fresnel rhombs, FR. As indicated above, this pumping scheme excites two orthogonally polarized, single-frequency, counterpropagating modes, D1 and D2, inside the ring-laser cavity.

In order to measure the beat frequency we must combine beams D1 and D2 on a detector, such that their polarizations are the same. The two outputs transmitted by the dichroic mirrors M1 and M2 retrace the path of the two argon pump-laser beams. Beam 1, which was vertically polarized in the ring cavity, becomes horizontally polarized and parallel to the polarization of beam 2 on passing through the Fresnel rhombs. The two beams are combined at the beam splitter and directed onto d, an RCA Model C30950E avalanche photodiode. This detector has a sensitivity of 0.2 V/μW and a bandwidth of 50 MHz. The frequency of the beat signal was measured with an HP5316A frequency counter interfaced to a computer.

The experimental apparatus just described would be complete if dispersion in the sodium gain medium could be neglected. However, fluctuations in the differential mode pulling by the gain medium limit the stability of the beat frequency and hence the sensitivity of the apparatus. Our estimates,[8] based on the data of Man and Brillet,[13] show that fluctuations in the differential pulling can cause variations in the beat frequency of some tens of kilohertz. These fluctuations are due to changes of the position of the modes within the gain profile and changes in the magnitude of the gain due to variations in the pump intensity. Several steps were taken to reduce the fluctuations in the beat frequency. Fluctuations caused by acoustic and thermal noise sources were reduced by enclosing both lasers in acoustically shielded boxes. The stability of the beat frequency was further enhanced by locking the frequency of the argon and the dimer lasers to external reference cavities, by stabilizing the intensity of the argon laser, and by matching the gains of the ring laser through the alignment of the pump-laser beams. Each reference cavity had a finesse of 30 and was housed in a pressure-tight, temperature-stabilized container. Alignment of the laser beams to the reference cavities was purposely off axis to reduce the coupling of the reflected beams to the rest of the system. Frequency fluctuations in both lasers were reduced to less then 1 MHz over a period of 10 sec by a feedback system and piezoelectric (PZ) crystals, which were used to adjust the lengths of the lasers. The polarizers P1–P3 and filters F1–F3 helped to isolate optically the different sections of the system and thus also contributed to the stability of the polarimeter.[7]

The pump source was a single-frequency Spectra-Physics Model 52 Ar^+ laser. Intensity fluctuations in the frequency range 0.01 to 10 Hz were less than 1%. In order to achieve the required frequency stability with our simple feedback system, we modified the laser. The plasma tube, along with its magnet and water-cooling supply, was suspended by wires. The mirrors and wavelength-selective elements were mounted directly on a vibration isolation table. This partly isolates the acoustic noise associated with the cooling water from the Invar rods that define the length of the cavity.

There are approximately 45 laser transitions in sodium when it is pumped with an argon laser. We find experimentally that it is easiest to obtain stable single-line, single-frequency operation of the ring laser by pumping with the 457.9-nm line of the argon laser while lasing on the 560-nm line of sodium. For a pump-laser power in the 10-mW range, the output power of the dimer laser is of the order of 10 μW. Operating with the widest possible wavelength separation between the pump and dimer lasers results in better stability because it lowers the finesse of the ring cavity at the pump frequency. A lower finesse means there is less resonant enhancement of the pump intensity inside the ring cavity. Thus a variation in the length of the ring cavity produces a smaller change in the stored pump power and concomitantly a smaller change in the amplitude of the gain.

We mentioned above that the largest source of anisotropy is the phase shift at the mirrors. This leads to a beat frequency $\nu_b{}^0$ of 2.7 MHz for the bare laser. It is advantageous to decrease this intrinsic beat frequency. We know from theory[14] that a smaller beat frequency should lead to a reduction in differential mode pulling. In addition, the frequency counter used in this experiment gave six digits of resolution. As a result, beat frequencies greater than 1 MHz would limit the resolution to 10 Hz or to a birefringence of 10^{-14} for our polarimeter. The beat frequency was reduced below 1 MHz by introducing an antireflection-coated optical flat into the cavity. The flat was squeezed to produce a phase shift due to strain birefringence, which partially compensated for the phase shift at the mirrors.

For Eq. (2) to remain valid, the intracavity cell must be aligned with the polarization modes of the cavity, and thus it is necessary to maintain stable and well-defined polarization modes. For this reason, care was taken not to overcompensate for the phase shift due to the mirrors. Overcompensation occurs when other sources of birefringence in the cavity become important enough significantly to alter the polarization modes of the cavity. If the beat frequency were reduced to zero, the laser would be effectively isotropic. Stephan and Hugon[15] have shown that in such lasers the polarization modes are extremely sensitive to stray effects. It was found that the stability of the beat frequency improved as the frequency was decreased to approximately 10 kHz, where the laser became unstable. As a compromise, the laser was operated with the bare-laser beat frequency reduced to ~200 kHz, which resulted in slightly improved stability and permitted a resolution of 1 Hz on the frequency counter.

Finally, it is known that the homogeneous width of the gain curve is ~10 MHz,[13] and thus the observed stability at values of the mode splitting much smaller than 10 MHz requires an explanation. We interpret the lack of mode competition as being due to velocity selection. For pump frequencies off the center of the absorption line, the resonant Raman condition preferentially selects a specific class of sodium dimers with nonzero velocity along the direction of the pump. For bidirectional pumping, therefore, the gain of each dimer laser mode is obtained from separate countermoving groups of molecules, and one expects little mode competition. Coupling of the modes by means of scattering at the mirrors does not appear to be a significant. This concludes the description of the intracavity polarimeter. In the next section we determine the sensitivity of the apparatus under the conditions described above.

BEAT-FREQUENCY STABILITY

In the experiment described below we wish to measure the change in the beat frequency on application of a static electric field to an intracavity sample. Thus it is the variation of the beat frequency that limits the precision with which the change in frequency may be measured. The optimum conditions under which to carry out a measurement will depend on the exact nature of the variations, and consequently an essential preliminary measurement is the characterization of the temporal behavior of the beat frequency.

Figure 3 shows the variation in the beat frequency, $\nu_b{}^0$, over a single 3-min time interval. We see that the variations are mostly deterministic,[16] i.e., $\nu_b{}^0$ is an irregular yet continuous function of time. Deterministic noise arises from variations in the length of the cavity, from fluctuations in the pump power, etc. Apart from drift, we see from the figure that $\nu_b{}^0$ has a mean excursion of ~1 kHz, which is much greater than the targeted 1-Hz precision. Because the noise is deterministic, for the Kerr effect one can employ rapid modulation of the field so that the variation in $\nu_b{}^0$ is small over one cycle of modulation. When differences are taken between successive field-on and field-off frequency measurements there is a large cancellation of the noise. The optimum modulation frequency can be chosen by considering the Allan variance[16] of $\nu_b{}^0$.

The Allan variance, σ^2, is often used to characterize frequency stability. For N consecutive measurements, each of duration τ, σ^2 is given by $\sigma^2(\tau) = \Sigma\{y[(n+1)\tau] - y(n\tau)\}^2/2(N-1)$, where $y(n\tau)$ is the deviation of the frequency of the nth measurement from the mean frequency, divided by the mean frequency. The Allan variance is a measure of the fractional average fluctuation from one reading to the next contiguous reading. The minimum value of $\sigma^2(\tau)$, as a function of τ, gives the most suitable modulation frequency $(1/\tau)$ for phase-sensitive detection. The Allan variance, as defined above, assumes zero dead time between measurements. We will neglect the presence of a nonzero dead time in our measurements, as we do not require a precise characterization of the noise.

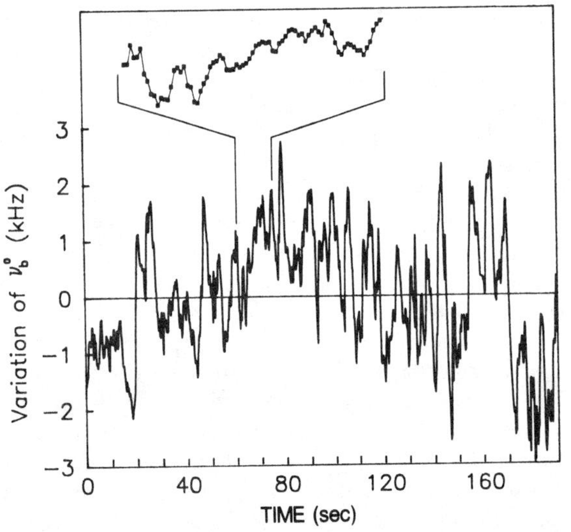

Fig. 3. Measured variation in the beat frequency of the bare laser, over a 3-min time interval. The mean beat frequency was approximately 400 kHz. Successive data points were measured for an integration time of 100 msec and separated by a 90-msec dead time. The expanded section shows the variation over a 15-sec period.

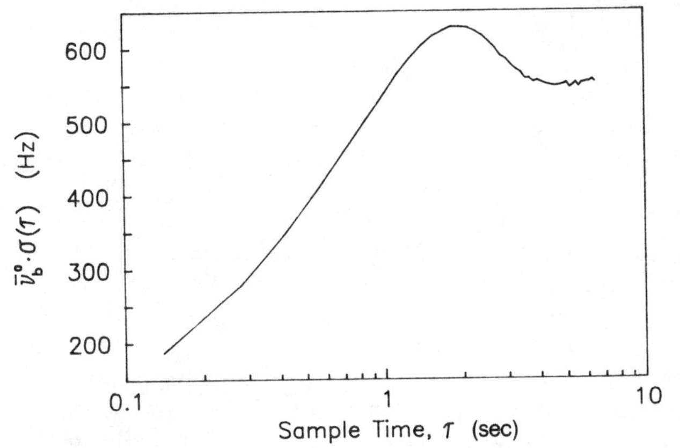

Fig. 4. Plot of the average beat frequency $\nu_b{}^0$ times the square root $\sigma(\tau)$ of the Allan variance as a function of sampling time. This is a measure of the stability or rate of change of the beat frequency at various sampling times.

The Allan variance as a function of time was computed for the data of Fig. 3. For each point the integration time, τ_c, was 100 msec, and the dead time was 90 msec. These data, consisting of 1000 consecutive measurements of the beat frequency, were lumped into sets of equal size, m. The average frequency of each of the $2000/m$ sets was used as a new set of frequencies corresponding to a measurement time of length $m\tau c$. The square root of the Allan variance, $\sigma(m\tau c)$, was then calculated for this new data set. This procedure was one of convenience, since a single record of the frequency could be used to construct an Allan variance for a large range of $\tau = m\tau_c$ simply by using a computer and altering the choice of m.

Figure 4 shows the results of these calculations, where we have plotted the mean frequency $\bar{\nu}_b{}^0$ times $\sigma(\tau)$. The approximate asymptotic value of 600 Hz at long times reflects the mean frequency excursions apparent in Fig. 3. We are interested in the short-time behavior of $\bar{\nu}_b{}^0\sigma(\tau)$, where the minimum of approximately 190 Hz occurs at the shortest sampling time. Other data sets showed various minimum values of $\bar{\nu}_b{}^0\sigma(\tau)$, but the rapidly decreasing value at short times was characteristic of all Allen variance plots. This indicates that it would be desirable to have still higher modulation frequencies because the minimum in $\sigma(\tau)$ has not been reached.

The potential sensitivity of the apparatus was calculated by analyzing data, like those of Fig. 3, without modulating the anisotropy of the intracavity cell. One simply pretends that consecutive readings correspond to field on, field off. As there is no real field, the change in the beat frequency will be zero within the noise or sensitivity of the apparatus. On a particularly good night, when the acoustic noise in the building was low, the average deviation was found to be 36 Hz for a data run consisting of 2000 points sampled at the maximum rate. This translates into a sensitivity of 1.2 Hz, approaching the limit of 1 Hz set by the frequency counter. A resolution of 1.2 Hz corresponds to a phase shift of 6×10^{-8} rad or, for a large filling factor, to a birefringence of 2×10^{-15}. For the modest filling factor used in our preliminary experiment, the sensitivity was quite sufficient to permit measurement of the Kerr constant of CO_2 at low pressures.

TEST OF THE APPARATUS: THE KERR CONSTANT OF CO_2

The Kerr effect was discovered by the Reverend John Kerr in 1875. While Kerr expressed his results in terms of a phase shift, the length of the sample, and the electric-field strength, it is now common practice to define a Kerr constant on a molecular basis. Like Buckingham and Orr,[17] we define the low-density molar constant A_K by

$$A_K = \frac{2}{27}\frac{\Delta n}{\rho E^2}, \tag{3}$$

where Δn is the difference between the index of refraction of light polarized parallel and perpendicular to the applied field E and ρ is the molar density. For our polarimeter, Eq. (2) directly relates a measured change in beat frequency to Δn, and thus we can determine A_K, provided that we also know or measure E, ρ, and the filling factor f.

The intracavity Kerr cell was a 26-cm stainless-steel cylinder closed on the ends by two antireflection-coated fused-silica windows 3 mm thick. It was designed for use up to 50 atm. A uniform electric field was supplied by two parallel metal plates separated by a distance of 3.80 ± 0.03 mm. The voltage applied to the Kerr cell was square-wave modulated with a peak voltage of 1.91 ± 0.01 kV, using a Kepco OPS2000 power supply under the control of the computer. This corresponds to an electric field E of 503 ± 4 kV/m.

Matheson Aquarator grade CO_2, with impurities less than 21 parts in 10^6, was used to fill the cell to pressures ranging from 1 to 50 atm. The cell was evacuated to 100 mTorr and filled twice to 50 atm to purge impurities. The temperature of the cell was 25.0 ± 0.5°C, and pressures were measured using a calibrated Bourdon gauge. The pressures were converted to molar densities, ρ, using thermodynamic data.[18]

To calculate f we must know the length of the Kerr cell and the optical length of the laser cavity. The index of refraction of air, CO_2, and the sodium gain medium was taken to be unity. The five fused-silica windows ($n = 1.46$) inside the cavity occupied a total length of 1.5 cm, while the physical length of the cavity was measured to be 322 ± 2 cm. This gives an optical length of 323 ± 2 cm for the cavity. The length of the electrodes in the Kerr cell was 20.00 ± 0.05 cm. The effective length over which the electric field was applied is somewhat longer than the plates, owing to the fringing fields at both ends. We solved Laplace's equation numerically and calculated the effective length of the cell by using the equation

$$l = \frac{2\int_0^\infty E^2(x)dx}{E^2(0)}. \tag{4}$$

Here E is the component of the electric field perpendicular to the axis of the cell and the field plates. The result is $l = 20.14 \pm 0.05$ cm, which yields a filling factor, f, equal to 0.0625 ± 0.0004.

The change in the beat frequency was measured at several pressures ranging from 1 to 50 atm. The modulation frequency was 3.5 Hz, i.e., field on for 1/7 sec, field off for 1/7 sec, a frequency determined by the maximum rate of data transmission to the computer. Figure 5 shows the change in the beat frequency, $\nu_b(E) - \nu_b^0$, due to the Kerr effect, as a function of molar density. An error-weighted least-squares procedure was used to fit the data to a straight line, giving a slope of 0.274 ± 0.002 kHz m^3/kg-mol and an intercept of 1.6 ± 3.4 Hz. The line passes through the origin within experimental error, as expected. Combining Eq. (2) and Eq. (3) with the slope of the line in Fig. 5 and the values of ν, f, and E yields the value of $(2.40 \pm 0.05) \times 10^{-27}$ C^2 m^5/J^2 for the molar Kerr constant of CO_2.

Previous measurements of the Kerr effect in CO_2 include the work of Quarles[19] and Buckingham *et al.*[20] We deduce from the early work of Quarles that the dispersion in the molar Kerr constant is small over the visible spectrum. We can therefore compare our results obtained at 560 nm with the measurements of Buckingham obtained at 633 nm. Interpolating Buckingham's measurements of A_K, which were made over a range of temperatures, we find that $A_K = 2.78 \times 10^{-27}$ C^2 m^5/J^2 at 298 K. This is 14% higher than our result. Buckingham claims an absolute accuracy of ±5% arising from difficulties in calibration, while our estimated random error is ±2%. There is a systematic difference between the two results, but we do not consider the discrepancy to be serious for an initial experiment.

The most probable source of systematic error is associated with matching the principal axes of the field-induced birefringence to the axes of the polarization modes of the cavity. Well after the experiment was performed we reexamined the polarization modes as the beat frequency was reduced with the optical flat. It was found that the polarization modes occasionally rotated as much as 10° when the beat frequency was reduced to values typical of those used in the experiment. We have estimated[7] the systematic error that such a rotation would have and find that it would decrease the change in the beat frequency or Kerr constant by some 15%. A misalignment of the Kerr cell would have a similar effect.

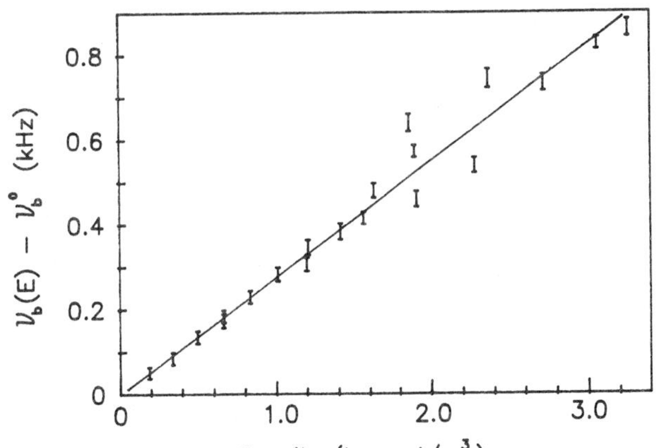

Fig. 5. Density dependence of the change in the beat frequency due to the Kerr effect in CO_2. The maximum density corresponds to a cell pressure of approximately 50 atm.

Simple experimental steps can be taken in future experiments to eliminate such systematic errors.

There are a number of possible ways to decrease the random error or to increase our signal-to-noise ratio. The small filling factor of 0.065 was a consequence of using a cell that was on hand from a previous experiment. Filling factors of close to 1/2 are possible and would increase the signal by an order of magnitude. The Allan variance measurements reported above indicated that higher modulation frequencies are needed to reduce the noise. We have made some preliminary noise measurements with a phase-locked loop to convert the beat frequency to a voltage. This permits modulation frequencies of up to 100 kHz, rather than the 3.5 Hz used here, and should increase the signal-to-noise ratio by another 2 orders of magnitude.

SUMMARY AND CONCLUSIONS

In this paper we have described the development and testing of a sensitive intracavity polarimeter based on a optically pumped sodium dimer laser. The sensitivity of the polarimeter is of the order of 6×10^{-8} rad. An analysis of the noise of the beat frequency indicates that higher modulation frequencies should permit us to increase the sensitivity significantly. The absolute Kerr constant of CO_2 was measured with an estimated random error of ±2%. Even at the present level of sensitivity, both the dc and the optical Kerr constants, including Raman resonances, should be measurable absolutely in dilute systems.

ACKNOWLEDGMENTS

We gratefully acknowledge the fine technical support of the physics department machine shop and the financial support of the Natural Sciences and Engineering Research Council of Canada.

A. D. May is also a principal investigator, Ontario Laser and Lightwave Research Centre, University of Toronto.

REFERENCES AND NOTES

1. C. G. Lefevre and R. J. W. Lefevre, "The Kerr effect," in *Techniques of Chemistry*, A. Weissberger, ed. (Wiley Interscience, New York, 1972), Vol. 1, Part 3C, pp. 399–427.
2. A. D. Buckingham and D. A. Dunmur, "Kerr effect in inert gases and sulphur hexafluoride," Trans. Faraday Soc., **64**, 1776 (1968).
3. A. K. Burnham, L. W. Buxton, and W. H. Flygare, "Kerr constants, depolarization ratios, and hyperpolarizabilities of substituted methanes," J. Chem. Phys. **67**, 4990 (1977).
4. A. D. Buckingham and H. Sutter, "Gas phase measurements of the Kerr effect in some *n*-alkanes and cyclohexane," J. Chem. Phys. **64**, 364 (1976).
5. One of us has recently shown that it is possible to achieve the shot-noise limit of the conventional technique [D. P. Shelton and R. E. Cameron, Rev. Sci. Instrum. (to be published)]. The fundamental sensitivity of the intracavity technique is better and will be discussed in a future publication.
6. W. M. Doyle and M. B. White, "Generation of combination tones by the interaction of orthogonal oscillations in a gas laser," Appl. Phys. Lett. **10**, 224 (1967).
7. S. C. Read, "Measurement of the Kerr effect in carbon dioxide using intracavity polarimetry," M.Sc. thesis (University of Toronto, Toronto, Canada, 1987).
8. Lai Ming, "Frequency stabilization of an argon ion laser and a sodium dimer ring laser for use in optical anisotropy measurements," M.Sc. thesis (University of Toronto, Toronto, Canada, 1985).
9. T. B. Cave, "An investigation of the sodium dimer ring laser for use in optical anisotropy measurements," M.Sc. thesis (University of Toronto, Toronto, Canada, 1983).
10. A. D. Guest, "Dispersion measurements with a dual-frequency infrared helium–neon laser for use in optical anisotropy measurements," M.Sc. Thesis (University of Toronto, Toronto, Canada, 1981).
11. B. Wellegehausen, "Optically pumped cw dimer lasers," IEEE J. Quantum Electron. **QE-15**, 1108 (1979).
12. H. Scheingraber and C. R. Vidal, "Heat pipe oven of well-defined column density," Rev. Sci. Instrum. **52**, 1010 (1981).
13. C. N. Man and Brillet, "Gain lineshapes in optically pumped dimer lasers," Opt. Commun. **45**, 95 (1983).
14. W. R. Bennett, "Gaseous optical masers," Phys. Rev. **126**, 580 (1962).
15. G. Stephan and D. Hugon, "Light polarization of a quasi-isotropic laser with optical feedback," Phys. Rev. Lett. **55**, 703 (1985).
16. J. A. Barnes, A. R. Chi, L. S. Cutler, D. J. Healy, D. B. Leeson, T. E. Mcgunigal, J. A. Mullen, Jr., W. L. Smith, R. L. Sydnor, R. F. C. Vessot, and G. M. R. Winkler, "Characterization of frequency stability," IEEE Trans. Instrum. Meas. **IM-20**, 105 (1971).
17. A. D. Buckingham and B. J. Orr, "Electric birefringence in molecular hydrogen," Proc. R. Soc. London Ser. A **305**, 259 (1968).
18. J. H. Dymond and E. B. Smith, *The Virial Coefficients of Gases: A Critical Compilation* (Clarendon, Oxford, 1980), p. 37.
19. G. G. Quarles, "The dispersion of the electro-optical Kerr effect in carbon dioxide," Phys. Rev. **46**, 692 (1934).
20. A. D. Buckingham, M. P. Bogaard, D. A. Dunmur, C. P. Hobbs, and B. J. Orr, "Kerr effect in some simple non-dipolar gases," Trans. Faraday Soc. **66**, 1548 (1970).

An Automatic Retardation Meter for Automatic Polarimetry by Means of an ADP Polarization Modulator

Hiroshi Takasaki, Mitsuo Isobe, Toshinobu Masaki, Atsuo Konda, Tadao Agatsuma, and Yasushi Watanabe

The construction of a new automatic retardation meter with an ADP polarization modulator and some results obtained with the apparatus are reported. An ADP solid Kerr cell which is subjected to an alternating voltage is inserted in an optical system which measures birefringence, and the retardation of a test piece is modulated. The modulated signal is amplified and fed to a servomotor, which controls the retardation compensation to get extinction. The retardation of the test piece is obtained from the amount of compensation. A practical apparatus, which is suited to study the characteristics of fibers, is made, and the effects of heat treatment and recrystallization of some synthetic fibers are measured.

I. Introduction

The birefringence measurement of some synthetic materials gives new means of studying properties of the materials.[1–4] The design of a new instrument for automatic measurement of retardation is based on a series of our studies,[5–9] which concern a new polarimetry with ADP (ammonium dihydrogen phosphate, $NH_4H_2PO_4$).

The ADP crystal has a strong electrooptical effect. Specifically, large retardation is induced when an electric field is applied on the crystal. Thus an ADP crystal unit subjected to an alternating electric field naturally makes a modulator of retardation. In addition to achieving the retardation modulation with the ADP crystal, we devised a method to vibrate the azimuth of polarized light with the crystal[5] and made an automatic polarimeter. The state of polarization can be modulated in two respects, retardation (or ellipticity) and azimuth.

The state of elliptically polarized light is defined by two parameters, ellipticity and azimuth, or retardation and amplitude ratio. Measurement of these two parameters has been usually made by controlling two polarization compensations to make the brightness of the field minimum. This operation is very tedious either by visual observation or by ordinary photoelectric observation.

H. Takasaki is with the Faculty of Engineering, University of Shizuoka, Hamamatsu, Japan. M. Isobe and T. Masaki are with Riken Keiki Fine Instrument Co. Azusawa, Itabashi-ku, Tokyo, Japan. A. Konda, T. Agatsuma, and Y. Watanabe are with the Textile Research Institution of the Japanese Government, Sawatari, Kanagawa-ku, Yokohama, Japan.

Received 5 August 1963.

We apply the two kinds of modulation simultaneously on the elliptically polarized light under observation and then separate the output signal into two signals corresponding to the modulations. In this way we made two kinds of universal polarimeters, the polarization goniophotometer which is useful for studying a thin film[6] and a universal birefringence meter which automatically determines retardation and orientation of a birefringent test piece.[7]

These methods of polarimetry were proved to be so accurate that some phenomena are automatically measured after conversion into polarization interference.[7–9] A shearing lens interferometer[7] and an instrument which measures the nutation of a carriage[9] are made following the idea. This paper discusses the simplest form of our new automatic polarimeter.

II. ADP and Its Applications

ADP is an artificially cultured transparent crystal. It is uniaxial under normal conditions but becomes biaxial when an electric field is applied.

Billings[10–12] and Carpenter[13] have made extensive studies on properties of the XH_2PO_4 type crystal and shown some of its applications. It is helpful to review here briefly the properties of ADP and its applications.

The optic axis of the ADP crystal is aligned with the Z-crystallographic axis in the normal state. It is split into two parts when an electric field is applied along the Z-crystallographic axis, and the crystal becomes biaxial. Each of the new optic axes makes the same angle to the Z-crystallographic axis and lies in a plane which bisects the X- and Y-crystallographic axes as shown in Fig. 1. If the polarity of the electric field

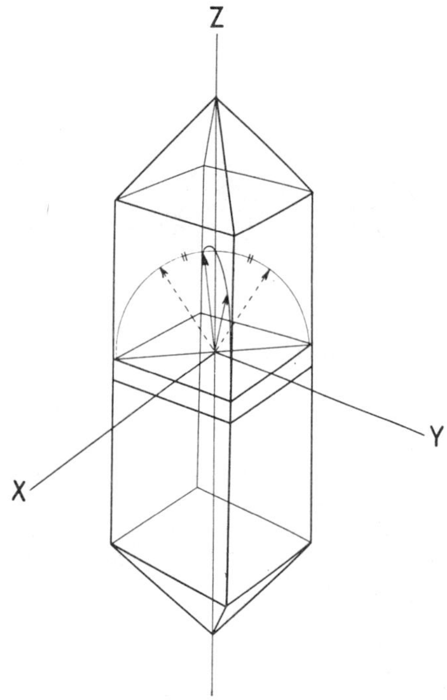

Fig. 1. The electrooptical characteristics of ADP crystal.

is reversed, the optic axis is split on another plane which is perpendicular to the former. This state is shown by the dashed arrows.

The conoscopic pattern of Z-cut ADP crystal viewed through crossed circular polarizers is the well-known concentric pattern as shown in Fig. 2(a). When the electrical field is applied along the Z direction, the crystal becomes biaxial, and the central dark spot is divided into two spots and a two-eyed pattern is observed as shown in Fig. 2(b). The direction of departure of the two spots rotates 90° with the reversal of the polarity of the electric field as shown in Fig. 2(c).

If the angle of view is restricted so as just to include the central dark spot of the pattern shown in Fig. 2(a), light is blocked in the normal state but is transmitted by applying electric voltage to the crystal.

The direction of the principal axis of the birefringence lies on the plane of the two axes at 45° to the X- and Y-crystallographic axes. The amount of retardation is proportional to the applied voltage (approximately a half-wavelength per 9.6 kV). Because of the reciprocal relation of thickness of crystal plate and the electric field, the amount of retardation is independent of the plate thickness and depends only on the voltage applied on the plate.

An electrical shutter and transmittance modulation using this crystal is thus possible. There are, however, two restrictions. First, since the electric field must be applied parallel to the direction in which the light travels, the electrodes must be transparent. The NESA glass or a metal grid satisfies this requirement. Second, strictly speaking, perfect extinction of light is obtained only for light collimated along the Z direction, consequently, the solid angle of light is limited.

Billings reported on the electrical tuning of the Lyot–Öhman filter with an ADP crystal.[14] Günter,[15] Billings,[11] ~~and~~ Fushimi[16] ~~studied an electrooptical shutter,~~ Namba and Wachi[17] studied an electrooptical shutter, Fujisawa,[18] and Namba and Ogawa[19] studied on the measurement of the high electric field, Fujimura and Ogawa[20,21] made a quite new sonagraph with an ADP piezooptical resonator, and light modulation was studied by Billings[11] and others.

We apply the crystal to automatic polarimetry. The application matched the properties of ADP as follows:

1. Polarimetric measurement is usually done using a parallel light beam, so that widening the effective aperture is unnecessary.

2. Small modulation depth is sufficient for automatic polarimetry. Consequently, the modulation voltage is small and there is no danger of electrical breakdown of the crystal.

3. Sinusoidal voltage of audio frequency is usually

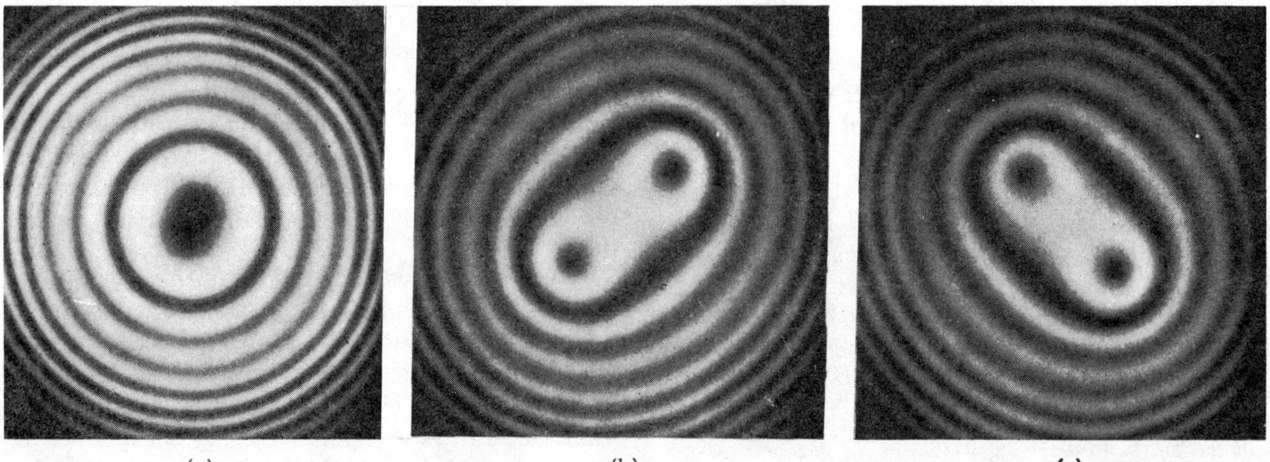

(a) (b) (c)

Fig. 2. Conoscopic patterns of ADP. (a) Without electric field; (b) +9.6 kV; (c) −9.6 kV.

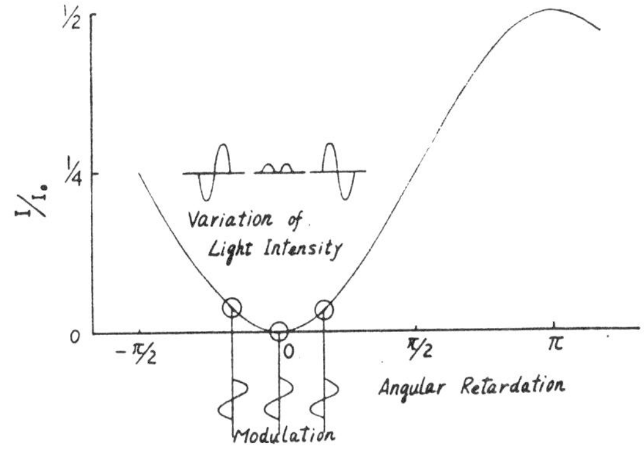

Fig. 3. The light transmittance with a retardation plate between a pair of crossed polarizers and the relationship between retardation modulation and error signal.

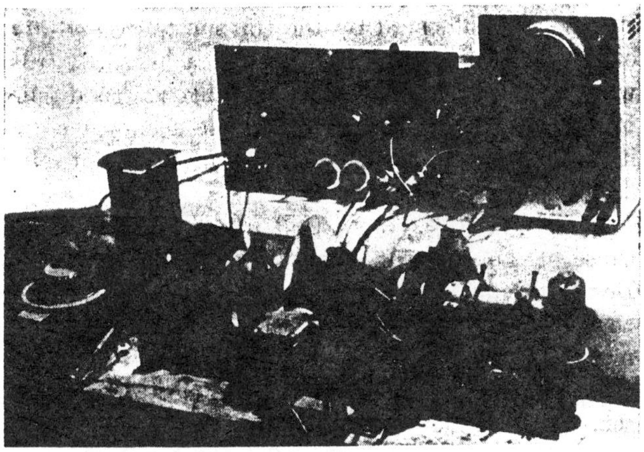

Fig. 4. General view of a tentative apparatus which measures birefringence of a test piece by direct compensation.

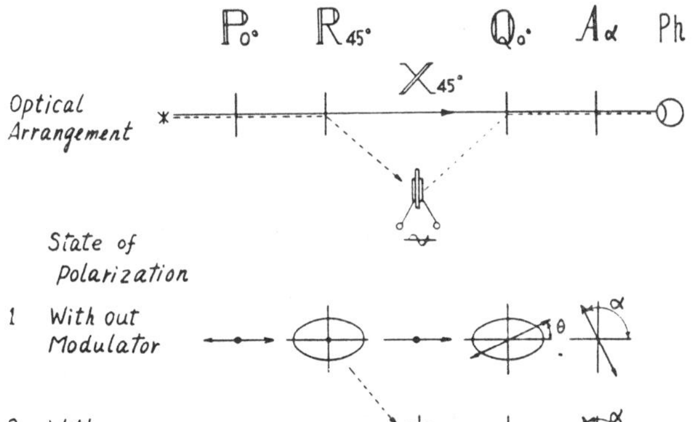

Fig. 5. The optical arrangement and state of polarization of a retardation meter of Sénarmont type.

adopted for modulation. This makes the electric circuit very simple.

III. Theory

A birefringent test piece is inserted, keeping the azimuth of birefringence at 45°, between a pair of crossed polarizers $P_{0°}$ and $A_{90°}$. The suffixes 0° and 90° stand for the azimuth angle of the variation plane of the polarization elements.

Light intensity transmitted through this optical arrangement is represented by:

$$I = {}^1/_4(1 - \cos 2\pi\Delta)I_0, \tag{1}$$

where I_0 and I are intensity of incident and transmitted light, and Δ is the retardation of the test piece. An ADP crystal unit is placed after the test piece, matching the principal axis of the electrically induced birefringence to the principal axis of birefringence of the test piece, and an alternating voltage of some hundred volts is applied.

The amount of retardation of the test piece is modulated with the ADP unit subjected to alternating voltage, and there occurs a cyclic variation of transmitted light as shown in Fig. 3. When the retardation of the test piece is compensated by a compensator such as a Babinet-Soleil, the fundamental component of the signal corresponding to the modulation disappears, and only a bit of second harmonic signal remains. When the birefringence is over-compensated, the fundamental signal occurs, but in this case the phase of the signal is inverted.

Such a signal that changes its phase at a balance point is most suited for automatic measurement. The signal is amplified and fed to the signal winding of a servomotor, and the output of the servomotor is linked to the retardation compensator so as to complete the compensation automatically. Thus the retardation of the test piece is measured from the amount of compensation. Our first instrument which is shown in Fig. 4 was made following this idea, but the practical apparatus was made using the Sénarmont's method, because the Babinet-Soleil compensator is rather expensive.

A pair of polarizers $P_{0°}$ and A_α, a quarter-wave plate $Q_{0°}$, and a test piece $R_{45°}$ are arranged as shown in Fig. 5.

The linearly polarized light is elliptically polarized after passing through the test piece $R_{45°}$. After passing through the quarter-wave plate $Q_{0°}$, the elliptically polarized light is brought back to the linearly polarized light, but the vibration plane is rotated to $\theta = \pi\Delta$ which is equal to the angle of ellipticity of the elliptically polarized light.

Now, the ADP unit is inserted in series with the test piece keeping its principal axis at 45°, and an alternating voltage is applied to the unit. The amount of retardation of the test piece is modulated, and the ellipticity of the elliptically polarized light varies in accordance with the modulation. This elliptically polarized light is brought back to vibrating linearly polarized light by passing through $Q_{0°}$. Since the modulating voltage is an alternating one and contains no dc component, the azimuth of the linearly polarized light vibrates around

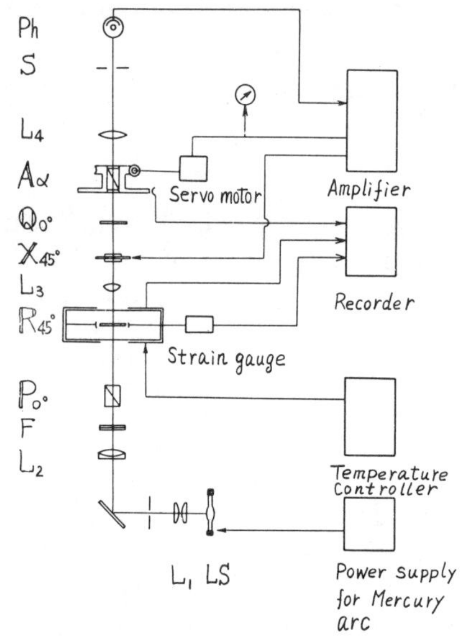

Fig. 6. Schematic diagram of an automatic retardation meter.

θ. When the angle of the analyzer does not lie crossed or parallel to θ, the signal is unbalanced and energizes the servomoter which rotates the analyzer so as to extinguish the signal. This signal vanishes at two points, extinction and maximum of the transmitted light. By adjusting the phase relation between the signal and field excitation of the servomotor, the direction of rotation is arranged to get extinction of light. The retardation Δ is measured from the analyzer angle $\alpha = \theta + {}^1/_2\pi$ at which the signal balances.

IV. Apparatus

A schematic diagram of the practical apparatus is shown in Fig. 6. This apparatus is specially designed for the purpose of measuring temperature-birefringence characteristics of plastic fibers.

Light from a high-pressure mercury arc LS is collimated by a lens L_2, is made into monochromatic light of 546 mμ by passing through an interference filter F, and is linearly polarized by a polarizer $P_{0°}$, and then it illuminates a test piece $R_{45°}$. The test piece is a sheet or a fiber of plastic material oriented to 45°. The sample is imaged by the lenses L_3 and L_4 on a plane with a small hole. The ADP modulator $X_{45°}$, quarter-wave plate $Q_{0°}$, and an analyzer A_α are located in the image-forming light path. These polarimetric elements form the birefringence meter of Sénarmont type shown in Fig. 5. Both polarizer and the analyzer are Glan–Thompson type. The illuminating light is collimated to 5′ and the beam diameter is 5 mm.

The tension applied to the test piece is accurately controlled from zero to 300 g, and recorded automatically with a resistance wire strain gauge. The test piece is enclosed in an electric oven, and its temperature is also controlled and recorded.

The magnification of the microscope consisting of L_3 and L_4 is 3 × 15, and the diameter of the pinhole S is 1 mm or less, so that the diameter of the piece under test is as small as 0.02 mm.

The ADP unit is composed as follows: 1 mm thick Z-cut ADP is sandwiched between a pair of carefully annealed NESA-coated, fused silica plates with oleic acid. An alternating voltage of 250 V, 60 cps is applied to the unit for modulation.

The phototube Ph is one equivalent to RCA 931A. The photoelectric signal is amplified by a three-stage amplifier and fed to a two-phase servomotor which rotates the analyzer A_α.

A potentiometer is mounted on the analyzer drum, and the analyzer rotation is converted into an electric voltage and recorded. The reproducibility of the instrument is influenced by the transmittance and area of the test piece under test. It reaches 1×10^{-4} wavelength of retardation under good conditions. However, it is safe to state that the accuracy of the instrument is of the order of 10^{-3} wavelength of retardation.

This accuracy is kept in all retardation ranges because of the optical null nature of the measurement. Besides, the instrument will follow such a retardation change that exceeds one wavelength keeping the aforementioned accuracy, if the speed of variation is not too rapid. However, owing to the electromechanical servo system, the response time of the record is rather slow. The full-scale traveling time is about 3 sec, so that the instrument misses the signal if the retardation abruptly changes over a half-wavelength.

Figure 7 shows the general view of the apparatus. Figure 8 shows the electric oven.

V. Application

In Fig. 9 curve a shows a retardation–temperature curve obtained from a commercial polyester "Mylar". Starting from room temperature, the retardation of the Mylar film increases very slowly with the temperature rise; then the rate of the variation abruptly changes at

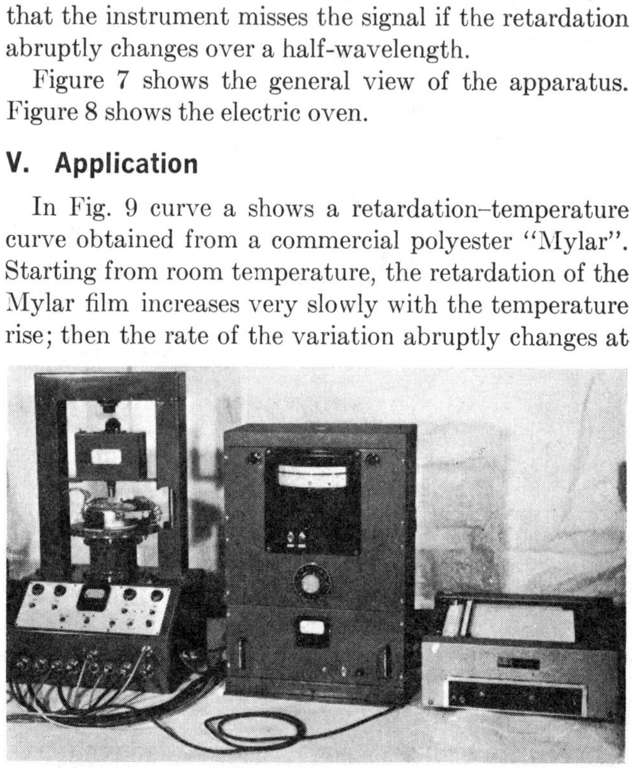

Fig. 7. General view of a practical retardation meter, together with thermocontroller and recorder.

Fig. 8. The test piece stage and the electric oven of the retardation meter.

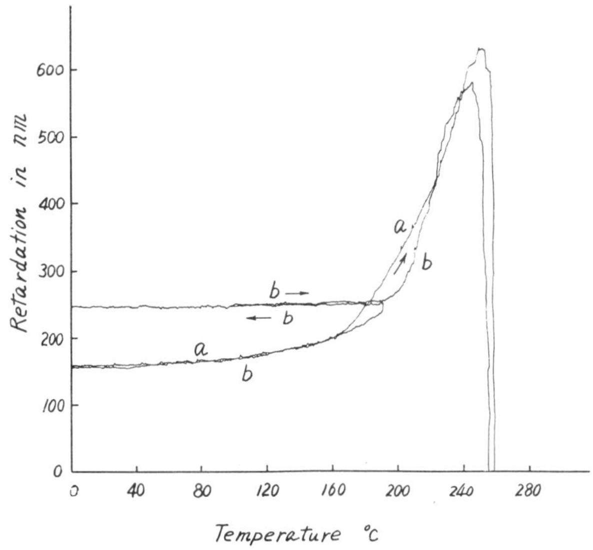

Fig. 9. Temperature–retardation characteristics of commercial polyester film "Mylar"

about 160°C. This sudden change seems to be caused by the heat treatment of the temperature which was previously applied on the film. This supposition is verified by the curve b, where the test piece is heated up to 190°C, then cooled down to the room temperature and heated again to 250°C. In case of the second heating, retardation of the test piece does not change up to 190°C.

Figure 10 shows the temperature–retardation characteristics of some stretched amorphous polyethylene terephthalate film "P.E.T." The test pieces are stretched in a water bath of 70°C at a rate of 0.5% per sec and fixed at the draw ratio of 2.0, 3.6, and 4.9. The rate of temperature rise of the oven is 3°C per min. All of the drawn test pieces show a sudden change of retardation at about 75°C. This is not because of the drawing process in a water bath of 70°C but because of properties of the material itself; that is, the molecular chains are disoriented at 75°C. This supposition is verified from the fact that the sudden change of retardation at 75°C is also apparent in case of draw ratio 1.0 or of the undrawn test piece.

The retardation increases again over 90°C, especially in the case of hard drawn test pieces. This is explained as follows: the molecular motion of the test piece becomes active at this high temperature. This causes the cold crystallization of the test pieces into a fiber structure, and the test pieces increase their birefringence. The cold crystallization is carried on until the tempera-

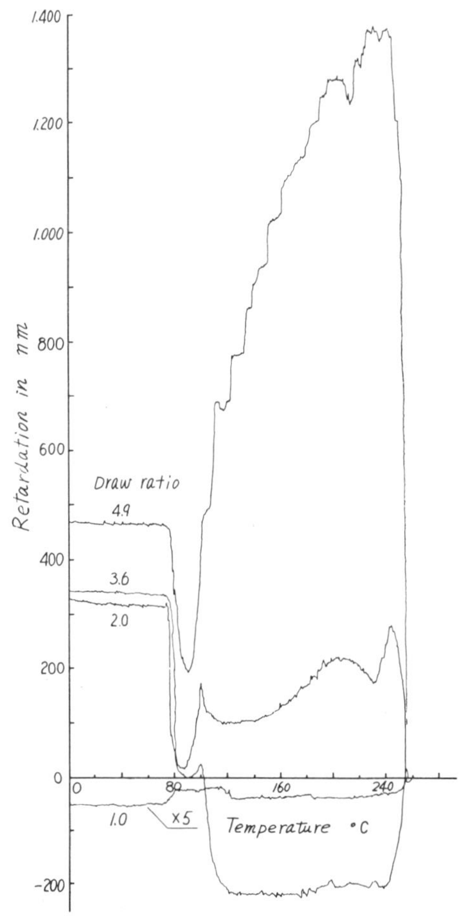

Fig. 10. Temperature–retardation characteristics of the drawn P.E.T. film.

tures of the test pieces come near the melting point, which is about 250°C.

VI. Conclusion

The accuracy of the instrument is some 10^{-3} wavelengths over all the retardation range. On the other hand, the adoption of an electromechanical servo system limits the speed of recording, so that the instrument is especially suited to pursue the slow retardation changes with variation of temperature, stress, or time.

Some results of temperature–retardation measurement of synthetic materials are shown. The results demonstrate that the instrument will be useful to study crystallization and relaxation characteristics of synthetic materials. Further study of the materials with this instrument will be reported in another paper.

The authors wish to express their sincere thanks to the chief members of the universities and companies concerned for their generous support to carry out this research and for their permission to publish the results.

References

1. L. R. G. Treloar, *The Physics of Rubber Elasticity* (Oxford Univ. Press, Oxford, 1949).
2. A. V. Tobolsky and R. S. Stein, Textile Res. J. **18**, 201 (1948).
3. A. V. Tobolsky and R. S. Stein, Textile Res. J. **18**, 302 (1948).
4. A. V. Tobolsky, R. S. Stein, and S. Krimm, Textile Res. J. **19**, 8 (1949).
5. H. Takasaki, J. Opt. Soc. Am. **51**, 462 (1961).
6. H. Takasaki, J. Opt. Soc. Am. **51**, 463 (1961).
7. H. Takasaki, J. Opt. Soc. Am. **51**, 1146 (1961).
8. H. Takasaki, J. Opt. Soc. Am. **52**, 718 (1962).
9. H. Takasaki, T. Shimodaira, and Y. Nagano, J. Opt. Soc. Am. **53**, 641 (1963).
10. B. H. Billings, J. Opt. Soc. Am. **39**, 797 (1949).
11. B. H. Billings, J. Opt. Soc. Am. **39**, 802 (1949).
12. B. H. Billings, J. Opt. Soc. Am. **42**, 12 (1952).
13. R. O'B. Carpenter, J. Opt. Soc. Am. **40**, 225 (1950).
14. B. H. Billings, J. Opt. Soc. Am. **37**, 738 (1947).
15. R. C. Günter, J. Opt. Soc. Am. **38**, 419 (1948).
16. K. Fushimi, Bull. Electro-Tech. Lab. Japan, No. 602 (1962).
17. Y. Wachi, Oyo Butsuri **27**, 670 (1958).
18. Y. Fujisawa, Oyo Butsuri **27**, 657 (1958).
19. S. Namba and T. Ogawa, Oyo Butsuri **26**, 498 (1957).
20. Y. Fujimura and T. Ogawa, Oyo Butsuri **26**, 509 (1957).
21. Y. Fujimura and T. Ogawa, Oyo Butsuri **26**, 520 (1957).

An Automatic Polarimeter. II. Automatic Polarimetry by Means of an ADP Polarization Modulator

Hiroshi Takasaki, Nobuo Okazaki, and Kōji Kida

An ADP solid Kerr cell subjected to an alternating voltage naturally makes a retardation modulator. An azimuth modulator with the ADP Kerr cell is easily made with the aid of simple polarization optics. Two types of azimuth modulators with an ADP Kerr cell, a Sénarmont-type and a circular polarizer-type respectively, are proposed and compared with each other. The construction of a tentative apparatus and a practical model is also reported.

I. Introduction

Among the various polarimetric instruments, those which measure the rotation of polarization plane are most common and are simply called polarimeters. Polarimeters are so useful for the sugar industry, and for the study of stereochemics, that many studies have been made to improve the instruments.

Ebert *et al.*[1] studied analytically the sensitivity and accuracy of the polarimeters. Rudolph[2] furnished a commercial photoelectric polarimeter which is based on the symmetrical angle method. Gates,[3] Gillham,[4] Levy,[5] and Landegren[6] studied automation of the instruments.

The means of the automation is grouped roughly into three classes: static matching of the brightness of two light paths, dynamic matching by means of the mechanical interchange of the two paths, and dynamic matching with a polarization plane vibrator.

Among these types, the last one, which uses only one light path, seems to promise the best results. Several instruments based on this principle are commercially available.[7] Most of them use a Faraday cell for the modulation, but, owing to small magnetooptical constants of the common Faraday cell, considerable electrical power is required for its operation, and troubles are likely to be caused by the large power dissipation of the cell. We replaced the Faraday cell with an optical system with the ADP Kerr cell. The ADP Kerr cell is naturally an electrooptical compensator of retardation, but this property can be converted into an electrooptical azimuth rotor with the aid of simple polarization optics. The crystal has a large electrooptical constant and is operated mostly by the electric voltage. Consequently, the electric power required is very small, and the troubles caused by the power dissipation in Faraday cells are minimized.

The paper deals with some comparative studies of two types of polarimeters with ADP polarization modulator and the construction of practical instruments.

Hiroshi Takasaki (presently with the National Bureau of Standards for a period of one year) is a member of the Faculty of Engineering, University of Shizuoka, Hammatsu, Japan. Nobuo Okazaki and Kōji Kida are with Scientific Instruments Works, Shimadzu Seisakusho Ltd., Kyoto, Japan.

Received 24 August 1963.

II. Theory

The key to making an electrooptically operated polarimeter is to make such an element that electrically rotates the polarization plane.[8,9]

First, the Sénarmont method which is commonly used to measure retardation is considered. In this method, retardation of the test piece is converted into azimuth rotation of linearly polarized light and measured by rotation of the analyzer.

If the test piece is replaced by the ADP Kerr cell subjected to an electric voltage, the polarization plane is electrically rotated. This is the principle of the polarization modulator of the Sénarmont-type.

By using this type of polarization modulator, we can make a polarimeter. The schematic diagram of a polarimeter is shown in Fig. 1.

The Mueller matrix[10,11] is helpful for calculating the intensity of transmitted light. Each polarization element is represented by a 4 × 4 matrix as follows:

a perfect polarizer, the azimuth of which is zero,

$$\mathbf{P}_0 = {}^1/_2 \begin{vmatrix} 1 & 1 & 0 & 0 \\ 1 & 1 & 0 & 0 \\ 0 & 0 & 0 & 0 \\ 0 & 0 & 0 & 0 \end{vmatrix}; \qquad (1)$$

a retardation plate, the azimuth of which is zero,

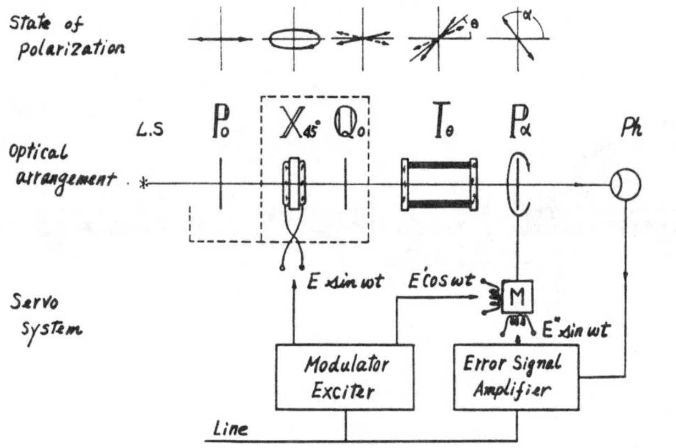

Fig. 1. Schematic diagram of an automatic polarimeter with the Sénarmont-type azimuth modulator.

$$\mathbf{R}_{0,\Delta} = \begin{vmatrix} 1 & 0 & 0 & 0 \\ 0 & 1 & 0 & 0 \\ 0 & 0 & \cos\Delta & -\sin\Delta \\ 0 & 0 & \sin\Delta & \cos\Delta \end{vmatrix}; \tag{2}$$

a rotator of rotation α,

$$\mathbf{T}_\alpha = \begin{vmatrix} 1 & 0 & 0 & 0 \\ 0 & \cos2\alpha & -\sin2\alpha & 0 \\ 0 & \sin2\alpha & \cos2\alpha & 0 \\ 0 & 0 & 0 & 1 \end{vmatrix}. \tag{3}$$

Then, applying rotators from both sides of an element of azimuth 0, the matrix of the rotated element is obtained.

A perfect polarizer, the azimuth of which is α,

$$\mathbf{P}_\alpha = {}^1/_2 \begin{vmatrix} 1 & \cos2\alpha & \sin2\alpha & 0 \\ \cos2\alpha & \cos^22\alpha & \sin2\alpha\cos2\alpha & 0 \\ \sin2\alpha & \sin2\alpha\cos2\alpha & \sin^22\alpha & 0 \\ 0 & 0 & 0 & 0 \end{vmatrix}. \tag{4}$$

A retardation plate of angular retardation Δ and of the azimuth of the principal axis α,

$$\mathbf{R}_{\alpha,\Delta} = \begin{vmatrix} 1 & 0 & 0 & 0 \\ 0 & 1-(1-\cos\Delta)\sin^22\alpha & \sin2\alpha\cos2\alpha(1-\cos\Delta) & \sin2\alpha\sin\Delta \\ 0 & \sin2\alpha\cos2\alpha(1-\cos\Delta) & 1-(1-\cos\Delta)\cos^22\alpha & -\cos2\alpha\sin\Delta \\ 0 & -\sin2\alpha\sin\Delta & \cos2\alpha\sin\Delta & \cos\Delta \end{vmatrix}. \tag{5}$$

The matrices are multiplied in the order of the optical system. The matrix representation of the optical arrangement shown in Fig. 1 is as follows:

$$\mathbf{S}' = \mathbf{P}_\alpha\cdot\mathbf{T}_\theta\cdot\mathbf{Q}_0\cdot\mathbf{X}_{45°\epsilon}\cdot\mathbf{P}_0\cdot\mathbf{S}, \tag{6}$$

where $\mathbf{S}$ and $\mathbf{S}'$ are incident and transmitted light which are represented by the Stokes' parameters.[10] The symbols $\mathbf{X}$ and $\mathbf{Q}$ denote an ADP unit and a quarter-wave plate, and the suffixes α, θ, 0, and 45° denote the azimuth of the principal axis of each element, and θ is the electrically induced birefringence.

The ADP unit subjected to an alternating voltage is expressed in the same form as Eq. (5), except that in this case Δ is represented by ϵ which is $KE\sin\omega t$, where K is the electrooptical constant of the ADP crystal and $E\sin\omega t$ is the alternating voltage for modulation. The electrooptical constant is as high as angular retardation of $\pi/9.6$ kV.

The 1:1 component of the matrix Eq. (6) gives the intensity of the transmitted light. This component is calculated as follows:

$$I = {}^1/_4\{1 + \cos2(\theta-\alpha)\cos\epsilon - \sin2(\theta-\alpha)\sin\epsilon\}I_0, \tag{7}$$

where I and I_0 denote the intensity of incident and transmitted light.

Since ϵ or $KE\sin\omega t$ is small, Eq. (7) is calculated as follows:

$$I = {}^1/_4\{1 + \cos2(\theta-\alpha) - {}^1/_2\cos2(\theta-\alpha)\sin^2(KE\sin\omega t) - \sin2(\theta-\alpha)\sin(KE\sin\omega t)\}I_0$$

$$\doteqdot {}^1/_4\left[\left\{1 + \cos2(\theta-\alpha) - \frac{K^2E^2}{4}\cos2(\theta-\alpha)\right\} + \frac{K^2E^2}{4}\cos2(\theta-\alpha)\cos2\omega t - KE\sin2(\theta-\alpha)\sin\omega t\right]I_0. \tag{8}$$

The first term of Eq. (8) represents the light leak which becomes a minimum at extinction. The second term is the second harmonic signal to the modulation. This signal remains even at extinction and overwhelms the fundamental signal. This unwanted second harmonic signal is choked by the interstage twin T filters in the amplifier. The third term corresponds to modulation and disappears at extinction. This signal increases with $\sin2(\theta-\alpha)$ and inverts the phase at the extinction. Thereupon, driving the polarizer $\mathbf{P}_\alpha$ with a servomotor driven by the signal, the analyzer is rotated to extinction, and the rotation by the test piece is read from the analyzer angle.

The aforementioned is for an ideal case. In practice, a setting error and a retardation error of the polarization components remain; therefore, the influence of such defects must be considered.

When the quarter-wave plate is imperfect both with respect to retardation and to azimuth setting, the intensity of the transmitted light is calculated as

$$I = {}^1/_4[1 + \{[1-(1-\cos q)\sin^22\gamma]\cos2(\theta-\alpha) - \sin2\gamma\cos2\gamma(1-\cos q)\sin2(\theta-\alpha)\}\cos\epsilon - \sin2(\gamma+\theta-\alpha)\sin q\cdot\sin\epsilon]I_0, \tag{9}$$

where γ is the azimuth angle of the wave plate and q is the angular retardation of the plate.

When the azimuth setting is perfect, Eq. (9) reduces to

$$I = {}^1/_4\{1 + \cos2(\theta-\alpha)\cos\epsilon - \sin q.\sin2(\theta-\alpha)\sin\epsilon\}I_0. \tag{10}$$

This means that the retardation error reduces the error signal in proportion to $\sin q$.

In the case of erratic azimuth setting of the quarter-wave plate, Eq. (9) is reduced to

$$I = {}^1\!/_4\{1 + \cos 2\gamma \cos 2(\gamma + \theta - \alpha)\cos\epsilon - \sin 2(\gamma + \theta - \alpha)\sin\epsilon\} I_0. \quad (11)$$

This means that the analyzer angle shifts by γ and the extinction is deteriorated.

The shift of analyzer angle does not cause trouble, because it is easily cancelled by additional measurement without the sample. On the other hand the extinction failure introduces serious troubles, because it increases the steady photocurrent at extinction and reduces the signal-to-noise ratio of the desired signal.

Now, suppose the polarizer of an ideal Sénarmont-type polarimeter is rotated. The transmitted intensity of light is then calculated as

$$I = {}^1\!/_4\{1 + \cos 2\beta \cos 2(\theta - \alpha)\cos\epsilon - \cos 2\beta \cdot \sin 2(\theta - \alpha)\sin\epsilon\} I_0, \quad (12)$$

where β is azimuth angle of the polarizer. This means that not only the error signal is reduced in proportion to $\cos 2\beta$ but also the extinction is deteriorated. It may sound strange, but the analyzer angle is not influenced by the polarizer rotation.

From the aforementioned, we can see that, in case of the Sénarmont-type modulator, there is no effective means of zero adjustment except the rotation of the whole modulator system around the optical axis of the instrument.

Now, we consider the azimuth modulator of the circular polarizer-type shown in Fig. 2.

The Mueller matrix of the system $\mathbf{Q}_{45°}\mathbf{X}_0\mathbf{Q}_{45°}$ is calculated as

$$\mathbf{Q}_{45°}\cdot\mathbf{X}_{0,\epsilon}\cdot\mathbf{Q}_{-45°} = \begin{vmatrix} 1 & 0 & 0 & 0 \\ 0 & 0 & 0 & 1 \\ 0 & 0 & 1 & 0 \\ 0 & -1 & 0 & 0 \end{vmatrix} \begin{vmatrix} 1 & 0 & 0 & 0 \\ 0 & 1 & 0 & 0 \\ 0 & 0 & \cos\epsilon & -\sin\epsilon \\ 0 & 0 & \sin\epsilon & \cos\epsilon \end{vmatrix} \begin{vmatrix} 1 & 0 & 0 & 0 \\ 0 & 0 & 0 & -1 \\ 0 & 0 & 1 & 0 \\ 0 & 1 & 0 & 0 \end{vmatrix}$$

$$= \begin{vmatrix} 1 & 0 & 0 & 0 \\ 0 & \cos\epsilon & \sin\epsilon & 0 \\ 0 & -\sin\epsilon & \cos\epsilon & 0 \\ 0 & 0 & 0 & 1 \end{vmatrix} = \mathbf{T}_{-(\epsilon/2)}. \quad (13)$$

The result is the matrix of a rotator of rotation $-\epsilon/2$. From this a perfect azimuth modulation, which is not affected by the azimuth of the incident linearly polarized light, is expected.

The intensity of the light transmitted through the optical system shown in Fig. 2 is calculated as

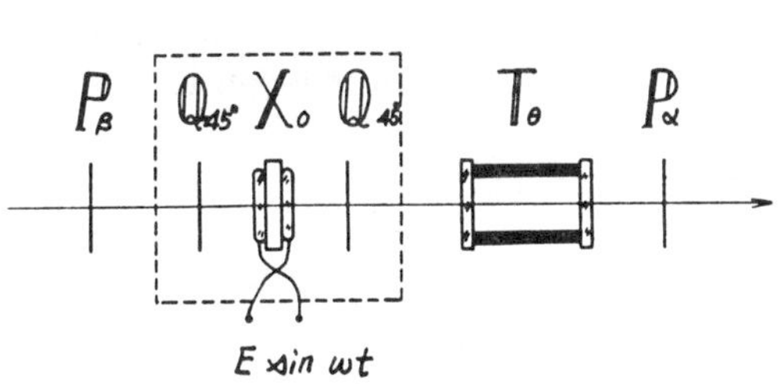

Fig. 2. Schematic diagram of the optical system of a polarimeter with the circular-type azimuth modulator.

$$I = {}^1\!/_4[1 + \cos 2(\theta - \alpha)\cos^2 q \cos 2\beta + \{\cos 2(\theta - \alpha)\sin^2 q \cos 2\beta - \sin 2(\theta - \alpha)\sin 2\beta\}\cos\epsilon + \sin 2(\beta + \theta - \alpha)\sin q \sin\epsilon] I_0, \quad (14)$$

where the angular retardation of the two wave plates are assumed to be equal, which is represented by q; the polarizer is rotated to β and the azimuth setting of other elements is assumed to be accurate.

When the angular retardation q is equal to $\pi/2$, Eq. (14) is reduced to

$$I = {}^1\!/_4\{1 + \cos 2(\beta + \theta - \alpha)\cos\epsilon + \sin 2(\beta + \theta - \alpha)\sin\epsilon\} I_0. \quad (15)$$

The result is essentially equal to that of the perfect Sénarmont-type and is not affected by the rotation of polarizer. Along with the rotation of polarizer the analyzer shifts by just the same angle of polarizer rotation, and no other disadvantages are introduced. This means that this optical system allows zero adjustment of the instrument with polarizer rotation.

When $\beta = 0$, Eq. (14) is reduced to

$$I = {}^1\!/_4\{1 + \cos 2(\theta - \alpha)\cos^2 q + \cos 2(\theta - \alpha)\sin^2 q \cos\epsilon + \sin 2(\theta - \alpha)\sin q \sin\epsilon\} I_0, \quad (16)$$

or in the corresponding form to Eq. (8), we get

$$I \doteqdot {}^1\!/_4 \left\{ 1 + \cos 2(\theta - \alpha) - \frac{K^2E^2}{4}\cos 2(\theta - \alpha)\sin^2 q + \frac{K^2E^2}{4}\cos 2(\theta - \alpha)\sin^2 q \cos 2\omega t + KE \sin 2(\theta - \alpha)\sin q \sin\omega t \right\} I_0. \quad (17)$$

From this, we can see that the extinction of the system is not deteriorated seriously by the retardation error of the quarter-wave plates. The error signal reduces with proportion to $\sin q$, but this reduction of signal does not make problems, because it is easily covered by increasing the gain of amplifier.

Thus the azimuth modulator of circular polarizer-type has several advantages compared with that of Sénarmont type. So, we planned to make an instrument using the circular-type azimuth modulator. But afterward, it was proved that the Sénarmont-type was much easier to make accurate adjustment and was sometimes even superior to the circular-type in practice because of its simplicity.

III. Apparatus

Our prototype apparatus which employs the circular-type azimuth modulator is shown in Fig. 3.

Light emitted from an incandescent lamp is filtered by a metal film interference filter. The bandwidth of the filtered light is about 6 mμ, and the peak wavelength is varied a little from 546 mμ by tilting the filter. This wavelength control serves to calibrate the instrument division with standard sucrose solution and eliminates

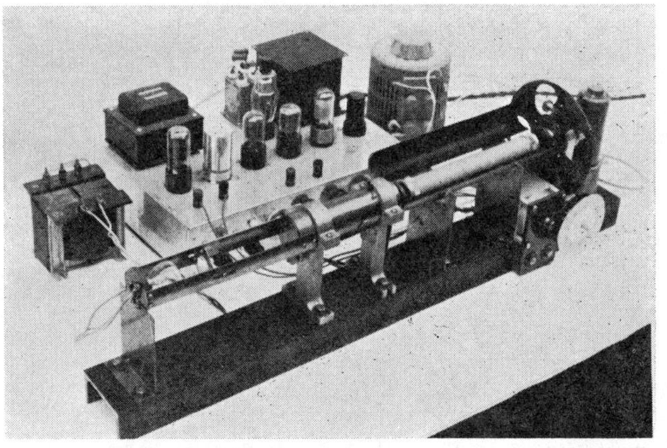

Fig. 3. General view of a polarimeter with the circular-type azimuth modulator.

Fig. 5. General view of a practical instrument.

the tube length error of the test tube. The zero point of the instrument is adjusted by rotating the polarizer.

The circular-type azimuth modulator allows such a design because its function is not seriously deteriorated by a wavelength change and by a rotation of polarizer.

We experienced some troubles with this instrument. The most serious one is the deterioration of extinction with sugar solution in the optical path. This is caused by the rotatory dispersion of the sugar solution under test and the bandpass characteristics of the metal interference filter. Light with some bandwidth is passed through the interference filter and is rotatory-dispersed by the sugar solution. This makes the perfect extinction impossible.

Furthermore, the modulator is tedious to adjust and it is hard to get good extinction of light even without sample. The light scattering caused by many optical surfaces and by incompleteness of polarization elements seems to be the main reason for the troubles.

A practical model which is shown schematically in Fig. 4 was planned, based on the above experience. In the new apparatus the light filtered from an incandescent lamp is replaced by the green line spectrum of a mercury arc, the azimuth modulator of the circular polarizer-type by the Sénarmont-type, and the zero adjustment is made by rotating the analyzer together with the modulator part. High accuracy of manufacture permitted this design.

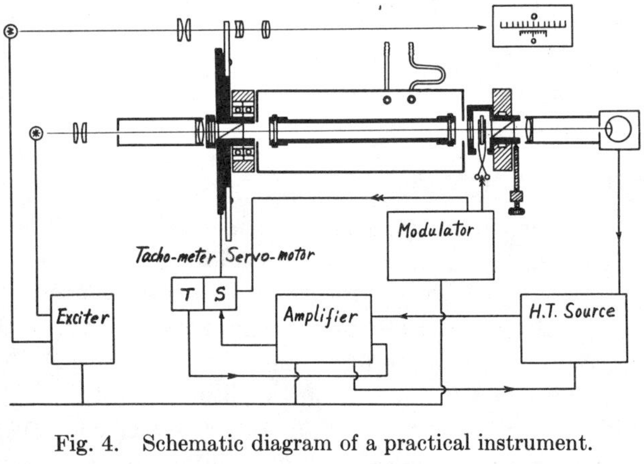

Fig. 4. Schematic diagram of a practical instrument.

A precision-divided circle is mounted on the polarizer drum, and the rotation angle is projected onto a screen on which 0.1° S division (the calibration of rotation in terms of sucrose concentration) is engraved in addition to the ordinary angular graduation.

The polarizer drum is driven with a steel belt by a two-phase motor with a tacho-generator, the output of which is fed back to the amplifier to improve the dynamic response of the instruments.

As to the electric circuit, the amplifier is tuned to 60 cps using the twin T feedback network. Two interstage twin T filters tuned to 120 cps are also involved—in the amplifiers to choke the second harmonic signal. The high voltage is supplied to the phototube from a source that consists of a 100 kc/sec oscillator and a rectifier. The output of the oscillator is controlled by the error signal. When the signal is large the high voltage is reduced from 1000 V dc to 200 V dc so as not to saturate the phototube and the amplifier. The modulation voltage is varied from 100 to 300 V ac. An appropriate voltage is chosen for the measuring conditions. Generally speaking, deep modulation is preferred when the test piece is turbid, but low-modulating voltage of 50 V ac is recommended when the sample condition is good. The general view of the instrument is shown in Fig. 5.

IV. Results

With this practical instrument, we encountered serious trouble of a large zero drift which had not been experienced before. The source of this trouble was found to be temperature rise of the instrument. Because the amplifier and the light source are mounted on the same base plate with the optical parts, heat is transferred to the optical parts. Among the optical parts the ADP unit is most sensitive to temperature rise. That is, the frozen stress in the NESA-coated transparent electrode is relaxed by small temperature rise and the polarization state is changed. This causes the serious zero drift shown in curve *a* of Fig. 6.

Several trials were made to decrease this drift, and it was found that the removal of the internal stress frozen in the transparent electrode was most effective. Upon replacing the transparent electrode with a carefully annealed one, the drift characteristics are diminished as shown in curve *b* of Fig. 6.

Another difficulty was an absolute error in the sucrose rotation. The cause of this was found to be impurity of the monochromatic light. The short wavelength spectrum light leaked from the filter increased the rotation about 10%. This trouble was eliminated by using a didymium filter together with a metal interference filter.

The reproducibility of the apparatus is of the order of 0.01°. The absolute precision of the instrument was checked using several Schmidt-Hensche quartz standard plates of 10.14°, 34.97°, and 40.61°. The readings of the instrument with these standard plates were 10.14°, 34.96°, and 40.58°, respectively.

We noticed some weak points in the electric circuit. Most of the troubles were caused from the fact that the gain control is operated solely by the error signal. The most serious trouble is reduced response speed at the start. The instant that the test piece is inserted, a considerable amount of light is transmitted by the analyzer, and this steady leak saturates the phototube.

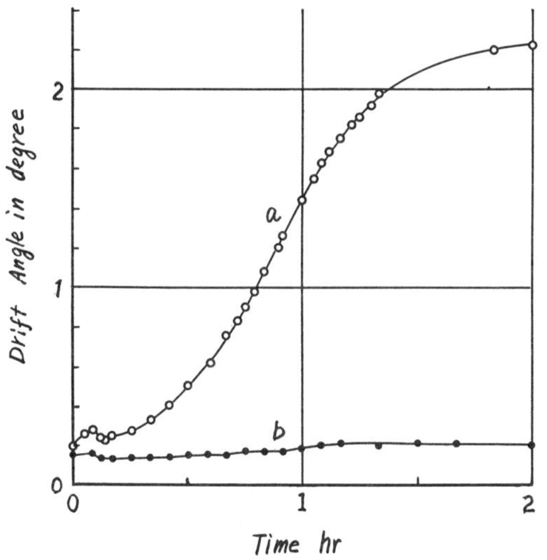

Fig. 6. Zero drift characteristics of a practical instrument.

On account of this saturation, the error signal is suppressed too much and the high voltage which is supplied to the phototube is raised. Thus the situation goes from bad to worse. Almost the same situation takes place when the test piece is turbid and the error signal is small compared with the steady leak. We are now planning to make a gain control referring to the steady leak of light and to make the modulation depth control by referring to the second harmonic signal.

V. Conclusion

The principle and the construction of a new automatic polarimeter with ADP polarization azimuth modulator are reported.

The instrument was proved to have an accuracy sufficient for practical use.

It is also possible to make a full electric polarimeter with ADP compensator subjected to dc voltage, as well as to the modulating ac voltage. However, applying dc voltage to the ADP units is apt to introduce troubles such as electrode damage and gas deposition by electrolysis, so some measure should be taken to eliminate these troubles.

The authors wish to express their deep appreciation to the Shimadzu Seisakusho Company for its generous support to the research and for its permission to publish the results. The authors would like to express their appreciation also to T. Furuhashi and I. Honda for their assiduous help throughout the whole course of the research.

References

1. L. Ebert and G. Kortüm, Z. physik. Chem. Abt. B **13,** 105 (1931).
2. H. Rudolph, J. Opt. Soc. Am. **45,** 50 (1955).
3. J. W. Gates, Chem. Ind. (London) **1958,** 190.
4. E. J. Gillham, J. Sci. Instr. **34,** 435 (1957).
5. G. B. Levy, Rev. Sci. Instr. **21,** 693 (1950).
6. G. F. Landegren, Rev. Sci. Instr. **26,** 578 (1955).
7. I. Becker, Zucker **15,** 356 (1959).
8. H. Takasaki, J. Opt. Soc. Am. **51,** 462 (1961).
9. H. Takasaki, Appl. Opt. **3,** 345 (1964).
10. B. H. Billings and E. H. Land, J. Opt. Soc. Am. **38,** 819 (1948).
11. R. C. Jones, J. Opt. Soc. Am. **37,** 107 (1947).

Automatic Ellipsometer. Automatic Polarimetry by Means of an ADP Polarization Modulator III

Hiroshi Takasaki

An analytical and experimental study of an automatic ellipsometer is reported. The polarized beam is modulated simultaneously by two ADP cells, and the corresponding two signals are separated and are used to adjust polarizer and analyzer settings in any of the four zones. The reproducibility of the setting has been demonstrated to be of the order of 0.01°.

I. Introduction

The state of polarization of light changes at reflection depending on the incidence angle, optical constants n and k of the reflecting material and, if any, thickness and optical constants of film on the surface. An ellipsometer is an instrument that measures the state of the elliptically polarized light, in particular, amplitude ratio and relative phase retardation of the p and s components of the reflected light. A measurement of the optical constants of the reflecting material or film formation on the surface is made possible from the reading. On the other hand, operation of a conventional ellipsometer is tedious because it requires at least two kinds of adjustments to determine the two parameters of an elliptically polarized light. Hence, there is a large demand for an automatic ellipsometer.

Several attempts have been made to design an automatic ellipsometer. Among these attempts, those based on a rotating polarizer[1–3] are very ingenious. Such a system gives large signals and is capable of being fully automated. This system promises to be particularly useful in the field of ultraviolet or infrared ellipsometry where there exist stringent limitations on the intensity of available light and availability of polarizing elements.

A new design for an automatic ellipsometer was developed by the author[4] and some results were reported.[5] Two parameters of polarized light are simultaneously modulated, the signal components separated electronically and used to adjust the optical compensating devices by means of a servosystem. In this work ADP (ammonium dihydrogen phosphate) cells were used as modulators. However, Faraday cells have also proven useful for this purpose.[6]

The author is with the National Bureau of Standards, Washington, D.C. 20234.

Received 20 August 1965.

In the earlier study,[5] the polarization optics were designed for easy construction. Because of the limitation of mechanical precision and the quality of the polarization optics of the instrument, the application was limited to those cases where the amount of change in the apparent optical constants is rather large.

The purpose of the present study is, first, to develop a method for automatically conducting measurements in the four zones, and, second, to achieve a higher accuracy using optics of considerably higher quality. The theoretical treatment will be confined to the ideal case.

II. Theory

One of the common arrangements of an ellipsometer consists of an adjustable polarizer, a test piece, a quarter-wave plate set to ±45°, and an adjustable analyzer.

This arrangement was modified slightly by the inclusion of two ADP cells, in which birefringence is produced by an alternating electric voltage and the transmittance of the optical system is modulated. The modified arrangement is shown schematically in Fig. 1.

The coordinate systems which are shown in the figure are defined as follows: (1) all angles are measured counterclockwise from the plane of incidence looking toward the light; (2) the angles of the polarizer and analyzer are determined by the angles of transmitted electric vectors; (3) the angle of the quarter-wave plate is determined by the angle of the slow axis; (4) the angle of the ADP cell is determined by the angle of the electrically induced slow axis.

The subscripts shown on the ADP cells in Fig. 1, $\alpha - 45°$ and $\beta + 45°$, mean that each cell is mounted on the polarizer or analyzer and rotates with it keeping the relative angle of 45°.

This polarizing optical system is best represented by the Mueller matrix:[7–9]

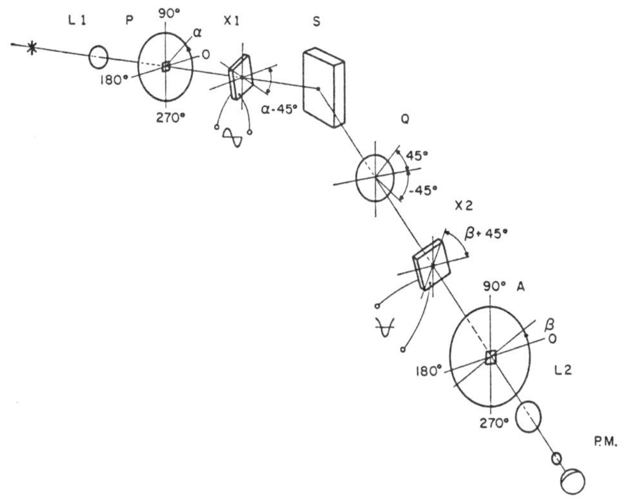

Fig. 1. Schematic diagram of the optical arrangement of an automatic ellipsometer. L_1 and L_2, collimating lenses; P, polarizer; A, analyzer; X_1 and X_2, ADP cells; Q, quarter-wave plate; S, test piece; and $P.M.$, photomultiplier tube.

$$\begin{vmatrix} I' \\ M' \\ C' \\ S' \end{vmatrix} = A_\beta \underset{\epsilon_2}{X_{\beta+45°}} Q_{\pm 45°} \underset{\Delta, R_p/R_s}{S_0} \underset{\epsilon_1}{X_{\alpha-45°}} P_\alpha \begin{vmatrix} I \\ M \\ C \\ S \end{vmatrix} \tag{1}$$

where P_α and A_β represent linear polarizers of azimuth angle α and β,[9] $\underset{\epsilon_1}{X_{\alpha-45°}}$, and $\underset{\epsilon_2}{X_{\beta+45°}}$ represent retardation plates of small retardation ϵ_1 and ϵ_2, respectively, with the azimuth angles $\alpha - 45°$ and $\beta + 45°$, and $Q_{\pm 45°}$ a quarter-wave plate with azimuth $\pm 45°$. These matrices as given are special cases of the matrices for general retardation plate.[9]

$\underset{\Delta, R_p/R_s}{S_0}$ represents a test piece with retardation Δ and reflectance ratio of R_p/R_s. This matrix is obtained by multiplying the matrix for a partial polarizer of zero azimuth[7] and the matrix for a retardation plate of zero azimuth, which is

$$\begin{aligned} \underset{\Delta, R_p/R_s}{S_0} &= \underset{R_p/R_s}{P_0} \underset{\Delta}{R_0} \\ &= \tfrac{1}{2}\begin{vmatrix} R_p^2 + R_s^2 & R_p^2 - R_s^2 & 0 & 0 \\ R_p^2 - R_s^2 & R_p^2 + R_s^2 & 0 & 0 \\ 0 & 0 & 2R_pR_s\cos\Delta & -2R_pR_s\sin\Delta \\ 0 & 0 & 2R_pR_s\sin\Delta & 2R_pR_s\cos\Delta \end{vmatrix}, \end{aligned} \tag{2}$$

where

$$\underset{R_p/R_s}{P_0} = \tfrac{1}{2}\begin{vmatrix} R_p^2 + R_s^2 & R_p^2 - R_s^2 & 0 & 0 \\ R_p^2 - R_s^2 & R_p^2 + R_s^2 & 0 & 0 \\ 0 & 0 & 2R_pR_s & 0 \\ 0 & 0 & 0 & 2R_pR_s \end{vmatrix}. \tag{3}$$

The transmittance of the system is calculated as

$$\begin{aligned} T &= (I'/I) \\ &= \tfrac{1}{8}\{[(R_p^2 + R_s^2) + (R_p^2 - R_s^2)\cos 2\alpha + 2R_pR_s \sin(2\beta \\ &\pm \Delta)\sin 2\alpha] \pm 2R_pR_s\cos(2\beta \pm \Delta)\epsilon_1 \\ &\pm [(R_p^2 + R_s^2)\cos 2\alpha + (R_p^2 - R_s^2)]\epsilon_2\}. \end{aligned} \tag{4}$$

The upper (lower) signs refer to the quarter-wave plate of azimuth $+45°$ ($-45°$).

ϵ_1 and ϵ_2 represent the electrooptically induced birefringence and are written as

$$\epsilon = KE\sin(\omega t + \theta), \tag{5}$$

where K is the electrooptical constant of the ADP cell and $E\sin(\omega\, t + \theta)$ is applied sinusoidal voltage. These two modulated signals can be characterized by modulation voltage of proper frequency or phase. The practical technique of separating these signals will be explained later.

The transmittance signal component associated with ϵ_1, which varies with $\cos(2\beta \pm \Delta)$, disappears and inverts its polarity at $\cos(2\beta \pm \Delta) = 0$, provided that neither R_p or R_s is zero. This signal component will be called the retardation signal and abbreviated *Sr*.

The signal component associated with ϵ_2 disappears and changes its polarity at $\tan\alpha = \pm R_p/R_s$. This component will be named the azimuth signal and abbreviated *Sa*.

The first term of Eq. (4), which is independent of the modulations, represents the steady transmittance of light through the optical system. This component will be named the transmittance term.

Any pair of values of α and β which make *Sa* and *Sr* vanish simultaneously causes the transmittance term to become zero or maximum. Thus, the adjustment of α and β which makes both of the signal components *Sa* and *Sr* vanish is equivalent to an adjustment that makes the transmittance of the optical system either zero or maximum.

There are two values for α which make *Sa* zero, and four values of β which make *Sr* zero; hence, eight combinations of α and β that extinguish both of the signals. Four of these eight combinations make the transmittance term zero, and another four make it maximum. The four combinations that make the transmittance maximum are usually discarded because of an adverse signal-to-noise ratio. The four remaining combina-

Table I. The Condition of Zero Retardation Signal

General		2β	$\sin(2\beta \pm \Delta)$
$\cos(2\beta \pm \Delta) = 0$	$Q_{45°}$	$2\beta = 90° - \Delta$	$\sin(2\beta + \Delta) = 1$
		$2\beta = -90° - \Delta$	$\sin(2\beta + \Delta) = -1$
	$Q_{-45°}$	$2\beta = 90° + \Delta$	$\sin(2\beta - \Delta) = 1$
		$2\beta = -90° + \Delta$	$\sin(2\beta - \Delta) = -1$

Table II. The Condition of Zero Azimuth Signal

General[a]		α
$\cos 2\alpha = -\dfrac{R_p^2 - R_s^2}{R_p^2 + R_s^2}$	$Q_{45°}$	$\alpha = \tan^{-1}R_p/R_s$
or $\tan\alpha = R_p/R_s$	$Q_{-45°}$	$\alpha = \tan^{-1}R_p/R_s$

[a] Neither R_p nor R_s is supposed to be zero.

tions of α and β that extinguish the transmittance are the four zones.

This reasoning is illustrated in the Tables I–III.

In the present treatment for the ideal case, the necessity for making measurements in the four zones is not apparent. In practice, there are differences in Δ and R_p/R_s obtained from measurements of α and β in each zone. These differences are cancelled by taking the average of Δ and R_p/R_s in the four zones. Thus, it becomes essential to make measurements in all four zones or at least two zones.[10,11] However, the polarities of the signals which adjust polarizer and analyzer angle are not necessarily consistent zone by zone, and unless control of the polarity of the servosystems is maintained, they fail to find extinction but obtain maximum transmission. In order to find their polarity, the retardation and azimuth signals are differentiated with respect to α and β, which gives $\partial Sr/\partial\alpha = 0$ and $\partial Sa/\partial\beta = 0$. This implies that the retardation signal is independent of the rotation of the polarizer and the azimuth signal is independent of the analyzer rotation.

The other differentiations are

$$\partial S_r/\partial\beta = \pm\tfrac{1}{2}R_pR_s\sin(2\beta\pm\Delta) \quad \text{for } Q\pm 45°, \tag{7}$$

$$\partial S_a/\partial\alpha = \pm\tfrac{1}{4}(R_p^2+R_s^2)\sin 2\alpha \quad \text{for } Q\pm 45°. \tag{8}$$

The signs of these differentiations, which indicate the polarity of the servosystems, are also tabulated in Table III for each zone.

Both of the derivatives of Sr and Sa are positive in the first and the fourth zones and negative in the second and the third zones. This means that in the second and the third zones the polarity of the servosystems must be opposite to the polarity used for the first zone. In the fourth zone the original polarities will work. It should be noted that this situation holds only for the particular optical arrangement shown in Fig. 1.

Because the four zones are defined by the combination of the two positions of polarizer and the two positions of the quarter-wave plate, the control of the polarities of the servosystems, as well as the indication

Table III. Combinations of α and β

α	Q	Δ	Transmittance	Zone	Sign of $\partial Sr/\partial\beta$	Sign of $\partial Sa/\partial\alpha$
$\alpha > 0$	$Q_{45°}$	90°-2β	$\frac{R_p^2R_s^2}{R_p^2+R_s^2}$			
		270°-2β	0	1	+	+
	$Q_{-45°}$	2β-90°	$\frac{R_p^2R_s^2}{R_p^2+R_s^2}$			
		2β-270°	0	2	−	−
$\alpha' < 0$	$Q_{45°}$	90°-2β	0	3	−	−
		270°-2β	$\frac{R_p^2R_s^2}{R_p^2+R_s^2}$			
	$Q_{-45°}$	2β-90°	0	4	+	+
		2β-270°	$\frac{R_p^2R_s^2}{R_p^2+R_s^2}$			

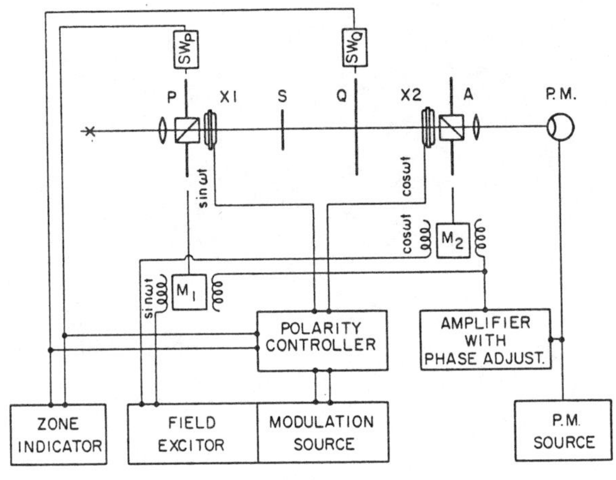

Fig. 2. A block diagram of an automatic ellipsometer. SW_P and SW_Q: *dp–dt* switches operated by cams that are mounted on the polarizer P, and the quarter-wave Q, and M_1 and M_2 are two-phase servomotors.

of the zone in which the measurement is taking place, is made by a relay circuit operated by two microswitches and cams mounted on the polarizer and the quarter-wave plate. The actual circuit will be given in the next section.

III. Experimental

A Gaertner ellipsometer, Model L119, is used as the basic instrument. The schematic diagram of the entire system is shown in Fig. 2.

The ADP cells with 12-mm aperture are mounted on the polarizer and the analyzer keeping the angle between the principal axis of the electrically induced birefringence and the principal axis of the polarizing prism exactly 45°.

The mounting for the crystal unit shown in Fig. 3 allows the independent adjustment of tilt and azimuth of the cell.

The light source is incandescent light monochromatized to approximately 546 nm by an interference filter B.L.H9546 and a gelatin filter Wratten No. 61.

The light source is located adjacent to fixed collimator, and a photomultiplier is attached to the telescope mounted on the swing arm. Because of the design of instrument, the direction of propagation of the light is opposite to that used in the analysis. Accordingly, some confusion in terminology occurs. But the names polarizer and analyzer that have been used in the above analysis will be employed in the customary manner.

A simple high-voltage source supplies about −70 V dc cathode voltage to the photomultiplier. Th signal is amplified and is fed to two servomotors, M and M_2.

The ADP cells X_1 and X_2 are subject to the same 60 cycle voltage of about 200 V, but with different phases X_1 is subject to voltage of $E\sin\omega t$, and X_2 is subject to E $\cos\omega t$.

Thus, the retardation signal and the azimuth signa in Eq. (4) are modulated by 60-cycle voltages wit

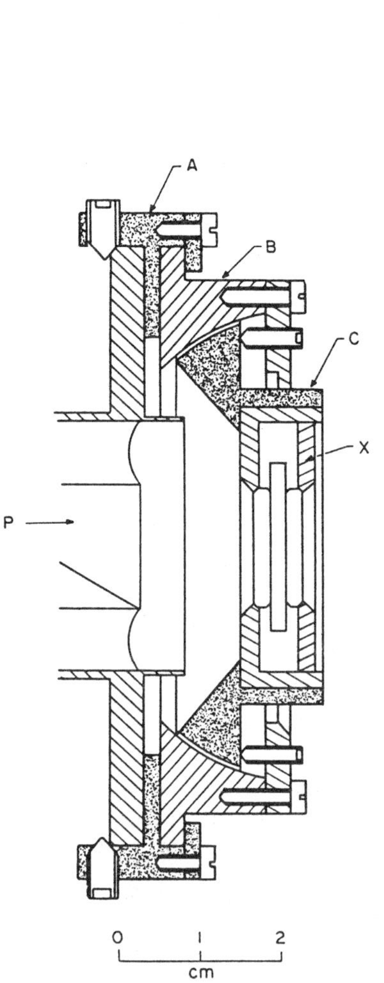

Fig. 3. A sectional diagram of the cell mounting. An ADP cell X is held in C, which a has spherical bottom and is accepted by a matching spherical recess of B. B is held in A, and azimuth adjustment can be made independent of the tilt adjustment.

orthogonal phase relation. From Eq. (4) it will be seen that the servomotor M_1 that controls the polarizer angle α must pick up the signal component modulated by X_2, i.e., the signal component with $\cos\omega t$ phase. The servomotor M_2 that controls analyzer angle β must pick up the signal component modulated by X_1, i.e., the signal component with phase $\sin\omega t$. By exciting the field windings of the M_1 with $\sin\omega t$ phase, signal component with $\sin\omega t$ phase is rejected, and only the signal component with $\cos\omega t$, the azimuth signal, is picked up by the motor and fed back to polarizer adjustment. The same is applied to the second servomotor M_2, which is excited with $\cos\omega t$ phase and is sensitive only to the signal component with $\sin\omega t$ phase, which is picked up by the motor and adjusts the analyzer angle to extinguish the retardation signal.

In practice, a modulator–exciter circuit provides two sets of orthogonal voltages for modulation of the ADP cells and the field excitation of the servomotors. The relative phase relation between the signals and the field excitation of the servomotors is adjusted by a phase shift circuit in the main amplifier.[5]

Two identical cams are mounted on the polarizer and the quarter-wave plate in the same way; that is, the cam throws an accompanying double-pole, double-throw microswitch to one way when the polarizer (quarter-wave plate) azimuth is between 0° and 90° or 180° and 270°, but throws the switch to the other way when the polarizer (quarter-wave plate) azimuth is between 90° and 180° or 270° and 360°. One of the double-throw poles is used to indicate the zone in which the measurement is being made. Another pole is used to control the relay circuit which adjusts the polarity of the servosystems. These circuits are shown in Fig. 4. A close-up of the modulator parts is shown in Fig. 5.

IV. Adjustments

A. ADP Cells

The ADP cells must be mounted so that their crystallographic optical axes coincide with the optical axis of the instrument. In addition, the principal axis of the electrically induced birefringence of the crystal must make a 45° angle with the principal axis of the polarizing prism on which it is mounted.

After the conventional alignment of a polarization goniophotometer, the polarizer and analyzer are adjusted to cross each other, preferably at 0° and 90°. One of the ADP cells is put in position; for instance, the X_1 on the polarizer. The tilt of the cell is adjusted so that the photometric field is covered by the central dark cross of the conoscopic pattern. Thus, the crystallographic optical axis of the cell is adjusted to coincide with the optical axis of the instrument.

Next, a birefringent sample, preferably a quarter-wave plate, is inserted between the cell and analyzer keeping the relative azimuth to the analyzer about 45°. Using a photomultiplier, a modulation voltage is applied to the cell. The amplified signal is displayed on a cathode ray oscilloscope or on an ac voltmeter, and the cell is rotated to extinguish the signal. Either of the principal axes of the electrically induced birefringence is now adjusted to coincide with the plane of polarization of the polarizer. By referring to the graduations on the cell mount, the cell is rotated by 45° and secured. The extinction is again checked without modulation, and the tilt of the cell is adjusted if necessary.

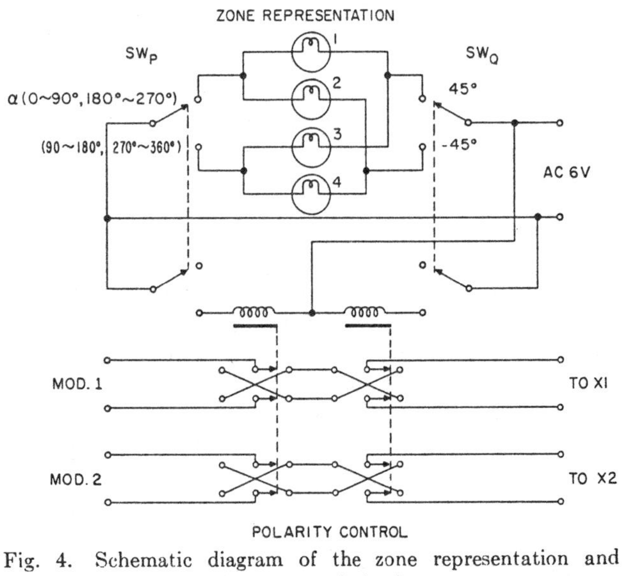

Fig. 4. Schematic diagram of the zone representation and polarity-control circuits.

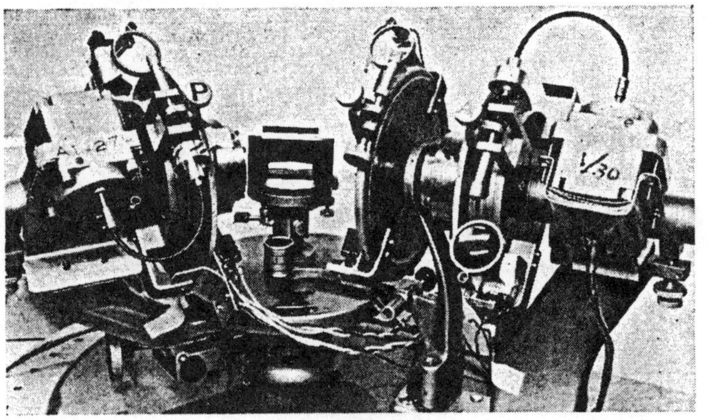

Fig. 5. Close-up view of the cell mounts and optical feedback system.

The same procedure is applied to the second ADP cell X_2 that is to be mounted on the analyzer. While setting the second cell, the first is left as is, without modulation.

B. Signal Separation and Optical Feedback

The orthogonality of the two modulating voltages and field excitations is adjusted. After setting the polarizer and the analyzer at extinction visually, the adjustments are displaced by small amounts, for example, 1° or 2°.

Using a photomultiplier and a modulation voltage applied to one of the cells—for instance, to the cell X_1 on the polarizer—amplified signal is supplied to the two servomotors. The servomotors are disconnected from both the polarizer and the analyzer. Both servomotors begin to rotate in response to the modulation. The phase-shift circuit in the amplifier is adjusted until servomotor M_1, which turns the polarizer, stops. Now the other servomotor M_2 is rotating at its maximum speed. If the analyzer angle is adjusted around the extinction point, M_2 changes its direction of rotation at extinction, and the direction of rotation should be such that it returns the analyzer to the extinction point when the motor is coupled to the analyzer. After the analyzer is set to stop the motor, the second cell X_2 on the analyzer is modulated together with X_1. The servomotor M_1 that controls the polarizer begins to rotate. The same procedure is applied to polarizer M_1 combination to check the rotational direction of M_1. If the direction is opposite, the polarity of either the modulation, the field excitation or the signal to the motor may be reversed to get proper rotational direction. If necessary, the phase may be adjusted to obtain the compromise so that the small displacement of the polarizer causes the least effect on the servomotor M_2, and at the same time analyzer displacement causes the least effect on the servomotor M_1. Next, the motor M_1 is coupled to the polarizer, and M_2 to the analyzer, to complete the optical feedback.

Referring to Eq. (4), when the polarizer and the analyzer are adjusted to their balance points by the motors and the motors stop running, the second and the third terms of the equation are set to zero. Polarizer setting α and analyzer setting β may be read using conventional Vernier scales on each of the polarizer and the analyzer drums. The retardation Δ and reflectance ratio R_p/R_s is obtained by referring to Tables II and III.

V. Performance

The results of the analysis have been demonstrated by experiments. When care is taken not to disturb the sample, the short period (of the order of 10 min) *reproducibility* of the angle settings of the polarizer and the analyzer has been shown to be 0.01° for a metallic sample.

The angle setting remains within ±0.01° for a light-source voltage drop from 120 V to 75 V ac (again for metallic sample), but begins to drift slowly when the light source voltage decreases below 75 V. This failure comes from the residual noise in the high voltage source and may be corrected by refinement of the electronic circuit. The response speed is rather slow. It takes 5 sec to arrive at balance from 1° away.

Figure 6 shows an example of application of the instrument, in which the polarizer and the analyzer settings with arbitrary original settings are plotted with length of time of microwave application on a gauge-block surface at the incidence angle 55°. The surface was cleaned using alcohol and was kept in position for more than 2 hr. With the application of microwave, the polarizer and the analyzer settings start to drift and, after 10 min, the settings become stable again.

This effect is considered as caused by boiling out a part of the vapor film adsorbed on the surface. After the surface arrives at this state, the polarizer and the analyzer settings respond quickly with off-and-on application of microwaves on the surface. The four measured points at the end of the curves show this situation. By turning off the microwave, the polarizer and the analyzer settings go back toward original values, and with application of microwave, both settings come back very fast to the former terminal values, as shown by the end of each curve. This experiment demonstrates that the instrument is suited to follow the change of both polarizer and analyzer settings simultaneously.

There is another point of improvement from the standpoint of instrumentation.

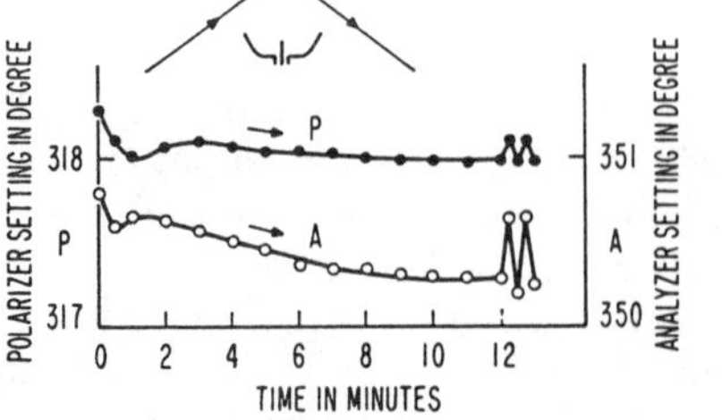

Fig. 6. Changes of angular settings of polarizer and analyzer as a function of time as a gauge-block surface is irradiated with microwave radiant energy. The power density and frequency of the microwave are not measured. Origin of angle is arbitrary.

The present instrument makes use of a lever and tangent-screw mechanism for fine adjustment. This mechanism is not efficient. In addition to limiting the response speed, it requires a coarse manual setting of the polarizer and analyzer before starting the automatic adjustment in order to keep the balance point in the limited adjustable range.

Additional problems now masked by the inefficiency of the fine-adjustment mechanism will probably become apparent when a better mechanism is used. With a system more efficient than the lever and screw arrangement, there may be an interacting hunting of the two servosystems. This can be corrected by adjusting the response time of one servosystem to a value of between two and ten times that of the other. If the fine-adjustment mechanism is changed to increase the range of adjustment, the automatic system will be started quite often at a position far away from the balance point, and the intense transmitted light and excess signals will saturate the photomultiplier and amplifier. In this case some refinements in the electronic circuit will be required.

VI. Discussion

The example is for demonstration of performance. The reproducibility of the settings under a given condition has been proven to be of the order of 0.01°, but this does not mean that the accuracy is of the same order. Errors that are estimated to be of the order of 0.1° may be introduced by the imperfection of elements.

From the theoretical point of view, the present design of polarization optics is considered as optimum. From the practical standpoint, however, extremely high precision is required for the rotating parts. The surfaces of the ADP cell must be ground accurately parallel to each other, and the cell must be adjusted so the crystallographic optic axis coincides with the axis of rotation as well as the optical axis of the instrument. The poorer of the two cells in the present instrument deflects light as much as 20 min of angle, causing nutation of light with rotation of the cell. This is serious when the required accuracy in the incident angle is high. In addition, if there is any residual error in alignment of the crystallographic axis and the optical axis of the instrument, it will result in change of birefringence with polarizer and analyzer settings and will make the correction difficult.

These considerations suggest a compromise between design and precision of manufacture. If the precision of manufacture can be improved sufficiently, the present optical arrangement will be used. If not, another optical arrangement, similar to the one reported earlier[4,5] which keeps the ADP cells stationary, must be used.

To summarize, an automatic ellipsometer system has been developed by the theoretical analysis, and the experimental instrument performed in a satisfactory manner as the analysis indicated.

Reproducibility of 0.01° has been obtained. The instrument is precise enough for comparative study and is especially suited to follow simultaneously the variations of both the polarizer and the analyzer settings.

Some troubles in absolute accuracy remain. Analysis of the error and further improvement of the instrument are left for the future.

The author wishes to express his sincere thanks to the staff of the National Bureau of Standards, especially T. R. Young for his guidance in the theoretical treatment of ellipsometry, J. S. Beers for arrangement of laboratory facilities, M. C. Sola for his help during the course of experimental work, and to O. N. Stavroudis and C. D. Patterson for their assistance in preparing this paper.

References

1. C. V. Kant and J. Lawson, J. Opt. Soc. Am. **27,** 117 (1937).
2. R. W. Ditchburn and G. A. J. Orchard, Proc. Phys. Soc. (London), **B67,** 608 (1954).
3. G. K. T. Conn and G. K. Eaton, J. Opt. Soc. Am. **44,** 484 (1954).
4. H. Takasaki, J. Opt. Soc. Am. **51,** 463 (1961).
5. H. Takasaki, Japan. Appl. Phys. **29,** 468 (1960).
6. A. B. Winterbottom, *Natl. Bur. Std. Misc. Publ. No. 256*, Government Printing Office, Washington, D. C., 1964), p. 97.
7. B. H. Billings, J. Opt. Soc. Am. **38,** 819 (1948).
8. W. A. Shurcliff, *Polarized Light* (Harvard University Press, Cambridge, 1962).
9. H. Takasaki, N. Okazaki, and K. Kida, Appl. Opt. **3,** 833 (1964).
10. A. B. Winterbottom, *Optical Studies of Metal Surfaces*, The Royal Norwegian Scientific Society Report No. 1 (F. Bruns, Trondheim, Norway, 1955).
11. F. L. McCrackin, E. Passaglia, R. R. Stromberg, and H. L. Steinberg, J. Res. Natl. Bur. Std. **67A,** 363 (1963).

Influence of the input state of polarization on interferometric and polarimetric measurements of birefringence

T. Tambosso and S. Donati

Department of Electronics, University of Pavia, 27100 Pavia, Italy

Received June 10, 1988; accepted February 24, 1989

Using a general model to schematize polarimetric and interferometric measurements, we derive the dependence of the output signal on the input state of polarization. We show that in any birefringent medium, in addition to the eigenvectors that maximize the interferometric phase signal, there exist dual vectors that maximize the amplitude of the polarimetric signal. The dual vectors lie on the maximum circle normal to the eigenvector axis.

The evolution of the state of polarization (SOP) is an important issue for coherent communications and optical fiber sensors. In interferometric sensors and coherent receivers, fluctuations in the output SOP induce the so-called polarization fading of the photodetected signal,[1–4] for which polarization controllers[2,4] and polarization diversity have been devised. In polarimetric fiber sensors[5] fluctuations of the SOP directly reflect themselves as a measurement error whose effect has not yet been analyzed in detail. Also, it is well known that interferometric and polarimetric measurements are complementary because they involve the phase and the amplitude, respectively, of the signal. As shown below, for those input SOP's maximizing the information in one measurement, the other measurement carries no information.

The aim of this Letter is to generalize the analysis of interferometric and polarimetric measurements in birefringent media so as to find which SOP is best suited to perform the measurement and which error is eventually generated.

First, let us model the *interferometric measurement* as in Fig. 1(A), where the input SOP is described by the Jones vector $\mathbf{V}_i$. After propagation the output vector $\mathbf{V}_u$ is combined onto the photodetector with the input delayed by an arbitrary phase ϕ_0. The current signal at the photodetector output is then

$$\begin{aligned} i \propto |\mathbf{V}_i \exp(i\phi_0) + \mathbf{V}_u|^2 \\ = |\mathbf{V}_i|^2 + |\mathbf{V}_u|^2 + 2\,\mathrm{Re}[\mathbf{V}_i \exp(i\phi_0)\mathbf{V}_u^*]. \quad (1) \end{aligned}$$

The interferometric signal, contained in the last term of relation (1), is the real part of the scalar product P between the input and output SOP's,

$$P = \mathbf{V}_i \exp(i\phi_0) \cdot \mathbf{V}_u^*. \quad (2)$$

Now we explicitly define $\mathbf{V}_i$ in terms of the eigenvectors (or polarization modes) of the medium $\mathbf{A}_1$ and $\mathbf{A}_2$ as

$$\mathbf{V}_i = R_1\mathbf{A}_1 + R_2\mathbf{A}_2, \quad (3)$$

where the coefficients R_1 and R_2 are complex numbers (given by the scalar product of $\mathbf{V}_i$ with $\mathbf{A}_1$ and $\mathbf{A}_2$). With $\mathbf{A}_1$ and $\mathbf{A}_2$ normalized ($|\mathbf{A}_1| = |\mathbf{A}_2| = 1$) and $\mathbf{A}_1 \cdot \mathbf{A}_2 = 0$, we have

$$|R_1|^2 + |R_2|^2 = 1. \quad (4)$$

Now let us write the output vector $\mathbf{V}_u$ as

$$\mathbf{V}_u = \mathbf{J} \cdot \mathbf{V}_i, \quad (5)$$

where $\mathbf{J}$ is the Jones matrix of the medium. By inserting Eq. (3) into Eq. (5), since $\mathbf{J} \cdot \mathbf{A}_1 = \lambda_1\mathbf{A}_1$ and $\mathbf{J} \cdot \mathbf{A}_2 = \lambda_2\mathbf{A}_2$ for the eigenvectors and $\lambda_1 = \lambda_2^* = \exp(i\Delta z)$ for the eigenvalues λ_1 and λ_2 of a Jones matrix $\mathbf{J}$ of a nondissipative medium (with the determinant equal to unity), we get easily for $\mathbf{V}_u$

$$\mathbf{V}_u = R_1\mathbf{A}_1 \exp(i\Delta z) + R_2\mathbf{A}_2 \exp(-i\Delta z), \quad (6)$$

where z is the path length and Δ is the birefringence coefficient of the medium. [Specifically, $\Delta = \beta_l/2$ for a linear birefringence, $\Delta = \beta_c/2$ for a circular birefringence, $\Delta = (1/2)(\beta_l^2 + \beta_c^2)^{1/2}$ for a coexisting linear and circular birefringence, etc.] Combining Eqs. (2), (3), and (6), one has

$$P = [\cos \Delta z + i(|R_2|^2 - |R_1|^2)\sin \Delta z]\exp(i\phi_0), \quad (7)$$

and the interferometric signal therefore reads

$$\mathrm{Re}\, P = \cos \Delta z \cos \phi_0 - (|R_2|^2 - |R_1|^2)\sin \Delta z \sin \phi_0. \quad (8)$$

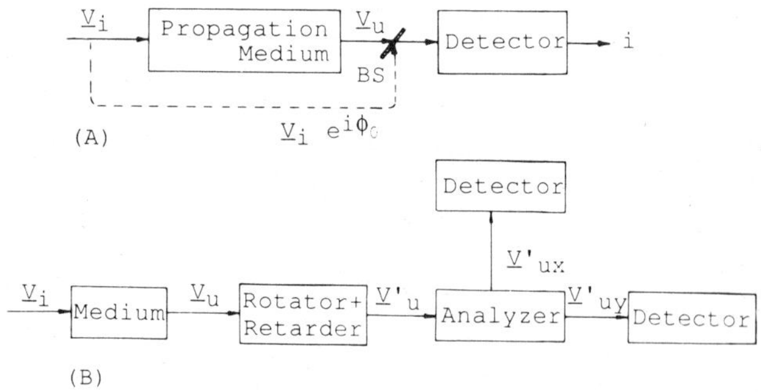

Fig. 1. Models of (A) interferometric and (B) polarimetric measurements. BS, beam splitter.

To discuss the dependence of the interferometric signal from the input SOP, we first note that if one of the eigenvectors of the medium is chosen as the input ($R_1 = 1$ and $R_2 = 0$ or vice versa) the result is

$$\mathrm{Re}\, P = \cos(\phi_0 \pm \Delta z), \tag{9}$$

as expected for a measurement of a phase shift $\pm\Delta z$ with an added reference phase ϕ_0. For all other input SOP's (R_1 and R_2 different from zero) the signal is smaller than the above value, and in the case of $|R_1| = |R_2|$ it becomes

$$\mathrm{Re}\, P = \cos \Delta z \cos \phi_0, \tag{10}$$

i.e., it is modulated in amplitude by the bias term cos ϕ_0. This is the extreme case of maximum fading of polarization.[1,4] Alternatively, Eqs. (9) and (10) can be interpreted in terms of phase- and amplitude-modulated signals. This is easily seen by letting $\phi_0 = \omega t$, so that the above results are also extended to the case of a heterodyne detection scheme. If an eigenvector $\mathbf{A}_1$ or $\mathbf{A}_2$ is used as the input, the beating (or interferometric) signal given by Eq. (9) contains the birefringence information Δz as a phase-modulated term of the carrier ωt, and there is no information in amplitude. The reverse is true if the SOP's with $|R_1| = |R_2|$ are used as input: the information on Δz is contained in the amplitude-modulation term cos Δz of the carrier cos ωt, and there is no information in phase.

We call these SOP's, obtained by equal weighting ($|R_1| = |R_2|$) of the eigenvectors $\mathbf{A}_1$ and $\mathbf{A}_2$, the *dual vectors* of the eigenvectors. On the Poincaré sphere the dual vectors lie on the maximum circle normal to the axis connecting the eigenvectors.

Since the propagation is described by a rotation[6–8] of the Poincaré sphere around the eigenvector axis, an input dual vector will change its SOP on propagation but still be a dual vector [this can also be seen by comparing Eqs. (3) and (6)]. The intersections of the dual-vector circle and of the Poincaré sphere equatorial circle give the two linearly polarized dual vectors of the medium. If $2\Phi_A$ and $2\Psi_A$ are the coordinates (inclination and ellipticity) of the eigenvectors, those of the two linear dual vectors are $2\Phi = \pm\pi/2 + 2\Phi_A$ and $2\Psi = 0$. For a medium presenting a distributed linear (β_l) and circular (β_c) birefringence, in which the rotation velocity[7,8] of the Poincaré sphere is $\mathbf{\Omega} = \mathbf{\Omega}_l + \mathbf{\Omega}_c$ with $|\mathbf{\Omega}| = (\beta_l^2 + \beta_c^2)^{1/2}$, we show in Fig. 2 the eigenvectors $\mathbf{A}_1$ and $\mathbf{A}_2$ and the locus of the dual vectors (circle D). Here the angular coordinates[6,7] of mode $\mathbf{A}_2$ are $2\Phi_A = 0$ (inclination) and $2\Psi_A = 2 \arctan \alpha$ (ellipticity), where $\alpha = (\beta_c/2)/(\beta_l/2 + \Delta)$ and $\Delta = (\beta_l^2 + \beta_c^2)^{1/2}/2$.

Further insight can be obtained by expressing the scalar product P [Eq. (2)] in the form

$$P = |P| \exp(i\phi), \tag{11}$$

so that from Eq. (7) one finds that

$$|P|^2 = 1 - 4|R_1|^2|R_2|^2 \sin^2 \Delta z, \tag{12}$$

$$\phi = \arctan[(|R_2|^2 - |R_1|^2)\tan \Delta z] + \phi_0. \tag{13}$$

For the eigenvectors ($R_1 = 0$ and $R_2 = 1$ or vice versa)

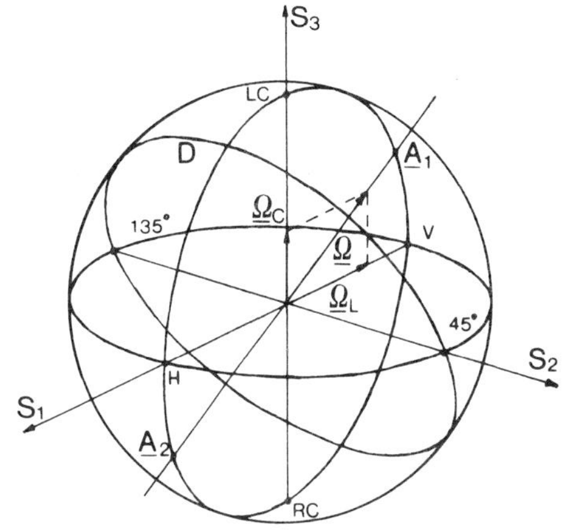

Fig. 2. Representation of eigenvectors $\mathbf{A}_1$ and $\mathbf{A}_2$ and dual vectors (circle D).

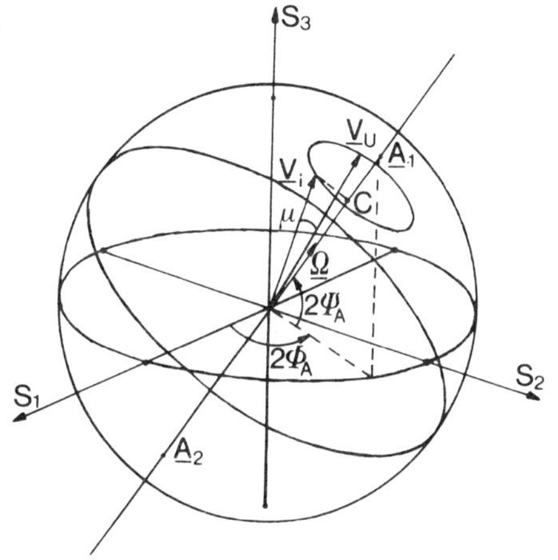

Fig. 3. Geometrical interpretation of the scalar product P and of weights R_1 and R_2.

we have $|P| = 1$ and $\phi = \pm\Delta z + \phi_0$, while for the dual vectors ($|R_1| = |R_2| = 1/\sqrt{2}$) we have $|P| = \cos \Delta z$ and $\phi = \phi_0$. On the Poincaré sphere the modulus $|P|$ of the scalar product has a simple geometrical interpretation related to the angle μ between the input and output SOP vectors (Fig. 3). It can be shown[9] that

$$|P| = |\mathbf{V}_i \cdot \mathbf{V}_u^*| = \cos \mu/2. \tag{14}$$

The projection of $\mathbf{V}_i$ on the eigenstates axis intercepts two segments $\overline{A_1C}$ and $\overline{A_2C}$ that are proportional to $|R_1|^2$ and $|R_2|^2$ (Ref. 10) (Fig. 3). The phase ϕ is connected to $|R_1|^2$ and $|R_2|^2$ through Eq. (13) but has no direct geometrical interpretation on the Poincaré sphere, because ϕ/z is in general dependent on z.

Both the modulus and the phase of the scalar product P depend in general on the phase shift variation Δz [Eqs. (12) and (13)]. These two quantities have the meaning of the information content carried by the amplitude and phase, respectively. We may maximize the information contained in the phase by requiring that the modulus is independent of Δz or, equivalently, that

$$\partial|P|/\partial\Delta z = 0. \tag{15}$$

By inserting Eq. (12) into Eq. (15) we obtain as a condition $|R_1|\,|R_2| = 0$, which is satisfied just by the eigenvectors; conversely, if we maximize the information contained in the amplitude by letting

$$\partial\phi/\partial\Delta z = 0, \tag{16}$$

we obtain the condition $|R_1| = |R_2|$, which is the one defining the dual vectors.

Last, we note that the above analysis is readily extended to the case of birefringence in both arms of the interferometer. For this case we can go back[4] to that of Fig. 1(A) by considering $\mathbf{J}\cdot\mathbf{J}_r^{-1}$ as the Jones matrix of the equivalent medium, where $\mathbf{J}_r^{-1}$ is the inverse of the Jones matrix of the reference-arm birefringence.

The general scheme of a *polarimetric measurement* can be modeled as in Fig. 1(B). Here $\mathbf{V}_i$ and $\mathbf{V}_u$ are the Jones vectors at the input and output of the birefringent medium, $\mathbf{V}_u$ may be modified in $\mathbf{V}_u'$ by means of a suitable rotator and retarder combination, and $\mathbf{V}_u'$ is analyzed in its components V_{ux}' and V_{uy}' with respect to axes S_1 and S_2.

Conceptually the polarimetric signal is maximized by applying a rotation and a retardance such that the input vector $\mathbf{V}_i$ is brought to coincide with the S_1 axis and the output vector $\mathbf{V}_u'$ is brought to lie on the equator of the Poincaré sphere ($S_3 = 0$). In view of the equivalence theorem,[6] this is always possible and physically realizable. Since any rotation of the Poincaré sphere leaves the distances unchanged,[6,8] we have

$$\overline{\mathbf{V}_u'S_1} = \overline{\mathbf{V}_u\mathbf{V}_i} = \mu, \tag{17}$$

where μ is obtained by the differential equation of polarization evolution,[6,8] i.e.,

$$d\mu = |\Omega \times \mathbf{V}|dz = \Omega \sin \overline{\Omega V}\, dz, \tag{18}$$

and the angle $\overline{\Omega V}$ can be related to longitudes Φ_Ω and Φ_V and latitudes Ψ_Ω and Ψ_V through spherical trigonometry as

$$\cos \overline{\Omega V} = \cos 2(\Phi_\Omega - \Phi_V)\cos 2(\Psi_\Omega - \Psi_V). \tag{19}$$

Limiting ourselves to the case of constant $\mathbf{\Omega}$, so that $\mathbf{V}$ rotates around $\mathbf{\Omega}$ with a constant angle $\overline{\Omega V}$, and in general letting $|\Omega| = 2\Delta$,[8] we have

$$\mu = 2\Delta z \sin \overline{\Omega V_i}. \tag{20}$$

Since the output $\mathbf{V}_u'$ has coordinates $\Phi = \mu$ and $\Psi = 0$, it can be written as the Jones vector,

$$\mathbf{V}_u' = \begin{bmatrix}\cos \mu/2\\ \sin \mu/2\end{bmatrix} = \begin{bmatrix}\cos \Delta' z\\ \sin \Delta' z\end{bmatrix}, \tag{21}$$

where the birefringence coefficient Δ' is defined explicitly, in terms of input vector angular coordinates Φ_i and Ψ_i, as

$$\Delta' = \Delta[1 - \cos^2 2(\Phi_\Omega - \Phi_i)\cos^2 2(\Psi_\Omega - \Psi_i)]^{1/2}. \tag{22}$$

As special cases of Eqs. (21) and (22), we consider the eigenvectors for which $\mathbf{V}_i \parallel \mathbf{\Omega}$, $\Delta' = 0$, and no polarimetric signal is developed, and the dual vectors for which $\mathbf{V}_i \perp \mathbf{\Omega}$, $\Delta' = \Delta$, and the output vector has the components $\cos \Delta z$ and $\sin \Delta z$.

Another useful expression can be obtained in terms of the weight coefficients R_1 and R_2 by writing $\mathbf{V}_i$ as in Eq. (3) and $\mathbf{V}_u$ as in Eq. (6). From the projection of the input vector on the eigenvector axis $\mathbf{\Omega}$ the two segments $\overline{A_1C} = 2|R_2|^2$ and $\overline{A_2C} = 2|R_1|^2$ are found, which allow us to evaluate $\sin \overline{\Omega V}$ as $(1 - \cos^2 \overline{\Omega V})^{1/2} = [1 - (1 - \overline{A_1C})^2]^{1/2} = 2|R_1|\,|R_2|$. Therefore we can rewrite Eq. (22) in the form

$$\Delta' = 2\Delta|R_1|\,|R_2|. \tag{23}$$

It is interesting to note that, with the rotator and retarder compensation, the output signal $\mathbf{V}_u'$ in a polarimetric measurement can always be brought to the form of cos and sin components of $\Delta' z$. The reduction factor $\Delta'/\Delta = 2|R_2|\,|R_1|$ is independent of the argument of R_2 and R_1 or, equivalently, from the longitude of $\mathbf{V}$ with respect to $\mathbf{\Omega}$; Δ'/Δ is unity for the dual vectors and decreases as it moves on the Poincaré sphere toward the eigenvectors, for which it is zero.

In conclusion, we have derived expressions regarding interferometric and polarimetric measurements in terms of input SOP's. We have shown that the dual vectors maximize the polarimetric amplitude and give a zero output for the interferometric output, as opposed to the eigenmode SOP's, for which the reverse is true.

References

1. D. W. Stowe, D. R. Moore, and R. G. Priest, IEEE J. Quantum Electron. **QE-18,** 1644 (1982).
2. T. Okoshi, IEEE J. Lightwave Technol. **LT-5,** 44 (1987).
3. K. H. Wanser and N. H. Safar, Opt. Lett. **12,** 217 (1987).
4. A. D. Kersey, A. Dandridge, and A. B. Tveten, Opt. Lett. **13,** 288 (1988).
5. T. G. Giallorenzi, J. Bucaro, A. Dandridge, G. Siegel, J. H. Cole, S. Rashleigh, and R. G. Priest, IEEE J. Quantum Electron. **QE-18,** 626 (1982).
6. P. S. Theocaris and E. E. Gdoutos, *Matrix Theory of Photoelasticity* (Springer-Verlag, New York, 1979).
7. S. C. Rashleigh, IEEE J. Lightwave Technol. **LT-1,** 312 (1983).
8. R. Ulrich and A. Simon, Appl. Opt. **18,** 2241 (1979).
9. R. Ulrich, Opt. Lett. **1,** 109 (1977).
10. Y. Yen and R. Ulrich, Appl. Opt. **20,** 2721 (1981).

Measuring Faraday rotation over extended wavelength ranges: a simple method

Marco Marinelli, Enrico Milani, and Paolo Paroli

Enrico Milani is with II University of Rome, Department of Mechanical Engineering, 6 Via O. Raimondo, I-00173 Rome, Italy; the other authors are with CNR Institute of Solid-State Electronics, 42 Via Cineto Romano, I-00156 Rome, Italy.

Received 3 February 1988.
0003-6975/88/214360-02$02.00/0.

Faraday rotation can be defined as the rotation undergone by the polarization plane of linearly polarized light after crossing a unit section of a sample magnetized parallel to the light propagation direction. The Faraday rotation is in general[1] proportional to the sample magnetization and, therefore, in diamagnetic and paramagnetic materials, to the externally applied field. The Faraday rotation has long been known as a sensitive tool to investigate the nature of optical transitions,[2] sublattice magnetization,[3] and magnetic phase transitions.[4] In recent years, the Faraday effect has also gained applicative importance: several magnetooptical devices[5] have been developed exploiting its nonreciprocity, namely, the fact that the sense of rotation changes sign if the sample magnetization is reversed.

The simplest way to measure the Faraday rotation is the extinction method: the sample is placed between linear polarizers, and the difference is measured between the angular settings of the analyzer yielding minimum transmitted intensity when the sample magnetization is parallel or antiparallel to the light propagation direction. This method is commonly used for visual measurements with polarizing microscopes. Its accuracy is rather poor, since it turns out[6] that the angular sensitivity is a minimum just close to the extinction angle to be measured. The angular sensitivity is instead a maximum[6] if the two polarizers are set 45° apart: this has been recently exploited by Hartmann *et al.*[7] for the readout devices of magnetooptical disks.

Much higher sensitivity is obtained with modulation methods and synchronous inphase detection. The light polarization state can be modulated making use of rotating polarizers[8] or photoelastic modulators.[9] However, the former method allows rather low modulation frequencies (typically 30 Hz) and (in our experience at least) introduces mechanical vibrations and instabilities. The latter method, on the contrary, operates at frequencies (typically 50 kHz) too high to be compatible with the PbS photoresistive detectors, which are very suitable because of their high sensitivity and wide operational wavelength range ($0.5\ \mu m \leqslant \lambda \leqslant 3.0\ \mu m$). Silicon and germanium photodiodes (or phototransistors) have to be used instead, which allow light detection only up to $\lambda = 1.2$ and $1.7\ \mu m$, respectively. This is a serious limitation both from the fundamental and applicative point of view: the wavelength range $1.8\ \mu m \leqslant \lambda \leqslant 3.0\ \mu m$ is of interest, on the one hand, to study the optical transitions of a number of transition metal[10] and rare-earth[11] compounds and, on the other hand, to realize integrated optical devices of future generation.

The above limitation is overcome with our measuring scheme (Fig. 1). The light beam passing the chopper is linearly polarized (say, at 45° from the vertical direction) and focused onto the magnetized sample. Emerging from the sample, it is divided into two beams, nominally of equal intensities, and directed 90° apart by means of a nonpolarizing beam splitter (e.g., a half-silvered mirror). Each of the two beams passes through a linear polarizer, set at +45 and −45°, respectively, relative to the original polarization plane and is finally focused onto a PbS detector. The output of each PbS detector is fed into the inverting and noninverting input, respectively, of a lock-in amplifier tuned to the chopper frequency. Of course, the beam splitter and the two Glan-Thomson prisms could be replaced (with a gain in the signal intensity) by a polarizing beam splitter (like Foster of Wollanstron prisms) that, however, are less common and more expensive pieces.

If the sample is absent, the scheme of Fig. 1 yields, in principle, zero signal from the lock-in, since the inverting and noninverting input receive a signal of identical amplitude. Small spurious signals due to inequalities between to two beams can be canceled by fine adjustments of the setting of the polarizer P_0.

When a platelet sample is inserted, magnetized parallel to the light beam, it will cause a rotation θ_F of the polarization plane of the light impinging on the beam splitter, so that the two detectors will receive signals of different amplitude. The lock-in amplifier will measure a signal proportional to

$$\sin^2(45 + \theta_F) - \sin^2(-45 + \theta_F) \propto \sin 2\theta_F.$$

To obtain again zero signal from the lock-in amplifier, the polarization state as in the absence of sample has to be

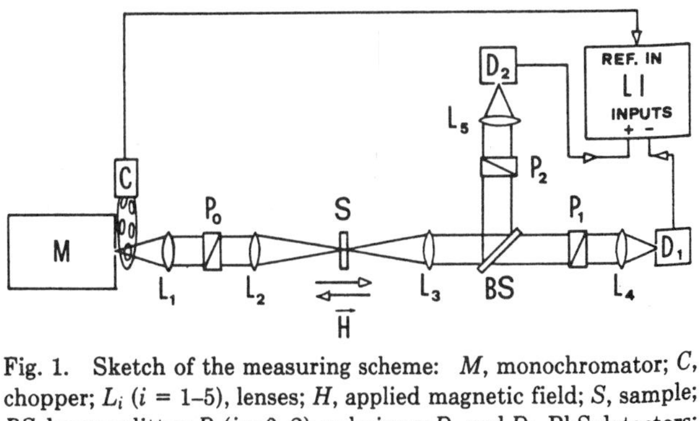

Fig. 1. Sketch of the measuring scheme: M, monochromator; C, chopper; L_i (i = 1–5), lenses; H, applied magnetic field; S, sample; BS, beam splitter; P_i (i = 0–2), polarizers; D_1 and D_2, PbS detectors; LI, lock-in amplifier.

restored for the light impinging on the beam splitter: it will suffice for that to change by $-\theta_F$, the angular setting of the first polarizer.

In the same way, if the sample is magnetized antiparallel to the light beam, it will rotate the polarization plane of the light by $-\theta_F$, and zero signal from the lock-in amplifier is obtained after rotating the polarizer P_0 by $+\theta_F$. Of course, then, a practical procedure consists in adjusting the first polarizer to have a zero signal for one magnetization direction, then inverting the magnetic field applied to the sample, finally taking the Faraday angle as one half of the angle by which the first polarizer has to be rotated to recover a zero signal from the lock-in amplifier. In principle, the sensitivity of the measurement is limited only by the reading accuracy of the angular setting of the polarizer P_0; the circle dividers with 10^{-2} or 10^{-3} deg reading accuracy are commercially available. The other two polarizers do not need such high accuracy mounts.

The above measurement can easily be made computer-controlled. A periodical switch of the magnetic field has to be accomplished, and between magnetization reversals the rotating mount of the first polarizer has to be driven by means of a step-motor to minimize the amplitude of the difference signal from the lock-in amplifier. Since the latter is used as a null-reading instrument, no adjustment of the setting of other optical pieces is required if the intensity of the light beam is changed, as when performing spectral scannings.

The spectral range of the method is limited only by the operational range of the polarizers (typically 0.3–3-μm wavelength with Glan-Thomson prisms) and of the detectors (0.5–3-μm wavelength with PbS cells), so that a wide portion of the visible and of the near-IR wavelength range is readily accessible.

As an example, we performed Faraday rotation measurements on a 170-μm thick single-crystal platelet of yttrium iron garnet ($Y_3Fe_5O_{12}$) doped with Si and Co. In this material, it has long been known[12] that Co gives rise to an optical transition at $\lambda = 2.2$ μm, but the associated magnetooptical activity had never been measured due to the above-mentioned limitations. Figure 2 shows the spectra of the Faraday rotation, measured with our method, and of the optical absorption, measured with a Cary 17D spectrophotometer, for 1.8 μm $\leqslant \lambda \leqslant$ 2.4 μm. (The accuracy of the angular measurement was $\sim$0.05°, which, for a sample thickness of 170 μm, corresponds to an accuracy of the Faraday rotation of $\sim$3° cm^{-1}.) It is clearly seen that the Faraday rotation curve has a dispersive shape centered at the maximum of the absorption curve, so that[1,2] the paramagnetic character of this transition is revealed. This indicates that both the ground and excited energy levels involved in this transition are singlets (i.e., unsplit); see, e.g., Ref. 1.

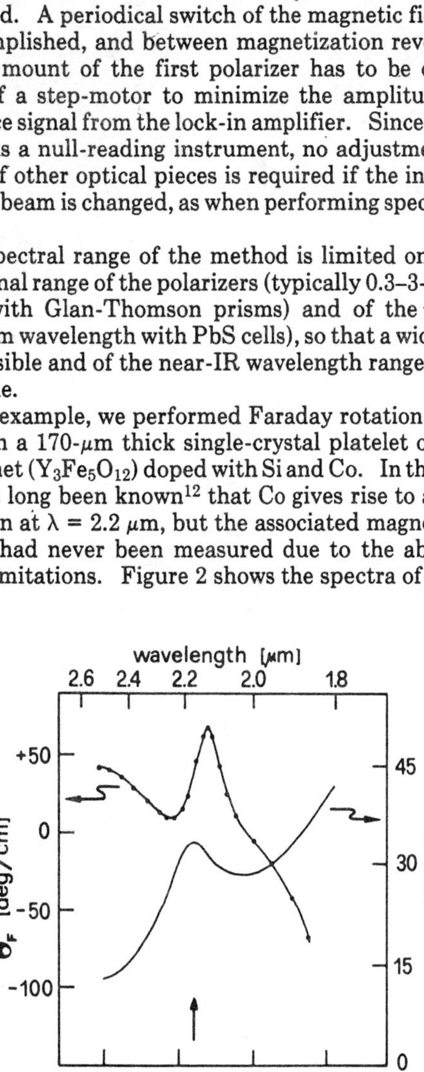

Fig. 2. Faraday rotation θ_F (dotted curve, left scale) and absorption α (continuous curve, right scale) spectra of Si- and Co-doped yttrium iron garnet in the medium infrared. The arrow marks the energy at which an optical transition occurs (see Ref. 11). The dispersive behavior of θ_F about the transition energy reveals the paramagnetic character of this transition; see Ref. 1.

If the magnetic circular dichroism Δ associated with the Faraday rotation (see, e.g., Refs. 1 and 6) is noticeable, the light emerging from the sample is elliptically polarized, and by our method (without any change in the measurement procedure) the Faraday rotation is correctly measured as one-half of the angular change of the major axis of the ellipse, obtained by reversing the sample magnetization. Of course, the ellipticity should not be such to make difficult to locate the major axis itself, namely, the sample thickness t should be small enough that $t\Delta < 1$. This condition was fulfilled in the case of Fig. 2, where $t = 170$ μm, and Δ ranged from 2 to 5 cm^{-1} around the transition, so that $t\Delta = 0.34$–0.68.

This work has been supported in part by Progetto Finalizzato MADESS of CNR.

Paolo Paroli also works in the Department of Mechanical Engineering the II University of Rome.

References

1. See, for example, W. A. Crossley, R. W. Cooper, J. L. Page, and R. P. van Stapele, "Faraday Rotation in Rare-Earth Iron Garnets," Phys. Rev. **181**, 896 (1969).
2. G. B. Scott, "The Optical Absorption and Magneto-Optical Spectra of YIG," in *Physics of Magnetic Garnets*, A. Paoletti, Ed. (North-Holland, Amsterdam, 1978), p. 445.
3. G. Abulafya and H. Le Gall, "Temperature Dependence of the Octahedral and Tetrahedral Magneto-Optical Coefficient in YIG," Solid State Commun. **11**, 629 (1972).
4. J. F. Dillon, "The Development of Magneto-Optical Research in Garnets and Magnetic Insulators," in *Ferrites*, H. Watanabe, S. Iida, and M. Sugimoto, Eds. (Reidel, Dordrecht, 1982).
5. P. Paroli, "Magneto-Optical Devices Based on Garnet Films," Thin Solid Films **114**, 187 (1984).
6. P. Paroli, "An Introduction to Magneto-Optics," in *Magnetic Properties of Matter*, F. Borsa and V. Tognetti, Eds. (World Scientific, Singapore, 1988).
7. M. Hartmann, B. A. J. Jacobs, and J. J. M. Braat, "Erasable Magneto-optical Recording," Philips Tech. Rev. **42**, 37 (1985).
8. T. K. Vien and H. LeGall, "Mesures des Birifringences Magnetiques Circulair et Lineaire par Analyseur et Lame Birefringente Tournants," Nouv. Rev. Opt. Appl. **4**, 115 (1973).
9. J. P. Krumme, V. Doorman, and C. P. Klages, "Measurement of the Magneto-Optical Properties of Bismuth-Substituted Iron Garnet Films Using Piezobirefringent Modulators," Appl. Opt. **23**, 1184 (1984).
10. D. S. McClure, *Electronic Spectra of Molecules and Ions in Crystals* (Academic, New York, 1979).
11. D. L. Wood and J. P. Remeika, "Effect of Impurities on the Optical Properties of Yttrium Iron Garnet," J. Appl. Phys. **38**, 1038 (1967).
12. D. L. Wood and J. P. Remeika, "Optical Absorption of Tetrahedral Co^{3+} and Co^{2+} in Garnets," J. Chem. Phys. **46**, 3595 (1967).

Verdet Constants of Rare-Earth Glasses measured with a New Polarimeter

IN some optical devices the Faraday effect is used with advantage as a shutter, and, as a modulator of a light beam, it is especially useful in polarimetry[1-3]. In such applications a material having a large value of the Verdet constant is usually sought, consistent with adequate transmission at the desired wavelengths. While numerous materials, especially thin metal films and ferrites, are known which have large absolute values of the Verdet constant[4], measurements reported here are limited to glasses.

From the data published by Cole[5] the Verdet constants of silicate glasses not containing rare-earths are seen to be approximately proportional to the reciprocal of their constringence, and positive. Therefore, a low-power magneto-optic modulator using the Faraday effect has often depended upon the use of a dense flint glass of low constringence.

It was discovered here that there is at least one readily available glass the Verdet constant of which is negative : didymium filter glass, type *ON*16 (Chance-Pilkington Optical Works, Ltd.). The negative Verdet constant of this glass raised the question of how the Verdet constant would depend upon the presence of a paramagnetic rare-earth in a glass.

A preliminary examination with a visual polarimeter of some experimental glasses, obtained from the research laboratories of Messrs. Pilkington Brothers, Ltd., was then made. These glasses contained a constant proportion of boric oxide, and samarium oxide replacing lanthanum oxide from 0 to 65 per cent by weight. It was found that their Verdet constants, while positive in sign, decreased with increasing samarium content.

Finally, the Verdet constants of a number of borate glasses specially made by the above Research Laboratories and containing graded proporti ons of the paramagnetic rare-earths gadolinium and dysprosium were measured using the recently available *ETL-NPL* automatic polarimeter.

The operation of this polarimeter has been described elsewhere[6], but it should be noted that the activity of a sample in this polarimeter appears as an analogue electric current. In these experiments this current was measured using an adjustable shunt and a potentiometric chart recorder.

The optical activity was measured using an incandescent source filtered in the neighbourhood of mercury green with a half-intensity width of 200 A. A reduced aperture was used so that the beam crossing the sample had a diameter of 2 mm.

The various samples were polished plane-parallel to thicknesses between 2 and 3 mm., and were left as fragments of less than $\frac{5}{8}$ in. in any other dimension.

An axial magnetic field of about 1,120 gauss was provided in a sample holder which used two axially polarized permanent magnets of an annular shape. The flux path of each magnet was nearly completed with turned soft-iron washers having a conical inside surface perforated with a central $\frac{1}{8}$-in. hole, and leaving an axial gap of $\frac{1}{8}$ in. to contain the sample. Two such assemblies with their magnetic fields in opposition were mounted within a plastic holder having an outside diameter of 30 mm. Only one magnet assembly was used to contain the sample, and the magnetic fields of the two tended to cancel, so that the effect of stray magnetic fields upon the polarimeter was reduced to less than 0·008° arc and could be corrected for. During the course of some fifty measurements where the sample holder was dismantled and re-assembled each time, the effective field-strength H was found to have decayed by less than 1 per cent.

The effective field-strength H in the perforated gap of the sample holder can be found by referring the observed magneto-optic rotation of fused quartz α_q to the known Verdet constant V_q of fused quartz[7] (16·64 min./kilogilbert) :

$$H = \alpha_q / V_q p_q \quad (1)$$

where p_q is the path-length. Measurements of the Verdet constants of the rare-earth glasses were then made referring to the Verdet constant of fused quartz. Thus, if the observed sample rotation α_s in a path-length p_s corresponds to a sample Verdet constant V_s :

$$V_s = \alpha_s / H_s p_s \quad (2)$$

Upon substituting the value of H from (1) then,

$$V_s = V_q(\alpha_s/\alpha_q)\ (p_q/p_s) \quad (3)$$

Chart records of the optical activity of the fused quartz reference and of the sample glasses were usually made alternately, and the ratios of optical activity taken from adjacent traces. The chart record was also convenient in observing the birefringence of the samples, as an oscillatory trace was obtained when the sample holder, together with the sample in it, was slowly turned about the optical axis. The activity of the sample was then taken as the average value of the trace over at least four cycles.

Table 1

Glass	V_s (min./kilogilbert) λ = 546 nm.	Mean deviation (per cent)	Amplitude of birefringence (per cent)
Didymium *ON* 16	−25·3	1·2	18·0
10 wt. per cent Gd_2O_3	21·2	5·6	43·0
20 ,, ,, ,,	20·5	4·6	64·0
30 ,, ,, ,,	19·8	2·2	25·2
10 ,, ,, Dy_2O_3	−35·2	3·7	22·0
20 ,, ,, ,,	−93·8	3·0	12·8
30 ,, ,, ,,	−150·0	2·7	6·7
030181	151·0	—	—

In all cases the birefringence greatly exceeded the inherent noise-level of the instrument, which was 0·001° arc in a 1 c./s. bandpass. The total excursion of the recorder trace due to residual birefringence is recorded in Table 1 as per cent of the mean optical activity.

The Verdet constants of the six new rare-earth glasses are listed in Table 1. The six samples all contained 37·5 per cent by weight boric oxide, and the balance both lanthanum and the paramagnetic rare-earth oxide.

For the comparison, the Verdet constant of a didymium filter glass, and the most dense flint glass available for construction of a Faraday cell, are included in Table 1.

These results show that certain glasses can have negative Verdet constants comparable in magnitude with that of the densest flint glasses.

It would be interesting to find out in more detail just how the presence of the paramagnetic rare-earth ion is related to the Verdet constant, and whether much greater Verdet constants could be obtained.

I am indebted to Mr. J. Kowal for making many of the measurements reported here, and to the Controller of Research, Dr. J. H. Mitchell, for permission to publish this paper.

NIELS O. YOUNG

Ericsson Telephones, Ltd.,
Instrument Division,
High Church Street,
New Basford, Nottingham.

June 9.

[1] Kudryavtsev, V. I., *Sakharnaya Prom.*, **27**, 14 (1953).
[2] Breton, J., *C.R. Acad. Sci., Paris*, **243**, 254 (1956).
[3] Gates, J. W., *Chem. and Indust.*, 190 (1958).
[4] Dillon, jun., J. F., *J. App. Phys.*, **29**, 539 (1958).
[5] Cole, H., *J. Soc. Glass Technol.*, **34**, 220 (1950).
[6] Gilham, E. J., *J. Sci. Instr.*, **34**, 435 (1957).
[7] Ramaseshan, S., *Proc. Indian Acad. Sci.*, **24**, 426 (1946).

Optical heterodyne measurement of the phase retardation of a quarter-wave plate

Lin Yao, Zhou Zhiyao, and Wang Runwen

Shanghai Institute of Optical and Fine Mechanics, Academia Sinica, Shanghai, China

Received October 26, 1987; accepted March 30, 1988

We propose and discuss an optical heterodyne technique to measure the phase retardation of a wave plate. The accuracy of this measurement technique in determining the optical path difference of a quarter-wave plate is of the order of 0.1 nm, thus exceeding the accuracy of other measurement techniques.

A wave plate is one of the most basic and common optical elements used in optical interferometers. Typical techniques[1–4] used to measure the optical path difference of a wave plate have accuracies ranging from λ/50 to λ/500, where λ is the wavelength of the light source.

An optical heterodyne technique to measure the phase retardation of a wave plate is proposed and discussed in this Letter. This technique uses a common-path interferometric system to measure the phase retardation of a wave plate and yields satisfying experimental results.

The principle of this experiment is shown in Fig. 1. The output of a frequency-stabilized longitudinal Zeeman He–Ne laser is a common-path beam consisting of left and right circularly polarized beams separated slightly in frequency. Passing through the wave plate to be measured, the two polarized beams interfere along the direction of the polarizer orientation. The resulting beam is incident upon a photodetector. The output of the photodetector is proportional to the intensity of the incident light, and a heterodyne signal containing information about the wave plate is displayed on the screen of an oscilloscope. By measuring the amplitude of the alternating component of the heterodyne signal, the phase retardation of the wave plate can be determined.

As the Cartesian coordinate system XOY, shown in Fig. 1, is fixed, the Jones matrices of the left and right circularly polarized beams are expressed as

$$E_+ = \frac{1}{\sqrt{2}}\begin{bmatrix}1\\-i\end{bmatrix} A \exp[-i(\omega_+ t + \varphi_{0+})], \quad (1)$$

$$E_- = \frac{1}{\sqrt{2}}\begin{bmatrix}1\\i\end{bmatrix} A \exp[-i(\omega_- t + \varphi_{0-})], \quad (2)$$

where the subscripts + and − correspond to the left and right circularly polarized beams, respectively, ω is the angular frequency, and φ_0 is the initial phase. When the polarizer is oriented along the Y direction, the Jones matrix of the polarizer P is given by[5]

$$P = \begin{bmatrix}0 & 0\\0 & 1\end{bmatrix}. \quad (3)$$

The angle between the rapid axis of the wave-plate crystal and the X direction is θ, and the phase retardation of the wave plate is δ. The Jones matrix of the wave plate is then given by[6]

$$M(\delta,\theta) = \begin{bmatrix}\cos\frac{\delta}{2} + i\sin\frac{\delta}{2}\cos^2\theta & -i\sin\frac{\delta}{2}\sin^2\theta\\ -i\sin\frac{\delta}{2}\sin^2\theta & \cos\frac{\delta}{2} - i\sin\frac{\delta}{2}\cos^2\theta\end{bmatrix}. \quad (4)$$

The optical fields $E_+(f_+)$ and $E_-(f_-)$ of the left and right circularly polarized beams at the photodetector are

$$\begin{aligned} E_+(f_+) &= PM(\delta,\theta)E_+ \\ &= \left(-i\sin\frac{\delta}{2}\sin^2\theta - i\cos\frac{\delta}{2} - \sin\frac{\delta}{2}\cos^2\theta\right) \\ &\quad\times \frac{1}{\sqrt{2}} A\exp[-i(\omega_+ t + \varphi_{0+})], \end{aligned} \quad (5)$$

$$\begin{aligned} E_-(f_- &= PM(\delta,\theta)E_- \\ &= \left(-i\sin\frac{\delta}{2}\sin^2\theta + i\cos\frac{\delta}{2} + \sin\frac{\delta}{2}\cos^2\theta\right) \\ &\quad\times \frac{1}{\sqrt{2}} A\exp[-i(\omega_- t + \varphi_{0-})]. \end{aligned} \quad (6)$$

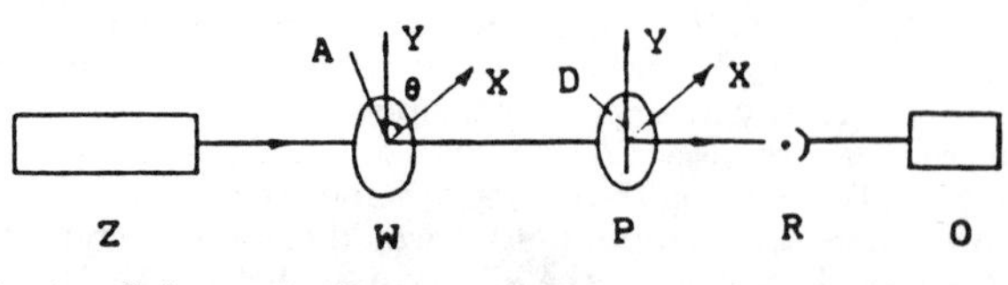

Fig. 1. Schematic diagram of the principle of the experiment. Z, Zeeman split He–Ne laser; W, wave plate; P, polarizer, R, photodetector; A, rapid axis; O, oscilloscope, D, direction of the polarizer orientation.

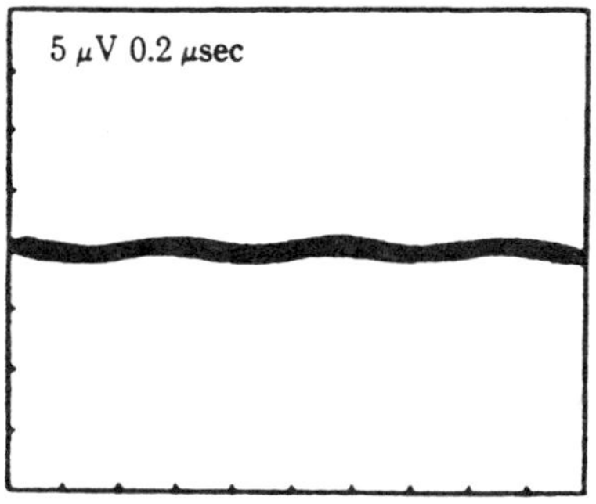

Fig. 2. Optical heterodyne wave when the angle between the rapid axis of the quarter-wave plate and the X direction is 45°.

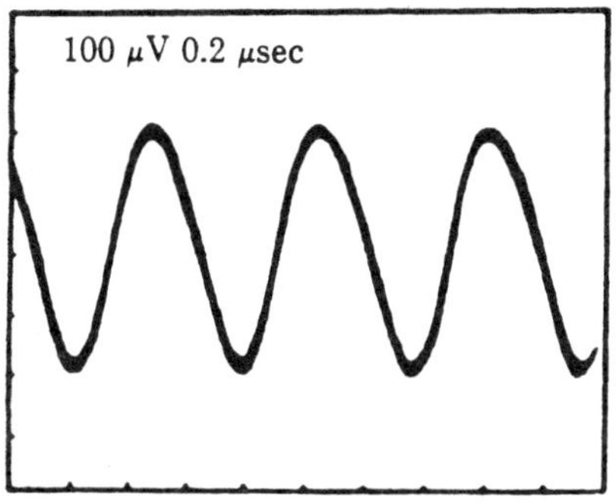

Fig. 3. Optical heterodyne wave when the angle between the rapid axis of the quarter-wave plate and the X direction is 90°.

The signal I from the photodetector is proportional to the irradiance $|E_+(f_+) + E_-(f_-)|^2$ and is given by

$$\begin{aligned} I &= |E_+(f_+) + E_-(f_-)|^2 \\ &= A^2 - A^2 \sin^2 \frac{\delta}{2} \cos(\Delta\omega t - 4\theta + \varphi_0) \\ &\quad - A^2 \cos^2 \frac{\delta}{2} \cos(\Delta\omega t + \varphi_0), \end{aligned} \tag{7}$$

where $\Delta\omega = \omega_+ - \omega_-$ is the angular Zeeman frequency and $\varphi_0 = \varphi_{0+} - \varphi_{0-}$ is the initial phase difference between the two polarized beams. The alternating component Ic is

$$\begin{aligned} Ic &= A^2 \sin^2 \frac{\delta}{2} \cos(\Delta\omega t - 4\theta + \varphi_0) \\ &\quad - A^2 \cos^2 \frac{\delta}{2} \cos(\Delta\omega t + \varphi_0). \end{aligned} \tag{8}$$

Equation (8) yields several interesting results, which are discussed below.

From Eq. (8), Ic is zero when $\delta = \pi/2$ and $\theta = (\pi/4)(2m + 1)$, where m is the nature number. The precise optical heterodyne measurement of wave-plate phase retardation relies on the accurate measurement of the microvariance of the amplitude of the heterodyne signal. Since a capacitance exists in the output electronic circuit of the photodetector, the signal shown on the screen of the oscilloscope is only relative to the alternating component Ic. Hence the measurement of the zero value of Ic offers a dependable foundation for the precise measurement of the phase retardation of the quarter-wave plate.

The minimum amplitude of Ic is zero, which follows from Eq. (8) when $\delta = \pi/2$ and $4\theta = (2m + 1)\pi$, where m is the nature number.

The maximum amplitude of Ic is A^2, which follows from Eq. (8) when $\delta = \pi/2$ and $4\theta = 2m\pi$.

It must be noted that some variation $\Delta\delta$ exists in the phase retardation of the quarter-wave plate being measured. Thus Ic cannot be exactly equal to zero when $\theta = (\pi/4)(2m + 1)$. The amplitude of Ic at that moment is actually the amplitude of ΔIc, which is due to the existence of $\Delta\delta$. ΔIc can be written as $(\mathrm{d}Ic/\mathrm{d}\delta)\Delta\delta$, hence the amplitude of ΔIc from Eq. (8) is given by

$$Ic \text{ amplitude} = \left.\frac{\mathrm{d}Ic}{\mathrm{d}\delta}\right|_{\theta=\frac{\pi}{4}(2m+1)} \text{amplitude} \cdot \Delta\delta = A^2\Delta\delta.$$

Thus the measuring accuracy of the phase retardation of a quarter-wave plate is given by

$$\Delta\delta = \Delta Ic\Big|_{\theta=\frac{\pi}{4}(2m+1)} \text{amplitude}/Ic \times \text{maximum amplitude}. \tag{9}$$

Since the ratio of the maximum and minimum amplitudes of the light-intensity signal received by the photodetector can be greater than 10^3, $\Delta\delta$ can be less than 10^{-3} rad. Thus, the measuring accuracy of the optical path difference of a quarter-wave plate is

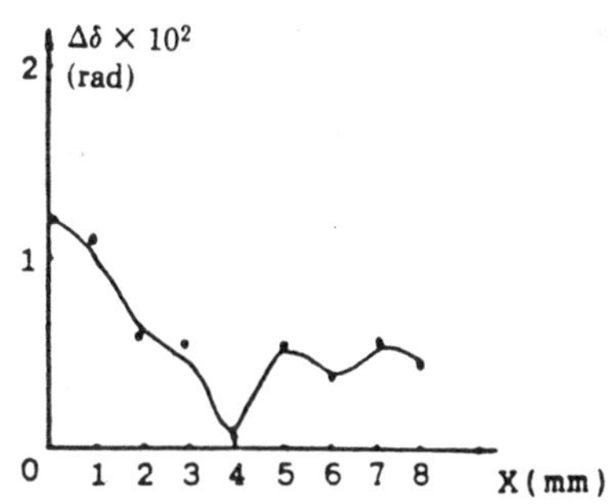

Fig. 4. Measurement of the phase retardation of the measured wave plate along the diameter in the X direction.

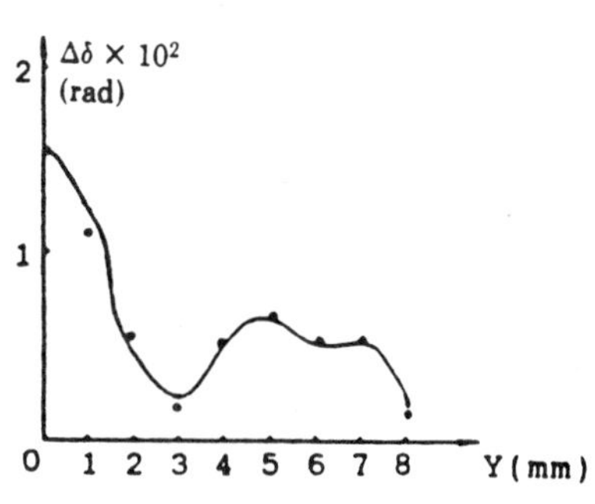

Fig. 5. Measurement of the phase retardation of the measured wave plate along the diameter in the Y direction.

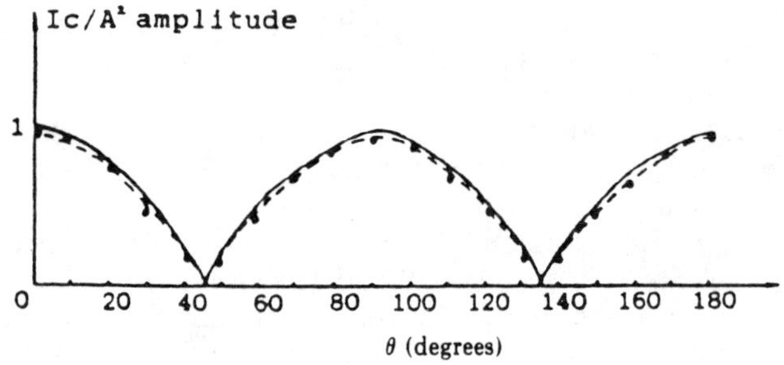

Fig. 6. Comparison of the experimental and theoretical results of the Ic amplitude.

$$\frac{\Delta\delta}{2\pi}\lambda < \frac{\lambda}{6000}, \qquad (10a)$$

and the measuring accuracy of the phase retardation of a quarter-wave plate is

$$\frac{\Delta\delta}{2\pi}360° < 0.06°. \qquad (10b)$$

The Ic minimum amplitude is $A^2|\cos\delta|$, which follows from Eq. (8) when $4\theta = (2m + 1)\pi$.

The Ic maximum amplitude is A^2, which follows from Eq. (8) when $4\theta = 2m\pi$. The expression for the phase retardation of a wave plate is given by $|\cos\delta| = Ic$ minimum amplitude/Ic maximum amplitude, where the sign of $\cos\delta$ is positive or negative, depending on whether the phase difference between the waves for which the amplitude of Ic is a maximum and minimum is 0 or π.

A frequency-stabilized longitudinal Zeeman He–Ne laser is used as the light source. Its wavelength is 633 nm, and its optical frequency stability is less than 2×10^{-8}. The frequency difference between the left and right circularly polarized beams is 1.64 MHz, with a frequency stability less than 1.3×10^{-4}. A photomultiplier is used as the photodetector.

A comparison is made between measurements made by using this optical heterodyne technique and those made by using accepted commercial instruments.

Figures 2 and 3 show the optical heterodyne waves for which the amplitude of Ic is a minimum and maximum, respectively. The amplitude of the heterodyne wave shown in Fig. 2 is not zero, owing to a difference between the actual phase retardation of the quarter-wave plate and $\pi/2$.

Figures 4 and 5 show the measurements of the phase retardation of the measured wave plate along the diameters in the X direction and the Y direction, respectively. It is necessary to measure the wave plate at many points and then to take the average. In this experiment, the ratio $\Delta\delta$ of the averages of the minimum and maximum amplitudes of Ic is 0.005 rad. Thus, the difference between the optical-path difference of the quarter-wave plate being measured and $\lambda/4$ is $(\Delta\delta/2\pi)\lambda = 0.503$ nm, and the difference between the phase retardation of the quarter-wave plate being measured and $\pi/4$ is $(\Delta\delta/2\pi)360° = 0.286°$.

By using a polarimeter with compensation, the difference between the optical path difference of the quarter-wave plate and $\lambda/4$ was estimated to be about 1 nm. The difference between these measurements is less than the stated accuracy of the polarimeter.

Figure 6 shows that the experimental variance of the amplitude of Ic agrees with theoretical calculation.

The experimental results and theoretical analysis prove that this optical heterodyne technique can precisely measure the phase retardation of a wave plate. This technique can be used to measure the phase retardations of different wave plates at the same optical wavelength or the same wave plate over a variety of optical wavelengths with a high degree of experimental accuracy.

References

1. C. M. McIntyre and S. E. Harris, J. Opt. Soc. Am. **58,** 1575 (1968).
2. H. G. Jerrard, J. Opt. Soc. Am. **44,** 634 (1954).
3. H. G. Berry, G. Gabrielse, and A. E. Livingston, Appl. Opt. **16,** 3200 (1977).
4. B. R. Grunstra and H. B. Perkins, Appl. Opt. **5,** 585 (1966).
5. L. Quanting, *Optics* (Mechanical Industry, China, 1980), p. 217.
6. H. Yuefeng, Opt. Tech. (China) **2,** 2 (1984).

Reprinted from *Engineering*, 213–219 (1912).

THE DESIGN AND CONSTRUCTION OF LARGE POLARISCOPES

By Professors E. G. Coker and S. P. Thompson, of the City and Guild of London Technical College, Finsbury

(*Monday, June 24th*, 1912.)

In optical investigations, for which a beam of polarized light is required, the best means yet devised for obtaining plane polarized light is the prism of Iceland spar first invented by Nicol and known by his name.

Such prisms are unsurpassed for the purity and intensity of the light transmitted through them; and they are almost always preferred for optical investigations wherever the conditions admit of their use.

Nicol's prisms have been made giving a clear view of an object as much as four inches in diameter, and this appears to be very nearly the greatest size for which spar has been found available, except in one or two isolated instances, where full advantage was not taken of the size and perfection of the crystal.

For most scientific purposes Nicol's prisms, of the size mentioned, would be more than ample if they could be readily obtained; but, at the present time, the number of large prisms in existence is very limited, and there is a complete dearth of spar of a suitable kind, owing to the stoppage of the supply from Iceland, and the lack of any other supply of sufficiently good optical quality.

Other means are, however, available for producing polarized light, and the object of the present paper is to describe certain forms of apparatus using sheets of glass and mica.

Although the instruments described are somewhat inferior in their polarizing and analysing properties to those using Nicol's prisms, yet they have many advantages of their own which make them almost indispensable for some kinds of investigation. The polarizing properties of glass plates are extremely useful in cases where a large doubly refracting object requires examination; and their behaviour has been examined by several physicists, and especially by Stokes,* who has obtained expressions for the intensities of the reflected and refracted rays for a pile of plates made of a glass which reduces the intensity of transmitted light by absorption.

If ρ is the fraction of the incident light of unit intensity reflected at the first surface of a single plate, as determined by Fresnel's laws, the intensity r of the reflected ray may be expressed by—

$$r = \rho + \frac{\rho(1-\rho)^2 g^2}{1-\rho^2 g^2};$$

and the intensity t of the refracted ray by—

$$t = \frac{(1-\rho)^2 g}{1-\rho^2 g^2},$$

where g is the fraction to which the intensity is reduced by absorption in a single transit.

If there is no absorption $g = 1$, and we have the sum of the intensities of the reflected and refracted rays equal to that of the incident beam; but in general the absorption causes a diminution in the intensity of both beams. If a pile of m plates is used, and $\phi(m)$, $\psi(m)$ are the intensities of the reflected and refracted rays respectively, the corresponding formulæ are shown to be—

$$\frac{\phi(m)}{\sin m\beta} = \frac{\psi(m)}{\sin \alpha} = \frac{1}{\sin(\alpha + m\beta)},$$

which reduces to—

$$\frac{r}{\sin \beta} = \frac{t}{\sin \alpha} = \frac{1}{\sin(\alpha + \beta)},$$

when $m = 1$, and α and β are imaginary since in general $r + t < 1$.

The intensities, corresponding to values of $g = 1$, 0·98 and 0·90 respectively, are shown in the accompanying diagram for reflection and refraction at the polarizing

* On the Intensity of the light reflected from or transmitted through a pile of plates. *Collected Papers*, vol. iv. p. 145.

angle : $\omega = \tan^{-1} \mu = \tan^{-1} 1 \cdot 52$.

The forms of the ϕ curves show that the intensity of the reflected light falls off very rapidly as the absorption increases, and that there is little advantage in using a very large number of plates; the ψ curves show similar results.

The refracted ray will, in general, consist of two components polarized in directions at right angles, and the ratio of these is a measure of the effectiveness of the polarization. These ratios are shown by the dotted curves; the one corresponding to perfect transparency coincides with the ψ curve for $g = 1$.

These ratios show the important result that although light transmitted through a pile of plates is materially reduced by absorption, the polarization effect is improved somewhat. If, therefore, the illumination is sufficiently intense it is not necessary to use glass that is of the clearest quality and therefore expensive.

The formulæ and curves form a general means of estimating the probable polarizing effect obtainable for a given design of polariscope, constructed with sheets of glass having definite values of ρ and g, ascertained by experiment.

In the construction of polariscopes we may make use of reflected or refracted rays, or both; and since plates of glass may be obtained of any convenient size at a low cost, they afford a cheap and easy means, when suitable illumination is provided, of obtaining, in polarized light, a field of view of almost any extent, thereby offering an advantage over Nicol's prisms, in which the field of view is always a very limited one.

In practice it is found that light which falls on a sheet of black glass at the polarizing angle is almost perfectly plane polarized, and although its intensity is only about one-seventh of the incident light, yet it is not as a rule worth while to modify this arrangement to produce a more intense reflected beam at the expense of the perfection of the plane polarization usually entailed, especially as the initial intensity can be made very great. It is, therefore, convenient to use such a sheet and illuminate it by a bank of lamps.

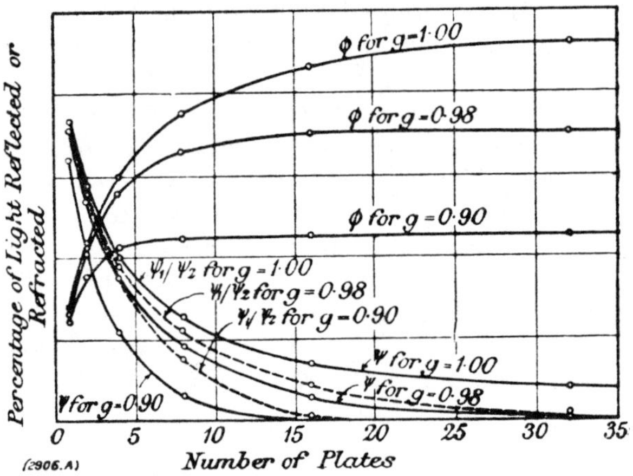

Fig. 1.—Reflection and Refraction of Light at the Polarising Angle.

To obtain equal illumination at all parts of the field the light must be of uniform intensity, and if filament lamps are used some means of diffusing the light must be employed. The interposition of a sheet of ground glass, or one or two sheets of white tissue paper, between the lamps and the reflecting surface gives an effect which is uniform enough for most purposes. The illumination may be intensified by reflectors behind the lamps, and also at the sides after the manner of a kaleidoscope.

An object, possessing doubly refracting properties, when viewed in this illuminated field by a suitable analyser, shows interference colours; and, since the analyser may be of moderate dimensions, it is possible to use a quite small Nicol's prism for this purpose. An example of a polariscope of this kind is shown by Fig. 2 in sectional elevation, in which the light from a bank of lamps, A, passes through a diffusing screen, B, of tissue paper, and is reflected at the polarizing angle by a sheet of black glass, C, to afford a horizontal beam for illuminating an object placed in the space, D, and viewed by a Nicol's prism, E.

An apparatus of this kind was used in an investigation for determining the distribution of stress in a plate subjected to shear * where a field of view of about 120 sq. ins. was required; and it replaced an earlier arrangement of Nicol's prisms giving a field of view of barely two square inches. This latter arrangement,

* "An Optical Determination of the Variation of Shear Stress in a Thin Rectangular Plate Subjected to Shear," by E. G. Coker, *Proc. Roy. Soc.*, **A**, vol. lxxxvi. (1912.)

on account of its small size, required to be constantly shifted to new positions for making observations, and this involved a serious waste of time and liability to error, owing to the frequent adjustments required. Its replacement by a large polariscope diminished the time of labour of observation to about one-quarter of that formerly needed.

For many investigations a circularly polarized beam of light is necessary, and if only a small field of view is required, the well-known Fresnel rhomb offers the great advantage with white light that waves of varying length are all nearly accurately circularly polarized. In the present case a glass prism of sufficient size would be

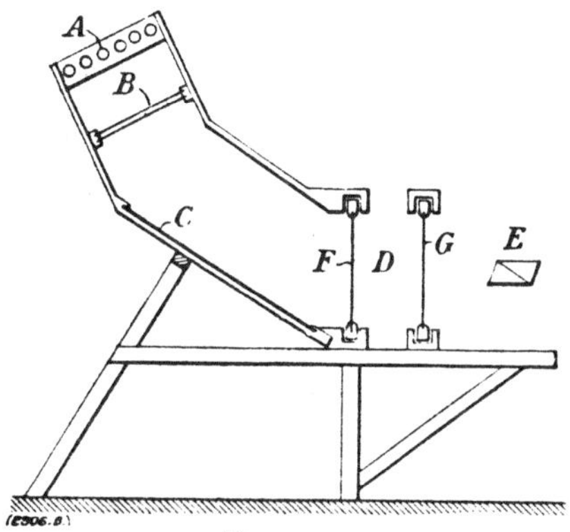

FIG. 2.

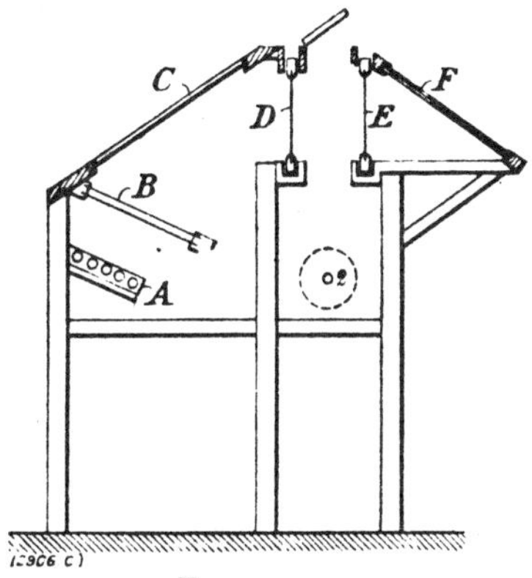

FIG. 3.

almost impossible to construct and use, but an admirable substitute is found in sheets of mica, which may, with some practice, be split into laminæ of large size to give approximately a quarter-wave phase difference to the two components into which a plane polarized beam is divided in passing through a lamina at perpendicular incidence.

Quarter-wave plates F and G are shown in this polariscope between the reflector and the analyser, and in the space D, therefore, we obtain circularly polarized light. An object set in this space will not, therefore, show a black cross effect when viewed by the analyser.

It may be interesting to state that quarter-wave plates up to 20 ins. in diameter, and also of rectangular form 40 ins. by 10 ins., have been constructed by one of us by joining two sheets of mica of the same thickness. It is believed that these dimensions have not been approached before.

For some experimental purposes it is convenient to construct of transparent materials, engineering and other models, and to load them in the way the actual structures and machines are loaded in practice; the stresses may then be obtained by methods which have already been described.*

A general optical examination of a model often shows the places where maximum stress occurs, and thereby affords information of value in the design of a large structure, especially in cases where calculations cannot be made owing to the complexity of the stress system. In many cases of construction, as in roofs, plate and lattice girders, ships and the like, the over-all dimensions are very great in some one direction, and a long and narrow field of view is required. Moreover, it may be necessary to make arrangements to apply and release the load at regular intervals.

A polariscope designed for these conditions is exhibited, and a cross-sectional view of it is shown, in the accompanying Fig. 3. Light obtained from a bank of filament lamps, A, is diffused by a tissue paper screen, B, before reflection from a black glass plate, C, 40 ins. by 21 ins., inclined at the polarizing angle. The light afterwards passes through two quarter-wave plates, D and E, each affording an unobstructed view 40 ins. wide and 10 ins. high. The analyser consists of several sheets, F, of clear glass set at a suitable angle. In this way an observer in front of the polariscope can view directly the interference effects produced by the application of load to the object under examination.

* "The Optical Determination of Stress," by E. G. Coker, *Phil. Mag.*, October 1910. "The Distribution of Stress in Plates of Variable Section," by E. G. Coker, *Trans. Inst. Naval Architects.* (1911.)

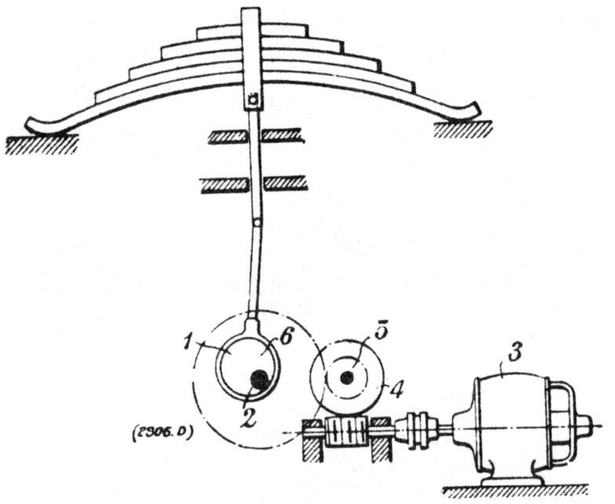

FIG. 4.

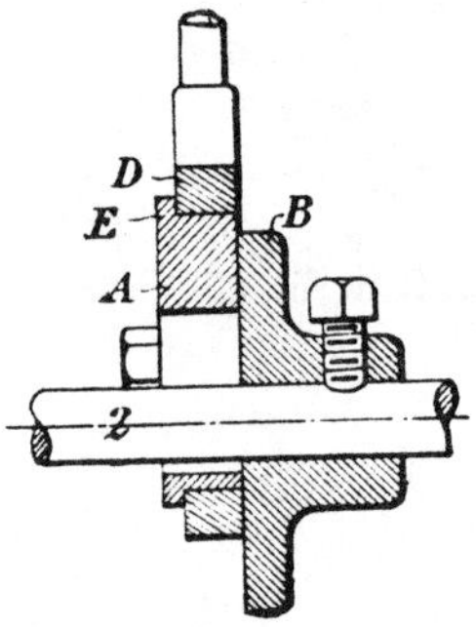

FIG. 5.

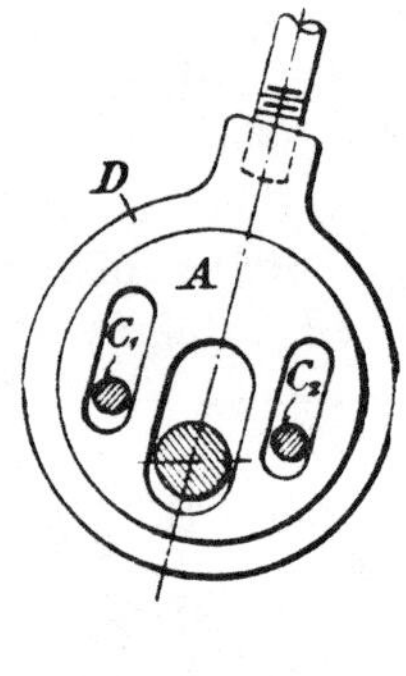

FIG. 6.

Cyclical loadings of the models, like the spring shown in Fig. 4, are obtained by means of eccentrics (1) keyed to a shaft (2), running the whole length of the polariscope. This shaft is driven from a motor (3) by means of a train of gearing comprising a worm reduction gear (4), driving a pinion (5), which latter gears with a spur wheel (6) on the main shaft; the reduction effected is from 2000 to ten revolutions per minute. A feature of interest is afforded by the eccentrics, which have a throw capable of variation between 0 and 1½ ins.

This is accomplished by the construction shown in section by Fig. 5 and in side elevation by Fig. 6. The main sheave (A) is not keyed to the shaft directly, but is bolted to a flange-plate (B) secured to the shaft. The sheave may be fixed at any definite eccentricity by means of a pair of studs (C_1, C_2) screwed into the flange-plate and clamping the sheave to it. Slotted holes are provided to accommodate the studs and shaft within the assigned limits of throw. The strap encircling the sheave is in one piece, and is prevented from moving sideways in one direction by a projecting collar (E) on the sheave, and in the opposite direction by the flange-plate. This arrangement has been found very useful for laboratory and demonstration purposes. We desire to express our warmest thanks to Messrs. Mould, Withycombe and several other members of the College staff, and to many students for the skill and care they have devoted to the construction of this apparatus.

Mr. T. SMITH said he had been much interested in watching the machine placed by the authors in the Exhibition. It appeared to him that in the models exhibited the strain was much greater than would usually be permissible in actual structures, and he would be glad to hear from the authors whether the distribution of the stresses in an actual structure, where the strain was proportionately much less, could be deduced from those of the experimental models, and if so, what the limits were within which such deduction might fairly be made.

Dr. W. ROSENHAIN asked whether calculations could be based on these qualitatively most interesting results.

Dr. A. S. PERCIVAL inquired how it was possible to make quarter-wave plates of such enormous sizes as shown (21 in. in diameter). He remembered spending an inconceivable time in making a quarter-wave plate one inch in diameter.

Dr. S. P. THOMPSON, in reply, said that it would be quite impossible now to obtain material of uniform thickness, if it were not that Messrs. Wiggins & Sons had been collecting specimens of exceptional quality. They had also helped him with their experience. A little dexterity and experience were all that was required, and he had no doubt that the quarter-wave plates could be repeated over again.

Prof. COKER stated in answer to Mr. T. Smith and Dr. Rosenhain that the points raised were, of course, very important. The strains were no doubt much greater in xylonite than in iron and steel; yet if the physical constants were taken into consideration—modulus of elasticity of xylonite 300,000, of steel 30,000,000; Poisson's ratio for xylonite 2·8, for steel 9·0—there was a fairly good agreement. The strains, moreover, appeared larger than was actually the case, owing to the movement of the colour bands which gave the eye a false impression of considerable strain, although measurement showed nothing abnormal. Quantitative results had been obtained; test cases like bending rectangular section beams, eccentrically loaded tie bars, and other specimens yielded values in transparent materials which were quantitatively correct within five per cent. or less, and agreed with similar experiments on steel. Glass was better for such experiments than xylonite. The neutral axis, generally in the middle of the beam, should go up as the stress from above increased. When shear was applied to a long plate, the quantitative results of the calculations based upon the optical examination were again fairly accurate. With regard to the production of the quarter-wave plate, Prof. Coker wished to add that his share in this had practically been nil, and that Prof. Thompson had hardly told the meeting all the skill that was required.

Vol. 44(12), 1753–1762 (December 1973).

A new polarization-modulated light scattering instrument*

Arlon J. Hunt and Donald R. Huffman

Department of Physics, University of Arizona, Tucson, Arizona 85721
(Received 25 June 1973; and in final form, 10 September 1973)

A new light scattering instrument is described for measuring, as a function of scattering angle, the elements of the matrix describing light scattered from small particle systems. The instrument uses a piezo-optical birefringence modulator to modulate the polarization state of the incident light beam, and uses lock-in amplifier detection of the scattered light. The first and second harmonics of the modulation frequency are used with various combinations of filters and orientations to give the separate matrix elements. A treatment of a simple scattering experiment by means of Mueller calculus is carried through to illustrate the interactions of the light with the scattering system and with the various components of the measuring instrument. Evaluation of the instrumental performance is shown through measurements on two systems of monodisperse polystyrene spheres of average radius 550 and 3940 Å, compared with Mie calculations. Further measurements on sulfur colloids are presented, and suggestions of the value of this type of instrument in future work are given.

I. INTRODUCTION

In order to completely characterize a system of small particles optically, it is necessary to determine the intensity and polarization of the scattered light for several different polarization states of the incoming light. Various techniques have been developed in the past to obtain the angular variation of intensity of a specific polarization of the scattered light. This was done in order to determine the size and/or size distribution of an assembly of spherical particles with known optical constants. In 1950, Kerker and La Mer[1] developed a method using either the polarization ratio or the phase angle of the scattered light as a function of the scattering angle to extract size information. Since that time, several techniques[2–4] have been used with greater or lesser success. However, few attempts have been made to experimentally characterize the scattered light completely enough to determine more than two elements of the scattering matrix.[5] To do this all four Stokes parameters of the scattered light must be measured at each angle for various states of polarization of the incoming beam. Holland and Gagne[6] recorded intensity information for 18 combinations of filters and subsequently took differences between the recorded values to obtain the matrix elements. The disadvantage of this technique is that the values needed to obtain the elements appear as small differences between relatively large quantities. Hence, small errors introduced by drift and instabilities are magnified.

In this paper a technique is reported which avoids these problems by the use of a polarization modulator.[7,8] Briefly, the polarization of the light is periodically varied in such a fashion that the synchronously detected components of the scattered light are proportional to the matrix elements. By proper choice of polarizers, orientations, and harmonics, selected normalized elements of the scattering matrix can be plotted automatically (by pairs or singly) on x-y recorders.

The system records the data in a relatively short time (a scan from 5° to 168° may be made in a few minutes) and produces an output in a convenient format for direct comparison with computer generated plots resulting from Mie calculations. The sensitivity of the system is such that it will respond to 1% differences in linear polarization, which is considerably better than in most subtractive techniques. The accuracy of the instrument is about ±3% presently but that figure should be improved with further development of calibration techniques. The angular resolution of the instrument is 1.5°. The normalized parameters obtained from the system are in a convenient form to obtain the degree of partial polarization which may be used to give a measure of the polydispersity in size.

Before describing the apparatus in detail, it will be useful to review, in Sec. II, the treatment of polarized light by Mueller calculus. This will serve to introduce the notation and matrix formalism as well as to discuss a simple case that is important in understanding the operation of the instrument. Section III contains a detailed discussion and analysis of the operation of the apparatus as well as calibration procedures. In Sec. IV, instrumental performance is evaluated by comparing the results of measurements with calculations for two model small-particle systems which have independently known characteristics. The final section contains a brief discussion of other applications of the technique.

II. LIGHT SCATTERING, BASIC CONSIDERATIONS

A. Formalism

The polarization state of any beam of noncoherent light may be characterized by a four element Stokes vector. This vector may be determined phenomenologically[9] or its elements may be expressed equivalently[10] in terms of quadratic functions of the electric fields as

$$\begin{aligned} I&=\langle E_lE_l^*+E_rE_r^*\rangle;\\ Q&=\langle E_lE_l^*-E_rE_r^*\rangle;\\ U&=\langle E_lE_r^*+E_rE_l^*\rangle;\\ V&=\langle i(E_lE_r^*-E_rE_l^*)\rangle. \end{aligned} \tag{1}$$

The subscripts refer to the components of the electric field parallel l, and perpendicular r, to the scattering plane which is defined by the direction of the incoming and scattered waves. The brackets refer to time averages and the asterisks denote the complex conjugates.

The matrix representing the linear transformation of the Stokes vector is referred to as the Mueller[11] matrix and is a useful tool for manipulating polarized light. Shurcliff[9] gives a list of matrices corresponding to various polarizing and retarding elements. The most general form of the Mueller matrix, one with no zero elements, allows for all possible mixing of polarization states; the general scattering matrix is of this form. This matrix may be derived from the amplitude transformation equations for the electric fields given by

$$\begin{aligned} E_l &= A_2E_{l0} + A_3E_{r0}, \\ E_r &= A_4E_{l0} + A_1E_{r0}, \end{aligned} \tag{2}$$

by substitution into Eq. (1) and solving for the elements of the 4×4 intensity transformation matrix.

The general form of the scattering matrix as given by van de Hulst[5] is

$$\begin{bmatrix} \frac{1}{2}(m_2+m_3+m_4+m_1) & \frac{1}{2}(m_2-m_3+m_4-m_1) & S_{23}+S_{41} & -D_{23}-D_{41} \\ \frac{1}{2}(m_2+m_3-m_4-m_1) & \frac{1}{2}(m_2-m_3-m_4+m_1) & S_{23}-S_{41} & -D_{23}+D_{41} \\ S_{24}+S_{31} & S_{24}-S_{31} & S_{21}+S_{34} & -D_{21}+D_{34} \\ D_{24}+D_{31} & D_{24}-D_{31} & D_{21}+D_{34} & S_{21}-S_{34} \end{bmatrix}.$$

All the elements are functions of the scattering angle. The relationships with the amplitude transformation coefficients are

$$\begin{aligned} M_i &= |A_i|^2, \\ D_{ij} &= i/2(A_iA_j^* - A_jA_i^*), \\ S_{ij} &= \tfrac{1}{2}(A_iA_j^* + A_jA_i^*). \end{aligned} \tag{3}$$

For the case of spherical particles, no mixing between the l and r channels occurs so that $A_3 = A_4 = 0$ and the matrix simplifies to

$$F_S = \begin{bmatrix} \frac{1}{2}(m_2+m_1) & \frac{1}{2}(m_2-m_1) & 0 & 0 \\ \frac{1}{2}(m_2-m_1) & \frac{1}{2}(m_2+m_1) & 0 & 0 \\ 0 & 0 & S_{21} & -D_{21} \\ 0 & 0 & D_{21} & S_{21} \end{bmatrix}. \tag{4}$$

B. Analysis of a Simple Scattering Experiment

In order to determine the results of a typical scattering experiment, it is necessary to multiply the Stokes vector of the incoming light by the matrix representing each optical element or scatterer in order of traversal. Consider a simple scattering experiment in which light, polarized at 45° to the scattering plane, is incident on a system of spherical scatterers. The resultant Stokes vector for this experiment is calculated to be

$$\begin{bmatrix} I \\ Q \\ U \\ V \end{bmatrix} = \frac{1}{2}\begin{bmatrix} \frac{1}{2}(m_1+m_2) \\ \frac{1}{2}(m_2-m_1) \\ S_{21} \\ D_{21} \end{bmatrix}. \tag{5}$$

The quantities m_1 and m_2 are proportional to the light scattered in the vertical and horizontal plane, respectively. Thus, for this polarization state of the incoming light, the four elements of the resultant vector are just the four independent elements of the scattering matrix for spheres. This result will be helpful in interpreting later results.

A useful property of Stokes vectors is that they are additive for combining beams of incoherent light. Thus, the scattering from a collection of randomly positioned spheres may be represented by the sum of the Stokes vectors for the individual spheres. The degree of partial polarization of the light is defined in terms of the Stokes vector elements as

$$F = \frac{(Q^2+U^2+V^2)^{\frac{1}{2}}}{I}. \tag{6}$$

In order to gain a clearer understanding of the scattering example above, it is worthwhile to examine the resultant Stokes vector for the degree of polarization. To do this one must consider the scattering medium more carefully: Deirmendjian[12] pointed out that the degree of partial polarization resulting from scattering by a collection of identical spheres is equal to unity (if the incoming light is completely polarized). However, if the spheres are not all of equal size, he also noted that the degree of partial polarization is less than one and may be used to help determine the amount of polydispersity in size. Thus in order to determine F, the problem must be solved in detail for the matrix elements or it may be determined directly from the experimental curves.

III. EXPERIMENTAL SYSTEM

A. The Instrument

In this section the experimental system is described in more detail, and examples are given of how the elements of the scattering matrix may be obtained. Figure 1 is a schematic diagram of the scattering apparatus and a complete block diagram of it is given in Fig. 2. Light from a high

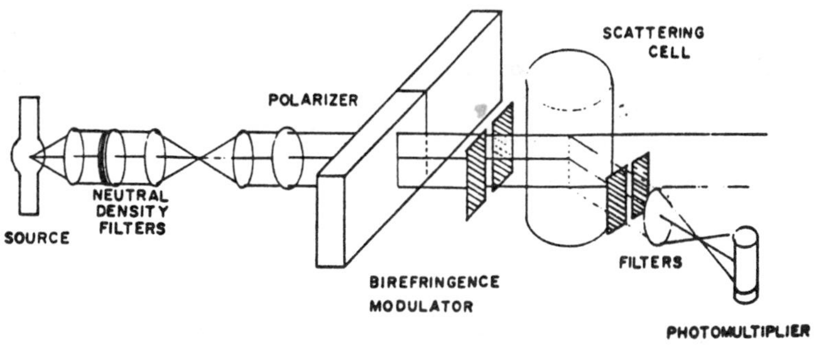

FIG. 1. Schematic diagram of the scattering apparatus.

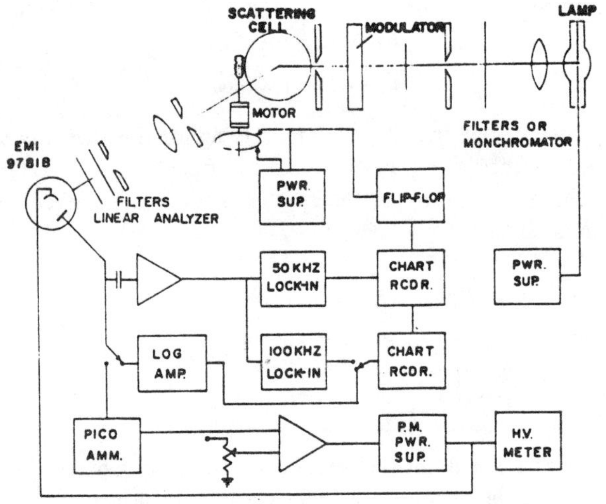

FIG. 2. Block diagram of the scattering apparatus.

pressure Hg lamp (Osram HBO 100) or a laser is collimated and passed through a polarization modulator before entering the scattering cell. The scattered light is detected by a photomultiplier tube (EMI 9781B) in an arm which may be rotated through an angle from 0° to 168° by a dc motor. Monochromatic filters and polarizers may be placed in front of the detector. The ac component of the photomultiplier signal is detected by a pair of lock-in amplifiers and displayed on x–y recorders. The dc component of the output is switched into one of two channels depending on the desired mode of operation. It may be measured directly by a logarithmic amplifier and recorder, or it may pass into a picoammeter which drives a control circuit. In the second mode the photomultiplier current is servoed to a constant value with an operational amplifier which controls the phototube high voltage.

The θ-drive for the recorders is produced by stepping motors which are triggered by a digitizer circuit. This consists of a flip-flop activated by a phototransistor which views a light emitting diode through an occulting disk. The disk is geared to the same shaft that drives the rotating arm so that the total number of holes that pass the phototransistor determines the x position of the recorder.

To understand the operation of the instrument in more detail first consider the birefringence modulator. The heart of the device (Fig. 3) is a vibrating block of amorphous quartz which is driven at its natural frequency (50 kHz) by a piece of crystalline quartz of similar proportions. The crystal forms the feedback element of an oscillator circuit which can be adjusted in amplitude. The applied stress in the amorphous quartz induces a periodic birefringence which, if of the correct amplitude, could modify the polarization state of the incoming light from linear (at 45° with respect to modulator axis) to alternately right and left circularly polarized.

To make the discussion more quantitative, Mueller calculus may be used to analyze the light as it progresses through the instrument. Before proceeding, the angular orientation of the polarizer and modulator must be specified. The two cases considered here are:

I. When the modulator axis is at 45° with respect to the scattering plane and the polarizer P oriented at 0°.

II. When the modulator axis is parallel to the scattering plane and the polarizer at 45°.

The modulator may be described in both cases as a sinusoidially varying phase plate. The eigenvector[9] of the phase plate is specified by the polarization state of light which remains unchanged when traversing the element and must be known to calculate the effect of a phase retarder. It depends on the azimuthal angle of the fast axis of the retarder and may be specified by the Stokes vector. The eigenvector is (1,0,1,0) and (1,1,0,0) for case I and II, respectively. Once the eigenvector is found, the phase retardance matrix may be calculated from the general form given by Shurcliff.

B. Linear Polarization and Intensity Measurements

For case I, the Stokes vector of the light as it emerges from the modulator is given by the matrix product

$$\frac{1}{2}\begin{pmatrix}1&0&0&0\\0&\cos\delta&0&-\sin\delta\\0&0&1&0\\0&\sin\delta&0&\cos\delta\end{pmatrix}\begin{pmatrix}1&1&0&0\\1&1&0&0\\0&0&0&0\\0&0&0&0\end{pmatrix}\begin{pmatrix}1\\0\\0\\0\end{pmatrix}=\frac{1}{2}\begin{pmatrix}1\\\cos\delta\\0\\\sin\delta\end{pmatrix},$$

where δ = instantaneous angle of retardation. The scattering from a system of spherical particles, given by the scattering matrix operating on the above Stokes vector is

$$\frac{1}{2}\begin{pmatrix}\frac{1}{2}(m_2+m_1)&\frac{1}{2}(m_2-m_1)&0&0\\\frac{1}{2}(m_2-m_1)&\frac{1}{2}(m_2+m_1)&0&0\\0&0&S_{21}&-D_{21}\\0&0&D_{21}&S_{21}\end{pmatrix}\begin{pmatrix}1\\\cos\delta\\0\\\sin\delta\end{pmatrix}=\frac{1}{2}\begin{pmatrix}\frac{1}{2}(m_1+m_2)+\frac{1}{2}(m_2-m_1)\cos\delta\\\frac{1}{2}(m_2-m_1)+\frac{1}{2}(m_2+m_1)\cos\delta\\-D_{21}\sin\delta\\S_{21}\cos\delta\end{pmatrix}.$$

If there are no further polarizers in the light path the photomultiplier will respond to the total intensity of the light, given by the first element of the Stokes vector,

$$I=\tfrac{1}{2}k[(m_1+m_2)+(m_2-m_1)\cos\delta], \tag{7}$$

where k is a constant depending on the phototube sensitivity, light collecting efficiency, etc. The photomultiplier current may be broken into two parts; (m_1+m_2) is a constant in time but varies with angle, $(m_2-m_1)\cos\delta$ is a complex time varying quantity. Consider how the time variation enters through the phase retardance δ,

$$\delta=\frac{2\pi d}{\lambda}s\sin wt=A\sin wt,$$

where d = thickness of the modulator element, λ = wavelength, s = stress coefficient. Hence the phase retardance is a sinusoidal function of time, but it occurs as an argument of a trigonometric function. Expanding $\sin\delta$ and $\cos\delta$ in terms of Bessel functions,[8]

$$\begin{aligned}\sin(A\sin wt)&=2J_1(A)\sin wt+2J_3(A)\sin 3wt+\cdots\\ \cos(A\sin wt)&=J_0(A)+2J_2(A)\cos 2wt+\cdots.\end{aligned}\tag{8}$$

The time averages of $\sin\delta$ and $\cos\delta$ are zero and $J_0(A)$, respectively. To simplify the data interpretation it would be convenient for $J_0(A)$ to be zero. This is accomplished by adjusting the amplitude of the vibration of the modulator so that $A=138°$, which corresponds to the first zero of the $J_0(A)$ function. This restriction fixes A for each wavelength used.

Another function of the electronics is to force the dc component of the phototube output to be a constant, thus effectively dividing by the total intensity. Normalizing the output in this way accomplishes two useful functions. The unnormalized signals vary over several orders of magnitude, (for example see the tables in Ref. 12) and are inconvenient to record. After normalization, the signals vary between the limits of ±1 (with one exception, see below). Secondly, random fluctuations in the intensity of the lamp are smoothed out, thus obviating the necessity of a second photomultiplier and subtractive circuits.

The circuit for accomplishing this normalization is shown in Fig. 4. The output of the photomultiplier is divided into two channels, the ac part is capacitively coupled to the lock-in amplifiers, while the dc component passes into a picoammeter. The output of the picoammeter is compared with an adjustable reference voltage by an operational amplifier. The difference signal drives a programmable phototube high voltage power supply. The effect of this servo circuit is to keep the dc phototube current at a constant value, thus effectively dividing the ac signal by the total light intensity. Provision is made for measuring the total light intensity by switching the dc signal directly to a logarithmic amplifier and x–y recorder while disabling the servo circuit.

After accomplishing these electronic functions, the output of the photomultiplier may be written

$$I=\tfrac{1}{2}k(m_1+m_2)\left[1+\frac{m_2-m_1}{m_1+m_2}\cos 2wt\right],$$

or since (m_1+m_2) is constant,

$$I\propto\text{const}\left[1+\frac{(m_2-m_1)}{m_1+m_2}\cos 2wt\right].\tag{9}$$

By examining the transformation equations [Eqs. (2), (3)], it can be seen that m_1 and m_2 are proportional to the intensities of the scattered light in the r and l directions. Hence, Eq. (9) states that for case I the signal proportional to the second harmonic of the oscillator frequency is just the degree of linear polarization of the scattered light, or

$$P=\frac{m_2-m_1}{m_2+m_1}.\tag{10}$$

It may be seen by reference to Eq. (4) that this is just the

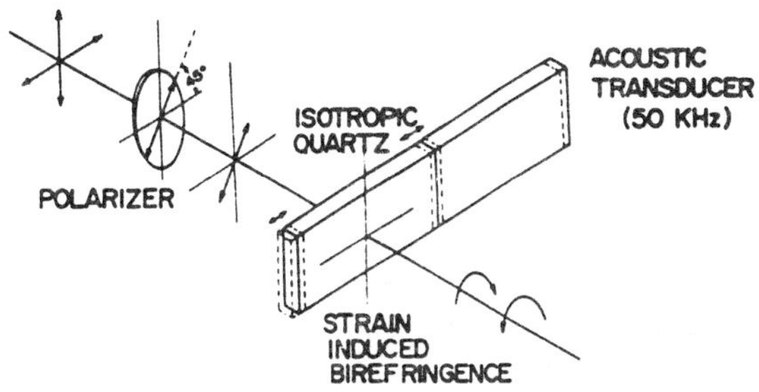

FIG. 3. Birefringence modulator element.

second element of the scattering matrix for spheres, normalized by the total light intensity.

The total direct current may be obtained by disabling the servo circuit, since the time average of $\cos 2wt$ is zero. In this case however, because the total intensity of the light is measured, there must be a correction for the observed scattering volume. This is done by dividing the total intensity term by the sine of the scattering angle. For convenience the logarithm of this variable is measured. Then the observed quantity corresponding to the matrix element m_1+m_2 is given by

$$\Sigma=\log\left[\frac{m_1(\theta)+m_2(\theta)}{\sin\theta}\times\frac{\sin\theta_0}{m_1(\theta_0)+m_2(\theta_0)}\right].\tag{11}$$

The second term in the logarithm normalizes the output so as to make Σ equal to zero for the minimum scan angle θ_0.

C. Circular and Oblique Polarization Measurements

In order to extract the other two elements of the scattering matrix for spheres, we must examine the operation of the system when the modulator axis is parallel to the scattering plane. Using the eigenvector for case II, the Mueller matrix is found to be

$$M=\begin{bmatrix}1&0&0&0\\0&1&0&0\\0&0&\cos\delta&\sin\delta\\0&0&-\sin\delta&\cos\delta\end{bmatrix}.$$

Following the same procedure as for case I, the Stokes vector

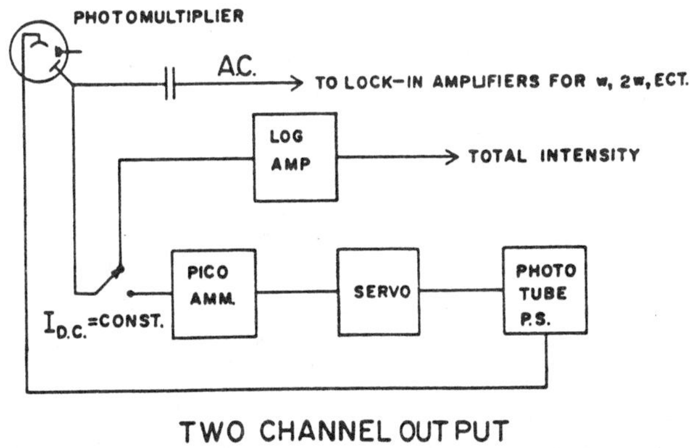

FIG. 4. Block diagram of normalization circuit.

for the scattered light is

$$\frac{1}{2}\begin{bmatrix} \frac{1}{2}(m_1+m_2) \\ \frac{1}{2}(m_2-m_1) \\ S_{21}\cos\delta+D_{21}\sin\delta \\ D_{21}\cos\delta-S_{21}\sin\delta \end{bmatrix}.$$

Now, the time varying components appear only in the third and fourth element. To obtain an ac signal, a linear polarizer is placed at 45° orientation in front of the phototube. This results in a new Stokes vector, which may be calculated in the usual fashion, given by

$$\begin{bmatrix} \frac{1}{2}(m_1+m_2)+S_{21}\cos\delta+D_{21}\sin\delta \\ 0 \\ \frac{1}{2}(m_1+m_2)+S_{21}\cos\delta+D_{21}\sin\delta \\ 0 \end{bmatrix}.$$

Again expanding the time varying components, taking account of $J_0(A)=0$, and normalizing, the output current is given by

$$I\propto[\tfrac{1}{2}+2J_1(A)D\sin wt+2J_2(A)S\cos 2wt)],$$

where

$$D=\frac{D_{21}}{m_1+m_2},\quad S=\frac{S_{21}}{m_1+m_2},\tag{12}$$

where D and S represent the normalized quantities corresponding to the measurements. It can be seen from the above that the signals proportional to the first and second harmonic of the driving frequency give the two remaining normalized elements of the scattering matrix for spheres.

The physical significance of these elements may be interpreted by examining the Stokes vector of the scattered light given in Eq. (5). This resulted from the case when the incident light is linearly polarized at 45°. Then S_{21} is proportional to the light scattered at ±45° to the scattering plane, and D_{21} gives the component of the light which is circularly polarized. The normalized quantity S should vary from +1 in the forward direction to −1 for backscatter if the scatterers exhibit no linear dichroism. Again D should be zero in the forward direction for systems exhibiting no circular dichroism. It is important to emphasize here that these quantities are inherent to the scattering matrix and their relation to the Stokes vector of the scattered light depends entirely on the particular experimental situation.

Until now, only spherical scatterers have been considered. The same method may be used to investigate the elements of more general scattering matrices. Perrin[13] showed that for isotropic scatterers with no optical activity the scattering matrix takes the following form:

$$\begin{bmatrix} s_{11} & s_{12} & 0 & 0 \\ s_{12} & s_{22} & 0 & 0 \\ 0 & 0 & s_{33} & s_{34} \\ 0 & 0 & -s_{34} & s_{44} \end{bmatrix}.$$

It can be shown by the same type of analysis that all the elements of this matrix may be obtained. In addition, they will be normalized by the total intensity and result from the same two cases with only the addition of a final circular polarizer. The elements of more complex matrices (see for example the 10 parameter matrix given by Van de Hulst on p. 50) may also be extracted, although not in such a direct manner.

D. Calibration

Calibration of the instrument is relatively easy since most of the measurements are normalized, in the case of total intensity to the intensity at some reference angle, and in the other three cases to unity. It is only necessary to insert a linear polarizer or phase plate of the proper retardance to produce a full scale (100%) calibration. For case I a polarizer may be inserted into the beam when the detector arm is set at 0°. (A set of attenuators reduces the intensity of the beam to normal operating levels in order to avoid damage to the photomultiplier tube.) From Eq. (9), it can be seen that aligning the polarizer in the r and l directions causes either m_1 or m_2 to be zero. This results in a positive or negative signal of equal value which may be adjusted to give full scale (100%) deflection. For case II, S is adjusted for 100% deflection at 0° for no intervening polarizers. The D signal may be calibrated by inserting a variable retarding stress plate into the beam and adjusting the phase plate for maximum deflection. The variable retarder assembly consists of a small clamp which can be rotated about an axis parallel to the beam that applies an adjustable force to a floating jaw. This produces a uniform stress in a carefully machined Plexiglas or quartz rectangle, thereby inducing a birefringent phase shift between light having a polarization parallel and perpendicular to the jaws.

The condition $J_0(A)=0$ may be obtained[14] by examining the resultant beam for case II with the scattering cell removed. By alternatively placing a polarizer in front of the photomultiplier tube at ±45°, the following current will result:

$$I=\frac{I_0}{1\pm J_0(A)}\left[1\pm\frac{2J_2(A)\cos 2wt}{1\pm J_0(A)}\right].$$

If $J_0(A)\neq 0$, switching from the +45° to the −45° polarizer will change the applied photomultiplier tube voltage, which will vary the gain of the tube, in turn causing unequal positive and negative second harmonic signals. Thus to obtain the necessary condition $J_0(A)=0$, the modulator amplitude is adjusted until the two signals have the same absolute magnitude.

IV. RESULTS FOR LATEX SPHERES

A. Mie Calculations

Mie[15] calculations were performed to obtain the elements of the scattering matrix for the spherical particles of two model systems. The computations were performed on a CDC 6400 with a computing routine developed by Dave.[16] The results of the calculations were graphed directly by a Cal Comp 665 plotter in the same format as the experimental output to facilitate direct comparisons. A Gaussian size distribution was used that matched the specifications given by the manufacturer of the test particles. The results of the Mie calculations were summed over 21 sizes, ranging more than two standard deviations from the center radius to obtain the curves for a given size distribution. Further

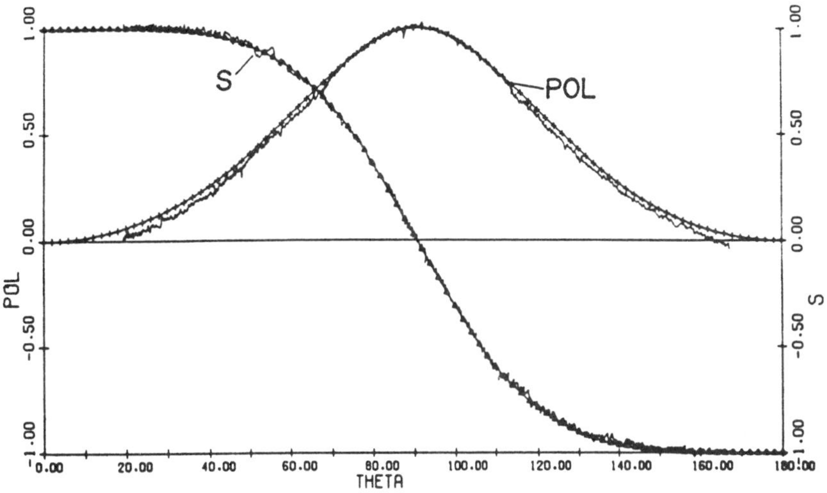

FIG. 5. Calculated and experimental POL and S signals for 550 Å radius latex particles for light of wavelength 4350 Å.

increasing the number of integration steps did not affect the results.

The effect of the spectral distribution of the Hg lamp was calculated by assuming that it had a Gaussian shape and performing numerically a double integration over the two distribution functions. That is, if $F(a/\lambda)$ represents the result of one Mie calculation for a given radius a, and wavelength λ, then the average result is given by

$$F\left(\frac{a}{\lambda}\right)=\int_0^\infty\int_0^\infty F\left(\frac{a}{\lambda}\right)\exp\{-[(\lambda-\lambda_0)/\sigma_1]^2\}\times\exp\{-[(a-a_0)/\sigma_2]^2\}da\,d\lambda,$$

where σ_1 and σ_2 give the Gaussian parameters for the distribution in wavelength and size, respectively. The spectral distribution, which was somewhat narrower than the size distribution, was summed over 11 wavelengths for the larger test particles. There was very little difference between the results that included both distribution functions and those with only the distribution in size.

B. Experimental Results and Comparison with Theory

In order to evaluate the performance of the instrument, a series of measurements was undertaken of two model systems. This acted as a test of the theory and also helped locate problem areas. The two systems used were prepared from Dow[17] polystyrene latex spheres dispersed in water. According to the manufacturer's specifications, one system had an average radius of 550 Å, the other 3940 Å radius. Measurements were made at three wavelengths, the 5461 and 4350 Å lines of mercury, and at 3250 Å using a He–Cd laser. The 550 Å spheres were small enough to produce Rayleigh-like scattering. This case is especially easy to visualize since the vertical component of the light is scattered isotropically and the horizontal component varies as the $\cos\theta$, where θ is the scattering angle. It is easy to see that the scattering amplitudes are given by

$$A_1=1$$
$$A_2=\cos\theta.$$

Substituting these values into Eqs. (3), (10), (11), and (12) and normalizing them so as to correspond to the signal output gives

$$P=\frac{1-\cos^2\theta}{1+\cos^2\theta},\quad S=\frac{\cos\theta}{1+\cos^2\theta},\quad D=0$$
$$\Sigma=\log_{10}\left\{\frac{1+\cos^2\theta}{\sin\theta}\,\frac{\sin\theta_0}{1+\cos^2\theta_0}\right\}. \tag{13}$$

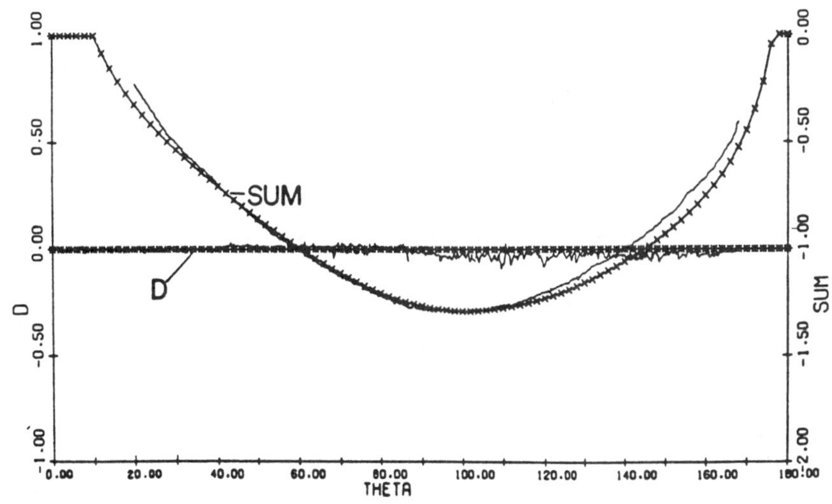

FIG. 6. Calculated and experimental D and SUM signals for 550 Å radius latex particles for light of wavelength 4350 Å.

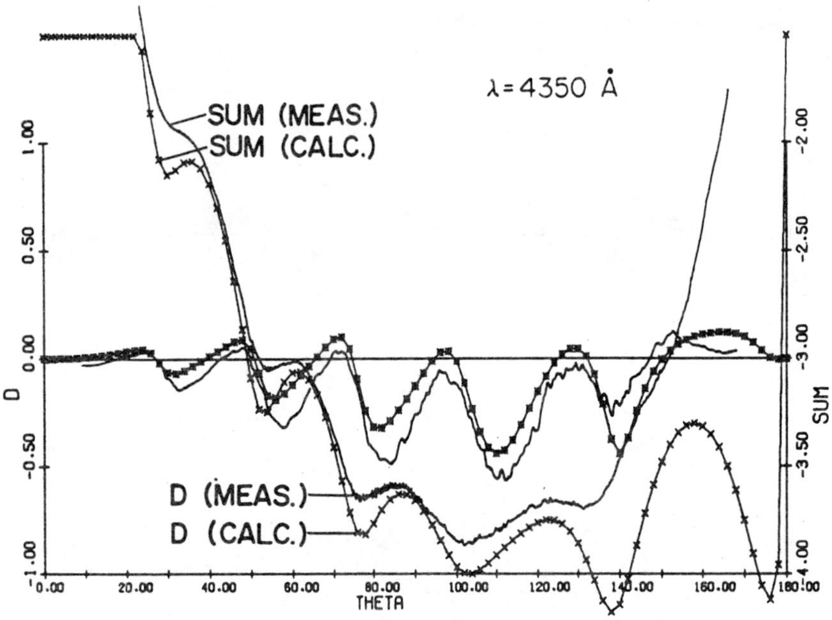

FIG. 7. Calculated and experimental D and SUM signals for 4060 Å radius latex particles for light of wavelength 4350 Å.

Figures 5 and 6 illustrate the computer-produced plots of these four variables ($\sum$=SUM, P=POL). The results are actually from the Mie calculation for spheres of real refractive index 1.592 in water, but correspond closely to the Rayleigh scattering case. The experimental curves for a wavelength of 4350 Å are also displayed in Fig. 5 and 6. It can be seen that the agreement between the calculated and experimental results is quite good for the three normalized quantities, with a slight deviation in magnitude at low and high angles for the total intensity (SUM) curve.

The second scattering system investigated was composed of the larger polystyrene spheres (Dow lot number 2F2J, diameter 0.794 μ). The Mie computation must be done for this case, and it predicts complex scattering curves characteristic of the higher order electromagnetic modes of a sphere. Figures 7 and 8 illustrate the results for all four measurements. It can be seen that the position of most of the theoretically predicted features are correctly given in the experimental curves. The angular position of these are indicative of the size of the spheres while the heights are more characteristic of the size distribution. The fact that the experimental positions of the peaks are given correctly is confirmation that the technique is working as predicted. The discrepancy in the height of the peaks can be attributed to multiple scattering, light scattered from the walls of the scattering cell, impurities, and clustering. The effect of multiple scattering on the SUM and D curves was experimentally investigated. For larger dilution factors, less light is multiply scattered into the acceptance angle of the detector. This trend may be seen in the sum curves for various dilutions in Fig. 9.

Another major factor which can lead to extraneous scattered light is the internal reflections of the scattered light within the cell.[18] This becomes more important for larger colloids which have strong scattering lobes. Several techniques were tried in order to reduce the amount of reflected

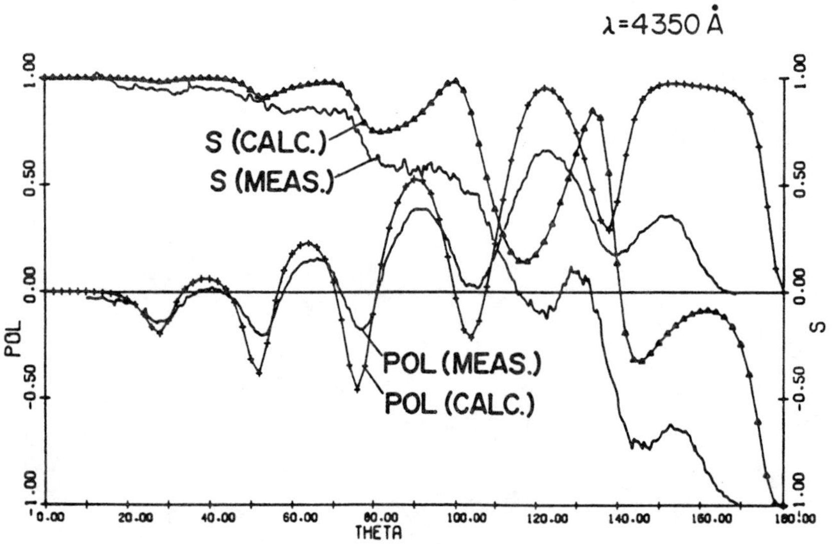

FIG. 8. Calculated and experimental POL and S signals for 4060 Å radius latex particles for light of wavelength 4350 Å.

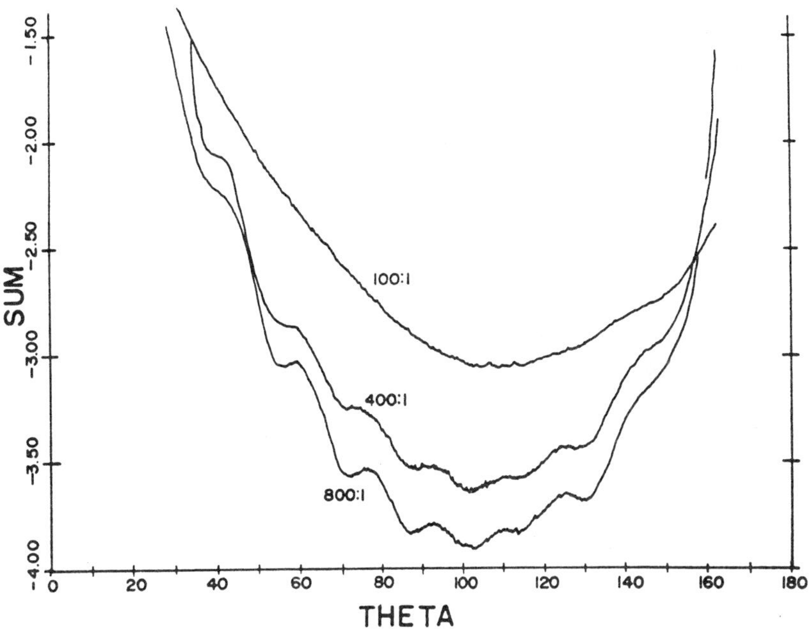

FIG. 9. Measurement of the effect of multiple scattering on the SUM signal for 4060 Å latex particles at 3250 Å.

scattered light. The most successful scheme was to coat the inside walls of the cell except the viewing area with a black, nonscattering paint.[19]

Recently it has been pointed out[20] that additional particles may be present in the latex sols not within the specified Gaussian size distribution. In addition, any impurities present including clusters of the sol particles,[21] would increase the general scattered light and cause a decrease in the size of the structure of the curves. The large latex sols were prepared with deionized water and centrifuged to decrease the number of two and three particle agglomerates. This produced a substantial increase in the peak height and brought the experimental results into better agreement with the calculations.

Figures 10 and 11 illustrate the experimental and theoretical results for the ultraviolet laser measurements at a wavelength of 3250 Å. The results are appreciably better with the laser source and allow the study of much more tenuous colloids and aerosols.

The index of refraction of bulk polystyrene, measured at 3250 Å, by use of a spectrophotometer and the method of minimum deviation was determined to be $1.672+\beta i$ where $\beta<0.0001$. The indices used at 4350 and 5461 Å were $1.61+0.0i$ and $1.60+0.0i$, respectively. With these values of the refractive index, it was found that the experimental results were in better agreement with calculations for slightly larger sphere sizes (at all wavelengths) than that specified by the manufacturer of the latex sols. The best fit was for a radius of 4060 Å. Since in this case, the method is able to measure the absolute size of the latex spheres to an accuracy of greater than 1%, it allows a new and sensitive determination of that parameter.

Now it is possible to reconsider the question of the degree of polarization raised in the last part of Sec. II. Equations (10) and (12) indicate that the quantities P, S, and D are just the original matrix elements divided by the total intensity of scattered light at that angle. However, in Eq. (5) they occur as the last three unnormalized elements of the

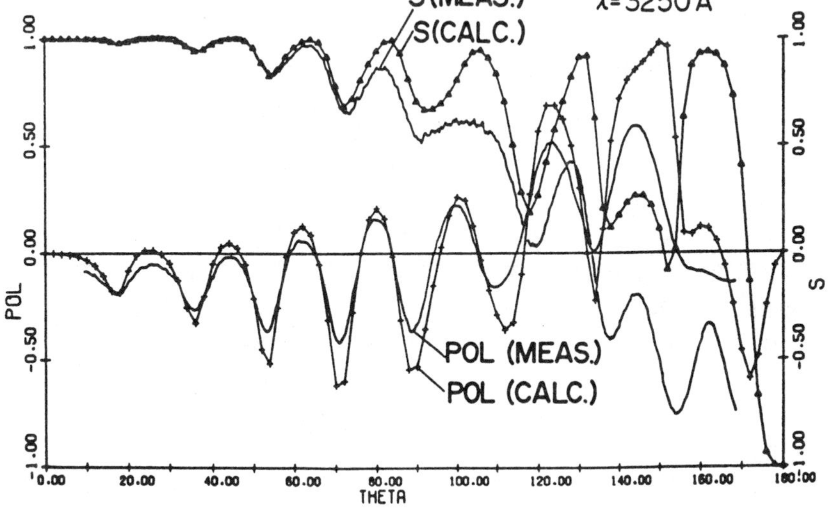

FIG. 10. Calculated and experimental POL and S signals for 4060 Å radius latex particles for laser light of wavelength 3250 Å.

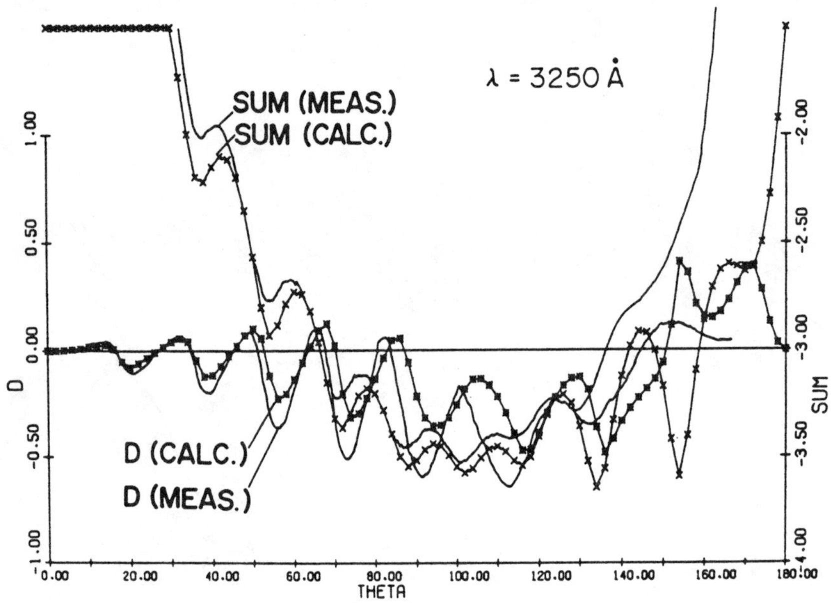

FIG. 11. Calculated and experimental D and SUM signals for 4060 Å radius latex particles for laser light of wavelength 3250 Å.

Stokes vector. Therefore, the degree of partial polarization defined in Eq. (6) for the experiment in Sec. II is given by

$$(P^2+S^2+D^2)^{\frac{1}{2}}=F,$$

hence the degree of partial polarization is easily determined from the measured quantities. For Rayleigh scattering it is equal to one for all angles. In more complex cases involving size distributions, it decreases nonuniformly from unity in the forward direction to a low at some intermediate angle. An analysis of both the theoretical and experimental results for the small spheres indicates that F is very nearly one for all angles. In the case of the larger latex spheres, F varies considerably with angle, approaching one for forward and backscatter directions while reflecting the complex variations in the other scattering parameters at intermediate angles. The experimental data follows most of the variations but is usually somewhat lower in value. This may be interpreted as due to either a size distribution wider than the manufacturer's specifications or the other nonideal factors mentioned before or both.

V. FURTHER INVESTIGATIONS

The instrument has been used for preliminary investigations of a number of other colloidal and aerosol systems. Rather extensive investigations of the growth of the La Mer sulfur sols[22] have already been carried out. Figures 12 and 13 illustrate the agreement between theory and experiment for determining the size of a sulphur sol at one stage in its growth. In general the experimental agreement with theory for the sulfur sols has been better than for the latex spheres. This is probably due to the fact that the sulfur has fewer particles outside the Gaussian size range than do the latex spheres. Small impurity particles might be effectively removed if they served as nucleation sites for the sulfur particles.

The instrument seems well suited to study multiple scattering effects experimentally by varying the concentration of a sol with known characteristics. The effect on the scattering matrix of alignment of ferromagnetic particles by an external field allows the variation of the symmetry of the

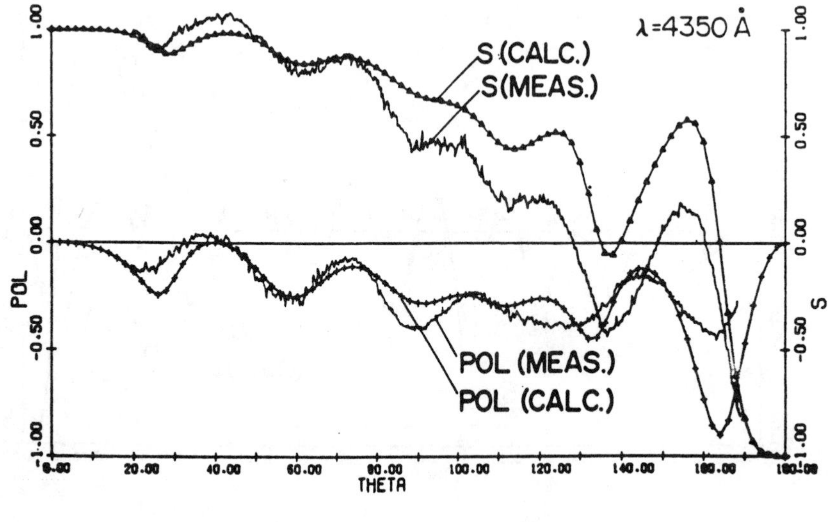

FIG. 12. Calculated and experimental POL and S signals for approximately 3000 Å spherical sulfur particles at 4350 Å.

FIG. 13. Calculated and experimental D and SUM signals for approximately 3000 Å spherical sulfur particles at 4350 Å.

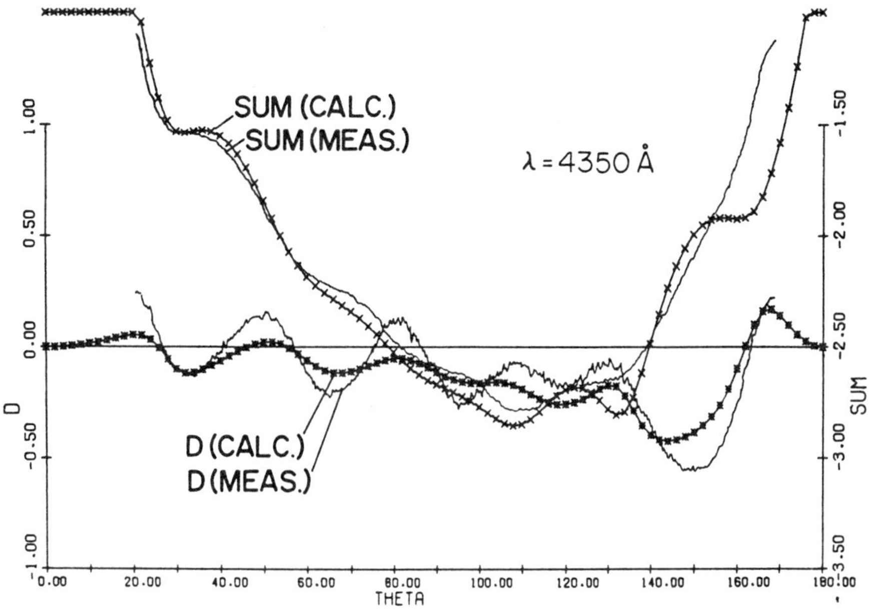

scattering system. Early measurements on smoke aerosols indicate they may have definite polarization signatures in spite of the wide particle size distribution.

Improvements are planned to the scattering cell to reduce the amount of scattered light from internal reflections within the cell by trapping it in an extended Rayleigh horn.[18] The possibility of suspending and performing measurements on a single particle will be pursued because the elimination of the smoothing effect produced by the particle size distribution would allow for much more precise measurements.

ACKNOWLEDGMENT

The authors would like to thank Professor James C. Kemp for the use of a polarization modulator, which made possible the development of the present light scattering instrument.

*Supported by the Research Corporation and by the Atmospheric Sciences Section, National Science Foundation.

[1]M. Kerker and V. K. La Mer, J. Am. Chem. Soc. **72,** 3516 (1950).
[2]W. B. Dandliker, J. Am. Chem. Soc. **72,** 5110 (1950).
[3]A. F. Stevenson, W. Heller, and M. L. Wallach, J. Chem. Phys. **34,** 1789 (1961).
[4]J. P. Kratohvil and C. Smart, J. Colloid Sci. **20,** 875 (1965).
[5]H. C. van de Hulst, *Light Scattering by Small Particles* (Wiley, New York, 1957).
[6]A. C. Holland and G. Gagne, Appl. Opt. **9,** 1113 (1970).
[7]J. C. Kemp, J. Opt. Soc. Am. **59,** 950 (1969).
[8]S. N. Jasperson and S. E. Schnatterly, Rev. Sci. Instrum. **40,** 761 (1969).
[9]W. A. Shurcliff, *Polarized Light* (Harvard University Press, Cambridge, 1966).
[10]M. J. Walker, Am. J. Phys. **22,** 170 (1954).
[11]H. Mueller, J. Opt. Soc. Am. **38,** 661 (1948).
[12]D. Deirmendjian, *Electromagnetic Scattering on Spherical Polydispersions* (American Elsevier, New York, 1969).
[13]F. Perrin, J. Chem. Phys. **10,** 415 (1942).
[14]S. N. Jasperson, private communication.
[15]G. Mie, Ann. Phys. (Leipz.) **25,** 377 (1908).
[16]J. V. Dave, IBM Scientific Center, Palo Alto, California, Report No. 320-3236, 1968.
[17]Dow Chemical Company, 1200 Madison Avenue, Indianapolis, Indiana 46225.
[18]W. Heller and J. Witeczek, J. Phys. Chem. **74,** 4241 (1970).
[19]3M brand velvet coating, 100 series No. 101-C10 black, Minnesota Mining and Manufacturing Co., St. Paul, Minnesota.
[20]D. D. Cooke and M. Kerker, J. Colloid Interface Sci. **42,** 150 (1973).
[21]J. A. Davidson, C. W. Macosko, and E. A. Collins, J. Colloid Interface Sci. **25,** 381 (1967).
[22]I. Johnson and V. K. La Mer, J. Am. Chem. Soc. **69,** 1184 (1947).

Section Four
Depolarizers

One of the difficulties of working in the field of polarization is the fact that most light sources are partially polarized and most optical equipment partially polarizes the light that it transmits.

Many of us at Polaroid Corporation were extremely conscious of this difficulty. I felt that not enough work had been done in the area of depolarizing systems and jokingly proposed to my friends that I would set up a company called "Depolaroid Corporation."

It is important to remember that no device exists that will convert polarized light to what is normally defined as unpolarized light. The devices that do exist convert polarized light to a state in which a particular detecting system will be unable to distinguish it from unpolarized light. In the Lyot depolarizer, each wavelength interval is polarized differently and a broadband detector will believe the light is unpolarized. In the case of my own device, the detector must integrate the signal over a long enough period of time to believe the light is unpolarized.

It is still disturbing to find advertisements for monochromatic depolarizers that, in fact, merely change one state of polarization to another. In the case of this device, the word "depolarizer" probably means conversion of linear polarized light to another state.

I have included one advertisement from the Melles Griot catalogue that covers the Cornu pseudodepolarizer. I wrote the first draft of the text and calculations for this article. It has been re-edited several times with the addition of the term "pseudodepolarizer" with which I am not too happy. The device is made by a company in Germany and I was excited by the thought that Cornu had invented the first depolarizer. This would have been long before Lyot and would indeed be a milestone. I wrote the company and found that they had invented the device and decided to name it the Cornu because of its relationship to the Cornu prism.

On a visit to Meudon, I had a lengthy discussion with Lyot, the inventor of the first depolarizer. I still believe this field has been neglected.

Reprinted from *Annales de l'Observatoire de Paris*, Vol. VIII, 100–104 (1929).

RECHERCHES SUR LA POLARISATION DE LA LUMIÈRE DES PLANÈTES ET DE QUELQUES SUBSTANCES TERRESTRES

Par B. LYOT

AIDE-ASTRONOME A L'OBSERVATOIRE

DISPOSITIFS EXPÉRIMENTAUX

Dispositif général

Construction. — Le dispositif utilisé pour la plupart de mes mesures est le suivant :

La substance S (fig. 26) est placée au centre d'un goniomètre dont l'un des bras porte la source lumineuse, l'autre le polarimètre.

La source est une lampe demi-watt A, de 6 volts, 100 bougies ; deux lentilles de 40 mm. de foyer chacune, B et C, projettent une image très agrandie de son filament sur la substance, par l'intermédiaire d'un prisme à réflexion totale D. Ce prisme est suivi d'une double lame cristalline E, étudiée de manière à supprimer la faible polarisation de la lampe et celle qu'introduisent les lentilles. (Cette lame sera décrite un peu plus loin.) Un diaphragme circulaire F, de 12 millimètres de diamètre, limite l'ouverture de la source.

Le polarimètre est muni d'une lentille biconvexe G, de 18 centimètres de foyer, qui forme l'image de la partie la plus éclairée de la substance sur un œilleton H, de 2 millimètres de diamètre, placé contre la bonnette polariscope (les verres de l'oculaire ont été enlevés). Le foyer antérieur de la lentille G est occupé par un diaphragme I, de $3^{mm},5$ de diamètre. Ce diaphragme est destiné à limiter le champ à une portion voisine de l'axe sous-tendant un diamètre apparent d'environ 70 minutes. Il facilite, en même temps, la vision des franges qui se forment à son intérieur, dans le même plan que lui.

La source est placée à 45 centimètres de la surface étudiée ; vue de celle-ci, elle sous-tend un diamètre apparent de 1°30′. Le diaphragme du polarimètre sous-tend un diamètre de 0°25′. Leur distance angulaire, ou angle de vision, est donnée par l'index J, mobile sur le cercle K. Elle peut descendre jusqu'à 1°30′. Au-dessous de cette valeur, les plans de vision relatifs aux divers points de la source sont très différents, ce qui donnerait une polarisation résultante notablement plus faible que si la source était ponctuelle.

Double lame dépolarisante. — Cette lame est constituée par deux lames de quartz taillées parallèlement à l'axe optique (à défaut de quartz, j ai utilisé deux lames de gypse clivé). Leurs épaisseurs sont dans le rapport de 2 à 1 et elles sont collées au baume de Canada de manière que leurs axes optiques fassent entre eux des angles de 45°.

Cet ensemble se comporte de la manière suivante :

Faisons tomber sur lui une lumière polarisée rectiligne. Soient (fig. 27) :

OX, OY et OX′, OY′, les axes des deux lames ;

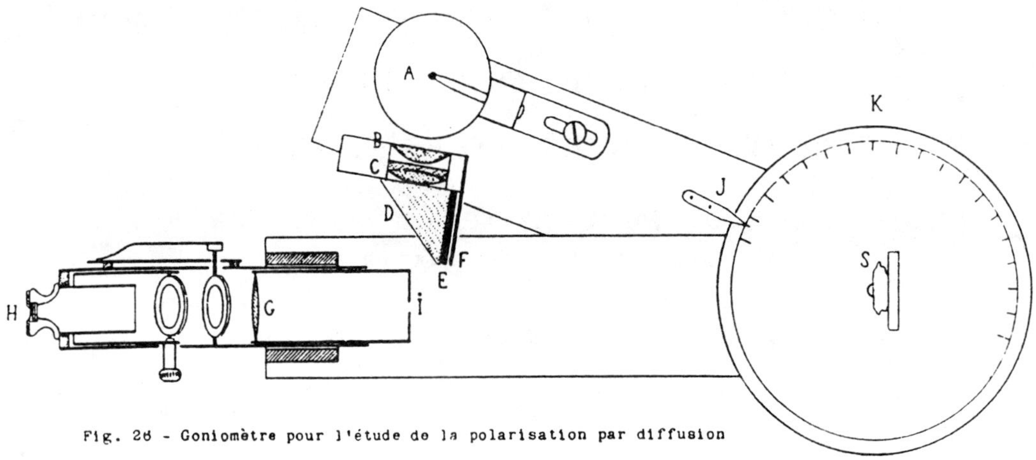

Fig. 26 - Goniomètre pour l'étude de la polarisation par diffusion

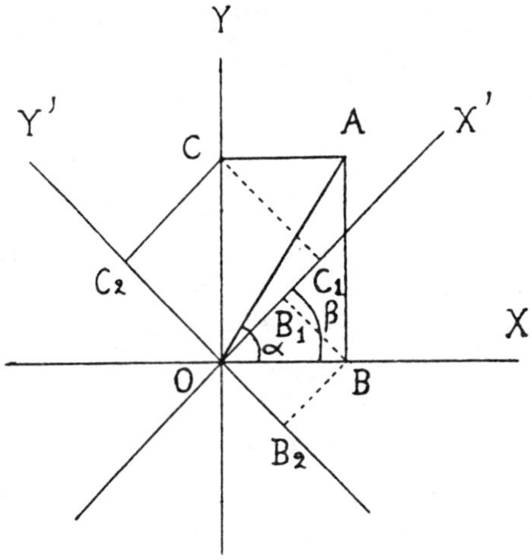

Fig. 27 - Double lame dépolarisante

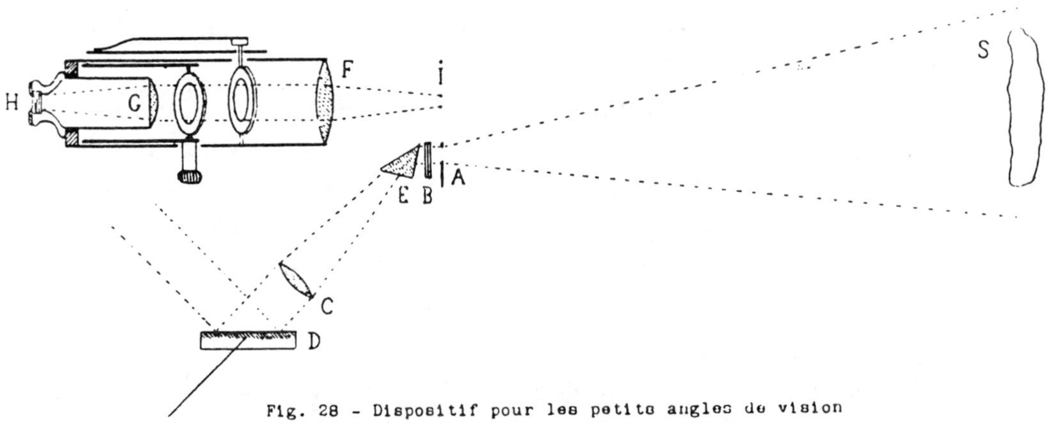

Fig. 28 - Dispositif pour les petits angles de vision

β l'angle que font OX′ et OX ;

α l'angle de la vibration rectiligne incidente OA et de OX.

Il sort de la seconde lame quatre vibrations confondues OB_1, OB_2, OC_1, OC_2.

Soient n et n' les indices principaux des lames, e et e' leurs épaisseurs. Les quatre vibrations ont subi des retards respectifs :

$$ne + ne' \qquad ne + n'e' \qquad n'e + ne' \qquad n'e + n'e'.$$

Dans le cas où $e' = 2e$, les retards des trois dernières par rapport à la première sont :

$$2(n' - n)e \qquad (n' - n)e \qquad 3(n' - n)e.$$

Si $(n' - n)e$ est assez grand (plus d'une dizaine de longueurs d'onde), ces vibrations ne donnent, en lumière blanche, aucune interférence observable.

Leurs intensités s'ajoutent.

Il sort donc deux faisceaux, polarisés respectivement suivant OX' et OY', dont les intensités sont :

$$OB_1^2 + OC_1^2$$

d'une part ;

$$OB_2^2 + OC_2^2$$

d'autre part.

Or, on a :

$$OB_1 = OA \cos \alpha \cos \beta \qquad OC_1 = OA \sin \alpha \sin \beta$$
$$OB_2 = OA \cos \alpha \sin \beta \qquad OC_2 = OA \sin \alpha \cos \beta$$

d'autre part :

$$\cos \alpha = \sin \beta = \frac{\sqrt{2}}{2}$$

d'où :

$$OB_1^2 + OC_1^2 = \frac{OA^2}{2} \qquad \text{et} \qquad OB_2^2 + OC_2^2 = \frac{OA^2}{2}.$$

Leur ensemble est équivalent à de la lumière naturelle d intensité OA, égale à celle de la lumière incidente.

Lorsqu'on examine une source de lumière très fortement polarisée, avec un polariscope de Savart, à travers cette double lame, on ne voit pas de franges au centre du champ : mais il apparaît, de chaque côté, d'autres systèmes de franges dans les directions pour lesquelles la biréfringence des lames de Savart compense, à quelques longueurs d'onde près, les différences de marche introduites par la double lame.

Comme les mesures sont effectuées au centre du champ, ces systèmes de franges n interviennent pas et la lumière se comporte comme si elle était parfaitement naturelle.

Disposition de la substance. — Pour étudier un minéral, il suffit de pratiquer dans celui-ci une cassure grossièrement plane et de le fixer, au moyen de cire molle, à un support placé au centre du goniomètre, de manière que la cassure soit parallèle à l'axe de celui-ci.

Lorsque le solide est très hétérogène, on le taille grossièrement sous la forme d'une sphère de 6 millimètres de diamètre portée par un fil de fer très fin. Un petit moteur électrique le fait tourner autour d'un axe confondu avec celui du goniomètre. On obtient ainsi la polarisation moyenne des divers constituants du minéral.

Pour étudier une poudre, on rend l'axe du goniomètre horizontal et on place la poudre sur une tablette parallèle à cet axe.

Pour examiner un jet de vapeur ténue, par exemple, il suffit de placer celui-ci au centre du goniomètre.

Etude de la lumière diffusée très près de la source

Mes observations astronomiques, en particulier celles effectuées sur l'anneau intérieur de Saturne, montrent que les rayons diffusés par certaines substances, à 12 minutes seulement de la source lumineuse, sont encore sensiblement polarisés.

L'appareil décrit ci-dessus ne permet pas d'effectuer des mesures au-dessous de 1°30′.

Pour atteindre des angles aussi petits que possible, il fallait réduire le diamètre apparent de l'ouverture du polarimètre et celui de la source, augmenter beaucoup la brillance de celle-ci, ainsi que la distance et les dimensions de la substance étudiée.

J'ai donc réalisé le dispositif suivant :

La substance étudiée S (fig. 28) est disposée sous forme d'une plaque de 15 à 20 centimètres de côté, au fond d'une pièce très obscure ; elle est éclairée par un diaphragme A, de 2 millimètres de diamètre, précédé d'une double lame cristalline B destinée à dépolariser la lumière, analogue à celle qui est décrite page 102. Une lentille C de 80 centimètres de foyer, précédée d'un sidérostat polaire D, forme, sur le diaphragme A, une image du centre du Soleil à travers un très petit prisme à réflexion totale E.

Le polarimètre a pour objectif une lentille F de 6 centimètres de foyer et pour oculaire une lentille semblable G ; l'ensemble donne, à l infini, une image de la substance S en, vraie grandeur.

La bonnette-polariscope est munie d'un diaphragme H, de 2 millimètres de diamètre, dont une image réelle I est formée, en vraie grandeur, par l'ensemble des lentilles, à l'extérieur du polarimètre.

Le polarimètre est porté par un support à hauteur variable qui permet de le pointer sur la substance et d'amener l'image I de son diaphragme à 4 millimètres du diaphragme A de la source.

L'angle de vision peut ainsi descendre à 2′48″. Pour le faire varier, il suffit de déplacer le polarimètre et son support. Il est facile de le calculer, en mesurant directement la distance du diaphragme de la source à l'image de celui du polarimètre.

A Monochromatic Depolarizer*†

BRUCE H. BILLINGS
Research Laboratory, Baird Associates, Inc., Cambridge 38, Massachusetts
(Received March 15, 1951)

Up to the present time no image-forming depolarizing filter has been described which will work for monochromatic light. This paper describes a simple monochromatic depolarizer. This depolarizer consists of two variable wave plates whose axes are separated by an angle of 45°. The retardation of the first wave plate varies from zero to 2π. The second wave plate varies from zero to 4π. This variation is made to occur in a time which is short compared to the time constant of the apparatus in which the depolarizer is used. Typical components are Z cut plates of ammonium-dihydrogen-phosphate. The retardation of these plates can be made to vary by applying an electric field along the optic axis.

A discussion is also given of the Lyot depolarizer which is effective for light which covers a wide wavelength range. Tolerances are derived for the behavior of the Lyot depolarizer and experimental measurements are shown of its behavior.

SINCE the time of Huygens there has been considerable attention paid to the study of polarized light. Many different varieties of light polarizers have been invented and many scientific and industrial applications have been found for polarized light. In contrast to this mass of material there appears to be only one paper devoted to depolarizers.[1] In particular, it is not possible to purchase a depolarizing light filter although it is possible to purchase a variety of polarizing light filters. No depolarizer seems to have been described which will work with monochromatic light. This paper describes a new depolarizer which is effective for either monochromatic light or light with any spectral characteristics.

In order to understand the operation of a depolarizer it is necessary to define exactly what is meant by unpolarized light. The definition of unpolarized light which leads most readily to the treatment of a depolarizer is that given by Perrin.[2] This definition has been discussed by Mueller[3] and Billings[4] using an optical calculus which was developed by Mueller.

According to this definition unpolarized light is light for which the time averages of the polarization com-

* Presented in part as an invited paper at the March, 1951, meeting of the Optical Society of America, Washington, D. C.
† Part of this paper is appearing in Spanish in *Ciencia e Investigacion*.

[1] Lyot, Ann. Observatoire Astron. Phys. de Paris (Meudon)—Tomi i Fasc. 1–2 8, 102 (1928).

[2] F. Perrin, J. Chem. Phys. 10, 415 (1942).

[3] H. Mueller, M.I.T. Course (8.26), Spring, 1945; J. Opt. Soc. Am. 38, 661(A) (1948).

[4] B. H. Billings and E. H. Land, J. Opt. Soc. Am. 38, 819–829 (1948).

ponents become equal to zero. In equation form this can be written:

$$\bar{M}=\frac{1}{T}\int_0^T Mdt=0,$$

$$\bar{C}=\frac{1}{T}\int_0^T Cdt=0,$$

$$\bar{S}=\frac{1}{T}\int^T Sdt=0, \tag{1}$$

where M, C, and S are the instantaneous Stokes vectors of the light. Another way of stating the definition is to say that unpolarized light is light in which the state of polarization changes more rapidly than can be followed by the particular detector which is being used. Operationally one might say that with unpolarized light there is no variation in intensity when it is passed through a rotating polarizer or wave plate. The vector for unpolarized light of unit intensity can be written:

$$\begin{vmatrix} I \\ M \\ C \\ S \end{vmatrix} = \begin{vmatrix} 1 \\ 0 \\ 0 \\ 0 \end{vmatrix}. \tag{2}$$

Unpolarized light can also be treated by the Jones[5] calculus, but it is necessary to use two sets of matrices.

A MONOCHROMATIC DEPOLARIZER

In treating the depolarizing filter we shall use the Mueller calculus. The equations of this calculus are found in reference 4. The equation connecting the incident light with the light which emerges from such a depolarizer can be written as follows:

$$\mathbf{D}\begin{vmatrix} I_1 \\ M_1 \\ C_1 \\ S_1 \end{vmatrix} = \begin{vmatrix} I_2 \\ M_2 \\ C_2 \\ S_2 \end{vmatrix}. \tag{3}$$

Here the vector I_1, M_1, C_1, S_1 represents the most general kind of polarized light incident upon the depolarizer. **D** is the matrix of the depolarizer. The vector I_2, M_2, C_2, S_2 represents the emergent light.

The instrument must see no change in intensity when a linear, circular, or elliptical polarizer is put in series with the system and when the incident light is considered to have any state of polarization.

Let us represent any analyzer by the matrix **A**. The equation showing the operation of this analyzer on the light emergent from the depolarizer can be written as follows:

$$\mathbf{A}\begin{vmatrix} I_2 \\ M_2 \\ C_2 \\ S_2 \end{vmatrix} = \begin{vmatrix} I_3 \\ M_3 \\ C_3 \\ S_3 \end{vmatrix}. \tag{4}$$

This can be expanded to give

$$I_3=A_{11}I_2+A_{12}M_2+A_{13}C_2+A_{14}S_2. \tag{5}$$

According to the definition given above for unpolarized light, I_3, the intensity of the light through the analyzer, must be independent of the characteristics of the incident light, I_2, M_2, C_2, S_2, and it must also be independent of the characteristics of the analyzer. In order to accomplish this M_2, C_2, and S_2 are made to vary rapidly with time. If this variation is rapid compared with the time used to measure I_3, the average of M_2, C_2, and S_2 can be made equal to zero.

From Eq. (3)

$$M_2=D_{21}I_1+D_{22}M_1+D_{23}C_1+D_{24}S_1=0 \tag{6}$$

with similar equations for C_2 and S_2. Here the D_{ij}'s are considered to be instantaneous values. When the observation time is τ, this equation is replaced by

$$\frac{1}{\tau}\int_0^\tau M_2dt=\frac{I}{\tau}\int^\tau D_{21}dt+\cdots=0. \tag{7}$$

The depolarizer D must then consist of a series of elements whose characteristics can be made to vary with time.

There are several possible mechanical arrangements of optical rotators and wave plates which can be made to give zero average for M_2, C_2, and S_2. The device described below has been chosen as one of the simplest and most flexible of these possible arrangements. This depolarizer consists of two variable wave plates whose axes are placed at an angle to each other. The matrix equation which describes the effect of such a combination can be written as follows:

$$\mathbf{R}_2(n\delta,\beta)\mathbf{R}_1(\delta,0)\begin{vmatrix} I_1 \\ M_1 \\ C_1 \\ S_1 \end{vmatrix} = \begin{vmatrix} I_2 \\ M_2 \\ C_2 \\ S_2 \end{vmatrix}. \tag{8}$$

$\mathbf{R}_1$ and $\mathbf{R}_2$ represent the matrices of the two retardation plates which have retardations δ and $n\delta$, respectively. Retardation plate 2 is inclined at an angle β to retardation plate 1. Retardation plate 1 is considered as having one of its axes parallel to the reference axis of the system. This can be done without any loss of generality.

[5] R. Clark Jones, J. Opt. Soc. Am. **31**, 488–503 (1941).

The matrix of the two plates is:

$$\begin{vmatrix} 1 & 0 & 0 & 0 \\ 0 & \cos^2 2\beta+\sin^2 2\beta\cos n\delta & \begin{matrix}[\cos 2\beta\sin 2\beta(1-\cos n\delta)]\cos\delta \\ +\sin 2\beta\sin n\delta\sin\delta\end{matrix} & \begin{matrix}-[\cos 2\beta\sin 2\beta(1-\cos n\delta)]\sin\delta \\ +\sin 2\beta\sin n\delta\cos\delta\end{matrix} \\ 0 & \sin 2\beta\cos 2\beta(1-\cos n\delta) & \begin{matrix}\cos\delta[\sin^2 2\beta \\ +\cos^2 2\beta\cos n\delta] \\ +\sin\delta[-\cos 2\beta\sin n\delta]\end{matrix} & \begin{matrix}-\sin\delta(\sin^2 2\beta+\cos^2 2\beta\cos n\delta) \\ -\cos\delta\cos 2\beta\sin n\delta\end{matrix} \\ 0 & -\sin 2\beta\sin n\delta & \begin{matrix}\cos\delta(\cos 2\beta\sin n\delta) \\ +\sin\delta\cos n\delta\end{matrix} & \begin{matrix}-\sin\delta(\cos 2\beta\sin n\delta) \\ +\cos\delta\cos n\delta\end{matrix} \end{vmatrix}$$

In this matrix δ is considered to be a periodic function of time. The average value of the matrix coefficients is obtained by integrating over one period of the variation of δ. From the resulting set of equations it is possible to obtain values for the parameters β and n which will make the combination effective as a depolarizer. Although there are twelve equations, the parameters can be determined with only a few integrals. The integral of D_{22} becomes:

$$\bar{D}_{22}=\frac{1}{T}\int_0^T(\cos^2 2\beta+\sin^2 2\beta\cos n\delta)dt=0$$

or

$$\bar{D}_{22}=\cos^2 2\beta+\frac{1}{T}\int_0^T\sin^2 2\beta\cos n\delta(t)dt=0. \quad (10)$$

From this equation it is intuitively clear that β should be 45°.

Before establishing values for the other parameters, it is necessary to select the function $\delta(t)$. The simplest mathematical solution is obtained when δ is a saw-toothed linear function of time and can be written

$$\delta=Kt \quad 0\leq t\leq\tau. \quad (11)$$

When $\delta=2\pi$ the function repeats. With this function the equation becomes

$$\bar{D}_{22}=\frac{1}{2\pi}\int_0^{2\pi}\cos n\delta d\delta=0. \quad (12)$$

From this equation it is clear that n should be an integer. The next integral can be evaluated to give

$$\bar{D}_{23}=\frac{1}{2\pi(n-1)}[\sin\pi(n-1)]+\frac{1}{2\pi(n+1)}[\sin\pi(n+1)]. \quad (13)$$

This equation shows that n cannot be unity. A suitable value for n is 2. With this value for n all the remaining integrals become zero and the matrix satisfies the conditions for a perfect depolarizer. The resulting instrument then consists of two wave plates which are mounted with an angle of 45° between their fast axes. The wave plates vary in retardation linearly with the time. Wave plate No. 1 varies periodically from zero to one wave retardation. Exactly the same effect will be realized if wave plate No. 2 varies from zero to one wave retardation at twice the frequency with which wave plate No. 1 is varied. With this combination all the elements of the depolarizer matrix become zero.

The simplest type of variable wave plate which can be operated at frequencies which are high compared with the frequencies of most physical measurements is a crystal electro-optic shutter.[6]

When voltage is applied to a Z-cut section of a uniaxial crystal of the type XH_2PO_4 the induced birefringence is much larger than the ordinary Kerr effect in solids or liquids. Research on these crystals has only been in progress for the past decade.[6–10]

The most available crystal is $NH_4H_2PO_4$ or ammonium-dihydrogen-phosphate (ADP). (This crystal is grown under the trademark "PN" by the Brush Development Company, 3405 Perkins Avenue, Cleveland, Ohio.) Along the axis the retardation is zero. When voltage is applied in this direction the retardation becomes:

$$\Gamma=d(n_e-n_0)/\lambda=pV, \quad (14)$$

where Γ is the retardation, p is a constant independent of the dimensions of the crystal plate, and V is the applied voltage. In ADP approximately 9500 volts is required to give a half-wave retardation at 5461A. When the voltage is applied the crystal becomes biaxial with an axial angle Ω whose value can be calculated from the equation:

$$\sin^2\Omega=pV/kd \quad (15)$$

where $k=(\omega-\epsilon)/\lambda$, $p=r_{63}\omega^3/\lambda$, ω and ϵ are the ordinary and extraordinary indices, and r_{63} is the electro-optic coefficient.

The crystal must be used in fairly well-collimated light as the retardation at angles to the optic axis is

[6] B. H. Billings, J. Opt. Soc. Am. **39**, 797–801 (1949); J. Opt. Soc. Am. **39**, 802–808 (1949); R. O'B. Carpenter, J. Opt. Soc. Am. **40**, 225–229 (1950).

[7] G. D. Gottschall, J. Soc. Motion Picture Engrs. **51**, 13 (1948).

[8] Hans Jaffe, Phys. Rev. **73**, 95 (1948).

[9] B. Zwicker and P. Scherrer, Helv. Phys. Acta **16**, 214 (1943).

[10] B. Zwicker and P. Scherrer, Helv. Phys. Acta **17**, 346 (1944).

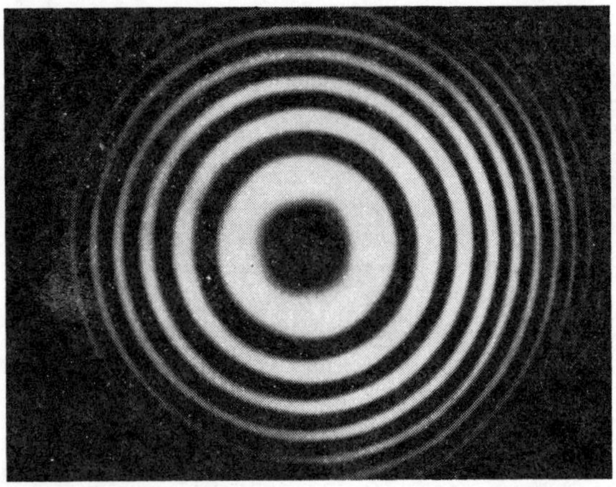

FIG. 1. Pattern observed at 546 mμ with basal section of ADP between crossed circular polarizers. Plate 0.31 cm thick.

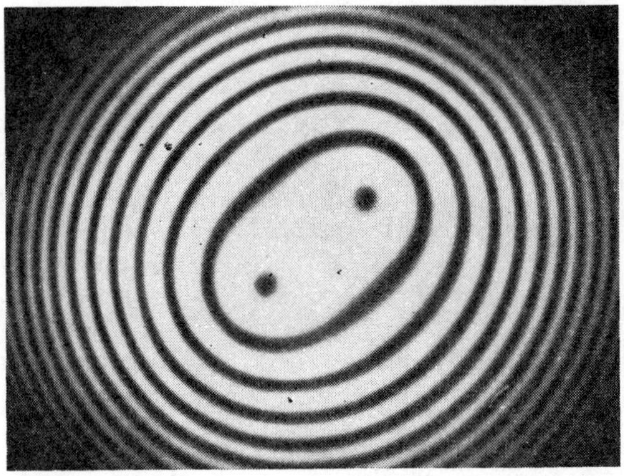

FIG. 2. Pattern observed at 546 mμ with basal section of ADP 0.31 cm thick between crossed circular polarizers. 4500 volts on plate.

not equal to zero. The useful angular field is given by the size of the dark central spot which appears when the crystal plate is examined between crossed circular polarizers. Such a pattern is shown in Fig. 1. In Fig. 2 is shown the pattern which results when voltage is applied. In this particular photograph the voltage was chosen to give a retardation of one-quarter wave.

THE POSSIBILITY OF SINUSOIDAL EXCITATION

A more desirable depolarizer would operate with the sinusoidally varying alternating current which is generally available. In the case of the crystal electro-optic shutter the retardation is directly proportional to the voltage which is supplied. Thus, with ordinary alternating current the retardation will be a sinusoidal function of time and can be written:

$$\delta = \delta_0 \sin\omega t. \tag{16}$$

The value for the average matrix element $\bar{D}_{22}$ can then be written as follows:

$$\bar{D}_{22} = (\omega/2\pi)\int_0^{2\pi/\omega} \cos(n\delta_0 \sin\omega t)dt. \tag{17}$$

This integral can be rewritten:

$$\bar{D}_{22} = (1/2\pi)\int_0^{2\pi} \cos(n\delta_0 \sin\phi)d\phi. \tag{18}$$

This turns out to be simply the Bessel function of zeroth order of the peak retardation of the second plate. It is thus easily possible to reduce the value of the matrix element $\bar{D}_{22}$ to zero by an appropriate choice for the peak retardation of the second plate. The integral for the expression $\bar{D}_{23}$ can be written:

$$\bar{D}_{23} = (1/2\pi)\int_0^{2\pi} \sin(n\delta_0 \sin\phi)\sin(\delta_0 \sin\phi)d\phi. \tag{19}$$

Let $\delta_0 \sin\phi = \alpha$ and use the relationship

$$\sin n\alpha \sin\alpha = \tfrac{1}{2}\cos(n\alpha-\alpha) - \tfrac{1}{2}\cos(n\alpha+\alpha). \tag{20}$$

The integral then becomes:

$$\bar{D}_{23} = \tfrac{1}{2}[J_0(n-1)\delta_0 - J_0(n+1)\delta_0]. \tag{21}$$

$\bar{D}_{33}$ becomes

$$\bar{D}_{33} = J_0(\delta_0), \tag{22}$$

and the remaining matrix elements can be similarly evaluated.

The simplest problem for a depolarizer is the depolarization of linear polarized light. A typical experiment might include a rotating linear analyzer in series with the depolarizer. The intensity of the light through the system should be independent of the position of the analyzer. If the incident light is polarized at an angle α and the analyzer is at an angle θ, the intensity through the combination is

$$I = 1 + \cos 2\theta(\bar{D}_{22}\cos 2\alpha + \bar{D}_{23}\sin 2\alpha) + \sin 2\theta(\bar{D}_{33}\sin 2\alpha). \tag{23}$$

This equation uses only the matrix elements which have been evaluated above. If these three elements can be simultaneously reduced to zero, the depolarizer will function in linear polarized light. This requires that values of n and δ_0 be found which will satisfy the three conditions.

$$J_0(\delta_0) = 0$$
$$J_0(n\delta_0) = 0$$
$$J_0(n-1)\delta_0 - J_0(n+1)\delta_0 = 0. \tag{24}$$

The first two conditions may be easily satisfied by making δ_0 and $n\delta_0$ equal to any of the roots of $J_0(x)$. There are evidently an infinite number of possible combinations. The third condition, however, is not satisfied by any of these values. I obtained a mathematical proof of this fact from Mr. Antonio Quesada of this laboratory. Unfortunately, it is too long to be included in this paper.

It is thus apparent that a perfect depolarizer cannot be made using sinusoidally varying voltage on the crystals.

THE LYOT DEPOLARIZER

The best of the previous depolarizing filters is probably that which was described by Lyot. The Lyot depolarizer is effective in an instrument which examines a large wavelength region simultaneously. It cannot be used in a spectrograph or any other device which looks at small wavelength regions at a time. The Lyot depolarizer consists of two plates of quartz whose thicknesses are in the ratio 2:1. They are X-cut plates. The angle between the x axes in the two plates is 45°. It is interesting to examine first the general case of a depolarizer made up of retardation sheets and rotators and then show how the simplest two plate depolarizer is the Lyot device.

In this type of depolarizer none of the elements vary with time, and the emergent light is, of course, not depolarized in the strict sense of the word. We can, however, make a definition for depolarized light or natural light similar to the Perrin definition. This system of retardation plates will act as a depolarizer provided the measuring instrument will satisfy the following conditions.

The instrument must see no change in intensity when a linear, circular, or elliptical polarizer is put in series with the system and when the incident light is considered to have any state of polarization.

In the case of monochromatic light this obviously will not take place. For white light it does take place and we must write a new definition of polarized light in which, instead of integrating over a period of time, the integral is carried over a range of wavelengths. Let us represent the depolarizer by the matrix D and any analyzer by the matrix A. The equation showing the operation of this analyzer on the light emergent from the depolarizer can be written as follows:

$$\mathbf{A}\begin{vmatrix} I_2 \\ M_2 \\ C_2 \\ S_2 \end{vmatrix} = \begin{vmatrix} I_3 \\ M_3 \\ C_3 \\ S_3 \end{vmatrix}. \tag{25}$$

This can be expanded to give

$$I_3 = A_{11}I_2 + A_{12}M_2 + A_{13}C_2 + A_{14}S_2. \tag{26}$$

According to the definition given above for unpolarized light, I_3, the intensity of the light through the analyzer must be independent of the characteristics of the incident light, I_2, M_2, C_2, S_2, and it must also be independent of the characteristics of the analyzer. Since we are operating in white light, the actual value of I_3, the intensity, is given by the expression

$$\bar{I}_3 = (1/\Delta\lambda)\int_{\lambda_1}^{\lambda_2} I_3 d\lambda = (1/\Delta\lambda)\int_{\lambda_1}^{\lambda_2} A_{11}I_2 d\lambda + (1/\Delta\lambda)\int_{\lambda_1}^{\lambda_2} A_{12}M_2 d\lambda + \cdots. \tag{27}$$

The simplest case is that in which the analyzer characteristics and, therefore, the A_{ij}'s are independent of wavelength. This yields a set of equations which are exactly analogous to those written by Perrin. The definition of unpolarized light can be written

$$\int_{\lambda_1}^{\lambda_2} M_2 d\lambda = \int_{\lambda_1}^{\lambda_2} C_2 d\lambda + \int_{\lambda_1}^{\lambda_2} S_2 d\lambda = 0 \tag{28}$$

or

$$\bar{M} = \bar{C} = \bar{S} = 0.$$

It is possible now to make a few more general statements about the depolarizer itself. In the most general case the incident light will vary in both intensity and state of polarization as a function of wavelength. It is useful to separate these two variations by rewriting the vector which represents the incident light in the following form:

$$\begin{vmatrix} I_1 \\ M_1 \\ C_1 \\ S_1 \end{vmatrix} = I_1 \begin{vmatrix} 1 \\ m_1 \\ c_1 \\ s_1 \end{vmatrix}, \tag{29}$$

where m_1, c_1, and s_1 now vary only with the state of polarization of the incident light.

We can now write a series of equations which must be satisfied if the depolarizer is to make $\bar{M}$, $\bar{C}$, and $\bar{S} = 0$. These equations will be of the type

$$\begin{aligned} &\int_{\lambda_1}^{\lambda_2} D_{21}I_1 d\lambda = 0, \quad m_1\int_{\lambda_1}^{\lambda_2} D_{22}I_1 d\lambda = 0, \\ &c_1\int_{\lambda_1}^{\lambda_2} D_{23}I_1 d\lambda = 0, \quad s_1\int_{\lambda_2}^{\lambda_2} D_{24}I_1 d\lambda = 0. \end{aligned} \tag{30}$$

There will be twelve more equations similar to these four. The general case of two retardation plates can now be examined to see what requirements they must meet in order to act as a depolarizer. Equation (31) becomes identical with Eq. (8).

$$R_2(n\delta, \beta)R_1(\delta, 0)\begin{vmatrix} I_1 \\ M_1 \\ C_1 \\ S_1 \end{vmatrix} = \begin{vmatrix} I_2 \\ M_2 \\ C_2 \\ S_2 \end{vmatrix}. \tag{31}$$

The matrix of this retardation plate combination is the same as the matrix of Eq. (8).

To find the values of the parameters which will make this retardation plate combination act as a depolarizer, it is convenient again to consider the intensity of the incident light as being completely independent of wavelength. Equation (30) above can then be rewritten

$$\bar{D}_{ij}=\int D_{ij}d\lambda=0. \tag{32}$$

The integrals of (32) can be evaluated very simply since the retardation plate characteristics repeat over wavelength intervals which cover a full wave of retardation. We will suppose that the spectral region covered by the incident light extends over an integral number of waves of retardation in the first plate. Since the light has equal intensity at all wavelengths, it is necessary then to integrate only over a range of one wave retardation. It is convenient to evaluate the integrals with respect to angular retardation rather than with respect to wavelength. The angular retardation is

$$\delta=[2\pi(n_e-n_0)d/\lambda] \tag{33}$$

where d is the thickness of the retardation plate, n_e and n_0 are the two indices of refraction. This gives the relationship

$$\bar{D}_{ij}=(1/\Delta\lambda)\int_{\lambda 1}^{\lambda_2} f(\delta)d\lambda$$

$$=(\delta_1\delta_2/\delta_1-\delta_2)\int_{\delta 1}^{\delta_2}(f(\delta)d\delta/\delta^2). \tag{34}$$

Since the integral is only being taken over a retardation range of 2π this becomes immediately

$$\bar{D}_{ij}=(\delta_1\delta_2/2\pi)\int_{\delta 1}^{\delta_2}(f(\delta)/\delta^2)d\delta. \tag{35}$$

The integrals in Eq. (32) can now be evaluated and will give a series of equations which will determine the retardation plate parameters. Although there are twelve equations, the parameters can be determined with only a few integrals. The integral of $\bar{D}_{22}$ becomes

$$\bar{D}_{22}=\frac{\delta_1\delta_2}{2\pi}\left[2\pi\cos^2 2\beta+\sin^2 2\beta\int_{\delta 1}^{\delta_2}\frac{\cos n\delta d\delta}{\delta^2}\right]. \tag{36}$$

From this one equation the following three intuitive requirements can be written down: n must be an integer, β must be 45°, δ must be very large. Even with δ large, the integral will not quite go to zero and there will be a slight residual polarized component for the light going through the instrument. Exact compensation would be realized if the light did not have equal intensity at all wavelengths but had an intensity distribution which was proportional to the inverse square of the wavelength

$$I_1=(Q/\lambda^2).$$

In this case the integrals of Eq. (30) are used and I_1 as assimilated into the operator $\bar{D}$. The $\bar{D}_{ij}$ becomes

$$\bar{D}_{ij}=(I_{Av}/2\pi)\int_{\delta 1}^{\delta_2} f(\delta)d, \tag{37}$$

where $I_{Av}=(Q/\lambda_1\lambda_2)$.

In the particular case of D_{22}, Eq. (36) then can be written:

$$\bar{D}_{22}=(I_{Av}/2\pi)\int_{\delta 1}^{\delta_2}\cos n\delta d\delta \tag{38}$$

and no restrictions need to be placed on the value of δ other than that it must vary over 2π in the wavelength range covered. For the rest of this section it will be assumed that δ is large, as this can be realized quite easily in practice. It will also be assumed that the light is uniformly distributed in wavelengths. For this case Eq. (38) can be written

$$D_{22}=(1/2\pi)\int_{\delta 1}^{\delta_1+2\pi}\cos n\delta d\delta$$

$$=(1/\pi n)\sin\pi n\cos n(\pi+\delta). \tag{39}$$

$\bar{D}_{23}$ can now be evaluated.

$$\bar{D}_{23}=[1/2\pi(n-1)]\sin\pi(n-1)\cos(n-1)(\delta+\pi)$$
$$+[1/2\pi(n+1)]\sin\pi(n+1)\cos(n+1)(\delta+\pi). \tag{40}$$

From this equation it is apparent that n cannot equal 1. Accordingly, the simplest white light depolarizer consists of two retardation plates with high retardation. One plate must have twice the retardation of the other and the axis of the first plate must be at 45° to the axis of the second. With this arrangement all of the matrix elements $\bar{D}_{ij}$ become zero and the device is effective for light of any state of polarization.

THE DETERMINATION OF DEPOLARIZER TOLERANCES

The values of the parameters for the two plate case were derived in a semi-intuitive fashion by Lyot. It is difficult in this fashion to define the tolerances which must be placed on the depolarizer to achieve a given amount of depolarization. The depolarizing ability of the unit can be characterized by the maximum percentage of change in light intensity which would be realized when the unit is set up in polarized light and the emergent light is passed through a rotating analyzer. For some orientation and state of polarization of the incident light, the emergent light will have a maximum in its percentage modulation. The relative modulation can be written

$$(\Delta I/I)=(I_{\max}-I_{\min})/I_{Av}. \tag{41}$$

To illustrate the calculation we shall consider the specific case of linear polarized light with a rotating linear analyzer. The general matrix equation for this case is

$$\mathbf{P}(\theta)\bar{D}\mathbf{P}(\alpha)\begin{vmatrix}1\\0\\0\\0\end{vmatrix}=\begin{vmatrix}I\\M\\C\\S\end{vmatrix}. \tag{42}$$

The intensity of the emergent light from this combination will be:

$$I=1+\cos 2\theta(\bar{D}_{22}\cos 2\alpha+\bar{D}_{23}\sin 2\alpha) +\sin 2\theta(\bar{D}_{32}\cos 2\alpha+\bar{D}_{33}\sin 2\alpha). \tag{43}$$

The two parameters whose variation will affect the performance of the depolarizer are the thickness ratio of the two plates and the wavelength region of the incident light. Here we shall calculate the effect of each parameter, assuming that the variable not being considerd is fixed at its best value. If the wavelength region of the incident light covers exactly an integral number of waves retardation, $\bar{D}_{32}$ and $\bar{D}_{33}$ are zero, and Eq. (3) can be rewritten:

$$I=1+\cos 2\theta[\bar{D}_{22}\cos 2\alpha+\bar{D}_{23}\sin 2\alpha]. \tag{44}$$

From this equation it is apparent that when the analyzer is oriented parallel to the axis of the second plate, the depolarizer acts perfectly no matter what is the value of n, the thickness ratio of the two plates. This takes place because the second plate is not active. At all other angles for the incident light there will be a difference in intensity as the polarizer in the incident beam is rotated. This maximum difference is a measure of the tolerances which need to be pinned on the value of n.

This difference will be greatest when the angle of the incident light is zero. At this angle, $\cos 2\theta=1$. The position of the polarizer for giving the maximum and minimum intensity under this condition is given by the equation:

$$\partial I/\partial\alpha=-2\bar{D}_{22}\sin 2\alpha+2\bar{D}_{23}\cos 2\alpha=0. \tag{45}$$

From this we find

$$\alpha_1=\tfrac{1}{2}\tan^{-1}(\bar{D}_{23}/\bar{D}_{22}) \tag{46}$$

and α can have two values separated by 90°. One gives maximum intensity and the other minimum intensity. With these values Eq. (41) becomes:

$$\Delta I/I_{\text{Av}}=2[\bar{D}_{22}\cos 2\alpha_1+\bar{D}_{23}\sin 2\alpha_1], \tag{47}$$

where I_{Av} is unity.

When values are substituted for $\bar{D}_{22}$ and $\bar{D}_{23}$, Eq. (46) can be written:

$$\alpha_1=\tfrac{1}{2}\tan^{-1} \times\frac{(n/n-1)\sin\pi(n-1)+(n/n+1)\sin\pi(n+1)}{\sin\pi n}. \tag{48}$$

The ratio of the plate thicknesses can be written:

$$n=n'+\epsilon,$$

where n' is an integer and ϵ is the error in preparing the plates. With this substitution the equation can be written

$$\alpha_1=\tfrac{1}{2}\tan^{-1}\frac{(n/n-1)\sin[\pi(n'-1)+\pi\epsilon]+(n/n+1)\sin[\pi(n'+1)+\pi\epsilon]}{2\sin(\pi n'+\pi\epsilon)}. \tag{49}$$

This can be reduced to:

$$\alpha=\tfrac{1}{2}\tan^{-1}(n^2/n^2-1). \tag{50}$$

For large values of n, α is $22\tfrac{1}{2}°$. If $n=2$, α is 26°33′. If $n=3$, α equals 24°11′.

It is possible to determine ΔI directly without determining α. From Eq. (46) the values for $\cos 2\alpha$ and $\sin 2\alpha$ are given by the expression:

$$\sin 2\alpha=\bar{D}_{23}/(\bar{D}_{22}{}^2+\bar{D}_{23}{}^2)^{\frac{1}{2}} \times\cos 2\alpha=\bar{D}_{22}/(\bar{D}_{22}{}^2+\bar{D}_{23}{}^2)^{\frac{1}{2}}. \tag{51}$$

When these are substituted in Eq. (47), we get:

$$\Delta I=2(\bar{D}_{22}{}^2+\bar{D}_{23}{}^2)^{\frac{1}{2}}. \tag{52}$$

With the values for $\bar{D}_{22}$ and $\bar{D}_{23}$ this equation can be reduced to:

$$\Delta I=(2\sin\pi\epsilon/\pi)\left[\left(\frac{n}{n^2-1}\right)^2+\frac{1}{n^2}\right]^{\frac{1}{2}}. \tag{53}$$

For small values of ϵ the intensity through the system is thus linear with ϵ. The proportionality constant is 5/3 with a retardation ratio of two and the percent modulation by the depolarizer failure will thus be slightly larger than the error in thickness ratio.

The effect of the width of the wavelength region which is to be depolarized is much less severe than the effect of an error in the value of n. An error in n affects the behavior of the depolarizer over each part of the spectrum which is passed. If the plates are perfect but the wavelength region covers a nonintegral number of waves of retardation, the only part of the depolarizer which is ineffective is that which operates on the last fraction of the wavelength region. By making the depolarizer plates very thick, it is possible to reduce this inefficiency to as small a value as is desired.

EXPERIMENTAL

The special optical system shown in Fig. 3 was set up to test the operation of the Lyot depolarizer. Light from a tungsten lamp was collimated with a lens and passed through a linear polarizer which could be rotated by hand. After leaving the polarizer the light passed through the depolarizer, which consisted of two

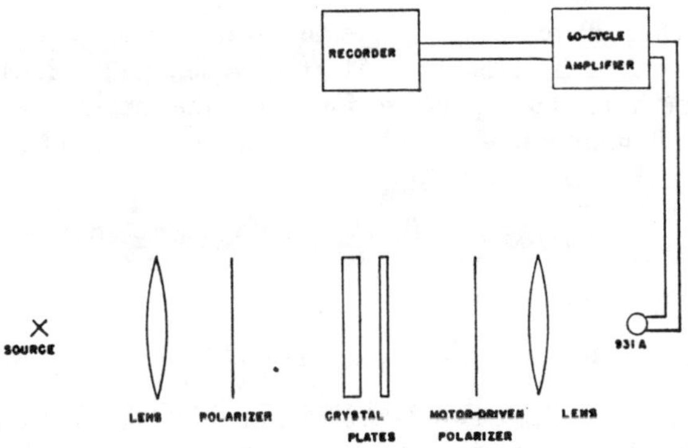

FIG. 3. Special optical system for measuring the efficiency of a white light depolarizer. The source was a tungsten filament. The motor driven linear polarizer was rotated at thirty revolutions per second.

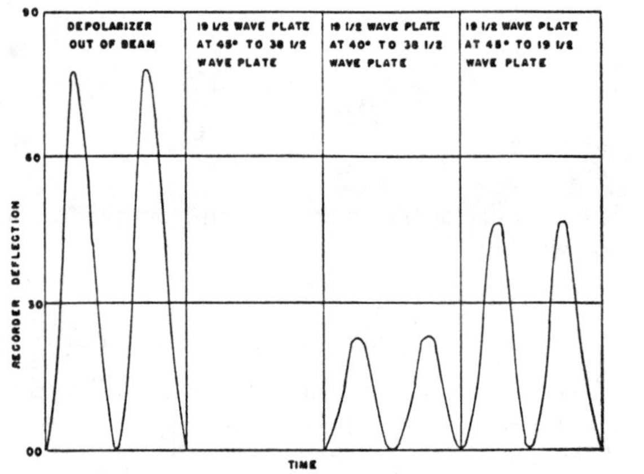

FIG. 4. Relative transmission measurements for different arrangements of the crystal plate in Fig. 3. In each section the first linear polarizer was slowly rotated by hand through one complete revolution.

plates of quartz cut parallel to the axis. These two plates could be rotated to assume any angle. From the depolarizer the light passed through a linear polarizer which was rotated at 30 rps by an electric motor. The light was finally focused onto a photomultiplier tube. The signal from this tube was passed through a 60-cycle amplifier, rectified and put on a Speedomax recorder. When the depolarizer was removed a 60-cycle signal was received on the phototube. This peak signal was the difference between the light transmitted when the linear polarizers were parallel and the light transmitted when the polarizers were crossed. It is apparent that when the light through the first polarizer is depolarized, the signal will be zero. For the first experiment one plate had a retardation of $19\frac{1}{2}$ waves at $H\alpha$. The second had a retardation of $38\frac{1}{2}$ waves. To orient the plates the linear polarizers were first set in a crossed position. The crystal plates were inserted and rotated to a position at which the light remained extinguished. At this point the crystal axes were parallel. One of the plates was then rotated to an angle of 45° with respect to the other. The motor on the second polarizer was then turned on and a record taken while the first polarizer was very slowly rotated. The motion of the pen was less than 1 percent of the signal which was measured without the depolarizer present. The angle between the two wave plates was then changed 5°. The linear polarizer was again rotated slowly. A final experiment was performed in which the two plates both had a retardation of $19\frac{1}{2}$ waves. The results of these experiments are summarized in Fig. 4. The importance of maintaining the 45° angle and the 2 to 1 thickness ratio is clearly shown.

ROTATING ELEMENTS AS DEPOLARIZERS

On discussing this paper with various scientists I was frequently asked why a simple rotating polarizer or wave plate would not function as a perfectly satisfactory depolarizer. It would seem at first glance that if a high order retardation plate were spun at a frequency which was short as compared with the observation time, the light passing through this plate would be completely depolarized. Actually, neither device alone will depolarize the incident light. Consider first the rotating polarizer at a particular instant of time when the polarizer is at an angle α. The equation for the effect of this polarizer on light of any state of polarization can be written:

$$p(\alpha)\begin{vmatrix} I_1 \\ M_1 \\ C_1 \\ S_1 \end{vmatrix} = \begin{vmatrix} I_2 \\ M_2 \\ C_2 \\ S_2 \end{vmatrix}. \tag{54}$$

The vector of the light which comes from this device can be expanded as:

$$\frac{1}{2}\begin{vmatrix} I_1+M_1\cos 2\alpha+C_1\sin 2\alpha \\ I_1\cos 2\alpha+M_1\cos^2 2\alpha+C_1\sin 2\alpha\cos 2\alpha \\ I_1\sin 2\alpha+M_1\sin 2\alpha\cos 2\alpha+C_1\sin^2 2\alpha \\ 0 \end{vmatrix}. \tag{55}$$

The value of the various vector components over a time which is long with the period of the rotation can be obtained by integrating each component with respect to α, the angle of the polarizer. The integral can be evaluated in this fashion since α is linear with time. In the case of M_2 the integral is simply:

$$M_2=\frac{M_1}{2\pi}\int_0^{2\pi}\cos^2 2\alpha d\alpha=M_1/2. \tag{56}$$

Similarly, the other integrals can be evaluated and the final expression for the emergent light can be written:

$$\begin{vmatrix} I_2 \\ M_2 \\ C_2 \\ S_2 \end{vmatrix} = \begin{vmatrix} I_1 \\ \frac{1}{2}M_1 \\ \frac{1}{2}C_1 \\ 0 \end{vmatrix}. \tag{57}$$

It is apparent that the emergent light is linearly polarized and has less intensity than the incident light.

For a rotating wave plate the same situation holds

true. The equation can be written:

$$R(\beta,\delta)\begin{vmatrix} I_1 \\ M_1 \\ C_1 \\ S_1 \end{vmatrix} = \begin{vmatrix} I_2 \\ M_2 \\ C_2 \\ S_2 \end{vmatrix}. \tag{58}$$

This equation relates the incident light to the emergent light when the wave plate is placed at an angle β to the system axis. If the wave plate is now uniformly rotated at a constant velocity, the matrix elements can be determined by integrating with respect to β. The final expression for the emergent light can be written

$$\begin{vmatrix} I_2 \\ M_2 \\ C_2 \\ S_2 \end{vmatrix} = \begin{vmatrix} I_1 \\ \frac{1}{2}(1+\cos\delta)M_1 \\ \frac{1}{2}(1+\cos\delta)C_1 \\ S_1\cos\delta \end{vmatrix}. \tag{59}$$

Again the device does not act as a depolarizer.

It is interesting to see exactly what is the characteristic of the emergent light for some specific rotating wave plates. In the case of a $\frac{1}{4}$ wave plate the emergent light has the vector

$$\begin{vmatrix} I_1 \\ \frac{1}{2}M_1 \\ \frac{1}{2}C_1 \\ 0 \end{vmatrix}. \tag{60}$$

This represents a mixture of linearly polarized light and depolarized light. If the incident light is linearly polarized, the polarized fraction of the emergent light has the same azimuth. If the incident light is circularly polarized the emergent light is completely unpolarized, and the device functions as a perfect depolarizer. For linearly polarized incident light the percentage polarization can be represented by the fraction

$$p=100(M_2^2+C_2^2)/(M_1^2+C_1^2)=25 \text{ percent}. \tag{61}$$

With a rotating $\frac{1}{2}$-wave plate the emergent light has the following vector:

$$\begin{vmatrix} I_1 \\ 0 \\ 0 \\ -S_1 \end{vmatrix}. \tag{62}$$

If the incident light is circularly polarized the emergent light is completely circularly polarized but with opposite sign. If the incident light is linearly polarized, the device acts as a perfect depolarizer.

By using the same general principle which was involved, both in the Lyot depolarizer and in the electro-optic depolarizer, a combination of rotating wave plates can be devised which will act as a complete depolarizer for monochromatic light.

From the analysis above it seems possible that a combination of a $\frac{1}{4}$-wave plate rotating at one speed followed by a $\frac{1}{2}$-wave plate rotating at twice the speed will give a device which is a perfect depolarizer and will operate on any state of polarization. Consider first the general case of two wave plates of retardation, δ_1 and δ_2, rotating at rate ω_1 and $n\omega_1$, respectively.

The matrix elements D_{ij} and D_{i1} for the combination are automatically equal to zero with the exception of D_{11} which is unity. For D_{22} and D_{23} the equation can be written:

$$\begin{aligned}\frac{1}{4}\Big\{&(1-\cos\theta_1-\cos\theta_2+\cos\theta_1\cos\theta_2)\cos[8\pi(n-1)t/T]\\ &+1+\cos\theta_1+\cos\theta_2+\cos\theta_1\cos\theta_2\\ &+(1-\cos\theta_1+\cos\theta_2-\cos\theta_1\cos\theta_2)\cos(8\pi t/T)\\ &+(1+\cos\theta_1-\cos\theta_2-\cos\theta_1\cos\theta_2)\cos(8\pi nt/T)\\ &+2\sin\theta_1\sin\theta_2\left[\cos\frac{4\pi(1+n)t}{T}-\cos\frac{4\pi(n-1)t}{T}\right]\Big\}\\ &=D_{22}\end{aligned} \tag{63}$$

$$\begin{aligned}D_{23}=\frac{1}{4}\Big\{&\sin[8\pi(n-1)t/T](\cos\theta_1+\cos\theta_2-1)\\ &-\sin[8\pi(n+1)t/T](\cos\theta_1\cos\theta_2)\\ &+\sin(8\pi t/T)(1+\cos\theta_2-\cos\theta_{\_}-\cos\theta_1\cos\theta_2)\\ &+\sin(8n\pi t/T)(1-\cos\theta_2+\cos\theta_1+\cos\theta_1\cos\theta_2)\\ &+2\sin\frac{4\pi(n+1)t}{T}\sin\theta_1\sin\theta_2\\ &+2\sin\frac{4\pi(n-1)t}{T}\sin\theta_1\sin\theta_2\Big\},\end{aligned} \tag{64}$$

where T is the period of revolution of the first retardation plate. The effect of these elements if found by integrating over one revolution of the first plate. By inspection of the various instantaneous matrix elements, it is apparent that n should be an integer. Under this condition the integrals of D_{22} and D_{23} become

$$\bar{D}_{22}=(1/T)\int_0^T D_{22}dt=\tfrac{1}{4}(1+\cos\delta_1)(1+\cos\delta_2) \tag{65}$$

$$\bar{D}_{23}=(1/2\pi)\int_0^{2\pi} D_{23}d\beta=0. \tag{66}$$

From these equations it is apparent that $\delta_1=\pi/2$ and $\delta_2=\pi$ are solutions. With these values all the $\bar{D}_{ij}$'s become equal to zero and the device becomes a perfect monochromatic depolarizer.

NOISE DRIVEN DEPOLARIZER

Although we have used an operational definition of polarized light in this paper, it is interesting to speculate on the possibility of making a depolarizer whose output conforms more closely to the quantum-mechanical definition of unpolarized light. According to this definition the electric vector should change in azimuth in a completely random fashion. For such a random motion the integrals which provide the empirical definition of unpolarized light will equal zero just as in the case of the light whose vector moves periodically but at a rate faster than can be detected by a measuring instrument.

A possible method of shifting the electric vector in a random fashion would be to apply random phase shift. Such a random phase shift could be applied with any variable wave plate. At first glance this seems particularly easy in the case of the electrically variable wave plate of ADP. Random variations in voltage are easily obtainable by amplifying thermal noise in a resistor or the shot noise from a device such as a photomultiplier tube. A simple depolarizer might consist of two plates of ADP with their fast axes at 45° just as was used with the sawtoothed voltage. These plates could be driven with the random voltage from a source of noise. Such a noise source can be described in many different ways. The noise voltage can be represented as a series of pulses of.equal length which are distributed randomly in sign and in time. The amplitude distribution can be represented by a Poisson function. The integrals to be evaluated are the same as those which occur in the previous section. A typical integral can be written:

$$\bar{D}_{33}=\frac{1}{T}\int_0^T \cos\delta dt. \tag{67}$$

To make this integral equal to zero it is necessary that the value of δ be greater than $\frac{1}{4}$ wave as often as it is less than $\frac{1}{4}$ wave. Since δ is linear with voltage for the electrical retardation plate this cannot be achieved with the noise source described above. This noise source has a distribution around zero in which positive values occur as often as negative values the values of δ between zero and $\pi/2$ are very much more predominant than the values of $\pi/2$ and π. Although the integral can be reduced by biasing the noise so that it is centered around $\pi/2$, when this is done the values of some of the other integrals are not equal to zero. To reduce all the integrals simultaneously to zero, it is necessary to have a noise source which consists of pulses which are all the same height and which are distributed at random in time and in sign. The height of the pulse will have to be adjusted to give a retardation exactly equal to π. This is not particularly easy to do. Accordingly, it seems easier to be satisfied with the purely operational definition of unpolarized light. Nearly all light sources are partially polarized. Nearly all light receivers including the eye behave slightly differently for light of different states of polarization. These characteristics frequently make precise polarization measurements extremely difficult. These difficulties are particularly apparent in the case of measurement of small percentages of polarization or dichroism. All the polarization phenomena in the source and the receiver must be carefully determined and subtracted from the final measurements. With the use of one or more depolarizers these difficulties can be largely eliminated.

Examples of some of the techniques of using depolarizers will be given in a later paper.

Depolarization of white light by a birefringent crystal

Shan-ling Lu* and A. P. Loeber

Department of Physics and Astronomy, Eastern Michigan University, Ypsilanti, Michigan 48197

(Received 10 July 1974)

The depolarization of white light upon passage through a birefringent crystal, due to the unequal phase retardation of different wavelengths is considered. Assuming the incident light to be completely linearly polarized with the spectral power distribution of a blackbody radiator, the degree of polarization remaining after passage through various thicknesses of birefringent material is calculated. The problem is simplified by assuming that there is no dispersion in the crystal. Results show that under the assumptions made, the light from a 3000 K source is less than 1% polarized after passage through 2.66×10^{-1} mm of quartz and 1.41×10^{-2} mm of calcite.

Index Headings: Polarization; Birefringence; Crystal.

It is well known that polarized white light tends to become depolarized upon passage through a birefringent crystal. This depolarization is due to the unequal phase retardation suffered by different incident wavelengths as they traverse the birefringent crystal. For example, a given crystal may serve as a quarter-wave plate for a particular wavelength. In this case, this crystal would have the effect of changing incident circularly polarized light of that particular wavelength to linearly polarized light when it emerges from the crystal, but incident light of the same polarization but different wavelength would emerge from the crystal with a different polarization.

If the incident light consists of a band of wavelengths all completely polarized with the same polarization, each wavelength will, in general, suffer a different phase retardation, with the result that the emerging light is less than completely polarized.

Using the terminology of Shurcliff,[1] we can think of the emerging light as consisting of that pair of completely and orthogonally polarized components that have the maximum difference of irradiance. If the irradiance of the dominant component is represented by I_d, and that of the inferior component is represented by I_i, the degree of polarization V is given by

$$V=(I_d-I_i)/(I_d+I_i)\ . \tag{1}$$

If we wish to express the emerging light in terms of the Stokes parameters I_e, M_e, C_e, S_e, the degree of polarization can be expressed as[2]

$$V=(M_e^2+C_e^2+S_e^2)^{1/2}/I_e\ . \tag{2}$$

STATEMENT OF THE PROBLEM

We wish to investigate the case of white light incident perpendicularly upon a uniaxial birefringent crystal. (See Fig. 1.) Let us consider the incident light to be completely linearly polarized at 45° to the horizontal. Assume the fast axis of the birefringent crystal to lie in the surface of the crystal, and to have a horizontal orientation. We are interested in determining the Stokes parameters (I_e, M_e, C_e, S_e) of the emerging light and using these to determine the degree of polarization, V, by use of Eq. (2). A knowledge of the Stokes parameters will also enable us to determine the ratios

$$V_m=M_e/I_e\ , \tag{3}$$

$$V_c=C_e/I_e\ , \tag{4}$$

$$V_s=S_e/I_e\ , \tag{5}$$

which we may call the degree of horizontally polarized light, the degree of +45° polarized light, and the degree of right-circularly polarized light, respectively. We wish to examine how the quantities V_m, V_c, V_s, and V vary with the thickness of the birefringent crystal through which the originally polarized light has traveled.

ASSUMPTIONS

To simplify the problem, we assume that (i) the incident white light has the spectrum of a blackbody radiator, (ii) all incident wavelengths are totally polarized with the same type of polarization (in this case +45° linearly polarized), (iii) we will ignore the dispersion with wavelength normally present in refracting materials, (iv) we will ignore absorption and scattering by the crystal, and (v) the crystal is regarded as an ideal homogeneous linear retarder.

MATHEMATICAL FORMULATION OF THE PROBLEM

We can express the problem in mathematical terms

$$[V_e]=[M_c][V_i]\ , \tag{6}$$

where $[V_e]$ is the Stokes vector of the emerging light, $[V_i]$ is the Stokes vector of the incident light, and $[M_c]$ is the Mueller matrix that represents the crystal.

For the case at hand,

$$[V_i]=\begin{bmatrix}1\\0\\1\\0\end{bmatrix} \tag{7}$$

because the incident light is +45° linearly polarized. The Mueller matrix[3] for the crystal $[M_c]$ for a homogeneous linear retarder, the fast axis of which is oriented horizontally is

$$[M_c] = \begin{bmatrix} 1 & 0 & 0 & 0 \\ 0 & D^2+G^2 & 0 & 0 \\ 0 & 0 & -D^2+G^2 & 2DG \\ 0 & 0 & -2DG & 2G^2-1 \end{bmatrix} , \quad (8)$$

where

$$\begin{aligned} D &= M \sin\tfrac{1}{2}\delta , \\ C &= \cos\tfrac{1}{2}\delta , \\ \delta &= \text{retardance in radians} , \\ M &= \text{orientation of fast eigenvector of crystal,} \end{aligned} \quad (9)$$

and in the case at hand, $M = 1$ (horizontal). Substituting from Eq. (9) into Eq. (8), and making use of simple trigonometric identities, we can simplify the matrix for the crystal to

$$[M_c] = \begin{bmatrix} 1 & 0 & 0 & 0 \\ 0 & 1 & 0 & 0 \\ 0 & 0 & (\cos^2\tfrac{1}{2}\delta - \sin^2\tfrac{1}{2}\delta) & (2\sin\tfrac{1}{2}\delta\cos\tfrac{1}{2}\delta) \\ 0 & 0 & (-2\sin\tfrac{1}{2}\delta\cos\tfrac{1}{2}\delta) & (\cos^2\tfrac{1}{2}\delta - \sin^2\tfrac{1}{2}\delta) \end{bmatrix} . \quad (10)$$

In Eq. (10) the retardance δ, which appears in some of the matrix elements, is a function of wavelength λ. Therefore, the value of the matrix $[M_c]$ is a function of λ.

In order, then, to evaluate the Stokes parameters of the light emerging from the crystal, it is necessary to know the power distribution with wavelength of the incident light. We will assume that the light source radiates like a blackbody. Thus, spectral power distribution of the source ($\psi_\lambda\, d\lambda$) belonging to the range of wavelengths $d\lambda$ is given by Planck's radiation law

$$\psi_\lambda d\lambda = \frac{8\pi hc}{\lambda^5} \cdot \frac{e^{-hc/\lambda kT}}{1 - e^{-hc/\lambda kT}} . \quad (11)$$

The total power density of the radiant source (ψ_T) will therefore be given by

$$\lambda_T = \int_0^\infty \psi_\lambda d\lambda = \int_0^\infty \frac{8\pi hc}{\lambda^5} \cdot \frac{e^{-hc/\lambda kT}}{1 - e^{-hc/\lambda kT}} d\lambda , \quad (12)$$

where the symbols have their usual meanings.

Because the irradiance of any beam of light taken from this source will be proportional to ψ_T, we may write the Stokes parameters for the incident light as

$$\begin{bmatrix} I_i \\ M_i \\ C_i \\ S_i \end{bmatrix} = \int_0^\infty \psi_\lambda d\lambda \begin{bmatrix} 1 \\ 0 \\ 1 \\ 0 \end{bmatrix} \quad (13)$$

and those of the emerging beam as

$$\begin{bmatrix} I_e \\ M_e \\ C_e \\ S_e \end{bmatrix} = \begin{bmatrix} 1 & 0 & 0 & 0 \\ 0 & 1 & 0 & 0 \\ 0 & 0 & (\cos^2\tfrac{1}{2}\delta - \sin^2\tfrac{1}{2}\delta) & (2\sin\tfrac{1}{2}\delta\cos\tfrac{1}{2}\delta) \\ 0 & 0 & (-2\sin\tfrac{1}{2}\delta\cos\tfrac{1}{2}\delta) & (\cos^2\tfrac{1}{2}\delta - \sin^2\tfrac{1}{2}\delta) \end{bmatrix} \cdot \int_0^\infty \psi_\lambda d\lambda \begin{bmatrix} 1 \\ 0 \\ 1 \\ 0 \end{bmatrix} . \quad (14)$$

Because δ in Eq. (14) is a function of λ, it is evident that the Stokes parameters of the emerging light will, in general, depend upon the power distribution with wavelength.

Carrying out the multiplication indicated in Eq. (14), we obtained the Stokes parameters for the emerging light,

$$I_e = \int_0^\infty \psi_\lambda d\lambda , \quad (15a)$$

$$M_e = 0 , \quad (15b)$$

$$C_e = \int_0^\infty \psi_\lambda d\lambda (\cos^2\tfrac{1}{2}\delta - \sin^2\tfrac{1}{2}\delta) , \quad (15c)$$

$$S_e = \int_0^\infty \psi_\lambda d\lambda (-2\sin\tfrac{1}{2}\delta\cos\tfrac{1}{2}\delta) . \quad (15d)$$

Our problem thus is to evaluate the four integrals of Eqs. (15) and then to evaluate the ratios given in Eqs. (3), (4), and (5).

The value of M_e is evidently zero (i.e., none of the emerging light is polarized horizontally or vertically). so we need evaluate only three integrals, I_e, C_e, and S_e.

EVALUATION OF I_e

From Eqs. (12) and (15a), we may write

$$I_e = \int_0^\infty \frac{8\pi hc}{\lambda^5} \frac{e^{-hc/\lambda kT}}{1 - e^{-hc/\lambda kT}} d\lambda . \quad (16)$$

This may be rewritten

$$I_e = C_1 \int_0^\infty \frac{1}{\lambda^5} \cdot \frac{e^{-C_2/\lambda}}{1 - e^{-C_2/\lambda}} d\lambda , \quad (17$$

where the constants $C_1 = 8\pi hc$ and $C_2 = hc/kT$ have replaced their equivalents. To facilitate the integration, we also change the variable. Let $x = C_2/\lambda$. Then $\lambda = C_2$ and $d\lambda = -(C_2/x^2)\,dx$. With these substitutions in Eq. (17) we obtain

$$I_e = \frac{C_1}{C_2^4} \int_0^\infty \frac{x^3}{e^x - 1} dx . \quad (18$$

This integral[4] has a value $(\pi^4/15)$. Therefore

$$I_e = (C_1/C_2^4)(\pi^4/15) . \quad (1$$

EVALUATION OF C_e

Because M_e is obviously zero, we proceed to the ev uation of C_e in Eq. (15c). With the same change of varia as before and substituting for the retardance δ its equivalent $2\pi t\Delta n/\lambda$, where t is the crystal thickness, Δn is the difference between the principal indices of refraction, and λ is the vacuum wavelength, we obtain

$$C_e = \frac{C_1}{C_2^4}\int_0^\infty \frac{x^3}{e^x - 1}\cos\beta x\,dx\,, \tag{20}$$

where $\beta = 2\pi t\Delta n/C_2$.

By integration[5] of the right-hand side of Eq. (20), we find

$$C_e = \frac{C_1\pi^4}{C_2^4}\left[\frac{2\cosh^2\beta\pi + 1}{\sinh^4\beta\pi} - \frac{3}{\beta^4\pi^4}\right]. \tag{21}$$

EVALUATION OF S_e

If we make the same substitutions for ψ_λ and δ in Eq. (15d) as we previously made in Eq. (15c), we obtain

$$S_e = -\frac{C_1}{C_2^4}\int_0^\infty \frac{x^3}{e^x - 1}\sin\beta x\,dx \tag{22}$$

where C_1, C_2, β, and x have the same meanings as before. Equation (22) can be rewritten as

$$S_e = (-C_1/C_2^4)\int_0^\infty \sin\beta x(x^3)(e^{-x} + e^{-2x} + e^{-3x} + \cdots)\,dx. \tag{23}$$

Thus, S_e consists of a series of integrals, the general term of which is

$$G_n = (-C_1/C_2^4)\int_0^\infty \sin\beta x(x^3)e^{-nx}\,dx\,, \quad n = 1, 2, 3, \ldots\,. \tag{24}$$

Upon integration,[6] this expression becomes

$$G_n = -\frac{C_1}{C_2^4}\left[\frac{(\beta^2 - n^2)24\beta n}{(\beta^2 + n^2)^4}\right], \tag{25}$$

so Eq. (23) can be rewritten as

$$S_e = \frac{C_1}{C_2^4}\int_0^\infty \sin\beta x\left(\frac{x^3}{e^x - 1}\right)dx$$

$$= \sum_{n=1}^\infty \frac{C_1}{C_2^4}\,\frac{24\beta n(\beta^2 - n^2)}{(\beta^2 - n^2)^4}\,, \quad n = 1, 2, 3, \ldots\,. \tag{26}$$

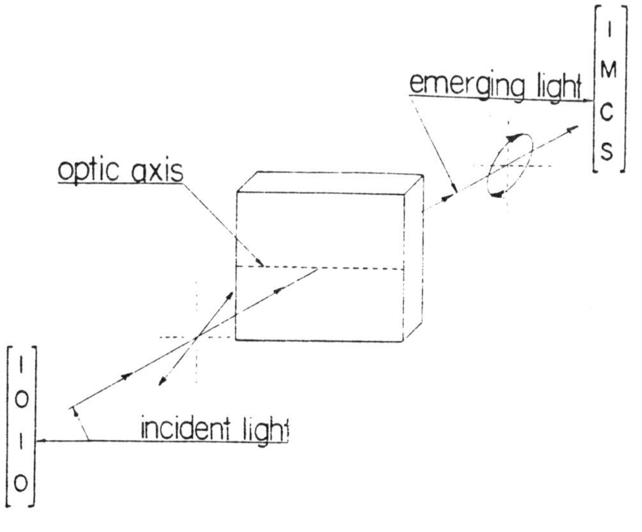

FIG. 1. Alteration of polarization (Stokes) parameters of light upon passage through a crystal.

TABLE I. Degree of polarization (V) remaining after passage of 45° linearly polarized light through crystal thickness (t) corresponding to thickness parameter (X). $X = 2\pi kT\Delta nt/hc$, where Δn = difference between principal refractive indices, T = Kelvin temperature of source, k = Boltzman's constant, h = Planck's constant, c = vacuum speed of light.

X	$V_e = C_e/I_e$ (%) 45° polarization	$V_s = S_e/I_e$ (%) circular polarization	$V = (V_e^2 + V_s^2)^{1/2}$ (%) Total polarization
0.01	99.99	−1.22	99.998
0.05	99.8	−6.09	99.9
0.10	99.1	−12.1	99.8
0.5	78.2	−54.0	95.0
1.0	31.4	−75.7	82.0
1.5	−8.3	−65.0	65.5
2.0	−27.2	−41.5	49.6
2.5	−29.9	−20.7	36.4
3.0	−25.2	−7.40	26.3
3.5	−19.0	−0.42	19.0
4.0	−13.5	2.64	13.8
4.5	−9.49	3.64	10.2
5.0	−6.66	3.69	7.61
5.5	−4.72	3.37	5.80
6.0	−3.46	2.94	4.55
6.5	−2.49	2.52	3.54
7.0	−1.86	2.14	2.84
7.5	−1.42	1.82	2.31
8.0	−1.10	1.55	1.90
8.5	−0.862	1.32	1.58
9.0	−0.686	1.14	1.33
9.5	−0.552	0.984	1.13
10.	−0.450	0.855	0.966
11.	−0.307	0.656	0.725
12.	−0.217	0.513	0.557
13.	−0.158	0.408	0.438
14.	−0.117	0.330	0.350
15.	−0.0889	0.270	0.284
16.	−0.0687	0.224	0.234
17.	−0.0539	0.188	0.195

DEGREE OF POLARIZATION

We are now in a position to calculate the degrees of the various types of polarization remaining after passage through a crystal, by using the definitions of Eqs. (3), (4), and (5) and by utilizing the expressions that we have obtained for I_e, C_e, and S_e. The degree of +45° polarized light, therefore, can be written

$$V_c = \frac{C_e}{I_e} = \left[15\,\frac{2\cosh^2\beta\pi + 1}{\sinh^4\beta\pi} - \frac{3}{\beta^4\pi^4}\right]. \tag{27}$$

Introducing the variable $X = \beta\pi$, we may write Eq. (27) in the form

$$V_c = 15\left[\frac{2\cosh^2 X + 1}{\sinh^4 X} - \frac{3}{X^4}\right]. \tag{28}$$

Similarly, the degree of right-circularly polarized light becomes, from Eqs. (5) and (26) ,

$$V_s = 15\sum_{n=1}^\infty \frac{24nX\pi(X^2 - n^2\pi^2)}{X^2 + n^2\pi^2}\,. \tag{29}$$

The degree of polarization (total) in this case (since $M_e = 0$) is, from Eq. (2),

$$V = \{(C_e/I_e)^2 + (S_e/I_e)^2\}^{1/2} = \{V_e^2 + V_s^2\}^{1/2}\,. \tag{30}$$

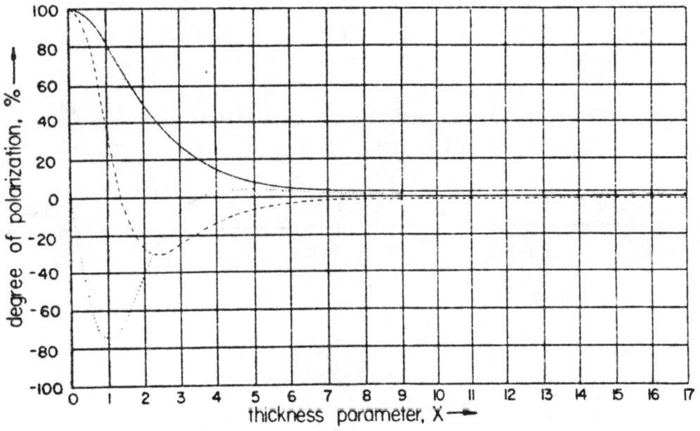

FIG. 2. Degree of polarization (V) remaining (expressed as %) after passage of 45° linearly polarized incident light through crystal thickness (t), corresponding to thickness parameter $X(x=2\pi k T\Delta n t/hc)$. V_c (dashed curve) is the degree of 45° linear polarization. V_s (dotted curve) is the degree of circular polarization. V (solid curve) is the total degree of polarization.

Equations (27), (29), and (30) now allow us to calculate the degree of +45° polarization and the degree of right-circularly-polarized light, as well as the total amount of polarization remaining after passage of the light through the crystal, in terms of the parameter X, which is a function of the crystal thickness and its principal indices of refraction. The relationship between X and t (the distance of travel through the crystal) is given by

$$t = \frac{X}{\Delta n}\,\frac{hc}{2\pi^2 kT} \ . \tag{31}$$

RESULTS

Calculations have been made for the value of V_c, V_s, and V corresponding to various values of the parameter X. The results are shown in Table I. In this table V_c, V_s, and V are shown as percentages. Thus, a given value of V, for example, gives the percentage of light still polarized after passing through the crystal thickness associated with the corresponding value of X. The values of X are taken as exact and the corresponding values for V_c, V_s, and V are calculated. The results are also plotted in Fig. 2.

COMMENTS AND CONCLUSIONS

If appropriate values for the constants h, c, and k ar inserted into Eq. (31), then, using 3000 K for the temperature of the source, we find that $t=(X/\Delta n)$ $(2.42\times10^{-4}$ mm). Using $|\Delta n| \simeq 0.0091$ for quartz and $|\Delta n| \simeq 0.172$ for calcite, we find $t_{\text{quartz}} = 2.66\times10^{-2}X$ m and $t_{\text{calcite}} = 1.41|\times10^{-3}X$ mm.

Thus, from the table, we see that the original completely polarized light is less than 1% polarized after passage through 2.66×10^{-1} mm of quartz and through 1.41×10^{-2} mm of calcite; these values correspond to $X=10$. Although these values are not completely realistic from a practical point of view, because of the assumptions employed, as well as the limits of integration used, nevertheless the results are of interest, in view of the small thickness required to yield a high degree of depolarization.

We should also point out that in spite of the high degree of depolarization that results after passage of lig through a comparatively thin crystal, it is not necessa ily enough depolarization for the light to be considere unpolarized. With one percent still polarized, there would be many applications for which this could not be considered completely depolarized.

We should also mention that, although a birefringent crystal is often considered to be a depolarizer of white light, a better term might be a scrambler of white ligh because the process is not random. The original stat of polarization could be restored by passage of the scrambled light through a second crystal that had identical thickness and other properties, with the fast and slow axes interchanged.

*Present address: Department of Industrial and Operations Engineering, University of Michigan, Ann Arbor, Mich. 48

[1]William A. Shurcliff, *Polarized Light, Production and Use* (Harvard U.P., Cambridge, Mass., 1962), pp. 11 and 12.

[2]Reference 1, p. 22.

[3]Reference 1, pp. 165 and 169.

[4]I. S. Gradshteyn and I. M. Ryzhik, *Tables of Integrals, Se and Products, Fourth ed.* (Translated from the Russian) (A ademic, New York, 1965), pp. 325 and 1079.

[5]Reference 4, p. 494.

[6]Reference 4, p. 490.

Depolarization of white light by a birefringent crystal. II. The Lyot depolarizer

A. P. Loeber

Department of Physics and Astronomy, Eastern Michigan University, Ypsilanti, Michigan 48197

Received September 26, 1981

An earlier paper [S. Lu and A. P. Loeber, J. Opt. Soc. Am. **65**, 248–251 (1975)] investigated the depolarization of white light on passage through a birefringent crystal because of the unequal phase retardation of different wavelengths. Assuming the incident light to be completely polarized with the spectral-power distribution of a blackbody, the degree of polarization remaining after passage through various crystal thicknesses was determined. This paper extends the result of the earlier study to an analysis of the Lyot depolarizer. Earlier work by Billings [J. Opt. Soc. Am. **41**, 966–975 (1941)] considered the general case of two crystal retardation plates of different retardances and orientations to determine what requirements the plates must meet in order to act as a depolarizer. Billings assumed the intensity of the incident light to be independent of wavelength. Despite the different assumptions regarding the spectral-power distribution of the incident light, our investigation agrees with Billings's conclusions that (1) the crystal retardances of both crystals must be large, (2) the crystal retardances cannot be equal, and (3) the angle between the crystal axes must be 45° for the combination of plates to act as a depolarizer. An additional conclusion by Billings, that the ratio of the two retardances must be an integral, is not indicated by our analysis. Results of our analysis are in qualitative agreement with the experimental work carried out by Billings on several modifications of the usual Lyot depolarizer configuration.

INTRODUCTION

An earlier paper[1] reported on the results of a study of the depolarization of white light passing through a birefringent crystal that occurs because of unequal phase retardation suffered by the different wavelengths traversing the crystal. In this paper we report on an extension of that study to include the Lyot depolarizer. We then compare results with earlier theoretical and experimental work of Billings.[2]

STATEMENT OF THE PROBLEM

The Lyot depolarizer consists of two X-cut quartz crystals of thickness ratio 2:1, whose optic axes are oriented at 45° to one another. We wish to investigate the case of white light incident perpendicularly upon this combination of crystals, which we will call the depolarizer (See Fig. 1). Let us consider the incident light to be completely polarized of polarization given by the Stokes parameters (i, m, c, s), where

$$i^2 = m^2 + c^2 + s^2 \tag{1}$$

because the light is completely polarized. We will assume the first crystal to be oriented with its fast axis horizontal, whereas that of the second crystal is oriented at +45° to the horizontal. We are interested in determining the Stokes parameters (I_e, M_e, C_e, S_e) of the emerging light and in using them to determine the degree of polarization

$$V = (M_e{}^2 + C_e{}^2 + S_e{}^2)^{1/2}/I_e, \tag{2}$$

which remains after the light traverses the depolarizer. Also the ratios

$$V_m = M_e/I_e, \tag{3}$$

$$V_c = C_e/I_e, \tag{4}$$

$$V_s = S_e/I_e \tag{5}$$

are of interest. We call them the degree of horizontally polarized light, the degree of +45° polarized light, and the degree of circularly polarized light, respectively.

ASSUMPTIONS

To simplify the problem, (1) we assume that all incident wavelengths are totally polarized with the same polarization, (2) we ignore the dispersion with wavelength normally present in refracting materials, (3) we ignore absorption and scattering in the crystal, and (4) we regard the crystal as an ideal homogeneous linear retarder.

MATHEMATICAL FORMULATION OF THE PROBLEM

If $[V_e]$ represents the Stokes vector of the emerging light and $[V_i]$ that of the incident light, the problem can be expressed in mathematical terms as follows:

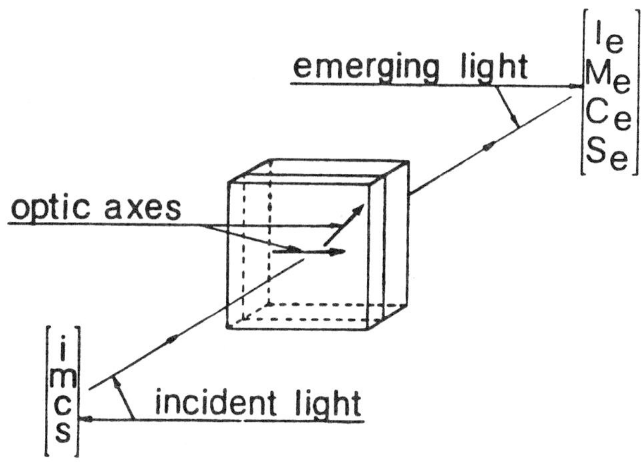

Fig. 1. The Lyot depolarizer.

$$[V_e] = [M_2][M_1][V_i] = [D][V_i], \quad (6)$$

where $[M_1]$ and $[M_2]$ are the Mueller matrices of the first and second crystals, respectively, of the depolarizer and $[D]$ is the matrix for the combination of the two crystals, i.e., the depolarizer matrix. These matrices can be written as follows[3]:

$$[M_1] = \begin{bmatrix} 1 & 0 & 0 & 0 \\ 0 & 1 & 0 & 0 \\ 0 & 0 & \cos\delta_1 & \sin\delta_1 \\ 0 & 0 & -\sin\delta_1 & \cos\delta_1 \end{bmatrix}, \quad (7)$$

$$[M_2] = \begin{bmatrix} 1 & 0 & 0 & 0 \\ 0 & \cos\delta_2 & 0 & -\sin\delta_2 \\ 0 & 0 & 1 & 0 \\ 0 & \sin\delta_2 & 0 & \cos\delta_2 \end{bmatrix}, \quad (8)$$

and

$$[D] = [M_1][M_2] = \begin{bmatrix} 1 & 0 & 0 & 0 \\ 0 & \cos\delta_2 & \sin\delta_1\sin\delta_2 & -\cos\delta_1\sin\delta_2 \\ 0 & 0 & \cos\delta_1 & \sin\delta_1 \\ 0 & \sin\delta_2 & -\sin\delta_1\cos\delta_2 & \cos\delta_1\cos\delta_2 \end{bmatrix}, \quad (9)$$

where δ_1 and δ_2 are the retardances of the first and second crystals, respectively.

The retardances δ_1 and δ_2, which appear in the matrix $[D]$, are functions of the wavelength λ. Therefore the matrix $[D]$ is a function of the wavelength. If we wish to evaluate the Stokes parameters of the light emerging from the depolarizer, we must know the power distribution with wavelength of the incident light. As before,[1] we will assume that the light source radiates like a blackbody. The total power density of the source ψ_T can then be written as

$$\psi_T = \int_0^\infty \psi_\lambda d\lambda = \int_0^\infty \frac{8\pi hc}{\lambda^5} \frac{\exp(-hc/kT)}{1 - e^{-hc/\lambda kT}}. \quad (10)$$

Since the irradiance of any beam of light taken from this source is proportional to ψ_T, we may write the Stokes parameters of the incident light as

$$\begin{bmatrix} I_i \\ M_i \\ C_i \\ S_i \end{bmatrix} = \int_0^\infty \psi_\lambda d\lambda \begin{bmatrix} 1 \\ m \\ c \\ s \end{bmatrix}, \quad (11)$$

where we are assuming identical polarization for all wavelengths and unit intensity for the incident light. The Stokes parameters of the emerging light can then be written as follows:

$$\begin{bmatrix} I_e \\ M_e \\ C_e \\ S_e \end{bmatrix} = \begin{bmatrix} 1 & 0 & 0 & 0 \\ 0 & \cos\delta_2 & \sin\delta_1\sin\delta_2 & -\cos\delta_1\sin\delta_2 \\ 0 & 0 & \cos\delta_1 & \sin\delta_1 \\ 0 & \sin\delta_2 & -\sin\delta_1\cos\delta_2 & \cos\delta_1\cos\delta_2 \end{bmatrix} \times \int_0^\infty \psi_\lambda d\lambda \begin{bmatrix} 1 \\ m \\ c \\ s \end{bmatrix}. \quad (12)$$

If we carry out the multiplication indicated in Eq. (12) we obtain, for the Stokes parameters of the emerging light,

$$I_e = \int_0^\infty \psi_\lambda d\lambda, \quad (13a)$$

$$M_e = \int_0^\infty (m\cos\delta_2 + c\sin\delta_1\sin\delta_2 - s\cos\delta_1\sin\delta_2)\psi_\lambda d\lambda, \quad (13b)$$

$$C_e = \int_0^\infty (c\cos\delta_1 + s\sin\delta_1)\psi_\lambda d\lambda, \quad (13c)$$

$$S_e = \int_0^\infty (m\sin\delta_2 - c\sin\delta_1\cos\delta_2 + s\cos\delta_1\cos\delta_2)\psi_\lambda d\lambda. \quad (13d)$$

Having evaluated the above Stokes parameters, we can determine the degree of polarization of the various types of light V_m, V_c, and V_s by use of Eqs. (3)–(5) and the total degree of polarization by Eq. (2).

The emerging light will be completely depolarized if all the integrals contained in Eqs. (13b)–(13d) are equal to zero. Billings[2] determined the conditions necessary to cause these integrals to go to zero assuming the intensity of the incident light to be independent of wavelength. In our analysis we have assumed the spectral-power distribution of a blackbody radiator.

Evaluation of I_e

In the previous study,[1] it was shown that I_i could be rewritten as follows:

$$I_e = \frac{C_1}{C_2^4} \int_0^\infty \frac{x^3}{e^x - 1} dx, \quad (14)$$

where $C_1 = 8\pi hc$, $C_2 = hc/kT$, and a change in variable $x = C_2/\lambda$ was employed. On integration, it was found that

$$I_e = (C_1/C_2^4)(\pi^4/15). \quad (15)$$

Evaluation of M_e, C_e, and S_e

Before proceeding with the evaluation of M_e, C_e, and S_e, let us for convenience rewrite Eqs. (13b)–(13d) as follows:

$$M_e = mI_1 + cI_2 - sI_3, \quad (16)$$

$$C_e = cI_4 + sI_5 \quad (17)$$

$$S_e = mI_6 - cI_7 + sI_8, \quad (18)$$

where

$$I_1 = \int_0^\infty \cos\delta_2\psi_\lambda d\lambda,$$

$$I_2 = \int_0^\infty \sin\delta_1\sin\delta_2\psi_\lambda d\lambda,$$

$$I_3 = \int_0^\infty \cos\delta_1\sin\delta_2\psi_\lambda d\lambda,$$

$$I_4 = \int_0^\infty \cos\delta_1\psi_\lambda d\lambda,$$

$$I_5 = \int_0^\infty \sin\delta_1\psi_\lambda d\lambda,$$

$$I_6 = \int_0^\infty \sin\delta_2\psi_\lambda d\lambda,$$

$$I_7 = \int_0^\infty \sin \delta_1 \cos \delta_2 \psi_\lambda d\lambda,$$

and

$$I_8 = \int_0^\infty \cos \delta_1 \cos \delta_2 \psi_\lambda d\lambda \quad (19)$$

Integrals I_1, I_4, I_5, and I_6 were already evaluated in the earlier study. If we use the same change of variable as was used in evaluating I_e, and if we substitute for the retardances δ_1 and δ_2 their equivalents $2\pi t_1 \Delta n/\lambda$ and $2\pi t_2 \Delta n/\lambda$, where t_1 and t_2 are the thicknesses, respectively, of the two crystals, Δn is the difference between the principal indices of refraction, and λ is the vacuum wavelength, we find that

$$I_1 = \int_0^\infty \psi_\lambda d\lambda \cos \delta_2 = \frac{C_1}{C_2^4} \int_0^\infty \frac{x^3}{e^x - 1} \cos \beta_2 x dx, \quad (20)$$

where $\beta_2 = 2\pi t_2 \Delta n/C_2$. On integration, this was shown to be

$$I_1 = \frac{C_1 \pi^4}{C_2^4} \left(\frac{2 \cosh^2 \beta_2 \pi + 1}{\sinh^4 \beta_2 \pi} - \frac{3}{\beta_2^4 \pi^4} \right). \quad (21)$$

Similarly,

$$I_4 = \int_0^\infty \psi_\lambda d\lambda \cos \delta_1 = \frac{C_1 \pi^4}{C_2^4} \left(\frac{2 \cosh^2 \beta_1 \pi + 1}{\sinh^4 \beta_1 \pi} - \frac{3}{\beta_1^4 \pi^4} \right), \quad (22)$$

where $\beta_1 = 2\pi t_1 \Delta n/C_2$.

Also, the integrals I_5 and I_6 were previously evaluated:

$$I_5 = \int_0^\infty \psi_\lambda d\lambda \sin \delta_1 = \frac{C_1}{C_2^4} \int_0^\infty \frac{x^3}{e^x - 1} \sin \beta_1 x dx, \quad (23)$$

which, on integration, becomes

$$I_5 = \sum_{n=1}^{\infty} \frac{C_1}{C_2^4} \frac{24\beta_1 n(\beta_1^2 - n^2)}{(\beta_1^2 + n^2)^4}, \qquad n = 1, 2, 3, \ldots \quad (24)$$

and similarly,

$$I_6 = \int_0^\infty \psi_\lambda d\lambda \sin \delta_2 = \sum_{n=1}^{\infty} \frac{C_1}{C_2^4} \times \frac{24\beta_2 n(\beta_2^2 - n^2)}{(\beta_2^2 + n^2)^4}, \qquad n = 1, 2, 3, \ldots, \quad (25)$$

where β_1 and β_2 have the same meanings as before.

In the previous study[1] we had considered the problem of plane-polarized light of +45° azimuth incident upon a single crystal, the fast axis of which was oriented horizontally. The degree of polarization remaining after passage through the crystal was given by

$$V = [(I_4/I_e)^2 + (I_5/I_e)^2]^{1/2} = (V_c^2 + V_s^2)^{1/2}, \quad (26)$$

where V_c and V_s represent the degree of +45° linearly and circularly polarized light, respectively. The ratios V_c and V_s were given by

$$V_c = 15 \left(\frac{2 \cosh^2 X + 1}{\sinh^4 X} - \frac{3}{X^4} \right) \quad (27)$$

and

$$V_s = 15 \sum_{n=1}^{\infty} \frac{24nX\pi(X^2 - n^2\pi^2)}{(X^2 + n^2\pi^2)^4}, \quad (28)$$

where $X = \beta_1 \pi$, which we call the thickness parameter, was introduced. Here

$$X = 2\pi^2 kT \Delta n t_1/hc. \quad (29)$$

Values of V_c and V_s and V were calculated and plotted in the previous study for values of X up to 17, assuming a source of temperature 3000 K. These calculations showed that the emerging light is less than 1% polarized after passage through a crystal for which $X = 10$, which corresponds to a quartz-crystal thickness of 2.66×10^{-1} mm. As the crystal thickness increases, the polarization remaining continues to decrease toward zero. Although it never actually reaches zero for any finite crystal thickness, the polarization can be reduced to any predetermined degree by choosing a sufficiently thick crystal.

In the present study the ratios

$$V_M = M_e/I_e = m(I_1/I_e) + c(I_2/I_e) - s(I_3 I_e), \quad (30a)$$

$$V_c = C_e/I_e = c(I_4/I_e) + s(I_5/I_e), \quad (30b)$$

and

$$V_s = S_e/I_e = m(I_6/I_e) - c(I_7/I_e) + s(I_8 I_e) \quad (30c)$$

are of interest [see Eqs. (3)–(5) and (16)–(18).]

We will now show that the integrals I_2, I_3, I_7, and I_8 contained in these ratios have similar characteristics to those of I_1, I_4, I_5, and I_6 already considered.

Evaluation of I_2, I_3, I_7, and I_8 for the Lyot Depolarizer

If we make the same change in variable and the same substitutions for the constant terms as we used in the previous integrals, and keep in mind that for the Lyot depolarizer $t_2 = 2t_1$, $\delta_2 = 2\delta_1$, and $\beta_2 = 2\beta_1$, we find that the integrals I_2, I_3, I_7, and I_8 can be written as follows:

$$I_2 = -\frac{C_1}{C_2^4} \int_0^\infty \frac{x^3}{e^x - 1} \sin \beta_1 x \sin 2\beta_1 x dx,$$

$$I_3 = -\frac{C_1}{C_2^4} \int_0^\infty \frac{x^3}{e^x - 1} \cos \beta_1 x \sin 2\beta_1 x dx,$$

$$I_7 = -\frac{C_1}{C_2^4} \int_0^\infty \frac{x^3}{e^x - 1} \sin \beta_1 x \cos 2\beta_1 x dx,$$

$$I_8 = -\frac{C_1}{C_2^4} \int_0^\infty \frac{x^3}{e^x - 1} \cos \beta_1 x \cos 2\beta_1 x dx. \quad (31)$$

Making use of common trignometric relationships involving sines and cosines of sums and differences, we obtain the following relationships:

$$\sin \beta_1 \cos \beta_2 = \frac{1}{2} \sin(\beta_1 + \beta_2) + \frac{1}{2} \sin(\beta_1 - \beta_2),$$

$$\cos \beta_1 \cos \beta_2 = \frac{1}{2} \cos(\beta_1 + \beta_2) + \frac{1}{2} \cos(\beta_1 - \beta_2),$$

$$\sin \beta_1 \sin \beta_2 = \frac{1}{2} \cos(\beta_1 - \beta_2) - \frac{1}{2} \cos(\beta_1 + \beta_2), \quad (32)$$

which, when applied to Eqs. (31), result in the following expressions for the integrals:

$$I_2 = -\frac{1}{2} \frac{C_1}{C_2^4} \int_0^\infty \frac{x^3}{e^x - 1} \cos \beta_1 x dx + \frac{1}{2} \frac{C_1}{C_2^4} \int_0^\infty \frac{x^3}{e^x - 1} \cos 3\beta_1 x dx,$$

$$I_3 = -\frac{1}{2}\frac{C_1}{C_2^4}\int_0^\infty \frac{x^3}{e^x - 1}\sin 3\beta_1 x dx$$

$$-\frac{1}{2}\frac{C_1}{C_2^4}\int_0^\infty \frac{x^3}{e^x - 1}\sin \beta_1 x dx,$$

$$I_7 = -\frac{1}{2}\frac{C_1}{C_2^4}\int_0^\infty \frac{x^3}{e^x - 1}\sin 3\beta_1 x dx$$

$$+\frac{1}{2}\frac{C_1}{C_2^4}\int_0^\infty \frac{x^3}{e^x - 1}\sin \beta_1 x dx,$$

$$I_8 = -\frac{1}{2}\frac{C_1}{C_2^4}\int_0^\infty \frac{x^3}{e^x - 1}\cos 3\beta_1 x dx$$

$$-\frac{1}{2}\frac{C_1}{C_2^4}\int_0^\infty \frac{x^3}{e^x - 1}\cos \beta_1 x dx. \quad (33)$$

So we see that the integrals I_2, I_3, I_7, and I_8 all have the form either of I_1 and I_4 or of I_5 and I_6 and can therefore be evaluated in a similar fashion. Therefore, as $\beta_1\pi = X$ becomes large (i.e., for t_1 large), all the integrals in the ratios V_m, V_c, and V_s of Eqs. (30a)–(30c) tend toward zero for large crystal thicknesses. Therefore the ratios V_m, V_c, and V_s also tend toward zero, and so, under the assumed conditions, the Lyot depolarizer will depolarize incident polarized light to any predetermined degree assuming sufficiently thick crystals.

Examination of Other Configurations

Billings[2] examined the general case of two crystal retardation plates of different retardances and orientations to determine what requirements the plates must meet in order to act as a depolarizer. He assumed that the intensity of incident light was completely independent of wavelength. Under this assumption, he concluded the following:

1. The ratio of the two retardances in the crystals δ_2/δ_1 must be an integer n.
2. The integer n cannot be unity, i.e. the retardances of the two crystals cannot be equal.
3. The retardances must be large.
4. The angle between the fast axes of the two crystals must be 45°.

It is interesting to examine the results obtained under the assumption of blackbody radiation that we have made and to compare them with the conclusions of Billings.

We already have shown that for a blackbody spectrum the retardances must be large, a conclusion that might be expected intuitively and that is in agreement with Billings's conclusion.

CRYSTALS WITH EQUAL RETARDANCES

Next, let us examine what our approach predicts with regard to Billings's conclusion that the retardances of the two crystals cannot be equal, a result that is perhaps not so obvious intuitively.

If the two crystals have equal retardances δ, the depolarizer matrix [D] (Eq. 9) becomes the following:

$$[D] = \begin{bmatrix} 1 & 0 & 0 & 0 \\ 0 & \cos\delta & \sin^2\delta & -\cos\delta\sin\delta \\ 0 & 0 & \cos\delta & \sin\delta \\ 0 & \sin\delta & -\sin\delta\cos\delta & \cos^2\delta \end{bmatrix}. \quad (34)$$

Using the above matrix in Eq. (12) leads to the following Stokes parameters for the emerging light in place of those given in Eqs. (13a)–(13d):

$$I_e = \int_0^\infty \psi_\lambda d\lambda, \quad (35a)$$

$$M_e = m\int_0^\infty \cos\delta\psi_\lambda d\lambda + c\int_0^\infty \sin^2\delta\psi_\lambda d\lambda$$

$$- s\int_0^\infty \cos\delta\sin\delta\psi_\lambda d\lambda, \quad (35b)$$

$$C_e = c\int_0^\infty \cos\delta\psi_\lambda d\lambda + s\int_0^\infty \sin\delta\psi_\lambda d\lambda, \quad (35c)$$

$$S_e = m\int_0^\infty \sin\delta\psi_\lambda d\lambda - c\int_0^\infty \sin\delta\cos\delta\psi_\lambda d\lambda$$

$$+ s\int_0^\infty \cos^2\delta\psi_\lambda d\lambda. \quad (35d)$$

All these integrals with the exception of the second integral of M_e [Eq. (35b)] and the third integral of S_e [Eq. (35d)] were examined previously and were found to lead to results like those of either Eq. (21) or Eq. (24) that tend toward zero as the crystal thickness and thus the retardances become large. The exceptions we will call

$$I_9 = \int_0^\infty \sin^2\delta\psi_\lambda d\lambda, \quad (36a)$$

$$I_{10} = \int_0^\infty \cos^2\delta\psi_\lambda d\lambda, \quad (36b)$$

which can be rewritten, using the same variable change as before,

$$I_9 = \frac{C_1}{C_2^4}\int_0^\infty \frac{x^3}{e^x - 1}\sin^2\beta x dx, \quad (37a)$$

$$I_{10} = \frac{C_1}{C_2^4}\int_0^\infty \frac{x^3}{e^x - 1}\cos^2\beta x dx. \quad (37b)$$

Using simple trigonometric identities, these can be rewritten as follows:

$$I_9 = \frac{C_1}{2C_2^4}\int_0^\infty \frac{x^3}{e^x - 1}dx + \frac{C_1}{2C_2^4}\int_0^\infty \frac{x^3}{e^x - 1}\cos 2\beta x dx, \quad (38a)$$

$$I_{10} = \frac{C_1}{2C_2^4}\int_0^\infty \frac{x^3}{e^x - 1}dx - \frac{C_1}{2C_2^4}\int_0^\infty \frac{x^3}{e^x - 1}\cos 2\beta x dx. \quad (38b)$$

The second integral in Eqs. (38a) and (38b) is once again similar to those previously examined and approaches zero as the crystal thickness becomes large. The first integral in each of these expressions is independent of the crystal thickness (contains no β) and is, except for the coefficient ½, identical to I_e as given in Eq. (14). The value of I_9 and I_{10} therefore approach, for large values of retardance, the constant value

$$I_9 = I_{10} = (C_1/2C_2^4)(\pi^4/15). \quad (39)$$

The conclusion is thus inescapable that the light emerging from such a depolarizer (of equally thick crystals) whose axes are oriented at 45° to one another can never be depolarized completely but will have large polarized components. Once again, this agrees with the result that Billings predicted for incident light of equal intensity for all wavelengths.

CRYSTALS WITH INTEGRAL RETARDANCE RATIOS

If we wish to examine how the integrals I_2, I_3, I_7, and I_8 behave with different retardance ratios, we can rewrite Eq. (31) using $\beta = 2\pi t\,\Delta n/C_2$ to represent the first crystal and $m\beta$ to represent the second crystal, where m is the ratio of the β's and also the retardance as well as thickness ratios. Then, using Eqs. (32) we obtain, for I_2,

$$I_2 = -\frac{C_1}{C_2{}^4}\left\{\int_0^\infty \frac{x^3}{e^x - 1}\left[\frac{1}{2}\cos(1-m)\beta x\,dx\right] - \int_0^\infty \frac{x^3}{e^x-1}\left[\frac{1}{2}\cos(1+m)\beta x\,dx\right]\right\}, \quad (40)$$

and similar equations for I_3, I_7, and I_8, some of which involve the sine rather than the cosine. These integrals once again are of the type previously evaluated and therefore will approach zero as β and therefore δ and t become large. There is no inconsistency with the requirement that m be an integer, integral values certainly allowing depolarization to occur, yet in our analysis there is no necessity for m to be an integer, and nonintegral values are also permissible.

CRYSTALS WITH OPTIC AXES AT ARBITRARY ANGLES

If we wish to examine the required angle between optic axes of the two crystals, we must first write the matrix representing the two crystals with an arbitrary angle α between their optic axes. (Until now, we have assumed that $\alpha = 45°$.) We assume the fast axis of the first crystal to be oriented horizontally. Then, using the conventions of Shurcliff,[3] we obtain for the depolarizer matrix the following:

$$[D] = \begin{bmatrix} 1 & 0 & 0 & 0 \\ 0 & \cos^2 2\alpha + \sin^2 2\alpha\cos\delta_2 & [\sin(2\alpha)\cos(2\alpha)(1-\cos\delta_2)]\cos\delta_1 + \sin(2\alpha)\sin\delta_2\sin\delta_1 & \sin(2\alpha)\cos(2\alpha)(1-\cos\delta_2)\sin\delta_1 - \sin(2\alpha)\sin\delta_2\cos\delta_1 \\ 0 & \sin(2\alpha)\cos(2\alpha)(1-\cos\delta_2) & \cos\delta_1[\sin^2\alpha + \cos^2(2\alpha)\cos\delta_2] - \sin\delta_1[\cos(2\alpha)\sin\delta_2] & \sin\delta_1[\sin^2 2\alpha + \cos^2(2\alpha)\cos\delta_2] + \cos\delta_1\cos(2\alpha)\sin\delta_2 \\ 0 & \sin(2\alpha)\sin\delta_2 & -\cos\delta_1[\cos(2\alpha)\sin\delta_2] - \sin\delta_1\cos\delta_2 & -\sin\delta_1[\cos(2\alpha)\sin\delta_2] + \cos\delta_1\cos\delta_2 \end{bmatrix}, \quad (41)$$

which is equivalent to the matrix given by Billings[2] except for certain sign conventions.

The Stokes vector $[V_i]$ for light from a blackbody source whose polarization is complete and constant with wavelength but otherwise arbitrary is given by Eq. (11). When this light is incident upon a depolarizer with optic axes at arbitrary angles represented by the matrix of Eq. (41), the Stokes parameters of the emerging light, I_e, M_e, C_e, and S_e, will consist of the sum of several integrals, each containing the factor $\int_0^\infty \psi_x d\lambda$ multiplied by a matrix element D_{ij} and one of the constant terms l, m, c, or s. For complete depolarization the integrals must all become zero.

We have already seen that integrals over the range zero to infinity, in which the integral consists of $\psi_\lambda d\lambda$ multiplied by terms such as $\sin\delta_1$, $\cos\delta_1$, $\sin\delta_1\sin\delta_2$, $\cos\delta_1\cos\delta_2$, and $\cos\delta_1\sin\delta_2$, are similar to integrals I_1 through I_8 [Eq. (19)], all of which approach zero as the deltas become large, indicating depolarization of the incident light. But, if the integrand consists of $\psi_\lambda d\lambda$ multiplied by $\cos^2\delta$, $\sin^2\delta$, or a constant, we have integrals similar to I_9, I_{10}, or I_e [Eq. (36) or Eqs. (13a) and (15)], which approach constant values that indicate incomplete depolarization of the incident light. Since all the terms of the Stokes parameters contain the factor $\int_0^\infty \psi_x d\lambda$, we need only to examine the matrix elements of Eq. (41) to determine the required angle α.

Matrix elements D_{23}, D_{24}, D_{33}, D_{34}, D_{42}, D_{43}, and D_{44} can be seen to be of the type that will result in integrals that approach zero, as the retardances increase, regardless of the values of α since the $\sin 2\alpha$ or $\cos 2\alpha$ that are multipliers of certain terms in these elements are merely constant multipliers and will not prevent the integrals from approaching zero. On the other hand, elements D_{22} and D_{32} do not approach zero as δ_1 or δ_2 increases.

Element D_{22}, for example, has the value

$$D_{22} = \cos^2 2\alpha + \sin^2(2\alpha)\cos\delta_2, \quad (42)$$

which leads to an integral of the type

$$I_{22} = \cos^2 2\alpha \int_0^\infty \psi_\lambda d\lambda + \sin^2 2\alpha \int_0^\infty \psi_\lambda d\lambda \cos\delta_2, \quad (43)$$

which, for a sufficiently thick crystal, becomes

$$I_{22} = \cos^2 2\alpha \int_0^\infty \psi_\lambda d\lambda = \cos^2 2\alpha (C_1/C_2{}^4)(\pi^4/15). \quad (44)$$

Similarly,

$$D_{32} = \sin(2\alpha)\cos(2\alpha)(1-\cos\delta_2), \quad (45)$$

which leads to an integral of the type

$$I_{32} = \sin(2\alpha)\cos(2\alpha)\int_0^\infty \psi_\lambda d\lambda - \sin 2\alpha\cos 2\alpha \int_0^\infty \psi_\lambda d\lambda\cos\delta_2, \quad (46)$$

which, for a sifficiently thick crystal, becomes

$$I_{32} = \sin(2\alpha)\cos(2\alpha)\int_0^\infty \psi_\lambda d\lambda = \sin(2\alpha)\cos(2\alpha)(C_1/C_2{}^4)(\pi^4/15). \quad (47)$$

So the integrals containing D_{22} and D_{32} will not tend toward zero as the retardances become large.

Integral I_{32} will equal zero for either $\alpha = 0$ or $\alpha = 45°$. However, both integrals I_{22} and I_{32} will be zero only if $\alpha = 45°$. Therefore the angle between the crystal axes must be 45° if complete depolarization is to be expected. This once again agrees with Billings's analysis as well as with his experimental work.

COMPARISON WITH EXPERIMENTAL WORK

It is interesting to compare the result of our analysis with Billings's exprimental work.

Billings allowed linearly polarized light to fall first on a Lyot depolarizer consisting of quartz crystals of thickness ratio nearly 2:1 (more exactly, 1.974:1), the fast axes of which were oriented at 45° to each other. He measured the intensity of the polarized component remaining after passage through the device and found it to be less than 1% of the original intensity. In his experiment, the thickness of the thinner crystal of his depolarizer had a value for the thickness parameter X of approximately 53, which should reduce the emerging polarization to well below 1%, according to our analysis.

Next he rotated one plate 5° so that the angles between the fast axes of the crystals were 40 and 50° rather than 45°. He again allowed linearly polarized light of arbitrary azimuth to impinge upon the crystals and measured the intensity of the polarized component emerging. Billings was merely demonstrating the necessity for the angle between axes to be 45°, and he did not give the precise values of his results. However, if one attempts to scale the values of his recorder curves as published in his paper, one finds that the maximum intensity of the polarized component remaining after passage through this depolarizer is about 29% of the intensity of the incident linearly polarized light.

Finally, linearly polarized light was impinged upon a depolarizer consisting of two quartz plates, identical in thickness but with axes oriented at 45° to one another. A similar measurement on the resulting curves yields an estimate of approximately 60% for the maximum polarized component remaining after passage through this depolarizer.

We have already considered the case of two crystals of equal retardance with the angle of 45° between their axes. The Stokes parameters of the light emerging from this depolarizer when light of parameters l, m, c, s, is incident are given by Eqs. (35a)–(35d). Let us now assume incident plane polarized light of arbitrary azimuth θ. (This follows the experimental arrangement of Billings.) Then $m = \cos 2\theta$, $c = \sin 2\theta$, and $s = 0$. Also, if we assume sufficiently thick crystals, most of the integrals in these equations approach zero, and the Stokes parameters become

$$\begin{aligned} I_e &= (C_1/C_2{}^4)(\pi^4/15), \\ M_e &= c \int_0^\infty \sin^2 \delta\psi_\lambda d\lambda = \sin 2\theta (C_1/2C_2{}^4)(\pi^4/15), \\ C_e &= 0, \\ S_e &= 0. \end{aligned} \tag{48}$$

So the degree of polarization remaining after passage through the depolarizer is

$$V = M_e/I_e = (1/2)\sin 2\theta, \tag{49}$$

the maximum value of which is 0.5, or 50% polarization at $\theta = 45°$. This compares with approximately 60%, which was the experimental result of Billings.

If we wish to investigate the effect of angles other than 45° between the crystal axes, we can use the matrix of Eq. (41). Then, for linearly polarized light of arbitrary azimuth θ incident upon this depolarizer, the Stokes parameters of the emerging light are

$$\begin{bmatrix} I_e \\ M_e \\ C_e \\ S_e \end{bmatrix} = [D] \int_0^\infty \psi_\lambda d\lambda \begin{bmatrix} 1 \\ \cos 2\theta, \\ \sin 2\theta \\ 0 \end{bmatrix}, \tag{50}$$

where $[D]$ is the matix of Eq. (41). Carrying out the indicated multiplication, we obtain

$$\begin{aligned} I_e &= \int_0^\infty \psi_\lambda d\lambda, \\ M_e &= \int_0^\infty D_{22} \cos 2\theta\psi_\lambda d\lambda + \int_0^\infty D_{23} \sin 2\theta\psi_\lambda d\lambda, \\ C_e &= \int_0^\infty D_{32} \cos 2\theta\psi_\lambda d\lambda + \int_0^\infty D_{33} \sin 2\theta\psi_\lambda d\lambda, \\ S_e &= \int_0^\infty D_{42} \cos 2\theta\psi_\lambda d\lambda + \int_0^\infty D_{43} \sin 2\theta\psi_\lambda d\lambda, \end{aligned} \tag{51}$$

where the D_{ij}'s are the elements of the matrix [Eq. (41)]. These matrix elements are functions of α, the angle between the crystal axes. It has already been shown, however, that regardless of the value of α all the above integrals approach zero if the crystal thickness is sufficiently large, with the exception of those integrals that contain D_{22} and D_{23}. Setting all other integrals equal to zero, therefore, we obtain, for the Stokes parameters of the emerging light,

$$\begin{aligned} I_e &= \int_0^\infty \psi_\lambda d\lambda = (C_1/C_2{}^4)(\pi^4/15), \\ M_e &= \int_0^\infty D_{22} \cos(2\theta)\psi_\lambda d\lambda = \cos(2\theta)\cos^2 2\alpha \int_0^\infty \psi_\lambda d\lambda \\ &= \cos(2\theta)\cos^2(2\alpha)(C_1/C_2{}^4)(\pi^4/15), \\ C_e &= \int_0^\infty D_{32} \cos(2\theta)\psi_\lambda d\lambda \\ &= \cos(2\theta)\sin(2\alpha)\cos(2\alpha) \int_0^\infty \psi_\lambda d\lambda \\ &= \cos(2\theta)\sin(2\alpha)\cos(2\alpha)(C_1/C_2{}^4)(\pi^4/15), \\ S_e &= 0 \end{aligned} \tag{52}$$

For $\alpha = 40°$ and $\theta = 0$, which is the incident azimuth that gives the maximum remaining polarization after passage through the crystals, we obtain, for the degree of polarization,

$$V = (M_e{}^2 + C_e{}^2)^{1/2}/I_e = 17.4\%, \tag{53}$$

which compares with about 29% measured from Billings's experimental curves.

SUMMARY AND DISCUSSION

We have extended our analysis of depolarization of white light by a birefringent crystal to include the Lyot depolarizer. The analysis of the Lyot depolarizer was previously carried out by Billings, who assumed the intensity of the incident light to be independent of wavelength. Our analysis assumes the spectral power distribution of a blackbody. Despite these differences in assumptions, the requirements regarding the de sign of the Lyot depolarizer as determined by Billings, namely, (1) large retardances by each crystal, (2) an angle between the fast axes of the crystals of 45°, and (3) the necessity for the two crystals to have unequal retardances, are also are obtained by our analysis. However, an additional requirement reached by Billings, namely, that the ratio of the retardances of the two crystals must be an integral, is not indicated in our analysis.

We have attempted to compare results of our analysis with Billings's experimental work, and although there is qualitative agreement, the quantitative agreement is not particularly good. The experimental arrangement described by Billings does not exactly fit either his or our assumptions, and the actual conditions probably lie somewhere between those assumed by either analysis. Although he used a tungsten source (probably approximating a blackbody radiator), the spectral characteristics of the polarizers used, the spectral response of the detector used, and any absorption by optical elements in the system are all unknown factors that have not been evaluated. Furthermore, since we have made a number of simplifying assumptions in our analysis, it is perhaps not surprising that quantitative agreement is not better.

Billings did not attempt to determine quantitative agreement between his analysis and experiment but did show experimentally the necessity for (1) large retardance, (2) crystals of unequal retardance, and (3) a 45° angle between the crystal axes. He did not experimentally demonstrate a requirement for an integral ratio of the retardance of the two crystals.

REFERENCES

1. S. Lu and A. P. Loeber, "Depolarization of white light by a birefringent crystal," J. Opt. Soc. Am. **65,** 248–251 (1975).
2. B. H. Billings, "A monochromatic depolarizer," J. Opt. Soc. Am. **41,** 966–975 (1941).
3. W. A. Shurcliff, *Polarized Light, Production and Use* (Harvard U. Press, Cambridge, Mass., 1962), pp. 165, 169.

Reprinted from *Journal of Lightwave Technology,* Vol. LT-1(3), 475–479 (September 1983).

Degree of Polarization in the Lyot Depolarizer

WILLIAM K. BURNS, MEMBER, IEEE

Abstract—**Analytic output equations for the degree of polarization of light exiting a Lyot depolarizer are derived from a coherency matrix representation. Design criteria are obtained for sources of different spectral shape.**

I. Introduction

LYOT DEPOLARZERS [1], [2] have recently attracted renewed attention due to the successful use of one in reducing polarization noise in a fiber-optic gyroscope [3]. Recently, such a device has been reported [4] which was constructed from birefringent single-mode fibers rather than the usual birefringent crystal plates.

The Lyot deploarizer classically consists of two birefringent crystal plates (X or Y cut), whose crystal axes are oriented at 45° with respect to each other, and whose thicknesses are in the ratio 1:2. The device is intended to be used with broadband light and may be viewed as performing a polarization conversion on the spectral components of polarized input light. At the output, each spectral component appears with a different polarization state, and upon averaging over bandwidth, the output is depolarized. The choice of a 45° angle between the plate axes is made so the depolarizer will be operative independent of the input polarization state. If depolarization is not achieved by the first plate, it will be achieved by the second.

Billings [2] analyzed the Lyot depolarizer by assuming sufficient birefringence to provide 2π of phase retardation over the source bandwidth in the first plate. His model for a rectangular-shaped source suggested that the plates should have an integer thickness ratio not equal to one. Loeber has treated the problem for a blackbody [5]. Recent approaches have focused on frequency decomposition of the output polarization states on the Poincare sphere [6], attempting to achieve a design that provides a uniform distribution of such output state [7], [8]. However, these approaches are not capable of providing quantitative design rules for Lyot depolarizers.

Rashleigh and Ulrich [9] have pointed out that polarization-mode dispersion in single-mode fibers leads to a group delay difference which causes the depolarization of broad-band light. The role of polarization dispersion as opposed to birefringence in depolarization has also been emphasized theoretically by Sakai *et al.* [10]. In this paper, we follow this approach to calculate analytic output equations for the degree of polarization of a Lyot depolarizer, as a function of the input polarization state, source bandwidth and band shape, and fiber or crystal group delay difference. This result will aid in the de-

Manuscript received April 18, 1983; revised June 6, 1983.

The author is with The Naval Research Laboratory, Washington, DC 20375.

sign of depolarization devices in applications. Our model has been presented qualitatively by Epworth [11]. We ignore mode coupling between polarization modes which has been treated in the paper by Böhm *et al.* [4].

II. THEORY

We assume a configuration for the Lyot depolarizer as shown in Fig. 1. The birefringent sections may be crystal plates or sections of briefringent fiber, although we will use fiber terminology. We assume fiber-mode propagation constants β_x and β_y, which are functions of optical frequency ω. The fiber birefringent axes are denoted by X and Y in the first section of length L_1, and X' and Y' in the second section of length $L_2 (L_2 \geqslant L_1)$. The second section is rotated relative to the first by 45°. We assume an optical source with spectral intensity $|\nu(\omega)|^2$ and center frequency ω_0. The optical input to the depolarizer is assumed to be linearly polarized with input azimuth at an angle θ_p from the mode axis X. For unity power input, the time-dependent electric-field input to the fiber may then be represented

$$\begin{bmatrix} E_{0x}(t) \\ E_{0y}(t) \end{bmatrix} = \begin{bmatrix} \cos\theta_p \\ \sin\theta_p \end{bmatrix} e(t)\, e^{i\omega_0 t} \tag{1a}$$

where

$$\langle e(t)\, e^*(t) \rangle = 1 \tag{1b}$$

and $\langle\ \rangle$ signifies a time average. At the joint between the two sections, polarization-mode coupling will occur. The transfer matrix for such coupling is given by

$$\begin{bmatrix} E_{x'}(L_1) \\ E_{y'}(L_1) \end{bmatrix} = \frac{1}{\sqrt{2}} \begin{bmatrix} 1 & 1 \\ -1 & 1 \end{bmatrix} \begin{bmatrix} E_x(L_1) \\ E_y(L_1) \end{bmatrix} \tag{2}$$

Propagation in the fiber sections is governed by the frequency-dependent propagation constants $\beta_x \equiv \beta_x(\omega)$ and $\beta_y \equiv \beta_y(\omega)$. Employing the transfer matrix in (2), we can write the frequency-dependent output fields at the end of section L_2 as

$$\begin{bmatrix} E_{x'}(\omega) \\ E_{y'}(\omega) \end{bmatrix} = \frac{1}{\sqrt{2}} \begin{bmatrix} e^{-i\beta_x(L_1+L_2)} & e^{-i(\beta_y L_1 + \beta_x L_2)} \\ -e^{-i(\beta_x L_1 + \beta_y L_2)} & e^{-i\beta_y(L_1+L_2)} \end{bmatrix} \begin{bmatrix} E_{0x}(\omega) \\ E_{0y}(\omega) \end{bmatrix} \tag{3}$$

where $E_{0x}(\omega)$ and $E_{0y}(\omega)$ are the input fields at the frequency component ω.

The degree of polarization is defined as the fraction of optical power that is polarized [10], [12]. It is calculated from

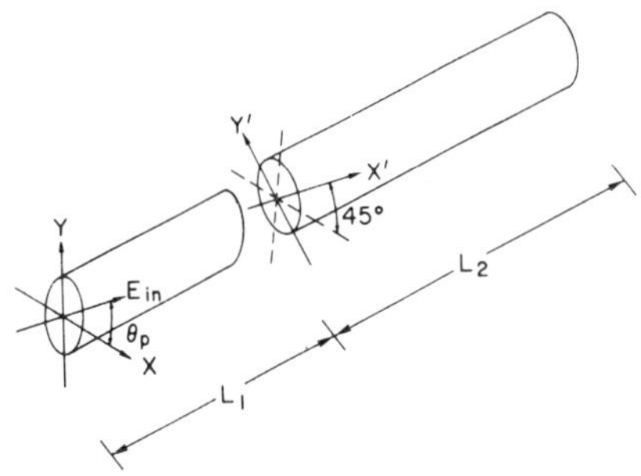

Fig. 1. Model for Lyot depolarizer formed with birefringent fibers.

the coherency matrix J

$$J = \langle E \cdot E^\dagger \rangle = \begin{bmatrix} \langle E_{x'} E_{x'}^* \rangle & \langle E_{x'} E_{y'}^* \rangle \\ \langle E_{y'} E_{x'}^* \rangle & \langle E_{y'} E_{y'}^* \rangle \end{bmatrix} \tag{4}$$

where † signifies the Hermitian transpose. To compute (4), we require time-varying electric-field outputs. We express time-varying electric fields in terms of a complex analytic signal in the usual way [10] by defining $e(t)$ in (1a) as

$$e(t) = 2\int_0^\infty \nu(\omega)\, e^{i(\omega-\omega_0)t}\, d\omega \tag{5}$$

where $\nu(\omega)$ is the (complex) amplitude spectrum of the source. The output time-varying fields from the depolarizer are then given by

$$\begin{aligned} E_{x'}(t) = \frac{1}{\sqrt{2}} \{ & \cos\theta_p\, e^{i[\omega_0 t - \beta_{x0}(L_1+L_2)]}\, e(t - \beta'_{x0}(L_1+L_2)) \\ & + \sin\theta_p\, e^{i(\omega_0 t - \beta_{x0} L_2 - \beta_{y0} L_1)} \\ & \cdot e(t - \beta'_{x0} L_2 - \beta'_{y0} L_1) \} \end{aligned} \tag{6a}$$

$$\begin{aligned} E_{y'}(t) = \frac{1}{\sqrt{2}} \{ & -\cos\theta_p\, e^{i(\omega_0 t - \beta_{x0} L_1 - \beta_{y0} L_2)} \\ & \cdot e(t - \beta'_{x0} L_1 - \beta'_{y0} L_2) + \sin\theta_p\, e^{i[\omega_0 t - \beta_{y0}(L_1+L_2)]} \\ & \cdot e(t - \beta'_{y0}(L_1+L_2)) \} \end{aligned} \tag{6b}$$

where $\beta_{x0} \equiv \beta_x(\omega_0)$, $\beta_{y0} \equiv \beta_y(\omega_0)$ and β' denotes differentiation of β by ω. To derive (6), we have used (3) and papers by Sakai *et al.* [10] and Burns *et al.* [13]. Using (6), we calculate the components of the coherency matrix by performing the necessary time averages

$$J_{x'x'} = \tfrac{1}{2}\,[S_0 + \sin 2\theta_p\, S_1(L_1) \cos\Delta\beta_0 L_1] \tag{7a}$$

$$J_{y'y'} = \tfrac{1}{2}\,[S_0 - \sin 2\theta_p\, S_1(L_1) \cos\Delta\beta_0 L_1] \tag{7b}$$

$$\begin{aligned} J_{x'y'} = \tfrac{1}{2}\, e^{i\Delta\beta_0 L_2} \{ & -\cos 2\theta_p\, S_1(L_2) \\ & + \tfrac{1}{2} \sin 2\theta_p\, e^{i\Delta\beta_0 L_1} [S_1(L_1 + L_2) \\ & - e^{-i2\Delta\beta_0 L_1} S_1(L_2 - L_1)] \} \end{aligned} \tag{7b}$$

$$J_{y'x'} = (J_{x'y'})^* \tag{7d}$$

where $\Delta\beta_0 = \beta_y(\omega_0) - \beta_x(\omega_0)$ and S_0 and $S_1(z)$ are defined by

$$S_0 = 8\pi \int_0^{\infty} |\nu(\omega)|^2 \, d\omega \tag{8a}$$

$$S_1(z) = 8\pi \int_0^{\infty} |\nu(\omega)|^2 \cos\,[(\omega - \omega_0)\delta\tau_g z]\, d\omega. \tag{8b}$$

To derive (7) and (8), we have assumed the source spectrum is symmetrical with respect to ω_0. In (8b), z is a length variable and $\delta\tau_g$ is the polarization-mode dispersion or group delay difference

$$\delta\tau_g = \frac{d(\beta_y - \beta_x)}{d\omega}. \tag{10}$$

The degree of polarization is obtained from the coherency matrix as [12]

$$P = \left[1 - \frac{4 \det J}{(trJ)^2}\right]^{1/2} \tag{10}$$

which yields $P(\theta_p)$ as a function of the input polarization azimuth. The result is

$$\begin{aligned} P(\theta_p)' = (&\cos^2 2\theta_p \gamma^2(L_2) \\ &+ \tfrac{1}{2}\sin^2 2\theta_p \{\gamma^2(L_1) + [\gamma^2(L_1) \\ &- \gamma(L_1 + L_2)\gamma(L_2 - L_1)] \cos 2\Delta\beta_0 L_1 \\ &+ \tfrac{1}{2}[\gamma^2(L_1 + L_2) + \gamma^2(L_2 - L_1)]\} \\ &- \sin 2\theta_p \cos 2\theta_p \gamma(L_2)[\gamma(L_1 + L_2) \\ &- \gamma(L_2 - L_1)] \cos \Delta\beta_0 L_1)^{1/2} \end{aligned} \tag{11}$$

where $\gamma(z) \equiv S_1(z)/S_0$ is the degree of coherence [12].

III. Discussion

Equation (11) gives the output degree of polarization for the Lyot depolarizer. It depends on the degree of coherence $\gamma(z)$

$$\gamma(z) = \frac{\int_0^{\infty} |\nu(\omega)|^2 \cos\,[(\omega - \omega_0)\delta\tau_g z]\, d\omega}{\int_0^{\infty} |\nu(\omega)|^2 \, d\omega} \tag{12}$$

which depends on the source parameters ω_0 and $|\nu(\omega)|^2$ and the fiber parameter $\delta\tau_g$. We note that $\gamma(z)$ is a real, even function of z. For band shapes with spectral width $2\delta\omega$, Sakai *et al.* [10] has calculated γ for the following spectral shapes

$$\text{Rectangular:} \quad \gamma = \frac{\sin\alpha}{\alpha} \tag{13a}$$

$$\text{Gaussian:} \quad \gamma = \exp\left[-\left(\frac{\alpha}{2\sqrt{\ln 2}}\right)^2\right] \tag{13b}$$

$$\text{Lorentzian:} \quad \gamma = \exp(-\alpha) \tag{13c}$$

where $\alpha \equiv \delta\omega\delta\tau_g z$. These functions are compared in Fig. 2. For the Gaussian and Lorentzian cases, γ is a monotonically decreasing function of α. Experimental results have been given for rectangular- [9] and Gaussian- [14] shaped sources.

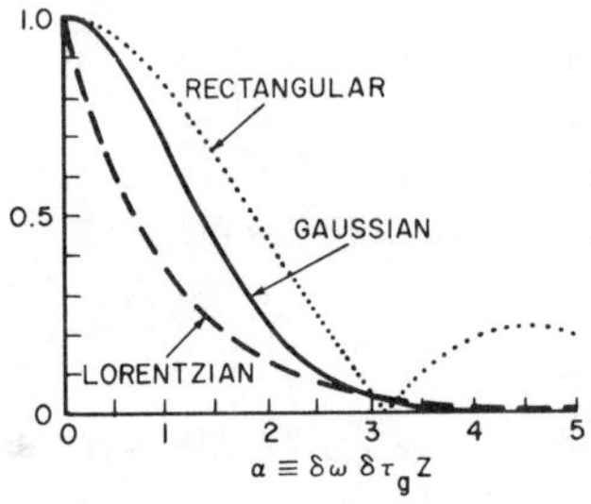

Fig. 2. Magnitude of the degree of coherence for sources with vari band shapes. $\delta\omega$ is the spectral half-width at half-power, $\delta\tau_g$ the fi group delay difference, and z the fiber length.

We first look at (11) for special cases of θ_p. For $\theta_p = 0$ t input is polarized parallel to the X-axis of the first section (L and no depolarization occurs in this section. The input to t second section (L_2) is at 45° to its birefringent axes and have $P(0) = |\gamma(L_2)|$. For $\theta_p = 45°$ depolarization will occur the first section and the second section becomes unnecessa We take $\theta_p = 45°$ and $L_2 = 0$ and obtain $P(45°) = |\gamma(L_1$ These results agree with the calculation for a single fiber len in the paper by Sakai *et al.* [10]. In general, we want $P(\theta_p) =$ for any θ_p. We assume $L_2 \geqslant L_1$ and a monotonically decre ing $\gamma(z)$. From inspection of (11), we require

$$\gamma(L_1) = 0 \tag{1}$$

and

$$\gamma(L_2) = 0 \tag{1}$$

which then implies $\gamma(L_1 + L_2) = 0$. We also require

$$\gamma(L_2 - L_1) = 0 \tag{1}$$

which is satisfied if (14a) is satisfied and if $L_2 - L_1 \geqslant L_1$ $L_2 \geqslant 2L_1$. The entire condition can then be stated as

$$\gamma(L_1) = 0 \tag{1}$$

$$L_2 \geqslant 2L_1 \tag{1}$$

to ensure $P(\theta_p) = 0$ for all θ_p with Lorentzian- or Gaussi shaped sources. This yields a mathematical explanation the original condition of $L_2 = 2L_1$ in a Lyot depolarizer. a rectangular shaped source, which was assumed by Billi [2], the situation is more complicated because γ does monotonically decrease with z. Following the argument abo we arrive at the condition

$$\gamma(L_1) = 0 \tag{1}$$

$$L_2 = nL_1, \quad n \text{ integer} \neq 1 \tag{1}$$

It is clear from Fig. 2 that precision in L_1 and L_2 is only quired with a rectangular-shaped source.

The condition for low output degree of polarization is t the quantity $\delta\omega\delta\tau_g z$ be large ($\geqslant\pi$). We may express the gr delay difference $\delta\tau_g$ in terms of the effective-index differe Δn_{eff} between the modes [9]. Since $\Delta\beta = \Delta n_{\text{eff}}\,\omega/c$, whe

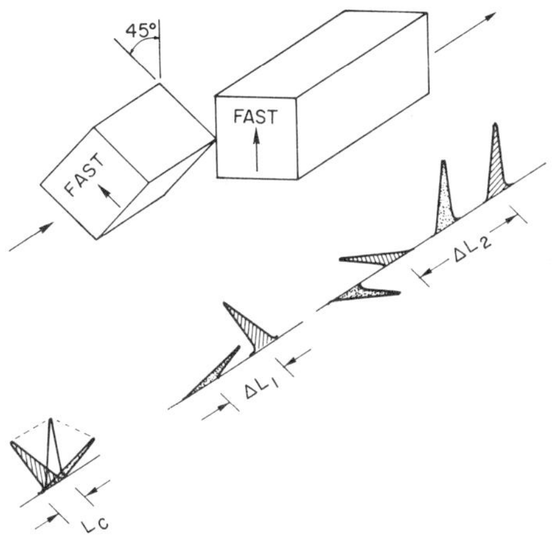

Fig. 3. Model for depolarization in a Lyot depolarizer of broad-band light with coherence length L_c. ΔL_1 and ΔL_2 are the group delay differences experienced in the first and second sections, respectively. For depolarization we require $L_c \ll \Delta L_1$ and $\Delta L_2 \sim 2\Delta L_1$. Depolarization is then independent of input azimuth.

is the light velocity, we have

$$\delta\tau_g = \frac{d\Delta\beta}{d\omega} = \frac{\Delta n_{\text{eff}}}{c} + \frac{\omega}{c}\frac{d(\Delta n_{\text{eff}})}{d\omega} \tag{17}$$

so that the dispersion of Δn_{eff}, when it is not negligible, contributes to depolarization as well as the index difference itself. This is the case, for example, in elliptically cored fibers.

Epworth [11] has given a useful physical picture of the mathematics presented here, following the original model of Rashleigh and Ulrich [9]. Broad-band light has a finite region of coherence [12]. Depolarization then occurs when a polarized input is evenly divided between two polarization modes, and the polarization-mode dispersion is sufficiently large that the resulting group delay is greater than the length of the coherent region. Using two fiber lengths with a relative axis rotation of 45° insures that this condition is met in one of the two sections for all linear inputs. An arbitrary input polarization state is then depolarized by superposition. The 1 : 2 length ratio of the sections avoids overlap of the coherent regions in the second section. We adapt this picture in Fig. 3, assuming a nonrectangular source.

Finally, we can compare our results here with the model used to analyze Lyot depolarizers by performing a spectral decomposition of output polarization states on the Poincare sphere [6]-[8]. The approach developed here points out the importance of spectral band shape and dispersion of Δn_{eff} to depolarization, factors which are not readily included in the Poincare sphere representation. The theoretical significance of a uniform distribution of output polarization states (spectrally decomposed) on the Poincare sphere, has not been rigorously defined. In constrast, depolarized light has been rigorously defined as light whose polarization state, viewed in a sequence of sufficiently short time intervals, does uniformly sample the surface of the Poincare sphere [15]; or as light whose Stokes parameters which describe polarization are zero [12], [15]. These criterion are directly related to the coherency matrix representation used above, and so must also be satisfied for any arrangement which gives zero degree of polarization as defined by (11).

Equations (11) and (15) predict that the degree of polarization with a given bandwidth source may be made as small as desired by using sufficient lengths of fiber with a large group delay difference. For a Gaussian source with half-width $\delta\lambda \simeq$ 7 nm and a fiber with $\delta\tau_g \simeq 1$ ns/km, a fiber length (L_1) of 0.5 m should be sufficient [14]. Of course the theory presented here will be limited by mode coupling between polarization modes, which was treated in the paper by Böhm *et al.* [4]. Experimentally, a value $P \sim 10^{-2}$ was achieved in a Lyot depolarizer [4], and a value $P \sim 10^{-3}$ in a single section of high-birefringence fiber [16].

IV. Conclusions

We have provided analytic output equations for the degree of polarization in Lyot depolarizers which can be applied directly to experimental arrangements. This derivation clearly points out the importance of source band shape and of group delay difference in the birefringent elements, and will allow quantitative device design. Our approach has been put in context with previous descriptions of the device. This analysis will be useful to device designers who will employ this important device in fiber gyroscopes.

Acknowledgment

The author appreciates helpful conversations with R. B. Smith on this subject.

References

[1] B. Lyot, Annales de l'observatoire d'astronomie de Paris (Mendon), vol. VIII, p. 102, 1929.

[2] B. H. Billings, "A monochromatic depolarizer," *J. Opt. Soc. Amer.*, vol. 41, pp. 966–975, 1951.

[3] K. Böhm, P. Marten, K. Peterman, E. Weidel, and R. Ulrich, "Low-drift fibre gyro using a superluminescent diode," *Electron. Lett.*, vol. 17, pp. 352–353, 1981.

[4] K. Böhm *et al.*, "Performance of Lyot-depolarizers with birefringent single-mode fibers," *J. Lightwave Technol.*, vol. 1, pp. 71–74, 1983.

[5] A. P. Loeber, "Depolarization of white light by a birefringent crystal. II The Lyot depolarizer," *J. Opt. Soc. Amer.*, vol. 72, pp. 650–656, 1982.

[6] R. Ulrich, "Polarization and depolarization in the fiber-optic gyroscope," in *Fiber Optic Rotation Sensors*, S. Ezekiel and H. J. Arditty, Eds. Berlin, Germany: Springer-Verlag, 1982, pp. 52–77.

[7] R. B. Smith, C. H. Anderson, and G. L. Mitchell, "Numerical modeling of dual polarization interferometric gyros and sensors," in *Fiber Optic Rotation Sensors*, S. Ezekiel and H. J. Arditty, Eds. Berlin, Germany: Springer-Verlag, 1982, pp. 93–101.

[8] R. B. Smith, G. L. Mitchell, and C. D. Anderson, "Fiber depolarizers using polarization preserving fibers," in *Proc.* OFC'83, New Orleans, 1983, paper TuG9.

[9] S. C. Rashleigh and R. Ulrich, "Polarization mode dispersion in single-mode fibers," *Opt. Lett.*, vol. 3, p. 60, 1978.

[10] J. I. Sakai, S. Machida, and T. Kimura, "Degree of polarization in anisotropic single mode optical fibers: Theory," *J. Quantum Electron.* vol. QE-18, p. 488, 1982.

[11] R. E. Epworth, "The temporal coherence of various semiconductor light sources used in optical fiber sensors," in *Fiber Optic Rotation Sensors*, S. Ezekiel and H. J. Arditty, Eds. Berlin, Germany: Springer-Verlag, 1982, pp. 237–244.

[12] M. Born and E. Wolf, *Principles of Optics*, 3rd ed. New York: Pergamon, 1965, pp. 316–323, 491–506, 544–554.

[13] W. K. Burns, R. P. Moeller, and C. L. Chen, "Depolarization in a single-mode optical fiber," *J. Lightwave Technol.*, vol. 1, pp. 44–50, 1983.

[14] W. K. Burns and R. P. Moeller, "Measurement of polarization mode dispersion in high-birefringence fibers," *Opt. Lett.*, vol. 8, pp. 195–197, 1983.

[15] R. M. A. Azzam and N. M. Bashara, *Ellipsometry and Polarized Light*. New York: North Holland, 1979, pp. 52–59.
[16] R. P. Moeller and W. K. Burns, "Depolarized broadband source," *Electron. Lett.*, vol. 19, pp. 187–188, 1983.

*

William K. Burns (M'80) was born in Philadelphia, PA in June 1943. He received the B.S. degree in engineering physics from Cornell University, Ithaca, NY, in 1965, and the M.S. and Ph.D. degrees in applied physics from Harvard University, Cambridge, MA, in 1967 and 1971, respectively. His thesis research was in nonlinear optics.

During 1971 he was a staff member at Arthur D. Little, Inc., Cambridge, MA, where he contributed to various laser-related studies and surveys. Since 1972 he has been a Research Physicist at the Naval Research Laboratory, Washington, DC. He is presently Head of the Optical Waveguide Section of the Optical Techniques Branch. His research interests include integrated optics, single-mode fiber optics, and the application of single-mode technology to communications, sensors, and signal processing.

Dr. Burns is a member of the Optical Society of America.

Performance of Lyot Depolarizers with Birefringent Single-Mode Fibers

KONRAD BÖHM, KLAUS PETERMANN, MEMBER, IEEE, AND EDGAR WEIDEL

Abstract—Fiber-optic Lyot depolarizers which are attractive for fiber-optic gyros have been studied theoretically and experimentally. It is shown that the amount of residual polarization at the depolarizer output is proportional to the square root of the polarization-maintaining parameter h, yielding a residual polarization of about 1 percent for a fiber-optic depolarizer with $h = 1.1 \cdot 10^{-4}$/m and a superluminescent diode.

I. Introduction

A LYOT DEPOLARIZER converts different spectral components of polarized input light to different polarization states at the output, so that the output light appears unpolarized if averaged over the spectrum [1], [2]. It usually consists of two birefringent plates with a length ratio of 2:1 with their principal axes at 45°, with respect to each other.

The use of such a depolarizer is very attractive in a fiber-optic gyroscope since then no polarization-control elements within the fiber loop are required [3]. In order to realize a compact fiber gyro, a fiber-optic solution for a depolarizer is desirable. Such a fiber-optic depolarizer may be realized by using a birefringent fiber, which is cut and then spliced again, after turning one end by an angle of 45°. In this letter, we will discuss the performance of such depolarizers both from the theoretical and the experimental point of view.

II. Theory

Before characterizing the overall performance of a fiber-optic depolarizer we will first describe a section of a birefringent fiber. In a practical birefringent fiber, there will be some random coupling between the two eigenstates of polarization [4], [5].

The fiber will be assumed to be linearly birefringent and the complex optical-field amplitudes of the two polarization eigenstates are denoted as A_x, A_y, with their propagation constants β_x, β_y yielding for a lossless fiber [6], [7]

$$dA_x/dz + j\beta_x A_x = j\kappa(z) A_y \tag{1}$$

$$dA_y/dz + j\beta_y A_y = j\kappa^*(z) A_x \tag{2}$$

with z denoting the coordinate in axial direction and $\kappa(z)$ the random coupling coefficient between the two eigenstates of polarization. The asterix * denotes complex conjugation. We will consider here only polarization-maintaining fibers where $|\kappa(z)| \ll \beta_x, \beta_y$. In the zeroth-order approximation ($\kappa = 0$) one obtains

$$A_x(z) = A_x(0) \exp(-j\beta_x z) \tag{3}$$

$$A_y(z) = A_y(0) \exp(-j\beta_y z). \tag{4}$$

Inserting this zeroth-order approximation into the right-hand side of (1), (2) yields as a first-order approximation [6]

$$A_x(z) = \exp(-j\beta_x z) \left(A_x(0) + jA_y(0) \int_0^z \kappa(z') \cdot \exp(j(\beta_x - \beta_y) z')\, dz'\right) \tag{5}$$

$$A_y(z) = \exp(-j\beta_y z) \left(A_y(0) + jA_x(0) \int_0^z \kappa^*(z') \cdot \exp(-j(\beta_x - \beta_y) z')\, dz'\right). \tag{6}$$

This first-order approximation will suffice for our analysis with only little polarization coupling. If we define a vector

$$A = \begin{pmatrix} A_x \\ A_y \end{pmatrix} \tag{7}$$

the output vector $A(z = L)$ at the end of a fiber with length L may be related to the input vector $A(z = 0)$ via a Jones matrix [8] $\underline{\underline{M}}$ as

$$A(z = L) = \underline{\underline{M}} A(z = 0) \tag{8}$$

with

$$\underline{\underline{M}} = \begin{pmatrix} \exp(-j\delta L) & j\int_0^L \kappa(z) \exp(2j\delta z - j\delta L)\, dz \\ j\int_0^L \kappa^*(z) \exp(-2j\delta z + j\delta L)\, dz & \exp(j\delta L) \end{pmatrix} \exp(-j\bar{\varphi}) \tag{9}$$

where

$$\delta = 0.5(\beta_x - \beta_y) \qquad \bar{\varphi} = 0.5(\beta_x + \beta_y) L.$$

Manuscript received August 26, 1982.
The authors are with AEG-Telefunken, Forschungsinstitut Ulm, 7900 West Germany.

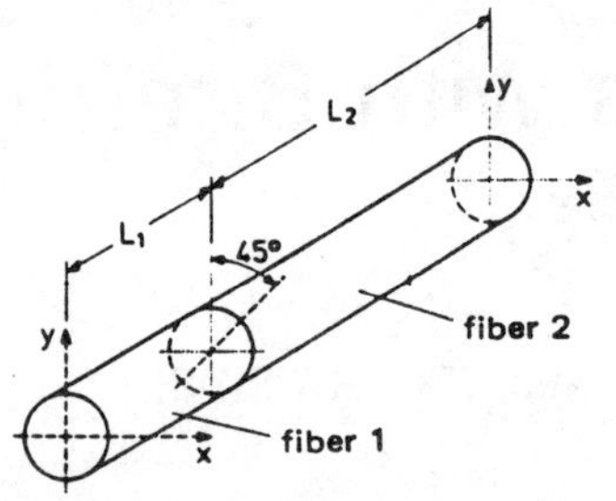

Fig. 1. Principle of a fiber-optic depolarizer.

The magnitude of random coupling between the polarizations is usually expressed by a polarization-holding parameter [4], [5]

$$h = \frac{1}{L}\left\langle\left|\int_0^L \kappa(z)\exp(2j\delta z)\,dz\right|^2\right\rangle \tag{10}$$

where $\langle\ \rangle$ denotes averaging over an ensemble of statistically equivalent fibers. Equations (5), (6), and (9) may be applied as long as $h \cdot L \ll 1$.

A fiber-optic depolarizer is sketched in Fig. 1, consisting of two birefringent fibers 1, 2 with different lengths L_1, L_2, which are turned at the joint by an angle of 45°. The principal axes of fibers 1, 2 are shown as broken and dotted lines, respectively. The Jones matrix of the depolarizer $\underline{\underline{M}}_D$ is obtained as

$$\underline{\underline{M}}_D = \begin{pmatrix} m_{11} & m_{12} \\ m_{21} & m_{22} \end{pmatrix} = \underline{\underline{M}}_2 \underline{\underline{R}}\, \underline{\underline{M}}_1 \tag{11}$$

with $\underline{\underline{M}}_1, \underline{\underline{M}}_2$ describing the matrices of the fibers 1, 2, respectively, and $\underline{\underline{R}}$ describing a rotation by 45°

$$\underline{\underline{R}} = \frac{1}{\sqrt{2}}\begin{pmatrix} 1 & 1 \\ -1 & 1 \end{pmatrix}.$$

$\underline{\underline{M}}_1, \underline{\underline{M}}_2$ corresponds to (9) with $\delta \to \delta_1 \delta_2$; $\kappa \to \kappa_1, \kappa_2$; $L \to L_1, L_2$ with the indices 1, 2 corresponding to fibers 1, 2 respectively. The matrix elements of the depolarizer according to (11) are then obtained as

$$\sqrt{2}\, m_{11} = \exp(-j\Delta^+) + j\exp(-j\Delta^-)K_1^* - j\exp(-j\Delta^+)K_2 \tag{12}$$

$$\sqrt{2}\, m_{12} = \exp(-j\Delta^-) + j\exp(-j\Delta^+)K_1 + j\exp(-j\Delta^-)K_2 \tag{13}$$

$$\sqrt{2}\, m_{21} = -\exp(j\Delta^-) + j\exp(j\Delta^+)K_1^* + j\exp(j\Delta^-)K_2^* \tag{14}$$

$$\sqrt{2}\, m_{22} = \exp(j\Delta^-) - j\exp(j\Delta^-)K_1 + j\exp(j\Delta^+)K_2^* \tag{15}$$

with

$$K_1 = \int_0^{L_1} \kappa_1(z)\exp(2j\delta_1 z)\,dz,$$

$$K_2 = \int_0^{L_2} \kappa_2(z)\exp(2j\delta_2 z)\,dz$$

and

$$\Delta^+ = \delta_2 L_2 + \delta_1 L_1$$

$$\Delta^- = \delta_2 L_2 - \delta_1 L_1.$$

In (12)–(15) the common phase factor has been omitted and terms of second order in κ like κ^2 have been neglected. A depolarizer acts properly only for a light source with a sufficiently broad spectrum. The normalized spectral intensity distribution of the light source will be denoted as $S(\omega)$ with $\int S(\omega)\,d\omega = 1$ and the circular optical frequency $\omega = 2\pi c/\lambda$ (c is velocity of light; λ is wavelength). In the following, a Lorentzian spectrum is assumed:

$$S(\omega) = \frac{2\tau_c/\pi}{1 + (\omega - \omega_c)^2 (2\tau_c)^2} \tag{16}$$

with the center emission frequency ω_c and the coherence time τ_c, which is defined as $\tau_c = 1/\Delta\omega$ with $\Delta\omega$ the spectral width (full width half maximum). The polarization state at the output appears as the average over the spectrum and this spectral average will be denoted by a bar as, for example $\bar{a} \triangleq \int a(\omega)\,S(\omega)\,d\omega$.

In order to get depolarized light for any input polarization the following requirements have to be met:

$$\overline{|m_{11}|^2} = \overline{|m_{12}|^2} = \overline{|m_{21}|^2} = \overline{|m_{22}|^2} = 1/2$$

and

$$\overline{m_{ij} m_{nm}^*} = 0, \qquad \text{for } ij \neq nm. \tag{17}$$

Without polarization coupling these requirements are fulfilled if

$$\left|\frac{d\Delta^+}{d\omega}\right|, \left|\frac{d\Delta^-}{d\omega}\right|, \left|\frac{d(\Delta^- \pm \Delta^+)}{d\omega}\right| \gg \tau_c \tag{18}$$

which means, that the polarization dispersion of any of the fiber sections 1, 2 and the difference between the polarization dispersions of fibers 1, 2 should significantly exceed τ_c.

But even if relation (18) is satisfied a fiber-optic depolarizer is not perfect because of polarization coupling. Somewhat arbitrarily, we will define a parameter ϵ

$$\epsilon = (\overline{|m_{11}|^2} - 1/2)/\overline{|m_{11}|^2} \simeq 2\overline{|m_{11}|^2} - 1 \tag{19}$$

which approximately corresponds to the relative amount of light, which is still polarized at the output of the depolarizer yielding

$$\epsilon = -2\,\mathrm{Re}\left\{ j \int_\omega \left[\int_0^{L_2} \kappa_2 \exp(2j\delta_2 z)\,dz - \int_0^{L_1} \kappa_1 \exp(-2j\delta_1 x)\,dx \right] S(\omega)\,d\omega \right\} \tag{20}$$

with $x = L_1 - z$. In (20) second-order terms in κ are neglected.

Usually, the fiber sections 1, 2 are made from the same fiber so that $\delta_1 = \delta_2 = \delta$. δ may be expanded around the cent nission frequency ω_c

$$\delta = \delta_0 + \frac{\Delta\tau}{2}(\omega - \omega_c) \tag{21}$$

ith the polarization dispersion per unit length $\Delta\tau$. Using the ectrum according to (16) and performing the integration er ω in (20) then yields

$$\epsilon = -2 \, \mathrm{Re}\left\{ j \int_0^{L_2} \kappa_2 \exp(2j\delta_0 z) \exp(-\Delta\tau z/2\tau_c) \, dz - j \int_0^{L_1} \kappa_1 \exp(-2j\delta_0 x) \exp(-\Delta\tau x/2\tau_c) \, dx \right\}. \tag{22}$$

The exact evaluation of (22) is difficult but if we use the asonable approximation

$$\exp(-\Delta\tau x/2\tau_c) \approx \begin{cases} 1 & \text{for } \Delta\tau x < 2\tau_c \\ 0 & \text{for } \Delta\tau x \geqslant 2\tau_c \end{cases} \tag{23}$$

e calculation considerably simplifies to

$$\epsilon = -2 \, \mathrm{Re}\left\{ j \int_0^{l_c} \kappa_2 \exp(2j\delta_0 z) \, dz - j \int_0^{l_c} \kappa_1 \exp(-2j\delta_0 x) \, dx \right\} \tag{24}$$

ith

$$l_c = 2\tau_c/\Delta\tau.$$

ecause of the random polarization coupling, ϵ is random as ell. Since κ_1 and κ_2 are uncorrelated one obtains

$$\epsilon_{\mathrm{rms}} = \sqrt{\langle \epsilon^2 \rangle} = 2\sqrt{h \cdot l_c} = 2\sqrt{2h\tau_c/\Delta\tau}. \tag{25}$$

The relative amount ϵ_{rms} of intensity which is still polarized the depolarizer output is thus proportional to the square ot of polarization coupling with respect to intensity during e length l_c. This situation is very similar to the performance quirements of a polarizer in a fiber gyroscope [9] where the uare root of the intensity extinction ratio for the rejected larization channel is proportional to the error in the detected tation rate.

Therefore, even for moderate quality of the depolarizer, tremely low values for the polarization-holding parameter are required.

III. Experiment

The single-mode fiber used for the experiments was a fiber ith a large linear birefringence supplied from Hitachi [10]. e fiber has a core diameter of 5 μm; the polarization-holding rameter was measured to be $h = 1.1 \cdot 10^{-4}$/m/4, 5/ and the at length $L = 2\pi/\beta_i$ was 4 mm at 633 nm. For the fabrica- n of the Lyot depolarizer two fiber pieces, fiber 1 and fiber with the length $L_1 = 5$ m and $L_2 = 10$ m were used. The rformance of the depolarizer depends critically on the angle tween the principal axes of fiber 1 and fiber 2. In order to

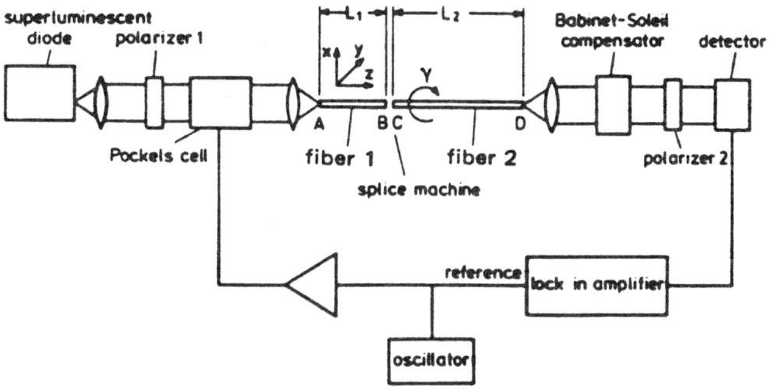

Fig. 2. Experimental setup.

adjust the optimum angle of 45° we used the experimental setup shown in Fig. 2.

The light from a superluminescent diode emitting at 811 nm with a full half-width of 14 nm is passing a polarizer 1, whose plane is adjusted with an angle of 45°, with respect to the axes of the following Pockels' cell [11]. By applying a sinusoidal voltage with a frequency of 1 kHz to the Pockels' cell the state of polarization is modulated. The amplitude of the ac voltage was chosen so that the state of polarization after the Pockels' cell moves on a complete meridian on the Poincaré sphere [8].

The light is then coupled into the end A of the fiber 1, which can be rotated around its axis. By applying the ac voltage to the Pockels' cell and with a step rotation of the end A any state of polarization with respect to the principal (i.e., fast) axis of fiber 1 can be coupled into the end A of fiber 1. From the end B of fiber 1, the light is coupled into the end C of fiber 2. The ends B and C are fixed in the mounting of a splicing unit, which was equipped with one rotatable jig allowing a precise rotation of the fiber end C of an angle γ with an accuracy of 1/60 deg. The light from the end D of fiber 2 is collimated, passes a Babinet Soleil compensator and a polarizer 2 and is then detected by a p-i-n photodiode. The signal from the photodiode is amplified by a lock-in-amplifier synchronized with the Pockels' cell-drive voltage.

In order to calibrate the setup, the end C of fiber 2 is rotated, until the principal axes of fiber 1 and 2 at B and C coincide. The principal axis of fiber 1 at A is then aligned parallel to the plane of the polarizer 1 and the principal axis of fiber 2 at D is aligned parallel to the plane of the polarizer 2, whereas the retardation of the Babinet Soleil is adjusted to zero. By applying the ac voltage to the Pockels' cell, the state of polarization is moving on a meridian on the Poincaré sphere; the two linear directions of polarization contained in this meridian are coupled into the fast and slow axes of fibers 1 and 2 and are maintained to the end D of fiber 2. The principal axis of the polarizer 2 is parallel to one of these polarizations and perpendicular to the other.

Therefore, full and zero light transmission through the polarizer 2 is achieved and a maximum ac-signal S is measured, which serves for calibration.

To obtain a depolarizer, the angle between the principal axes of fiber 1 at B and of fiber 2 at C is changed by turning the fiber end C by 45°. Now the depolarization of the two fibers 1 and 2 causes a reduction of the ac signal to a minimum. To determine the maximum residual ac signal s for a fixed angle

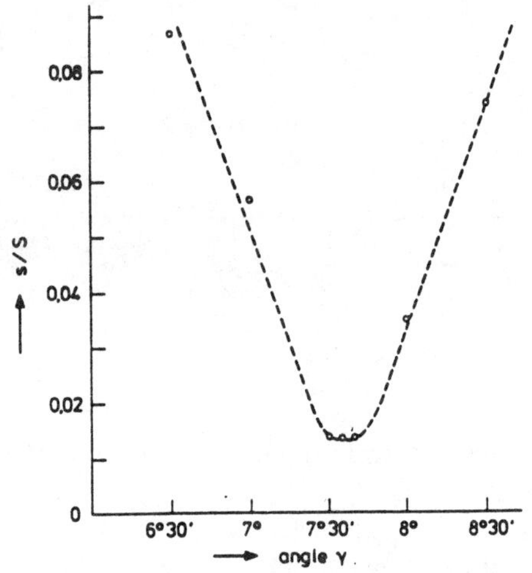

Fig. 3. Normalized residual ac signal s/S versus the angle γ of the rotatable jig.

γ, the fiber end A of fiber 1 was turned in steps of about 10° and the retardation of the Babinet Soleil compensator was changed arbitrarily until a maximum signal s was obtained. The normalized ac signal s/S is shown in Fig. 3 as a function of the angle γ of the rotatable jig of the splicing unit. As can be seen, s/S decreases to a minimum value of $1.4 \cdot 10^{-2}$. This value has to be compared with the quantity ϵ_{rms} of the preceeding chapter. From the above given data of the superluminescent diode and the fiber ϵ_{rms} can be calculated to $\epsilon_{rms} = 0.7 \cdot 10^{-2}$. This value is of the same order as the experimental value s/S.

To determine the sensitivity of the measuring setup, the ac signal of the setup without polarizer 2 was also measured. In this case the ratio s/S was about 10^{-3}.

For practical use, the fiber ends B and C were spliced together in the optimum position. The depolarizer was measured after the splicing process and no change could be measured.

IV. Conclusions

We have demonstrated that fiber-optic depolarizers can be made with similar quality as usual calcite depolarizers. However, since the polarization holding with respect to the optical-field amplitude is the limiting factor for the performance of the depolarizer, low values for the polarization-holding parameter h are required. The best value $h = 5 \cdot 10^{-6}$/m reported until now [12] would yield a depolarizer with $\epsilon_{rms} = 1.4 \cdot 10^{-3}$ for the superluminescent diode of our experiment, which is an excellent value.

Acknowledgment

The authors wish to thank L. Staudigel for assistance in the measurements.

References

[1] B. Lyot, "Annales de l'observatoire d'astronomie de Paris (Mendon), vol. VIII, p. 102, 1929.
[2] B. H. Billings, "A monochromatic depolarizer," *J. Opt. Soc. Amer.*, vol. 41, pp. 966–975, Dec. 1951.
[3] K. Böhm, P. Marten, K. Petermann, E. Weidel, and R. Ulrich, "Low-drift fibre gyro using a superluminescent diode," *Electron. Lett.*, vol. 17, pp. 352–353, May 14, 1981.
[4] I. P. Kaminow, "Polarization in optical fibers," *IEEE J. Quantum Electron.*, vol. QE-17, pp. 15–22, Jan. 1981.
[5] S. C. Rashleigh, W. K. Burns, R. P. Moeller, and R. Ulrich, "Polarization-holding in birefringent single-mode fibers," *Opt. Lett.*, vol. 7, pp. 40–42, Jan. 1982.
[6] D. Marcuse, *Theory of Dielectric Optical Waveguides.* New York: Academic, 1974.
[7] R. Ulrich, "Representation of codirectional coupled waves," *Opt. Lett.*, vol. 1, pp. 109–111, Sept. 1977.
[8] G. N. Ramachandran and S. Ramaseshan, "Polarization of light" in "Crystal optics," *Handbuch der Physik*, vol. XXV/1, S. Flügge, Ed. Berlin-Göttingen-Heidelberg, Springer, 1961, pp. 1–54.
[9] E. C. Kintner, "Polarization control in optical-fiber gyroscopes," *Opt. Lett.*, vol. 6, pp. 154–156, Mar. 1981.
[10] T. Katsuyama, H. Matsumura, and T. Suganuma, "Low loss single polarization fibre," *Electron. Lett.*, vol. 17, pp. 473–474, 1981.
[11] *Handbook of Optics*, W. D. Driscoll and W. Vaugham, Eds. New York: McGraw-Hill, 1978, pp. 10-(141–146).
[12] S. C. Rashleigh and M. J. Marrone, "Polarization-holding in high-birefringence fibre," *Electron. Lett.*, vol. 18, pp. 326–32 April 15, 1982.

*

K. Böhm, photograph and biography not available at the time of publication.

*

Klaus Petermann (M'76) was born in Mannheim, Germany, on October 2, 1951. He received the Dipl.-Ing. degree in 1974 and the Dr. Ing. degree in 1976, both in electrical engineering from the Technische Universität, Braunschweig, Germany.

From 1974 to 1976 he was a Research Associate at the Institut für Hochfrequenztechnik, Technische Universität, Braunschweig, where he worked on optical waveguide theory. Since 1977 he has been with AET-Telefunken Forschungsinstitut, Ulm, Germany, where he is engaged in research work on semiconductor lasers and optical fibers.

Dr. Petermann is a member of the Nachrichtentechnische Gesellschaft in the Verein Deutscher Elektrotechniker (VDE).

*

E. Weidel, photograph and biography not available at the time of publication.

©1984 Optical Society of America. Reprinted from *Applied Optics*, Vol. 23(19), 3284–3288 (October 1984).

Degree of polarization in jointed fibers: the Lyot depolarizer

Kiyofumi Mochizuki

Analytic equations for the degree of polarization of the output lightwave from jointed fibers are derived taking into account the coherence time of light and polarization mode dispersion in the fibers. On the basis of the results, the depolarizing conditions and the principles of optics of the Lyot depolarizer are discussed.

I. Introduction

The Lyot depolarizer originally consisted of two birefringent crystals of 2:1 thickness ratio, whose optic axes are oriented at 45° to each other. The performance of the Lyot depolarizer and the requirement regarding the design of the Lyot depolarizer have been discussed by Billings[1] and Loeber.[2] In their papers, depolarization was considered to be caused by the unequal phase retardation suffered by the different wavelengths traversing the crystal, and they showed that the retardance by each crystal should be large and unequal to the other for depolarization. However, they could not quantitatively induce the length of each crystal.

Recently, Böhm *et al.* introduced the concept of the coherence time of light to analyze the performance of the Lyot depolarizer with birefringent single-mode fibers and showed the quantitative requirements of the length to get complete depolarization.[3] Meanwhile, Burns derived the degree of polarization of light exiting the Lyot depolarizer with birefringent fibers and investigated the effect of the degree of coherence for sources with various shapes on the degree of polarization.[4] As a result, Burns clarified that the integer relation of the ratio of the two retardances obtained by Billings is only required with a rectangular-shaped source.[4]

As mentioned above, so far the requirements regarding the design of the Lyot depolarizer with birefringent fibers and the performance have been discussed. However, in their treatment, the angle between the birefringent axes of two fibers was fixed at 45°, and the angular dependence of the degree of polarization has not been discussed. Furthermore, the principles of optics constructing the Lyot depolarizer have not been much discussed yet either.

In this paper, we derive the degree of polarization of light exiting jointed fibers with their birefringent axes at all possible angles taking into account the coherence time of light and polarization mode dispersion in the fibers. On the basis of the theoretical results, the conditions of depolarization for lightwaves exiting the birefringent fibers are shown, and the principles of optics in the Lyot depolarizer are discussed.

II. Theory

We assume a configuration for jointed large birefringent fibers as shown in Fig. 1 and the model without mode coupling between polarization modes. The fiber birefringent axes are indicated by X_1 and Y_1 in fiber I and by X_2 and Y_2 in fiber II. We assume that an incident lightwave consists of two orthogonally polarized waves, $E_X(t,\Psi)$ and $E_Y(t,\Psi)$, which originate from a lightwave, and that one of the waves, say $E_Y(t,\Psi)$, is delayed by τ. Any state of polarization for incident lightwaves can be represented by suitably selecting the values of τ; i.e., $\lambda/2c$ for linear polarization and $\lambda/4c$ for circular polarization (λ = wavelength, c = speed of light). Here, we assume that the waves operate in the multilongitudinal mode and that the analytic signals are represented by[5]

$$E_X(t,\Psi) = \frac{1}{\sqrt{2}} \cdot \Sigma_n E_n \cdot \exp i[2\pi(\nu + n\delta\nu)\, t + \Psi_n], \quad (1)$$

$$E_Y(t,\Psi) = \frac{1}{\sqrt{2}} \cdot \Sigma_n E_n \cdot \exp i\, [2\pi(\nu + n\delta\nu)(t - \tau) + \Psi_n], \quad (2)$$

where $E_n = E_0 \exp[-2(n\delta\nu/\Delta\nu)^2]$. A Gaussian distribution is assumed for the mode amplitude E_n. Here ν is its center frequency, $\Delta\nu$ is its full width between half-peak intensity points, and $\delta\nu$ is the mode spacing. The relationships are shown in Fig. 2. Ψ_n represents

The author is with KDD Research & Development Laboratories, 1-23, Nakameguro 2-chome, Meguro-ku, Tokyo 153, Japan.

Received 10 March 1984.
0003-6935/84/193284-05$02.00/0.

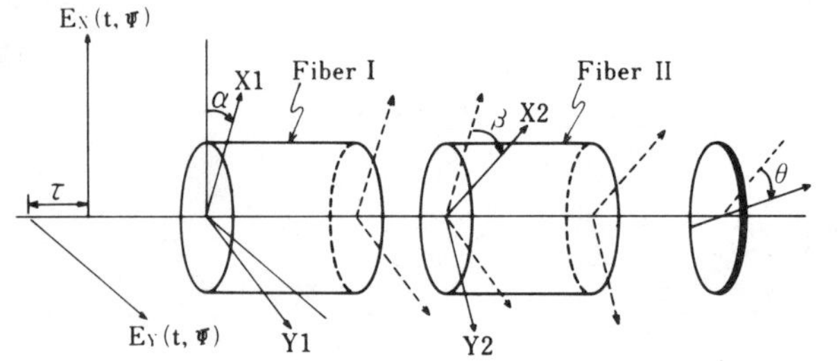

Fig. 1. Schematic configuration of jointed birefringent fibers.

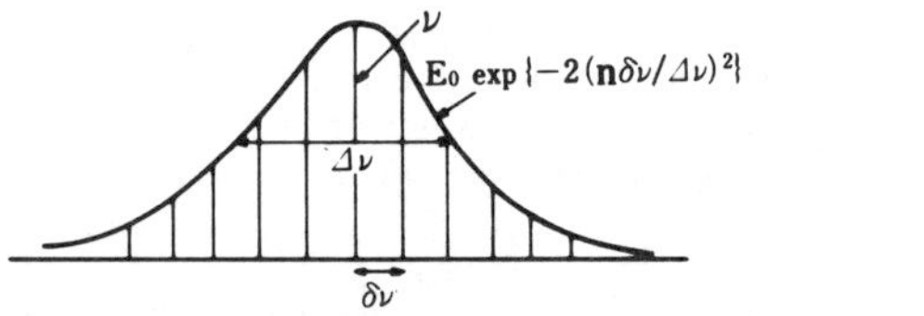

Fig. 2. Schematic spectrum of a laser diode operating in the multi-longitudinal mode.

the phases of the individual modes. If the phases Ψ_n are statistically independent, Eqs. (1) and (2) are, within one repetition period $1/\delta\nu$, Gaussian fluctuating lightwaves with center frequency ν and full width $\Delta\nu$ between half-peak intensity points.[5] We will assume that $\delta\nu \ll \Delta\nu$, so that we can replace the sum by an integral in the following calculations.

When the waves are sent into fiber I forming the angle α with respect to the birefringent axes as shown in Fig. 1, the waves propagating along the birefringent axes are given by

$$\begin{bmatrix} E^i_{X1}(t,\Psi) \\ E^i_{Y1}(t,\Psi) \end{bmatrix} = \begin{pmatrix} \cos\alpha & \sin\alpha \\ -\sin\alpha & \cos\alpha \end{pmatrix} \begin{bmatrix} E_X(t,\Psi) \\ E_Y(t,\Psi) \end{bmatrix}. \tag{3}$$

In fiber I, the spectral components of $E^i_{X1}(t,\Psi)$ and $E^i_{Y1}(t,\Psi)$ experience phase changes, Φ_{nX1} and Φ_{nY1}. These Φ_{nX1} and Φ_{nY1} can be calculated by expanding the propagation constant, β_{X1} or β_{Y1}, around the center frequency $\nu(\omega/2\pi)$ with respect to the frequency difference $(n\delta\nu)$ of the nth mode. Since third- and higher-order terms in the expansion can be neglected if $\Delta\nu/\nu < 10^{-2}$,[5] the phase changes can be represented by

$$\Phi_{nX1} = -Z_1\left[\beta_{X1} + 2\pi\delta\nu\cdot\frac{d\beta_{X1}}{d\omega}\cdot n + \tfrac{1}{2}(2\pi\delta\nu)^2\cdot\frac{d^2\beta_{X1}}{d\omega^2}\cdot n^2\right], \tag{4}$$

$$\Phi_{nY1} = -Z_1\left[\beta_{Y1} + 2\pi\delta\nu\cdot\frac{d\beta_{Y1}}{d\omega}\cdot n + \tfrac{1}{2}(2\pi\delta\nu)^2\cdot\frac{d^2\beta_{Y1}}{d\omega^2}\cdot n^2\right], \tag{5}$$

where β_{X1} and β_{Y1} are propagation constants for two orthogonal polarization modes in fiber I, Z_1 is the length of fiber I, and ω is the radian frequency. Accordingly, without loss of generality two waves emerging from fiber I are expressed by

$$E^o_{X1}(t,\Psi) = E^i_{X1}(t,\Psi + \Phi_{X1}), \tag{6}$$

$$E^o_{Y1}(t,\Psi) = E^i_{Y1}(t,\Psi + \Phi_{Y1}). \tag{7}$$

When fiber I is jointed with fiber II forming the angle β between each birefringent axis as shown in Fig. 1, two lightwaves propagating along the birefringent axes of fiber II are given by

$$\begin{bmatrix} E^i_{X2}(t,\Psi) \\ E^i_{Y2}(t,\Psi) \end{bmatrix} = \begin{pmatrix} \cos\beta & \sin\beta \\ -\sin\beta & \cos\beta \end{pmatrix} \begin{bmatrix} E^o_{X1}(t,\Psi) \\ E^o_{Y1}(t,\Psi) \end{bmatrix}. \tag{8}$$

The spectral components of the two lightwaves also experience the phase changes given by

$$\Phi_{nX2} = -Z_2\left[\beta_{X2} + 2\pi\delta\nu\cdot\frac{d\beta_{X2}}{d\omega}\cdot n + \tfrac{1}{2}(2\pi\delta\nu)^2\cdot\frac{d^2\beta_{X2}}{d\omega^2}\cdot n^2\right], \tag{9}$$

$$\Phi_{nY2} = -Z_2\left[\beta_{Y2} + 2\pi\delta\nu\cdot\frac{d\beta_{Y2}}{d\omega}\cdot n + \tfrac{1}{2}(2\pi\delta\nu)^2\cdot\frac{d^2\beta_{Y2}}{d\omega^2}\cdot n^2\right], \tag{10}$$

where β_{X2} and β_{Y2} are propagation constants for two orthogonal polarization modes in fiber II and Z_2 is the length of fiber II. Accordingly, the output lightwaves from fiber II without consideration of transmission loss are given by

$$E^o_{X2}(t,\Psi) = E^i_{X2}(t,\Psi + \Phi_{X2}), \tag{11}$$

$$E^o_{Y2}(t,\Psi) = E^i_{Y2}(t,\Psi + \Phi_{Y2}). \tag{12}$$

Using Eqs. (3) and (6)–(8), Eqs. (11) and (12) are recast as

$$\begin{aligned} E^o_{X2}(t,\Psi) = {} & \cos\beta[\cos\alpha\cdot E_X(t,\Psi + \Phi_{X1} + \Phi_{X2}) \\ & + \sin\alpha\cdot E_Y(t,\Psi + \Phi_{X1} + \Phi_{X2})] \\ & + \sin\beta[-\sin\alpha\cdot E_X(t,\Psi + \Phi_{Y1} + \Phi_{X2}) \\ & + \cos\alpha\cdot E_Y(t,\Psi + \Phi_{Y1} + \Phi_{X2})], \end{aligned} \tag{13}$$

$$\begin{aligned} E^o_{Y2}(t,\Psi) = {} & -\sin\beta[\cos\alpha\cdot E_X(t,\Psi + \Phi_{X1} + \Phi_{Y2}) \\ & + \sin\alpha\cdot E_Y(t,\Psi + \Phi_{Y1} + \Phi_{Y2})] \\ & + \cos\beta[-\sin\alpha\cdot E_X(t,\Psi + \Phi_{X1} + \Phi_{Y2}) \\ & + \cos\alpha\cdot E_Y(t,\Psi + \Phi_{Y1} + \Phi_{Y2})]. \end{aligned} \tag{14}$$

Using Eqs. (13) and (14), the elements of the coherency matrix[6] can be obtained as follows (see the Appendix):

$$\begin{aligned} J_{xx} = {} & 1 + \sin 2\alpha\cdot\cos 2\beta\cdot f(\tau) - \sin^2\alpha\cdot\sin 2\beta\cdot f(\tau + \tau_1) \\ & + \cos^2\alpha\cdot\sin 2\beta\cdot f(\tau - \tau_1), \end{aligned} \tag{15}$$

$$\begin{aligned} J_{yy} = {} & 1 - \sin 2\alpha\cdot\cos 2\beta\cdot f(\tau) + \sin^2\alpha\cdot\sin 2\beta\cdot f(\tau + \tau_1) \\ & - \cos^2\alpha\cdot\sin 2\beta\cdot f(\tau - \tau_1) \end{aligned} \tag{16}$$

$$\begin{aligned} J_{xy} = {} & -\tfrac{1}{2}\sin 2\alpha\cdot\sin 2\beta\cdot\Gamma(\tau + \tau_2)\cdot\exp[-j(2\pi\nu\tau + Z_2\Delta\beta_2)] \\ & - \tfrac{1}{2}\sin 2\alpha\cdot\sin 2\beta\cdot\Gamma(\tau - \tau_2)\cdot\exp[j(2\pi\nu\tau - Z_2\Delta\beta_2)] \\ & + \sin^2\alpha\cdot\sin^2\beta\cdot\Gamma(\tau + \tau_1 - \tau_2) \\ & \cdot\exp[j(2\pi\nu\tau + Z_1\Delta\beta_1 - Z_2\Delta\beta_2)] \\ & - \cos^2\alpha\cdot\sin^2\beta\cdot\Gamma(\tau - \tau_1 + \tau_2) \\ & \cdot\exp[-j(2\pi\nu\tau - Z_1\Delta\beta_1 + Z_2\Delta\beta_2)] \\ & + \cos^2\alpha\cdot\cos^2\beta\cdot\Gamma(\tau - \tau_1 - \tau_2) \\ & \cdot\exp[j(2\pi\nu\tau - Z_1\Delta\beta_1 - Z_2\Delta\beta_2)] \\ & - \sin^2\alpha\cdot\cos^2\beta\cdot\Gamma(\tau + \tau_1 + \tau_2) \\ & \cdot\exp[-j(2\pi\nu\tau + Z_1\Delta\beta_1 + Z_2\Delta\beta_2)], \end{aligned} \tag{17}$$

and

$$\begin{aligned}J_{yx} = &-\tfrac{1}{2}\sin 2\alpha\cdot\sin 2\beta\cdot\Gamma(\tau+\tau_2)\cdot\exp[j(2\pi\nu\tau+Z_2\Delta\beta_2)]\\&-\tfrac{1}{2}\sin 2\alpha\cdot\sin 2\beta\cdot\Gamma(\tau-\tau_2)\cdot\exp[-j(2\pi\nu\tau-Z_2\Delta\beta_2)]\\&+\sin^2\alpha\cdot\sin^2\beta\cdot\Gamma(\tau+\tau_1-\tau_2)\\&\cdot\exp[-j(2\pi\nu\tau+Z_1\Delta\beta_1-Z_2\Delta\beta_2)]\\&-\cos^2\alpha\cdot\sin^2\beta\cdot\Gamma(\tau-\tau_1+\tau_2)\\&\cdot\exp[j(2\pi\nu\tau-Z_1\Delta\beta_1+Z_2\Delta\beta_2)]\\&+\cos^2\alpha\cdot\cos^2\beta\cdot\Gamma(\tau-\tau_1-\tau_2)\\&\cdot\exp[-j(2\pi\nu\tau-Z_1\Delta\beta_1-Z_2\Delta\beta_2)]\\&-\sin^2\alpha\cdot\cos^2\beta\cdot\Gamma(\tau+\tau_1+\tau_2)\\&\cdot\exp[j(2\pi\nu\tau+Z_1\Delta\beta_1+Z_2\Delta\beta_2)],\end{aligned} \tag{18}$$

where

$$f(\tau-\tau_i) = \exp\left[-\left(\frac{\Delta\nu\pi}{2}\right)^2\cdot(\tau-\tau_i)^2\right]\cdot\cos(2\pi\nu\tau-\Delta\beta_i Z_i)$$

$$(i = 1,2) = \Gamma(\tau-\tau_i)\cdot\cos(2\pi\nu\tau-\Delta\beta_i Z_i), \tag{19}$$

$$\tau_i = Z_i\left(\frac{d\beta_{Xi}}{d\omega}-\frac{d\beta_{Yi}}{d\omega}\right), \tag{20}$$

$$\Delta\beta_i = (\beta_{Xi}-\beta_{Yi})Z_i. \tag{21}$$

Here, the approximation shown in the Appendix is used, and J_{xx}, J_{yy}, J_{xy}, and J_{yx} are normalized by $E_0^2\cdot(\pi^{1/2}/4)\cdot(\Delta\nu/\delta\nu)$. Here also τ_1 and τ_2 are the total polarization mode dispersion arising in fiber I and fiber II, respectively.

In Eq. (19), $\cos(2\pi\nu\tau-\Delta\beta_i Z_i)$ shows fringes as τ is varied and the envelope, $\Gamma(\tau-\tau_i)$, represents the degree of coherence. Here, the coherence time τ_c is defined by $\Gamma(\tau_c) = 1/e$. One example for Eq. (19) is illustrated in Fig. 3.

Using the elements of the coherency matrix, the degree of polarization is given by[6]

$$P = \sqrt{1-\frac{4|\mathbf{J}|}{(J_{xx}+J_{yy})^2}}, \tag{22}$$

where $|\mathbf{J}| = J_{xx}\cdot J_{yy}-J_{xy}\cdot J_{yx}$. Then, on substituting from Eqs. (15)–(18) into Eq. (22), the degree of polarization for all possible incident polarization states and all possible angles between the birefringent axes of the two fibers can be obtained. When the incident polarization state is linear, $\tau = 0$. Then the degree of polarization is obtained as

$$\begin{aligned}P = &[\cos^2 2\eta\{\cos^2 2\beta+\sin^2 2\beta[\Gamma(\tau_2)]^2\}\\&+\sin^2 2\eta\{\tfrac{1}{2}\sin^2 2\beta\cdot[\Gamma(\tau_1)]^2+\cos^4\beta[\Gamma(\tau_1+\tau_2)]^2\\&+\sin^4\beta[\Gamma(\tau_1-\tau_2)]^2\}\\&-\sin 4\eta\cdot\sin 2\beta\cdot\Gamma(\tau_2)\\&\times[\cos^2\beta\cdot\Gamma(\tau_1+\tau_2)-\sin^2\beta\cdot\Gamma(\tau_1-\tau_2)]\cos(Z_1\Delta\beta_1)\\&+\tfrac{1}{2}\sin^2 2\eta\cdot\sin^2 2\beta\cdot\{[\Gamma(\tau_1)]^2\\&-\Gamma(\tau_1-\tau_2)\cdot\Gamma(\tau_1+\tau_2)\}\cos(2Z_1\Delta\beta_1)]^{1/2},\end{aligned} \tag{23}$$

where the relation, $\eta = \alpha + 45°$, is used. η is the angle between the incident linearly polarized lightwave and the birefringent axis in fiber I.

III. Discussion

Equation (23) gives the degree of polarization of the lightwave exiting jointed fibers when a linearly polarized lightwave is coupled. When $\beta = 45°$, which is the case of the Lyot depolarizer, Eq. (23) is the same as the results obtained by Burns.[4]

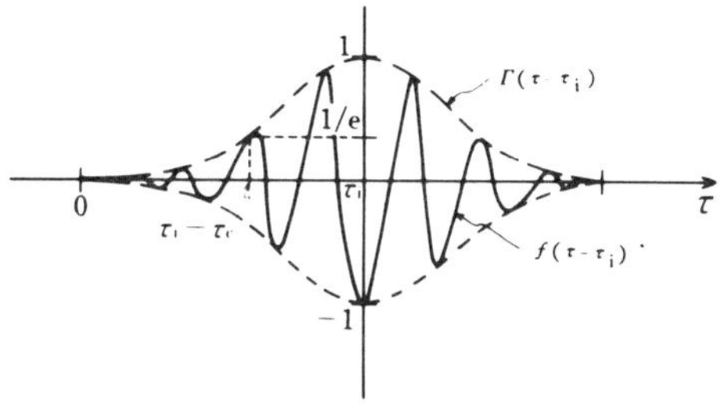

Fig. 3. Illustration of the equation $f(\tau-\tau_i)$. If $\tau_i \gg \tau_c$, $f(\tau-\tau_i)$ becomes zero at around $\tau = 0$.

As described earlier,[3,4] it will be expected from Eq. (23) that at least the conditions

$$\tau_1,\ \tau_2,\ |\tau_2-\tau_1| \gg \tau_C \tag{24}$$

are needed for complete depolarization. In this case, the elements of the coherency matrix are given by

$$J_{xx} = 1+\sin 2\alpha\cdot\cos 2\beta\cdot f(\tau),$$

$$J_{yy} = 1-\sin 2\alpha\cdot\cos 2\beta\cdot f(\tau),$$

$$J_{xy} = J_{yx} = 0.$$

Hence, the degree of polarization is given by

$$P = |\sin 2\alpha\cdot\cos 2\beta\cdot f(\tau)|. \tag{25}$$

Equation (25) shows the degree of polarization for all possible incident polarization states and all possible angles between the birefringent axes of the two fibers in the conditions given by Eq. (24). It is found that complete depolarization can be achieved by satisfying the following:

$$\sin 2\alpha\cdot\cos 2\beta\cdot f(\tau) = 0. \tag{26}$$

Equation (26) is fulfilled by one of the following conditions:

(i) $\sin 2\alpha = 0$

This condition signifies that two lightwaves propagating along two birefringent axes have the same power. This condition is easily created by sending a linearly polarized lightwave into the fiber forming the 45° angle with respect to the birefringent axes [see Fig. 4(a)]. In this case, it is not necessary for the fibers to be jointed and a depolarized lightwave can be obtained using only one fiber whose total polarization mode dispersion significantly exceeds the coherence time of the lightwave. This agrees with the results obtained in Refs. 7 and 8.

(ii) $\cos 2\beta = 0$

This condition means that two birefringent fibers are jointed forming the 45° angle between each birefringent axis [see Fig. 4(b)]. If fibers I and II have the same propagation constants, the conditions given by Eq. (24) are satisfied by $Z_2 = 2Z_1$ when $\tau_1 \gg \tau_c$. This is the Lyot depolarizer. Figure 5 shows the angle (β) dependence of the degree of polarization for linearly polarized incident light in the conditions given by Eq. (24).

(iii) $f(\tau) = 0$

From Eq. (19), it is found that this condition is satisfied when $\tau = \tfrac{1}{4}\nu = \lambda/4c$. This means that the incident lightwave is circularly polarized. In this case, this

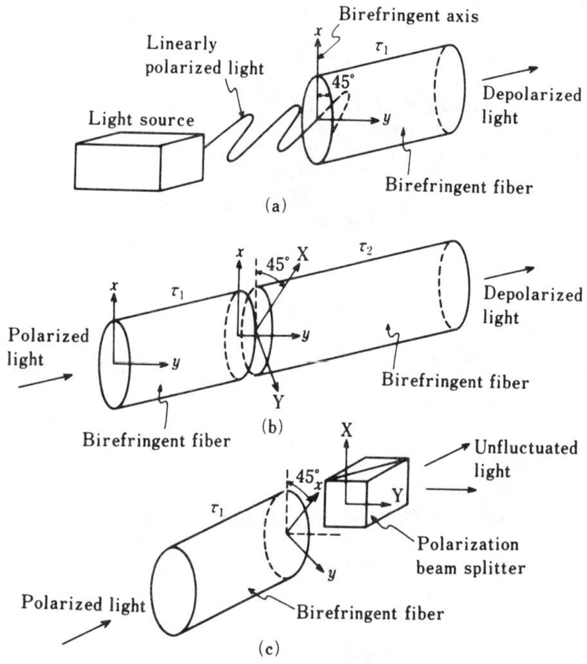

$\tau_1,\ \tau_2,\ |\tau_1-\tau_2|>\tau_C,$

Fig. 4. Schematic configuration to obtain the depolarization of light exiting birefringent fibers. (a) and (c) are examples of principles (I) and (II), respectively; (b) shows the Lyot depolarizer with birefringent fibers.

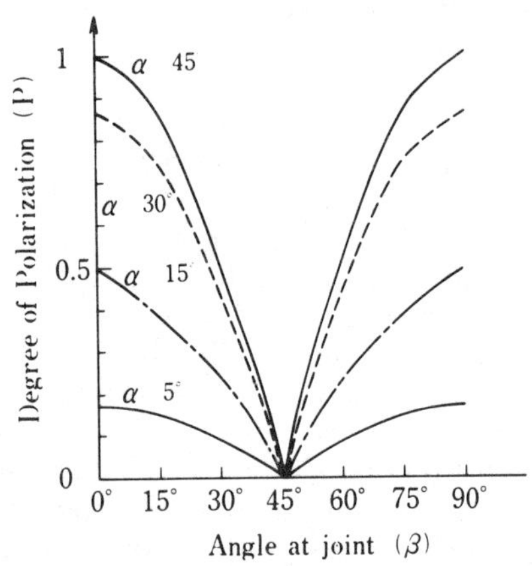

Fig. 5. Variation of the degree of polarization for linearly polarized incident light as a function of β for several values of α.

condition is also satisfied by using only one fiber as condition (i).

Meanwhile, in the conditions of Eq. (24), the output power is given by

$$I = J_{xx} + (J_{yy} - J_{xx})\sin^2\theta + \tfrac{1}{2}(J_{xy} + J_{yx})\sin 2\theta$$
$$= E_0^2 \cdot \frac{\pi^{1/2}}{4} \cdot \frac{\Delta\nu}{\delta\nu} \cdot [1 + \sin 2\alpha \cdot \cos 2\beta \cdot \cos 2\theta \cdot f(\tau)], \qquad (27)$$

where $E_0^2 \cdot (\pi^{1/2}/4) \cdot (\Delta\nu/\delta\nu)$ appears because the elements of the coherency matrix are normalized by it. Here, it is interesting to note that, when $\cos 2\theta = 0$, the output power from analyzer I becomes constant for all

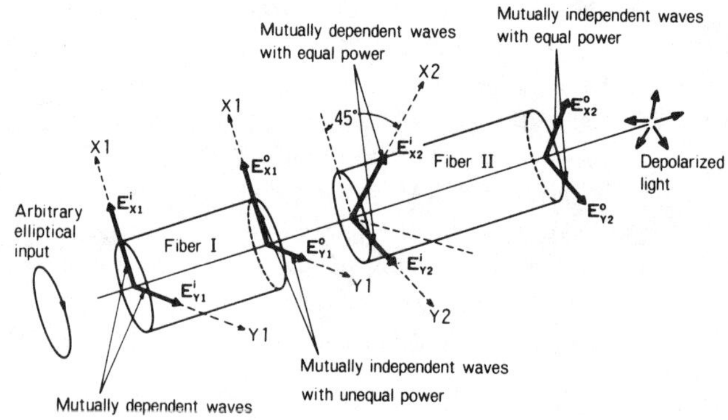

Fig. 6. Explanatory illustration of the Lyot depolarizer consisting of two principles of optics.

possible incident polarized states. This means that the output power from the analyzer, whose polarization axes form the 45° angle with respect to the fiber birefringent axes, is not affected by the polarization state of incident lightwaves to the fiber as long as the total power is constant, although the output power drops to a half. In this case, it is found that output power I is constant regardless of the angle β. Hence, this condition is obtained by only one fiber whose total polarization mode dispersion much exceeds the coherence time of the lightwave [see Fig. 4(c)].

The above-mentioned phenomena are based on the following principles of optics:

(I) The lightwave, of intensity I_0, consisting of two independent linearly polarized waves, each of intensity $\tfrac{1}{2}I_0$, with their electric vectors vibrating in two mutually perpendicular directions, is a depolarized lightwave.[6]

(II) The power in the direction which makes a 45° angle with respect to two independent linearly polarized waves with their electric vectors vibrating in two mutually perpendicular directions is half of the total power regardless of the power ratio of the two waves.

Condition (i) for depolarization is based on principle (I), and the phenomenon shown in Fig. 4(c) indicates principle (II).

Next, we show that the Lyot depolarizer consists of the above principles (see Fig. 6).

The first fiber constructing the Lyot depolarizer offers two independent linearly polarized waves with their electric vectors vibrating in two mutually perpendicular directions at the fiber end because the total polarization mode dispersion in the first fiber is much larger than the coherence time of the incident lightwave. Consequently, according to principle (II), the lightwaves propagating along two birefringent axes in the second fiber have the same power regardless of the state of polarization of the incident lightwave to the first fiber. When the incident lightwave is linearly polarized and launched parallel to one of the birefringent axes in the first fiber, two independent linearly polarized waves cannot be obtained at the end of the first fiber. However, it is obvious that two lightwaves propagating along the two birefringent axes in the second fiber have the same power because the birefringent axes of the two fibers make a 45° angle with each other.

The two lightwaves with the same power propagating in the second fiber become independent at the end of the fiber because of the condition $|\tau_2 - \tau_1| \gg \tau_c$. The mechanism of the independence is described in Ref. 4. This indicates that the output lightwave from the second fiber consists of two independent linearly polarized waves—of the same intensity—with their electric vectors vibrating in two mutually perpendicular directions, which becomes the depolarized lightwave according to principle (I). Hence, the output lightwaves from the Lyot depolarizer are depolarized regardless of the state of polarization of the incident lightwaves when the condition given by Eq. (24) is satisfied.

IV. Conclusion

We analytically derived the degree of polarization of light exiting jointed fibers. On the basis of the results, the depolarization conditions are clarified, and it was shown that the Lyot depolarizer consists of two principles of optics.

Appendix

The elements of the coherency matrix are given by

$$J_{xx} = \langle E^0_{X2}(t,\Psi) \cdot E^0_{X2}(t,\Psi)^* \rangle,$$
$$J_{yy} = \langle E^0_{Y2}(t,\Psi) \cdot E^0_{Y2}(t,\Psi)^* \rangle,$$
$$J_{xy} = \langle E^0_{X2}(t,\Psi) \cdot E^0_{Y2}(t,\Psi)^* \rangle,$$
$$J_{yx} = \langle E^0_{Y2}(t,\Psi) \cdot E^0_{X2}(t,\Psi)^* \rangle.$$

Since the time average of the optical power is simply obtained by calculating the intensity profile of each spectral component separately and then integrating the overall intensity profile, after some algebra[9] Eqs. (15)–(18) are obtained.

Where, respectively,

$$f(\tau - \tau_i) = \frac{\{\gamma(\tau_i)\}^{-1/4}}{\Delta\nu} \cdot \exp\left[\frac{-\left(\frac{\pi}{\Delta\nu}\right)^2}{4\gamma(\tau_i)}(\tau - \tau_i)^2\right] \cdot \cos[\phi(\tau - \tau_i)], \tag{A1}$$

$$\phi(\tau - \tau_i) = 2\pi\nu\tau - \Delta\beta_i z_i + \frac{\pi^4 \cdot \frac{d\tau_i}{d\omega} \cdot (\tau - \tau_i)^2}{8\gamma(\tau_i)} - \tfrac{1}{2}\arctan\left[\tfrac{1}{2} \cdot (2\pi\Delta\nu)^2 \cdot \frac{d\tau_i}{d\omega}\right], \tag{A2}$$

where

$$\gamma(\tau_i) = \frac{\pi^4}{4} \cdot \left(\frac{d\tau_i}{d\omega}\right)^2 + \left(\frac{1}{\Delta\nu}\right)^4. \tag{A3}$$

Since $[(d\tau_i)/(d\omega)] < 1 \times 10^{-33}(\mathrm{s}^2)$ in elliptical core fibers[10] and $[(d\tau_i)/(d\omega)] < 1 \times 10^{-28}(\mathrm{s}^2)$ in stress-induced fibers,[11]

$$\gamma(\tau_i) = \left(\frac{1}{\Delta\nu}\right)^4.$$

Then, Eqs. (A1) and (A2) are approximately rewritten as

$$f(\tau - \tau_i) = \exp\left\{-\left(\frac{\Delta\nu\pi}{2}\right)^2 \cdot (\tau - \tau_i)^2\right\} \cdot \cos(2\pi\nu\tau - \Delta\beta_i Z_i)$$
$$(i = 1,2) = \Gamma(\tau - \tau_i) \cdot \cos(2\pi\nu\tau - \Delta\beta_i Z_i),$$
$$\phi(\tau - \tau_i) = 2\pi\nu\tau - \Delta\beta_i Z_i.$$

The author wishes to thank Y. Namihira for his discussion and C. Ota and Y. Iwamoto for their encouragement.

References

1. B. H. Billings, "A Monochromatic Depolarizer," J. Opt. Soc. Am. **41**, 966 (1951).
2. A. P. Loeber, "Depolarization of White Light by a Birefringent Crystal. II. The Lyot Depolarizer," J. Opt. Soc. Am. 72, 650 (1982).
3. K. Böhm, K. Petermann, and E. Weidel, "Performance of Lyot Depolarizers with Birefringent Single-Mode Fibers," IEEE/OSA J. Lightwave Technol. **LT-1**, 71 (1983).
4. W. K. Burns, "Degree of Polarization in the Lyot Depolarizer," IEEE/OSA J. Lightwave Technol. **LT-1**, 475 (1983).
5. H. G. Danielmeyer and H. P. Weber, "Direct Measurement of the Group Velocity of Light," Phys. Rev. A **3**, 1708 (1971).
6. M. Born, and E. Wolf, *Principles of Optics* (Pergamon, New York, 1975).
7. J. Sakai, S. Machida, and T. Kimura, "Degree of Polarization in Anisotropic Single-Mode Optical Fibers: Theory," IEEE J. Quantum Electron. **QE-18**, 488 (1982).
8. R. P. Moeller and W. K. Burns, "Depolarised Broadband Source," Electron. Lett. **19**, 187 (1983).
9. R. Cubeddu and O. Svelto, "Effect of Dispersion on Laser Self-Locking," Phys. Lett. A **29**, 78 (1969).
10. K. Okamoto, T. Hosaka, and Y. Sasaki, "Linearly Single Polarization Fibers with Zero Polarization Mode Dispersion," IEEE J. Quantum Electron. **QE-18**, 496 (1982).
11. K. Okamoto, T. Edahiro, and N. Shibata, "Polarization Properties of Single-Polarization Fibers," Opt. Lett. **7**, 569 (1982).

New diagrammatical method for calculation of fiber-optic Lyot depolarizer performance

Kazumasa Takada, Kazunori Chida, and Juichi Noda

Electrical Communications Laboratories, Nippon Telegraph and Telephone Corporation, Tokai, Ibaraki-ken, 319-11, Japan

Received June 10, 1987; accepted November 30, 1987

We present a calculation for the performance of a fiber-optic Lyot depolarizer in the first-order approximation. To ensure a straightforward calculation, we developed a new procedure with diagrams representing light propagation. The complicated expression for the degree of polarization of the depolarizer output is expressed by the propagation diagrams in a simple form by using a few formulas about the diagrams. The degree of polarization of linearly polarized incident light is calculated with an arbitrary incidence angle and with an arbitrary rotation angle between the first (shorter) fiber and the second (longer) fiber constructing the depolarizer. Especially for an incidence of nearly 45°, it is clarified by calculation that the residual polarization depends solely on the polarization coupling existing in the longer fiber and does not depend on that existing in the shorter fiber. Comparisons of calculated and measured results are also presented.

1. INTRODUCTION

A fiber-optic Lyot depolarizer consists of two birefringent fibers with a 2:1 length ratio and with their principal axes at a 45° angle to each other. This kind of depolarizer is attractive for fiber-optic gyroscopes because no polarization elements are required within the fiber loop.[1,2]

Several authors[2,3] have calculated depolarizer performance, i.e., the residual polarization obtained with birefringent fibers. Böhm *et al.*,[2] for example, calculated performance by taking the distributed polarization-coupling effect in birefringent fibers into consideration. However, they did not calculate the degree of polarization strictly according to its definition. The reason may be that the calculation of the degree of polarization is complicated and tedious even in the first-order approximation.

With the perturbation method, many interference terms appear in the mathematical representation of the degree of polarization. There are two main origins of error in calculation: (1) The signs of individual terms are frequently mistaken, leading to inadvertent cancellation of comparable terms. (2) Almost all terms depend on the coherence length of the light. It is rather tedious and difficult to find the terms that are negligible under the short-coherence-length condition.

In this paper we present a calculation of the performance of a fiber-optic Lyot depolarizer in the first-order approximation. To ensure a straightforward calculation, we developed a new procedure with diagrams representing light propagation. The degree of polarization of the depolarizer output is expressed by the propagation diagrams in a simple form by using a few formulas around the diagrams. Furthermore, the optical path difference between two light waves constituting the interference terms is obtained easily by using the diagrams. As a result, the terms that are negligible under the low-coherence condition are also found.

2. CALCULATION OF DEGREE OF POLARIZATION

A. Formulation of Light Propagation in a Birefringent Fiber by a Diagrammatic Procedure

It is assumed that the dielectric tensor in a birefringent fiber is changed randomly along the fiber by internal and/or external perturbations. When the tensor change is relatively minor, the electromagnetic fields $\{\tilde{E}(r, t), \tilde{H}(r, t)\}$ in the perturbed system can be represented by the unperturbed principal modes $\{E_\mu(r), H_\mu(r)\}$ as

$$\tilde{E}(r, t) = \sum_{\mu=x,y} C_\mu(z, t)E_\mu(r),$$

$$\tilde{H}(r, t) = \sum_{\mu=x,y} C_\mu(z, t)H_\mu(r), \tag{1}$$

where $C_\mu(z, t)$ denotes the analytic signal of the μ (=x or y) mode at a fiber length z. Since the electromagnetic field $\{\tilde{E}(r, t), \tilde{H}(r, t)\}$ in perturbed fiber is assumed to be quasi-monochromatic, the analytic signal $C_\mu(z, t)$ is expressed by means of the amplitude spectrum $V_\mu(\omega, z)$ as

$$C_\mu(z, t) = \int_0^\infty V_\mu(\omega, z)e^{j\omega t}d\omega. \tag{2}$$

By substituting Eqs. (1) into the Maxwell equations, we obtain[4]

$$\frac{dV_x(\omega, z)}{dz} + j\tilde{\beta}_x(\omega)V_x(\omega, z) = -j\Gamma(z)V_y(\omega, z),$$

$$\frac{dV_y(\omega, z)}{dz} + j\tilde{\beta}_y(\omega)V_y(\omega, z) = -j\Gamma^*(z)V_x(\omega, z), \tag{3}$$

where $\tilde{\beta}_{x,y}(\omega)$ are the propagation constants of the two principal $HE_{11}{}^{x,y}$ modes and $\Gamma(z)$ is the coupling coefficient between the two principal modes. We consider only a birefringent fiber with low polarization cross talk, i.e., $|\Gamma| \ll \tilde{\beta}_{x,y}(\omega)$; that is, the first-order approximation will suffice for our analysis in the minimal-polarization-coupling scenario.

Equations (3) can be solved by using the Born approximation of general perturbation theory to give the solution with the zeroth- and first-order perturbations of the coupling coefficient $\Gamma(z)$:

$$V_\mu(\omega, z) = V_\mu^{(0)}(\omega, z) + V_\mu^{(1)}(\omega, z) + \ldots \quad (\mu = x, y). \quad (4)$$

Here $V_\mu^{(0)}(\omega, z)$ and $V_\mu^{(1)}(\omega, z)$ are the zeroth- and first-order solutions of the perturbation. Substitution of Eq. (4) into Eqs. (3) yields

$$\frac{dV_\mu^{(0)}(\omega, z)}{dz} + j\tilde{\beta}_\mu(\omega)V_\mu^{(0)}(\omega, z) = 0 \quad (\mu = x, y) \quad (5)$$

and

$$\frac{dV_x^{(1)}(\omega, z)}{dz} + j\tilde{\beta}_x(\omega)V_x^{(1)}(\omega, z) = -j\Gamma(z)V_y^{(0)}(\omega, z),$$

$$\frac{dV_y^{(1)}(\omega, z)}{dz} + j\tilde{\beta}_y(\omega)V_y^{(1)}(\omega, z) = -j\Gamma^*(z)V_x^{(0)}(\omega, z). \quad (6)$$

From Eq. (5) we obtain

$$V_\mu^{(0)}(\omega, z) = V_\mu^{(0)}(\omega)\exp[-j\tilde{\beta}_\mu(\omega)z] \quad (\mu = x, y), \quad (7)$$

where $V_\mu^{(0)}(\omega)$ is the Fourier coefficient of the input electrical field. After substituting Eq. (7) into Eq. (2), we obtain the zeroth-order analytic signal at the output end of a fiber of length L:

$$C_\mu^{(0)}(L, t) = \hat{C}_\mu \exp[j(\omega_0 t - \beta_\mu L)]F\left(t - \frac{L}{v_\mu}\right) \quad (\mu = x, y), \quad (8)$$

where $\hat{C}_\mu$ is the amplitude constant and ω_0 is the central angular light frequency. The propagation constant $\tilde{\beta}_\mu(\omega)$ is expanded into a first-order derivation as $\tilde{\beta}_\mu(\omega) = \beta_\mu + (\omega - \omega_0)/v_\mu$, with v_μ as the group velocity of the μ mode. $F(t)\exp(j\omega_0 t)$ is the analytic signal of the incident electrical field, and $\langle F(t)F^*(t + \tau)\rangle$ is normalized to unity at $\tau = 0$, where $\langle\ \rangle$ represents the time average.

Schematic diagrams of light propagation represented by zeroth-order solutions are given as Figs. 1(a) and 1(b) for the principal $HE_{11}{}^x$ and $HE_{11}{}^y$ modes, respectively, excited at the input end of the fiber. The principal $HE_{11}{}^x$ and $HE_{11}{}^y$ modes are represented by two transmission lines, and light propagation is indicated by the arrow. As light propagates through the fiber in the $HE_{11}{}^x$ and $HE_{11}{}^y$ modes, the phases $\beta_x L$ and $\beta_y L$ and the transit times L/v_x and L/v_y, respectively, are accumulated.

By inserting Eq. (7) into the right-hand side of Eq. (6), we obtain the first order solutions to the analytic signals:

$$C_x^{(1)}(L, t) = -j\hat{C}_y \int_0^L \Gamma(z)\exp\{j\omega_0 t - j[\beta_y z + \beta_x(L - z)]\} \times F\left(t - \frac{z}{v_y} - \frac{L - z}{v_x}\right)dz,$$

$$C_y^{(1)}(L, t) = -j\hat{C}_x \int_0^L \Gamma^*(z)\exp\{j\omega_0 t - j[\beta_x z + \beta_y(L - z)]\} \times F\left(t - \frac{z}{v_x} - \frac{L - z}{v_y}\right)dz. \quad (9)$$

Figures 1(c) and 1(d) are schematic diagrams of light propagation represented by first-order solutions when the principal $HE_{11}{}^y$ and $HE_{11}{}^x$ modes, respectively, are excited at the input end of the fiber. At any point z_i in the fiber, a small portion of the propagating light in the $HE_{11}{}^y$ and $HE_{11}{}^x$ modes is coupled with cross-polarized $HE_{11}{}^x$ and $HE_{11}{}^y$ modes at amplitude ratios of $-j\Gamma(z)$ and $-j\Gamma^*(z)$. The coupled light from the $HE_{11}{}^y$ to the $HE_{11}{}^x$ mode and vice versa accumulates phases of $\beta_y z_i$ and $\beta_x z_i$ after propagating from the input end to the perturbed point z_i and accumulates the phases $\beta_x(L - z_i)$ and $\beta_y(L - z_i)$, respectively, after

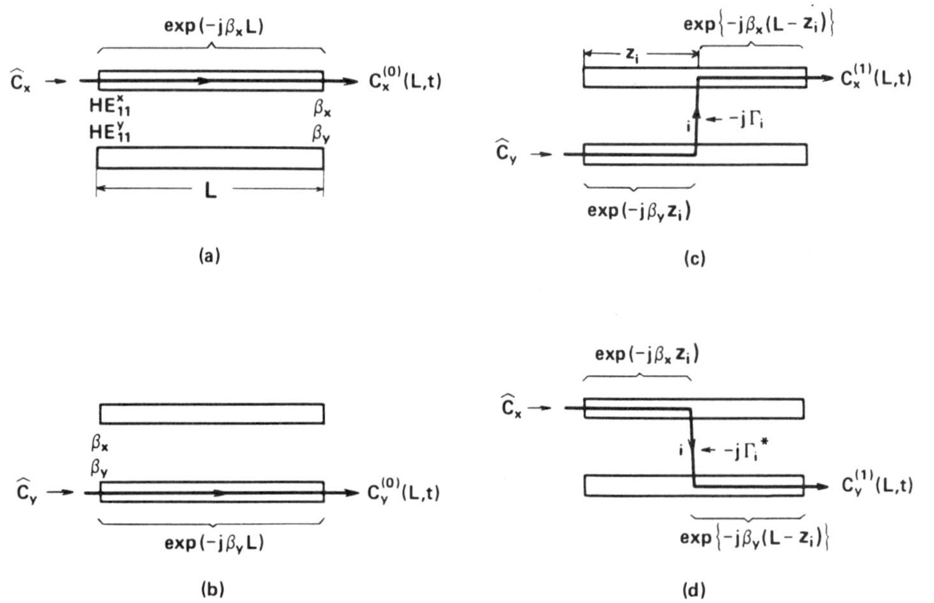

Fig. 1. Models of light propagation represented by the zeroth-order solution when principal (a) $HE_{11}{}^x$ and (b) $HE_{11}{}^y$ modes are excited at the input end of the fiber. (c), (d) Are models of polarization coupling represented by the first-order solution when the principal $HE_{11}{}^y$ and $HE_{11}{}^x$ modes, respectively, are excited at the input end of the fiber.

$$\equiv \exp(-j\beta_x L)F\left(t-\frac{L}{v_x}\right),$$

$$\equiv \exp(-j\beta_y L)F\left(t-\frac{L}{v_y}\right), \qquad (10)$$

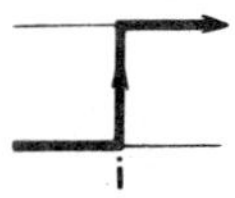

$$\equiv -j\int_0^L \Gamma(z_i)\exp[-j\{\beta_y z_i+\beta_x(L-z_i)\}]F\left(t-\frac{z_i}{v_y}-\frac{L-z_i}{v_x}\right)dz_i$$

$$=-j\sum_i \Gamma_i \exp[-j\{\beta_y z_i+\beta_x(L-z_i)\}]F\left(t-\frac{z_i}{v_y}-\frac{L-z_i}{v_x}\right),$$

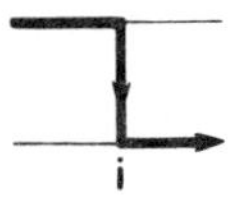

$$\equiv -j\int_0^L \Gamma^*(z_i)\exp[-j\{\beta_x z_i+\beta_y(L-z_i)\}]F\left(t-\frac{z_i}{v_x}-\frac{L-z_i}{v_y}\right)dz_i$$

$$=-j\sum_i \Gamma_i^* \exp[-j\{\beta_x z_i+\beta_y(L-z_i)\}]F\left(t-\frac{z_i}{v_x}-\frac{L-z_i}{v_y}\right). \qquad (11)$$

Fig. 2. Diagrammatic representations of zeroth- and first-order solutions.

propagating from the perturbed point z_i to the output end. Together with the phases, the coupled light also accumulates the transit times through the fiber and from the input end to the perturbed point z_i (z_i/v_y and z_i/v_x) and from the point z_i to the output end [$(L - z_i)/v_x$ and $(L - z_i)/v_y$]. The output light represented by the first-order solution is only the summation of all the cross-coupled lights at any perturbed points in the fiber.

On the basis of this intuitive understanding, we shall illustrate the zeroth- and first-order solutions [Eqs. (8) and (9)] by the diagrams shown in Fig. 2, in which the common phase factor $\exp(-j\omega_0 t)$ is omitted for simplicity. The subscript i of Eqs. (11) in Fig. 2 denotes the summation over the perturbed points causing polarization coupling. The diagrams of Eqs. (10) in Fig. 2 denote the analytic signal transmitted through the principal $HE_{11}{}^x$ and $HE_{11}{}^y$ modes when a signal with unit amplitude is excited at the input end of the fiber in the $HE_{11}{}^x$ and $HE_{11}{}^y$ modes, respectively. Also, Eqs. (11) in Fig. 2 denote coupled analytic signals from the $HE_{11}{}^y$ mode to the $HE_{11}{}^x$ mode and from the $HE_{11}{}^x$ mode to the $HE_{11}{}^y$ mode when a signal with unit amplitude is excited at the input end in the $HE_{11}{}^y$ and $HE_{11}{}^x$ modes, respectively. Here it should be pointed out that the boldface arrow is used in the propagation diagrams only so that the reader can clearly understand the light propagation. When we perform calculations, a different representation of light propagation can be used to simplify our diagrammatic description.

By using these diagrams, analytic signals of two principal modes at the output end of the fiber can be represented by Eqs. (12) in Fig. 3, in which the coefficient shown in the upper part of each diagram is multiplied by the value represented by the diagram, as illustrated by Eq. (13a) in Fig. 4. The degree of polarization can be calculated by the coherence matrix whose elements consist of products of the orthogonal components of the electrical fields.[5] Therefore the scalar product of the terms Ω_1 and Ω_2 is defined as

$$(\Omega_1, \Omega_2) = \langle \Omega_1 \times \Omega_2{}^* \rangle,$$

and a few formulas about the scalar product can be derived. The formula (13b) in Fig. 4 is obtained initially. From the definition, Eq. (14a) in Fig. 4 is established, where $R(\tau)$ is $\langle F(t)F^*(t+\tau)\rangle$ and $D = 1/v_x - 1/v_y$ is the polarization mode dispersion per unit length of fiber. On the other hand, Eq. (14b) in Fig. 4 is also obtained; here the relation $R(-\tau) = R^*(\tau)$ is used.[5] Therefore formula (14c) in Fig. 4 is obtained. Equation (14c) indicates that the complex conjugate of the scalar product of the zeroth- and first-order solutions is equivalent to the scalar product multiplied by -1 in which the two principal modes in the propagation diagrams are interchanged.

A schematic diagram of a depolarizer consisting of two birefringent fibers (1 and 2) with their principal axes at an angle $\beta = 45°$ with respect to each other is shown in Fig. 5. It is assumed that linearly polarized light with unit amplitude is launched into the input end of fiber 1 at an angle α with respect to the principal $HE_{11}{}^x$ axis. Transmission lines in the depolarizer are denoted by $\boxed{X}^\beta$, where $\boxed{X}^\beta$ repre-

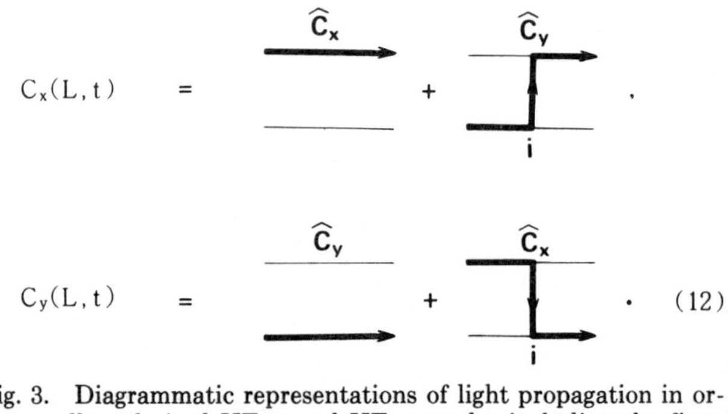

Fig. 3. Diagrammatic representations of light propagation in orthogonally polarized $HE_{11}{}^x$ and $HE_{11}{}^y$ modes including the first-order solution.

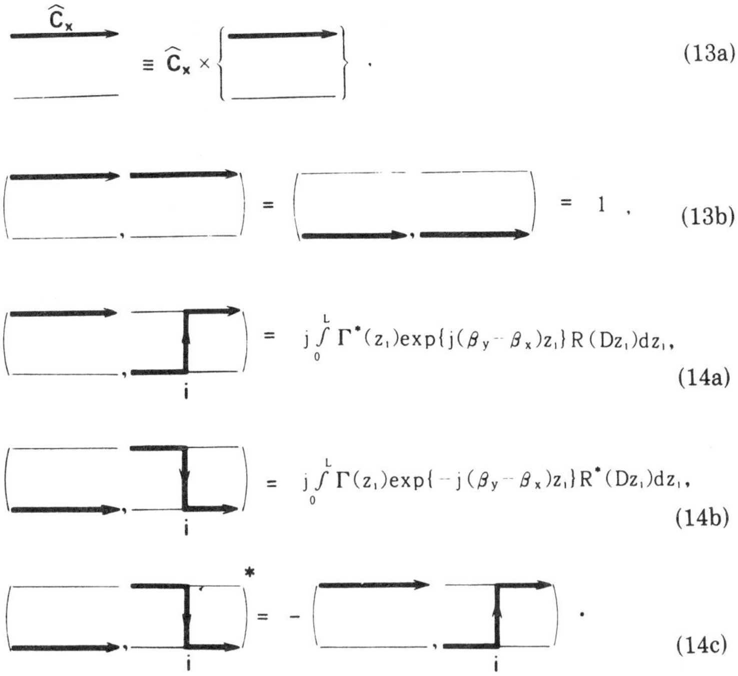

Fig. 4. Relations between diagrammatic scalar products [Eqs. (13) and (14)].

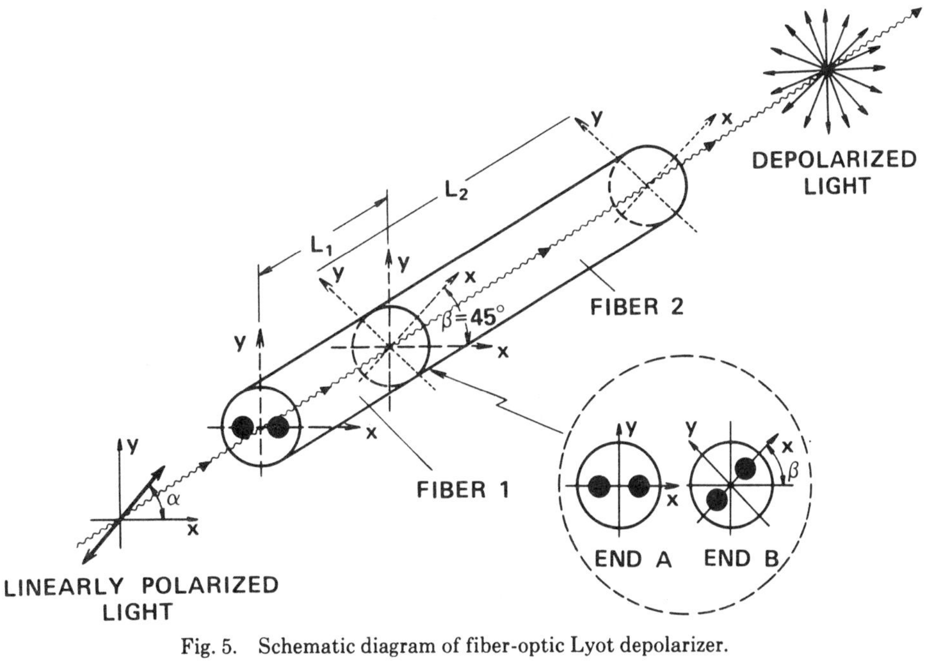

Fig. 5. Schematic diagram of fiber-optic Lyot depolarizer.

sents the joint of two fibers with their principal axes rotating at an angle β from each other. The analytic signal at the output end of the fiber can be determined easily as follows. In the case of the zeroth-order solution of the $HE_{11}{}^x$ mode, there are two kinds of light propagation in which the output field is in the $HE_{11}{}^x$ mode. One is the propagation in the $HE_{11}{}^x$ mode in both fiber 1 and fiber 2. The other is the propagation in the $HE_{11}{}^y$ mode of fiber 1 and then in the $HE_{11}{}^x$ mode of fiber 2, where the polarization is changed from the $HE_{11}{}^y$ mode to the $HE_{11}{}^x$ mode at the joint.

Similarly, for the first-order solution, there are four kinds of light propagation in which the output field is in the $HE_{11}{}^x$ mode. Two of them correspond to propagation in which polarization coupling occurs (1) from the $HE_{11}{}^y$ mode to the $HE_{11}{}^x$ mode and (2) from the $HE_{11}{}^x$ mode to the $HE_{11}{}^y$ mode at any points in fiber 1 while polarization coupling does not occur in fiber 2. In the latter propagation, the polarization is changed from the $HE_{11}{}^y$ mode to the $HE_{11}{}^x$ mode at the joint. The other kinds of propagation are (3) through the $HE_{11}{}^x$ mode and (4) through the $HE_{11}{}^y$ mode in fiber 1 without polarization coupling and with polarization coupling from the $HE_{11}{}^y$ mode to the $HE_{11}{}^x$ mode in fiber 2. In propagation (3) the polarization is changed from the $HE_{11}{}^x$ mode to the $HE_{11}{}^y$ mode at the joint. Therefore the analytic signal $C_x(L, t)$ can be represented easily by Eq. (15), shown in Fig. 6. In Eq. (15), polarization couplings at the joint are denoted by ⊻ and ⊼ , which correspond to coupling from the $HE_{11}{}^x$ mode to the $HE_{11}{}^y$ mode and from the $HE_{11}{}^y$

mode to the $HE_{11}{}^x$ mode, respectively. Because the principal $HE_{11}{}^x$ axis of fiber 2 is rotated by an angle β with respect to that of fiber 1, as shown in Fig. 5, coupling coefficients at the joint are $\cos\beta$, $\sin\beta$, $-\sin\beta$, and $\cos\beta$ for couplings from $HE_{11}{}^x$ to $HE_{11}{}^y$, from $HE_{11}{}^y$ to $HE_{11}{}^x$, from $HE_{11}{}^x$ to $HE_{11}{}^y$, and from $HE_{11}{}^y$ to $HE_{11}{}^y$, respectively. These coefficients are elements of the two-dimensional rotation matrix

$$\begin{bmatrix} \cos\beta & \sin\beta \\ -\sin\beta & \cos\beta \end{bmatrix}.$$

On the other hand, the principal $HE_{11}{}^x$ axis is rotated α de relative to the polarization direction of the linearly polarize light so that the amplitudes at the input end are $\cos\alpha$ an $\sin\alpha$ for the $HE_{11}{}^x$ and $HE_{11}{}^y$ modes, respectively. There fore the coefficients shown in the upper parts of the indivic ual diagrams of Eq. (15) are determined easily by the propa gation diagram shown in Fig. 6.

Similarly, the analytic signal $C_y(L, t)$ at the output en can be transcribed as the diagrammatic representation [E (16)] of Fig. 6. Clearly, the formulas given in Eqs. (13) an

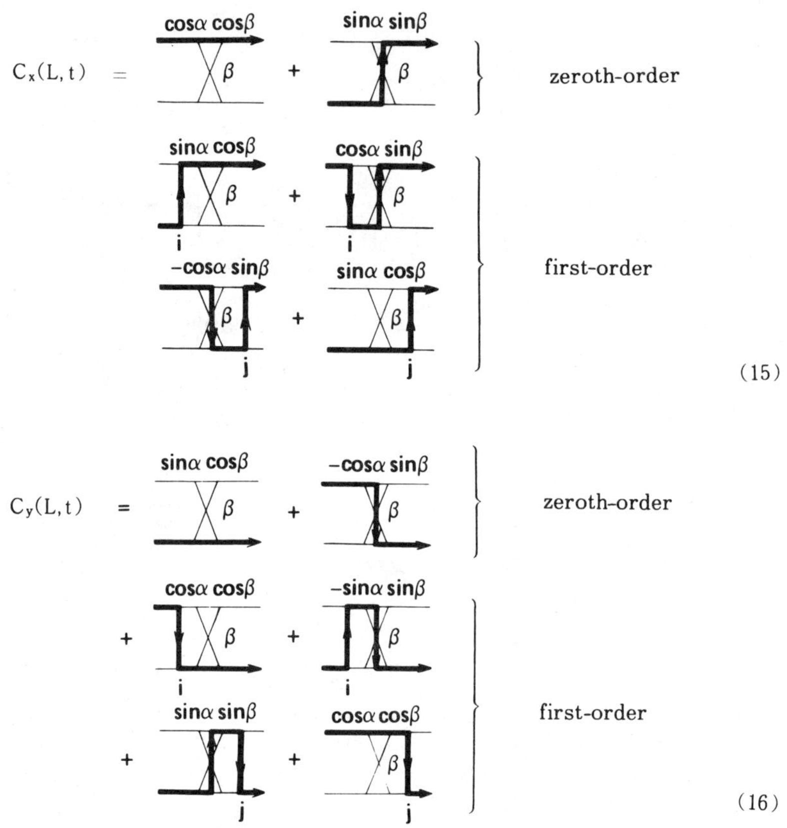

Fig. 6. Diagrammatic representations of analytical signals $C_x(L, t)$ and $C_y(L, t)$, including zeroth- and first-order solutions at the output end the fiber.

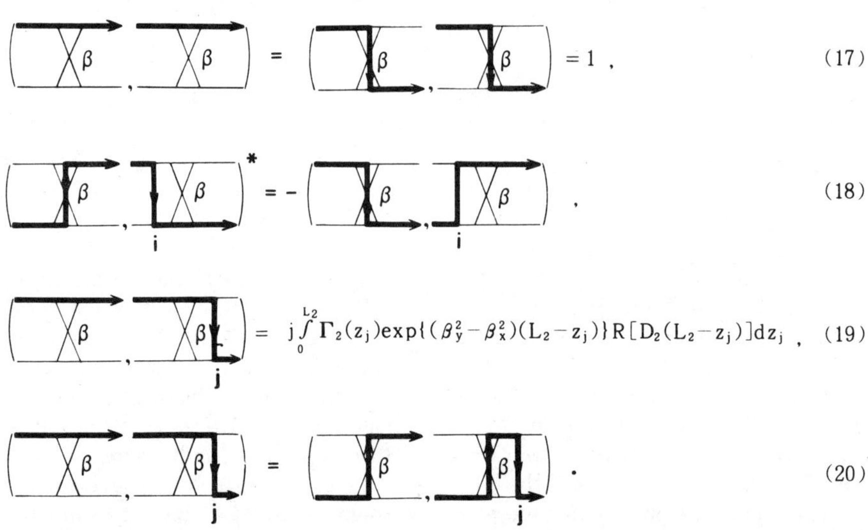

Fig. 7. Relations between diagrammatic scalar products [Eqs. (17)–(20)].

(14) are also extended in the depolarizer construction. For example, Eqs. (17) and (18) in Fig. 7 are established. On the other hand, Eq. (19) in Fig. 7 is obtained, where L_k, $\beta_{x,y}{}^k$, and D_k are the length, the propagation constant, and the polarization mode dispersion per unit length of fiber k ($k = 1, 2$), respectively. $\Gamma_k(z)$ is the coupling coefficient in fiber k. Equation (19) indicates that the scalar product between the zeroth- and first-order solutions transmitted through the same optical path in fiber 1 is independent of the parameters of fiber 1 (e.g., fiber length, propagation constant). It is dependent only on the extent to which each light propagates in fiber 2. Therefore Eq. (20) in Fig. 7 is established.

B. Degree of Polarization in One Birefringent Fiber

We calculate the degree of polarization of the fiber output by the proposed diagrammatic method. The degree of polarization is denoted by[5]

$$\hat{P} = \left[\frac{(J_{xx} - J_{yy})^2 + 4|J_{xy}|^2}{(J_{xx} + J_{yy})^2}\right]^{1/2}, \tag{21}$$

where $J_{\mu\nu}$ ($\mu, \nu = x, y$) are elements of the coherence matrix J:

$$J = \begin{bmatrix} \langle C_x C_x{}^* \rangle & \langle C_x C_y{}^* \rangle \\ \langle C_y C_x{}^* \rangle & \langle C_y C_y{}^* \rangle \end{bmatrix}.$$

We introduce $J_{\mu\nu}{}^{(0)}$ and $J_{\mu\nu}{}^{(1)}$, defined by the zeroth- and first-order solutions of coherence matrix elements $J_{\mu\nu}$, i.e., $J_{\mu\nu} = J_{\mu\nu}{}^{(0)} + J_{\mu\nu}{}^{(1)} + \ldots$, to calculate the zeroth- and first-order solutions of the degree of polarization $\hat{P}^{(0)}$ and $\hat{P}^{(1)}$. From Eqs. (12) we obtain Eqs. (22) in Fig. 8. In deriving Eqs. (22), Eq. (13) was used. $\hat{C}_x$ and $\hat{C}_y$ are replaced by cos α and sin α, respectively. This is because linearly polarized light with unit amplitude is launched into the fiber with a polarizer direction set at an angle α relative to the principal $HE_{11}{}^x$ mode of the fiber. The appearance of a term for the degree of coherence $R(DL)$ in Eqs. (22) is understood easily because the optical path difference between two light waves expressed by the diagrams in Eqs. (22) is DL. From Eqs. (22), the zeroth-order solution is obtained by

$$\hat{P}^{(0)} = |R(DL)|,$$

with $\alpha \cong 45°$. To obtain the degree of polarization limited by polarization coupling, the zeroth-order solution of the degree of polarization $\hat{P}^{(0)}$ should be less than the first-order solution $\hat{P}^{(1)}$; that is,

$$|R(DL)| \ll \hat{P}^{(1)}. \tag{23}$$

This is the condition required for the fiber.

Under the condition represented by relation (23), the first-order solution is dominant and can be obtained as follows. From Eqs. (12), Eqs. (24) and (25) in Fig. 9 are established, in which C.C. denotes the complex conjugate of (,). Since Eq. (26a) in Fig. 10 is obtained from Eq. (14), Eq. (26b) in Fig. 10 is established. Equation (24) then leads to Eq. (27) in Fig. 11. By means of Eq. (14), Eq. (25) is transformed into Eq. (28) in Fig. 11, whose value becomes zero with $\alpha \cong 45°$. Equation (27) and relation (28) indicate that polarization coupling close to the output end of the fiber contributes to the degree of polarization only because of the relatively short coherence length of the light, as represented by relation (23).

When $\Gamma(z)$ changes rapidly along the fiber [compared with $R(Dz)$], as may occur in perturbations such as microbending, $J_{xx}{}^{(1)} - J_{yy}{}^{(1)}$ can be approximated as

$$J_{xx}{}^{(1)} - J_{yy}{}^{(1)} = j\sum_{i=1}^{N} R(Dz_i)\int_{L_{i-1}}^{L_i} \Gamma^*(z_i)\exp[j(\beta_y - \beta_x)z_i]dz_i + \text{c.c.},$$

where the total fiber length L is divided into regions (L_{i-1}, L_i) ($i = 1, N$) and where the value $R(Dz)$ is assumed to be constant in the individual region. Therefore, by averaging statistically equivalent fibers, we can represent a first-order solution of the degree of polarization by

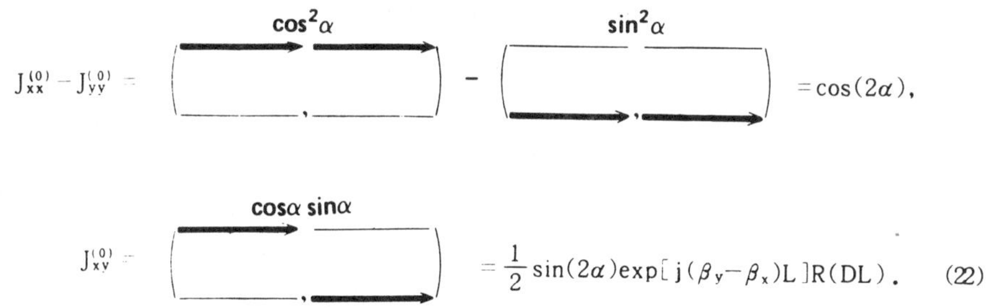

Fig. 8. Diagrammatic representations of $J_{xx}{}^{(0)} - J_{yy}{}^{(0)}$ and $J_{xy}{}^{(0)}$.

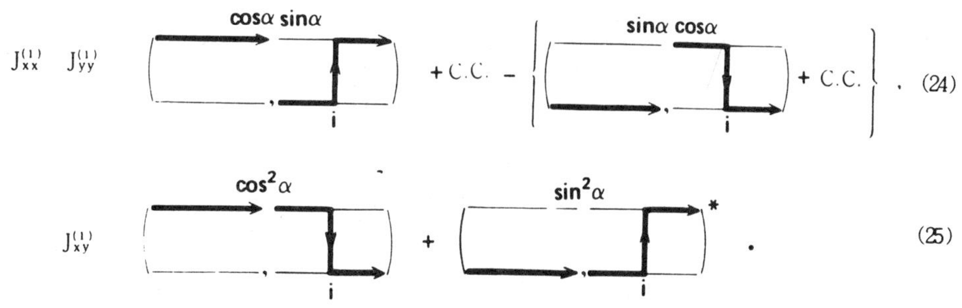

Fig. 9. Diagrammatic representations of $J_{xx}{}^{(1)} - J_{yy}{}^{(1)}$ and $J_{xy}{}^{(1)}$.

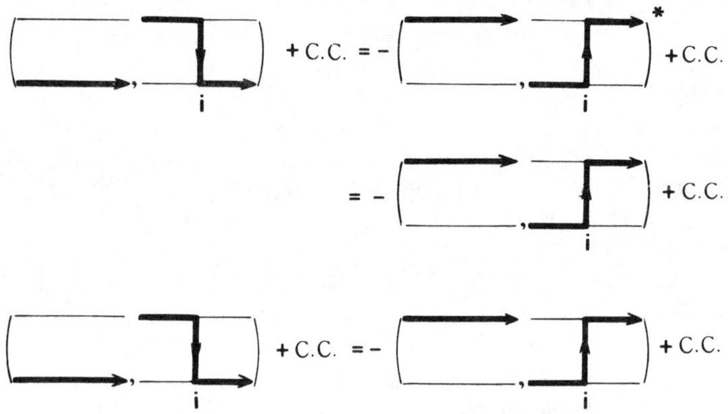

(26a)

(26b)

Fig. 10. Relations between diagrammatic scalar products [Eqs. (26)].

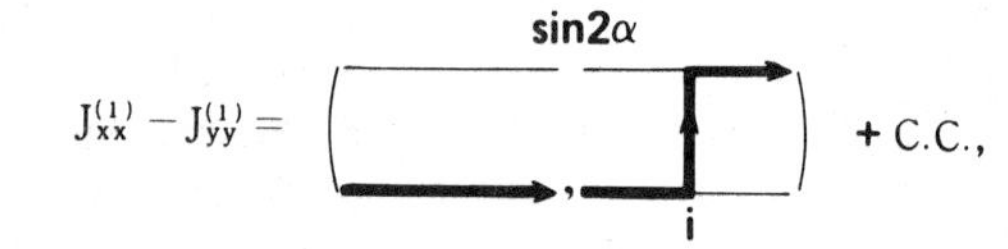

$$= j\int_0^L \Gamma^*(z_i)\exp[j(\beta_y-\beta_x)z_i]R(Dz_i)dz_i + \mathrm{C.C.} \tag{27}$$

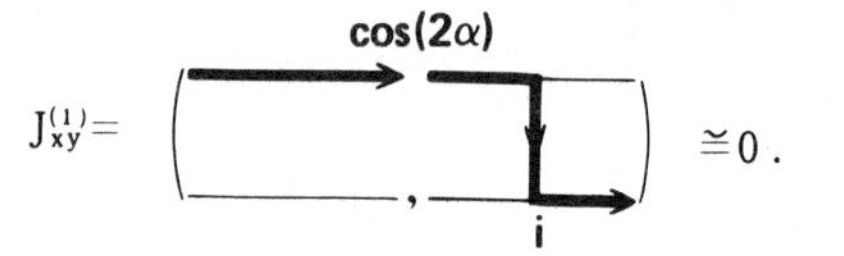

(28)

Fig. 11. Exact representations of $J_{xx}{}^{(1)} - J_{yy}{}^{(1)}$ and $J_{xy}{}^{(1)}$ obtained by diagrammatic transformation.

$$\hat{P}^{(1)} = \overline{[(J_{xx} - J_{yy})^2]}^{1/2}$$

$$= \left\{2\sum_{i=1}^{N} R^2(Dz_i)\overline{\left|\int_{L_{i-1}}^{L_i} \Gamma(z)\exp[j(\beta_x-\beta_y)z]dz\right|^2}\right\}^{1/2}$$

$$= \left[2\sum_{i=1}^{N} R^2(Dz_i)h\Delta z_i\right]^{1/2} = \left[2h\int_0^{\infty} R^2(Dz)dz\right]^{1/2}, \tag{29}$$

where the ensemble average is denoted by an overbar and h is a polarization-coupling parameter of the birefringent fiber defined by[6]

$$h = \frac{1}{L}\overline{\left|\int_0^L \Gamma(z)\exp[(\beta_x-\beta_y)]dz\right|^2}.$$

C. Degree of Polarization in a Lyot Depolarizer

From Eqs. (15) and (16), $J_{xx}{}^{(0)}$ and $J_{yy}{}^{(0)}$ can be represented by Eqs. (30a) and (30b), respectively, in Fig. 12. Since the equation shown in Fig. 13 is found by analogy with Eq. (26), the last terms in Eqs. (30a) and (30b) are equal. Therefore, with Eq. (17), $J_{xx}{}^{(0)} - J_{yy}{}^{(0)}$ leads to Eq. (31) in Fig. 14. $J_{xy}{}^{(0)}$ is denoted by Eq. (32) in Fig. 14. To attain the degree of polarization limited by fiber polarization coupling, the zeroth-order solution $\hat{P}^{(0)}$ should be less than the first-order solution $\hat{P}^{(1)}$; that is, $\hat{P}^{(0)} \ll \hat{P}^{(1)}$. Because of the technical limit of $\beta = 45°$ alignment accuracy with a splicing unit, it is rather difficult to decrease the value of $J_{xx}{}^{(0)} - J_{yy}{}^{(0)}$ represented by Eq. (31) by less than $\hat{P}^{(1)}$ ($\ll 1$). Therefore it is also necessary to align the incident angle α to 45° as precisely as possible to minimize the degree of polarization.

From Eqs. (31) and (32), the zeroth-order solution satisfies

$$\hat{P}^{(0)} = 2|J_{xy}{}^{(0)}| < \tfrac{1}{2}[|R(D_2L_2 + D_1L_1)| + 2|R(D_2L_2)| + |R(D_2L_2 - D_1L_1)|],$$

with $\alpha \cong \beta \cong 45°$, where the condition $D_2L_2 = 2D_1L_1$ is required for Lyot depolarizers. The conditions $|R(D_2L_2 + D_1L_1)| = |R(3D_1L_1)| < |R(D_1L_1)|$ and $|R(D_2L_2)| < |R(D_1L_1)|$ are then satisfied because $R(\tau)$ decreases as the time delay τ increases. Therefore the condition required for the fiber is

$$2|R(D_1L_1)| \ll \hat{P}^{(1)}. \tag{33}$$

Under this condition, the first-order solution is dominant.

We now derive $J_{xx}{}^{(1)} - J_{yy}{}^{(1)}$ and $J_{xy}{}^{(1)}$. First, $J_{xx}{}^{(1)}$ is expressed by Eq. (34) in Fig. 15. This complicated equation is transformed into a simple form by using Eqs. (18) and

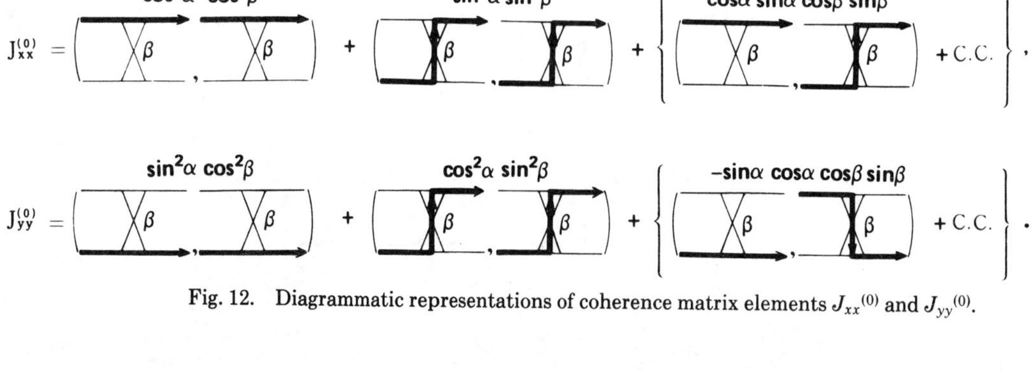

(30a)

(30b)

Fig. 12. Diagrammatic representations of coherence matrix elements $J_{xx}{}^{(0)}$ and $J_{yy}{}^{(0)}$.

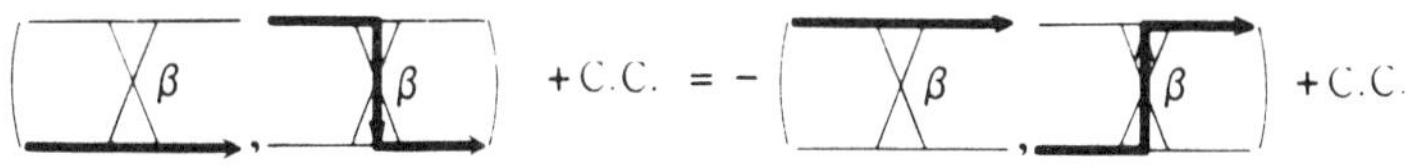

Fig. 13. Relation between diagrammatic scalar products.

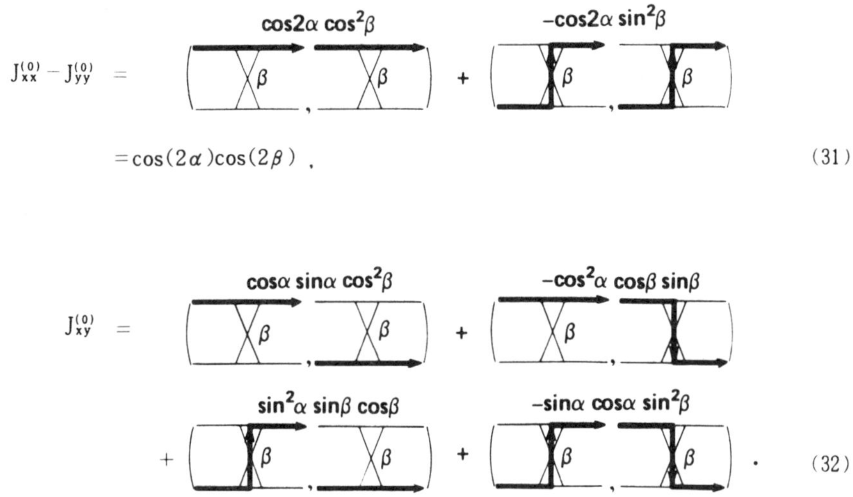

Fig. 14. Diagrammatic representations of $J_{xx}^{(0)} - J_{yy}^{(0)}$ and $J_{xy}^{(0)}$.

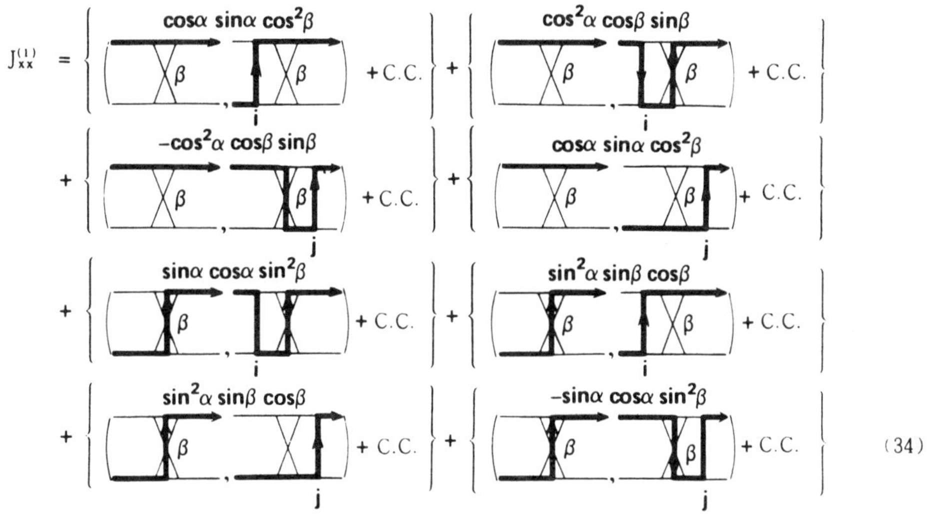

Fig. 15. Diagrammatic representation of coherence matrix $J_{xx}^{(1)}$.

(20); that is, Eq. (35b) in Fig. 16 is derived because the first term in line 3 of Eq. (34) can be transformed into Eq. (35a) in Fig. 16. In the same way, we obtain Eq. (35c) in Fig. 16. Furthermore, with Eq. (20), Eq. (35d) in Fig. 16 is found. Therefore, in Eq. (34), the first terms in lines 1 and 3 and in lines 2 and 4 and the second terms in lines 1 and 3 are subtracted from each other. $J_{xx}^{(1)}$ finally becomes Eq. (36) in Fig. 17. Similarly, we obtain Eq. (37) in Fig. 17.

By analogy with Eqs. (35b) and (35d), the relations represented by the diagrams in Fig. 18 are determined. Therefore, by subtracting Eq. (37) from Eq. (36), we obtain Eq. (38) in Fig. 19. Because the group-delay time difference between the two light waves involved in the fourth term in Eq. (38) is larger than D_1L_1, as is clearly seen from the propagation diagram itself, the fourth term can be neglected for condition (33). Equation (38) ultimately leads to Eq. (39) of Fig. 19, in which the first term is neglected for the depolarizer condition of $\beta \cong 45°$.

$J_{xy}^{(1)}$ is denoted by Eq. (40) in Fig. 20. From Eq. (18), individual terms in lines 1 and 2 and in lines 3 and 4 in Eq. (40) can be subtracted to give Eq. (41) in Fig. 20. Because group-delay time differences between light waves involved in the first and second terms in lines 1 and 2 and in the fourth term in line 1 in Eq. (41) are all larger than D_1L_1, as is evident from the propagation diagrams themselves, these terms can be neglected according to condition (33). Furthermore, the third term in line 1 and the fourth term in line 2 in Eq. (41) can be subtracted to give Eq. (42) in Fig. 20, in which the first term is neglected for the depolarizer condition $\beta \cong 45°$.

For $\alpha \cong 45°$ together with $\beta \cong 45°$, we obtain Eqs. (43) as shown in Fig. 21. Only polarization coupling that is close to $z_j = (D_1/D_2)L_1$ and $z_j = L_2 - (D_1/D_2)L_1$ in fiber 2 is dominant for $\alpha \cong 45°$ incidence as explained by Eq. (43). This result is not described in Ref. 2 and is therefore (to our knowledge) a new result. By a method similar to the derivation of Eq. (29), the residual polarization of a Lyot depolarizer for nearly 45° ($\alpha \cong 45°$) incidence can be obtained by

$$\hat{P}^{(1)} = \left[\frac{3}{2} h_2 \int_0^{\infty} R^2(D_2 z)\mathrm{d}z\right]^{1/2}, \tag{44}$$

where h_2 is the polarization-coupling parameter of fiber 2.

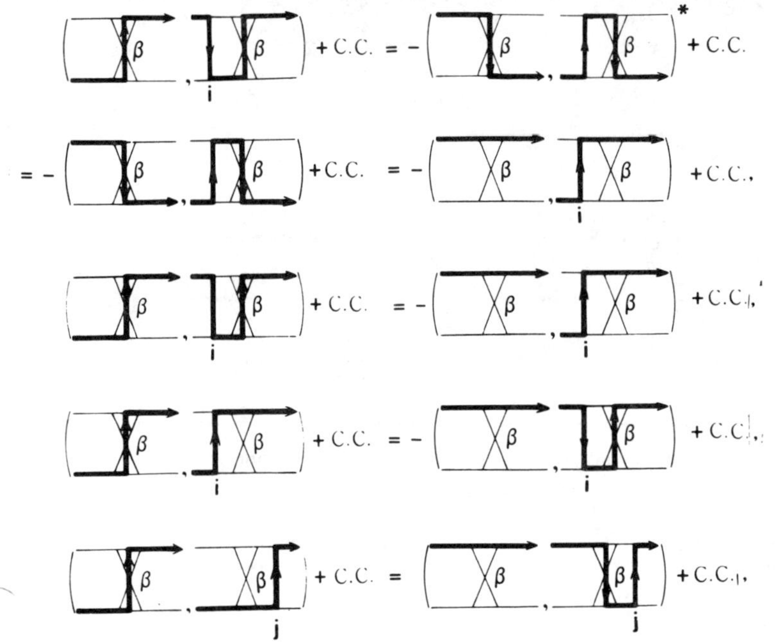

(35a)

(35b)

(35c)

(35d)

Fig. 16. Relations between diagrammatic scalar products [Eqs. (35)].

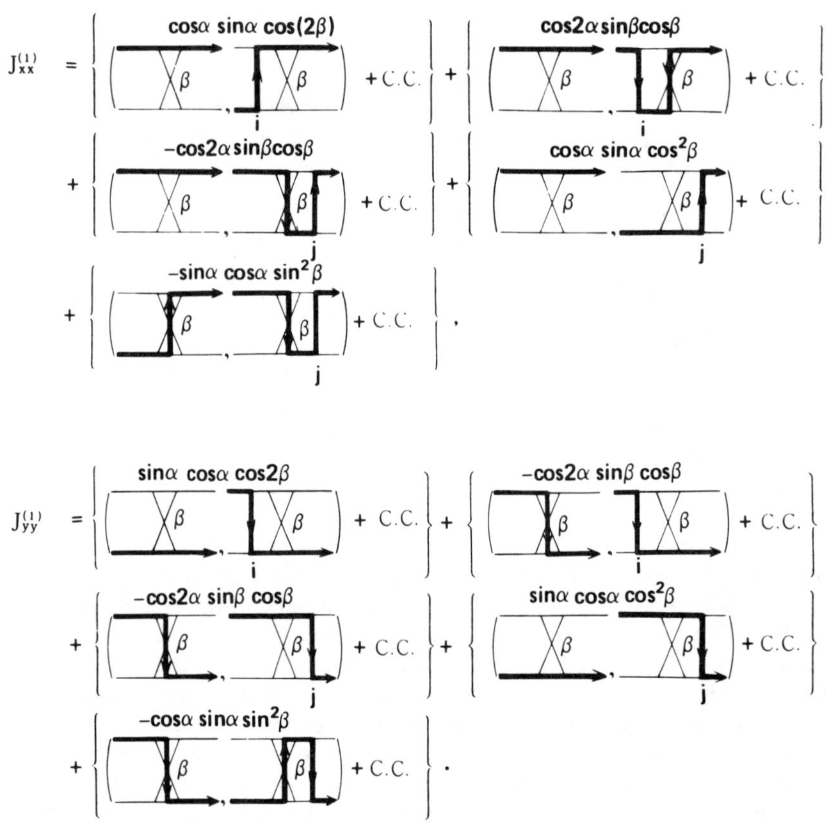

(36)

(37)

Fig. 17. Diagrammatic representations of coherence matrix elements $J_{xx}^{(1)}$ and $J_{yy}^{(1)}$.

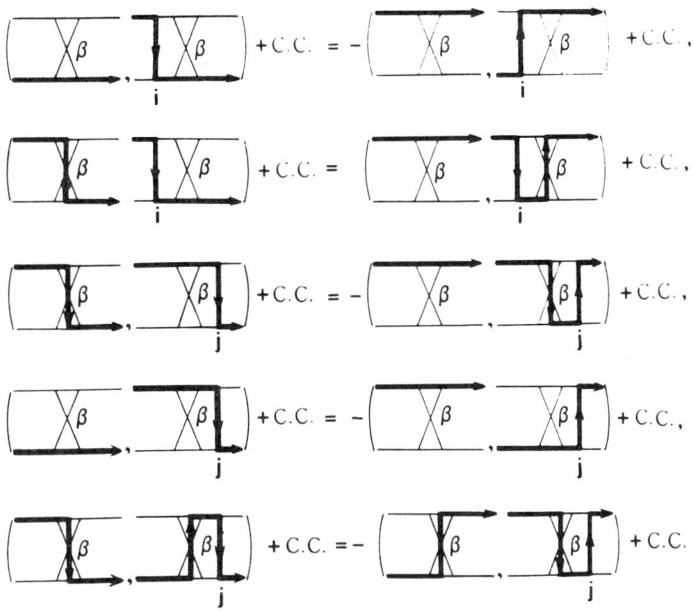

Fig. 18. Relations between diagrammatic scalar products.

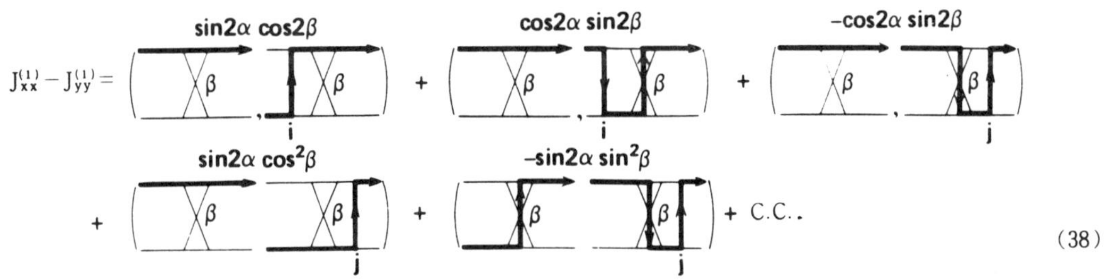

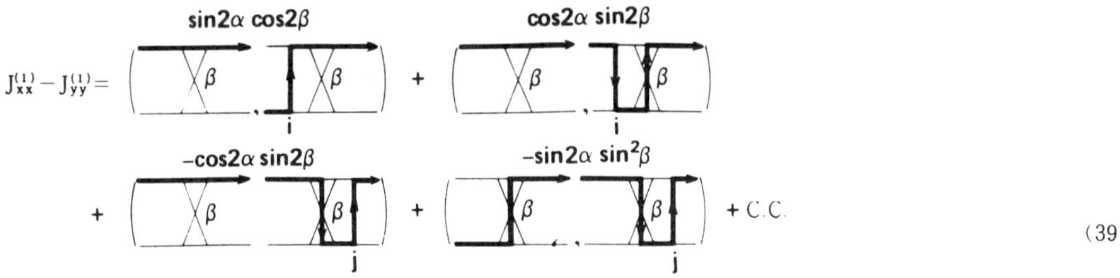

Fig. 19. Diagrammatic representations of $J_{xx}^{(1)} - J_{yy}^{(1)}$ [Eq. (38)] and the representation [Eq. (39)] simplified by neglecting the fourth term in the original representation [Eq. (38)].

3. RESULTS AND DISCUSSION

The degree of polarization obtained by the Lyot depolarizer was calculated precisely in Section 2 by the diagrammatic method. An important result is that for nearly 45° incidence relative to a principal axis, only the polarization coupling in the longer fiber (i.e., fiber 2 in Fig. 2) contributes to the residual polarization, independently of that in the shorter fiber (i.e., fiber 1 in Fig. 2) so far as the first-order approximation is concerned. In order to confirm this theoretical result, we performed two experiments. In the first one we compared the effects of lateral force on the residual polarization measured when a pointlike lateral force is applied to the center of the second fiber and to a point on the first fiber that is close to the joint. In the other experiment we measured the residual polarization obtained when fibers with different cross talk were used as the first fiber for a given second fiber.

In the experiments, an edge-emitting diode was used as a light source because there was little spectral modulation in the output. The measured visibility versus the time delay, $|R(\tau)|$, for the edge-emitting diode is shown in Fig. 22. $|R(\tau)|$ was measured with a Michelson interferometer at individual time delays. From the visibility curve, the coherence time (defined by a full half-width of the visibility) is $\tau_c = 0.07$ psec, corresponding to the $l_c = 21\ \mu$m coherence length. The visibility $|R(\tau)|$ decreases from 1 at zero delay to 10^{-3} at $\tau = \tau_0$ (= 0.16 psec). Although visibility for $\tau > \tau_0$ could not be measured because of the electrical noise, $|R(\tau)|$ is expected to decrease as the time delay increases because of the absence of spectral modulation. As is shown below, the residual polarization is $\hat{P} = 2 \times 10^{-3}$ for the test fibers. From relation (33) therefore the dispersion D_1L_1 should satisfy the condition of $|R(D_1L_1)| < \hat{P}/2 = 10^{-3}$; that is, $D_1L_1 < \tau_0 = 0.16$ psec. In our experiment, however, we chose $D_1L_1 = 19\tau_0 = 3.0$ psec, enough to decrease the cladding mode effect in the fibers. The cross talk in fibers of length L_1 were all less than 10^{-3}, so the cladding mode effect can be ignored when the degree of polarization is larger than 10^{-3}.

The experimental setup for measuring the degree of polarization of the depolarizer output is shown in Fig. 23. Light from the ELED with a 1300-nm central wavelength and a 50-nm full width passes through a polarizer and is launched into fiber 1 of the depolarizer. From end A of fiber 1, the light is coupled into end B of fiber 2. End B is equipped with one rotatable jig in a splicing unit so that it can be rotated precisely. Light from the output end of fiber 2 is collimated, passes through a quarter-wave plate (QW) and a half-wave

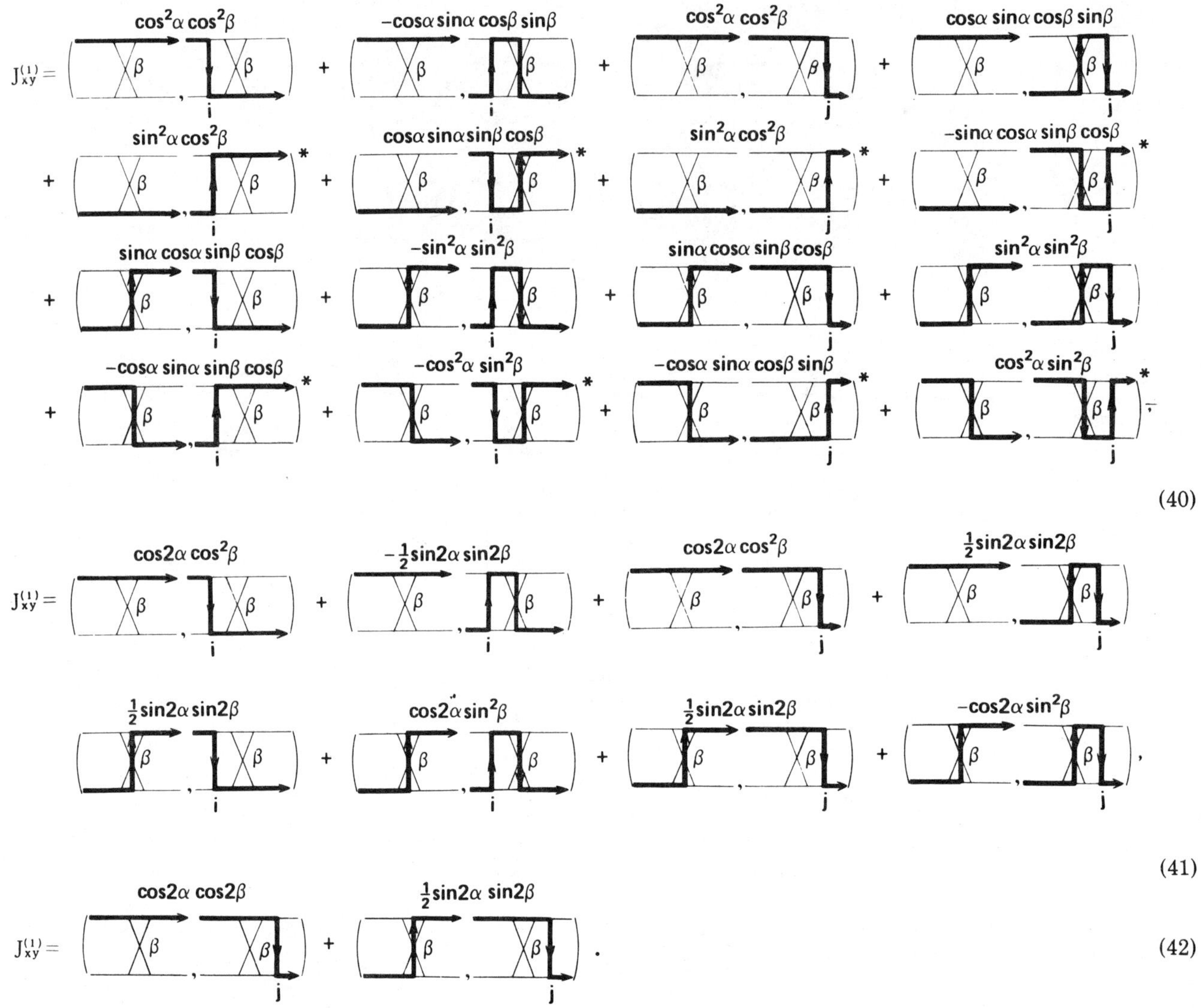

Fig. 20. Original diagrammatic representation [Eq. (40)] of coherence matrix element $J_{xy}^{(1)}$ and its simplified representations [Eqs. (41) and (42)].

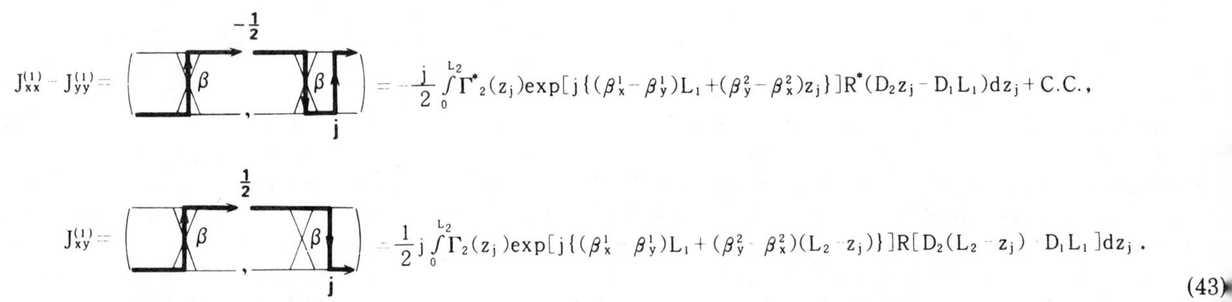

Fig. 21. Exact representations of $J_{xx}^{(1)} - J_{yy}^{(1)}$ and $J_{xy}^{(1)}$ obtained by using diagrammatic transformation

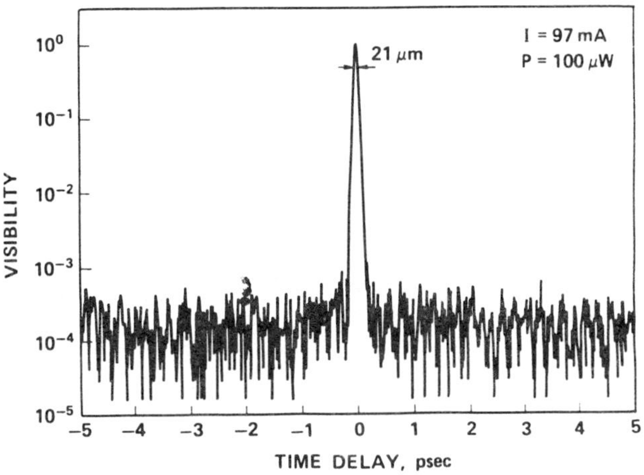

Fig. 22. Visibility curve of the edge-emitting diode output.

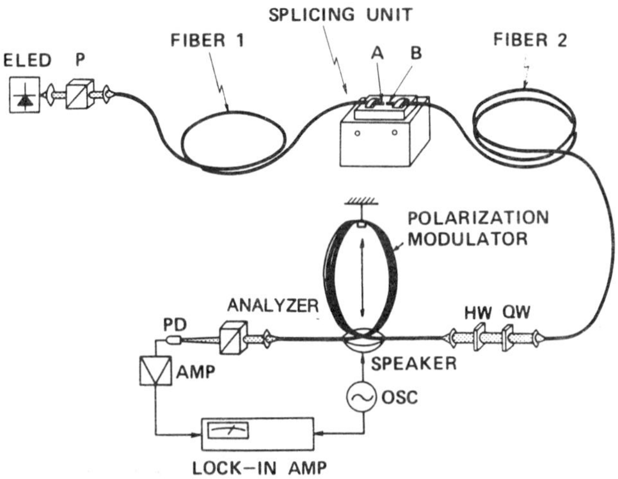

Fig. 23. Experimental setup for measuring the degree of polarization of the depolarizer output. P, Polarizer; OSC, oscillator; PD, photodiode.

plate (HW), and is launched into a single-mode fiber loop with a total length of 20 m.

The cone of a loudspeaker is attached to one side of this loop. By application of a sinusoidal voltage with a resonance frequency of 30 Hz to the fiber loop, the state of polarization in the single-mode fiber loop can be modulated.[7] The relative power of the polarized-light component existing in the depolarizer output is determined by measuring the optical intensity change after the light passes through an analyzer because the state of polarization of the polarized light is changed periodically by the fiber-loop polarization modulator, while that of the depolarized light is unchanged. The polarized light in the depolarizer output is generally elliptically polarized. In the measurement setup, therefore, the quarter-wave plate is used to transform the residual (elliptically) polarized light to linear polarization. The direction of the linearly polarized light is then adjusted to a 45° angle relative to the bending-induced principal axis of the fiber loop by rotating the half-wave plate. Output light from the fiber loop is launched into a Ge photodiode after passing through an analyzer. The signal from the photodiode is amplified by a lock-in amplifier synchronized with the loudspeaker drive voltage.

At a wavelength λ, the polarization coupling of bare fiber that is due to a pointlike lateral force is expressed by[8]

$$\Gamma_0 \Delta z_0 = 1.58\pi n^3 F_0 \frac{(P_{11} - P_{12})(1 + \nu)}{2aE\lambda} \sin(2\phi), \qquad (45)$$

where Γ_0 and Δz_0, respectively, denote the coupling coefficient and the coupling length of the lateral force. In Eq. (45), F_0 denotes the applied force, n denotes the mean refractive index, E and ν, respectively, denote Young's modulus and Poisson's ratio of fiber core material, P_{11} and P_{12} denote components of the stress-elastic tensor, and $2a$ is the outer diameter of the fiber. The dependence of the coupling on the azimuth angle ϕ is contained only in the factor $\sin(2\phi)$ in Eq. (45). Thus the coupling $\Gamma_0 \Delta z_0$ is maximized when a pointlike lateral force is applied with a force direction of $\phi = 45°$ relative to a principal fiber axis. By inserting $\lambda = 1.3$ μm here and also inserting the parameters of a typical silica fiber ($n = 1.46$, $2a = 125$ μm, $E = 7.3 \times 10^{10}$ N/m^2, $\nu = 0.17$, $P_{11} - P_{12} = -0.15$), we find that a force of $F = 100$ (gram force) (gf) produces a polarization coupling of $\Gamma_0 \Delta z_0 = 0.22$ for $\phi = 45°$.

Figure 24 shows the degree of polarization of the depolarizer output versus the applied force when a pointlike lateral force is applied to the center of fiber 2 [curve (a)] and to the point on fiber 1 that is close to the joint [curve (b)]. The test fiber was a bare polarization-maintaining and absorption-reducing fiber[9] of $L_1 = 4$ m without a silicon coating. The angle α incident upon fiber 1 was $\alpha = 45°$. Figure 24 shows the results obtained when a lateral force was applied in the direction $\phi = 45°$, which was experimentally determined to be optimal individually for curves (a) and (b) by sequentially rotating the force-applied point until the measured degree of polarization maximized at a relatively large force ($F = 100$ gf).

When a force was applied to the center of fiber 2, the degree of polarization increased almost linearly with the applied force, as shown in curve (a) of Fig. 24. One the other hand, when force was applied to the point on fiber 1 that is close to the joint, the degree of polarization remained constant at a minimum value of $\hat{P}^{(1)} = 2 \times 10^{-3}$ and then increased from $F = 20$ gf on, as shown in curve (b) of Fig. 24.

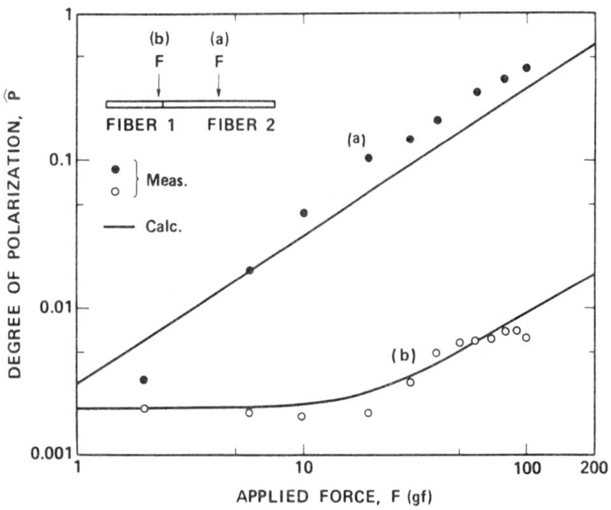

Fig. 24. Degree of polarization versus applied force. A pointlike lateral force is applied to (a) the center of fiber 2 and (b) the point on fiber 1 that is close to the joint.

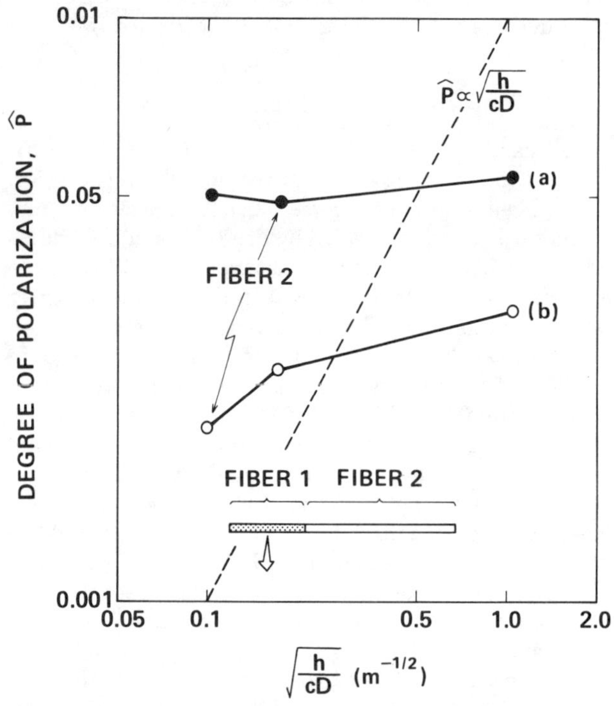

Fig. 25. Degree of polarization of the depolarizer output versus the parameter $[h/(cD)]^{1/2}$ for fiber 1 when fiber 2 is at (a) $[h/(cD)]^{1/2} = 0.18\ m^{-1/2}$ and (b) $[h/(cD)]^{1/2} = 0.1\ m^{-1/2}$. The dashed line shows the degree of polarization proportional to $[h/(cD)]^{1/2}$.

Compared with these two curves in the $F > 20$ gf region, each degree of polarization shown in curve (b) is less than that shown in curve (a) by at least a factor of $\gamma = 0.03$.

For a pointlike lateral force applied to the center of fiber 2, Eq. (43) leads to $J_{xx} - J_{yy} = \Gamma_0 \Delta z_0 \sin \varphi_1$ and $J_{xy} = (j/2) \Gamma_0 \Delta z_0 \exp(j\varphi_2)$ with phase constants of φ_1 and φ_2. The degree of polarization is then expressed by

$$\hat{P} = [2(\Gamma_0 \Delta z_0)^2 + (\hat{P}^{(1)})^2]^{1/2}, \tag{46}$$

with $\varphi_1 = \pi$, where $\Gamma_0 \Delta z_0$ is calculated from Eq. (45), and the minimum degree of polarization of $\hat{P}^{(1)} = 2 \times 10^{-3}$ is added in Eq. (46). On the other hand, when force is applied to the point on fiber 1 that is close to the joint, Eqs. (39) and (42) lead to $J_{xx} - J_{yy} = -2\cos(2\alpha)\Gamma_0 \Delta z_0$ with $\varphi_1 = \pi$ and $J_{xy} = 0$, so that

$$\hat{P} = [4(\cos 2\alpha)^2(\Gamma_0 \Delta z_0)^2 + (\hat{P}^{(1)})^2]^{1/2} \tag{47}$$

is obtained. Therefore the degree of polarization in curve (b) is expected to be smaller than that in curve (a) for a large F by a factor of $\gamma = 2^{1/2}|\cos(2\alpha)|$. Thus the angular misalignment of α is estimated to be $\Delta\alpha = 0.5°$ by using the measured value $\gamma = 0.03$. Also shown in Fig. 24 are degrees of polarization calculated with Eqs. (46) and (47) at $\alpha = 45° \pm 0.5°$. The agreement of calculated and measured values shows the precision of our diagrammatically obtained results.

The polarization of the depolarizer output was sequentially measured when fibers with different cross talk were used as fiber 1 for a given fiber 2. Figure 25 shows the measured degrees of polarization versus the $[h/(cD)]^{1/2}$ for fiber 1, where h denotes the polarization-coupling parameter, D denotes the polarization mode dispersion per unit length of fiber 1, and c denotes the velocity of light in vacuum. According to the calculation,[2] the residual polarization is proportional to $[h\tau_c/(cD)]^{1/2}$, where τ_c denotes the coherence time of the light. Therefore the parameter $[h/(cD)]^{1/2}$ was used to distinguish individual birefringent fibers. In Fig. 25 the value of this parameter for fiber 2 is $[h/(cD)]^{1/2} = 0.18\ m^{-1/2}$ for curve (a) and $[h/(cD)]^{1/2} = 0.1\ m^{-1/2}$ for curve (b). It is clear from Fig. 25 that the residual polarization is almost constant as the parameter $[h/(cD)]^{1/2}$ for fiber 1 changes; otherwise the slopes of curves (a) and (b) would be much smaller than that of the curve proportional to $[h/(cD)]^{1/2}$ (dashed line). This fact is also consistent with the result represented by Eq. (43).

4. CONCLUSION

In conclusion, we have presented a calculation of the performance of a fiber-optic Lyot depolarizer in the first-order approximation. To ensure a straightforward calculation, we have developed a new procedure with diagrams representing light propagation. The complicated expression for the degree of polarization of the depolarizer output is expressed by the propagation diagrams in a simple form with only a few formulas around the diagrams. The degree of polarization for linearly polarized incident light is calculated with an arbitrary incidence angle and with an arbitrary rotation angle between the first (shorter) fiber and the second (longer) fiber constructing the depolarizer. Especially for nearly 45° incidence, it is clarified by calculation that the residual polarization depends solely on the polarization coupling existing in the longer fiber. This result has been confirmed in two experiments, in which the experimental results agreed well with the theoretically obtained result. Therefore the proposed diagrammatic procedure represents a powerful technique for easily calculating polarization-coupling effects on systems with birefringent fibers.

ACKNOWLEDGMENTS

The authors would like to express their sincere thanks to N. Inagaki and H. Takata for their continual encouragement.

REFERENCES

1. K. Böhm, P. Marten, K. Petermann, E. Weidel, and R. Ulrich, "Low drift fiber gyro using a superluminescent diode," Electron. Lett. **17,** 352–353 (1981).
2. K. Böhm, K. Petermann, and E. Weidel, "Performance of Lyot depolarizers with birefringent single-mode fibers," IEEE J. Lightwave Technol. **LT-1,** 71–74 (1983).
3. W. K. Burns and R. P. Moeller, "Depolarization in single-mode optical fiber," IEEE J. Lightwave Technol. **LT-1,** 44–50 (1983).
4. K. Takada, K. Okamoto, and J. Noda, "Polarization coupling with a broadband source in birefringent polarization-maintaining fibers," J. Opt. Soc. Am. A **2,** 753–758 (1985).
5. M. Born and E. Wolf, *Principles of Optics*, 3rd ed. (Pergamon, New York, 1965), pp. 491–506, 544–554.
6. I. P. Kaminow, "Polarization in optical fibers," IEEE J. Quantum Electron. **QE-17,** 15–22 (1981).
7. K. Takada, K. Okamoto, and J. Noda, "New fiber-optic depolarizer," IEEE J. Lightwave Technol. **LT-4,** 213–219 (1986).
8. N. Chinone and R. Ulrich, "Elasto-optic polarization measurement in optical fiber," Opt. Lett. **6,** 16–18 (1981).
9. Y. Sasaki, T. Hosaka, K. Takada, and J. Noda, "8-km-long polarization-maintaining fiber with highly stable polarization state," Electron. Lett. **19,** 792–793 (1983).

CORNU PSEUDODEPOLARIZER

In laser power measurement, radiometry, and spectral radiometry, polarization properties of the source often interact with detection apparatus to produce significant measurement error. This happens because oblique reflection from either dielectric or metallic surfaces, oblique transmission by dielectric surfaces, dispersion by either gratings or prisms, and passage through slits may modify the polarization state.

True depolarizers do not exist. Psuedodepolarizers transform one polarization state into a continuum of states which under suitable circumstances may mimic the behavior of unpolarized radiation. Properly employed in radiometric measurements, pseudodepolarizers can dramatically improve both the repeatability and absolute accuracy of measurement.

The Cornu pseudodepolarizer (CPD) employs optical activity in crystalline quartz (see earlier description of optical activity). It operates on collimated beams of light, transforming linear polarization states into a complicated and spatially changing continua of linear polarization states. The CPD transmits circularly polarized light without change, and CPD effectiveness for elliptically polarized light is greatly diminished. In general it can be said that the Cornu pseudodepolarizer is more effective in polychromatic than in monochromatic light, since the mix of output polarizations is more complex in the polychromatic case. The discussion which follows concentrates on the more difficult monochromatic case.

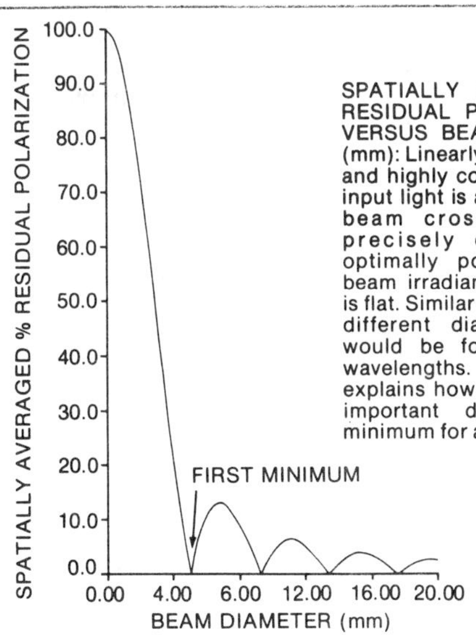

SPATIALLY AVERAGED % RESIDUAL POLARIZATION VERSUS BEAM DIAMETER (mm): Linearly polarized axial and highly collimated yellow input light is assummed. The beam cross section is precisely circular and optimally positioned. The beam irradiance distribution is flat. Similar curves, but with different diameter scales, would be found for other wavelengths. The text explains how to estimate the important diameter at first minimum for any wavelength.

Assume that linearly polarized monochromatic light propagates from left to right through the CPD. It does not matter which end of the CPD is made of left or right-handed quartz. What is important is that the two halves be made of the opposite kinds of quartz, as is the case in the CPD. In both halves the optic axis is normal to the XY plane of the figure (and hence parallels the direction of propagation). A well collimated axial beam is presumed. This beam experiences circular retardation of opposite sign in the two halves, and hence the plane of polarization rotates about the direction of propagation. The amount of rotation is a continuous function of the y-coordinate on the exit face, but is independent of the x-coordinate. This structure of the CPD assures that the collimated beam is transmitted without deviation, and further has the result that the output polarization plane is twice as sensitive to the y-coordinate as it would have been if ordinary glass had been used for one half of the CPD.

The residual "degree of polarization" at each point (x,y) in the output beam has a strict mathematical definition in terms of the Stokes parameters (See Shurcliff & Ballard, *Polarized Light*, p. 78). The effectiveness of the CPD is conveniently measured by weighted averaging of the residual degree of polarization over the output beam irradiance distribution at the exit face. For monochromatic input, CPD effectiveness demands well-collimated axial input, and is additionally very sensitive to input polarization state, beam size, beam position, and beam irradiance distribution. Beam position is subject to mechanical control. Beam size can be controlled with iris diaphragms and/or beam expanders. For any beam shape or irradiance distribution, beam position and beam size can be experimentally optimized to achieve minimum spatially averaged residual polarization. In special cases, assuming optimum positioning, it is possible to calculate optimum beam size as a guide to such experimentation.

The graph is the numerical result for the special case of a circular beam of constant irradiance. The calculation presumed linear input polarization, yellow light (589.5nm), and that the beam was optimally positioned. The graph is of spatially averaged residual polarization as a function of beam diameter. The calculation was continued up to beam diameters twice as large as the CPD can accommodate. It is apparent that there are several different beam diameters for which the spatially averaged residual polarization vanishes. The first minimum, at a diameter of 5.05mm for this wavelength, is of special interest because it is attainable with a real CPD. The beam diameter at first minimum is a function of wavelength, and for many wavelengths only the first minimum is attainable with a CPD of 10mm aperture. For a beam of constant irradiance the diameter at first minimum bears a special relationship to a characteristic distance ΔY_{180} on the exit face. Specifically, the beam diameter at first minimum is approximately 3/4 of this characteristic distance. For a beam of constant irradiance and rectangular shape the optimum beam width in the y-direction is approximately 2/3 of this same characteristic distance ΔY_{180}.

The significance of the characteristic distance ΔY_{180} is that, for linearly polarized input to the CPD, points on the exit face whose y-coordinates differ by this distance have identical states of linear polarization. The plane of output polarization rotates smoothly by 180 degrees as the y-coordinate is changed by this distance. This characteristic distance can easily be calculated for any wavelength. Specifically,

$$\Delta Y_{180} = \frac{90^\circ}{\alpha} + \frac{w}{4} ,$$

where w is the 10mm clear aperture width of the CPD, and α is the specific rotation or optical rotatory power of crystalline quartz in degrees per millimeter. T. Martin Lowry, in his book *Optical Rotatory Power* (Dover, 1964, page 258), publishes the following dispersion formula for specific rotation:

$$\alpha = \frac{9.5639}{\lambda^2 - 0.0127493} - \frac{2.3113}{\lambda^2 - 0.000974} - 0.1905 .$$

In this formula the wavelength λ must be expressed in microns. With the preceding formulas it is possible to estimate the optimum beam diameter for any wavelength.

Two CPD's may be used in tandem, with the y-direction of the first paralleling the x-direction of the second, to "scramble" the output polarization more effectively than is possible with one alone. For any beam diameter, whether optimum or not, the beam may be focused to a spot which more closely mimics the ideal of unpolarized radiation.

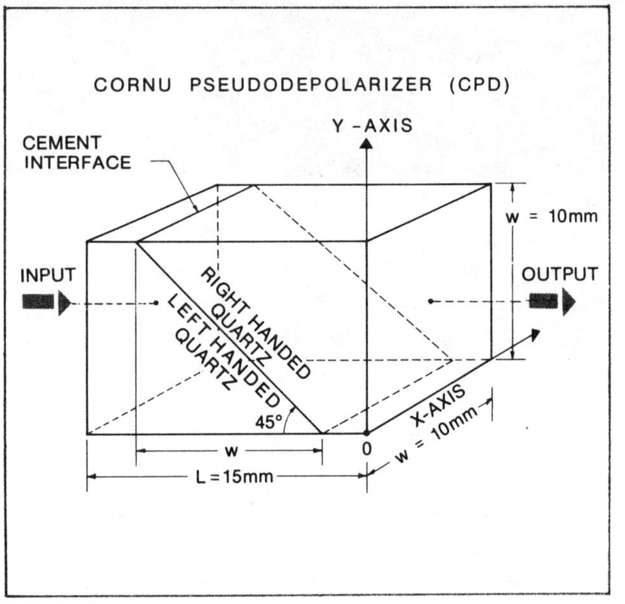

SPECIFICATIONS: CORNU PSEUDODEPOLARIZER

Aperture: 10mm square
Length: 15mm
Dimensional Tolerance: ±0.5mm
Material: Optical Grade Crystalline Quartz
Cement: Optical Grade Epoxy
Useful Spectral Range: 0.35 to 2.5 Microns

Cornu Pseudodepolarizer

	PRODUCT NUMBER
One	04 LDP 001

New Fiber-Optic Depolarizer

KAZUMASA TAKADA, KATSUNARI OKAMOTO, AND JUICHI NODA

Abstract—A new type of single mode fiber-optic depolarizer is proposed and demonstrated. The new depolarizer consists of a Mach–Zehnder interferometer with one arm used as a delay line with respect to the other. Light having a 100-MHz spectral width, for example, can be depolarized with high coupling efficiency to single mode fibers by using a polarization-maintaining fiber-delay line of a few meters in length.

I. Introduction

When polarized light having a narrow spectrum is launched into a single mode fiber, a high degree of polarization is maintained for a fiber length of 21 km [1]. The state of polarization (SOP), however, fluctuates randomly due to temperature and external force-induced birefringences. If polarization-dependent optical elements such as a polarizer or a blazed grating are inserted in a single mode fiber line, SOP fluctuations due to external birefringence give rise to intensity noise.

Depolarizing devices to reduce this kind of noise have been proposed and demonstrated [2], [3]. Among them, a crystal depolarizer consisting of two pairs of birefringent wedges reduces the polarization noise to less than -20 dB, which is applicable even for single mode laser radiation with its narrow spectrum [2]. The birefringent crystal, however, spatially produces a beam divergence which degrades the efficiency of coupling to single mode fibers. It cannot therefore be employed in a single mode fiber line such as an optical time domain reflectometer [4]. This problem also applies to a graded index fiber-optic depolarizer [2].

A depolarizer with high coupling efficiency to single mode fibers has been realized by constructing two birefringent plates or fibers with a length ratio of 2 : 1 and their principal axes at 45° to each other [3]. This device is called a Lyot depolarizer, which utilizes the light depolarization due to the group-delay time difference between the two principal modes.

To depolarize light having a narrow spectrum, for example, a spectral width of 100 MHz, it is necessary to use a 2-km birefringent fiber because the group-delay time difference is at most 5 ns/km. Similarly, a depolarizer consisting of a birefringent crystal requires at least a few meters for a Lyot depolarizer. Therefore, Lyot-type depolarizers are neither compact nor practical for depolarizing narrow spectral light.

To solve this problem, a new type of single mode fiber-optic depolarizer is presented. One advantage of this depolarizer is that light having a narrow spectrum can be depolarized with high coupling efficiency to single mode fibers by using a birefringent fiber of short length.

Manuscript received March 4, 1985; revised August 14, 1985.

The authors are with Ibaraki Electrical Communication Laboratories, Nippon Telegraph and Telephone Corporation, Tokai, Ibaraki-ken, 319-11, Japan.

IEEE Log Number 8406254.

II. Depolarizer Structure

With a coherence matrix, completely depolarized light can be represented by [5]

$$J_0 = \frac{I_0}{2}\begin{pmatrix} 1 & 0 \\ 0 & 1 \end{pmatrix} \tag{1}$$

where I_0 is the total power of the light. This coherence matrix is satisfied under two conditions, that is, 1) that the two orthogonal components of the electric fields are completely uncorrelated, and 2) that the field intensities (i.e., powers) of the two components are equal.

The schematic diagram of the new depolarizer, which consists of a Mach–Zehnder interferometer, is presented in Fig. 1.

The light passes through the half-wave-plate HW1 and is divided into P and S components (called P and S waves, respectively) at the polarization beam splitter BS1. The S-wave is launched into a polarization-maintaining PANDA fiber [6] where one principal axis is parallel to the polarization of the S wave. Its output light from the PANDA fiber is then coupled orthogonally with the P wave at the second beam splitter BS2 after passing through the half-wave-plate HW2 and the polarizer PO1. By employing the fiber as a delay line of the S wave, it is easy to achieve a large group-delay time difference of 5 ns/m, which is three orders of magnitude larger than that of the conventional Lyot depolarizer.

When the group-delay time difference due to the fiber delay line is longer than the coherence time of the light, condition 1) is satisfied. Furthermore, by adjusting the rotation angles of the two half-wave-plates HW1 and HW2 the P and S waves are superimposed with an equal power at the second beam splitter BS2. As a result, condition 2) can also be satisfied. When the SOP in the fiber delay line changes due to external birefringences, the light power of the S wave will fluctuate randomly after passing through the polarizer PO1. Consequently, the birefringent fiber for maintaining the linear polarization, such as a PANDA fiber, is necessary as a fiber delay line.

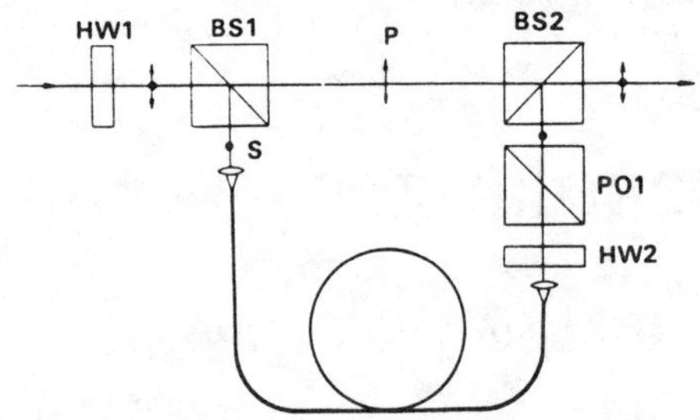

Fig. 1. Setup for proposed single mode fiber-optic depolarizer.

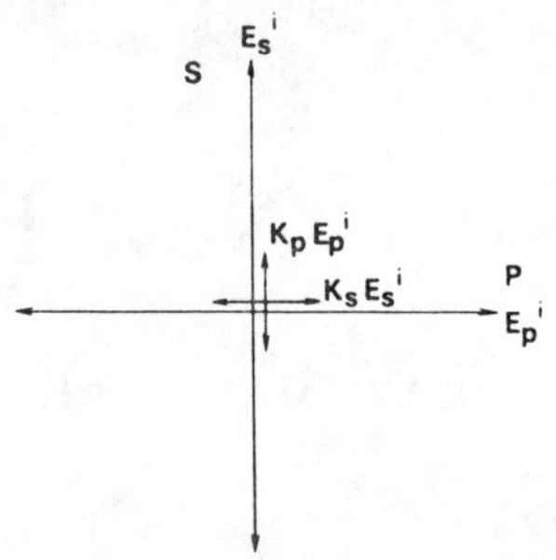

Fig. 2. Mode coupling between P and S waves at polarization beam spli ters BS2. $E_p{}'$ and $E_s{}'$ are incident P and S waves, respectively, and and k_s are coupling coefficients.

III. Performance of the Depolarizer

We will now present the design for the new depolarizer. When a mode coupling exists between the P and S waves at the second beam splitter BS2 the superimposed P and S waves are no longer uncorrelated. The mode coupling causes limitations to depolarizer performance. The electric fields E_p^o and E_s^o of the superimposed P and S waves are described by the incident electric fields E_p^i and E_s^i as

$$E_p^o = E_p^i + k_s E_s^i \tag{2.1}$$

$$E_s^o = E_s^i + k_p E_p^i. \tag{2.2}$$

Here, k_m $(m = p, s)$ are the coupling coefficients of the P and S waves at the second beam splitter BS2 as shown in Fig. 2. Additionally, the extra losses of the P and S waves at BS2 are neglected.

The degree of polarization $\hat{P}$ which is defined as the ratio of the polarized power I_{pol} to the total power I_{tot}, i.e., $\hat{P} = I_{\text{pol}}/I_{\text{tot}}$, is defined by the coherence matrix J of the superimposed field as

$$\hat{P} = \sqrt{1 - 4\frac{\det(J)}{(\mathrm{Tr}(J))^2}} = \sqrt{\frac{(J_{pp} - J_{ss})^2 + 4|J_{ps}|^2}{(J_{pp} + J_{ss})^2}} \tag{3}$$

where J_{mn} $(m, n = p, s)$ are matrix elements of J and are represented by $J_{mn} = \langle E_m^o E_n^{o*} \rangle$ with the symbol $\langle\,\rangle$ representing the time average and the asterisk * representing the complex conjugate.

From (2.1) and (2.2), the elements J_{mn} are denoted by the coherence matrix elements $j_{mn} = \langle E_m^i E_n^{i*} \rangle$, of the incident P and S waves as

$$J_{pp} = j_{pp} + |k_s|^2 j_{ss} + k_s j_{sp} + k_s^* j_{ps} \tag{4.1}$$

$$J_{ps} = J_{sp}^* = k_p^* j_{pp} + k_s j_{ss} + j_{ps} + k_s k_p^* j_{sp} \tag{4.2}$$

and

$$J_{ss} = j_{ss} + |k_p|^2 j_{pp} + k_p j_{ps} + k_p^* j_{sp}. \tag{4.3}$$

By substituting (4.1) ~ (4.3) into (3), and neglecting terms of higher order than $k_m k_n$ $(m, n = p, s)$ because of $|k_{p,s}| << 1$, the degree of polarization $\hat{P}$ is simply represented by

$$\hat{P} = \sqrt{|2K - 1|^2 + |k_s^* + k_p + R(L)|^2}. \tag{5}$$

Here, the power ratio $K = j_{ss}/(j_{ss} + j_{pp})$ is approximated as $K = \frac{1}{2}$, i.e., $|2K - 1| = 0$, and $R = j_{sp}/j_{ss}$ is the degree of coherence [5] of the incident fields E_p^i and E_s^i, respec-

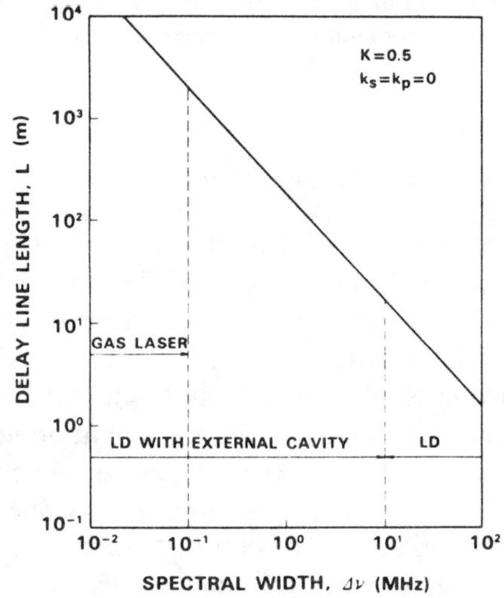

Fig. 3. Spectral width dependence of required delay line length for rea izing the degree of polarization $\hat{P} = -20$ dB. Power ratio is $K = 0$ ar coupling coefficients are $k_p = k_s = 0$.

tively. The degree of coherence in laser diodes whos emitting spectrum is Lorentzian [7] is described by

$$R(L) = \exp(-2\Delta\nu n\pi L/c) \tag{6}$$

where $\Delta\nu$ is a half width at half maximum of the Loren zian power spectrum, L is fiber length of the delay line n is the refractive index of the core, and c is the veloci of light in a vacuum.

Fig. 3 demonstrates the required delay line length as function of the spectral width $\Delta\nu$ for realizing the loga rithmic degree of polarization $\hat{P} = -20$ dB, where th coupling coefficients are assumed to be $k_p = k_s = 0$ an the power ratio $K = 0.5$. Even if the spectral width of th lasers such as a He-Ne laser or a laser diode with an ex ternal cavity is 100 kHz, a fiber length of 2 km is lon enough to depolarize the linearly polarized light of thes lasers. It cannot, however, be depolarized in the case o a Lyot-type depolarizer [3] consisting of a birefringe fiber even with 10-km length. The spectral width of co ventional laser diodes extends from 10 MHz to 10^3 GH It is clear from Fig. 3 that only a fiber length of 20 m sufficient as a delay line for the depolarization of the LD'

The relationship between the degree of polarization and the spectral width $\Delta\nu$ is plotted in Fig. 4, where th fiber delay line is 20 m and η represents the percentage

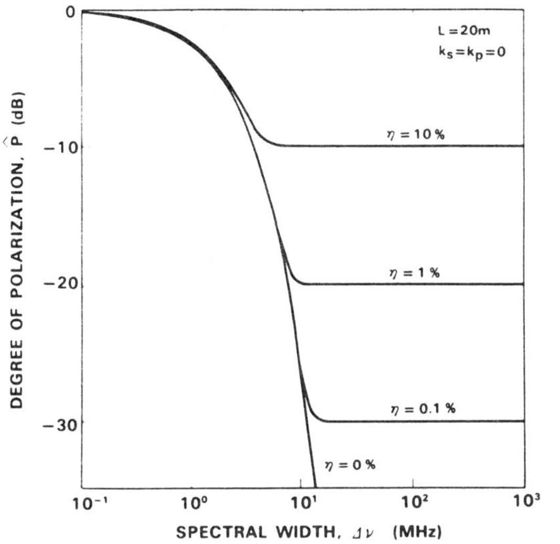

Fig. 4. Degree of polarization versus spectral width. Fiber delay line is 20 m and coupling coefficients are $k_p = k_s = 0$.

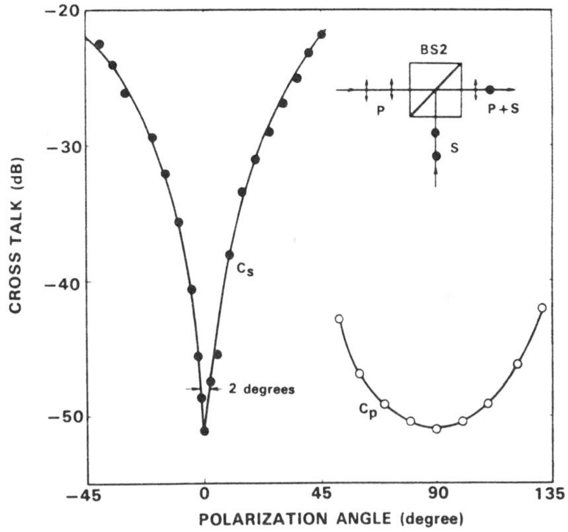

Fig. 5. Crosstalks $C_{p,s}$ of output P and S waves from polarization beam splitter BS2 versus incident polarization angles of P and S waves.

the power ratio-deviation from 0.5, i.e., $\eta = 100|K - 0.5|/0.5$. To decrease the degree of polarization to less than $\hat{P} = -20$ dB, the power-ratio deviation must be set at $\eta = 1$ percent, i.e., $K = 0.5 \pm 0.005$. If the power-ratio deviation is $\eta > 1$ percent, even the linearly polarized light of a broad-band light source with a 10^3 MHz spectral width cannot be depolarized to less than $\hat{P} = -20$ dB.

One limiting depolarizer factor is the mode coupling between the P and S waves at the beam splitter, BS2 as indicated in (5). To examine the limit of mode coupling produced by the misalignment of the incident linearly polarized light, the incident polarization angle dependences of the power crosstalk $C_{p,s} = 10 \cdot \log_{10} |k_{p,s}|^2$, were measured by rotating the polarizer PO1 and the polarization beam splitter BS1 as shown in Fig. 5. The minimum crosstalks were $C_p = C_s = -51$ dB which corresponded to $|k_p| = |k_s| = 2.8 \times 10^{-3}$.

The coupling coefficients k_p and k_s are generally complex numbers, although their phases are not known. By considering the phase changes of the incident waves at the conventional beam splitters or half mirrors, it can be assumed that k_s is a purely imaginary and k_p is real, i.e., $k_s = |k_s| \exp(i\pi/2)$ and $k_p = |k_p|$. From (5), the minimum degree of polarization is $\hat{P} = 10 \cdot \log_{10} (\sqrt{|k_p|^2 + |k_s|^2}) = -24$ dB with $K = 0.5$ and $R = 0$. In Fig. 5, the half width of the polarization angle relative to the minimum crosstalk is 2° for the S wave. It is easy to align and maintain the polarization angle of PO1 within ±1°.

IV. Depolarization Measurement System

As a method to measure the degree of polarization, the electrooptic modulators such as a $LiNbO_3$ modulator have been employed [3]. The relative power of the polarized light component existing in the incompletely depolarized light is determined by measuring the optical intensity change through a polarizer because the SOP of the polarized light is periodically changed by the modulator, while

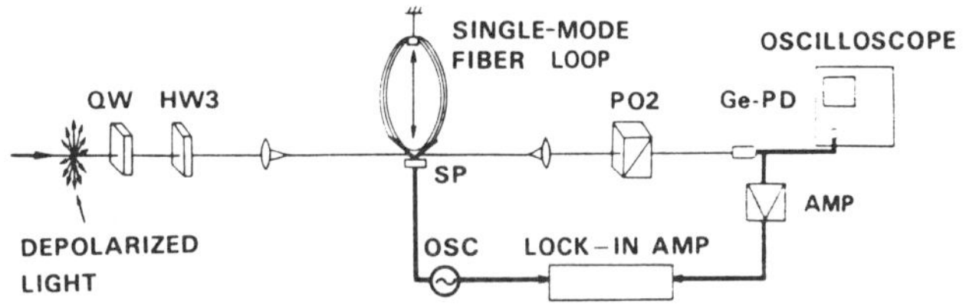

Fig. 6. Setup for measuring degree of polarization of depolarizer output.

the SOP of the completely depolarized light is unchanged. We used here a polarization modulator with the single mode fiber loop instead of the electrooptic modulators as shown in Fig. 6.

The output light from the depolarizer passing through the quarter-wave-plate QW and the half-waveplate HW3 is launched into the single mode fiber loop with the total length of 20 m and with a diameter of 10 cm. The cone of a loudspeaker SP which could be forced to vibrate, was attached to one side of this loop. By applying a sinusoidal voltage with a resonance frequency of 55.6 Hz to the fiber loop, the SOP in the single mode fiber could be modulated [2]. The ac voltage amplitude was selected so that the output polarization through the fiber loop could move completely along the meridian of the Poincaré sphere. This condition was optimized when the incident plane of polarization was adjusted to an angle of 45° with respect to the principal axis of the bending-induced birefringent fiber due to the fiber loop.

The polarized light existing in the depolarizer output light is, generally, elliptically polarized due to the phase difference between the P and S waves passing from the first to the second beam splitters BS1 and BS2 at different paths. In the measurement system, therefore, the quarter-wave-plate QW was used to transform the SOP of the residual (elliptically) polarized light to the linearly polarized light, and then the direction of the linearly polarized light was adjusted to the angle of 45° relative to the bending-induced principal axis of the fiber loop by rotating the half-wave-plate HW3. The light from the end of the fiber loop passing through the polarizer PO2 is detected by the Ge-photodiode. The signal from the photodiode is monitored by the oscilloscope and is amplified by a lock-in amplifier synchronized with the loudspeaker drive voltage.

V. Results and Discussion

Two kinds of InGaAsP-LD's were used for the depolarization experiment. LD-1 had a single longitudinal mode as shown in Fig. 7(a), LD-2 had about 20 longitudinal modes as shown in Fig. 7(b). A 2-m polarization maintaining fiber called PANDA fiber was used as the fiber delay line. To calibrate the setup shown in Fig. 6, the optical power levels of the P and S waves passing through the second beam splitter BS2 were first investigated. Small birefringence existed in the looped single mode fiber due to bending. Therefore, the polarization planes of the waves were adjusted to be 45° with respect to the bending-induced principal axis of the fiber loop by rotating the half-wave-plate HW3 whereas the retardation of the quarter-wave-plate QW was aligned at zero.

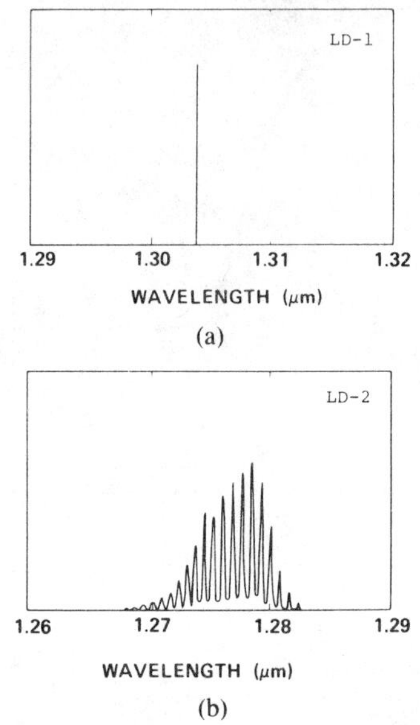

Fig. 7. Light spectrums of InGaAsP-laser diodes, (a) LD-1, (b) LD-2 with single- and multi-longitudinal modes, respectively. Resolution of spectrum was 0.1 nm.

By applying sinusoidal vibration to the fiber loop with the loud speaker SP the SOP moved almost completely along a meridian of the Poincaré sphere. The principal axis of the polarizer PO2 was parallel to one of the two linear polarizations contained in the meridian. When monitored by the oscilloscope, the transmitted power of the P wave of the depolarizer through the PO2, which is obtained by interrupting the optical path of the S wave in Fig. 1, exhibited sinusoidal variations as shown in Fig. 8(a), while the S wave as shown in Fig. 8(b), where the LD-2 was used. It should be noted that the phases of the P and S wave variations are opposite each other.

Measurement of the total power of the depolarizer output light was accomplished by measuring the signal amplitudes V_p and V_s obtained from the lock-in amplifier. When the P and S waves were superimposed, the resultant ac signal became small due to depolarization as shown in Fig. 8(c). To determine the maximum signal amplitude v from the lock-in amplifier which corresponds to the power of the residual polarized light, the quarter-waveplate QW and the half-wave-plate HW3 were arbitrarily rotated until the maximum signal was obtained. The logarithmic degree of polarization was then determined to be $\hat{P} = 10 \cdot \log_{10} (v/(V_p + V_s))$. The polarizer PO2 was next removed to obtain a minimum detectable degree of polarization in the measurement setup. The measurement results showed that the noise was mainly due to power loss in the fiber loop caused by the sinusoidal modulation and that the minimum detectable signal level would be less than $\hat{P} = -30$ dB.

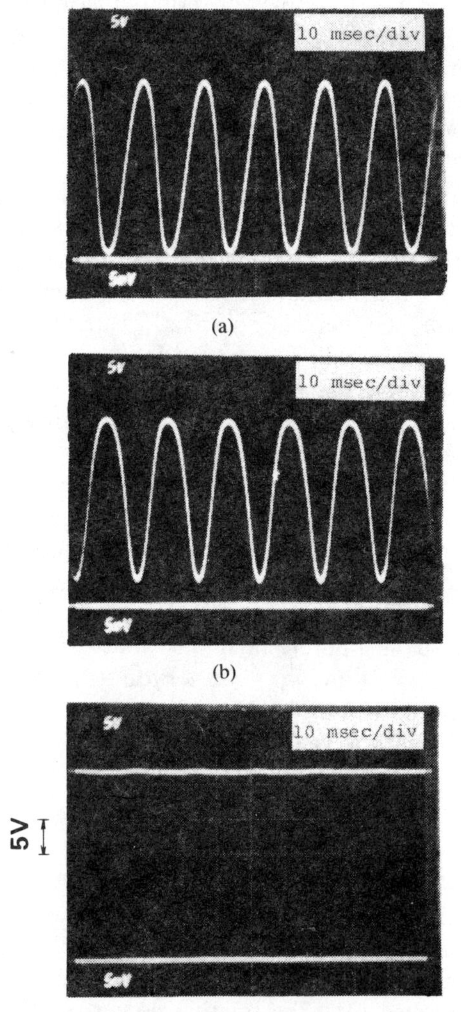

Fig. 8. Oscilloscope traces of polarization variations with sinusoidal vibrations when (a) P wave, (b) S wave, and (c) superimposed P and S waves are launched into fiber loop setup for measuring degree of polarization.

To obtain perfectly depolarized light, the power ratio K of the P and S waves was adjusted to 0.5 by rotating the half-wave-plates HW1 and HW2. Fig. 9 shows the degree of polarization versus the rotation angle θ of the half-wave-plate HW2 for two light sources LD-1 and LD-2. The crosstalk of the PANDA fiber-delay line was set at -43 dB by aligning one of the principal axis of the fiber

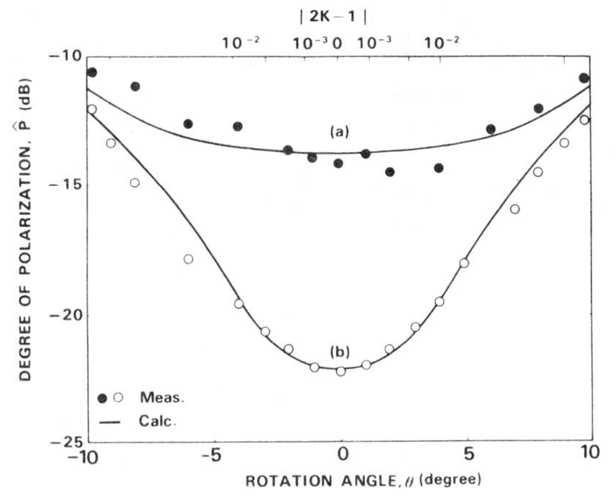

Fig. 9. Degree of polarization of depolarizer output versus rotation angle of half-wave-plate HW2 for (a) LD-1, and (b) LD-2 shown in Fig. 7. Degree of polarization is plotted as logarithmic scale.

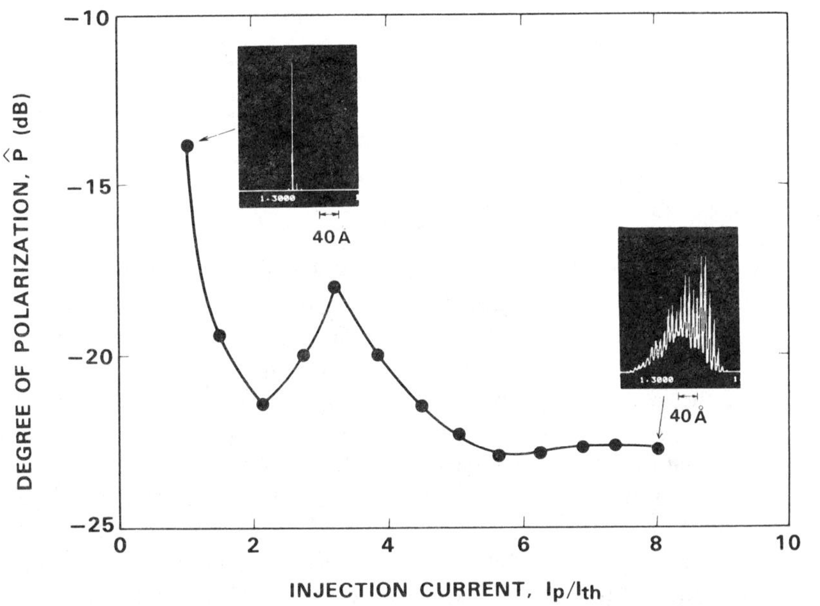

Fig. 10. Degree of polarization of depolarizer output versus relative peak current I_p/I_{th} of LD-1. I_p is peak value of the rectangular modulation current and I_{th} is threshold current of the LD-1.

with the linear polarization direction of the output light from the polarization beam splitter BS1. The fluctuations of the light power passing through the polarizer PO1 was then less than 0.01 percent. The measurement of the degree of polarization was made after roughly adjusting the power ratio K to 0.5 with the half-wave-plate HW1 while the axis of HW2 was set to be parallel to the plane of polarization of PO1. The change of the power ratio on the rotation angle θ of the HW2 is, in this condition, expressed by $K = 1/(1 + \cos^2(2\theta))$.

From Fig. 9, the linearly polarized light of the LD-1 was depolarized within −14 dB using only a 2-m PANDA fiber. This means that the polarization noise can be reduced to −14 dB, which represents a good value for improving the SNR of polarization noise in a single mode fiber line. On the other hand, the linearly polarized light of the LD-2 with the multilongitudinal modes was depolarized to −22 dB, which is a similar value to that obtained by the Lyot-depolarizers. Around the rotation angle of $\theta = 0°$, the adjustment of HW2 for maintaining the depolarization of less than −20 dB is not critical when the LD-2 is used. Curves fitting to the degree of polarization, $\hat{P}$ versus rotation angle θ were performed using the relationship of $|2K - 1| = \sin^2(2\theta)/(1 + \cos^2(2\theta))$ and the values of $|k_p| = |k_s| = 2.8 \times 10^{-3}$. The degrees of coherence of the sources LD-1 and LD-2 are then estimated to be $R_1 = 4.0 \times 10^{-2}$ and $R_2 = 6.3 \times 10^{-3}$, respectively. In this way, the degree of coherence having a definite group-delay time difference can also be measured by this depolarizer.

To clarify the limitation of the depolarizer performance, the rectangular modulation with a high peak current was used to broaden the LD-1 spectrum. Fig. 10 plots the degree of polarization of the depolarizer output versus the relative peak current I_p/I_{th} where I_{th} is the laser threshold current. As the peak current increases, the degree of

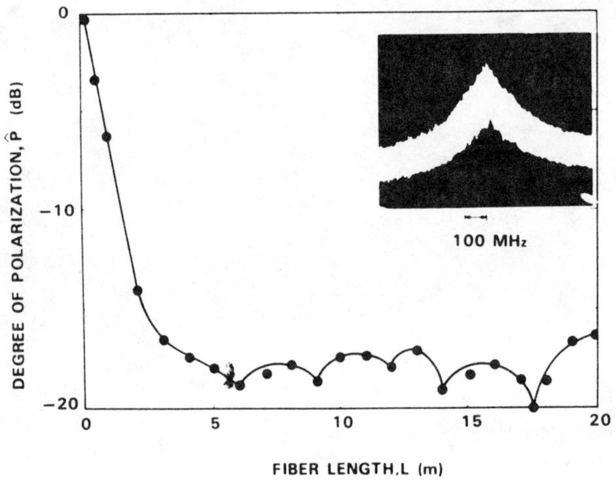

Fig. 11. Degree of polarization of depolarizer output versus fiber length of the delay line.

polarization decreased from $\hat{P} = -14$ dB at $I_p = 1.1I_{th}$ to $\hat{P} = -23$ dB at $I_p = 6I_{th}$ through the subpeak of the degree of polarization at $I_p = 3.2I_{th}$. The observed peak was due to decreasing the number of multilongitudinal modes of the LD-1.

In the $I_p > 6I_{th}$ range, the light spectrum became almost continuous. Since the minimum detectable signal level is less than -30 dB, the degree of polarization of $\hat{P} = -23$ dB represents the limit of this depolarizer. Inserting $\hat{P} = 5.0 \times 10^{-3}$, $|2K - 1| = 0$, and $R = 0$ into (5), the coupling coefficients were estimated to be $|K_p| = |K_s| = 2.5 \times 10^{-3}$, i.e., $C_p = C_s = -52$ dB, which agreed well with the experimental results shown in Fig. 5. Therefore, it was experimentally confirmed that the depolarizer performance is limited by the mode coupling between the P and S waves at the beam splitter BS2.

In order to achieve the depolarized light of the single longitudinal mode LD-1, the length of the fiber delay line was changed by steps of 1 m as shown in Fig. 11. Within the fiber length $L = 2$ m, the logarithmic degree of polarization linearly decreased with the fiber length L, i.e., $\hat{P} = -AL$ with a constant of $A = 6.8$ dB/m. The constant A is described by $A = (2\Delta\nu n\pi/c)\ (10 \cdot \log_{10}(e))$ under the condition of $|2K - 1| = 0$ and $|k_p|, |k_s| << 1$. Therefore, using the values of $n = 1.46$, $c = 3.0 \times 10^8$ m/s, and $A = 6.8$ dB/m, the spectral width is estimated to be $\Delta\nu = 51$ MHz.

Beyond the fiber length of $L = 2$ m, the degree of polarization did not decrease to the minimum value of $\hat{P} = -23$ dB. The degree of polarization actually fluctuated with about a $\Delta L = 3$ m period. As the delay line length dependence of the degree of coherence corresponds to the autocorrelation function $R(L)$ it is clear that the light spectrum of the LD has side bands around the carrier frequency. The spectral interval of the side bands is estimated to be $\Omega = c/(n\Delta L) = 68$ MHz. These side bands are probably caused by the formation of an external cavity due to the reflection of the light at the input end of the fiber loop in the setup shown in Fig. 6. This is because the distance between the LD and the input end of the fiber loop almost completely agrees with the calculated spa ing, $c/(2\Omega) = 2$ m, of the external cavity.

To verify the measured spectral width of $\Delta\nu = 51$ MH we constructed a self-heterodyne system [8] which analogous to the depolarizer system. The measurement that the polarization planes of the superimposed waves the beam splitter BS2 are parallel so as to interfere ea other. A polarizer was placed in front of the beam splitt BS2 at an angle of 45° with respect to these polarizati planes. Light then passing through the polarizer consist of the parallel components of the P and S waves project in the direction of the principal axis of the polarizer. T light was detected by a Ge-APD and its electrical sigr was monitored by a microwave spectrum analyzer.

The fiber length of the delay line was set at 500 which corresponded to the spectral resolution of 1.5 MH Theoretically, the spectral width monitored in the spe trum analyzer is twice that of the original Lorentzi spectral width of the LD. It was therefore confirmed tl the spectral width of the LD was 50 MHz as shown Fig. 11, which agreed well with that obtained by the c polarizer.

VI. Conclusion

We have presented a new type of single mode fib optic depolarizer. It is advantageous in that the fiber leng required to depolarize light is three orders of magnitu shorter than that for conventional fiber-optic Lyot dep larizers. As a result, light having a 51-MHz spect width, for example, can be depolarized to within -14 using only a 2-m fiber. The best depolarization val achieved is -23 dB, which is comparable with that conventional calcite or fiber-optic depolarizers. It was a ditionally found that depolarizer performance was limit by the mode coupling between the P and S waves at t polarization beam splitter.

Acknowledgment

The authors would like to express their sincere thar to Dr. Y. Sasaki for supplying the PANDA fiber and I. Yokohama for his useful discussions.

References

[1] K. Okamoto, Y. Sasaki, T. Miya, M. Kawachi, and T. Edahiro, " larisation characteristics in long-length V.A.D. single mode fiber *Electron. Lett.*, vol. 16, pp. 768–769, 1980.

[2] B. Hillerich and E. Weidel, "Polarization noise in single mode fil and its reduction by depolarizers," *Opt. Quantum Electron.*, vol. pp. 281–287, 1983.

[3] K. Böhm, K. Petermann and E. Weidel, "Performance of Lyot de larizers with birefringent single mode fibers," *J. Lightwave Techn* vol. LT-1, pp. 71–74, 1983.

[4] M. Nakazawa, M. Tokuda, and N. Uchida, "Measurement and a ysis of polarization properties of Rayleigh scattering for single-m optical fibers," *J. Quantum Electron.*, vol. QE-17, pp. 2326–2 1981.

[5] M. Born and E. Wolf, *Principles of Optics*, 3rd ed. New York: gamon, 1965, ch. 5.

[6] Y. Sasaki, T. Hosaka, K. Takada, and J. Noda, "8 km-long polarization-maintaining fibre with highly stable polarization state," *Electron. Lett.*, vol. 19, pp. 792–793, 1983.
[7] E. D. Hinkley and C. Freed, "Direct observation of the Lorentzian line shape as limited by quantum phase noise in a laser above threshold," *Phys. Rev. Lett.*, vol. 23, pp. 277–281, 1969.
[8] T. Okoshi, K. Kikuchi, and A. Nakazawa, "Novel method for high resolution measurement of laser output spectrum," *Electron. Lett.*, vol. 16, pp. 630–631, 1980.

*

Kazumasa Takada was born in Saitama Prefecture, Japan, on May 14, 1955. He received the B.S. degree in physics from Saitama University, and M.S. degree in physics from University of Tokyo, Japan, in 1979 and 1981, respectively.

In 1981, he joined the Ibaraki Electrical Communication Laboratory, Nippon Telegraph and Telephone Public Corporation, Ibaraki-ken Japan, where he has been engaged in research on transmission characteristics of the birefringent fibers and on the optical fiber gyroscopes.

Mr. Takada is a member of the Institute of Electronics and Communication Engineers of Japan and the Japan Society of Applied Physics.

Katsunari Okamoto was born in Hiroshima Prefecture, Japan, on October 19, 1949. He received the B.S., M.S., and Ph.D. degrees in electronics engineering from Tokyo University, Tokyo, Japan, in 1972, 1974, and 1977, respectively.

He joined Ibaraki Electrical Communication Laboratory, Nippon Telegraph and Telephone Public Corporation, Ibaraki, Japan, in 1977. He has been engaged in research on transmission characteristics of multimode, single-mode, and single-polarization optical fibers.

Dr. Okamoto received the Niwa Memorial Award in 1977, and the Achievement Awards and the Book Awards in 1978 and 1984, respectively, from the IECE. He is a member of the Japan Society of Applied Physics.

*

Juichi Noda was born in Yokohama, Japan, on February 6, 1943. He received the B.S., M.S., and Ph.D. degrees in electronics engineering from Hokkaido University, in 1965, 1967, and 1979, respectively. He joined the Musashino Electrical Communication Laboratory, Nippon Telegraph and Telephone Public Corporation, Tokyo, Japan, in 1967. He was engaged in research on electrooptic bulk devices, $LiNbO_3$-$LiTaO_3$ waveguides, and their integrated optical devices. Since 1981, he has been engaged in research on waveguide devices, polarization-maintaining fibers, and fiber applications at the Ibaraki Electrical Communication.

Dr. Noda is a member of the Institute of Electronics and Communication Engineers of Japan and the Japan Society of Applied Physics.

Effectiveness of a Quartz Rod as a Light Depolarizer

D. L. Portigal

Sperry Rand Corporation, Univac Division, Philadelphia, Pennsylvania 19101.
Received 25 November 1968.

It is often necessary to depolarize a light beam in an optical system. Translucent diffusers such as opal glass have been used for this purpose,[1] but in many situations, use of such diffusers leads to unacceptable loss of intensity. Various depolarizers involving birefringent material have also been used.[2–4] F. Cerdeira *et al.* have reported[5] the use of an ordinary quartz rod. However, these authors[5] have not published results showing the effectiveness of such a depolarizer. By using such a rod to depolarize the output of a monochromator, we have been able to decrease the polarization by at least two orders of magnitude with minimal intensity loss.

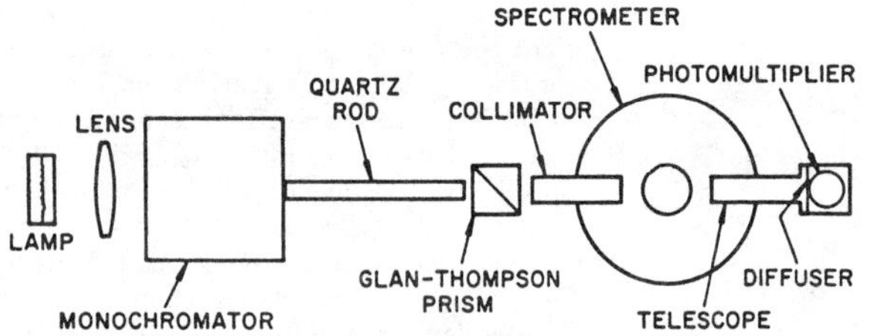

Fig. 1. Schematic diagram of apparatus used to measure depolarization of quartz rods.

The apparatus used in these measurements is shown in Fig. 1. A Hilger & Watts D–330 monochromator, with a 600 line/mm grating blazed at 300 mμ in first order is used. This has an angular aperture $f/5.8$. The measurements were taken using 100-μ wide slits. The 1P28 detecting photomultiplier is mounted on the telescope of a Gaertner L–114 spectrometer. A diffuser was mounted in front of the photomultiplier to eliminate its polarization dependence. A quartz–iodine light source was used. The end of the quartz rod was placed at the collimator focus, and its image was focused just in front of the photomultiplier diffuser. The objective lenses on the spectrometer are 28 mm in diameter. When this full aperture was used, the angular aperture of the detector assembly was $f/8.9$. Some measurements were taken with an 18-mm aperture. In that case, the angular aperture was reduced to $f/13.9$. The quartz rods tested were commercial grade, nominally 5 mm in diameter. In Table I, we show, as a function of rod length and objective aperture, the maximum relative variation in light intensity as the polarizer is rotated as well as the apparent polarization, i.e., the difference between the intensities of horizontally and vertically polarized light, divided by the mean intensity. Measurements were taken at 435 mμ. Note that the total variation and the polarization increase as the angular aperture decreases. This is the result that we would expect if depolarization results from multiple internal reflection, since the number of reflections which a ray undergoes increases with the angle that it makes with the rod axis.

In Table II, the maximum variation and apparent polarization are tabulated as a function of wavelength, with the optics aligned for minimum total variation at 435 mμ. A 0.6-m long rod and the full objective aperture were used in those measurements. A maximum variation of 0.5%, with a polarization of 0.2% w measured at 510 mμ with a 0.9-m rod. Better depolarizat was obtained at 350 mμ by adjusting the optics for that wa length and inserting a filter that rejects light whose wavelen is longer than 420 mμ.

The measurements of the polarization of the monochroma

Table I. Intensity Variation and Apparent Polarization Under Various Conditions λ = 435 mμ

Rod length (m)	Objective aperture (mm)	Maximum variation (%)	Apparent polarizatioı (%)
0.6	28	0.2	0.2
0.9	28	0.1	0.0
0.9	18	0.4	0.1

Table II. Wavelength Variation of Intensity Change and Apparent Polarization

λ (mμ)	Maximum variation (%)	Apparent polarization (%)
435	0.2	0.2
400	0.2	0.0
360	0.8	0.0
350	0.6	0.1

output without the quartz rod are not as precise, since measurements were made without a diffuser in front of photomultiplier. Also, a low pressure mercury arc lamp, wh has higher polarization than the quartz–iodine lamp, was us However, the polarization dependence of the photomultiplie only about 1%, and, at 400 mμ and 435 mμ, the polarizat with and without a diffuser in front of the light source diffe by 2% or less. Taking these factors into account, we infer t the monochromator output polarization at these waveleng was 19 ± 1%. It is somewhat larger at 510 mμ. At 365 r the polarization was found to be 10% when the light fr mercury lamp was depolarized, and 20% otherwise. We h also determined that the polarization of the quartz–iodine la was about one-third that of the mercury lamp. Taking t and the photomultiplier polarization into account we may in that the monochromator output has a polarization of about 1 at 360 mμ.

The light losses were not explicitly measured. No apprecia variation with rod length was noted. A review of our d indicates an overall loss of about 50%. In contrast, opal gl reduced the output by three to four orders of magnitude

From the foregoing results, we see that the light polarizat can be decreased by two orders of magnitude or more using commercial grade quartz rod. This result is obtained w reasonable loss of light, over a wide wavelength range, wit low cost, readily available material.

References

1. A. C. Hardy and F. H. Perrin, *The Principles of Op* (McGraw-Hill Book Company, Inc., 1930), p. 605.
2. B. H. Billings, J. Op. Soc. Amer. **44,** 66 (1951).
3. R. H. Hughes, Rev. Sci. Instrum. **31,** 1156 (1960).
4. C. J. Peters, Appl. Opt. **12,** 1502 (1964).
5. F. Cerdeira *et al.*, Bull Amer. Phys. Soc. **12,** 1049 (1967).

©1986 IEEE. Reprinted from *Journal of Lightwave Technology*, Vol. LT-4(2), 220–227 (February 1986).

Depolarization in Channel Glass Waveguides

ELSA M. GARMIRE, FELLOW, IEEE, AND KAZUO HONDA

***Abstract*—Depolarization was observed in channel waveguides fabricated by ion exchange in glass. This depolarization, as much as 10 percent in distances the order of 5 mm, was apparently due to edge roughness of the channels. A theory is introduced which analyzes this depolarization in terms of coupling between the longitudinal field components in linearly polarized modes of rectangular waveguides. In multimode channel guides it is shown that some depolarization will always be present whenever coupling between modes is large enough to create "*m*" lines. It is also shown that comparable depolarization may occur in single mode channel guides.**

I. Introduction

We REPORT a small conversion to the opposite polarization in channel glass waveguides formed by ion exchange in soda lime silicate substrates. This depolarization occurs as a result of deviations of the waveguide from perfect linear geometry. Such deviations may be due to inhomogeneities within the ion exchanged glass, sidewall roughness, or transitions between different waveguide geometries. The case analyzed here is for sidewall roughness. Depolarizations of the order of 10 percent for path lengths of guided light less than 1 cm long were observed in both single mode and multimode channel waveguides with widths from 6 to 150 μm. The possibility of depolarization is of concern for integrated optics devices such as beam splitters in isotropic materials for applications requiring polarization preservation. An analysis is presented to show that some depolarization is always present whenever there is mode-mixing or loss to radiation modes due to waveguide imperfections.

Any mechanism which causes mode mixing or radiation loss will also cause depolarization. The theory presented in this paper considers edge roughness of the waveguide sidewalls, but inhomogeneities within the waveguide will cause comparable depolarization. Although such depolarization has normally been considered small in the past, the purpose of this paper is to show that it may be large enough to degrade performance of devices requiring polarization preservation.

A. *Experimental Evidence of Depolarization*

Channel waveguides were fabricated by potassium ion exchange in soda lime silicate glass substrates using an aluminum mask to define channels. Because of previous observation of depolarization in ring resonators [1], and the hypothesis that this was due to mask roughness, we fabricated a mask with purposely rough walls in a series of channels of varying width but same wall roughness (by double exposure), as shown in Fig. 1(a). This made it possible to measure the waveguide loss as a function of width of the channel. Typical experimental results are plotted in Fig. 1(b), as a function of inverse waveguide width squared. This dependence will be explained in the theoretical analysis below.

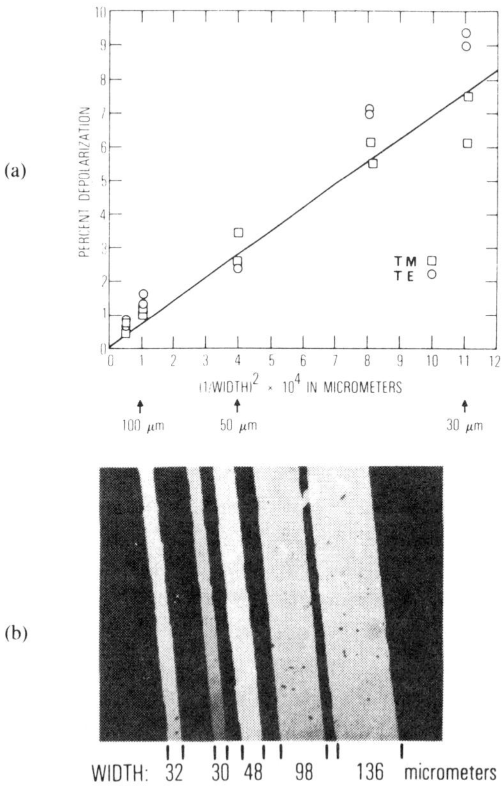

Fig. 1. Channel widths dependence of depolarization using the purposely rough mask shown in Fig. 1(b). Depolarization is defined as the ratio of the cross polarization intensity to the parallel polarization intensity at the output. Measurements were for a 3.8-mm propagation length in the channels.

For comparison, a high quality electron-beam-written single mode mask was used which was 6 μm wide. Even this waveguide showed depolarization, as shown in Fig. 2, because of slight lifting of the aluminum mask during the ion exchange, which led to edge roughness in the finished waveguides. This *e*-beam-written narrow single mode guide had considerably less loss than a linear extrapolation of the data in Fig. 1 to a waveguide width 6 μm would predict. At 6-μm width, the data from the pur-

Manuscript received April 12, 1985; revised August 23, 1985. This work was partially supported by the Air Force Office of Scientific Research and the Dupont Corporation.

E. Garmire is with the Center for Laser Studies, University of Southern California, Los Angeles, CA 90089-1112.

K. Honda was with the Center for Laser Studies, University of Southern California, Los Angeles, CA 90089-1112. He is now with Sony Corporation, Yokohama, 240 Japan.

IEEE Log Number 8406255.

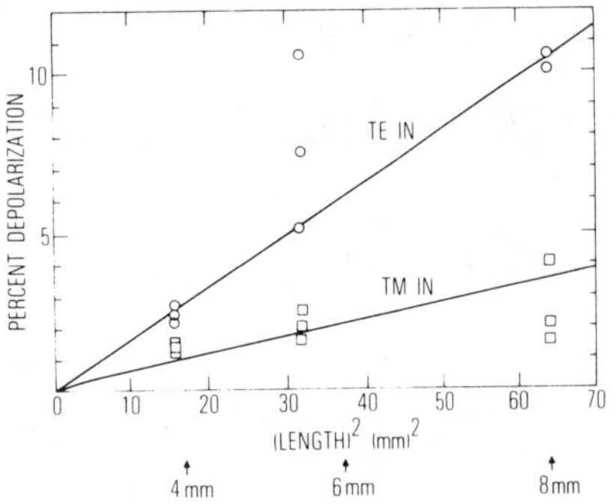

Fig. 2. Measure length dependence of depolarization in waveguides of 6-μm width, using prism couplers. The length is defined as the distance between the prisms plus two coupling distances of 1.5 mm each.

posely rough waveguides predicts complete depolarization, while in the *e*-beam-written guide, for the same length, only 2-percent depolarization was measured. This shows the advantage of the *e*-beam-written mask over the purposely rough mask. The dependence of the loss on the square of the waveguide length is characteristic of coherent coupling phenomena and was also observed with the waveguides shown in Fig. 1.

The experiments reported here were made by prism input and output coupling of polarized HeNe light. By collimating the input beam and properly adjusting the prism angle, only the lowest order mode was originally excited in the waveguide. The prism output coupler analyzed the single mode and multimode content of the guided waves. Experiments were carried out to determine that this small depolarization was not due to other causes. As a check on the polarizers and analyzers, planar waveguides were investigated and no depolarization was seen to be less than 0.1 percent. As a check on prism coupling, endfire input and output coupling were used and found to result in comparable depolarization. Finally, to eliminate the possibility that induced strain was the cause of the depolarization, the waveguides were measured after a series of anneals, with essentially the same amount of depolarization, independent of anneal time, up to the point at which the waveguides were annealed away. Further experimental data will be presented in Section IV.

B. Theoretical Approach

An estimate of the depolarization which is caused by waveguide imperfections is made by expanding the perturbed channel waveguide modes in linearly polarized modes of the "perfect" waveguide (which assumes small refractive index changes in an isotropic medium). In channel waveguides these modes are no longer truly TM and TE, but both have small components of polarization in the direction of propagation. It is these longitudinal components which are coupled by waveguide imperfections and cause conversion in the isotropic medium to the opposite polarization.

In this paper, the simplest possible analytic theory will be presented, using a step-index approximation for the ion-exchanged waveguides and the effective index method for the channels. Extension to graded index waveguides is straightforward but algebraically complex and will no change the nature of the results. The theoretical discus sion will assume edge roughness on the waveguid sidewalls. It is important to point out, however, that th approach of this analysis is also valid for imperfection due to inhomogeneities within the refractive index of th waveguide itself.

Since the longitudinal components of the polarizatio are very much less than the transverse components, might be expected that depolarization is always very muc smaller than polarized mode mixing. If it is assumed tha the edges of the waveguide are rough, with a roughnes independent of the correlation length, we will show tha an approximate expression for the ratio of depolarizatio to mode mixing is (36)

$$R = \sqrt{\frac{2\Delta n}{n}} \cdot \frac{\lambda}{2\pi n A}$$

where A is the thickness of the waveguide, and n is it refractive index, here 1.5. The wavelength of light λ = 0.63 μm in all experiments. For $A = 2$ μm, $R = 0.00$ This hypothesis means that even when there is 50-percen mode mixing, there should be only 0.15-percent depola ization, much smaller than we observed.

We will show in this section that the near-degenerac of the TE and TM polarizations for the same order moc in these channel waveguide allows for a much larger de polarization due to the rms surface roughness than the ca culations above. Under a series of simplifying assum tions we will estimate that this ratio is (see (37)):

$$R = (\tfrac{3}{4})(A/2B)^2$$

where the width of the waveguide is $2B$. For a typic single mode waveguide reported here this is $R \sim 0.08$ In this case, even a single mode waveguide with 10-pe cent radiation loss will have on the order of 1-percent d polarization. These numbers are consistent with exper mental results.

While mode mixing may occur between all confined ar radiation modes, the above argument suggests that th largest depolarization will occur from scattering betwe the nearly degenerate TE and TM modes of the same o der. Thus only this depolarization will be considered her

II. Theoretical Analysis for Imperfect Channel Waveguides

Modes in a waveguide are coupled only through wa roughness whose periodicity causes coherent coupling power between modes with different propagation co stants. Since the TE_0 and TM_0 modes are nearly dege erate, we will show that depolarization depends on t rms surface roughness, while polarized mode-mixing c pends on the Fourier transform of the surface roughne at a period of a few hundred micrometers, which is us ally much smaller than the rms surface roughness. In o der to study the relative strengths of depolarization a

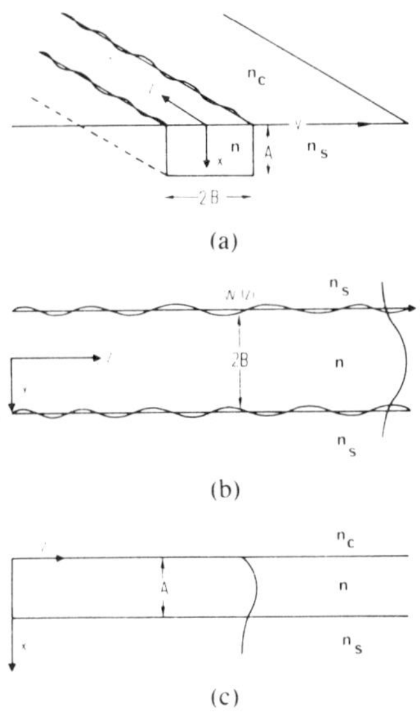

Fig. 3. Geometry of mode conversion analysis: (a) three-dimensional geometry, (b) y–z-plane, highlighting the wall roughness, and (c) x–z-plane and the asymetric guide profile in the z dimension.

mode mixing, therefore, it is first necessary to determine the relevant correlation lengths.

A. Calculation of Periodicity for Mode Conversion

The correlation length required to produce coherent coupling between two waveguide modes is given by

$$\Lambda = 2\pi/|\beta_i - \beta_f| \tag{1}$$

where β_i and β_f are the propagation constants for the initial and final modes, respectively. Since $\beta_i \sim \beta_f$, we may determine the correlation length through the expression

$$\beta_f - \beta_i \approx [\beta_f^2 - \beta_i^2]/2\beta_i. \tag{2}$$

Consider a rectangular waveguide of the geometry shown in Fig. 3. Using the effective index method described by Kogelnik [2, pp. 64–66], the propagation constant for the guided mode traveling in the z direction is given by

$$\beta^2 = n^2k^2 - k_x^2 - k_y^2 \tag{3}$$

where k_x is the transverse propagation constant of a planar waveguide in the x direction and k_y is the propagation constant of a planar waveguide in the y direction whose refractive index discontinuity is reduced by the effective refractive index calculated in the x-waveguide; k is the propagation constant in free space.

An approximate expression for k_x can be determined using arguments similar to those of Marcuse (leading up to [3, eq. 1.3-91], under the assumption that the mode is far from cutoff. Assume that the transverse propagation constant k_x is smaller than that for a perfectly confined mode by an amount e. Squaring Marcuse's equation [3, eq. 1.3-88] gives (under the assumption that e is small)

$$k_x^2 = [(N + 1)\ \pi/A]^2\ [1 - 2e_x] \tag{4}$$

where

$$e_x = \frac{m_c^2\sqrt{n^2 - n_s^2}\,k + m_s^2\sqrt{n^2 - n_c^2}\,k}{m^2\sqrt{n^2 - n_s^2}\sqrt{n^2 - n_c^2}\,k^2A}$$

where the m's = 1 for TE polarization, the m's = the refractive index of their respective medium (shown in Fig. 3) for TM polarization and N is the mode order. Marcuse included a small correction in each of the square roots which will be ignored here since we assume the modes are well confined.

There is an equivalent expression for the y component. For well-confined modes

$$k_y^2 = [(M + 1)\ \pi/2B]^2\ [1 - 2e_y] \tag{5}$$

where

$$e_y = \frac{m_c^2\sqrt{n^2 - n_s^2}\,k + m_s^2\sqrt{n^2 - n_c^2}\,k}{m^2\sqrt{n^2 - n_s^2}\sqrt{n^2 - n_c^2}\,k^2 2B}.$$

and M is the mode order. In this case the full width of the guide is defined as $2B$.

In a rectangular waveguide the proper values for m are chosen by the polarizations for the equivalent planar guides, shown in Fig. 3. That is, when the input polarization is along the y axis, e_x is defined for the TE polarization, since the electric field is parallel to the top and bottom walls. We call this the TE-like polarization. However, e_y is defined for the TM polarization, since the electric field is perpendicular to the rectangular waveguide sidewalls, which are the top and bottom walls of the effective planar y-waveguide. Furthermore, we make the simplification that the refractive index discontinuity from the glass to the air is much larger than from the waveguide to the substrate. That is

$$n_s^2 - n_c^2 \gg n^2 - n_s^2.$$

Then we obtain an approximate expression for the TE polarized mode:

$$e_{xTE} = \frac{1}{\sqrt{n^2 - n_s^2}\,kA}. \tag{6}$$

Similiarly, using the fact that $n_s = n_c$ in the y direction.

$$e_{yTE} = \frac{(n_s/n)^2}{\sqrt{n^2 - n_s^2}\,kB}. \tag{7}$$

The expressions for rectangular waveguides with incident light polarized along x are the same as (4) and (5), with a new definition for the e's. We have

$$e_{xTM} = \frac{(n_s/n)^2}{\sqrt{n^2 - n_s^2}\,kA} \tag{8}$$

and

$$e_{yTM} = \frac{1}{\sqrt{n^2 - n_s^2}\,kB}. \tag{9}$$

B. Periodicity for Depolarization

Consider the correlation length for conversion from one polarization to another of the same order (same values of

N and M). Combining (2)–(9) we determine, approximately

$$\beta_2 - \beta_1 = \left[\frac{\lambda^2 \sqrt{n^2 - n_s^2}}{4n^3}\right] \cdot \left[\frac{(N+1)^2}{A^3} - \frac{(M+1)^2}{B^3}\right]. \tag{10}$$

Using experimental numbers (justified in Section IV) for the lowest order mode ($N = 0$, $M = 0$), $n - n_s = 0.006$, $A = 2$ μm and $2B = 6$ μm, the periodicity which causes depolarization is

$$\Lambda_{dp} = 2\pi/|\beta_2 - \beta_1| = 1.8 \text{ cm}.$$

This means that it takes a relatively long distance for the TE and TM modes to get out of phase and it will be shown that coupling occurs via the rms deviation of edge roughness.

C. Correlation Length for Polarized Mode Mixing

The correlation length for modes which change values of N and M can be determined from the lowest order approximations in (4) and (5), without evaluating the e's. Mode-mixing from the NMth mode to the SPth mode has a correlation length given by

$$\frac{2\pi}{\Lambda_{\text{pol}}} = \left[\frac{\lambda}{4n}\right]\left[\frac{|(N+1)^2 - (S+1)^2|}{A^2} + \frac{|(M+1)^2 - (P+1)^2|}{B^2}\right]. \tag{11}$$

For mode mixing between the two lowest order modes, which gives the longest possible correlation length, $N = S = 0$ and $M = 0$, $P = 1$. In this case

$$\Lambda_{\text{pol}} = 32nB^2/3\lambda. \tag{12}$$

For the experimental device, with $2B = 6$ μm, $\Lambda_{\text{pol}} = 406$ μm. This is almost 50 times smaller than the correlation length for depolarization. This shows that depolarization and polarized mode mixing occur from very different periodicities in wall roughness.

D. Coupled Mode Theory in Imperfect Guides

In order to compare mode conversion between modes of the same number but different polarization modes, and mode conversion between different mode numbers, it is necessary to observe that in glass waveguides the former are very nearly degenerate while the latter are not. This means that in a typical propagation distance of a few millimeters, different polarizations remain in phase and light can couple coherently between them while modes of different mode number experience many phase reversals. To analyze both cases, we use coupled mode theory, beginning with [2, eqs. 2.6.23] in Kogelnik's chapter in *Integrated Optics*, which describe the growth of the mode amplitudes. The electric field of the guided mode can be expressed as an expansion in terms of forward going modes

$$E(x, y, z, t) = \sum_m A_m(z) \exp(i\beta_m z)\, E_m(x, y) \exp(-i\omega t).$$

When only two forward-going modes A and B are considered they will be coupled through the following equations:

$$dA(z)/dz = -iK(z)\, B(z)\, e^{-iDz}$$

and

$$dB(z)/dz = -iK(z)\, A(z)\, e^{iDz} \tag{13}$$

where $D = \beta_B - \beta_A$.

The coupling coefficient K is also given by Kogelnik [2, eqs. 2.6.21, 2.6.22]. When $\epsilon - \epsilon_s \equiv [n^2 - n_s^2]\,\epsilon_0 \equiv \Delta\epsilon << \epsilon$, then

$$K(z) = \iint_{-\infty}^{\infty} dx\, dy\, \Delta\epsilon E_i(x, y) \cdot E_f(x, y)/P. \tag{14}$$

The factor ω is the radian frequency and P is the normalizing power given by [2, eqs. 2.3.12 and 2.3.46]:

$$P = 2 \iint_{-\infty}^{\infty} E_T \times H_T\, dx\, dy$$

where the subscripts T represent the transverse components of the field.

Depolarization requires E_i and E_f to have the same mode number but opposite polarization. Because of their orthogonality, in the isotropic medium the transverse components of the field cannot contribute to (14) and to depolarization. Only the longitudinal components contribute to coupling between TE and TM modes. The transverse components of the field, on the other hand, contribute to polarized mode conversion (between TE_0 and TE_1 modes, for example).

In general, $\Delta\epsilon$ is considered to be a function of x, y, and is given by the perturbation of the guide from perfection. Consider the case shown in Fig. 3 in which the perturbations are variations in width of the guide. We can write

$$\Delta\epsilon = \epsilon - \epsilon_s, \qquad \text{for } W(z) > 0$$

and

$$\Delta\epsilon = -(\epsilon - \epsilon_s), \qquad \text{for } W(z) < 0 \tag{15}$$

where $W(z)$ is the deviation of the waveguide width from its "perfect" width. If the width deviations are small, the magnitude of the field does not change much over the integration and, following a reasoning similar to the derivation by Kogelnik of [2, eq. 2.6.44], (14) can be written as

$$K(z) = W(z)\, C \tag{16}$$

where C is a constant given by

$$C = 2\omega\Delta\epsilon \int dx\, E_i(x, B) \cdot E_f(x, B)/P. \tag{17}$$

The fields are expressed at the boundary of the waveguide, given by $y = B$, and a factor of 2 is introduced to include roughness on both boundaries of the waveguide. Calculations of C will be made later.

Including the z variation of W, (13) can be separated by differentiating and combining terms to obtain:

$$d^2A/dz^2 - [(1/W)(dW/dz) + D/i]\,(dA/dz) + W^2C^2A = 0 \tag{18}$$

and an identical equation for $B(z)$. In general, we need to include all possible modes, both confined and radiation. However, here only two modes are analyzed for simplicity. Attention will be focused on the TE and TM modes of the same order, as well as on mode mixing between only two modes. Extension to multimode analysis can be made following the treatment by Marcuse [3]. Such an analysis is not expected to change the character of the results.

Now let us separately analyze the two cases, corresponding to depolarization and polarized mode-mixing.

E. Depolarization

In this case, since the modes are nearly degenerate and the correlation length was shown to be so long, assume that $D = 0$. Furthermore, consider that perturbations in waveguide width W along the z direction occur at a rate which is much faster than any mode conversion. In that case, it is possible to average (18) over rapid variations in W, and the last term becomes $\langle W^2 \rangle$. Let us consider the middle term by referring to Fig. 3(b). When the "perfect" waveguide is chosen as the center of the perturbation, it can be seen that for each value of dW/dz, there are regions in which $W(z)$ is equal and opposite in sign. Thus it can be seen that $\langle (1/W)(dW/dz) \rangle = 0$. This allows us to ignore the second term in (18), which can then be expressed as a growth of A in terms of the rms value of W, $W_{rms} = \sqrt{\langle W^2 \rangle}$:

$$d^2A/dz^2 + (W_{rms})^2\,C^2A = 0. \tag{19}$$

This shows that depolarization is determined by the rms value of the height deviations.

F. Nondegenerate Mode Mixing

For nondegenerate mode mixing, $D \neq 0$ so that coherent coupling and growth of mode A occur only when the perturbation has a periodicity close to the difference in propagation constants D. That is, if

$$W(z) = W_{ft}(e^{iDz} + e^{-iDz}) \tag{20}$$

then the first term has a derivative which gives

$$(1/W)(dW/dz) = iD$$

and cancels the middle term in (18). The other term in (20) leads to an oscillatory term in (18) which varies rapidly with long enough guides and will average to zero. Thus nondegenerate mode mixing obeys the same equation (19), with the understanding that $(W_{rms})^2$ is now replaced by $(W_{ft})^2$, the square of the Fourier component of the perturbation at the spatial frequency given by the difference in propagation vectors between the two modes, $\Lambda = 2\pi/D$. This is an approximate model, of course, since for finite lengths a band of Fourier components should be included.

G. Growth of New Modes

Equation (19) and its companion identical equation for $B(z)$ can be solved, under the boundary conditions that at the entrance all power is in mode B. For small distances, the growth of the opposite polarization or higher order mode is

$$I(z) = C^2w^2z^2. \tag{21}$$

Note the quadratic dependence on z, which is characteristic of all coherent coupling.

III. Estimate for Imperfections Due to Edge Roughness

The exact nature of the waveguide imperfections sidewall roughness is not known. so that only the simplest possible step-index model will be used for this analysis. Comparisons between experiment and theory should be considered only qualitative. In fact, the use of diffused guides may tend to decrease the importance of roughness [4]. What will be shown here is that the theoretical results are not unreasonable.

The coupling coefficient necessitates calculating (17), where the fields may be longitudinal or transverse, depending on the modes being coupled. When the components of polarization transverse to the direction of propagation are considered, this expression describes coupling between modes of the same polarization, since E_x and E_y are orthogonal. As far as depolarization is concerned, the only nonzero terms are the longitudinal components of the electric field $E_{zi}E_{zf}$, which are introduced by the fact that the waveguides are rectangular rather than planar.

There are two polarizations to consider:

TE-like:	$E_T \| y$	$H_T \| x$
TM-like:	$E_T \| x$	$H_T \| y$.

Using separation of variables we express the magnetic field in the waveguide as

$$H(x, y, z) = H_0 \sin(k_x x + \phi_x) \cos(k_y y + \phi_y) \exp(ik\beta z) \tag{22}$$

where the ϕ's are phases. Since the refractive index of the cladding (air) is much smaller than of the glass substrate, the coordinate system is chosen so that $\phi_x = 0$. Furthermore, if only modes symmetric in y are considered, then ϕ_y also $= 0$. For simplicity consider only this case. Thus the magnetic field is:

$$H(x, y, z) = H_0 \sin(k_x x) \cos(k_y y) \exp(i\beta z). \tag{23}$$

The electric field obeys a similar relationship:

$$E(x, y, z) = E_0 \sin(k_x x) \cos(k_y y) \exp(i\beta z). \quad (24)$$

A. Depolarization

Using Maxwell's equations the longitudinal component of the field may be expressed in terms of the transverse components. Consider an incident TE-like wave. Then

$$E_{zi} = (-1/\omega\epsilon)\, dH_{xi}/dy = (1/\omega\epsilon)\, H_0 k_{yi} \sin k_{xi}x \sin k_{yi}y$$

and

$$E_{zf} = (1/\omega\epsilon)\, dH_{yf}/dx = (1/\omega\epsilon)\, H_0 k_{xf} \cos k_{xf}x \cos k_{yf}y. \quad (25)$$

This means that the coupling constant (17) will be

$$C_{dp} = [\Delta\epsilon/\omega P\epsilon^2]\, H_0^2 I k_{yi} k_{xf} \sin k_{yi}B \cos k_{yf}B \quad (26)$$

where

$$I = 2 \int_0^A \sin k_{xi}x \cos k_{xf}x \, dx.$$

The integral I can be evaluated for any particular final mode. The near-degenerate case under consideration here is scattering from one polarization to another within the same mode. Under these conditions $k_{xi} = k_{xf} = k_x$ and the integral is easily evaluated to obtain

$$I = (1/k_x) \sin^2 k_x A.$$

In order to evaluate (26), the normalization must be included. According to Kogelnik [2, eq. (2.3.46)]

$$P = A_e B_e H_o^2 (\beta/\omega\epsilon) \quad (27)$$

where A_e and $2B_e$ are the effective waveguide height and width, respectively. This gives the coupling constant for depolarization

$$C_{dp} = [\Delta\epsilon/\epsilon][k_y/\beta][1/A_e B_e] \cdot [\sin^2(k_x A) \cos(k_y B) \sin(k_y B)]. \quad (28)$$

While this equation was derived for TE polarization, the symmetry of (17) means that TM input has the same depolarization as TE input, given by (28).

The maximum value of C_{dp} is for $\sin^2(k_x A) = 1$ (a waveguide near cutoff) and for $\cos(k_y B) \sin(k_y B) = \frac{1}{2}$. In addition, k_y can be evaluated using (3). Since $k_x < k_y$, we can approximately say

$$k_y/\beta = \sqrt{n^2k^2 - \beta^2}/\beta \le \sqrt{n^2 - n_s^2}. \quad (29)$$

Thus

$$C_{dp} \le [\Delta n/n][\sqrt{2n\,\Delta n}][1/A_e B_e]. \quad (30)$$

Equation (30) shows an inverse dependence on the waveguide width, so that the depolarization is proportional to the inverse width squared, by (22), in agreement with the data of Fig. 1.

In order to make numerical comparison with Fig. 1, we use experimentally determined waveguide parameters, as described in Section IV: $\Delta n = 0.006$, $n = 1.5$, $A_e = 1.8$ μm. The data shown in Fig. 1 can be numerically fit b assuming that $W_{rms} = 3.75$ μm. As an example, for guide of width $2B = 30$ μm, $C = 0.20\ \text{cm}^{-1} \cdot \mu\text{m}^{-1}$, an an rms roughness of 3.75 μm will show 8-percent mod conversion in 3.8 mm. This large roughness is consister with the roughness of the mask shown in Fig. 1(a).

The data in Fig. 2 shows the quadratic dependence o distance for depolarization. In this case $2B = 6$ μm. Thi gives $C_{dp} = 1.0\ \text{cm}^{-1} \cdot \mu\text{m}^{-1}$. The average of the TE an TM data can be fit by an rms surface roughness of 0.2 μm, which gives 5-percent depolarization in 8 mm. In th experiments reported here, there was a tendancy of th aluminum mask to lift during ion exchange, so that thi relatively large roughness at the mask edges is not unrea sonable.

The different depolarization for TE compared to TM i the 6-μm wide guides occurs because there is more los for the TE than the TM modes. It is well known that i planar guides wall roughness causes more coupling to ra diation modes for the TM than for the TE modes, sinc the TM fields at the walls are slightly larger than the T fields. For channel waveguides with rough sidewalls, a input TE-like polarization corresponds to a TM polariza tion in the effective planar waveguide confined by th sidewalls, as shown in Fig. 3(b). Thus it is this polar zation which will experience the most loss. If the TE-lik polarization of the channel waveguide always experience more loss than the TM-like polarization, then the appare depolarization for TE modes will be more and the appa ent depolarization for TM modes will be less. This can b seen by assuming that TM modes are lossless and T modes experience an exponential loss e^{-aL}. Then the rati of depolarization for TE to TM modes is

$$I_{dp}(\text{TE})/I_{dp}(\text{TM}) = \exp(2aL). \quad (3$$

Thus the effective TE depolarization is expected to b larger than the effective TM depolarization.

We have shown numerically that a large edge roughne due to imperfect masks and the lifting of masks durir fabrication can lead to the observed depolarization. As th guides become narrower, the tolerances on rms roughne to avoid mode conversion increase quadratically. Goc masks and fabrication procedures will be required to avo mode conversion.

B. Polarized Mode Conversion

Let us contrast the above result with polarized moc conversion in which the mode number changes. In th case (17) will involve transverse field components rath than the much smaller longitudinal field components ar mode conversion would be expected to be much mo likely than depolarization. However, mode conversic occurs with much shorter correlation lengths than dep larization, and the Fourier transform of edge roughness these correlation lengths is much smaller than the rms su face roughness. Thus polarized mode conversion and d polarization effects may be within an order of magnitud

The analysis begins with the polarized mode conversion coefficient which is given by (16) and (17) with E_i and E_f representing transverse field components. Under the same assumptions as in the last section the y integral can be immediately evaluated to obtain

$$C_{mc} = (\omega/P)\Delta\epsilon\, E_o^2 \cos(k_{iy}B)\cos(k_{fy}B)\, l \qquad (32)$$

where

$$l = 2\int_0^A [\sin(k_{ix}x)\sin(k_{fx}x)\, dx.$$

To evaluate the integral we make use of the fact that during mode conversion the value of k_x is unchanged, only k_y changes. Thus $l = A$. To evaluate (32) we need the proper normalization [2, eq. (2.3.12)]:

$$P = [G_o\beta\omega/k^2][E_o^2 A_e B_e]. \qquad (33)$$

Then (including also the fact that $\beta \sim k$ and $A \sim A_e$):

$$C_{mc} = [\Delta\epsilon/\epsilon_o][k/B_e]\cos(k_{iy}B)\cos(k_{fy}B). \qquad (34)$$

Maximum values for the cosine functions will be 1, although this is near cutoff and both modes cannot have this value. Thus assume that the product of the cosines is less than about $\frac{1}{2}$. This leads to

$$C_{mc} \leq [n\Delta n][k/B_e]. \qquad (35)$$

We may put numbers into this expression to determine the amount of mode mixing which will occur from a roughness of $W_{ft}\cos Dz$. From an assumption of $B_e = 3$ μm, $\lambda = 0.63$ μm, $\Delta n = 0.006$, and $n = 1.5$, we obtain $C_{mc} = 300$ cm$^{-1}\cdot\mu$m^{-1}. It is difficult to know what to assume for the Fourier transform of the edge roughness at a correlation length of 400 μm. As a reasonable model, consider the ratio of edge roughness to correlation length as a constant. Since the rms surface deviation of 0.28 μm was 3×10^{-5} of the depolarization correlation length, the same fraction of the mode conversion length leads to $W_{ft} = 0.012$ μm. This will cause 50-percent mode conversion in 2 mm, consistent with experimental observations in multimode guides. This demonstrates that whenever mode mixing is observed, as witnessed by the observation of "m" lines, it is expected that there will be some degree of depolarization.

C. Ratio of Depolarized to Polarized Scattering

The ratio of the intensities for depolarization and mode conversion is given by

$$R = I_{dp}/I_{mc} = [C_{dp}/C_{mc}]^2[W_{rms}/W_{ft}]^2$$

$$= \sqrt{\frac{2\Delta n}{n}}\left[\frac{\lambda}{2\pi n A_{eff}}\right]\left[\frac{W_{rms}}{W_{ft}}\right]^2 \qquad (36)$$

Notice that for constant values of W, this is independent of the width of the guide, depending only on its height. If, however, it is assumed that W is a constant fraction of the correlation length, we can determine (W_{rms}/W_{ft}) from (10) and (11) (assuming $B >> A$):

$$\frac{W_{rms}}{W_{ft}} = \frac{3\pi n^2}{4\sqrt{n^2 - n_s^2}}\frac{A^3}{\lambda B^2}.$$

Inserting into R, we obtain

$$R = (\tfrac{3}{8})(A^2/B^2). \qquad (37)$$

It is interesting to note that this is independent of everything but geometry and predicts that the fraction of depolarization goes up quadratically with decrease in waveguide width. That this is reasonable can be seen from the fact that as the waveguides become more square, longitudinal components become larger. Inserting numbers for the 6-μm-wide guide, we expect 6 percent of the mode conversion to be depolarized.

IV. Further Experimental Discussions

The substrates used for these glass waveguides were Kodak lustra glass, which are flat soda lime silicate blanks used as the basis for photoresist masks. The waveguides were fabricated by immersing the substrates into molten KNO_3 at 400°C for periods between 1 and 6 h. The refractive index profile was determined by experiments with multimode planar waveguides. Propagation constants were measured and the waveguide parameters were determined from the WKB method. Fig. 4 shows a typical refractive index profile. From this data the surface index increase of 0.006 was determined.

Single mode waveguides 6 μm wide were fabricated. The depth of the channel waveguides was determined from the effective index method [2]. The symmetric equivalent planar guide in the y dimension of width 6 μm and $\Delta n = 0.006$ has a V parameter of 8. Using the graph in Kogelnik [2, fig. 2.8], the two lowest order modes have b parameters = 0.9 and 0.6, giving an effective index of 0.0054 and 0.0036, respectively. These effective indices are used in the equivalent asymmetric planar guide in the x dimension. The single mode regime has a V parameter between 2 and 5, so that the waveguide will be single mode for 1.6 μm $< A <$ 2.0 μm. As a model for the single mode regime, we take $A = 1.8$ μm.

The mask used to protect against KNO_3 ion exchange was aluminum, with photoresist lift-off as the procedure to define the channels. When the aluminum film was too thin, there was a tendancy for it to lift during the ion exchange procedure. This is the source of the roughness which caused depolarization. Use of thicker aluminum reduces the roughness and depolarization. The nonuniform lifting of the edges of the mask caused nonuniform depolarization. Fig. 5 shows a histogram of sixteen "identical" 6-μm-wide single mode channel waveguides. It can be seen that roughly a third have very low depolarization, roughly a third have the highest possible depolarization, and a third have depolarizations lying between the two extremes. It can also be seen that the depolarization of TE

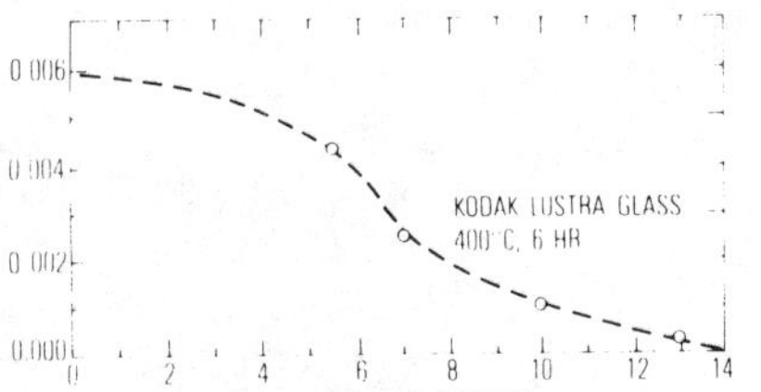

Fig. 4. Refractive index profile measured for potassium ion exchange in Kodak lustra grass.

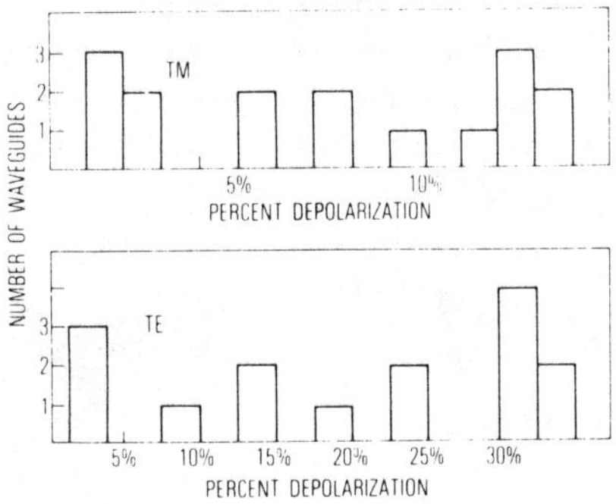

Fig. 5. Histograms showing the number of channel waveguides with a given percent depolarization in a number of nominally identical channels, demonstrating that nonuniform mask lifting was the cause of depolarizing edge roughness.

modes is almost three times as large as for TM modes. This was discussed above as due to the fact that the TE polarization experienced much more mode conversion loss.

V. Conclusions

We have demonstrated that depolarization can occur in glass channel waveguides when the sidewalls are rough, due to coupling between the longitudinal field components. Numerical estimates show that the roughness must be significant before substantial mode conversion takes place. However, depolarization increases as the aspect ratio of the rectangular waveguide decreases. Whenever there is enough mode conversion to obtain "*m*" lines some depolarization is expected. This is even true in single-mode guides. In the case reported here, approximately 6 percent of the mode conversion was depolarized.

Any deviation of a waveguide from a perfect channel will lead to some mode conversion. Discussed here is the case of surface roughness. Other possible sources of mode conversion are *Y* branches and curved guides. Numerical analysis of mode conversion in these cases awaits further investigation.

Acknowledgment

The authors wish to thank Dr. K. Wilson and Dr. C. Mueller for assistance in portions of the experimental program.

References

[1] K. Honda, E. M. Garmire, and K. E. Wilson, "Characteristics of an integrated optics ring resonator in glass," *J. Lightwave Technol.*, vol. LT-2, no. 5, pp. 714–718, Oct. 1984.

[2] H. Kogelnik, "Theory of dielectric waveguides," in *Integrated Optics*, T. Tamir, Ed. New York: Springer Verlag, 1975.

[3] D. Marcuse, *Theory of Dielectric Optical Waveguides*. New York: Academic, 1974.

[4] T. Findakly, E. Garmire, and H. T. Moon, "Comparison of losses in imperfect surface-diffused and buried optical waveguides," *Opt. Lett.*, vol. 4, no. 5, pp. 149–151, May 1979.

Section Five
Polarization Filters

Polarization narrow-passband filters have applications in most fields of science. Lyot was the inventor of the first narrow-band polarization filter. This first filter made it possible to look directly at the sun with the light of hydrogen alpha. This let astronomers get a glimpse of the tremendous complexity of the solar surface and spawned an increasing number of filters based on his technique.

The transmission of Lyot's original filter was quite low because it required a large number of polarizers. A tremendous improvement was made by Šolc. His paper was written in Czechoslovakian and I had planned to include it and rely on Evans's paper. Fortunately, I did find in my own archives a translation, which I have included.

Reprinted from *Comptes Rendus*, Vol. 197, 1593–1595 (1933).

ASTRONOMIE PHYSIQUE. — *Un monochromateur à grand champ utilisant les interférences en lumière polarisée.* Note de M. **Bernard Lyot**, présentée par M. Charles Fabry.

L'appareil que nous allons décrire très sommairement permet d'observer simultanément les différents points d'une source lumineuse étendue, en laissant passer une région étroite quelconque de son spectre et sans perdre beaucoup de lumière. Ces propriétés le rendent supérieur aux meilleurs écrans colorés et permettent de le substituer, avec avantage, au spectroscope à plusieurs fentes, au spectrohélioscope et au spectrohéliographe.

Le principe de cet appareil est le suivant :

La lumière de la source traverse successivement une série de polariseurs dont les sections principales sont croisées. Entre chaque polariseur et le suivant est placée une lame cristalline de spath ou de quartz, par exemple, taillée parallèlement à l'axe optique. Les sections principales de ces lames forment des angles de 45° avec celles des polariseurs. Chacune d'elles a une épaisseur double de la précédente ([1]).

Négligeons d'abord la dispersion des lames cristallines et les pertes de lumière par réflexion. La vibration d'amplitude A, qui sort du 1er polariseur, est décomposée par la lame la plus mince en deux vibrations rectangulaires d'amplitude $A/\sqrt{2}$, présentant une différence de marche δ. Le 2e polariseur les rend parallèles et réduit leur amplitude à $A/2$. Après la 2e lame et le 3e polarisateur nous avons, de même, 4 vibrations d'amplitude $A/4$, présentant des retards 0, δ, 2δ, 3δ. Soit n le nombre des lames; à la sortie du monochromateur nous aurons 2^n vibrations parallèles d'amplitude $A/2^n$ et présentant des retards 0, δ, 2δ, $\ldots$, $(2^n - 1)\delta$. Le spectre de la lumière résultant de leurs interférences consiste en un petit nombre de bandes dont l'amplitude au centre est A et la largeur totale $2\lambda/2^n\delta$. Elles sont distantes de λ/δ et encadrées par quelques maxima secondaires faibles. La dispersion des lames a pour effet de rapprocher les maxima et de les rendre plus fins dans le rapport $1 - (\lambda/n_e - n_0)\, d(n_e - n_0/d\lambda$.

L'épaisseur de la lame la plus mince sera choisie assez faible pour qu'on puisse isoler facilement une seule bande avec un écran coloré; le nombre de lames dépendra de la largeur de la portion de spectre qu'on veut isoler.

Réalisé sans précautions spéciales, cet appareil présenterait de graves inconvénients que nous éviterons de la manière suivante :

Lames biréfringentes. — La biréfringence des lames cristallines varie très rapidement lorsque l'on s'écarte de la normale aux faces, soit dans le plan de l'axe optique, soit dans le plan perpendiculaire. Dans les deux cas, les variations sont inverses. Ces variations qui limiteraient beaucoup le champ du monochromateur seront compensées, en grande partie, en remplaçant la lame d'épaisseur e par 2 lames d'épaisseur $e/2$ superposées de telle manière que leurs axes optiques soient rectangulaires. Pour que leurs

---

([1]) Mascart, puis Fabry et Perot avaient utilisé le spectre cannelé produit par une seule lame de quartz parallèle à l'axe, pour éteindre à volonté chacune des deux raies D du sodium (Mascart, *Annales scientifiques de l'École Normale supérieure*, **3**, 1874, p. 395; Fabry et Perot, *Comptes rendus*, **130**, 1900, p. 653).

biréfringences s'ajoutent au lieu de se retrancher, on intercale une lame demi-onde dont la section principale forme, avec leurs axes optiques, des angles de 45°. Les courbes isochromatiques de cet ensemble sont des cercles beaucoup plus grands que les cercles inscrits dans les hyperboles données par la simple lame. On obtient un champ encore plus grand en taillant les lames dans un cristal biaxe, parallèlement aux axes optiques. Soient n_1, n_2, n_3 ses indices principaux. La biréfringence δ de la double lame sous l'incidence i est liée à la biréfringence normale δ_n par la formule $\delta = \delta_n(1 + Ki^2)$, avec $K = (n_2^2 - n_1 n_3) / n_2^2 n_3$.

Pour le spath, $K = +0{,}0104$; pour le quarts K, $= -0{,}0006$.

Une deuxième solution intéressante consiste â tailler les deux lames dans deux cristaux, uniaxes par exemple, de signes différents. Leurs axes optiques doivent être parallèles et la lame demi-onde doit être supprimée. Le rapport des épaisseurs est tel que les surfaces isochromatiques soient circulaires. L'ensemble convient dans un domaine spectral étendu. Dans le cas d'une lame spath et quartz, l'épaisseur du quartz doit être 17,5 fois celle du spath et le coefficient K est 0,0055.

La biréfringence de l'ensemble sera rendue variable dans les limites de quelques longueurs d'onde en taillant, par exemple, les deux lames légèrement prismatiques et en déplaçant l'une d'elles. Un index et une règle graduée donneront la biréfringence.

Polariseurs. — Les polariseurs pourront être du type Glazebrook. Le fort astigmatisme qu'ils introduisent sur des faisceaux de lumière convergente sera éliminé en croisant leurs sections principales.

Disposition générale de l'appareil. — Pour diminuer les pertes de lumière par réflexion et la diffusion par les nombreuses surfaces optiques, les polariseurs et les lames biréfringentes sont immergés dans un liquide transparent contenu dans deux tubes placés à la suite l'un de l'autre. L'objectif d'une lunette donne une image de la source sur une lentille convergente qui précède le premier tube. Cette lentille forme une image de l'objectif entre les deux tubes. Un objectif symétrique, placé à cet endroit donne une deuxième image de la source au centre du second tube, sur une lentille convergente qui fait passer tous les rayons à travers un objectif placé à la suite du dernier tube. Celui-ci forme l'image monochromatique définitive. Le faisceau lumineux garde ainsi un diamètre constant d'un bout à l'autre du monochromateur et le champ est très étendu.

Tout l'appareil est contenu dans un thermostat pour éviter les variations de la longueur d'onde du monochromateur avec la température qui sont, par degré centigrade, de 0,4 Å pour le spath et 0,6 Å pour le quartz.

Un appareil de ce type est en construction. Il comportera 7 lames spath-quartz et 2 lames quart-quartz. La plus épaisse sera constituée par 3mm,63 de spath e 63mm,76 de quartz. Il isolera dans le vert, une bande spectrale de largeur moyenne 1Å en utilisant environ, le quart de la lumière. Il aura une ouverture utile de 18mm et donnera un champ de 6°. Il sera utilisable sans modification entre 3800 et 20000 Å.

Ce monochromateur permettra, par exemple, d'obtenir des photographies détaillées de la couronne solaire en dehors des éclipses, avec la lumière de ses raies brillantes. Il permettra également d'étudier visuellement ou au moyen de films cinématographiques accélérés les mouvements de l'atmosphère solaire : plages faculaires, filaments et protubérances.

A Tunable Narrow-Band Optical Filter

Bruce H. Billings*
Research Laboratories, Polaroid Corporation, Cambridge, Massachusetts

(Received July 18, 1947)

Polarization interference filters are described which can have a pass band ranging from a fraction of an angstrom to several hundred angstroms in width. The pass band can be shifted to any desired region of the spectrum. These tunable filters are based on the fixed filters discussed by Lyot and Evans. The transmission band is formed by the superposition of the polarized channel spectra, produced by x-cut plates of quartz or other birefringent media placed between parallel polarizers. The birefringent plates have thicknesses in the ratio 1:2:4 etc. The tuning is accomplished by changing the retardation of successive elements so that transmission maxima in the various channel spectra coincide at the desired wave-length. The retardation change can be made mechanically, for example, by stretching supplemental plastic sheets in series with the filter elements, or can be made electrically by using Kerr cells or crystals with high electro-optic coefficients, such as ammonium dihydrogen phosphate. The additional retardation never has to exceed a full wave at the wave-length of peak transmission. The measured transmission of an experimental filter is shown. The electrical tuning method is particularly adapted to cathode-ray oscillograph presentation of spectra. The filter also has possible application in color reproduction and colorimetry. With a pass band of a half-angstrom line of sight motion of solar prominences could be determined by the use of the Doppler shift of the prominence radiation.

In 1933 Lyot[1] discussed a filter which was planned to have an aperture of 25 mm and a pass band in the green of only 1A. Using the technique described in his paper, various investigators[2] have built filters of wide apertures and narrow transmission bands. The filters of this type have been fairly expensive to make, and have been used so far in the study of solar prominences. An improvement in such a filter would be to make it variable so that the pass band could be shifted at will over a considerable range of wave-lengths. Such an improvement has been described by Lyot[3] and by Evans[4] but is difficult to manipulate.

In this paper a simple tuning method is described which makes it possible to set the pass band of the filter at any point over a considerable range.

Both the fixed and, more especially, the tunable filter represent new scientific tools which have many possibilities for exploitation. The filters can be made with any pass band from a fraction of an angstrom to hundreds of angstroms. They can also be made to function in the ultraviolet and in the infra-red through the quartz-transmission region. In colorimetry and color reproduction there are many applications for a filter of perhaps 100-angstroms band width which could be tuned in discrete jumps rapidly through the visible spectrum. With the high speed provided by electrical tuning methods it should be possible to make spectral analyses of very short-lived phenomena, such as the discharge of an Edgerton lamp. These spectra could be displayed on a cathode-ray oscillograph. Presentation on cathode-ray oscillographs of other absorption or emission spectra seems quite straightforward with electrical tuning methods. With very narrow band filters it should be possible to make an analysis of the fine structure of Raman lines. Another application for the tunable filter is for the analysis of three-dimensional motion of solar prominences. A more detailed description of the astronomical applications will be made in another paper.

The peak transmission of the filters will depend very much on the method of construction. With a many-element filter using sheet polarizers the peak transmission may be of the order of 10 percent. With quartz or calcite polarizers and a laminated construction, the peak transmission might perhaps be as high as 35 or 40 percent. The angular aperture of the filter will also depend on its construction and on the narrowness of the desired pass band. For filters of several hundred-

* Now with Baird Associates Inc. Cambridge, Massachusetts.

[1] B. Lyot, Comptes rendus **197**, 1593 (1933).
[2] Y. Öhman, Nature **141**, 291 (1938); J. Evans, Pub. Astro. Soc. Pacific **52**, 305 (1940).
[3] B. Lyot, Ann. Astrophys. **7**, No. 1–2, (1944).
[4] J. Evans, Cientia e Investigacione, to be published.

angstroms band width it might be of the order of 20 degrees total field, whereas for a one-angstrom band filter the angular field might be of the order of a degree.

THEORY OF THE FILTER

In the basic filter the narrow band is formed by the superposition of the polarized channel spectra produced by many retardation plates. An x-cut plate of quartz or other birefringent crystal, when placed between crossed or parallel polarizers, transmits light at periodic intervals through the spectrum. In a spectroscope such a plate shows a series of dark bands across the spectrum—hence, the term "channel spectra." In the pile of alternate plates and polarizers making the filter, the thickest plate is chosen so that the width between a pair of black bands is the width of the desired pass band. The second plate is chosen half the thickness of the first plate. This second plate will cancel out every other pass band of the second plate. This is continued until the final pass bands of the system are far enough apart to be removed by auxiliary dye filters or filters of other types.

The transmission of a four-component filter and the transmission of the individual elements is shown in Fig. 1. Associated with the pass bands in this figure are some low intensity secondary bands. The location and height of these secondary pass bands is shown by a mathematical analysis of the filter transmission.

For the transmission of an N plate filter we can write:

$$T=(A'\cos\theta\cos2\theta\cos4\theta\cdots\cos2^{N-1}\theta)^2, \quad (1)$$

where $\theta=\pi(n_\epsilon-n_0)d/\lambda$, d is the thickness of the thinnest plate, n_e and n_0 are the ordinary and extraordinary indices of the crystal, and λ is the wave-length. We can replace $\cos n\theta$ with its equivalent:

$$\cos n\theta=(e^{ni\theta}+e^{-ni\theta})/2. \quad (2)$$

This can be rewritten:

$$\cos n\theta=(e^{ni\theta}/2)(1+e^{-2ni\theta}). \quad (3)$$

If this substitution is made, Eq. (1) becomes:

$$T^{\frac{1}{2}}=A'\frac{e^{i\theta}}{2}(1+e^{-2i\theta})\frac{e^{2i\theta}}{2}(1+e^{-4i\theta})\frac{e^{4i\theta}}{2}(1+e^{-8i\theta})\cdots$$
$$\times\frac{\exp[2^{N-1}i\theta]}{2}(1+\exp[-2^Ni\theta]). \quad (4)$$

This can be expanded to give:

$$T^{\frac{1}{2}}=(A'/2^N)e^{i\theta\sum_{N=1}^{N}2^{N-1}}$$
$$\times[1+e^{-2i\theta}+e^{-4i\theta}+e^{-6i\theta}+\cdots e^{-i\theta\sum_{N=1}^{N}2^N}]. \quad (5)$$

Now:

$$\sum_{N=1}^{N}2^{N-1}=2^N-1;\quad \sum_{N=1}^{N}2^N=2^{N+1}-2, \quad (6)$$

$$T^{\frac{1}{2}}=\frac{A'}{2^N}(\exp[i\theta(2^N-1)])\frac{1-\exp[-2i\theta(2^N)]}{1-\exp[-2i\theta]}; \quad (7)$$

or:

$$T^{\frac{1}{2}}=\frac{A'}{2^N}\frac{\exp[2^Ni\theta]-\exp[-2^Ni\theta]}{2i}\cdot\frac{2i}{e^{i\theta}-e^{-i\theta}}, \quad (8)$$

which becomes:

$$T=[A'\sin2^N\theta/2^N\sin\theta]^2. \quad (9)$$

For comparison with this equation, consider the expression for the transmission of a diffraction grating:

$$T=A(\sin^2\alpha/\alpha^2)\cdot(\sin^2n\varphi/\sin^2\varphi), \quad (10)$$

where

$$\alpha=(\pi/\lambda)d\sin\theta,\qquad \varphi=(\pi/\lambda)(d+b)\sin\theta. \quad (11)$$

In these equations n is the number of lines in the grating, d is the width of the transparent line of the grating, b is the width of the opaque lines in the grating, and θ is the angle from the grating normal. It is apparent from these last

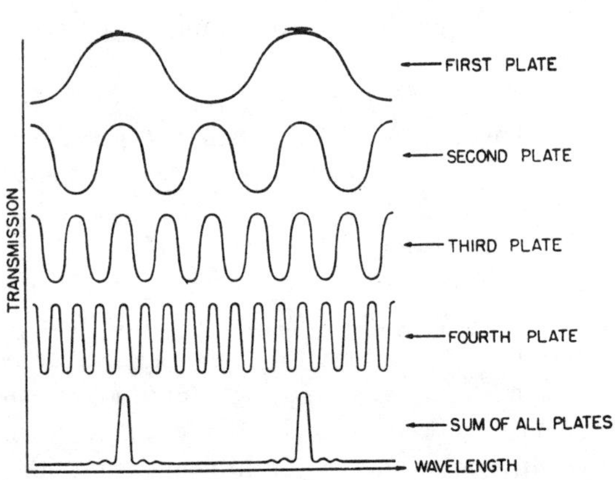

FIG. 1. Transmission between parallel polarizers of the individual retardation plates of a four-element filter and the transmission of the assembled filter.

equations that the N plate filter is analogous to a grating with 2^N lines, and the secondary pass bands are the equivalent of the secondary lines associated with the main lines of a diffraction grating.

ANALYSIS OF TUNING METHOD

If the retardation or thickness of the plates making up the pile were changed in such a way that the 2, 4, 8 ratio in retardation could be maintained, the tuning would be easily accomplished. Since some plates may have retardations of several thousand waves, the 2, 4, 8 ratio cannot be easily kept in a tuning system which uses variable retardation plates in series with the fixed elements. Most variable plates, such as Kerr calls, stretched sheets of plastic, or rotating birefringent plates have a retardation of only one or two waves. A scheme was devised by Lyot[3] in which the individual plates of the filter were cut into two wedges which could be slid across each other so as to vary the length of the light path through the combination. Since each plate has to be so tuned, the adjustment is very difficult and the filter is not considered variable.

It will be shown below that the 2, 4, 8 ratio can be considerably upset without introducing significant distortion of the filter pass band. In fact, the ratio can be departed from to such an extent that tuning can be accomplished by including with each element a low order variable retardation plate, such as a Kerr cell.

The retardation of the Kerr cell is a function of the voltage applied to it. Retardations up to full wave in the yellow are not too difficult to attain. The problem is to tune the filter and yet keep the change in retardation of the individual plates under a definite limit—in particular, under one half-wave.

The wave-length at which the filter transmission is to be peaked will be expressed in terms of the corresponding retardation of the thickest plate. The technique used in tuning will be to vary the retardation of each plate so that its nearest transmission peak is shifted to the point at which the over-all transmission is to be a maximum. In order to do this, the argument of the cosine for each term in the expression for the filter transmission must be a multiple of π at this point. We will operate in the region $p\pi<\theta<(p+1)\pi$, i.e., in the region where the first plate has a retardation Γ, where $p\lambda_1<\Gamma<(p+1)\lambda_2$ and λ_1 is the wave-length at which the angular retardation is $2p\pi$, and λ_2 the wave-length at which it is $2(p+1)\pi$. We will suppose that we wish to shift the pass band to a point where the angular retardation of the last plate is $2\pi(2^{N-1}p+x)$. At this point the argument of the cosine of the last plate can be written:

$$\theta_{Nx}=2^{N-1}\theta_{1x}=x\pi+2^{N-1}p\pi. \qquad (12)$$

At the transmission peak of the Nth plate which is nearest to x:

$$\theta_{Nn}=(n_N+2^{N-1}p)\pi, \qquad (13)$$

where n_N is an integer.

n_N is chosen so that $|x-n_N|$ is a minimum. If the transmission peaks of the Nth plate are numbered starting with the filter transmission peak as zero, n_N is the number of the peak nearest x.

We wish to multiply θ_{Nx} by a number "a" such that θ'_{Nx}, the value for the shifted plate, will equal θ_{Nn}.

$$\theta'_{Nx}=a\theta_{Nx}. \qquad (14)$$

From this equation we find for a:

$$a=(2^{N-1}p+n_N)/(2^{N-1}p+x). \qquad (15)$$

For the Kth plate:

$$a=[2^{N-1}p+(2^{N-K})n_K]/[2^{N-1}p+x], \qquad (16)$$

where n_K is the number of the peak of the Kth plate nearest to the point at which maximum transmission is desired, and x refers again to the Nth plate.

Since the produce $(2^{N-K})n_K$ may be independent of K over quite a range of K, this same value of a_K holds for several plates. In this case we can write a single symbol, a_i, for a series of plates where i is the number of the thickest plate which has a_i as a factor:

$$a_i=a_K=a_{K+1}\cdots=a_i. \qquad (17)$$

For the transmission of a filter tuned to a point x:

$$T'=A^2[\cos a_1\theta\cos 2a_2\theta\cos 4a_3\theta\cdots]^2. \qquad (18)$$

Because of the equality of many of the a_K's, this can be rewritten in the form of Eq. (9):

$$T' = \prod_{i=1}^{p} (\sin 2^{n_i} a_i \theta / 2^{n_i} \sin a_i \theta)^2, \tag{19}$$

where

$$\sum n_i = N. \tag{20}$$

If, instead of shifting to the nearest peak of each plate, we shift to the next highest peak, the equations are unchanged in form but n_K becomes the next highest peak to $x/2^{N-i}$.

CONSTRUCTION AND CHARACTERISTICS OF A DEMONSTRATION FILTER

To demonstrate the properties of the filter, a four-element demonstration model was built by Mr. Edgar Barr. The thinnest element of this filter had a retardation of two waves in the green. Both the fixed and variable elements of the filter consisted of plates of plastic. For the fixed plates the plastic was polyvinyl alcohol which had been stretched to the desired orientation and laminated between glass plates with Acryloid F–10.[5] The variable plates consisted of sheets of polyvinyl butyrate plasticized with 15 percent butyl sebacate. These sheets were mounted in a clamp which could be adjusted with a screw so as to stretch the sheet to the desired retardation.

TABLE I.

Plate number	Retardation
1	8880 mu
2	4480 mu
3	2250 mu
4	1050 mu

Each element of the filter was mounted on an optical bench in such a way that it could be swung out of the system.

The measured retardation of the filter components is shown in Table I. With a slight adjustment of the tuning elements associated with these plates, their retardation is made to have the 2, 4, 8 ratio at which they have a single pass band.

To operate the filter, light from a ribbon-filament lamp was projected through the system and through a carbon disulfide prism onto a screen. All the plates except the thickest were removed from the system. On the screen there appeared the channel spectrum of this first plate. The screw on the corresponding variable sheet was then adjusted until a pass band of this plate corresponded with the desired transmission band of the whole filter. The second plate was then thrown into the system and the same adjustment made. When all the plates

FIG. 2. Four-element plastic tunable filter. The circular filter elements are laminated to polarizers. The tuning elements are mounted with screw clamps. Examination of the behavior of the filter is made with the prism.

[5] This material is made by the Resinous Products Company in Philadelphia, Pennsylvania.

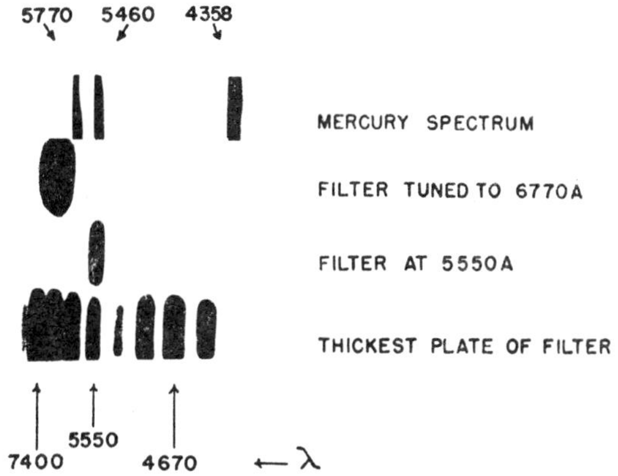

FIG. 3. Carbon disulfide prism spectrum of the light transmitted by an untuned four-element filter and spectrum of the light transmitted by the same filter tuned to a different wave-length region. The source is a tungsten lamp. A mercury arc spectrum is shown for comparison.

had been adjusted, a single band was transmitted by the filter. The complete filter is shown in Fig. 2.

In Fig. 3 is shown a series of photographs of the filter spectra made with the carbon disulfide prism. In the top line is the spectrum of the thickest plate between parallel polarizers. In the next line is the green line transmitted by the completely assembled filter. The next line shows the filter tuned to one of the red lines transmitted by the thickest plate. A mercury arc spectrum is shown for comparison.

It is apparent that the filter can only be tuned over a range which is equivalent to the separation between two successive pass bands in the fixed filter. As the filter pass band is shifted farther from the fixed-filter position, the 2, 4, 8 ratio of retardation in the individual elements is more and more upset. The experiments in the preceding section seem to indicate that this upsetting of the 2, 4, 8 retardation ratio does not give rise to any extraneous pass bands. To test this further a series of transmission curves has been plotted for a five-element filter whose first four elements are identical with the experimental filter described above. The transmission of the filter with no retardation in the variable plates is given in Fig. 4.

To go to the first peak of the thickest plate:

$$x=1. \tag{21}$$

In Eq. (16):

$$p=2, \quad n_K=0, \quad a_1=32/33. \tag{22}$$

Since $2^4\times(1/32)<1$, the same a may be used all the way through to the last plate for which $a_N=1$. The transmission is given by the equation:

$$T_1=[\cos a\theta \cos 2a\theta \cos 4a\theta \cos 8a\theta \cos 16\theta]^2. \tag{23}$$

At the second peak $a_1=16/17$, but only the first three plates need to be altered. This transmission is plotted in Fig. 5. To go to the third peak requires two a's: $a_1=32/35$ for the first three plates and $a_2=34/35$ for the fourth plate. The transmission is given by the equation:

$$T_3=[\cos a_1\theta \cos 2a_1\theta \cos 4a_1\theta \cos 8a_2\theta \cos 16\theta]^2. \tag{24}$$

This transmission is plotted in Fig. 6.

In Fig. 7 is shown the transmission for the shift to the fourth peak.

It is apparent that over the range shown the filters have not degenerated as far as band width and signal-to-noise are concerned.

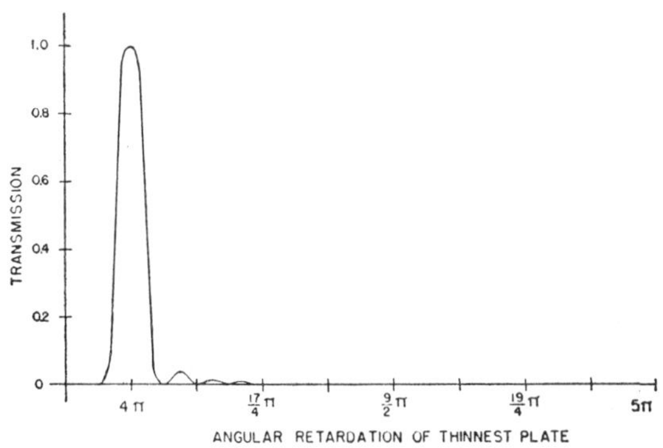

FIG. 4. Calculated transmission of a five-element filter with two waves retardation for the thinnest element.

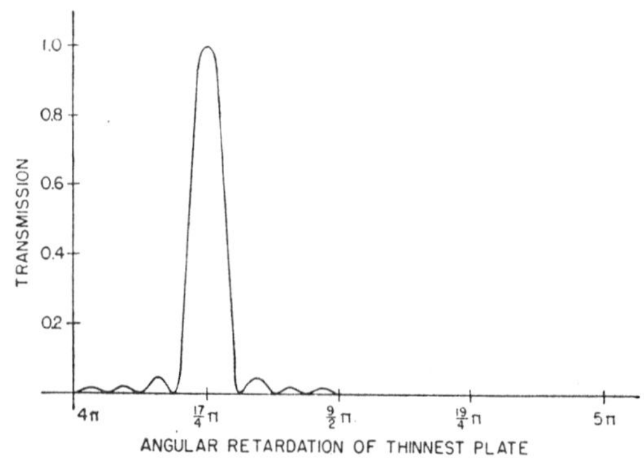

FIG. 5. Calculated transmission of a five-element filter tuned to transmit a wave-length at which the angular retardation of the thinnest plate is $17\pi/4$.

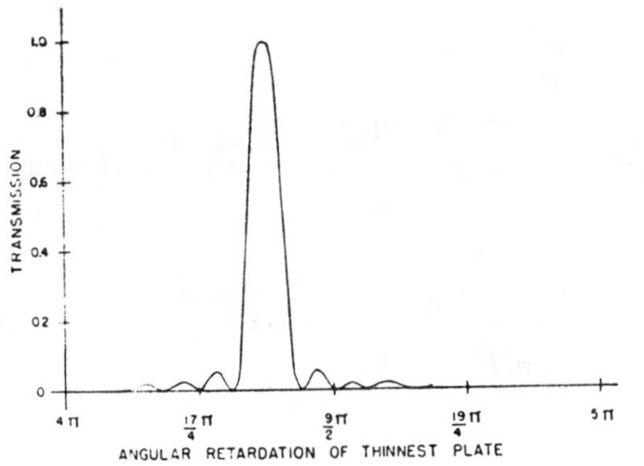

FIG. 6. Calculated transmission of a five-element filter tuned to transmit a wave-length at which the angular retardation of the thinnest plate is $35\pi/8$.

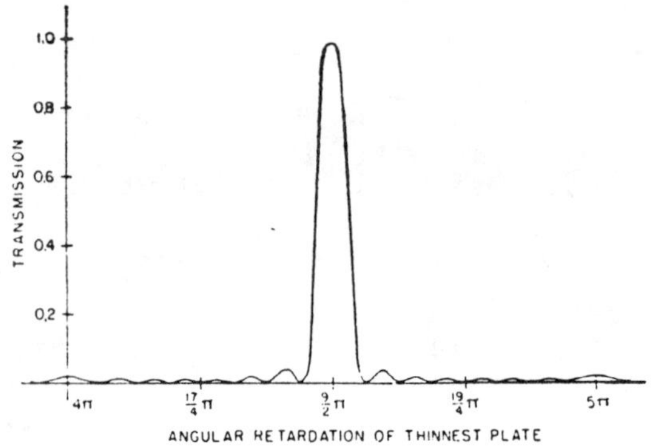

FIG. 7. Calculated transmission of a five-element filter tuned to transmit a wave-length at which the angular retardation of the thinnest plate is shifted to $9\pi/2$.

The a_i calculated on the basis of a maximum shift to the next highest n_K, gives more distortion to the undisturbed filter curve than those arising from a shift to the nearest n_K. The best factor will be the one which gives the longest wave-length interval with no secondary peaks.

To compare the effect of shifting to the nearest peak with a shift to the next highest peak, the peak at 2π was shifted up to $11\pi/4$. The equation for the transmission can be written:

$$T_{12} = [\cos(8/11)\theta \cos(20/11)\theta \cos 4\theta \cos 8\theta \cos 16\theta]^2. \quad (25)$$

In this curve (see Fig. 8) there is a strong secondary maximum at $9\pi/4$. In Fig. 9 are shown the results of shifting the peak at 3π down to $11\pi/4$. The equation for the transmission is:

$$T_{12} = [\cos(12/11)\theta\ (\cos 24/11)\theta \cos 4\theta \cos 8\theta \cos 16\theta]. \quad (26)$$

The secondary maximum at $9\pi/4$ is tremendously reduced, and the effective usefulness of this filter is increased by a total range of $\pi/2$.

If, instead of tuning the filter by adding an ordinary retardation plate, the filter is tuned by adding achromatic phase plates, there is no change in the secondary maxima even at wave-lengths far removed from the point to which the filter is tuned. The transmission of a single element is:

$$T = A \cos^2 a\theta. \quad (27)$$

With the addition of an achromatic plate this becomes:

$$T = A \cos^2(a\theta + \delta), \quad (28)$$

where δ is independent of θ. Tuning in this case does not change the spacing of the peaks in the curves but merely slides them along the wave-length scale.

In the non-achromatic case which has just

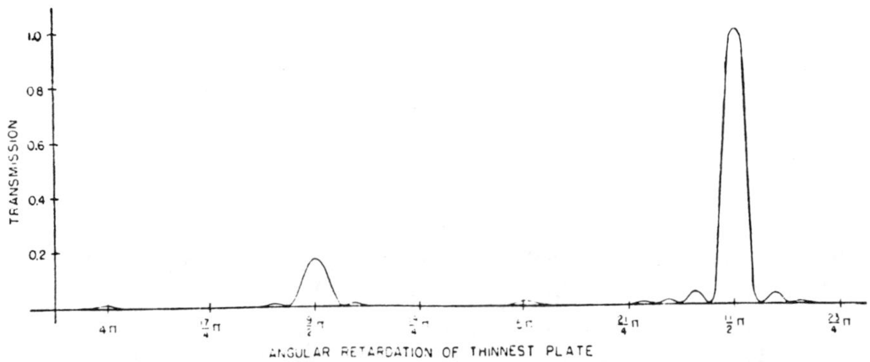

FIG. 8. Calculated transmission of a five-element filter tuned to transmit a wave-length at which the angular retardation of the thinnest plate is $11\pi/2$. In the calculation retardation was subtracted from each plate.

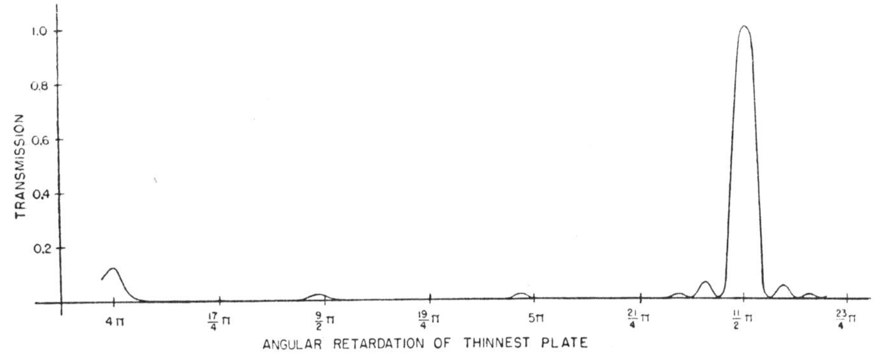

FIG. 9. Calculated transmission of a five-element filter tuned to transmit a wave-length at which the angular retardation of the thinnest plate is $11\pi/2$. In this calculation retardation was added to each plate.

been treated, the spacing between the peaks changes and it is the change in separation which changes the character of the secondary spectra. The maximum change in peak separation occurs when the θ is small, e.g., for a two-wave retardation plate. Since we have shown that the secondary maxima are not particularly distorting with a two-wave plate, it is apparent that any filter whose first plate has higher retardation than two waves can be successfully tuned over an interval of 2π in angular retardation.

In many of the applications of the tunable filter it is desirable to know how much retardation must be added to each plate to shift the pass band by a given amount. This information is especially necessary in building filters where the variable plates are adjusted by a single control. The retardation to be added to the Kth plate to produce a given shift can be written:

$$\Delta\delta_K = 2(\theta_{K0} - \theta'_{K0}), \tag{29}$$

where $2\theta_{K0}$ is the retardation of the Kth plate at λ_0. $2\theta'_{K0}$ is the retardation of the Kth plate at λ_0 after the transmission peak has been shifted. λ_0 is a reference wave-length. By substituting the value of θ'_{K0}, one gets

$$\Delta\delta_K = 2^K(a_K - 1)\theta_{1,0}. \tag{30}$$

From Eqs. (16) and (30) one finds

$$\Delta\delta_K = (2^N n_K - 2^K x/2^{N-1}p + x)\theta_{1,0}. \tag{31}$$

The formula for x can be written:

$$x = 2^{N-1}p(\lambda_0 - \lambda)/\lambda, \tag{32}$$

where λ is the wave-length at which peak transmission is desired. Substitution of Eq. (32) into Eq. (31) gives:

$$\Delta\delta_K = \frac{\lambda(n_K + 2^{K-1}p) - 2^{K-1}p\lambda_0}{p\lambda_0} \cdot 2\pi, \tag{33}$$

where this retardation is added in the region:

$$(2^K p\lambda_0/2n_K + 1 + 2^K p) < \lambda < (2^K p\lambda_0/2n_K - 1 + 2^K p). \tag{34}$$

For a filter which is to cover a continuous range between angular retardation $2(p-1)\pi$ and $2(p+1)\pi$, one can plot the retardation change for a given plate as a function of wave-length. This plot is shown in Fig. 10 for the first three plates of the filter described above.

TYPES OF TUNING AND APPLICATIONS OF THE FILTER

In the first part of this paper it was suggested that the variable retardation plates be Kerr

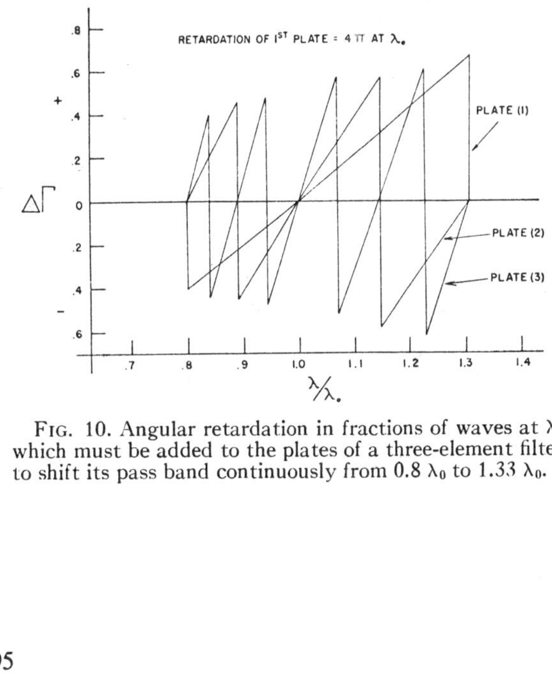

FIG. 10. Angular retardation in fractions of waves at λ_0 which must be added to the plates of a three-element filter to shift its pass band continuously from 0.8 λ_0 to 1.33 λ_0.

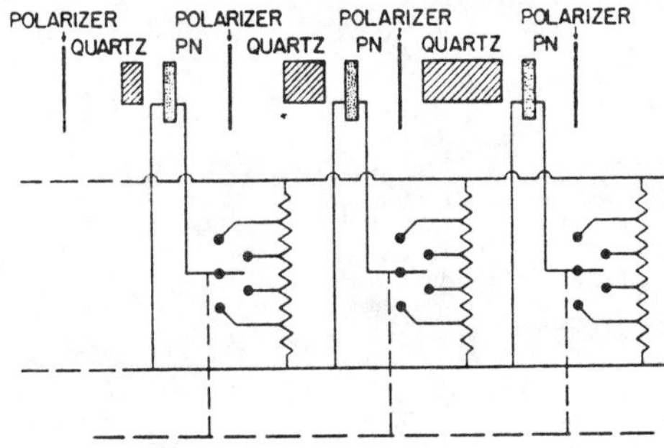

FIG. 11. Schematic diagram of a four-element tunable filter with an associated electrical network used to set the pass band to four discrete wave-lengths.

cells. The Kerr cell might be used rather loosely to indicate any material whose birefringence changes in an electric field. A material which seems to be especially adapted to the construction of these filters is crystalline $NH_4H_2PO_4$ or PN.[6] This uniaxial crystal has a high electro-optic effect and in particular it exhibits a strong longitudinal effect. For fields applied parallel to the direction of light propagation and at the same time parallel to the optical axis, there is a strong induced birefringence. In this crystal the effect is linear with field strength; and, since the light path and field strength are in the same direction, the effect becomes linear with voltage and independent of thickness. Because of this linearity a resistance network can be easily designed to give the appropriate voltages and, hence, retardation to the PN tuning element in a filter.

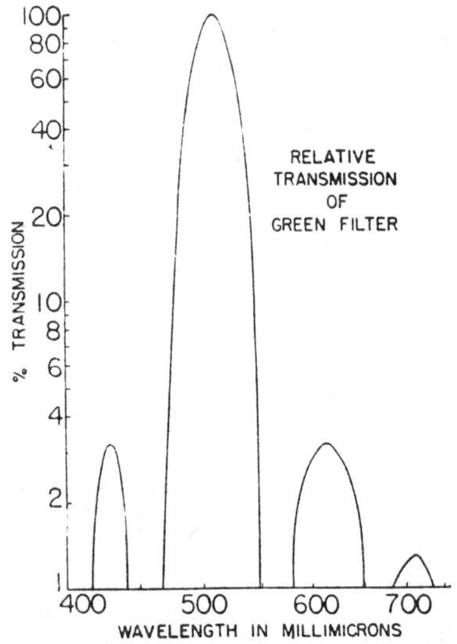

FIG. 12. Green transmission of a three-element tunable filter to be used for transmitting red, green, and blue for color television.

[6] This crystal is grown by the Brush Development Company in Cleveland, Ohio.

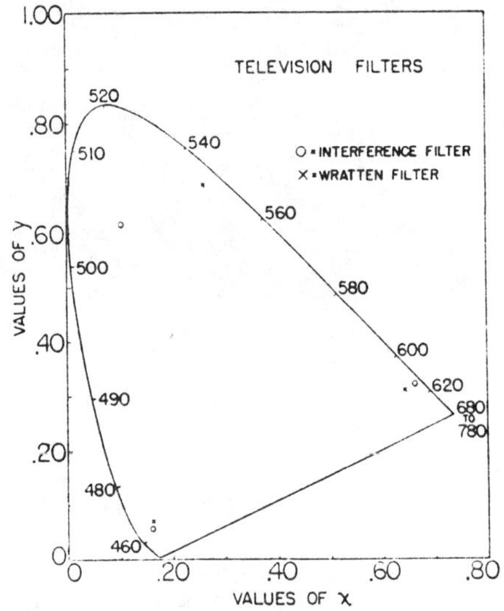

FIG. 13. Color of three possible positions for the pass band of a three-element tunable filter which might be used in color television. The Wratten *A*, *B*, and *C* are shown for comparison.

Figure 11 is the schematic diagram of a filter designed to be set to four discrete wave-lengths.

The electrical tuning method lends itself particularly to continuous scanning of a given spectral region. The curves of Fig. 10 can be considered as a set of sawtooth-voltage waves to be applied to the successive tuning elements of a filter. The basic cycle could be used as the sweep on a cathode-ray oscillograph, and the output of a photo-cell would be presented on the screen. This would give an instantaneous spectral curve of the light shining on the filter. Such cathode-ray tube spectral analysis might be useful for quick investigation of series of dyes or for process control which involved spectral-transmission curves.

The electrical tuning method is also interesting because of its possible application to sequential color television. The first three elements of the filter described in the previous section might be used in such an application.

By applying suitable voltages to the tuning elements the filter can be made to pass either red, green, or blue light. The transmission curve of the green band is shown in Fig. 12. In Fig. 13 is shown the position of the three pass bands on a color triangle for equal energy.

The author is particularly indebted to Professor Donald H. Menzel and to Dr. E. H. Land for many stimulating discussions, and to the Polaroid Corporation in whose laboratory most of the experimental work was carried out.

DETAILED THEORY OF THE $\frac{1}{8}$ Å BIREFRINGENT FILTER

By W. H. Steel*

[Manuscript received November 21, 1960]

Summary

A method for studying the passage of light through birefringent plates is developed and applied to derive the on-axis transmission of a birefringent filter with $\frac{1}{8}$ Å pass band. Some possible variations of such filters are discussed.

I. Introduction

The filter (Steel, Smartt, and Giovanelli 1961), designed and constructed in the Division of Physics, is a development of Lyot's (1933, 1944) original design; it also incorporates refinements by Evans (1949). A filter of this type consists of a series of "elements", composed of plates of a birefringent crystal, set between polarizers. The simplest form of element is a single plate of uniaxial crystal, cut with its optic axis normal to the direction of passage of the light and placed between polarizers whose axes are at an angle of $\frac{1}{4}\pi$ to the crystal axes. The transmittance of such an element is a sinusoidal function of wavelength: a "channel spectrum".

At a fixed temperature, the simplest element has a fixed transmittance at any wavelength, but this is constant over a small angular field only. In the filter considered here, the elements have been modified in three respects:

(i) Some elements are made "tunable"; their transmission function can be moved along the wavelength scale,

(ii) Lyot's first type of wide-field element is used to reduce the variation of transmittance with angular field,

(iii) by placing two elements between each pair of polarizers to form Evans' "split elements", the number of polarizers in the filter is reduced and the light transmittance increased.

The details of the filter are given in Table 1. At a wavelength of 6563 Å, the channel spectra obtained from each element have spacings $\frac{1}{4}$, $\frac{1}{2}$, 1, . . ., 512 Å so that the final transmission of the filter is in bands of half-width $\frac{1}{8}$ Å spaced at intervals of 512 Å.

II. Birefringent Plates

Consider a block of birefringent crystal cut with parallel faces normal to one of its three principal axes. Polarized light passing through the crystal in the direction of this axis may be decomposed into two components polarized in the directions of the other two axes. In a biaxial crystal, these two components travel with different velocities due to the different refractive indices for polarizations in these directions. The same applies in a uniaxial crystal if the light travels normally to its optic axis. The two components emerge from the plate with a relative "retardation"

* Division of Physics, C.S.I.R.O., University Grounds, Sydney.

TABLE 1

COMPONENTS* OF $\frac{1}{8}$ Å FILTER

The quarter- and half-wave plates were made of pairs of crossed quartz plates. However, mica could have been used, as in the last half-wave plate

Element and Component No.	Material	Retardation (λ) at 6563 Å	Orientation (given in range $0 \rightarrow \pi$)	Element and Component No.	Material	Retardation (λ) at 6563 Å	Orientation (given in range $0 \rightarrow \pi$)	Element and Component No.	Material	Retardation (λ) at 6563 Å	Orientation (given in range $0 \rightarrow \pi$)
First				Second				Third			
1	"Polaroid"†	—	$\frac{1}{4}\pi$	9	"Polaroid"	—	$\frac{1}{4}\pi$	17	"Polaroid"	—	$\frac{1}{4}\pi$
2	Quartz	$12\frac{1}{4}$	0	10	Quartz	$24\frac{1}{4}$	0	18	Quartz	$48\frac{1}{4}$	0
3	Quartz	192	$\frac{1}{4}\pi$	12	Quartz	384	$\frac{1}{4}\pi$	19	Calcite	743	$\frac{1}{4}\pi$
4	Quartz	$\frac{1}{4}$	0	12	Quartz	$\frac{1}{4}$	0	20	Quartz	$\frac{1}{4}$	0
5	Quartz	$\frac{1}{2}$	θ	13	Quartz	$\frac{1}{2}$	2θ	21	Quartz	$\frac{1}{2}$	4θ
6	Quartz	$\frac{1}{4}$	$\frac{1}{2}\pi$	14	Quartz	$\frac{1}{4}$	$\frac{1}{2}\pi$	22	Quartz	$\frac{1}{4}$	$\frac{1}{2}\pi$
7	Quartz	192	$\frac{3}{4}\pi$	15	Quartz	384	$\frac{3}{4}\pi$	23	Calcite	743	$\frac{3}{4}\pi$
8	Quartz	$12\frac{1}{4}$	0	16	Quartz	$24\frac{1}{4}$	0	24	Quartz	$48\frac{1}{4}$	0

* A scale drawing of these components is given by Steel, Smartt, and Giovanelli (1961).

† KN40.

TABLE 1 (*Continued*)

Element and Component No.	Material	Retardation (λ) at 6563 Å	Orientation (given in range $0 \to \pi$)	Element and Component No.	Material	Retardation (λ) at 6563 Å	Orientation (given in range $0 \to \pi$)	Element and Component No.	Material	Retardation (λ) at 6563 Å	Orientation (given in range $0 \to \pi$)
Fourth				Fifth				Seventh			
25	"Polaroid"	—	$\frac{1}{4}\pi$	37	"Polaroid"	—	$\frac{1}{4}\pi$	51	"Polaroid"	—	$\frac{1}{4}\pi$
				38	Quartz	$6\frac{1}{4}$	0				
26	Quartz	96	$\frac{1}{2}\pi$	39	Calcite	2972	$\frac{1}{4}\pi$	52	Calcite	11888	0
				40	Quartz	$\frac{1}{4}$	0				
27	Quartz	$\frac{1}{4}$	$\frac{3}{4}\pi$	41	Quartz	$\frac{1}{2}$	$2^4\theta$	53	Quartz	$\frac{1}{2}$	$\frac{1}{4}\pi$
				42	Quartz	$\frac{1}{4}$	$\frac{1}{2}\pi$				
28	Quartz	$\frac{1}{2}$	$\frac{1}{8}\pi-\frac{1}{4}\theta$	43	Calcite	2972	$\frac{3}{4}\pi$	54	Calcite	11888	$\frac{1}{2}\pi$
				44	Quartz	$6\frac{1}{4}$	0				
29	Calcite	$1486\frac{1}{4}$	$\frac{3}{4}\pi$					55	Quartz	$\frac{1}{4}$	$\frac{1}{2}\pi$
30	Quartz	$\frac{1}{4}$	$\frac{1}{2}\pi$	Sixth	"Polaroid"	—	$\frac{1}{4}\pi$	56 (first design)	"Polaroid"	—	$\frac{1}{4}\pi+2^7\theta$
31	Quartz	$\frac{1}{2}$	$2^3\theta$	45	Calcite	5944	0	56 (final)	Mica	$\frac{1}{2}$	$\frac{1}{8}\pi+\frac{1}{2}\beta+2^6\theta$ (one band) $\frac{1}{2}\beta+2^6\theta$ (two bands)
32	Quartz	$\frac{1}{4}$	0	46	Quartz	$\frac{1}{4}$	$\frac{3}{4}\pi$				
33	Calcite	1486	$\frac{1}{4}\pi$	47	Quartz	$\frac{1}{2}$	$\frac{1}{4}\pi+2^5\theta$				
34	Quartz	96	$\frac{1}{2}\pi$	48	Quartz	$\frac{1}{4}$	$\frac{1}{4}\pi$	67 (final)	Polarizing beam-splitter	—	β (one band) β and $\beta+\frac{1}{2}\pi$ (two bands)
35	Quartz	$\frac{1}{4}$	$\frac{3}{4}\pi$	49	Calcite	5944	$\frac{1}{2}\pi$				
36	Quartz	$\frac{1}{2}$	$\frac{1}{8}\pi-\frac{1}{4}\theta$	50							

between them that depends on the refractive index difference for the crystal and the thickness of the plate.

To avoid the necessity of distinguishing between uniaxial and biaxial crystals, and between ordinary and extraordinary rays in the former case, we shall refer only to the fast and slow axes of the two crystal axes that are normal to the direction of the light. The orientation of the crystal plate around this direction will be specified by the direction of the fast axis.

If the plate has a thickness d and refractive indices n_f and n_s for light polarized in the fast and slow directions ($n_s > n_f$), the optical paths through the crystal for these polarizations are $n_f d$ and $n_s d$. The path difference, or retardation due to the plate, is $(n_s - n_f)d$. The corresponding phases are

$$p_f = k n_f d,$$

$$p_s = k n_s d,$$

where

$$k = 2\pi/\lambda.$$

The phase difference is

$$p = p_s - p_f. \tag{2.1}$$

The final results in this paper will be expressed in terms of q, the retardation of a quartz plate that gives channel spectra at 512 Å spacing around 6563 Å, and of c, the retardation of a calcite plate giving 4 Å spacing. These retardations are functions of wavelength and have the values of 12λ and 1486λ at 6563 Å; these values are not in the ratio of an integral power of 2 since quartz and calcite have different dispersions of birefringence. The channel spectra spacing $\Delta\lambda$ depends not only on the birefringence but also its dispersion,

$$\frac{\Delta\lambda}{\lambda} = \frac{1}{d}\left\{\frac{\partial(n_s - n_f)}{\partial\lambda} - \frac{n_s - n_f}{\lambda}\right\}^{-1},$$

and, around 6563 Å and at a temperature of 40°C, the spacings for the quartz and calcite plates are in the ratio of an integral power of 2.

III. Basic Theory

We shall use, as a mathematical representation of the effects of birefringent plates on collimated polarized light, a modification of the method due to Jones (1941). It will also be compared with Poincaré's (1892) representation.

Any fully polarized light, in general elliptically polarized, can be represented either by the orientation and eccentricity of the polarization ellipse or by the resolved components along some conveniently chosen x- and y-axes. In this work the latter system is the more convenient. The two resolved components, A_x and A_y, differ in general in amplitude and phase; in the usual complex notation

$$A_x = u_1\, e^{j\psi_1}, \qquad A_y = u_2\, e^{j\psi_2},$$

and Poincaré's "state of polarization" is given by the ratio

$$\begin{aligned} Q &= A_y/A_x \\ &= (u_2/u_1)\, e^{j(\psi_2-\psi_1)} \\ &= \tan \tfrac{1}{2}\gamma\, e^{j\beta}, \end{aligned} \tag{3.1}$$

where β and γ are defined by this equation. Poincaré has shown that Q may be represented in terms of the angles γ and β as a point on the surface of a sphere, and the passage of the light through birefringent plates is represented by simple geometrical constructions on the sphere. Other angles on the sphere give the parameters of the polarization ellipse.

This treatment, which can be expressed in terms of spherical trigonometry, is not as simple analytically as a treatment in terms of the two components A_x and A_y separately. The extra information retained is however redundant for fully polarized light of constant intensity—it is required only for the treatment of partially polarized light—but, when a description of the final state of polarization is required, the redundancy is removed by calculating, at the end of the analysis, Poincaré's ratio Q. This redundant information is not important in the simple analysis of birefringent filters, but it can cause confusion in the study of certain types of wave plates considered in Section VI of the present paper.

Jones (1941) has given, in matrix form, an analysis in terms of the two resolved components. He defines a "vector polarization"

$$\mathbf{A} = \begin{pmatrix} A_x \\ A_y \end{pmatrix},$$

and shows that the vector polarization $\mathbf{A}'$ after a plate with its fast axis orientated at an angle α to the x-axis is related to the polarization $\mathbf{A}$ before the plate by the matrix formula

$$\mathbf{A}' = \mathbf{S}(\alpha)\, \mathbf{N}\mathbf{S}(-\alpha)\mathbf{A}, \tag{3.2}$$

where

$$\mathbf{S}(\alpha) = \begin{pmatrix} \cos\alpha & -\sin\alpha \\ \sin\alpha & \cos\alpha \end{pmatrix},$$

$$\mathbf{N} = \begin{pmatrix} e^{-jp_f} & 0 \\ 0 & e^{-jp_s} \end{pmatrix}.$$

This matrix treatment has proved very powerful in Evans' (1958) discussion of the Šolc filter.

In the present paper, a modification of Jones' method has been used in which the vector polarization is represented by a complex number

$$\mathbf{A} = A_x + iA_y. \tag{3.3}$$

This representation involves two sets of complex numbers since A_x and A_y have already been shown to include phase terms represented by the complex number j. The two numbers i and j both when squared give -1, for a reversal in direction is equivalent to a phase change of π. Otherwise they can be treated as completely independent, one representing an angle in a plane, the other a phase.

It has been found more convenient to express results, not in terms of A_x and A_y as in Jones' method, but in terms of **A** and **A***, the complex conjugate in i. The conjugate in j will be denoted by $\mathbf{A}^\dagger$. As usual

$$A_x = \frac{1}{2}(\mathbf{A}+\mathbf{A}^*),$$

$$A_y = \frac{1}{2\mathrm{i}}(\mathbf{A}-\mathbf{A}^*). \tag{3.4}$$

(a) Effect of a Birefringent Plate

The expression in complex notation that corresponds to (3.2) will now be derived in detail. If a plate has its fast axis at an angle α to the x-direction and gives phase changes p_f and p_s to light polarized in the fast and slow directions, the fast component of **A** as it enters the plate is given by

$$A_x \cos\alpha + A_y \sin\alpha,$$

and as it leaves by

$$(A_x \cos\alpha + A_y \sin\alpha)\, \mathrm{e}^{-\mathrm{j}p_f},$$

where the convention is as usual that the phase decreases as a wave advances. Analogous expressions apply to the slow component.

The two emergent components may now be resolved along the original axes to give A_x' and A_y' and these combined to give the emergent polarization,

$$\begin{aligned}\mathbf{A}' &= A_x' + \mathrm{i}A_y',\\ &= A_x[\cos^2\alpha\, \mathrm{e}^{-\mathrm{j}p_f} + \sin^2\alpha\, \mathrm{e}^{-\mathrm{j}p_s} + \mathrm{i}\sin\alpha\cos\alpha\{\mathrm{e}^{-\mathrm{j}p_f} - \mathrm{e}^{-\mathrm{j}p_s}\}]\\ &\quad + A_y[\sin\alpha\cos\alpha\{\mathrm{e}^{-\mathrm{j}p_f} - \mathrm{e}^{-\mathrm{j}p_s}\} + \mathrm{i}\sin^2\alpha\, \mathrm{e}^{-\mathrm{j}p_f} + \mathrm{i}\cos^2\alpha\, \mathrm{e}^{-\mathrm{j}p_s}]\\ &= \mathrm{e}^{-\frac{1}{2}\mathrm{j}(p_f+p_s)}[A_x\{\cos\tfrac{1}{2}p - \mathrm{j}\sin\tfrac{1}{2}p\, \mathrm{e}^{2\mathrm{i}\alpha}\} + A_y\{\mathrm{i}\cos\tfrac{1}{2}p + \mathrm{ij}\sin\tfrac{1}{2}p\, \mathrm{e}^{2\mathrm{i}\alpha}\}]\\ &\doteq \mathrm{e}^{-\frac{1}{2}\mathrm{j}(p_f+p_s)}[\mathbf{A}\cos\tfrac{1}{2}p - \mathrm{j}\,\mathbf{A}^*\sin\tfrac{1}{2}p\, \mathrm{e}^{2\mathrm{i}\alpha}],\end{aligned}$$

where $p = p_s - p_f$.

In this paper, the first term, which gives the absolute phase, is omitted as only relative phases are important. Then

$$\mathbf{A}' = \mathbf{A}\cos\tfrac{1}{2}p - \mathrm{j}\,\mathbf{A}^*\sin\tfrac{1}{2}p\, \mathrm{e}^{2\mathrm{i}\alpha}. \tag{3.5}$$

The corresponding expression in terms of Poincaré's Q is

$$Q' = \frac{Q \cos \frac{1}{2}p - \mathrm{j} \sin \frac{1}{2}p(\sin 2\alpha - Q \cos 2\alpha)}{\cos \frac{1}{2}p - \mathrm{j} \sin \frac{1}{2}p(\cos 2\alpha + Q \sin 2\alpha)},$$

a less convenient form. But, at any stage, the state of polarization Q may be calculated from the vector polarization $\mathbf{A}$ using

$$Q = -\mathrm{i}\,\frac{\mathbf{A}-\mathbf{A}^*}{\mathbf{A}+\mathbf{A}^*}. \tag{3.6}$$

(b) *Effect of Polarizers*

Plane polarized light of unit intensity from a polarizer at an orientation β has a vector polarization

$$\mathbf{A}' = \mathrm{e}^{\mathrm{i}\beta}. \tag{3.7}$$

If light of polarization $\mathbf{A}$ is incident on a polarizer of orientation β, only the component at this angle is transmitted. The complex amplitude A' of the transmitted light is given by the resolved component of the polarization vector,

$$\begin{aligned} A' &= A_\beta \\ &= A_x \cos \beta + A_y \sin \beta \\ &= \tfrac{1}{2}(\mathbf{A}\,\mathrm{e}^{-\mathrm{i}\beta} + \mathbf{A}^*\,\mathrm{e}^{\mathrm{i}\beta}). \end{aligned} \tag{3.8}$$

(i) *Element between Parallel Polarizers.*—The action of any element of the filter is found by applying equation (3.5) to each crystal plate. The result obtained is in the form

$$\mathbf{A}' = c_1\mathbf{A} + c_2\mathbf{A}^*.$$

If this element is placed between a pair of polarizers of orientation β, equations (3.7) and (3.8) give the final amplitude as

$$A' = \mathrm{Re}(c_1) + \mathrm{Re}(c_2) \cos 2\beta + \mathrm{Im}(c_2) \sin 2\beta, \tag{3.9}$$

where $\mathrm{Re}(c)$ and $\mathrm{Im}(c)$ are the real and imaginary parts of c in terms of i;

$$c = \mathrm{Re}(c) + \mathrm{i}\,\mathrm{Im}(c).$$

(ii) *Final Intensity.*—The final intensity of the emergent plane polarized light is given, as usual, by the squared modulus of its complex amplitude. This may be written as

$$I' = A'A'^{\dagger}, \tag{3.10}$$

where $A'^{\dagger}$ is the complex conjugate of A' in j.

IV. Analysis of Basic Units

As the filter is built up of a repetition of standard combinations of plates, the theory of these standard combinations will be developed in turn.

(a) *Wave Plates*

(i) For a quarter-wave plate, $p = \frac{1}{2}\pi$. If its fast axis is at α,

$$\mathbf{A}' = \frac{1}{\sqrt{2}}(\mathbf{A} - \mathrm{j}\mathbf{A}^{*}\,\mathrm{e}^{2\mathrm{i}\alpha}). \tag{4.1}$$

(ii) For a half-wave plate, $p = \pi$ and

$$\mathbf{A}' = -\mathrm{j}\mathbf{A}^{*}\,\mathrm{e}^{2\mathrm{i}\alpha}. \tag{4.2}$$

(b) *Wide-field and Tuning Units*

Both of these units consist of a half-wave plate between two crossed plates (axes perpendicular) of equal retardations. For the first plate, where $p = r$ and $\alpha = \theta$,

$$\mathbf{A}_2 = \mathbf{A}_1 \cos \tfrac{1}{2}r - \mathrm{j}\mathbf{A}_1^{*} \sin \tfrac{1}{2}r\, \mathrm{e}^{2\mathrm{i}\theta}.$$

For the half-wave plate, where $p = \pi$ and $\alpha = \phi$,

$$\begin{aligned}\mathbf{A}_3 &= -\mathrm{j}\mathbf{A}_2^{*}\,\mathrm{e}^{2\mathrm{i}\phi} \\ &= -\mathbf{A}_1 \sin \tfrac{1}{2}r\, \mathrm{e}^{2\mathrm{i}(\phi-\theta)} - \mathrm{j}\mathbf{A}_1^{*} \cos \tfrac{1}{2}r\, \mathrm{e}^{2\mathrm{i}\phi}.\end{aligned}$$

For the final plate, where $p = r$ and $\alpha = \theta + \frac{1}{2}\pi$,

$$\begin{aligned}\mathbf{A}_4 &= \mathbf{A}_3 \cos \tfrac{1}{2}r + \mathrm{j}\mathbf{A}_3^{*} \sin \tfrac{1}{2}r\, \mathrm{e}^{2\mathrm{i}\theta} \\ &= \mathrm{i}\mathbf{A}_1 \sin r \sin 2(\theta-\phi) - \mathrm{j}\mathbf{A}_1^{*}\,\mathrm{e}^{2\mathrm{i}\theta}[\cos 2(\theta-\phi) - \mathrm{i} \cos r \sin 2(\theta-\phi)].\end{aligned}$$

(i) For Lyot's first type of wide-field element, the half-wave plate is set at $\frac{1}{4}\pi$ to the two retardation plates. The general expression for this element in terms of p and α, where α is the orientation of the first plate, is

$$\mathbf{A}' = -\mathrm{i}\mathbf{A} \sin p - \mathrm{i}\mathrm{j}\mathbf{A}^{*} \cos p\, \mathrm{e}^{2\mathrm{i}\alpha}. \tag{4.3}$$

If this unit is set at $\alpha = 0$ between parallel polarizers at $\frac{1}{4}\pi$,

$$A' = -\mathrm{j} \cos p.$$

(ii) A tuning unit consists of two crossed quarter-wave plates with a half-wave plate at a variable orientation between them; it is the rotation of the half-wave plate that tunes the element in which the unit is incorporated. For the quarter-wave plates, $r = \frac{1}{2}\pi$ and

$$\mathbf{A}' = \mathrm{i}\mathbf{A} \sin 2(\alpha-\phi) - \mathrm{j}\mathbf{A}^{*}\,\mathrm{e}^{2\mathrm{i}\alpha} \cos 2(\alpha-\phi). \tag{4.4}$$

(c) *Tunable Wide-field Element*

If the half-wave plate in the Lyot wide-field element is replaced by a tuning unit, a tunable wide-field element is obtained. For the first plate, where $p = r$ and $\alpha = \theta$,

$$\mathbf{A}_2 = \mathbf{A}_1 \cos \tfrac{1}{2}r - \mathrm{j}\mathbf{A}_1^* \sin \tfrac{1}{2}r \, \mathrm{e}^{2\mathrm{i}\theta}.$$

The tuning unit, represented by equation (4.4), is set with $\alpha = \theta - \frac{1}{4}\pi$, so that

$$\begin{aligned}\mathbf{A}_3 &= -\mathrm{i}\mathbf{A}_2 \cos 2(\theta-\phi) + \mathrm{ij}\mathbf{A}_2^* \,\mathrm{e}^{2\mathrm{i}\theta} \sin 2(\theta-\phi) \\ &= -\mathrm{i}\mathbf{A}_1 \cos(\tfrac{1}{2}r+2\theta-2\phi) + \mathrm{ij}\mathbf{A}_1^* \,\mathrm{e}^{2\mathrm{i}\theta} \sin(\tfrac{1}{2}r+2\theta-2\phi). \qquad (4.5)\end{aligned}$$

For the last plate, where $p = r$ and $\alpha = \theta + \frac{1}{2}\pi$,

$$\begin{aligned}\mathbf{A}_4 &= \mathbf{A}_3 \cos \tfrac{1}{2}r + \mathrm{j}\mathbf{A}_3^* \sin \tfrac{1}{2}r \,\mathrm{e}^{2\mathrm{i}\theta} \\ &= -\mathrm{i}\mathbf{A}_1 \cos(r+2\theta-2\phi) + \mathrm{ij}\mathbf{A}_1^* \,\mathrm{e}^{2\mathrm{i}\theta} \sin(r+2\theta-2\phi),\end{aligned}$$

or, if subscripts are omitted and p and α used,

$$\mathbf{A}' = \mathrm{i}\mathbf{A} \cos(p+2\alpha-2\phi) + \mathrm{ij}\mathbf{A}^* \,\mathrm{e}^{2\mathrm{i}\alpha} \sin(p+2\alpha-2\phi). \qquad (4.6)$$

It should be noted that equation (4.5) has the same form as (4.6); the tuning unit can also follow the plate of a simple (not wide-field) element.

When set at $\alpha = 0$ between polarizers at $\frac{1}{4}\pi$,

$$A' = \mathrm{j} \sin(p-2\phi);$$

this gives the required cosine form when tuned to $\phi = \frac{1}{4}\pi$.

(d) *Evans' Split Element with Tuned Central Component*

In the split element, an element of the types considered above is set between a pair of parallel plates of equal retardations. If these plates have orientation θ and phase difference p

$$\mathbf{A}_2 = \mathbf{A}_1 \cos \tfrac{1}{2}p - \mathrm{j}\mathbf{A}_1^* \sin \tfrac{1}{2}p \,\mathrm{e}^{2\mathrm{i}\theta}.$$

If the central section is a tunable wide-field element, described by equation (4.6), set with $\alpha = \theta + \frac{1}{4}\pi$ and $p = r$,

$$\begin{aligned}\mathbf{A}_3 &= \mathrm{i}\mathbf{A}_2 \sin(r+2\theta-2\phi) - \mathrm{j}\mathbf{A}_2^* \,\mathrm{e}^{2\mathrm{i}\theta} \cos(r+2\theta-2\phi) \\ &= \mathrm{i}\mathbf{A}_1 \cos \tfrac{1}{2}p \sin(r+2\theta-2\phi) - \mathrm{ij}\mathbf{A}_1^* \,\mathrm{e}^{2\mathrm{i}\theta} \sin \tfrac{1}{2}p \sin(r+2\theta-2\phi) \\ &\quad - \mathbf{A}_1 \sin \tfrac{1}{2}p \cos(r+2\theta-2\phi) - \mathrm{j}\mathbf{A}_1^* \,\mathrm{e}^{2\mathrm{i}\theta} \cos \tfrac{1}{2}p \cos(r+2\theta-2\phi).\end{aligned}$$

The final plate is identical with the first, and

$$\mathbf{A}_4 = \mathbf{A}_1[-\sin p \cos(r+2\theta-2\phi) + \mathrm{i} \sin(r+2\theta-2\phi)] - \mathrm{j}\mathbf{A}_1^* \cos p \cos(r+2\theta-2\phi).$$

If the whole unit is set at $\theta = 0$ between parallel polarizers at $\frac{1}{4}\pi$,

$$A' = -\sin p \cos(r-2\phi).$$

Thus, to obtain cosine terms for both retardations, the outer plates must have phase retardations $p = n+\frac{1}{2}\pi$, where p is an integral multiple of 2π. Then in terms of an orientation α of the first plate,

$$\mathbf{A}' = -\mathbf{A}\cos n\cos(r+2\alpha-2\phi)+\mathrm{i}\mathbf{A}\sin(r+2\alpha-2\phi)-\mathrm{j}\mathbf{A}^{*}\sin n\cos(r+2\alpha-2\phi). \quad (4.7)$$

It should be noted that, just as equation (4.3) for a fixed wide-field element can be obtained from equation (4.6) by putting $\phi = \alpha+\frac{1}{4}\pi$, the equation for a fixed split element can be obtained from equation (4.7) by the same substitution.

(e) *Doubly Tunable Split Element*

The split element considered above has tuning on the central component only. In one element of the filter, the two outside components are also tuned.

The first component of this combined element is a plate of phase difference n at $\alpha = \theta$ followed by a tuning unit in which the quarter-wave plates are parallel at $\theta+\frac{1}{4}\pi$. As this does not correspond directly to equation (4.5), the detailed analysis is given. It is:

$$\mathbf{A}_2 = \mathbf{A}_1\cos\tfrac{1}{2}n-\mathrm{j}\mathbf{A}_1^{*}\sin\tfrac{1}{2}n\,\mathrm{e}^{2\mathrm{i}\theta},$$

$$\mathbf{A}_3 = \frac{1}{\sqrt{2}}\mathrm{e}^{-\frac{1}{2}\mathrm{i}n}(\mathbf{A}_1-\mathrm{ij}\mathbf{A}_1^{*}\mathrm{e}^{2\mathrm{i}\theta}),$$

$$\mathbf{A}_4 = \frac{1}{\sqrt{2}}\mathrm{e}^{\mathrm{i}x}(\mathrm{i}\mathbf{A}_1-\mathrm{j}\mathbf{A}_1^{*}\mathrm{e}^{2\mathrm{i}\theta}), \quad (4.8)$$

where $x = \frac{1}{2}n-2\theta+2\psi$, ψ being the angle of the half-wave plate, and

$$\mathbf{A}_5 = -\mathbf{A}_1\sin x-\mathrm{j}\mathbf{A}_1^{*}\mathrm{e}^{2\mathrm{i}\theta}\cos x. \quad (4.9)$$

The next component is a wide-field tunable element, represented by equation (4.6) with $\alpha = \theta+\frac{1}{4}\pi$. Then

$$\begin{aligned}\mathbf{A}_6 &= \mathrm{i}\mathbf{A}_5\sin y-\mathrm{j}\mathbf{A}_5^{*}\mathrm{e}^{2\mathrm{i}\theta}\cos y\\ &= \mathbf{A}_1(-\mathrm{i}\sin x\sin y-\cos x\cos y)+\mathrm{j}\mathbf{A}_1^{*}\mathrm{e}^{2\mathrm{i}\theta}(-\mathrm{i}\cos x\sin y+\sin x\cos y),\end{aligned}$$

where $y = p+2\theta-2\phi$.

The last component is the same as the first but with the last quarter-wave plate omitted. It is thus represented by equation (4.8)

$$\begin{aligned}\mathbf{A}_7 &= \frac{1}{\sqrt{2}}\mathrm{e}^{\mathrm{i}x}(\mathrm{i}\mathbf{A}_6-\mathrm{j}\mathbf{A}_6^{*}\mathrm{e}^{2\mathrm{i}\theta})\\ &= \frac{1}{\sqrt{2}}[\mathrm{i}\mathbf{A}_1\{\sin y-\mathrm{e}^{2\mathrm{i}x}\cos y\}+\mathrm{j}\mathbf{A}_1^{*}\mathrm{e}^{2\mathrm{i}\theta}\{\sin y+\mathrm{e}^{2\mathrm{i}x}\cos y\}]. \quad (4.10)\end{aligned}$$

If this whole element is set with $\theta = \frac{1}{2}\pi$ between parallel polarizers at $\frac{1}{4}\pi$,

$$A' = \frac{1}{\sqrt{2}}(\mathrm{j}-1)\sin(n+4\psi)\cos(p-2\phi),$$

giving a cosine function for both retardations when tuned so that $\psi = \frac{1}{8}\pi$ and $\phi = 0$.

V. Transmittance of Filter

The general expressions above can now be applied to each element of the filter.

(a) *Split Elements*

The first, second, third, and fifth elements are split elements with the central component tunable. The first plate is set at $\alpha = 0$ and the element is between polarizers at $\beta = \frac{1}{4}\pi$. Equations (3.9), (4.7), and (3.10) give

$$I' = \cos^2 n \cos^2(r-2\phi).$$

The phase differences n and r for each element can be expressed in terms of q and c of the thinnest quartz and calcite plates of equivalent single-plate elements. The tuning angle ϕ can be expressed in terms of θ, the angle of the half-wave plate in the first element.

(i) *First Element*:

$$n = 2\pi q, \quad r = 2^5\pi q, \quad \phi = \theta,$$

and

$$I_1' = \cos^2 2\pi q \cos^2(2^5\pi q-2\theta). \tag{5.1}$$

(ii) *Second Element*:

$$n = 2^2\pi q, \quad r = 2^6\pi q, \quad \phi = 2\theta,$$

and

$$I_2' = \cos^2 2^2\pi q \cos^2(2^6\pi q-2^2\theta). \tag{5.2}$$

(iii) *Third Element*:

$$n = 2^3\pi q, \quad r = \pi c, \quad \phi = 4\theta,$$

and

$$I_3' = \cos^2 2^3\pi q \cos^2(\pi c-2^3\theta). \tag{5.3}$$

(iv) *Fifth Element*:

$$n = \pi q, \quad r = 2^2\pi c, \quad \phi = 16\theta,$$

and

$$I_5' = \cos^2\pi q \cos^2(2^2\pi c-2^5\theta). \tag{5.4}$$

(b) *Doubly Tunable Element*

The fourth element is represented by equation (4.10) with $\alpha = \frac{1}{2}\pi$ and

$$I' = \sin^2(n+4\psi)\cos(p-2\phi).$$

For this element,

$$n = 2^4\pi q, \quad p = 2\pi c, \quad \psi = \tfrac{1}{8}\pi-\tfrac{1}{4}\theta, \quad \phi = 2^3\theta,$$

and

$$I_4' = \cos^2(2^4\pi q-\theta)\cos^2(2\pi c-2^4\theta). \tag{5.5}$$

(c) *Wide-field Element*

The sixth element is a tunable, wide-field element, and is represented by equation (4.6) and $\alpha = 0$. For this element,

$$p = 2^3\pi c, \quad \phi = \tfrac{1}{4}\pi + 2^5\theta,$$

and

$$I'_6 = \cos^2(2^3\pi c - 2^6\theta). \tag{5.6}$$

(d) *Last Element*

The last element is a wide-field element tuned at the end. Initially the tuning unit was a quarter-wave plate at $\frac{1}{4}\pi$ followed by a polarizer at β, rotation of the polarizer giving the tuning. For a fixed wide-field element at $\alpha = 0$ following a polarizer at $\frac{1}{4}\pi$ and followed by this tuning unit

$$A' = -\frac{1}{\sqrt{2}}(1+\mathrm{j})\cos(p + \tfrac{1}{4}\pi - \beta),$$

or, with $p = 2^4\pi c$, $\beta = \frac{1}{4}\pi + 2^7\theta$,

$$I'_7 = \cos^2(2^4\pi c - 2^7\theta). \tag{5.7}$$

Later the tuning was changed to a rotating half-wave plate at ϕ followed by a polarizing beam-splitter. If the direction of polarization β from the beam-splitter is taken,

$$I' = \cos^2(p + \tfrac{1}{4}\pi - 2\phi + \beta),$$

corresponding to equation (5.7) when $\phi = 2^6\theta + \frac{1}{2}\beta + \frac{1}{8}\pi$.

The other beam from the beam-splitter is polarized at $\beta + \frac{1}{2}\pi$. If the half-wave plate is turned an angle $\frac{1}{8}\pi$ from the position given above, instead of a band tuned centrally to the rest of the filter, two bands

$$I' = \cos^2(2^4\pi c - 2^7\theta \pm \tfrac{1}{4}\pi)$$

are obtained, one from each beam, and these are symmetrically placed within the band given by the rest of the filter.

A second half-wave plate could also be added. Then

$$I' = \cos^2(p + \tfrac{1}{4}\pi - 2\phi_1 + 2\phi_2 - \beta);$$

one plate could then be used to tune with the rest of the filter while the other could give the $\frac{1}{8}\pi$ rotation to change from a single to a double pass band.

(e) *Whole Filter*

Each element provides plane polarized light for the succeeding element and the final intensity is the product of the intensities given above, namely,

$$I' = \cos^2\pi q \ldots \cos^2 2^3\pi q \cos^2(2^4\pi q - \theta) \ldots \cos^2(2^6\pi q - 2^2\theta)\cos^2(\pi c - 2^3\theta)$$
$$\ldots . \cos^2(2^4\pi c - 2^7\theta). \tag{5.8}$$

In this expression, q and c are functions of wavelength and, over a small range of wavelengths around 6563 Å for which q and c are matched, a change of θ moves the transmission band along the wavelength scale.

Equation (5.8) is the theoretical transmittance on-axis for a perfect filter with perfect polarizers. The transmittance off-axis and the effects of imperfect polarizers are considered by Lyot and by Evans and also by Giovanelli and Jefferies (1954).

VI. Extension of Design

A further extension of the split element will be studied now, although this is not included in the $\frac{1}{8}$ Å filter. The design is then considered of a filter that could be tuned to any wavelength within the range covered by the transmission of the birefringent materials and the polarizers.

(a) *Fully Tunable Wide-field Split Element*

Both types of split element considered earlier have only wide-field correction on the central component. It is therefore of interest to consider a split element where both components have a wide field and are tunable.

Such an element would consist of two identical tunable wide-field elements with the same orientation with a third set at $\frac{1}{4}\pi$ between them. Successive applications of equation (4.6) give:

(i) For the first component, where $p = n$ and $\alpha = \theta$,

$$\mathbf{A}_2 = -\mathrm{i}\mathbf{A}_1 \cos x + \mathrm{ij}\mathbf{A}_1^* \, \mathrm{e}^{2\mathrm{i}\theta} \sin x,$$

where $x = n+2\theta-2\psi$.

(ii) For the second component, where $p = r$ and $\alpha = \theta+\frac{1}{4}\pi$,

$$\begin{aligned}\mathbf{A}_3 &= \mathrm{i}\mathbf{A}_2 \sin y - \mathrm{j}\mathbf{A}_2^* \, \mathrm{e}^{2\mathrm{i}\theta} \cos y \\ &= \mathbf{A}_1(\cos x \sin y - \mathrm{i} \sin x \cos y) - \mathrm{j}\mathbf{A}_1^* \, \mathrm{e}^{2\mathrm{i}\theta}(\sin x \sin y + \mathrm{i} \cos x \cos y),\end{aligned}$$

where $y = r+2\alpha-2\phi$.

(iii) For the third component, where $p = n$ and $\alpha = \theta$,

$$\begin{aligned}\mathbf{A}_4 &= -\mathrm{i}\mathbf{A}_3 \cos x + \mathrm{ij}\mathbf{A}_3^* \, \mathrm{e}^{2\mathrm{i}\theta} \sin x \\ &= -\mathrm{i}\mathbf{A}_1 \cos 2x \sin y - \mathrm{j}\mathbf{A}_1^* \, \mathrm{e}^{2\mathrm{i}\theta}(-\mathrm{i} \sin 2x \sin y - \cos y).\end{aligned}$$

Thus, in full,

$$\begin{aligned}\mathbf{A}' = &-\mathrm{i}\mathbf{A} \cos 2(n+2\alpha-2\psi) \sin(r+2\alpha-2\phi) \\ &-\mathrm{j}\mathbf{A}^* \, \mathrm{e}^{2\mathrm{i}\alpha}[\mathrm{i} \sin 2(n+2\alpha-2\psi) \sin(r+2\alpha-2\phi) - \cos(r+2\alpha-2\phi)]. \qquad (6.1)\end{aligned}$$

When set at $\alpha = 0$ between parallel polarizers at $\beta = \frac{1}{4}\pi$, the complex amplitude obtained is

$$A' = -\mathrm{j} \sin 2(n-2\psi) \cos(r-2\phi).$$

(b) *Adjustable Wave Plates*

At present the tuning range of the filter is limited by the fixed elements; the band can only be tuned over a fraction of the bandwidth of the thickest fixed element.

If all the elements were made tunable, the tuning range would not be greatly increased if the tuning ratio remained fixed between the quartz and calcite components, for these materials have different dispersions of birefringence. But this drawback could be overcome by having the quartz and calcite elements tuned separately in proportion to their dispersions. Such a modification to the filter would have several advantages: the tuning range would be increased considerably and the filter could operate at any temperature; at present it can operate only at 40°C, the temperature at which the two materials are matched at 6563 Å. Finally, there would be the practical advantage that errors in any component could be corrected by adjusting the angle of the corresponding tuning plate. In the present filter, the residual errors in retardation of the plates in the fixed elements lead to some stray light outside the transmission band.

The range of tuning would then be limited by the variation with wavelength of the retardations of the quarter- and half-wave plates. These variations could be reduced by using composite wave plates of the types described by Pancharatnam (1955) or Koester (1959). However, a more attractive alternative is to consider tunable wave plates that are corrected as the filter is tuned and which then have, in theory at least, no residual errors. Possible tunable quarter- and half-wave plates will therefore be considered.

If two plates of retardation p are orientated at α_1 and α_2 the combination gives

$$\mathbf{A}' = \mathbf{A}\{1-2\sin^2\tfrac{1}{2}p\cos(\alpha_1-\alpha_2)\mathrm{e}^{-\mathrm{i}(\alpha_1-\alpha_2)}\}-\mathrm{j}\mathbf{A}^*\sin p\cos(\alpha_1-\alpha_2)\mathrm{e}^{\mathrm{i}(\alpha_1+\alpha_2)}.$$

This reduces to

$$\mathbf{A}' = \frac{1}{\sqrt{2}}\mathrm{e}^{\mathrm{i}\delta}(\mathbf{A}-\mathrm{j}\mathbf{A}^*\mathrm{e}^{2\mathrm{i}\theta}), \tag{6.2}$$

where

$$\tan\delta = \frac{(-\cos 2p)^{\frac{1}{2}}}{2\cos p+1},$$

provided that

$$\left.\begin{aligned} p &\geqslant \tfrac{1}{4}\pi, \\ \cos(\alpha_1-\alpha_2) &= \frac{1}{\sqrt{2}\sin p}, \\ \alpha_1+\alpha_2 &= 2\theta+\delta. \end{aligned}\right\} \tag{6.3}$$

Equation (6.2) is the same as that for a quarter-wave plate, as given by (4.1), apart from the factor $\mathrm{e}^{\mathrm{i}\delta}$. But, if plane polarized light is passed through the combination of plates and the state of polarization Q is calculated, the factor $\mathrm{e}^{\mathrm{i}\delta}$ disappears and the expression for circularly polarized light is obtained. Also, if circularly polarized light is sent through the plates, plane polarized light is obtained but not orientated at an angle $\theta-\frac{1}{4}\pi$ but at $\theta-\frac{1}{4}\pi+\delta$. Thus it appears that the factor $e^{i\delta}$ has no significance; the combination of plates can be used to replace a simple quarter-wave plate.

At the same time, as the phase p changes with wavelength, the angles α_1 and α_2 can be altered, following equation (6.3), to tune the plate.

A half-wave plate can be formed by a combination of two such composite quarter-wave plates. If two plates of phase p are set at α_1 with a plate of $2p$ at α_2 between them, then

$$\mathbf{A}' = \mathbf{A}[\cos^2 p-\sin^2 p \cos 2(\alpha_1-\alpha_2)]$$

$$-j\mathbf{A}^* e^{2i\alpha_1}[\sin p \cos p\{1+\cos 2(\alpha_1-\alpha_2)\}-i \sin p \sin 2(\alpha_1-\alpha_2)].$$

This is equivalent to a standard half-wave plate

$$\mathbf{A}' = -j\mathbf{A}^* e^{2i\theta},$$

provided that

$$\left.\begin{aligned} p &\geqslant \tfrac{1}{4}\pi, \\ \alpha_1 &= \theta-\tfrac{1}{2}\psi, \\ \cos(\alpha_1-\alpha_2) &= \frac{1}{\sqrt{2}\sin p}, \end{aligned}\right\} \tag{6.4}$$

where

$$\tan\psi = -\frac{(-\cos 2p)^{\frac{1}{2}}}{\cos p}.$$

To show how these composite plates could be used in a filter, the tuning element represented by equation (4.4) is recalculated as a combination of two composite quarter-wave plates at α_1 and α_2 with a composite half-wave plate at ϕ_1 between them. The tuning conditions for the wave plates (6.3) and (6.4) are assumed to be satisfied. Then

$$\mathbf{A}' = -\mathbf{A}\, e^{i(-\alpha_1+\alpha_2+\delta)} \cos(2\phi_1-\alpha_1-\alpha_2-\delta)+ij\mathbf{A}^* e^{i(\alpha_1+\alpha_2+\delta)} \sin(\alpha_1+\alpha_2-2\phi_1+\delta).$$

This is equivalent to equation (4.4) if

$$\left.\begin{aligned} \alpha_1 &= \alpha, \\ \alpha_2 &= \alpha-\delta-\tfrac{1}{2}\pi, \\ \phi_1 &= \phi. \end{aligned}\right\} \tag{6.5}$$

It should be noted that this is only an example of an adjustable tuning unit; it may not however be the simplest.

(c) *Rotary Dispersion Units*

The transmission bands of the present filter are 512 Å apart in red light. To separate them further, plates of lower retardation than the thinnest quartz plates would be required. Although not practical as single quartz plates, they could be made as crossed pairs of plates with the required difference of retardation, or of other materials of lower birefringence.

A simpler method would be the use of elements based on the rotary dispersion of quartz. Plane polarized light passing through a plate of quartz in the direction of the optic axis has its plane of polarization rotated by an amount depending on the thickness of the plate. The rotation also depends on wavelength, and a plate of this type followed by a second polarizer gives again a channel spectrum. The tuning now is simply done by rotating either polarizer.

The advantage of this system, apart from its simplicity of tuning, is that the extra elements required to extend the band separation of the filter from 512 Å to several thousand ångströms can be made of quartz plates of convenient thicknesses.

VII. References

Evans, J. W. (1949).—*J. Opt. Soc. Amer.* **39**: 229.

Evans, J. W. (1958).—*J. Opt. Soc. Amer.* **48**: 142.

Giovanelli, R. G., and Jefferies, J. T. (1954).—*Aust. J. Phys.* **7**: 254.

Jones, R. C. (1941).—*J. Opt. Soc. Amer.* **31**: 488.

Koester, C. J. (1959).—*J. Opt. Soc. Amer.* **49**: 405.

Lyot, B. (1933).—*C. R. Acad. Sci. Paris* **197**: 1593.

Lyot, B. (1944).—*Ann. Astrophys.* **7**: 31.

Pancharatnam, S. (1955).—*Proc. Indian Acad. Sci.* A **41**: 130, 137.

Poincaré, H. (1892).—"Théorie mathématique de la lumière." Vol. 2. Ch. 12. (Gauthiers-Villars: Paris.)

Steel, W. H., Smartt, R. N., and Giovanelli, R. G. (1961).—*Aust. J. Phys.* **14**(2): (in press).

A Narrow Passband Polarization Interference Filter for Hydrogen Alpha*†

BRUCE H. BILLINGS, S. SAGE, AND W. DRAISIN
Research Laboratory, Baird Associates, Inc., Cambridge 38, Massachusetts
(Received April 23, 1951)

This paper discusses the construction and testing of a Lyot-Öhman type polarization interference filter. This filter has a band width of approximately 1A peaked at H alpha. Previous filters have been fabricated, mainly of quartz. For a 1A passband filter a quartz thickness of approximately 24 cm would be needed for the thickest plate. This filter has been constructed using ammonium dihydrogen phosphate. Because the birefringence of this material is much higher than that of quartz, the tolerances associated with its construction are severe. By the use of mica corrector plates or optical shims, the tolerances can be made less critical. The final filter is equipped with a Senarmont compensator at each end so that the passband can be adjusted over a range of about 3A.

DURING the last two decades a great deal of research has been done on the development of narrow band filters. The narrowest of these are of the polarization type.[1–3] These have been built particularly for use in studies of the sun. In a previous paper a series of methods was shown whereby a Lyot-Öhman polarization filter could be tuned over a range of wavelengths.[4] As part of a program to develop such a tunable filter, it was necessary to manufacture a series of fixed filters. This paper discusses the problem of constructing such a fixed polarization filter with a band width of approximately 1A.

The polarization filters which have been constructed previously were made by individuals who were interested in the filter as a tool for study of a particular problem. Their concern was not so much with the device as with the answer to the problem. As a result, the materials used and the manufacturing methods were frequently more suitable for a one-shot process rather than for construction of several devices. Also, since each scientist in general constructed only one filter, there was not much opportunity to experiment with varieties of materials or varieties of elements as the cost would be too high.

The basic filter consists of a series of birefringent plates which are separated by polarizers. A simple filter might have x-cut plates of quartz as the birefringent elements and these plates of quartz might be separated by parallel polarizing sheets. When a single birefringent plate is placed between parallel polarizers the transmission is given by the expression:

$$I=\cos^2[\pi d(n_e-n_0)/\lambda], \quad (1)$$

where d is the plate thickness, n_0 and n_e are the ordinary and extraordinary indices of refraction, and λ is the wavelength.

When light is shone through such a combination into a spectrograph the spectrum is seen to be crossed with a series of dark bands. In the polarization filter this first plate is placed in series with another plate whose thickness is half as great. This other plate will have black bands which are spaced at twice the distance of the bands from the first plate. The black bands from the second plate can thus be made to fall on every other passband of the first plate. The resultant of the combination will be a series of transmission bands which are separated by twice their width. This process can be repeated by adding more plates until the final transmission is a series of bands whose separation is so many times their width that all the bands but one can be removed with a supplementary filter such as a dye filter or a Fabry-Perot interference filter.

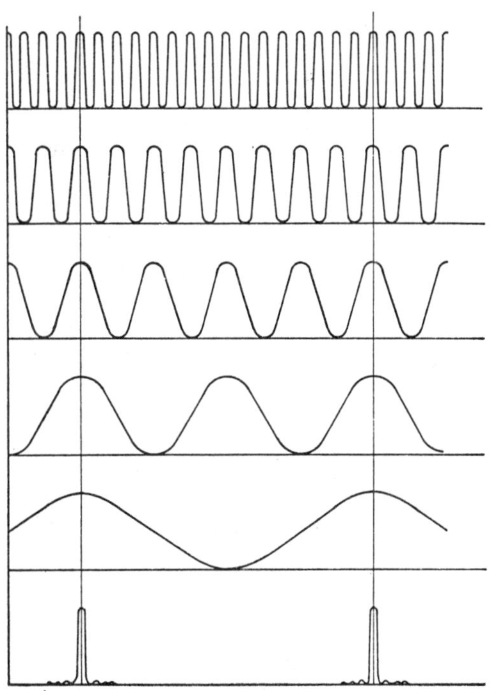

FIG. 1. Transmission between parallel polarizers of the individual retardation plates of a five element filter and the transmission of the assembled filter.

* This work has been supported by the Air Force, Cambridge Research Laboratories and the U. S. Navy, Bureau of Aeronautics, Bureau of Ordnance, Bureau of Ships, and ONR.

† Presented at the March, 1950 meeting of the Optical Society of America.

[1] B. Lyot, Compt. rend. **197**, 1593 (1933).
[2] Y. Öhman, Nature **141**, 291 (1938).
[3] J. Evans, Pub. Astron. Soc. Pacific **52**, 305 (1940).
[4] B. H. Billings, J. Opt. Soc. Am. **37**, 738 (1947).

TABLE I.

Crystal	Class	Indexes	$2V$
ADP	Tetragonal	$\epsilon=1.4792$ $\omega=1.5246$	...
KDP	Tetragonal	$\epsilon=1.4684$ $\omega=1.5095$	...
Rochelle salt	Orthorhombic	$\alpha=1.492$ $\beta=1.493$ $\gamma=1.496$	72°
EDT	Monoclinic	$\alpha=1.5086$ $\beta=1.5893$ $\gamma=1.5930$	21°
Quartz	Hexagonal (Trig)	$\epsilon=1.553$ $\omega=1.544$	...
Calcite	Hexagonal (Trig)	$\epsilon=1.486$ $\omega=1.658$	...
$NaNO_3$	Hexagonal (Trig)	$\epsilon=1.3361$ $\omega=1.5874$	...
Rutile	Tetragonal	$\epsilon=2.903$ $\omega=2.616$	...
Mica	Monoclinic	$\alpha=1.503$ to 1.623 $\beta=1.545$ to 1.685 $\gamma=1.545$ to 1.704	0 to 45°

In Fig. 1 is the transmission curve of the individual members of such a filter and also the transmission curve of a complete assembly of plates. The band width of such a filter will be determined by the thickest plate.

In the case of a 1A passband filter, the actual thicknesses involved can become quite large. For quartz and a simple filter the thickest plate will be 23.84 cm. This follows from the formula

$$W=0.5\lambda_0^2/[d(n_e-n_0)]. \quad (2)$$

Here W is the passband of the filter in angstroms and λ_0 is the position of the passband. Most of the possible applications for narrow band filters of this type require linear apertures of the order of 2 in. There are few pieces of optical quartz in this country which are large enough to be used for making these thick plates. Another possible material for a birefringent filter is calcite. Here the difference between the ordinary and extraordinary index of refraction is considerably larger than in the case of quartz and as a result a piece thick enough for a 1A passband filter is very much thinner. From Eq. (2) the actual thickness for a filter at H alpha can be calculated to be 1.268 cm. However, even in calcite a 2-in. diameter x-cut slab of this thickness, which will be of sufficient optical quality to make the last plate of a birefringent filter, is essentially unobtainable. Furthermore, the price of such material is quite high.

During the last few years a whole family of new crystals has been appearing. These crystals are being grown synthetically for the use of the electronic industry. One such crystal is ammonium dihydrogen phosphate.[5] These crystals are used in large Langevin plates for underwater signalling at supersonic frequencies. Another crystal which has been grown synthetically in the last few years is ethylenediamine tartrate. This crystal is being used as a piezoelectric oscillator for narrow band electrical filters and is being grown in large quantities by the Western Electric Company. The optical characteristics of a group of these crystals are shown in Table I. All the crystals mentioned above, with the exception of quartz, are exceedingly difficult to handle. They are soft and most of them are water soluble. However, the synthetic crystals are available in tremendous quantities. Ton lots can be obtained without too much difficulty. They also are much less expensive than either quartz or calcite. Because of their availability and cheapness, these crystals were used in the experiments which were performed in this laboratory. The particular crystal which was used in fabricating the filters described below was ammonium dihydrogen phosphate. This crystal had been grown in sizes such that 2-in. disks could be cut and also showed unusual optical uniformity. The optical quality of the material used in a polarization filter is extremely important since the device depends for its action on the phase shift which polarized light undergoes when it traverses a birefringent material. It is essential that the polarized light not have its state of polarization changed by any other means except the crystal birefringence. Any small speck or cloud or strain in the crystal will change the polarization state and render the filter less efficient. If at any stage through such a multi plate polarization filter a small amount of depolarization is introduced, the plate where this depolarization takes place will lose part of its action. At

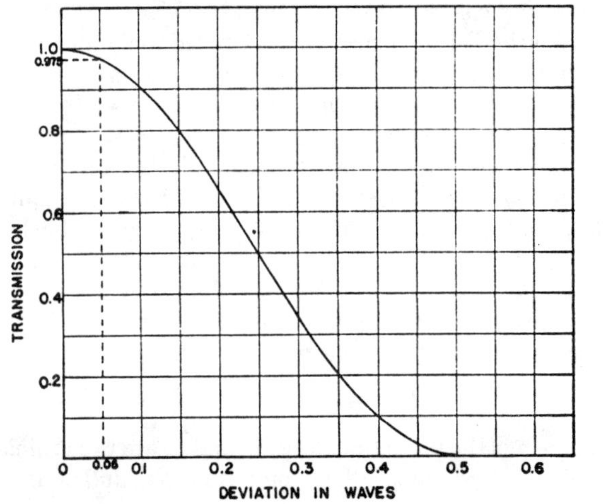

FIG. 2. Theoretical curve showing the effect on transmission of a polarization filter of a variation in retardation of the thickest crystal plate. The source is monochromatic light whose wavelength corresponds to the peak transmission of the filter.

[5] This crystal is grown by the Brush Development Company of Cleveland, Ohio under the trademark "PN."

those wavelengths where this plate should completely obstruct the light, there will be a leak and wavelengths will be passed by the filter which normally would be rejected.

MANUFACTURING TOLERANCES ON FILTER ELEMENTS

One of the first problems to be faced in constructing a filter is that of the tolerances which must be applied to the various filter elements. Tolerance must be placed on the thickness, plane parallelism, polish, and axial position of the ADP slabs. The thickness tolerance is determined by two factors. First, in the case of the thickest plate, by the wavelength which is to be passed. The retardation of this plate must be such that there is a transmission peak at the wavelength which is desired. Next, in the case of the plates which are thinner, the tolerance is determined more by the rejection which is achieved. Each plate is supposed to have a transmission zero at a position where each successive plate has a transmission maximum. If the thickness is not exactly correct, the zero will not coincide with the maximum and there will be a slight leak which will impair the wavelength purity of the light transmitted by the filter. In Fig. 2 is a curve of the transmission of a filter for monochromatic light as the thickest plate is changed so that the passband does not exactly coincide with the desired wavelength. From this curve a thickness tolerance of approximately 1/20 of a wave seems to be allowable. At this tolerance the transmission of the desired wavelength is reduced by only 4 percent. The thickness equivalent to a 1/20 wave retardation can be obtained from the following formula

$$d=(\Gamma\times\lambda)/(n_e-n_0), \tag{3}$$

where Γ is the number of waves of retardation. The thickness tolerance corresponding to 1/20 of a wave retardation turns out to be extremely small, both for quartz and for ADP. The tolerance has been deter-

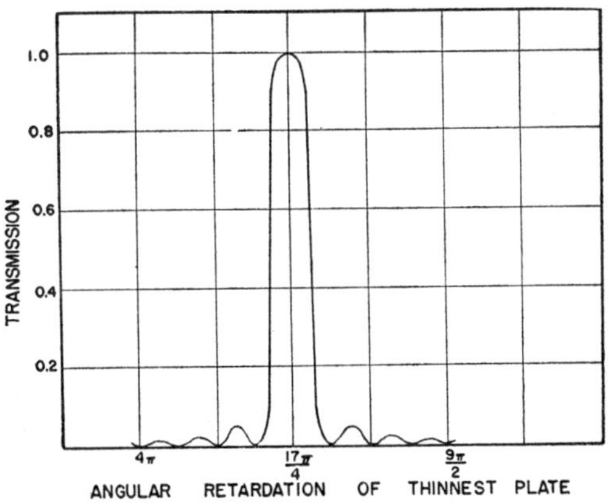

FIG. 3. Calculated transmission of a five element filter in which the first four elements have the 2, 4, 8 relationship, but the retardation of the thickest element is displaced one wave from this relationship. The thinnest plate has a retardation of two waves.

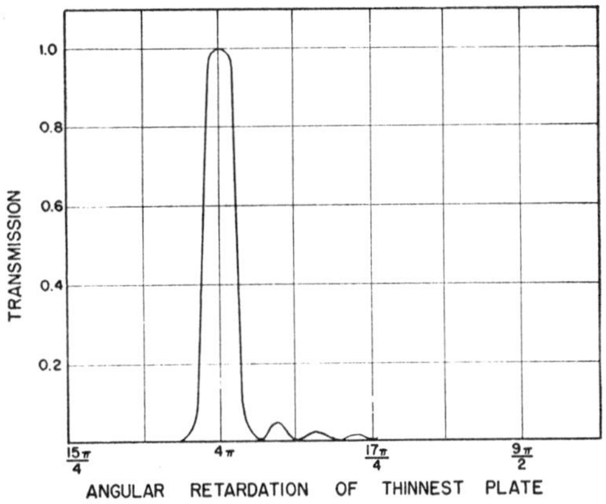

FIG. 4. Calculated transmission of a five element filter in which each element has a retardation exactly twice that of the preceding element.

mined at a wavelength of 5400A. At this wavelength the difference in refractive indices for ADP is 0.045, for the quartz the refractive index difference is 0.009. When these figures are put in the equation the thickness tolerance for ADP can be calculated to be ± 0.6 micron and for quartz ± 3 microns. Although this tolerance can be achieved in quartz, it seems well beyond the limit of accuracy in the case of ADP. This is particularly true since the blocks must have a particular flatness and a particular freedom from wedge angle at the same time. Whereas it might be possible to control the absolute thickness to this very fine tolerance, it seems quite impossible to control thickness, wedge angle, and flatness all at the same time. Accordingly, it was necessary to look for some trick which would enable the transmission peak to be located with the required accuracy but would not require the maintenance of such a fine tolerance in thickness. Such a trick was found in examining the tuning method described in a previous publication.[4] In this paper it was shown that the 2, 4, 8 relationship between the thickness of successive plates in the filter could be relaxed quite radically without introducing spurious passbands which could not be tolerated. This relaxation was based on the assumption that the filter had to operate only over a region as wide as the separation between two successive orders. Any secondary maxima which were introduced an order beyond the principle maxima which was being used would be removed by a supplementary filter. The tuning method simply required that a transmission peak be placed at the wavelength which was desired to be transmitted. It was not particularly important whether this transmission peak was twice the order of the transmission peak in the preceding plate. In Fig. 3 is a curve of the transmission of a filter in which the last plate has been displaced 1 peak from the 2, 4, 8 relationship. The retardation of the last plate, instead of being m waves, is $m-1$ wave at the wavelength to be transmitted. In Fig. 4 is a transmission curve of the same filter in which the last plate is precisely adjusted to m waves of retardation. The difference in secondary maxima between the two curves is barely distinguishable. Accordingly, it seems possible to change the thickness tolerance mentioned above from 1/20 of a wave to a whole number of waves +1/20 where the number must be a small percentage of the total number of waves retardation of the plate. Thus the thickness tolerance can be made very much larger than the tolerance of 1/20 of a wave. However, the gain has only been superficial so far since the transmission peak must be adjusted within 1/20 of a wave of the desired wavelength. This is accomplished by putting in series with the crystal a retardation plate which is of such value that the total retardation of the thick plate becomes correct ±1 wave. If the original thick plate is made to have the correct thickness within 1 wave, then the corrector plate or optical shim must have a thickness which will give a retardation less than 1 wave. At first glance it might appear that an enormous collection of retardation plates would be required in order to adjust the retardation of the thick elements to the correct value. This, however, is not the case as a tolerance has been set of 1/20 of a wave on the partial retardation of the thick plates. Accordingly, it is only necessary to have ten categories of retardation plates in steps of 1/20 of a wave. These retardation plates can then be either placed parallel or crossed with the thick element which is to be corrected. Because of the need for permanent and high quality in the optical corrector plates, it was decided to use high grade optical mica. Mica plates were cleaved and laminated and the resulting selection was catalogued into categories at 1/20 wave intervals. During the course of this project a library of several hundred retardation plates has been so collected and laminated.

The parallelism tolerance is determined from the same consideration which governs the choice of the thickness tolerance. In this case the wedge between the two surfaces must be no greater than that which would produce a 1/20 wave change in retardation across the surface of the filter. For the 2-in. aperture which was chosen, the parallelism tolerance in ADP is $\pm 1.2\times 10^{-5}$ radian or 2.5 seconds of arc. For quartz the figure is five times as great. For this tolerance there seemed to be no trick which would enable it to be reduced. Accordingly, the crystals had to be polished by a brute force method which would maintain the tolerance.

The tolerance on the orientation of the optic axis is determined by the angular field of the crystal. This information is obtained from the following formula which describes the retardation of a polarized light wave passing through a uniaxial x-cut crystal plate at various angles

$$\frac{\Gamma\lambda}{d}=(\epsilon-\omega)\left\{1-\frac{1}{2\omega}\left(\frac{\sin^2 i\cos^2\zeta}{\omega}-\frac{\sin^2 i\sin^2\zeta}{\epsilon}\right)\right\}, \quad (4)$$

where i is the angle of incidence, ζ is the angle between plane of incidence and the principal plane of the crystal, ϵ and ω are the extraordinary and ordinary refractive indexes. This is the formula for duplicate sets of hyperbolas. For an angle of incidence 0° from the normal

$$\Gamma_0\lambda/d=(\epsilon-\omega). \quad (5)$$

If Γ_0 is 1/20 wave greater

$$\frac{(\Gamma_0+1/20)\lambda}{d}=(\epsilon-\omega)\left\{1-\frac{\sin^2 i}{2\omega}\left(\frac{\cos^2\zeta}{\omega}-\frac{\sin^2\zeta}{\epsilon}\right)\right\} \quad (6)$$

subtracting (5) from (6)

$$\frac{\lambda}{20d}=(\omega-\epsilon)\frac{\sin^2 i}{2\omega}\left(\frac{\cos^2\zeta}{\omega}-\frac{\sin^2\zeta}{\epsilon}\right), \quad (7)$$

which may then be written

$$\sin^2 i=\frac{4\lambda\omega^2\epsilon}{20(\omega-\epsilon)[(\epsilon-\omega)+\cos^2\zeta(\omega+\epsilon)]d}. \quad (8)$$

Since $\omega>\epsilon$ for ADP the minimum angle i for which the retardation varies by 1/20 wave will occur at an azimuth of 90°.

$$\sin^2 i=\lambda\omega\epsilon/[10(\epsilon-\omega)d]. \quad (9)$$

For the crystal plates used in the 1A filter 5′ of arc seemed to be a reasonable tolerance. For a 50-mm thick slab of ADP at 5400A the maximum uniform angular field of view is approximately 26′ of arc in radius. If two crystals are set in series, the angular field of view is then the mutual field of view of each. If the orientation of optic axes of the two differ by an angle α, then the relation of the resultant angular field of view to the original angular field of view will be proportional to

$$[\tan^2(i-\alpha)]/\tan^2 i. \quad (10)$$

Thus, the angular field of view will be reduced by a factor of 0.6 for a variation in angle of 5′ of arc.

ANGULAR FIELD OF VIEW

The angular field of view of the simple filter is directly related to its bandwidth as may be seen by combining Eqs. (2) and (9). The resulting expression is

$$\sin^2 i=W\epsilon\omega/5\lambda. \quad (11)$$

This gives approximately 1° as the total field of a 1A filter at H alpha. For much of the work involving such a filter this value is much too small. Both Lyot[6] and Evans[7] have discussed several methods of increasing this low angular field in the polarization filter. The only successful technique which does not involve the use of thick plates of two different materials is Lyot's type I

[6] B. Lyot, *Annales D'Astrophysique,* Tome 7, fascicule 1–2, Janvier-Avril (1944).

[7] J. W. Evans, J. Opt. Soc. Am. **39**, 229 (1949).

system. In this system the thick elements are split into two equal halves. The slow axes of these two halves are then put at 90°. The two thick retardation plates are then effectively crossed and the retardation along the normal to the perpendicular is zero. Between the two plates is placed $\frac{1}{2}$ wave plate with its slow axis at 45° to the axes of the thick retardation plates. By this means the birefringences add while their variations with angle which are of opposite sign subtract. The retardation of the combination can be written

$$\Gamma=\frac{2(\epsilon-\omega)d}{\lambda}-\frac{(\epsilon-\omega)^2 d\sin^2 i}{2\omega^2\epsilon\lambda}. \qquad (12)$$

Along the normal this is the same retardation which was given by the thick single plate. Away from the normal, however, this retardation varies much more slowly with angle than in the case of the thick single plate. The loci of constant retardation are circles instead of hyperbolas as in the case of a single plate. The angular field of view is increased by a factor

$$(\epsilon/\epsilon-\omega)^{\frac{1}{2}}. \qquad (13)$$

For the 1A filter this technique was used for the four thickest ADP elements.

MECHANICAL PROCESSING OF AMMONIUM DIHYDROGEN PHOSPHATE

ADP has a hardness of about 3 according to Mohs scale of hardness. This is about equivalent to calcite and is accordingly very difficult to polish. The material also cleaves quite readily and what is more important fractures in all directions producing a typical conchoidal surface at the least provocation. Fracturing occurs if the material is subjected to any excessive mechanical stress or to sudden changes in temperature. Finally, and perhaps its worst property from the viewpoint of the optician, is its extreme solubility in water. This means that finished surfaces cannot be touched by ungloved hands lest a set of fingerprints be indelibly dissolved in the surface. Close visual inspection requires that the inspector must not breathe upon the surface, and fine polished surfaces must be protected by immediate lamination to prevent the deterioration of the surface even under conditions of low relative humidity.

In the latter half of 1948 a considerable amount of time was spent in attempts to hand polish ADP blanks to desired thicknesses. Handled one at a time it became more and more obvious that many plates could not be produced by this method as the final finished elements were often out of parallel by 10 or more fringes and the surfaces more nearly resembled a well-used skating rink. On the basis of the results of this preliminary experimentation a mechanical solution was sought in which a series of elements could be handled at one time and each would fall within the narrow allowable tolerances at the completion of the process.

Preliminary grinding runs were made early in 1949 at Frank Cooke's optical laboratory in North Brookfield, Massachusetts. At first no attempt was made to maintain crystallographic orientation of the plates, the prime purpose of these first tests was to determine the feasibility of grinding a group at one time, working on both sides of the elements at the same time, and the evaluation of the surfaces produced with respect to flatness and parallelism. Preliminary runs were sufficiently encouraging to warrant continuing this "mass production" approach to the problem.

The raw ADP blanks as obtained from the Brush Development Company in Cleveland have surfaces which, as a result of abrasive disk cutting, may be as much as 1° 30′ out of parallel with each other, and out of orientation with the crystallographic axes. By means of an optical lever arm and extinction of the blanks between oriented polarizers, the angular correction of one major face was first determined. Since the physical properties of this material preclude the usual methods of holding the elements (mechanical clamping, heating and cementing to a glass plate with beeswax, etc.) a special vacuum jig was fabricated which would rigidly hold the blanks. The vacuum jig was mounted on a Bemesine unit and the angular correction required was set on the latter. Once a new surface was produced on the raw blank, the bemesine was set to zero. The blank was then removed from the vacuum chuck, turned over and a surface was produced on the second major face plane parallel to the first.

Several methods were used to produce these new surfaces. The first used was to hold the oriented blank up against a kerosene saturated grinding belt. Apparently as a result of localized heating by the abrasive particles in the belt, fracturing occurred in the blanks which required excessive grinding in the later operations. This method was abandoned in favor of using a shaper which would produce much less localized heating and thus reduce the possibility of fracturing. This produced a surface relatively free of large fractures but containing to some depth a myriad of small check marks due to the action of the tool moving across the surface. It was finally found that the best results were obtained when the Bemesine unit was mounted on the face plate of a lathe and the surface turned by means of a round nosed cutting tool. A large number of crystals have now been processed by this method with exceptionally low mortality.

The actual grinding of the plates to specific thicknesses was performed after the parallel, oriented surfaces were prepared as described above. Techniques were developed which ultimately led to production so closely controlled that wafers can now be produced flat and parallel across two inches to less than 0.000010 inch.

ASSEMBLY AND TESTING OF BIREFRINGENT FILTERS

As demonstrated in previous sections the tolerance on the thickness and orientation of the axis of the

Table II.

Element	Orientation	Retardation
1. External rotating polarizer	Rotating	
2. Quarter wave mica plate	45° left	165 mμ
3. Mica corrector plate	Horizontal	55 mμ
4. ADP plate	Horizontal	608 ¼ waves
5. Half-wave plate	45° left	323 mμ
6. Mica corrector plate	Horizontal	195 mμ
7. Mica corrector plate	Vertical	166 mμ
8. ADP plate	Vertical	608 ¼ waves
9. Polarizer	45° right	
10. Mica corrector plate	Vertical	273 mμ
11. Mica corrector plate	Vertical	170 mμ
12. ADP plate	Vertical	304 ¼ waves
13. Half-wave mica plate	45° left	332 mμ
14. Mica corrector plate	Vertical	72 mμ
15. ADP plate	Horizontal	304 ¼ waves
16. Polarizer	45° left	
17. Mica corrector plate	Horizontal	161 mμ
18. ADP plate	Vertical	152 ¼ waves
19. Mica plate } half-wave	45° right	186 mμ
20. Mica plate } half-wave	45° right	144 mμ
21. Mica corrector plate	Vertical	146 mμ
22. ADP plate	Horizontal	152 ¼ waves
23. Polarizer	45° right	
24. Mica corrector plate	Horizontal	312 mμ
25. ADP	Vertical	152 ½ waves
26. Polarizer	45° left	
27. Mica corrector plate	Vertical	308 mμ
28. Quartz plate	Vertical	76 ½ waves
29. Polarizer	45° right	
30. Mica corrector plate	Vertical	233 mμ
31. Quartz plate	Vertical	38 ½ waves
32. Polarizer	45° left	
33. Mica corrector plate	Horizontal	164 mμ
34. ADP plate	Vertical	1216 ¼ waves
35. Half-wave mica plate	45° right	323 mμ
36. Mica corrector plate	Vertical	234 mμ
37. ADP plate	Horizontal	1216 ¼ waves
38. Quarter wave mica plate	45° right	163 mμ
39. External rotating polarizer	Rotating	

crystal plate is extremely small. Similarly the orientation of the plates and polarizers in an assembled filter must be precisely controlled. To assemble and test the filter a holder had to be devised which would maintain this precise orientation. Because of the extreme accuracy of polishing and laminating the crystals, it was not necessary to construct an orienting clamp for each crystal. Instead, it was possible to stack the crystals on a *V* block. The crystal surfaces provided orientation in one plane and friction with the *V* block provided orientation in the other plane.

Since the quartz and ADP plates were made with a tolerance in thickness of ±0.0002 in., the retardations they produce at λ6563 differed from the desired value by various amounts. The retardation (*n*) produced by a given thickness (*d*) may be calculated from the formula

$$n = d\mu/\lambda, \tag{14}$$

where λ = wavelength and μ = birefringence of the crystalline material at λ.

The variation in retardation produced by 0.0002 in. then becomes 0.35 wave for ADP and 0.07 wave for quartz.

To produce satisfactory isolation of a wavelength region by the filter, it is necessary that the retardation of the plates differ from the desired values by no more than ±0.05 wave. This was accomplished with the mica plates described above. The adjustment of the retardation of the various filter elements is done as the filter is assembled. Because of the high temperature coefficient of birefringence for ADP, it is necessary to do this assembly and correction at a constant temperature. Once this temperature is selected, the filter must always be operated at this point. This is necessary even for filters which are made of only one variety of crystal. As the birefringence changes with temperature the passband of the filter changes. It can be shown that in a filter whose elements are cut in the 2, 4, 8 ratio and in which the plates are all of the same material, a temperature change does not introduce any secondary maxima and the passband shifts smoothly through the spectrum. For filters which are made of two or more different materials, a temperature change not only shifts the passband but also introduces secondary maxima. This takes place because the 2, 4, 8 ratio between the retardations of the various plates is no longer maintained.

The temperature chosen for the operation of these filters was 40°C. Accordingly, the assembly and correction was done in a rectangular tank of mineral oil which was constructed with optical glass windows at the ends. This tank was mounted on an optical bench and a collimated incandescent light beam sent through the tank and focused on the entrance slit of a two meter grating spectrograph. This instrument is described below in connection with the testing of the assembled filter.

A fixed polarizer oriented vertically was mounted on the optical bench between the tank and the spectrograph. All polarizers in the filter were KN36C sheet polarizers made by the Polaroid Corporation. A mica plate having a retardation of exactly ¼ wave at λ6563 was mounted with its slow axis perpendicular to the axis of the fixed polarizer between the tank and the light source. A polarizer on a calibrated mount was then placed on the optical bench between the ¼ wave plate and the light source. The thickest ADP plate to be corrected was mounted then on the *V* block within the tank and oriented so that its axes were at 45° to the axis of the fixed polarizer.

This combination of a fixed polarizer, a retardation plate at 45°, a ¼ wave plate, and a rotating polarizer constitutes a Senarmont compensator.[8] The intensity of the light transmitted through this assembly can be written

$$I = \cos^2(\pi\Gamma - \phi) \tag{15}$$

where Γ = the retardation of the ADP plate, φ = the angle of the rotating polarizer measured from the vertical.

---

[8] H. deSenarmont, Ann. Chim. Phys. **73**, 337 (1840).

This is equivalent to the intensity produced by a single ADP plate with retardation $\Gamma-\phi\pi$.

The intensity peaks produced by the ADP plate were compared in the spectrograph with the emission line at λ6563 produced by a hydrogen lamp. The rotating polarizer was adjusted until an intensity peak produced by the ADP plate coincided with the emission line of the hydrogen lamp. The angle ϕ which could be read directly from the calibrated polarizer mount represented the variation in retardation of the ADP plate from the desired value. The retardation in $m\mu$ of the mica plate needed to correct this deviation can be determined from the ratio

$$m\mu/656=\phi/\pi.$$

The mica plate with the determined value was then oriented with its axes either parallel or perpendicular to those of the ADP plate and placed next to the ADP plate in the tank on the side nearer the light source, and a fixed polarizer was added next to the mica plate. The rotating polarizer was removed from the system and the intensity peak produced by the ADP plate and the mica plate between fixed polarizers was again compared with the emission peak of the hydrogen lamp. The second thickest ADP plate to be corrected was then added next to the fixed polarizer in the tank oriented with its axes at 45° to the axis of this polarizer. The rotating polarizer was then replaced in the system and rotated until the peak of the second plate coincided with that of the first plate and the secondary peaks were minimized. A mica plate having retardation determined by the angle ϕ was then added to the second ADP plate and the process continued until the whole filters, except for the two thickest plates, had been assembled. These plates needed no correction since they were mounted at opposite ends of the filter with this combination of a mica $\frac{1}{4}$ wave plate and a rotating plate so that they could be corrected by rotating the external polarizers on the filter.

FIG. 5. Brass cradle containing the filter elements adjusted to the proper angle.

The final mount for the filter consisted of two plane parallel brass bars affixed to the 120° and 240° positions of two ring-shaped brass end pieces. The oriented crystal plate stack was transferred from the *V* block onto the brass bars. With the elements in place a Neoprene gasketed brass washer ring was positioned in the rack and held in place against the end quarter wave plate by three set screws passing through the brass end piece. The top unit in the 0° position with respect to the end piece was a brass backed strip of Neoprene sponge separated from a rigid brass section by adjusting screws. This unit was bolted onto the end pieces to form a top support to prevent the filter components from slipping upward out of the *V* type supporting unit.

FIG. 6. Assembled filter with heating coil and the bellows which allows for expansion of the immersion liquid.

The first field trial filter was composed of the 39 elements listed in Table II. All of the components with the exception of the external rotating polarizers were held in the cradle and, therefore, were in the temperature controlled section of the filter. The retardation values listed in the table for the quartz and ADP elements represent the total retardation of the plate plus its mica corrector. The exact retardation of the latter plate is individually noted. The orientation of the slow crystallographic axis of each element and the polarizing axes of the polarizers is noted as well.

The entire cradle (Fig. 5) containing the filter elements was then placed in a brass tube which was threaded at each end. The brass tube was sealed by placing Neoprene gaskets on each side of a crystal-x strain free glass window and tightening the pair of brass end nuts. Oil was introduced into the system

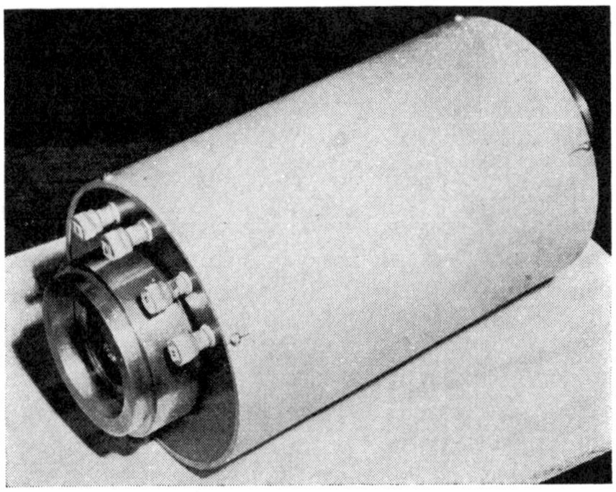

FIG. 7. Completed filter.

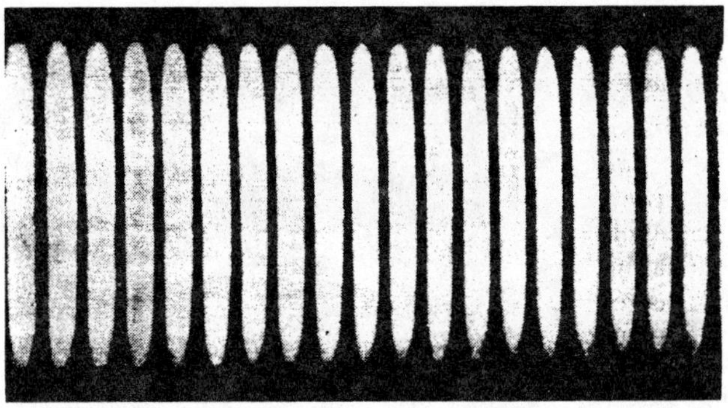

Fig. 8. Channel spectrum of an ammonium dihydrogen phosphate crystal between crossed polarizers. This photograph was taken with a 2-meter grating spectrograph.

prior to positioning the second nut. The final complete filling with oil was accomplished through a plug in the housing so that a minimum of air space remained within the unit. Since the filter was assembled at room temperature but is operated at 40.6°C, an expansion bellows was incorporated into the design to allow for an increase in volume of the immersion liquid. Both the filling plug and expansion bellows are shown in Fig. 6.

A pair of studs were soldered to the tubular brass housing at each end, just inside the threaded area to which were attached the winding terminals. A layer of glyptal was baked onto the center portion of the brass housing, and a 25-ohm looped winding of 0.0028-in. enameled copper wire was wound from one pair of terminals in close contact with the brass to act as a temperature sensitive element. This in turn was covered by another layer of glyptal and then by a layer of insulating mica tape. The latter is approximately 0.007 in. thick and is composed of a layer of woven glass fibers, a layer of mica, and a layer of Japanese rice paper. Number 22 nichrome wire, the heating element of the filter, was wound directly on the mica tape and connected to the other pair of terminals. Successive layers of mica-glass tape, cambric tape, and asbestos tape overlay the nichrome.

The brass end nuts are of sufficient size that a large radiating surface is present and, therefore, it was decided to enclose the entire metal housing in an insulated tube. A series of 8 blind holes were drilled and tapped into each end nut. An adapter ring, threaded to fit the external rotating polarizer mount, was fashioned out of polystyrene so that no direct metal to metal contact would aid in the conduction of heat outward from the filter. This ring was bolted to the end nut and between them was sandwiched a micarta end plate $\frac{1}{4}$ in. thick and 6 in. in diameter. Figures 6 and 7 show this assembly as well as the polarizer mount. Four terminals were placed in one of the micarta end plates, two for the heating leads and two for the sensitive element. Number 14 copper wire was used to connect the terminals attached to the filter tube to those in the micarta end plate so that all connections could be reached from the outside once the insulation was completed. A micarta tube was cut to slip over the end plates and extend to the outer face of these plates. All existing space between the micarta tube and the inner metal parts were filled with exploded vermiculite which provided excellent insulation around the filter.

To produce a 1A passband with a stable peak wavelength in a filter of this type the temperature must be controlled to within 0.08°C. To maintain this accuracy the previously described extensive thermal insulation was a "must." Further, it was imperative that a controller of extreme sensitivity be obtained.

In the case of most semimechanical temperature controllers (i.e., bimetallic units), the sensitive elements are somewhat bulky. Since most of these are activated by a mechanically operated contact, the orientation of the filter housing might possibly affect the control point. Further, these are the "on-off" types of control, producing full heat until the upper limit is reached and then none while the unit drops to the lower temperature limit. Of necessity, this method does not produce a fixed temperature, but an oscillation above and below a mean temperature and has little sensitivity with respect to changes in heat requirements brought on by external temperature changes. Due to the inadequacies of this variety, it appeared that the electronic proportional controller was more desirable. In this device the amount of heat introduced is proportional to the deviation in temperature from the desired temperature so that as the temperature of the unit controlled approaches a set value, the amount of heat introduced decreases until just enough heat is added to equal the amount dissipated by the unit. A proportional controller was designed and built by the W. S. Macdonald Company, Inc. of Cambridge, Massachusetts.

The temperature sensitive element consists of a resistance wire wrapped around the inner metal housing of the filter. As the temperature changes, small changes occur in the resistance of this element. These changes are measured by making the resistance wire one arm of a Wheatstone bridge. Two of the other arms are fixed resistors while the fourth is variable, to set the controller to a desired temperature. With this device it is possible to maintain the operating temperature to within 0.05°C over a wide range of external temperature conditions.

FIG. 9. Photograph showing the transmission of the completed filter for tungsten light. The spectrum lines at the top and bottom of the picture are hydrogen alpha from a hydrogen discharge tube. The light transmitted through the filter was from a tungsten lamp.

OPTICAL TESTS WITH THE ASSEMBLED FILTER

In order to study the behavior of the filter, light which has passed through it is examined spectrographically with a high dispersion spectrograph. It is necessary that the resolution and slit width of the spectrograph be such that about one-tenth of the passband of the filter is passed by the spectrograph. In a filter with 1A passband this requires a grating spectrograph. The particular test instrument used on this project was a two meter concave grating spectrograph made by Baird Associates, Inc. The grating was a 15,000 line per inch, four-inch grating ruled on the Baird Associates, Inc. ruling engine. In Fig. 8 is a typical channel spectrum of an ammonium dihydrogen phosphate crystal plate taken with this spectrograph.

In order to determine whether the initial 1A passband filter constructed in accordance with the technique described above would perform suitably, several types of tests were made. The first experiment consisted of elevating the complete filter to operating temperature, mounting it in the path of a collimated light beam from an incandescent line filament lamp and observing the transmitted results with a 2-meter Baird Associates, Inc. grating spectrograph. To determine the accuracy with which the filter was tuned to the H alpha-band λ6563 the emission spectrum of an H type hydrogen discharge lamp was used for comparison. Results of these tests shown in Fig. 9 indicate that the passband transmitted by the filter is peaked at the proper wavelength. Further corroboration was obtained by observing the emission of the hydrogen discharge tube directly through the filter. When properly oriented the filter transmitted the passband rather brightly. When the external polarizers were rotated to change the wavelength of peak transmission, the H alpha-radiation from the discharge tube grew dim and then disappeared.

The optical resolution of the filter was determined by photographing a standard National Bureau of Standards' resolution chart with a properly positioned camera first without the filter and then with the filter directly in front of the camera lens. The film using the camera alone showed a resolution of around 56 lines per millimeter; a film using the camera and a density 3 optical grade filter in front of the camera lens showed a resolution of about 40 lines per millimeter. With the birefringent filter the film showed a resolution of just over 28 lines per mm.

Electro-optically tuned spectral filters: a review

W. J. Gunning
Rockwell International Science Center
1049 Camino Dos Rios
Thousand Oaks, California 91360

Abstract. Electro-optically tunable spectral filters have been developed over the years and now include a rather diverse collection of devices. These devices, which are of increased importance for remote sensing applications, employ linear electro-optic (Pockels) materials such that tuning is accomplished via externally controlled voltages. A brief review of the properties of linear electro-optic materials is presented followed by descriptions of tunable birefringence interference filters, Fabry-Perot devices, and other novel filter structures Recent developments in electro-optic tunable filters are included. Issues relating to the further development of infrared tunable spectral filters are discussed

Keywords: electronically tunable optical spectral filters; electro-optics; optical filters birefringent filters; Fabry-Perot filters; tunable filters.

Optical Engineering 20(6), 837-845 (November/December 1981).

CONTENTS

1. INTRODUCTION

Recently there has been considerable interest in solid-state spectral filters whose transmission characteristics can be controlled solely by the application of electric fields. Development of such devices began with the realization that the electro-optic effect in certain crystals could be employed to replace mechanical tuning sections of birefringence interference filters.[1] The initial motivation for tunable filter development was for applications ranging from color projection systems to astronomical telescopes for which simultaneous spatial and spectral resolution was a requirement. Since then, interest in remote sensing and, in particular, space-based platforms has motivated the development of devices which were both compact and free of the difficulties inherent in mechanically tuned spectral filters. In many of these applications restrictions on size, weight, and power consumption have made the use of compact solid-state tunable filters highly desirable. Also, while early efforts were concentrated on visible applications, current interest is directed increasingly toward infrared filters.

Invited Paper TF-105 received July 30, 1981; accepted for publication Aug. 1, 1981; received by Managing Editor Aug. 5, 1981.

This article reviews the development of electro-optically tunab spectral filters that can be controlled by the application of a dc low-frequency electric field. In Sec. 2, a description of electro-opt materials provides background for the discussion of specific device Sections 3, 4, and 5 discuss electro-optic birefringent interferen filters, Fabry-Perot devices, and electro-optic coupled-wave filter respectively. Section 6 describes new electro-optic filter concep which have not yet been experimentally demonstrated, and Sec. presents a summary of the present status of electrically tuned spe tral filters. Acousto-optic tunable filters which rely on the intera tion of light with an rf acoustic wave propagating in a suitable cryst have been discussed elsewhere[2] and will not be reviewed here.

2. ELECTRO-OPTIC MATERIALS

As discussed in the introduction, this paper deals with optical filte that can be spectrally tuned through the use of optical materia whose refractive indices are modified by the application of an ele tric field. The materials include those that exhibit linear (Pockels) quadratic (Kerr) electro-optic effects in which index changes res from the nonlinear response of the crystalline material. In additic liquid crystal materials, in which large index changes result from t electric-field-induced rotation of optically anisotropic molecul may be employed. The discussion of electro-optic effects whi follows will concentrate on the Pockels effect since, in most cas the magnitude of the linear effect is greater than that observed Kerr effect materials. The object of this discussion is to acquaint t reader with those effects that are important for the device appli tions that are discussed later. For a more in-depth discussion electro-optic materials, the reader is directed to several excell review articles and reference works.[3-10] Descriptions of electro-op effects of liquid crystals will be deferred to the discussions of speci devices.

The effect of an applied electric field on the optical properties of a crystal can be represented as[3]

$$\Delta\left(\frac{1}{n_{ij}^2}\right) = \sum_{kl}\left(r_{ijk}E_k + R_{ijkl}E_k E_l\right) \tag{1}$$

where r_{ijk}, R_{ijkl} are the linear and quadratic electro-optic coefficients respectively, and all subscripts run from 1 to 3. Strictly speaking, these coefficients represent the direct effects, that is, ndependent of crystal strain. However, at all but the highest driving requencies, the electric-field-induced strain contributes to the index hange via the elasto-optic effect. This may be of either the same or pposite sign as the direct effect, and in some materials, can reduce he magnitude of the observed effect by as much as 50%. At high requencies, such that the material cannot strain macroscopically, nly the direct effect is observed.

By invoking simple symmetry arguments it is easy to see that the inear effect will not be observed in materials which possess a center f symmetry. In a centrosymmetric crystal, the change of sign of an pplied field must result in a physically indistinguishable response, ince no preferred crystallographic orientation with respect to an nversion operation exists. Therefore, the linear effect must be idenically zero. For noncentrosymmetric materials, the quadratic ffects are typically orders-of-magnitude smaller than the linear ffects. We will therefore take the R_{ijkl} to be zero for the remainder of he discussion.

In order to illustrate the effects of an applied field, it is useful to troduce the concept of the index ellipsoid or indicatrix. This is a epresentation of the three-dimensional index-of-refraction surface, hich, in the principal axis system of the crystal, is

$$\frac{x_1^2}{n_1^2} + \frac{x_2^2}{n_2^2} + \frac{x_3^2}{n_3^2} = 1\ . \tag{2}$$

order to determine the refractive indices for light propagating in arbitrary direction, a section is taken through the ellipsoid, ormal to the direction of propagation. The semimajor axes of the n general) elliptical cross section are the refractive indices for the o orthogonal polarization modes.

The effect of an externally applied electric field on the index lipsoid of a noncentrosymmetric crystal is represented by

$$\sum_{ij}\left(\frac{1}{n_{ij}^2} + r_{ijk}E_k\right)x_i x_j = 1\ . \tag{3}$$

symmetry, $r_{ijk} = r_{jik}$, and therefore the ij subscripts are usually ntracted using the convention 1 = 1 1, 2 = 2 2, 3 = 3 3, 4 = 2 3, 5 = , and 6 = 1 2.

In general, many of the r_{mk} are identically zero, as dictated by stal symmetry. Tables illustrating the general form of the r_{mk} sors for all allowed crystal symmetries, and tables of linear ctro-optic materials, may be found in several sources.[3,5,8,10] lues of the r_{mk} are typically given in units of 10^{-12} m/V.

The choice of a particular material for a specific application will determined by several factors. First, a material possessing the propriate crystal symmetry is required in order to produce the alitatively desired electro-optic response. In most cases, a material ving the highest possible value of the required r_{mk} is chosen in ler to keep the required voltages to a minimum. However, as is the e with materials such as SBN, KTN, and KDP and its isomorphs, maximum values of the r_{mk} occur at temperatures near their roelectric phase transitions. In this regime, the r_{mk} are quite perature dependent and their dielectric constants are quite large, s requiring increased electrical power for high frequency operation.

We will now describe, with several examples, several of the ctro-optic effects which are of major importance for device appliions. These fall qualitatively into three categories. First, and simplest, is an isotropic change in index as may be observed in crystals of symmetry groups 4, 4 mm, 3, 3 m, 6 or 6 mm. In the case of $LiNbO_3$[11] (point group 3 m), the nonvanishing coefficients are $r_{13} = r_{23}$, r_{33}, and $r_{42} = r_{51}$. If we consider an electric field applied along the optic axis (E_3), the index ellipsoid becomes

$$(x_1^2 + x_2^2)\left(\frac{1}{n_1^2} + r_{13}E_3\right) + x_3^2\left(\frac{1}{n_3^2} + r_{33}E_3\right) = 1\ . \tag{4}$$

For light incident along the optic axis and polarized at any direction in the x-y plane, we take the $X_3 = 0$ section of the ellipsoid and obtain

$$\left(\frac{1}{n_1^2} + r_{13}E_3\right)\left(x_1^2 + x_2^2\right) = 1\ . \tag{5}$$

Assuming the index change, Δn, is small, we find

$$\Delta n = -\frac{n_1^3 r_{13} E_3}{2}\ , \tag{6}$$

with the result that the use of nonstatic fields produces a phase modulation. Since this is a longitudinal field effect, no advantage is realized by increasing the optical path length. The half-wave voltage may be defined as that voltage required to induce a half-wave phase shift of the transmitted light resulting in

$$V_{1/2} = \pm\frac{\lambda}{2n_1^3 r_{13}}\ . \tag{7}$$

For $LiNbO_3$, $r_{13} = 8.6 \times 10^{-12}$ m/V[11] resulting in $V_{1/2} = \pm 3.7$ kV at 6328 Å and ±23.5 kV at 4 μm. A few observations may be made regarding the form of Eq. (7). First, since little dispersion is observed in the electro-optic coefficients, higher voltages are required at longer wavelengths which imposes practical limitations in many device applications. This will also be the case for other geometries, and indeed several filter geometries result in an even stronger wavelength dependence. Second, the dependence of Δn on the sign of the voltage, a requirement for noncentrosymmetric crystals, will only hold as long as, in the case of ferroelectric materials, a critical (coercive) field is not exceeded. In that event, the polarity of the ferroelectric domain is reversed and the advantage of sign reversal of the applied field is lost, resulting in a doubling of the maximum voltage. $LiNbO_3$ is not observed to exhibit domain reversal at room temperature for any value of applied field up to its dielectric breakdown.[12] On the other hand $Sr_6Ba_4Nb_2O_6$ (SBN) goes through domain reversal for field strengths of only 2 kV cm^{-1}.

The second and third major categories of electro-optic effects result in a change in the birefringence and/or a rotation of the optical axes of the crystal. These effects are by far the most common, since the application of an electric field to a crystal will introduce a new symmetry direction and thereby reduce the symmetry of the crystal. This forms the basis for the development of tunable birefringent network filters, since such electro-optic elements may be employed as tunable or rotatable waveplates.

We will illustrate the above effects with several examples. First, consider $LiNbO_3$ again, but now propagate along the X_2-axis, with the electric field still directed along the optic axis. If the incident light is polarized at 45° with respect to the optic axis, we obtain

$$\left(\frac{1}{n_1^2} + r_{13}E_3\right)x_1^2 + \left(\frac{1}{n_3^2} + r_{33}E_3\right)x_3^2 = 1\ , \tag{8}$$

resulting in

$$\Delta n_1 = -\frac{n_1^3 r_{13} E_3}{2} \qquad \Delta n_3 = -\frac{n_3^3 r_{33} E_3}{2}\ . \tag{9}$$

Since no cross terms appear, the axes are not rotated, but we find that we have a transverse electro-optic waveplate whose field dependent birefringence is

$$B = (n_1 - n_3) + \frac{1}{2} n_o^2 r_c E_3 \,, \tag{10}$$

where

$$r_c = r_{13} - \left(\frac{n_3}{n_1}\right)^3 r_{33} \,. \tag{11}$$

In the transverse geometry, a voltage reduction from the full half-wave voltage may be realized in the ratio of the optical path length to the dimension along which the field is applied.

Next, consider the case in which the applied field introduces a cross term into the index ellipsoid, thereby causing a rotation. Two possibilities exist, as illustrated in Fig. 1, depending upon whether or not the crystal is initially isotropic for the direction of propagation. To illustrate, we now turn our attention to the KH_2PO_4, $NH_4H_2PO_4$ (KDP, ADP)[13,14] type crystals which are of group $\bar{4}2m$ above their phase transition temperatures. These have the coefficients $r_{41} = r_{52}$ and r_{63}. With the electric field along x_1, the index ellipsoid becomes

$$\frac{x_1^2}{n_1^2} + \frac{x_2^2}{n_1^2} + \frac{x_3^2}{n_3^2} + 2 r_{41} E_1 x_2 x_3 = 1 \,. \tag{12}$$

The $x_1 = 0$ section is an ellipse in the absence of E_1. Consequently, the small cross term causes a small rotation of the ellipse by angle α, given by

$$\tan 2\alpha = \frac{2 r_{41} E_1}{\dfrac{1}{n_1^2} - \dfrac{1}{n_3^2}} \,, \tag{13}$$

in addition to a change in the birefringence.

For fields along x_3 with light propagating in that same direction, we obtain a result similar to Eq. (12). However, the $x_3 = 0$ section is initially circular and Eq. (13) yields $\alpha = 45°$ for all finite values of applied field. Light initially polarized along x_1 or x_2 experiences a phase retardation of $\Gamma = 2\pi n_1^3 r_{63} E_3 \ell/\lambda$ where ℓ is the optical path length. The half-wave voltage required to obtain a retardation of $\Gamma = \pi$ is $V_{1/2} = \lambda/2 n_1^3 r_{63}$.

Additional effects are observed in crystals of other symmetry classes. However, the above examples should provide sufficient background for the discussion of tunable filter applications which follows. Additional information, as well as a discussion of experimental techniques for determining the values of electro-optic coefficients, will be found in the rather complete literature on the subject.[3-10]

3. ELECTRO-OPTIC BIREFRINGENT INTERFERENCE FILTERS

Birefringence interference filters have been used for many years for astronomical spectroscopy, and are now finding applications in laser stabilization and tuning. As a class, these filters are linear combinations of birefringent elements and linear polarizers.[15] Typical materials employed for birefringent filters are quartz, calcite, and ADP. The two simplest configurations are the Lyot-Ohman[16,17] and Solc[18,19] filters. Of the two, the Lyot-Ohman[16,17] is conceptually the simplest, as illustrated in Fig. 2. Each filter stage, consisting of one birefringent element between parallel polarizers, has a transmission spectrum given by

$$\tau = \cos^2 \frac{\Gamma}{2} \tag{14}$$

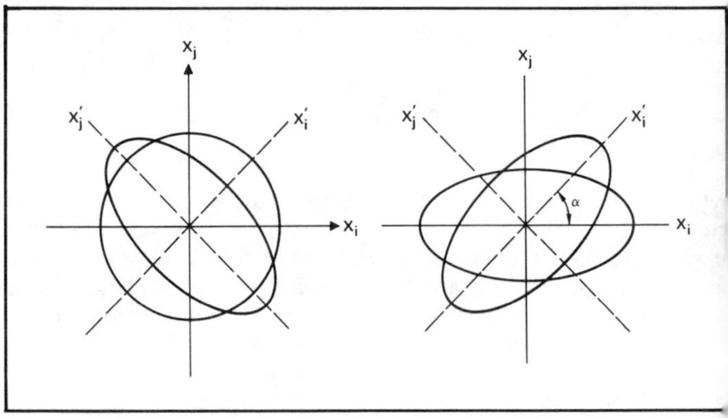

Fig. 1. Effect of an applied transverse field on the indicatrix (a) with an initially circular cross section and (b) with an initially elliptical cross section for crystals with off diagonal electro-optic coefficients r_{63}, r_{41}, or r_{52}.

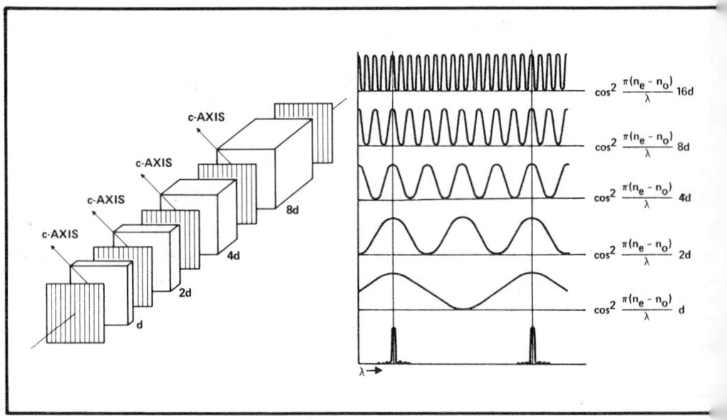

Fig. 2. Lyot-Ohman birefringent filter geometry and spectral transmission.

where $\Gamma = 2\pi\, \Delta n d/\lambda$. Here, Δn is the birefringence of the birefringent element, d is its thickness, and λ is the wavelength. Whe[n] several such stages are placed in tandem, with a constant pla[te] thickness ratio of 2:1, the overall transmission is given by

$$T = \prod_{i=1}^{N} \cos^2 \pi \frac{\Delta n}{\lambda} d_o 2^{N-1} \,, \qquad (1$$

where d_o is the thickness of the thinnest plate. This reduces to

$$T = \frac{1}{4^N} \frac{\sin^2 \dfrac{2^N \pi \Delta n d_o}{\lambda}}{\sin^2 \dfrac{\pi \Delta n d_o}{\lambda}} \,. \qquad (1$$

Physically, the thickest plate determines the bandwidth of the filt[er] while the addition of thinner plates increases the free spectral ran[ge] (FSR) by causing extinction of every other maximum of the pr[e]viously thicker stage. The bandwidth of a Lyot-Ohman filter is

$$\mathrm{FWHM} = 0.44 \frac{\lambda_o^2}{2^{N-1} \Delta n d} \,, \qquad ($$

while the free spectral range is approximately

$$\mathrm{FSR} = \frac{\lambda_o^2}{\Delta n d_o} \,. \qquad ($$

The opposite end of the spectrum in birefringent filters is the one nvented by Solc.[18,19] This configuration, which is not as physically ntuitive as the Lyot-Ohman, consists of a series of birefringent plates of equal thickness between a single pair of polarizers. In the Solc fan geometry, shown in Fig. 3, the angular orientations of the optic axes of the individual plates are arranged in an increasing sequence be-ween parallel polarizers. For an N-element filter, the orientation of he kth plate is $\phi_k = (2k - 1)\,\rho$ referenced to the input polarizer axis, vith $\rho = 45°/N$. The transmission of the Solc fan filter is found, via he Jones calculus,[20] to be

$$T = \left[\frac{\sin N\chi}{\sin\chi}\cos\chi\tan\rho\right]^2 , \tag{19}$$

here

$$\cos\chi = \cos\rho\cos\frac{\pi\Delta nd}{\lambda} . \tag{20}$$

urprisingly, the Solc folded filter with plate orientations $\phi_k = (-1)^k\rho$ nd crossed polarizers has a transmission profile identical to the fan nfiguration, but with its passband shifted by one half of its FSR.

Many variations of each of the above structures are possible, and owerful techniques have been developed to generate structures aving desired transmission characteristics.[21-30] However, in every se, the basic physical mechanism is the relative phase shift of the o polarizations, which is caused by the birefringence of the ele-ents and the relative intensities in each polarization, which is termined by the orientation of the elements. The ability to vary her of these properties introduces the possibility of tuning of the ssband of the filter. A number of mechanical tuning schemes have en developed which function by rotating the plane of polarization the light via rotating waveplates.[17,28,31] Figure 4 illustrates this heme for a general Solc element. A tunable Lyot element would nply omit the second quarter-wave plate. With the use of electro-tic elements, it becomes possible to replace the rotating half-wave ate with its electro-optic equivalent, a rotating-field phase shift-[32,33] In addition, fully electro-optic birefringent plates may be ployed,[4,29,33-43] in which case the birefringence is directly tunable h no additional tuning elements required. Each of these above proaches has been reported in the literature and will be briefly cussed below.

Billings[1] was the first to suggest the use of electro-optic retarda-n plates as tuning elements. The basic filter was a fixed wave-gth Lyot-Ohman filter with quartz birefringent elements. Each ge includes, in tandem with the quartz element, a longitudinal d z-cut ADP retardation plate. With no voltage applied, the ADP te is optically isotropic, and hence the spectrum of the fixed rtz filter is observed. As voltage is applied along the z-axis, efringence is induced which adds to that of the quartz element and ilts in a shift of the passband. A practical filter was not produced Billings, however, because of the lack of electro-optic crystals of quate quality with low half-wave voltage.

With the advent of large high quality crystals of the ADP, KDP s, electrically tunable birefringent filters were fabricated by a nber of workers. Babits and Hicks[34] were the first to demonstrate lor filter based on the model proposed by Billings. Ahmed and [36] demonstrated a switched-voltage tuning concept which reduced maximum voltage required for continuous tuning of the filter. In original concept discussed by Billings, larger voltages were ired for the electro-optic elements in the thicker filter stages. was necessary in order to produce the same wavelength shift of igh order stages as in the low order stages. The design of Ahmed Ley employed odd half-order waveplates in the sequence $\lambda/2$, 2, $5\lambda/2$, $7\lambda/2$, and $19\lambda/2$ with their accompanying ADP ro-optic tuning plates. One such complete stage is illustrated in 5. Each stage was delineated with crossed polarizers. Wave-s from $9\lambda/2$ through $17\lambda/2$ were omitted because of their igible effect on filter bandwidth. To tune the filter through its

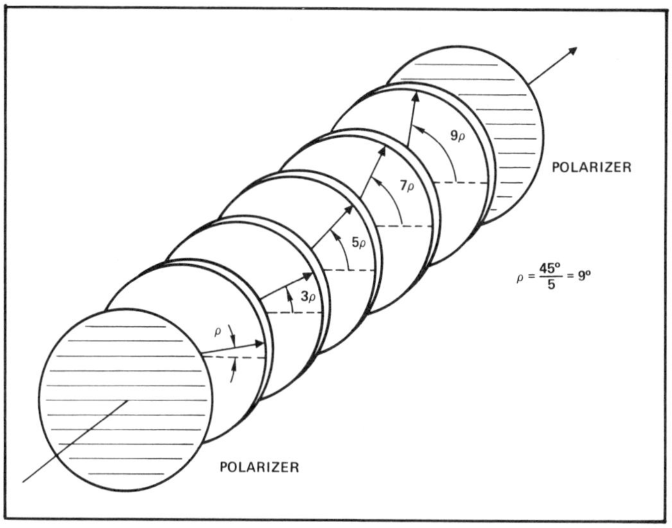

Fig. 3. Solc filter in the fan geometry.

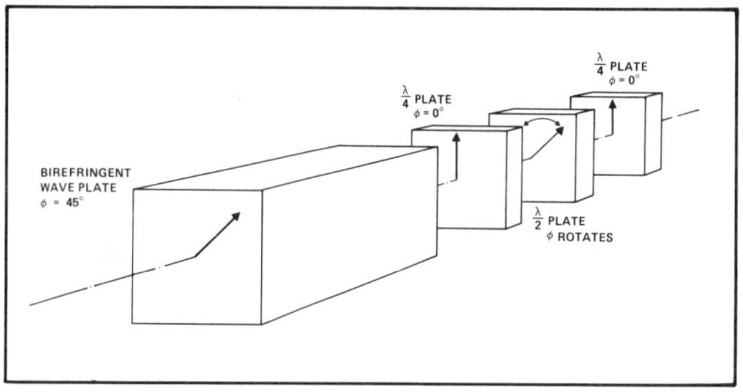

Fig. 4. General Solc type mechanically tuned stage. The second $\lambda/4$ plate is removed for use in a Lyot type filter.

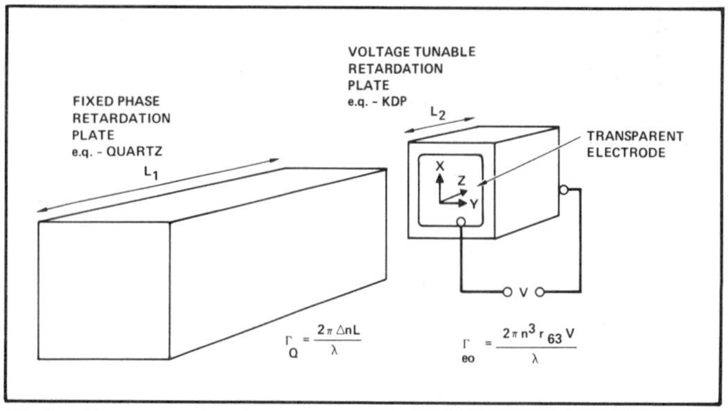

Fig. 5. Electro-optically tuned birefringent filter stage using a longitudinal field tuning element.

free spectral range would, in principle, require a shift of the spectrum of the highest order plate through 19 FSRs. Consequently, while the ADP tuning element in the $\lambda/2$ stage requires ± 5 kV to scan from $\lambda = 2500$ Å to $\lambda = 7500$ Å, 19 times the phase shift, and hence, voltage is required in the $19\lambda/2$ stage. The switched voltage approach allows the voltage in the higher order stages to drop back to zero whenever that stage has scanned through its free spectral range once. This approach results in some discontinuous changes in filter bandwidth and sideband structure, however, voltages were limited to ± 30 kV rather than the original ± 95 kV. In this manner, a filter was demonstrated which was continuously tunable from $\lambda = 2500$ Å with FWHM = 80 Å through $\lambda = 7500$ Å with FWHM = 480 Å.

Several other workers have designed and built filters, based on the above concepts, for applications as diverse as imaging solar spectrophotometry[29,37,39] and laser frequency stabilization.[38,40,42] Drichko[35] incorporated thermo-optic compensation into various electro-optic birefringent filter designs in order to reduce the temperature sensitivity of the passband characteristics inherent in these devices. Thermo-optic compensation is achieved by combining birefringent elements having different temperature dependencies. The condition for thermo-optic compensation is

$$N_1A_1 - N_2A_2 = 0\ , \tag{21}$$

where

$$A = \frac{1}{\Delta n}\frac{\partial \Delta n}{\partial T} + \frac{1}{\ell}\frac{\partial \ell}{\partial T} \tag{22}$$

and

$$N = \frac{\ell \Delta n}{T}\ . \tag{23}$$

Here, ℓ is the crystal length and T is the temperature. Drichko, et al.[29] demonstrated such a filter in which the KDP tuning plates also served as thermo-optic compensation plates for the fixed calcite elements. In this way, a 0.12 Å wide filter, which would normally have required temperature control of ±0.02°C, required no thermal stabilization over a range of 2300 Å in the visible. Different thickness KDP plates were required in each stage of the filter in order to achieve this result. However, through use of a switched voltage technique, peak voltages were limited to 9 kV. Development of a longitudinal field Solc type filter is presently being pursued for applications in the infrared.[44] This structure would consist of many thin, independently voltage addressable, electro-optic plates.

Thus far we have described tuning elements which have longitudinal field geometries. This is principally due to the ready availability of large crystals of the KDP, ADP class. However,transparent electrodes are required making extention to longer wavelengths difficult. In addition, the full half-wave voltage is required to produce one half-wave of phase retardation. Both of these potential difficulties are overcome through the use of transverse field elements. As discussed earlier, the tuning voltage may be significantly reduced by simply increasing the optical path length while maintaining a constant cross section as in Fig. 6. Such transverse field waveplates have been routinely employed as optical modulators,[45-48] which are simply tunable low-order retardation plates between polarizers, a simple birefringent filter stage.

The alternate approach to electrically tunable birefringent filters was proposed and demonstrated by von Willisen.[32] This employed electro-optic rotating-field phase shifters in direct analogy with mechanical tuning methods. As illustrated in Fig. 7, transverse field electrodes are employed in such a manner that the applied electric field may be oriented in any direction in the crystal. The electro-optic material is therefore required to be isotropic in the direction of light propagation in the absence of applied fields. In the design proposed by von Willisen, crossed rf fields are applied at angular frequency ω in order to continuously scan the filter elements. For crystals belonging to symmetry group $\bar{4}3$ m, the fields are directed along the [1, 1, 2] direction and the axes of the induced birefringence rotate in the opposite direction at half the angular velocity of the rf fields. The field required to produce a half-wave phase shift is

$$E = \frac{3}{8}\frac{\lambda}{\ell n_o^3 r_{41}}\ , \tag{24}$$

where ℓ is the length of the crystal. Quadratic, Kerr effect materials may also be used in this application. Indeed, von Willisen demonstrated the principle of rotating field phase shifters by using nitrobenzene Kerr cells. This tuning approach has the advantage that,

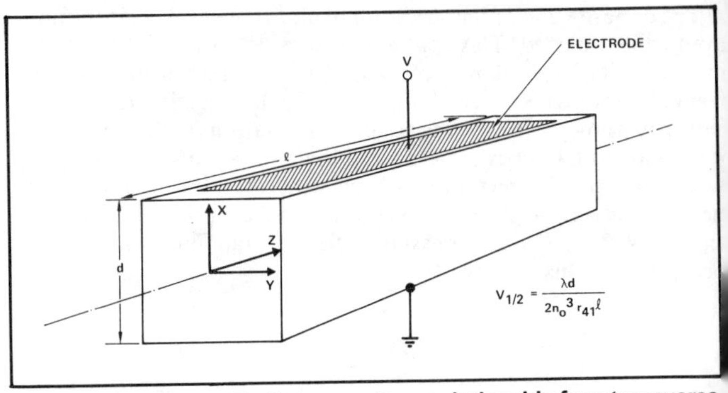

Fig. 6. Geometry and half-wave voltage relationship for a transverse field electro-optic phase retardation element. A material of symmetry group $\bar{4}3$ m (for example CdTe) has been used for illustration.

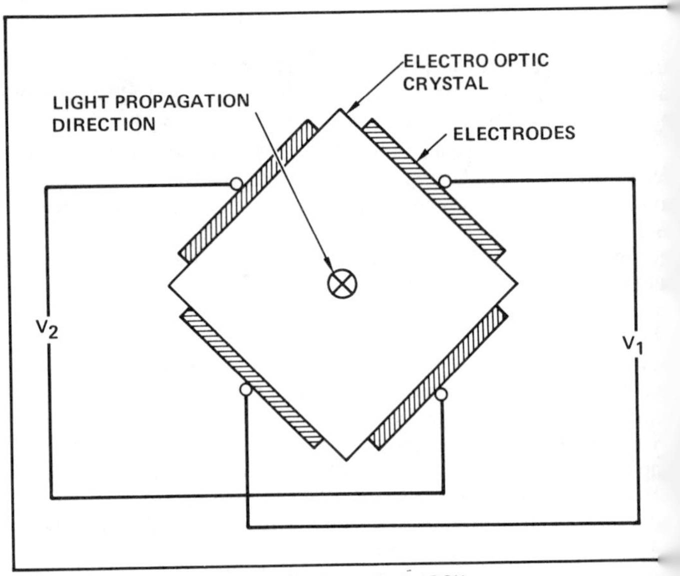

Fig. 7. Rotating field phase shifter. (Ref. (32)).

being a transverse field structure, no electrodes are introduced in t optical path, and identical voltages are required for each element the filter. Only the phasing of the voltage applied to the two electro pairs is critical.

Recently, Pinnow et al.[49,50] developed a radically differe approach to electro-optic birefringent filters. In their device, a sp tially periodic electric field is applied across a thin plate of Y-c $LiTaO_3$ (see Fig. 8). This single plate is positioned between cross polarizers, and light is transmitted along the length of the very t crystal. By suitably selecting the voltage profile, the optic axis of crystal is rotated locally in such a manner as to approximate Solc-type filter. Different wavelengths may be selected by vary the period of the periodic voltage waveform. In the device describ a peak-to-peak voltage of 50 V was applied across the 140 μm th crystal via 100 finger electrodes. Typical Solc filter transmiss spectra were obtained over the range 4700-6000 Å with bandwid of 80 Å to 170 Å. Tuning to infrared wavelengths was mention however, the λ^2 dependence of voltage on wavelength leads difficulties in extension of this physical structure to longer wa lengths. In addition, as a result of the restricted aperture, implem tation of this filter in an imaging system would not be straightforw

A current area of interest is the development of integrated opt electrically tuned filters and modulators. The design of these devic conceptually similar to that of the filter described by Pinnow et al. This growing field will not be discussed here and the interested re is directed to the accumulating literature on the subject.[51]

As an alternative to solid electro-optic crystal materials, Ta

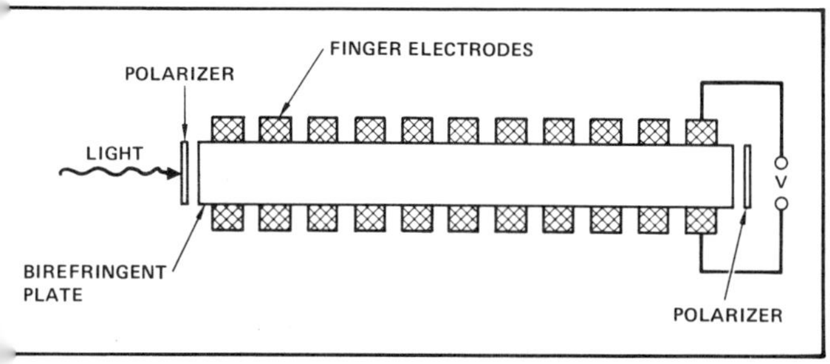

ig. 8. Single crystal transverse field programmable Solc filter (Refs. 9 and 50).

emonstrated a filter which employs thin liquid crystal cells oriented the Solc geometry. Here, the extremely large optical anisotropy of quid crystal molecules is exploited to tune the filter. Nematic liquid ystals may have a birefringence as large as 0.2 and are readily oriented by small electric fields. For example, a nematic liquid ystal with positive dielectric anisotropy is initially oriented by rface treatments such that all of the molecules lie parallel to the cell alls. In this orientation, the maximum birefringence is expe- enced. Then, as a dc voltage is applied, the molecules begin to tate until, at some saturation voltage, they are all aligned parallel ith the electric field. Therefore, for light propagating along the ectric field direction, the liquid crystal appears isotropic. Tarry monstrated tuning through the visible in a 6 stage filter with ltages ranging from 1.7 to 2.5 V. Each liquid crystal cell was μm thick.

Liquid crystal devices, on the other hand, have severe limitations. particular, achievement of uniform liquid crystal layers is rather fficult, leading to nonuniformity of device response. Environmen- factors are also an important consideration. Most liquid crystals ve a very narrow range of operating temperatures, some requiring eration within 10°C of the designed operating point. Low temper- ıre operation, as is required for many infrared applications, is not ossibility because of phase changes to the solid crystalline state. cessive temperature may irreversibly degrade the liquid crystal. hile crystalline electro-optic materials do exhibit some tempera- e dependence in their properties, this is a reversible and usually a all effect for all but the most exotic materials. Finally, the infrared sorptions of liquid crystals may be troublesome for some applica- ns, but certainly do not rule out infrared applications, as will be monstrated later.

ELECTRO-OPTICALLY TUNED FABRY-PEROT EVICES

bry-Perot (FP) filters represent a different class of devices from se discussed above. Here the passband is determined by the ical boundary conditions of an optically resonant cavity. As a up, Fabry-Perots are capable of very narrow bandwidths (a few r less), exhibit a narrow field of view, and usually have a small spectral range. The transmission of an FP is given by[53]

$$\lambda, \theta) = \frac{1}{1 + F \sin^2 \delta/2}\ , \qquad (25)$$

n

$$F = \frac{4R}{(1-R)^2} \qquad (26)$$

$$\delta = \frac{4\pi n d \cos\theta'}{\lambda}\ . \qquad (27)$$

Here, R is the reflectance at the cavity boundary, n is the index of the cavity medium, d is its thickness, and θ' is the angle of incidence within the cavity medium.

The condition for a transmission maximum is

$$2nd \cos\theta' = m\lambda\ , \qquad (28)$$

where m is an integer. Obviously a change in the optical path length, nd, will change the tuning point of the filter. In the simplest manifes- tation of a tunable FP, an optical cavity is defined by two highly reflecting parallel surfaces, while the cavity medium is air. Tuning is achieved by either varying the air pressure and hence the index or, more commonly, by varying the plate spacing via a piezoelectric transducer. In both cases, a mechanical support structure fabricated from low thermal expansion materials is required, and careful temperature control is quite critical for operation. Consequently, in many applications, for reasons of mechanical stability, tuning speed, and simplicity of design, a completely electrically-tunable FP would be desirable.

Electro-optic Fabry-Perots are direct analogs of solid cavity Fabry-Perot etalons and, in the limit, dielectric interference filters. Here, the optical cavity is an optically flat and parallel plate of electro-optic material. High efficiency reflective coatings are applied to the optical faces, and transparent electrodes are applied. Tuning is achieved through the linear electro-optic effect by which the refrac- tive index, and therefore the optical path length, is modified by the applied voltage. It is desirable to select a uniaxial material of appro- priate symmetry such that the index changes are identical in both polarizations. This eliminates the need for polarizers and the accompanying 50% insertion loss. Transverse field FPs have been demonstrated but exhibit extremely small free spectral range (sub Angstrom in the visible), since the optical path lengths of the cavities are quite long. This configuration also typically requires polarized light. Such a device has been demonstrated as an opto-electronically controlled modulator, capable of pulse shaping and optical switching.[54,55]

In order to achieve a larger free spectral range and simultane- ously maintain a uniform electric field distribution, the longitudinal geometry is required. The electrodes may be thin layers of semicon- ducting material which form an integral part of the dielectric reflec- tor structure. By choosing electro-optic materials from symmetry classes 4, 4 mm, 3, 3 m, 6, or 6 mm, we see that a voltage applied along the c-axis, corresponding to the optical propagation direction, induces a change in refractive index given by Eq. (6). If $r_{13} = r_{23}$, then the effect is independent of polarization, and therefore the need for an input polarizer is eliminated. Such a device has been demon- strated as an electro-optic laser modulator.[56]

The FSR of a longitudinal device is still quite limited for many filtering applications. In order to increase the FSR to the point that conventional interference filters may be used for blocking, the mul- tiple cavity structure may be implemented. This concept has been employed in piezoelectrically tuned FPs for astronomical applica- tions.[57,58] An examination of the tuning principles of this filter will convince the reader that the voltage required to tune a multiple stage filter over the extended FSR is the same as that required to tune the thickest stage over its FSR. Indeed, the filter may be tuned over several discrete wavelengths by tuning one stage through its FSR while the rest are passive.

Gunning and Yeh[43,59] describe a multiple cavity electro-optic filter of this type using $LiNbO_3$. A two cavity filter is shown schemat- ically in Fig. 9 with a difference of cavity thickness of 10%. Conse- quently, the FSR will be ten times that of the thicker cavity. High transmission through the center electrode, which is a thin metal film, in the infrared is possible via the phenomenon of induced transmis- sion. This multiple cavity design has two modes of tuning, both discrete and continuous. In the discrete mode, one cavity is tuned while the other is passive. Transmission maxima appear at discrete wavelengths corresponding to the passbands of the passive stage. For continuous tuning, both cavities are simultaneously tuned, there- by reaching those wavelengths not accessible by discrete tuning.

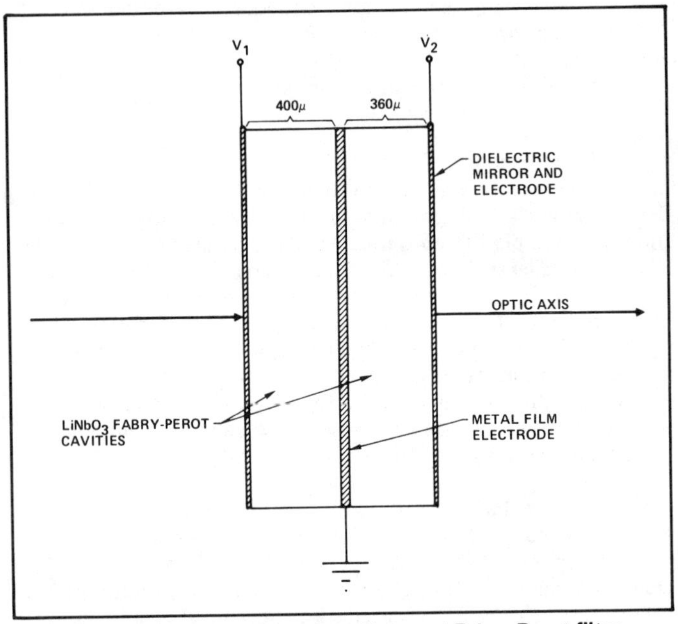

Fig. 9. Double-cavity electro-optically tuned Fabry-Perot filter.

Both modes of tuning have been experimentally demonstrated by Gunning.[60]

The difficulty of high-tuning voltages for infrared applications remains. For many applications which require continuous tuning, the voltages are prohibitive. Development of more efficient electro-optic materials is possible and several promising possibilities exist.[61] However, infrared applications would still require voltages in the kV range. (Visible devices could require only 500 V using these materials.) In addition, no material candidates currently exist for operation at wavelengths longer than 5 μm.

The possible use of liquid crystals for tunable FP filters has been pursued. Ito[62] demonstrated optical modulation at 6328 Å by using a device which is essentially a FP filter. A negative nematic liquid crystal was introduced into a 8.5 μm thick cell. Surface treatment resultedin a homeotropic alignment in the voltage off state, which is optically isotropic for light at normal incidence. As a longitudinal voltage is applied, the molecules rotate toward the parallel, homogeneous orientation. The resulting change of index caused a shift through 2 1/2 FSRs. In this configuration, an output polarizer is required because of the birefringence of the liquid crystal. This approach has the disadvantage that no preferred parallel orientation exists and therefore, as the field is applied, the molecules may initially rotate in any direction, thus requiring a reorientation of the polarizer after each return to the zero voltage condition.

Recently Gunning et al.[63] reported the operation of a tunable liquid crystal FP filter in both the visible, at 6328 Å, and in the 3 to 5 μm band. Positive nematic material was used, such that in the zero voltage state the molecules were aligned parallel to the cell walls. Surface treatments of the cell resulted in a unique orientation and thus eliminated the ambiguity of polarizer orientation. Only 12 volts were required to shift the transmission spectrum through 75% of one FSR at 4 μm. As a result of the small fields required to induce large index changes, thinner cells than those required for solid electro-optic Fabry-Perots may be used. Consequently, larger FSRs and bandwidths are possible. It is speculated that operation in other regions in the infrared is possible where regions of high transparency of the liquid crystal exist.

5. ELECTRO-OPTIC COUPLED WAVE FILTERS

A novel filter which has recently been developed is the electro-optic coupled wave filter.[64,65] This device is characterized as having both a narrow passband and an extremely large field of view. Strictly speaking, this is not an electro-optic tunable filter, since the applied field does not result in a spectral shift but is the means by which the filter is activated. The filter consists of a single crystal of a birefringent material which possesses an iso-index point (wavelength at which the birefringence crosses zero), positioned between crossed polarizers. When the crystal's optic axis is oriented parallel to the input polarizer, no light is transmitted in the absence of some coupling field. Henry[66] and Laurenti et al.[67] demonstrated an optical filter of this type, using CdS, in which coupling from one polarization mode to the other at the iso-index point was induced by applied strain or magnetic fields. Indeed, since the modes are degenerate at the iso-index wavelength, residual strain in the crystal is often sufficient to produce mode coupling.

Lotspeich[64,65] describes a coupled wave filter which uses the electric-field rotation of the indicatrix to achieve mode coupling. Using a material such as $AgGaS_2$ with symmetry $\bar{4}2$ m, either longitudinal or transverse geometries are possible. As discussed earlier with a field applied along the x_1-direction, the optic axis of the birefringent material is rotated by α, which is given by Eq. (13). α is an extremely small angle for wavelengths at which the birefringence is finite, but abruptly becomes 45° at the iso-index point for any value of applied field. Therefore, light at wavelengths different from the iso-index point will remain linearly polarized and will therefore be extinguished by the analyzer; at the iso-index point, however, the linearly polarized light will become, in general, elliptically polarized. The transmitted intensity depends on the magnitude of the applied voltage, with maximum transmission being obtained for

$$V = \frac{\lambda}{2n^3 r_{41}} , \quad (29$$

the same condition as for a classic Pockels cell.

The bandwidth of this filter depends on the rate of birefringenc dispersion. Lotspeich[64,65] calculates the dispersion for $AgGaS_2$ to b

$$\frac{d\Delta n}{d\lambda} = 1.87\ \mu m^{-1} \text{ at } 4970\ Å , \quad (3C$$

which obtains a passband of 0.2 Å for a 10 mm thick crystal. In ord to achieve peak transmission, the required voltage is found to b 3.9 kV. Varying the applied voltage will permit control of filt transmission.

An extremely important feature of this filter is its field of vie Since the passband is determined solely by the iso-index point of t material, there will be no shift of the passband position with angle incidence. This is a radical departure from other types of bandpa filters which typically exhibit sizeable shifts of passband positi with incident angle. Iso-index coupled wave filters exhibit so broadening of their passband and will be limited ultimately by t off-axis performance of the polarizers. Lotspeich[64,65] finds a ma mum line broadening of 20% in $AgGaS_2$ for extreme rays ($\pi/2$ ang of incidence) at the worst-case azimuth angle. An important cons eration is the occurrence of optical rotation in $AgGaS_2$ and oth chalcopyrites and materials of symmetry $\bar{4}2$ m. This will lead increased out-of-band transmission for rays incident in the pla containing the c-axis. Iso-index materials of class 6 mm, such CdS, CdS_xSe_{1-x}, and $Zn_xCd_{1-x}S$[67] are not optically active and therefore promising candidate materials.

The lack of tunability of the iso-index coupled wave filter (f tuning may be accomplished by temperature variation or appl strain) restricts its use to the material-determined iso-index wa length. Recently, Yeh[68] described a dispersive birefringent fi which, by using the basic birefringent filter geometries, exploits birefringence dispersion of materials at wavelengths removed fr their iso-index points. This filter, which exhibits narrow bandwi and extremely large field of view,[69] may be designed to operate at wavelength for which the birefringence dispersion is large. Henc is not restricted by the point spectral properties of materials. Eff are now underway to develop electronic tuning mechanisms for filter.

6. FUTURE ELECTRO-OPTIC FILTER DEVELOPMENTS

Thus far we have considered electrically tuned filters which, with the exception of the coupled-wave filter, are direct analogs of classical mechanically tuned devices. Recently, new device concepts have been proposed which have yet to be experimentally demonstrated. Among these is the programmable Bragg[43] electro-optic tunable filter.

The programmable Bragg filter consists of a periodic structure of alternating layers of different refractive index. By using the formalism developed by Yeh et al.,[70,71] the properties of an idealized Bragg reflector have been modeled. The reflection bandwidth of a Bragg reflector consisting of alternating quarter-wave layers[72] of different refractive indices is found to be[73]

$$\frac{\Delta\lambda}{\lambda} = \frac{4}{\pi}\sin^{-1} r \,, \tag{31}$$

where

$$r = \left| \frac{n_2 - n_1}{n_2 + n_1} \right| \tag{32}$$

and n_1 and n_2 are the refractive indices of the alternating layers. Layer thicknesses are simply

$$= \frac{\lambda}{4n_2}$$

and

$$= \frac{\lambda}{4n_2} \,. \tag{33}$$

The peak reflectivity of a Bragg reflector with N pairs of quarter-wave layers is given by[74]

$$= \tanh^2 \left| N \log\left(\frac{n_2}{n_1}\right) \right| . \tag{34}$$

For $n_2 \approx n_1$, Eq. (32) can be written as

$$= \tanh^2 [N\, \Delta n / n_o] \,, \tag{35}$$

where $\Delta n = n_2 - n_1$, $n_o = n_1 \approx n_2$. From Eq. (35), a Bragg reflector is found to have a band rejection ratio of $1/\cosh^2(N\Delta n/n)$ or approximately $2\exp(-N\Delta n/n)$.

An electro-optic Bragg reflector may be realized by applying a spatially periodic dc electric field to an electro-optic material. Consider a single electro-optic crystal embedded with a periodic array of planar electrodes separated at a distance of one quarter-wave, $\lambda/4n$. This crystal will become a Bragg reflector when a dc voltage is applied to every other electrode. The incident radiation will be reflected from the crystal at $\lambda = 4nd$. Therefore, the structure becomes an active Bragg reflector that can be switched on and off electronically.

The wavelength of the rejection band depends on the thickness of the crystal layer. In order to build a filter structure with a tunable rejection band and a variable bandwidth, a chirped multilayer stack is considered. This layered structure is similar to that discussed above, except that the layers are graduated in thickness (chirping). This structure becomes a broadband reflector when voltage is applied to every other electrode. This occurs since, at any wavelength, some part of the structure satisfies the Bragg reflection condition. By controlling the electrodes selectively, the radiation of any desired band can be rejected. Therefore, this structure is fully voltage programmable. In addition, it can be operated in a tunable bandpass mode. For example, if only a few of the electrodes are switched off while the rest are on, then only a narrowband of radiation will be transmitted.

Physical realization of the programmable Bragg filter will require significant developments in optical layered structure fabrication techniques. Optical quality single crystal layers of electro-optic materials separated by thin, optically transparent electrode layers will be required. Chemical-vapor deposition[75,76] and molecular-beam epitaxy[77] are two new powerful techniques which may prove useful in realizing this and other layered optical structures. Electrical addressing of individual electrodes in such a thin structure will require photolithographic techniques to be applied in a rather difficult geometry. However, the versatility of such a conceptually simple structure for both bandpass and band-rejection applications makes it a very interesting device for future development.

This discussion of a programmable Bragg filter is included as an indication of future directions in tunable spectral filter development. Clearly, significant progress in materials and fabrication techniques will be required before this and other advanced longitudinal structures can be experimentally demonstrated. Other tunable filter structures which fall into this same category include tunable versions of the Christiansen-Bragg filter[78] and tunable dispersive birefringent filters.[68]

7. SUMMARY

In this review we have discussed the major developments which have occurred in the development of electro-optically tunable spectral filters. Most of these devices are direct analogs of mechanically tuned filter structures, as is the case with birefringent filters and Fabry-Perot devices.

In the case of the electro-optic coupled wave filter and the programmable Bragg filter, the electro-optic effect leads to the development of new filter concepts. At this writing there is still considerable room for development. No truly compact solid-state electro-optically tunable filter, having moderate bandwidths (10s to 100s of Å), has been developed which may be employed for remote sensing applications. The situation in the infrared is particularly pressing. ADP and KDP type materials which are commonly used for visible applications are not transparent past 1.4 μm. Other materials, including the numerous ferroelectric oxides, adequately cover the middle infrared out to 5 μm. However, these crystals are quite expensive, especially in large sizes. Very few electro-optic materials are transparent in the 8 to 12 μm band. For those materials which are available, the operating voltages are prohibitively high for their use in standard filter configurations. Consequently, the development of new more efficient electro-optic materials, combined with innovative filter structures, will likely have the greatest impact on the future development of electro-optic tunable filters.

8. ACKNOWLEDGMENTS

The author would like to thank P. Yeh, J. Tracy, and the many other people who have contributed their valuable insight during the preparation of this paper.

9. REFERENCES

1. B. H. Billings, "A Tunable Narrow-Band Optical Filter," J. Opt. Soc. Am. 37, 738(1947).
2. I. C. Chang, "Development of an Infrared Tunable Acousto-Optic Filter," Proc. SPIE 131, 2(1978) and references cited therein; M. Khoshnevisan, E. Sovero, P. R. Newman, and J. Tracy, "Development of a Cryogenic Infrared Acousto-Optic Tunable Spectral Filter," Proc. SPIE 245, 63(1980); and E. A. Sovero and M. Khoshnevisan, " A Generalized Method for Designing Acousto-Optic Tunable Filters," in Proc. 1980 Ultrasonics Symposium, B. R. McAuoy, ed. pp. 492(1980).
3. I. P. Kaminow and E. H. Turner, "Electro-optic Light Modulators," IEEE Proc. 54, 1374(1966).
4. I. S. Zheludev and O. G. Vlokh, "The Electro-optic Effect in Crystals," Sov. Phys. Crystallogr. 3, 647(1959).
5. I. P. Kaminow and E. H. Turner, "Linear Electro-optical Materials," in *Handbook of Lasers*, R. J. Pressley, ed. CRC Press, Cleveland, 447-459(1971).

6. I. P. Kaminow, *An Introduction to Electro-optic Devices,* Academic Press, New York, (1974).
7. J. T. Milek and M. Neuberge, "Linear Electro-optic Modulator Materials," in *Handbook of Electronic Materials,* Vol. 8 IFI/Plenum, New York, Washington, London, (1972).
8. W. R. Cook, R. F. S. Hearmon, H. Jaffe, and D. F. Nelson, "Piezooptic and Electro-optic Constants," in Landolt Bornstein—New Series III/11 *Elastic, Piezoelectric, Pyroelectric, Piezooptic, Electro-optic Constants and Nonlinear Dielectric Susceptibilities of Crystals,* Springer-Verlag, Heidelberg-New York, (1979) pp. 495-606.
9. T. S. Narasimhanurty, *Photoelastic and Electro-Optic Properties of Crystals,* Plenum Press, New York-London, pp. 345-421(1981).
10. J. F. Nye, *Physical Properties of Crystals,* Clarendon Press, Oxford, pp. 235-259(1957).
11. E. H. Turner, "High Frequency Electro-Optic Coefficients of Lithium Niobate," Appl. Phys. Lett. 8, 303(1966).
12. K. Nassau, H. J. Levinstein, and G. M. Loiacono, "Ferroelectric Lithium Niobate 2. Preparation of Single Domain Crystals." J. Phys. Chem. Solids 27, 989(1966).
13. B. H. Billings, "The Electro-optic Effect in Uniaxial Crystals of the Type XH_2PO_4. I. Theoretical," J. Opt. Soc. Am. 39, 797(1949).
14. B. H. Billings, "The Electro-optic Effect in Uniaxial Crystals of the Type XH_2PO_4. II. Experimental," J. Opt. Soc. Am. 39, 802(1949).
15. W. G. Driscoll, *Handbook of Optics,* McGraw-Hill, New York, 8-111 to 8-115 (1978).
16. B. Lyot, "The Birefringent Filter and its Application in Solar Physics," Ann. Astrophys. 7, 31(1944).
17. J. W. Evans, "The Birefringent Filter," J. Opt. Soc. Am. 39, 229(1940).
18. I. Solc, "A new Type of Birefringent Filter," Czech. J. Phys. 4, 53(1954).
19. J. W. Evans, "Solc Birefringent Filter," J. Opt. Soc. Am. 48, 142(1958).
20. R. C. Jones, J. Opt. Soc. Am. 31, 488(1941).
21. S. E. Harris, E. O. Ammann, and I. C. Chang, "Optical Network Synthesis using Birefringent Crystals. I. Synthesis of Lossless Networks of Equal Length Crystals," J. Opt. Soc. Am. 54, 1267(1964).
22. E. O. Ammann and I. C. Chang, "Optical Network Synthesis using Birefringent Crystals. II. Synthesis of Networks Containing one Crystal, Optical Compensator, and Polarization Per Stage," J. Opt. Soc. Am. 55, 835(1965).
23. E. O. Ammann, "Optical Network Synthesis Using Birefringent Crystals. III. Some general properties of lossless birefringent networks," J. Opt. Soc. Am. 56, 943(1966); and "Optical network synthesis using birefringent crystals. IV. Synthesis of lossless double-pass networks," J. Opt. Soc. Am. 56, 952(1966).
24. E. O. Ammann and J. M. Yarborough, "Optical Network Synthesis using Birefringent Crystals. V. Synthesis of Lossless Networks Containing Equal-length crystals and compensators," J. Opt. Soc. Am. 56, 1746(1966).
25. J. M. Yarborough and E. O. Ammann, "Experiments on Single-pass and Double-pass Birefringent Networks," J. Opt. Soc. Am. 58,776(1967).
26. J. M. Yarborough and E. O. Ammann, "Experiments on a Multistage Electro-optic Amplitude modulator," IEEE J. Quant. Elect. QE-4, 230(1968).
27. A. M. Title and S. A. Schoolman, "Report Advancer in Birefringent Filters," Proc. SPIE 88, 23(1976); and "Optical Network Synthesis using Birefringent Crystals. VI. Additional Techniques for the Synthesis of Lossless Double-pass Networks," J. Opt. Soc. Am. 57, 349(1967).
28. A. M. Title and W. J. Rosenberg, "Spectral Management," Proc. SPIE 268, 178(1980).
29. N. M. Drichko, S. B. Ioffe, and S. G. Podkorytova, "Thermooptic Compensation of Wide-Angle Controllable Stages of Interference-Type Polarization Filters," Sov. J. Opt. Technol. 39, 727(1972).
30. J. Katzenstein, "New Type of Birefringent Filter," J. Opt. Soc. Am. 58, 1348(1968).
31. A. M. Title and W. J. Rosenberg, "Tunable Birefringent Networks," Proc. SPIE 202, 47(1979).
32. F. K. von Willisen, "A Tunable Birefringent Filter," Appl. Opt. 5, 97(1966).
33. N. M. Drichko and S. B. Ioffe, "Interference-Polarizing Filter Stage with Dual Control," Sov. J. Opt. Tech. 36, 300(1969).
34. B. R. Babits and H. R. Hicks, "Electrical Color Filters," Electronics 23, 112(1950).
35. N. M. Drichko, "Narrow-band Electrically-Controlled Interference-Polarization Filters," Sov. J. Opt. Tech., 36, 386(1969).
36. M. S. Ahmed and J. M. Ley, "A Switched-Voltage Tunable Birefringent Filter," J. Phys. D. 3, 1737(1970).
37. N. M. Drichko, S. B. Ioffe, and E. I. Mogilevskii, "Narrow-band Electrically Controlled Interference-Polarizing Filter for Solar Physics Studies," Opt. & Spectrosk 31, 241(1970).
38. S. G. Varnado and W. D. Smith, "Electro-optic Ceramics as Wavelength Selection Devices in Dye Lasers," IEEE J. Quant. Electr. QE-8, 88(1972).
39. N. M. Drichko, "Spectrally Controllable Electrooptical Device," Sov. J. Opt. Technol. 42, 623(1975).
40. V. P. Lapasov and M. M. Makogan, "Laser Frequency Control with Birefringent Polarizing Filters," Sov. Phys. Tech. Phys. 20, 216(1975).
41. J. P. Auton, M. J. Brookfield, and Jaqueline Dawson, "Non-mechanically Tuned Optical Bandpass Filters, a Survey and Assessment," Final Report ESRO Contract 2275/74 PP. (1975).
42. M. M. Makogan, "Ruby Laser with High-Frequency Control of Lasing Conditions," Radio Eng. and Electron. Phys. 24, 97(1979).
43. P. Yeh and W. Gunning, "Electro-optic Tunable Filter Structures," Proc. SPIE 202, 2(1979).
44. D. M. Henderson, Private communication.
45. I. V. Nikolaev and M. M. Koblova, "Modulation of Optical Radiation of $\lambda = 10.6$ micron," Sov. J. Quant-Elect. 1, 158(1971).
46. L. F. Champagne, F. O'Neill, and W. T. Whitney, "Reliable Halfwave Operation of a GaAs Pockels Cell," Opt. Commun. 11, 11(1974).
47. J. F. Figueira, "Extinction ratios of GaAs and CdTe Electro-optical Modulators," IEEE J. Quant. Electr. QE-10, 572(1974).
48. L. F. Champagne, F. O'Neill, and W. T. Whitney, "A Large Aperature CdTe Pockels Cell for CO_2 Lasers," Opt. Commun. 13, 282(1975).
49. D. A. Pinnow, R. L. Abrams, J. F. Lotspeich, D. M. Henderson, T. K. Plant, R. R. Stephens, and C. M. Walker, "An Electro-optic Tunable Filter," Appl. Phys. Lett. 34, 391(1979).
50. D. M. Henderson, J. F. Lotspeich, R. R. Stephens, "Programmable Electro-optic Tunable Filters," Proc. SPIE 202, 16(1979).
51. R. C. Alferness, "Efficient Waveguide Electro-optic TE $\rightleftarrows$ TM Mode Converter/Wavelength Filter," Appl. Phys. Lett. 36, 513 (1980).
52. H. A. Tarry, "Electrically Tunable Narrowband Optical Filter," Electronics Lett. 18, 471(1975).
53. M. Born and E. Wolf, *Principles of Optics,* Macmillan, New York, pp. 322-340(1970).
54. W. Sohler, "Optical Bistable Device as Electro-optical Multivibrator," Appl. Phys. Lett. 36, 351(1980).
55. P. W. Smith, E. H. Turner, and P. J. Maloney, "Electro-optic Nonlinear Fabry-Perot Devices," IEEE. J. Quant. Electr. QE-14, 207(1978).
56. X. DeAngelis and W. Niblack, "Electro-optic Interference Filter Light modulator," Proc. IEEE 51, 1258(1963).
57. L. A. Marscall and L. M. Hobbs, "Interferometric Studies of Interstellar Calcium Lines," Astr. Phys. Journ. 173, 43(1972).
58. J. V. Ramsay, H. Hobler, and E. G. V. Mugridge, "A New Tunable Filter with a very Narrow Passband," Sol. Phys. 12, 492(1970).
59. W. J. Gunning and P. Yeh, "Multiple-Cavity Infrared Electro-optic Tunable Filter," Proc. SPIE 202, 21(1979).
60. W. J. Gunning, "An Electro-Optically Tunable Multiple Cavity Fabry-Perot," in prep.
61. See for example, Y. Itoh and H. Iwasaki, "Ferroelectric and Optical Properties of $Ba_6Ti_2Nb_8O_{30}$ Single Crystals," J. Phys. Chem. Solids 34, 1639(1973).
62. Y. Ito, "Light Modulation with Multilayer Thin Films," Appl. Opt. 13, 2464(1974).
63. W. Gunning, J. Pasko, and J. Tracy, "A Liquid Crystal Tunable Spectral Filter; Visible and Infrared Operation," Proc. SPIE 268, 190(1980).
64. J. F. Lotspeich, "Iso-Index Coupled-Wave Electro-optic Filter," IEEE J Quantum Electr. QE-15, 47(1979).
65. J. F. Lotspeich and D. M. Henderson, "Blue/Green Coupled Wave Filter Study," Final Report ONR Contract N00014-79-C-0491 (1980).
66. C. H. Henry, "Coupling of Electromagnetic Waves in CdS," Phys. Rev 143, 627(1966).
67. J. P. Laurenti, K. C. Rustagi, and M. Rouzeyre, "Optical Filters using Coupled Lightwaves in Mixed Crystals," Appl. Phys. Lett. 28, 212(1976).
68. P. Yeh, "Dispersive Birefringent Filters," Opt. Commun. 37, 153(1981).
69. W. Gunning, R. Hall, and J. Tracy, "Dispersive Birefringent Filters Experimental Development," in prep.
70. P. Yeh, A. Yariv, and C. S. Hong, "Electromagnetic Propagation in Periodic Stratified Media. I. General Theory," J. Opt. Soc. Am. 67, 423(1977).
71. A. Yariv and P. Yeh, "Electromagnetic Propagation in Periodic Stratified Media. II. Birefringence, Phase Matching, and X-ray Lasers," J. Opt. Soc. Am. 67 438(1977).
72. P. Yeh, Ph. D. Thesis (Caltech, 1978) p. 167.
73. W. G. Driscoll, *Handbook of Optics,* McGraw-Hill, New York, pp. 8-61(1978).
74. See for example M. Born and E. Wolf, *Principles of Optics,* Macmillan, New York, p. 69 (1970).
75. H. M. Manaseuit and W. I. Simpson, "The use of Metal-Organics in the Preparation of Semiconductor Materials. I. Epitaxial Gallium Compounds," J. Electrochem. Soc. 116, 1725(1969).
76. R. D. Dupis and P. D. Dapkus, "Room-Temperature Operation of $Ga_{(1-x)}Al_xAs/GaAs$ Double-Heterostructure Layers Grown by Metal Organic Chemical Vapor Deposition," Appl. Phys. Lett. 31, 466(1977); "$Ga_{(1-x)}Al_xAs/Ga_{(1-y)}Al_yAs$," Appl. Phys. Lett. 31, 839(1977).
77. A. Y. Cho and J. R. Arthur, *Progress in Solid State Chemistry,* Vol. 1[illegible], Pergamon Press, New York, 157(1975).
78. W. G. Driscoll, *Handbook of Optics,* McGraw-Hill, New York, 8-03 to 8-106; 8-111 to 8-115(1978).

Ferroelectric liquid-crystal tunable filter

H. J. Masterson, G. D. Sharp, and K. M. Johnson

Center for Optoelectronic and Computing Systems, Department of Electrical and Computer Engineering, University of Colorado at Boulder, Boulder, Colorado 80309-425

Received February 23, 1989; accepted August 23, 1989

A new discrete-channel tunable Lyot filter that uses ferroelectric liquid-crystal (FLC) smectic C^* wave plates is described and demonstrated. The device is tuned by changing the retardation of each birefringent element. This is achieved by electrically rotating the FLC wave plates in each stage by 45°, which effectively changes the design wavelength of the Lyot filter. The transmission characteristics of such a device are demonstrated in a three-stage filter and show good agreement with theory. Advantages of the FLC tunable filter over existing filter structures include low switching voltages (±10 V), rapid tunability (~100 kHz), potentially high transmission and wide field of view, and large entrance aperture.

Birefringent filters were first used in solar research, where subangstrom spectral resolution is required to observe solar prominences. The first birefringent filter was invented by Lyot[1] in 1933. The basic Lyot filter[2] can be decomposed into a series of individual filter stages. Each stage consists of a birefringent element placed between parallel polarizers. The exit polarizer for a particular stage acts as the input polarizer for the following stage. The birefringent elements are oriented with optic axes parallel to the interface and rotated 45° from the direction of the input polarization. The thickness, and therefore the retardation of the birefringent elements, increases geometrically in powers of two for each successive stage in the Lyot geometry. Multiple-stage devices have been demonstrated with high resolution (0.01 nm) and a broad free spectral range.[3]

Another birefringent filter, proposed by Solc,[4] uses only two polarizers but requires more elements to achieve the same bandwidth as a Lyot filter.

More recently, research in optical filters has focused on tuning the wavelength of peak transmission. An optical filter that can be rapidly tuned has applications in remote sensing, signal processing, displays, and wavelength-division demultiplexing. Tunability of otherwise fixed-frequency Lyot filters has been suggested and implemented using various techniques, including mechanical methods,[5,6] electromechanical techniques,[3] and temperature or electro-optic tuning the birefringence of the elements.[5] Temperature-tuning and mechanical-tuning methods are inherently slow. Electro-optic tuning, while potentially rapid, requires large drive voltages and is limited in tunability by the breakdown voltages of the elements.[7]

Other electronically tunable filters that have been demonstrated include acousto-optic tunable filters,[8] electro-optic tunable filters,[9] multiple-cavity Fabry–Perot devices,[10] and hybrid filters such as the Fabry–Perot electro-optic Solc filter.[7] Kaye patented a tunable filter using twisted nematic liquid crystals.[11] Other researchers have also demonstrated twisted nematic liquid-crystal tunable filters.[12–14] The main disadvantage of such devices is their slow tuning speeds (~100 msec).

In this Letter a new type of discretely tunable Lyot filter is demonstrated that utilizes microsecond-switching ferroelectric liquid-crystal (FLC) smectic C^* wave plates. The physics and operation of the surface-stabilized FLC have been extensively described elsewhere.[15] Briefly, the optic axis can be electronically rotated through a net angle of 45°. The voltage requirements for switching the devices are modest (±10 V), and the power consumption is low because the voltage need not be applied to maintain the FLC in the switched state (the devices are bistable[16]). Typical switching times are <20 μsec at room temperature.[17] Furthermore, the FLC tunable filter (FLCTF) is not hampered by small entrance apertures, as FLC cells can be fabricated on large substrates.

The FLC wave plates function as electronically controlled phase retarders that are added to each stage of a fixed-frequency Lyot filter. The addition of a retarder to a filter stage, oriented with a crystal axis along the direction of polarization, does not change the spectral transmission characteristics of the filter. However, rotation of the FLC by 45° is equivalent to an increase in the thickness of the birefringent element. This effectively changes the design wavelength of the Lyot filter, allowing discrete tuning between wavelengths. To maintain the Lyot geometry the retardation of the FLC wave plates must increase in the same ratio as that of the birefringent elements. Therefore, the number of FLC cells required in the nth stage is 2^{n-1}. For the three-stage filter demonstrated in this Letter, seven FLC cells were necessary.

The operation of a Lyot filter can be understood by analyzing a single stage, as shown in Fig. 1. We assume that the direction of propagation of light is along the z axis of the Cartesian coordinate system. The faces of the birefringent plates and the FLC's are normal to the z axis, and the electric vector of light transmitted by the polarizers lies along the y axis. The optic axes of the wave plates are in the plane normal to the z axis. To describe the operation of a typical fixed-frequency Lyot filter, it is initially assumed that the FLC is oriented along the y axis, transmitting the field with no retardation ($\alpha = 0°$). Since the fixed

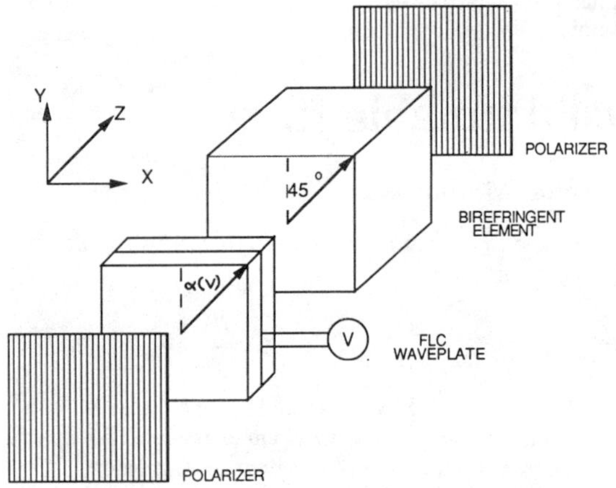

Fig. 1. Single-stage of the FLC tunable Lyot filter. The net retardation of the stage can be modulated by electronically rotating the crystal axes [$\alpha(V)$] of the FLC wave plate.

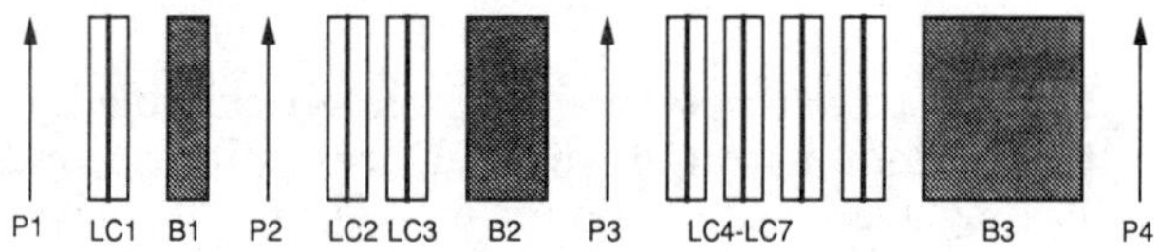

Fig. 2. Proof-of-principle three-stage Lyot filter. This device requires four polarizers (P1–P4), seven FLC wave plates (LC1–LC7), and three birefringent elements (B1–B3), which are one-wave, two-wave, and four-wave retarders at the design wavelength.

birefringent element is rotated by 45° about the z axis, the incident linearly polarized light is divided into two equal-amplitude eigenwaves that travel at different phase velocities through the material. The retardation between the two waves at the exit of the crystal is given by

$$\Gamma(\lambda) = \frac{2\pi\Delta n d}{\lambda}, \tag{1}$$

where Δn is the birefringence of the material, d is the material thickness, and λ is the free-space wavelength. The two waves interefere at the exit polarizer such that only wavelengths that are in phase achieve unity transmission. The transmission spectrum for the nth stage of a Lyot filter, as shown below, is given by

$$T_n(\lambda) = \cos^2[\Gamma_n(\lambda)/2]. \tag{2}$$

The transmission of a multiple-stage filter is the product of the intensity transmittances of the individual filter stages. For the Lyot geometry, the thickness of each birefringent element is always that of the previous stage. Each subsequent stage exhibits a transmission spectrum with half the spectral period of the previous stage and therefore provides blocking for the following stage. The transmission spectrum of an N-stage filter can be written in the form of a replicated sinc function.[2] The spectral period of the filter, or free spectral range, is determined by the stage with the thinnest birefringent element (the first stage). The resolution of the filter is determined by the thickest element (the final stage).

The transmission spectrum of a single Lyot filter stage can be determined using the 2 × 2 Jones calculus.[18] Referring to Fig. 1, the output of the nth stage is the matrix equation

$$E_n'(\lambda) = P_y W_n(\lambda) P_y E_n(\lambda), \tag{3}$$

where $E_n(\lambda)$ and $E_n'(\lambda)$ are the column vectors giving the x and y components of the input and transmitted electric fields, respectively, P_y is the matrix for polarizers oriented along the y axis, and $W_n(\lambda)$ is the matrix for a retarder with crystal axes rotated 45° about the z axis. These matrices are expressed as[2]

$$P_y = \begin{bmatrix} 0 & 0 \\ 0 & 1 \end{bmatrix},$$

$$W_n(\lambda) = \begin{bmatrix} \cos[\Gamma_n(\lambda)/2] & -i\sin[\Gamma_n(\lambda)/2] \\ -i\sin[\Gamma_n(\lambda)/2] & \cos[\Gamma_n(\lambda)/2] \end{bmatrix}, \tag{4}$$

where the total retardation, $\Gamma_n(\lambda)$, is given by

$$\Gamma_n(\lambda) = \Gamma_n^F(\lambda) + \Gamma_n^C(\lambda). \tag{5}$$

Here, $\Gamma_n^F(\lambda)$ is the retardation of the fixed birefringent plate given by Eq. (1) as

$$\Gamma_n^F(\lambda) = 2^n\pi\frac{\lambda_A}{\lambda}, \tag{6}$$

where λ_A ($= \Delta n d$) is the design wavelength of the filter in the unswitched state. This is the wavelength at which the birefringent element in the first stage is a full-wave plate. We assume negligible dispersion of the birefringent elements throughout the tuning range. $\Gamma_n^C(\lambda)$ is the net additional retardation due to the 2^{n-1} FLC's. In the unswitched state, this retardation is zero. In the switched state ($\alpha = 45°$), the filter is tuned to the new design wavelength, λ_B, owing to the additional retardation. This retardation can therefore be written as

$$\Gamma_n^C(\lambda) = \begin{cases} 0 & \text{unswitched} \\ 2^n\pi\dfrac{\Delta\lambda}{\lambda}\dfrac{\Delta n(\lambda)}{\Delta n(\lambda_B)} & \text{switched} \end{cases}, \tag{7}$$

where $\Delta n(\lambda)$ is the wavelength-dependent birefringence of the FLC's and $\Delta\lambda = (\lambda_B - \lambda_A)$. Owing to the highly dispersive nature of liquid crystals, this expression includes the effect of dispersion of the FLC birefringence. Use of Eqs. (3) and (4) and the relation $T(\lambda) = |E_y'(\lambda)/E_y(\lambda)|^2$ yields the intensity transmission given by Eq. (2), where

$$\Gamma_n(\lambda) = \begin{cases} 2^n\pi\dfrac{\lambda_A}{\lambda} & \text{unswitched} \\ \dfrac{2^n\pi}{\lambda}\left[\lambda_A + \Delta\lambda\dfrac{\Delta n(\lambda)}{\Delta n(\lambda_B)}\right] & \text{switched} \end{cases}. \tag{8}$$

The FLCTF was experimentally demonstrated using the arrangement shown in Fig. 2. Three birefringent elements, which retard light at 475 nm by one wave (B1), two waves (B2), and four waves (B3), respectively, are sandwiched between vertical dichroic sheet polarizers (P1–P4). Each of these stages in the FLCTF is then modulated by one (LC1), two (LC2, LC3), and four (LC4–LC7) FLC's, respectively. These seven FLC devices (fabricated by Displaytech

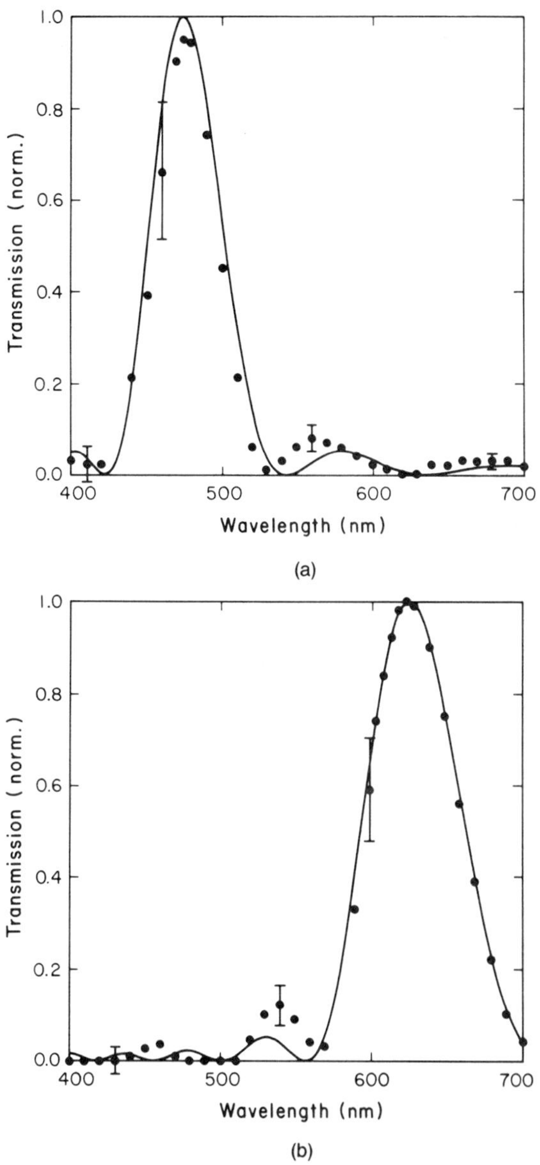

Fig. 3. Measured transmission spectra of the three-stage Lyot filter in (a) the unswitched state and (b) the switched state shown with simulation results (solid curves).

Inc.) are half-wave at 400 nm. The FLC's were actively switched using a Hewlett-Packard model 8116A function generator. The light source used was a 280-W tungsten lamp. The filter output was analyzed with a photodiode, a Hewlett-Packard 1726A oscilloscope, and a monochromator.

The experimental results are plotted in Fig. 3 along with the numerical solutions of the theoretical curves. The transmission spectra of the three-stage Lyot filter with the FLC's in the unswitched state ($\alpha = 0°$) and the switched state ($\alpha = 45°$) are shown in Figs. 3(a) and 3(b), respectively. The transmission is maximum at 475 and 625 nm, respectively, which agrees well with the theoretical curves (the solid curves). This was achieved by taking into account the FLC dispersion when calculating Δn. A model describing the birefringence of liquid crystals[19] allows us to express Δn as

$$\Delta n(T, \lambda) = G(T)\, \frac{\lambda^2(\lambda^*)^2}{\lambda^2 - (\lambda^*)^2}\,, \qquad (9)$$

where $G(T)$ is a temperature-dependent parameter in inverse squared nanometers and λ^* is the mean UV resonance wavelength. Experimentally measured values for these parameters are $G(T)d = 2.08 \times 10^{-3}$ nm^{-1} and $\lambda^* = 245.0$ nm.

High-quality fixed-frequency Lyot filters are capable of transmitting 35–40% of incident unpolarized light.[6] Additional transmission losses owing to absorption, scattering, and Fresnel reflections with the addition of FLC's to a fixed-frequency Lyot filter can be estimated given the transmission of a single FLC device. This was measured to be typically 0.94 with broadband antireflection coating on the substrates. The aperture of the FLC devices was 2.5 cm.

Computer simulations indicate that a five-stage FLCTF, which employs three FLC's in the first stage, each with a retardation of $\pi/4$ at a wavelength of λ = 400 nm, would allow selection between wavelengths separated by nearly 50 nm with a 10-nm bandwidth. The use of FLC's of different thicknesses in the FLCTF would significantly increase the number of possible wavelength channels.

In conclusion, we have demonstrated what is to our knowledge the first FLCTF. The FLCTF has advantages over many other tunable filters with respect to switching voltages, power consumption, entrance aperture, field of view, and switching speeds.

We gratefully acknowledge the support of National Science Foundation award EET 8451485, Pete Peterson, and AT&T Information Systems. We also acknowledge stimulating discussions with H. John Caulfield, Bruce Billings, Joseph W. Goodman, and Thomas and Joanna Jannson.

References

1. B. Lyot, C. R. Acad. Sci. **197,** 1593 (1933).
2. A. Yariv and P. Yeh, *Optical Waves in Crystals* (Wiley, New York, 1984), Chap 5.
3. A. M. Title and W. J. Rosenberg, Opt. Eng. **20,** 815 (1981).
4. I. Solc, J. Opt. Soc. Am. **55,** 621 (1965).
5. B. H. Billings, J. Opt. Soc. Am. **37,** 738 (1947).
6. J. W. Evans, J. Opt. Soc. Am. **39,** 229 (1949).
7. R. S. Weis and T. K. Gaylord, J. Opt. Soc. Am. A **4,** 1720 (1987).
8. I. C. Chang, Opt. Eng. **20,** 824 (1981).
9. J. F. Lotspeich, R. R. Stephens, and D. M. Henderson, Opt. Eng. **20,** 830 (1981).
10. W. Gunning, Appl. Opt. **21,** 3129 (1982).
11. W. I. Kay, U.S. patent 4,394,069 (June 5, 1979).
12. H. A. Tarry, Electron. Lett. **18,** 47 (1975).
13. W. Gunning, Proc. Soc. Photo-Opt. Instrum. Eng. **268,** 190 (1980).
14. S. Wu, Appl. Opt. **28,** 48 (1989).
15. N. A. Clark, M. A. Handschy, and S. T. Lagerwall, Mol. Cryst. Liq. Cryst. **94,** 213 (1983).
16. N. A. Clark and S. T. Lagerwall, Appl. Phys. Lett. **36,** 899 (1980).
17. ZLI-4003 mixture available from Merck (Frankfurterstrasse 250, D-6100 Darmstadt 1, Federal Republic of Germany).
18. R. C. Jones, J. Opt. Soc. Am. **31,** 488 (1941).
19. S. Wu, Phys. Rev. A **33,** 1270 (1986).

Czechoslovak Journal of Physics A, Vol. 10(1), 16–34 (1960).

CHAIN BIREFRINGENT FILTERS

by

Ivan Šolc

INTRODUCTION

We use the term chain birefringent filter to designate the arrangement of birefringent plates in which no plate is separated from another by polaroids and in which the path difference of the individual filter elements is repeated periodically. Hence, the simplest arrangement answering this definition is a system of plates of equal thickness, ground from the same material and with the same orientation of the angle of cut. Filters are either asymmetrical (the filter effect can be changed by changing the direction of the rays) or symmetrical (the filter effect is independent of the direction of the rays), depending on the arrangement of the azimuths of the individual filter elements. For practical reasons, symmetrical filters are the most convenient, so we shall concentrate on them in this paper. Birefringent filters are generally used to make light monochromatic, and thus it is our object to attain the optimum effect along these lines.

BASIC ASSUMPTIONS

A ray may not change its path direction while passing through the filter. Hence it follows that only one cut of a crystal is suitable, namely the cut parallel to the plane of the optic axes (uniaxial crystals are cut parallel with the optic axis). Huygen's construction of the refraction of light in crystals shows clearly that all other cuts are unsuitable. I described the simplest type of chain filter in an article published in 1953 [1]. A more detailed analysis of such filters can be found in [2]. The system of equations derived by Hsien-Yü Hsü, M. Richartz and Yüng-Kang Liang in 1947 [5] from the general formula for computing the intensity of polarized light passing through a series of birefringent plates has proved to be effective for analysis of the filter function

$$J_n = J_0 \cdot \left[\prod_{s=1}^{n} \cos\gamma_s\right]^2 \cdot \left\{\left[\sum_{l=0}^{n/2} (-)^l \sum_{j_{2l}>\dots>j_1=1}^{2l} \prod_{k=1}^{2l} x_{jk} \cos\left(\Phi - 2\sum_{k=1}^{2l} (-)^k \varrho_{jk}\right)\right]^2 + \left[\sum_{l=0}^{n/2} (-)^l \sum_{j_{2l}>\dots>j_1=1}^{2l+1} \prod_{k=1}^{2l+1} x_{jk} \cos\left(\Phi - 2\sum_{k=1}^{2l+1} (-)^{k-1} \varrho_{jk}\right)\right]^2\right\}. \qquad (1)$$

In this formula, Φ is the azimuth angle of the analyzer with respect to the azimuth of the polaroid, $x_i = \tan \delta_s$, $\gamma = \delta_s/2$, δ is the phase difference, ρ_i the azimuth of the i-th plate, J_o the intensity of the light after the first polaroid. The phase difference and the path difference are related by the familiar relationship

$$\delta = \frac{2\pi}{\lambda} \cdot d \cdot D \tag{2}$$

in which λ is the wavelength of light, d is the plate thickness and D is the index of refraction.

To make satisfactory birefringent filters, one must use suitable, highly homogeneous birefringent crystal material, maintain a strictly accurate angle of cut, and control the grinding such that the plate thickness remains strictly within the assigned tolerances. A condition for the use of birefringent filters is that they must fit into an optical system in which the allowed angle of convergence of the rays is not exceeded.

GENERAL RELATIONSHIPS FOR SIMPLE CHAIN FILTERS

For simple chain filters, with which we are concerned here, the individual elements have the same path difference, which is expressed by the condition

$$\gamma_1 = \gamma_2 = \gamma_3 = \dots = \gamma_s = \gamma,$$

then, also,

$$x_1 = x_2 = x_3 = \dots = x_s = x. \tag{3}$$

In further computations, intensity J_o will be taken as 1.

We will use X_n for the double brackets in general expression (1). Then, for example,

$$\begin{aligned}
X_1 &= \cos^2 \Phi + [x_1 \cdot \cos(\Phi - 2\varrho_1)]^2, \\
X_2 &= [\cos \Phi - x_2 x_1 \cos(\Phi - 2\varrho_2 + 2\varrho_1)]^2 + \\
&\quad + [x_2 \cos(\Phi - 2\varrho_2) + x_1 \cos(\Phi - 2\varrho_1)]^2, \\
X_3 &= [\cos \Phi - x_3 x_2 \cos(\Phi - 2\varrho_3 + 2\varrho_2) - x_3 x_1 \cos(\Phi - 2\varrho_3 + 2\varrho_1) - \\
&\quad - x_2 x_1 \cdot \cos(\Phi - 2\varrho_2 + 2\varrho_1)]^2 + [x_3 \cos(\Phi - 2\varrho_3) + \\
&\quad + x_2 \cos(\Phi - 2\varrho_2) + x_1 \cos(\Phi - 2\varrho_1) - x_3 x_2 x_1 \cdot \\
&\quad \cdot \cos(\Phi - 2\varrho_3 + 2\varrho_2 - 2\varrho_1)]^2,
\end{aligned}$$

etc.

Next, let us introduce the expression $L_{u,v}$:

$$L_{u,v} = \sum_{j_u > \dots > j_1 = 1}^{u} \cos\left(\Phi - 2\sum_{k=1}^{v} (-1)^{k-1} \varrho_{jk}\right). \tag{4}$$

By successive substitution, we get

$$\begin{aligned}
L_{1,1} &= \cos(\Phi - 2\varrho_1),\\
L_{2,1} &= \cos(\Phi - 2\varrho_2) + \cos(\Phi - 2\varrho_1),\\
L_{2,2} &= \cos(\Phi - 2\varrho_2 + 2\varrho_1),\\
L_{3,1} &= \cos(\Phi - 2\varrho_3) + \cos(\Phi - 2\varrho_2) + \cos(\Phi - 2\varrho_1),\\
L_{3,2} &= \cos(\Phi - 2\varrho_3 + 2\varrho_2) + \cos(\Phi - 2\varrho_3 + 2\varrho_1) + \cos(\Phi - 2\varrho_2 + 2\varrho_1),\\
L_{3,3} &= \cos(\Phi - 2\varrho_3 + 2\varrho_2 - 2\varrho_1),
\end{aligned}$$

etc.

For the sake of clarity, we may express these relationships by the following numerical scheme:

$L_{1,1}$ 1
$L_{2,1}$ 2, 1
$L_{2,2}$ 21
$L_{3,1}$ 3, 2, 1
$L_{3,2}$ 32, 31, 21
$L_{3,3}$ 321
$L_{4,1}$ 4, 3, 2, 1
$L_{4,2}$ 43, 42, 41, 32, 31, 21
$L_{4,3}$ 432, 431, 421, 321
$L_{4,4}$ 4321

etc.

By successive substitution of $L_{u,v}$ and relationships (3) for X_n, we arrive at equations

$$\begin{aligned}
X_1 &= \cos^2\Phi + [xL_{1,1}]^2,\\
X_2 &= [\cos\Phi - x^2L_{2,2}]^2 + [xL_{2,1}]^2,\\
X_3 &= [\cos\Phi - x^2L_{3,2}]^2 + [xL_{3,1} - x^3L_{3,3}]^2,\\
X_4 &= [\cos\Phi - x^2L_{4,2} + x^4L_{4,4}]^2 + [xL_{4,1} - x^3L_{4,3}]^2,\\
X_5 &= [\cos\Phi - x^2L_{5,2} + x^4L_{5,4}]^2 + [xL_{5,1} - x^3L_{5,3} + x^5L_{5,5}]^2,
\end{aligned}$$

$$X_n = [\cos\Phi - x^2L_{n,2} + x^4L_{n,4} \dots \pm x^sL_{n,s}]^2 + [xL_{n,1} - x^3L_{n,3} + \dots \pm \\ \pm x^tL_{n,t}]^2.$$

In this equation, s is the largest even number that fulfills the condition $s \leqq n$, and t is the largest odd number that fulfills the similar condition $t \leqq n$.

It is convenient to arrange these expressions by decreasing powers of x:

$$\begin{aligned}
X_1 &= x^2L_{1,1}^2 + \cos^2\Phi,\\
X_2 &= x^4L_{2,2}^2 + x^2(L_{2,1}^2 - 2L_{2,2}\cos\Phi) + \cos^2\Phi,\\
X_3 &= x^6L_{3,3}^2 + x^4(L_{3,2}^2 - 2L_{3,1}L_{3,3}) + x^2(L_{3,1}^2 - 2L_{3,2}\cos\Phi) + \cos^2\Phi,\\
X_4 &= x^8L_{4,4}^2 + x^6(L_{4,3}^2 - 2L_{4,2}L_{4,4}) + x^4(L_{4,2}^2 - 2L_{4,1}L_{4,3} + 2L_{4,4}\cos\Phi) +\\
&\quad + x^2(L_{4,1}^2 - 2L_{4,2}\cos\Phi) + \cos^2\Phi,\\
X_5 &= x^{10}L_{5,5}^2 + x^8(L_{5,4}^2 - 2L_{5,3}L_{5,5}) + x^6(L_{5,3}^2 - 2L_{5,2}L_{5,4} + 2L_{5,1}L_{5,5}) +\\
&\quad + x^4(L_{5,2}^2 - 2L_{5,1}L_{5,3} + 2L_{5,4}\cos\Phi) + x^2(L_{5,1}^2 - 2L_{5,2}\cos\Phi) +\\
&\quad + \cos^2\Phi,
\end{aligned}$$

etc.

The brackets with functions $L_{u,v}$ with values of x are of the nature of coefficients and express the characteristic properties of a chain filter. Let us call them $A_{k,\ell}$. Then,

$$\begin{aligned}
A_{1,1} &= L^2_{1,1}\,,\\
A_{2,1} &= L^2_{2,2}\,,\\
A_{3,1} &= L^2_{3,3}\,,\\
A_{4,1} &= L^2_{4,4}\,,\\
&\vdots\\
A_{k,1} &= L^2_{k,k}\,;
\end{aligned}$$

$$\begin{aligned}
A_{1,2} &= \cos\Phi\,,\\
A_{2,2} &= L^2_{2,1} - 2L_{2,2}\cos\Phi\,,\\
A_{3,2} &= L^2_{3,2} - 2L_{3,1}L_{3,3}\,,\\
A_{4,2} &= L^2_{4,3} - 2L_{4,2}L_{4,4}\,,\\
A_{5,2} &= L^2_{5,4} - 2L_{5,3}L_{5,5}\,,
\end{aligned}$$

$$A_{k,2} = L^2_{k,k-1} - 2L_{k,k-2}\cdot L_{k,k}\,;$$

$$\begin{aligned}
A_{1,3} &= 0\,,\\
A_{2,3} &= \cos\Phi\,,\\
A_{3,3} &= L^2_{3,1} - 2L_{3,2}\cos\Phi\,,\\
A_{4,3} &= L^2_{4,2} - 2L_{4,1}L_{4,3} - 2L_{4,4}\cos\Phi\,,\\
A_{5,3} &= L^2_{5,3} - 2L_{5,2}L_{5,4} + 2L_{5,1}L_{5,5}\,,\\
A_{6,3} &= L^2_{6,4} - 2L_{6,3}L_{6,5} + 2L_{6,2}L_{6,6}\,,
\end{aligned}$$

$$A_{k,3} = L^2_{k,k-2} - 2L_{k,k-3}L_{k,k-1} + 2L_{k,k-4}L_{k,k}\,;$$

$$\begin{aligned}
A_{1,4} &= 0\,,\\
A_{2,4} &= 0\,,\\
A_{3,4} &= \cos\Phi\,,\\
A_{4,4} &= L^2_{4,1} - 2L_{4,2}\cos\Phi\,,\\
A_{5,4} &= L^2_{5,2} - 2L_{5,1}L_{5,3} + 2L_{5,4}\cos\Phi\,,\\
A_{6,4} &= L^2_{6,3} - 2L_{6,2}L_{6,4} + 2L_{6,1}L_{6,5} - 2L_{6,6}\cos\Theta\,,\\
A_{7,4} &= L^2_{7,4} - 2L_{7,3}L_{7,5} + L2_{7,2}L_{7,6} + 2L_{7,1}L_{7,7}\,,
\end{aligned}$$

$$A_{k,4} = L^2_{k,k-3} - 2L_{k,k-4}L_{k,k-2} + 2L_{k,k-5}L_{k,k-1} - 2L_{k,k-6}L_{k,k}\,;$$

$$\begin{aligned}
A_{1,5} &= 0\,,\\
A_{2,5} &= 0\,,\\
A_{3,5} &= 0\,,\\
A_{4,5} &= \cos\Phi\,,\\
A_{5,5} &= L^2_{5,1} - 2L_{5,2}\cos\Phi\,,\\
A_{6,5} &= L^2_{6,2} - 2L_{6,1}L_{6,3} + 2L_{6,4}\cos\Phi\,,\\
A_{7,5} &= L^2_{7,3} - 2L_{7,2}L_{7,4} + 2L_{7,1}L_{7,5} - 2L_{7,6}\cos\Phi\,,
\end{aligned}$$

$$\begin{aligned}
A_{8,5} &= L^2_{8,4} - 2L_{8,3}L_{8,5} + 2L_{8,2}L_{8,6} - 2L_{8,1}L_{8,7} + 2L_{8,8}\cos\Phi\,,\\
A_{9,5} &= L^2_{9,5} - 2L_{9,4}L_{9,6} + 2L_{9,3}L_{9,7} - 2L_{9,2}L_{9,8} + 2L_{9,1}L_{9,9}\,,
\end{aligned}$$

$$A_{k,5} = L^2_{k,k-4} - 2L_{k,k-5}L_{k,k-3} + 2L_{k,k-6}L_{k,k-2} - 2L_{k,k-7}L_{k,k-1} + 2L_{k,k-8}L_{k,k}\,,$$

etc.

The formulas combine formally if we introduce the symbol $\cos\Phi = L_{k,0}$. Then the general relationship for expression $A_{k,\ell}$ has the form

$$A_{k,l} = L^2_{k,k-l+1} - 2L_{k,k-l}L_{k,k-l+2} + 2L_{k,k-l-1}L_{k,k-l+3} - \\ - 2L_{k,k-l-2}L_{k,k-l+4} \dots \pm 2L_{k,1}L_{k,k}. \tag{5}$$

If we use coefficients of $A_{k,\ell}$, the formula for X_n assumes the form

$$\begin{aligned} X_1 &= x^2A_{1,1} + \cos^2\Phi, \\ X_2 &= x^4A_{2,1} + x^2A_{2,2} + \cos^2\Phi, \\ X_3 &= x^6A_{3,1} + x^4A_{3,2} + x^2A_{3,3} + \cos^2\Phi, \\ X_4 &= x^8A_{4,1} + x^6A_{4,2} + x^4A_{4,3} + x^2A_{4,4} + \cos^2\Phi, \\ X_5 &= x^{10}A_{5,1} + x^8A_{5,2} + x^6A_{5,3} + x^4A_{5,4} + x^2A_{5,5} + \cos^2\Phi, \\ &\vdots \\ X_n &= x^{2n}A_{n,1} + x^{2n-2}A_{n,2} + \dots + x^2A_{n,n} + \cos^2\Phi. \end{aligned} \tag{6}$$

With the aid of formula (1) we may convert equation (6) into expressions suitable for direct calculation of intensity:

$$\begin{aligned} J_1 &= \sin^2\gamma \,.\, A_{1,1} + \cos^2\gamma\cos^2\Phi, \\ J_2 &= \sin^4\gamma \,.\, A_{2,1} + \sin^2\gamma\cos^2\gamma \,.\, A_{2,2} + \cos^4\gamma\cos^2\Phi, \\ J_3 &= \sin^6\gamma \,.\, A_{3,1} + \sin^4\gamma\cos^2\gamma \,.\, A_{3,2} + \sin^2\gamma\cos^4\gamma \,.\, A_{3,3} + \cos^6\gamma\cos^2\Phi, \\ J_4 &= \sin^8\gamma \,.\, A_{4,1} + \sin^6\gamma\cos^2\gamma \,.\, A_{4,2} + \sin^4\gamma\cos^4\gamma \,.\, A_{4,3} + \sin^2\gamma\cos^6\gamma \,.\, \\ &\quad A_{4,4} + \cos^8\gamma\cos^2\Phi, \\ &\vdots \\ J_n &= \sin^{2n}\gamma \,.\, A_{n,1} + \sin^{2n-2}\gamma\cos^2\gamma \,.\, A_{n,2} + \dots + \sin^2\gamma\cos^{2n-2}\gamma \,.\, A_{n,n} + \\ &\quad + \cos^{2n}\gamma\cos^2\Phi. \end{aligned} \tag{7}$$

Equations (4), (5) and (7) are general relationships describing simple chain filters, both symmetrical and asymmetrical.

In what follows, we shall concentrate on symmetrical filters. We shall use the term "ideal" to define the filter assembly with the simplest arrangement of azimuths of the individual plates. We shall use the term "optimum" to define the filter assembly in which the attempt is made to reduce the secondary maxima as much as possible through arrangement of the azimuths of the individual plates. We shall also distinguish two different cases described by angle Φ. In the first case, angle Φ will be either 0° or 90° (basic positions), in the second case angle Φ will be general (general positions). Of course, all cases will have to be divided into two modifications, the first and the second, which are mutually complementary. We shall take up all the aforementioned cases systematically but separately.

THE FIRST MODIFICATION OF THE CHAIN FILTER IN THE BASIC POSITION, IDEAL ASSEMBLY [OF PLATES]

In this case,

$$\Phi = 90^\circ, \quad \varrho_1 = \varrho_3 = \varrho_5 = \varrho_7 = \dots = \varrho_{2k+1} = \varrho, \\ \varrho_2 = \varrho_4 = \varrho_6 = \varrho_8 = \dots = \varrho_{2k} = -\varrho. \tag{8}$$

Through substitution, equation (4) takes on the form

$$
\begin{aligned}
L_{1,1} &= \sin 2\varrho\,, \\
L_{2,1} &= 0\,, \\
L_{2,2} &= -\sin 4\varrho\,, \\
L_{3,1} &= \sin 2\varrho\,, \\
L_{3,2} &= 0\,, \\
L_{3,3} &= \sin 6\varrho\,, \\
L_{4,1} &= 0\,, \\
L_{4,2} &= -2\sin 4\varrho\,, \\
L_{4,3} &= 0\,, \\
L_{4,4} &= -\sin 8\varrho\,, \\
L_{5,1} &= \sin 2\varrho\,, \\
L_{5,2} &= 0\,, \\
L_{5,3} &= 3\sin 6\varrho + \sin 2\varrho\,, \\
L_{5,4} &= 0\,, \\
L_{5,5} &= \sin 10\varrho\,, \\
&\text{etc.}
\end{aligned}
$$

Equation (5) simplifies as follows:

$$
\begin{aligned}
A_{1,1} &= \sin^2 2\varrho\,, \\
A_{2,1} &= \sin^2 4\varrho\,, \\
A_{3,1} &= \sin^2 6\varrho\,, \\
A_{4,1} &= \sin^2 8\varrho\,, \\
A_{5,1} &= \sin^2 10\varrho\,; \\
\\
A_{1,2} &= 0\,, \\
A_{2,2} &= 0\,, \\
A_{3,2} &= -2\sin 6\varrho \sin 2\varrho\,, \\
A_{4,2} &= 4\sin 8\varrho \sin 4\varrho\,, \\
A_{5,2} &= -6\sin 10\varrho \sin 6\varrho - 2\sin 10\varrho \sin 2\varrho\,, \\
\\
A_{1,3} &= 0\,, \\
A_{2,3} &= 0\,, \\
A_{3,3} &= \sin^2 2\varrho\,, \\
A_{4,3} &= 4\sin^4 4\varrho\,, \\
A_{5,3} &= 9\sin^2 6\varrho + \sin^2 2\varrho + 6\sin 6\varrho \sin 2\varrho + 2\sin 10\varrho \sin 2\varrho \\
\\
A_{1,4} &= 0\,, \\
A_{2,4} &= 0\,, \\
A_{3,4} &= 0\,, \\
A_{4,4} &= 0\,, \\
A_{5,4} &= -6\sin 6\varrho \sin 2\varrho - 2\sin^2 2\varrho\,, \\
&\text{etc.}
\end{aligned}
\tag{5a}
$$

Due to substitution in intensity formulas (7), a condition is created by which the intensity is maximum in the main maxima. This is the case when $A_{1,1} = A_{2,1} = A_{3,1} = \ldots = A_{n,1} = 1$, or

$$\text{for 1 plate} \quad L_{1,1}^2 = \sin^2 2\varrho \ = 1\,,$$
$$\text{for 2 plates} \quad L_{2,2}^2 = \sin^2 4\varrho \ = 1\,,$$
$$\text{for 3 plates} \quad L_{3,3}^2 = \sin^2 6\varrho \ = 1\,,$$
$$\text{for k plates} \quad L_{k,k}^2 = \sin^2 2k\varrho = 1\,.$$

The fundamental relationship for attaining maximum transmission

$$\rho = \frac{45^\circ}{\text{number of plates}} \tag{9}$$

follows from the above equations. By substitution of condition (9) in equation (4), we can construct the following table of numerical values of $L_{n,v}$.

Table 1

n \ v	1	2	3	4	5	6	7	8	9
1	1								
2	0	−1							
3	0,500	0	1						
4	0	−1,414	0	−1					
5	0,309	0	2,736	0	1				
6	0	−1,500	0	−4,464	0	−1			
7	0,222	0	4,418	0	6,375	0	1		
8	0	−1,530	0	−9,758	0	−9,137	0	−1	
9	0,174	0	6,04	0	16,50	0	12,082	0	1

Values of $L_{n,v}$ with higher indexes can be approximated roughly, e.g., by extrapolation on graph paper.

Using table 1, we can construct table 2 for $A_{k,\ell}$, on the basis of relationship (5).

Table 2

k \ l	1	2	3	4	5	6	7	8	9
1	1								
2	1	0							
3	1	− 1	0,250						
4	1	− 2,818	2,000	0					
5	1	− 5,472	8,104	− 1,091	0,0055				
6		− 8,928	22,927	− 13,302	2,250	0			
7	1	−12,750	49,477	− 56,774	22,350	− 1,962	0,049		
8	1	−18,274	103,001	−181,38	123,179	− 29,859	2,341	0	
9	1	−24,164	178,97	−410,79	411,124	−203,525	42,224	−2,102	0,020

Substituting numerical values in formula (7), we establish the regularity of the variation of intensity of an ideal assembly of the first-modification chain filter:

$$J_1 = \sin^2 \gamma ,$$
$$J_2 = \sin^4 \gamma ,$$
$$J_3 = \sin^6 \gamma - \sin^4 \gamma \cos^2 \gamma + 0{,}25 \sin^2 \gamma \cos^4 \gamma ,$$
$$J_4 = \sin^8 \gamma - 2{,}818 \sin^6 \gamma \cos^2 \gamma + 2 \sin^4 \gamma \cos^4 \gamma ,$$
$$J_5 = \sin^{10} \gamma - 5{,}472 \sin^8 \gamma \cos^2 \gamma + 8{,}104 \sin^6 \gamma \cos^4 \gamma - 1{,}691 . \cdot \sin^4 \gamma \cos^6 \gamma + 0{,}095 \sin^2 \gamma \cos^8 \gamma , \tag{7a}$$

etc.

By successive substitution of the path differences, we find the curve of the transmission capacity of the filter. The expressions for intensity are polynomial in form. The main maximum is characterized by the first term, the subsequent terms pertain to the secondary maxima and minima. Their number also determines the number of secondary extremes.

By analyzing the first term, we can arrive at the position of the main maximum, which becomes

$$\gamma = \frac{\delta}{2} = \frac{\pi}{2} , \frac{3}{2}\pi, \ldots, \frac{2k + 1}{2}\pi .$$

By substituting this condition in equation (2), we get the familiar relationship for the wavelengths of maxima,

$$d \cdot D = \frac{2k + 1}{2} \cdot \lambda .$$

The position of the secondary maxima can also be defined roughly as the extremes of the subsequent individual terms of the polynomial, which have the form

$$y = \sin^{2k} \gamma \cdot \cos^{2\ell} \gamma .$$

The extremes of this function are given by the equation

$$\sin \gamma = \sqrt{\frac{k}{k + \ell}} .$$

In these extremes, the function has the value

$$y_{max} = \frac{k^k \cdot \ell^\ell}{(k + \ell)^{k+\ell}} .$$

The true position of the secondary maxima is always somewhat different, since the extremes of the entire polynomial must be considered. Thus, only half the extremes act as maxima, the other half correspond approximately to minima.

The curves of the transmission [of light] for filters with one to nine

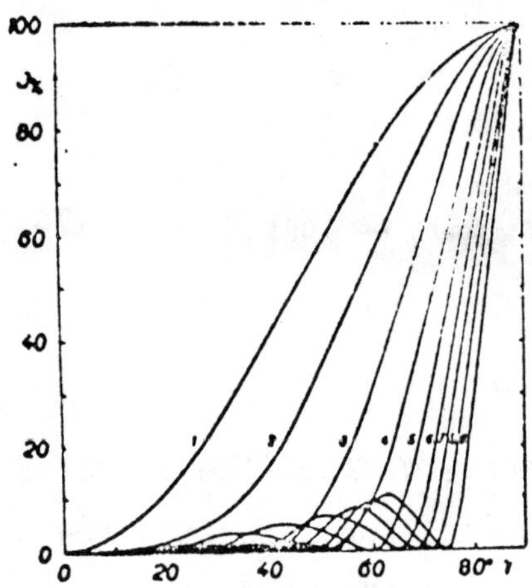

Figure 1. The variation of intensity of first-modification chain filters with one to nine plates, ideal assembly in the basic position.

plates are computed on the basis of equations (7). This is illustrated in figure 1. An interesting fact becomes apparent from equations (7), namely that a filter with two plates has no secondary maximum. This could be used in constructing a Lyot filter. Further, computation shows that the transmission curve is considerably steeper near the main maximum than one would expect from the first term of the polynomial itself. The main maximum is always somewhat broader than the case of a single plate which is as thick as the total number of plates of a chain filter.

THE SECOND MODIFICATION OF A CHAIN FILTER IN THE BASIC POSITION WITH AN IDEAL ASSEMBLY [OF PLATES]

In this case,

$$\Phi = 0^\circ, \quad \varrho_1 = \varrho,\ \varrho_2 = 3\varrho,\ \varrho_3 = 5\varrho,\ \ldots,\ \varrho_n = (2n-1)\,\varrho .$$

By substitution in equations (4) we get

$$\begin{aligned}
L_{1,1} &= 0, \\
L_{2,1} &= 0, \\
L_{2,2} &= -\cos 4\varrho, \\
L_{3,1} &= 0, \\
L_{3,2} &= 2\sin 2\varrho - \cos 4\varrho, \\
L_{3,3} &= 0, \\
L_{4,1} &= 0, \\
L_{4,2} &= \cos 4\varrho - \cos 12\varrho, \\
L_{4,3} &= 0, \\
L_{4,4} &= \cos 8\varrho, \\
L_{5,1} &= 0, \\
L_{5,2} &= 2\sin 6\varrho + 2\sin 2\varrho - 2\cos 8\varrho - \cos 12\varrho + \cos 4\varrho, \\
L_{5,3} &= 0, \\
L_{5,4} &= \cos 8\varrho, \\
L_{5,5} &= 0,
\end{aligned} \qquad (4b)$$

etc.

Equation (5) simplifies as follows:

$$\begin{aligned}
A_{1,1} &= 0, \\
A_{2,1} &= \cos^2 4\varrho, \\
A_{3,1} &= 0, \\
A_{4,1} &= \cos^2 8\varrho, \\
A_{5,1} &= 0;
\end{aligned}$$

$$
\begin{aligned}
A_{1,2} &= 1\,, \\
A_{2,2} &= 2\cos 4\varrho\,, \\
A_{3,2} &= 4\sin^2 2\varrho + \cos^2 4\varrho - 4\sin 2\varrho\cos 4\varrho\,, \\
A_{4,2} &= \cos 12\varrho\cos 8\varrho - \cos 8\varrho\cos 4\varrho\,, \\
A_{5,2} &= \cos^2 8\varrho\,;
\end{aligned}
$$

$$
\begin{aligned}
A_{1,3} &= 0\,, \\
A_{2,3} &= 1\,, \\
A_{3,3} &= 2\cos 4\varrho - 4\sin 2\varrho\,, \\
A_{4,3} &= \cos^2 4\varrho + \cos^2 12\varrho - 2\cos 4\varrho\cos 12\varrho + 2\cos 8\varrho\,, \\
A_{5,3} &= 4\cos^2 8\varrho + 2\cos 12\varrho\cos 8\varrho - 4\cos 8\varrho\sin 6\varrho - 4\cos 8\varrho\sin 6\varrho - \\
&\quad - 2\cos 8\varrho\cos 4\varrho\,;
\end{aligned}
$$

$$
\begin{aligned}
A_{1,4} &= 0\,, \\
A_{2,4} &= 0\,, \\
A_{3,4} &= 1\,, \\
A_{4,4} &= 2\cos 12\varrho - 2\cos 4\varrho\,, \\
A_{5,4} &= 4\sin^2 6\varrho + 4\sin^2 2\varrho + 4\cos^2 8\varrho + \cos^2 12\varrho + \cos^2 4\varrho + \\
&\quad + 8\sin 6\varrho\sin 2\varrho - 8\sin 6\varrho\cos 8\varrho - 4\sin 6\varrho\cos 12\varrho + \\
&\quad + 4\sin 6\varrho\cos 4\varrho + 4\sin 2\varrho\cos 4\varrho - 4\sin 2\varrho\cos 12\varrho - \\
&\quad - 8\sin 2\varrho\cos 8\varrho + 4\cos 12\varrho\cos 8\varrho - 4\cos 8\varrho\cos 4\varrho - \\
&\quad - 2\cos 12\varrho\cos 4\varrho + 2\cos 8\varrho\,,
\end{aligned}
$$

etc.

The maximum intensity in the main maxima of the second modification need not be verified separately, since the last term of equations (7) has the coefficient $\cos^2\Phi = 1$.

However, the second modification can be made complementary to the first by proper choice of angle ρ. This angle is given by relationship (9). Through substitution of its numerical values in equations (4), we can again construct a table for $L_{u,v}$.

Table 3

$n \backslash v$	1	2	3	4	5	6	7	8	9
1	0								
2	0	0							
3	0	0,500	0						
4	0	1,414	0	0					
5	0	2,736	0	0,309	0				
6	0	4,464	0	1,500	0	0			
7	0	6,375	0	4,418	0	0,222	0		
8	0	9,137	0	9,758	0	1,530	0	0	
9	0	12,082	0	16,50	0	6,04	0	0,174	0

A comparison of this table with table 1 shows the mutual relationship of coefficients $L_{u,v}$ for both filter modifications.

Again we can construct a table for coefficients of $A_{k,\ell}$ on the basis of relationships (5).

Table 4

k \ l	1	2	3	4	5	6	7	8	9
1	0								
2	0	0							
3	0	0,250	−1						
4	0	0	+2	− 2,818					
5	0	0,0955	−1,691	+ 8,104	− 5,472				
6	0	0	+2,250	−13,392	+ 22,927	− 8,928			
7	0	0,049	−1,962	+22,350	− 56,774	+ 49,477	− 12,750		
8	0	0	+2,341	−29,859	+123,179	−181,38	+103,00	− 18,274	
9	0	0,030	−2,102	+42,224	−203,525	+411,124	−410,79	+178,97	−24,164

This time we construct the equation for intensity in the reverse order of terms of the polynomial. This shows the complementary nature of the two modifications more clearly:

$$
\begin{aligned}
J_1 &= \cos^2 \gamma\,, \\
J_2 &= \cos^4 \gamma\,, \\
J_3 &= \cos^6 \gamma - \cos^6 \gamma \sin^2 \gamma + 0{,}25 \cos^2 \gamma \sin^4 \gamma\,, \\
J_4 &= \cos^8 \gamma - 2{,}818 \cos^4 \gamma \sin^2 \gamma + 2 \cos^4 \sin^4 \gamma\,, \\
J_5 &= \cos^{10} \gamma - 5{,}472 \cos^8 \gamma \sin^2 \gamma + 8{,}104 \cos^6 \gamma \sin^4 \gamma - 1{,}691 \cos^4 \sin^6 \gamma + \\
&\quad + 0{,}0955 \cos^2 \gamma \sin^8 \gamma\,,
\end{aligned}
\tag{7b}
$$

etc.

The system of equations (7b) is formally identical with system (7a), except that the functions of sin γ and cos γ are interchanged. We get complete agreement by introducing a complementary angle. In other words, the transmission curves of the second-modification filter are shifted 90° with respect to those of the first-modification filter, but otherwise are in agreement. The condition for the position of the main maxima of the second-modification filter, $\gamma = \pi, 2\pi, \ldots, k\pi$, follows from equations (7b), which leads to relationship

$$d \cdot D = k \cdot \ell\,.$$

Otherwise, the analysis of the intensity variation is the same as that discussed for the first filter modification.

THE FIRST MODIFICATION OF A SYMMETRICAL CHAIN FILTER IN THE BASIC POSITION, OPTIMUM ASSEMBLY [OF PLATES]

It is quite difficult to define the optimum assembly, for it is difficult to define this case mathematically, since the variation of the main maximum deteriorates somewhat due to the condition that the secondary maxima be suppressed. Before we begin the main analysis of the requirements of an optimum assembly, let us derive the general equation for a symmetrical first-modification filter in the basic position.

In this case,

$$\Phi = \frac{\pi}{2}, \quad |\varrho_1| = |\varrho_n|, \ |\varrho_2| = |\varrho_{n-1}|, \ \ldots, \ |\varrho_i| = |\varrho_{n-i+1}| .$$

Again the signs of the azimuths alternate. The condition for maximum transmission of the filter follows from the general equation

$$A_{1,1} = A_{2,1} = A_{3,1} = \ldots = A_{n,1} = 1,$$

or else

$$L^2_{1,1} = L^2_{2,2} = L^2_{3,3} = \ldots = L^2_{n,n} = 1.$$

According to equation (4), this relationship leads to the result

$$|\varrho_1| + |\varrho_2| + |\varrho_3| + \ldots + |\varrho_n| = 45^\circ , \qquad (10)$$

which is the generalized relationship (9). Then, in equations (4), we will substitute, successively,

for 1 plate	ϱ_1,	$\varrho_1 = 45^\circ$,
for 2 plates	$\varrho_2 = -\varrho_1$,	$\varrho_1 - \varrho_2 = 45^\circ$,
for 3 plates	$\varrho_3 = \varrho_1$, ϱ_2,	$2\varrho_1 - \varrho_2 = 45^\circ$,
for 4 plates	$\varrho_4 = -\varrho_1$, $\varrho_3 = -\varrho_2$,	$2\varrho_1 - 2\varrho_2 = 45^\circ$,
for 5 plates	$\varrho_5 = \varrho_1$, $\varrho_4 = \varrho_2$, ϱ_3,	$2\varrho_1 - 2\varrho_2 + \varrho_3 = 45^\circ$,
for 6 plates	$\varrho_6 = -\varrho_1$, $\varrho_5 = -\varrho_2$, $\varrho_4 = -\varrho_3$,	$2\varrho_1 - 2\varrho_2 + 2\varrho_3 = 45^\circ$,

etc.

The general formulas (4) specialize as follows:

$$L_{1,1} = \sin 2\varrho_1 \ (= 1), \qquad (4c)$$
$$L_{2,1} = 0,$$
$$L_{2,2} = \sin(2\varrho_2 - 2\varrho_1) \ (= -1),$$

$$
\begin{aligned}
L_{3,1} &= 2\sin^2\varrho_1 - \sin 2\varrho_2\,, \\
L_{3,2} &= 0\,, \\
L_{3,3} &= \sin(4\varrho_1 - 2\varrho_2)\ (=1)\,, \\
L_{4,1} &= 0\,, \\
L_{4,2} &= -\sin 4\varrho_1 - \sin 4\varrho_2 - 2\sin(2\varrho_2 + 2\varrho_1) + 2\sin(2\varrho_2 - 2\varrho_1)\,, \\
L_{4,3} &= 0\,, \\
L_{4,4} &= \sin(4\varrho_2 - 4\varrho_1)\ (= -1)\,, \\
L_{5,1} &= 2\sin 2\varrho_1 + 2\sin 2\varrho_2 + \sin 2\varrho_3\,, \\
L_{5,2} &= 0\,, \\
L_{5,3} &= 2\sin(2\varrho_1 - 2\varrho_2 + 2\varrho_3) + 2\sin 2\varrho_1 + 2\sin(4\varrho_1 - 2\varrho_2) + \\
&\quad + 2\sin(2\varrho_1 + 2\varrho_2 - 2\varrho_3) + \sin(4\varrho_1 - 2\varrho_3) + \sin(4\varrho_2 - 2\varrho_3)\,, \\
L_{5,4} &= 0\,, \\
L_{5,5} &= \sin(4\varrho_1 - 4\varrho_2 + 2\varrho_3)\ (=1)\,,
\end{aligned}
\qquad \text{(4c) cont.}
$$

etc.

According to equations (7a), one- and two-plate filters in the ideal assembly do not have secondary maxima; therefore, they are, simultaneously, filters in the optimum assembly. The three-plate filter has secondary maxima characterized by the coefficients $A_{3,2}$ and $A_{3,3}$, according to equation (7). Now let us seek such a solution for azimuths ρ_1 and ρ_2 in which these coefficients are minimum. According to formulas (5),

$$
\begin{aligned}
A_{3,2} &= L_{3,2}^2 - 2L_{3,1}L_{3,3}\,, \\
A_{3,3} &= L_{3,1}^2 - 2L_{3,2}\cos\Phi
\end{aligned}
$$

According to (4c), the expressions reduce to

$$
\begin{aligned}
A_{3,2} &= -2L_{3,1}\,, \\
A_{3,3} &= L_{3,1}^2\,.
\end{aligned}
$$

Then we need concern ourselves only with the expression $L_{3,1}$. Let us seek the condition for its minimum, which can be used in this case to solve the equations

$$
\begin{aligned}
L_{3,1} &= 2\sin 2\varrho_1 + \sin 2\varrho_2 = 0\,, \\
2\varrho_1 - \varrho_2 &= 45^\circ\,.
\end{aligned}
$$

The quadratic equation

$$
2\sin^2 2\varrho_1 + 2\sin 2\varrho_1 - 1 = 0\,,
$$

stems from these equations; its solution gives the angles

$$
\rho_1 = 10.7^\circ, \quad \rho_2 = -23.6^\circ.
$$

With these plate azimuths, the three-plate filter has no secondary maximum. The equation for the variation of intensity has the form

$$
J = \sin^6 \gamma\,.
$$

Proceeding in a similar manner, let us seek the solution of the equation $L_{4,2} = 0$ for a four-plate filter, or

$$\sin 4\varrho_1 + \sin 4\varrho_2 + 2 \sin (2\varrho_1 + 2\varrho_2) + 2 \sin (2\varrho_1 - 2\varrho_2) = 0 ,$$

$$2\varrho_1 - 2\varrho_2 = 45° .$$

By a simple operation we arrive at the equation

$$\sin 4\varrho_1 . (1 + 2 \cos 45°) - \cos 4\varrho_1(1 + 2 \sin 45°) + 2 \sin 45° = 0 ,$$

whose solution gives the angles

$$\rho_1 = 7.6°, \quad \rho_2 = - 14.9°.$$

A filter with these azimuths of the plates again has only a main maximum and its intensity variation is given by the equation

$$J = \sin^2 \gamma.$$

We can continue thus, but with ever-increasing computational difficulties. For instance, the solution of a single trigonometric equation no longer suffices. The distinction between the ideal assembly and the same assembly without secondary maxima is an important one. The two types of filters can best be judged on the basis of the intensity variation, which is shown in figure 2 for a four-plate filter.

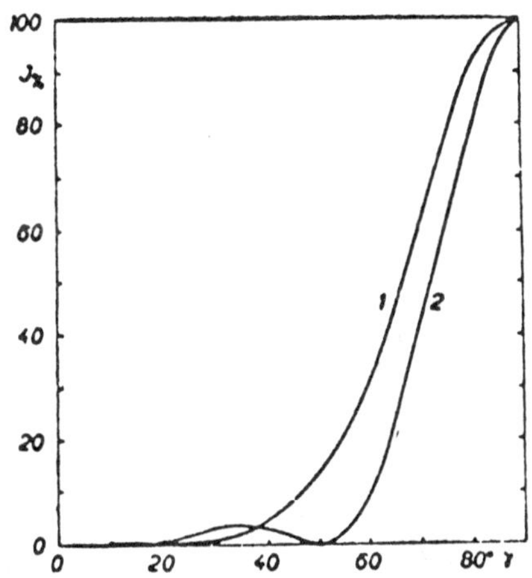

Figure 2. Variation of the intensity of a first-modification chain filter with four plates in an ideal assembly (the steeper curve with the secondary maximum) and in an optimum assembly (the flatter curve without a secondary maximum)

It is easy to see that when the secondary maxima are eliminated, the main maximum broadens perceptibly. Therefore we need not concern ourselves with an exact analysis of the possibility of eliminating the secondary maxima in filters with a large number of plates, but we can be satisfied with some compromise solution in the choice of the azimuths of the plates. Furthermore, we are concerned with a considerable reduction of the intensity of the secondary maxima without essentially increasing the width of the main maximum. The following method has proved best: Select the azimuth angle of the first plate; for filters with more than two plates this angle is less than that indicated by formula (9) and, according to circumstances, reaches 50-75% of the value of the angles for an ideal assembly. The absolute values of the azimuth angles of the plates themselves increase progressively toward the center of the filter. The growth functions can be chosen differently, and the arithmetic progression will also serve the purpose well.

$$\varrho_1, \quad \varrho_2 = -(\varrho_1 + \alpha), \quad \varrho_3 = \varrho_1 + 2\alpha, \quad \varrho_4 = -(\varrho_1 + 3\alpha),$$
$$\varrho_5 = \varrho_1 + 5\alpha \quad \text{etc.}$$

The azimuths of all the plates must then fulfill condition (10) [+].

As an example of such an optimum assembly, let us analyze a filter with seven plates. We shall choose the coefficient of reduction of the first azimuth as 70%, the azimuths increase in arithmetic progression. Let us compute the angles

$$\begin{aligned} \varrho_1 &= 4{,}5^\circ, & \varrho_4 &= -9^\circ, \\ \varrho_2 &= -6{,}0^\circ, & \varrho_5 &= +7{,}5^\circ, \\ \varrho_3 &= 7{,}5^\circ, & \varrho_6 &= -6^\circ, \\ & & \varrho_7 &= 4{,}5^\circ. \end{aligned}$$

For seven plates, equations (4) have the form

$$\begin{aligned}
L_{7,1} = {} & 2\sin 2\varrho_1 + 2\sin 2\varrho_2 + 2\sin 2\varrho_3 + \sin 2\varrho_4, \\
L_{7,2} = {} & 0, \\
L_{7,3} = {} & 4\sin 2\varrho_1 + 2\sin 2\varrho_2 + 4\sin(2\varrho_1 - 2\varrho_2 + 2\varrho_3) + \\
& + 2\sin(2\varrho_1 - 2\varrho_2 + 2\varrho_4) + 4\sin(2\varrho_1 + 2\varrho_2 - 2\varrho_3) + \\
& + 2\sin(4\varrho_1 - 2\varrho_2) + 2\sin(4\varrho_1 - 2\varrho_3) + 2\sin(4\varrho_2 - 2\varrho_3) + \\
& + 2\sin(2\varrho_1 - 2\varrho_3 + 2\varrho_4) + 2\sin(2\varrho_2 + 2\varrho_3 - 2\varrho_4) + \\
& + 2\sin(2\varrho_1 + 2\varrho_3 - 2\varrho_4) + 2\sin(2\varrho_1 + 2\varrho_2 - 2\varrho_4) + \\
& + 2\sin(2\varrho_2 - 2\varrho_3 + 2\varrho_4) + \sin(4\varrho_1 - 2\varrho_4) + \\
& + \sin(4\varrho_2 - 2\varrho_4) + \sin(4\varrho_3 - 2\varrho_4), \\
L_{7,4} = {} & 0, \\
L_{7,5} = {} & 2\sin 2\varrho_1 + 2\sin(4\varrho_1 - 2\varrho_2) + 2\sin(2\varrho_1 + 2\varrho_3 - 2\varrho_4) + \\
& + 2\sin(2\varrho_1 - 2\varrho_3 + 2\varrho_4) + 2\sin(4\varrho_1 - 2\varrho_2 + 2\varrho_3 - 2\varrho_4) + \\
& + 2\sin(4\varrho_1 - 4\varrho_2 + 2\varrho_3) + 2\sin(2\varrho_1 - 4\varrho_3 + 2\varrho_2 + 2\varrho_4) + \\
& + 2\sin(4\varrho_1 - 2\varrho_2 - 2\varrho_3 + 2\varrho_4) + \sin(4\varrho_1 - 4\varrho_3 + 2\varrho_4) + \\
& + \sin(4\varrho_1 - 4\varrho_2 + 2\varrho_4) + \sin(4\varrho_2 - 4\varrho_3 + 2\varrho_4) + \\
& + 2\sin(2\varrho_1 - 2\varrho_2 + 4\varrho_3 - 2\varrho_4), \\
L_{7,6} = {} & 0, \\
L_{7,7} = {} & \sin(4\varrho_1 - 4\varrho_2 + 4\varrho_3 - 2\varrho_4).
\end{aligned}$$

From these formulas we get the following values of $L_{u,v}$

$$\begin{aligned} L_{7,1} &= +0{,}106, \\ L_{7,2} &= 0, \\ L_{7,3} &= +2{,}851, \\ L_{7,4} &= 0, \\ L_{7,5} &= 4{,}714, \\ L_{7,6} &= 0, \\ L_{7,7} &= 1. \end{aligned}$$

[+] The azimuth of the first plate can even be chosen as 0°; however, in this case the first and last plates will not function and one must reckon with a filter which has two more plates. The first and last plates are omitted in constructing the filter.

The corresponding coefficients of $A_{k,\ell}$ for equations (5) are

$$A_{7,1} = 1,$$
$$A_{7,2} = -9{,}428,$$
$$A_{7,3} = 27{,}924,$$
$$A_{7,4} = -27{,}092,$$
$$A_{7,5} = 9{,}127,$$
$$A_{7,6} = -0{,}6064,$$
$$A_{7,7} = 0{,}0123.$$

Then the variation of intensity is given by the equation

$$J_7 = \sin^{14}\gamma - 9{,}428\sin^{12}\gamma\cos^2\gamma + 27{,}924\sin^{10}\gamma\cos^4\gamma - 27{,}092\sin^8\gamma\cos^6\gamma + 9{,}127\sin^6\gamma\cos^8\gamma - 0{,}606\sin^4\gamma\cos^{10}\gamma + 0{,}012\sin^2\gamma\cos^{12}\gamma.$$

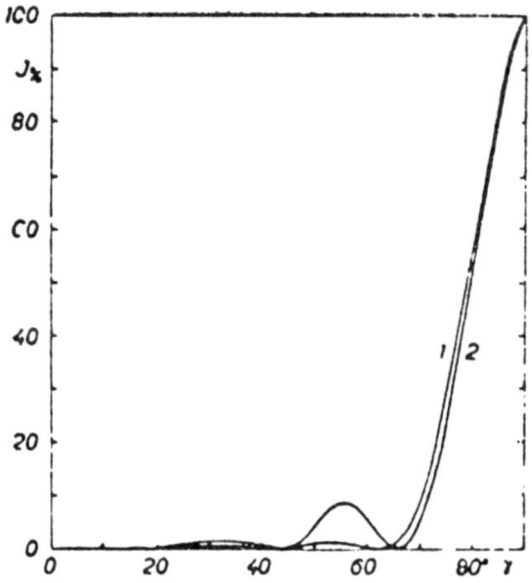

Figure 3. The variation of intensity of a first-modification chain filter with seven plates, in an ideal assembly (the curve with the higher secondary maxima) and in the optimum assembly (the curve with the lower secondary maxima)

A comparison of the intensity variation of the "optimum" assembly of a chain birefringent filter with the "ideal" assembly shows a marked suppression of the secondary maxima (figure 3). It should also be mentioned that when the azimuth of the individual plates is changed, not only the intensity but also the position of the secondary maxima change.

Analysis of the optimum filter assembly shows that the concept of the optimum can be adapted as needed. For example, one may wish to find the sharpest main maximum without considering the secondary maxima. Most often, however, it is important to suppress the secondary maxima. Then the relationship of the azimuths based on arithmetic progression is of use, and this applies not only to filters with a small number of plates but also to filters with tens or hundreds of plates.

THE SECOND MODIFICATION OF A SYMMETRICAL CHAIN FILTER IN THE BASIC POSITION, OPTIMUM ASSEMBLY [OF PLATES]

This example again yields similar results, so we will treat it but briefly. Here

$$\Phi = 0, \quad \varrho_n = 90^\circ - \varrho_1, \; \varrho_{n-1} = 90^\circ - \varrho_2, \; \ldots, \; \varrho_{n-i} = 90^\circ - \varrho_{i+1}.$$

When there is an odd number of plates, due to symmetry the middle plate will have an azimuth of 45°. The condition of maximum transmission need not be determined separately.

Formulas (4) specialize as follows:

$$
\begin{aligned}
L_{1,1} &= 0\,, \\
L_{2,1} &= 0\,, \\
L_{2,2} &= -\cos 4\varrho_1\,, \\
L_{3,1} &= 0\,, \\
L_{3,2} &= 2\sin 2\varrho_1 - \cos 4\varrho_1\,, \\
L_{3,3} &= 0\,, \\
L_{4,1} &= 0\,, \\
L_{4,2} &= -\cos 4\varrho_1 - \cos 4\varrho_2 - 2\cos(2\varrho_2 + 2\varrho_1) + 2\cos(2\varrho_2 - 2\varrho_1)\,, \\
L_{4,3} &= 0\,, \\
L_{4,4} &= \cos(4\varrho_1 - 4\varrho_2)\,, \\
L_{5,1} &= 0\,, \\
L_{5,2} &= 2\cos(2\varrho_1 - 2\varrho_2) - 2\cos(2\varrho_1 + 2\varrho_2) - \cos 4\varrho_1 - \cos 4\varrho_2 + \\
&\quad + 2\sin 2\varrho_1 + 2\sin 2\varrho_2\,, \\
L_{5,3} &= 0\,, \\
L_{5,4} &= \cos(4\varrho_1 - 4\varrho_2) + 2\sin 2\varrho_1(4\varrho_1 - 2\varrho_2) - 2\sin 2\varrho_1\,, \\
L_{5,5} &= 0\,,
\end{aligned}
$$

etc.

According to equations (7b), one- and two-plate filters do not have secondary maxima. Again the elimination of a secondary maximum in three-plate filters is formulated by the condition $L_{3,2} = 0$, or

$$2\sin 2\varrho_1 - \cos 4\varrho_1 = 0\,.$$

This equation has the same solution as in the case of the first filter modification; only angle ρ_2 need be computed with respect to symmetry, hence it is different:

$$\varrho_1 = 10{,}7^\circ\,, \quad \varrho_2 = 79{,}3^\circ\,.$$

The curve of the transmission [of light] is the same as for the first filter modification, except that it is shifted by $\pi/2$.

Once again, the solution is similar for a four-plate, second-modification filter without secondary maxima. Conditions $L_{4,2} = 0$ and $L_{4,4} = 0$ must be fulfilled. Then the following equations must be solved:

$$
\begin{aligned}
-\cos 4\varrho_1 - \cos 4\varrho_2 - 2\cos(2\varrho_2 + 2\varrho_1) + 2\cos(2\varrho_2 - 2\varrho_1) &= 0\,, \\
4\varrho_1 - 4\varrho_2 &= \pm 90^\circ\,.
\end{aligned}
$$

This system leads to equation

$$-\cos 4\varrho_1(1 + 2\cos 45^\circ) + \sin 4\varrho_1(1 + 2\sin 45^\circ) + 2\cos 45^\circ = 0\,,$$

which is equivalent to the equations derived for the first-modification filter. The solution gives the values

$$\varrho_1 = 7{,}6^\circ , \quad \varrho_3 = 59{,}9^\circ ,$$
$$\varrho_2 = 30{,}1^\circ , \quad \varrho_4 = 82{,}4^\circ .$$

Again, this filter does not have secondary maxima. Solutions can be reached for multi-plate filters, but greater difficulties are involved; however, what has been said about the first modification also applies to the second modification, so let us be satisfied with the analogy of the arithmetic increase of azimuths applicable to the first modification. In the second modification, the increase of the azimuths from the edge toward the center of the filter can be expressed as follows:

$$\varrho_1 , \quad \varrho_2 = 3\varrho_1 + \alpha , \quad \varrho_3 = 5\varrho_1 + (1 + 3)\,\alpha , \quad \varrho_4 = 7\varrho_1 + (1 + 3 + 5)\,\alpha ,$$
$$\varrho_5 = 9\varrho_1 + (1 + 3 + 5 + 7)\,\alpha \ \ldots \ \varrho_{n-2} = [2(n-2) - 1]\,\varrho_1 + (1 + 3 + 5 +$$
$$+ 7 + \ldots + 7 + 5)\,\alpha , \quad \varrho_{n-1} = [2(n-1) - 1]\,\varrho_1 + (1 + 3 + 5 + 7 + \ldots +$$
$$+ 7 + 5 + 3)\,\alpha , \quad \varrho_n = (2n - 1)\,\varrho_1 + (1 + 3 + 5 + 7 + \ldots + 7 + 5 + 3 +$$
$$+ 1)\,\alpha .$$

In selecting the azimuths, ρ_1 and ϑ can be defined as in the case of the first modification. Thus, the seven-plate first-modification filter of which we spoke in the preceding section has the same azimuths ($\rho_1 = 4.5^\circ$, $\vartheta = 1.5^\circ$) for the second modification:

$$4.5^\circ,\ 15^\circ,\ 28.5^\circ,\ 45^\circ,\ 61.5^\circ,\ 75^\circ,\ 85.5^\circ.$$

Further analysis of this filter shows that the transmission curve is shifted by $\pi/2$, as in the case of the first filter modification.

CHAIN FILTERS IN THE GENERAL POSITION

Up to this point, all types of birefringent filters have been calculated and constructed for the basic position of the polarizing elements only, whether for parallel or crossed polaroids. However, these two fundamental positions are not a necessary condition for good filter function. Therefore, at this point we will indicate (albeit briefly) a method of solving for chain filters in a general position.

To explain the fundamental relationships, let us begin with one plate. For intensity, according to (7):

$$J_1 = \sin^2 \gamma \cos^2 (\Phi - 2\varrho_1) + \cos^2 \gamma \cos^2 \Phi .$$

Angle Φ is arbitrary in the general position. The first step in the analysis is to find angles ρ_1 that will correspond to the first and second modification of the filter. The first modification shows secondary maxima for $\gamma = \pi/2$. Then,

$$J_{max} = \cos^2(\Phi - 2\varrho_1),$$

or

$$\varrho_1 = \frac{\Phi}{2} = \Omega_1 . \tag{11}$$

Here, too, the maxima reach full values. The minima have the intensity

$$J_{min} = \cos^2 \Phi .$$

The minima are not pure; their intensities increase as the distance of angles Φ from $\pi/2$. The minima vanish when $\Phi = 0^\circ$.

The second filter modification has its maxima at $\gamma = 0$, its minima at $\gamma = \pi/2$. For one plate,

$$J_{min} = \cos^2(\Phi - 2\varrho_1),$$

or

$$\Phi - 2\varrho_1 = \frac{\pi}{2},$$

$$\varrho_1 = \frac{\pi}{4} - \frac{\Phi}{2} = \Omega_2 . \tag{12}$$

The maxima reach values of

$$J_{max} = \cos^2 \Phi$$

Therefore, the analogy of the second modification has pure minima; however, the maxima do not reach full values and decrease as angle Φ increases. The maxima vanish when $\Phi = \pi/2$.

This analysis indicates that two distinct azimuths of a birefringent plate, defined by equations (11) and (12), can also be established for a common angle Φ. In the case of equation (11), the plate behaves analogously with the first filter modification; the maxima and minima have the same wavelengths as when $\Phi = \pi/2$. The intensity variation is then given by the relationship

$$J = \sin^2 \gamma + \cos^2 \gamma \cos^2 \Phi .$$

When the azimuth of the plate agrees with equation (12), the result is analogous to the second filter modification and the intensity variation is given by the equation

$$J = \cos^2 \gamma \cos^2 \Phi .$$

If there are more plates, the general equations can be used in the analysis. For the first filter modification we get the following result: the azimuths of the individual plates deviate alternately to the left and right of the mean angle Γ,

$$\Gamma = -\Omega_2 .$$

The following relationship holds for the azimuth angles measured from the mean angle Γ,

$$|\bar{\varrho}_1| + |\bar{\varrho}_2| + |\bar{\varrho}_3| + \cdots + |\bar{\varrho}_n| = \Omega_1 . \tag{13}$$

In filters with a large number of plates, there are also clear minima for angles Φ which differ considerably from $\pi/2$. By selection of angles $\bar{\rho}_i$, a filter with the ideal assembly of plates and one with the optimum assembly can be constructed in a manner completely analogous to that used for the basic position, except that equation (13) will apply instead of equation (8).

In the case of the second filter modification, again we arrange the azimuths of the plates in the shape of a fan. The center of the fan must agree with angle Ω_2, according to equation (12). The azimuth difference of the first and last plate is determined by the equation

$$\varrho_n - \varrho_1 \gtreqless \Omega_2 . \tag{14}$$

One can construct a filter with either the ideal or the optimum assembly, depending on the arrangement of azimuths. When the number of plates is increased, the intensity in the maxima also increases and approaches full value. The special property of filters that operate at a general angle of the polaroids and analyzers is that they offer the possibility of tuning filters in the basic position. Thus, through slight departures from the parallel or crossed positions of the polaroids and analyzer, the filter can be tuned still more. This generalization of the chain filters is also important from the theoretical standpoint.

SUMMARY

In their application, chain birefringent filters fall into the same category as multilayered interference filters prepared by steaming in vacuum. The most selective multilayered interference filters have a transmission band of about 20 Å in the visible region and it can be assumed that, for technical reasons, the selectivity of these filters will not be increased. Birefringent filters are suitable for a range of band widths of about 50 Å to 0.5 Å. Therefore it is desirable to develop birefringent filters with high transmission and, simultaneously, with the purest possible spectral characteristic. Apropos, the numerical analysis performed here shows a method for producing filters with high transmission and nearly pure maxima without secondary transmissions.

Submitted 11 May 1959

LITERATURE CITED

1. Šolc, I. Československý Časopis pro Fysiku, 3: 366, 1953.

2. Šolc, I. Československý Časopis pro Fysiku, 4: 669, 1954.

3. Šolc, I. Československý Časopis pro Fysiku, 4: 607, 1954.

4. Šolc, I. Československý Časopis pro Fysiku, 5: 114, 1955.

5. Hsü, Hsien-Yü, Yüng-Kang Liang, and M. Richartz. Optical Society of America, Journal, 37: 99, 1947.

6. Šolc, I. Czechoslovak Journal of Physics, 9: 237, 1959.

7. Dollfus, A. Revue d'Optique, 35: 539, 1956.

8. Dollfus, A. Revue d'Optique, 35: 625, 1956.

9. Rozenberg, G. V. Uspekhi Fizicheskikh Nauk, 17: 3, 1952.

10. Rozenberg, G. V. Uspekhi Fizicheskikh Nauk, 17: 173, 1952.

11. Mie, K. Optik, 14: 1, 1957.

Solc Birefringent Filter

John W. Evans
Sacramento Peak Observatory, Geophysics Research Directorate, AFCRC, Sunspot, New Mexico
(Received August 26, 1957)

A new birefringent filter composed of a series of retardation plates between a single pair of polarizers has been described by Solc. The filter is analyzed and compared with the Lyot filter. It is found that the Solc filter is inferior to the Lyot filter in the suppression of parasitic light in secondary maxima near the primary transmission bands, although its band width is slightly less, and its transparency is very much better if both filters use Polaroid film for polarizers.

Birefringent filters of the type invented by Lyot, with transmission bands ranging from 0.5 to 5 A in width have become standard tools of solar research. While they are not excessively difficult to construct, and their performance is excellent, any possible simplification or improvement would be welcomed. A new form of birefringent filter recently developed in Czechoslovakia is therefore worth examination, and comparison with the Lyot filter.

In 1953 to 1955 Ivan Solc[1] published three papers on a new form of birefringent filter which he apparently developed and investigated purely by experiment. He does not give a general expression for the transmission of his filter as a function of wavelength because of its mathematical complexity. He did, however, succeed in determining its more important characteristics experimentally with considerable accuracy. The purpose of this paper is to derive an expression for the transmission of the Solc filter, and to compare its performance with that of the Lyot filter.

The Solc filter consists of a pile of identical retardation plates of birefringent material, with only two linear polarizers, one at each end. Figure 1 shows the arrangement. The plates are cut, as in the Lyot filter, with the crystal optic axis parallel to the surfaces. Solc describes two possible arrangements of the orientations of the axes of successive plates, which we shall term the fan and folded filters, respectively.

Let the electric vector of light traversing the first polarizer be the reference direction from which the orientation, ω_j, of the optic axis of the jth plate is measured. In the fan filter the two polarizers are parallel,

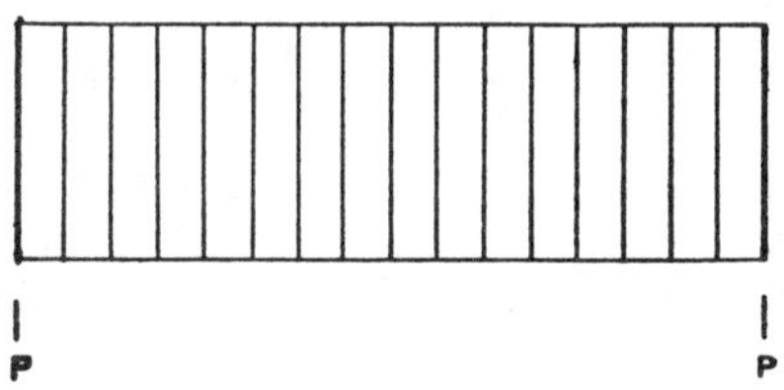

Fig. 1. Section of a Solc filter, consisting of 16 retardation plates between a pair of polarizers, P.

[1] I. Solc, Czechoslov. Cosopis pro Fysiku, **3**, 366 (1953); **4**, 607, 669 (1954); **5**, 114 (1955).

and ω progresses monotonically. The angle ω_j is given by

$$\omega_j=(\alpha/2)+(j-1)\alpha, \tag{1}$$

where α is a small angle which we shall determine.

In the folded filter the polarizers are crossed, and ω alternates between $(\alpha/2)$ and $-(\alpha/2)$. The angle ω_j is given by

$$\omega_j=(-1)^{j+1}(\alpha/2). \tag{2}$$

Solc determined experimentally that both forms of the filter appeared to work best when $\alpha\approx\pi/2n$, where n= the total number of plates. He states further that the curve of transmission as a function of wavelength closely resembles that of a Lyot filter in which the thinnest element (which fixes the spacing of the bands) has the same thickness as one of his retardation plates,

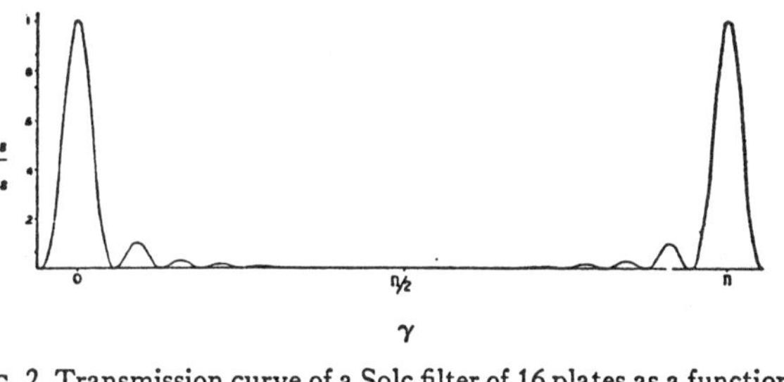

Fig. 2. Transmission curve of a Solc filter of 16 plates as a function of the retardation, γ, of a single plate.

and the thickest element (which determines the band width) has the thickness of the whole pile of Solc plates.

It is difficult to see intuitively how either form of the Solc filter accomplishes its purpose. The problem of deriving an expression for the transmission by the usual methods does not appear to have an easy solution. However, the use of the matrix calculus developed by Jones[2] proves to be effective. The reader is referred to his papers for a very clear explanation of the method. In the following application, we retain his notation for convenience of reference.

Let the light travel along the z axis of a rectangular coordinate system. The faces of the n retardation plates of a Solc filter are normal to the z axis, and the electric vector of light transmitted by the first polarizer is parallel to the x axis. Let the electric vectors of light

[2] R. C. Jones, J. Opt. Soc. Am. **31**, 488 (1941); **31**, 500 (1941).

entering and emerging from the system be represented by the one column matrices,

$$\epsilon_0=\begin{pmatrix}E_{x0}\\E_{y0}\end{pmatrix}\quad\text{and}\quad\epsilon_n=\begin{pmatrix}E_{xn}\\E_{yn}\end{pmatrix}. \tag{3}$$

Then the initial matrix equations relating ϵ_0 and ϵ_n in the fan and folded filters are,

$$\begin{pmatrix}E_{xn}\\E_{yn}\end{pmatrix}_{\text{Fan}}=\mathbf{P}_x\mathbf{S}(\alpha/2)\mathbf{S}(\pi/2)[\mathbf{S}(-\alpha)\mathbf{G}]^n \times\mathbf{S}(-\alpha)\mathbf{P}_x\begin{pmatrix}E_{x0}\\E_{y0}\end{pmatrix}, \tag{4}$$

$$\begin{pmatrix}E_{xn}\\E_{yn}\end{pmatrix}_{\substack{\text{Folded}\\ n\text{ even}}}=(-1)^{(n/2)}\mathbf{P}_y\mathbf{S}(\alpha/2)[\mathbf{S}(-\alpha)\mathbf{G}\mathbf{S}(\alpha)\mathbf{G}]^{(n/2)} \times\mathbf{S}(-\alpha)\mathbf{P}_x\begin{pmatrix}E_{x0}\\E_{y0}\end{pmatrix}, \tag{5}$$

$$\begin{pmatrix}E_{xn}\\E_{yn}\end{pmatrix}_{\substack{\text{Folded}\\ n\text{ odd}}}=i(-1)^{(n-1)/2}\mathbf{P}_y\mathbf{S}(\alpha/2) \times\mathbf{G}[\mathbf{S}(-\alpha)\mathbf{G}\mathbf{S}(\alpha)\mathbf{G}]^{(n-1)/2} \times\mathbf{S}(-\alpha)\mathbf{P}_x\begin{pmatrix}E_{x0}\\E_{y0}\end{pmatrix}. \tag{6}$$

Here $\mathbf{S}(\alpha)$ and $\mathbf{G}$ are matrices representing rotation through the angle α, and the retardation of a single plate, respectively. They are defined as follows:

$$\mathbf{S}(\alpha)=\begin{pmatrix}\cos\alpha & -\sin\alpha\\ \sin\alpha & \cos\alpha\end{pmatrix}, \tag{7}$$

and

$$\mathbf{G}=\begin{pmatrix}e^{-i\gamma} & 0\\ 0 & e^{-i\gamma}\end{pmatrix}, \tag{8}$$

where $\gamma=\pi(d/\lambda)(\epsilon-\omega)$ in the fan filter, $\gamma=\pi(d/\lambda)(\epsilon-\omega)+(\pi/2)$ in the folded filter, d = thickness of a single retardation plate, λ = wavelength of the light, and ϵ and ω = refractive indexes of the birefringent material.

The matrices $\mathbf{P}_x$ and $\mathbf{P}_y$ represent polarizers transmitting the electric vector parallel to the x and y axes, respectively, and are defined by

$$\mathbf{P}_x=\begin{pmatrix}1 & 0\\ 0 & 0\end{pmatrix}\quad\text{and}\quad\mathbf{P}_y=\begin{pmatrix}0 & 0\\ 0 & 1\end{pmatrix}. \tag{9}$$

The transmission of the filter, τ_S, is the square of the ratio of the amplitudes of the entering and emerging electric vectors, ϵ_0 and ϵ_n. The problem of deriving a useful expression for τ_S is basically the problem of raising the factors in square brackets in Eqs. (4), (5) and (6) to the indicated powers without multiplying them out explicitly, an impossibly laborious task when n is large. These factors are two by two matrices. Jones has devised an ingenious method for raising such matrices to any desired power. His method leads to a single expression for the transmission of initially unpolarized light through the fan and the even and odd folded filters as follows,

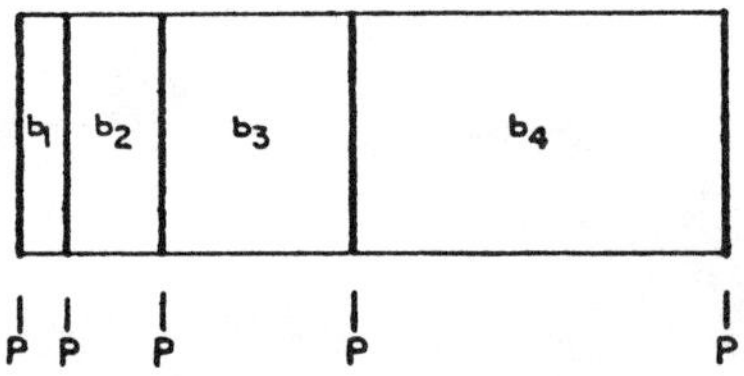

FIG. 3. Section of a Lyot filter of four birefringent elements, b_1–b_4, sandwiched between polarizers, P.

$$\tau_S=\frac{t_S}{2}\left[\frac{\sin n\chi}{\sin\chi}\cos\chi\tan\alpha\right]^2. \tag{10}$$

The parameter χ is related to the retardation, γ, by the following expression:

$$\cos\chi=\cos\gamma\cos\alpha. \tag{11}$$

The factor t_S represents absorption and reflection losses in the filter.

Examination of Eq. (10) shows that Solc was correct in his experimental finding that $\alpha=(1/n)(\pi/2)$ is an optimum condition. Adopting this value for α, we obtain the transmission curve of Fig. 2, and derive the following conclusions:

(a) At the principal transmission bands, $\tau_S=(t_S/2)$, when $\chi=\alpha$ or $\chi=\pi-\alpha$. The corresponding values of the retardation are $\gamma=k\pi$, where k is any integer.

(b) $\tau_S=0$ when $\chi=(l/n)\pi=2l\alpha$, where $l=1, 2, 3\cdots(n-1)$.

(c) Secondary maxima occur between the zeros. Their exact positions are not readily determined, although they are very near the midpoints between successive zeros. At these points, $\chi=(2l+1)\alpha$, where $l=1, 2, 3, \cdots(n-2)$. The transmission at a midpoint between zeros is, then,

$$\tau_{Sm}=\frac{t_S}{2}\left[\frac{\sin n(2l+1)\alpha}{\sin(2l+1)\alpha}\cos(2l+1)\alpha\tan\alpha\right]^2. \tag{12}$$

For purposes of comparison, the reader will find the theory of the Lyot filter and its split-element modification in papers by Lyot,[3] Evans,[4] and Dollfus.[5] To review briefly, the Lyot filter consists of a multiple sandwich of birefringent crystal layers and polarizers. The crystal layers, termed b-elements, are cut with the crystal optic axis parallel to the surfaces, perpendicular to the instrumental optical axis. In the simplest form, the thicknesses of the b-elements form a series in powers of 2. Polarizers, usually sheets of Polaroid film, are placed between successive b-elements and at each end, with their axes parallel. The arrangement is shown in Fig. 3. The b-elements are also oriented with their axes parallel, at an angle of 45° to the electric vector trans-

[3] B. Lyot, Ann. astrophys. **7(1)**, 2 (1944).
[4] J. W. Evans, J. Opt. Soc. Am. **39**, 229 (1949).
[5] A. Dollfus, Rev. opt. **35**, 625 (1956).

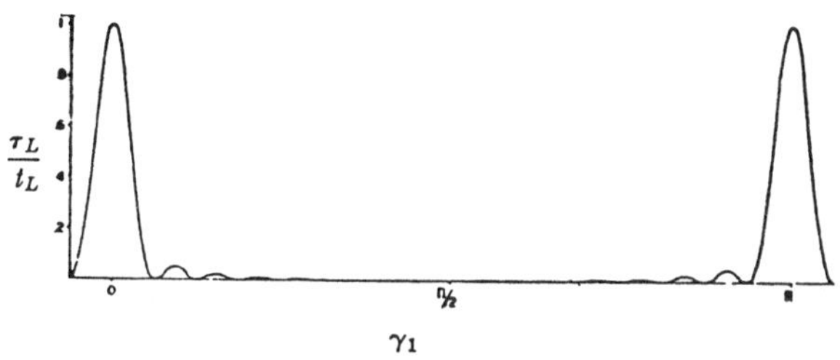

FIG. 4. Transmission curve of a Lyot filter of four elements as a function of the retardation, γ_1, of the thinnest element.

mitted by the polarizers. The transmission, τ_L, of the assembly is given by the equation,

$$\tau_L = (t_L/2)[\cos\gamma_1 \cos 2\gamma_1 \cos 4\gamma_1 \cdots \cos 2^{N-1}\gamma_1]^2, \quad (13)$$

where N = number of b-elements, γ_1 = retardation of the thinnest element $=\pi(d_1/\lambda)(\epsilon-\omega)$, d_1 = thickness of the thinnest element, and t_L = a factor representing absorption and interface reflection losses in the filter.

Lyot[3] shows that Eq. (13) can be written as follows

$$\tau_L = \frac{t_L}{2}\left[\frac{\sin 2^N\gamma_1}{2^N \sin\gamma_1}\right]^2. \quad (14)$$

He points out that this is identical to the expression for the intensity curve of a diffraction grating of 2^N rulings at fixed angles of incidence and diffraction. The curve, shown in Fig. 4, consists of widely spaced sharp principal maxima or transmission bands. Between successive bands are 2^N-1 zeros interspersed with 2^N-2 secondary maxima. The separation of successive bands is inversely proportional to γ_1, and the band width to $2^{N-1}\gamma_1$. These are the retardations in the thinnest and thickest b-elements, respectively.

Lyot filters, in both the simple and split element forms, made of quartz and calcite, less than 20 cm long, are now widely used for observation of the sun in the light of the Hα line of hydrogen and the 5303 line of Fe XIV in the corona. The band widths range from 0.5 to 5 A.

In assessing the practical value of a birefringent filter, the relevant characteristics are the transmission at the centers of the transmission bands; the band width between the first zeros on each side of the band; the transmission of parasitic light in the secondary maxima, particularly those near the primary bands; and finally, the cost of the filter. It is therefore of interest to compare the Solc and Lyot filters in these particulars. For convenience, the curves of τ_S and τ_L are superposed in Fig. 5.

A comparison of Eqs. (10) and (14) suggests that a Lyot and Solc filter are equivalent if $\gamma_S = \gamma_{1L}$ and $n = 2^N$. Then the spacing of the transmission bands in γ (or λ), and the number of zeros and secondary maxima between bands is the same for the two. We therefore compare a Solc filter composed of 2^N identical retardation plates with a Lyot filter of N b-elements, the thinnest of which is identical with one of the Solc retardation plates. The total thickness of birefringent material in the Solc filter exceeds that in the Lyot filter by the thickness of one retardation plate. This is not in accord with Solc's experimental finding that the two are equivalent when the thickness of his filter is approximately half that of a Lyot filter, an item of some economic importance.

In transmission, the Solc filter is definitely superior to the Lyot filter as usually constructed. The polarizers are usually Polaroid film, which has a transmission in the visible spectrum of about 0.75 for light polarized parallel to the transmission direction. Thus in the Solc filter $t_S=0.56$, while in the Lyot filter $t_L=(0.75)^{N+1}$. The split element form of the Lyot filter is considerably better, with $t_L=(0.75)^{(N/2)+1}$. In most practical filters, with transmission band widths between 0.5 and 5 angstroms, N ranges from 6 to 10, and t_L is between 0.14 and 0.05 in the simple Lyot filter, or 0.32 and 0.16 in the split element filter. These losses in the Lyot filter can be avoided, however, by the use of more transparent (and much more expensive) polarizers like Rochon prisms. A Lyot filter of this construction is very nearly as transparent as the equivalent Solc filter, but with the practical disadvantage of some added optical length which vignettes the field unless the aperture is enlarged.

The band widths of the two filters may be most readily compared in terms of $\Delta\gamma$, the increment in retardation of a single Solc plate (or the thinnest Lyot b-element) between the center of a transmission band and the first zero on either side. For the Lyot filter, $\Delta\gamma_L = (\pi/2N)$. For the Solc filter, $\Delta\chi = 2\alpha$. If n is large (≥ 16, say), α is small, and $\Delta\gamma_S$ approaches $(\sqrt{3}/2)(\pi/n)$. Since $n=2^N$, we conclude that the Solc filter has the sharper transmission band by a factor of $(\sqrt{3}/2)=0.87$.

The transmitted parasitic light in the secondary maxima is approximately proportional to the transmission at the midpoint between zeros. This is given in Eq. (12) for the Solc filter. For the Lyot filter,

$$\tau_{Lm} = \frac{t_L}{2}\left[\frac{\sin(2l+1)\frac{\pi}{2}}{2^N \sin\frac{\pi}{2}(2l+1)/2^N}\right]^2. \quad (15)$$

We are interested in the ratio $(\tau_{Sm}/t_S)/(\tau_{Lm}/t_L) = \tau_{Sm}'/\tau_{Lm}'$ at the midpoints between corresponding zeros in the two filters (i.e., for the same values of l). This is

$$\frac{\tau_{Sm}'}{\tau_{Lm}'} = \left[2^N \cos\left(\frac{2l+1}{2^N}\right)(\pi/2)\tan\frac{1}{2^N}\frac{\pi}{2}\right]^2. \quad (16)$$

If n is large, the secondaries near the transmission

bands (l small) approach the ratio

$$(\tau_{Sm}'/\tau_{Lm}') = (\pi/2)^2 = 2.47. \tag{17}$$

Thus the parasitic light in the immediate neighborhood of the primary transmission bands is roughly 2.5 times as great in the Solc filter as in the Lyot filter. Since the first secondaries on either side of the primary band of a Lyot filter transmit together about 0.05 as much light as the band itself, it is evident that the parasitic light of the Solc filter may be very serious in certain applications. A textbook example would be the observation of solar flares in the light of the Hα line of hydrogen. Here we must isolate the light at the center of an absorption line roughly one angstrom broad. The light intensity at the center of the line is 0.16. that of the neighboring continuum. Thus a Lyot filter with an interval of 0.5 A between the center of the band and the first zeros on either side has its first secondaries well out in the continuum, and the diluting parasitic light from these secondaries is roughly $0.05/0.16 = 0.3$ times the light in the transmission band. This reduces the contrast of the solar Hα features very appreciably. With the Solc filter the diluting light would be nearly 2.5 times as great, and only the most contrasty of the solar features would de detectable. For the observation of emission line objects against a continuous background, however, like the prominences at the solar limb, the Solc filter should function reasonably well, although it is still definitely inferior to the equivalent Lyot filter in contrast.

One device for reducing the parasitic light in the Solc filter would be the addition of a secondary suppressor plate, in the form of a single thick retardation plate with a transmission maximum coinciding with the passband, and zeros coinciding with the first secondaries. This calls for one additional polarizer. The same device applies equally to the Lyot filter, and in practice it improves the performance quite substantially. The suppressor plate does not change the relative merits of the two filters.

The off-axis performance of the Solc filter has not been investigated analytically. However, we should expect that when n is large, the folded form should have very nearly the same off-axis characteristics as the simple Lyot filter. The field characteristics of the fan filter, however, are not so readily apparent. Solc states that the two forms have the same off-axis characteristics, which is surprising. He further states that the wavelengths of the transmission bands can be shifted over a broad range by tilting the filter. This finding is correct in the same sense that it is true for a Lyot filter. As the angle of inclination increases, however, the angular field over which the wavelength of the transmission band is uniform within a given tolerance (0.1 of the band width, for instance) decreases approximately with the reciprocal of the angle of inclination.

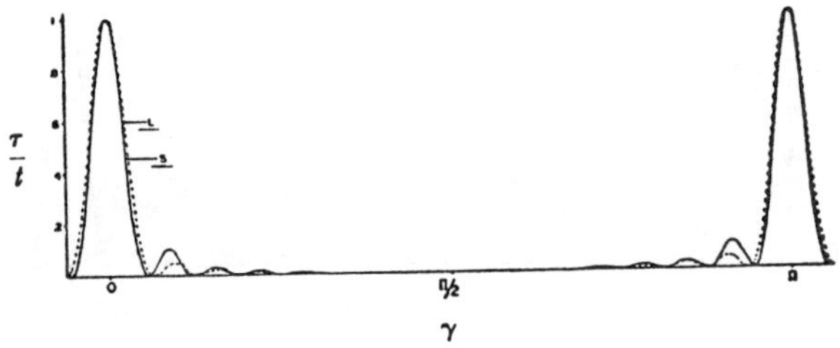

FIG. 5. Superposed transmission curves of a 16-plate Solc filter (solid curve) and a 4-element Lyot filter (dotted curve).

Thus this type of tuning would be useful for the observation of a point source at infinity, but is of little use in the observation of an extended field, or in a converging beam of light.

The useful field of the Solc filter can be greatly extended by the same devices that are applicable to the Lyot filter. However, this calls for a compound construction of each retardation plate, and multiplies the complexity of construction several fold.

A Solc filter with only two polarizers is probably more expensive than the equivalent Lyot filter for band widths greater than 3 A, and most certainly so for sharper bands. A filter with a 5-A band width calls for 60 or 70 retardation plates, and the number increases with the reciprocal of band width. Since they are identical, many such plates can be ground and polished in one operation. However, the loss in birefringent material in the sawing process would be very considerable, and the mechanical problem of mounting and cementing so many elements in their proper orientation might be expensive. If the material is quartz, the thicknesses of individual retardation plates presumably would be between 1 and 2 mm. For band widths of less than 2 A it is almost necessary to use calcite to avoid excessively thick filters. The thickness of a single plate is then of the order of 0.06 to 0.12 mm, and the problem of working this rather difficult material to the requisite accuracy in thickness and flatness (about 0.2 micron) in several hundred plates becomes really formidable.

In practice, one would avoid excessively thin plates in the Solc filter by compounding it. For example, a $\frac{1}{2}$-A filter could be made with a first stage of 25 quartz retardation plates about 1.1 mm thick and a second stage of 25 calcite plates 1.4 mm thick (the exact thicknesses depend upon the desired wavelengths of the transmission bands). An additional polarizer must, of course, be inserted between the two stages. I suspect that such a filter would be comparable in cost with an equivalent Lyot filter equipped with Rochon prism polarizers. Of the two the Lyot filter is preferable because of its superior suppression of parasitic light.

In summary, then, the Lyot filter performs better than the Solc filter except for possible applications where the parasitic light of the latter near the transmission bands is of no importance.

ELECTRICALLY TUNABLE NARROWBAND OPTICAL FILTER

Indexing terms: Liquid-crystal devices, Optical filters, Tuning

A narrowband optical filter of the Solc type is made from a twisted stack of birefringent plates between polarisers. If the plates are replaced by homogeneously aligned liquid-crystal cells whose retardation is controllable, the transmitted wavelength of the filter can be varied over the visible spectrum. The characteristics of a rudimentary device of this type agree with predictions.

An optical filter with a fixed, narrow passband can be made in several ways by combining birefringent retardation plates and polarisers. In the Solc filter structure[1] a series of M identical retardation plates is used and these are arranged with their optic axes perpendicular to the light beam (Fig. 1). The optic axis of each plate is at an angle α $(= \pi/2M)$ from the optic axis of the plate preceding it, and at each end of the stack are polarisers with their axes $\alpha/2$ and $90° - \alpha/2$ from the optic axes of the plates adjacent to them. The polariser axes are therefore parallel.

Evans[1] has analysed the action of this filter and has shown that for incident unpolarised light a maximum transmission of 0·5 is obtained in a series of narrow bands centred on wavelengths λ_T, given by

$$\lambda_T = \frac{\Delta n d}{m} \qquad (1)$$

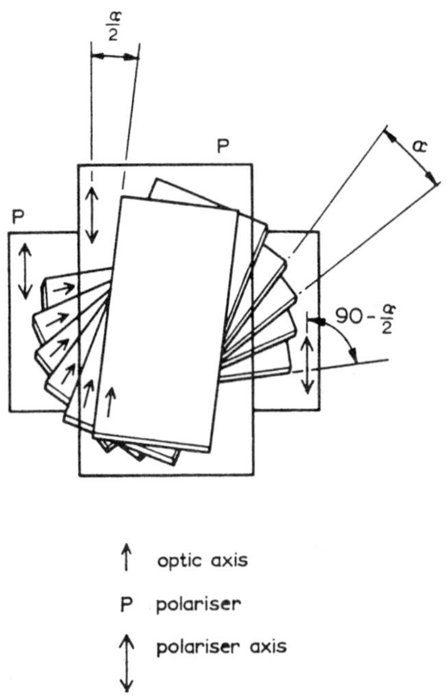

Fig. 1 *Solc filter structure viewed along light beam*

where m is the order number of the bands (0, 1, 2, ...) Δn, is the difference between the principal refractive indices of the plates (n_e and n_0) and d is their thickness. The transmission is limited to 0·5 by the action of the first polariser. The bandwidth $\Delta\lambda_T$ measured at half the maximum transmission is, for the 1st-order ($m = 1$) peak, given by

$$\frac{\Delta\lambda_T}{\lambda_T} \simeq \frac{1}{M} \qquad (2)$$

It is interesting to note that if the number of plates is made large, while keeping the total thickness of birefringent material (Md) constant, the predicted spectrum has only one peak of 0·5 transmission at $m = 0$, and is identical to the spectrum predicted for a twisted-nematic liquid-crystal device.[2]

An electrically tunable version of this filter may be envisaged if Δn can be electrically controlled. This characteristic can be obtained if each plate is replaced by a film of nematic liquid crystal of positive dielectric anisotropy between transparent electrodes, with its optic axis aligned in the required direction by suitable coatings on the elctrodes. Δn is now the birefringence of the liquid crystal, typically 0·2. When an electric field is applied between the electrodes, the molecules of the liquid crystal are realigned, and the effective birefringence of the film Δn_{eff} falls from Δn towards zero as the field is increased. Eqn. 1 shows that λ_T will therefore fall, and in this way the wavelengths transmitted by the Solc structure can be varied electrically. One of the principal transmission maxima can be moved across the visible spectrum with a suitable choise of d.

A rudimentary liquid-crystal Solc filter has been constructed to verify these predictions. It was made from six liquid-crystal cells, connected in parallel, nominally 12 μm thick, containing a eutectic mixture of cyanobiphenyl liquid crystals[3] having a nematic phase between −5 and 45°C. An obliquely evaporated film of silicon monoxide on each electrode was used to align the molecules.[4] On one surface, the angle of incidence was 60° giving an alignment of the optic axis of the liquid crystal parallel to the electrode. On the other surface, the angle of incidence was 85°, giving an alignment

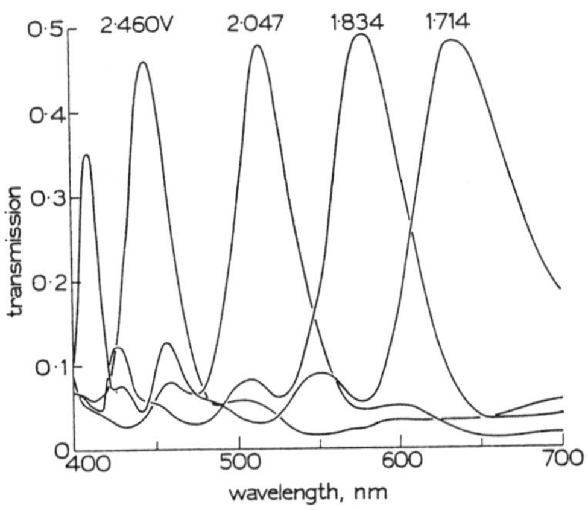

Fig. 2 *Transmission spectra of 6-cell liquid-crystal filter for four values of applied voltage*

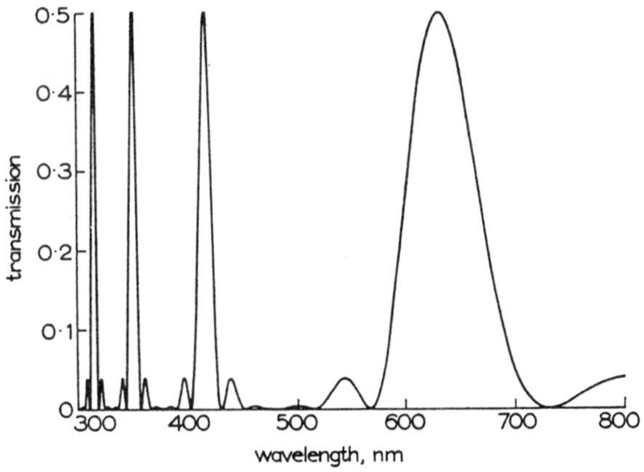

Fig. 3 *Predicted transmission spectrum 6-cell filter with* $\Delta n_{eff} = 0{\cdot}27\Delta n$

tilted slightly from the plane of the electrode. This prevented ambiguity in the direction in which the electric field realigns the molecules, so preventing the formation of disclinations which would scatter light.

Transmission spectra for this structure were obtained using a spectrophotometer.* After compensation for the losses due to absorption by the cells and polarisers,† these are shown in Fig. 2 as a function of applied voltage. As the voltage applied to the six cells was increased, a transmission peak moved from the red towards the blue end of the visible spectrum, and the maximum transmission of the peak remained close to the predicted value of 0·5. The spectrum obtained with no applied field contained many maxima and is not included. The presence of a second peak at 410 nm of order $m+1$ on the 1·714 V trace enables the order number m of the principal peak to be calculated, provided that Δn is known. Values of n_e and n_0 were therefore measured using an Abbé refractometer, and, at 25°C, were 1·712 and 1·520 at 644 nm, rising to 1·791 and 1·549 at 436 nm. Inserting the appropriate values into eqn. 1 shows that the peak traversing the visible spectrum is of first order ($m = 1$).

The refractive-index measurements also made it possible to calculate the detailed transmission spectrum of a 6-plate Solc structure. For its 1st-order peak to coincide with the 635 nm peak of the 1·714 V trace in Fig. 2, a value of $\Delta n_{eff}/\Delta n = 0{\cdot}27$ is required. This calculated spectrum is shown in Fig. 3. It predicts a value of 0·115 for $\Delta\lambda_T/\lambda_T$ of the 1st-order peak, and this compares well with the experimental value of 0·12. The subsidiary peaks in the experimental spectra are larger and more asymmetric than those predicted, but this is almost certainly because the cells used were not matched in thickness sufficiently closely to have an identical dependence of Δn_{eff} on voltage.

Several precautions must be taken to give the liquid-crystal device an optical performance comparable to that of the solid filter. The films must be parallel and matched in thickness to an accuracy of $\gtrsim d/M$. Both Δn and d will depend on temperature, and so some degree of temperature stabilisation will be necessary. Losses due to reflections at interfaces and absorption will be greater than those in the solid filter because of the greater number of interfaces created by the use of transparent electrodes, but the application of anti-reflection coatings can considerably reduce this problem. The scattering of light by thermal fluctuations in the liquid crystal will increase the transmission of the device between its principal peaks, but such an effect will not normally cause a transmission of more than a few percent. However, it has been shown that a narrowband electrically tunable filter is practicable and that its response agrees well with predictions.

H. A. TARRY *20th August 1975*

Services Electronics Research Laboratory
Baldock, Herts. SG7 6NG, England

References

1 EVANS, J. W.: 'Solc birefringent filter', *J. Opt. Soc. Am.*, 1958, **48**, pp. 142–145

2 GOOCH, C. H., and TARRY, H. A.: 'Optical characteristics of twisted nematic liquid-crystal films', *Electron. Lett.*, 1974, **10**, pp. 2–4

3 HULME, D., RAYNES, E., and HARRISON, K.: 'Eutectic mixtures of 4' substituted 4-cyanobiphenyls', *Chem. Commun.*, 1974, pp. 98–99

4 GUYON, E., PIERANSKI, P., and BOIX, M.: 'On different boundary conditions of nematic films deposited on obliquely evaporated plates', *Appl. & Eng. Sci.*, 1973, **1**, pp. 19–24

* Beckman DK–2A

† Polaroid HN32

©1988 Optical Society of America. Reprinted from *Journal of the Optical Society of America A*, Vol. 5(9), 1565–1570 (September 1988).

Fabry–Perot/Šolc filter with distributed Bragg reflectors: a narrow-band electro-optically tunable spectral filter

R. S. Weis

Department of Electrical Engineering, U.S. Naval Academy, Annapolis, Maryland 21402

T. K. Gaylord

School of Electrical Engineering, Georgia Institute of Technology, Atlanta, Georgia 30332

Received December 16, 1987; accepted February 26, 1988

A new narrow-band (typically 0.01-nm) spectral filter called the Fabry–Perot/Šolc filter, which combines the characteristics of the Fabry–Perot étalon and the folded Šolc filter, was introduced recently [J. Opt. Soc. Am. A **4**, 1720 (1987)]. The design of the Fabry–Perot/Šolc filter is extended to include narrow-band end reflectors to produce high-finesse spectral filters. The narrow-band end reflectors can be implemented as distributed Bragg reflectors (DBR's), thus permitting tailoring of the spectral characteristics of the Fabry–Perot/Šolc filter. Sample designs are presented, including one filter with an effective finesse of 5420. Possible uses for the Fabry–Perot/Šolc filter with DBR's are discussed, and it is compared with the narrow-band electro-optically tunable notch filter proposed by Lytel and Lipscomb [Appl. Opt. **25**, 3889 (1986)].

1. INTRODUCTION

Narrow-band optical spectral filters are useful in a variety of applications including spectroscopy, astronomy,[1] and optical communications.[2] For example, one of the proposed solutions for the ever-increasing need for high-volume, high-speed communications is the wavelength-division multiplexing of many signals over single-mode optical-fiber networks. This in turn has created a requirement for narrow-band spectral filters for optical multiplexing and demultiplexing. These spectral filters can be realized by using isotropic multilayered interference filters.[2,3] The spectral-filtering characteristics of multilayered interference filters can be enhanced greatly with the inclusion of anisotropic layers, as demonstrated by the anisotropic multilayered Fabry–Perot/Šolc filter.

A new narrow-band (typically 0.01-nm) spectral filter called the Fabry–Perot/Šolc filter, which combines the characteristics of the Fabry–Perot étalon and the folded Šolc filter, was introduced recently.[4] Several versions of the Fabry–Perot/Šolc filter, each composed of different materials, were analyzed.[4] However, all the previously analyzed versions included end reflectors that were assumed to have a constant high reflectance over a region much broader than the total tuning range (TTR) of the Fabry–Perot/Šolc filter.

In this paper the design of the Fabry–Perot/Šolc filter is extended to include end reflectors whose high reflectance regions are not constant over the TTR of the filter. It is shown here that the narrow-band end reflectors can be used as distributed Bragg reflectors (DBR's) to design high-finesse spectral filters.

In Section 2 the concepts of the Fabry–Perot/Šolc filter and the DBR are reviewed briefly. In Section 3 a few representative designs of Fabry–Perot/Šolc filters with DBR's are presented. In Section 4 possible uses for the Fabry–Perot/Šolc filter with DBR's are discussed, and the Fabry–Perot/Šolc filter with DBR's is compared with the narrow-band electro-optically tunable notch filter proposed by Lytel and Lipscomb.[5]

2. CONCEPTS OF OPERATION

A. Fabry–Perot/Šolc Filter

The Fabry–Perot/Šolc filter is a hybrid device that is designed so that the center of the passband that is due to the multiple-pass folded Šolc filter coincides with the center of the narrow passband for the Fabry–Perot cavity formed between two high-reflectance mirrors. The Fabry–Perot/Šolc filter is configured by satisfying all the following conditions at the center wavelength of the passband: (1) the first layer is a birefringent quarter-wave plate; (2) the second layer is a birefringent half-wave plate; (3) the third layer is a birefringent quarter-wave plate; and (4) the fourth layer is optically isotropic (for normal incidence), and its thickness is chosen such that the optical path length of the device corresponding to the input polarization is equal to an integer number of free-space half-wavelengths.

Thus the Fabry–Perot/Šolc filter consists of four plates and two end reflectors. The first three plates are oriented so that they form the multiple-pass Šolc filter. The fourth layer is oriented so that it is optically isotropic at normal incidence; it is used to make the optical path length for the input polarization equal to an integer number of free-space half-wavelengths (for the center wavelength of the pass band) after the three wave-plate conditions have been satisfield. The amplitudes of the Fabry–Perot peaks are determined by the multiple-pass folded Šolc filter passband and the stopband of the end reflectors.

Interference filters are classified typically by the following characteristics of their transmittance spectra: the minimum and maximum transmittances, the minimum and maximum rejections, the free-space center wavelength, the full width at half-maximum (FWHM) of the passband, and the spectral distance between successive peaks [the free spectral range (FSR)]. These characteristics of the Fabry–Perot/Šolc filter can be tailored for specific applications as described below.

Consider, as an example of the tailoring of filter characteristics for a specific application, the design of a filter with an extremely large effective finesse. The finesse of an interference filter ($\mathcal{F}$) is defined as $\mathcal{F}$ = FSR/FWHM, where FSR is defined as the spectral separation between adjacent Fabry–Perot peaks. For an isotropic interference filter, the FSR is determined by the optical thickness of the filter. The effective finesse ($\mathcal{F}_{\text{eff}}$) is defined similarly as $\mathcal{F}_{\text{eff}} = \text{FSR}_{\text{eff}}/\text{FWHM}$, where FSR_{eff} is the spectral separation between adjacent nonzero peaks when these neighboring peaks do not correspond to adjacent Fabry–Perot maxima. Thus FSR_{eff} is not determined simply by the optical thickness of the filter. In order for a Fabry–Perot/Šolc filter with an extremely large effective finesse to be obtained, the FSR must be greater than half of the width of the multiple-pass Šolc filter envelope. With this condition satisfied, the Fabry–Perot peaks adjacent to the center peak are outside the Šolc envelope; thus the effective finesse of this filter is determined by the spacing between the Šolc envelope peaks, not by the spacing between Fabry–Perot peaks. For constant mirror reflectance and flatness, the filter's FWHM could be narrowed by increasing its total length. For tunable versions of the Fabry–Perot/Šolc filter, effective finesses of the order of 1000 are possible,[4] whereas, for nontunable versions of the Fabry–Perot/Šolc filter, effective finesses greater than 5000 are possible.

The angle that the optic axis should be tilted from the vertical axis in the plane of the layer surface to obtain the maximum transmitted power at the center wavelength of the passband is determined by the end-mirror reflectance at the center wavelength. It was calculated by iteration. With end-mirror amplitude reflectances equal to 0.98 the optimum center-element tilt angle (α) is α = 0.592 deg. The rigorously determined value of α and the folded Šolc filter relation $\alpha = \pi/2N$, where N is the number of plates, indicate that the filter acts as a 152-plate folded Šolc filter within a Fabry–Perot cavity; however, the filter also acts as nonoptimal . . . , 144-, 148-, 156-, 160-, . . .-plate folded Šolc filters, and therefore the envelope of the Fabry–Perot/Šolc filter's transmittance spectrum does not exhibit the nulls expected of a 152-element folded Šolc filter.

A linearly polarized incident wave whose wavelength is equal to the center wavelength produces a circularly polarized wave between the exit mirror and the exit polarizer. Conversely, a circularly polarized incident wave whose wavelength is equal to the center wavelength produces a linearly polarized wave between the exit mirror and the exit polarizer. Thus the throughput of the filter at the center wavelength can be increased by a factor of 2 (to 50%) either by converting the linearly polarized light emerging from the entrance polarizer to circularly polarized light or by converting the circularly polarized light emerging from the exit mirror to linearly polarized light. This can be accomplished by adding an antireflection-coated quarter-wave plate between either polarizer and its corresponding mirror. The throughput of the filter can be increased to 100% at the center wavelength by placing interferometrically flat quarter-wave plates between both polarizers and their corresponding mirrors and suitably adjusting the orientation of the exit polarizer. However, since the relative orientation of the polarizers is no longer orthogonal, the rejection of the filter is decreased. The throughput of a single birefringent plate between crossed polarizers is 100% when the birefringent plate is a half-wave plate because both the input wave and the output wave are linearly polarized. Thus another conceivable means of increasing the multilayered filter's throughput would be the substitution of Faraday rotators of the two birefringent quarter-wave plates.

The Fabry–Perot/Šolc filter can be tuned electro-optically by using longitudinal transparent electrodes. The tuning is characterized by the fundamental tuning range (FTR) and the TTR. The FTR, as depicted in Fig. 1, is the spectral range around each Fabry–Perot peak through which the filter may be tuned, provided that the required electric field is less than 10^7 V/m (the value above which breakdown effects start to occur). The FTR is determined by the material properties (piezoelectric and electro-optic) and thickness of the spacer plate. Similarly, the TTR (the spectral range through which the envelope can be turned), as depicted in Fig. 2, is determined by the material parameters and thicknesses of the three birefringent plates.

B. Distributed Bragg Reflector

A DBR is a structure whose refractive index varies periodically with thickness. The direction of periodicity is perpendicular to the surfaces bounding the structure. The reflector is designed so that an optical wave with a given free-space wavelength normally incident upon the bounding surface is diffracted in a direction antiparallel to the incident wave. The DBR is a type of planar pure-reflection grating. A thin-film structure with a sinusoidal refractive-index profile is referred to sometimes as a rugate filter.[6] A review of the analysis and applications of optical diffraction by periodic structures (gratings) in isotropic materials was presented in Ref. 7.

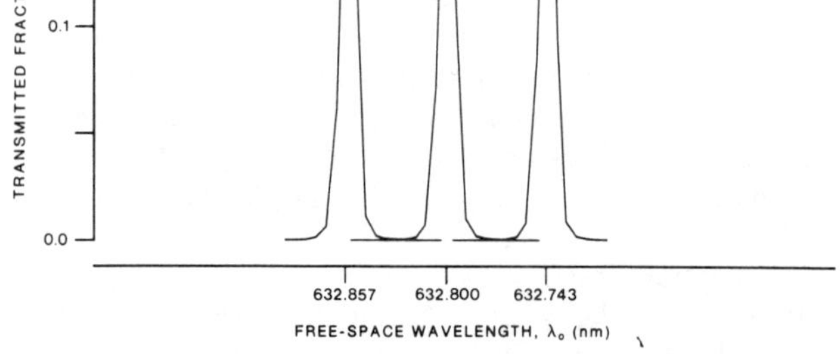

Fig. 1. Detailed view of the central FTR of the all-$LiNbO_3$ Fabry–Perot/Šolc filter with broadband end reflectors. The magnitudes of the electric field in the spacer plate (E_4) are shown above each corresponding peak.

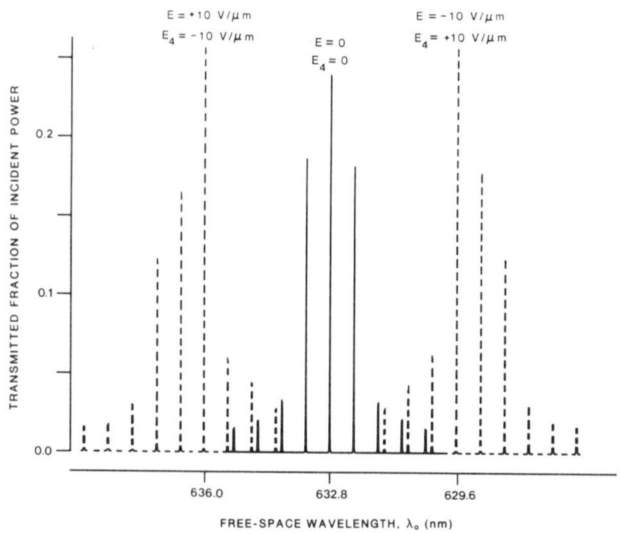

Fig. 2. Transmitted power as a function of the phase retardance (γ) of the center wave plate for all-$LiNbO_3$ Fabry–Perot/Šolc filter with broadband end reflectors over its TTR. The spectral separation between two adjacent peaks is defined as the FSR. The required electric field strength magnitudes (E and E_4) are noted at the top of each set of peaks.

Pure-reflection gratings are not strictly periodic, since they have only a finite number of periods. Frequently the gratings are composed of a large number of layers, but in order for a theory based on the Floquet theorem to be employed rigorously, there must be an infinite number of layers. Isotropic pure-reflection gratings can be analyzed rigorously by using a chain-matrix formulation,[8,9] a modal formulation,[10] or a characteristic-matrix formulation.[11] The matrix-based formulations can be applied to pure-reflection gratings with continuous refractive-index profiles (refractive index as a function of thickness) by modeling the grating as a large stack of thin homogeneous layers.[12] By making the layers sufficiently thin the analysis can be made arbitrarily accurate.[8] The calculation method used in this paper[4] can be used for the rigorous analysis of anisotropic periodic structures with a finite number of periods.

3. FABRY-PEROT/ŠOLC FILTERS WITH NARROW-BAND END REFLECTORS

A. End-Reflector Characteristics

All the versions of the Fabry–Perot/Šolc filter presented in this paper are the same as the all-lithium-niobate Fabry–Perot/Šolc filter presented in Ref. 4 except that the versions introduced in this paper have narrow-band end reflectors. In this context, a narrow-band reflector is one in which the spectral width of the high-reflectance region is either equal to or less than the TTR of the Fabry–Perot/Šolc filter. The largest TTR presented in this section is 6.1 nm.

Tailoring the spectral width of the end reflectors is another means of adjusting the amplitudes of the adjacent Fabry–Perot peaks, which in turn makes it possible to alter the minimum and maximum rejections of the filter. The narrow-band end reflectors used are DBR's. The spectral width of the high-reflectance region of a DBR is defined as the DBR's stopband.

For a large number of periods and a small refractive-index modulation, the transmission and reflection characteristics of an alternating high- and low-refractive-index quarter-wave multilayered stack and a DBR with alternating high- and low-refractive-index layers are nearly identical.[13] Thus the well-known relations for the reflectance characteristics of multilayered quarter-wave stacks can be used directly to calculate the reflectance characteristics of the DBR's. For a quarter-wave stack with alternating high- and low-refractive-index layers, the spectral width of the stopband is determined by the ratio (n_H/n_L), where n_H is the high refractive index and n_L is the low refractive index in the multilayered stack and the reflectance within the stopband is determined by the number of layers in the stack.[11] The refractive-index modulation required to achieve a desired stopband width for a DBR can be derived from the following formula for the fractional stopband width of an alternating multilayered quarter-wave stack[11]:

$$\Delta\left(\frac{\lambda_{oc}}{\lambda_o}\right) = \left(\frac{2}{\pi}\right)\arcsin\left(\frac{n_H - n_L}{n_H + n_L}\right), \tag{1}$$

where λ_{oc} is the free-space center wavelength of the stopband and λ_o is the nearest free-space wavelength less than λ_{oc} that corresponds to a reflectance that is one-half of the reflectance at λ_{oc}. The spatially weighted refractive-index average (n_0) is $n_0 = \bar{n} = (n_H + n_L)/2$. With Δn defined as $\Delta n = (n_H - n_L)/2$, Eq. (1) can be reexpressed as

$$\frac{\Delta n}{n} = \frac{\pi}{2}\sin\left(\frac{\lambda_{oc}}{\lambda_o} - 1\right). \tag{2}$$

The quantity $\Delta n/\bar{n}$ is defined as the refractive-index modulation of the DBR.

The thicknesses of the high- and low-refractive-index layers (t_H and t_L, respectively) are equal and are given by the first Bragg condition for normal incidence:

$$t_H = t_L = \frac{\lambda_{oc}}{4n_0}. \tag{3}$$

The number of layers required for a given stopband reflectance to be achieved was determined by calculating the reflectance for the periodic structure as a function of the number of periods. The stopband power reflectance for all the versions of the Fabry–Perot/Šolc filter presented in this paper is 0.96.

Diffraction gratings with the required number of layers and refractive-index modulation for distributed Bragg reflectors were fabricated previously by using ultraviolet beams intersecting in organic crystals.[5] For the versions of the Fabry–Perot/Šolc filter presented in this section the organic crystal 2-methyl,4-nitroaniline (MNA) is used for the end reflectors. The refractive index of MNA is 2.0.[5]

B. Tunable $LiNbO_3$:$LiNbO_3$ Fabry-Perot/Šolc Filter with 6-nm Stopband Distributed Bragg Reflectors as End Reflectors

The required refractive-index modulation for a 6-nm-wide stopband, calculated by using Eq. (2), is 0.0048. Thus, for $\bar{n} = 2.0$ (MNA), $n_H = 2.01$ and $n_L = 1.99$. The required layer thicknesses, calculated by using Eq. (3), are $t_H = t_L = 79.1$ nm. The required number of periods is 230. The power reflectance as a function of free-space wavelength for a wave

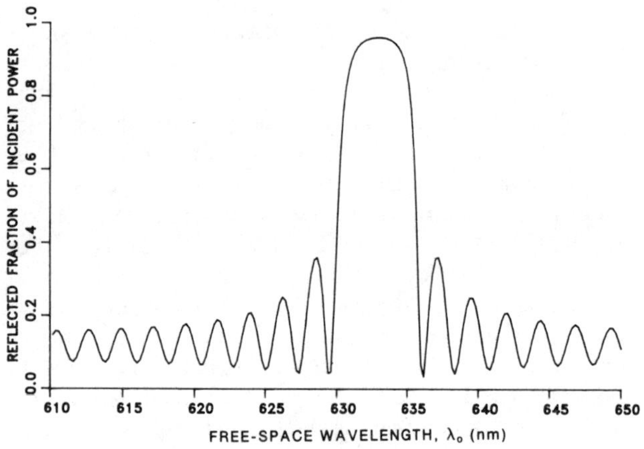

Fig. 3. Spectral reflectance characteristics of a 271-period, 6-nm stopband DBR centered at λ_o = 632.8 nm.

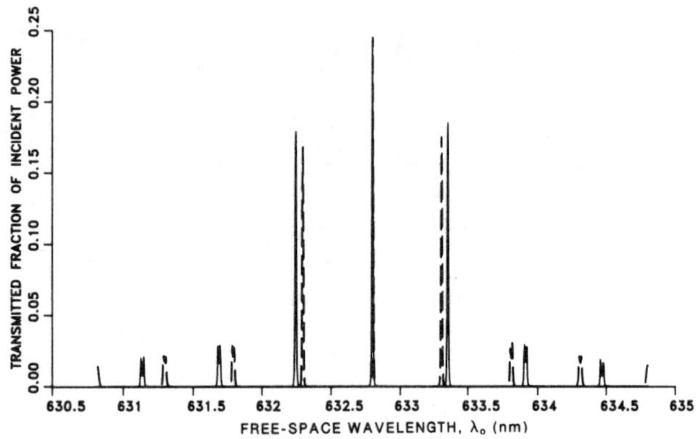

Fig. 4. Comparison of the transmittance spectra of the all-$LiNbO_3$ Fabry–Perot/Šolc filter with infinite stopband reflectors (solid line) with those of the all-$LiNbO_3$ Fabry–Perot/Šolc filter with 6-nm stopband end reflectors (dashed line).

incident from a region with n = 2.24 (average refractive index for $LiNbO_3$) of this DBR is presented in Fig. 3.

Since the TTR (and hence the spectral region of interest) of the same Fabry–Perot/Šolc filter with broadband end reflectors is equal to the stopband width for this version (6mm), the two transmittance spectra over this spectral region are approximately the same, as shown in Fig. 4. The Fabry–Perot peaks for the version with the narrow-band end reflectors are shifted slightly closer to the center wavelength because the effective length of the Fabry–Perot cavity formed between the two end reflectors is slightly greater in the narrow-band-reflector version than in the broadband-reflector version.

C. Tunable $LiNbO_3$:$LiNbO_3$ Fabry-Perot/Šolc Filter with 1-nm Stopband Distributed Bragg Reflectors as End Reflectors

The required refractive-index modulation for a 1-nm-wide stopband, calculated by using Eq. (2), is 0.00124. Thus, for $\bar{n}$ = 2.0 (MNA), n_H = 2.0025 and n_L = 1.9975. The required layer thicknesses, calculated by using Eq. (3), are $t_H = t_L$ = 79.1 nm. The required number of periods is 1081. The power reflectance as a function of free-space wavelength for a wave incident from a region with n = 2.24 (average refractive index for $LiNbO_3$) of this DBR is presented in Fig. 5.

The improved rejection of this version compared with the Fabry–Perot/Šolc filter with 6-nm stopband-width DBR's as end reflectors is evident in Fig. 6. The amplitudes of the Fabry–Perot peaks adjacent to the central peak are reduced from 0.16 to 0.11.

D. Nontunable $LiNbO_3$:$LiNbO_3$ Fabry-Perot/Šolc Filter with 1-nm Stopband Distributed Bragg Reflectors as End Reflectors

Except for the thickness of the spacer plate, the design parameters of the filter presented here are identical to those presented in the Subsection 3.C. As discussed above (in Subsection 3.A), a Fabry–Perot/Šolc filter can exhibit an extremely large effective finesse, $\mathcal{F}_{\text{eff}}$, if the Fabry–Perot peaks adjacent to the center wavelength lie outside the envelope formed by the spectral transmittance characteristics of the multiple-pass Šolc filter and the end reflectors. The Hg_2Cl_2:$LiNbO_3$ Fabry–Perot/Šolc filter presented in Ref. 4 shifted the adjacent Fabry–Perot peaks away from the central peak, by using a material with a large natural birefringence (Hg_2Cl_2) for the first three layers, to reduce the total optical path length of the structure. Another filter design that has an even larger effective finesse is presented here.

Reducing the spacer-plate thickness to a minimum (and thereby reducing the tunability within the FTR to a minimum) maximizes the separation between adjacent Fabry–

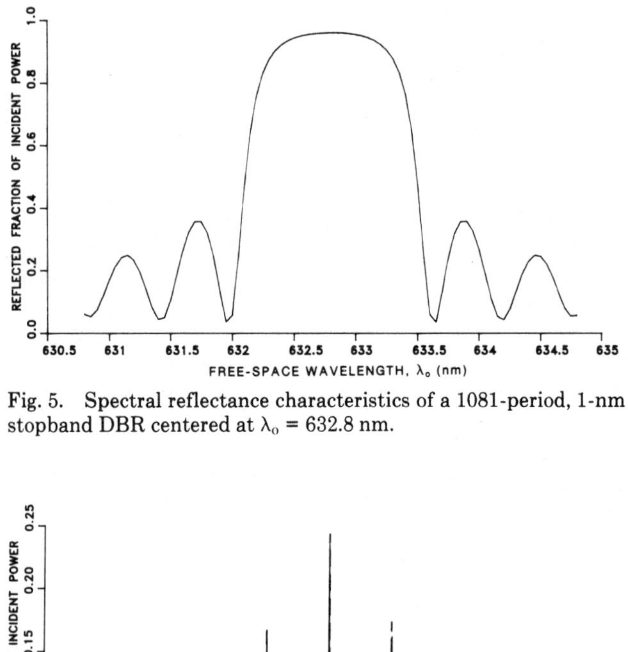

Fig. 5. Spectral reflectance characteristics of a 1081-period, 1-nm stopband DBR centered at λ_o = 632.8 nm.

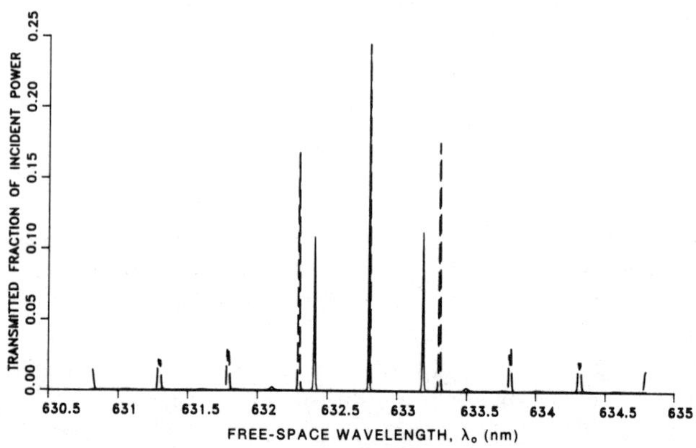

Fig. 6. Comparison of the transmittance spectra of the all-$LiNbO_3$ Fabry–Perot/Šolc filter with 6-nm-wide stopband DBR end reflectors (dashed line) and the all-$LiNbO_3$ Fabry–Perot/Šolc filter with 1-nm wide stopband DBR end reflectors (solid line).

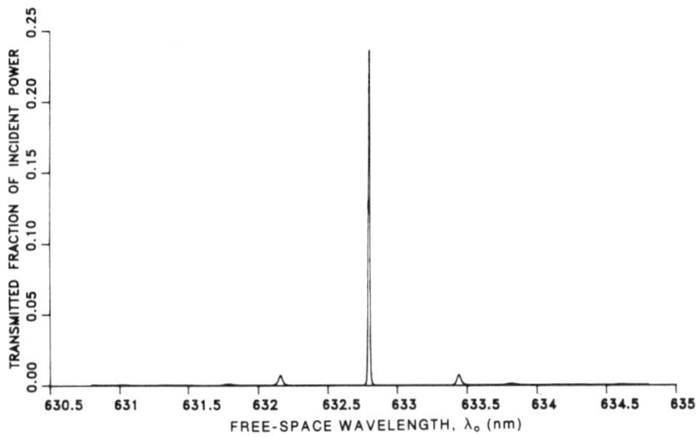

Fig. 7. Transmittance spectrum of an all-$LiNbO_3$ nontunable Fabry–Perot/Šolc filter with 1-nm-wide stopband narrow-band end reflectors.

Perot peaks. The separation between adjacent Fabry–Perot peaks for the filter with minimum spacer-plate thickness is 0.6 nm. Since the stopband of the DBR end reflectors extends 0.5 nm on both sides of λ_{oc}, the Fabry–Perot peaks immediately adjacent to the center wavelength lie outside the envelope formed by the spectral transmittance characteristics of the multiple-pass Šolc filter and the end reflectors, and thus their amplitudes are greatly reduced, as shown in Fig. 7.

The nearest wavelength at which another peak occurs is that at which a folded Šolc filter envelope and a high-reflectance region overlap. The folded Šolc filter envelopes are centered at free-space wavelengths that satisfy the condition

$$\lambda_o = \frac{m'}{m} \lambda_{oc}, \tag{4}$$

where m is the arbitrary odd integer chosen in the design of the folded Šolc filter (see App. B of Ref. 4) and m' is an odd integer. The DBR's high-reflectance regions are centered at free-space wavelengths that satisfy the condition

$$\lambda_o = \frac{1}{m''} \lambda_{oc}, \tag{5}$$

where m'' is an odd integer less than m. For this particular version $m = 7$, and the nearest Fabry–Perot peak with a significant amplitude is centered at $\lambda_o = (1/7)\lambda_{oc} = 90.4$ nm. The FSR_{eff} is 542 nm, and the FWHM = 0.1 nm; thus $\mathcal{F}_{eff} = 5420$. Although dispersion and other wavelength-dependent properties were neglected in this calculation, it is nevertheless important to realize that this filter's FWHM = 0.1 nm, while its blocking range extends from the ultraviolet to the far infrared.

4. DISCUSSION AND SUMMARY

Although all the versions of the Fabry–Perot/Šolc filter presented here were tuned electro-optically by using longitudinal transparent electrodes, the Fabry–Perot/Šolc filter also could be implemented by using transverse electrodes similar to those used by Pinnow *et al.*[14] and others[15–18] in the construction of their electro-optically tunable filter. The longitudinal version of the Fabry–Perot/Šolc filter could be fabricated by using liquid-phase epitaxy. A longitudinal electro-optically tunable filter for the infrared portion of the spectrum, composed of alternating layers of low- and high-resistivity silver thiogallate ($AgGaS_2$), was constructed recently.[19] The layers were grown sequentially by liquid-phase epitaxy using a vertical dipping technique.

A single-mode-optical-fiber analog of the Šolc filter was proposed by Johnson[20] and fabricated by Yen and Ulrich,[21] who replaced the birefringent crystal plates with sections of a continuous single-mode birefringent fiber. The required birefringence was induced in the fiber by bending it.[22] Sorin *et al.*[23] recently demonstrated a tunable single-mode-fiber DBR made by inserting a fan-shaped grating into the evanescent field of a side-polished fiber. Thus a single-mode-fiber Fabry–Perot/Šolc filter could be fabricated as sketched in Fig. 8. Since there are many different ways to induce birefringence in a single-mode optical fiber,[24] many other fiber implementations of the Fabry–Perot/Šolc filter are possible.

The Fabry–Perot/Šolc filter with narrow-band end reflectors is similar to the narrow-band electro-optically tunable notch filter of Lytel and Lipscomb.[5] The Lytel–Lipscomb filter is a Fabry–Perot étalon formed by placing an electro-optic material between two DBR's. It is tuned by electro-optically changing the optical path length of the étalon. The principal limitation of the Lytel–Lipscomb design is that the light incident upon the filter must be prefiltered to

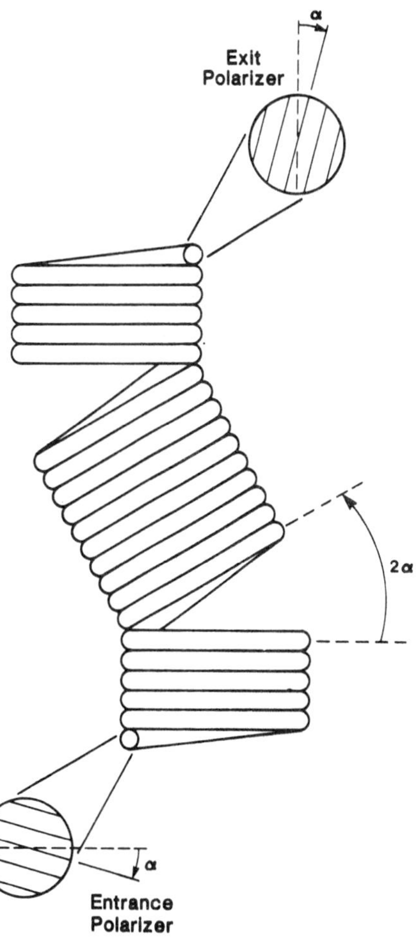

Fig. 8. Single-mode-fiber implementation of the Fabry–Perot/Šolc filter. The center-element tilt angle (α) is exaggerated for illustration purposes. The end reflectors are not shown.

be within the stopband of the DBR's. By contrast, the Fabry–Perot/Šolc filter with DBR's has an extremely large blocking range and a FWHM equal to that of the Lytel–Lipscomb filter. However, the maximum throughput of the Fabry–Perot/Šolc filter (50%) is less than that of the Lytel–Lipscomb filter (100%).

In summary, the design of the Fabry–Perot/Šolc filter introduced in Ref. 4 is extended in this paper to include end reflectors whose high-reflectance regions are not constant over the TTR of the filter in order to produce high-finesse spectral filters. The narrow-band end reflectors may be used as DBR's and can be used to tailor the spectral characteristics of the Fabry–Perot/Šolc filter. Example designs are presented, including one filter with an effective finesse of 5420. Possible uses for the filter are discussed, and the Fabry–Perot/Šolc filter with DBR's is compared with the narrow-band electro-optically tunable notch filter proposed by Lytel and Lipscomb.[5]

ACKNOWLEDGMENTS

This research was supported by a National Science Foundation Fellowship (to R. S. Weis) and by the Joint Services Electronics Program under contract no. DAAG29-84-K-0024.

REFERENCES

1. C. H. Burton, A. J. Leistner, and D. M. Rust, "Electrooptic Fabry–Perot filter: development for the study of solar oscillations," Appl. Opt. **26,** 2637–2642 (1987).
2. A. Fornier, R. Richier, and E. Pelletier, "Realization of Fabry–Perot filters for wavelength demultiplexing," in *Thin Film Technologies II*, J. R. Jacobsson, ed., Proc. Soc. Photo-Opt. Instrum. Eng. **652,** 27–32 (1986).
3. J. Minowa and Y. Fujii, "Dielectric multilayer thin-film filters for WDM transmission systems," IEEE J. Lightwave Technol. **LT-1,** 116–121 (1983).
4. R. S. Weis and T. K. Gaylord, "Electromagnetic transmission and reflection characteristics of anisotropic multilayered structures," J. Opt. Soc. Am. A **4,** 1720–1740 (1987).
5. R. Lytel and G. F. Lipscomb, "Narrowband electrooptic tunable notch filter," Appl. Opt. **25,** 3889–3895 (1986).
6. P. Baumeister, "Simulation of a rugate filter via a stepped-index dielectric multilayer," Appl. Opt. **25,** 2644–2645 (1986).
7. T. K. Gaylord and M. G. Moharam, "Analysis and applications of optical diffraction by gratings," Proc. IEEE **73,** 894–937 (1985).
8. M. G. Moharam and T. K. Gaylord, "Chain-matrix analysis of arbitrary-thickness dielectric reflection gratings," J. Opt. Soc. Am. **72,** 187–190 (1982).
9. M. G. Moharam and T. K. Gaylord, "Comments on analyses of reflection gratings," J. Opt. Soc. Am. **73,** 399–401 (1983).
10. R. S. Chu and J. A. Kong, "Modal theory of spatially periodic media," IEEE Trans. Microwave Theory Tech. **MTT-25,** 18–24 (1977).
11. H. A. Macleod, *Thin-Film Optical Filters* (Macmillan, New York, 1986).
12. W. W. Rigrod, "Diffraction efficiency of nonsinusoidal Bragg reflection gratings," J. Opt. Soc. Am. **64,** 97–99 (1974); erratum, J. Opt. Soc. Am. **64,** 895 (1974).
13. R. S. Weis and T. K. Gaylord, "Periodic alternating-layer reflective structures: relationship between quarter-wavelength stacks and Bragg pure reflection gratings," submitted to Appl. Opt.
14. D. A. Pinnow, R. L. Abrams, J. F. Lotspeich, D. M. Henderson, T. K. Plant, R. R. Stephens, and C. M. Walker, "An electro-optic tunable filter," Appl. Phys. Lett. **34,** 391–393 (1979).
15. J. F. Lotspeich, R. R. Stephens, and D. M. Henderson, "Electrooptic tunable filters for infrared wavelengths," IEEE J. Quantum Electron. **QE-18,** 1253–1258 (1982).
16. D. M. Henderson, J. F. Lotspeich, and R. R. Stephens, "Programmable electro-optic tunable filters," in *Active Optical Devices*, J. Tracy, ed., Proc. Soc. Photo-Opt. Instrum. Eng. **202,** 16–20 (1979).
17. J. F. Lotspeich, R. R. Stephens, and D. M. Henderson, "Electro-optic tunable filter," Opt. Eng. **20,** 830–836 (1981).
18. J. F. Lotspeich, R. R. Stephens, and D. M. Henderson, "Harmonic response of a tunable Šolc filter," Appl. Opt. **22,** 2743–2746 (1983).
19. S. R. Sashital, R. R. Stephens, and J. F. Lotspeich, "A multilayer $AgGaS_2$ structure for the infrared (2–10 μm) electro-optic tunable filters: fabrication and performance," J. Appl. Phys. **59,** 757–760 (1986).
20. M. Johnson, "Single-mode-fiber birefringent filters," Opt. Lett. **5,** 142–144 (1980).
21. Y. Yen and R. Ulrich, "Birefringent optical filters in single-mode fiber," Opt. Lett. **6,** 278–280 (1981).
22. R. Ulrich, S. C. Rashleigh, and W. Eickhoff, "Bending-induced birefringence in single-mode fibers," Opt. Lett. **5,** 273–275 (1980).
23. W. V. Sorin, P. Zorabedian, and S. A. Newton, "Tunable single-mode fiber reflective grating filter," IEEE J. Lightwave Technol. **LT-5,** 1199–1202 (1987).
24. S. C. Rashleigh, "Origins and control of polarization effects in single-mode fibers," IEEE J. Lightwave Technol. **LT-1,** 312–331 (1983).

©1989 Optical Society of America. Reprinted from *Applied Optics*, Vol. 28(1), 48–52 (January 1989).

Design of a liquid crystal based tunable electrooptic filter

Shin-Tson Wu

Procedures for designing a tunable liquid crystal (LC) electrooptic filter are illustrated. Selective filtering of one of the two mixed laser lines is used as an example for calculation. It is demonstrated that a LC filter can select a desired beam to pass while rejecting the other one, or vice versa, by switching the voltage biased on the LC cell. The switching times are determined by the liquid crystal properties, thickness, and biased voltage. A multilayered structure is proposed to improve the response times and power loading capability.

I. Introduction

An optical filter is a useful device for selecting a desired wavelength from a broadband light source or a multiple-line laser source. For example, a narrow bandpass filter allows a specific wavelength to pass and meanwhile rejects (by absorption or reflection) all the others. Normally, this narrow bandpass filter is designed to match a specific wavelength and is not electronically tunable. Thus, to select some desired wavelengths (one at a time) from a multiline laser (such as Ar^+ and He–Ne lasers) or from a fiber delivering more than one laser wavelength (e.g., 1.3- and 1.5-μm laser beams multiplexed in a fiber), several filters are needed.

Wavelength tunable electrooptic filters employing birefringent crystalline[1] or nematic liquid crystal (LC) materials[2,3] have been realized some time ago. A LC-based electrooptic filter would exhibit many attractive features: (1) Its applicable spectral range is extraordinarily broad. Many nematic LCs show very large birefringence and good transparency in both near-UV, visible, and IR regions (except in the vicinities of some resonant electronic and molecular vibrational absorption bands).[4] (2) Its wavelength tunability is determined by the controlled birefringence of the LC used. Usually, a large birefringence change (~0.2) can be achieved by a relatively small voltage (~ a few volts, depending on the LC used).[5] (3) No deviation in the outgoing beams takes place. (4) The LC thickness requirement for a filter is rather flexible. (5) The temperature effect on the filter's performance can be compensated by the applied voltage. (6) A large area device is possible. (7) The device fabrication is simple and cost is low.

A detailed analysis of the performance of a tunable LC electrooptic filter is described in Sec. II. An example of practical design is also given. The electrooptic response of a LC filter is analyzed in Sec. III. Also included are the confirming experimental results.

II. Theoretical Analysis

The operation principle of a tunable LC filter is based on the phase retardation experienced by each linearly polarized light while traversing through a birefringent LC plate. These phase retardations are then analyzed by an analyzer resulting in amplitude modulation of each beam. A small voltage is applied to the LC cell to control the phase retardation of each beam so that the filter transmits only one beam with high efficiency and, in the mean time, reflects completely the others. The optical transmittance of a parallel-aligned LC cell under crossed polarizers (the director's axis of the LC is at 45° to the axis of the polarizer) can be described as[6]

$$T_j = \sin^2(\delta_j/2), \tag{1a}$$

where

$$\delta_j = 2\pi d \Delta n_j/\lambda_j \tag{1b}$$

is the phase retardation, d is the LC layer thickness, and Δn_j is the birefringence of the jth beam. The birefringence of a LC depends on wavelength, temperature T, and voltage V as

$$\Delta n_j(T,\lambda_j,V) = \Delta n_0(T,\lambda_j)f(V), \tag{2}$$

where $\Delta n_0(T,\lambda_j)$ represents the birefringence of a LC at temperature T, wavelength λ_j, and $V = 0$, and $f(V)$ is

The author is with Hughes Research Laboratories, 3011 Malibu Canyon Road, Malibu, California 90265.

Received 18 April 1988.
0003-6935/89/010045-05$02.00/0.

the functional form of the voltage dependent portion of birefringence. In general, $f(V)$ is a nonlinear function of the applied voltage. However, at a voltage which is not too far above the threshold voltage (V_{th}), $f(V)$ is a linear function of V: $f(V) = a - bV$.[7,8] For example, $(a,b) = (1.651,0.635)$ and $(2.044,0.889)$ for BDH-E7 and ZLI-1132 LCs at room temperature, and $1 \lesssim V \lesssim 2$ V, respectively.[8] Beyond this region, $f(V)$ becomes nonlinear, and at $V \gg V_{th}$, $f(V)$ is found to be inversely proportional to V.

From the phenomenological theory of birefringence,[4]

$$\Delta n_0(T,\lambda_j) = G(T)\frac{\lambda_j^2\lambda^{*2}}{\lambda_j^2 - \lambda^{*2}}, \tag{3}$$

where $G(T)$ is related to the microscopic structure of a LC, and its temperature dependence is contributed primarily from the orientational order parameter of the LC, and λ^* is the mean electronic resonant wavelength, which depends on the LC composition and is usually in the UV region.

For the purpose of illustrating the calculation procedures, we shall consider the operation of a LC filter at room temperature ($T \cong 24°C$) for two mixed laser lines with wavelength λ_1 and λ_2. It should be kept in mind that an ideal filter would block λ_1 completely and transmit λ_2 with high efficiency, or vice versa, so that high extinction between these two beams can be achieved. First, let us consider the case in which the LC filter is tuned to reject λ_1 and pass λ_2. Correspondingly, we set $T_1 = 0$ and then optimize T_2. When $T_1 = 0$, we find from Eqs. (1a) and (1b) that

$$d\Delta n_1(V,\lambda_1) = N\lambda_1, \tag{4}$$

where N is a positive integer. In this circumstance,

$$T_2 = \sin^2\left[N\pi\left(\frac{\lambda_2}{\lambda_1}\right)\left(\frac{\lambda_1^2 - \lambda^{*2}}{\lambda_2^2 - \lambda^{*2}}\right)\right]. \tag{5}$$

From Eq. (5), T_2 is strongly affected by N for given λ_1 and λ_2 (λ^* is determined by the LC used). Although the choice of N is quite flexible, it is by no means arbitrary. From Eq. (4) and knowing that $\Delta n_1(V,\lambda_1) \le \Delta n_0(V = 0,\lambda_1)$, we find that n is related to d as

$$1 \le N \le d\Delta n_0(\lambda_1)/\lambda_1. \tag{6}$$

The larger the $d\Delta n_0(\lambda_1)/\lambda_1$, the higher the N_{max} (the maximum value of N). A large N_{max} means a higher probability for selecting an appropriate N to obtain a high T_2, as illustrated later. In the visible region, the wavelength is short and birefringence is high, so that a thin LC layer can be used for designing a high performance LC filter. However, as the wavelength increases, the birefringence of LC declines, which implies that a thicker LC layer is required to obtain a large N_{max}. The use of a thick LC layer would result in a slow response time, as shown in Sec. III. Therefore, a trade-off between fast response times and high transmission efficiency should be taken into consideration in the practical design.

To proceed with the calculation, we select $\lambda_1 = 0.5145$ μm and $\lambda_2 = 0.488$ μm (the two strong Ar^+ laser lines) as an example. From experimental results,[4] $G = 3.06 \times 10^{-6}$ nm^{-2} and $\lambda^* = 0.25$ μm for an E-7 LC mixture at $T \cong 24°C$. Thus Δn_0 is found to be 0.25 at $\lambda_1 = 0.5145$ μm. Substituting this Δn_0 into Eq. (6), we find $N_{max} = 9$ for a 20-μm E-7 layer at $\lambda_1 = 0.5145$ μm. The calculated T_2 as a function of N ($1 \le N \le N_{max}$) is tabulated in Table I.

The choice of the LC layer thickness is also flexible because it can always be compensated by an applied voltage (i.e., bias voltage V_b). In the case of $d = 20$ μm (E-7 LC) and $N = 5$ ($T_2 = 98.4\%$), V_b is calculated to be 1.79 V_{rms} from Eqs. (4), (3), and (2). If we choose $d = 15$ μm and $N = 5$, then V_b turns out to be 1.52 V_{rms}.

On the other hand, we want to tune the LC filter to reject λ_2 completely and simultaneously transmit λ_1 with high efficiency. By setting $T_2 = 0$, we have

$$d\Delta n_2(V,\lambda_2) = M\lambda_2. \tag{7}$$

In this situation,

$$T_1 = \sin^2\left[M\pi\left(\frac{\lambda_1}{\lambda_2}\right)\left(\frac{\lambda_2^2 - \lambda^{*2}}{\lambda_1^2 - \lambda^{*2}}\right)\right]. \tag{8}$$

Similarly, M should satisfy the following boundary condition:

$$1 \le M \le d\Delta n_0(\lambda_2)/\lambda_2. \tag{9}$$

For a 20-μm E-7 LC layer, $1 \le M \le 10$ is found for $\lambda_2 = 0.488$ μm. Again, T_1 should be evaluated as a function of integer M. Results are shown in Table II. From Table II, T_1 approaches 100% at $M = 6$, which is below M_{max} and is acceptable. Substituting $M = 6$, $d = 20$ μm, and $\Delta n_0(\lambda_2) = 0.259$ into Eqs. (7) and (2), we find $V = 1.71$ V_{rms}. Table III summarizes the results we have

Table I. Calculated T_2 as a Function of N Using Eq. (5) for E-7 LC when $T_1 = 0$. The Wavelengths Used for Calculation are $\lambda_1 = 0.5145$ μm and $\lambda_2 = 0.488$ μm

N	1	2	3	4	5	6	7	8	9
T_2	0.081	0.298	0.581	0.837	0.984	0.974	0.810	0.545	0.266

Table II. Calculated T_1 as a function of M using Eq. (8) for E-7 LC when $T_2 = 0$. The Wavelengths Used for Calculation are $\lambda_1 = 0.5145$ μm and $\lambda_2 = 0.488$ μm

M	1	2	3	4	5	6	7	8	9	10
T_1	0.068	0.255	0.508	0.760	0.940	0.999	0.923	0.730	0.474	0.226

Table III. Summary of the Calculated Results Using a 20-μm-thick E-7 LC Filter for $\lambda_1 = 0.5145$ μm and $\lambda_2 = 0.488$ μm[a]

V_b	T_1	T_2
1.79	0	0.984
1.71	0.999	0

[a] V_b is the bias voltage on the LC cell, and T_1 and T_2 are the optical transmittances of the LC filter at those biased voltages for λ_1 and λ_2, respectively.

obtained so far. It is quite impressive to note that such a LC filter (a 20-μm thick E-7 cell sandwiched between two crossed polarizers) transmits λ_1 and λ_2 alternately with very high efficiency and extinction ratio when the bias voltage is switched between 1.71 and 1.79 V_{rms}.

To examine whether such a LC filter can also be used for separating the 1.30- and 1.55-μm laser beams (assume same polarization) from a fiber, we perform similar calculations as demonstrated above. Results are listed in Table IV. From Table IV, the same LC filter (20-μm E-7) can switch these two beams (λ_1 = 1.30 and λ_2 = 1.55 μm) ON and OFF alternately at V_b = 1.36 and 1.05 V_{rms}, respectively.

To operate a LC filter in a broad spectral range requires two broadband polarizers with high transmittance and high extinction ratio. The substrate materials and transparent conductive coatings are also dependent on the spectral range of application. For example, quartz and glass substrates with indium–tin–oxide (ITO) coatings are useful for visible and near-IR applications. However, in the IR region (say 10 μm), germanium substrates with antireflection coatings could be considered.[9] ITO is not very transparent in the IR region due to its free carrier absorption, which is proportional to λ^2.

III. LC Response Times with Bias Voltage

The switching times of the LC director under a small bias voltage (not too large compared to V_{th}) condition and single elastic constant approximation can still be described by a simplified Leslie-Ericksen equation[10]:

$$K_{11}\frac{\partial^2\phi}{\partial Z^2} + \epsilon_a E^2 \sin\phi \cos\phi = \gamma_1 \frac{\partial\phi}{\partial t} \quad \text{(rise)}, \tag{10a}$$

$$K_{11}\frac{\partial^2\phi}{\partial Z^2} + \epsilon_a E_b^2 \sin\phi \cos\phi = \gamma_1 \frac{\partial\phi}{\partial t} \quad \text{(decay)}, \tag{10b}$$

where $\phi(Z,t)$ represents the deformed angle of the LC directors, K_{11} is the splay elastic constant for the parallel-aligned cell, ϵ_a is the dielectric anisotropy of the LC, E and E_b are the field strength of the applied and bias voltages to the cell, respectively, and γ_1 is the rotational viscosity. Solutions of Eqs. (10a) and (10b) should have the same form. To solve Eqs. (10a) and (10b), we expand $\sin\phi \cos\phi$ to the ϕ^3 term (assuming ϕ is small) and follow the same procedures as described in Ref. 10. We obtain

$$\tau_{\text{rise}} = \frac{\gamma_1 d^2/K_{11}\pi^2}{\left(\frac{V}{V_{\text{th}}}\right)^2 - 1}, \tag{11a}$$

$$\tau_{\text{decay}} = \frac{\gamma_1 d^2/K_{11}\pi^2}{\left|\left(\frac{V_b}{V_{\text{th}}}\right)^2 - 1\right|}. \tag{11b}$$

Equation (11a) is a well-known rise time expression for $V > V_{th}$. However, Eq. (11b) has some interesting features. The absolute value in the denominator of Eq. (11b) takes both $V_b < V_{th}$ and $V_b > V_{th}$ into consideration. At $V_b = 0$, $\tau_{decay} = \tau_0 = \gamma_1 d^2/K_{11}\pi^2$ is called the free relaxation time of the LC director. As V_b increases, τ_{decay} increases and finally diverges when V_b approaches V_{th}. This phenomenon can be understood as follows: the relaxation process of the LC directors in the bias voltage condition is governed by two torques, the elastic torque and the electric field-induced torque. However, these two torques are always working against each other for a parallel-aligned cell filled with LC molecules of positive dielectric anisotropy. At $0 < V_b < V_{th}$, the elastic torque exceeds the electric torque and dominates the relaxation process. However, in this region, the relaxation time is slower than τ_0 because part of the restoring elastic torque is canceled out by the electric torque. At $V_b \approx V_{th}$, these two torques become approximately equal and a very sluggish decay results.

On the other hand, when $V_b > V_{th}$ the electric torque exceeds the elastic torque and dominates the decay process. However, due to the partial cancellation from the elastic torque the decay time is still slower than τ_0 at $V_b < \sqrt{2}V_{th}$. At $V_b = \sqrt{2}V_{th}$, the net torque (i.e., the difference between electric torque and elastic torque) is just equal to the magnitude of the elastic torque so that $\tau = \tau_0$ as described by Eq. (11b). Once V_b exceeds $\sqrt{2}V_{th}$, τ_{decay} is less than τ_0, and improvement in the decay time is observed. Further increase in V_b results in a larger improvement in τ_{decay}.

Experimental confirmation of Eq. (11b) was conducted under small signal excitations using an ~10-μm LC cell filled with 5CB LC. The director's response times were measured by the optical probing method[11] at some bias voltages. Results are shown in Fig. 1, in which τ_0/τ_{decay} is plotted as a function of V_b^2. Dots in Fig. 1 represent the experimental results, and solid lines represent the theoretical calculations using Eq. (11b) and the V_{th} obtained from Fig. 1. Agreement between experiment and theory is very good except in the neighborhood of threshold voltage, where theory predicts a divergence in the decay time. In addition, the obtained threshold voltage (V_{th} = 0.72 V) from Fig. 1 agrees well with that determined from the voltage dependent transmission curve. Very similar results to those shown in Fig. 1 were obtained for some other LCs including E-7. The only difference is in the slopes of the lines, which are determined by the threshold voltage. For E-7, $V_{th} \cong$ 0.93 V at $T \cong$ 24°C and 10 kHz sine wave excitation. Caution has to be taken that Eqs. (11a) and (b) are valid only for small signal excitations. For large signal excitations, some deviation from these expressions is expected.

Knowing the viscoelastic coefficient (γ_1/K_{11} = 19 ms/μm^2 at $T \cong$ 24°C) of E-7,[11] we can calculate the switching times of the tunable LC filter. From Table III and Eqs. (11a) and (11b), the LC filter (20-μm E-7)

Table IV. Same as Table III but for λ_1 = 1.3 μm and λ_2 = 1.55 μm

V_b	T_1	T_2
1.05	0	0.999
1.39	0.925	0

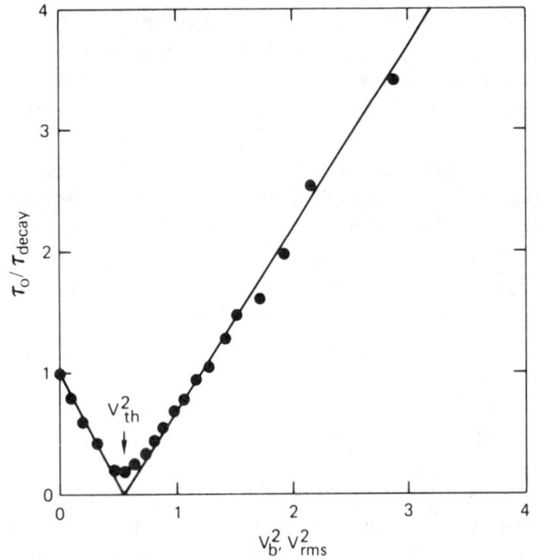

Fig. 1. Bias voltage dependent director's decay time τ_{decay} of 5CB LC (4-cyano-4′-n-pentylbiphenyl) under small signal excitations. $\tau_0 = \gamma_1 d^2/K_{11}\pi^2$ is the free relaxation time of the LC. Dots represent experimental results on τ_0/τ_{decay} and solid lines theoretical calculations using Eq. (11 b). The threshold voltage determined by the intersection of these two lines is equal to 0.72 V_{rms}, which agrees well with that measured by the optical transmission method. The temperature used for these measurements was ~24°C. The electric field used for excitation has a frequency of 10 kHz; sine wave.

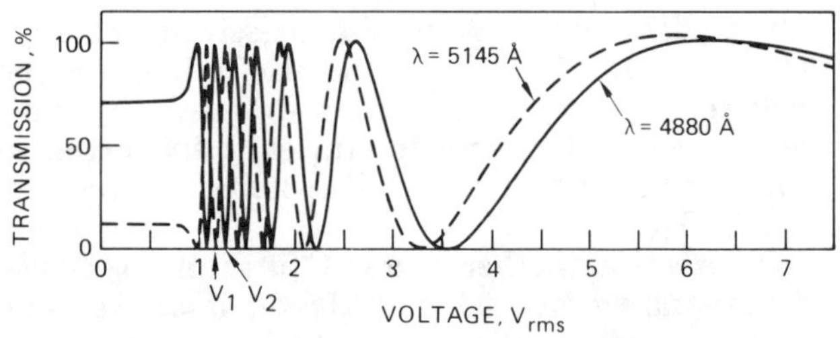

Fig. 2. Voltage dependent optical transmission of a 12.6-μm E-7 LC cell under crossed polarizers for two Ar^+ laser lines, $\lambda_1 = 0.5145$ and $\lambda_2 = 0.488$ μm. The experimental conditions are: excitation frequency = 10 kHz, $T \cong 24$°C. V_1 and V_2 are 1.19 and 1.30 V_{rms}, respectively.

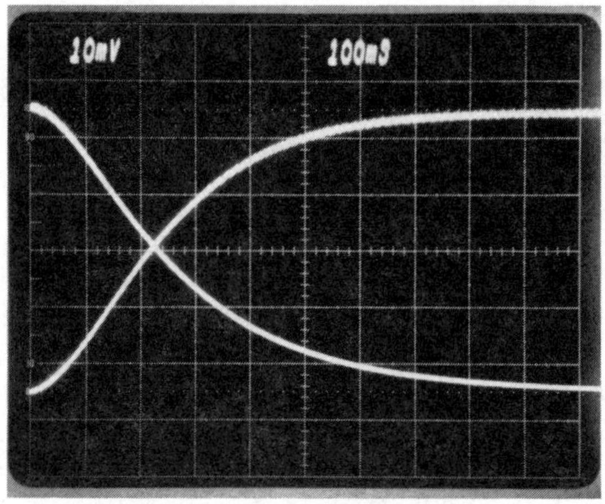

Fig. 3. Switching times of the 12.6-μm LC filter. The rise time (from V_1 to V_2) is measured to be ~275 ms and decay time (from V_2 to V_1) ~ 300 ms.

will take ~284 and ~323 ms to switch from λ_1 to λ_2 and from λ_2 to λ_1, respectively. These switching times can be improved substantially by using a thin LC layer with high Δn and small γ_1/K_{11} (Ref. 9) or by using a cell consisting of multiple thin LC layers. It is worth mentioning that the multilayer approach has several attractive features. Suppose we divide a LC layer of thickness d into N equally spaced thin layers, each having thickness d' so that $d = Nd'$. The total phase of the multilayered LC cell is the same as that of the single thick LC cell. However, the response time of the multilayered LC cell is determined by d' instead of d and is improved by a factor of N^2. Furthermore, the power loading capability for the multilayered LC cell is N^2 times higher than a single thick LC cell. This is because the threshold power for the optical field-induced Freedericksz transition[11,12] is inversely proportional to the square of the LC layer thickness. The occurrence of this nonlinear optical effect will result in diffraction patterns in the outgoing beam which are not desirable. Therefore, the multiple-thin-layered LC cell is particularly suitable for high power laser applications.

IV. Experimental Results

In experiment, a parallel-aligned E-7 LC cell was used. Its thickness was measured to be 12.6 μm by the interferometric method. The cell was sandwiched between two crossed polarizers, and the LC director's axis was aligned at 45° to the polarization axis of the incident beam. Voltage dependent transmissions of two Ar^+ laser lines, $\lambda_1 = 0.5145$ and $\lambda_2 = 0.488$ μm, are shown in Fig. 2. As indicated in the figure, V_1 ($\cong$1.19 V_{rms}) is the voltage which gives null transmission for λ_1, while keeping maximum transmission for λ_2, and V_2 (=1.30 V_{rms}) is for the reverse. V_1 and V_2, calculated following the procedures illustrated in Sec. III are found to be 1.19 and 1.31 V_{rms}, respectively. The agreement is fairly good.

The switching times of this 12.6-μm E-7 LC filter were measured between V_1 and V_2. Results are shown in Fig. 3. The 0–100% trace indicates the rise time from V_1 to V_2, and the 100–0% trace represents the decay time from V_2 to V_1. From these curves, the rise and decay time constants (at $1/e$ intensity) are measured to be ~275 and ~300 ms, respectively. These response times can also be calculated from Eq. (11). Using $V_{th} \cong 0.9$ V_{rms} and $\gamma_1/K_{11} \cong 19$ ms/μm^2 for E-7 LC mixture, the rise and decay times are found to be 278 and 388 ms, respectively. The rise time is in close agreement with the experimental value. However, the decay time is ~30% slower than the experimental result. This phenomenon is also observed in Fig. 1. This discrepancy can be explained as follows: Eq. (11b) is derived based on the very simplified small angle and single elastic constant approximations. When the bias voltage of a LC cell is close to V_{th}, a small voltage increase can cause a large increase in the deformation angle of the LC directors. In the experimental condition, the decay from V_2 to V_1 corresponds to about a π-radian phase change at $\lambda = 0.5145$ μm so that the small angle approximation may not be valid. The larger the deformation angle the larger the elastic torque and, therefore, the faster the decay time.

V. Discussion

One limitation of the LC-based filter arises when $\lambda_1 \approx m\lambda_2$ (m = a positive integer) occurs. This problem can be better demonstrated by looking at Eq. (5). If both λ_1^2 and λ_2^2 are much greater than λ^{*2} (this condition is met for long wavelength, compared to λ^*, applications of a LC filter), then Eq. (5) is reduced to

$$T_2 \approx \sin^2\left[N\left(\frac{\lambda_1}{\lambda_2}\right)\pi\right]. \tag{12a}$$

For $\lambda_1 \approx m\lambda_2$, $N(\lambda_1/\lambda_2)$ is close to an integer. So T_2 approaches zero, as does T_1. That means both beams are rejected by the filter. On the other hand, when $T_2 = 0$, T_1 [Eq. (8)] can be simplified to

$$T_1 \approx \sin^2\left[M\left(\frac{\lambda_2}{\lambda_1}\right)\pi\right] \approx \sin^2[(M/m)\pi]. \tag{12b}$$

Since both M and m are integers, T_1 can have either a zero or nonzero value depending on the selection of M, which can be accomplished by varying the biased voltage of the cell. Therefore, a filter in this operating condition (i.e., $\lambda_1 \approx m\lambda_2$) cannot switch λ_1 OFF completely while maintaining a high T_2 for λ_2. An example which has $m \approx 2$ follows. The same 20-μm E-7 LC filter is used to separate the two He–Ne laser lines ($\lambda_1 = 1.152$ and $\lambda_2 = 0.633$ μm). Results (in the calculation, both N and M are chosen to be 3) are summarized in Table V. It is not surprising that when T_1 is zero, the maximum T_2 is only 24.3%. In this example, λ_1/λ_2 is not exactly an integer. Thus we still can improve T_2 by using a thicker LC layer but with some sacrifice in the switching times.

When the incident beam consists of more than two discrete laser lines with the same polarization direction, more than one LC filter is needed. Let us assume that we use two filters in series for separating three beams. The first filter is tuned to reject λ_1 and pass λ_2 and λ_3 with high transmission. The second one is then tuned to reject either λ_2 or λ_3 and transmit one of the two beams. The detailed analyses are, in principle, similar to those we demonstrate in this paper except for the increasing complexity in algebra.

Finally, the extinction ratio of the LC based electrooptic filter is governed by the following factors: (1) light scattering due to the long-range director's orientational fluctuations[13]; (2) alignment uniformity of the LC layer; and (3) extinction ratio of the crossed polarizers. In the nematic phase, the forward light scattering is not negligible[14] and may ultimately limit the extinction ratio of the filter. This light scattering increases drastically as wavelength decreases ($\sim\lambda^{-4}$). A typical extinction ratio for a LC display device using a monochromatic light source is about several hundreds to one.

V. Conclusion

A simple tunable electrooptic filter using LC as the active medium is analyzed. This filter can be used in a broad spectral region. The switching times depend on the LC layer thickness, viscoelastic coefficient of the LC used, and bias voltage. A multilayered structure is found to improve the response times significantly and improve the power loading capability, which is important for high power laser applications.

The author would like to thank L. D. Hess and T. Y. Hsu for useful discussions.

References

1. D. A. Pinnow, R. L. Abrams, and J. F. Lotspeich, "Electro-optic Tunable Optical Filter," U.S. Patent 4,197,008 (1980).
2. H. A. Tarry, "Electrically Tunable Narrowband Optical Filter," Electron. Lett. **11**, 471 (1975).
3. W. I. Kaye, "Liquid Crystal Tuned Birefringent Filter," U.S. Patent 4,394,069 (1983).
4. S. T. Wu, "Birefringence Dispersions of Liquid Crystals," Phys. Rev. A **33**, 1270 (1986).
5. F. J. Kahn, "Electric-Field-Induced Orientational Deformation of Nematic Liquid Crystals: Tunable Birefringence," Appl. Phys. Lett. **20**, 199 (1972).
6. M. Born and E. Wolf, *Principles of Optics* (Pergamon, New York, 1980).
7. V. G. Chigrinov and M. F. Grebenkin, "Determination of the Elastic Constants and Coefficient of Viscosity of Nematic Liquid Crystals from Oriented Electro-optical Effects," Sov. Phys. Crystallogr. **20**, 747 (1976).
8. S. T. Wu, U. Efron, and L. D. Hess, "Birefringence Measurements of Liquid Crystals," Appl. Opt. **23**, 3911 (1984).
9. S. T. Wu, "Infrared Properties of Liquid Crystals: an Overview," Opt. Eng. **26**, 120 (1987).
10. For example, see L. M. Blinov, *Electro-optical and Magneto-optical Effects of Liquid Crystals* (Wiley, New York, 1983).
11. N. V. Tabiryan, A. V. Sukhov, and B. Ya.Zel'dovich, "The Orientational Optical Nonlinearity of Liquid Crystals," Mol. Cryst. Liq. Cryst. **136**, 1 (1986).
12. I. C. Khoo and Y. R. Shen, "Liquid Crystals: Nonlinear Optical Properties and Processes," Opt. Eng. **24**, 579 (1985).
13. P. G. deGennes, *The Physics of Liquid Crystals* (Clarendon, Oxford, 1974).
14. S. T. Wu and K. C. Lim, "Absorption and Scattering Measurements of Nematic Liquid Crystals," Appl. Opt. **26**, 1722 (1987).

Table V. Same as Table III but for $\lambda_1 = 1.152$ μm and $\lambda_2 = 0.633$ μm

V_b	T_1	T_2
1.25	0.984	0
1.94	0	0.243

An electro-optic tunable filter

D. A. Pinnow, R. L. Abrams, J. F. Lotspeich, D. M. Henderson, T. K. Plant,[a] R. R. Stephens, and C. M. Walker

Hughes Research Laboratories, Malibu, California 90265
(Received 4 December 1978; accepted for publication 9 January 1979)

This letter describes a tunable optical filter that couples light energy between ordinary and extraordinary polarizations in a birefringent electro-optic crystal at the phase-matched wavelength by means of a spatially periodic dc electric field. The specific embodiment described here is a $LiTaO_3$ platelet operating with transverse electric field and located between cross polarizers. Analogies to both the Šolc birefringent filter and the Harris acousto-optic filter are cited. The electro-optic filter is tuned by varying the spatial period of the electric field via an array of separately addressable finger electrodes. We have demonstrated tunability from 4700 Å to 4.5 μm. Experimental results of transmittance versus applied voltage and of bandwidth versus pass wavelength agree well with theory. The flexibility of this type of electronic filter in allowing synthesis of arbitrary transfer functions through nonperiodic voltage distributions is emphasized.

PACS numbers: 42.80.Cj, 78.20.Jq, 78.20.Fm

An optical filter whose characteristics can be electronically controlled has been recognized as an important element in optical sensors and signal processors. Until now, the device which could most closely provide this function was the acousto-optic tunable filter (AOTF).[1] In this type of filter, the index-of-refraction perturbations that selectively-couple light from the ordinary to extraordinary axis of a birefringent crystal are generated by acoustic waves imposed on the crystal. Since the required acoustic power is excessive for many applications, the use of the AOTF—particulary in the infrared portion of the spectrum—has been limited. In this letter we describe an electro-optic tunable filter (EOTF) in which the selective coupling is induced electro-optically by dc voltages. In this case, the electrical power dissipation results solely from leakage currents in the electro-optic crystal and, with the selection of high-resistivity materials, can be many orders of magnitude smaller than that of the AOTF. The specific configuration that is described in this paper—a thin platelet of $LiTaO_3$ within which a spatially periodic electric field is imposed—has been used previously in optical mode control[2] and optical modulation.[3] This is the first time, to our knowledge, that such an interaction has been used for optical filtering.

[a]Present address: Department of Electrical and Computer Engineering, Oregon State University, Corvallis, Ore. 97331.

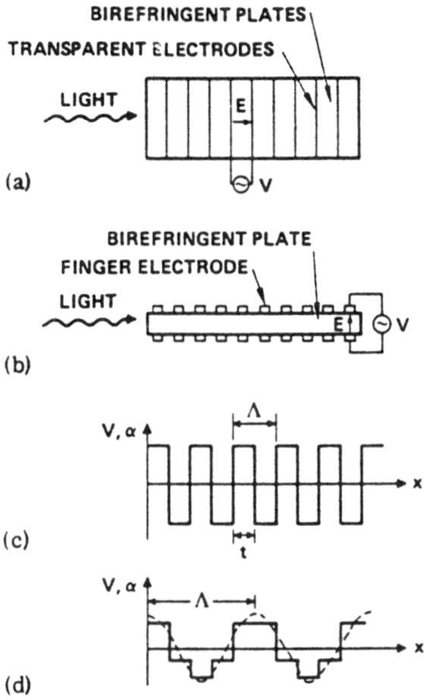

FIG. 1. Two possible filter configurations are shown in (a) and (b). The Šolc-type filter waveform of (c) determines the shortest operating wavelength. Operation at longer wavelength is accomplished by sampling longer voltage periods as seen in (d).

While the performance of the EOTF can be described in terms of the coupled-wave theory used for the acoustic filter, better insight into the filter operation is gained by recognizing that the EOTF is an extension of the Šolc birefringent filter.[4] A Šolc filter consists of a series of half-wave plates contained between crossed polarizers with the optic axes of the plates alternately aligned at angles $+\alpha$ and $-\alpha$ with respect to the plane of polarization of the input light. This angle is commonly called the rocking angle. At the fundamental wavelength of the filter, each $\frac{1}{2}\lambda$ plate rotates the plane of polarization symmetrically through its optic axis. This wavelength is simply given by $\lambda_0 = 2(n_0 - n_e)t$, where n_0 and n_e are refractive indices of the ordinary and extraordinary axes, respectively, and t is the plate thickness. In passing through the stack of $\frac{1}{2}\lambda$ plates, the optical plane of polarization experiences a continually growing angle with respect to the input plane of polarization, emerging finally at an angle $2N\alpha$, where N is the number of plates. By assuring that this angle is 90° at the filter output, light at wavelength λ_0 encounters no loss in passing through the crossed output analyzer. Light at all other wavelengths, excluding odd harmonics, does not experience a complete 90° polarization rotation and therefore incurs loss at the crossed output polarizer. The basic filter transfer function exhibits a $\sin x/x$ dependence which peaks at the wavelength λ_0.

Since the Šolc filter utilizes static birefringence, tuning of the device is not easily achieved. We have developed, however, an easily tunable filter based on the concepts outlined above by relying on the electro-optic interaction to control the rocking angle of the optic axis. The basic concept is shown in Fig. 1(a). In this case each plate in the stack is comprised of an electro-optic crystal whose optic axis is rotated through an angle α by the application of an electric field across the crystal. A specific example of a crystal that exhibits such dependence is Y-cut KDP with the electric field applied in the Y direction. Since the rocking angle of each plate can be controlled by its applied voltage, an electrically active Šolc filter can be synthesized through application of equal and opposite voltages to this plate, as shown in Fig. 1(c). With this configuration, however, one can easily apply voltage distributions other than the simple Šolc square-wave distribution. It is this feature of the EOTF that leads to its tunability and agility.

Filter tuning is achieved by imposing a spatially sampled sine wave on the filter, as shown in Fig. 1(d). For such a distribution with spatial period Λ, the pseudomomentum of the field can couple light from the ordinary to the extraordinary axis. Phase matching is achieved at the wavelength λ satisfying

$$\frac{2\pi(n_e - n_0)}{\lambda} = \frac{2\pi}{\Lambda}, \tag{1}$$

which leads to the tuning criterion $\lambda = (n_e - n_0)\Lambda$. Note that this is the identical condition for the Šolc filter where $\Lambda = 2t$. Using coupled-wave theory it can be shown that the fractional linewidth $\Delta\lambda/\lambda$ of a filter of length L is given by

$$\frac{\Delta\lambda}{\lambda} = \frac{\pi}{4}\frac{\Lambda}{L} = \frac{\pi}{4}\frac{\lambda}{(n_e - n_0)L}. \tag{2}$$

This result shows that the fractional linewidth is governed by the number of voltage periods the light encounters in passing through the filter and that the actual linewidth grows quadratically with wavelength for a filter of fixed length. Identical expressions hold for the AOTF, in which case Λ represents the wavelength of the acoustic wave. However, the EOTF offers greatly reduced drive power and a greater flexibility in applying other than simple sine-wave spatial distributions to the filter. Examples of such flexibility include variable filter bandpass, simultaneous operation at multiple wavelengths, and sidelobe suppression.

We have demonstrated these filter concepts in the visible portion of the spectrum with a Y-cut $LiTaO_3$ single-plate filter, oriented between crossed polarizers and operating with transverse electrodes, as shown in Fig. 1(b). This configuration has been selected for the initial device demonstra-

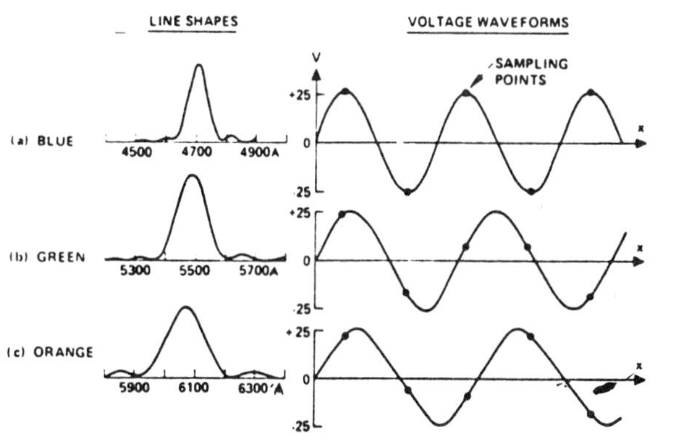

FIG. 2. Measured line shapes accompanied by sampled spatial voltage waveforms. The exact wavelengths are (a) 4700 Å, (b) 5480 Å, and (c) 6060 Å.

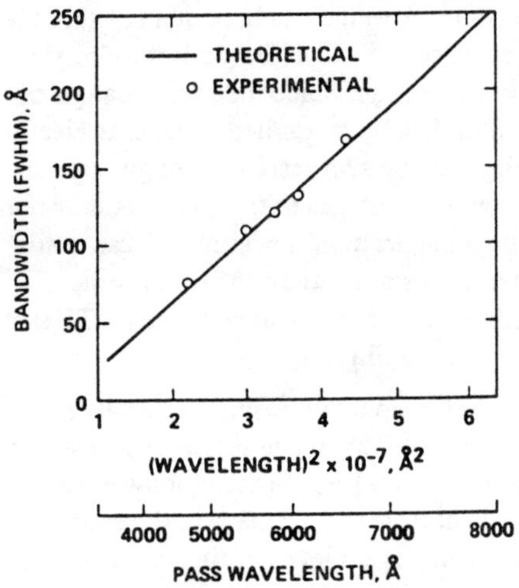

FIG. 3. Bandwidth characteristics of electro-optic tunable filter.

tion because it avoids the need for transparent electrodes required in the longitudinal electrode design of Fig. 1(a). The crystal, which is 140 μm thick, has 100 pairs of finger electrodes deposited on its top and bottom surfaces. The voltage to each electrode pair can be independently set, which allows a variety of voltage distributions to be imposed on the filter.

Filter tuning is shown in Fig. 2 where the sampled voltage waveforms as well as the measured filter response are presented. At the fundamental wavelength of the filter, $\lambda_0 = 4700$ Å, the sampled spatial sine wave is merely the square-wave voltage distribution of the Šolc filter. By sampling larger-wavelength spatial periods, the filter tunes to longer wavelengths, in agreement with Eq. (1). The increasing linewidth with wavelength seen in Fig. 2 results from the quadratic wavelength dependence of Eqs. (2). By comparing the theoretical linewidths with the measured values, excellent agreement is found, as shown in Fig. 3. The maximum transmittance of the filter at the fundamental wavelength was 41% with a peak-to-peak voltage of 50 V, again in good agreement with theoretical predictions.

We have tuned this filter through the near-infrared portion of the spectrum out to 4.5 μm — the point where the $LiTaO_3$ transmission cuts off. We are presently addressing birefringent materials with longer-wavelength transmission in order to take advantage of the EOTF concept described in this letter in the infrared portion of the spectrum.

[1]S.E. Harris and R.M. Wallace, J. Opt. Soc. Am. **59**, 744 (1969).
[2]Y. Okamura, S. Yamamoto, and T. Makimoto, Appl. Phys. Lett. **32**, 161 (1978).
[3]M. Kondo, Y. Ohta, M. Sakaguchi, and Y. Fugino, Jpn. J. Appl. Phys. **18**, 107 (1977).
[4]I. Šolc, J. Opt. Soc. Am. **55**, 621 (1965).

5. Frustrated Total Reflection Interference Filters. PAUL LEURGANS AND A. F. TURNER, *Bausch & Lomb Optical Company.*

An interference filter can be made by replacing the customary metallic films with non-absorbing films at which frustrated total reflection takes place. Frustrated total reflection occurs when light is incident upon a thin film between two media of higher index such that the angle of incidence exceeds the critical angle. The transmittance and reflectance depend on the angle of incidence, the relative indices of the film and the adjacent media, and the thickness of the film. An interference filter embodying this principle may be made, for instance, by filming the hypotenuse of a 30°–60° high index prism with a layer of low index material (e.g., cryolite), a spacer layer of higher index material (e.g., zinc sulfide), and a second layer of the low index material. A second high index prism is cemented on so that the light is incident on the films at a 60° angle. For a given order of interference there are two separate transmitted peaks, identified with the p and s vibrational directions. The transmittance of each peak is 100 percent for the respective vibrational direction. The wave-length difference between the peaks arises from the difference in phase shift at the frustrated total reflection (low index) films. The transmission characteristics of the peaks have two main features: (1) the peak transmittance is always independent of the half-width, and (2), the half-width can be made arbitrarily small, evaporation techniques permitting. The properties of the filter have been investigated theoretically as a boundary value problem and also experimentally.

A Frustrated Total Reflection Filter for the Infra-Red

B. H. Billings
Research Laboratory, Baird Associates, Inc., Cambridge 38, Massachusetts

and M. A. Pittman
Madison College, Harrisonburg, Virginia

(Received August 18, 1949)

The Frustrated Total Reflection filter is a transmission band pass filter. It can be constructed for any part of the spectrum for which there exist transparent materials of different indices of refraction. Measurements are shown of a filter which was designed for the rocksalt infra-red region. This filter has two oppositely polarized pass bands which, at a particular angle of incidence, are at $4\frac{1}{2}$ and 5 microns, respectively. Both the pass bands are approximately $\frac{1}{10}$ of a micron wide. They can be shifted over a range of approximately half a micron by tilting the filter.

The isolation of a narrow band of wave-lengths is often a necessary part of a scientific experiment or of an industrial instrument. If the source of the band of wave-lengths is continuous, it is necessary to use either a monochromator or a narrow band filter. Until ten years ago, only a few special filters and techniques have been available. These include reststrahlen plates,[1] the Lyot polarization filter,[2] the Christiansen filter,[3] the reflection from potassium chlorate crystals,[4] and Lippmann plates.[5] Of these, the Lyot filter, the Lippmann plate, and the Christiansen filter provide a pass band anywhere in the visible spectrum. Only the Christiansen filter has ever been manufactured for use in the infra-red. The reststrahlen plates can only be used at those few wave-lengths where nature has provided materials with a narrow metallic reflection band. The potassium chlorate crystal has narrow pass bands whose location is determined by chance. Because of this and because of their small size they are not particularly useful as filters. The Christiansen filter is not image-forming and is not particularly useful either in the infra-red or ultra-violet.

During the last decade, an increasing number of filters have been developed which can be used to isolate a narrow band and which are image-forming. The solid Fabry-Perot filter is the most popular of these filters.[6] The filter is essentially a Fabry-Perot etalon operating in the first and second order with a solid dielectric spacer. The operation of the etalon is discussed in most textbooks on optics.[7] It consists simply of two parallel plates separated by an air gap. The surfaces of the plates which face the gap are partially silvered. The action of the interferometer may be explained by assuming that at the first silvered surface the light is partly reflected and partly transmitted. The transmitted part continues to the second silvered surface where it is again partly reflected and partly transmitted. The reflected light passes back to the first surface. Part of this light is transmitted and joins with the light that was previously reflected. If these two beams are out of phase the light will be transmitted through the plates. If they are in phase, it will be reflected. For a quantitative treatment of the instrument it is necessary to consider multiple reflections between the plates.

The standard treatment ignores absorption in the film and leads to the equation

$$\frac{I}{I_0}=\frac{t^2}{t^2+4r\sin^2\delta}, \tag{1}$$

where I is the intensity of the transmitted light, I_0 the intensity of the incident light, t the transmission of the reflecting film, r its reflectivity, and δ the phase shift through dielectric layer. This phase shift is equal to $(2\pi nd/\lambda)$, where n is the index of the layer between the plates and d is its thickness. This equation has a maximum at all wave-lengths at which the optical thickness, "nd," is a whole number of half-waves.

In Fig. 1 is a family of curves of the transmission of the filter represented by Eq. (1). As the reflectivity of the boundary layers increases, the band width of the filter narrows. The peak transmission, however, stays constant. The exact behavior of a metal layer must include absorption. It has been carried through by several workers.[8] The exact treatment of the interference filter has been given by Hadley.[6]

Figure 2 shows the curves which are derived from his formulas. It is clear that the band width cannot be narrowed indefinitely without losing peak transmission to the extent that the filter is no longer useful. Similarly,

[1] John Strong, *Procedures in Experimental Physics* (Prentice-Hall, Inc., New York, 1943). Maurice Parodi, *Recherches dans L'Infrarouge Lointain par la Methode des Rayons Restants* (Hermann & Cie, Paris, 1938).
[2] B. Lyot, Comptes Rendus **197**, 1593 (1933).
[3] C. Christiansen, Ann. Physik u. Chemie **23**, 298 (1884); **24**, 439 (1885). E. D. McAlister, Smithsonian Misc. **93**, No. 7 (1935).
[4] R. W. Wood, *Physical Optics* (The Macmillan Company, New York, 1934), p. 199.
[5] See reference 4, p. 214.
[6] L. N. Hadley and D. M. Dennison, J. Opt. Soc. Am. **37**, 451–465 (1947); **38**, 483–496 (1948).
[7] See reference 4. Also, F. A. Jenkins and H. E. White, *Fundamentals of Physical Optics* (McGraw-Hill Book Company, Inc., New York, 1937).
[8] Thornton Fry, J. Opt. Soc. Am. **22**, 307 (1932).

if the reflectivity is lowered the filter loses its value because of the increased transmission away from the peak. Another limitation on these filters is that they are only useful in the visible part of the spectrum. The characteristics of metals are such that the filters cannot be made to operate either in the ultraviolet or infra-red spectral regions. In both these regions the absorption associated with a film of given reflectivity is much higher than in the visible spectrum.

The results shown in Fig. 2 are rather disappointing compared with the curves of Fig. 1. The tremendous gain to be realized by having materials of high reflectivity with low absorption makes the search for such materials particularly keen. There are actually no metals which will give any better results than those shown in Fig. 2. There are, however, three methods which have been used to achieve surfaces with high reflectivity and low absorption. All these methods give results which are better than those shown in Fig. 2.

These methods are: (1) The Frustrated Total Reflection filter which was described by Turner[9] at the Cincinnati meeting of the Optical Society. (2) The use of multiple layers of dielectric. (3) The use of multilayers of dielectric in combination with layers of silver.

THEORY OF THE FRUSTRATED TOTAL REFLECTION FILTER

This filter uses the principle of the penetration into the second medium which occurs when light is incident on a surface beyond the angle of total internal reflection. When light is sent through the base of a right angle prism it is reflected at the hypotenuse. This reflection is 100 percent if the light strikes the hypotenuse of the prism at an angle of incidence greater than the critical angle. If now a convex lens is put in contact with the hypotenuse, the light at the point of contact will not be reflected but instead will be transmitted into the lens. When the light reflected from the prism hypotenuse is observed, a black spot is seen where this penetration takes place. The size of this black spot is found to be much larger than one would expect if the light went through only where actual optical contact were achieved. In other words, the light appears to penetrate into the air space between the lens and the prism and where that air space is sufficiently thin, it is able to reach the lens surface and be transmitted. Examination of Maxwell's equations at the prism hypotenuse shows that at angles bigger than the critical angle, the light actually does penetrate into the low index medium. In this medium the amplitude drops off exponentially with the distance. If a high index surface is placed close enough to the hypotenuse so that the light amplitude has not dropped to zero, then a fraction of the incident light will be transmitted. No light will be lost in the low index layer. The surface of this low index layer thus constitutes a semitransparent reflector with no absorption loss.

The Frustrated Total Reflection filter consists of two such semitransparent reflectors. The prism hypotenuse is covered with a low index layer whose thickness is determined by the desired reflectivity. This is followed by a high index layer whose thickness determines the wave-length which will be transmitted. Another low index layer is then deposited so as to give another frustrated total reflection. Another prism is then laminated to the first prism with a material whose index is the same as the prism material. The resulting rhomb constitutes the interference filter.

The Frustrated Total Reflection filter, unlike the regular solid Fabry-Perot filter, has no inherent limitations which make it unsatisfactory for wave-lengths in the ultraviolet or infra-red. The only requirements are that the materials of which the filter is made be transparent in the desired spectral region and that the indices have the proper relationship.

In studying the Frustrated Total Reflection filter the quickest insight into its behavior is obtained by con-

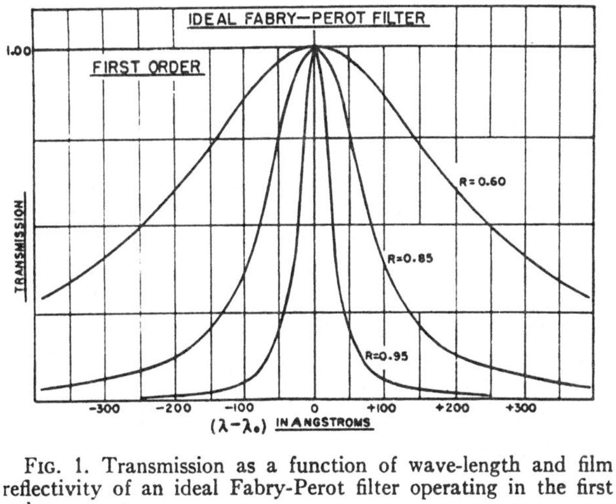

FIG. 1. Transmission as a function of wave-length and film reflectivity of an ideal Fabry-Perot filter operating in the first order.

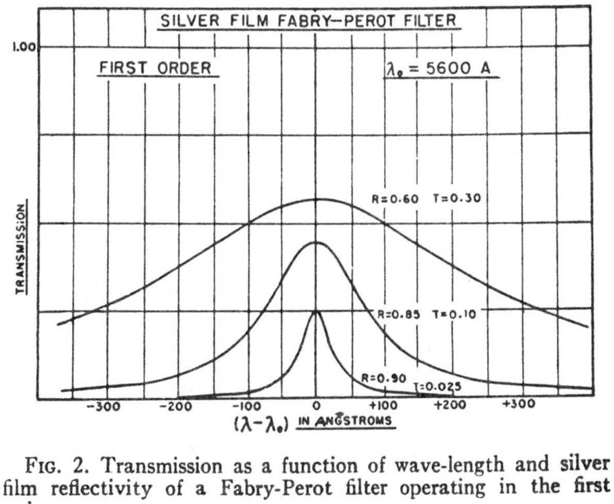

FIG. 2. Transmission as a function of wave-length and silver film reflectivity of a Fabry-Perot filter operating in the first order.

[9] Paul Leurgans and A. F. Turner, J. Opt. Soc. Am. **37**, 983 (1947).

sidering it as a simple Fabry-Perot filter. The wave-length of peak transmission is determined by the thickness of the high index layer and the band width by the thickness of the low index layer. Equation (1) gives a good approximation to the behavior of the Fabry-Perot filter at normal incidence. When the filter is tilted the equivalent thickness of the high index layer changes and the peak transmission accordingly changes.

The change of peak transmission with angle of incidence is most easily derived by considering the behavior of the filter as arising from the interference of the light reflected from the first silver surface with light reflected from the second silver surface.

The retardation through a Fabry-Perot interference filter can be written

$$\Gamma=[2d(n^2-\sin^2\theta)^{\frac{1}{2}}]/\lambda, \tag{2}$$

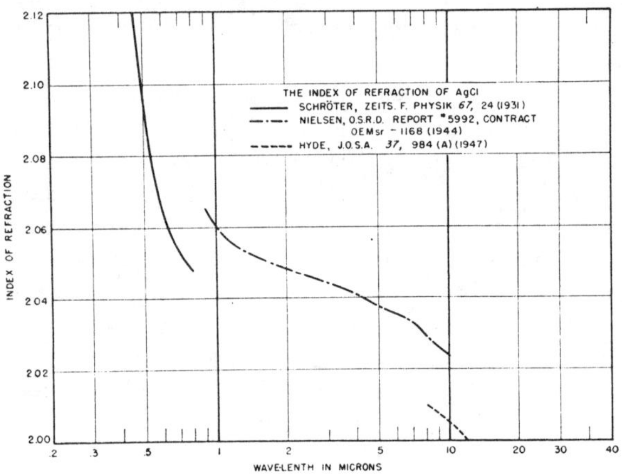

FIG. 3. Refractive index in the infra-red of silver chloride.

where d is the thickness of the middle dielectric layer, n is the index of this layer, θ is the angle of incidence, and λ is the wave-length. The filter has maximum transmission when Γ is an integer. The wave-length for peak transmission can be written

$$\lambda=(2d/\Gamma)(n^2-\sin^2\theta)^{\frac{1}{2}}. \tag{3}$$

For the transmission peak at normal incidence

$$\lambda_0=(2d/\Gamma)n. \tag{4}$$

The wave-length for peak transmission at an angle θ can be written in terms of the peak at normal incidence

$$\lambda=(\lambda_0/n)(n^2-\sin^2\theta)^{\frac{1}{2}}. \tag{5}$$

It is interesting to note that the relationship between the wave-length for peak transmission and the angle of incidence is independent of the order of interference.

Since the wave-length of peak transmission varies with the angle of incidence, it is apparent that the angular field of view of the filter, for monochromatic light, is directly related to the width of the pass band. The angular field can be considered as twice the angle through which the filter must be turned to drop the transmission by 50 percent. From Eq. (5) this can be written

$$\Theta=2\sin^{-1}\left[n\left(\frac{\lambda_0^2-\lambda_{I/2}^2}{\lambda_0^2}\right)^{\frac{1}{2}}\right], \tag{6}$$

where Θ is the angular field of view and $\lambda_{I/2}$ is the wave-length at which the transmission through the filter is one-half the transmission at the peak.

If the transmission of the filter for monochromatic light is measured in a system whose angular field of view is wider than the angular field of the filter, it will appear lower than the transmission for a collimated beam. In other words, in monochromatic light, the filter passes light over an angle Θ. Light incident at angles outside Θ is not transmitted. In a beam in which the intensity is uniformly distributed over an angle $\alpha>\Theta$, the transmission will be approximately

$$T=T_N(\Theta/\alpha), \tag{7}$$

when T_N is the transmission at normal incidence.

By the same token, the band width of a filter will appear wider if it is examined in a system of greater angular aperture than the angular field of the filter. The beam can be split into angular sections of width Θ. Each section will have a transmission peak at a different wave-length. At each wave-length the transmission will be given by Eq. (7). The apparent band width will be approximately

$$\Delta\lambda=(\alpha/\Theta)\Delta\lambda_N, \tag{8}$$

where $\Delta\lambda_N$ is the band width at normal incidence.

The same criteria which establish the angular field of view of the filter also establish the tolerance which must be placed on the homogeneity and plane parallelism of the various layers. If the low index layer is of such thickness that the filter band width is 1/1000 of the wave-length of the pass band, the high index layer must be plane parallel to within better than 1/1000 of the thickness.

In deriving the equations which show the behavior of the filter, it was considered that the phase shift through the filter depended entirely on the thickness of the high index layer. Actually, there is another additional phase difference between the light which goes directly through the high index layer and the light which is reflected from the opposite sides of this layer. This is caused by the phase change which occurs when light is reflected at a boundary. At normal incidence and with silver boundary films this additional phase shift merely means that the peak wave-length is slightly displaced from the position calculated on the basis of Eq. (1). When light goes through the filter obliquely, the phase shift on reflection becomes different for the two states of polarization. This results in splitting of the filter pass bands. This splitting can be demonstrated in the ordinary Fabry-Perot filter and has been discussed by Hadley.[6] In the Frustrated Total Reflection filter there

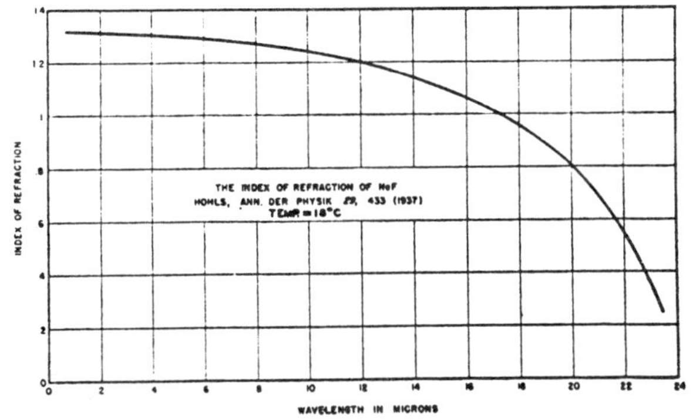

FIG. 4. Infra-red refractive index of sodium fluoride.

is also a phase shift on reflection and, as in the case of the Fabry-Perot filter, this phase shift is different for the two states of polarization. The angular shift can be calculated from the equations

$$\tan\frac{\delta_p}{2}=\frac{(\sin^2\theta-(n_1/n_0)^2)^{\frac{1}{2}}}{(n_1/n_0)^2\cos\theta};$$

$$\tan\frac{\delta_s}{2}=\frac{(\sin^2\theta-(n_1/n_0)^2)^{\frac{1}{2}}}{\cos\theta},\qquad(9)$$

where δ_s is the phase shift for light polarized perpendicularly to the plane of incidence, δ_p is the phase shift for light polarized parallel to the plane of incidence, n_1 is the index of the low index layer, n_0 the index of the prism, and θ the angle of incidence. This difference in phase which takes place on total internal reflection is used in the Fresnel rhomb to convert linearly polarized light to circularly polarized light.

As a result of this phase difference all the pass bands of the Frustrated Total Reflection filter are doubled. The long wave pass band is for light polarized perpendicularly to the plane of incidence, the short wave pass band for the light polarized in the plane of incidence. Because the pass bands are thus polarized, the maximum transmission in any band is 50 percent for unpolarized light.

AN INFRA-RED FRUSTRATED TOTAL REFLECTION FILTER

It seemed interesting to attempt to construct a very narrow band filter to operate in the rocksalt infra-red region. One problem was the choice of materials for the various layers in the filter. These layers had to be of high and low index, transparent in the infra-red and easily evaporated. This last requirement was fairly difficult because few materials can be evaporate tod make smooth, thick layers. Figures 3 and 4 show the dispersion curves of AgCl and NaF which were the two materials finally selected. The indices given in these curves are for bulk materials. For evaporated layers the indices are probably different. Unfortunately, very little work has been done in determining the indices of evaporated layers, either in the visible or infra-red.

A first attempt to make the filter was made using prisms of KRS-5. Since this material has a very high index it was not possible to find a laminating liquid of the same index; accordingly, it was decided to use a laminating liquid for the low index layer. The filter was constructed with an evaporated layer of sodium fluoride of an optical thickness of 11/4 waves at 6500A and of silver chloride of an optical thickness of 11/4 waves at 6500A, followed by a layer of mineral oil for the next low index layer and as a laminating fluid. No energy was transmitted by the filter at any wave-length. It was decided that the control of the uniformity and thickness of the low index liquid layer was going to be too difficult to make this particular prism work. A prism of rocksalt was located. This was a 60° prism and was sawed in two so as to make two 30, 60, 90 prisms. The hypotenuses of these prisms were polished. A low index layer of sodium fluoride was evaporated on one of these surfaces to an optical thickness of 17/2 waves at 6500A. This was followed by a layer of silver chloride evaporated to a thickness of 23/4 waves at 6500A and a second layer of sodium fluoride. The prisms were then laminated with mineral oil. A sketch of the resulting filter is shown in Fig. 5. The predicted location of the peaks can be calculated approximately from the Fabry-Perot formulas. From Eq. (4) the wave-length peak at normal incidence for the first order is

$$\lambda_0=2dn=2\times\frac{23}{4}\times\frac{6500\times n'}{n_{6500}}.\qquad(10)$$

Here n' is the index at the wave-length of the peak. Since the dispersion of AgCl is small, a good approximation is obtained simply by assuming $n'=n_{6500}$ and then getting a more exact value by using the value of n' corresponding to resulting peak wave-length. The value of n_{6500} for AgCl is found from Schröter[10] to be 2.06. The final calculation gives $\lambda_0=7.3\mu$.

From Eq. (5) the wave-length for peak transmission in the 60° filter is

$$\lambda=\frac{7.3\times n_{\mathrm{NaCl}}}{n_{\mathrm{AgCl}}}\left[\left(\frac{n_{\mathrm{AgCl}}}{n_{\mathrm{NaCl}}}\right)^2-\sin^2 60°\right]^{\frac{1}{2}}=5.6\mu.\qquad(11)$$

Here all the high layer indices are divided by the index

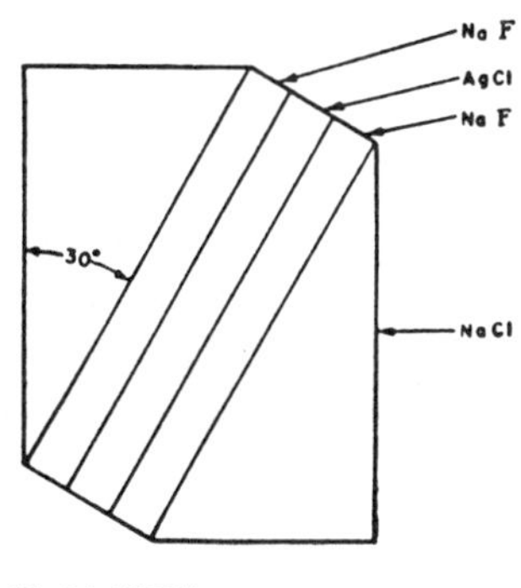

FIG. 5. Sketch of infra-red Frustrated Total Reflection filter. The rhomb has dimensions approximately 1¼ in.×1¼ in.×2 in.

[10] Schröter, Zeits. f. Physik **67**, 24 (1931).

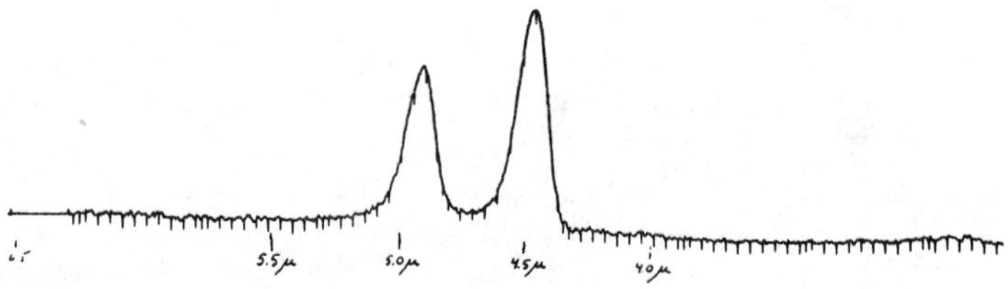

FIG. 6. Recorded transmission curve of the filter shown in Fig. 5. This curve was run on a Perkin-Elmer spectrometer with the source mirror covered with a half-inch wide slit.

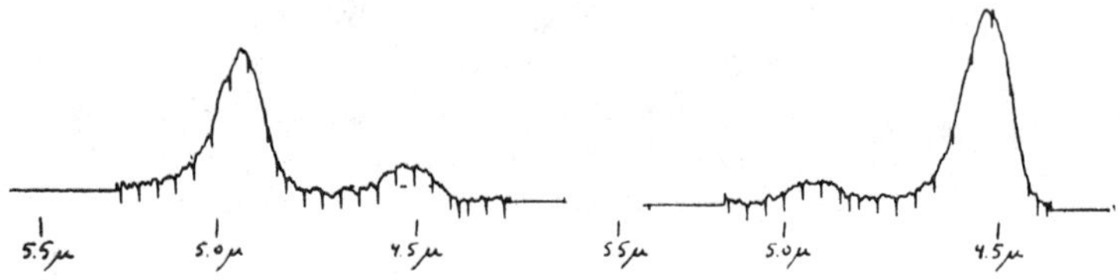

FIG. 7. Transmission of the filter shown in Fig. 5 in series with a two-plate silver chloride polarizer.

of NaCl since the filter is effectively embedded in this material. To locate the pass bands, transmission curves were run of the resulting rhomb with a Baird Associates double beam infra-red spectrometer.[11] A curve was also run to show the transmission of the mineral oil.

As predicted from the values of the thickness of the high index layer, the filter had a first-order peak at approximately 5.6μ. The prediction could probably have been made closer if the exact formula for the filter transmission had been used.

The filter was then studied with a Perkin-Elmer spectrometer in the Massachusetts Institute of Technology spectroscopy laboratory. The problem of measuring the spectral transmission of the finished filter was complicated by the dependence of band width and peak transmission on angular field of view. As a result of preliminary measurements it became apparent that the angular size of the beam was greater than the field of the filter. Accordingly, the angular size of the beam was diminished by putting a stop over one of the source mirrors. In order to avoid cutting down the energy too much, the stop was made in the form of a slit with its axis parallel to the plane of the prism hypotenuse. The slit reduced the beam width to approximately 3°.

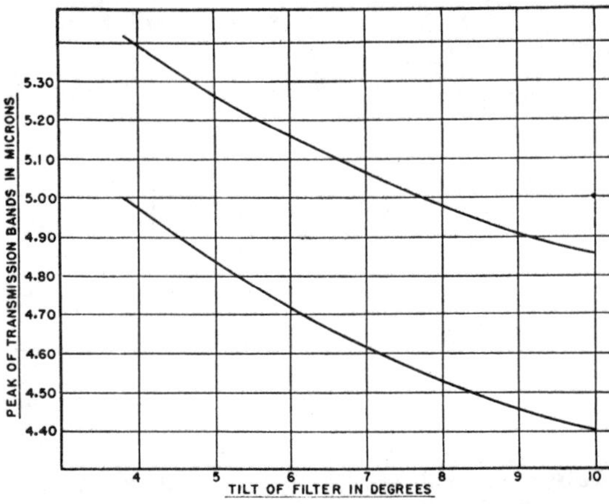

FIG. 8. Wave-length of peak transmission of filter as a function of the angle of incidence. The upper curve is for the peak which is polarized in the plane of incidence. The lower curve is for the peak polarized at right angles to the plane of incidence.

In Fig. 6 is a transmission curve of the filter. The band width of the two peaks is about 0.1μ. This seems wide for the particular thickness of the low index layer. This large width may very well be caused by the large angular beam in which the prism was examined.

To demonstrate the polarization of the bands a transmission polarizer was made of two sheets of silver chloride[12] at 63° to the beam. The use of such polarizers in the infra-red has been described by Wright[13] and by Halford.[14] Since two sheets are not sufficient to give complete polarization of the beam, neither transmission band of the filter can be completely extinguished. In Fig. 7 is shown the transmission of the filter with the polarizer first parallel and then perpendicular to the plane of incidence on the prism hypotenuse.

The variation of the transmission peak with angle of incidence was measured for both states of polarization and is shown in Fig. 8. The small size of the angular field of view is immediately apparent from this curve.

APPLICATION OF THE INFRA-RED FILTER

A most important use of the filter in the infra-red should be in problems of gas analysis. In this application one of the two pass bands of the filter is made to coincide with an absorption band of the gas to be detected. The other pass band is placed in a region where the gas has no absorption band. A polarizer is rotated in front of the filter. A beam of radiation is passed through the polarizer, the filter, a gas cell, and on to a detector such as a bolometer or thermocouple. With no gas present in the gas cell, an alternating current signal will be present in the detector because of the slight difference in

[11] Baird, O'Bryan, Ogden, and Lee, "An automatic recording infra-red spectrophotometer," J. Opt. Soc. Am. **37**, 754–761 (1947).

[12] Silver chloride sheets are provided by the Harshaw Chemical Company, Cleveland, Ohio.

[13] N. Wright, J. Opt. Soc. Am. **38**, 69 (1948).

[14] R. Newman and R. S. Halford, Rev. Sci. Inst. **19**, 270 (1948).

radiation intensity at the two wave-lengths. This zero signal can be removed optically as with a third polarizer in the system or can be removed electronically. If the absorbing gas is present in the gas cell, an additional alternating current will be produced in the receiver. For small concentrations, this will be proportional to the amount of gas in the cell. Anything which affects both beams will not change the alternating current in the receiver.

Another use for the infra-red filter will, perhaps, be as a means for doing spectral analysis in the long wave infra-red. It should be possible with these filters to get resolutions as high as 1000. At 300μ this is difficult to achieve by other means.

Since the pass band of the filter can be tuned over a range, perhaps 10 percent of the wave-length of the pass band, a fairly large spectral range could be covered with just a few filters.

The authors are especially indebted to Professor Richard Lord of the Massachusetts Institute of Technology for permission to use the equipment of the Massachusetts Institute of Technology spectroscopy laboratory. They are also indebted to Mr. Robert MacDonald for his help in making the measurements.

A Birefringent Frustrated Total Reflection Filter

B. H. Billings
Research Laboratory, Baird Associates, Inc., Cambridge, Massachusetts

(Received January 18, 1950)

In the standard Turner Frustrated Total Reflection filter the pass bands occur in pairs which are oppositely polarized. The maximum theoretical transmission in any band is thus 50 percent. In this paper it is shown that if the high index layer of the filter is made of a birefringent material the oppositely polarized pass bands can be made to coincide so that the maximum theoretical transmission becomes 100 percent and the bands occur singly. Measurements are shown on a filter made with such a birefringent layer.

During the last few years an increasing amount of attention has been paid to the development of methods of isolating narrow spectral regions. A large part of this effort has been spent in the investigation of the possibilities of interference filters. At any given stage of development the complexity of these filters seems to vary inversely with the width of the pass band desired. For a band width of 15 mμ a simple Fabry-Perot interference filter[1] is perfectly satisfactory. With its three evaporated layers this filter certainly is the least complex of the interference family.

A pass band of 6 mμ requires a much more complex structure than that of the solid Fabry-Perot filter. These band widths were produced several years ago with a multilayer technique by Fogelsanger at Evaporated Metal Films. They have recently been described by Polster.[2] Here the two silver layers of the simple Fabry-Perot filter are each replaced by seven layers of dielectric. These narrow pass bands can also be achieved by the use of a mixture of dielectric layers and silver layers. This has been demonstrated by DuFour.[3]

Pass bands of 0.6 mμ or less require the still more complex structure of the polarization interference filter.[4–6] Although this filter pre-dates the Fabry-Perot and other interference filters by several years, it is so very difficult to manufacture that there are less than a dozen of these filters in existence. The Turner Frustrated Total Reflection filter[7–8] considerably reduces the complexity required for a given band width. With this filter good transmission can be achieved in the visible spectrum with band widths of only 1 mμ. The filter, however, is not much more complex than the simple Fabry-Perot filter and has only three interfering layers. In its basic form, the Frustrated Total Reflection filter has pass bands which occur in pairs and which are oppositely polarized. As a result, the theoretical maximum transmission of any one pass band is 50 percent. In this paper a method is described whereby the polarized double bands are replaced by a single isotropic band of twice the transmission.

THEORY OF THE BASIC FRUSTRATED TOTAL REFLECTION FILTER

The operation of the Frustrated Total Reflection filter is most easily understood by considering it as an analog to the solid Fabry-Perot filter. This filter consists of a layer of dielectric between two semitransparent layers of silver. Its operation may be explained by assuming that at the first silvered surface the light is partly reflected and partly transmitted. The transmitted part continues to the second silvered surface where it is again partly reflected and partly transmitted. The reflected light passes back to the first surface, part of this light is transmitted, and joins with the light that was previously reflected. If these two beams are out of phase the light will be transmitted through the filter. If they are in phase the light will be reflected. The pass band of the filter depends on the number of times the light can be reflected between the boundaries of the dielectric layer of the filter. The number of reflections will depend on the reflectivity of the silver layers. By summing the various transmitted amplitudes of the rays passing through the filter one obtains the formula:

$$\frac{I}{I_0}=\frac{t^2}{t^2+4r\sin^2[(\varphi-2\delta)/2]},\tag{1}$$

where t is the transmission of the reflecting film, r its reflectivity, φ the phase shift through the dielectric layer is equal to $4\pi nd/\lambda$ and δ is the phase shift at the boundary of the dielectric layer. This equation ignores the absorption in the film and shows that the peak transmission is independent of the silver reflectivity. In the exact formula which considers the absorption of the metallic layers, the peak transmission is no longer independent of the band width and, as the reflectivity is increased and the band width decreased, the peak transmission is also decreased, so that for very narrow band widths the resulting filter is not particularly useful because of extremely low peak transmission.

The Frustrated Total Reflection filter uses the principle of the penetration into the second medium which occurs when light is incident on a surface beyond the angle of total internal reflection. When light is sent

[1] Walter Geffcken, German Patent No. 716153 (filed December 8, 1939).
[2] H. D. Polster, J. Opt. Soc. Am. **39**, 1054A (1949).
[3] C. DuFour, Le Vide **16**, 3 (1948).
[4] B. Lyot, Comptes Rendus **197**, 1593 (1933).
[5] J. Evans, Pubs. Astron. Soc. Pacific **52**, 305 (1940).
[6] B. Billings, J. Opt. Soc. Am. **37**, 738 (1947).
[7] P. Leurgans and A. F. Turner, J. Opt. Soc. Am. **37**, 983 (1947).
[8] B. Billings and M. A. Pittman, J. Opt. Soc. Am. **39**, 978 (1949).

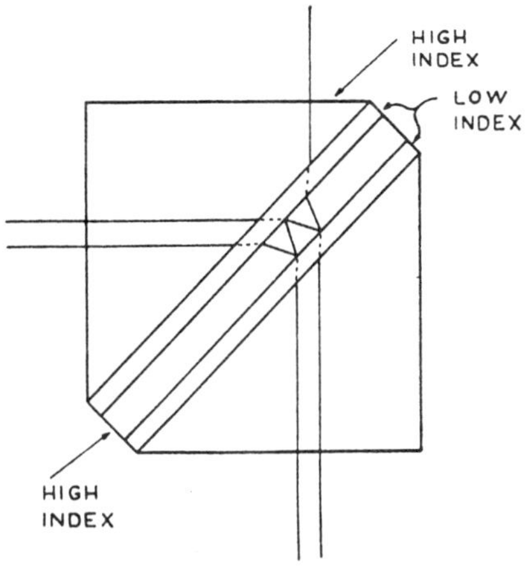

FIG. 1. Arrangement of the elements and path of the rays in the Turner Frustrated Total Reflection filter.

through the base of a right-angle prism, it is reflected at the hypotenuse. This reflection is 100 percent if the light strikes the hypotenuse of the prism at an angle of incidence greater than the critical angle. If now a convex lens is put in contact with the hypotenuse, the light at the point of contact will not be reflected but instead will be transmitted into the lens. When the light reflected from the prism hypotenuse is observed, a black spot is seen where this penetration takes place. The size of this black spot is found to be much larger than one would expect if the light went through only where actual optical contact was achieved. In other words, the light appears to penetrate into the air space between the lens and the prism, and where that air space is sufficiently thin it is able to reach the lens surface and be transmitted. Examination of Maxwell's equations at the prism hypotenuse shows that at angles bigger than the critical angle the light actually does penetrate into the low index medium. In this medium the amplitude drops off exponentially with the distance. If a high index surface is placed close enough to the hypotenuse so that the light amplitude has not dropped to zero, then a fraction of the light will be transmitted. No light will be lost in the low index layer. The surface of this low index layer thus constitutes a semitransparent reflector with no absorption loss.

The Frustrated Total Reflection filter consists of two such semitransparent reflectors. The prism hypotenuse is covered with a low index layer whose thickness is determined by the desired reflectivity. This is followed by a high index layer whose thickness determines the wave-length which will be transmitted. Another low index layer is then deposited so as to give another frustrated total reflection. Another prism is then laminated to the first prism with a material whose index is the same as the prism material. The resulting rhomb constitutes the interference filter.

This filter, unlike the regular solid Fabry-Perot filter, has no inherent limitations which make it unsatisfactory for wave-lengths in the ultraviolet or infra-red. The only requirements are that the materials of which the filter is made be transparent in the desired spectral region and that the indices have the proper relationship.

The wave-length of peak transmission is determined by the thickness of the high index layer and the band width by the thickness of the low index layer. Equation (1) gives a good approximation to the behavior of the filter at normal incidence. When the filter is tilted, the equivalent thickness of the high index layer changes and the peak transmission accordingly changes.

The change of peak transmission with angle of incidence is most easily derived by considering the behavior of the filter as arising from the interference of the light reflected from the boundary of the first low index layer with light reflected from the second low index layer.

In Fig. 1 is a sketch showing the path of the rays through the Frustrated Total Reflection filter. Since the same multiple reflections take place that occur in a solid Fabry-Perot filter, we can use the Fabry-Perot formula to describe the operation of this filter. The expressions for r, t, δ and φ are, however, quite different. By applying Maxwell's equations to reflection at a prism hypotenuse, they can be shown for the perpendicular component to be:

$$r_s=\frac{(\sinh^2 u)(n_1^2-n_2^2)^2}{D_s}, \tag{2}$$

$$t_s=\frac{4(n_1^2\sin^2\theta-n_2^2)n_1^2\cos^2\theta}{D_s}, \tag{3}$$

$$D_s=\sinh^2 u(n_1^2\cos^2\theta-n_1^2\sin^2\theta+n_2^2)^2 +(n_1^2\sin^2\theta-n_2^2)(2n_1\cos\theta)^2\cosh^2 u, \tag{4}$$

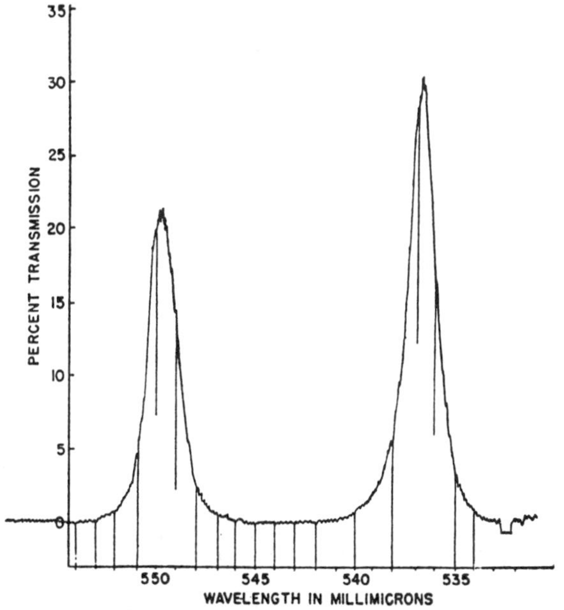

FIG. 2. Transmission as a function of wave-length of a simple Frustrated Total Reflection filter. The filter is made with 67.5° glass prisms. The low index layer has $n=1.38$ and an optical thickness of $\frac{3}{4}$ of a wave at 560 mμ. A high index layer has $n=2.38$ and an optical thickness of 6 waves at 600 mμ.

where

$$u=(2\pi d)/\lambda(n_1^2\sin^2\theta-n_2^2)^{\frac{1}{2}}. \tag{5}$$

n_1 is the low index, n_2 the high index, and θ the angle at which light is incident on the low index boundary. Similar expressions exist for the parallel component. For the various phase shifts

$$\tan\frac{\delta_s}{2}=\frac{(\sin^2\theta-(n_1/n_0)^2)^{\frac{1}{2}}}{\cos\theta}, \tag{6}$$

$$\tan\frac{\delta_p}{2}=\frac{(\sin^2\theta-(n_1/n_0)^2)^{\frac{1}{2}}}{(n_1/n_0)^2\cos\theta}, \tag{7}$$

$$\varphi=(4\pi nd/\lambda)\cos\theta, \tag{8}$$

where n_0 is the prism index. The subscript s refers to the perpendicular component and p to the parallel component. Note that there are now two values for the quantity δ, the phase shift at the boundary. This is because the light is incident at an oblique angle on the interfering surfaces. In Fig. 2 is shown the transmission curve of a typical Frustrated Total Reflection filter. This filter has the following characteristics:

$$\theta_0=67.5°$$
$$n_{\text{low}}=1.38$$
$$n_{\text{high}}=2.38$$
$$nd_{\text{low}}=3/4\lambda \text{ at } 560\text{ m}\mu$$
$$nd_{\text{high}}=6\lambda \text{ at } 600\text{ m}\mu.$$

The prism material had an index of 1.63.

One of the problems in making measurements on filters of this type is the high resolution required in the

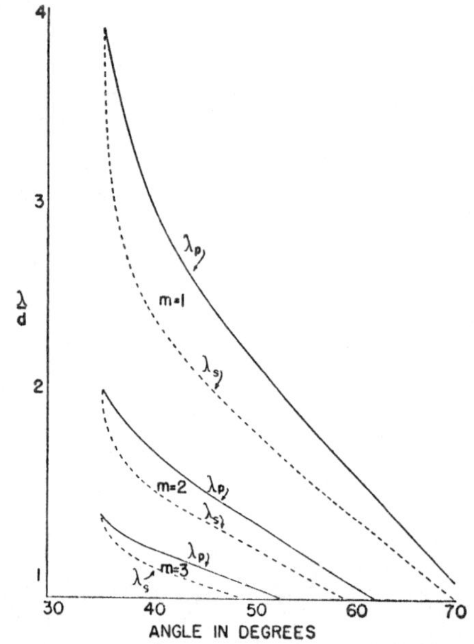

FIG. 3. Variation of wave-length of peak transmission with angle of incidence for a Frustrated Total Reflection filter. λ_s is the wave-length for light polarized perpendicular to the plane of incidence; λ_p for light polarized parallel to the plane of incidence. d is the thickness of the high index layer. The indices are the same as in Fig. 2.

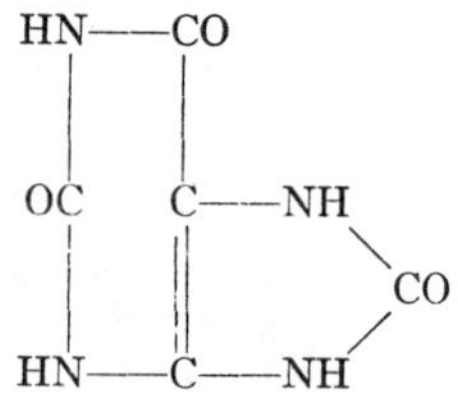

FIG. 4. Formula for uric acid.

measuring instrument. To measure accurately the percent transmission of a filter whose band width is 1 mμ requires a spectrometer which can resolve at least 0.2 mμ, and for really accurate work even higher resolution is necessary. Most of the conventional recording spectrophotometers do not have this high resolution. A grating instrument has been assembled at Baird Associates, Inc. for the purpose of working at this high resolving power. The instrument uses a 15,000 line per inch grating and for these measurements was operated with a photo-multiplier tube as the receiver. The light falling on the entrance slit is interrupted by a mechanical chopper. The output of the photo-multiplier tube is fed through conventional a.c. electronics to a Leeds and Northrup recorder. With this instrument it is possible to achieve resolution of a fraction of an angstrom and thus make accurate and rapid transmission measurements of these narrow band filters. Complete description of the grating monochromator will appear in a forthcoming paper.

The spacing of the pass bands is a function of the angle of the filter, of the indices of the high and low index layers, and of the order number. The equation for the wave-lengths of the peaks is:

$$\frac{4\pi nd\cos\theta}{\lambda}-2\delta=2m\pi \tag{9}$$

or

$$\lambda=\frac{2\pi nd\cos\theta}{m\pi+2\delta}, \tag{10}$$

where m is the order of the pass band. The band separation in cm^{-1} is expressed by the equation:

$$\Delta\nu=\frac{\delta_s-\delta_p}{2\pi nd\cos\theta}. \tag{11}$$

In Fig. 3 is a series of curves showing the way the wave-length of peak transmission changes as a function of the angle of incidence. In order to make the pass bands come together, it is apparent by inspection of Eq. (1) that the quantity $\varphi=2\delta$ must be made to be the same for each state of polarization. This can only be accomplished by making some layer of the filter birefringent. If the central layer is made birefringent the sum of the phase shift of traversal and the phase shift of reflection can be made equal for both states of polarization. The angular dependence of these phase shifts for birefringent layers becomes more complex than that given in Eqs. (6) to (8) because the birefringence depends on angle. This dependence is a function of the orientation of the crystallographic axes of the layer. It is only recently that research at the Polaroid Corporation made it possible to prepare birefringent layers by evaporation.

EVAPORATED BIREFRINGENT LAYERS

Although the preparation of evaporated birefringent layers has not yet appeared in scientific literature, it is covered fairly well in a series of patents[9–12] owned by Polaroid Corporation. The substances which are strongly birefringent after evaporation are all, in general, organic. They are usually planar and have one or more benzene rings. Distributed around the rings are groups which will exert force between the molecule and a surface or between the molecules and each other. Layers of such molecules are symmetrical about an axis normal to the surface on which they are evaporated. The resulting medium is birefringent with its optic axis normal to the film. At a given angle the indices of such a film are:

$$\omega(\theta)=\omega$$

$$\epsilon'(\theta)=\frac{\omega\epsilon}{(\epsilon^2\cos^2\theta+\omega^2\sin^2\theta)^{\frac{1}{2}}}, \tag{12}$$

where ω is the ordinary index, ϵ the extraordinary index and θ is the angle between the wave front of the light

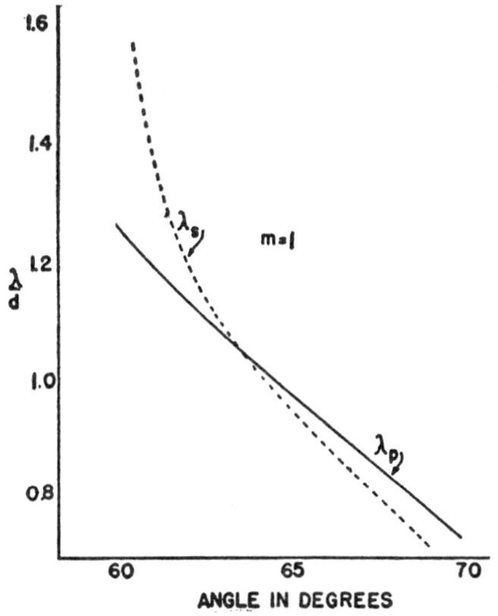

FIG. 5. Variation of wave-length of peak transmission with angle of incidence for a Frustrated Total Reflection filter. In this filter the central layer is birefringent with ω 1.72, ϵ 1.55, and the low index layer 1.38.

[9] E. E. Barr and C. D. West, U. S. Patent No. 2,447,790 (filed April 11, 1945).
[10] E. R. Blout, U. S. Patent No. 2,447,792 (filed March 24, 1945).
[11] M. Hyman, Jr., U. S. Patent No. 2,447,805 (filed April 11, 1945).
[12] R. B. Woodward, U. S. Patent No. 2,447,831 (filed March 3, 1945).

and the optic axes. It is possible to achieve quite large values for the birefringence of these evaporated materials. A birefringence of 0.17 is obtainable with uric acid. This formula is given in Fig. 4.

BIREFRINGENT LAYERS FOR THE FRUSTRATED TOTAL REFLECTION FILTER

For the Frustrated Total Reflection filter the birefringent layers must have certain, very specific characteristics in order to have a single pass band for both states of polarization. In case the central layer is birefringent both indices must be higher than the low index material. If one index is too low then for one state of polarization the filter transmits all wavelengths, whereas for the opposite state of polarization, the filter will have single narrow pass bands. If one of the lower index layers is made birefringent, both indices must be lower than the index of the high central layer, otherwise again a continuum will be passed for one state of polarization.

Consider now the case where the central layer is birefringent. Both indices must be high. It is apparent from Eq. (1) and Eqs. (6), (7), and (8) that a filter can be made which will have a single pass band at any desired wave-length. The two parameters which can be adjusted to achieve this result are the angle of the filter and the thickness of the central layer. For an anisotropic central layer Eq. (10) for the wave-length of peak transmission becomes:

$$\frac{4\pi\epsilon'(\theta)\omega d\cos\theta}{\omega\lambda}-2\delta_s(\theta)=2m\pi, \tag{13}$$

$$\frac{4\pi\omega d\cos\theta}{\lambda}-2\delta_p(\theta)=2m\pi. \tag{14}$$

By eliminating $\omega d/\lambda$ from these two equations one obtains the following expression:

$$\epsilon'(\theta)=\omega\frac{m\pi+2\delta_s(\theta)}{m\pi+2\delta_p(\theta)}. \tag{15}$$

As shown above, δ_s is larger than δ_p in the isotropic case. If this were true in the anisotropic case, ϵ would have to be bigger than ω and the central layer would thus need to be positively birefringent. When these experiments were first conceived, the false conclusion was reached that a positive layer would be necessary to bring the pass bands together. The intuitive reasoning behind this error seemed quite logical. The pass band of the light vibrating in the plane of incidence was at a shorter wave-length than the pass band for light vibrating perpendicular to the plane of incidence. To compensate for this difference it seemed obvious that the optical thickness of the filter for light vibrating in the plane of incidence should be greater than the optical thickness for the light vibrating perpendicular to the plane of incidence. Since these evaporated birefringent layers are optically equivalent to a uniaxial crystal with its axes perpendicular to the plane of the layers, a positive crystal thus seemed necessary. Positive crystals or positive evaporated layers are relatively rare. This is because such materials are generally formed from rod like structures where the rods are arranged parallel to each other. Oil films or films of paraffin are positive but they cannot be made very thick as they tend to form aggregates which make the layers quite hazy.

An experiment was performed using a negatively birefringent material to see if perhaps the peak from one order could be made to coincide with the peak from the next higher order. When the experiment was carried out it was discovered, of course, that the bands of a single order could be brought together. The reason for this was that δ_s and δ_p, the two phase shifts, are quite strongly dependent on index and for a negatively birefringent layer, these phase shifts change in such a way that δ_p becomes larger than δ_s. Equation (16) can then be solved to find the value of θ at which the bands can be brought together in a given order. The birefringent layer must be negative. This is most fortunate as negative materials are easily prepared and a filter with a single pass band thus becomes a practical possibility.

When m is very large $\epsilon'(\theta)$ approaches ω. This is true only at normal incidence. Since the filter will only operate at angles larger than the critical angle, coincidence cannot be achieved at high orders. In Fig. 5 is shown a calculated curve for the first order showing the way in which the bands shift position with angle of incidence. The characteristics of this filter were:

$$\theta_0=67.5°$$
$$n_{\text{low}}=1.38$$
$$\omega=1.72$$
$$\epsilon=1.55.$$

The coincidences for the different orders occur at angles and optical thicknesses shown in Table I. A filter was made up using magnesium fluoride for the low index layer and uric acid for the high index layer. The characteristics of this filter were as follows:

$$\theta_0=67°$$
$$(nd)_{\text{low}}=3/4\lambda \text{ at } 520\ \text{m}\mu$$
$$(nd)_{\text{high}}=9/4\lambda \text{ at } 600\ \text{m}\mu.$$

In Fig. 6 is shown the transmission of this filter when rotated through an angle of 10°. This is rotated so that the angle of incidence on the hypotenuse is smaller. In Fig. 7 the filter is tilted still farther so that the two bands coincide. In Fig. 8 the filter has been tilted so that the angle of incidence is still smaller and the bands have crossed over. The parallel band is now at a larger wave-length than the perpendicular band.

For very narrow transmission bands the tolerances in the thickness and uniformity of the evaporated layers are extremely small. The lower transmission of the birefringent filter shown in the figures is probably caused by roughness and non-uniformity in the organic film. To achieve high transmission it will be necessary to reduce this non-uniformity.

TABLE I. Values of $\omega d/\lambda$ and θ at which coincidence is achieved between bands of a given order. This is calculated for a filter with $n=1.38$ for the low index layer and with $\omega=1.72$ and $\epsilon=1.58$ for the high index layer.

$\omega d/\lambda$	θ	n
0.69	66.75°	0
1.68	64°	1
2.59	62°	2
3.47	60.75°	3

If the low index layer is made birefringent the crossover will depend only on the phase shift at reflection. In this case, all bands in all orders coincide at the same angle. This is quite different from the previous case in which coincidence was achieved at a different angle in each order. The wave-length of peak transmission is given simply by:

$$1/\lambda=\frac{m\pi+\delta(\theta)}{2\pi nd\cos\theta}, \tag{16}$$

where θ is that angle at which $\delta_s(\theta)=\delta_p(\theta)=\delta(\theta)$.

This case is somewhat more difficult to realize in practice because of the unavailability of birefringent substances in which both indices are low. A possible method of making such layers is the technique described by Wiener.[13] In this method birefringent layers are made by piling up alternate layers of high and low index material. The individual layer thicknesses are a small fraction of a wave-length in the region where the sandwich is to be birefringent.

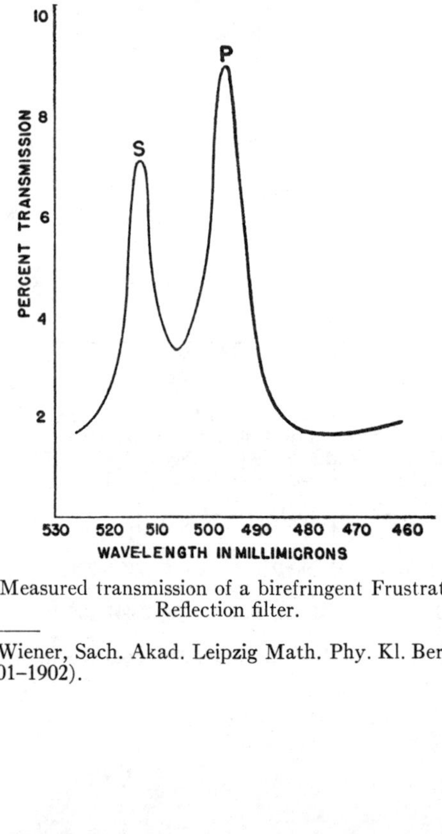

FIG. 6. Measured transmission of a birefringent Frustrated Total Reflection filter.

[13] O. Wiener, Sach. Akad. Leipzig Math. Phy. Kl. Berichte **53, 54,** (1901–1902).

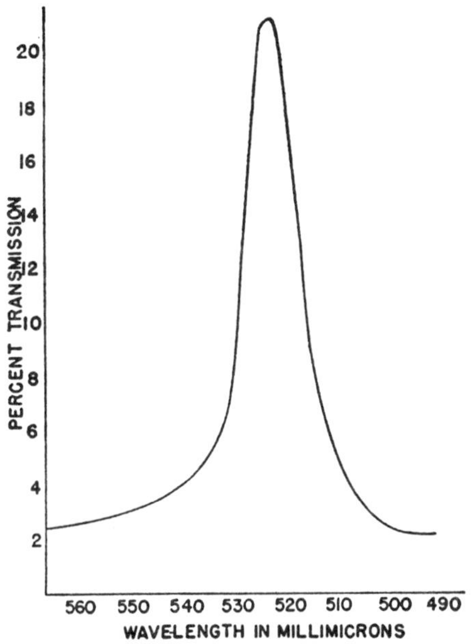

FIG. 7. Measured transmission of the filter shown in Fig. 6 tilted to the angle at which the pass bands coincide.

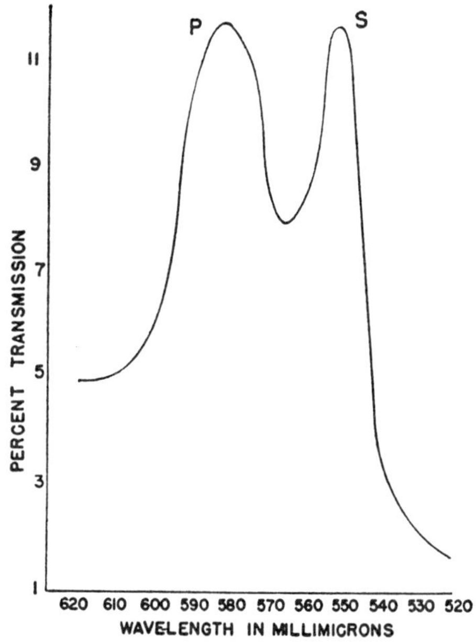

FIG. 8. Measured transmission of the birefringent filter of Figs. 6 and 7 tilted until the two bands have reversed their positions.

A treatment of this low index case will be given in a later paper.

CONCLUSION

Experimental measurements have been shown for the transmission of a birefringent Frustrated Total Reflection filter. It should be possible, using the technique demonstrated in this paper, to make filters of band widths of the order of 1 mμ and transmissions of perhaps 70 to 80 percent. There are many places where such filters could be most useful. The resolution is sufficiently high so that in many situations they become competition for an ordinary monochromator. Their main use, however, probably lies in image forming systems. There are occasions when it is desirable to take a photograph of a flame or of some luminous phenomenon in monochromatic light. A typical problem is the photography of solar prominences. If no occulting disk is used good measurements of the prominences require a band width of an angstrom or so. This is normally achieved with polarization filters which require many components. The Frustrated Total Reflection filter can be used as a fore filter for such a device. In this fashion it will save the use of a large part of the crystalline material used in the construction of polarization filters.

It is a pleasure to acknowledge the help of Edgar E. Barr, Grace E. Griffith, and Antonio F. Quesada in the preparation of this manuscript.

Metal Grid Interference Filter*

EDWARD A. LEWIS
U.S.A.F. Cambridge Research Laboratories, Cambridge, Massachusetts

AND

JOSEPH P. CASEY
U.S.A.F. Cambridge Research Laboratories and Boston University, Boston, Massachusetts

(Received January 26, 1951)

AN interference filter has been constructed and tested using metal grids as the reflecting elements in place of the customary highly absorbing silver films. The results indicate, despite the crudeness of the experimental microwave model, that the absorption is far less than would be expected had continuous metal films of the same reflection coefficient been used.

The theory of the reflection and transmission of a single grating composed of lossless parallel wires has been worked out by Wessel.[1] The reflection coefficient when the E-vector is parallel to the wires is dependent in both amplitude and phase upon the two ratios λ/d and d/a, where λ is the wavelength of the incident radiation, d is the center-to-center spacing of the individual wires, and a the wire radius. It has been shown numerically from Wessel's data that, for gratings of high reflectivity, the modulus of the amplitude reflection coefficient is approximately equal to minus the cosine of the phase change on reflection. At a large distance from the wires, the reflected and transmitted waves are therefore practically the same as would be expected from hypothetical, continuous, lossless, and very thin reflecting layer.

The properties of an interferometer consisting of two such surfaces can be deduced by the rigorous application of the ray method. (Alternately, the same result is obtained by setting up the boundary value problem of two infinite sets of wires, and using asymptotic formulas for the bessel functions which are encountered.) If $e^{i\theta}$ is the amplitude coefficient for reflection for a single surface, and $\theta=\cos^{-1}(-\rho)$, it is found that the reflected and transmitted intensities for two surfaces are, respectively,

$$R=[4\rho^2\sin^2(\psi-\theta)]/[(1-\rho^2)^2+4\rho^2\sin^2(\psi-\theta)],$$
$$T=[(1-\rho^2)^2]/[(1-\rho^2)^2+4\rho^2\sin^2(\psi-\theta)],$$

where $\psi=2\pi s/\lambda$, s being the distance separating the surfaces.

It will be seen that, for proper choice of ψ, the transmission becomes 100 percent, while the reflection vanishes. Assuming ρ does not change rapidly with wavelength, it follows that the half-power points have a fractional frequency separation of

$$\Delta f/f=[\lambda/\pi s]\sin^{-1}[(1-\rho^2)/2\rho].$$

From the calculated current in the ideal wire grid, the loss was estimated for a typical silver or copper wire grating with high reflection coefficient, taking into account the skin depth. This turns out to be much less than 1 percent at $\lambda=1.25$ cm.

A very crude microwave model was constructed of commercially available screening. This screening had cross wires, of course, and was therefore equivalent to two crossed gratings, rendering the screen insensitive to polarization. The cross wires increased the reflection coefficient slightly over that computed for a single grating. The measured value was 94 percent (intensity). The double screen filter had a band width within 3 percent of that computed for a grating of the same reflection coefficient and passed 60 percent of the incident radiation at the desired wavelength. The calculated transmission for a filter using continuous metal films of the same reflection coefficient was about 0.02 percent.

The possibility of constructing metal grid surfaces for use in the millimeter and infrared regions is now under consideration. The shorter the wavelength the finer the wires and their separation, so ultimately it may be necessary to resort to silver filaments or meshes deposited on flat surfaces.

* This work was done under the sponsorship of the Radio Physics Research Directorate, AF Cambridge Research Laboratories, AMC.

[1] W. Wessel, Hochfrequenztechnik 54, 62 (1939).

©1987 Optical Society of America. Reprinted from *Applied Optics*, Vol. 26(9), 1693–1695 (May 1987).

High frequency Fabry-Perot phase modulator

David W. Sesko and Carl E. Wieman

We have constructed an efficient high frequency phase modulator using a KDP crystal modulator in a Fabry-Perot interferometer. We present the theoretical analysis for the performance of such a modulator and compare this with experimental results. We have obtained a modulation index of 0.6 at 2.7 GHz with 0.3 W of microwave power. This implies that the use of a Fabry-Perot cavity can reduce the amount of drive power needed to achieve a given modulation index by 3 orders of magnitude or more. Such a device is particularly useful in the ultraviolet region of the spectrum where one must use relatively inefficient modulator crystals.

The technique of FM or frequency modulation spectroscopy has made possible optical absorption measurements of unprecedented sensitivity.[1,2] One of the many exciting applications of this technique is the detection of trace species in the atmosphere.[3] For this application, it is necessary to have an optical phase modulator which produces a modulation index of the order of one, at a frequency higher than the pressure broadened transition linewidth of the species of interest. This is typically several gigahertz. There has been a great deal of work done on high frequency modulation.[4–8] Recently, impressive advances in modulator technology have attained a high modulation index at microwave frequencies using $LiTaO_3$ as a modulator material.[7] This material works very well in the visible portion of the spectrum. Unfortunately, it is opaque in the ultraviolet where the absorption lines of some particularly important species such as OH are found. The modulator materials which are transparent in this spectral region, such as KDP, are far less efficient and thus would require unacceptably large amounts of drive power to achieve a satisfactory modulation index. A possible solution to this problem is to modulate a visible laser and then double the laser frequency. This involves substantial problems for cw laser sources, however, because the doubling efficiency is low and boosting the efficiency by intracavity doubling can introduce large amounts of undesirable amplitude modulation. We have explored the alternative approach of enhancing the frequency modulation efficiency in the UV by using an optical cavity. The realization that a resonant optical cavity will provide an enhancement of the modulation index is quite old. It was demonstrated qualitatively by Ruscio at low modulation frequencies,[5] and a number of people including Gallagher *et al.*[8] have proposed it could be done at microwave frequencies. We have demonstrated this by placing a high frequency KDP modulator inside a resonant Fabry-Perot interferometer. In this paper we will provide a theoretical analysis of our microwave modulator and then a qualitative discussion of the experimental performance.

Phase modulation is accomplished by orienting the optic axis (Z) of a nonlinear crystal along the direction of propagation of a light beam. The light beam is polarized along the X axis of the crystal so that the modulating field along the Y axis produces a phase modulation index of

$$\delta = \omega n_O^3 r_{41} E_m l/2c, \tag{1}$$

where ω is the optical frequency, n_O is the refractive index of the ordinary ray, r_{41} is the electrooptic coefficient, E_m is the strength of the modulation field, and l is the length of the crystal. Equation (1) is valid if l is short enough so that the optical wave and the modulating field remain in phase throughout the crystal.[4] The optical field, as it leaves the phase modulator, is described by a sum of Bessel functions:

$$E_{\text{out}} = A\{J_0(\delta)\cos\omega t + J_1(\delta)[\cos(\omega + \omega_m)t - \cos(\omega - \omega_m)t] + (\text{higher-order terms})\}, \tag{2}$$

where A is the amplitude of the incident field, and ω_m is the frequency of modulation. Equation (2) shows the

---

The authors are with Joint Institute for Laboratory Astrophysics, University of Colorado and U.S. National Bureau of Standards, Boulder, Colorado 80309-0440.

Received 20 October 1986.
0003-6935/87/091693-03$02.00/0.

field amplitude of the carrier and the sidebands as a function of the modulation index.

We will now calculate the enhancement of power in the sidebands when the crystal is inside the Fabry-Perot interferometer. As we will show, the resonant frequency of the Fabry-Perot and the modulation frequency must be set so that the optical carrier and the sidebands all fall within the transmission of the interferometer. Also, the position of the KDP crystal in the interferometer must be chosen to insure that the sidebands produced before and after reflection of the optical wave are in phase. We analyze this system by first decomposing the microwave and optical standing waves into forward and backward traveling waves. The traveling microwave field propagating in the same direction as one of the optical waves produces a modulation on this wave. If the modulator is an integral number of microwave half-wavelengths long, there is no modulation due to the counterpropagating microwave field.[4] This condition was true for our experiment and will be assumed in the following analysis.

To get an explicit expression for the modulation, we consider an input optical field of magnitude A into the interferometer. The steady-state optical field may be obtained by summing the multiple round-trip passes of the fields,[9] keeping track of the phases of the carrier and the sidebands in each of the passes through the modulator. To first order in $J_1(\delta)$ the fields of the first pass, the next round trip, and the nth round trip through the crystal are, respectively,

$$E_0 = At_1(1-\gamma)(J_0(\delta)\exp(i\omega_0 t) + J_1(\delta)\{\exp[i(\omega_0+\omega_m)t] - \exp[i(\omega_0-\omega_m)t]\}), \tag{3}$$

$$\begin{aligned} E_1 = {} & At_1r_1r_2(1-\gamma)^3\exp\left(i\omega_0\frac{2L}{c}\right)\Bigg(J_0^3(\delta)\exp(i\omega_0 t) \\ & + J_1(\delta)J_0^2(\delta)\Bigg\{\left[1+\exp\left(i\omega_m\frac{2L}{c}\right)+\exp\left(i\omega_m\frac{2x}{c}\right)\right] \\ & \times\exp[i(\omega_0+\omega_m)t] \\ & -\left[1+\exp\left(-i\omega_m\frac{2L}{c}\right)+\exp\left(-i\omega_m\frac{2x}{c}\right)\right]\exp[i(\omega_0-\omega_m)t]\Bigg\}\Bigg), \end{aligned} \tag{4}$$

$$\begin{aligned} E_n = {} & At_1\left[r_1r_2(1-\gamma)^2\exp\left(i\omega_0\frac{2L}{c}\right)\right]^n\Bigg[J_0^{2n+1}(\delta)\exp(i\omega_0 t) \\ & + J_1(\delta)J_0^{2n}(\delta)\Bigg(\sum_{j=0}^{n}\Bigg\{\exp\left(i\omega_m\frac{2jL}{c}\right)\exp[i(\omega_0+\omega_m)t] \\ & - \exp\left(-i\omega_m\frac{2jL}{c}\right)\exp[i(\omega_0-\omega_m)t]\Bigg\} \\ & + \sum_{k=0}^{n-1}\Bigg\{\exp\left(i\omega_m\frac{2x}{c}\right)\exp\left(i\omega_m\frac{2kL}{c}\right)\exp[i(\omega_0+\omega_m)t] \\ & - \exp\left(-i\omega_m\frac{2x}{c}\right)\exp\left(-i\omega_m\frac{2kL}{c}\right)\exp[i(\omega_0-\omega_m)t]\Bigg\}\Bigg)\Bigg], \end{aligned} \tag{5}$$

where t_j and r_j are the amplitude transmission and reflection coefficients of mirror j, γ is the loss through the crystal, L is the optical path length of the interferometer, and x is the optical path length from the center of the crystal to the input mirror.

By requiring the phases of the carrier wave terms in Eqs. (3)–(5) to match, we obtain the usual Fabry-Perot condition that

$$\omega_0\frac{2L}{c} = 2M\pi. \tag{6}$$

Similarly, for the sideband terms to also match in phase,

$$\omega_m\frac{2L}{c} = 2N\pi, \tag{7}$$

where N and M are integers. This states that both the carrier and sideband frequencies must be an integral number of the free spectral range of the interferometer.

By summing all the components of the carrier from one to an infinite number of passes, we find that the field transmitted by the Fabry-Perot cavity is

$$E_c = At_1t_2(1-\gamma)J_0(\delta)/[1-r_1r_2(1-\gamma)^2J_0^2(\delta)]. \tag{8}$$

The field of the sideband may be obtained by summing in a similar manner and taking the real part:

$$E_s = \frac{At_1t_2(1-\gamma)J_1(\delta)\left[1+r_1r_2(1-\gamma)^2J_0^2(\delta)\cos\frac{\omega_m 2x}{c}\right]}{[1-r_1r_2(1-\gamma)^2J_0^2(\delta)]^2}. \tag{9}$$

The quantity of interest is the ratio of sideband to carrier field amplitudes which we represent by an effective modulation index defined as $2(E_s/E_c)$. For $\delta \ll 1$, we can make the approximations that $J_0(\delta) = 1$ and $J_1(\delta) = \delta/2$ and from Eqs. (8) and (9) we obtain

$$\delta_{\text{eff}} = 2(E_s/E_c) = \frac{\delta F[1+r_1r_2(1-\gamma)^2\cos(\omega_m 2x/c)]}{\pi(r_1r_2)^{1/2}(1-\gamma)}, \tag{10}$$

where F is the finesse of the cavity given by

$$F = \pi(r_1r_2)^{1/2}(1-\gamma)/[1-r_1r_2(1-\gamma)^2]. \tag{11}$$

From Eq. (10) we see the effective modulation index will be largest when the crystal is positioned so that

$$\frac{2x}{c}\omega_m = 2p\pi \text{ or } x = \frac{pL}{N}, \tag{12}$$

where p is an integer less than N. Thus for $N = 4$, there will be three locations in the interferometer where the crystal can produce maximum modulation.

Equations (10) and (11) can be simplified by assuming the plausible conditions that the crystal is positioned properly, $r_{1,2}$ are equal and close to one, and γ is small. This gives the simple result that

$$\delta_{\text{eff}} = \frac{2\delta F}{\pi}, \tag{13}$$

and $F = \pi/1 - R$, where R is the mirror reflectivity. One other relevant factor is the maximum transmission through the interferometer which is[10] $T_{\text{int}} = (1 - A/1 - R)^2$, where A is the single pass absorption. Dielectric mirrors typically have losses of <0.1% and

KDP crystals can be obtained with insertion losses of <1%. Thus for a mirror reflectivity of 95%, from Eq. (13) we see the modulation index will be increased by a factor of ~40 with less than a 30% loss in the total optical power. Equivalently, to obtain a given modulation index this means the microwave power required is reduced by a factor of 1600.

We have demonstrated the predicted enhancement of the modulation index at microwave frequencies using the setup in Fig. 1. The output of a polarized He–Ne laser was mode matched into an optical resonator which contained a microwave modulator (commercial Sylvania modulator, $Q \simeq 1000$). The resonator was a Fabry-Perot interferometer with one mirror having a radius of curvature of 15 cm and $R = 0.99$ and the other being flat with $R = 0.995$. The curved mirror was mounted on a piezoelectric transducer which allowed it to be translated by a few microns and thereby vary the resonant frequency of the interferometer. The mirror separation was ~20 cm. The phase modulation was done by a 3.53-cm long KDP crystal contained in a cylindrical resonant microwave cavity. The TE_{111} mode of the cavity was excited to produce a standing wave with its field perpendicular to the optic axis of the crystal. The resonant frequency of the modulator was 2.7 GHz and could be tuned by changing the length of the cavity. The cavity was excited with 0.3 W of microwave power measured just before entering the modulator. With the modulator in place, the free spectral range of the optical cavity was very close to one-quarter of 2.7 GHz.

It should be noted that the purpose of this experiment was only to demonstrate the enhanced modulation. Therefore many of the parameters of this setup were determined by what equipment was readily available and hence are far from optimum for a real modulator. In particular the KDP crystal had antireflection coatings for the wrong wavelength which caused an insertion loss of 12% per pass. This limited the finesse of the cavity, and in combination with the excessively high reflectivity of the cavity mirrors, gave a low optical throughput.

The resonant frequency of the optical cavity was held at the frequency of the He–Ne laser using a servo system based on the Hansch-Couillaud approach.[11] In this scheme an error signal is derived by analyzing the circular polarization of the light reflected off the optical cavity. This error signal was sent through a high voltage integrating amplifier whose output drove the piezoelectric transducer.

We examined the light transmitted through the optical cavity with a scanning Fabry-Perot interferometer optical spectrum analyzer. This allows us to measure the relative power in the carrier and the modulation sidebands. With the microwave frequency adjusted to satisfy the phase-matching condition the ratio of sideband to carrier height was 0.10(1), or a modulation index of 0.60(4). With only a single pass through the modulator (mirrors removed) the ratio was $1.7(3) \times 10^{-4}$, giving a modulation index of 0.026(4). We determined that the finesse of the optical cavity was 25(5) by measuring the linewidth and free spectral range. Using this value and Eq. (9), we calculate that the optical cavity should increase the modulation index by a factor of 18(3). This is in reasonable agreement with the measured value of 23(4).

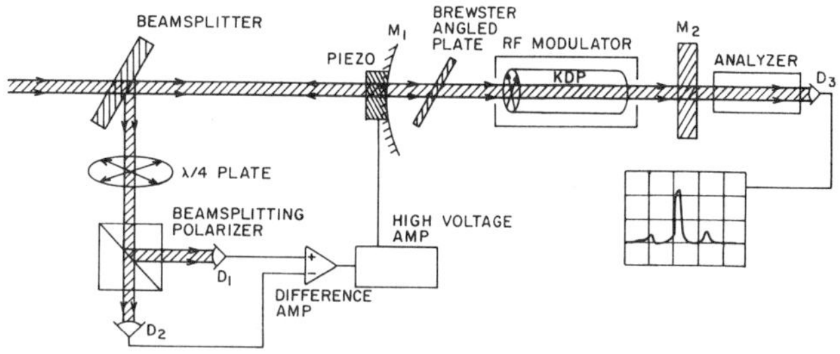

Fig. 1. Diagram of optical servo system phase modulator and analyzer.

Thus we have demonstrated that a resonant optical cavity can give an enormous improvement in the efficiency of a high frequency optical modulator. This allows a large modulation index to be obtained at microwave frequencies using relatively inefficient materials such as KDP and modest modulation powers. The primary advantage to such a device is that it will work in the ultraviolet region of the spectrum where traditional $LiTaO_3$ modulators are opaque.

We are happy to acknowledge the loan of the microwave modulator by J. Magyar and valuable discussions with J. Ward. This work was supported by the National Science Foundation. One of us (CEW) is an Alfred P. Sloan fellow.

References

1. G. C. Bjorklund, "Frequency-Modulation Spectroscopy: A New Method for Measuring Weak Absorptions and Dispersions," Opt. Lett. **5**, 15 (1980).
2. N. C. Wong and J. L. Hall, "Servo Control of Amplitude Modulation in Frequency-Modulation Spectroscopy: Demonstration of Shot-Noise-Limited Detection," J. Opt. Soc. Am. B **2**, 1527 (1985).
3. J. F. Ward, C. C. Wang, and C. E. Wieman, "A Proposal for Improved OH Detection Using Frequency Modulation Spectroscopy," Bull. Am. Phys. Soc. **30**, 643 (1985).
4. For a comprehensive overview and collection of reprints, see I. P. Kaminow, *An Introduction to Electrooptic Devices* (Academic, New York, 1974).
5. J. T. Ruscio, "A Coherent Light Modulator," IEEE J. Quantum Electron. **QE-1**, 182 (1965).
6. I. P. Kaminow and W. M. Sharpless, "Performance of $LiTaO_3$ and $LiNbO_3$ Modulators at 4 GHz," Appl. Opt. **6**, 351 (1967).
7. N. H. Tran, T. F. Gallagher, J. P. Watjen, G. R. Janik, and C. B. Carlisle, "High Efficiency Resonant Cavity Microwave Optical Modulator," Appl. Opt. **24**, 4282 (1985), and references therein.
8. T. F. Gallagher, N. H. Tran, and J. P. Watjen, "Principles of a Resonant Cavity Optical Modulator," Appl. Opt. **25**, 510 (1986).
9. M. Born and E. Wolf, *Principles of Optics*, (Pergamon, New York, 1980), pp. 327–328.
10. E. Hecht and A. Zajac, *Optics* (Addison-Wesley, Reading, MA, 1976), p. 308.
11. T. W. Hänsch and B. Couillaud, "Laser Frequency Stabilization by Polarization Spectroscopy of a Reflecting Reference Cavity," Opt. Commun. **35**, 441 (1980).

Section Six
Polarization and Nature

We are of course surrounded by polarized light. Perhaps most startling to me was the discovery that only a small number of optical scientists are aware that the eye can distinguish polarized light and have taken the trouble to look at Haidinger's brush. I was most tempted to put Land's polarized headlights in this section because the reduction in polarized glare is the principal use of this invention.

The paper by Rinne written in 1926 complains that too few documents on the rainbow mention that it is polarized. This is still true today. There is almost no textbook that, in describing the optics of the rainbow, mentions the light is polarized. Fortunately, Minnaert[2] gives a good description and fortunately for me the reference to Rinne's paper.

---

[2] Minnaert, *Light and Colour in the Open Air*, Dover Publications Inc., 1954.

Reprinted from *Mem. Phys. Chim Soc. D'Arcueil*, Vol. 2, 143–158 (1809).

*Sur une propriété de la lumière réfléchie**

PAR M. MALUS.

LORSQU'UN rayon solaire est réfléchi ou réfracté, il conserve en général ses propriétés physiques; et, soumis à de nouvelles épreuves, il se comporte comme s'il émanoit directement du corps lumineux : le prisme, en dispersant les rayons colorés, ne fait que changer leur direction respective, sans altérer leur nature. Il y a cependant des circonstances où l'influence de certains corps imprime aux rayons qu'ils réfléchissent, ou qu'ils réfractent, des caractères et des propriétés qu'ils transportent avec eux, et qui les distinguent essentiellement de la lumière directe.

La propriété de la lumière que je vais décrire est une modification de ce genre. Elle avoit déja été apperçue dans une circonstance particulière de la duplication des images, offerte par le spath calcaire; mais le phénomène qui en résultoit étant attribué aux propriétés de ce cristal, on ne soupçonnoit pas qu'il pût être produit non-seulement par tous les corps cristallisés qui donnent une double réfraction, mais encore par toutes les autres substances diaphanes solides ou liquides et même par les corps opaques.

Si on reçoit un rayon de lumière perpendiculairement à la face d'un rhomboïde de spath calcaire, ce rayon se divise en deux faisceaux, l'un qui se prolonge dans la direction du rayon incident, et l'autre qui fait avec celui-ci un angle de quelques degrés. Le plan qui passe par ces deux rayons jouit de plusieurs propriétés particulières, et on le nomme plan de la section principale. Il est toujours parallèle à l'axe des molécules intégrantes du cristal et perpendiculaire à la face réfringente naturelle ou artificielle. Lorsque le rayon incident est incliné à la face réfringente il se divise également en deux faisceaux, l'un qui est réfracté suivant la loi ordinaire, et l'autre suivant une loi extraordinaire qui dépend des angles que le rayon incident forme avec la surface réfringente et la section principale. Lorsque la face d'émergence est parallèle à la face d'incidence, les deux rayons émergens sont parallèles au rayon incident, parce que chaque rayon éprouve aux deux faces opposées le même genre de réfraction.

Si actuellement on reçoit sur un second rhomboïde dont la section principale soit parallèle à celle du premier, les deux rayons qui ont déja traversé celui-ci, ils ne seront plus divisés en deux faisceaux comme l'eussent été des rayons de lumière directe. Le faisceau provenant de la réfraction ordinaire du premier cristal sera réfracté par le second suivant la loi de la réfraction ordinaire, comme si celui-ci avoit perdu la faculté de doubler les images. De même le faisceau provenant de la réfraction extraordinaire du premier cristal sera réfracté par le second suivant la loi de la réfraction extraordinaire.

Si le premier cristal restant immobile on fait tourner le second de manière que la face d'incidence reste parallèle à elle-même, chacun des deux rayons provenant de la réfraction du premier cristal commence à se diviser en deux faisceaux; en sorte, par exemple, qu'une partie du rayon provenant de la réfraction ordinaire, commence à se réfracter extraordinairement, ce qui produit quatre images. Enfin, après un quart de révolution le faisceau provenant de la réfraction ordinaire du premier cristal est en entier réfracté extraordinairement par le second; et réciproquement, le faisceau provenant de la réfraction extraordinaire du premier cristal est en entier réfracté par le second suivant la loi ordinaire, ce qui réduit de nouveau à deux le nombre des images. Ce phénomène est indépendant des angles d'incidence, puisque dans le mouvement du second cristal les faces réfringentes des deux rhomboïdes conservent entre elles la même inclinaison.

Ainsi le caractère qui distingue la lumière directe de celle qui a été soumise à l'action d'un premier cristal, c'est que l'une a constamment la faculté d'être divisée en deux faisceaux, tandis que dans l'autre cette faculté dépend de l'angle compris entre le plan d'incidence et celui de la section principale.

Cette faculté de changer le caractère de la lumière et de lui imprimer une nouvelle propriété qu'elle transporte avec elle, n'est pas particulière au spath d'Islande; je l'ai retrouvée dans toutes les substances connues qui doublent les images, et ce qu'il y a de remarquable dans ce phénomène c'est qu'il n'est pas nécessaire pour le produire d'employer deux cristaux d'une même espèce. Ainsi le second cristal, par exemple, pourroit être de carbonate de plomb ou de sulfate de barite : le premier pourroit être un cristal de soufre et le second un cristal de roche Toutes ces substances se comportent entre elles de la même manière que deux rhomboïdes de spath calcaire. En général cette disposition de la lumière à se réfracter en deux faisceaux ou en un seul, ne dépend que de la position respective de l'axe des molécules intégrantes des cristaux qu'on emploie, quels que soient d'ailleurs leurs principes chimiques et les faces naturelles ou artificielles sur lesquelles s'opère la réfraction. Ce résultat prouve que la modification que la lumière reçoit de ces différens corps est parfaitement identique.

Pour rendre plus sensibles les phénomènes que je viens de décrire, on peut regarder la flamme d'une bougie à travers deux prismes de matières différentes donnant la double réfraction et posés l'un sur l'autre. On aura en général quatre images de la flamme; mais si on fait tourner lentement un des prismes, autour du rayon visuel comme axe, les quatre images se réduiront à deux, toutes les fois que les sections principales des faces contigues seront parallèles ou rectangulaires. Les deux images qui disparoissent ne se confondent pas avec les deux autres, on les voit s'éteindre peu-à-peu tandis que les autres augmentent d'intensité. Lorsque les deux sections principales sont parallèles, une des images est formée par des rayons réfractés ordinairement par les deux prismes, et la seconde par des rayons réfractés extraordinairement. Lorsque les deux sections principales sont rectangulaires, une des images est formée par des rayons réfractés ordinairement par le premier cristal, et extraordinairement par le second, et l'autre image par des rayons réfractés extraordinairement par le premier cristal et ordinairement par le second.

Non-seulement tous les cristaux qui doublent les images peuvent donner à la lumière cette faculté d'être réfractée en deux faisceaux ou en un seul, suivant la position du cristal réfringent, mais tous les corps diaphanes solides ou liquides, et les corps opaques eux-mêmes, peuvent imprimer aux molécules lumineuses cette singulière disposition qui sembloit être un des effets de la double réfraction.

Lorsqu'un faisceau de lumière traverse une substance diaphane, une partie des rayons est réfléchie par la surface réfringente, et une autre partie par la surface d'émergence. La cause de cette réflexion partielle qui a jusqu'ici échappé aux recherches des physiciens, semble avoir, dans plusieurs circonstances, quelque analogie avec les forces qui produisent la double réfraction.

Par exemple, la lumière réfléchie par la surface de l'eau sous un angle de 52°. 45′ avec la verticale, a tous les caractères d'un des faisceaux produits par la double réfraction d'un cristal de spath calcaire dont la section principale seroit parallèle ou perpendiculaire au plan qui passe par le rayon incident et le rayon réfléchi que nous nommerons plan de réflexion.

Si on reçoit ce rayon réfléchi sur un cristal quelconque, ayant la propriété de doubler les images, et dont la section principale soit parallèle au plan de réflexion, il ne sera pas divisé en deux faisceaux comme l'eût été un rayon de lumière directe, mais il sera réfracté tout entier suivant la loi ordinaire, comme si ce cristal avoit perdu la faculté de doubler les images. Si, au contraire, la section principale du cristal est perpendiculaire au plan de réflexion, le rayon réfléchi sera réfracté tout entier suivant la loi extraordinaire. Dans les positions intermédiaires il sera divisé en deux faisceaux suivant la même loi et dans la même proportion que s'il avoit acquis son nouveau caractère par l'influence de la double réfraction. Le rayon réfléchi par la surface du liquide a donc, dans cette circonstance, tous les caractères d'un rayon ordinaire formé par un cristal dont la section principale seroit perpendiculaire au plan de réflexion.

Pour analyser complètement ce phénomène j'ai disposé verticalement la section principale d'un cristal, et après avoir divisé un rayon lumineux, à l'aide de la double réfraction, j'ai reçu les deux faisceaux qui en provenoient sur la surface de l'eau et sous l'angle de 52°. 45′. Le rayon ordinaire, en se réfractant, a abandonné à la réflexion partielle une partie de ses molécules comme l'eût fait un faisceau de lumière directe, mais le rayon extraordinaire a pénétré en entier le liquide; aucune de ses molécules n'a échappé à la réfraction. Au contraire, quand la section principale du cristal étoit perpendiculaire au plan d'incidence, le rayon extraordinaire produisoit seul une réflexion partielle, et le rayon ordinaire étoit réfracté en entier.

L'angle sous lequel la lumière éprouve cette modification en se réfléchissant à la surface des corps diaphanes, est variable pour chacun d'eux, il est, en général, plus grand pour les corps qui réfractent davantage la lumière. Au-delà et en deça de cet angle une partie du rayon est plus ou moins modifiée, et d'une manière analogue à ce qui se passe entre deux cristaux dont les sections principales cessent d'être parallèles ou rectangulaires.

Lorsqu'on veut simplement prendre connoissance de ce phénomène sans le mesurer avec exactitude, il faut placer en avant d'une bougie ou le corps diaphane ou le vase contenant le liquide qu'on veut soumettre à l'expérience. On examine à travers un prisme de cristal l'image de la flamme réfléchie à la surface du corps ou du liquide, on voit généralement deux images; mais en tournant le cristal autour du rayon visuel comme axe, on s'apperçoit qu'une des images s'affoiblit à mesure que l'autre augmente d'intensité. Au-delà d'une certaine limite, l'image qui s'étoit affoiblie recommence à augmenter d'intensité aux dépens de la seconde. Il faut saisir à-peu-près le point où l'intensité de lumière est au *minimum*, et rapprocher ou éloigner de la bougie le corps réfléchissant, jusqu'à ce que l'angle d'incidence soit tel qu'une des deux images disparoisse totalement; cette distance déterminée, si on continue à faire tourner lentement le cristal, on s'appercevra qu'une des deux images s'éteindra alternativement à chaque quart de révolution.

Le phénomène que nous avons remarqué dans les rayons qui se réfléchissent sous un certain angle à la surface d'un corps diaphane, a lieu aussi sous un autre angle dans les faisceaux réfléchis intérieurement par la surface d'émergence, et le sinus du premier angle est au sinus du second, dans le même rapport que les sinus d'incidence et de réfraction; ainsi, en supposant la face d'incidence et la face d'émergence parallèles, et l'angle d'incidence tel que le rayon réfléchi à la première surface présente le phénomène que nous avons décrit, le rayon réfléchi à la seconde surface sera modifié de la même manière. Si le rayon incident est tel que toutes ses molécules échappent à la réflexion partielle en traversant la face d'entrée, elles y échapperont de même en traversant la face de sortie. Cette nouvelle propriété de la lumière offre un moyen de mesurer d'une manière précise la quantité de rayons absorbés à la surface des corps diaphanes, problême que la réflexion partielle rendoit presque impossible à résoudre.

Lorsqu'un corps, qui donne la double réfraction, réfléchit la lumière à sa première surface, il se comporte comme une substance diaphane ordinaire. La lumière réfléchie sous un certain angle d'incidence acquiert la propriété que j'ai décrite ; et cet angle est indépendant de la position de la section principale qui n'influe que sur la double réfraction, ou sur les réflexions qui ont lieu dans l'intérieur du cristal.

En effet, les rayons qui se réfléchissent intérieurement à la seconde surface, présentent des phénomènes particuliers qui dépendent à la fois des forces réfringentes, et des propriétés de la lumière réfléchie que j'ai déja exposées.

Lorsqu'un faisceau lumineux a été divisé en deux rayons à la première surface d'un rhomboïde de spath calcaire ; ces deux rayons sortent par la seconde face en deux faisceaux parallèles au rayon incident, parce que chacun d'eux éprouve, à cette face, le même genre de réfraction qu'à la première. Il n'en est pas de même de la lumière réfléchie. Quoique le rayon réfracté ordinairement à la première face, soit réfracté ordinairement à la seconde, il est néanmoins réfléchi à cette surface en deux faisceaux, l'un ordinaire, l'autre extraordinaire. De même le rayon réfracté extraordinairement se réfléchit en deux autres ; en sorte qu'il y a quatre rayons réfléchis, tandis qu'il n'y en a que deux émergens. Ces quatre rayons revenant à la première face du cristal en sortent par quatre faisceaux parallèles, qui font, avec cette surface, mais en sens contraire, le même angle que le rayon incident, et qui sont parallèles au plan d'incidence. Pour lier ce genre de réflexion à celui de la double réfraction, il faut concevoir, par les deux points d'émergence de la seconde face, deux rayons incidens, faisant, avec cette surface, mais en sens contraire, le même angle que les rayons émergens. Ces deux rayons, par leur réfraction à travers le cristal, produiront quatre faisceaux qui suivront exactement la route des rayons réfléchis. Ainsi, la loi de la double réfraction étant connue, celle de la double réflexion peut s'en déduire facilement.

Nous allons passer actuellement au genre de phénomène qui fait l'objet de ce Mémoire, et qui est relatif non à la loi suivant laquelle se dirigent les rayons, mais à la quantité et aux propriétés de la lumière qu'ils contiennent.

Supposons l'angle d'incidence constant et le cristal posé horisontalement. Si on fait tourner le rhomboïde autour de la verticale de manière à rapprocher sa section principale du rayon incident, on voit diminuer peu-à-peu l'intensité du rayon ordinaire réfléchi extraordinairement, et du rayon extraordinaire réfléchi ordinairement. Enfin, lorsque le plan de la section principale passe par le rayon incident, ces deux rayons réfléchis disparoissent totalement, et il ne reste que le rayon ordinaire réfléchi ordinairement, et le rayon extraordinaire réfléchi extraordinairement. Ce dernier a néanmoins une intensité beaucoup moindre que le premier.

Si actuellement le rayon incident continuant à être compris dans la section principale, on augmente ou on diminue l'angle d'incidence, jusqu'à ce qu'il soit égal à 56°.30′, alors le dernier rayon réfléchi disparoît totalement, et il ne reste que celui qui a été réfracté ordinairement et réfléchi ordinairement. Audelà et en deça de cet angle, le rayon extraordinaire réfléchi extraordinairement, reparoît avec d'autant plus d'intensité qu'on s'éloigne davantage de cette limite L'angle d'incidence dont je viens de parler est celui sous lequel un rayon réfléchi à la première surface, auroit acquis la propriété de se diviser en deux faisceaux ou en un seul, comme cela à lieu à la surface de tout autre corps diaphane. Le phénomène précédent se lie facilement à l'expérience dans laquelle nous avons pris l'eau pour exemple ; car si on fait tomber sur la surface du rhomboïde, et sous l'angle d'environ 56°.30′, un rayon disposé à ne se réfracter qu'en un seul faisceau extraordinaire, ce rayon ne produit pas de réflexion partielle à la première surface, ce qui semble expliquer pourquoi il n'en produit pas à la seconde.

Cependant, il n'en est pas de même lorsque le plan d'incidence fait un angle sensible avec la section principale. Si on fait tomber dans ce plan et sous l'angle d'environ 56°.30', le rayon dont nous venons de parler, il se comporte à la première surface comme dans le cas précédent ; il la traverse sans se réfléchir, mais à la seconde surface il est réfléchi en deux faisceaux qui parviennent à leur *maximum* d'intensité lorsque le plan d'incidence est perpendiculaire à la section principale.

On sent que la lumière réfléchie à la seconde face ne se comporte pas ici comme dans le cas précédent, parce que dans la première expérience le rayon incident réfracté et réfléchi est toujours dans un même plan, au lieu que dans le dernier cas la force répulsive qui produit la réfraction extraordinaire, détourne la lumière du plan d'incidence, en sorte qu'elle cesse d'être dans les mêmes circonstances par rapport aux forces qui agissent sur elle.

Si on examine la lumière qui provient de la réflexion partielle des corps opaques, tels que le marbre noir, le bois d'ébène, etc., on trouve également un angle pour lequel cette lumière jouit des propriétés de celle qui a traversé un cristal de spath d'Islande. Les substances métalliques polies sont les seules qui ne semblent pas susceptibles de fournir ce phénomène, mais si elles n'impriment pas aux rayons lumineux cette disposition particulière, elles ne l'altèrent pas lorsque la lumière l'a déja acquise par l'influence d'un autre corps.

Cette propriété se conserve aussi dans les faisceaux qui traversent les corps qui réfractent simplement la lumière.

J'exposerai dans la seconde partie de ce Mémoire les circonstances, où à l'aide de la réflexion, sur les miroirs métalliques, on peut changer la disposition mutuelle des molécules d'un même rayon ordinaire ou extraordinaire, de manière que les unes se réfractent toujours ordinairement, tandis que les autres se réfractent extraordinairement. L'examen de ces diverses circonstances nous conduira à la loi de ces phénomènes, qui dépend d'une propriété générale des forces répulsives qui agissent sur la lumière.

Reprinted from a special reproduction produced by Itek Corporation on the occasion of the dedication of the Rayleigh collection at the Air Force Cambridge Research Laboratories, Bedford, Massachusetts, on March 30, 1966.

On the Light from the Sky, its Polarization and Colour

by the Hon. J.W. Strutt

On the Light from
the Sky —
Its' Polarization & Colour
by the Hon J. W Strutt,
Fellow of Trinity College Cambridge

1

It is now, I believe, generally admitted that the light which we receive from the clear sky is due in some way or another to small suspended particles, which divert the light from its regular course. On this point the experiments of Tyndall with precipitated clouds seem quite decisive. Whenever the particles of the foreign matter are sufficiently fine, the light emitted laterally is blue in colour, & in a direction perpendicular to that of the incident beam is completely polarized.

About the colour there is no *primâ facie* difficulty. For as soon as the question is raised it is seen that the standard of linear dimension with reference to which the particles are called small is the wave length of light & that a given set of

2

particles ~~p~~ would (on any conceivable view as to their mode of action) produce a ~~greater~~ continually increasing disturbance as we pass along the spectrum towards the more refrangible end; and there seems no reason why the colour of the compound light thus scattered laterally should not agree with that of the sky.

On the other hand the direction of polarization (perpendicular to the path of the primary light) seems to ~~be~~ have been felt as a difficulty. Tyndall says "...... the polarization of the beam by the incipient cloud has thus far proved itself to be absolutely independent of the polarizing angle. The law of Brewster does not apply to matter in this condition; & it rests with the

3

undulatory theory to explain why whenever the precipitated particles are sufficiently fine, no matter what the substance forming the particles may be, the direction of maximum polarization is at right angles to the illuminating beam, the polarizing angle for matter in this condition being invariably 45°. This I consider to be a point of capital importance with reference to the present question."× As to the importance there will not be two opinions; but I venture to ~~say~~ think that the difficulty is entirely imaginary, & is caused mainly by ~~the~~ misuse of the word reflection. Of course there is nothing in the etymology of reflection or refraction to forbid their application in this sense, but the words have acquired technical meanings & ~~have~~

× Phil Mag vol 37 p 304

become associated with certain well known laws called after them. Now a moments' consideration of the principles according to which reflection & refraction are explained in the wave theory is sufficient to shew that they have no application unless the surface of the disturbing body is larger than many square wave lengths, whereas the particles to which the sky is supposed to owe its illumination must be ~~much~~ smaller than the wave length, or else the explanation of the colour breaks down. The idea of polarization by reflection is therefore quite out of place, & that "the law of Brewster does not apply to matter in this condition" (of extreme fineness) is only what might have been inferred from the principles of the wave theory.

Nor is there any difficulty in foreseeing what according to the wave theory the direction of polarization ought to be. Conceive a beam of plane polarized light to move among a number of ~~small~~ particles all small compared to any of the wave lengths. The foreign matter if ~~be~~ optically denser than air may be supposed to *load* the aether so as to increase its *inertia* without altering its resistance to distortion, provided that we agree to neglect effects analogous to chromatic dispersion. If the particles were away the wave would pass on unbroken & no light would be emitted laterally. Even with the particles retarding the motion of the ether the same will be true if to counterbalance the increased inertia we imagine suitable forces to act on the ether at all points where the inertia is altered. These forces have the same period & direction as the undisturbed luminous vibrations themselves. The light actually emitted laterally is thus the same as would be caused by forces exactly the opposite of these acting on the medium otherwise free from disturbance, & it only remains to see what the effect of such forces would be.

On account of the smallness of the particles the forces acting throughout the volume of any one are all of the same intensity & direction & may be considered as a whole. The determination of the motion in the ether due to the action of a periodic force at a given point requires of course the aid of mathematical analysis, but very simple considerations will lead us to a conclusion on the particular point now under discussion. In the first place there is a complete symmetry round the direction of the force. The disturbance consisting of transverse vibrations is propagated outwards in all directions from the centre, & in consequence of the symmetry the direction of vibration in any ray lies in ~~plan~~ the plane containing the ray & the axis, that is to say, the direction of vibration in the scattered or diffracted ray makes with the direction of vibration in the incident or primary ray the least possible angle. The symmetry

8

also requires that the intensity of the scattered light should vanish for the ray which would be propagated along the axis; for there is nothing to distinguish one direction transverse to the ray from another — we have now got what we want. Suppose for distinctness of statement that the primary ray is vertical, & that the plane of vibration is that of the meridian. The intensity of the light scattered by a small particle is constant & a maximum for rays which lie in the vertical plane running east & west, while there is no scattered ray along the north & south line. If the primary ray is unpolarized the light scattered north & south is entirely due to that component ~~of~~ which vibrates east & west & is therefore *perfectly polarized*, the direction of its vibration being also east & west. Similarly ~~in any direc-tion~~ any other ray scattered horizontally is perfectly polarized & the vibration is performed in the horizontal plane. In other directions the polarization becomes less & less complete as we approach the vertical, & in the vertical direction itself altogether disappears.

So far then as disturbance by very small particles is concerned theory appears to be in complete accordance with the experiments of Tyndall & others. At the same time if the above reasoning be valid the question as to the direction of the vibrations in polarized light is decided in accordance with the views of Fresnel. Indeed the observations on the plane of polar-zation of the scattered light is virtually only another form of Prof Stokes' original test with the diffraction grating. In its present shape however it is free from certain difficulties both of theory & experiment which have led different physicists who have used the other method to contradictory conclusions. I confess I cannot see any room for doubt as to the result it leads to. ×

The argument used is apparently open to a serious objection which I ought to notice. It seems to prove too much. For if one disturbing particle is unable to ~~scatter light in the~~ send out a scattered ray in the direction of original vibration, it would appear that no com-

Footnote 10'

[I only mean ~~that~~ if light, ~~consists~~ as is generally supposed, consists of transversal vibrations similar to those which take place in an elastic solid, the vibration must be normal to the plane of polarization. ~~Over a very wide range of phenomena the vibrations of light are unquestionably analogous~~ There is unquestionably a formal analogy between the two sets of phenomena extending over a very wide range, but it is another thing to assert that the vibrations of light are really & truly to & fro motions of a medium having mechanical properties (with reference to small vibrations) like those of ordinary solids. The fact that ~~O identically~~ the ~~same~~

[~~form of wave theory~~ theory of elastic solid led Green 10"
to Fresnel's formulae for the
reflection & refraction of polar-
ized light seems amply suf-
ficient to warrant its em-
ployment here, while the
question whether the ~~su~~ ana-
logy is more than formal
is still left open.]

bination of them (such as a small body may be supposed to be) could do so ~~either~~, at least at such a distance that the body subtends only a small solid angle. Now we know that when light vibrating in the plane of incidence falls on a reflecting surface at an angle of 45°, light is send out according to the law of ordinary reflection whose direction of vibration is perpendicular to that ~~of~~ in the incident ray. And not only is this so in experiment, but it has been proved by Green× to be a consequence of the very same view as to the nature of the difference between media of various refrangibility as has been adopted in this paper.

× Camb Phil Trans vol VII

The apparent contradiction is however easily explained. It is true that the disturbance due to a foreign body of any size is the same as would be caused by forces acting through the space it fills in a direction parallel to that in which the primary light vibrates, *but these forces must be supposed to act on the medium as it actually is, that is with the variable density.* Only on the supposition of complete uniformity would it follow that no ray could be emitted parallel to the line in which the forces act. When however the sphere of disturbance is small compared to the wave length, this want of uniformity is of little account & cannot alter the law regulating the intensity of the vibration propagated in different directions.

Having disposed of the polarization let us now consider how the intensity of the scattered light varies from one part of the spectrum to another, still supposing that all the particles are many times smaller than the wave length even of violet light. The whole question admits of ~~exact~~ analytical ~~mathematical~~ treatment, but before entering upon that it may be worth while to shew how the principal result may be anticipated from a consideration of the dimensions of the quantities concerned.

The object is to compare the intensities ~~of the scattered ra~~

of the incident & scattered rays 14
for these will clearly be propor-
tional. The number (i) expressing
the ratio of the two amplitudes
is a function of the following quanti-
ties ~~which are the only elements~~
~~of the problem~~: — T the
volume of the disturbing particle,
r the distance of the point
under consideration from it,
λ the wave length, b the
velocity of propagation of light,
D & D′ the original & altered den-
sities; of which the first three
depend only on space, the fourth
on space & time, while the
fifth & sixth introduce the
consideration of mass. Other
elements of the problem there
are none, except mere numbers
& angles which do not de-

pend on the fundamental measurements of space time & mass — Since the ratio i whose expression we seek is of no dimensions in mass it follows at once that D & D' only occur under the form $D : D'$ which is a simple number & may therefore be omitted. It remains to find how i varies with T r λ b. Now of these quantities b is the only one ~~which~~ depending on time & therefore as i is of no dimensions in time, b cannot occur in its expression. We are left then with T r & λ, and from what we know of the dynamics of the question we may be sure that i varies directly as T & inversely as r, & must therefore be proportional to $T \div r\lambda^2$,

16

T being of three dimensions in space. In passing from one part of the spectrum to another λ is the only quantity which varies, & we have the important law:—

When light is scattered by particles which are very small compared to ~~the~~ any of the wave lengths, the ratio of the amplitudes of the vibrations of the scattered and incident light varies inversely as the square of the wave lengths, & the intensity of the lights themselves as the inverse fourth power.

I will now investigate the mathematical expression for the disturbance propagated in any direction from a small particle ~~on~~ which a beam of light strikes.

Let the vibration corresponding to the incident light be expressed by $A\cos\frac{2\pi}{\lambda}bt$. The acceleration ~~at~~ is ~~therefore~~

$$-A\left(\frac{2\pi}{\lambda}b\right)^2\cos\frac{2\pi}{\lambda}bt\,,$$

so that the force which would have to be applied to the parts where the density is D' in order that the wave might pass on undisturbed is per unit of volume

$$-(D'-D)\,A\left(\frac{2\pi b}{\lambda}\right)^2\cos\frac{2\pi}{\lambda}bt\,.$$

To obtain the ~~Thus the~~ total force which must be supposed to ~~total accelerating force which~~ act over the space occupied by the particle ~~must act over the volume~~ the factor T must be introduced.

~~$\frac{D'-D}{D'}\,AT\left(\frac{2\pi b}{\lambda}\right)\cos\frac{2\pi}{\lambda}bt$.~~

The opposite of this conceived ~~supposed~~ to act at O (the position of the particle) gives the same disturbance in the medium as is actually caused by the presence of the particle. Suppose now that the ray is incident along OY, and that

Let the particle be situated 17
at the origin O of rectangular co-
coordinates,

~~Let the vibration~~ 18 ~~17~~

the direction of vibration makes an angle α with the axis of x which is the ~~direction~~ line of the scattered ray under consideration, a supposition wh ~~this~~ involves no loss of generality because of the symmetry which we have shewn to exist round the line of action of the force. The question is now entirely reduced to the discovery of the disturbance produced in the ether by a given periodic force acting at a fixed point in it. In his valuable paper on the Dynamical Theory of Diffraction[×] Prof Stokes has given a complete investigation of this problem, & I might assume the result at once. The method there used is however for this particular purpose

× Camb Phil Trans. vol IX

very indirect & accordingly I (19

have thought it advisable to give a comparatively short cut to the result which will be found at the end of the present paper. It is proved that if the total ~~accelerating~~ force acting at O in the manner supposed is $F \cos \frac{2\pi}{\lambda} bt$, the resulting disturbance in the ray propagated along Ox is

$$\zeta = \frac{F \sin\alpha}{4\pi b^2 D r} \cos \frac{2\pi}{\lambda}(bt - r)$$

Substituting for F its value we have

$$\zeta = A \frac{D' - D}{D} \frac{T}{r\lambda^2} \cos \frac{2\pi}{\lambda}(bt - r) \sin\alpha$$

an ~~expression~~ equation which includes all our previous results & more.

One reservation however must not be omitted. Since we have supposed the medium uniform throughout, whereas

20

it really has a different density at the place where the force acts, our investigation ~~is~~ does not absolutely correspond to the actual circumstances of the case. As before remarked no error is on that account to be feared in the law determining the intensity of the vibration in different directions ~~in consequence~~, but it is probable that the coefficient so far as it depends on $D : D'$ may be changed, & there will be a change in the phase comparable with $\frac{2\pi}{\lambda}$ × the linear dimension of the particles, which is of importance ~~in combining the~~ when the scattered & primary waves have to be compounded.

So much for a single particle. In actual experiments as for instance with Prof Tyndall's 'clouds', we have to deal with an immense number of such particles & the question now is to ~~determine~~ deduce what their effect must be from the results already obtained. Were the particles absolutely motionless, the partial waves sent out in any direction from them would have ~~a~~ permanent relations as to phase & the total disturbance would have to be found by compounding the vibrations due to all the particles. ~~But~~ Such a supposition would however be very wide of the mark, for in consequence of ~~the~~ extreme smallness of λ

22

the slightest motion of any particle will cause an alteration of phase passing through many periods in a less time than the eye could appreciate. Our particles are then to be treated as so many <u>unconnected</u> sources of light, & instead of adding the <u>vibrations</u> we must take the <u>intensities</u> represented by their squares. Only in one direction is a different treatment necessary, namely along the course of the primary light. I mention this because it would not otherwise appear how the reduction in the intensity of the transmitted light is effected, But we do not require to follow the details of the process, because when once we know the intensity of the light emitted laterally the principle of energy will tell us what the primary wave has lost.

~~Let m be the numbe~~

The intensity of the light scattered from a cloud is thus equal to

$$A^2 \left(\frac{D'-D}{D}\right)^2 \sin^2\alpha \; \frac{\Sigma T^2}{\lambda^4 r^2}$$

where ΣT^2 is the sum of all the squares of T. If $\overline{T}^2$ be understood to denote the mean square of T (not the square of the mean value of T), & m be the number of particles

$$\Sigma T^2 = m \, \overline{T}^2$$

If the primary light be unpolarized, the intensity in a direction making an angle β with its

24

course becomes

$$A^2 \frac{(D'-D)^2}{D^2}(1+\cos^2\beta)\frac{m\,T^2}{\lambda^4 r^2}$$

Backwards from the cloud the light is thus twice as bright as normally. To the light scattered nearly in the direction of the primary ray our expression does not apply.

I will now endeavour to trace some of the consequences of the law

Scattered light $\propto \lambda^{-4}$, & first as to the effect on the light which passes on. Since there is no increase of energy on the whole the light emitted laterally is to be regarded as merely diverted from the main stream, & so if I represent the intensity

Fig 1 shews the curve representing the intensity of the scattered light for each part of the spectrum referred to the intensity in the primary light as a standard. The abscissa being proportional to λ, the base line represents the diffraction spectrum with the principal fixed lines. ~~With~~ over the brighter portion of the spectrum from B to G the curve differs but little from a straight line, while the small curvature ~~which there~~ ~~is~~, is turned downwards indicating a deficiency in the green & yellow.

Before making out the theory I had endeavoured to ascertain by observation the actual prismatic composition of the blue of the sky, & had ob-

26

tained preliminary results. The experimental method (whose description I must reserve for another opportunity) was fully adequate to the comparison of two given lights, but the difficulty was to find something to compare the blue light with. In the only complete set of observations that I have hitherto been able to make, the blue of the sky (apparently a very good one) taken from the neighbourhood of the zenith was compared with sun light diffused through white paper. About 30 consistent comparisons were made ranging over the spectrum from C to beyond F, & a curve drawn on the plan of Fig 1. I do not give the complete curve, because I hope before long to complate & confirm the observations, but the following numbers will give an idea of the result. 27

C	D	b_3	F	~~from Fig 1~~
25	40	63	80	from Fig 1
25	41	71	90	observed

The upper line ~~repres~~ gives the theoretical intensities for the fixed lines C, D, b_3, F, while the lower gives the observed ratios between the lights (sky & diffused sun lights) the two sets of numbers being made to agree at C. Considering the difficulties & uncertainties of the case the two curves agree very well, & it should be noticed that the sky compared to diffused light was even bluer than theory makes it on the supposition that the diffused light through the paper may be taken as similar to that whose scattering illuminates the sky. It is ~~not improbable~~ possible that the paper was slightly yellow, or the cause may lie in the yellowness of ~~the~~ sunlight as it reaches us compared to the colour it possesses in the upper regions of the atmosphere. It would be a mistake to lay any great stress on the observations in their present incomplete form, but at any rate they shew that a colour more or less like that of the sky would result from taking the elements of white light in quantities proportional to λ^{-4}. I do not know how it may strike others but individually I was not prepared for so great a difference as the observations shew, the ratio for F being more than 3 times greater than for C. 29

There is one point in which our calculations do not exactly meet the case of the sky. In the experiments with precipitated clouds the total quantity of light scattered is quite insignificant compared with the incident beam, but it is by no means so clear that ~~this~~ the same is the case with the sky. Each particle is thus struck not only by the direct light of the sun but also by that scattered from others. It does not seem that the chromatic effects would be much affected by this consideration, but it is worth notice that the conclusion as to complete polarization perpendicular to the incident ray would have to be modified. To see this imagine as before the (unpolarized) light incident along Oy ~~or~~ upon a particle O; we have seen that the ray diffracted along Ox contains no vibration parallel to Oy. By the aid however of another particle P in the xy plane such a vibration may be communicated to it. For in the ray diffracted from P to O there is a component vibration in the xy plane perpendicular to PO which when again diffracted along Ox will ~~have~~ give a component parallel to Oy. This is perhaps the explanation of the incomplete polarization of sky light at right angles to the solar beams, but it must be re-

31

re-membered that an insufficient fineness in some of the particles of foreign matter would have a like result.

By many physicists from Newton downwards the light of the sky has been supposed to be reflected from thin plates, & the colour to be the blue of the first order in Newton's scale. Such a view is fundamentally different from that adopted in this paper, though it might not at first seem so. In support of this assertion it may be sufficient to notice that the two theories are at variance as to the law connecting the intensity with wave length. By an argument from dimensions similar to that already used it is easy to find how the intensity of the light reflected from a thin plate (that is thin compared to any of the wave lengths) varies with λ. Instead of our former quantities T & λ, we now have merely λ & δ the thickness of the plate. Since the reflected vibration necessarily varies as δ it must also be proportional to λ^{-1}, & so the intensity of the reflected light $\propto \lambda^{-2}$ instead of λ^{-4}. The ordinary analytical expression for the reflected light leads readily to the same ~~law~~ conclusion (Airy's Tracts p 297). There can I think be no question that the composition of the light of the sky agrees more nearly with the latter than with the former law.

The principle of energy makes it clear that the ~~[illegible]~~ light emitted laterally is not a new creation but only diverted from the main stream. If I represent the intensity of the primary light after traversing a thickness x of the turbid medium, we have

$$dI = -k I \lambda^{-4} dx$$

where k is a constant independent ~~of λ~~. On integration

$$I = I_0 \, \varepsilon^{-k\lambda^{-4}x}$$

~~if~~ if I_0 correspond to $x = 0$ — a law altogether similar to that of absorption, & shewing how the light tends to become yellow & finally red as the thickness of the medium increases. Fig 2 shews a series of curves representing the composition of the originally white light after passing through thicknesses in the ratio of 1, 2, 4, 8, 16, 32. The reader will observe how little of the violet light remains when the red is still in nearly its original force. I cannot but think that this rapid diversion of the rays of short wave length has a good deal to ~~to~~ do with the absence of light of the highest refrangibility, from the direct ~~light~~ rays of the Sun. For the line A at the extreme red & R ~~at~~ near the upper limits of the photographic spectrum the wavelengths are 7617 & 3108. The ratio of the fourth powers is about 36 : 1, so that whatever the fraction representing the transmission of A may be, its 36th power will give the ~~percentage~~ ~~tra~~ transmission of R. To take an instance, if .9 of the ray A get through, only .018 of

35

R would be able to penetrate. For the rays of still higher refrangibility which Prof Stokes found abundant in the electric light but missing in the solar rays the fraction would be smaller still, but I am not aware of any measurements of smaller wavelength on which to found a calculation.

We have hitherto supposed that the light scattered by the finely divided matter reaches the eye without modification, & we have taken no account of any change in the composition of the primary light before diffraction. If x be the total ~~distance~~ length of the path of the ray through the turbid medium we may express the quality of the light in terms of x, for it makes no difference whether the absorption takes effect before diffraction or after. In fact

$$I \propto \lambda^{-4} \varepsilon^{-k\lambda^{-4}x}$$

an expression which shews that I vanishes for very small as well as for very large values of λ while for some definite value (say Λ) it rises to a maximum (I_0). Expressing I in terms of I_0 & Λ we have

$$\frac{I}{I_0} = \frac{\Lambda^4}{\lambda^4} \varepsilon^{1 - \frac{\Lambda^4}{\lambda^4}}$$

from which we may fall back on our original law by supposing Λ indefinitely small & replacing $\Lambda^4 I_0$ by a finite constant. An approximate idea of the character of these lights may be obtained by

Subtracting the successive curves of Fig 2. Thus the difference of the curves marked 2 & 4 represents a light having its maximum brightness (of course relatively to the primary light) in the blue green portion of the spectrum. I find by calculation that if the maximum intensity be at b & be taken as unity the intensities at G & C are given by the numbers ·713, ·710 respectively. The colour of would be greenish, but whether the green of the sky is to be accounted for in this way I am not able to say. Some I believe consider it to be entirely a contrast effect.

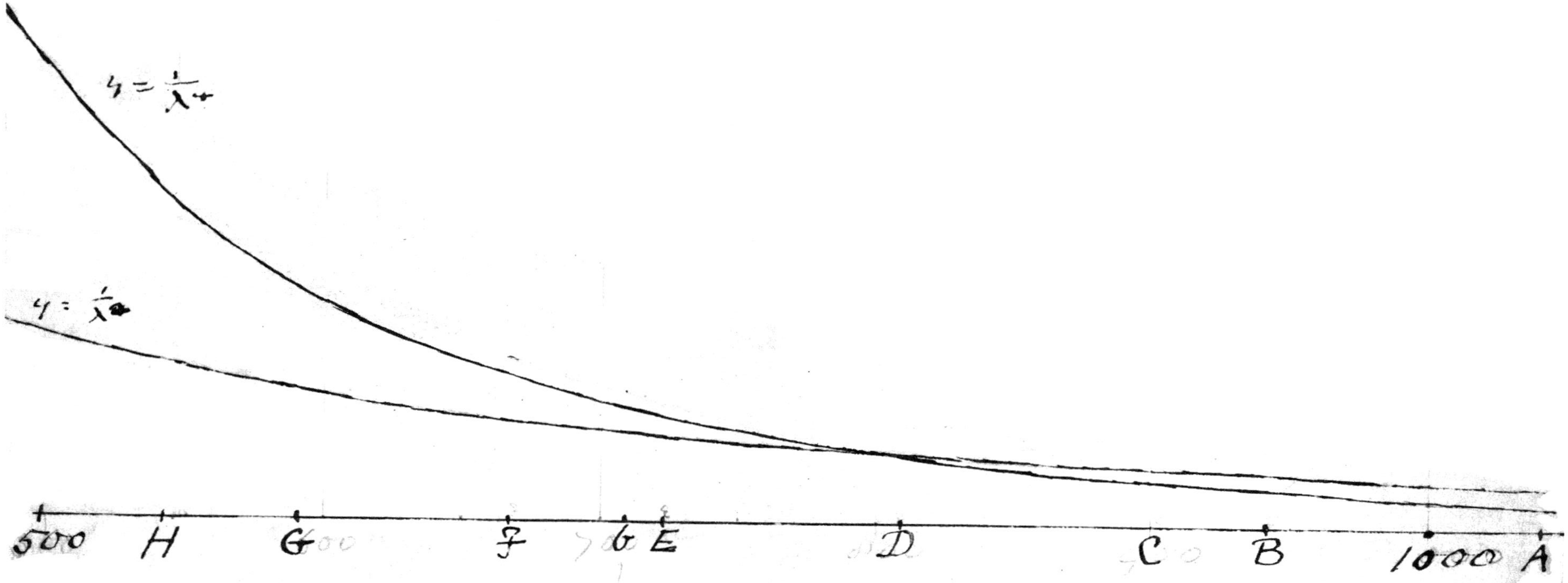

Curve representing theoretical composition of blue sky light supposed to be scattered from particles small compared to any of the wave lengths

Fig 1

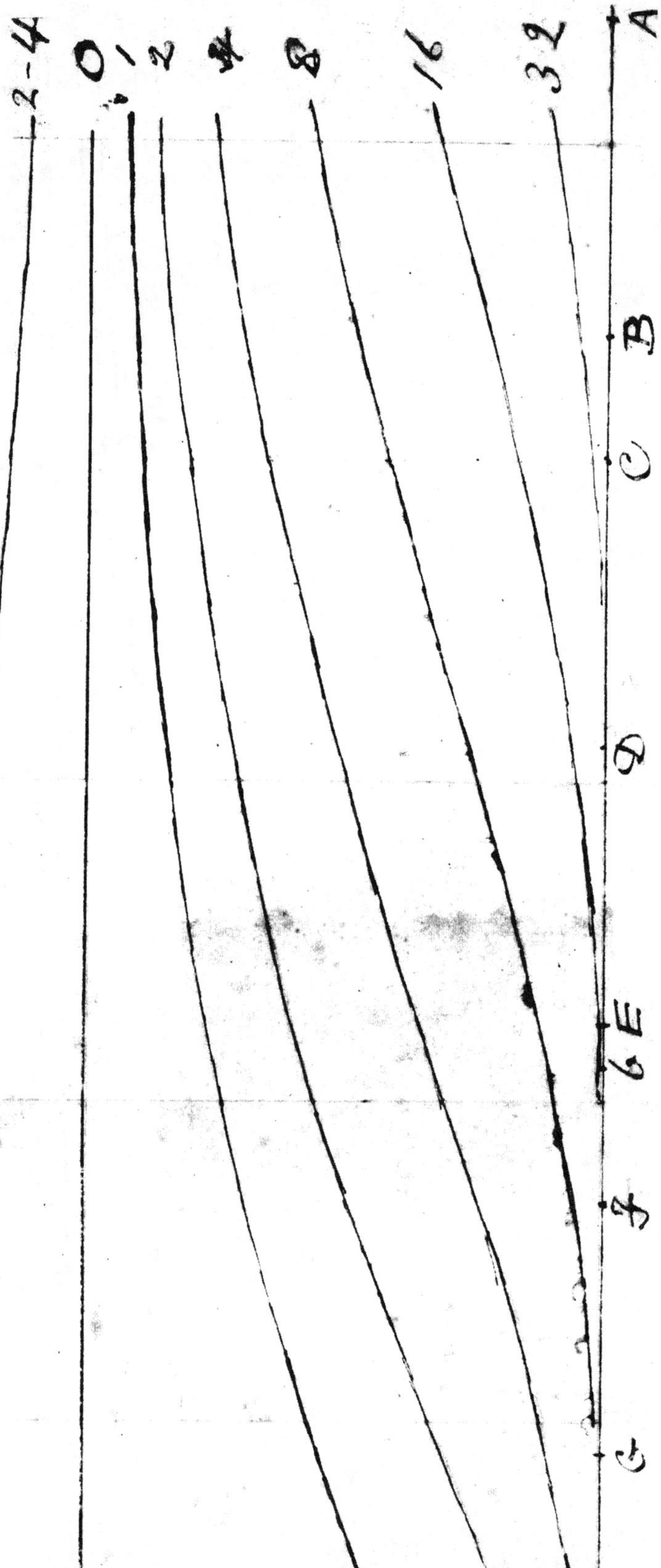

Fig 2

Appendix 1

Within a space T small in all its dimensions against λ and situated at the origin of coordinates let a force parallel to OZ and so far as it depends upon the time expressed by a simple circular function act on the medium.

If $\xi\ \eta\ \zeta$ denote the displacements parallel to the axes at the point xyz, and $\delta = \frac{d\xi}{dx} + \frac{d\eta}{dy} + \frac{d\zeta}{dz}$

$$\left.\begin{aligned} \frac{d^2\xi}{dt^2} &= b^2\nabla\xi + (a^2-b^2)\frac{d\delta}{dx} \\ \frac{d^2\eta}{dt^2} &= b^2\nabla\eta + (a^2-b^2)\frac{d\delta}{dy} \\ \frac{d^2\zeta}{dt^2} &= b^2\nabla\zeta + (a^2-b^2)\frac{d\delta}{dz} + Z \end{aligned}\right\} \cdots \overset{\times}{(A)}$$

× (Thomson & Tait's Natural Philosophy p 530)

where ∇ stands for $\frac{d^2}{dx^2} + \frac{d^2}{dy^2} + \frac{d^2}{dz^2}$

a^2 and b^2 are constants depending on the nature of the medium supposed to be isotropic. For the luminiferous ether Green has shewn that a is to be regarded as indefinitely great.×

× Camb Phil Trans vol. VII

2.

To represent the periodic force write for $Z - Z\varepsilon^{int}$. Similar transformations will then apply to $\xi\ \eta\ \zeta$ and δ; so that on substitution in (A) and dividing out the common factor ε^{int} there results

$$\left.\begin{aligned}(b^2\nabla + n^2)\xi + (a^2 - b^2)\frac{d\delta}{dx} &= 0\\ (b^2\nabla + n^2)\eta + (a^2 - b^2)\frac{d\delta}{dy} &= 0\\ (b^2\nabla + n^2)\zeta + (a^2 - b^2)\frac{d\delta}{dz} &= -Z\end{aligned}\right\} \text{---(B)}$$

Writing $\frac{d\xi}{dy} - \frac{d\eta}{dx} = \varpi_3$ &c

we obtain from (B) by differentiation and subtraction

$$\left.\begin{aligned}(b^2\nabla + n^2)\varpi_3 &= 0\\ (b^2\nabla + n^2)\varpi_1 &= \frac{dZ}{dy}\\ (b^2\nabla + n^2)\varpi_2 &= -\frac{dZ}{dx}\end{aligned}\right\} \text{---(C)}$$

$\varpi_1\ \varpi_2\ \varpi_3$ are the rotations of the elements of the medium round axes parallel to those of coordinates

The disturbance which we are investigating is that caused and maintained by the force Z acting within the space T. Accordingly*

*(Helmholtz, Crelle's Journal 1860)

$$\varpi_3 = 0$$

$$\varpi_1 = -\frac{1}{4\pi b^2}\iiint \frac{dZ}{dy}\,\frac{\varepsilon^{\pm ikr}}{r}\,dx\,dy\,dz$$

r being the distance between the element $dx\,dy\,dz$ and the point where ϖ_1 is estimated, and

$$k = \frac{2\pi}{\lambda} = \frac{n}{b} \quad \ldots\ldots \quad (D)$$

~~Since the $\varepsilon^{\pm}$~~

Since $\varepsilon^{\pm ikr}$ will be finally multiplied by ε^{int}, and the disturbance which ~~are~~ we are dealing with is propagated <u>outwards</u> from T, it is evident that the <u>lower</u> sign is to be employed.

Now $\int \frac{\varepsilon^{-ikr}}{r}\frac{dZ}{dy}\,dy = \left[Z\,\frac{\varepsilon^{-ikr}}{r}\right]$

$$-\int Z\,\frac{d}{dy}\left(\frac{\varepsilon^{-ikr}}{r}\right)dy\,,$$ of which the term within brackets vanishes

4

because the value of Z is only finite within the space T. Thus

$$\varpi_1 = \frac{1}{4\pi b^2}\iiint Z \frac{d}{dy}\left(\frac{\varepsilon^{-ikr}}{r}\right) dx\,dy\,dz$$

The factor $\frac{d}{dy}\left(\frac{\varepsilon^{-ikr}}{r}\right)$ within the space T is sensibly constant, so that if Z stand for the mean value of Z over the volume T

$$\varpi_1 = \frac{TZ}{4\pi b^2}\frac{d}{dy}\left(\frac{\varepsilon^{-ikr}}{r}\right)$$

$$\varpi_2 = \frac{-TZ}{4\pi b^2}\frac{d}{dx}\left(\frac{\varepsilon^{-ikr}}{r}\right)$$

$$x\varpi_1 + y\varpi_2 \propto \left(x\frac{d}{dy} - y\frac{d}{dx}\right)\cdot\frac{\varepsilon^{-ikr}}{r} = 0$$

and if $R = \sqrt{x^2+y^2}$

$$\varpi = \frac{x\varpi_2 - y\varpi_1}{R} = \frac{-TZ}{4\pi b^2 R}\left(x\frac{d}{dx}+y\frac{d}{dy}\right)\cdot\frac{\varepsilon^{-ikr}}{r}$$

$$= \frac{-TZ}{4\pi b^2}\frac{d}{dR}\cdot\frac{\varepsilon^{-ikr}}{r} = \frac{TZ\sin\alpha}{4\pi b^2}\frac{d}{dr}\cdot\frac{\varepsilon^{-ikr}}{r}$$

where α denotes the angle between r and Z.

~~In differentiating the $\frac{\varepsilon^{-ikr}}{r}$ with respect to r, we may neglect~~

646

The resultant rotation at any point is thus about an axis perpendicular to the plane passing through the point and the axis of Z, & its magnitude is given by ϖ. In differentiating $\frac{\varepsilon^{-ikr}}{r}$ with respect to r we may neglect the term divided by r^2 as altogether insensible, kr being an exceedingly great quantity at any ordinary distance from the origin of disturbance. Thus

$$\varpi = \frac{-ik.\,TZ\sin\alpha}{4\pi b^2}\cdot\frac{\varepsilon^{-ikr}}{r}\quad\ldots\ldots(E)$$

which completely determines the rotation at any point. For a given disturbance it is seen to be everywhere about an axis perpendicular to r & the direction of the force, & in magnitude dependent only on the angle between these two directions (α) and on the distance (r).

The intensity of the light is however more usually expressed in terms of the actual displacement in the plane of the wave. In order to find the connection between the two quantities, it will be more convenient to suppose the scattered ray parallel to x & that the force F (for Z is no longer appropriate) acts in the plane of zx at an angle α with Ox. ϖ becomes identical with ϖ_2 that is with $\frac{d\zeta}{dx}$, for ξ as well as η is zero; so that

$$\zeta = \int \varpi \, dx^{r} = \frac{T F \sin \alpha}{4\pi b^2} \cdot \frac{\varepsilon^{-ikr}}{r}$$

Restoring the factor ε^{int}, we have

$$\zeta = \frac{T F \sin \alpha}{4\pi b^2} \cdot \frac{\varepsilon^{i(nt-kr)}}{r}$$

or throwing away the imaginary part

7

$$S = \frac{T \mathfrak{F} \sin\alpha}{4\pi b^2 r} \cos\frac{2\pi}{\lambda}(bt - r) \quad \ldots\ldots (\mathfrak{F})$$

This corresponds to a total accelerating force ~~$T\mathfrak{F}$ given by~~

~~$\mathfrak{F}$~~

equal to

$$\mathfrak{F} T \cos\frac{2\pi}{\lambda} bt$$

results which ~~are~~ agrees with that of Prof Stokes more complete investigation with the exception of a slight difference of notation.

~~Dec 30~~ Tenby Place
William

Dec 30 – 1870.

Haidinger's Brushes and Circularly Polarized Light

W. A. SHURCLIFF
Research Laboratory, Polaroid Corporation, Cambridge, Massachusetts
(Received December 20, 1954)

HAIDINGER'S brushes, as seen by linearly polarized light, have been described rather fully.[1,2] I have been unable to find any mention, however, of observation of the brushes by circularly polarized light, which is surprising since the brushes appear as prominent by circularly polarized light as by linearly polarized light.

The phenomenon is studied conveniently with the aid of commercially available circular polarizers of sheet type. On looking through a right circular polarizer towards a clear or cloud-covered sky the observer sees a brush of the usual size, shape, and (yellowish) color. The direction of the brush is an upward-to-the-right, downward-to-the-left direction, and rotating the polarizer in its own plane has no effect on the direction. A sheet of left circular polarizer leads to an upward-to-the-left, downward-to-the-right direction of the brushes. (These conclusions were confirmed by all five persons who were invited to verify the observations.)

For a given type of circular polarizer, the observer's right eye sees the brush at approximately the same angle that the left eye sees it at,—indicating what is perhaps an unexpected lack of bilateral symmetry in the pair of eyes. (As judged by the author, the brush direction is not quite the same for both eyes but differs by approximately 10 deg.)

Looking first through the right circular polarizer and then the left tends to accentuate the brushes, due to fatigue effects having a time constant of the order of a few seconds. The fatigue effects found using linearly polarized light and circularly polarized light are not independent but rather are essentially identical, provided the linear polarizer's azimuth is adjusted so that its brush direction is the same as that of the circular polarizer under consideration.

The present observation perhaps suggests the need of revising present theories[1,2] as to the cause of the general phenomenon.

The aforementioned brush direction rule permits quick visual determination of the class (right or left) of a circular polarizer.

[1] H. von Helmholtz, *Physiological Optics*, edited by Southall (Optical Society of America, New York, 1924).
[2] O. Cords, Optik 2, 423 (1947).

THE ROLE OF THE MACULAR PIGMENT IN THE DETECTION OF POLARIZED LIGHT

RICHARD A. BONE

Department of Physics, University of the West Indies, Kingston 7, Jamaica

(*Received* 15 *February* 1979)

Abstract—A direct comparison technique for measuring the dichroism responsible for the Haidinger brush effect is described, together with a subsidiary method for assessing the birefringence of the cornea. It is shown that when this birefringence is taken into account, the results are consistent with the theory of a symmetrical arrangement of a fraction of the macular pigment molecules. Tests on a number of subjects indicated that they all possessed similar areal densities of such symmetrically arranged molecules.

INTRODUCTION

If a white surface, illuminated with short wavelength light is viewed through a polarizing filter, Haidinger's brushes are seen, centred on the fixation point (Haidinger, 1844). Former theories to account for this phenomenon have been reviewed by Stanworth and Naylor (1950). Experiments, subsequently carried out by DeVries *et al.* (1953) and Naylor and Stanworth (1954), provided strong evidence that some degree of symmetry in the orientation of dichroic macular pigment molecules in the retina was responsible for the brushes. This explanation is largely in keeping with that proposed by Helmholtz (1866).

An alternative model suggested by Summers *et al.* (1970) ascribes to the retina the properties of an anisotropic absorbing crystal. In view of the lack of supporting experimental evidence however, this suggestion has gained little recognition. More recently, Hochheimer (1978), using polarized light, photographed a "Maltese cross" figure in the rhesus monkey eye, which he attributed to the same structure that produces Haidinger's brushes in the human eye. In contrast to the visual phenomenon, however, this figure appeared not only when using short wavelength light, but also on photographs taken with red light where the macular pigment has zero absorption. Hochheimer concluded that the Maltese cross, and therefore also Haidinger's brushes, could not be related to dichroism of the macular pigment. Shute (1978), on the other hand, has pointed out that there is little, if any, evidence that they are more than indirectly related and has suggested an alternative explanation of the Maltese cross.

Certainly the results of the present investigation are compatible with the macular pigment theory of Haidinger's brushes and may be summarized as follows: Firstly, in keeping with DeVries *et al.* (1953) and Naylor and Stanworth (1954), the dichroism responsible for Haidinger's brushes has a spectral distribution virtually indistinguishable from the optical density spectrum of the macular pigment. Theory, presented below, shows that this result requires the involvement of a long chain pigment molecule. Secondly, apparent variations in the dichroism with orientation of the polarization plane of the incident light follow a trend to be expected as a result of birefringence of the cornea. Thirdly, a comparison between the strength of the dichroism and the macular pigment density for a number of observers showed that the number of preferentially orientated pigment molecules per unit area of retina did not vary much among them.

THEORY

DeVries *et al.* (1953) proposed that Haidinger's brushes were due to a fraction of the macular pigment molecules possessing symmetry. Lutein, which is probably the pigment involved (Wald, 1945; Bone, 1976) has a long chain molecule which will show greater absorption if the incident light is polarized with the **e** vector parallel to the molecule axis than with it perpendicular* (Dewar, 1949). To account for the brushes, a fraction of the otherwise randomly aligned molecules must be preferentially arranged tangentially with respect to the foveal centre. The image of a centrally fixated illuminated slit will then fall on an area of the retina where the orientated molecules are essentially perpendicular to the slit. This is shown schematically in Fig. 1. The "dichroic ratio" (G) of the macular pigment as a whole may then be defined as the ratio of transmittances of such an area to light polarized perpendicular and parallel to the orientated molecules.

Thus
$$G = T_{\perp}/T_{\parallel}. \qquad (1)$$

In the present investigation, subjects matched the brightness of a polarized slit with that of an unpolarized surrounding field, this leading directly to an estimate of G. The following derivation establishes the

* Hereafter, "polarized parallel to ..." will mean "polarized with the **e** vector parallel to ..." etc.

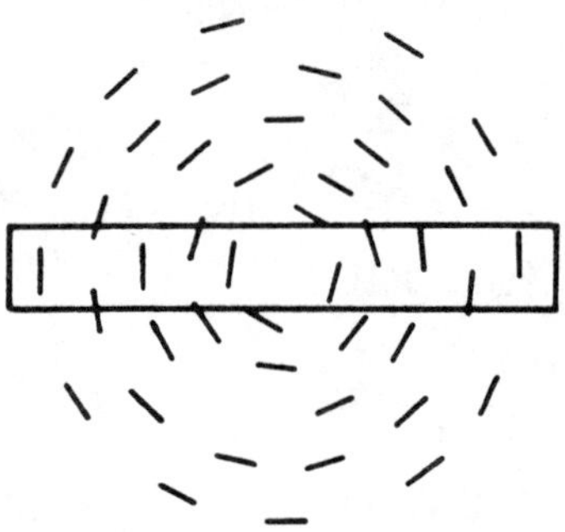

Fig. 1. Schematic representation of the preferentially arranged macular pigment molecules. The rectangle represents the image of a centrally fixated slit, within which the molecules are seen to be approximately parallel to one another.

expected relationship between G and the optical density D of the macular pigment.

Consider the pigment within the retinal area defined by the slit image as a medium of uniformly spaced molecules of concentration c per unit volume forming a layer of thickness d. If a fraction x of these molecules is preferentially orientated tangentially with respect to the foveal centre, they form, in effect, a layer of thickness xd on top of another layer of thickness $(1 - x)d$ of randomly arranged molecules. (It is assumed, for simplicity, that all molecules lie in planes parallel to the surfaces of the layers.) Now the random layer will affect light in the same way as two superimposed half layers of parallel molecules, which are crossed with respect to each other. Furthermore, we may align one of these half layers such that its molecules are parallel to those in the layer of thickness xd. We then have, in effect, one layer (layer 1) of thickness $\frac{1}{2}(1 + x)d$ whose molecules are perpendicular to the axis of the slit and another (layer 2) of thickness $\frac{1}{2}(1 - x)d$ whose molecules lie parallel to the slit axis.

Let the intrinsic extinction coefficients of a molecule be $\varepsilon_{\parallel}$ and $\varepsilon_{\perp}$ for light polarized parallel and perpendicular respectively to its main axis. Consider then the passage of a beam of monochromatic light through the combined layers. If this is polarized parallel to the molecules of layer 1, the transmittance $T_{\parallel}$ is given by:

$$T_{\parallel} = 10^{-\varepsilon_{\parallel}\frac{1}{2}cd(1+x)} \cdot 10^{-\varepsilon_{\perp}\frac{1}{2}cd(1-x)}$$

Similarly if it is polarized at right angles to the molecules of layer 1, we obtain:

$$T_{\perp} = 10^{-\varepsilon_{\perp}\frac{1}{2}cd(1+x)} \cdot 10^{-\varepsilon_{\parallel}\frac{1}{2}cd(1-x)}$$

Substituting in equation (1),

$$\log_{10}G = (\varepsilon_{\parallel} - \varepsilon_{\perp})cdx \tag{2}$$

Consider now the case for unpolarized light. This may be represented by two components of equal intensity I polarized parallel and perpendicular to the molecules of layer 1 (and with a random phase difference between them). The intensities on emerging from the combined layers will be reduced to $I\,T_{\parallel}$ and $I\,T_{\perp}$ respectively. Since the incident intensity was $2I$, the effective transmittance (T) of the combined layers is given by

$$T = (T_{\parallel} + T_{\perp})/2$$

with a corresponding optical density D given by

$$D = \log_{10}\frac{2}{T_{\parallel} + T_{\perp}}$$

Substituting for $T_{\parallel}$ and $T_{\perp}$ and incorporating equation (2),

$$D = \frac{cd}{2}(\varepsilon_{\parallel} + \varepsilon_{\perp}) + \log_{10}\left[\frac{2G^{1/2}}{1 + G}\right]$$

Typical values of G lie in the range 1.05 to 1.15. Substituting the latter value into the above equation gives

$$D = \frac{cd}{2}(\varepsilon_{\parallel} + \varepsilon_{\perp}) - 0.001$$

With typical values of D lying in the range 0.1 to 1.0, the second term is clearly insignificant and we may write

$$D = \frac{cd}{2}(\varepsilon_{\parallel} + \varepsilon_{\perp}) \tag{3}$$

D, however, is usually defined in terms of an extinction coefficient, ε, without reference to particular directions, i.e.

$$D = cd\varepsilon$$

Comparison with equation (3) shows that ε is therefore the mean of $\varepsilon_{\parallel}$ and $\varepsilon_{\perp}$.

Eliminating the product cd between equations (2) and (3) gives the required relationship between D and $\log_{10}G$:

$$\log_{10}G = 2Dx\left[\frac{\varepsilon_{\parallel} - \varepsilon_{\perp}}{\varepsilon_{\parallel} + \varepsilon_{\perp}}\right] \tag{4}$$

Similarity between the spectral distributions of D and $\log_{10}G$, as discovered by DeVries *et al.* (1953) and Naylor and Stanworth (1954), would imply the necessary and sufficient condition that $\varepsilon_{\perp}$ was zero over the visible spectrum, i.e. the molecules had only one direction of absorption. For then

$$\log_{10}G = 2Dx. \tag{5}$$

This essentially was the relationship derived by DeVries *et al.* (1953) from the initial assumption that $\varepsilon_{\perp}$ was zero. In their case, however, similarity between the two distributions implied that this had been a sufficient assumption, but not also a necessary one as is shown by equation (4).

Naylor and Stanworth (1954), on the other hand, obtained

$$\log_{10}G = Dx$$

having assumed that

$$T_{\parallel} = 10^{-\frac{1}{2}(1+x)D}$$

and

$$T_{\perp} = 10^{-\frac{1}{2}(1-x)D}$$

but this seems incorrect since it leads to

$$T_{\parallel} = T_{\perp} = 10^{-\frac{1}{2}D} \text{ and not } 10^{-D}$$

for a completely random arrangement of molecules.

In the measurement of G, as briefly outlined earlier, a complication arises from the birefringence of the optical media of the eye (Boehm, 1940). The relative retardation associated with this birefringence has been measured and quite a wide range of values reported (Boehm, 1940; DeVries *et al.*, 1953; Naylor and Stanworth, 1954; Shute, 1974). The contribution of the lens to the overall birefringence (Weale, 1978) is negligibly small, amounting to an absolute phase difference between the ordinary and extraordinary rays of less than 3° at 500 nm. The only other medium with a suitable structure for producing birefringence is the cornea (Maurice, 1969) and this must therefore be the main contributor.

Its effect is to transform initially plane polarized light into elliptically polarized light except when the former is parallel to either of its privileged directions. Thus the slit intensities, with the emerging light polarized first parallel and then perpendicular to the slit, which match the unpolarized surrounding field, only lead to the true dichroic ratio (G) under this condition. At other angles (ϕ) between the polarization plane of the incident light and either privileged direction, we measure an "apparent dichroic ratio" (G'), related to G by the formula:

$$G' = \frac{G + (1 - G)\sin^2 2\phi \sin^2 \psi/2}{1 - (1 - G)\sin^2 2\phi \sin^2 \psi/2} \quad (6)$$

where ψ is the relative retardation produced by the cornea. (The derivation of this formula is given in the appendix.)

It is easily seen that G' is equal to G when ϕ is equal to 0 or 90°. Except during the experimental verification of equation (6), this condition was always met by carrying out a preliminary experiment to determine the orientations of the subject's privileged directions.

METHODS

Measurement of G *and* G′

The layout of the apparatus is shown schematically in Fig. 2. Monochromatic light was provided by a 250 W tungsten halogen lamp H in conjunction with a graded spectrum filter F, slit K and lens L_1. This monochromator, which has been described elsewhere (Bone and Sparrock, 1971), gave a bandwidth of approx. 12 nm which could be centred anywhere in the range 400–700 nm. Lenses L_2 and L_3 (L_3 being cylindrical) directed the monochromatic beam towards a magnesium oxide coated screen S_1 where it formed a uniform patch of light. The lower half of this beam, however, was intercepted by a mirror M_1 and, via mirror M_2, provided adequately uniform illumination of a second white screen S_2. In the middle of S_2 was cut a slit J which, when viewed through the 6 mm exit pupil E and the magnifying lens L_4, subtended a visual angle of 2.5° lengthwise. Preliminary experiments revealed that more repeatable results could be obtained if, instead of J being rectangular, it consisted of two narrow angles (30°) sectors (see inset). This shape was selected as a compromise between obtaining a large area and still remaining within the boundaries of the Haidinger brush pattern as would be seen at the normal observation distance.

Polaroid filter P_2 determined the plane of polarization of the light emerging through J from S_1 whereas rotation relative to P_2 of a second Polaroid filter P_1 determined its intensity. P_1, P_2 and S_2 were all mounted in calibrated, rotatable holders.

To facilitate accurate measurements, a dental bite was provided for the subject, together with a small fixation mark at the centre of J.

Depending on the nature of the test, the monochromator was first set to the desired wavelength and the screen S_2 rotated to produce the required orientation of J relative to the subject's eye. P_2 was then rotated to align the plane of polarization either parallel or perpendicular to the main axis of J. The subject fixated voluntarily on the fixation mark and adjusted P_1, via a remote control, to secure a luminance match between J and the unpolarized surround.

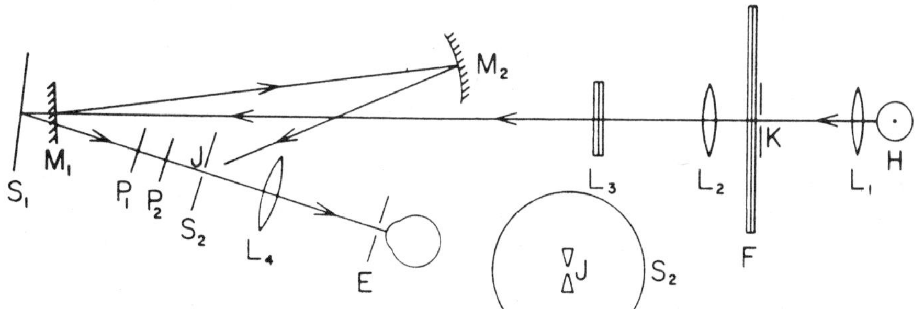

Fig. 2. Plan view of the apparatus for determining G and G'. The light source H, together with the graded spectrum filter F, provide illumination of screens S_1 and S_2. Inset is the subject's view of S_2. Light emerging from the aperture J is polarized by Polaroid filter P_2 and its intensity controlled by rotation of P_1.

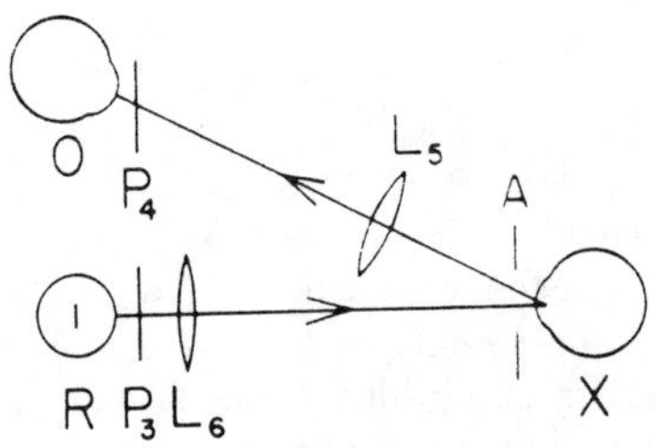

Fig. 3. Plan view of the apparatus for determining the cornea's privileged directions. Rotatable Polaroid filters P_3 and P_4 are coupled so as to be always crossed. The operator O makes observations on the Purkyně-Sanson images of the ribbon filament of lamp R, produced by the subject's eye X.

Generally about six settings were recorded for each orientation of P_2. The ratio of slit luminances, determined by the settings of P_1 relative to P_2, led directly to an estimate of G (or G').

Birefringence measurements

Observations of the Purkyně-Sanson images, which are formed by reflections at the anterior and posterior surfaces of the lens and cornea, were found to provide a quick, convenient means of locating the fast and slow axes of the cornea. The principal component of the apparatus, shown in Fig. 3, consisted of two Polaroid filters P_3 and P_4 mounted in the inner races of two ball bearings and linked by a gear train. By this means, the filters could be rotated together in the same sense and at the same rate, and were mounted in such a way as to be always crossed. The operator O made observations through a magnifying lens L_5 of the images in the subject's eye X of the ribbon filament of lamp R (magnified by L_6).

Assuming birefringence in the lens is negligible, extinction of either the second, third or fourth image, as seen by the operator, will occur when the incident light is polarized parallel to the fast or slow axis of the subject's cornea. At any other angle, the reflected light will be elliptically polarized after its double passage through the cornea and will not be extinguished by P_4. The relatively intense first image, on the other hand, should always be extinguished. In practice however, crossed Polaroid filters transmit a small fraction of light, and the first image was always faintly visible. Due to the close proximity of this and the second and fourth images, difficulty was experienced in finding the angle at which the latter two were extinguished. Observations were therefore restricted to the relatively distant third image, which is formed by reflection from the aqueous humor–lens interface.

Strictly speaking however, the foregoing arguments apply only in the case of light rays incident normally on the lens, and complications may arise for large angles of incidence. It is then possible for the components of the incident light, whose planes of polarization are parallel and perpendicular to the plane of incidence, to undergo significantly different degrees of reflection, resulting in a change in the polarization state of the reflected light. If, for example, the light incident on the eye is plane polarized parallel to the fast axis of the cornea, it will still be plane polarized at the lens, but upon reflection there will be an effective rotation of the plane of polarization relative to the fast axis. Thus the light re-emerging from the cornea will be elliptically polarized and will not be totally extinguished by P_4.

Under the experimental conditions, the maximum angle of incidence was about 10°. The corresponding ratios of the reflected to incident amplitudes were calculated to be −0.029 and −0.031 for the components parallel and perpendicular respectively to the plane of incidence (the negative signs indicating phase changes of 180°). These data, together with typical values for the relative retardation of the cornea and the orientations of its privileged directions (obtained from later results), were used to calculate the relative intensity of the light transmitted by P_4 for different directions of polarization of the light entering the eye. Using only one value of relative retardation, however, is equivalent to using monochromatic light, whereas white light was used experimentally to obtain a sufficiently high intensity. Since there is no simple relationship between the relative retardation of a birefringent medium and wavelength, it was decided to try to simulate the effect of white light by assuming that a large number of relative retardations were present simultaneously. Hence the above calculations were repeated, varying the relative retardation from 0 to 350° in increments of 10°. The superposition of the sets of results, with an equal weighting for each set is shown in Fig. 4, where the relative intensity of the light transmitted by P_4 is plotted for a small range of angles between the polarization plane of the incident light and a privileged direction of the cornea.

This graph shows clearly that for angles of incidence up to at least 10° (any undesirable effects increasing with angle) there are no noticeable complications, almost complete extinction of the third Pur-

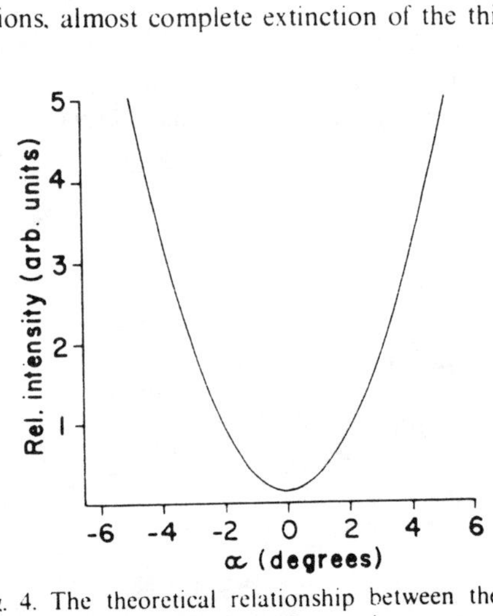

Fig. 4. The theoretical relationship between the relative intensity of the third Purkyně-Sanson image as seen by the operator and the angle α between the polarization plane of the incident light and either of the privileged directions of the cornea.

kynē–Sanson image, as seen by the operator, occurring when the incident light is polarized parallel to either privileged direction of the cornea.

Thus, with the aid of a 0.5″ aperture A and a dental bite, the subject placed his eye in the required position and was asked to fixate on the ribbon filament of R. The operator then rotated the polarizing filters, noting the angles at which the third image disappeared. These angles, measured relative to the dental bite, were found, as expected, to be separated by 90°. From these measurements it was possible to orientate J and P_2 (Fig. 2) to the required angle which would ensure that light reaching the subject's retina from J remained plane polarized.

This experimental technique may be compared favorably with earlier methods (e.g. Naylor and Stanworth, 1954; Smith and Weale, 1974; Shute, 1974). The main advantage is that it relies for accuracy entirely on the skill of the operator with no demands being made on the subject. This was reflected in the repeatability of results, which indicated an accuracy of the order of 1° or 2°. It will be seen later (Fig. 5) that this is more than sufficient for the purpose of obtaining the true dichroic ratio G. The second advantage is that unlike the earlier methods, it does not rely on observations of Haidinger's brushes and is therefore suitable for subjects who have difficulty in seeing them. Finally there is the advantage of the speed with which readings could be taken, a set of six occupying little more than 1 min.

Optical density measurements

In an earlier investigation (Bone and Sparrock, 1971), flicker photometry was successfully employed to obtain a subject's macular pigment density spectrum from foveal and extra-foveal sensitivity measurements. The same apparatus was used in the present investigation, and consisted, in part, of the monochromator described above. The apparatus shown in Fig. 2 could, in fact, be rapidly converted for such a measurement of D, thus allowing measurements of both D and G to be made in a single session. The conversion consisted mainly in removing lenses L_2 and L_3 from the monochromator beam and placing

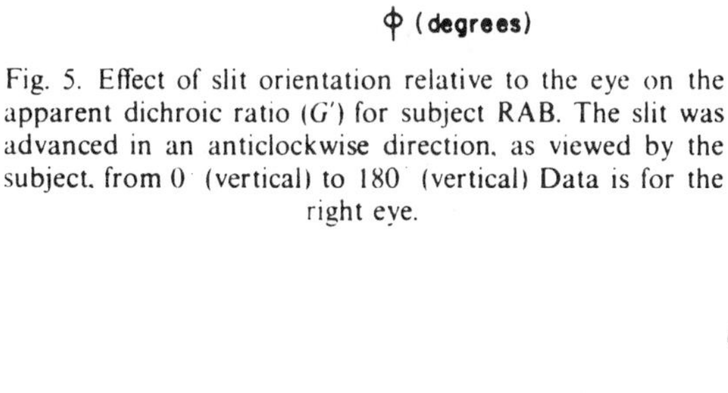

Fig. 5. Effect of slit orientation relative to the eye on the apparent dichroic ratio (G') for subject RAB. The slit was advanced in an anticlockwise direction, as viewed by the subject, from 0° (vertical) to 180° (vertical) Data is for the right eye.

the flicker photometer head in their former position. A neutral density wedge to the right of slit K allowed the subject to control the beam intensity.

With the monochromator set to the wavelength under consideration, the subject formed a brightness match, first foveally and then 8° extra-foveally against the light from a comparison lamp-interference filter combination. This combination limited the light in the comparison field to wavelengths greater than 530 nm, above which the macular pigment has zero absorption. From the wedge settings obtained, the pigment density could easily be calculated (see Bone and Sparrock, 1971).

RESULTS AND DISCUSSIONS

Birefringence of the cornea

The effect of slit orientation, relative to the eye, on the apparent dichroic ratio G' at 480 nm, was assessed for subject RAB. This wavelength was selected as one for which the value of G was high (see later results) and the light level sufficient for reasonably repeatable measurements.

The results, which are for the subject's right eye, are plotted in Fig. 5 (solid circles). The vertical bars represent the standard deviations of the mean values derived from six readings for each setting of P_2 (i.e. parallel and perpendicular to J). The continuous curve was given by equation (6) with the parameters G, ϕ and ψ adjusted to give the best fit between the curve and the experimental points.

The correspondence is very encouraging. It must, after all, be borne in mind that equation (6) assumes an ideal condition where the true dichroic ratio G is uniform over the retinal area under consideration. In fact Naylor and Stanworth (1954), having corrected for birefringence in the cornea, found that G was generally higher along the horizontal meridian as compared with the vertical meridian.

The continuous curve of Fig. 5 yielded a value of G of 1.145 and a relative retardation of the cornea of 51°. This retardation may be compared with the 40° at 480 nm obtained by Naylor and Stanworth (1954), 0 and 65° at 460 nm reported for two observers by DeVries *et al.* (1953) and 45° using blue light in the range 430–485 nm obtained by Boehm (1940). These variations can be easily understood if the absolute phase difference produced by the cornea between the ordinary and extraordinary rays is of the order of many wavelengths and not merely a fraction of a wavelength. Then small variations in corneal thickness, or the use of different wavelengths, would result in seemingly random variations in relative retardation.

The positions of the maxima of the curve indicate that one of the cornea's privileged directions was inclined downwards at 19° below the horizontal in the nasal direction. This was later established, by means of retardation plates, to be the direction of the slow axis. The value of 19° was less than the 32° obtained

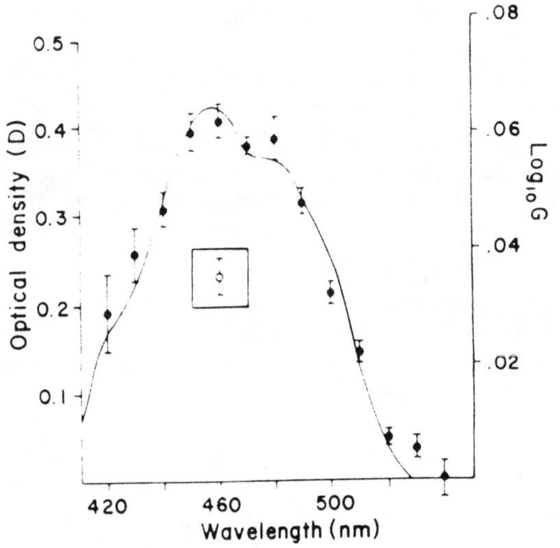

Fig. 6. Comparison between the spectral distributions of D (continuous curve and left-hand ordinate scale) and $\log_{10}G$ (solid circles and right-hand ordinate scale) for subject RAB. Standard deviations of the mean have been omitted for the D curve, but an average value is shown in the inset.

by DeVries *et al.* (1953) and the 30° of Naylor and Stanworth (1954). It was, however, completely in keeping with the value of 20° measured directly using the apparatus of Fig. 3.

Spectrum of G *and* D

Tests were then conducted on the same subject in order to obtain the dichroic ratio of the pigment as a function of wavelength. To remove the effects of birefringence in the cornea, slit J was rotated to a position parallel to the subject's slow axis, and appropriate measurements made at 10 nm wavelength intervals.

The results, again for the subject's right eye, are shown in Fig. 6 (solid circles), the vertical bars being as in Fig. 5. Also shown in Fig. 6 is the optical density spectrum of the pigment (continuous curve) for the same subject, this too being based on measurements every 10 nm. For the sake of clarity, the standard deviations for this curve have been omitted, but an average value is indicated in the inset (centre). To emphasize the expected similarity between the two spectral distributions, as suggested by equation (5), a log scale was used for G and the scale divisions deliberately chosen to give the best fit between the two curves. Such similarity is clearly present, and the assumption of DeVries *et al.* (1953), that the pigment molecules have only one direction of absorption over the visible spectrum, is well supported by these results.

Fraction of orientated molecules

Fifteen subjects participated in the final experiment, designed to bring to light any variations in x, the fraction of preferentially orientated molecules. Having accepted the validity of equation (5), it was therefore only necessary to determine G and D at one particular wavelength, and this was chosen, as before, to be 480 nm. It was also necessary to determine the privileged directions of each subject's cornea and to align the slit J accordingly. It might be remarked in passing that a considerable range of values was encountered here. One of the directions was found to be inclined downwards towards the nose at anywhere between 0 and 77° below the horizontal, with a mean value of 22°. No attempt, however, was made to determine whether it was the fast or the slow direction.

Values of x were found to range from about 2 to 10% and an interesting trend which emerged indicated that subjects with a high pigment density had a low x value and vice versa. This is evident from the positive slope in Fig. 7, a graph of pigment density at 480 nm versus $1/x$. The error bars once again represent the standard deviations of the mean values.

GENERAL CONCLUSIONS

The results described here reinforce, but also go beyond those of the earlier investigations which have been frequently quoted (DeVries *et al.*, 1953; Naylor and Stanworth, 1954). Figure 6, for example, shows a slight extension of the wavelength range over which G has been measured, but, more importantly, we have here the first direct comparison between the D and $\log G$ spectral distributions not only for the same subject but obtained using the same monochromator. The similarity between the distributions implies two things: firstly, and more apparently, they have their origins in the same retinal pigment. Secondly, the pigment must consist of molecules possessing only one direction of absorption over the visible spectrum. Hence lutein, with its long chain molecule still stands as the most likely candidate and we may rule out any contender with at least two directions of absorption, for example, Dartnall and Thomson's (1949) suggested cytochrome.

The birefringence of the cornea is well illustrated by the curve in Fig. 5. DeVries *et al.* (1953) also

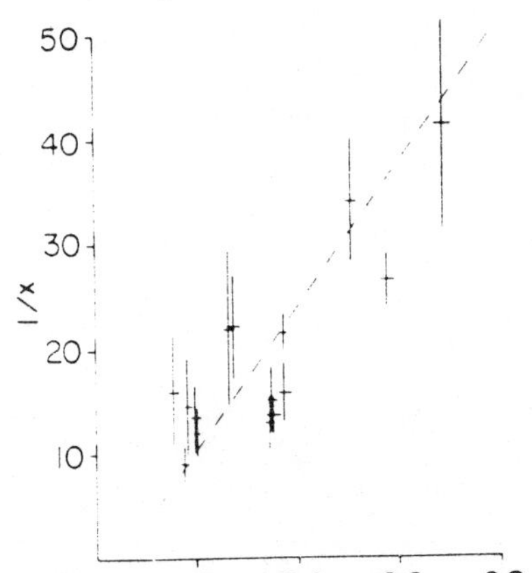

Fig. 7. The reasonably linear relationship between the optical density of the macular pigment and the reciprocal of the fraction (x) of preferentially orientated pigment molecules for 15 subjects. The best straight line associated with the data points is indicated.

measured variations in G', but only presented a few values over a limited range of angles ($\sim 70°$). As Naylor and Stanworth (1954) pointed out, their technique of tilting the observer's head relative to the test field, in order to vary the angle, could result in inaccuracies. Certainly their curve does not exhibit as clearly as Fig. 5 the trend to be expected on the basis of equation (6).

The birefringence of the cornea, as explained by Shute (1974), is consistent with a preferential orientation of corneal collagen. From his observations using Haidinger's brushes, Shute noted that the direction of orientation appeared to be upwards-and-outwards, coinciding therefore with the line of pull of the superior and inferior oblique muscles of the eye, a result of possible significance. However, the wide range of orientations of the cornea's privileged directions, measured accurately by the Purkyně–Sanson image technique in the present study, must cast doubt on any suggestion of a link between the direction of the muscles and that of the collagen.

With regard to the relationship between D and x (Fig. 7), it would appear that the areal density of preferentially orientated molecules did not vary considerably among the subjects tested. For then the subject with the high pigment density, caused possibly by a high concentration of pigment molecules, would have a low fraction which were preferentially aligned. Conversely, the low pigment density observer would have a relatively large x value. Generally we may assume that variations in D among subjects are due to variations in the product cd (equation 3), D being therefore proportional to the total number of pigment molecules per unit area of retina (N). If N_0 is the number of preferentially orientated molecules per unit area, and this is the same for all subjects, then since $x = N_0/N$, D will be inversely proportional to x; hence the choice of axes in Fig. 7.

Now the macular pigment is found between the outer and inner limiting membranes of the retina (Polyak, 1941), a region containing the radially arranged nerve fibres. It is conceivable that these fibres may be responsible for producing the alignment of the molecules, in much the same way that a stretched sheet of polyvinyl alcohol acts as the aligning matrix for polymeric iodine in the Polaroid *H* polarizer (Land, 1951). To account for Fig. 7, it would then be necessary to assume that the fibres are only capable of aligning a certain number of molecules, a not unreasonable hypothesis if indeed the molecules are membrane-bound to the fibres. Thereafter, the remaining molecules would be in a free state between the inner and outer limiting membranes and their orientations would certainly be expected to be random.

REFERENCES

Boehm G. (1940) Über maculare (Haidinger'sche) Polarisationsbüschel und über einen polarisationsoptischen Fehler der Auges. *Acta ophth.* **18,** 109–142.

Bone R.A. (1976) Computer-enhanced resolution as an aid to identifying the macular pigment. *Vision Res.* **16,** 1191–1192.

Bone R. A. and Sparrock J. M. B. (1971) Comparison of macular pigment densities in human eyes. *Vision Res.* **11,** 1057–1064.

Dartnall H. J. A. and Thomson L. C. (1949) Retinal oxygen supply and macular pigmentation. *Nature* **164,** 876.

DeVries H., Spoor A. and Jielof R. (1953) Properties of the eye with respect to polarized light. *Physica* **19,** 419–432.

Dewar M. J. S. (1949) *The Electronic Theory of Organic Chemistry.* Oxford University Press, London.

Haidinger W. (1844) Über das direkte Eriennen des polarisierten Lichts. *Annln Phys.* **63,** 29.

Helmholtz H. (1866) *Handbuch der Physiologischen Optik.* Leopold Voss, Hamburg.

Hochheimer B. F. (1978) Polarized light retinal photography of a monkey eye. *Vision Res.* **18,** 19–23.

Land E. H. (1951) Some aspects of the development of sheet polarizers. *J. opt. Soc. Am.* **41,** 957–963.

Maurice D. M. (1969) The cornea and sclera. In *The Eye,* 2nd edn, Vol. I (Edited by Davson H.). Academic Press, London.

Naylor E. J. and Stanworth A. (1954) Retinal pigment and the Haidinger effect. *J. Physiol.* **124,** 543–552.

Polyak S. L. (1941) *The Retina.* University of Chicago Press, Chicago.

Shute C. C. D. (1974) Haidinger's brushes and predominant orientation of collagen in corneal stroma. *Nature* **250,** 163–164.

Shute C. C. D. (1978) Haidinger's brushes. *Vision Res.* **18,** 1467.

Smith R. W. and Weale R. A. (1974) A new method for the determination of the birefringence of the intact human pre-retinal media. *J. Physiol.* **246,** 37P–38P.

Stanworth A. and Naylor E. J. (1950) Haidinger's brushes and the retinal receptors. *Br. J. Ophthal.* **34,** 282–291.

Summers D. M., Friedmann G. B. and Clements R. M. (1970) Physical model for Haidinger's brush. *J. opt. Soc. Am.* **60,** 271–272.

Wald G. (1945) Human Vision and the spectrum. *Science* **101,** 653–658.

Weale R. A. (1978) On the birefringence of the human crystalline lens. *J. Physiol.* **284,** 112P–113P.

APPENDIX

Derivation of equation (6)

Consider a plane polarized wave, represented by $a \cos \theta'$, incident on the cornea at an angle ϕ with the slow axis OS as shown in Fig. A-1. OF is the fast axis. The components

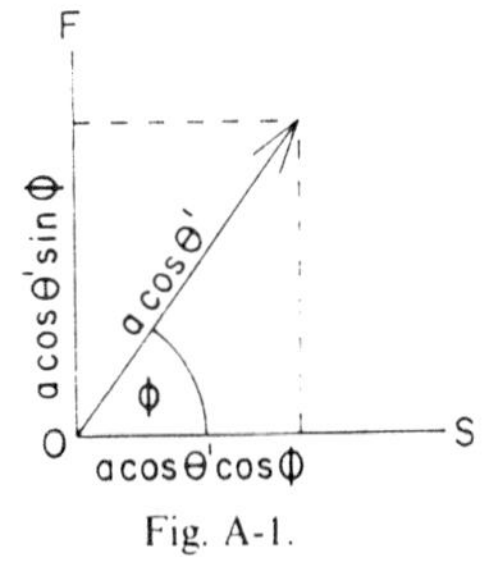

Fig. A-1.

along OS and OF will be as indicated. On emerging from the cornea, the components are, say,

$$a \cos \theta \cos \phi \text{ along } OS$$

and

$$a \cos (\theta + \psi) \sin \phi \text{ along } OF$$

where ψ represents the relative retardation. Each of these may be further resolved into components parallel and perpendicular to the original wave. The first becomes

$$a \cos\theta \cos^2\phi \text{ parallel}$$

and

$$a \cos\theta \cos\phi \sin\phi \text{ perpendicular.}$$

The second becomes

$$a \cos(\theta + \psi) \sin^2\phi \text{ parallel}$$

and

$$-a \cos(\theta + \psi) \sin\phi \cos\phi \text{ perpendicular.}$$

The resultant intensity I_1 of the two parallel components is given by summing the squares of the coefficients of $\cos\theta$ and $\sin\theta$.

Thus

$$I_1 = a^2[(\cos^2\phi + \cos\psi \sin^2\phi)^2 + (-\sin\psi \sin^2\phi)^2] = a^2(1 - \sin^2 2\phi \sin^2\psi/2).$$

Similarly the resultant intensity I_2 of the perpendicular components is given by:

$$I_2 = a^2 \sin^2 2\phi \sin^2\psi/2.$$

Consider now the passage through the dichroic pigment layer. If the original wave was polarized parallel to the slit axis, the component I_1 will be perpendicular to the preferentially aligned molecules and the component I_2 will be parallel. The total intensity incident on the receptors will then be given by:

$$T_\perp I_1 + T_\parallel I_2 = a^2 T_\perp + a^2(T_\parallel - T_\perp)\sin^2 2\phi \sin^2\psi/2.$$

Consequently the "apparent transmission" $T'_\perp$ of the pigment is given by:

$$T'_\perp = T_\perp + (T_\parallel - T_\perp)\sin^2 2\phi \sin^2\psi/2.$$

Similarly if the original wave was polarized perpendicular to the slit, the apparent transmission $T'_\parallel$ is given by:

$$T'_\parallel = T_\parallel - (T_\parallel - T_\perp)\sin^2 2\phi \sin^2\psi/2.$$

Thus the "apparent dichroic ratio" $G'(= T'_\perp/T'_\parallel)$ is given by:

$$G' = \frac{G + (1 - G)\sin^2 2\phi \sin^2\psi/2}{1 - (1 - G)\sin^2 2\phi \sin^2\psi/2}$$

where $G = T_\perp/T_\parallel$ as before.

Vol. 14(6), 1395–1405 (June 1975).

Polarization of light reflected from rough planetary surface

Milo Wolff

A calculation is made of the luminance and polarization of light due to single and double reflections from the faces of particles in a surface composed of random, irregular particles using equations of electromagnetic waves and materials with a complex index of refraction. Some geometric properties of shadows are derived and used. Good agreement is obtained between these results and measurements of polarized light from Mars, Mercury, and the moon, including the phenomenon of negative polarization at small phase angles. Negative polarization is found to be caused by shade and shadows affecting the double-reflected rays. Graphical results are provided for materials of varied real and complex indices of refraction. The model can be used to calculate polarization and luminance of rough astronomical bodies and surfaces as a function of the viewing angle. Calculated ratios of single-reflected, double-reflected, and randomly diffused light can be related to the surface structure and optical properties of the material.

I. Introduction

Measurements of polarization vs lunar phase angle for small regions of the moon were obtained by Gehrels *et al.*,[1] Dollfus and Bowell,[2] and Pellicori[3] at wavelengths ranging from 0.327 μm to 1.05 μm. A feature that has attracted much attention is the existence of negative polarization, a 90° shift of the plane of polarization at small phase angles as shown in Fig. 1. No explanation has accounted for it. Dollfus[4] performed an experiment to show that a pile of powdered sand will generate a negative polarization branch, but if the grains are widely separated by pouring them in front of the polarimeter, the negative branch almost disappears. Bowell *et al.*[5] observed blue $CuSO_4$ crystals in red light, equivalent to opaque crystals, and found a pronounced negative branch. However, when the same crystals were observed in blue light, making the crystals semitransparent, the negative branch was absent. In general, the negative branch is always found to be most pronounced when the reflecting material is very dark, extremely rough, porous, and pitted. Dark astronomical bodies with slight atmospheres such as the moon, Mars, and asteroids show maximum negative polarization of about −1.0%, and some laboratory samples such as carbon black[4] reach −1.9%.

Returned lunar samples have been examined to compare their polarization properties with microscopic appearance. Lunar and terrestrial rocks polarize more strongly than lunar fines.[5] However, thin coatings of microscopic fines reduce the polarization of rocks and breccias to telescopically observed values. Lunar particles have agglomerated small particle coatings that appear essential to produce polarization curves with maxima at lower angles than do smooth, opaque particles like sand. The particle sizes of lunar fines vary widely, most falling between 6 μm and 800 μm with a mean size of about 20 μm.[6,7] The positive branch of polarization resembles the curve easily calculated for single reflection from any plane dielectric; however, the maxima occurs at lower phase angles than predicted by simple particle models, and the amplitude is much less.

Many suggestions have been offered to explain the negative polarization from the moon and other dark, rough surfaces. Lyot (p. 111)[8] speculated, "... must come in greater part from the interior ...," apparently emphasizing the porosity of such surfaces. Ohman[9] pointed out that light reflected from retroreflectors can be negatively polarized, and if tiny (~0.01-mm) crystals with such properties existed on the moon, the diffraction broadening could explain the negative peak at small phase angles. McCoyd[10] calculated the polarization caused by a jagged dielectric surface layer on another dielectric with a plane interface. He found that if the upper layer has a higher index, negative polarization occurs because of internal reflection. Hopfield[11] has pointed out that significant negative polarization occurs in the diffraction edges of shadows around electrically conducting particles of sizes about 10 μm and smaller. Since some lunar dust is metallic and of this size, it is reasonable that a very small fraction of the negative polarization is due to this effect. None of these suggestions agree with actual measurements or with properties of lunar samples.

The author was with the Aerospace Corporation Los Angeles; he is now at 1600 Nelson Avenue, Manhattan Beach, California 90266.

Received 11 December 1974.

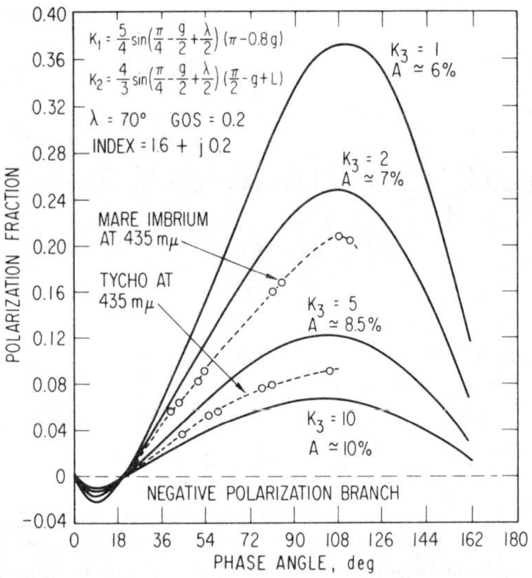

Fig. 1. Simulated curves of lunar polarization obtained by using surface structure parameters. Measurements from Dollfus and Bowell.[2]

An unusual feature of lunar photometry, also found to some degree in any dark, rough surface, is the property of backscattering light so that the curve of brightness vs phase angle is very peaked around zero phase angle. Hapke[12] analyzed the phenomenon and derived a peaked photometric function that accounts for single reflections. According to Gehrels *et al.*,[1] the half-width of peaking is about 3–5°, although this is difficult to determine since no measurement can be made at exactly 0° phase angle.

The possibility of deducing surface properties from polarization measurements at small phase angles is of great interest, especially in the case of the outer planets and satellites that cannot be observed at large angles from earth. Pollack and Sagan[13] have reviewed martian photometry and polarimetry. Dollfus[14] made a study of planetary polarimetric properties. Polarization of minor planets and satellites have been measured by Provin,[15] Veverka,[16] and Zellner.[17,18] Gehrels, *et al.*[19] use the method of Widorn[20] to find albedo from the slope of polarization-phase angle curves. Pellicori[3] made qualitative deductions of surface properties.

In this paper I treat the general problem of the interactions of incident light with a rough surface and show that the negative polarization phenomenon as well as various other characteristics of light from planetary bodies are completely predicted.

A significant feature of this problem is the varied ways reflection can take place. To calculate these, a computer has been used with a model consisting of a cloud of random crystals with faces that obey Fresnel's laws of reflection for complex indices. Then the cloud is converted to a surface by adding effects of structure and shadows when the particles are closely spaced.

This paper deals mainly with dark, rough surfaces that appear to be found on astronomical bodies with slight atmospheres, such as Mars, Mercury, the moon, and asteroids and terrestial surfaces such as cinders, dark coral growth, mosses, and volcanic rock.[8]

II. Light Processes in a Surface

Light rays emitted from a rough surface can be classified according to the processes that they have undergone. These processes depend on the shape and substance of the particles constituting the surface. The more important ones are

(1) single reflection from particle faces; (2) shadowing; (3) double reflection from particle faces; (4) absorption after refraction inside particles; (5) triple, quadruple, and higher order reflections; (6) resonant scattering by small particles and corners; and (7) internal reflections after refraction inside particles.

The first three processes produce significant polarization of emitted light, whereas the last three yield mostly unpolarized light. The preceding list is in the approximate order of importance for a dark, rough surface, and this ordering implies a definition for such surfaces.

In this paper all these processes are taken into account. Single and double reflection are calculated for a random cloud of particles using laws of electromagnetic waves. Absorption is included by using a complex index of refraction and by varying the amount of emergent unpolarized light. Shadowing for single reflections uses the backscattering results of Hapke.[12] Shadowing for double reflections uses geometric results derived here. And finally, processes 5, 6, and 7 are lumped together and added to other calculations as unpolarized light.

III. Gray Crystal Cloud Model

This model of a surface consists of a cloud of dark crystals of random shape and position. The majority of crystals in the model have dimensions larger than the wavelength of light so that the reflection laws may be used. Light reflected from the crystal faces obeys Fresnel's equations. Light that enters the crystals is either absorbed or emitted unpolarized because of many reflections and refractions.

In the simplest model, there is no preferential alignment of crystal faces. A face at any angle is equally probable. In other models, gravity could be taken into account by giving greater probability to face angles near the horizontal; however, microscope studies of lunar samples suggest this is not important for the moon.[5]

The calculations to follow will determine the intensity and polarization of light that undergoes single and double reflection and light emitted from higher order events by assuming it to be unpolarized. The computer program sums up all the probabilities of single and double reflections at all possible angles in a complete sphere by integration over small intervals of light path and angle. Finally, the effects of shadowing and structure will be included by eliminating shadowed and forbidden paths in the computation.

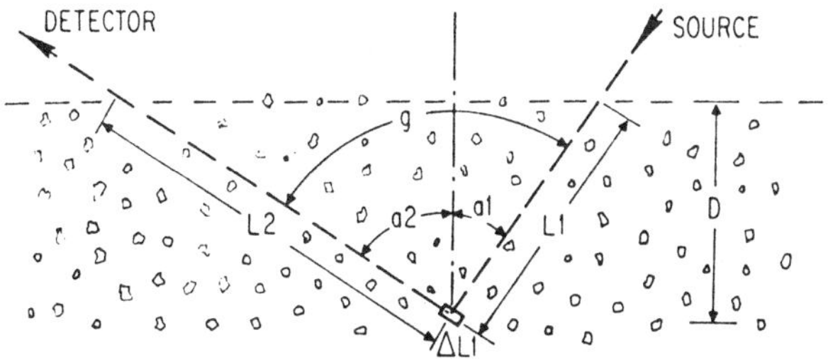

Fig. 2. Light enters a particulate surface material, is reflected once, and escapes.

IV. Single Reflection

A portion of the light that enters the surface and leaves at the phase angle g will be reflected only once. Before and after the reflection, the intensity will be attenuated. The intensity will also be affected according to the reflection laws at the crystal face and by the probabilities of a light ray striking a particle at an incident angle between i and $i + \Delta i$, per steradian of solid angle into which it is reflected. In a random collection of particles, all face angles are equally likely, so that the normalized probability used here is $\frac{1}{4}\pi$.

As the light rays travel through the crystal cloud, their intensity is diminished proportional to the distance X

$$H = H_0 \exp(-X/K) \qquad \mathrm{W/m^2},$$

where H_0 is the intensity before attentuation.

The observer is located at an angle g away from the source of light as in Fig. 2. Taking account of attenuation on the way in, the probability of finding a properly angled face, the length of the differential path element $\Delta L1/K$ and the reflection coefficient FF, we obtain the differential radiance leaving the point of reflection

$$\Delta B = H_0(1/4\pi)\,[\exp(-L1/K)](\Delta L1/K)FF \qquad \mathrm{W/(m^2\ sr)}.$$

The reflection coefficient of a surface is also termed a Fresnel factor FF and must be computed for each angle. K is the unit of distance.

Light will then be attenuated on the way out and received at the observer's detector. The differential intensity now becomes

$$\Delta B = H_0(1/4\pi)\,[\exp(-L1/K)](FF)\,[\exp(-L2/K)](\Delta L1/K).$$

The total intensity arriving at the detector is the integral of the above expression from a surface depth of zero to infinity. The pathlengths $L1$ and $L2$ can be written in terms of the depth D, using the triangles of Fig. 2. The resulting integral for the radiance of the surface due to single reflection is

$$\begin{aligned} B1 &= [H_0/\cos(a2)](1/4\pi)FF \int_0^\infty \{\exp[-D/K\cos(a1)]\} \\ &\quad \{\exp[-D/K\cos(a2)]\}(dD/K) \\ &= \frac{H_0(1/4\pi)FF\,[\cos(a1)]}{\cos(a1)+\cos(a2)} \qquad \mathrm{W/(m^2\ sr)}. \end{aligned}$$

Except for the inclusion of $\frac{1}{4}\pi$ and FF, this is the classical Lommel-Seeliger law.[12] If the surface being observed is on a moon or planet, the angles $a1$ and $a2$ in Fig. 3 can also be written as

$$a1 = g - \lambda \qquad a2 = \lambda,$$

where λ is the colatitude and g is the phase angle in earth-centered coordinates of the observed point.

V. Double Reflection

The calculation of double reflections is complicated by the many directions in which the rays may go. The problem will be considered in two parts: (1) the calculation of attentuation along the ray paths and (2) calculation of the Fresnel factors for the components of polarization at each of the crystal faces. Both will be later combined for a complete result.

A. Attenuation Along the Path

The path of a typical double-reflected ray is shown in Fig. 3. Initially, a light ray goes along path $D0$ in a direction specified by angle $a1$. Its radiance leaving $P1$ the first point of reflection is

$$\Delta B = H_0 \exp(-D0/K)\mathrm{PROB1}(\Delta D0/K) \qquad \mathrm{W/(m^2\ sr)},$$

where **D0, D1, D2** are unit vectors along paths $D0$, $D1$, $D2$. The light may now go in any direction in a 4π sphere and may eventually strike another crystal at point $P2$.

The irradiance into a solid angle, $\Delta\phi\Delta\theta \sin\phi$, arriving at $P2$, as shown in Fig. 4, is

$$\Delta H = H_0 \exp(-D0/K)\exp(-D1/K)(1/4\pi)(\Delta D0/K)(\Delta\phi \sin\phi\Delta\theta),$$

where θ, ϕ are the spherical angles of the path between $P1$ and $P2$ with point $P1$ as the center. The rays reflect at $P2$ with probability $\frac{1}{4}\pi$ and, finally, travel outward in a direction specified by $a2$. They are attenuated along $D2$ and detected with radiance

$$\begin{aligned} \Delta B2 &= H_0 \exp[(-D0 - D1 - D2)/K](1/4\pi)^2 \\ &\quad \Delta\phi \sin\phi\Delta\theta(\Delta D0)\Delta D2/K^2. \end{aligned}$$

This result must now be combined with the proper Fresnel factors at $P1$ and $P2$ to obtain the final emitted radiance.

B. Fresnel Factors for Double Reflection

Fresnel's equations contain the laws of reflection for the two components of light polarized perpendic-

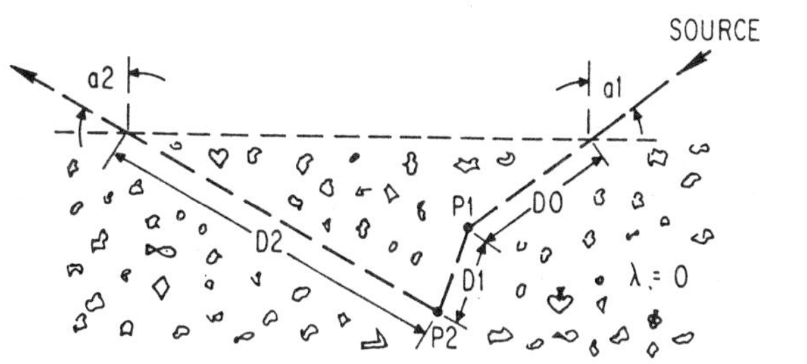

Fig. 3. A light ray enters the particulate surface material, is reflected twice at $P1$ and $P2$, and escapes.

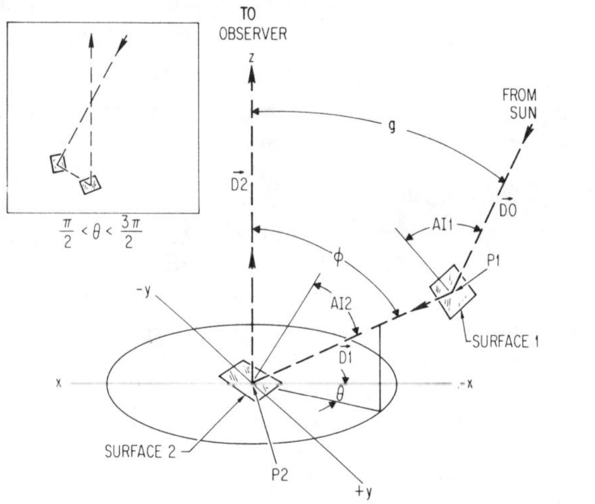

Fig. 4. The geometry used to compute the Fresnel factors for doubly reflected rays. Inset: the path crosses when angle θ is in the forward hemisphere.

ular and parallel to the plane of incidence. They are used to compute the fraction of each component, which is reflected from each particle surface, which is a number between 0 and 1. The remainder of the light is refracted into the crystal and absorbed in the black model. A gray model of random crystal shapes will emit some unpolarized light after refraction.

C. Integration along the Ray Path

In order to sum all possible ray intensities, the differential radiance $\Delta B2$ must be integrated as follows:

(1) along $D0$ (from 0 to ∞) to $P1$, as in Fig. 3; (2) along $D1$ (from 0 to ∞ or to the limits of the model surface) to $P2$; (3) over a sphere of 4π, with angles ϕ and θ as variables of integration, as in Fig. 4.

The Fresnel factors must be computed at each angle during the integration. The expression for the emitted radiance is then

$$B2 = \int H_0 \exp[(-D0 - D1 - D2)/K] \sin\phi (FM) \, dD0 dD1 d\phi d\theta / 4\pi K^2 .$$

Choosing the unit of length $K = 1$, without loss of generality, inserting limits, and writing integrals as sums, we obtain

$$B2 = H_0 \sum_{D0=0}^{\infty} \sum_{D1=0}^{LIM} \sum_{\theta=0}^{2\pi} \sum_{\Phi=0}^{\pi} \exp(-D0) \exp(-D1) \exp(-D2) (FM) \sin\phi \Delta\theta \Delta\phi \Delta D0 \Delta D1/4\pi^2 ,$$

FM stands for either $FM1$ or $FM2$ depending on which polarization component is being computed. FM represents a series of matrixlike computations for the whole ray path and is explained later. LIM is either infinity or the distance to the limit of the model surface in Fig. 3.

Figure 4 illustrates the geometry of the mathematical problem to be solved. It shows a typical ray of light that enters from infinity parallel to the X-Z plane and is reflected twice before leaving. An observer at infinity on the Z axis detects the light.

We wish to find the net polarization and intensity factors of the outgoing ray for every value of the angles ϕ and θ. The calculation for one complete ray path consists of four steps:

(1) Find the angles of incidence and refraction and the Fresnel factors at the first surface to obtain polarization components that are parallel and normal to the plane of incidence of the first surface.

(2) Convert (rotate) the above components into new components parallel and normal to the plane of incidence of the second surface.

(3) Find the angles of incidence and refraction and the Fresnel factors at the second surface to obtain the polarization components that leave the second surface.

(4) Convert (rotate) the above components into two components parallel and normal to the plane containing the observer, the surface, and the sun.

The ray-path relationships can be seen more clearly by visualizing three planes, shown in Fig. 5, each defined by two of the three ray vectors as shown in Table I.

The unit vectors **D0, D1, D2** lie along paths $D0$, $D1$, $D2$. The four steps can be regarded as matrix operations on the initial components of light intensity, $I1$ and $I2$, to obtain the final components, $FM1$ and $FM2$. The entire process is written

$$\begin{pmatrix} FM1 \\ FM2 \end{pmatrix} = |M20| \cdot (C2) \cdot |M12| \cdot (C1) \cdot \begin{pmatrix} I1 \\ I2 \end{pmatrix} = |FM| \cdot \begin{pmatrix} I1 \\ I2 \end{pmatrix} .$$

$C1$ contains the two coefficients of reflection for light

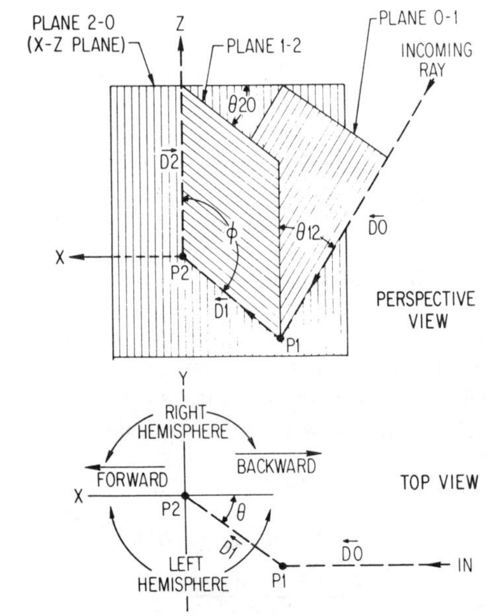

Fig. 5. The three planes of incidence are determined by three ray vectors: **D0, D1,** and **D2.**

Table I. Planes of Figure 4

Plane	Defined by vectors	Plane contains	Plane is
0–1	D0 and D1	Incident rays.	Plane of incidence at surface one.
1–2	D1 and D2	Rays traveling from the first particle to the second particle.	Plane of incidence at surface two. It connects the other two planes.
2–0	D2 and D0 or X and Z	Rays going to the observer.	Identical to the $X - Z$ plane.

intensity components at the first surface. They are a function of the angle of incidence $AI1$.

$$C1 = \begin{bmatrix} FF1\ (AI1) \\ FF2\ (AI1) \end{bmatrix}.$$

$M12$ is a 2×2 matrix that operates (row by column) to obtain components of intensity rotated an angle $\theta 12$ from the first to the second reflection surface:

$$M12 = \begin{vmatrix} \cos^2(\theta 12) & \sin^2(\theta 12) \\ \sin^2(\theta 12) & \cos^2(\theta 12) \end{vmatrix}.$$

$C2$ contains the coefficients of reflection for components at the second surface.

$$C2 = \begin{bmatrix} FF1\ (AI2) \\ FF2\ (AI2) \end{bmatrix}.$$

$M20$ is a 2×2 matrix that operates (row by column) to obtain components rotated an angle $\theta 20$ from the second surface to the observer.

$$M20 = \begin{vmatrix} \cos^2(\theta 20) & \sin^2(\theta 20) \\ \sin^2(\theta 20) & \cos^2(\theta 20) \end{vmatrix}.$$

The angles $\theta 12$ and $\theta 20$ are just those between the planes 0–1, and 1–2, and 1–2 and 2–0, respectively, in Fig. 5. All the angles can be obtained from the vector relations

$$AI1 = \frac{1}{2}[\pi - \arccos(\mathbf{D0} \cdot \mathbf{D1})],$$

$$AI2 = \phi/2,$$

$$\theta 12 = \arccos\left(\frac{\mathbf{D0} \times \mathbf{D1}}{|\mathbf{D0} \times \mathbf{D1}|}\right) \cdot \left(\frac{\mathbf{D1} \times \mathbf{D2}}{|\mathbf{D1} \times \mathbf{D2}|}\right),$$

$$\theta 20 = \theta.$$

The rectangular components of the three unit vectors in the coordinates of Fig. 4 are

$$\mathbf{D0} = [\sin(g), 0, -\cos(g)],$$

$$\mathbf{D1} = (\sin\phi \cos\theta, -\sin\phi \sin\theta, -\cos\phi),$$

$$\mathbf{D2} = (0, 0, 1).$$

After carrying through the vector operations, we get

$$AI1 = \frac{1}{2}\{\pi - \arccos[\sin(g)\sin(\phi)\cos(\theta) + \cos(g)\cos(\phi)]\},$$

$$AI2 = \phi/2,$$

$$\theta 20 = \theta,$$

$$\cos(\theta 12) = \frac{\cos(g)\sin(\phi) - \sin(g)\cos(\phi)\cos(\theta)}{\{1 - [\sin(g)\sin(\phi)\cos(\theta) + \cos(g)\cos(\phi)]^2\}^{1/2}},$$

which completes the mathematics needed for the matrix elements.

D. Complex Index of Refraction

The model so far has only accounted for absorption by the naive device of ignoring the rays that enter the crystals. Properly, absorption should be accounted for by the use of a complex index of refraction. In general, a complex index will increase the reflectivity of the surface and change the dependence on the angle of incidence.

The reflection coefficient for the parallel component of intensity is

$$FF1 = (E_{0p}'/E_{0p})^2 = \frac{|m^2 - A|^2}{|m^2 + A|^2}$$

$$= \frac{|\mathrm{Re}(m^2) - \mathrm{Re}(A)|^2 + |\mathrm{Im}(m^2) - \mathrm{Im}(A)|^2}{|\mathrm{Re}(m^2) + \mathrm{Re}(A)| + |\mathrm{Im}(m^2) + \mathrm{Im}(A)|^2}$$

and for the normal component is

$$FF2 = (E_{0n}'/E_{0n})^2 = |1 - A|^2/|1 + A|^2$$

$$= \frac{|1 - \mathrm{Re}(A)|^2 + |\mathrm{Im}(A)|^2}{|1 + \mathrm{Re}(A)|^2 + |\mathrm{Im}(A)|^2},$$

where m is the complex index, i is the angle of incidence either $AI1$ or $AI2$, and E_0 is the electric field, and

$$A = \left(\frac{m^2 - \sin^2 i}{1 - \sin^2 i}\right)^{1/2}.$$

The preceding formalism was used to compute components of single-reflected light, $B11$ and $B12$, and double-reflected light, $FM1$ and $FM2$, and then the observers polarization

$$\text{Singles polarization} = \frac{B11 - B12}{B(\text{Total})},$$

$$\text{Doubles polarization} = \frac{FM1 - FM2}{B(\text{Total})}.$$

Results are shown in Fig. 6. Note the absence of a negative polarization branch. Shadow and surface effects have not yet been included.

VI. Converting the Cloud to a Surface

The major fraction of a rough surface is covered with shadows. Clearly, the optical radiation from the surface depends on the way shadows are formed. This section examines the nature of a surface and makes a mathematical conversion of the cloud model to a shadowed surface, which results in the appearance of negative polarization.

To make a surface of the gray cloud model, particles are brought together and joined. Bringing parti-

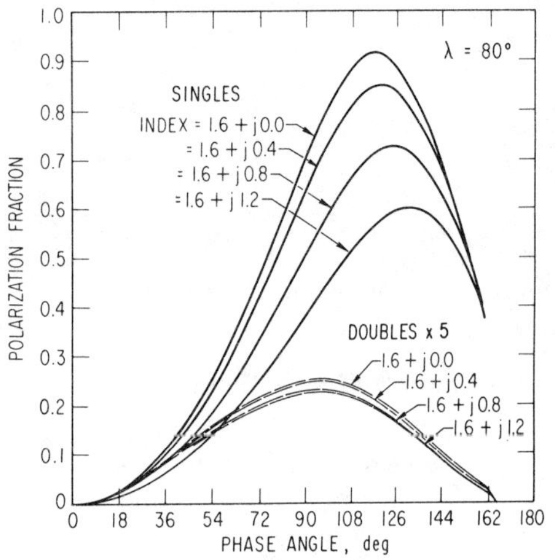

Fig. 6. Computed polarization from a particle cloud due to single and double reflection without shadowing or structure effects.

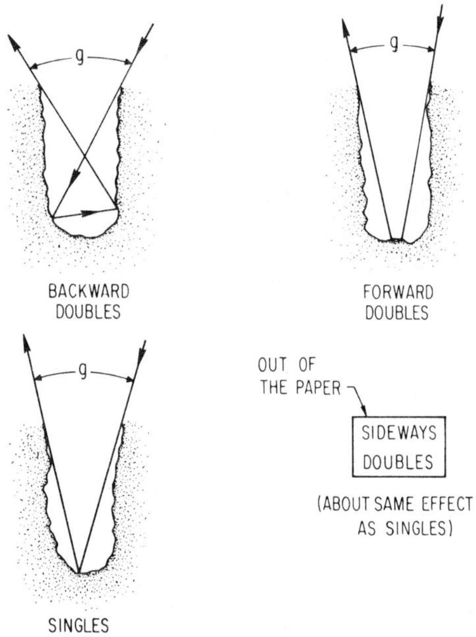

Fig. 7. The retrodirected light from a rough surface has different angular widths for different types of rays.

cles together creates shadows and surface structure. The effects of shadows and structure were incorporated into the computer program by enhancing or attenuating those ray paths of the cloud model that were affected.

A. Retrodirective Effect

The extreme brightness of the full moon is due to a retrodirective effect from light, which only at that period is able to escape from the bottom of surface pits toward the observer. Most of the intensity is single reflection, but it also affects double reflections, as shown in Fig. 7, which shows how forward-going doubles are restricted to the smallest angle. To account for peaking of the lunar brightness, the function SHAG was used as a multiplier of intensity in the singles computation:

$$\text{SHAG} = \{1 + \exp(-g/GOS)^2\},$$

where GOS is the width of the brightness peak in radians.

B. Effect on Double-Reflected Light

To understand the effects of structure and shadows on polarization, it is useful to picture the mechanism of polarization by double reflection. This is illustrated in Figs. 8(A) and 8(B). When the intermediate path of a double reflection goes into a forward or backward quadrant as in Fig. 8(A), the light is polarized positive, similar to single reflection. If the paths are in quadrants left or right of the plane of incidence, as in Fig. 8(B), the polarization is negative because the local planes of incidence at each particle are perpendicular to the plane of incidence formed by the observer, surface, and sun.

In the unmodified particle cloud model, the contributions of polarization from all the intermediate paths in each quadrant are summed. The paths in the forward quadrant contribute the greatest amount of polarization because the angles at each reflection are closest to the Brewster angle of total polarization. The left and right quadrants each contribute about 10% less (near zero phase angle), and the backward quadrant contributes the least (about 20% less near zero phase angles). The reader can verify this by calculating angles for a few cases and referring to a graph of reflectivity vs angle such as in Ref. 21. At zero phase angle all contributions sum to zero polarization as required by symmetry. At phase angles around 90° the polarization from rays in the forward quadrant becomes even stronger because of favorable

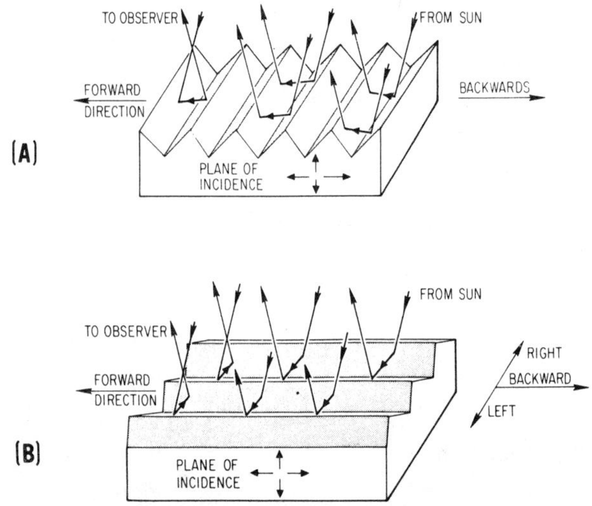

Fig. 8. (A) Double reflections with the intermediate path parallel to the plane of incidence always produces positively polarized light. (B) Reflections with the intermediate path perpendicular to the plane of incidence always produces negatively polarized light.

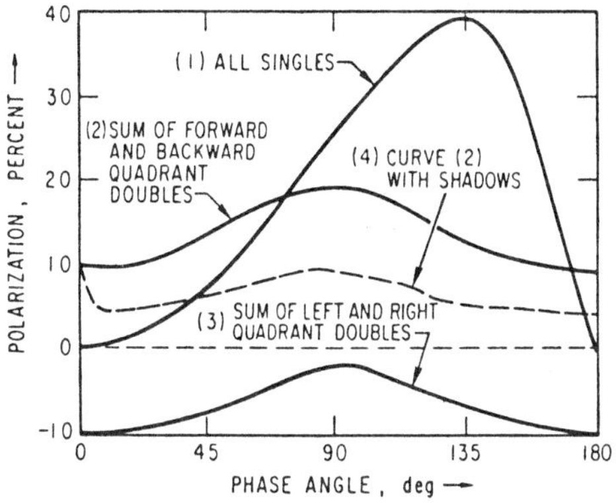

Fig. 9. Typical contributions to polarization by single and double reflected light paths.

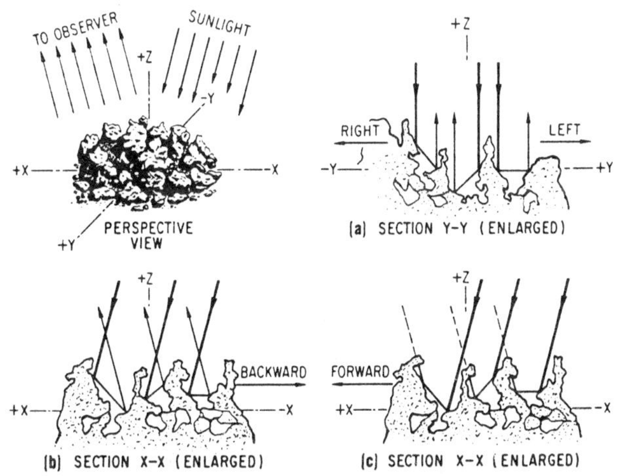

Fig. 10. A pitted backscattering surface showing doubly reflected rays: (a) left and right paths are permitted; (b) backward paths are permitted; (c) forward paths are blocked.

angles at each reflection, while that from left and right quadrants is weaker, as in Fig. 9.

When the particle cloud model is modified to simulate a rough surface, the rays with forward-going intermediate paths are partly blocked, thus diminishing the contribution of positive polarization. Negatively polarized light from left-right quadrants can then dominate at small and very large phase angles and produce a negative branch in the polarization curve. This is the principal cause.

C. Blocking of Forward-Going Doubles

There are two mechanisms that block double-reflected light with forward-going intermediate paths. One of these, illustrated in Figs. 7 and 10, occurs because of all the rays that enter and leave a hole by two reflections, the forward-going rays are restricted to the smallest subtended angle. Figure 10 shows four views of double-reflected light paths in a pitted, backscattering surface such as the moon. View (a) shows typical reflection paths for rays whose intermediate path is in the left or right quadrant. View (b) shows reflection paths of rays whose intermediate path is backward, and view (c) shows how rays with forward-going intermediate paths are blocked by the deep backscattering pits of a rugged surface.

The second mechanism is more easily explained with logic than by a drawing. At zero phase angle the observer sees no shadow because the light source is behind him. As phase angle is increased, shadowing increases until, at angles greater than the width of the backscatter peak, the majority of a rugged surface appears in shade or shadow. By shade is meant the dark side of a body that is intercepting light, and shadow means the dark image of the intercepting body on another surface. No single-reflected light comes from regions of shade or shadow. These regions can, however, doubly reflect light with an important geometric constraint. Regions of shade cannot emit rays with forward-going intermediate paths because in order for the second reflection to strike on the dark side, the intermediate path must go backward, which obviously is not forward. There is no constraint on rays with left, right, or backward intermediate paths.

A mathematical method to describe the blocking of rays with forward intermediate paths is to introduce as an intensity factor, the function SHAD;

$$\text{if } \cos\theta > -0.707,\ \text{SHAD} = 1,$$

$$\text{if } \cos\theta < -0.707,\ \text{SHAD} = 1 + \cos\theta\{1 - \exp[-(2g/GOS)^2]\}.$$

Cosθ is minus in the forward direction. The use of this function to modify the particle cloud model results in decreased positive polarization of curve (4) compared with curve (2) of Fig. 9. This is the most important effect.

Figure 11 shows the computed polarization with the functions SHAG (for singles) and SHAD (for doubles) included to account for backscattering (of singles) and blocking of paths (of doubles) due to

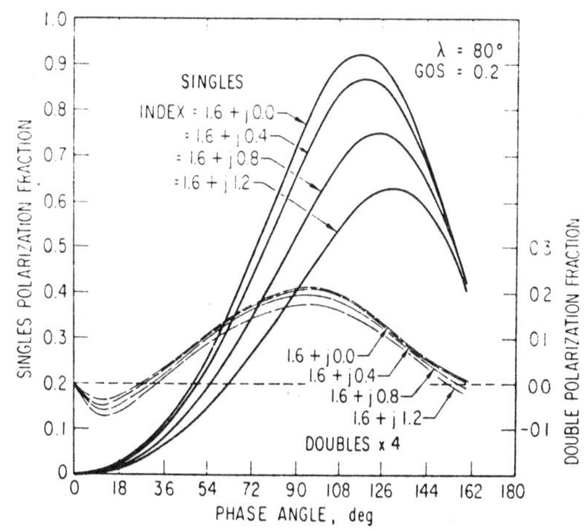

Fig. 11. Computed polarization of light from a particle cloud with shadowing effects included.

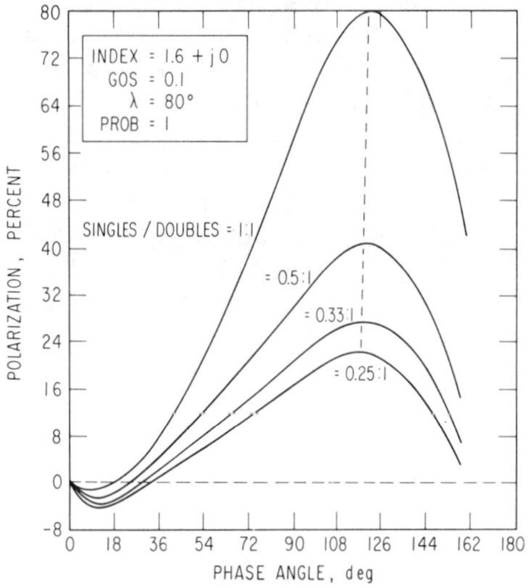

Fig 12. The effect on polarization of assuming different ratios of single and double reflected light.

structure and shade in a surface. The main result of the SHAD function is to add a negative branch to the polarization curve. SHAG has an almost negligible effect on polarization.

Figure 12 shows the result of assuming smaller ratios of single and double reflected light. Smaller relative amounts of single reflections would occur, for example, in a foamlike surface having pits with sharp edges and closed bottoms. A ratio of 1:1 is the same as the original particle cloud, a ratio ½:1 has half as many singles as doubles etc. This ratio would roughly correspond to the ratio of specular area to hole area of the surface. Note how the peak of polarization is slightly shifted to lower phase angles. These curves do not correspond to any real surface, except possibly a totally black sponge or moss, because no unpolarized light has been added yet.

D. Albedo and Polarization

Observations indicate a close relation between albedo and polarization. For regions of the moon, the angle of peak polarization increases with decreasing geometric albedo, and the peak value of polarization decreases with albedo increase. The geometric albedo due to single reflections is only of the order of $(m - 1)^2/(m + 1)^2 = 0.05$ if $m = 1.6$. Hence, the observed higher albedo must be accounted for by other events, particularly refraction and emission of light involving three or more reflections and refractions. Such light is assumed randomized and nonpolarized.

Postulating that the observed photometric function of the moon is accounted for by the sum of light from singles $B1$ plus doubles $B2$ plus nonpolarized events, UNPOL, a boundary condition can be written

$$K_1 B1(g) + K_2 B2(g) + K_3\,\text{UNPOL}\,(g, A) = \text{Observed } B,$$

where $B1$ and $B2$ are the intensity of light from a random particle cloud including shadowing effects. $B1$ and $B2$ are functions of the phase angle g, and the unpolarized light is a function of albedo A as well. The parameters K_1, K_2, and K_3 indicate the relative quantity of each type.

Computer experiments were done using the above relations with the goal of simultaneously matching observed lunar polarization and intensity as a function of phase angle. The parameters K_1, K_2, and K_3 were restricted to smooth functions of g. In general, the polarization due to $B1$ and $B2$ alone resembles the moon's, and large changes are not needed. Thus, the task was to match up small details. A significant fact emerged. In order to replicate the observed decrease of angle of peak polarization with increasing albedo, it was necessary that the slope of the UNPOL (g) function in the region $g = 90°$ to $130°$ be less than that for singles and doubles, or

$$\text{require: } \frac{d(\text{UNPOL})}{dg} < \frac{d}{dg}(B1 + B2),$$

or, in words, the unpolarized light varies more smoothly with lunar phase angle than the light from singles and doubles. This illustrates how the angular distributions of emitted light have an important effect on the angular position of polarization-curve features. For example, this effect dominates the effect shown dotted in Fig. 12.

It is to be expected that the angular distribution functions for the three components are quite different because the functions depend on the number of reflections or refractions involved. It is qualitatively obvious that

(1) single reflected rays are confined mainly to the small angles of the opposition effect; (2) double reflected rays occur at larger phase angles because the second reflection allows some rays to escape; and (3) unpolarized light, caused by multiple events, is least angle dependent, since refracted light can travel through the particles independent of the surface structure.

The derived functions are illustrated in Fig. 13 which shows typical decreasing angular dependence of $K_1 \cdot B1$, $K_2 \cdot B2$, and $K_3 \cdot \text{UNPOL}$ together with algebraic expressions that represent them.

The UNPOL function might have been made constant as a first approximation, but because the ray paths are random, attenuation occurs proportional to the distance traveled, and path distances are known as a function of phase, a form can be derived. With these conditions given, the mathematical derivation is identical to Sec. IV and Fig. 2, with the result that

$$\text{UNPOL} = (A - b)\cos(g - \lambda)/[\cos(g - \lambda) + \cos\lambda],$$

where the parameter b is the contribution of $B1 + B2$ to the albedo A.

The polarization and intensity that result from the functions of Fig. 13 are shown in Figs. 1 and 14 which also shows (dotted curves) two typical measurements taken from the extensive work by Dollfus and Bowell.[2] The agreement is excellent. It is possible to fit quantitatively these and all other measurements of the moon, Mars, Mercury, and the asteroids by ad-

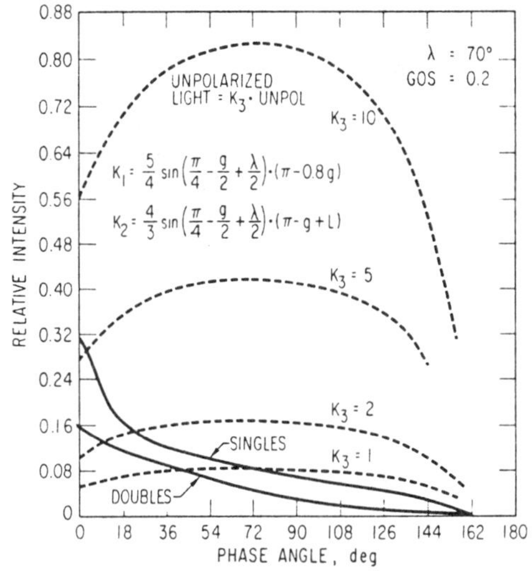

Fig. 13. The single-reflected, double-reflected, and unpolarized light components that make up the lunar polarization curves of Fig. 1 and the photofunction of Fig. 14.

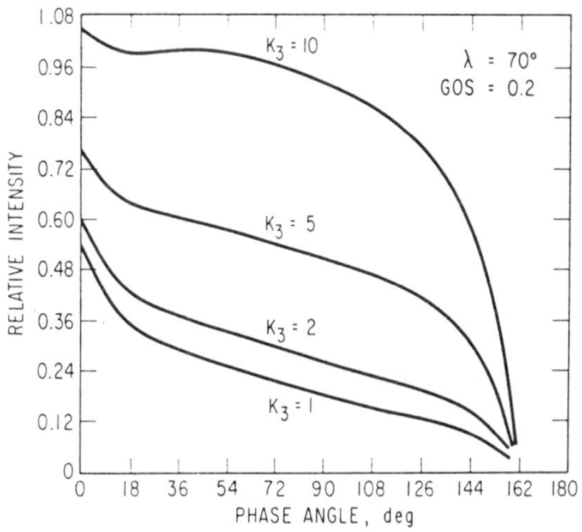

Fig. 14. Photometric functions corresponding to the lunar polarization curves of Fig. 1.

justing the relative quantities of singles, doubles, and unpolarized light. The resulting relative quantities then have simple interpretations in terms of surface structure, surface perforations, and albedo, as discussed in Sec. VIII. The results of this surface model provide a theoretical basis for the empirical Umov principle, which states that polarization is inversely correlated with albedo for surfaces of similar structure, as is apparently true for these planets and satellites.

We may ask, "Why should a model mainly using singles + doubles + nonpolarized light give such good results for the polarization of surfaces with widely varying properties?" The answer is that the main effect of each of the three kinds of light occurs where the other two do not interfere. The negative polarization due to doubles occurs at small angles where polarization due to singles is tiny. The peak positive polarization due to the singles easily dominates at around 90–100°. The main effect of nonpolarized light is even reduction of polarization at all angles. Hence we can understand why simple additive mixtures result in good modeling.

VII. Other Effects

During this work, other properties of surfaces were examined for relevance to polarization. Not all were significant, but the two below are of some importance.

A. Surface Geometry and Tunnels

An interesting geometric effect of changing from a cloud to a real surface is to prevent some double reflecting rays from passing between two particles as they do in the cloud model. Figure 15 illustrates this situation by showing the four possible variations of ray path in the plane of incidence: types A, B, C, and D.

Note that in types *A* and *B* (which correspond to $\theta = 0°$, $\phi > g$, and $\theta = 180°$, $\phi < (\pi - g)$, both reflecting particles lie below the incoming and outgoing rays. These types of paths occur in the interstices and pits of typical surfaces.

In types B and C, one reflecting particle is between the rays. This type of path occurs only if there exist tunnels or overhangs. The probability of tunnels is lower in a compacted surface than in a cloud. Hence, calculations should attenuate or exclude the tunnel rays as a first approximation in all coplanar paths. The forward and backward rays are coplanar, but the left-right rays are not. Hence, the exclusion of tunnel rays tends to favor negative polarization.

The ray paths to be excluded from calculations of real surfaces can be defined, as a first approximation, as

$$\text{EXCLUDE, } IF\ \phi < g(\cos\theta) \text{ and, } IF\ \cos\theta = +,$$

$$\text{EXCLUDE, } IF\ \phi > (\pi + g\cos\theta) \text{ and, } IF\ \cos\theta = -.$$

The result on computer plots is to contribute about one-tenth of the amplitude of the negative branch near $g = 0°$. However, near grazing angles, $g = 175–180°$, a large negative branch is predicted. This

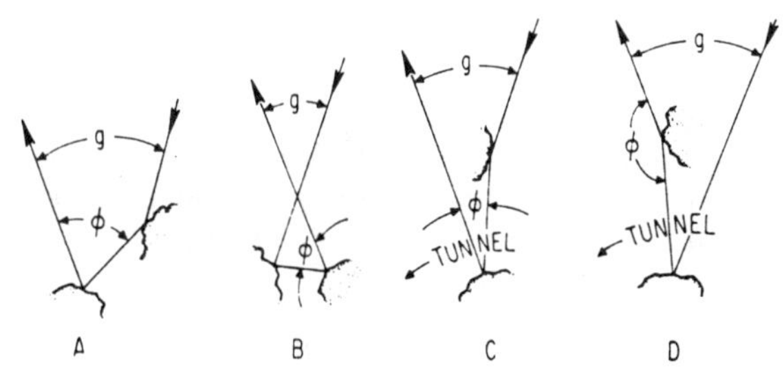

Fig. 15. The four types of coplanar paths for double-reflected light shown with the same phase angle *g*. Types *C* and *D* lead to tunnel paths.

has never been observed because of the difficulty of pointing a telescope toward the sun.

B. Gravity Vector and Its Effect

It is possible that gravity could cause a preferential alignment of particles in a surface. For example, bits of mica might all tend to lie flat. Visual microscopic examination[5] has shown this is not the case for the moon. Some computations using gravity factors showed a tiny polarization at small phase angles and a larger negative polarization at grazing angles.

VIII. Conclusions

This model can be used to calculate the polarization and intensity curves of the moon and other astronomical bodies and laboratory samples by adjustment of major parameters: (1) fraction of single-reflected light with respect to double-reflected; (2) fraction of unpolarized light; (3) amount of shadowing effect; and (4) scattering functions of surface structure. These parameters correspond to characteristics of the surface structure which are, respectively, (1) ratio of surface specular area to hole area, (2) transmissivity of surface material, (3) roughness of surface structure, and (4) detailed shape of surface structure.

A. Location of Positive Peak

The computer experiments showed that the position of the positive polarization peak varies with the relative quantities of light from single, double, or nonpolarizing events. The peak of singles polarization is near 120–130°, and the doubles peak is near 90°. Depending on the mixture ratio, the combined peak will fall somewhere between these two extremes.

B. Shift of Positive Peak Angle with Albedo

This model explains the observed decrease of angular position of peak polarization with increasing albedo as an effect of increasing amounts of nonpolarized light. The phase function of single-reflected light, the main contributor to the peak, has a much steeper slope than the nonpolarized light. Thus, as the fraction of unpolarized light is increased, the peak is forced to smaller angles. This can also be seen by analysis of the polarization equation as a function of phase angle, g,

$$\text{Polarization} = \frac{B_{\perp}(g) - B_{\parallel}(g)}{B_{\perp}(g) + B_{\parallel}(g) + 1_{\text{unpol}}(g)} .$$

Although the possibility exists that the relative quantities of single and double-reflected light vary with albedo change, there is little evidence for this. On the other hand, if translucency should change with wavelength,[23] this mode could be important to explain the wavelength dependence.

C. Generation of Negative Polarization

This is a property of light that has been doubly reflected out of the plane of vision, thus shifting the polarization vector 90°. The out-of-plane (left-right) components dominate at small phase angles because the in-plane (forward-going) component is severely attenuated by shadow and shade in the complex surface structure. At zero phase angle, no selective shadowing exists, and all components cancel each other, as required by symmetry.

The reflection processes of the model only permit a net polarization, which is either perpendicular or parallel to the plane of symmetry, that is, the plane of vision. Exceptions could occur when observing surface areas so small that local features such as a rock or edges of a mountain range create an asymmetrical, nonrandom distribution of reflecting particles.

D. Location of Negative Peak

The location of the peak of negative polarization depends mainly on the width of the photometric peak, since that width is a measure of the onset of shadowing and its resultant blocking of forward-going double events, which create the negative polarization branch.

If the photometric peak is very sharp, the negative polarization peak occurs at small phase angles. If the photometric peak is very broad, the onset of shadowing begins at a large phase angle where singles polarization is already strongly positive, and negative polarization is very weak or does not occur at all. Some confirmation of this conclusion can be deduced from the laboratory results of Lyot[8] and Dollfus.[4]

These results do not explain the constant amplitude of the negative branch of the moon's polarization despite changing wavelength or the increase of the crossover angle with increasing wavelength, both reported by Bowell and Dollfus.[2] However, the negative branch amplitude varies among different asteroids.[22]

E. Index of Refraction

The real index (1.6) used here has been chosen to agree with measurements of lunar samples. The proper imaginary index for astronomical surfaces is not clear. Lunar samples show only a small fraction of metallic material, and hence a small imaginary index seems appropriate, but it has been pointed out that particle bombardment could cause surface darkening. This would imply a very thin layer of material of high imaginary index and possibly wavelength-dependent indices. Further measurements would be valuable to enable more refined calculations.

F. Predictions

This model would predict that negative polarization should disappear if particles are widely separated because double reflections and shadowing are almost nil. This agrees with the loosely poured sand experiment of Dollfus.[4] It predicts that a retrodirective photometric function and negative polarization occur together, as observed. It also predicts that a negative branch may exist at grazing angles, which is a condition hard to observe since direct light enters the detector. Furthermore, given the photometric and polarimetric characteristics of an unknown surface, curves may be fitted to them from which the porosity or roughness and the index of the particles may be deduced.

References

1. T. Gehrels, T. Coffeen, and D. Owings, Astron J. **69,** 826 (1964).
2. A. Dollfus and E. Bowell, Astron. Astrophys. **10,** 29 (1971).
3. S. F. Pellicori, Appl. Opt. **10,** 270 (1971).
4. A. Dollfus, Ann. Astrophys. **19,** 83 (1956).
5. E. Bowell, A. Dollfus, and J. E. Geake, "Polarimetric Properties of the Lunar Surface and Its Interpretation," in *Proc. of Third Lunar Sci. Conf.* (MIT Press, Cambridge, Mass., 1972).
6. S. F. Pellicori, Technical Report 42, Optical Sci. Center, Univ. of Arizona (1969).
7. E. Bowell, *Polarimetric Studies—in Geology and Physics of the Moon* (American Elsevier, New York, 1972), Chap. 9.
8. B. Lyot, NASA Technical Translation TTF-187, Washington, D.C. (1964) and/or Annales de L'Observatoire de Paris, section Meudon **8,** No. 1 (1929).
9. Y. Ohman, Ann. Stockholm Obs. **18,** no. 8 (1955).
10. T. McCoyd, J. Opt. Soc. Am. **57,** 1345 (1967).
11. J. Hopfield, Science **151,** 1380 (1966).
12. B. Hapke, J. Geophys. Res. **68,** 4571 (1963).
13. J. B. Pollack and C. Sagan, Space Sci. Rev. **9,** 243 (1969).
14. A. Dollfus, Nasa Technical Translation TTF-188, Washington, D.C. (1964); Thesis, Univ. of Paris (1955).
15. S. Provin, Publ. Astron. Soc. Pac. **67,** 115 (1955).
16. J. Veverka, Icarus **15,** 454 (1971).
17. B. Zellner, Astron. J. **77,** 183.5 (1972).
18. B. Zellner, Astron. J. **174,** L107 (1972).
19. T. Gehrels, E. Roemer, E. Taylor, and B. Zellner, Astron. J. **75,** 186 (1970).
20. T. Widorn, Ann. Univ. Stern W. Wien **27,** 112 (1967).
21. F. W. Sears, *Optics* (Addison-Wesley, Reading, Mass., 1949), p. 174.
22. B. Zellner, T. Gehrels, and J. Gradie, Astron. J. **79,** 1100 (1974).
23. S. F. Pellicori, Astron. J. **74,** 1066 (1969).

©1980 Academic Press, Inc. Reprinted from *Icarus,* Vol. 44, 780–792 (1980).

Theory and Application of the Polarization–Albedo Rules

MILO WOLFF

Technotran Group, 1600 Nelson Avenue, Manhattan Beach, California 90266

Received March 24, 1980; revised November 14, 1980

Results are reported of a computer calculation of the scattering of light from porous, particulate surfaces such as those of the Moon, asteroids, and airless planets. The scattered-light intensity and polarization are computed as a function of phase angle, and plots are made corresponding to the observed empirical rules for deducing albedo from the slope, maximum, and minimum of polarization curves. The approximate linearity of these rules on a log–log plot is confirmed for most solar system objects. The linearity is due to the similarity of the real indices of refraction of solar system soils. The theory shows that the albedo vs polarization plots correspond directly to a plot, in two dimensions, of the real vs imaginary index of refraction of these soils, using curved axes. This new information provides a straightforward method of measuring optical and structural properties of the major soil constituents leading to mineral identification. The theory is based upon geometric optics with an estimated error of 5 to 10% which agrees reasonably well with available measurements. A comparison is made with 30 asteroids and some moons. The soils of these bodies appear similar to that of the Earth, with some interesting exceptions.

INTRODUCTION

The polarized light reflected from airless planets has a unique dependence on phase angle which identifies their surfaces as rough and particulate—a condition attributed to meteoroid bombardment. Polarization is known to be very sensitive to both microstructure and albedo of the surface, making it one of the powerful techniques for exploration of the optically unresolvable minor planets. Before this work, no theoretical means of calculating optical properties of these surfaces had been available. However, the realization of the value of the empirical rules has been a strong motivation to accumulate observations. The pioneering work on polarimetry, begun by Lyot (1929) is well summarized by Bowell and Zellner (1974), and Dollfus and Zellner (1979).

The first rule to estimate albedo from polarization was devised by KenKnight (1967) and Widorn (1967) who noted an empirical relationship of the form $d(\log h) = -d(\log p)$ between the polarization slope, h, and the albedo, p, of the planet. A team, Zellner *et al.* (1977), at the University of Arizona has calibrated and made extensive use of the albedo rule to estimate albedos and diameters of about 100 asteroids.

Two other rules to approximate albedo of planets utilize the maximum and minimum values of polarization as a function of phase angle. They are reviewed and illustrated by Dollfus and Geake (1977). These two rules are less often used to determine albedo because the scatter of measurements has appeared to be much greater; also, maximum polarization cannot be observed from Earth for the outer planets. All three rules are displayed in Fig. 1.

Wolff (1975) used a computer model of a particulate planetary surface to compute polarization as a function of phase angle. The results provided a theoretical reason—double reflections—for the negative branch, which was previously not understood, but the accuracy was insufficient for quantitative results. The shortcoming was a lack of a theoretical means to calculate the unpolarized diffuse light, a condition rectified in this paper.

This paper presents the results of an improved theoretical model of the

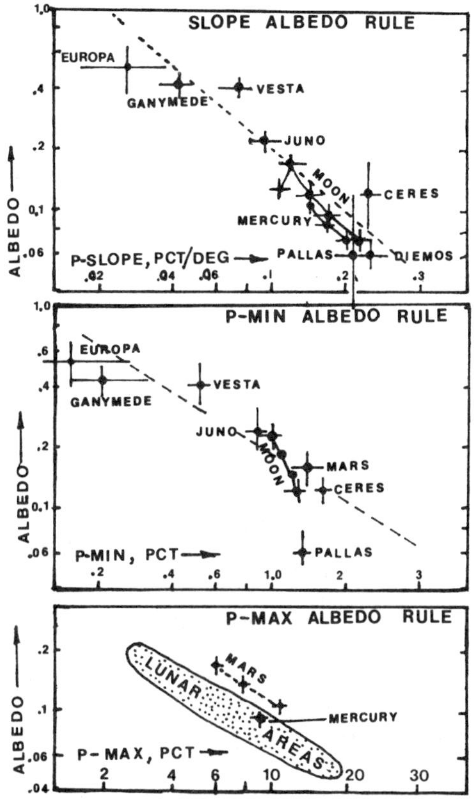

FIG. 1. These plots of the three albedo rules suggest that some regularity of nature causes the plotted planets to lie on an approximate straight line. This has enabled astronomers to obtain an approximate albedo from a measurement of polarization. The range of values of polarization and albedo on these plots is restricted to that of planetary observations. This is only part of the complete theoretical range shown in Figs. 3 to 5.

reflection and polarization processes in a particulate surface. The accuracy now corresponds with the accuracy of state-of-the-art measurements of polarization and albedo. An important result of this model is the ability to calcualte the indices of refraction of soil minerals distributed through the solar system on the airless planets.

The reason for the occurrence of the empirical albedo rules is now understood from the results of these calculations. The rules depend on the ranges of the two indices of refraction; the real index has a small range in nature, and the imaginary index has a wide range in nature. These two facts lead to the observed band of points plotted on polarization-albedo curves.

In this paper, the symbolism which has been adopted is that of FORTRAN variables. This has been done so that there is a direct correspondence with the computer programs used to make the calculations. This may be of value to readers who wish to obtain the programs. As an example, the symbols for the real and imaginary indices are *MR* and *MI*.

PROOF OF THE GENERAL FORM OF THE ALBEDO RULES

Before discussing details, there is an enlightening argument to demonstrate the general shape required of the curves on a polarization-albedo plot which covers a complete range of the imaginary index from zero to infinity.

The argument makes use of two optical laws: (1) Fresnel's equations for real and imaginary indices of refraction, and (2) The law of exponential attenuation of light in an absorbing dielectric material. It is presumed that the reader is familiar with them, particularly the dependence on the imaginary index, which varies widely.

The following assumptions identify a surface model:

(a) The surface is a thick layer of irregular particles.

(b) The particle sizes are larger than the wavelength of light, so that the laws of geometric optics apply. (The model can be improved by including small particles which cause a significant component of diffracted light, but it does not change the results of the argument.)

(c) The optical properties of the particle material are described by an average index of refraction, $N = MR - iMI$.

With the above assumptions, a good description of the scattered light can use the following three components:

(1) Diffuse light *BD*, which has entered particles, undergone random internal

reflection, been partly absorbed, and then emitted. It is unpolarized.

(2) Single reflected light $B1$. It is partly polarized.

(3) Doubly reflected light $B2$. It is partly polarized.

Triple and higher reflections are of negligible amplitude if the particles are not metallic (have small MI). In the metallic region of MI, the model can be improved by considering triples and higher, but it does not affect this argument.

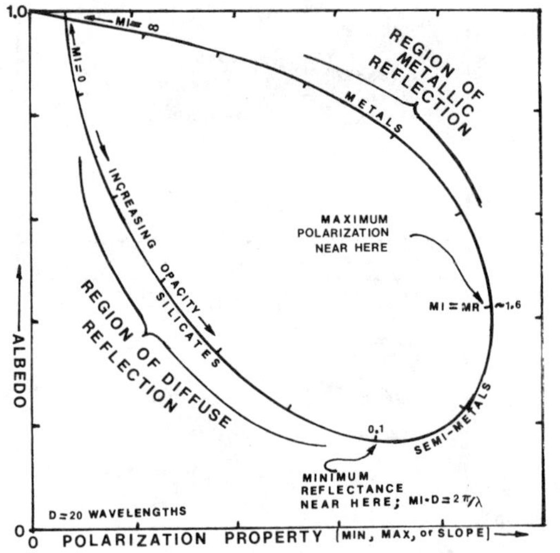

FIG. 2. This diagram illustrates the derivation of a general polarization vs albedo plot. The general shape of this curve and other information shown on it can all be deduced from Fresnel's Equations and the law of light absorption. If the plot uses log–log scales, then Fresnel's Equations predict that the descending branch is almost linear.

We need not know the detailed dependence of $B1$, $B2$, and BD. It is sufficient to know from Fresnel's equations that polarization will decrease to zero as MI approaches infinity, but this is not significant until MI describes metals; also, that light refracted into particles will decrease as MI increases; also, that absorption will increase as MI increases, reaching a preponderance for an MI midway between dielectric particles and metallic particles.

The albedo will be proportional to the sum of $B1$, $B2$, and BD. The polarization will arise from reflected light $B1$ and $B2$.

The albedo rules are plotted using albedo versus a desired polarization function; P-max, P-min, or P-slope, as shown in Fig. 2. This plot can be mentally constructed by varying MI between zero and infinity, while holding MR constant. A small MI corresponds to bright objects with crystalline particles. A larger MI corresponds to dark objects, and a very large MI corresponds to bright objects with very metallic particles. All planets must fall somewhere in this range—most of them in the small MI (crystalline) range.

Consider how the properties of the light components are used to construct Fig. 2. The important property of the diffuse light BD is that it must decrease as MI increases. This is because light is absorbed inside the grains, as

$$I = I_0 \exp(-kx), \qquad k = 4\pi MI/\lambda,$$

where x is distance traveled and λ is wavelength. When $MI = 0$, no light is absorbed, albedo is a maximum, and the diffuse light dominates the other components. About 90% of incident light enters and leaves the grains. When $MI \simeq 0.1$, and if the particles are about 20 wavelengths of size (Moon), most of the diffuse light will be absorbed, so that now the reflected light, $B1$ and $B2$, dominates the light seen by an observer. A minimum of albedo occurs near here.

It is a direct result of Fresnel's equations that the polarized light is almost constant for values of MI between 0.0 and 0.1, when MR is that of typical minerals. This theoretical fact has not been recognized, but has been proposed as an empirical result by Bowell and Zellner (1974), who recognized that it leads to a polarization-slope albedo rule. That is, the left-hand part of the curve in Fig. 2 can be drawn linearly on a log–log plot, as albedo decreases. For the same reason, the calculations of this paper also produce an almost linear plot in this region.

After the diffuse light has become nil, forming an albedo minimum nearby, the right-hand side of the curve depends *only on Fresnel's equations*. This enables us to easily deduce the shape of the remaining curve. As *MI* increases, the particle reflectivity increases asymptotically towards 100%—this is the *MI* region of commercial mirrors. But polarization, on the other hand, decreases, because both of the Fresnel polarization coefficients tend to become equal. Good metal mirrors polarize only slightly. Accordingly, the curve of Fig. 2 turns back towards the upper left corner.

SOME GENERAL CONCLUSIONS FROM THE PROOF

(1) If $MI = 0$, albedo = 100%, and *P*-max, *P*-min, and *P*-slope are a minimum.

(2) If *MI* approaches ∞, albedo approaches 100%, and polarization approaches zero.

(3) A minimum of albedo occurs near $MI = 0.1$, midway between crystalline and metallic particles.

The first two conclusions are boundary conditions on the general problem of light scattering from the particulate surface of an airless planet. They have been incorporated into the model of this paper.

It is instructive to compare Fig. 2 with Figs. 3 to 5, where the actual polarization-albedo rules have been computed. Note that the descending branches are almost linear on a log-log plot as surmised by Widorn, Zellner, and others.

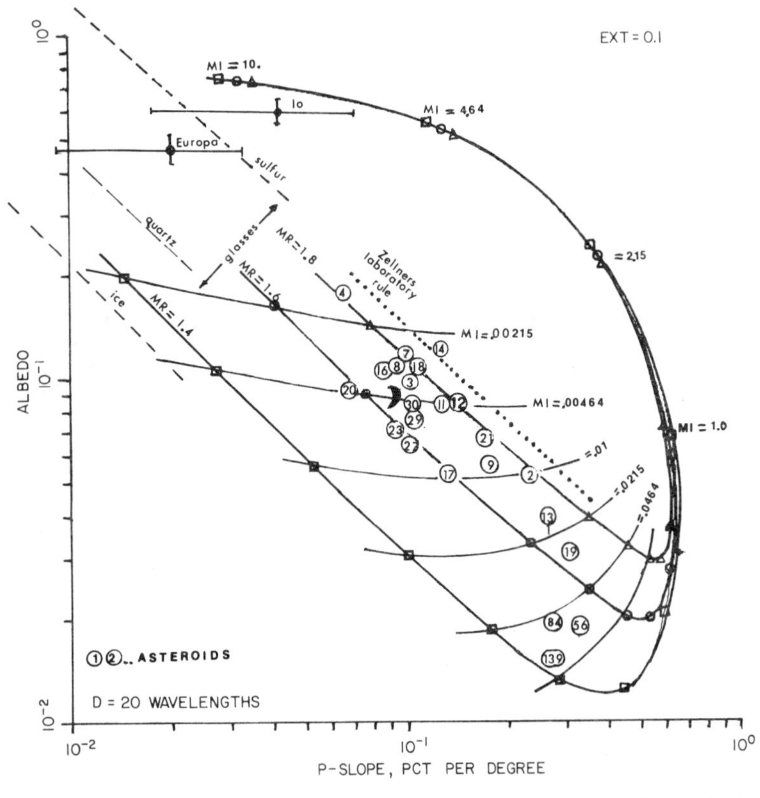

FIG. 3. This theoretical slope–albedo rule is compared to representative observations of asteroids and moons. The dotted line is an experimental rule obtained from crushed meteorites in the laboratory. Asteroids 1 to 30 are shown. The theoretical model indicates that a point on the plot is determined by values of the real and imaginary index of refraction for the surface observed. The physical meaning of the rule is revealed clearly here: *the broad band of points formed by observations of airless planets displays the typical range of indices of average solar system soils*. The range of *MR* is narrow and the range of *MI* is broad, thus appearing like a line—quite similar to Earth soil.

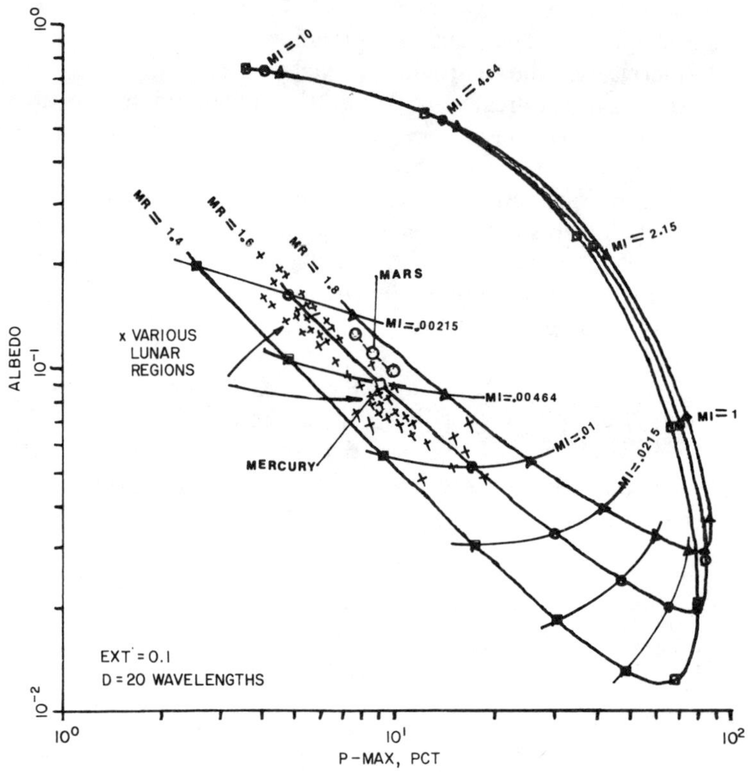

FIG. 4. The *P*-max Albedo Rule. This theoretical plot is compared to observations of many lunar areas, Mars, and Mercury. Data on asteroids and other planets is not available because of limited observing angles. Exploration by spacecraft is not so constrained so that maximum polarization data could be very useful. If diffracted light from small particles were included, the lower right of these curves would be shifted towards the upper left—more in agreement with lab measurements of metallic particles. Both *P*-min and *P*-max are strongly affected by the fraction, *EXT*, of smooth surface. Comparison of this figure with Fig. 3 indicates that the most probable value of *EXT* lies between 0.1 and 0.2.

CHARACTERISTICS OF THE COMPUTER MODEL

The computer model uses the basic ideas of Wolff (1975) to calculate the light scattered from a particulate surface. As above, the emergent light is considered to arise from three components: diffuse, singly reflected, and doubly reflected. It is assumed that the process of meteoroid bombardment produces a particulate regolith on all solar system objects, which are sufficiently similar to each other so that differences can be accurately described by variations in (a) the real index *MR*, (b) the imaginary index, *MI*, (c) the average width/depth ratio of grain interstices *WID*, (d) average particle size *D*, and (e) the fraction *EXT*, of planetary surface which is not covered with particles.

The improved computer model differs in three important respects from the 1975 model. (1) A quantative equation for diffuse light has been derived (Wolff, 1979) to replace the former empirical approximation. Importantly, the diffuse light equation is based upon optical laws so that the diffuse light is a function of the indices of refraction and particle size. (2) Esoteric approximations such as particle clouds have been abandoned and replaced by as exacting a description of a real particle surface as possible. (3) The boundary conditions found in the previous section have been incorporated.

The result of these improvements has been to produce a quantitative theory functionally dependent on refractive indices and surface structure which agrees with experimental observations of a broad range of photopolarimetry.

The equation for diffuse light was obtained by using Fresnel's Equations to calculate the fraction of light refracted into the surface particles. This light bounces around inside zero, one, or more times by internal reflection. The probability of eventual escape is calculated proportional to the angular aperture of a cone formed by the critical angle around a perpendicular at the point of striking. Absorption is proportional to the total path traveled during reflections and the imaginary index of refraction. Finally, the total light emitted is equal to the sum of probabilities at each striking point. The result, for vertical incidence and observation, is

$$BD = \text{diffuse light} = I(\text{sun}) \cdot REF/[1 + (4\pi)^2 \cdot MI \cdot D \cdot CONE/\lambda], \quad (1)$$

where REF is the fraction of light refracted into particles (typically 90%), $CONE$ is the escape cone solid angle divided by 4π, and D is the average particle diameter. At phase angles other than zero, two additional factors, the Lommel–Seeliger function and a backscattering function, are used to account for shadowing and the opposition effect. In the backscattering function $GP0$,

$$GP0 = A(1 - EXT)\exp(-G/W). \quad (2)$$

The constant A must be set equal to the amplitude of the observed opposition effect, and W must be the width. Therefore these are not adjustable constants.

The singly reflected light is obtained from

$$B1 = I(\text{sun}) \cdot FF \cdot SHAD \cdot [EXT + (1 - EXT) \cdot GP1, \quad (3)$$

where FF is either Fresnel coefficient; EXT is the fraction of planetary surface not covered with particles; $GP1$ is the geometric probability (~15%) of single reflctions occurring in a pitted particulate surface; and $SHAD$ is a function to account for the changing illumination on a particle, or the entire planet, as a function of the phase angle.

$$SHAD = \cos\{\pi G/(2L - \pi)\}. \quad (4)$$

$SHAD$ is constrained to cause a terminator at the proper longitude L and phase angle, G. The meaning of geometric probability is shown in Fig. 6.

The doubly reflected light is computed in four parts, where the intermediate ray travels in quadrants which are: forward, left, right, and backward. This is done, because for nonzero phase angles, the probability of each type of ray is different due to unsymmetrical shadowing and escape opportunities from the interstices between the particles. It is this asymmetry, favoring the left–right-going rays which creates the negative polarization branch. Following the method of Wolff (1975), the reflection coefficients in the four quandrants are computed using a binary matrix $FM_n{}^m$ where m has 2 values, one for each polarization, and n has 4 values, one for each quadrant. Thus

$$\text{doubles light} = B2_n = I(\text{sun}) \cdot FM_n{}^m GP2_n \cdot (1 - EXT). \quad (5)$$

All values of $GP2_n$ are identical at zero phase angle as required by symmetry and have a value of about $\frac{5}{8}$. All are exponentially decreasing functions of phase angle because of the decreased probability of escaping from the deep interstices. Doubles with forward-going intermediate paths decrease most rapidly and backward ones least rapidly. This asymmetry of the $GP2$'s is caused by the asymmetrical geometry of doubles reflection.

The use of the parameter EXT essentially divides the whole surface into two kinds: that part which is covered with particles and that which is not.

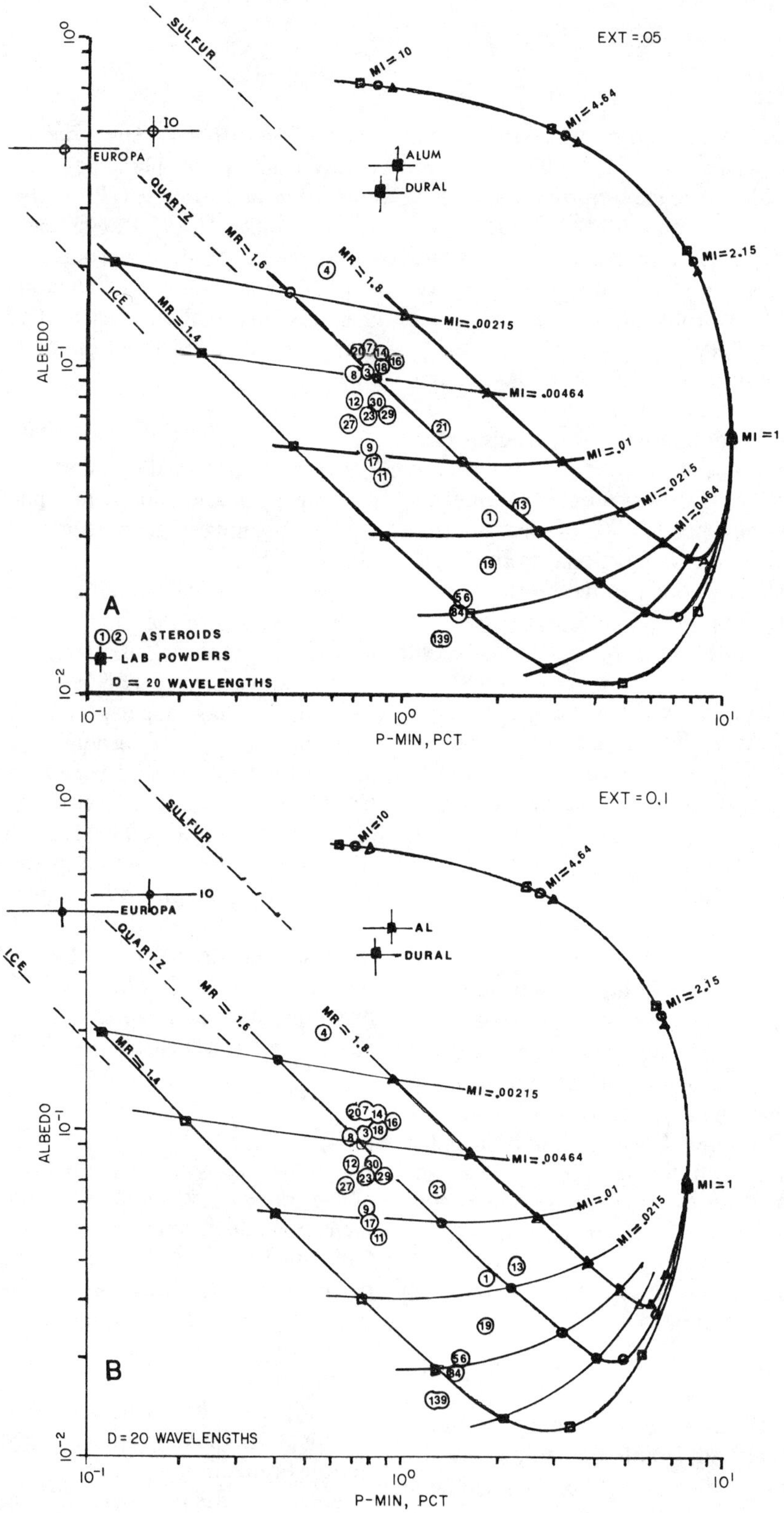

FIGS. 5A–C. This theoretical plot of minimum polarization is compared to observations of solar system objects including asteroids 1 to 30, plus the brightest and darkest known objects. These three diagrams also illustrate how *P*-min depends on the fraction *EXT*, of smooth specular surface on an observed planet. Comparison of these three diagrams with Fig. 3 suggests that the most probable value for *EXT*, averaged for all the objects plotted, is about 0.2.

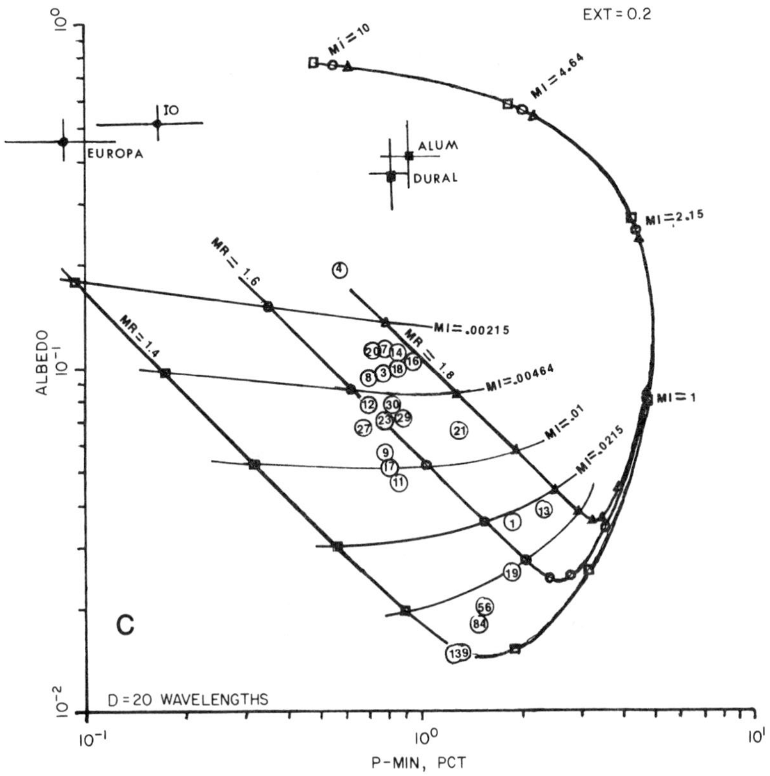

FIG. 5—Continued.

The fraction *EXT* leads to singly reflected light, and the fraction 1-*EXT* leads mostly to doubly reflected light. For example in Eq. (5) for doubles, 1-*EXT* must be a factor of the equation. Some single reflection can occur in particle-covered area, hence in Eq. (3) the area (1 − *EXT*) times the probability *GP*1 is a term of the equation.

The geometric probability functions are all related to each other because they all describe the effect of the same particulate surface on the probability of light-reflection events in the surface. They are:

$$
\begin{aligned}
GP1 &= \tfrac{1}{8} \cdot \exp(-G/W) && \text{(singles)},\\
GP2_1 &= \tfrac{5}{8} \cdot \exp(-G/\tfrac{1}{2}W) \cdot \tfrac{1}{4} && \text{(forward doubles)},\\
GP2_3 &= \tfrac{5}{8} \cdot \exp(-G/4W) \cdot \tfrac{1}{4} && \text{(sideways doubles)},\\
GP2_4 &= \tfrac{5}{8} \cdot \exp(-g/10W) \cdot \tfrac{1}{4} && \text{(backward doubles)}. \qquad (6)
\end{aligned}
$$

The factor of $\frac{1}{4}$ is due to the division of the doubles intensity into four equal parts, one in each quadrant. The factors $\frac{1}{8}$ and $\frac{5}{8}$ are the estimated probabilities of single reflection or double reflection from particle interstices at vertical incidence. They were determined by pencil and paper experiments with drawings of lunar fines as seen in a stereo microscope.

Note that there are no adjustable constants in this model. The parameters *A*, *W*, and the functions *GP1* and $GP2_n$ are

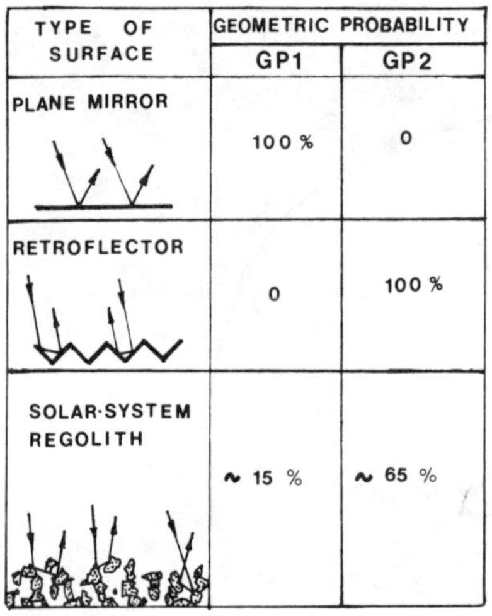

FIG. 6. This chart illustrates the meaning of "geometric probability." This is the probability of a ray path undergoing one (*GP1*) or two (*GP2*) reflections. The probabilities depend on the structure of the surface. A plane mirror and a retroflector illustrate the extreme cases. The last diagram illustrates the surface assumed for airless planets of the solar system. This chart only shows the situation at vertical incidence. In general, both *GP1* and *GP2* are a maximum at zero phase angle and decrease with increasing phase resulting in the opposition effect.

all determined by measurement. The variables *MR, MI, D,* and *EXT* are descriptions of the unknown surface which are sought using observations of albedo and polarization as a function of phase angle and wavelength.

CALCULATING POLARIZATION AND REFLECTANCE

The net polarization is obtained by summing each polarization component and finding the difference divided by the total light intensity

$$P = \{B1_x + \sum_{n=1}^{4} B2_{nx} - B1_y - \sum_{n=1}^{4} B2_{ny}\}/(B1 + B2 + BD), \quad (7)$$

where the x subscript indicates the electric vector polarized perpendicular to the plane of vision, and the subscript y indicates the parallel vector.

The reflectance of the surface is given by

$$\text{Reflectance} = (B1 + B2 + BD)/I(\text{sun}). \quad (8)$$

RESULTS

Polarization and reflectance were calculated for a wide range of the imaginary index and for several values of the real index and the bare surface fraction *EXT*. Figures 7 and 8 show polarization and reflectance for a single point on a planet. Figures 3 to 5 show various examples of the three albedo laws. For the albedo, reflectance is integrated over the illuminated portion of the planet and the value at $G = 0$ used for plotting.

Experimental data is also plotted in these diagrams. In order to provide well-known examples, every one of the asteroids from 1 to 30 for which data is available was plotted. Asteroid albedo was taken from the radiometric measurements of Hansen (1976) and polarization from Zellner and Gradie (1976). In order to provide the perspective of a broad range, the three darkest asteroids and the two brightest moons were plotted using data from Dollfus (1975). These authors did not give measurement errors, so the error bars shown were estimated from their comments and the scatter of the data. The dotted line shown is the slope–albedo rule obtained by Zellner *et al.* (1977) using selected crushed meteorites.

The most important feature of these plots is the realization that the field of the slope–albedo diagram corresponds to specific values of the real and imaginary indices of refraction of the observed soils. Similarly, if the proper value of *EXT* is used in a *P*-min or *P*-max diagram, then a planet point will also correspond to the indices of the soil particles on that planet.

A comparison of the slope–albedo diagram with the *P*-min diagrams shows that

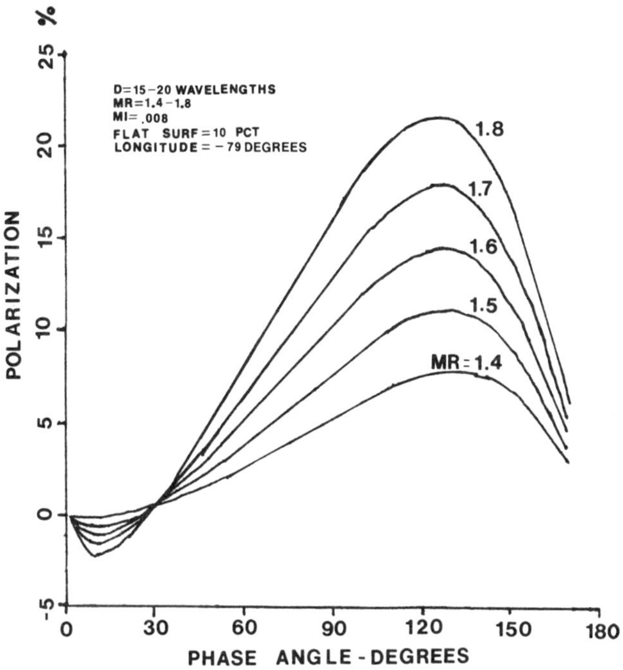

FIG. 7. The polarization of light from a planet, shown here as a function of phase angle, was computed using Eq. (7). Polarization is produced by Fresnel reflections, once or twice, and is diluted by the unpolarized diffuse light.

the relative position of these asteroids and moons match best in the (EXT = 0.2) P-min diagram. Positions in the (EXT = 0.5 or 0.1) P-min diagram are generally lower. One concludes that EXT = 0.2 is the most probable value for most of these objects.

One should be mindful of the errors present in these plots: about 10–20% for the observations, and about 5–10% in the theory for the left-hand side. The right-hand side is too far right because diffraction and higher reflections are not in.

MEANING OF THE SLOPE–ALBEDO RULE

The origin of this rule is now clear. The straight band into which most objects fall expresses the well-known fact that most soil minerals have a very narrow range of values of the real index, typically 1.4 to 2.2. On the other hand, the range of values of opacity or imaginary index is extremely wide—about two decades (0.001 to 0.1 for asteroids). When many different minerals are mixed to produce soils, the range of real indices are even narrower, but the imaginary indices still range widely. As a result,

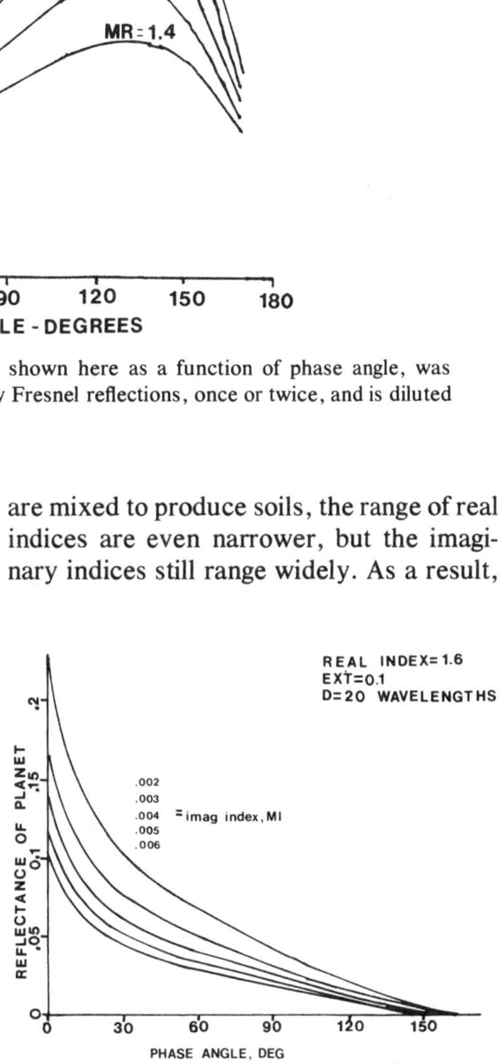

FIG. 8. The computed albedo of a planet, shown here as a function of phase angle, was obtained by integrating the three components of light (singly reflected, doubly reflected, and diffuse light) over the illuminated portion of the planet.

the slope–albedo plots form a long narrow band. The ensemble of 30 points in Fig. 3 displays a panorama of soils which may be fairly representative of the solar system. Most of the soils of these asteroids and moons are not too far different from the Earth and the Moon.

This new understanding of the slope-albedo plot has application to planetary science. From measurements of polarization and albedo at varying wavelengths, spectra of refractive indices can be computed as a function of wavelength. The spectra of imaginary indices should be similar to that of reflection spectroscopy, but the spectra of the real indices can be obtained in no other way from astronomical observations. Together, the tools of mineral identification are strengthened. With a reasonable number of measurements, it should be possible to distinguish major surface constituents such as ice, sulfur, silicates, and iron.

USE OF THE MAXIMUM AND MINIMUM POLARIZATION RULES

These two rules display more observational scatter than the slope–albedo rule. This is because they depend on the surface roughness through the parameter *EXT,* the fraction of surface not covered by pits or particles. Determining this fraction appears to be the main use of the rules.

The dependence on *EXT* is *opposite* for each rule and the reason is found in the opposite origins of positive and negative polarization. For the *P*-max rule, positive polarization arises from specular reflection and is thus proportional to *EXT*. On the other hand, negative polarization, which produces the *P*-min rule, arises from pits and particles, and is thus proportional to $1 - EXT$.

The different dependence is quite useful because either rule can be used to find *EXT* after the slope rule has been used to find *MR* and *MI*. One rule can check the other. Finally, a third check is the position of the inversion angle, mentioned in corollary II below.

THE POLARIZATION COMPLEMENTARITY PRINCIPLE

A principle useful for interpretation of polarimetry can be deduced from the geometric meaning of the parameter *EXT*. The principle explains why *P*-slope is almost independent of *EXT*. The principle can be stated, "Any increase of positive polarization, due to a change of the smooth-surface fraction *EXT,* will be accompanied by a decrease of negative polarization and vice versa." Or more briefly, "Negative polarization changes are inverse to positive polarization changes."

The principle follows from these three facts:

(1) Surface pits produce double reflections and negative polarization.

(2) Flat area (or nonpits) produce single reflections and positive polarization.

(3) The area of pits plus the area of nonpits is a constant equal to the total area. (Incidentally, this is a boundary condition of the surface.) Thus if the ratio of pits and nonpits should differ on some planet from a norm characterizing the interplanetary environment, then the negative and positive polarizations for that planet will be found to have changed inversely.

Corollary I. *Somewhere between the points of maximum and minimum polarization, the slope of a polarization curve is unaffected by variation of EXT.*

This corollary explains the observed smaller amount of scatter of points on a slope-albedo rule plot. Paper and pencil experiments with polarization curves quickly demonstrate this corollary. It is quite convincing.

Corollary II. *If EXT increases, the angle of the inversion point must decrease and vice versa.*

This idea, too, is quickly demonstrated by pencil and paper experiments. In principle, this is a useful means to ascertain the value of *EXT,* but the variation must be separated out from the affects of changing *MR* and *MI* indices.

CONCLUSIONS

The basic phenomena underlying the albedo rules has been revealed and this new understanding should lead to a reorientation of our thinking about them. Their value as a means of estimating albedo has diminished although it is still useful if no other means is at hand and we are willing to presume that the objects involved have average indices of refraction. In return for this loss, a method has been found to determine the mineral properties and structure of airless planets.

Already, a panoramic view of solar system soils has been achieved and it is seen that they are generally not very different from the soils of the Earth and the Moon. Still, a few asteroids possess properties far from the average and their further investigation is a future challenge.

CONSEQUENCES FOR SOLAR SYSTEM MEASUREMENT AND EXPLORATION

Often the investigation of situations which appear to deviate strongly from the expected norm lead to revelations and new knowledge. Among the asteroids and moons of major planets, there are many objects with unusual properties of an unknown cause. This new theory provides basic understanding of the deviations from albedo rules. It is realized that the rules are only averages of the physical and optical properties of surfaces which have been subjected to similar bombardment by meteoroids. Accurate measurement of the deviations can now provide accurate determinations of those intriguing objects which have been observed but not understood.

The theory opens up new areas of investigation. After verifying that the slope–albedo rule is log–log linear where it described type *C* and *S* asteroids, it enables calculations of metallic and semimetallic particle surfaces. This second branch is very nonlinear and measurements of the particles which it described were formerly not understood—even empirically. Now all measurements of all materials can be fitted into one logical pattern of behavior.

An interesting situation of planetary evolution involves the assumption of uniform meteoroid bombardment in the solar system—a process which requires about 10^7 years to reach equilibrium. If an equilibrium condition of bombardment is not found on an object (by measurement of *EXT*), there is an immediate presumption that faster or more recent events are occurring, such as volcanic activity or ice formation.

With currently available data and theory, it seems possible to determine an index, say 1.6, with an accuracy of ± 0.05, but this falls short for quantitative mineralogy. Actually present instrumentation is probably several times better but other errors are present. The only fundamental way to measure albedo is the radiometric method of Morrison (1973) and Hansen (1976). Further improvements of this ingenious scheme will produce immediate dividends.

One problem is the lack of uniformity in the definitions of measurables. The same-named quantity will often have a different working definition. Especially important are albedo, opposition width, color, and reddening with phase angle.

Most of the theory is based upon laws of optics, but there is considerable error, say 5–10%, due to uncertainty in the geometric probability functions that describe the structure of the particulate surface. Also, the neglect of diffraction could cause polarization to be perhaps 50% too large for very low albedo objects. However, there is very little error in the type *S* object range. For metallic particles, neglect of triple and higher reflections also causes too large a polarization. As a result, the accuracy of Figs. 3 to 5 decreases across the page, left to right. These improvements of the model are under development.

ACKNOWLEDGMENTS

Part of this work was supported by the National Aeronautics and Space Administration under grant NASW-3201. I am indebted to John Geake and Leonard Dougherty of the University of Manchester Institute of Science and Technology for fruitful discussions. Auduoin Dollfus and the Observatoire de Paris, Meudon, generously provided opportunity for study there in November–December 1979. Ben Zellner has given valuable criticism.

REFERENCES

Bowell, E., and Zellner, B. (1974). Polarizations of asteroids and satellites. In *Planets, Stars and Nebulae Studied with Photopolarimetry* (T. Gehrels, Ed.), pp. 381–404. Univ. of Arizona Press, Tucson.

Dollfus, A. (1975). Optical polarimetry of the Galilean satellites of Jupiter. *Icarus* **25,** 416–431.

Dollfus, A., and Geake, J. (1977). Polarimetric and photometric studies of lunar samples. *Philos. Trans. Roy. Soc. London, Ser. A.* **285,** 397–402.

Dollfus, A., and Zellner, B. (1979). Optical polarimetry of asteroids and laboratory samples. In *Asteroids* (T. Gehrels, Ed.), Univ. of Arizona Press, Tucson.

Hansen. O. (1976). Radii and albedos of 84 asteroids from visual and infrared photometry. *Astron. J.* **81,** 74–84.

KenKnight, C. E., Rosenberg, D. L., and Wehner, G. K. (1967). Parameters of the optical properties of the lunar surface powder in relation to solar wind bombardment. *J. Geophys. Res.* **72,** 3105–3129.

Lyot, B. (1929). *Recherches sur la polarisation de la lumiere des planetes et de quelques substances terrestres.* Ph.D. Thesis, Université Paris; NASA TT F-187, 1964.

Morrison, D. (1973). Determination of radii of satellites and asteroids from radiometry and photometry. *Icarus* **19,** 1–14.

Widorn, T. (1967). Zur Photometrischen Bestimmung der Durchmesser der kleinen Planeten. *Ann. Univ. Sternw. Wein* **27,** 112–119.

Wolff, M. (1975). Polarization of light reflected from rough planetary surface. *Appl. Opt.* **14,** 1395–1405.

Wolff, M. (1979). *Photopolarimetry of Scattering Surfaces and Their Interpretation by Computer Model.* Final Report, NASA Contract NASW-3201, Washington, D.C.

Zellner, B., and Gradie, J. (1976). Minor Planets and related objects. XX. Polarimetric evidence for the albedos and compositions of 94 asteroids. *Astron. J.* **81,** 262–280.

Zellner, B., Leake, M., Lebentre, T., Duseaux, M., and Dollfus, A. (1977). The asteroid albedo scale. I. Laboratory polarimetry of meteorites. *Proc. Lunar Sci. Conf. 8th,* 1091–1110.

POLARISATION DES RADIATIONS MONOCHROMATIQUES DU CIEL NOCTURNE ET DE LA RAIE D CRÉPUSCULAIRE

J. Bricard et Kastler

Institut de Physique du Globe de Paris et École Normale Supérieure.

Sommaire. — *Description d'un appareillage permettant la détection des radiations du Ciel Nocturne et Crépusculaire sans spectrographe, en photographiant les franges d'interférences données par des cristaux en lumière monochromatique. (On utilise un ensemble convenable filtre émulsion.)*

Si la raie est initialement polarisée, les franges apparaissent même en l'absence du polariseur, et l'appareil fonctionne alors comme un polariscope. Convenablement agencé, il permet d'apprécier de très faibles polarisations ; on montre ainsi que les raies rouges et verte du Ciel Nocturne ne sont pas sensiblement polarisées.

Si l'on veut étudier la polarisation de la raie jaune crépusculaire, on est gêné par la présence du fond continu, lui-même polarisé, qui l'accompagne. Ce fond est éliminé en projetant les franges sur la fente d'un spectrographe dispersif ; on trouve ainsi que la raie présente une faible polarisation dans le sens correspondant à une excitation par résonance optique.

Summary. — *In this paper, we give the description of an equipment detecting photographically Night Sky and Twilight radiations. Interferences given in monochromatic light by cristals are used ; no spectrograph is necessary, but instead we have taken filters and proper plates.*

If the incident light is polarized, fringes are visible even without polarizor ; this apparatus properly disposed can detect small polarizations. We found out this way that the green and red rays of the Night sky light are practically not polarized.

Some difficulties are encountered in the study of the polarization of the D line present in the Twilight radiations because of the light coming from the background, and which is polarized too. This extra light is eliminated by receiving the fringes on the slit of a very dispersive spectrograph. We found out this way that the D rays present in the twilight is feably polarized, which is in favour of an optical resonance excitation origin for this rays.

Détection des radiations du ciel nocturne par la méthode du polariscope.

Pour étudier les radiations monochromatiques du ciel nocturne, on peut utiliser, à la place d'un spectrographe, une méthode interférentielle. Babcock [1], le premier, a employé ce procédé, et a étudié la raie verte du ciel nocturne avec un interféromètre Fabry-Pérot. Le même appareil a été employé par Vegard et Har ng [2] pour l'étude de la raie verte et du doublet rouge de l'Oxygène, et par Cabannes, Dufay et Gauzit [3] d'une part, Bernard [4] d'autre part, pour l'identification du doublet du sodium dans le spectre du ciel nocturne et du crépuscule.

L'emploi de l'interféromètre présente quelques inconvénients : sa luminosité est faible, à cause des pertes de lumière par la première réflexion et par absorption dans les couches métallisées ; son réglage est difficile, et surtout sa stabilité défectueuse ; il est sensible aux variations de température et demande, pour des poses de longue durée, à être thermostaté. Ces difficultés s'éliminent si l'on emploie les franges d'interférence données par les cristaux entre polariseurs en lumière monochromatique.

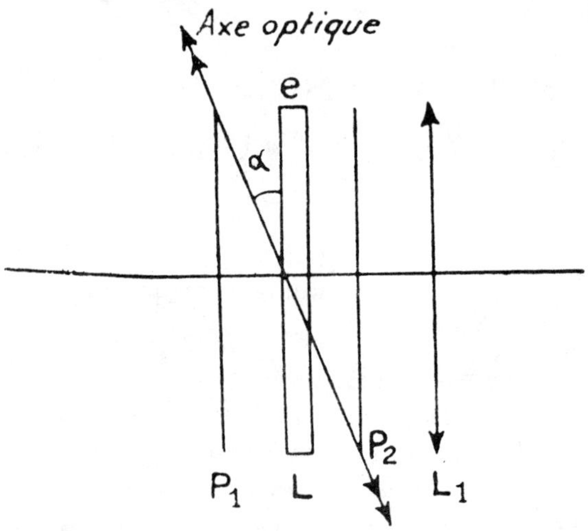

Fig. 1.

Considérons une lame de quartz à faces parallèles d'épaisseur e, taillée obliquement par rapport à son axe optique, dont la normale fait un angle $\frac{\pi}{2} - \alpha$ avec celui-ci. Plaçons-la entre deux polaroïds P_1 et P_2 croisés ou parallèles, et éclairons-la par un faisceau de lumière monochromatique de grande ouverture (fig. 1). En plaçant derrière la lame une lentille L_1, on observe photographiquement dans son plan focal des franges à l'infini, équidistantes et de très faible courbure, qui sont les

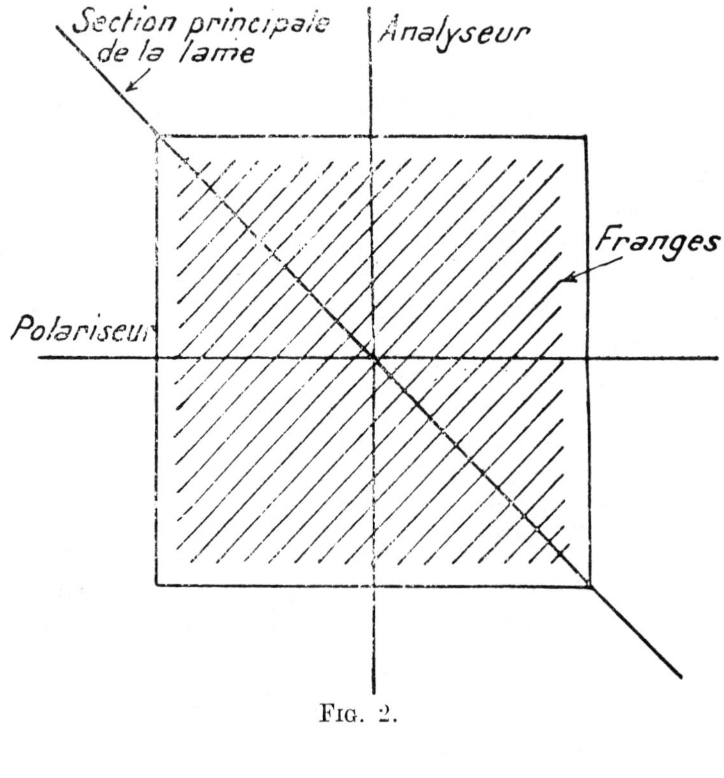

Fig. 2.

franges de Savart. Elles sont perpendiculaires au plan contenant l'axe optique et la normale à la lame.

Ces franges ont une netteté maxima lorsque la vibration transmise par le polariseur est à 45° de ce plan (fig. 2). Les deux systèmes de franges produits lorsque la vibration transmise par l'analyseur est respectivement perpendiculaire ou parallèle à la direction de la vibration incidente sont tels que les franges brillantes de l'un des systèmes coïncident avec les franges sombres de l'autre.

Il est facile de montrer que les deux indices principaux de la lame sont :

$$\left\{\begin{array}{l} n' = n_0 \\ n'' = n_0 + (n_e - n_0)\cos^2\alpha = n_0 + \Delta n \cos^2\alpha. \end{array}\right. \tag{1}$$

En incidence normale (ce qui correspond au centre du champ de la lentille), la différence de marche entre les deux vibrations transmises par la lame suivant ses deux axes principaux est :

$$\Delta_0 = e(n'' - n') = e\,\Delta n \cos^2\alpha. \tag{2}$$

Si les rayons incidents ne sont plus normaux, mais frappent la lame sous une faible incidence i, étant donné que la biréfringence est peu élevée, on peut admettre qu'à l'intérieur de la lame, les deux rayons ordinaire et extraordinaire sont confondus, et qu'ils font avec la normale un angle $r = i/n$, n désignant l'indice moyen du cristal. Il existe entre les deux vibrations rectangulaires transmises dans cette direction une différence de marche :

$$\Delta = e\Delta n \cos^2\left(\alpha + \frac{i}{n}\right) \sim e\Delta n\left(\cos^2\alpha - \frac{i}{n}\sin 2\alpha\right) = \Delta_0 - e\frac{\Delta n}{n} i \sin 2\alpha. \tag{3}$$

λ désignant la longueur d'onde de la lumière monochromatique incidente, on obtiendra l'intervalle angulaire Δi séparant deux franges successives en écrivant que $\Delta - \Delta_0 = \lambda$, d'où il résulte que :

$$\Delta i = \frac{\lambda}{e}\frac{n}{\Delta n}\frac{1}{\sin 2\alpha} = \frac{n\lambda}{2\Delta_0 \operatorname{tg}\alpha}. \tag{4}$$

Pour avoir un temps de pose acceptable avec les radiations du ciel nocturne, il faut que la lentille L_1 soit très lumineuse. Nous avons employé des objectifs photographiques de 50 mm de foyer (Saphir Boyer F/1,4 et F/1, objectif Penciolelli F/1), toutes les surfaces optiques, lames, objectifs etc., étant traitées à l'antiréflex. Les caractéristiques de nos lames de quartz ont été déterminées pour qu'avec de tels objectifs, les franges sur la plaque photographique soient convenablement séparées.

Pour la raie verte ($\lambda = 5\,577$ Å) et une lame d'épaisseur $e = 13$ mm, taillée à $\alpha = 30°$ de l'axe, les franges sont distantes de 0,42 mm. Le problème est plus compliqué dans le jaune et dans le rouge étant donné la présence simultanée de deux composantes, dont les systèmes de franges ne doivent pas se détruire mutuellement. L'ordre d'interférence au centre du champ est :

$$K = \frac{e \Delta n \cos^2 \alpha}{\lambda}. \tag{5}$$

Dans le jaune ($\lambda = 5\,890$ Å), il est de l'ordre de 150 ; il diffère d'une unité pour deux radiations λ et $\lambda + \Delta\lambda$ si :

$$K\lambda = (K + 1)(\lambda - \Delta\lambda) \tag{6}$$

soit

$$K = \frac{\lambda}{\Delta\lambda}. \tag{7}$$

Pour $\lambda = 5\,890$ Å, on trouve $\Delta\lambda = 39$ Å. L'intervalle correspondant au doublet du sodium, soit $\delta\lambda = 6$ Å sera $\frac{\delta\lambda}{\Delta\lambda} \sim 0{,}15$. Pour les composantes D_1 et D_2, les franges sont donc pratiquement confondues. Cette lame pourra ainsi convenir pour la raie verte et la raie jaune.

Pour l'étude du doublet rouge 6 300-6 364 Å du ciel, nous avons utilisé une lame de quartz calculée pour obtenir la coïncidence exacte des franges correspondant à chacune des deux composantes. $\Delta\lambda = 64$ Å dans la relation (7) donne $K = 100$, et la relation (5) fournit alors une lame d'épaisseur $e = 13{,}9$ mm taillée à 45° de son axe. Avec le même objectif que précédemment, l'interfrange correspondant est de 0,34 mm pour cette lame, qui peut ainsi convenir pour les trois radiations verte, jaune et rouge du ciel nocturne.

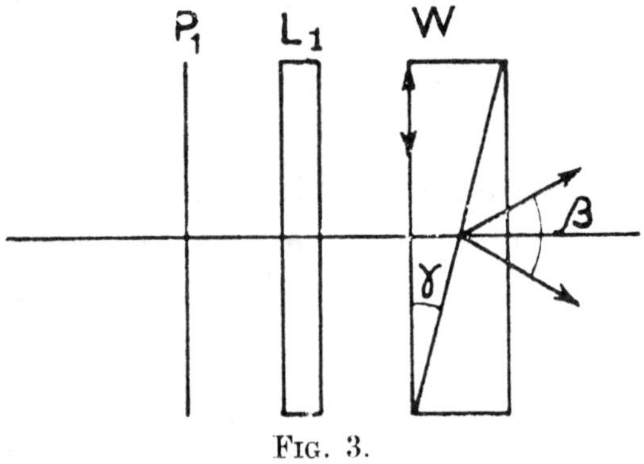

Fig. 3.

La luminosité des franges peut être améliorée en utilisant le procédé de Lyot, qui consiste à remplacer l'analyseur par un Wollaston (fig. 3), dont les sections principales sont orientées à 45° de celles de la lame, et donnant une séparation angulaire telle que les franges brillantes correspondant à l'une des images coïncident avec celles de l'autre image. Soit γ l'angle au sommet de l'un des prismes constituant le Wollaston W. La séparation angulaire qu'il donne en incidence normale est :

$$\beta = 2(n_0 - n_e) \operatorname{tg} \gamma. \tag{8}$$

D'autre part, les vibrations se trouvant à 45° de la direction des franges, l'arête doit être elle-même à 45° des franges. Si Δi désigne l'interfrange, pour que les deux systèmes coïncident, le déplacement donné par le Wollaston devra être :

$$\beta = \Delta i \frac{\sqrt{2}}{2}. \tag{9}$$

On trouve ainsi que l'angle au sommet γ doit être, pour la première lame, de 19°,4, et pour la seconde, de 17°,15.

Ce système est particulièrement commode pour l'étude des trois radiations ci-dessus, un étroit domaine spectral de chacune d'elles étant d'abord isolé par une combinaison convenable de plaques photographiques et d'écrans Wratten. Nous avons employé les combinaisons suivantes : Pour 5 577 Å de OI ($^1S_0$-$^1D_2$) : Plaques Ilford chromatic avec écran Wratten n° 15 ou 16 ;

Pour le doublet jaune du Sodium 5 890-5 896 Å : Plaques Crumière Panchro ou Kodak T avec écran Wratten n° 22 ou 23.

Pour le doublet rouge 6 300-6364 Å de O I ($^1D_2$-^3P), plaques Ilford HP3 avec écran Wratten n° 26 (*).

Dans ces conditions, avec un polaroïd très peu absorbant (High transmission polaroïd), il suffit d'une pose nocturne d'une demi-heure pour obtenir nettement les franges données par la raie verte, et d'une heure environ pour les raies jaunes et rouge. En remplaçant le polaroïd par une lame de verre incliné, on obtient des franges moins contrastées, mais il suffit alors d'un quart d'heure de pose pour la raie verte. La figure *4a* représente un spectre du Ciel nocturne obtenu avec un spectrographe Cojan, d'ouverture F /0,7 et des plaques Ilford chromatic ; le spectre montre uniquement la raie 5 577 Å. La figure *4b* représente les franges données par le dispositif précédemment décrit, que nous appellerons Polariscope Savart-Lyot, les deux poses étant faites simultanément. La figure 5 représente le ciel et la chaîne des Pyrénées photographiés avec la raie verte, à travers le polariscope. Les franges apparaissent non seulement sur le ciel, mais sur les montagnes elles-mêmes, grâce à la lumière diffusée par la neige. Cette dernière photographie a été faite avec un objectif photographique ordinaire ouvert à F /4,5 avec 6 heures de pose.

Au cours de ces essais, nous avons obtenu des clichés homogènes, qui ne permettaient en aucun cas de penser que la brillance de la raie pouvait varier beaucoup dans le champ de l'appareil. Il faut donc penser que les variations de brillance mises en évidence par Abadie, A et E Vassy [5] et par Elvey, Swings et Linke [6] avec des récepteurs photo-électriques entre différents points du ciel, doivent se compenser, étant donné la pose assez longue que nécessite la faible luminosité du ciel nocturne à l'état normal. (Nous n'avons pas eu l'occasion d'essayer notre appareil en période aurorale.)

Bien entendu, la structure fine des raies ne peut être mise en évidence avec cet appareillage comme avec l'étalon de Fabry et Pérot, mais le polariscope peut servir à l'enregistrement permanent des trois radiations principales du ciel noc-

---

(*) Cet appareillage a été présenté au colloque de Londres, juillet 1947. Voir *Physical Society Gassiot Committe Report* 1948, p. 70.

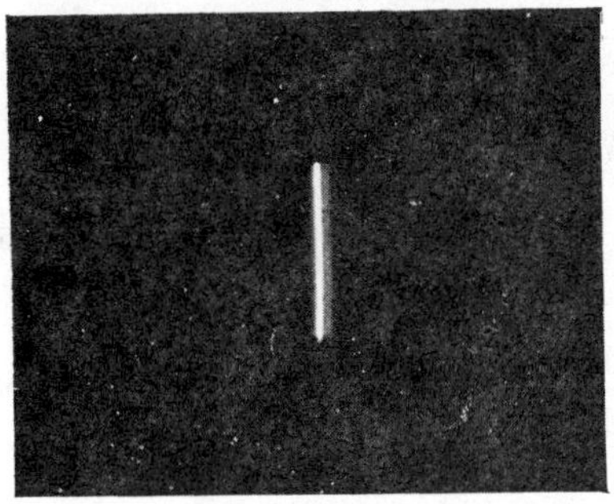

FIG. 4 *a*.

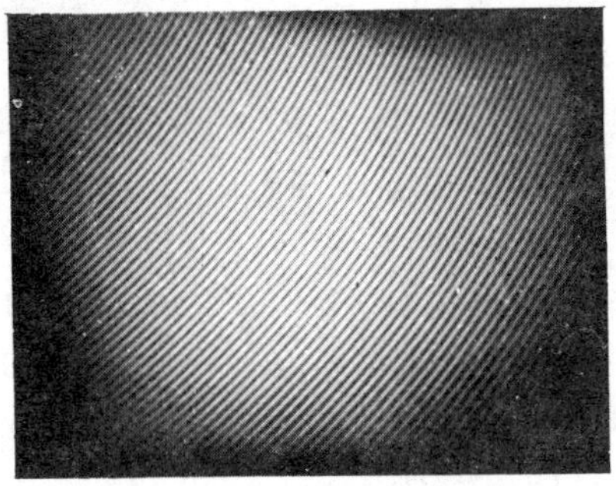

FIG. 4 *b*.

FIG. 5.

FIG. 6.

FIG. 7.

FIG. 11 *a*.

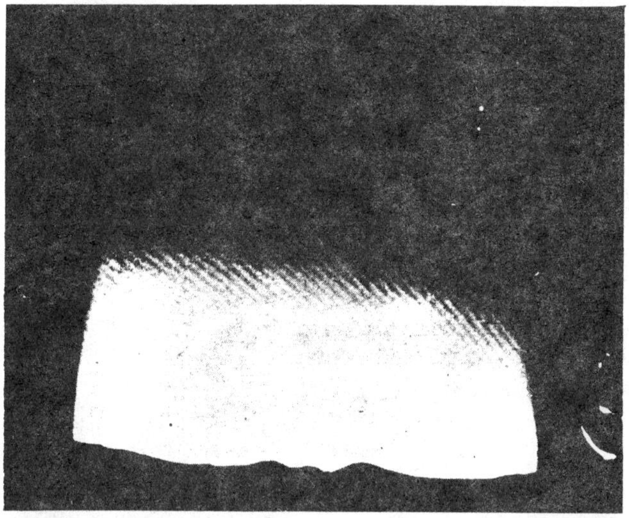

FIG. 11 *b*.

FIG. 11 *c*.

Fig. 8.

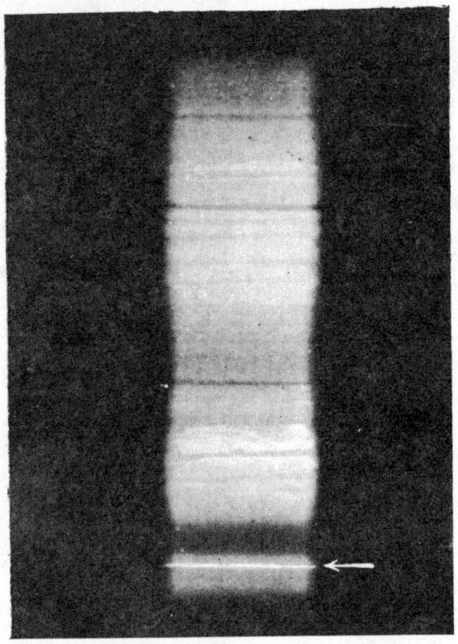

Fig. 9.

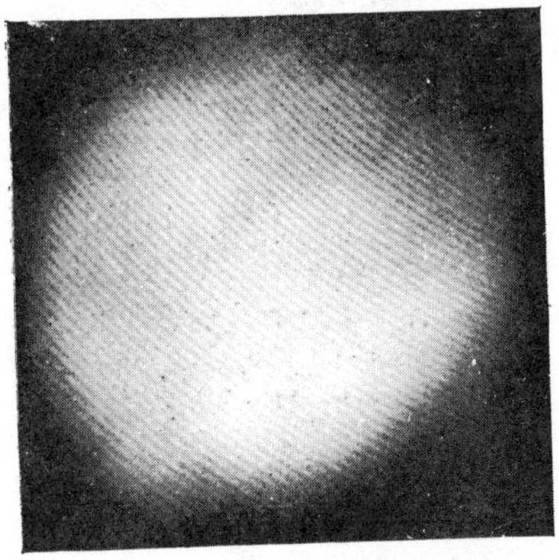

Fig. 10.

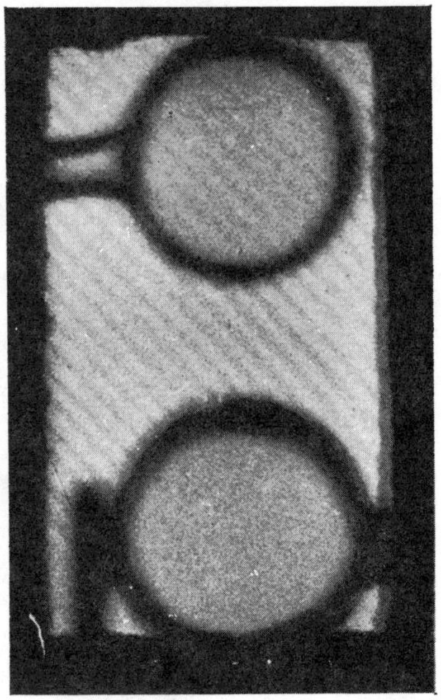

Fig. 12.

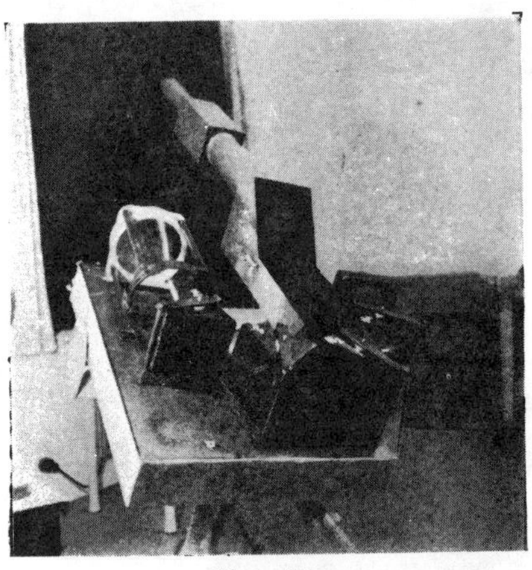

Fig. 13.

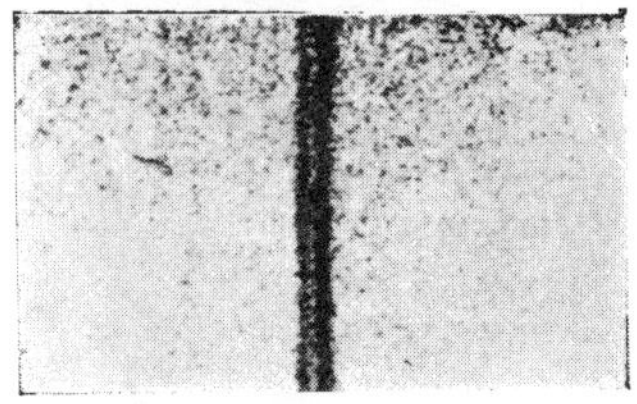

Fig. 14.

turne, et à l'étude statistique de leurs variations. La figure 6 représente un appareil à franges muni d'un dispositif permettant l'avance automatique et discontinue de la plaque photographique.

ÉTAT DE POLARISATION DES RAIES DU CIEL NOCTURNE

Cet appareillage peut être utilisé pour étudier la polarisation des raies du ciel nocturne : une raie initialement polarisée, photographiée à travers l'ensemble quartz biréfringent convenablement orienté, peut donner des franges, même si le polariseur placé devant la lame de quartz a été enlevé. On obtient au laboratoire des franges parfaitement nettes avec les émulsions employées ci-dessus, lorsque la lumière incidente est constituée par une raie spectrale pure, présentant une polarisation d'au moins 3 %. Le contraste des franges peut être modifié par interposition d'une lame de verre inclinée. Il est renforcé si la vibration incidente prédominante est dans le plan d'incidence de cette lame, et affaibli si la vibration incidente est perpendiculaire à ce plan. En utilisant côte à côte deux lames de verre inclinées G_1 et G_2 dont les plans d'incidence sont croisés, toute polarisation initiale de la lumière incidente doit se manifester par une dissymétrie du contraste des franges données par les deux lames.

Pour étudier la polarisation des radiations monochromatiques du ciel nocturne, l'axe de visée de l'appareil est orienté perpendiculairement à la direction des rayons solaires (pratiquement, on vise autour de minuit solaire dans la direction de l'étoile polaire). Nous appellerons plan de diffusion D celui qui passe par la direction de visée et le soleil. L'axe de rotation de G_1 étant situé dans ce plan, et celui de G_2 lui étant perpendiculaire, les deux lames de verre sont disposées dans le plan focal objet d'une lentille de projection placée devant le polariscope, et l'objectif forme ainsi l'image de G_1 et G_2 sur la plaque photographique. En inclinant convenablement les glaces (une quarantaine de degrés sur le plan des axes), on introduit une polarisation suffisante de la lumière incidente, qui entraîne l'apparition des franges. La figure 7 représente les résultats obtenus avec la raie verte du ciel nocturne ; le coin supérieur droit P représente les franges données par un polaroïd ; aucune dissymétrie de contraste n'a pu être constatée entre les franges données par les deux glaces, ni pour la raie verte, ni pour le doublet rouge du ciel nocturne ; les résultats que nous avons obtenus pour la raie jaune nocturne ne sont pas encore assez nombreux pour que nous puissions en tirer une conclusion.

On peut se rendre compte de la sensibilité d'un tel dispositif en ajoutant une 3e glace G_3 recouvrant partiellement chacune des glaces G_1 et G_2, et inclinée autour d'un axe parallèle à celui de G_1 par exemple, la même inclinaison donnée à G_1 et G_2 étant tout juste suffisante pour faire apparaître des franges en A et B présentant le même contraste. (Voir fig. 8 correspondant à la raie verte du ciel nocturne.) En C, les effets de G_1 et G_3 se renforcent, et le contraste augmente, alors qu'en D, les effets de G_2 et G_3 sont en opposition, ce qui produit la disparition presque totale des franges. On voit sur la figure 8 que le contraste des franges diminue dans l'ordre des plages C, A = B, D. Nous avons montré [7] que dans ces conditions, en tenant compte de l'inclinaison des trois glaces sur le rayon incident, et en évaluant le contraste sur les trois plages, la polarisation minima que l'on peut déceler est inférieure à 1,5 %.

Pour les raies verte et rouge du ciel nocturne, on peut conclure des poses faites que la polarisation est inférieure à cette quantité, c'est-à-dire pratiquement inexistante. En ce qui concerne la raie jaune du ciel nocturne, nous pouvons affirmer que sa polarisation est très inférieure à celle qui fut trouvée par Khvostikov [8] par une autre technique.

Détection de la raie crépusculaire

Le polariscope peut être également employé pour mettre en évidence la raie D crépusculaire. On orientera son polariseur de façon à affaiblir au maximum le spectre continu très polarisé diffusé par l'atmosphère, afin d'augmenter le contraste entre les franges et le fond continu qui les accompagne dans ce cas. La figure 9 représente un spectre du ciel crépusculaire montrant l'existence de la raie D, et la figure 10 représente les franges obtenues au crépuscule par le polariscope muni du même équipement que pour l'étude de la raie jaune nocturne, les deux poses ayant été effectuées simultanément.

L'enregistreur employé pour étudier l'évolution du ciel nocturne, peut être utilisé pour suivre celle du ciel crépusculaire. Les figures 11 *a b c* représentent trois photographies prises à l'aube, l'appareil visant à l'horizon dans le plan vertical du soleil. Les durées de pose étaient 1 minute 30 secondes, et les clichés étaient pris à 4 minutes d'intervalles. On voit nettement l'aube monochromatique précéder l'aube générale, se développer vers le haut et envahir tout le ciel à mesure que le soleil se rapproche de l'horizon.

Les spectres que l'on obtient au crépuscule sont caractérisés sur les négatifs par une large bande claire dans le jaune orangé due à l'absorption par l'ozone et par la vapeur d'eau. Cette bande a deux maxima à 6 100 Å et 5 750 Å, séparés par un minimum [9] vers 5 872 Å, qui se traduit par une bande noire étroite de 100 à 150 Å de largeur. Nous nous sommes assurés que les franges obtenues au crépuscule sont bien dues à la présence de la raie D, et non pas à celle de cette bande particulièrement intense pour les faibles dépressions solaires, en procédant de la façon suivante : on place devant le polariscope une lentille L_2 dans le plan focal de laquelle on a disposé deux cuves à faces parallèles contenant de la vapeur de sodium ; l'image de ces deux cuves est au point sur la plaque. L'une d'elles est froide, l'autre est maintenue à une température suffisante pour assurer l'absorption totale de la raie crépusculaire. [12]. La figure 12 représente le cliché obtenu dans ces conditions, et montre que les franges correspondant aux rayons ayant traversé la cuve chaude ont disparu.

Nous avons essayé d'étudier la polarisation de la raie jaune crépusculaire en utilisant la méthode décrite ci-dessus et appliquée aux raies du ciel nocturne. Nous n'avons pas obtenu de résultats satisfaisants : en effet la raie est toujours accompagnée d'un fond continu très fortement polarisé, qui se manifeste de façon particulièrement gênante, étant donné la largeur de la bande spectrale à laquelle est sensible l'ensemble filtre-émulsion. L'inégal affaiblissement du fond par les deux glaces, dû à sa polarisation, crée des noircissements dissymétriques, qui modifient le contraste des franges. Il est donc nécessaire d'éliminer ce fond continu, si l'on veut obtenir un résultat net concernant la polarisation de la raie crépusculaire.

Adaptation du polariscope a un spectrographe ouvert et dispersif

Le montage que nous avons employé consiste à projeter les franges de Savart du polariscope sur la fente d'un spectrographe. Pour réduire le fond continu, il faut employer un spectrographe à grande dispersion, mais aussi lumineux que possible, afin de ne pas entraîner des temps de pose prohibitifs. Nous avons utilisé un appareil de type Arnulf-Lyot [13] à 4 prismes de 60° d'angle, en flint, de 14 cm de hauteur et de longueur de côté. Le collimateur a une distance focale de 1 m 30, l'objectif de chambre est constitué par un miroir sphérique de 28 cm de distance focale, ouvert à F /2 ; le rapport de réduction est de 4,6, et la dispersion de 60 Å /mm au voisinage de la raie D. La fente, de largeur réglable, ouvrant au maximum à 3 mm,

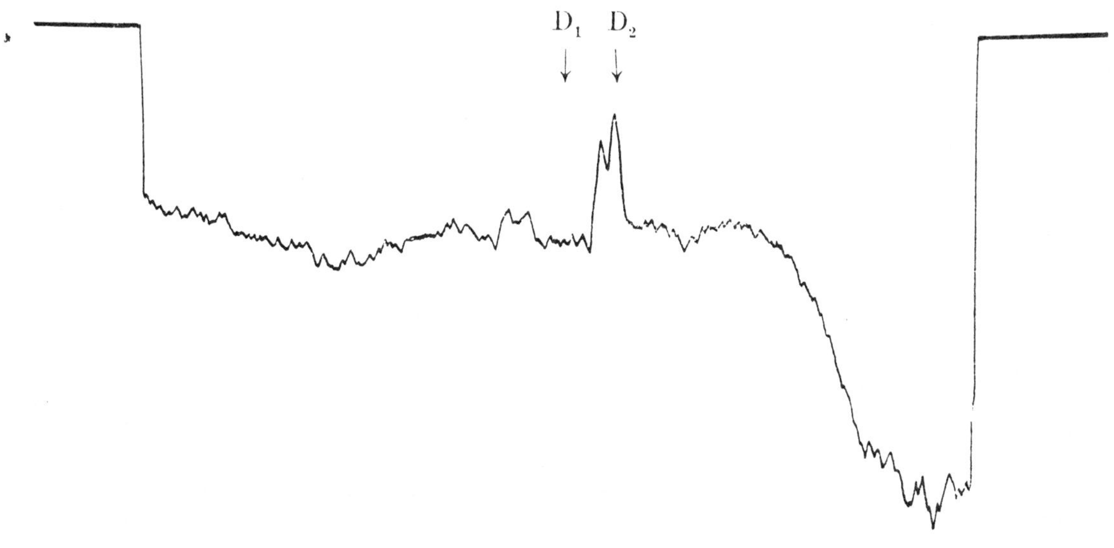

Fig. 15.

a une hauteur de 22 mm, et la hauteur de son image sur la plaque est de 4,8 mm. Les photographies sont faites sur des petites plaques carrées de 1 cm de côté. La figure 13 représente ce spectrographe lorsque les prismes sont à découvert (*). Les figures 14 et 15 représentent le doublet du sodium du ciel crépusculaire pris en fente fine avec cet appareil, et son enregistrement microphotométrique.

La figure 16 représente le schéma du montage employé pour l'étude de la polarisation. La lentille L_1 projette les franges sur la fente large du spectrographe ; W et L représentent le Wollaston et la lame de quartz. Les glaces G_1 et G_2 se trouvent dans le plan focal objet de la lentille de projection L_2. Les images de G_1 et G_2 recouvrent respectivement les deux moitiés supérieure et inférieure de la fente F,

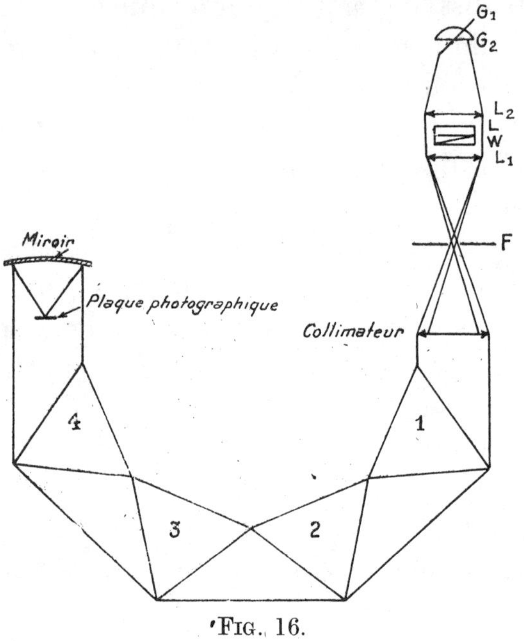

FIG. 16.

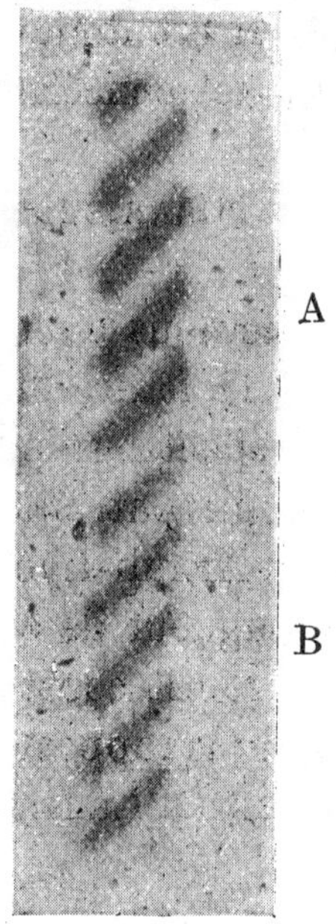

FIG. 17. — A plus contrasté que B.

et la lame de quartz est orientée de sorte que les franges qu'elle donne avec la raie D soient inclinées à 45° sur le fente. Avec les lentilles employées, et compte tenu du rapport de réduction du spectrographe, on trouve que les franges données par la raie D sont écartées de 0,46 mm sur la plaque ; la dispersion du spectrographe est telle que l'on peut opérer en fente large (D_1 et D_2 ne sont plus séparées), ce qui donne des franges plus nettement visibles qu'en fente fine, le fond continu conservant une intensité suffisamment faible.

La figure 17 représente une pose obtenue par M. Lecamp avec ce dispositif, en éclairant avec de la lumière provenant d'une lampe à sodium, complètement

(*) Les caractéristiques de cet appareil ont été calculées par MM. ARNULF et PANCIOLELLI, de l'Institut d'Optique. Le collimateur a été réalisé par la Société Générale d'Optique, et le miroir nous a été offert gracieusement par l'Office d'études et de recherches aéronautiques, grâce à la bienveillance de M. M. ROULLEAU. La partie mécanique a été réalisée par les établissements ANGELÉ, d'après les indications de M.M. LAFFINEUR.

polarisée par un polaroïd. Il faut remarquer que les maxima des deux systèmes de franges ne sont pas aussi intenses de part et d'autre du centre de la fente, ce qui tient au fait que les glaces non seulement modifient le taux de polarisation d'une lumière déjà polarisée, mais de plus affaiblissent différemment les diverses composantes ; les noircissements de la partie inférieure et supérieure sont donc différents. Pour les poses effectuées sur le ciel, l'inclinaison des glaces est réglée pour produire des franges de même contraste en lumière monochromatique naturelle (lumière d'une lampe à sodium ayant traversé deux verres dépolis successifs normaux au faisceau lumineux).

Avec une telle dispersion, le spectre continu du ciel est relativement faible au crépuscule, mais nous devons en tenir compte, et éliminer complètement son influence : avec une lame de quartz épaisse donnant un ordre d'interférence élevé, les franges de Savart ne sont visibles qu'en lumière monochromatique. Éclairé en lumière blanche, le polariscope donne un fond continu, constitué par du blanc d'ordre supérieur, l'ordre d'interférence variant très rapidement avec la longueur d'onde de la lumière. Lorsqu'on projette ce fond sur la fente fine du spectrographe, on observe un spectre cannelé. Les cannelures sont perpendiculaires à la direction de dispersion lorsque les franges de Savart sont parallèles à la fente. Lorsqu'on incline les franges de Savart par rapport à la fente, les cannelures s'inclinent également mais il est possible d'orienter le polariscope de telle façon que les cannelures données en fente fine par le fond continu soient inclinées en sens opposé des franges de Savart observées en fente large avec de la lumière monochromatique. Nous avons adopté une telle orientation pour éviter toute confusion sur la plaque entre les franges données par la raie D, inclinées à 45° sur la fente et les cannelures du fond. En élargissant la fente, les cannelures du fond s'estompent dans le spectre, et en choisissant par tâtonnements des largeurs de fente convenables, on arrive à les effacer complètement. Grâce à cette double précaution, on obtient ainsi du ciel crépusculaire un fond spectral faible et absolument continu, sur lequel se détache l'image monochromatique rectangulaire de la fente, sillonnée des franges de Savart. Il subsiste une petite difficulté lorsqu'on photographie ainsi à travers le polariscope précédé des deux glaces G_1 et G_2, le spectre du ciel. A cause de la très forte polarisation de la lumière solaire diffusée par l'atmosphère, les fonds continus des deux spectres correspondant aux images des deux glaces G_1 et G_2 n'ont pas la même intensité. Mais après quelques tâtonnements, on arrive à égaliser les fonds des deux spectres, en éclairant pendant un temps convenable la moitié de la fente qui correspond au fond le plus faible par une source de spectre continu auxiliaire (écran éclairé à travers un filtre bleu « day light » par une petite lampe à incandescence 4 volts).

Avant de photographier dans ces conditions le spectre du ciel crépusculaire, nous avons vérifié qu'une source de lumière monochromatique non polarisée (lampe à sodium visée à travers deux verres dépolis perpendiculaires au faisceau lumineux) donne sur les spectres correspondant aux glaces G_1 et G_2 inclinées à 50°, des franges de Savart de même contraste. Visant ensuite le ciel crépusculaire à angle droit du soleil, nous avons obtenu sur l'image de la raie D crépusculaire une dissymétrie de contraste nette. Le sens de cette dissymétrie indique une polarisation correspondant à un vecteur électrique prédominant perpendiculaire au plan de diffusion. C'est le sens à prévoir en cas d'excitation par résonance optique (dissymétrie positive) (*).

Pour nous assurer que cette dissymétrie n'est pas d'origine instrumentale, nous avons créé une faible dissymétrie en sens contraire en inclinant davantage (de 55°) l'une des deux glaces, dissymétrie qui est contrôlée par les spectres de la lampe à sodium (dissymétrie instrumentale négative). Dans ces conditions encore,

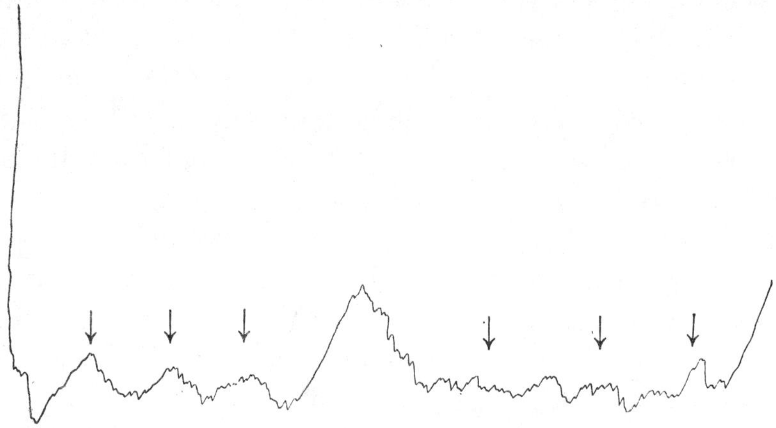

Fig. 18. — Enregistrement microphotométrique des franges projetées sur la fente du spectrographe en lumière crépusculaire. L'image de la fente du spectrographe est explorée dans le sens de la longueur, la fente du microphotomètre étant inclinée parallèlement aux franges. Les flèches indiquent les emplacements des minima de noircissement.

la raie D du ciel nous a fourni sur les plaques une dissymétrie positive. La polarisation de la raie D du ciel existe donc certainement.

Afin de pouvoir faire des poses crépusculaires de longue durée nous avons fait un séjour à haute latitude (observatoire météorologique d'Abisko en Laponie Suédoise à 68,2° de latitude) durant l'hiver 1948-1949.

Le ciel ayant été fréquemment couvert de nuages durant ce séjour, nous n'avons pu enregistrer qu'une demi-douzaine de poses mettant qualitativement en évidence la polarisation de la raie D du ciel.

La figure 18 représente un microphotogramme des franges de Savart obtenu à Abisko en février 1949. Il met nettement en évidence la dissymétrie des franges.

La méthode de la compensation progressive de la dissymétrie des franges par une dissymétrie instrumentale de sens inverse, obtenue en inclinant inégalement

(*) Ces résultats ont été exposés dans une note préliminaire. *C. R.* 228, 1601, 1949.

les deux glaces G_1 et G_2, doit permettre une évaluation quantitative de la polarisation. Nous nous proposons de poursuivre nos recherches dans ce sens.

A titre de garantie supplémentaire, nous avons tenu à étudier avec le même appareillage (même spectrographe et même montage) la polarisation de la résonance optique de la vapeur de sodium au laboratoire. M. M. Lecamp [14] s'est chargé à notre demande de cette étude. Il a montré que la lumière de résonance optique de la vapeur de sodium sous une pression de vapeur correspondant à une température de 120 à 130° C et éclairée latéralement par la lumière naturelle d'une lampe à sodium, produit des franges de Savart dont la dissymétrie de contraste est très nette. Lorsqu'on superpose à la lumière de résonance optique un spectre continu pour réaliser des conditions semblables à celles de la pose sur le ciel, cette dissymétrie est comparable à celle donnée par la raie D du ciel. M. Lecamp a également vérifié, selon les prévisions théoriques, que la polarisation de la lumière excitatrice renforce ou annule la dissymétrie des franges suivant que le vecteur électrique excitateur est perpendiculaire, ou parallèle au plan de diffusion, et que l'application à la cuve de résonance d'un champ magnétique de quelques centaines de gauss dirigé suivant l'observateur annule également la polarisation.

Conclusion

Nous avions déjà montré antérieurement que la raie D du ciel crépusculaire possède une finesse de l'ordre de celle des raies de résonance en étudiant l'absorption de cette raie par la vapeur de sodium [15] et [12]. La mise en évidence d'une faible polarisation de cette raie à angle droit du soleil met hors de doute le mécanisme d'excitation dans la haute atmosphère : il s'agit d'une résonance optique excitée par les radiations jaunes des rayons solaires. Une luminescence produite par photodissociation de composés sodiques sous l'influence de l'ultraviolet solaire produirait en effet une raie entièrement dépolarisée [16].

La connaissance du mécanisme d'excitation et des radiations solaires efficaces permet, à partir de l'évolution de la luminosité dans le temps, de calculer l'altitude du centre de gravité de la couche luminescente qui se trouve ainsi située vers 80-90 km d'altitude. Nous estimons que les calculs d'altitude à partir de l'évolution de la luminosité par la visée près du zénith et près de l'horizon sont peu précis. Ils ont donné en effet, pour différentes mesures d'un même auteur et d'un auteur à l'autre, des résultats considérablement dispersés [17]. Nos conclusions n'excluent pas l'existence de couches de sodium moins denses mais plus élevées, responsables de la luminosité post-crépusculaire beaucoup plus faible, mise récemment en évidence par Barbier et Roach [18].

Pour qu'une excitation par résonance optique soit possible, il faut que le sodium existe dans l'atmosphère à l'état d'atomes libres. Or à l'altitude de 80 à 90 km, ces atomes entrent fréquemment en collision avec des atomes et molécules d'oxygène qui oxydent le sodium. Il faut donc admettre l'existence d'un mécanisme tel que celui qui a été proposé par S. Chapman [19], qui régénère continuellement le sodium à l'état atomique.

Il se pourrait que les rayons ultraviolets du soleil interviennent par un mécanisme de photo-dissociation. Mais étant donné la symétrie de la luminescence à l'aube et au crépuscule, tout effet d'accumulation est à exclure ; il faudrait que ce mécanisme soit équilibré instantanément par les chocs destructifs pour que l'évolution de la luminescence soit conditionnée par l'illumination de l'atmosphère par l'ultraviolet solaire comme le pensent certains auteurs [20], et non pas par les radiations jaunes du soleil. En tout cas, l'origine du sodium atmosphérique et le maintien de celui-ci à l'état d'atomes de sodium libre constituent des énigmes non encore résolues.

Une autre propriété caractéristique de la luminescence crépusculaire du sodium est sa forte variation saisonnière d'intensité, qui présente un maximum en hiver, et un minimum prononcé en été. Sur notre demande, la mission antarctique française a commencé l'étude de la variation saisonnière d'intensité dans l'hémisphère austral. Il semble résulter d'une première séries de poses faites par M. Pommier, que la variation y soit commandée par la saison australe, c'est-à-dire que les variations saisonnières d'intensité dans l'hémisphère austral soient en opposition de phase avec celles de l'hémisphère boréal. Si ces résultats se confirment, ils prouveront que les variations d'intensité ont une cause terrestre, et non pas une cause cosmique annuelle telle que la chute de poussières météoriques.

Notre mission à Abisko a pu être réalisée grâce à une subvention de l'Union Internationale de Physique et grâce à l'aide financière du C. N. R. S., organismes que nous sommes heureux de remercier.

Nous remercions également les autorités suédoises qui ont aimablement mis à notre disposition l'Observatoire d'Abisko et qui ont facilité par tous les moyens l'accomplissement de notre travail.

Manuscrit reçu le 11 *octobre* 1950.

BIBLIOGRAPHIE

[1] Babcock H. D., *Astrophys. Journ.*, **57**, 1923, p. 209.

[2] Vegard L. et Harang L., *Geofys. Pub.*, II, n° I, 1934, II; n° 15, 1937 ; **12**, n° 5, 1938.

[3] Cabannes, Dufay et Gauzit, *C. R. Acad. Sc. Paris*, **206**, 1938, p. 870 et 1525 ; *Astrophys. Journ.*, **88**, 1938, p. 164.

[4] Bernard R., *C. R. Acad. Sc. Paris*, **206**, 1938, p. 928.

[5] P. Abadie, A. Vassy et E. Vassy, *Ann. de Geoph.*, **1**, 1945, p. 189 ; *C. R. Acad. Sc. Paris.*, **222**, 1946, p. 99.

[6] Elvey C. T. Swings P. et Linke W., *Astrophys. Journ.*, **93**, 1941, p. 337.

[7] J. Bricard et A. Kastler., *Ann. de Geophys.*, **3**, 1947, p. 308 et *C. R. Acad. Sc. Paris*, **224**, 1947, p. 1555.

[8] I. A. Khvostikov, *C. R. Acad. Sc. U. R. S. S.*, **21**, n° 7, 1938, p. 322 ; **27**, n° 3, 1940, p. 219.

[9] Mme A. Vassy, *C. R. Acad. Sc. Paris*, **206**, 1938, p. 1368 ; *Ann. de Phys*, **16**, 1941, p. 155.

[10] J. Gauzit, *Ann. de Phys.*, **4**, 1935, p. 460.

[11] R. Bernard, *C. R. Acad. Sc. Paris*, **206**, 1938, p. 448 ; *Zeits. für Phys.*, **110**, 1938, p. 291.

[12] J. Bricard et A. Kastler, *Ann. de Geophys.*, **1**, 1944, p. 53.

[13] A. Arnulf et B. Lyot, *C. R. Acad. Sc. Paris*, **201**, 1935, p. 1480 ; A. Arnulf, *Ann. d'Astrophys.* **6**, 1943, p. 21.

[14] M. Lecamp, Diplôme d'études supérieures, Paris, 1949.

[15] A. Kastler, *C. R. Acad. Sc. Paris*, **210**, 1940, p. 530.

[16] T. R. Hogness et J. Frank, *Zeits. für Phys*, **44**, 1927, p. 26.

[17] L. Vegard et E. Tonsberg, *Geof. Pub.*, **13**, n° 1, 1941.
L. Vegard et G. Kwifte, *Geof. Pub.*, **16**, n° 7, 1945.
L. Vegard, *Nature*, **145**, 1940, p. 623 ; R. Bernard, Compte Rendu *Colloque Int. du C. N. R. S.*, Lyon, p. 249, 1947. L. Vegard, *ibid.*, p. 254.
J. Dufay, *C. R. Acad. Sc., Paris*, **225**, 1947, p. 690.

[18] B. Barbier et E. Roach, *Trans. Amer. Geophys. Union*, **31**, 1950, p. 13.

[19] S. Chapman., *Astro. Jour.*, **90**, 1939, p. 309.

[20] R. Penndorf., *Jour. of Met. U. S. A.*, 1948, p. 152 ; *Phys. Rev.*, **78**, 1950, p. 66.
Voir également références L. Vegard et collab.

Addition à la correction des épreuves :

Nous signalons sans pouvoir les analyser les articles récents sur le problème du sodium atmosphérique :

F. D. Kahn, *Phys. Rev.* **79**, 1950, p. 167.

G. Cario et U Stille, *Abhandl. Brannschweig. Wiss. Ges.*, **2**, 1950, p. 40.

D. R. Bates et M. Nicolet, *J. of Geophys. Research.*, **55**, 1950, p. 235.

©1926 Springer-Verlag Heidelberg. Reprinted from *Die Naturwissenschaften,* Vol. XIV, 1283 (1926).

Notiz über die Art des Regenbogenlichtes.

Beobachtet man die schöne Himmelserscheinung des Regenbogens durch ein vor das Auge gehaltenes NICOLsches Prisma, so erkennt man beim Drehen des Nicols ein bogenstückweises Verschwinden des Phänomens. Das Regenbogenlicht ist somit sehr ausgeprägt linear polarisiert, und zwar verlaufen seine Schwingungen tangential zum Bogen. Das gilt sowohl für den Haupt- als auch für den Nebenregenbogen. Eine einfache hier nicht näher zu erörternde Überschlagsrechnung liefert die damit harmonierenden zahlenmäßigen Verhältnisse.

Der erwähnte, immerhin doch sehr drastische Umstand ist, wie es scheint, im Kreise der naturkundlich Interessierten sehr wenig bekannt. In den mir zugänglichen physikalischen Lehrbüchern, ja selbst in dem großen meteorologischen Werke von PERNTNER-EXNER, das dem Regenbogen 80 Seiten widmet, fehlt eine entsprechende Darlegung. Indes ist in der Abhandlung von CHR. WIENER über die Helligkeit des klaren Himmels der in Rede stehende Umstand in allgemeineren Erörterungen mit enthalten, und in den theoretischen Studien von W. MÖBIUS über den Regenbogen findet man eine einschlägige Berechnung, indes auch nicht für Wassertropfen, sondern für Glaskugeln.

Es ist der Zweck obiger Notiz, die erwähnte, mir seit vielen Jahren bekannte Erscheinung mehr als jetzt der Fall ist in den Interessenkreis, besonders auch der Lehrer an Schulen, zu rücken.

Nach einer von meinem Assistenten Dr. S. RÖSCH ausgeführten Berechnung ist das Verhältnis der Intensitäten des senkrecht zur Einfallsebene schwingenden Lichtes zu der des in der Einfallsebene schwingenden $I^s : I^p$ beim Hauptregenbogen etwa 21 : 1. Für den Nebenregenbogen ergibt sich $I^s : I^p$ zu etwa 8,5 : 1.

Der Gang der Rechnung und ihre Ergebnisse hinsichtlich der Regenbögen höherer Ordnung sollen im Zentralbl. für Mineralogie veröffentlicht werden.

Leipzig, den 22. November 1926.

FRIEDRICH RINNE.

Gelöste und ungelöste Rätsel der Bienensprache.

Von K. v. Frisch (Graz).

(Fortsetzung und Schluß.)

II. Die Tänze auf horizontaler Unterlage.

1. Richtungsweisung.

Wenn man aus einem Bienenkasten eine Wabe mit tanzenden Bienen vorsichtig heraushebt und horizontal legt, so werden häufig die Tänze fortgesetzt. Die Tänzerin kann aber unter diesen Umständen den Winkel zum Sonnenstand, der beim Flug zum Futterplatz einzuhalten ist, nicht mit Bezug auf die Richtung der Schwerkraft angeben (wie in Fig. 6, S. 16), denn ihre Einstellung zu dieser bleibt ja nun dauernd dieselbe. Sie tanzt vielmehr in horizontaler Lage so, *daß der geradlinige Schwänzellauf unmittelbar die Richtung zum Futterplatz weist*. Dreht man die Wabe in ihrer Ebene, wie die Drehscheibe der Eisenbahn, so sucht der Tanz seine Orientierung im Raume beizubehalten und spielt sich immer wieder — einer Kompaßnadel vergleichbar — auf die alte Richtung ein. Daß die Bienen für diese scheinbar unnatürliche Lage sofort eine ihr angemessene andere Form der Richtungsweisung bereit haben, wirkt überraschend. Man muß aber bedenken, daß Tänze auf horizontaler Unterlage in Wirklichkeit nicht unnatürlich sind. Man sieht sie nicht selten auf dem horizontalen Anflugbrettchen vor dem Flugloch, wenn daselbst an warmen Tagen viele Bienen versammelt sind und den Nektar, wie auch die Botschaften der Heimkehrer schon unter freiem Himmel in Empfang nehmen.

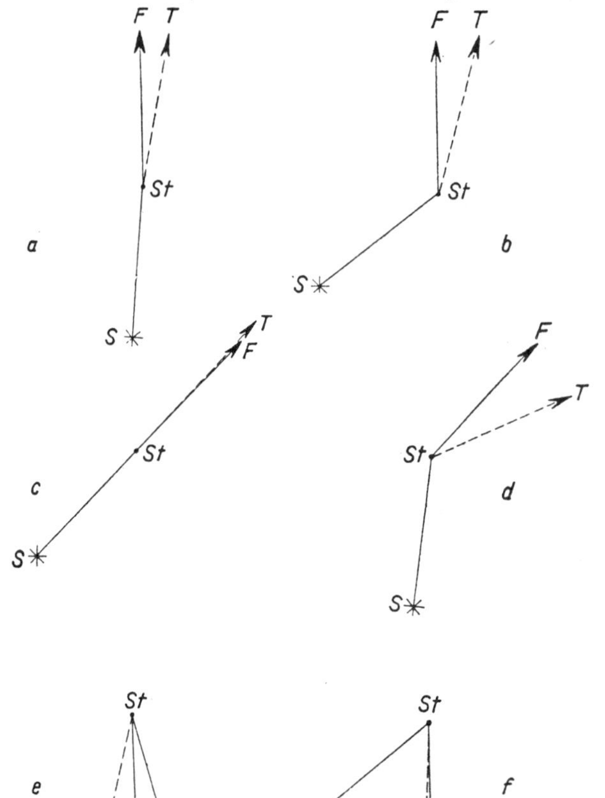

Fig. 21. Sechs Versuche über Richtungsweisung auf horizontaler Wabenfläche. *St* = Beobachtungsstock, *S* = Sonnenstand, *F* = Futterplatz, *T* = Tanzrichtung (Mittelwert aus 10—20 Beobachtungen).

Man kann die Tänze auf horizontaler Wabenfläche in aller Ruhe studieren, wenn man einfach den ganzen Beobachtungsstock horizontal legt. Die Bienen lassen sich dadurch in ihrer Tätigkeit nicht stören. Es ist für das Folgende wichtig, sich zu vergegenwärtigen, daß bei der Beobachtung der hölzerne Schutzdeckel des Bienenkastens abgehoben ist, so daß durch die Glasscheiben Licht auf die Waben fällt. Doch war direktes Sonnenlicht stets durch ein Schutzdach abgehalten.

Um eine Vorstellung vom Grad der Genauigkeit dieser Richtungsweisung zu vermitteln, bringe ich in Tabelle 6 das Ergebnis von 6 Versuchen, bei welchen der Futterplatz nacheinander in drei verschiedene Richtungen gelegt wurde. Die graphische Darstellung Fig. 21 zeigt in anschaulicherer Weise, was dabei herausgekommen ist. Während auf der lotrechten Wabenfläche der Schwänzeltanz mit dem Lauf der Sonne seine Richtung ändert (vgl. S. 15), bleibt er auf der horizontalen Fläche zu jeder Tageszeit gleichgerichtet und zeigt direkt nach dem Futterplatz. Wie können die Tänzerinnen das machen, und wie können es ihre Kameraden verstehen?

Man ist versucht, an einen *magnetischen Sinn* zu denken, der den Bienen die Orientierung über die Himmelsrichtungen auch im Inneren des Stockes gestattet. Aber so ist es nicht. Denn die Tänze werden durch Annäherung eines starken Magneten aus beliebiger Richtung nicht gestört.

Es war ferner mit der Möglichkeit zu rechnen, daß die Bienen auch im Schatten des Daches die Sonne wahrnehmen, indem sie ein besonderes Sinnesorgan für *langwellige, das hölzerne Schutzdach durchdringende Strahlen* hätten. Wenn sie vom Stock aus die Sonne sehen, brauchen sie ja nur beim Tanz auf der Wabe denselben Winkel zum Sonnenstand einzuhalten wie beim Flug zum Futterplatz, um die Richtung zur Futterstelle zutreffend zu weisen. Bei den Versuchen, diese Hypothese durch verschiedenartige Abschirmungen zu prüfen, fiel auf, daß die Tänze der Bienen unter Umständen schon gestört werden, wenn man das einfallende Licht durch ein Blatt Papier abschirmt. Dies paßte schlecht zu der Vorstellung von einer durchdringenden Strahlung und führte auf die Frage, ob denn die Bienen auch im geschlossenen, finsteren Stock bei horizontaler Lage die Richtung weisen. Ich hatte es bisher angenommen. Doch wie sollte man sich überzeugen? In den geschlossenen Stock kann man nicht hineinsehen.

Um die Frage zu entscheiden, bauten wir um den Bienenstock aus *Dämmplatten*, einer Preßmasse aus Holzfasern, ein kleines Haus, das geräumig genug war, um den Beobachter in sein Inneres aufzunehmen (Fig. 22)[14]. Nun konnte man, gleichsam in einem vergrößerten, geschlossenen Bienenkasten, bei dem roten — für Bienen nicht wahrnehmbaren — Licht einer Dunkelkammerlampe oder bei schwachem, diffusem weißen Licht beobachten. Das Ergebnis war eindeutig und gegen die Erwartung. Die Bienen tanzten zwar mit unverminderter Lebhaftigkeit, aber *die Tänze waren völlig desorientiert*. So hatten z. B. die 11 Schwänzelläufe eines einzigen Schwänzeltanzes die in Fig. 23 durch Pfeile angegebenen Richtungen. Manchmal trippelten die Tänzerinnen auch, wie unschlüssig, im Kreise und schwänzelten im Rundlauf oder gar nicht. Dies gilt aber nur für die Tänze auf *horizontaler* Wabenfläche. Stellt man den Stock, der normalen Lage entsprechend, vertikal, so weisen die Bienen auch im geschlossenen Dämmplattenzelt zutreffend durch ihre nach der Schwerkraft orientierten Tänze die Richtung zum Futterplatz mit Bezug auf den Sonnenstand (vgl. Fig. 6, S. 16).

Tabelle 6. *Richtungsweisung auf horizontaler Wabenfläche, Kegelbahnvolk 1945.*

Versuch Nr.	Tag MEZ.	Sonnen-azimut	Azimut des Futterplatzes	Von den Bienen gewiesene Richtung (n° östl. v. Nord)	Differenz (Fehler)	Entfernung des Futterplatzes	n	Wetter
a) 295	22. 10. 12.03	183°	357° (3° westl. v. N.)	8,5°	**11,5°**	135 m	20	teilw. bewölkt
b) 297	22. 10. 15.15	232°	357° (3° westl. v. N.)	13°	**16°**	135 m	10	bedeckt
c) 311	24. 10. 14.38	224°	42° (42° östl. v. N.)	41,5°	**0,5°**	125 m	10	wolkenlos
d) 325	27. 10. 12.10	186°	42° (42° östl. v. N.)	65°	**23°**	125 m	11	„
e) 336	29. 10. 10.43	162°	177° (3° östl. v. S.)	191°	**14°**	100 m	20	sonnig
f) 344	29. 10. 15.08	230°	177° (3° östl. v. S.)	179,5°	**2,5°**	100 m	10	„

Fig. 22. Das Zelt aus Dämmplatten. Auf der Bank ist der Flugtrichter des Beobachtungsstockes zu erkennen. Die im Bilde sichtbare, sonnenbeschienene Südwand läßt sich zum Betreten des Zeltes öffnen.

Die seitlichen Holzdeckel des Beobachtungsstockes waren aus 10 mm dicken Sperrholzplatten gemacht. Um dessen sicher zu sein, daß auch diese Abschirmung die Orientierung der Tänze aufhebt, beobachtete ich im horizontal liegenden Stock die Tänze *unter diesem Holzdeckel*. Es war dazu nötig, den Deckel durch einen aus Sperrholz gearbeiteten Aufsatz in solchen Abstand von den Glasscheiben zu bringen, daß ich durch ein seitlich angebrachtes Loch den Kopf hineinstecken und durch ein zweites Loch eine rote Taschenlampe bedienen konnte. Die Tänze waren ebenso desorientiert wie im Dämmplattenzelt, obwohl der Bienenkasten beim Versuch vollem Sonnenschein ausgesetzt war.

Daß die horizontalen Tänze im geschlossenen Stock desorientiert sind, läßt sich mit Hilfe einer anderen Methode auch ohne die erwähnten Umbauten und Abänderungen nachweisen: indem man ihre *Wirkung auf die Stockgenossen* prüft. Auch auf der horizontalen Wabe erwecken die Tänzerinnen das Interesse der Kameraden und werden von einem Schwanz nachtrippelnder Bienen verfolgt. Das gilt für des-

[14]) Die 7 mm dicken Dämmplatten lassen sich mit einem scharfen Messer leicht in beliebiger Größe zurechtschneiden. Da sie nicht naß werden sollen, habe ich sie über Nacht oder bei Regen stets ins Haus genommen. Bei ihrem geringen Gewicht ist das Abbauen und Aufstellen des Häuschens rasch und mühelos zu machen.

orientierte Tänze in gleicher Weise. Aber wenn die Sammlerinnen keine bestimmte Richtung angeben, dann werden die anderen Bienen — wenn überhaupt — *ungerichtet* ausschwärmen. Wir haben nun einige numerierte Bienen des Beobachtungsvolkes 240 m westlich vom Stock auf duftender Unterlage gefüttert und die Verteilung der Neulinge durch Auslegen von Duftplatten in verschiedenen Richtungen geprüft. Dabei zeigte sich, daß sie ungerichtet suchen, wenn der Stock horizontal liegt und geschlossen ist. Entfernt man den Deckel, so daß durch die Glasscheiben Licht auf die horizontal liegenden Waben fällt, oder läßt man den Stock geschlossen und bringt ihn in die normale, vertikale Stellung, so bewegt sich der Strom der weiterhin ausschwärmenden Neulinge in der Richtung des Futterplatzes, und bei den Duftplatten in anderer Richtung wird es still. Dieser Versuch ist in mehrfacher Hinsicht lehrreich: er zeigt, daß *auch die horizontalen Tänze alarmieren*, und daß die Richtungsweisung *auch in dieser anderen Art verstanden und befolgt* wird; er bestätigt ferner, daß *bei Lichtabschluß die Richtungsweisung auf der vertikalen Wabenfläche funktioniert, auf der horizontalen aber versagt.*

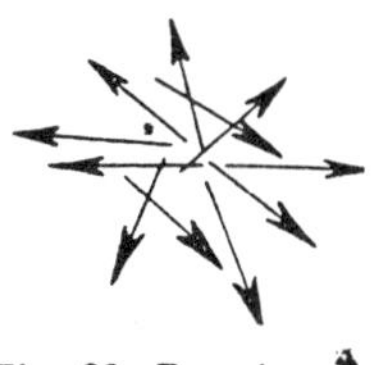

Fig. 23. Desorientierter Schwänzeltanz. Die Pfeile geben die Richtungen der einzelnen Schwänzelläufe an.

Wenn das Licht der maßgebende Faktor ist, der es den Bienen ermöglicht, auf horizontaler Fläche durch die Richtung des Schwänzellaufes unmittelbar die Richtung zum Futterplatz anzugeben, so kann man ihnen vielleicht im Dunkelzelt die Sonne durch eine künstliche Lichtquelle ersetzen? Mit diesem Gedanken ließ ich im geschlossenen Dämmzelt das Licht einer Taschenlampe aus der Richtung, in der zur Zeit die Sonne stand, auf meine desorientierten Bienen scheinen. Sofort kam Ordnung in ihre Tänze und der geradlinige Schwänzellauf wies nach Westen, wo sich der Futterplatz befand. Wenn ich aber das Licht der Taschenlampe von der entgegengesetzten Seite auf die Wabe fallen ließ, dann richteten sie ihren Schwänzellauf nach Osten. Ich konnte sie durch entsprechende Lampenstellung tanzen lassen, wohin ich wollte. *Sie hielten bei ihrem Schwänzellauf stets zur Stellung der Lampe denselben Winkel ein, den sie bei ihrem Flug nach dem Futterplatz zur Sonne einhalten mußten.* Benützt man eine schwache Taschenlampe, so wird dieser Winkel nicht sehr genau innegehalten. Als ich aber eine elektrische Lampe von 50 Kerzen verwendete und sie in der geschlossenen Metallglocke einer Mikroskopierlampe[15]) anbrachte, aus der das Licht durch eine kreisrunde Öffnung in der Glocke gut gerichtet auf die Tänzerinnen fiel, liefen die Bienen diesen Winkel mit einer erstaunlichen Genauigkeit. Als Beleg bringt Tabelle 7 das Ergebnis einer Versuchsreihe, in der ich das Licht dieser Lampe nacheinander aus allen vier Himmelsrichtungen auf die tanzenden Bienen fallen ließ, und zwar zunächst am Vormittag und dann, bei ganz anderem Sonnenstand, am Nachmittag. Unter dem *Tanzwinkel* ist hier der Winkel zu verstehen, den die tanzende Biene beim Schwänzellauf zur Himmelsrichtung des Lampenlichtes einhält. Mit dem *Sonnenwinkel* ist, wie immer, der Winkel gemeint zwischen der Himmelsrichtung der Sonne zur mittleren Beobachtungszeit und der Richtung vom Stock zum Futterplatz. Die größte Differenz zwischen den Durchschnittswerten der Tanzwinkel und dem Sonnenwinkel beträgt 12° in einem Versuch, bei dem wegen einer Stromstörung statt 10 Tänzen nur 4 beobachtet wurden. In den 7

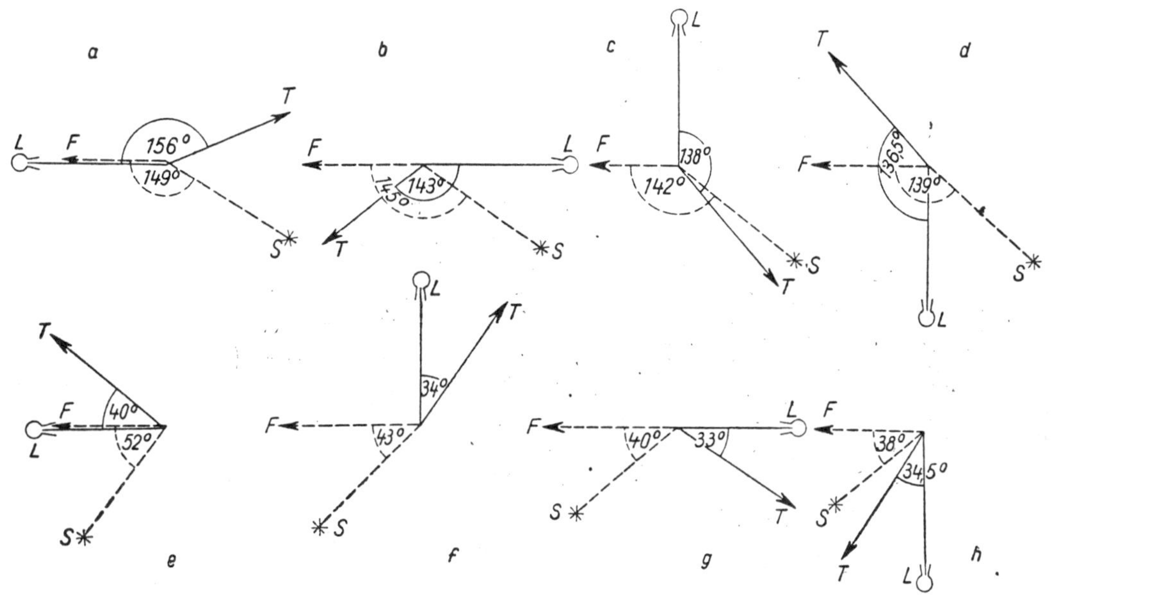

Fig. 24. Richtungsweisung auf horizontaler Wabe im Dunkelzelt nach einer künstlichen Lichtquelle. L = Lampe, T = Tanzrichtung, S = Sonnenstand, F = Futterplatz. Die Biene hält mit Bezug auf die Lampe die Richtung ein, die mit Bezug auf die Sonne zum Futterplatz führt.

Tabelle 7. *Richtungsweisung auf horizontaler Wabe im Dämmzelt nach dem gerichteten Licht einer Lampe, 14. 8. 1947. Futterplatz 240 m vom Stock, genau westlich.*

Protokoll Nr.	MEZ.	Sonnenazimut	Sonnenwinkel	Lampenazimut	Tanzwinkel (zur Lampe)	Diff. d. Winkel (Fehler)	n
a) 230	9.30	121°	149°	270° (West)	156°	7°	10
b) 231	9.42	125°	145°	90° (Ost)	143°	2°	10
c) 232	9.53	128°	142°	0° (Nord)	138°	4°	10
d) 233	10.04	131°	139°	180° (Süd)	136,5°	2,5°	10
e) 241	13.43	218°	52°	270° (West)	40°	12°	4
f) 245	14.10	227°	43°	0° (Nord)	34°	9°	10
g) 246	14.20	230°	40°	90° (Ost)	33°	7°	10
h) 247	14.27	232°	38°	180° (Süd)	34,5°	3,5°	10

[15]) „Frisch-Lampe" der Firma Lautenschläger, München.

anderen Versuchen war die Abweichung im Durchschnitt 5°. Anschaulicher als die Tabelle führt Fig. 24 das Ergebnis vor Augen. Die punktierten Linien zeigen für jeden Versuch die Stellung der Sonne und die Richtung zum Futterplatz. Die ausgezogenen Linien geben die Richtung des einfallenden Lampenlichtes und die Richtung des Schwänzellaufes wieder. Man sieht, daß *mit Bezug auf die Lampe die Richtung angezeigt wird, die mit Bezug auf die Sonne zum Futterplatz führt.*

Hierbei ist offenbar das für uns sichtbare Licht der wirksame Faktor, oder — vorsichtiger ausgedrückt — es genügt das für uns sichtbare Licht für die Orientierung der Bienen. Denn die Richtungsweisung war nicht gestört, wenn die ultravioletten Strahlen durch ein Glasfilter (GG_{13}) von Schott, Jena, ausgeschaltet wurden; es sind anderseits auch nicht die Wärmestrahlen maßgebend, denn die Richtungsweisung wurde durch Vorschalten einer 2,4 cm dicken Wasserschicht in einer planparallelen Kuvette nicht gestört, aber durch Vorschalten einer Rotscheibe aufgehoben

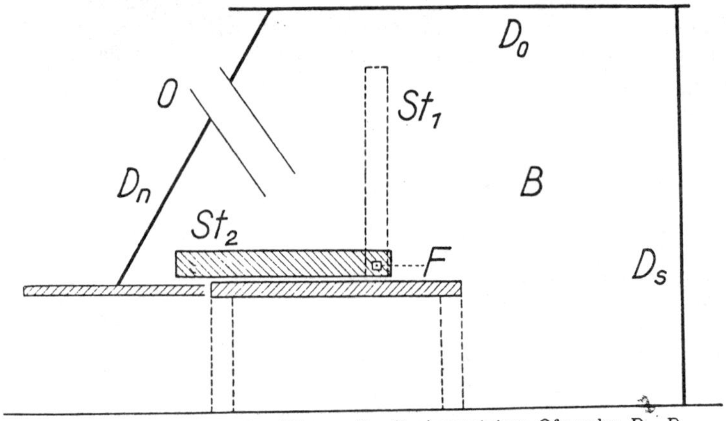

Fig. 25. Schnitt durch das Dämmzelt mit eingesetztem Ofenrohr. D_s, D_n, D_0 = südliche, nördliche und obere Dämmplatte des Zeltes. St_1 = Beobachtungsstock in normaler (vertikaler) Stellung. St_2 = Beobachtungsstock, horizontal gelegt. F = Lage des Flugloches. O = Ofenrohr. B = Raum für den Beobachter.

(das Bienenauge ist für Rot unempfindlich); die Annäherung eines heißen Lötkolbens, dessen Wärmestrahlung viel stärker war als jene der Lampe, hatte nicht die geringste orientierende Wirkung.

Nach diesen Erfahrungen mit der Lampe liegt die Deutung nahe, daß die Bienen auf der horizontalen Fläche die Richtung zum Futterplatz anzeigen, indem sie beim Schwänzeltanz denselben Winkel zum Sonnenstand einhalten wie bei ihrem Flug zum Futterplatz. Die Frage ist nur: *wie können sie unter dem schattenspendenden Schutzdach den Sonnenstand wahrnehmen?*

Vielleicht erkennen sie ein Helligkeitsmaximum am Himmel in der Richtung, in der zur Zeit die Sonne steht? Diese Annahme wurde durch folgende überraschende Erfahrung umgestoßen: der Stock lag horizontal im geschlossenen Dämmzelt (Fig. 22), die Tänze waren desorientiert; nun entfernte ich nur die Dämmplatte der Nordseite. *Das Helligkeitsmaximum lag im Norden*, also in ganz anderer Richtung als die Sonne. Trotzdem *waren die Tänze sofort richtig orientiert* und wiesen nach Westen, wo der Futterplatz lag.

Die Bienen konnten sich in diesem Falle nicht nach der Einfallsrichtung des Lichtes orientiert haben. Bei der Suche nach einer anderen Erklärung war an die Möglichkeit *kinästhetischer Wahrnehmungen* zu denken. Zwar spricht das Versagen der Orientierung bei Lichtabschluß gegen eine solche Deutung — aber vielleicht ist eine gewisse Helligkeit nötig, damit die Biene alle ihre Drehungen und Wendungen richtig erkennt und im Gedächtnis behält. So unwahrscheinlich dies klingt, ich wollte es prüfen, und bespannte zu diesem Zweck das Gerüst um den Bienenstock statt mit den Dämmplatten mit weißen Leintüchern. In dem sonnenbeschienenen Leinenzelt war es sehr hell. Trotzdem waren die Tänze vollkommen desorientiert. Es kam offenbar nicht auf die Helligkeit an, sondern die Bienen *mußten nach irgendeiner Seite freien Ausblick haben*, wenn sie ihre Tänze richtig nach dem Sonnenstand einstellen sollten.

Dieser Ausblick muß nicht groß sein. Es war nicht nötig, eine Seitenwand des Dämmzeltes ganz wegzunehmen. Es genügte schon ein 10 cm breiter Spalt, um die Orientierung der Tänze herbeizuführen. Im Bestreben, den wirksamen Bereich weiter einzuengen, kam ich zu folgender Versuchsanordnung: In die nördliche, schräg gestellte Seitenwand des Dämmzeltes wurde ein Ofenrohr von 15 cm Durchmesser und 40 cm Länge so eingesetzt, wie Fig. 25 zeigt. Die innere Öffnung war auf die Stelle der Wabe gerichtet, wo die Bienen tanzten. Von da aus war durch das Ofenrohr nichts anderes zu sehen als *ein Stück blauen Himmels*. Dies genügte für die Orientierung der Tänzerinnen. Sie richteten ihren Schwänzellauf nach Westen und zeigten hiermit die Richtung zum Futterplatz. Wurde die Öffnung des Ofenrohres durch einen Deckel verschlossen, so waren die Tänze ungerichtet. Nach Entfernung des Deckels war die Richtungsweisung wieder da. Ich schließe daraus, daß die Bienen an dem Ausschnitt blauen Himmels eine nach der Sonne ausgerichtete Erscheinung wahrnehmen, nach der sie sich zum Sonnenstand orientieren können. Vielleicht ist es *die Polarisation des Himmelslichtes*, die ja sehr bestimmte Beziehungen zum Sonnenstand aufweist.

Für diese Annahme spricht, daß die Tänzerinnen unsicher wurden, wenn der Himmel vor dem Rohr verschleiert war, und überhaupt nicht mehr in der Richtung des Futterplatzes tanzten, wenn das Rohr nach einer Wolke blickte. Wolken depolarisieren das Licht. Die Tänze waren allerdings auch unter diesen Umständen gerichtet, aber in anderer Weise. Das helle Loch wirkte so, wie im geschlossenen Dämmzelt die seitliche Lampe: die Tänzerin hielt nun beim Schwänzellauf den Winkel zur Lichtquelle ein, den sie beim Flug zum Futterplatz gegenüber der Sonne eingenommen hatte. Sobald die Wolke wich und blauer Himmel vor dem Rohr erschien, wiesen die Tänze wieder direkt die Richtung zum Futterplatz. Ob die Sonne selbst, vom Standort des Bienenvolkes gesehen, zur Zeit der Beobachtung frei oder durch Wolken verdeckt ist, ob also der Stock im Sonnenschein oder im Wolkenschatten liegt, spielt keine Rolle.

Die Annahme, daß sich die Bienen nach einer gerichteten Erscheinung am blauen Himmel orientieren, wird ferner durch folgenden Versuch gestützt: ich brachte vor dem Ofenrohr einen Spiegel an, so daß auf der Wabe statt eines nördlichen Himmelsausschnittes das *Spiegelbild* eines südlich gelegenen Himmelsteiles sichtbar wurde. Die Tänze verliefen nun zwar merklich gehemmt, aber eindeutig nach der spiegelbildlichen Richtung und *wiesen nach Osten*. Das war offenbar nicht dem Spiegel an sich, sondern der Bildumkehr zuzuschreiben. Denn wenn durch einen zweiten Spiegel *doppelt gespiegeltes* Nordlicht durch das Rohr geschickt wurde, wiesen die Tänze wieder nach der richtigen Seite, *nach Westen*.

Da die Bienen an einem blauen Himmelsfleck die Stellung der Sonne erkennen, an einem Wolkenfleck aber nicht, und doch auch bei geschlossener Wolkendecke den Sonnenstand wahrnehmen (vgl. S. 18), muß man wohl eine Orientierung nach verschiedenen optischen Methoden annehmen, gemäß dem Prinzip der doppelten Sicherung. So ist die Richtungsweisung ein größeres Problem, als es zuerst den Anschein hatte. Anderseits können wir als eine Klärung und Vereinfachung die Erkenntnis buchen, daß die Bienen die Richtung zum Futterplatz sowohl auf der vertikalen wie auf der horizontalen Wabe nach dem *Sonnenstand* weisen — nur daß diese Nachrichtenübermittlung auf der vertikalen Wabe auf die Richtung zur Schwerkraft transponiert wird, während sie bei horizontalem Tanz[16]) ohne diesen Umweg vonstatten gehen kann.

Es ist zunächst nicht abzusehen, ob und wie weit die hypothetische Wahrnehmung der Polarisation des Himmelslichtes auch beim freien Flug und bei der Orientierung im Gelände mitspielt. Das starr befestigte Facettenauge mit seinen nach allen Seiten gerichteten Ommatidien könnte, bei entsprechenden optischen Eigenschaften der Kristallkegel, einen geradezu idealen Apparat darstellen, um die Stärke und Richtung der Polarisation am ganzen Himmelsgewölbe gleichzeitig zu überblicken und dauernd im Auge zu behalten. Aber es wäre verfrüht, dem Leser einen duftigen Strauß aus verlockenden Blüten der Phantasie zu binden. Man muß warten, bis das herbere Kraut der Tatsachen herangewachsen ist.

2. *Die Entfernungsweisung.*

Über den *Rhythmus des Schwänzeltanzes auf horizontaler Wabenfläche* habe ich bisher nur wenige Aufzeichnungen.

Wenn die Tänze auf horizontaler Unterlage *orientiert* sind, also nach einer bestimmten Richtung weisen, dann steht ihr Rhythmus mit den auf vertikaler Wabe gefundenen Werten in Übereinstimmung. In zwei Versuchen war der Mittelwert aus 7 bzw. 9 beobachteten Tänzen 7,5 und 7,55 Wendungen je ¼ Minute. Die Entfernung des Futterplatzes betrug 240 m. Für diese Entfernung ist aus der Kurve Fig. 3 (S. 13), der die Beobachtungen bei normaler Wabenstellung zugrunde liegen, ein Durchschnittswert von 7,4 Wendungen abzulesen.

Wenn aber die Tänze *desorientiert* sind, somit nicht regelmäßig auf jeden Halbkreis ein Schwänzellauf folgt, dann versagt auch die Entfernungsweisung. Bei einem Versuch im diffusen Licht des Leinenzeltes fand ich als Durchschnitt 6,9 Wendungen je ¼ Minute bei der gleichen Entfernung von 240 m. Wie auf der vertikalen Wabe, ist offenbar auch hier eine zutreffende Entfernungsweisung an die regelmäßige Form des Schwänzeltanzes gebunden.

III. Die Tänze auf schräg gestellter Wabenfläche.

Auf der vertikalen Wabe weist der Schwänzeltanz den Winkel mit der Richtung nach oben, der beim Flug zum Futterplatz mit der Sonnenrichtung einzuhalten ist; auf der horizontalen Wabe weist der Schwänzeltanz direkt nach dem Futterplatz; nun wollen wir wissen, bei welcher Neigung der Wabenfläche die eine Art der Richtungsweisung von der anderen abgelöst wird, und ob der Übergang plötzlich oder allmählich erfolgt. Tabelle 8 bringt das Ergebnis einer darauf bezüglichen Versuchsreihe.

[16]) Unter normalen Umständen auf dem Anflugbrettchen vor dem Flugloch, also unter freiem Himmel.

Der *Neigungswinkel* des Stockes geht von der Normallage aus, so daß bei 0° die Wabenfläche vertikal steht, bei 90° horizontal liegt. Der *Tanzwinkel* ist hier, zur besseren Vergleichbarkeit der Zahlen, durchgehend von 1 bis 359° im Sinne des Uhrzeigers gezählt: „0°" bedeutet einen Schwänzellauf nach oben (bezogen auf die vertikale Wabenstellung) oder nach Norden (bezogen auf die horizontale Lage); „260°" bedeutet einen Schwänzellauf 100° nach links von der Richtung nach oben, oder (auf die horizontale Lage bezogen) nach 10° südlich von West. Unter „*Fehler*" ist im linken Stab die Differenz zwischen Tanzwinkel und Sonnenwinkel angegeben, also der Fehler nach der Richtungsweisung auf vertikaler Fläche, im rechten Stab die Abweichung der Richtung des Schwänzellaufes von der Richtung zum Futterplatz, also der Fehler nach der Richtungsweisung auf horizontaler Fläche[17]). Der Stock stand mit dem Flugloch nach Westen und wurde für die Schräglagen nach Norden geneigt. Der Futterplatz lag 240 m westlich vom Stock.

Tabelle 8. *Richtungsweisung auf schräger Wabenfläche, Versuchsreihe vom 11. 8. 1947. Verschleierter Sonnenschein.*

Versuch Nr.	Neigungswinkel	MEZ.	Sonnenazimut	Sonnenwinkel	Tanzwinkel	Fehler Abweichung v. Sonnenwinkel	Fehler Abweichung v. Richtg. z. Futterpl.	*n*
181	0°	9.14	116°	154°	165°	11°	—	10
182	45°	9.29	120°	150°	185°	35°	85°	10
183	60°	9.40	123°	147°	218°	71°	52°	10
184	75°	9.53	127°	143°	260°	117°	10°	10
185	90°	10.04	130°	140°	268°	—	2°	10
187	75°	10.20	135°	135°	199°	64°	71°	10
188	60°	10.36	140°	130°	140°	10°	130°	10
189	45°	10.48	144°	126°	143,5°	17,5°	126,5°	10
190	0°	10.58	148°	122°	123°	1°	—	10

Bei mehrmaligen Wiederholungen waren die Befunde recht verschieden. Es kam vor, daß die Bienen noch bei einem Neigungswinkel von 60°, ja sogar von 75° wie auf der vertikalen Wabe die Richtung mit Bezug auf die Schwerkraft wiesen; es kam anderseits vor, daß sie schon bei einer Neigung von 70° die Richtung direkt zum Futterplatz anzeigten. Auch in Versuchen, die kurz aufeinanderfolgen, kann das Ergebnis bei gleichem Neigungswinkel wechseln (vgl. Tabelle 8). Wodurch diese Verschiedenheiten bedingt sind, weiß ich nicht.

Wichtig scheint mir, daß kein plötzlicher Umschlag erfolgt, sondern daß sich in einem Übergangsbereich, der mehr oder weniger breit sein kann, die Richtung des Schwänzellaufes mit zunehmender Neigung der direkten Richtung zum Futterplatz allmählich nähert. Die Orientierung nach dem Licht setzt sich gegenüber der Orientierung nach der Schwerkraft immer stärker durch. Dies legt den Gedanken nahe, daß die bei vertikaler Wabenstellung beobachteten Mißweisungen in ähnlichem Sinne zustande kommen, indem die Richtungsweisung nach der Schwerkraft, bei geöffnetem Stock, durch die Belichtung der Bienen gestört wird.

[17]) Der auf der schrägen Wabe eingehaltene Winkel ist hierbei unverändert auf die horizontale Wabe übertragen, gibt also zutreffend den auf der Wabenfläche gemessenen Winkel an, um den die Biene von der zum Futterplatz weisenden Richtung abwich. Um die Differenz der Himmelsrichtungen wiederzugeben, wäre die Projektion der auf der schrägen Wabe eingehaltenen Richtung auf die horizontale Fläche zugrunde zu legen. Für das, worauf es hier ankommt, genügt aber die in der Tabelle gewählte Darstellungsweise.

IV. Deutung der „Mißweisung".

Es war im 4. Abschnitt des I. Teiles (S. 19) davon die Rede, daß die Bienen bei der üblichen, also lotrechten Wabenstellung oft einen erheblich anderen Winkel anzeigen, als nach unserem Schlüssel (Fig. 6, S. 16) zu erwarten ist, und daß in diesen „Mißweisungen" ein gewisses System liegt. Im II. Teil hat sich bei den Versuchen auf horizontaler Wabenfläche ergeben, daß die Bienen schon an einem kleinen Himmelsausschnitt, der weitab von der Sonne gelegen sein kann, den Sonnenstand zu erkennen vermögen. Schließlich haben wir soeben besprochen, daß sie auf schräg gestellter Wabenfläche, innerhalb gewisser Grenzen des Neigungswinkels, gleichsam einen Kompromiß schließen zwischen dem Tanzwinkel, den sie auf lotrechter Wabe mit der Richtung nach oben einzuhalten hätten, und der Tanzrichtung, die auf horizontaler Fläche direkt nach dem Futterplatz weist. Daraus ergibt sich folgende Deutung für die Mißweisungen: Beim Öffnen des Bienenstockes zum Zweck der Beobachtung fällt direktes Himmelslicht auf die Bienen und hiermit wird, auch wenn die Wabenfläche etwa nach Norden gerichtet ist, der Sonnenstand für sie erkennbar. Vielleicht schließt die Biene unter diesen Umständen auch auf der lotrechten Wabenfläche einen Kompromiß zwischen der Richtungsweisung mit Bezug auf die Schwerkraft, wie sie im geschlossenen, finsteren Stock geboten ist, und der Richtungsweisung mit unmittelbarem Bezug auf den Sonnenstand, wie auf der belichteten horizontalen Wabe.

Um zu prüfen, ob Sinn und Ausmaß der Mißweisungen mit dieser Deutung in Einklang stehen, war es nötig zu klären, welche Richtung und welchen Winkel zur Sonne die auf der lotrechten Wabe tanzenden Bienen bei jedem Versuch eingenommen haben. Dies wurde durch einen einfachen Apparat ermöglicht, der es gestattet, die Stellung der Wabenfläche im Raum sowie das Azimut und die Höhe der Sonne für jeden Versuch mit wenigen Griffen einzustellen und sodann die auf die Wabenfläche projizierte Richtung der Sonne durch den vom „Sonnenzeiger" entworfenen Schatten an einem Teilkreis abzulesen[18]).

Das Ergebnis einer Überprüfung der 163 Versuche über Richtungsweisung, die für unsere Frage in Betracht kommen, war, daß die Bienen tatsächlich bei geöffnetem Stock auf lotrechter Wabe *angenähert in der Resultierenden tanzen zwischen den beiden Richtungen, die sie beim Transponieren des Sonnenwinkels auf die Richtung nach oben, und die sie bei unmittelbarem Einhalten des Sonnenwinkels* — wie auf horizontaler Fläche — *einzuschlagen hätten.*

Es würde über den hier gesteckten Rahmen hinausgehen, dies im einzelnen darzulegen. Es sind ja auch diese Dinge aus Versuchen, die mit anderer Fragestellung angelegt waren, nachträglich herausgelesen und müssen nun mit Anordnungen, die unmittelbar aufs Ziel gerichtet sind, neuerlich untersucht werden. Darum möchte ich mich hier auf den Hinweis beschränken, daß die im Abschnitt über die „Mißweisung" erwähnten Regeln, die zunächst so unverständlich, ja sinnlos schienen, durch diese Deutung ihre ungezwungene Erklärung finden.

Für die Zukunft ergibt sich die Notwendigkeit, bei der Beobachtung der Tänze das Himmelslicht auszuschließen, wenn man die Richtungsanzeige von der „Mißweisung" befreien will. Erweist sich so die Erscheinung im Grunde als bedingt durch einen Versuchsfehler bei der Beobachtung, so bringt sie uns doch eine Bestätigung größten Umfanges für den so wenig glaubhaft scheinenden Befund, daß die Biene wirklich schon aus einem Stück blauen Himmels den Sonnenstand entnehmen und sich danach richten kann.

Eingegangen am 16. März 1948.

[18]) Herrn Prof. H. Benndorf, Graz, bin ich für die Konstruktion und Herstellung des Apparates nach meinen Wünschen zu Dank verpflichtet.

©1986 Macmillan Magazines Ltd. Reprinted from *Nature,* Vol. 323, 128–131 (September 1986).

Polarization vision in bees

Samuel Rossel & Rüdiger Wehner

Department of Zoology, University of Zürich, Winterthurerstrasse 190, CH-8057 Zürich, Switzerland

A solution to the long-standing problem of how honey bees detect polarized light from the sky is presented. The mechanism involves the transformation of polarization information into modulations of perceived brightness while the bee scans the sky by rotating its field of view. Using this technique the bee needs only a very simple strategy to read compass information from the polarization patterns in the sky.

KARL VON FRISCH discovered that honey bees can detect linearly polarized light and can use this ability to derive compass information from the Sun-linked polarization patterns in the sky[1-3]. Subsequently, the basis of polarization sensitivity in the retina of bees and other invertebrates was shown to lie in the oriented arrays of pigment molecules packed in the membranes of the layered microvillar structure of the rhabdomeric photoreceptors, such that the sensitivity of the receptors is greatest when light is polarized parallel to the microvilli (see ref. 4 for review). Furthermore, it is known for bees that a small, dorsal region of the retina (the POL area) is both essential and sufficient for navigation by means of polarized sky light[5] and that this specialized retinal area contains an array of polarization-sensitive ultraviolet photoreceptors which equip it for this task[6-9]. One outstanding problem in the elucidation of polarization vision is how bees use information from the receptor array to analyse and orient by the patterns of polarized sky light.

Here we present a novel and comprehensive solution to this problem by showing how the array of receptors forms a template which the bee uses to scan and match the polarization patterns in the sky. We begin with a brief review of our recent model for the detection of polarized light and go on to report experiments designed to test the model. We conclude that the mechanisms uncovered may be generally operative in polarization-sensitive invertebrates.

Polarized light detection model

Imagine that in each eye of the bee the orientation of microvilli of polarization-sensitive photoreceptors varies in a graded manner through 180°, being horizontal in the anterior and posterior part of the eyes and oriented at all other angles between these positions (Fig. 1). Consider next what happens to the output of such an array when the bee, when presented with a small stationary patch of polarized light, rotates about its vertical body axis. The patch is positioned at 60° above the horizon and stimulates successively a ring of receptors as the bee rotates. The angle between microvilli and the *e*-vector varies gradually during the turn. As a consequence, the output of the array is a modulating signal passing through a peak as the microvilli become parallel to the *e*-vector in the patch. A similar modulation is detected by the bee when the *e*-vector direction is changed, except that the modulation now rises to a peak in a different position on the retina. Thus, we assume that for the analysis of *e*-vector directions in the sky, the bee sweeps its retinal analysers across the *e*-vectors and records which part of the retina is maximally active during the sweep.

This simple model of *e*-vector detection arose from behavioural experiments in which we analysed orientation performance of the bee to small spots of polarized sky light[10-13]. The implications of these experiments are most easily explained if we consider how the scanning mechanism might be used by bees to obtain compass bearings from the polarization patterns in the sky. Consider first a polarization pattern as it is realized

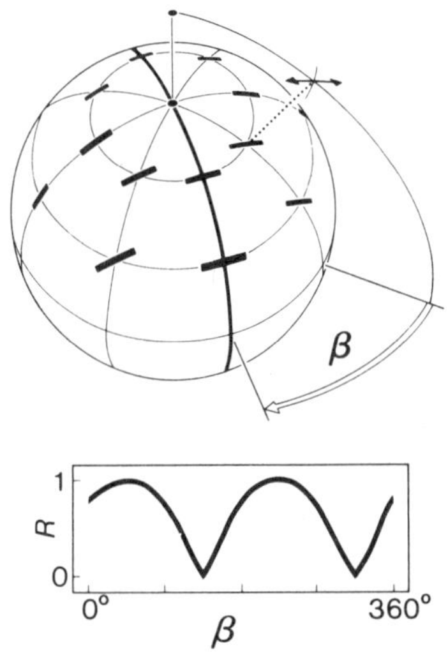

Fig. 1 Mechanism of detecting polarized light by bees. Sphere, visual fields of the left and the right eye (separated by thick continuous lines). The systematic variation of microvillar (analyser) directions of polarization-sensitive photoreceptors is shown for two elevations above the horizon (30 and 60°). Note that the dichroic visual pigments are oriented parallel to the microvilli so the sensitivity of the receptors is greatest when the *e*-vector of polarized light is parallel to the microvilli and least when it is at right angles to them. The operational principles of the model are demonstrated for a situation in which the bee views a patch of polarized light exhibiting a single *e*-vector direction (double arrow). The bee is first aligned with the patch ($\beta = 0°$) and then performs a full turn about its vertical axis ($\beta = 0$–$360°$), thus sweeping a ring of receptors with different microvilli directions across the *e*-vector. The relative output (*R*, log scale) of successive receptors within the ring is plotted as a function of β. A peak output is induced by the patch of polarized light when its *e*-vector direction is matched by a suitably arranged microvillar direction. The rationale behind the scanning model is that each *e*-vector direction has its own location on the retina, specified by a peak output of a corresponding photoreceptor.

at dawn and dusk (Fig. 2*a*). When the model bee is aligned with the solar and the antisolar meridian of such a pattern, the layout of polarization analysers over the eyes matches closely the distribution of *e*-vectors in the sky. Given this coincidence, the bee can rotate about its vertical body axis and assume that whenever there is a peak in the output of the *e*-vector detectors, the body is aligned with the solar and the antisolar meridian. It can then turn from this reference direction until it faces the desired compass course.

The situation changes during the day when the Sun is high in the sky (Fig. 2*b*). In this case there are substantial discrepancies between the *e*-vector pattern and the array of polarization analysers. Thus when the bee, while scanning a small patch of

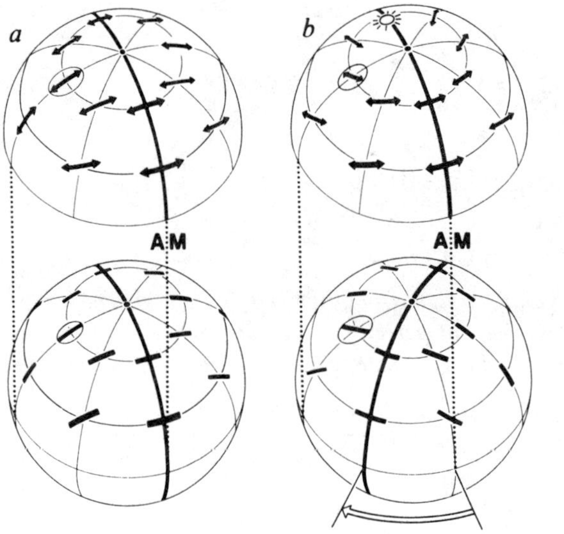

Fig. 2 Use of the scanning model in *e*-vector pattern orientation. ***a*,** Upper hemisphere, distribution of *e*-vectors (arrows) across the sky when the Sun is at the horizon. The celestial sphere is viewed from the back (AM, antisolar meridian). The pattern of polarized light depends strictly on the elevation of the Sun. As a general rule the angle of polarization (*e*-vector direction) is always perpendicular to the plane containing the Sun, the patch in the sky and the observer in the centre of the celestial hemisphere. The lower sphere shows the layout of analysers within the eyes of the bee. Note that the analysers match the *e*-vectors when the bee is aligned with the solar and the antisolar meridian. Thus, to determine the position of these meridians from one or another *e*-vector in the sky (in the figure exemplified for a vertical *e*-vector) the bee sweeps its field of view across the *e*-vector and assumes that it is aligned with the reference meridians when it detects a peak output from its analyser array. ***b*,** A different polarization pattern is realized in the sky when the Sun is at an elevation of 65°. When now presented with an isolated *e*-vector (examplified here by an oblique *e*-vector), the bee again assumes that it is aligned with the solar and the antisolar meridian when it perceives a peak intensity from the patch of polarized light. But, in this case, it is mistaken (open arrow) and ultimately steers an incorrect course.

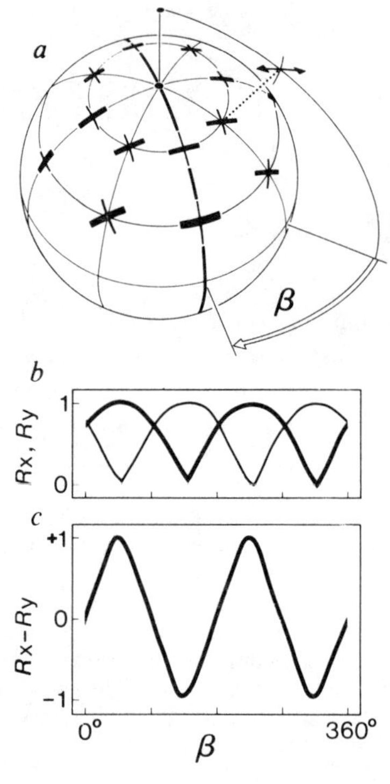

Fig. 3 Extended version of the scanning model. *a*, In addition to the array of receptors shown in Fig. 1 (here called x-type receptors) there is an array of y-type receptors with orthogonally arranged microvillar directions. Equipped with this array of crossed analysers, the bee scans a patch of polarized light displaying a single *e*-vector direction (double arrow). *b*, Output of the x- and y-type receptors (Rx and Ry, log scale) as a function of the horizontal orientation relative to the aximuthal angle β of the patch of polarized light. *c*, Output of the crossed analyser array as a function of β, when the outputs of the x- and y-type receptors are compared differentially by the visual system (Rx − Ry, log scale).

a daytime pattern, applies its simple method of determining the reference meridians, it may arrive at an incorrect estimate of the position of the meridians. This in turn should lead to errors of orientation expressing the discrepancy between the layout of analysers and the *e*-vector pattern. Our findings show that bees do indeed make consistent orientation errors, depending on the elevation of the Sun and the *e*-vector direction which is viewed by the bee in the sky. Therefore, the interpretation of the orientation data in terms of a scanning mechanism of *e*-vector detection allows us to infer the spatial layout of retinal analysers shown in Fig. 1.

These behavioural interferences are supported by anatomical and optical observations on the POL areas of the bees, which show that the microvillar directions of photoreceptors vary across the retina in the way predicted. However, the details of this correspondence have not been completely checked[7,9].

Finally, an important interference derives from combined anatomical and physiological studies, showing that each ommatidium of the POL area contains a set of orthogonally arranged microvillar directions[7–9]. To accommodate this, we have extended our model by adding a second array of receptors to the one already described (Fig. 3). We then suppose that the visual system compares the outputs of perpendicularly oriented receptors such that the differential response will be maximum when the microvilli of one receptor array are oriented parallel to the *e*-vector and least when the *e*-vector is parallel to the microvilli of the other[9,14]. Clearly, such antagonistic interactions of crossed analysers enhance the specificity of *e*-vector detection, but they are not essential in principle for the scanning model to work. In fact, it has already been shown theoretically that only one analyser is needed per ommatidium, if polarized light is to be analysed by a scanning method[15].

Our model can be tested by exploiting the ability of the bee to orient by means of small patches of polarized skylight. As can be deduced from Fig. 2, bees invariably treat a small patch of polarized light of a given *e*-vector as though it were at a fixed azimuthal angle relative to the solar and the antisolar meridian, although this is in fact only true of the pattern of polarization across the sky at dawn and dusk. For instance, bees orient to a patch of light with a horizontal *e*-vector as it lies along the antisolar meridian and to a patch with a vertical *e*-vector as though it is at right angles to the antisolar meridian. Thus we know both were a particular *e*-vector direction is represented within the retina and also what azimuthal angle the bee associates with a patch of light of that *e*-vector direction. The scanning model then predicts that a patch of unpolarized light arranged to evoke a peak output within a particular part of the retina will cause the bee to orient as though the light were polarized in the appropriate direction. Although this prediction is simple, a complex experiment is necessary to test it.

Experimental evidence

The standard method for analysing *e*-vector orientation is to see how polarized light influences the 'waggle dance' bees perform when they return to the hive after foraging. The orientation of this dance on the surface of the comb signals the direction

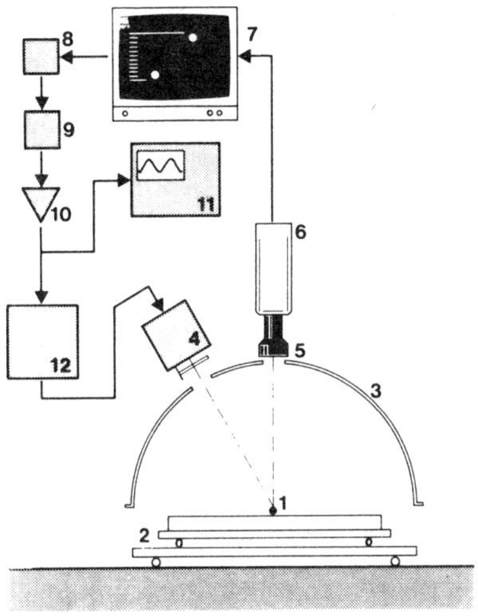

Fig. 4 Experimental device for testing the scanning model. The bee (1) performs its directional dances on horizontal combs (2) covered with a black Plexiglass hemisphere (3). Light from a Xenon arc (4), equipped with an ultraviolet filter (Schott; UG 11), is the only cue for dance orientation. The light stimulus subtends an angle of 10° and is at an elevation of 60°. The horizontal body-rotation of the dancing been controls the intensity of the Xenon arc. For this purpose two circular pieces (1 mm) of a reflecting foil (3M; 7610) are stuck onto the thorax and the abdomen of the bee, exactly 8 mm apart. The ring lamp (5), which surrounds the lens of a television camera (6), is then switched on and the aperture of the lens closed until the image on the television monitor (7) fades out completely, except for the two light spots reflected from the body of the bee. In each frame the number of lines between the first and the last image signals is counted. Because the distance between the two reflecting spots is constant, the number of lines provides a direct measure of the direction of the medial plane of the bee and hence can be used to control the intensity of the Xenon arc (8, line counter; 9, digital-to-analogue converter; 10, amplifier; 11, storage oscilloscope; 12, power supply of Xenon arc). Maximal and minimal intensities are produced when the bee is aligned either parallel or at right angles, respectively, to the vertical axis of the monitor. With respect to the medial plane of the bee, maximal intensities can be induced at any angle by rotating the camera relative to the aximuthal angle of the Xenon arc. The intensity between maximal and minimal intensity varies by a factor of three. All waggle dances are videotaped and later analysed by measuring the direction of individual waggle runs.

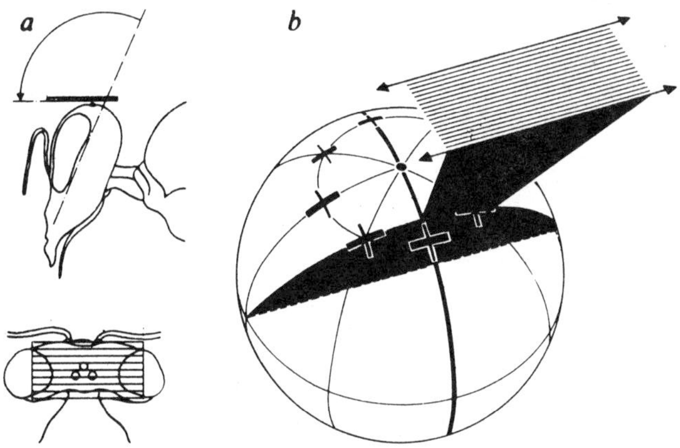

Fig. 5 Method for stimulating the receptors selectively. *a*, Side and dorsal view of the head of the bee that contains a piece of polarizing filter (Polaroid, HNP'B). The transmission axis of the filter is parallel to the transverse axis of the head. The inclination of the filter is defined relative to the back plane of the eyes. *b*, Model eyes with crossed analysers (*x*- and *y*-type receptors; Fig. 3) at an elevation of 60°, and part of the overhead filter. Throughout the array the microvillar directions of the x-type receptors are defined by the set of planes which include the transverse axis of the eyes (here one plane is shaded). In addition, these planes include the transmission axis of the filter. Therefore, light is polarized by the filter to stimulate predominantly the x-type receptors.

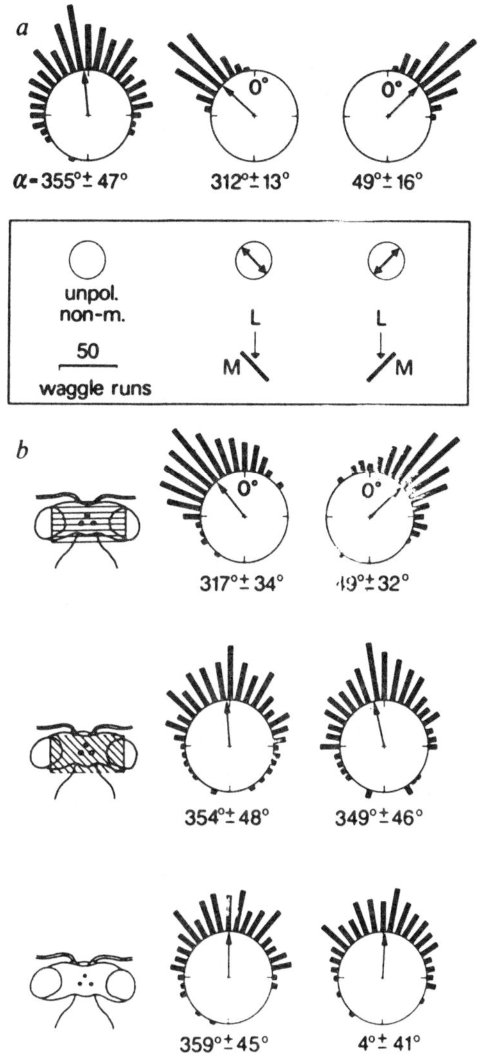

Fig. 6 a, Dance directions of bees when presented with a patch of unpolarized non-modulating ultraviolet light and a patch of polarized light with one of two oblique vectors (stimulus conditions are indicated in the upper half of the inset; *e*-vector directions are marked by double arrows). Mean dance directions (α, arrow) are calculated to indicate where the navigating bee expects the light stimulus to occur in the sky (0° marks the antisolar meridian Fig. 2). Data are based on 931 waggle runs recorded from 16 bees. *b*, Dance directions of bees when presented with a patch of unpolarized modulating ultraviolet light. The modulations are arranged to simulate the two *e*-vector directions presented to the bees in *a* (in the inset M is the orientation of the bees relative to the azimuthal angle of the light stimulus L when its intensity passes through a peak; compare with Fig. 1 for retinal positions of oblique *e*-vectors). In the upper and middle parts of *b* the dancing bees are equipped with an overhead polarization filter (the transmission axis of the filter is oriented parallel and at 45° to the transverse axis of the head). In the lower part of *b* the bees are not carrying overhead filters. Data based on 2,760 waggle runs arecorded from 24 bees.

of the foraging site from the hive. On vertical combs in the dark, bees orient with respect to the gravitational vector. On horizontal combs they will orient with respect to the Sun or the polarization patterns across the sky[3].

In the present experiments, several individually labelled bees were trained to a feeding station several hundred metres from the hive. The hive had horizontal combs and an artificial light source provided the trained bees with the only directional cue by which they could orient their dance. When given a single patch of polarized light in such a situation, the orientation of the dance varies with the *e*-vector direction. We sought to mimic this behaviour by means of a beam of unpolarized non-modulating ultraviolet light which is modulated in intensity according to the horizontal orientation of the bee, so that the intensity of the beam peaks when it stimulates a preditermined part of the retina. Details of the servo system are given in Fig. 4.

As we described above, each ommatidium within the POL area is equipped with pairs of orthogonally oriented polarization detectors which we assume interact to give a differential output. At first sight a modulated beam of unpolarized light appears to be an inappropriate stimulus because pairs of crossed analysers will be stimulated equally, so that the differential output will be zero throughout the modulation cycle. To circumvent this effect we selectively stimulated one of the two crossed receptor arrays on its own by covering the eyes with a small piece of polarizing filter oriented to transmit *e*-vectors parallel to the transverse axis of the eyes (Fig. 5). This filter transmits the set of *e*-vector planes which contain lines that are parallel to the transverse axis. Microvilli are oriented over the retinal region of interest so that light transmitted by the filter will stimulate one of the two receptor arrays more than the other. As a control, some bees are equipped with a filter with its transmission axis oriented 45° from the transverse axis, in which case the two crossed analysers should again be stimulated equally.

The light was modulated so that the beam was of peak intensity when it fell on one of two different retinal positions which according to the model contain detectors for one of two oblique *e*-vectors. The orientation of the dances under these conditions is shown in Fig. 6. Bees indeed behave as the scanning model predicts. With the transmission axis of the filter parallel to the transverse axis of the eyes, dances orient as though the bees are presented with one of the two oblique *e*-vector directions. In contrast, with the filter oriented at 45° from the transverse axis of the eyes, the effect disappears and the bees behave as though they are stimulated by a patch of unpolarized, non-modulating ultraviolet light.

Before these results can be considered to provide conclusive evidence in support of our model, we asked whether there is an alternative mechanism of *e*-vector detection which could produce the same results. Previous accounts assumed that each individual ommatidium of the compound eyes is able to assess and report independently on the *e*-vector direction of light falling within its receptive field of view (see refs 3, 16 for review). Thus one might argue that polarization orientation in our experiments is not affected by the intensity modulations of the ultraviolet beam, but is caused by the *e*-vectors projected onto the eyes by the overhead polarization filter. This possibility, however, can be excluded. Because the direction of the *e*-vectors projecting onto the eyes varies as both the bee and filter rotate, different ommatidia would give the bee conflicting and uninterpretable information about *e*-vector directions. Similarly, we would have expected the dance directions to be the same whether the filter was parallel to the transverse axis of the eyes or oriented obliquely to it.

Conclusions

We can now explain the principles underlying polarization vision in bees in terms of the arrangement of dichroic photoreceptors, the visual cues abstracted from these receptors and the influence this processing strategy has on the orientation behaviour.

Specifically, our data show that bees have an array of retinal analysers, each of which is maximally sensitive to a different *e*-vector direction. When a patch of polarized sky light is swept across the array, the perceived direction of its *e*-vector is determined by which analysers within the array produce the largest output. We also deduce from our experiments that *e*-vector detection by the array of analysers exploits antagonistic interactions between receptors with orthogonally oriented microvilli. We believe that such interactions provide the bee with a high-contrast *e*-vector signal.

The spatial layout of retinal analysers forms the structural basis of *e*-vector detection. In addition, it matches the distribution of *e*-vectors in the sky, when the bee is aligned with the solar and the antisolar meridian (Fig. 2). Thus to determine the position of these reference meridians, the bee turns about its vertical axis and detects when the array of analysers is maximally active.

This simple strategy of reading compass information from the polarization patterns explains the orientation performance of bees whose vision of the sky is restricted to small areas containing a very small range of *e*-vector directions. At dawn and dusk there is a good match between the sky pattern and the array of detectors, but during the day substantial discrepancies occur. Thus, when the bee applies its simple compass strategy it may arrive at an incorrect course, as we have previously described[10-13]. But this shortcoming is not as serious as it might first appear. Under normal circumstances the navigating bee has large areas of the polarization patterns at its disposal. Therefore, it can benefit from the fact that *e*-vectors in each celestial half (defined by the solar and the antisolar meridian) are mirror images of each other (Fig. 2). This implies that error angles induced by *e*-vectors from the left and the right half of the sky are always opposite in sign, and thus cancel each other out[12,13]. Furthermore, during free flight the navigating bee can resort to powerful backup systems such as the Sun and landmarks[3,17].

Finally, note that the structural pecularities that characterize the array of polarization analysers within the retina of bees seem to be common in invertebrates. The retinas of cephalopds, crustacens and insects have sets of receptors with orthogonally arranged microvilli (see refs 16, 18 for review). Furthermore, there is increasing evidence that the directions of microvilli vary in fan-shaped arrays across the retina in various groups of insects[9,19,20]. It seems likely, therefore, that the principles discussed here are generally applicable to invertebrates that detect and use polarized sky light for orientation.

We thank H. Baumann, K. Feller and T. Grob for design and construction of electronic equipment, H. Baumann, K. Cheng, S. Gut and A. Rossel-Jaeckle for assistance, D. Rigoli for the drawings, T. S. Collett and S. B. Laughlin for reading and improving the manuscript. The work was supported by Swiss NSF grant 3.665-0.84.

Received 26 March; accepted 10 July 1986.

1. von Frisch, K. *Naturwissenshaften* **35,** 38-43 (1948).
2. von Frisch, K. *Experientia* **5,** 142-148 (1949).
3. von Frisch, K. *The Dance Language and Orientation of Bees* (Harvard University Press, Cambridge, Massachusetts, 1967).
4. Laughlin, S. B., Menzel, R. & Snyder, A. W. in *Photoreceptor Optics* (eds Snyder, A. W., Menzel, R. 237-259 (Springer, Berlin, 1975).
5. Wehner, R. & Strasser, S. *Physiol. Ent.* **10,** 337-349 (1985).
6. Helversen, O. & Edrich, W. *J. comp. Physiol.* **94,** 33-47 (1974).
7. Sommer, E. W. Thesis, Univ. Zürich (1979).
8. Labhart, T. *J. comp. Physiol.* **141,** 19-30 (1980).
9. Wehner, R. *Himmelsnavigation bei Insekten. Neurophysiologie und Verhalten, Neujahrsbl. Naturforsch. Ges. Zuerich* **184** (1982).
10. Rossel, S., Wehner, R. & Lindauer, M. *J. comp. Physiol.* **125,** 1-12 (1978).
11. Rossel, S. & Wehner, R. *Proc. natn. Acad. Sci. U.S.A.* **79,** 4451-4455 (1982).
12. Rossel, S., Wehner, R. *J. comp. Physiol.* **A154,** 607-615 (1984).
13. Wehner, R. & Rossel, S. in *Experimental Behavioural Ecology* (eds Hoelldobler, B. & Lindauer, M.) 11-53 (Fischer, Stuttgart, 1985).
14. Waterman, T. H. & Horch, K. W. *Science* **154,** 467-475 (1966).
15. Kirschfeld, K. *Z. Naturforsch.* **27b,** 578-579 (1972).
16. Waterman, T. H. in *Handbook of Sensory Physiology, Vol. VII/6B* (ed. Autrum, H.) 281-469 (Springer, Berlin, 1981).
17. Dyer, F. C. & Gould, J. L. *Science* **214,** 1041-1042 (1981).
18. Waterman, T. H. In *Photoreception and Vision in Invertebrates* (ed. Ali, M. A.) 63-114 (Plenum, New York, 1984).
19. Burghause, F. *Zool. Jb. Physiol.* **83,** 502-525 (1979).
20. Wunderer, H. & Smola, U. *Int. J. Insect Morph. Embryol.* **11,** 25-38 (1982).

Polarization Patterns in Submarine Illumination

Talbot H. Waterman

Osborn Zoological Laboratory, Yale University, New Haven, Connecticut and *Bermuda Biological Station*, St. George's West, Bermuda*

ALTHOUGH the polarization of sky light has been known since 1809 and widely studied (*1, 2*), the comparable phenomenon in natural waters has scarcely been considered. The possible importance of this unexplored field has recently been emphasized by the discovery that polarized light has significant behavioral and physiological effects on a variety of arthropods. First demonstrated by von Frisch (*3*) in the honeybee, striking sensitivity to polarized light has since been found in a variety of insects and their larvae, crustaceans, and chelicerates.

Not only have these animals been shown capable of detecting polarized light, but also some of them have been found to use the natural polarization of the blue sky in navigating from one place to another. These facts have opened up a whole new field of visual physiology and behavior (*4*).

Among aquatic animals also polarized light sensitivity has proved to be widespread in the arthropods. The compound eye of *Limulus* was discovered to be sensitive to plane-of-light polarization, although so far the behavioral or ecological significance of such ability in this animal is unknown (*5*).

Behavioral reactions to polarized light have, however, been found by Baylor and Smith (*6*) in a number of fresh-water aquatic forms. Most interesting was their discovery that the swimming of cladoceran crustaceans like *Daphnia* could be predictably oriented by vertically or horizontally directed beams of plane-polarized light. Swimming was predominantly at right angles to the direction of the *e* vector in a vertically incident beam and could be repeatedly changed when the polarization plane was rotated. In a horizontal beam of polarized light, the crustacean's swimming direction was similarly related to the plane of polarization, being upward in a horizontally polarized beam and lateral, with the animals swimming on their sides, in vertically polarized illumination.

That polarized light of natural origin can have similar orienting effects on populations of these crustaceans was also demonstrated by these same workers, who briefly reported appropriate orientation of Cladocera exposed to the polarized light of the blue sky at sunset or sunrise. Homing on polarized shafts of penetrating light around shadows was predicted, too. The present data, however, indicate the importance for underwater polarization of the directionality of daylight illumination everywhere in the water. This presumably applies not only to the sea, where these observations were made, but also to any other extensive body of water, fresh or salt.

In view of these facts, then, the occurrence and pattern of light polarization in aquatic environments becomes a subject of wide potential relevance to various problems in oceanography and limnology. The basic principles of light polarization by scattering in gases and liquids have been well known since the pioneer work of Tyndall and Lord Rayleigh. Brief mention of the occurrence of polarized light in natural waters has been made (*7–10*), yet few if any, actual data are available.

The present observations on submarine illumination were, therefore, undertaken to outline the major characteristics of its polarization. The results indicate that to an even greater degree than terrestrial animals, aquatic forms of the photic zone are surrounded by a complex pattern of polarized light. For those that have eyes sensitive to it, this pattern may provide significant cues for orientation and migration.

The observations made so far have been carried out visually by the author, skin diving in the sea around Bermuda. Most of the data were taken in the surface 5 to 6 m, but a few measurements were made down to 15 m. Stations investigated included shallow turbid water in Ferry Reach, deeper more transparent areas in Castle Roads and the North Shore of St. George's Island, very clear oceanic water beyond the 2000-m contour of the South Shore, and cave waters of Harrington Sound. A face plate was employed for clear vision, and a special hand-held polarization analyzer was employed in examining the light from all directions. This device indicates by inspection the presence and plane of polarized light and at the same time allows an estimate of the degree of its polarization.

The polarized light detector utilized makes use of an interference pattern well known to crystallographers. The light being examined first passes in a convergent beam through a suitably thick plate of a uniaxial crystal cut perpendicularly to its axis. The light emerging from this is then circularly analyzed through a quarter wave plate and a disk of Polaroid film. The whole optic sandwich is cemented together and protected with glass covers.

If unpolarized light is viewed through this instrument, nothing is observed except a decrease in light intensity. If, however, plane-polarized white light is examined, a brightly colored interference pattern consisting of concentric, broken rings appears at the far point of vision. This figure subtends a visual angle of about 15° in the unit used. In partially polarized light, the intensity of this pattern and the number of concentric rings observable vary with the percentage polarization. Thus with strongly polarized, moderately intense light, 30 or more rings can be seen. With weakly polarized light of the same intensity, only a few rings, perhaps 6 to 8, are visible. This device, then, provides a rough but ready estimate of the degree of polarization. So far no precise method of calibration has been attempted.

The interference pattern further indicates the planes of the *e* and *h* vectors of the polarization. As mentioned in the preceding paragraph, the concentric rings are broken. The interruptions occur in quadrants and establish a right cross, the arms of which define the orientation of the polarized light traversing the analyzer. At the center of the pattern two diametrical quadrants are dark, and two are light, This relationship provides a ready means of discriminating the *e* and *h* vectors, since the position of these quadrants is specifically related to the plane of polarization and the kind of crystal plate used.

Most of the time no close measure of the polarization angle was made. However, in a few cases, the analyzer was used with a plumb line for zenith reference and a protractor for measuring the angle. Even so, accuracies greater than ±2° to 3° would not be expected, because sights were all taken while swimming freely. At least for exploratory purposes the superiority of the present analyzer to a complicated prismatic polarimeter or a Savart plate should be obvious. Rapid, convenient, reproducible estimates of degree and plane of polarization were thereby accessible for inspection or photography.

In general the daylight reaching any observed submarine point from almost all possible directions was found to be partially polarized. The pattern of this polarization, described in detail in the following paragraphs, is shown diagrammatically in Fig. 1. There were at least two major components of this polarized light.

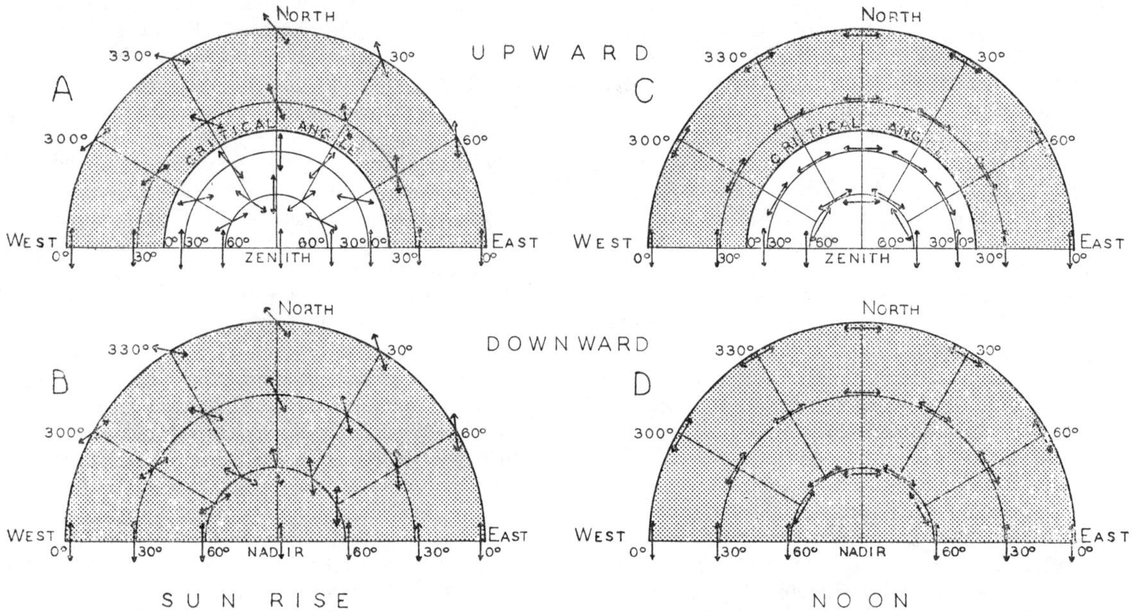

Fig. 1. Polarization planes of submarine illumination as seen from a point a few meters under water at two different times of a clear cloudless day (diagrammatic). In *A* and *B* the pattern shown is that observed at sunrise with the sun on the horizon directly east. In *C* and *D* the pattern is that observed at noon with the sun directly overhead in the zenith. The diagrams are plotted on polar coordinates with azimuths represented by radii, and elevation or depression of the observer's line of sight as concentric circles. The outermost circle (0°) in each case is the underwater horizontal—that is, the edge of a plane passing through the observer and normal to the vertical. Diagrams *A* and *C* show polarization appearing when the line of sight is raised upward toward the zenith; *B* and *D* show polarization planes seen when it is depressed downward to the nadir. In looking upward, when the critical angle (about 42° elevation) is reached, the sky and its polarization are seen, as they are at all greater elevations. The areas showing sky polarization as seen underwater are unshaded; those where the polarization is produced by sunlight in the water itself are gray. The numbers labeling angles between the critical angle and the zenith are altitudes in the sky measured from the water-level horizon (0° inside the semicircles). These are distorted by refraction, as the plot shows. In all cases the polarization planes are represented by vectors whose angular relationship to the azimuthal radius that they intersect is the same as that between the *e* vector of the polarized light and the vertical at the point plotted. The pattern of light polarization in the observer's other visual hemisphere would be a mirror image of that in the hemisphere diagramed. Intensities of polarization are not represented in this figure but vary markedly as described in the text.

1) One of the components originated in the blue sky and was readily visible at depths of 5 to 6 m or less. As is well known, the surface of the sea acts as a wide-angle lens which provides about a 180° aerial field of view to an underwater observer. The whole water-level horizon and the celestial hemisphere are seen restricted within a circular area subtending an angle of just over 96° centered around the vertical. Within this area, determined by the angle of total reflection, any atmospheric polarization was readily detected and found to be the same as when observed in air except, of course, that apparent elevation was distorted by refraction and surface roughness.

For this reason then the pattern of polarization reaching any submarine point (at least within a few meters of the surface and from directions included by a 96° to 97° angle symmetrical around the zenith point) will be determined by factors the same as those influencing the sky's polarization. These primarily are relative position of the sun and amount of overcast but secondarily include distance of the point observed from the zenith, the earth's albedo, amount of atmospheric dust, multiple scattering and depolarization owing to anisotropy of air molecules (*2, 11*). In addition, these rays may be modified by scattering and absorption at the surface and by the water itself. However, in the present measurements, none of the numerous secondary factors had sufficient effect to be noted.

In the few observations made at 15 m, neither the sky nor its polarization was visible in the turbid water concerned.

2) The other major component of polarized submarine light, under the conditions stated, was found to arise outside of the angle of total reflection and to originate from the water itself. In depths of 5 to 6 m, this underwater polarization could be observed in all directions not included within the critical angle and a small transition region immediately outside this where some elliptical polarization appeared to be present. At 15 m this component of the polarization was seen in all directions.

This strictly aquatic polarized light was found in turbid shallow water containing much suspended matter and in the far clearer oceanic water, where the absorption coefficient would be intermediate between that of the clearest known sea water and that typical of western North Atlantic continental slope regions (*12*). In these highly transparent waters the degree of polarization was found to be considerably greater than in the shore areas under the same conditions of illumination. With the sun close to the zenith at the deep-water oceanic station, the water's polarization in a horizontal direction was even appreciably greater than the polarization of the sky near the horizon in the same azimuth (*13*). The blue of the oceanic water, however, appeared far more luminous and brilliant than that of the sky, so that quantitative comparison was not feasible by the visual method described.

Besides being influenced by the clarity of the sea water, the intensity of submarine polarization was found to be affected by the position of the sun relative to the line of sight. When the sun was close to the zenith maximum polarization was found in the water observed horizontally. Depressing the direction of vision downward toward the antisun decreased the amount of polarization to zero at the latter point which lay within the observer's underwater shadow directly below. Raising the line of sight toward the angle of total reflection also decreased the polarization to a fraction of the total light.

When, on the other hand, the sun was close to the horizon at sunrise or sunset, the distribution of polarized light intensities was quite different. Maximum polarization then was found directly downward and along an arc outside the critical angle and at all points 90° from the bearing of the sun. Minimum polarization with level lines of sight occurred in the sun's azimuth and also toward the antisolar point. Therefore in sweeping the direction of view around horizontally, the percentage of polarization varied continuously passing through minima in the bearings mentioned and maxima at azimuths 90° from the sun.

In addition to its intensity and its degree of polarization, the plane of the light's polarization was discovered also to vary systematically with the relative position of the sun and the point observed underwater. When the sun was near the zenith, the electric vector of the polarized light was horizontal and the same at all azimuths.

At sunset, on the other hand, the plane of polarization in the sun's azimuth was horizontal. As the direction of view made an increasingly large angle with the sun's azimuth the plane of polarization tilted more and more toward the west. When it was observed 90° from the sun this tilt was about 45°. Turning still farther from the sun, the westward tilt decreased to zero 180° from the sun's bearing. At sunrise a similar pattern of polarization was found, except that the slant of the polarization plane was toward the east. At either sunrise or sunset, when the observer looked directly down in the water, the magnetic vector of the polarization coincided with the great circle through the sun, the point observed and the antisolar point.

On the basis of the data reported, it seems obvious that the polarization of light underwater is primarily determined, as one might expect (*8*), by factors similar to those that determine the underlying pattern of sky polarization (*14, 15*).

In this case the degree of polarization, assuming it all to be due to isotropic molecular scattering, is given by the relationship, $P = \frac{\sin^2 \theta}{1 + \cos^2 \theta}$, where P is the fraction of total light polarized and θ is the angle between the sun's direction and the line of sight. Obviously P varies from 0, when $\theta = 0°$, to 1, when $\theta = 90°$. Similarly, the *e* vector of the scattered light is at all points normal to the direction of the impinging light rays.

Note, however, that superimposed on these fundamental similarities to sky polarization the pattern of polarized light in natural waters will show an important difference—namely, as long as the light originates out of the water, the direction of the primary rays, which are to be scattered, will be altered by refraction. The resultant amount of deflection varies with the angle of incidence. Thus with a flat calm the refractive effect at the sea surface is nil for vertical incidence. It is maximum near sunrise and sunset when the angle of incidence is 90° and, according to Snell's law, the angle of refraction is only 48°20′ (estimated at Bermuda salinities and temperatures).

There are a number of specific observations, in addition to the general facts already presented, which give further support to the scattering theory for the origin of the primary pattern of polarized submarine illumination. Most instructive was the examination of polarized light distribution in caves and under deep overhanging ledges, Here it was possible to compare the plane of polarization in water that was illuminated by direct sunlight and water that was illuminated by an aperture lighted generally by sky and clouds but not directly by the sun. In suitably chosen areas, it could be demonstrated dramatically that the direction of the penetrating illumination was the main determinant of the plane of underwater light polarization.

Thus, in one particular case, this was shown directly by my swimming a few feet from open water into water overhung by a large rock ledge. In the open sunlit water the plane of polarization, as seen looking north and horizontally, was tilted about 35° toward the west. This tilt corresponded with the angle of refraction of the sun's rays and the direction of view. In the shaded water, where the rock overhead cut out all illumination except from the eastern third to half of the sky, there was a tilting of the *e* vector toward the east by 30° while the line of sight was still level and to the north. Therefore, the plane of polarization differed by 65° in these differently illuminated areas, although the sun's position and the line of sight were the same. This is a crucial observation in support of the scattering hypothesis.

Somewhat similar effects have been seen underwater on several instances when heavy overcast of limited extent obscured the direct rays of the sun. For example, on one occasion in moderately clear water, the following was observed in a horizontal direction 90° from the sun's azimuth. The sun was close to the horizon, and the plane of submarine polarization was tilted west by about 40°. While this was being observed a dense cloud drifted in front of the sun, although most of the rest of the sky was clear. At this time the tilting of the *e* vector became less pronounced by 10° to 15°. A return to the initial condition, when the sun again came out from behind the cloud, provided a good control.

Clearly the alteration in net direction of the total incident illumination produced a concomitant shift in the polarization. This is also supported by observations under a thick general overcast when the plane of submarine polarization was found to be horizontal in all azimuths. When the cloud cover was thin or incomplete, the obliquity of the polarization plane depended on the position of the sun and the contribution of its direct rays to the over-all underwater illumination.

Finally, some more quantitative measurements of the actual orientation of the polarization plane in the water under different specific conditions should be mentioned. At sunset the deviation of the polarization from horizontal was 48° toward the west, 90° from the sun. This is well within the observational error of the theoretical value obtained with the foregoing assumptions. Similarly, when the sun's zenith distance was 35°, the measured displacement of the interference pattern was 28°, while the predicted value on the basis of simple refraction would be 25°, again within the experimental error.

The observations described in foregoing paragraphs demonstrate that submarine illumination, at least at the depths observed, is highly polarized in patterns primarily related to the directional incident light from the sun and sky. Several additional points of interest arise from these findings. These relate, on the one hand, to problems in physical oceanography and, on the other, to the biological implications of the data.

At first glance one of the most surprising aspects of the pattern of submarine illumination is the lack of evidence that surface reflection plays any primary part in it. On the basis of simple refraction and reflection at the air-water interface, one would expect strong vertically polarized components, particularly when the sunlight is incident at Brewster's angle (53°15′ at Bermuda salinities and temperatures).

Vertical polarization has, in fact, never been observed underwater. The only possible part of the polarization pattern that does not seem directly related in its origin to scattering of the incident light occurred, as mentioned earlier, in an annular area near the angle of complete reflection. Here the interference pattern gave evidence of elliptical polarization of the light. This may have been related to polarization by refraction and reflection at the surface.

The absence of a marked effect of the water surface on the submarine polarization may be the result of the complex nature of the reflection, refraction, and scattering that occur there. In the first place, the surface is not usually smooth, so that simple Brewsterian reflection may not be prominent. This kind of rough surface effect may be observed in model form by comparing the polarization of reflected light by a smooth glass surface and by a scattering-reflecting surface such as the cloth binding of a book. In the former case, as in a flat calm at sea, light reflected is highly polarized horizontally and follows Fresnel's formulas. In the case of the scattering surface the polarization plane may be tilted far from the horizontal and toward the source in highly directional illumination if this is not normal to the surface in the line of sight.

Furthermore, the light reflected from the sea surface is partly polarized by the blue sky which it reflects (*7, 16*). Thus the surface polarization of the total reflected light and, hence, that of the penetrating light will have at least three major sources whose relative importance will vary under different conditions and angles of incidence. The present observation that the surface polarization ordinarily has a small effect on the over-all submarine polarized light is consistent with the more general finding that surface scattering and reflection, except at high angles of incidence, have negligible effects on light penetration as a whole (*8, 17*).

If these surface phenomena, or any other sources of polarized light than Rayleigh scattering, are significantly involved in the polarization of underwater illumination, they should be revealed by a more detailed study of the pattern of polarized light. Interaction between light rays polarized in different planes would produce idiosyncrasies in the pattern analogous to the neutral points and lines in the blue sky (*18*). As mentioned earlier, such have not yet been observed underwater with the present technique.

Another matter of considerable interest is the question of how penetration into deep layers affects the polarization of submarine light. If, as the present data indicate, polarized light underwater originates mainly by Rayleigh scattering of beams of penetrating light, one would expect polarization to occur as deep as significant directionality is present. Although scattering tends to decrease directionality, most of the scattered light in sea water is directed forward (*19*). Photographic measurements at 400 m in very clear water demonstrated that the ratio of total horizontal to vertical illumination was still 60 percent (*20*). It seems clear then that directionality of illumination is maintained to great depths, in terms of the total distance of submarine penetration of light. On this basis one would predict the presence of polarization produced by Rayleigh scattering at all depths where significant light from the surface is present.

From the known data on the directionality of light penetration in general, one would also predict that the polarization pattern would become simpler and increasingly stable in time in deep water. These differences from the polarized light patterns near the surface would arise from the antagonistic interaction of scattering and absorption which eventually should produce an equilibrium pattern of illumination centering around the vertical (*21*).

Actual observations have shown that, while directionality is maintained, average obliquity is reduced and the angular spread of the radiation is broadened with depth (*22*). Thus at some depth characteristic for the optical properties of the water column concerned, the directional pattern will center about the zenith and be independent of the sun's altitude. In the clearest water this equilibrium point, which is reached gradually, has been estimated by Jerlov (*22*) to occur at about 300 m. In less transparent water it would occur nearer the surface. One would expect, therefore, that as depth increases, the light polarization pattern would become, at all times when there was adequate illumination, increasingly like the pattern in the upper layers, when the sun is in the zenith, or even more like that with a completely overcast sky.

The physical optic problem of the precise origin of submarine polarization by scattering is also an interesting matter that needs further study. If the scattering is molecular, then the polarization and the brilliant blue color of very clear oceanic waters are linked together as they are in the sky (*9, 23*). Upward scattering of daylight in highly transparent oceanic water has been found to have a maximum at 425 mμ (*22*). This is the value one would predict on the basis of molecular scattering by the water. On the other hand, the color of clear ocean water corresponds with the wavelength of greatest transparency, 467 mμ (*22*), and at least in inshore waters the scattering is predominantly the effect of transparent mineral particles (*19*). Perhaps the present findings on submarine light polarization may stimulate further work in this interesting and still somewhat controversial field (*7, 8, 24*).

Another area where further research is required relates to the whole question of the biological effects of a complex pattern of polarization such as actually occurs in natural waters. Not only is the region within the critical angle affected by sky polarization, but the water itself has a complicated pattern dependent on the obliquity of incident light. This pattern is omnipresent during the day in surface waters, and, as outlined in foregoing paragraphs, at least the part originating from scattering in the water itself probably extends with modifications into the lowest layers of the photic zone.

Since bathypelagic animals are orienting to light and undertaking extensive diurnal vertical migrations down to 600 to 800 m (*25*), the possibility that light polarization may be involved in cuing such activity should not be overlooked. Above the level of the equilibrium directional distribution of light the movement of the sun through the sky and changes in sky overcast would be reflected by characteristic changes in the polarization pattern. At the present time we know that the polarization patterns are there. We also know that polarized light sensitivity and corresponding behavior patterns also occur widely in one phylum at least. To what extent the natural behavior of such animals may be modulated or initiated by these factors is still largely unknown.

To answer this question, more must be known about the physiology of polarized light sensitivity and arthropod vision in general (*26*). Further data on polarized light behavior patterns must be gathered, and their relationship to the distribution and migrations of planktonic animals determined. When such biological knowledge is correlated in detail with the physical optical facts, we may then look forward to an understanding of the importance of polarization patterns in submarine illumination. Meanwhile, the prospect of such correlation and evaluation should act as a stimulus to workers in the many pertinent fields of science.

This report may be briefly summarized as follows. Polarization patterns of submarine illumination in the upper layers of the sea near Bermuda have been observed by means of a sensitive detector and analyzer, which produces an oriented interference figure when traversed by linearly polarized light. Underwater polarization was found to consist of two primary components, one directly transmitted through the surface from the blue sky, the other originating in the water itself and being nearly everywhere oriented at right angles to the direction of penetrating light. As a result of their ultimate dependence on the sun's position, both of these components were found to undergo marked changes during the course of the day. The polarization arising within the water would seem to be largely accountable on the basis of Rayleigh scattering of light in the water. If this is so, the deep water pattern of polarization would most likely be similar to that near the surface on a heavily overcast day. Since a number of aquatic animals are known to be visually sensitive to polarized light, the possible relationship of the patterns described to the orientation and migration of zooplankton offers a promising area for future research.

References and Notes

* Contribution No. 213.

1. H. C. van de Hulst, *Atmospheres of the Earth and Planets,* G. P. Kuiper, Ed. (Univ. of Chicago Press, Chicago, 1949), pp. 49–111.
2. Z. Sekera, in *Compendium of Meteorology*, T. F. Malone, Ed. (American Meteorological Society, Boston, 1951), pp. 79–90.
3. K. von Frisch, *Naturwiss.* **35**, 38 (1948).
4. T. H. Waterman, *Trans. N.Y. Acad. Sci.* **14**, 11 (1951).
5. ———, *Science* **111**, 252 (1950).
6. E. R. Baylor and F. E. Smith, *Am. Naturalist* **87**, 97 (1953).
7. C. V. Raman, *Proc. Roy. Soc. (London)* **A101**, 64 (1922).
8. Y. LeGrand, *Ann. Inst. Océanog. Monaco* **19**, 393 (1939).
9. K. Kalle, *Rapports et procès-verbaux des réunions. Conseil permanent international pour l'exploration de la mer* **109**, pt. 3, 98 (1939).
10. H. V. N. Schenck and H. W. Kendall, *Underwater Photography* (Cornell Maritime Press, Cambridge, Md., 1954), pp. 1–110.
11. J. Cabannes, *La Diffusion moléculaire de la Lumière* (Presses Univ., Paris, 1929), pp. 1–326.
12. G. L. Clarke, *J. Marine Research* **4**, 221 (1941).
13. Molecular scattering by a given volume of water, although less per molecule, is 159 times (at 30°C) that produced by an equal volume of dust-free air (7).
14. Lord Rayleigh (J. W. Strutt), *Phil. Mag.* Ser. 4, **41**, 107, 274 (1871).
15. The present data relate only to daylight-illuminated water. The moonlit sky is polarized in a pattern essentially like that produced by the sun except, of course, that the light intensities, even with a bright full moon, are much lower. Theoretically, moonbeams penetrating natural waters would undergo scattering that would polarize them in a plane normal to the axis of the beam. Evidence for this lunar polarization pattern was sought at full moon, swimming in the turbid waters of Ferry Reach, but none was detected by the present method. The negative result was probably the effect of the overall underwater light intensity. Even though this was bright enough to illuminate bottom detail a meter or so away, the light through the analyzer was of extremely low intensity. Clearer water or more sensitive measurements would undoubtedly show the presence of a pattern of polarized light of lunar origin. Although a light compass reaction in certain insects has been shown to depend on the moon's position, no evidence is available at present for any biological importance of nocturnal polarization.
16. J. Ahlgrimm, *Zur Theorie der atmosphärischen Polarisation* (Inaug. Diss., Univ. of Kiel, 1915), pp. 1–66; E. O. Hulburt, *J. Opt. Soc. Amer.* **24**, 35 (1934).
17. W. M. Powell and G. L. Clarke, *J. Opt. Soc. Amer.* **26**, 111 (1936).
18. S. Chandrasekhar and D. Elbert, *Nature* **167**, 51 (1951).
19. W. R. G. Atkins and H. H. Poole, *Proc. Roy. Soc. (London)* **B140**, 321 (1952).
20. N. G. Jerlov and F. Koczy, *Rep. Swed. Deep-Sea Exped.* **3**, 63 (1951).
21. L. V. Whitney, *J. Marine Research* **4**, 122 (1941); H. H. Poole, *Sci. Proc. Roy. Dublin Soc.* **24**, 29 (1945).
22. N. G. Jerlov, *Rep. Swed. Deep-Sea Exped.* **3**, 1 (1951).
23. K. Kalle, *Ann. der Hydrog. Mar. Meteor.* **66**, 1 (1939).
24. Lord Rayleigh (J. W. Strutt), *Nature* **83**, 48 (1910).
25. T. H. Waterman *et al.*, *Biol. Bull.* **76**, 256 (1939).
26. T. H. Waterman, *Proc. Nat. Acad. Sci. (U.S.)* **40**, 252, 258 (1954); T. H. Waterman and C. A. G Wiersma, *J. Exptl. Zool.* **126**, 59 (1954).

Author Index

Subject Index